STIMSON'S

Introduction to Airborne Radar

Third Edition

STIMSON'S

Introduction to Airborne Radar

Third Edition

George W. Stimson

Hugh D. Griffiths
Chris J. Baker
Dave Adamy

Published by SciTech Publishing, an imprint of the IET.
www.scitechpub.com
www.theiet.org

Editor: Dudley R. Kay
Production Coordinator: Rachel Williams
Graphics: Antonio Shyam Reyes and Techset
Cover Design: Brent Beckley

10 9 8 7 6 5 4 3 2 1

ISBN 978-1-61353-022-1 (hardback)

Typeset in India by Techset
Printed in the USA by Sheridan Books, Inc.

Contents

PART III Fundamentals of Radar

In Memory of George W. Stimson

George W. Stimson became fascinated with radio waves as a teenage amateur radio enthusiast, designing and building transmitters and receivers.

His first brush with radar, which came in the early years of World War II, was bouncing echoes off Navy blimps in between experiments outside the ultra-high frequency lab at Stanford University. Upon receiving his bachelor's degree in electrical engineering, he did some additional course work at Caltech, went through the Navy's radar schools at Bowdoin and MIT, and wound up as an electronics officer on an attack transport.

Following the war, he served as an engineer on Southern California Edison's frequency-change project and at its completion joined Northrop's Snark Missile project. There quite by chance he became involved in technical publications and motion pictures.

In 1951, he was hired by Hughes Aircraft Company to write a widely circulated technical periodical called the *Radar Interceptor*. Working closely with the Company's top designers, in the ensuing years he observed at first hand the fascinating evolution of airborne radar from the simple systems for the first all-weather interceptors to the advanced pulsed doppler systems of today. He witnessed the development of the first radar-guided air-to-air missiles, the first incorporation of digital computers in small airborne radars, the birth of laser radar, SAR, and the programmable digital signal processor; and he saw the extension of airborne radar technology to space applications.

Following his retirement in 1990, he remained active in the field, teaching a short course in modern radar at the National Test Pilots School in Mojave, writing a technical brochure on Hughes antenna radiation-pattern and RCS measurement facilities, producing a fully narrated interactive multimedia presentation on the new HYSAR radar, and writing the article on radar for the 1998 edition of the *Encyclopedia Americana*.

Preface

Introducing the topic of radar is no easy undertaking. George W. Stimson's original text set new standards for clarity of explanation, wonderfully crafted illustrations, and boxed special topics. The writing, editing, and refining of this third edition has been both a daunting and extremely rewarding task, and we hope the result is every bit as clear as the first and second editions.

It is over 15 years since the last updates were made, and much has happened in the intervening period. Perhaps the most notable has been the remarkably rapid development of digital technology. Not only is this all-pervasive in our everyday lives, but it has also had a huge impact on how radar systems are designed and how echoes are processed. You will see the impact of advances in digital technology and radar signal processing throughout the Third Edition. Furthermore, airborne electronically scanned antennas, or AESAs, have firmly moved from the realm of new and advanced concepts and are routinely appearing in new systems. We recognize this by moving the ESA chapters from the "Advanced Concepts" of the 2nd Edition to the "Radar Fundamentals" early section of the book. Wholly new concepts such as bistatic and multistatic radar systems have begun to appear. Also, the ability to detect faint echoes has improved so much that even tiny objects such as bees can be confused with real target. Thus now the problem is shifting away from just detection to one of discrimination.

In order to address these and other exciting modern radar topics, we enlisted subject area experts, much as George W. Stimson drew upon the expertise of the many outstanding radar engineers at Hughes Aircraft. Drafts of chapters underwent rigorous editing and peer reviews with chapter versions reaching up to ten iterations as we endeavored to maintain a consistent voice, as much in line with Stimson's as we could. Additionally, all the art was reevaluated with some figures being redrawn, the colors adjusted for modern and consistent hues, and many new figures and photographs added. We never lost sight of the special Stimson genius for combining text and graphics to engage and enlighten the reader.

What's New

If you're familiar with the first and second editions, you may be wondering what's new in the third. Prompted by the advent of "stealth," the daunting prospect of ever more sophisticated radar countermeasures, and the explosive growth of digital-processing throughput, 12 entirely new chapters have been added. Additionally, a like number of chapters were extensively revised. Briefly, they reflect the following:

- Electronically scanned array antennas (ESAs) – besides providing extreme beam agility and resistance to sources of interference, they are also consistent with making the whole aircraft stealthy.
- Low-probability of intercept techniques (LPI) – besides greatly reducing vulnerability to counter-measures, they amazingly enable a radar to detect targets without its signals being usefully detected by an enemy.
- The entire Electronic Warfare section expanded and updated in text and figures to reflect both 21st Century EW terminology and technology.
- Multi-frequency operation and small-signal target detection—also essential in the era of stealth—plus space-time adaptive processing, true-time-delay beam steering, and 3D SAR.
- New modes and approaches to mode control that take advantage of the ESA's versatility.
- Advanced airborne digital processing architectures, key to most of the above capabilities.
- Detection and tracking of low-speed moving targets on the ground, an important topic missed in the previous edition.
- Bistatic, networked, and cognitive radar techniques, arguably the future of airborne radar.
- Some emerging concepts are presented as a peek into the future in new Chapter 48, and an update has been given to the dozen or so airborne systems currently in service.

These are the major content revisions that were necessary to bring the book into a thoroughly modern presentation of radar. However, from reviews and previous readers, certain additional enhancements to format were needed:

Test Your Understanding – each chapter now concludes with questions that test a reader's comprehension of the material. They are intentionally limited in mathematical challenge. We have the self-learning reader and short course practitioner in mind rather than the academic graduate student.

Further Reading – we offer books and tutorial articles worth exploring for extending the material for each chapter rather than a list at the end of the book.

SI Units – in keeping with today's scientific and technical books for an international audience, we have employed SI units throughout.

What's Not So New

Although the content has been updated and several enhancements introduced, we have striven to retain the same style and features that have made this book unique. Beginning from scratch, the book presents the wide range of airborne radar techniques in the form of an unfolding saga, not of individuals, but of radar concepts and principles. Each chapter tells a story, and the story flows naturally from chapter to chapter. The book remains designed to fulfill the needs of all who want to learn about radar, regardless of their technical backgrounds. It has sufficient technical depth and mathematical rigor to satisfy the instructor, the engineer, the professor. Yet, as long as a reader has a basic understanding of algebra and knows a little trigonometry and physics, the text takes the reader in bite-sized increments to the point of being able to talk on a sound footing with the radar experts.

Every technical concept is illustrated with a simple diagram immediately next to the text it relates to. Nearly every illustration has a descriptive caption, which we have striven to improve upon to assist reader understanding of the key point of the graphic. Where additional detail may be desired by some readers, additional material is conveniently placed in a "blue panel," which one may skip on a first reading and come back to later. We have added a number of new blue panels on modern topics. Exceptions, caveats, and clarifications are presented in side notes. One can follow the development of each chapter by reading just the text, just the illustrations and captions, or by moving along between text and illustrations. Lastly, recognizing that people interested in airborne radar love airplanes, dispersed through the book are photos and renderings of radar-bearing aircraft from all over the world, spanning the history of airborne radar from the first radar patent in 1903, the very first experiments in 1936, and then on through time to the aircraft, unmanned air vehicles, and satellites of the present and future.

The third edition has similarly benefitted from the insights and input provided by radar engineers from all over the world. It is a testament to the affection in which this book is held that so many people have given their time so freely. We must also acknowledge Brent Beckley for seeking out images from all over the world, Shyam Reyes for producing the wonderful illustrations, and all those who have reviewed and commented on early drafts of the chapters. Last, but by no means least, a special word for Dudley Kay, not just for his insistence on perfection and his ability to bring order out of chaos, but also for unstinting lifetime service in which he has been instrumental in educating the next generation of radar engineers.

Finally, we pay tribute to the vision of George W. Stimson in seeing the need for this book. We trust that we have done justice to that vision.

For corrections and suggestions to improve future printings and edition, please write to **stimson3@scitechpub.com**

Hugh Griffiths, Chris Baker and Dave Adamy
May, 2014

Publisher's Note

That there was a Stimson2 and now a Stimson3 has been a series of coincidences and serendipity. Most know the book was conceived at Hughes Aircraft Company by the Radar System Group's President as a private publication, given away or sold at a nominal charge to Hughes' customers and friends, mostly within the Department of Defense. George Stimson was Manager of Special Publications Projects and was given the task of pulling together a radar book that most people with a technical background could understand. No doubt Stimson had earned his reputation as a fine writer, but his zeal for clear graphics and meticulous formatting of pages emerged with this book. The radar engineers and advertising department of industry giant Hughes – 65,000 employees at the time – were put at his disposal, and Stimson took full advantage of these resources. Its release in 1983 underscored Hughes' leadership in radar, but it was never publicly promoted and sold; it did not even carry an ISBN, a requirement in the book trade for retail sales. Eventually the book was made available to those who wanted a copy, but only directly from Hughes, for the sum of $50.00.

With over 20,000 copies in circulation, tragedy struck in 1992. Reprinting was halted when a fire set off the sprinkler system in the hangar that housed the film printing plates. The plates were ruined, so no more books could be made. "The coffee table book of radar" appeared to have died, but its legendary reputation lived on. Lucky owners put their books under lock and key, lest they "walk away."

The second edition came about when editorial advisor Joe Brewer called from his Westinghouse (now Northrop Grumman) office in Baltimore. Joe reported his radar colleagues were always raving about this "greatest technical book ever published" but that it was no longer available. Might I consider obtaining the rights from Hughes to put it back into print, if just in a black and white edition? Joe Brewer is the unsung hero of the Stimson revival.

Two trips to Hughes showed me they would support a reprinting, but at the second meeting George Stimson, retired since 1987, was brought in for his opinions and advice. He stunned the gathering with two pronouncements: 1) he had been working on a revision for years and 2) he had been promised all rights to any future editions. Hughes, then in a "due diligence" acquisition period with Raytheon, capitulated. A "dead book" was a mere annoyance to the lawyers and management.

Right off Stimson declared he was not interested in either a reissue or a black and white version, and he demanded first class graphics to accompany his updates. SciTech, being a fledgling company with shallow pockets, sought one or more publishing partners while Stimson waited, but nobody else, whether commercial publisher or professional association, wanted the financial risk of a full color book, either. This all changed with two fortuitous events. First, Hughes offered to send me two file drawers packed with letters from companies and individuals that typically began, "If this book ever becomes available again, we would like X copies." Secondly, we hired a printer's representative to check the damaged printing plates to see if anything was salvageable for use in a revision. The report was that the film was in perfect condition! The deep secret, unbeknownst even to Stimson, was that Hughes had fabricated the fire story in order to end the bothersome small orders in checks and money orders that nobody wanted to handle.

George Stimson relied upon a small subset of his contacts still at Hughes/Raytheon, plus his own research, to accomplish the revision. SciTech supplied a very talented artist for new figures and page layout, a former Jet Propulsion Lab graphics artist, Shyam Reyes. Much was picked up from the first edition so that it was a hybrid of the timeless fundamentals of early chapters and what Stimson termed "advanced concepts" in the later new chapters. The revision proved to be a monumental success, even though material that had been termed "advanced" was fast becoming operational and commonplace in modern aircraft. Our radar authors and advisors would later comment, "It's an exceptional book but still built upon the technology of the 1980s."

Stimson's advancing age and death in 2009 left a vacuum in revision planning. The making of the 3rd edition set the challenge of finding the leadership and authorship to see it through. Who would have George's drive, knowledge, and genius? Perhaps no one person, but I had met a rising radar star named Hugh Griffiths many years previously and had published a bistatic radar book he had co-edited with Nick Willis. Moreover, Hugh has been editor of the journal *IET Radar, Sonar, and Navigation* for over 15 years, Chair of the IEEE Aerospace and Electronic Systems Society's Radar Systems Panel, and was then President of the Society itself. Most importantly, he really *wanted* the challenge. We were fortunate to pick up Hugh's good friend Chris Baker at Ohio State (and frequent consultant to Wright-Patterson AFB). I had also been fortunate to befriend and publish books by electronic warfare guru Dave Adamy, so Dave was enlisted to oversee the major updating and expansion of the EW section. Hugh, Chris, and Dave enlisted additional contributors from industry, military, and academia. All understood the Stimson tutorial level and the importance of graphics to support and illuminate the text. Like Stimson at Hughes, our subject experts were numerous and diverse, both the contributors and over 55 volunteer reviewers. How fortunate it was, also, to track down artist Shyam Reyes to render new art and a fresh page design. In our own way, we very nearly replicated the support team that had been put at George Stimson's disposal for the first edition. Can 65 editors, authors, and reviewers work together harmoniously to equal *one* George Stimson? Perhaps not, but we have given it our best effort and invite your comments and criticisms for refinements to future printings and editions. Thanks to the IET, a co-publisher of the 2nd Edition and now parent to SciTech Publishing, for supporting our efforts to keep the dream alive.

Dudley Kay

Founder of SciTech Publishing

Senior Commissioning Editor, SciTech Publishing – An Imprint of the IET

Technical Editors

Hugh D. Griffiths – Hugh Griffiths holds the Thales/Royal Academy of Engineering Chair of RF Sensors at University College London. His research interests include radar and sonar systems, signal processing (particularly synthetic aperture radar and bistatic and multistatic radar), as well as antennas and antenna measurements. Professor Griffiths has received numerous awards and he served as President of the IEEE Aerospace and Electronic Systems Society for 2012/13. He is a member of the IEEE AES Radar Systems Panel, serving as Chair from 2006–2009, and is Editor-in-Chief of *IET Proceedings on Radar, Sonar and Navigation*. He served as Chairman of the IEE International Radar Conference RADAR 2002 in Edinburgh, UK and he has advisory roles for the UK Ministry of Defence. He is a Fellow of the IEE and the IEEE and, in 1997, he was elected to Fellowship of the Royal Academy of Engineering. In 2013, he won the prestigious AF Harvey prize from the IET, honoring an exceptional individual researcher for their outstanding achievements and promising future research. He holds PhD and DSc(Eng) degrees from the University of London.

Christopher J. Baker – Chris Baker is the Ohio Research Scholar in Integrated Sensor Systems at The Ohio State University. Until June 2011 he was the Dean and Director of the College of Engineering and Computer Science at the Australian National University (ANU). Prior to this he held the Thales-Royal Academy of Engineering Chair of intelligent radar systems based at University College London. Professor Baker is the recipient of the IEE Mountbatten premium (twice), the IEE Institute premium and is a Fellow of the IET. He is a visiting Professor at the University of Cape Town, Cranfield University, University College London and Adelaide University. He has been actively engaged in radar system research since 1984 and is the author of over 200 publications. His research interests include: coherent radar techniques, radar signal processing, radar signal interpretation, electronically scanned radar systems, natural echo locating systems and radar imaging.

David L. Adamy – Dave Adamy is an internationally recognized expert in electronic warfare. He has 47 years' experience as a systems engineer and program technical director, developing EW systems from DC to Light, deployed on platforms from submarines to space, with specifications from QRC to high reliability. For the last 26 years, he has run his own company, performing studies for the US Government and defense contractors. He has also presented dozens of courses in the US, Europe, and Australia on Electronic Warfare and related subjects. He has published over 180 professional articles on Electronic Warfare, receiver system design, and closely related subjects, including the popular EW101 column in the *Journal of Electronic Defense*. He has eleven books in print and is a past National President of the Association of Old Crows.

Expert Contributors

Christopher D. Bailey – Chris Bailey is Associate Chief of the Radar Systems Division of the Georgia Tech Research Institute (GTRI). He has 15 years of experience with phased array design, analysis and modeling, and architecture trades. His current research efforts include digital beamforming, overlapped subarray architectures, and low-cost/low power arrays. Bailey has written numerous reports on phased array technology and regularly teaches phased array courses with the Georgia Tech Defense Technology Education Program. He holds an MSEE from The Johns Hopkins University and an MBA from the Georgia Institute of Technology.

Subject expert and contributor to Chapters 9, 10, and 42

David Blacknell – Dave Blacknell is Fellow at the UK Defence Science and Technology Laboratory (Dstl) and a visiting professor at University College London. He has worked on a large variety of topics in radar signal and image processing including image formation techniques, clutter modeling, and automatic target recognition. He is a recognized international expert on radar image exploitation and has published over 90 journal and conference papers. He has worked at the GEC-Marconi Research Centre (1984–1991), at QinetiQ (1991–2007), and held the Chair in Radar Systems at Cranfield University (2007–2009). He received a B.A. in Mathematics from Cambridge University, an M.Sc. from University College London, a Ph.D. from Sheffield University, and is a Fellow of the Institute of Physics.

Subject expert and contributor to Chapter 46

Shannon D. Blunt – Shannon Blunt is Associate Professor in the Electrical and Computer Engineering Department at the University of Kansas, where he is Director of the Radar Systems & Remote Sensing Lab. He began his career in the Radar Division of the Naval Research Laboratory. He was recipient of the IEEE/AESS Nathanson Memorial Radar Award, AFOSR Young Investigator Award, and multiple teaching awards at KU. He is a member of the IEEE/AESS Radar Systems Panel, Associate Editor for *IEEE Transactions on Aerospace & Electronic Systems*, and is on the Editorial Board for IET *Radar, Sonar, and Navigation*. He was General Chair of the 2011 IEEE Radar Conference and is Chair of the NATO SET-179 research task group on Dynamic Waveform Diversity & Design. He holds a PhD degree from the University of Missouri.

Subject Expert and contributor to chapters 16 and 46

Eli Brookner – Eli Brookner is a global radar authority known for his contributions to airborne, intelligence, space, air-traffic control and defense mission systems. Retired Principal Engineering Fellow from Raytheon Company's Integrated Defense Systems, Sudbury, Massachusetts, Dr. Brookner has played a key role in many major radar and phased-array radar systems developed during the past 40 years. Over 10,000 engineers and students have attended his lectures.

Subject expert and contributor to chapters 9, 10 and 42

Anthony M. Kinghorn – Tony Kinghorn is Chief Technical Officer (RF Systems) at Selex ES and is responsible for advanced radar research and development programs. He joined industry initially to work on radar signal processing. He has extensive experience in radar systems engineering and phased array systems and has amassed over 35 years' experience in the field of airborne radar. He has played a leading role in the technical research and development of a number of key systems and technologies, including the Captor multimode radar for Eurofighter Typhoon and a range of AESA radar systems both for fighter aircraft and airborne surveillance platforms. He graduated from the University of Cambridge in 1978.

Subject expert and contributor to chapters 14 and 15

Ronald McDivitt – Ronnie McDivitt is Chief Engineer in the Advanced Projects Group at Selex ES and is responsible for engineering innovative solutions for numerous sensor research and development programs. He has developed airborne radar signal and data processors for a wide variety of airborne radars. He has been a principal contributor to the architecture and design of the radar processor for the Captor multimode radar for Eurofighter Typhoon and a number of other AESA radars. He graduated from the University of Glasgow in 1979.

Subject expert and contributor to chapter 43

James M. Stiles – Jim Stiles is an Associate Professor within the Electrical Engineering and Computer Science Department at the University of Kansas and currently Associate Director of the KU Information and Telecommunication Technology Center, the unit that houses the Radar Systems & Remote Sensing Lab (RSL). He previously was a design engineer in the Defense Systems and Electronics Group at Texas Instruments (Raytheon). He is author of more than 50 refereed papers and winner of both Distinguished Research and Teaching awards. His PhD was earned at the University of Michigan.

Subject expert and contributor to Part VII – Radar Imaging, Chapters 32–35

Simon Watts – Simon Watts was Deputy Scientific Director at Thales UK and a Visiting Professor at University College London, in the Electronics and Electrical Engineering Department. He has undertaken extensive research into the modeling of radar sea clutter and the development of operational detection signal processing methods. He was appointed MBE in 2000 for services to the defense industry, is a Fellow of the Royal Academy of Engineering and of the IET, IEEE and IMA. He received the MA from the University of Oxford, PhD from the CNAA and MSc and DSc degrees from the University of Birmingham.

Subject expert and contributor to Part V – Clutter, Chapters 23–26

Richard G. Wiley – Dick Wiley is an internationally known expert in Electronic Intelligence (ELINT). He has written several books on the interception and analysis of radar signals. Several thousand participants have attended his lectures. He is vice-president and a founder of Research Associates of Syracuse, Inc. (RAS). He received BSEE and MSEE degrees from Carnegie Mellon University, the PhD from Syracuse University, and was elected a Fellow of the IEEE.

Subject expert and contributor to Part VIII, specifically Chapter 41

Reviewer Acknowledgements

The technical editors and publisher are deeply indebted to the volunteer reviewers who lent their expertise and gave their time so unselfishly to benefit the radar community with this updated edition. If and when you see any of these outstanding engineers, you can thank them, too.

Carly Aderton, Engineer Systems Architect, Northrop Grumman Corporation, USA

Christopher Allen, PhD, Professor, University of Kansas, USA

Larry Altshuler, Senior Engineering Fellow, Raytheon Space & Airborne Systems (Retired), USA

Andrea Antonini, PhD, Researcher, Consorzio LAMMA, Italy

Paul Antonik, PhD, Senior Scientist, USA

Bevan D. Bates, Adjunct Professor, University of Adelaide, Australia

Daniel A. Bernabei, Weapon System Engineer, Department of Defense, USA

Lee Blanton, Radar Engineer, General Atomics Aeronautical Systems, Inc., USA

Robin Blasberg, U.S. Naval Research Laboratory – Radar Division, USA

JJ Campbell, Lt Col USAF (Ret), JJ Campbell, LLC, USA

Kernan Chaisson, Captain, USAF (retired), USA

Carmine Clemente, Research Associate, University of Strathclyde, UK

Bill Correll, Jr, Research Scientist, General Dynamics Advanced Information Systems, USA

Gregory E. Coxson, PhD, Naval Research Laboratory – Radar Division, USA

G. Richard Curry, Consultant in Radar System Applications, USA

Antonio De Maio, PhD, DIETI University of Napoli, Italy

Manohar Deshpande, PhD, Microwave & Radar Engineer, NASA Goddard Space Flight Center, USA

John Erickson, PhD, United States Air Force, USA

Karl Erik Olsen, PhD, Principal Scientist, Norwegian Defence Research Establishment, Norway

Christopher N. Folley, PhD, Engineering Specialist, The Aerospace Corporation, USA

Mark Frank, Principal Engineer, Rohde & Schwarz, USA

Tony Gillespie, Defence Science & Technology Laboratory, UK

James Gitre, Engineering Manager, RF Design and Development, Intel Mobile Communications, USA

Paul J. Hannen, Senior Systems Engineer, Leidos, USA

Michael Inggs, PhD, Professor, University of Cape Town, South Africa

M. Jankiraman, PhD, Senior Radar Advisor, Larsen & Toubro Ltd, India

Randy Jost, PhD, Ball Aerospace and Technologies Corporation, USA

Joel Johnson, PhD, Professor, Ohio State University, USA

Seong-Hwoon Kim, PhD, Raytheon, USA

Stéphane Kemkemian, Senior Expert, Radar, Thales Airborne Systems, France

Theodoros G. Kostis, PhD, Lecturer (adj), Hellenic Military Academy, Greece

Jeff Lange, PhD, Senior Defence Scientist, Defence Research & Development Canada, Canada

Andon D. Lazarov, DrSc., Professor, Burgas Free University, Bulgaria

Anthony D. Leotta, President, ADL Associates, USA

Yasser M. Madany, PhD, Professor, Alexandria University, Egypt

Bob McShea, Director, Avionic Systems, National Test Pilot School, USA

John M. Milan, PhD, Consultant, ITT Gilfillan (retired), USA

Richard E. Miller, PhD, Senior Project Engineer, The Aerospace Corporation, USA

Douglas Moody, Mercer University, USA

Brian Mork, PhD, Professor and Research Engineer, USAF Test Pilot School, USA

Lee R. Moyer, Radar Subject Matter Expert, EOIR Technologies, Inc., USA

Mark A. Richards, Principal Research Engineer, Georgia Institute of Technology, USA

John Roulston, PhD, Scimus Solutions, UK

Earl Sager, Consultant, USA

Necmi Serkan Tezel, Postdoctoral Researcher, Stellenbosch University, South Africa

Graeme E. Smith, PhD, The Ohio State University, USA

Dr. Michael J. Staniforth, ESL Defence Ltd, UK

John J. SantaPietro, PhD, Principal Sensor Systems Engineer, The MITRE Corporation, USA

Craig Stringham, Brigham Young University, USA

Margaret Swassing, Sensors Integration Engineer, 412 Test Wing, 775 Test Squadron, Edwards Air Force Base, USA

Børge Torvik, Senior Scientist, Norwegian Defence Research Establishment (FFI), Norway

Jay Virts, Senior Systems Engineer, Exelis Inc, USA

Bradley A. Wilson, PhD, Associate Professor, Lakehead University, Canada

Sevgi Zübeyde Gürbüz, Assistant Professor, TOBB University of Economics and Technology, Turkey

David M. Zasada, PhD, Senior Principal Engineer, The MITRE Corporation, USA

List of Acronyms

AESA	Airborne Electronically Scanned Antenna
AEW	Airborne Early Warning
AGC	Automatic Gain Control
AI	Airborne Interception
ALU	Arithmetic Logic Unit
AMRAAM	Advanced Medium Range Air-to-Air Missile
ARM	Anti Radiation Missile
ASIC	Application Specific Integrated Circuit
ASV	Air-to-Surface Vessel
ATR	Automatic Target Recognition
AWACS	Airborne Control and Warning System
AWG	Arbitrary Waveform Generator
BSC	Beam Steering Controller
CBS	Cavity Backed Spiral
CCD	Coherent Change Detection
CEP	Circular Error Probable
CFAR	Constant False Alarm Rate
CMOS	Complementary Metal Oxide Semiconductor
COR	Coherent On Receive
COSRO	Conical Scan on Receive Only
COTS	Commercial Off-The-Shelf
CPI	Coherent Processing Interval
CRT	Cathode Ray Tube
CVR	Crystal Video Receiver
CW	Continuous Wave
DBS	Doppler Beam Sharpening
DBF	Digital Beam Forming
DD	Differential Doppler
DDS	Direct Digital Synthesis
DFT	Discrete Fourier Transform
DOA	Direction of Arrival
DPCA	Displaced Phase Center Antenna
DRFM	Digital Radio Frequency Memory
DSP	Digital Signal Processor
EA	Electronic Attack
ECM	Electronic Counter Measures
ECCM	Electronic Counter Counter Measures
ELINT	Electronic Intelligence
EMI	Electromagnetic Interference
ENR	Excess Noise Ratio
EP	Electronic Protection
ES	Electronic Support
ESA	Electronically Scanned Array
ESM	Electronic Support Measures
EW	Electronic Warfare
FAR	False Alarm Rate
FDOA	Frequency Difference Of Arrival
FFT	Fast Fourier Transform
FIFO	First In, First Out
FLOPS	Floating Point Operations per Second
FMCW	Frequency Modulation Continuous Wave
FOPEN	Foliage Penetration
FPGA	Field Programmable Gate Array
GMT	Ground Moving Target
GMTI	Ground Moving Target Indication
GPU	Graphic Processing Unit
HARM	High-speed Anti-Radiation Missile
IF	Intermediate Frequency
IFM	Instantaneous Frequency Measurement
InSAR	Interferometric Synthetic Aperture Radar
IRST	InfraRed Search and Track
ISAR	Inverse Synthetic Aperture Radar
JEM	Jet Engine Modulation
LFM	Linear Frequency Modulation
LO	Local Oscillator
LORO	Lobe on Receive Only
LP	Log Periodic
LPI	Low Probability of Intercept
LSB	Least Significant Bit
LSI	Large Scale Integration
MFR	Multi Function Radar
MIMO	Multiple Input, Multiple Output
MISO	Multiple Input, Single Output
MLC	Main Lobe Clutter
MLE	Maximum Likelihood Estimate
MMIC	Monolithic Microwave Integrated Circuit
MOTS	Military Off-The-Shelf
MSA	Mechanically Steered Antenna
MTI	Moving Target Indication
NCTR	Non Cooperative Target Recognition
NLFM	Nonlinear Frequency Modulation
NRCS	Normalized Radar Cross Section
OSU	Ohio State University
PBR	Passive Bistatic Radar
PCB	Printed Circuit Board
PDI	Postdetection Integration
PRBS	Pseudo Random Binary Sequence
PRF	Pulse Repetition Frequency
PRI	Pulse Repetition Interval
PSL	Peak Sidelobe Level
RCS	Radar Cross Section
RGPO	Range Gate Pull Off
rms	root mean square
RWR	Radar Warning Receiver
SAM	Surface-to-Air Missile
SAR	Synthetic Aperture Radar
	Search And Rescue
SAW	Surface Acoustic Wave
SCR	Signal-to-Clutter Ratio
SIMO	Single Input, Multiple Output
SLAR	Sideways Looking Airborne Radar
SLC	Sidelobe Canceler
SNR	Signal-to-Noise Ratio
SOJ	Stand Off Jamming
SPJ	Self Protection Jamming
STAP	Space Time Adaptive Processing
STC	Sensitivity Time Control
SWT	Search While Track
TBJ	Terrain Bounce Jamming
TDOA	Time Difference Of Arrival
TTD	True Time Delay
TWS	Track While Scan
TWT	Traveling Wave Tube
UCL	University College London
UHF	Ultra High Frequency
ULA	Uncommitted Logic Array
VCO	Voltage Controlled Oscillator
VHF	Very High Frequency

Avro Anson Mk.I (1936)

The Anson was the first aircraft in the world to fly a complete airborne radar (RDF-2) in August 1937. It was equipped with a modified EMI receiver, together with a lightweight transmitter and 1 kW portable petrol generator unit. Alan Blumlein, a gifted inventor and designer, working within the EMI Research Department in the 1930s, was responsible for taking airborne radar experiments to a stage where operational systems were sufficiently reliable to give RAF squadrons a true night-fighting capability.

"The Big Echo" - an original painting by Anthony Cowland

PART I

Overview of Airborne Radar

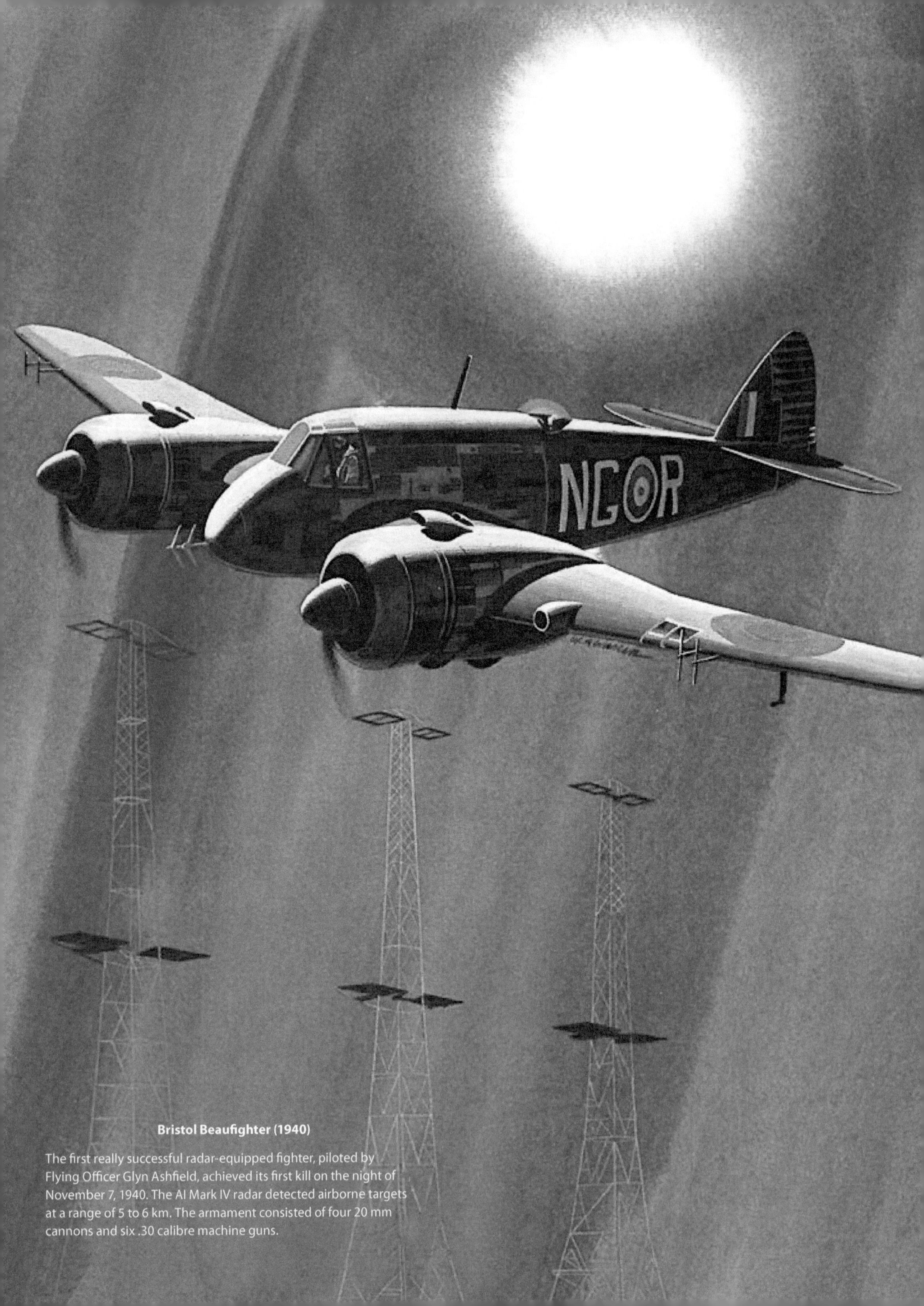

Bristol Beaufighter (1940)

The first really successful radar-equipped fighter, piloted by Flying Officer Glyn Ashfield, achieved its first kill on the night of November 7, 1940. The AI Mark IV radar detected airborne targets at a range of 5 to 6 km. The armament consisted of four 20 mm cannons and six .30 calibre machine guns.

1

Basic Concepts

from Hülsmeyer's original radar patent, 1904

1.1 Echolocation

Across a period of more than 50 million years, the echolocating bat has perfected the technique of transmitting a sequence of pulses and interrogating echoes. It depends on success for its very survival, intercepting prey on the wing and identifying sources of food in a complex background of unwanted reflections from debris such as leaves (or clutter). It is often accomplished in the presence of "jamming" from insects attempting to avoid capture. This is truly a remarkable capability, honed over time, enabling the bat to be one of the most widespread species of mammal on Earth.

Less well-known is the inherent ability of humans to echolocate, exploited so brilliantly by those who are sightless. These extraordinary people can literally *see with sound*. Not only can they dispense with the long cane but can even go cycling on- or off-road! Try an Internet search for Daniel Kish and be prepared to be amazed.

Using the same principle of echolocation, the pilot of a supersonic fighter is able to close in unerringly on a possible enemy intruder, hidden behind cloud cover, perhaps 200 km away (Fig. 1-1). How is it done?

Figure 1-1. Looking out through a streamlined fairing in the nose of a supersonic fighter, a small but powerful radar enables the pilot to home in on an intruder hidden behind or in a cloud bank 200 km away. (Courtesy of US Navy.)

Underlying all of these remarkable feats is a very simple principle: detecting objects and determining their distances (range) from reflected echoes. The chief difference is that, in the cases of the bat and the blind man, the echoes are those of sound waves, whereas in the case of the fighter they are echoes from radio waves. This chapter briefly introduces the fundamentals of the radar[1] concept and shows how radar is put to practical uses, such as detecting targets and measuring their ranges and locations.

A second important concept is also examined: determining the relative speed (or range-rate) of a reflecting object. This is achieved by measuring the shift in the radio frequency of the reflected waves relative to that of the transmitted waves, a phenomenon known as the *Doppler effect*. By sensing Doppler shifts,

1. Radar = **RA**dio **D**etection **A**nd **R**anging.

Figure 1-2. Rather than rejecting echoes from the ground, as when searching for moving targets, the radar may use them to produce real-time high-resolution maps of the terrain, such as this image of Wright-Patterson Air Force Base. (Image courtesy of Air Force Research Laboratory, Public Release Number: 88 ABW-12-0578.)

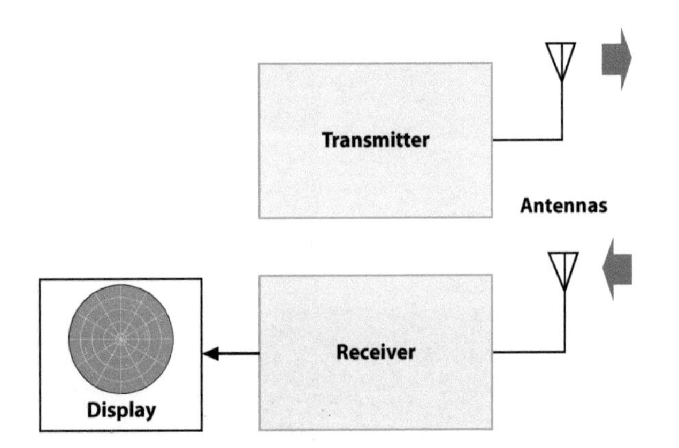

Figure 1-3. In rudimentary form, a radar system consists of five basic elements: a radio transmitter; a radio receiver; two antennas; and a display.

radar not only can measure range-rates but also can differentiate between echoes from moving targets and the unwanted echoes (clutter) caused by reflections from the ground and other stationary objects. It will be further learned that, rather than rejecting the echoes from the ground, the radar can alternatively use them to produce high-resolution map-like images (Fig. 1-2) using a technique known as synthetic aperture radar (SAR).

1.2 Radio Detection

Most objects such as aircraft, ships, vehicles, buildings, and terrain features will reflect radio waves much as they do light waves. Both types of waves are, in fact, the same thing: the flow of electromagnetic energy. The sole difference is that the frequencies of light are much higher. The reflected energy is scattered in all directions, but a detectable portion of it is generally scattered back in the direction from which it originally came.

At the longer wavelengths (lower frequencies) that many shipboard and ground-based radars use and even at short wavelengths of airborne radars, the atmosphere is almost completely transparent. By detecting the reflected radio waves, it is possible to "see" objects at night, in the daytime, and through haze, fog, or clouds. This is a major reason that radar is so widely employed.

In its most rudimentary form, a radar system consists of five elements: a *radio transmitter*; a *radio receiver* tuned to the transmitter's frequency; *two antennas*; and a *display* (Fig. 1-3). To detect the presence of an object (target), the transmitter generates radio waves. These are radiated by one of the antennas in the form of a collimated or focused beam. The receiver, meanwhile, listens for the "echoes" of these radio waves, which are picked up by the other antenna. If a target is detected, a blip indicating its location appears on the display.

In practice, the transmitter and receiver generally share a common antenna (Fig. 1-4). This antenna can be rotated through 360°. If a target is detected, it is displayed as a function of range and bearing on a *plan position indicator* (PPI) display. The radar is usually located at its center, and each angle bearing represents the antenna beam as it scans through all 360°.

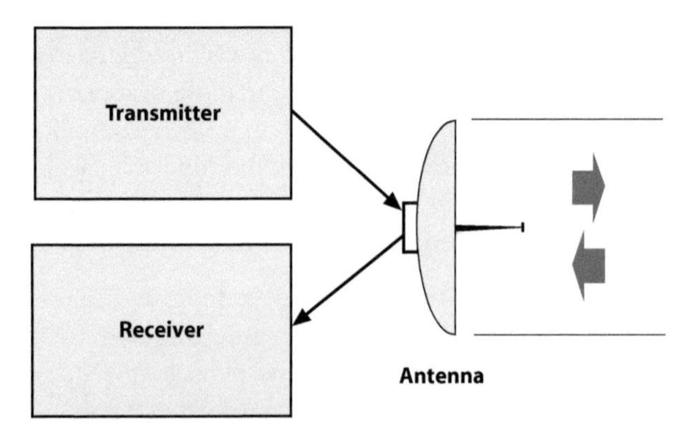

Figure 1-4. In practice, a single antenna is typically time shared by the transmitter and the receiver.

To avoid problems of the transmitter interfering with reception, the radio waves are usually transmitted in pulses and the receiver is turned off (*blanked*) during transmission. When using a single antenna for transmitting and receiving, a device known as a *circulator* or *transmit/receive switch* is used to reduce transmitter interference. The rate at which the pulses are transmitted is called the *pulse repetition frequency* (PRF), and the time between the pulse transmissions is the *pulse repetition interval* (PRI). Thus, the PRF is the inverse of the PRI (i.e., PRF = 1/PRI) (Fig. 1-5).

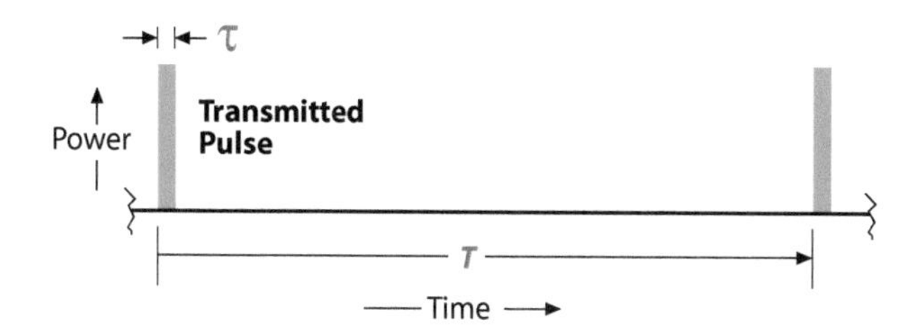

Figure 1-5. To keep transmission separate from reception, the radar usually transmits radio waves in a sequence of pulses and listens for the echoes in the time between them, *T*.

The term *target* is broadly used to refer to almost anything that is to be detected: an aircraft, a ship, a vehicle, a man-made structure on the ground, a specific point in the terrain, rain (weather radars), aerosols, even free electrons. Like light, radio wave frequencies used by most airborne radars essentially travel in straight lines. Consequently, for a radar system to receive echoes from a target, the target must be within the line of sight (Fig. 1-6).

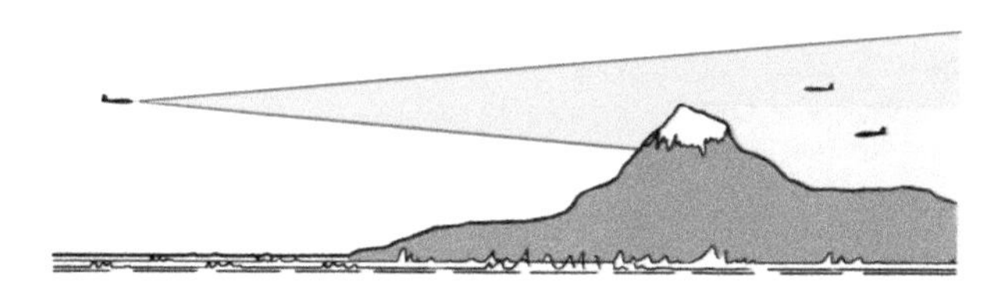

Figure 1-6. To be seen by most radar systems, a target must be within the line of sight. This is not the case for the target in the blue "shadow" zone, and this target will not be seen by the radar.

Even then, the target will not be detected unless its echoes are strong enough to be discerned above the background either of electrical noise that always exists in the receiver or of simultaneously received echoes from the ground (*ground clutter*). In some situations ground clutter may be substantially stronger than the noise in a receiver.

The strength of a target's echoes is inversely proportional to the target's range raised to the fourth power ($1/R^4$). Therefore, as a distant target approaches, its echoes rapidly grow stronger (Fig. 1-7).

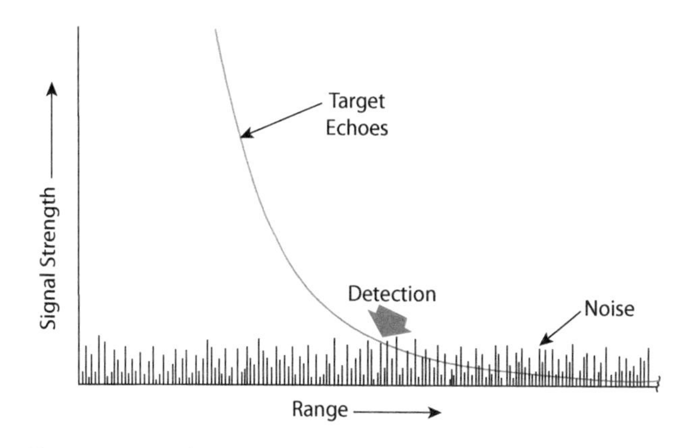

Figure 1-7. As a distant target approaches, its echoes rapidly grow stronger. Only when they emerge from the background of noise or ground clutter are they strong enough to be detected.

The range at which they become strong enough to be detected depends on a number of factors, the most important of which include:

- The power of the transmitted waves
- The fraction of time during which power is transmitted (τ/T, where τ is the pulse duration and T is the time between emitted pulses, the PRI)
- The size of the antenna (the bigger the antenna, the narrower and more intense the beam)
- The reflecting characteristics of the target (generally, the bigger the target, the bigger the reflection)
- The length of time the target is in the antenna beam (more echo pulses can be received)
- The number of scans in which the target appears (even more echo pulses can be received)
- The wavelength of the radio waves
- The strength of background noise or clutter

Much as sunlight reflected from a car on a distant highway scintillates and fades, the strength of the echoes scattered in the radar's direction varies more or less at random (Fig. 1-8).

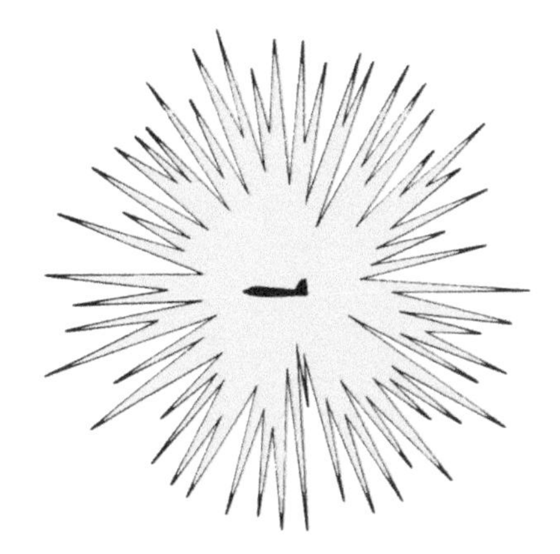

Figure 1-8. Since the target return scintillates and fades, and noise varies randomly, detection ranges must be expressed in terms of probabilities.

Figure 1-9. Radars in larger aircraft like airborne warning and control systems (AWACS) can detect small aircraft at ranges in excess of 500 km. (Courtesy of US Air Force.)

$$
\begin{aligned}
R &= \frac{1}{2}\ \text{(Round-Trip Time)} \times \text{(Speed of Light)} \\
&= \frac{1}{2} \times \frac{10}{1{,}000{,}000}\ \text{s} \times 300{,}000{,}000\ \text{m/s} \\
&= 1.5\ \text{km}
\end{aligned}
$$

Figure 1-10. The transit time is measured in millionths of a second (µs). A transit time of 10 µs corresponds to a range of 1.5 km.

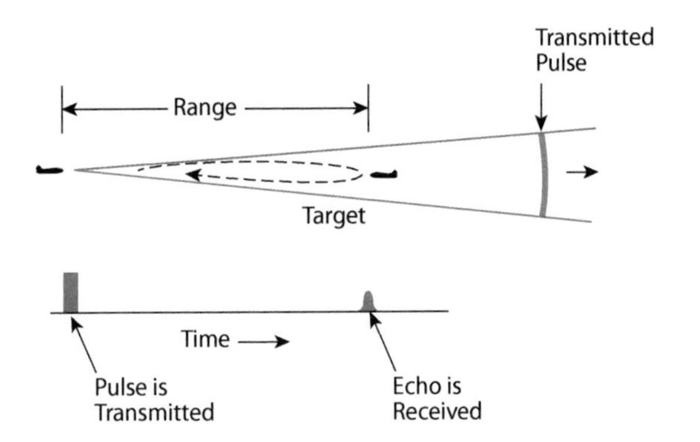

Figure 1-11. Usually a target's range may be most easily determined by measuring the time between transmission of a pulse and the reception of its echo.

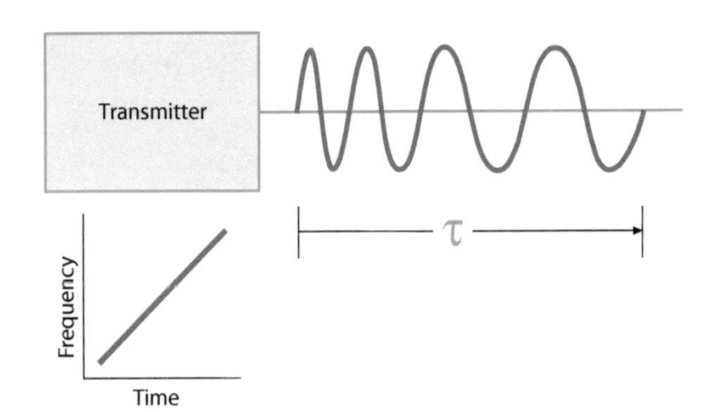

Figure 1-12. In a chirp pulse compression modulation, the transmitter's frequency increases linearly throughout the duration, τ, of each pulse.

Because of both this and the randomness of the background noise, the range at which a given target is detected by the radar will not always be identical. Nevertheless, the probability of its being detected at any particular range can be predicted with considerable certainty.

By optimizing the parameters that can be controlled, a radar can be made small enough to fit in the nose of a fighter yet detect small targets at ranges of the order of 200 km. Larger radar systems on larger aircraft (Fig. 1-9) can detect targets at even greater ranges.

1.3 Determining Target Position

In most applications, it is not enough merely to know that a target is present. It is also necessary to know the target's location, that is, its distance (range) and direction (angle).

Measuring Range. Range may be determined by measuring the time it takes the radio waves to reach the target and return. Radio waves travel at essentially the speed of light, which is a constant. A target's range therefore is half the transit time (because the pulse has to travel to the target and back to the receiver) times the speed of light (Fig. 1-10). Since the speed of light is high, 300 million meters per second, ranging times are generally expressed in millionths of a second (microseconds). For example, a round-trip transit time of 10 microseconds corresponds to a range of 1.5 km.

The transit time is measured by observing the time delay between transmission of a pulse and reception of the echo of that pulse (Fig. 1-11), a technique called *pulse-delay ranging*. To avoid echoes overlapping from targets closely spaced in range, and hence appearing to be the return from a single target, the width of the pulse, τ, must be made sufficiently small. However, this may be insufficient to radiate enough energy to detect distant targets. Thus, to detect closely spaced targets at long ranges, pulses must be made wider. This dilemma is resolved by *compressing* the echoes after they are received.

One method of compression, called *chirping,* is to linearly increase the frequency of each transmitted pulse throughout its duration (Fig. 1-12). The received echoes are passed through a filter, introducing a *delay* that decreases with increasing frequency and thereby compresses the received energy into a narrow pulse. A resolution of 30 cm or so may be obtained without limiting detection range, and this corresponds to a *chirp* frequency span of approximately 500 MHz. This is discussed in more detail in Chapter 16.

Radars that transmit continuously (continuous wave, or CW) measure range with a technique called *frequency modulation* (FM). Here the frequency of the transmitted wave is varied as

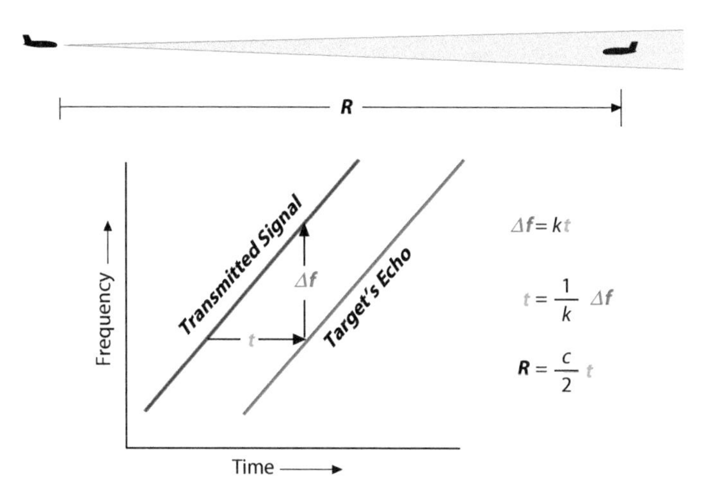

Figure 1-13. In FM ranging, the frequency of the transmitted signal is varied linearly, and the instantaneous difference, Δf, between the transmitter's frequency and the target echo frequency is sensed. The round-trip transit time, t, to the target (the target's range, R) is proportional to this difference.

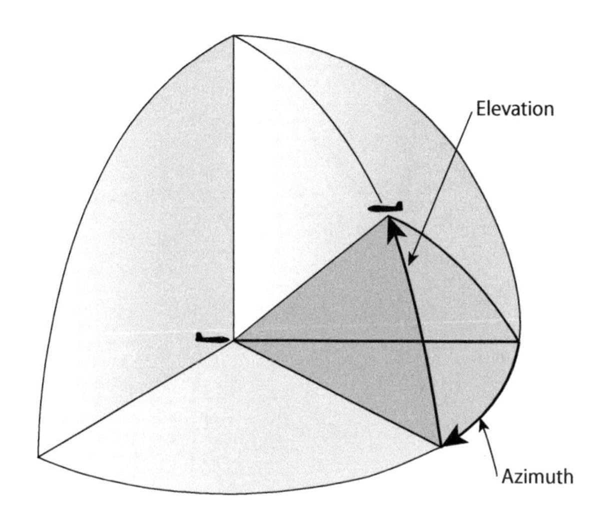

Figure 1-14. Angle between the fuselage reference axis and the line of sight to a target is usually resolved into azimuth and elevation components.

a function of time, and range is determined by observing the lag in time between this modulation and the corresponding modulation of the received echoes (Fig. 1-13).

Measuring Direction. In most airborne radars, direction is measured in terms of the angle between the line of sight to the target and a reference direction, such as north, or the longitudinal reference axis of the aircraft's fuselage. This angle is usually resolved into a horizontal component called the *azimuth* and a vertical component called the *elevation* (Fig. 1-14).

Where both azimuth and elevation are required, the beam is given a more or less conical shape, called a *pencil beam* (Fig. 1-15a). Where only azimuth is required, as for long-range surveillance, mapping, or detecting targets on the ground, the beam may be given a *fan shape* (Fig. 1-15b).

Automatic Tracking. The goal is often to follow the movements of one or more targets while continuing to search for others. This may be done in a mode of operation called *track-while-scan*. In this mode, the position of each target of interest is tracked on the basis of the periodic samples of its range, range rate, and direction obtained whenever the antenna beam sweeps across it (Fig. 1-16).

Track-while-scan is ideal for maintaining situation awareness. It provides sufficiently accurate target data for launching guided missiles, which after departure can correct their trajectories. It is particularly useful for launching missiles in rapid succession against several widely separated targets. However, track-while-scan does not provide accurate enough data for predicting the flight path of a target for a fighter's

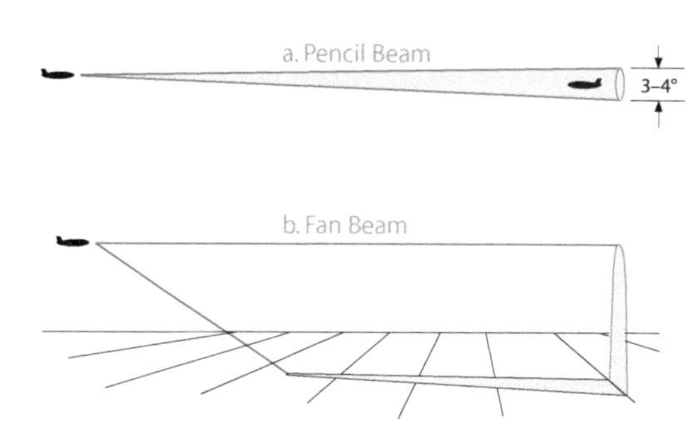

Figure 1-15. A pencil beam (a) is used for detecting and tracking aircraft, and a fan beam (b) is employed for long-range surveillance, mapping, or detecting targets on the ground.

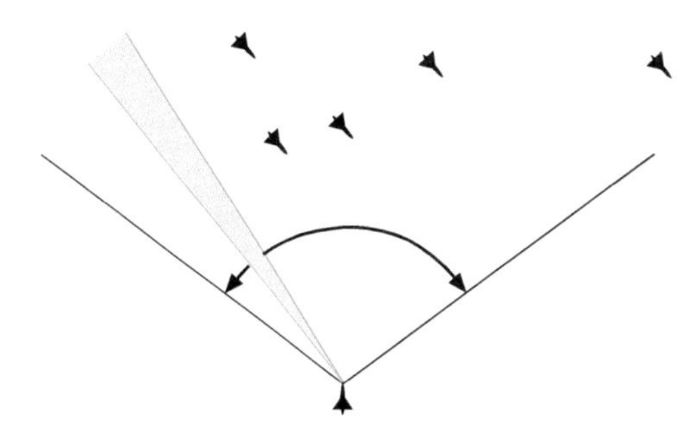

Figure 1-16. In track-while-scan, any number of targets may be tracked simultaneously on the basis of samples of each target's range, range rate, and direction obtained when the beam sweeps across it in the course of the search scan.

Figure 1-17. For tasks requiring precision, such as predicting the flight path of a tanker in preparation for refueling, a single-target tracking mode is generally provided.

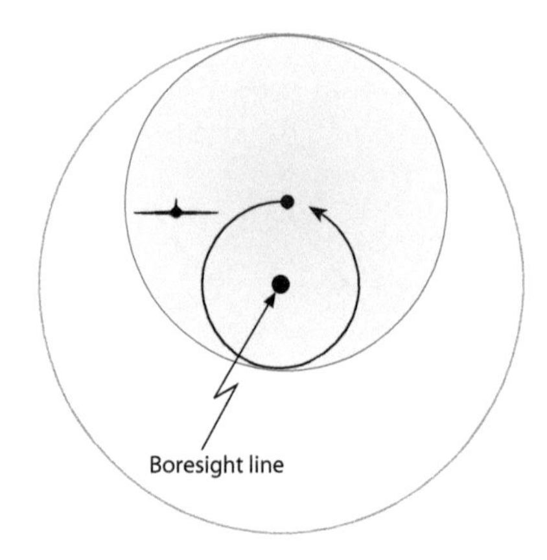

Figure 1-18. In a conical scan, rotating the antenna's beam about the boresight line and sensing the resulting modulation of the received echoes allows angle tracking errors to be sensed.

guns or for tanker refueling (Fig. 1-17). In such cases, the antenna must be trained on the target continuously in a *single-target track* mode.

To keep the antenna trained on a target in this mode, the radar must be able to measure its own pointing errors. This may be done in several ways. Older radar systems use a technique called *conical scanning* in which the beam is rotated so that its central axis sweeps a small cone about the pointing axis (the boresight line) of the antenna (Fig. 1-18). If the target is on the boresight line (i.e., no error exists), its distance from the center of the beam will be the same throughout the conical scan and the amplitude of the received echoes will be unaffected by the scan. However, since the strength of the beam falls off toward its edges, if a tracking error exists the conical scan will modulate the echoes. The amplitude of the modulation indicates the magnitude of the tracking error, and the point in the scan at which the amplitude reaches its minimum indicates the direction of the error.

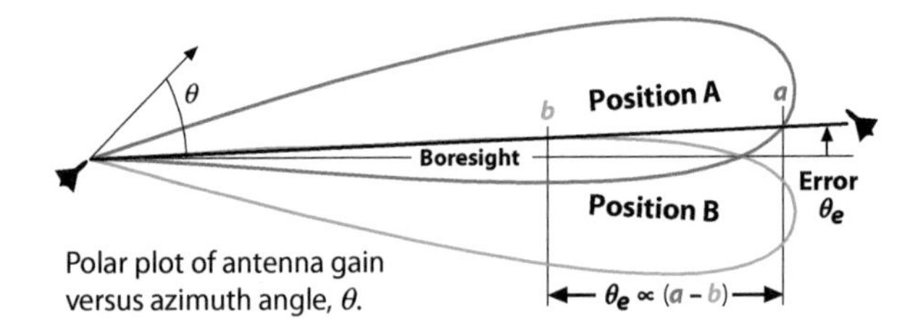

Figure 1-19. In lobing, the antenna lobe is alternately deflected to the right and left of the boresight line to measure the angle-tracking error, θ_e.

Alternatively, in a technique called *sequential lobing* the pointing error may be measured by sequentially placing the center of the beam on one side and then the other side of the boresight line during reception only (Fig. 1-19).

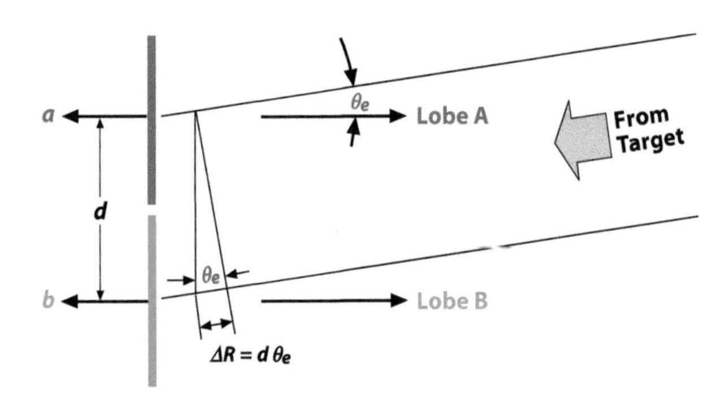

Figure 1-20. In phase-comparison monopulse, the difference in the distance from a target to the antenna's two halves is ΔR. Hence, for small angles the difference in phases of the outputs a and *b* is proportional to the tracking error, θ_e.

However, both conical scanning and sequential lobing suffer from inaccuracies due to pulse-to-pulse fluctuations in the echo strength from a target. To overcome this, most modern radars form lobes simultaneously, thereby enabling the pointing error to be sensed with a single pulse, called *monopulse*. In one such version, called *amplitude-comparison monopulse*, the antenna is divided into halves that produce overlapping lobes. In another, called *phase-comparison monopulse*, both halves of the antenna produce beams pointing in the boresight direction. In phase-comparison monopulse, if a tracking error exists, the distance from the target to each half will differ slightly in proportion to the tracking error, θ_e. Consequently, the error can be determined by sensing the resulting difference in the radio frequency phase of the signals as received by the two halves of the antenna (Fig. 1-20).

By continuously sensing the tracking error with either of these techniques and correcting the antenna's pointing direction to minimize the error, the antenna can be made to follow the target's movement very precisely.

While the target is being tracked in angle, its range and direction may be continuously measured. Its range rate may then be computed from the continuously measured range. Its angular rate—the rate of rotation of the line of sight to the target—may be computed from the continuously measured direction. The target's range, range rate, direction, and angular rate allows its velocity and acceleration to be calculated, as illustrated in Fig. 1-21.

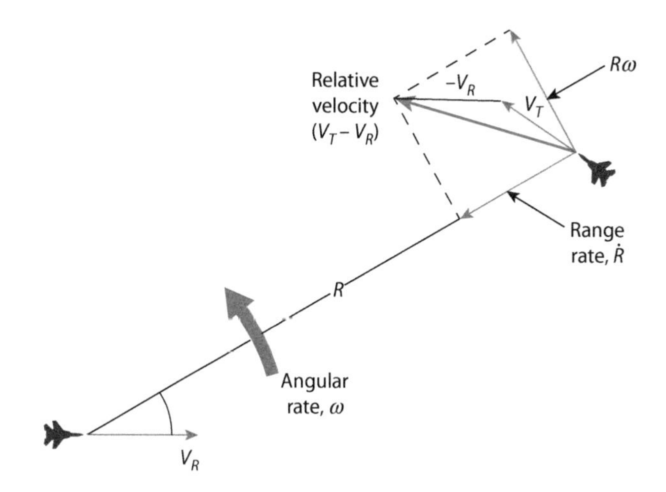

Figure 1-21. A target's relative radial velocity may be computed from measured values of range, range rate, and angular rate of line of sight.

Most radar systems now use electronic scanning. This blurs the distinction between track-while-scan and single-target tracking because the radar beam can be made to dwell for chosen periods, in contrast to mechanical scanning in which the dwell is for a fixed period determined by the scan rate.

1.4 The Doppler Effect

A classic example of the Doppler effect is the change in pitch of a vehicle as it passes by on the highway. As the vehicle approaches the pitch is increased but as it passes and moves away the pitch is reduced. This is because the apparent wavelength is shortened when the vehicle is approaching and lengthened when receding (Fig. 1-22).

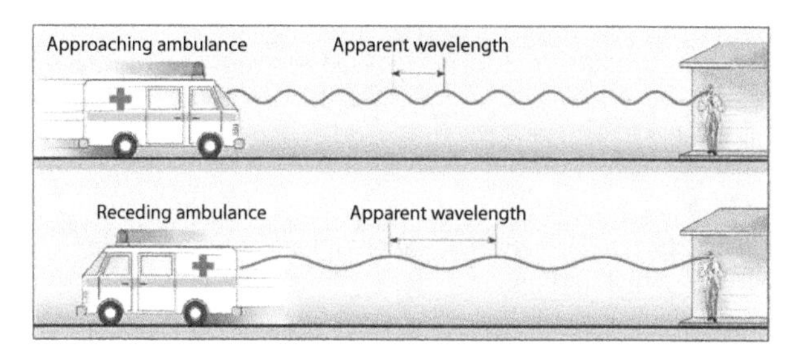

Figure 1-22. In this common example of the Doppler shift, the motion of a vehicle condenses the sound waves propagated ahead (increasing their apparent frequency) and spreads the waves propagated behind (reducing their apparent frequency).

Because of the Doppler effect, the radio frequency of an echo received from an object is shifted relative to the frequency of the transmitter in proportion to the object's range rate. (Note: this is the component of range-rate projected in the direction of the radars position, more commonly called *radial velocity*). Since the range-rates encountered by radar are a minuscule fraction of the speed of radio waves (i.e., the speed of light), the Doppler shift (or *Doppler frequency*) of even the most rapidly closing target is so extremely slight that it shows up simply as a pulse-to-pulse shift in the radio frequency phase of the target's echoes. However, like a laser, radar is a coherent sensor, and this allows the phase shift imparted on an echo to be measured. This phase shift is measured by cutting the radar's transmitted pulses from the same position on a continuous signal (Fig. 1-23). The phase of the echo is referenced to the phase of the transmitted pulse, which allows phase changes due to target motion to be measured for only each transmitted pulse. The rate of change of phase from a sequence of pulses provides a direct measure of the Doppler frequency or radial velocity of a target.

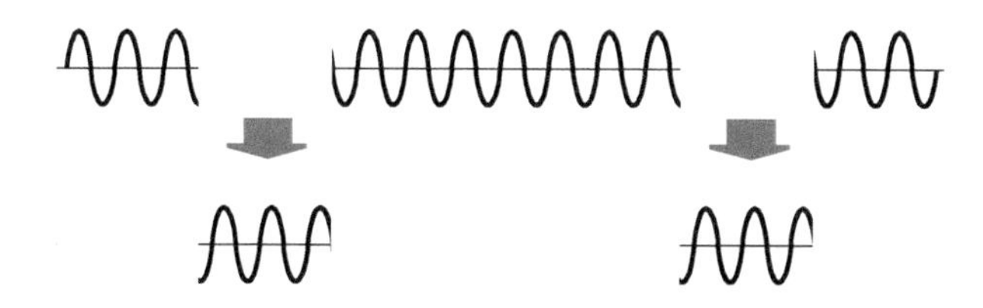

Figure 1-23. By cutting transmitted pulses from a continuous wave, the radio frequency phase of successive echoes from the same target will be coherent, enabling their Doppler frequency to be readily measured.

By sensing phase shifts associated with successive echoes, a radar system not only can measure range rates directly but also can expand its capabilities in other respects. Chief among these is the substantial reduction, or complete elimination in some cases, of clutter. The range rates of targets are generally quite different from the range rates of most points on the ground and other stationary or slow-moving sources of unwanted return. By sensing Doppler frequencies, a radar

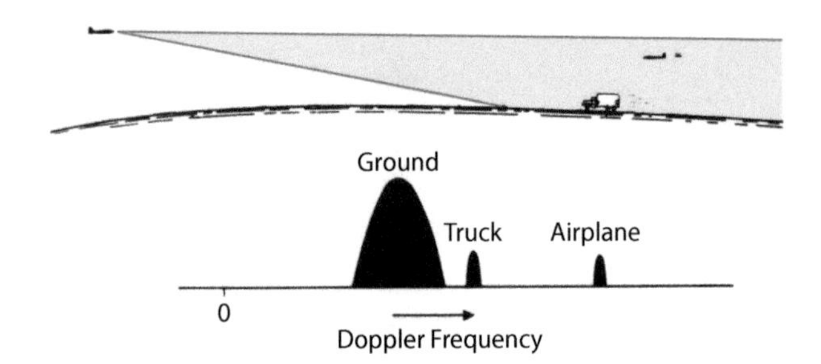

Figure 1-24. With MTI, echoes from targets and moving vehicles on the ground are separated from ground clutter on the basis of the differences in their Doppler frequencies. Generally, echoes from aircraft and echoes from moving vehicles on the ground may be differentiated as a result of the ground vehicles' lower speed.

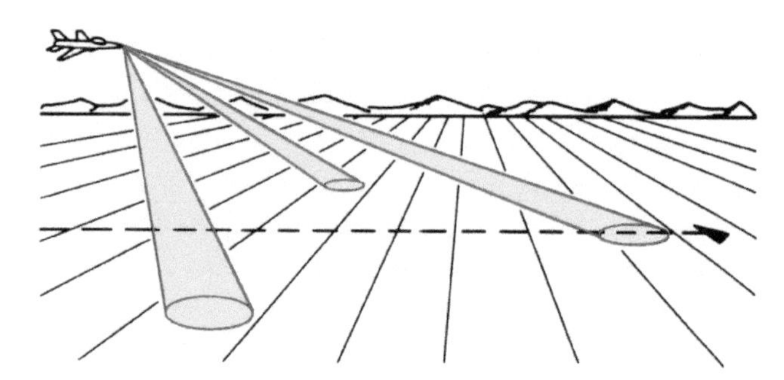

Figure 1-25. A radar's velocity may be computed from Doppler frequencies of three or more points on the ground at known angles.

2. This depends on the lookdown angle. Water and flat ground directly below a radar system produce very strong returns.

system can differentiate echoes of moving targets from clutter. This is called *moving target indication* (MTI). Often MTI is separated into *airborne moving target indication* (AMTI) and *ground moving target indication* (GMTI).

MTI is of inestimable value in airborne radars that must operate at low altitudes or look down in search of aircraft flying below them. The antenna beam commonly intercepts the ground at the target's range. Without MTI, the target echoes would be lost in the ground return (Fig. 1-24). MTI can also be immeasurably important when aircraft must fly at higher altitudes and look straight ahead. Even then, the lower edge of the beam may intercept the ground at long ranges.

Where desired, by sensing the Doppler shift a radar system can measure its own velocity. For this the antenna beam is generally pointed ahead and down at a shallow angle. The echoes from the point at which the beam intercepts the ground are then isolated and their Doppler shift is measured. By sequentially making several such measurements at different azimuth and elevation angles, the aircraft's horizontal ground speed can be computed accurately (Fig. 1-25).

1.5 Imaging

The radio waves transmitted by a radar system are scattered back in the direction of the radar in different amounts by different objects. Not much is scattered back from smooth surfaces such as lakes[2] and roads. More comes from farmland, brush, and trees, with strongest scattering tending to come from man-made structures. Thus, by displaying the differences in the intensities of the received echoes when the antenna beam is swept across the ground, it is possible to produce a pictorial map of the terrain, called a *ground map* or *clutter map*. These are examples of lower resolution imagery, usually with range resolution being different from cross range resolution (which is determined by the antenna beamwidth).

Radar maps differ from aerial photographs and road maps in several fundamental respects. In the first place, because of the difference in wavelengths, the relative reflectivity of the various features of the terrain may be quite different for radio waves than for visible light. Consequently, what is bright in a photograph may not be bright in a radar map and vice versa. In addition, unlike road maps, radar maps contain shadows, may be distorted, and, unless special measures are taken to improve azimuth resolution, may show only large-scale features.

Shadows are produced whenever the transmitted waves are intercepted by hills, mountains, or other obstructions and there is no line of sight to the ground beyond. The effect can be visualized by imagining that you are looking directly down on a relief map illuminated by a single light source at the radar's

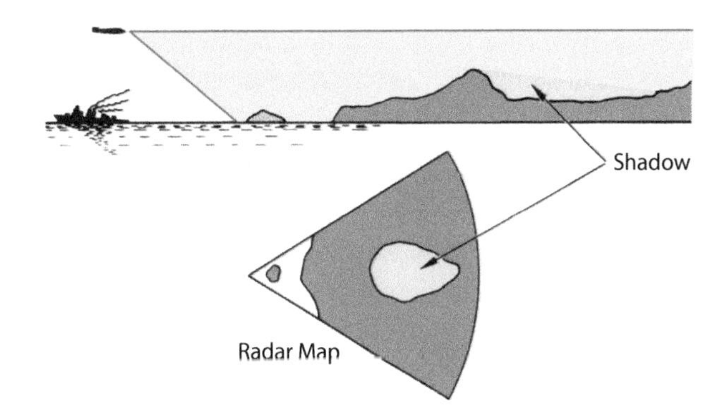

Figure 1-26. Shadows leave holes in radar maps. At steep lookdown angles, shadowing is minimized.

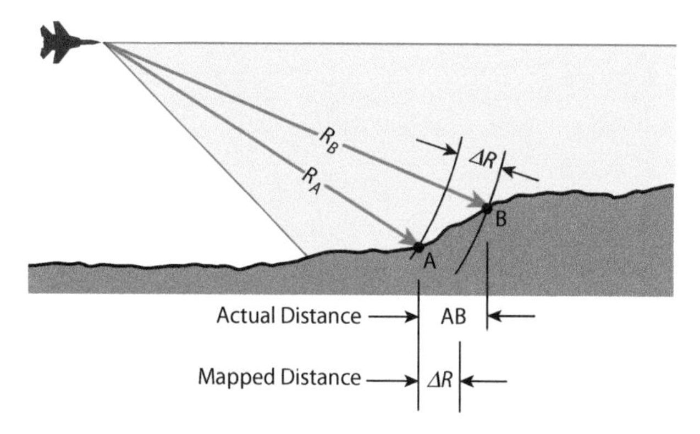

Figure 1-27. At steep lookdown angles, mapped distances are foreshortened. Except for distortion due to slope of the ground, foreshortening may be corrected before the map is displayed.

location (Fig. 1-26). Shadowing is minimal if the terrain is reasonably flat or if the radar is looking down at a fairly steep angle.

Distortion arises if the lookdown angle is large. Since the radar measures distance in terms of slant range, *layover* occurs when the apparent horizontal distance between two points at the same azimuth is foreshortened (Fig. 1-27). If the terrain is sloping, two points separated by a small horizontal distance can, in the extreme, be mapped as a single point. Usually, the foreshortening can be corrected on the basis of the lookdown angle before the map is displayed.

Because radar is a coherent sensor, reflections contain *speckle noise* similar to lasers. This arises because of the coherent addition and subtraction of reflections from targets composed of multiple scatters that create constructive and destructive interference. This causes target reflections to vary greatly with small changes in look angle.

The degree of detail a radar map provides depends on the ability of the radar to separate or resolve closely spaced objects in range and azimuth. Range resolution is limited primarily by the width of the radar's pulses.

SAR Imaging. By transmitting long duration pulses and employing pulse compression, the radar may obtain strong returns even from very long ranges and achieve range resolution as fine as a 30 cm.

Fine azimuth resolution is not so easily obtained. In conventional (real-beam) ground mapping, azimuth resolution is determined by the width of the antenna beam (Fig. 1-28). With a beamwidth of 3°, for example, at a range of 10 km the azimuth resolution of a real-beam map may be no finer than 0.5 km (Fig. 1-29).

Azimuth resolution may be improved by operating at higher frequencies or by making the antenna larger. But if exceptionally high frequencies are used, detection ranges are reduced by atmospheric attenuation, and there are practical limitations

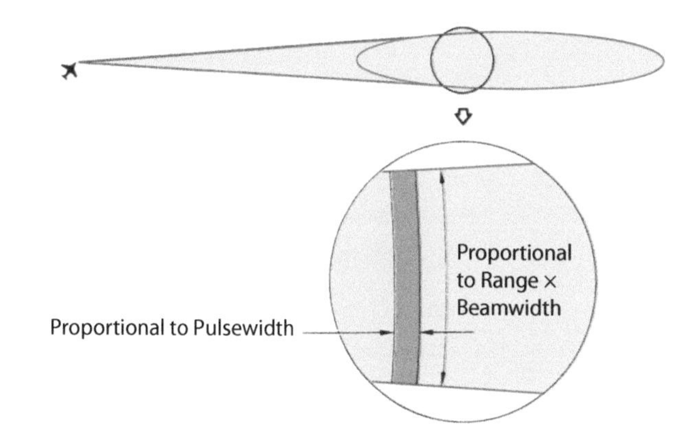

Figure 1-28. With conventional mapping, the dimensions of a resolution cell are determined by the pulse width and antenna beamwidth.

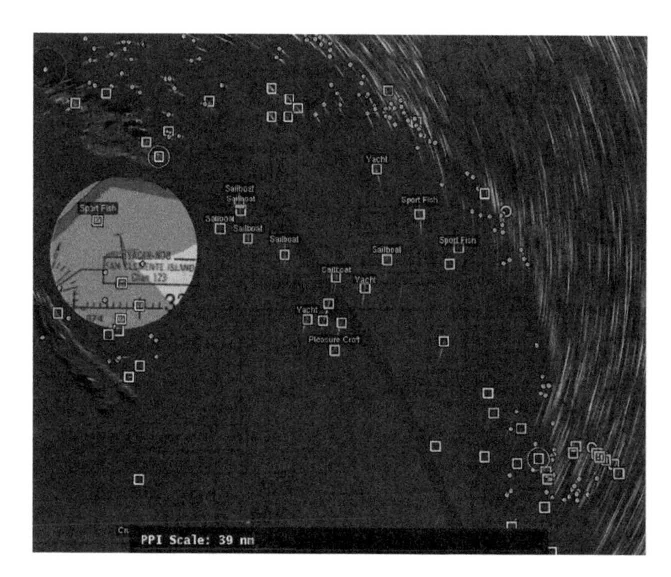

Figure 1-29. Raytheon's SeaVue XMC Maritime Surveillance Radar detects small maritime targets in high seas and provides search in ISAR and SAR modes. (Courtesy Raytheon Company.)

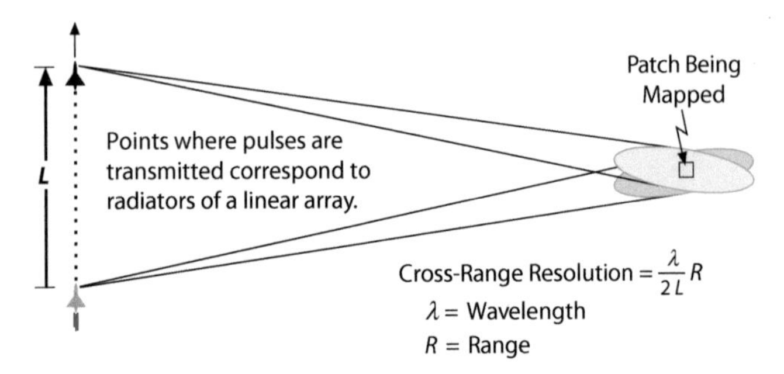

Figure 1-30. With its antenna trained on a patch to be mapped, each time the radar transmits a pulse it assumes the role of a single radiator. When the returns of a great many pulses are added up, the results are essentially the same as would have been obtained with a linear array antenna of length, *L*. The mode illustrated here is called spotlight SAR.

on how large an antenna an aircraft can accommodate. However, an antenna of almost any length can be synthesized using SAR.

Rather than scanning the terrain with a rotating antenna, SAR maintains the radar beam in a direction orthogonal to the trajectory of the aircraft. Each time the radar radiates, the pulse assumes the role of a single radiating element of the synthetic aperture. Because of the aircraft's velocity, each element is a little farther along on the flight path (Fig. 1-30). By storing the returns of a great many pulses and combining them (much as a feed system combines the returns received by the radiating elements of a real antenna), the radar can synthesize the equivalent of a linear array long enough to provide azimuth resolution as fine as 15 cm (Fig. 1-31).

Moreover, by increasing the length of the synthesized array in proportion to the range of the area being mapped, the same fine resolution can be obtained at a range of 100 km as can be obtained at a range of only a few km. In fact, the resolution is effectively independent of both range and wavelength. This is one of the main reasons that SAR has become such an invaluable remote sensing and surveillance tool and that systems are able to operate from spacecraft orbiting the Earth at distances on the order of 750 km.

1.6 Summary

By transmitting radio waves and listening for their echoes, a radar can detect objects day or night, and in all kinds of weather. By concentrating the waves into a narrow beam, radars can determine direction, and by measuring the transit time of the waves they can measure range.

To find a target, the radar beam is repeatedly swept through a scan. Once detected, the target may be automatically tracked and its relative velocity computed on the basis of either (1) periodic samples of its range and direction obtained during the scan or (2) continuous data obtained by training the antenna on the target. In the latter case, the target's echoes must be singled out in range or Doppler frequency, and some means such as lobing must be provided to sense angular tracking errors.

Because of the Doppler effect, the radio frequencies of the radar echoes are shifted in proportion to the reflecting object's range rates. By sensing these shifts, the radar can measure target radial velocity, reject clutter, and differentiate between ground return and moving vehicles on the ground. It can even measure its own velocity.

Since radio waves are scattered in different amounts by different features of the terrain, a radar system can map the ground. With SAR, high-resolution map-like images can be made (Fig. 1-31).

Figure 1-31. This is a 15 cm, spotlight SAR image, of Stonehenge in the UK, one of the best known prehistoric monuments in Europe. Note, the long shadows cast by the ancient arrangement of stones is due to illumination from the radar. The image is oriented with North pointing downward. (Copyright QinetiQ Ltd.)

Further Reading

Historical Background

S. S. Swords, *Technical History of the Beginnings of Radar*, Peter Peregrinus, 1986.

E. G. Bowen, *Radar Days*, Adam Hilger, 1987.

L. Brown, *Technical and Military Imperatives: A Radar History of World War II*, Taylor & Francis, 1999.

J. B. McKinney, "Radar: A Case History of an Invention," *IEEE Aerospace and Electronic Systems Magazine*, Vol. 21, No. 8, Part II, August 2006.

Technical Background

S. Kingsley and S. Quegan, *Understanding Radar Systems*, SciTech-IET, 1999.

G. R. Curry, *Radar Essentials*, SciTech-IET, 2012.

P. Hannen, *Principles of Radar and Electronic Warfare for the Non-Specialist*, 4th Edition, SciTech-IET, 2014.

Test your understanding

1. What are the five elements comprising a radar system?
2. What is meant by the terms PRF and PRI, and what is the relationship between them?
3. The round-trip transit time to detect a target is measured to be 666 microseconds. How far away is the target from the radar?
4. When should track-while-scan be used, and when should single-target tracking be used?
5. Name three techniques used to improve angular accuracy in tracking radar systems.
6. How is Doppler frequency (or radial velocity) measured by a radar system?
7. What is the technique used to generate high cross-range resolution in radar imaging?

Messerschmitt Bf-110 G4 (1941)

The Bf-110 was the first Luftwaffe radar-equipped fighter. It was equipped with a bulky nose-mounted antenna system which cut its top speed by 25 mph. The Telefunken FuG 212 'Lichtenstein' radar was able to detect targets from ranges of 200 m out to 5 km. The armament consisted of four 20 mm and four 7.9 mm guns firing forward and a 20 mm cannon firing vertically upward.

2

Approaches to Implementation

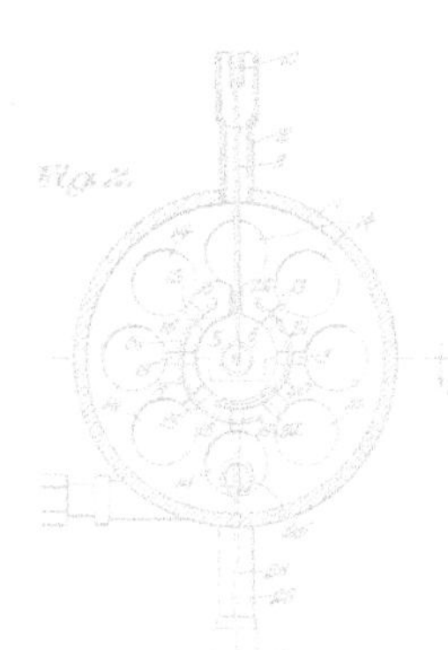

Original cavity magnetron, 1940

Now that we have reviewed the basic concepts of radar, this chapter examines their practical considerations and implementation. Although radar design varies greatly, it is possible to get a rough idea of what is involved by considering two different types: (1) *noncoherent pulsed radar,* where only the amplitude of echoes is measured; and (2) *coherent pulsed radar,* where both the amplitude and the phase of echoes are measured.

Noncoherent pulsed radar is associated with older radar forms like the all-weather interceptors in the 1950s and 1960s, which were often referred to as simple pulsed radar. Noncoherent pulsed radar is still very widely used in marine navigation and surveillance applications and in other configurations.

Coherent radar systems are far more capable—albeit more complex and expensive. They are often referred to as *pulse-Doppler* radars and are used for almost all military and an increasing number of civil functions. Because both echo amplitude and phase are measured, coherent radars are able to determine Doppler velocity, which both improves performance and supports a much wider variety of applications.

One other technology development that improves performance and capability still further is *electronically scanned array* (ESA) *antennas* (though they add even more complexity and expense). Electronic scanning, in effect, allows the radar beam to be pointed anywhere at any time. It also enables adaptation to local conditions, thereby improving detection of targets against clutter and other sources of interference.

2.1 Noncoherent Pulsed Radar

This form of radar (Fig. 2-1) is capable of automatic searching, single-target tracking, and real-beam ground mapping. It is the workhorse of older military systems and, in ground-based

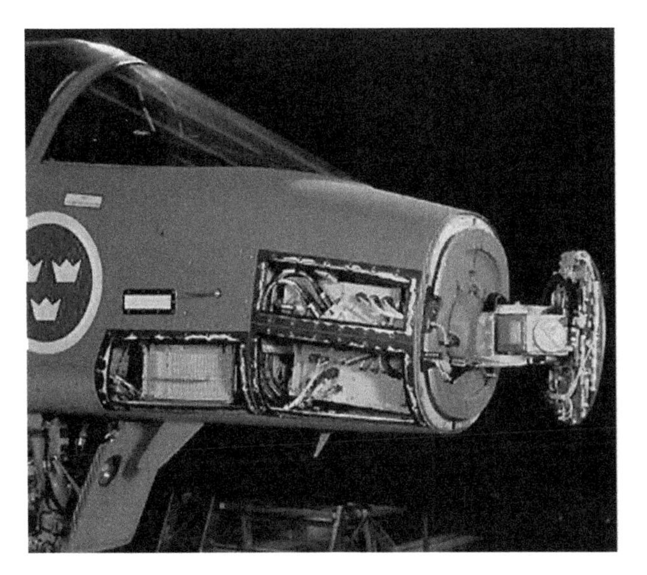

Figure 2-1. This is a Saab Microwave Systems PS-05 multipurpose pulse Doppler radar in a Gripen aircraft.

PULSED RADAR

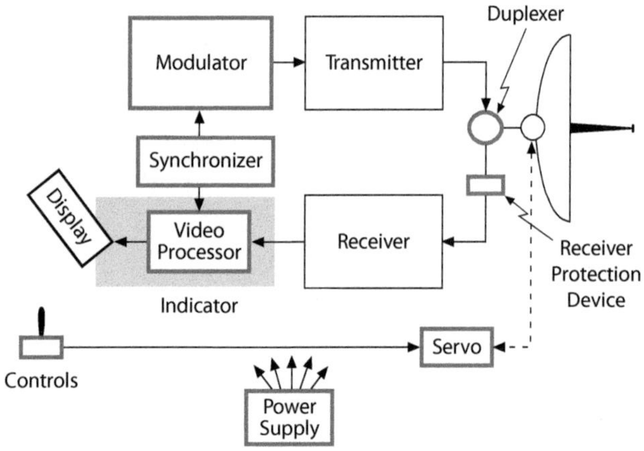

Figure 2-2. Elements outlined in blue must be added to the transmitter, receiver, antenna, and display for even a simple generic noncoherent pulsed radar.

form, is still widely used for maritime surveillance from both commercial and leisure craft. Indeed, it is the most common form of radar. The transmitter is called a *magnetron* (whose development is said to have been responsible for helping the Allies win World War II). The magnetron has become even more ubiquitous as the power source in the microwave oven.

Remember from Chapter 1 that a pulsed radar system consists of four basic functional elements: transmitter; receiver; time-shared antenna; and display. However, to implement even a simple practical noncoherent radar system, several other elements are also required, the most important of which are included in Figure 2-2.

Synchronizer. This unit enables time synchronization of the transmitter and indicator by generating a continuous stream of very short, evenly spaced pulses. Through being fed to the modulator and indicator, these pulses designate the times at which successive radar pulses are to be transmitted.

Modulator. Upon receipt of each timing pulse, the modulator produces a high-power pulse of direct current (DC) energy and supplies it to the transmitter, acting in effect as a switch to turn the transmitter on and off.

Transmitter. This high-power oscillator, generally a *magnetron* (Fig. 2-3), generates a high-power radio frequency wave for the duration of the input pulse from the modulator. In effect, this converts the DC pulse to a pulse of radio frequency energy. How it does this is illustrated in the blue panel on pages 18–19. The wavelength of the energy is typically between 3 and 10 cm. The exact value either may be fixed by the design of the magnetron or may be tunable by an operator over a range of about 10%. The wave is radiated into a metal pipe (Fig. 2-4) called a *waveguide*, which conveys it to the *duplexer*.

Figure 2-3. The magnetron transmitter tube converts pulses of direct current (DC) power to pulses of microwave energy.

Duplexer. Essentially a waveguide switch (Fig. 2-5), the duplexer connects the transmitter and the receiver to the antenna while maintaining a separation between the transmitter and receiver. This is to avoid the high-power transmitted signal from damaging the delicate and highly sensitive components of the receiver. The duplexer is usually a passive device that is sensitive to the direction of flow of the radio waves. Thus, it allows the waves coming from the transmitter to pass with negligible attenuation to the antenna while blocking their flow to the receiver. Similarly, when echo

Figure 2-4. A waveguide is a metal pipe down which radio waves may be ducted. The width is usually about three-quarters of the wavelength, and the height is roughly half the wavelength.

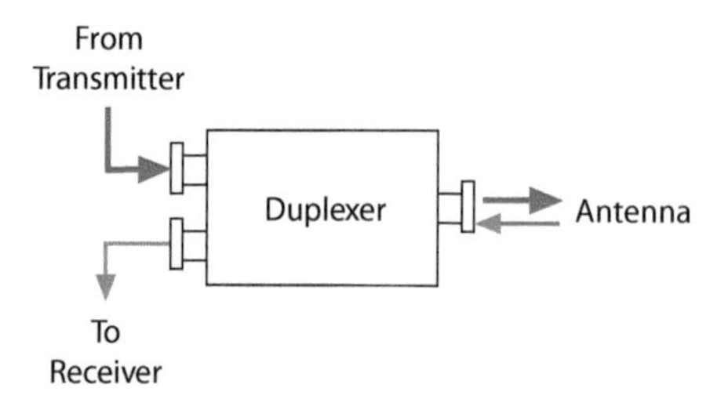

Figure 2-5. A duplexer is a device that passes the transmitter's high-power pulses to the antenna and the received echoes from the antenna to the receiver. It separates the transmitter and receiver sides of the radar system.

signals return to the radar, the duplexer allows the waves coming from the antenna to pass with negligible attenuation to the receiver while blocking their way to the transmitter.

Antenna. In simple radars, the antenna generally consists of a *radiator* and a parabolic-shaped *reflector* (often called a *dish*), which are mounted on a common support. In its most rudimentary form, the radiator is little more than a horn-shaped nozzle on the end of the waveguide coming from the duplexer. The horn directs the radio wave arriving from the transmitter onto the dish, which reflects the wave in the form of a narrow beam (Fig. 2-6). Echoes intercepted by the dish are reflected into the horn and conveyed by the same waveguide to the duplexer and then to the receiver.

Generally, the antenna is mounted on gimbals, which allow it to be pivoted about both the azimuth and elevation axes. In some cases, a third gimbal may be provided to isolate the antenna from the roll of the platform (e.g., aircraft, ship). Transducers on the gimbals may provide the indicator with signals proportional to the displacement of the antenna about each axis.

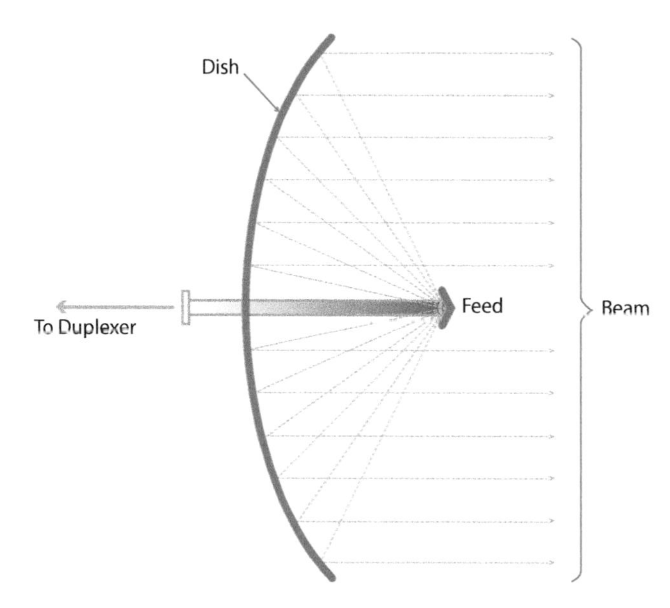

Figure 2-6. The antenna for a simple pulsed radar consists of a single feed and a parabolic dish reflector. The parabolic reflector forms the transmitted beam and also reflects the returned echoes into the feed.

Receiver Protection. Because of electrical discontinuities (mismatch of impedances) between the antenna and the waveguide, some of the energy of the radio waves ends up being reflected from the antenna back to the duplexer. Since the duplexer performs its switching function purely on the basis of direction of flow, there is nothing to prevent this reflected energy from flowing onto the receiver (just as the radar echoes do). The reflected energy amounts to only a very small fraction of the transmitter's output. However, because of its high power, the reflections are strong enough to potentially damage the receiver. To prevent the reflections from reaching the receiver and to block any of the transmitter's energy that has leaked through the duplexer, an additional *receiver protection device* is employed.

The receiver protection device (Fig. 2-7) is a high-speed microwave switch, which automatically blocks any radio waves strong enough to damage the receiver. Besides leakage and energy reflected by the antenna, the device also stops any exceptionally strong signals that may be received from outside the radar—echoes received when the radar is inadvertently fired up in a lab or is operated while facing, say, a wall at point-blank range.

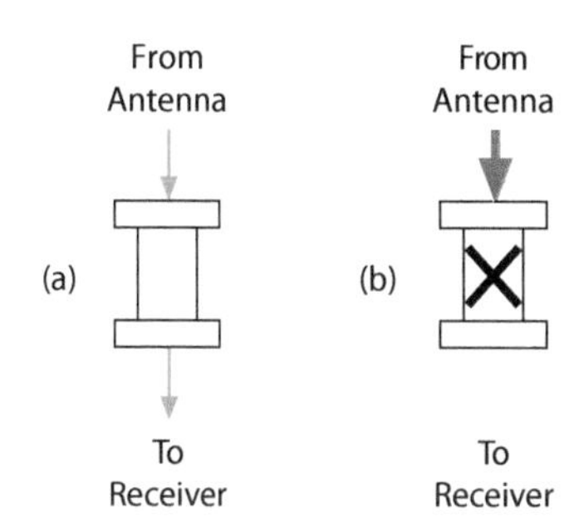

Figure 2-7. The receiver protection device (a) allows the weak echoes to pass from the duplexer to the receiver with negligible attenuation and (b) blocks any signals strong enough to damage the receiver.

Receiver. Typically, the receiver is of a type called a *superheterodyne* (Fig. 2-8). It translates the received signals to a lower frequency at which they can be filtered and amplified more using simpler and less costly equipment. Translation is accomplished by "beating" the received signals against the output of a low-power oscillator (called the *local oscillator,* or LO) in a device called a mixer. The frequency of the resulting signal, called the *intermediate frequency* (IF), equals the difference between the signal's original frequency and the LO frequency.

The output of the mixer is amplified by a tuned circuit (IF amplifier). It filters out any interfering signals as well as the electrical background noise lying outside the band of frequencies occupied by the received signal.

RECEIVER

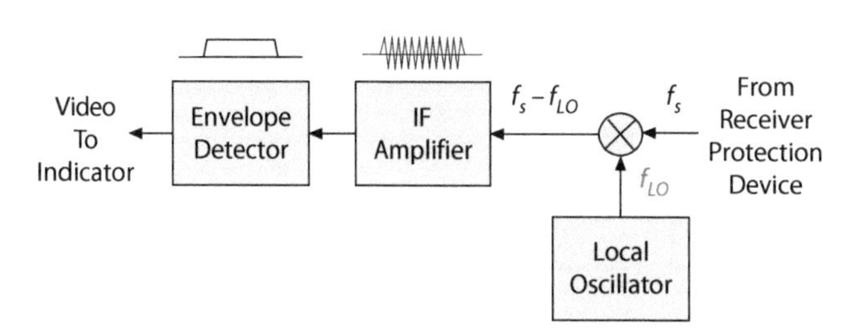

Figure 2-8. The receiver translates the received radio waves (signal) to a lower frequency (IF), amplifies them, filters out signals of other frequencies, and produces a video output proportional to the received signal's amplitude.

The Venerable Magnetron

Developed in the early years of World War II, the magnetron was the breakthrough that first made high-power microwave radars practical. Because of its comparatively low cost, small size, light weight, high efficiency, and rugged simplicity (plus its ability to produce high output powers with moderate input voltages), the magnetron has been widely used in radar transmitters ever since. It is the workhorse of commercial and leisure craft civil marine radar systems, both of which are used for surveillance and navigation. It is highly reliability and long-lasting and because of continuous development is now very affordable.

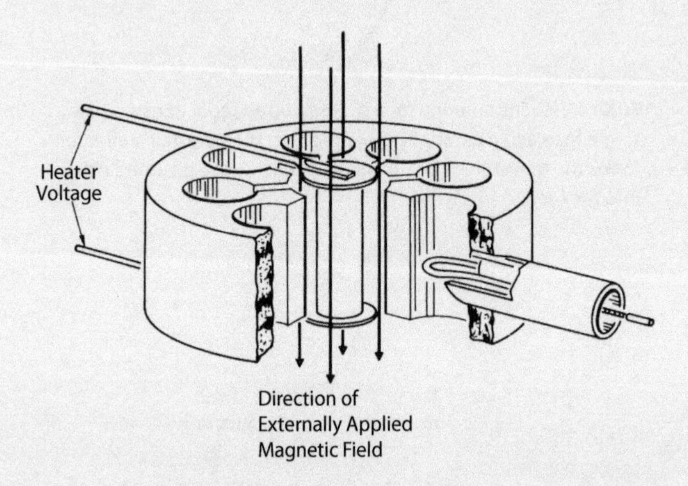

The magnetron is one of a family of vacuum tube oscillators and amplifiers that utilizes the force exerted by a magnetic field whose direction is normal to the electron's velocity when the electron moves through it and causes its path to curve. The greater the electron's speed, the greater the curvature.

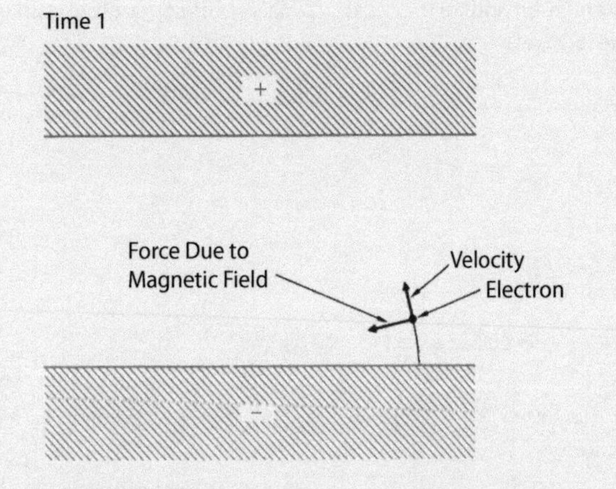

These tubes are called *cross-field* tubes because the electric field that produces the electrons' motion is normal to the magnetic field. Slicing a magnetron in two would reveal a cylindrical central electrode (cathode) ringed by a second

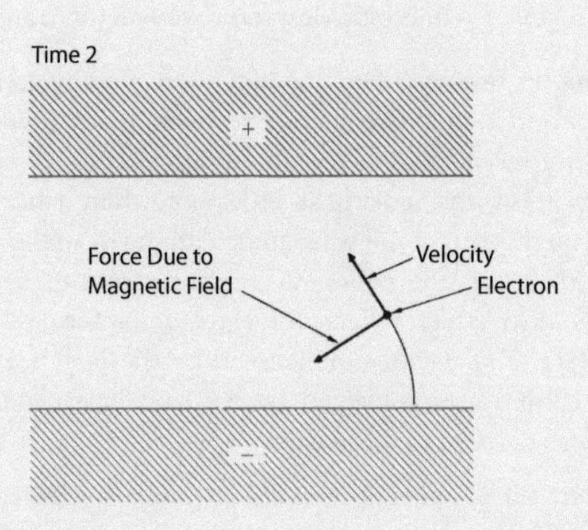

cylindrical electrode (anode), with a gap (called the interaction space) in between.

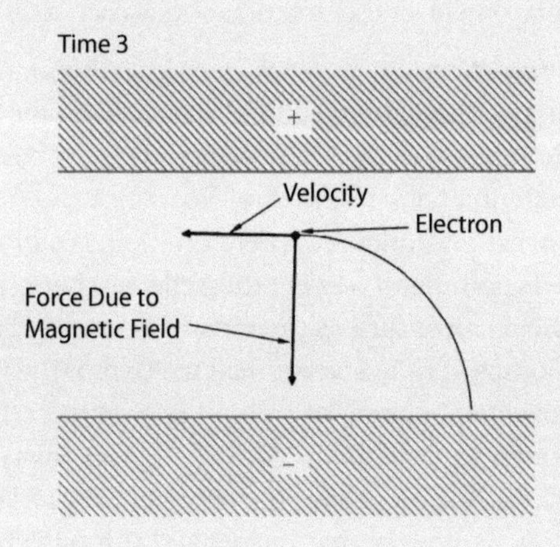

Evenly distributed around the inner circumference of the anode are resonant cavities opening into the interaction space. The cathode is heated so that it emits electrons, which form a dense "cloud" around it. An externally mounted permanent magnet produces a strong magnetic field within the interaction space, normal to the axis of the electrodes.

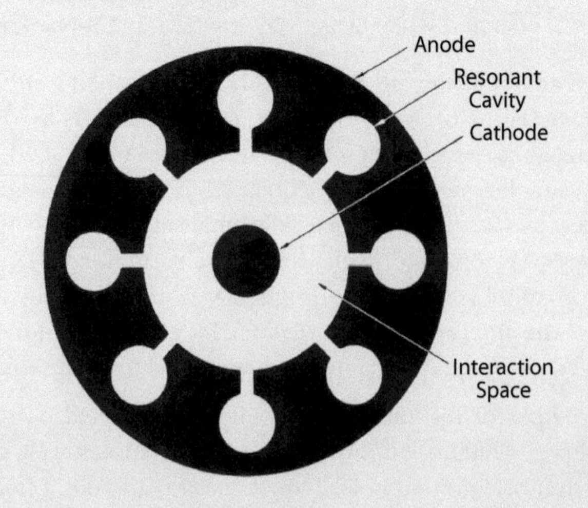

The Venerable Magnetron *continued*

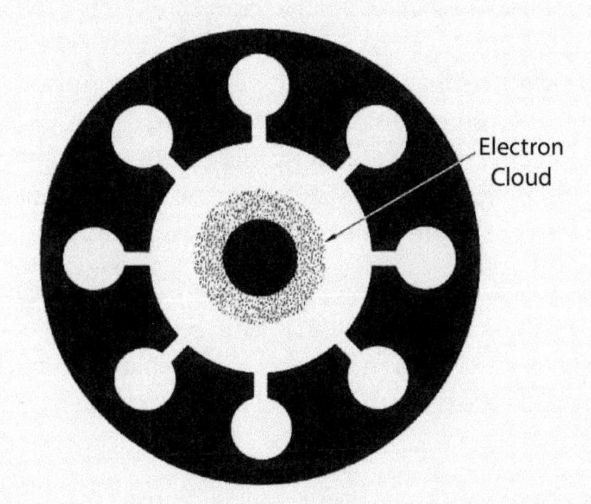

To cause the tube to generate radio waves, a strong DC voltage is applied between the electrodes (cathode negative, anode positive). Attracted by the positive voltage, the electrons accelerate toward the anode. But as the velocity of each electron increases, the magnetic field produces an increasingly strong force on the electrons, causing them to follow curved paths that carry them past the openings of the cavities.

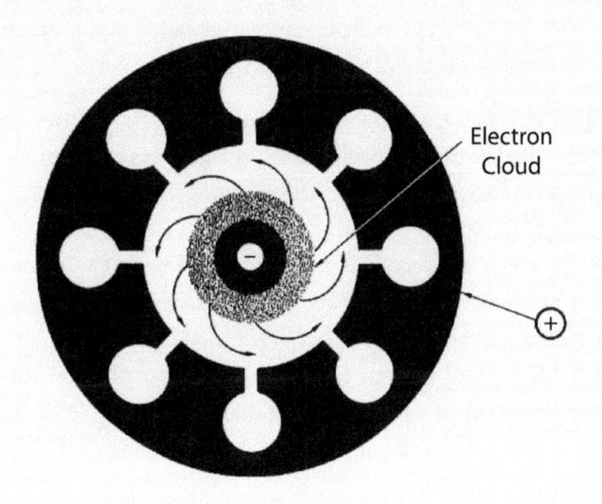

Much as a sound wave builds up in a bottle when you blow air across its opening, an oscillating electromagnetic field (radio wave) builds up as a result of the electrons sweeping past the cavity openings. As with the sound wave, the frequency of the radio wave is the resonant frequency of the cavities.

It all starts with a minute, random disturbance that initiates an electromagnetic oscillation in one of the cavities. This oscillation propagates from cavity to cavity via the interaction space. The electric field of this radio wave causes the electrons sweeping past the cavity openings during one peak of each cycle to slow down and move out toward the anode and those sweeping past during the other to speed up and move in toward the cathode. Consequently, the electrons quickly bunch up and form swirling spokes whose rotation is synchronized with the travel of the radio wave around the interaction space.

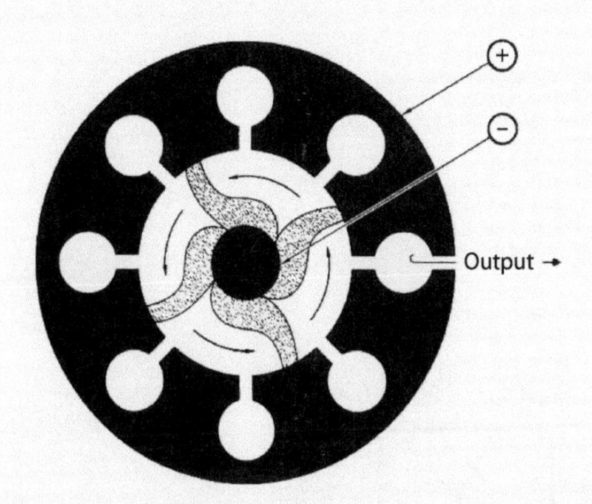

The electrons forming the spokes are gradually slowed down by their interaction with the traveling wave and in the process give up energy to the wave, thereby increasing its power. The slowing, of course, reduces the curvature of each electron's path, with the result that the electron soon reaches the anode. By the time it does, however, it has transferred to the radio wave up to 70 percent of the energy it acquired in being accelerated by the inter-electrode voltage. (What remains of the energy is absorbed as heat in the anode and must be carried away by a cooling system.) The spent electrons are returned to the cathode by the external power source, so the transfer of energy from the power source to the radio wave continues as long as the DC power is supplied.

Meanwhile, a tiny antenna inserted in one of the cavities bleeds the energy of the radio wave off into a waveguide. This is the output "port" of the tube. A magnetron's frequency may be varied over a limited range by changing the resonant frequency of the cavities through such techniques as lowering plungers into them.

Over the years a number of refinements have been made to the basic magnetron design. In one, a coaxial resonant output cavity is added.

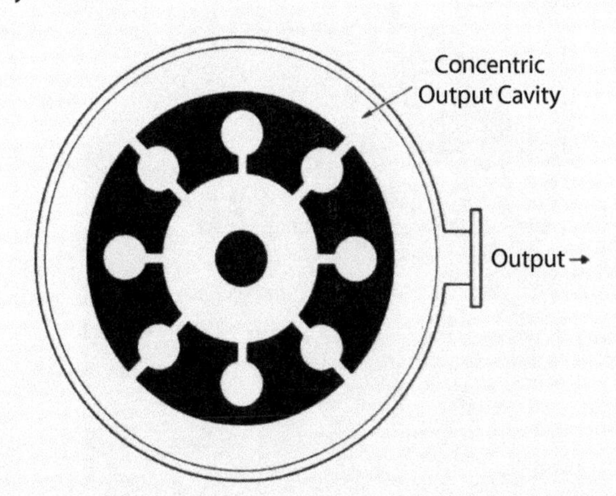

Energy is bled into it through slots in alternate cavities. The magnetron is tuned by changing the output cavity's resonant frequency.

Finally, the amplified signal is applied to an envelope detector that produces an output voltage corresponding to the peak amplitude (or envelope) of the signal. This signal is supplied to the indicator so that its magnitude can be portrayed on a screen and the presence or absence of a target can be determined by an operator. Alternatively, the output of the envelope detector can be further processed such that only automatically detected targets are presented to the radar operator.

Display Systems. These systems are used to (1) display the received echoes in a format that will satisfy the operator's requirements; (2) control the automatic searching and tracking functions; and (3) extract the desired target data when tracking a target.

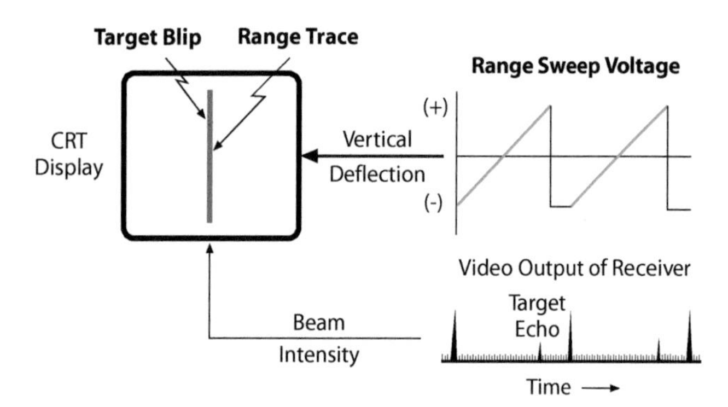

Figure 2-9. Triggered by timing pulses from synchronizer, a linear increase in the vertical deflection voltage produces a range sweep. The video output of the receiver intensifies the beam, producing a target blip. The strong video spikes are leakage of transmitted pulse through duplexer.

Any of a variety of display formats may be used, only one of which (the B display, or B-scope) will be described here (see the blue panel for details of the most common types of display).

In the B form of display, targets are displayed as bright spots, or *blips,* on a rectangular grid representing range and azimuth angle. In early systems a video amplifier raised the receiver output to a level suitable for controlling the intensity of the display tube's cathode ray beam. The operator generally sets the gain of the amplifier so that noise spikes make the beam barely visible (Fig. 2-9). Target echoes strong enough to be detected above the noise will then produce a blip.

To control the vertical and horizontal positions of the beam, each timing pulse from the synchronizer triggers the generation of a linearly increasing voltage that causes the beam to trace a vertical path from the bottom of the display to the top. Since the start of each trace is thus synchronized with the transmission of a radar pulse, if a target echo is received the distance from the start of the trace to the point at which the target blip appears will correspond to the round-trip transit time for the echo or target range. For this reason the trace is called the *range trace* and the vertical motion of the beam is the *range sweep.*

Figure 2-10. In the cockpit of a Typhoon aircraft, the combining glass for the head-up display is in the center of the windscreen.

Meanwhile, the azimuth signal from the antenna is used to control the horizontal position of the range trace, and the elevation signal may be used to control the vertical position of a marker on the edge of the display, where an elevation scale is provided.

As the antenna executes its search scan, the range trace sweeps back and forth across the display in unison with the azimuth scan of the antenna. Each time the antenna beam sweeps across a target, a blip appears on the range trace, providing the operator with a plot of the range versus the azimuth of the target. The typical location of the displays in a cockpit is shown in Figure 2-10.

Modern systems use a digital display, and targets are painted onto the display by deriving the positional data from the synchronizer and antenna. Processing is applied to the

Common Radar Displays

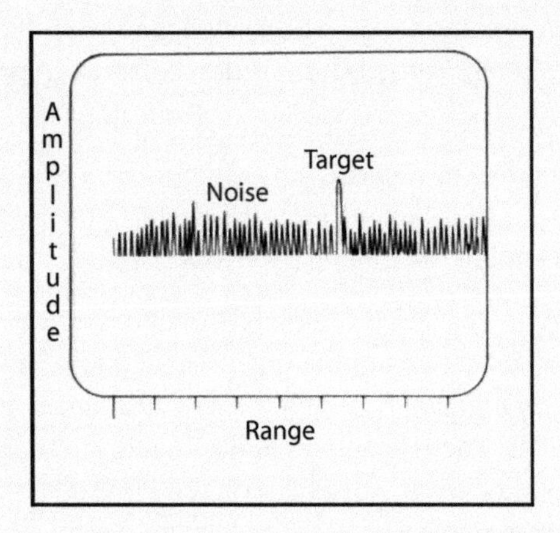

Display. Plots amplitude of receiver output versus range on horizontal line, called a range trace. Simplest of all displays but rarely used because it does not indicate azimuth.

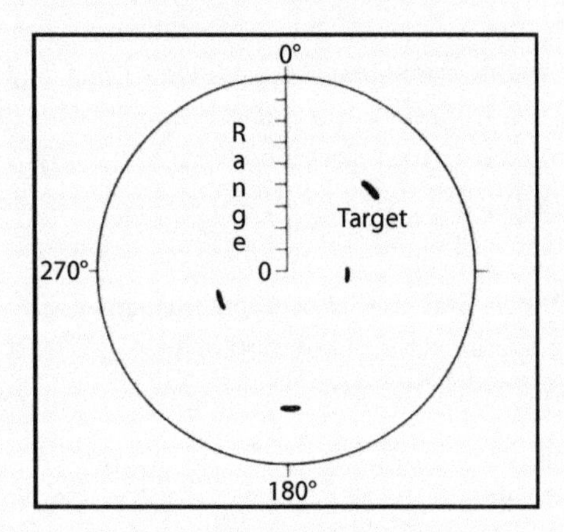

PPI Display. Targets displayed in polar plot centered on radar's position. Ideal for radars that provide 360° azimuth coverage.

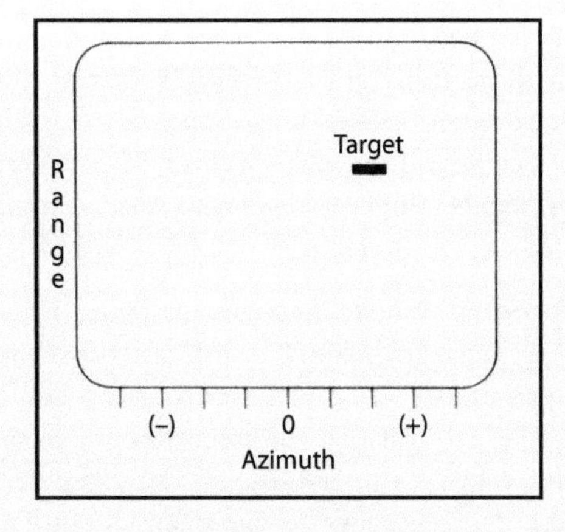

Display. Targets displayed as blips on a rectangular plot of range versus azimuth. Widely used in fighter applications, where horizontal distortion near zero range is of little concern.

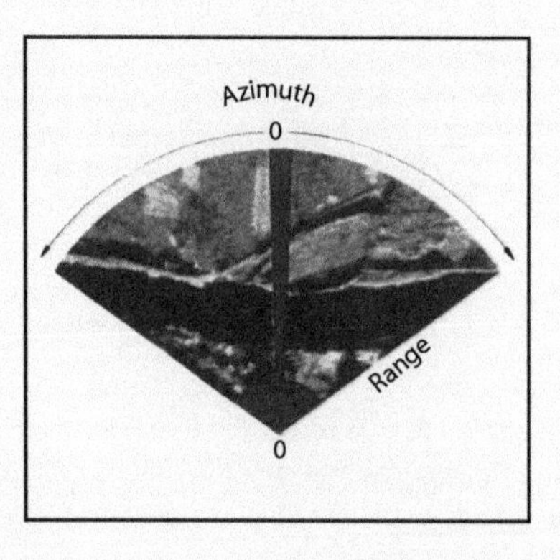

Sector PPI Display. Gives an undistorted picture of region being scanned in azimuth. Commonly used for sector ground mapping.

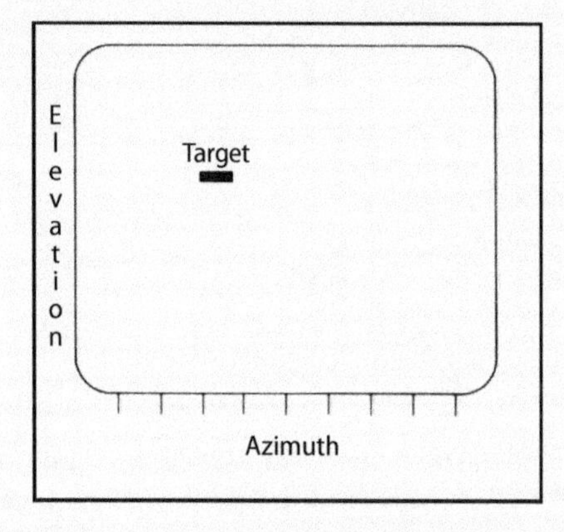

Display. Shows target position on plot of elevation angle versus azimuth. Useful in pursuit attacks since display corresponds to pilot's view through windshield. Commonly projected on windshield as "Head-Up Display."

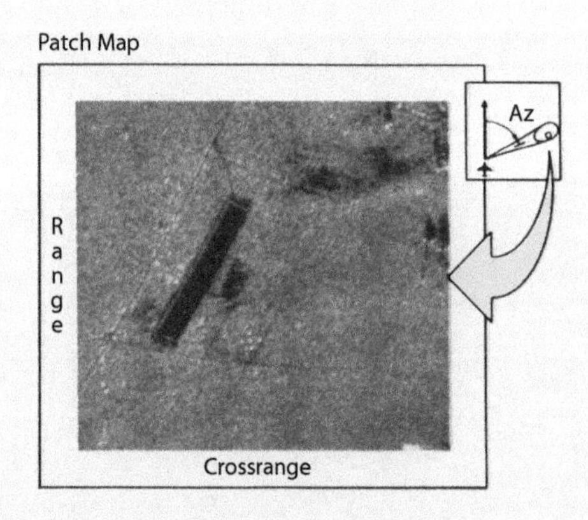

Imaging. In high-resolution *synthetic aperture radar* (SAR) ground mapping, a rectangular patch map may be displayed. This is a detailed image of a specific area of interest at a given range and azimuth angle. The range dimension of the image is displayed vertically, and the cross-range dimension (i.e., dimension normal to the line of sight to the patch) appears horizontally. Alternatively, a continuous scrolling image can be generated as the radar flies its trajectory, allowing surveillance of a much broader area.

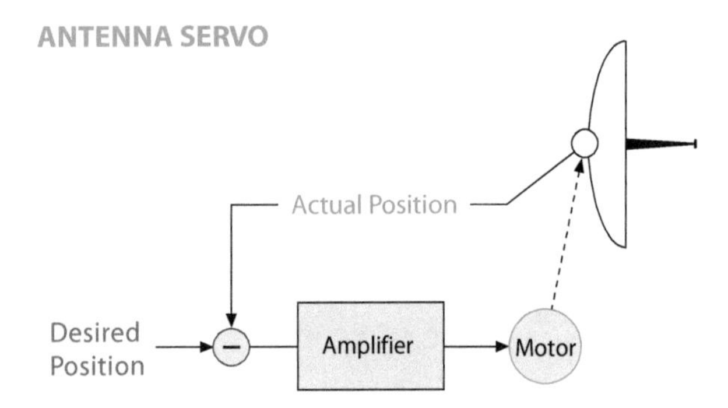

Figure 2-11. An antenna servo compares actual position of the antenna with the desired position, amplifies the resulting error signal, and uses it to drive the antenna in the direction to reduce error to zero.

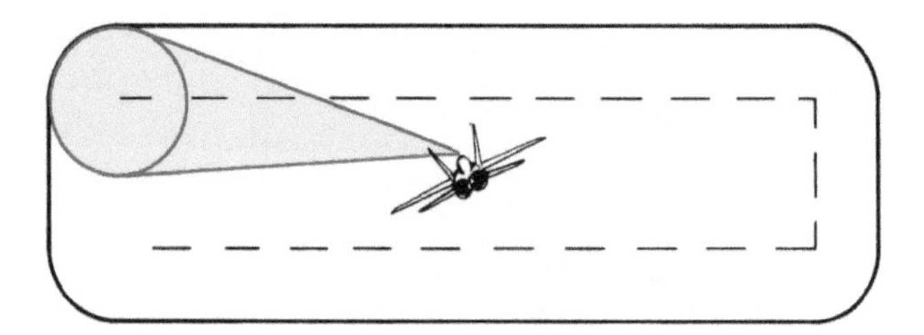

Figure 2-12. The antenna's search scan is stabilized in pitch and roll so that the region searched will be unaffected by changes in aircraft attitude.

Figure 2-13. With a pistol grip hand control, the operator gains control of the antenna by pressing a trigger. Fore and aft motion controls the position of the range maker. Right and left motion controls antenna azimuth. A tilt switch controls elevation.

envelope-detected signals, which identify targets and reject noise and clutter automatically. Digital displays also allow additional information such as speed and heading to be easily incorporated and routinely use color to improve the readability.

Of the other forms of display, the *plan position indicator* (PPI) is by far the most common, but all are able to exploit digital displays and exploit the inherently greater flexibility this offers.

Antenna Servo. This unit positions the antenna in response to control signals generated by angle tracking systems such as *lobing* and *monopulse*. A separate servo channel is provided for each gimbal. Their operation is illustrated in Fig. 2-11. The voltage obtained from a transducer on the gimbal is subtracted from the control signal, thereby producing an error signal proportional to the error in the antenna's position. This signal is then amplified and applied to a motor that rotates the antenna about the gimbal axis to reduce the error to zero.

So that the search scan, which is usually much wider in azimuth than in elevation, will be unaffected by the attitude of the aircraft, stabilization may be provided (Fig. 2-12). If the antenna has a roll gimbal, the roll position of the antenna is compared with a reference provided by a vertical gyro and the resulting error signal is used to correct the roll position of the antenna.

Otherwise, the azimuth and elevation error signals are resolved into horizontal and vertical components on the basis of the reference provided by the gyro.

Power Supply. In airborne radar, power is typically supplied as a 115 volt, 400 Hz primary power source and has to be converted to the various DC forms required by the radar. This is the job of the power supply. It first transforms the 400 Hz power to the standard voltages required and then converts them to DC and smooths them. When necessary, the power supply "regulates" the DC voltages so they remain constant, regardless of changes in current or voltage drawn by the system. Though these tasks are superficially mundane, elegant techniques have been devised to accomplish them at a minimum cost in weight and dissipated power. The antenna servo is generally operated directly off the 400 Hz supply and the relays operated off the aircraft's 28 volt DC supply.

Automatic Tracking. Where automatic tracking is required, three additions must be made to the system to provide the necessary feed signals to the antenna servo. First, some means must be provided for isolating the target echoes in time (i.e., range). Second, lobing or monopulse, as described in the preceding chapter, must be added to the antenna. Third, controls must be provided with which the operator can *lock* the radar onto the target's echoes.

For lock-on, the operator uses a handheld control such as pistol grip (Fig. 2-13) to position a marker at any desired point on the range trace. Depressing a switch or button tells the system that the operator has aligned the marker with the target he wishes to track.

To lock onto a target, the operator takes control of the antenna with the hand control, aligns the antenna in azimuth to center the range trace on the target blip, adjusts the elevation of the antenna to maximize the brightness of the blip, runs the marker up the trace until it is just under the blip, and presses the lock-on button.

Automatic Range Tracking

To CONTROL THE TIMING OF A RANGE GATE SO IT AUTOMATICALLY FOLLOWS (tracks) the changes in a target's range, a range-tracking servo is employed.

Target Return
Tracking Gate
Early Gate
Late Gate
Time →

Typically, it samples the returns passed by the tracking gate with two secondary gates—the early and late gates—each of which remains open only half as long as the tracking gate. The early gate opens when the tracking gate opens and samples the return passed by the first half of the tracking gate. The late gate opens when the early gate closes and thus samples the returns passed by the second half of the tracking gate.

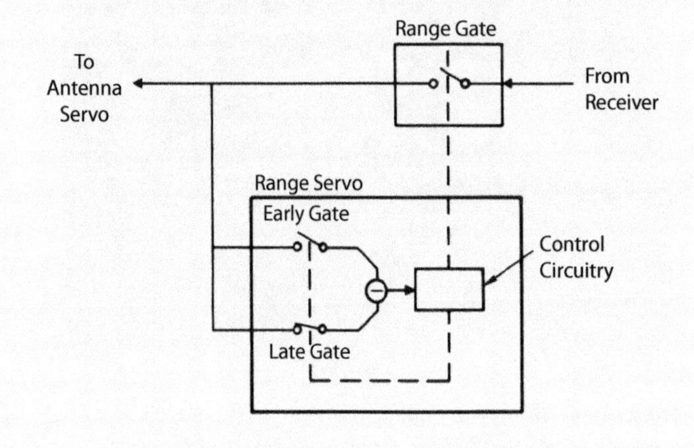

The range servo continuously adjusts the timing of the tracking gate to equalize the outputs of the early and late gates, thereby keeping the tracking gate centered on the target. In this way the operation of the early–late-gate technique is very similar in concept to that of monopulse except that it operates in the range domain rather than the angle domain.

After processing to smooth erratic disturbances, signals proportional to the azimuth, elevation, and range tracking errors are supplied to the antenna servo. Subsequently, the antenna is adjusted to maintain the peak of the beam on the target being tracked (Fig. 2-14).

Where extremely precise tracking is desired, rate-integrating gyros may be mounted on the antenna. They establish azimuth and elevation axes to which the antenna servo is slaved, thereby holding the antenna solidly in the same position, regardless of disturbances due to the aircraft's maneuvers. This feature is called *space stabilization*.

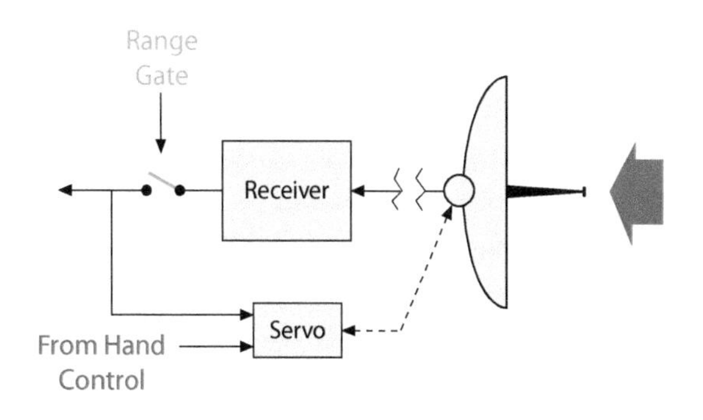

Figure 2-14. For automatically tracking a target, echoes are isolated by closing an electronic switch (called the range gate) at the exact time each echo will be received.

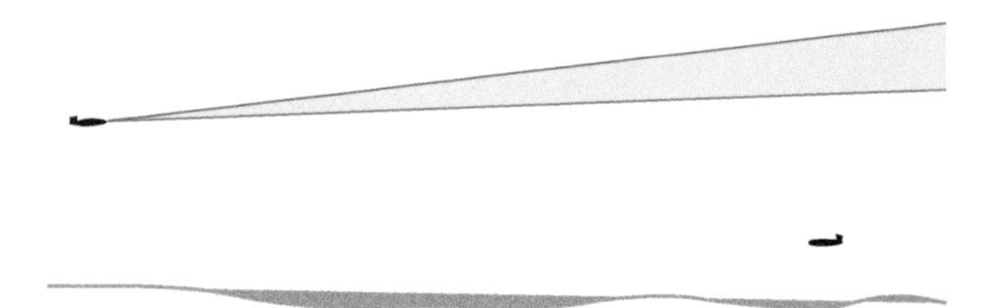

Figure 2-15. In early radars, ground clutter was avoided by keeping the radar beam from striking the ground, but this limited the radar's tactical capability.

Figure 2-16. The pulse Doppler radar on this F-15 is the mechanically steered APG-63. (Courtesy of Raytheon Company.)

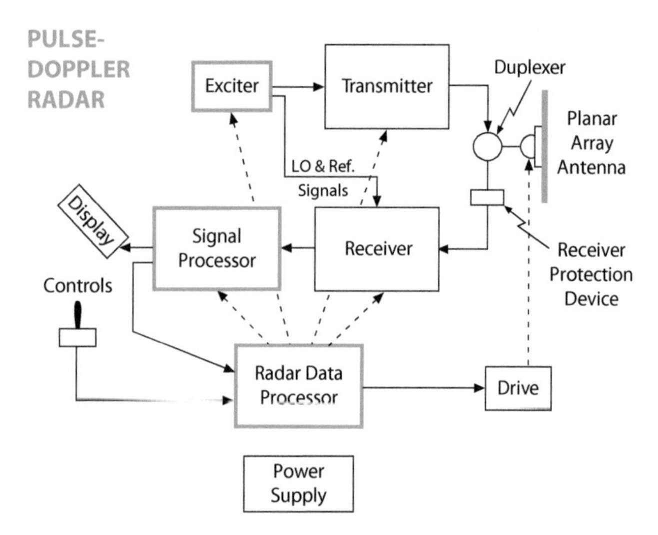

Figure 2-17. This shows the principal elements of a pulse-Doppler radar. Boxes with blue borders were introduced with this generic system. Data processor controls all elements, verifies their operation, and isolates faults.

The tracking error signals are smoothed and have corrections added to them to anticipate the effect of the aircraft's acceleration with respect to the target's relative position. They are then applied to torque motors, which precess the gyros, thereby changing the directions of the reference axes to reduce the tracking errors to zero.

The principal shortcoming of the simple noncoherent pulsed radar is that it cannot easily differentiate between targets and clutter. In early radars (Fig. 2-15), clutter was avoided by keeping the radar beam from striking the ground. As can be imagined, this seriously limits the radar's ability. To overcome these limitations, coherent forms of radar were developed to enable retention of echo signal phase and amplitude.

2.2 Coherent Pulse-Doppler Radar

This form of radar (Fig. 2-16) can be identical in size or even smaller than its noncoherent counterpart yet provides a radical improvement in performance. It can detect small targets at long ranges (even when their echoes are buried in strong ground clutter) by exploiting phase to compute Doppler velocities. If the Doppler velocity of a moving target is substantially different from that of stationary clutter, the two may easily be differentiated.

Coherent pulse-Doppler radar can track targets either singly or several at a time while continuing to search for more. It can detect and track moving targets on the ground or in the air, and it can make real-time high-resolution SAR ground images where the resolution at long range is the same as at short range. A simplified functional diagram of a pulse-Doppler radar is shown in Figure 2-17.

Coherence is achieved by providing a LO signal that synchronizes the transmitter and the receiver such that the phase of the transmitted signal is always the same and therefore the relative phase between the transmitted signal and the received echo is due only to movement of the target. In conjunction with phase coherent radars there has been and continues to be rapid improvements in digital signal processing. Together these permit received echoes to be processed into detections and tracks in a manner that has much greater automation, allowing the operator to concentrate more on ensuring mission success. Comparing the diagram of the noncoherent pulsed radar (Fig. 2-2) with Figure 2-17 reveals the following obvious differences in the latter:

- Addition of a computer called the radar data processor
- Addition of a unit called the exciter
- Elimination of the synchronizer (its function is absorbed partly by the exciter but mostly by the data processor)
- Elimination of the modulator (its task is reduced to the point where it can be performed in the transmitter)

- Addition of a digital signal processor
- Elimination of the indicator (its functions are absorbed partly by the signal processor and partly by the data processor)

These elements, although typical of the class of radar, by no means represent the only form of configuration. In fact, digital technology can enable direct synthesis of waveforms for transmission and direct digitization at radio frequency. This provides an ever-increasing flexibility and range of implementable designs. The aforementioned elements, together with some important differences in the transmitter, antenna, and receiver, are briefly described in the following paragraphs to show how they work together to enable coherent pulse-Doppler radar.

The Exciter. This element generates a continuous, highly stable, low-power signal of the desired frequency[1] and phase for the transmitter, the LO signals, and a reference-frequency signal for the receiver.

1. The frequency is selectable over a fairly wide range by the operator.

The Remarkable Gridded Traveling Wave Tube

THE TRAVELING WAVE TUBE (TWT) AMPLIFIER IS ONE OF THE KEY developments of the 1960s that made it possible to have truly versatile multimode radar systems. With the TWT, for the first time both the width and repetition frequency of radar's high-power transmitted pulses could be controlled precisely. Further, they could also be readily changed, almost instantaneously, to a wide variety of values within the power handling capacity of the tube. Added to these capabilities were those of the basic TWT: the high degree of coherence required for Doppler operation; versatile, precise control of radio frequency; and the ability to conveniently code the pulse's radio frequency or phase for pulse compression.

The Basic TWT. The TWT is one of a family of "linear beam" vacuum tube amplifiers (including the klystron) that convert the kinetic energy of an electron beam into microwave energy. In simplest form, a TWT consists of four elements:

- An electron gun, which produces the high-energy electron beam
- A helix, which guides the signal to be amplified
- A collector, which absorbs the unspent energy of the electrons that are returned to the gun by a DC power supply
- An electromagnet (solenoid), which keeps the beam from spreading because of the repulsive forces between electrons. (Alternatively, a chain of periodic permanent magnets (PPMs) is often used, called thus because polarities of adjacent magnets are reversed.)

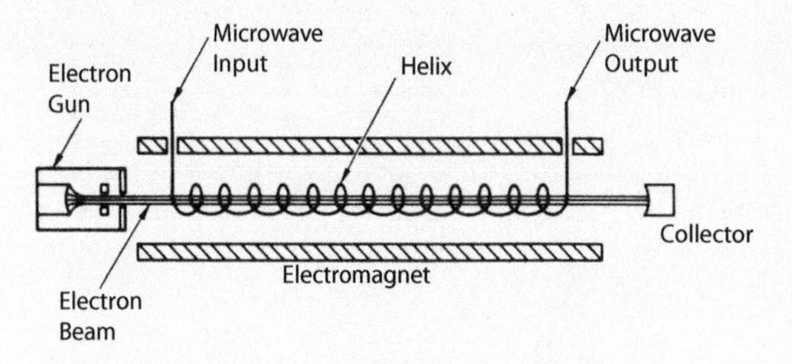

The microwave input signal is introduced at one end of the helix. Although the speed of the signal is essentially that of light, because of the greater distance the signal must cover in spiraling down the helix its linear speed is slowed to the point where it travels slightly slower than the electrons in the beam. For this reason, the helix is called a slow-wave structure.

As the signal progresses, it forms a sinusoidal electric field that travels down the axis of the beam. This field speeds up the electrons that happen to be in positive nodes and slows down those in the negative nodes. The electrons therefore tend to form bunches around the electrons at the nulls whose speed is unchanged.

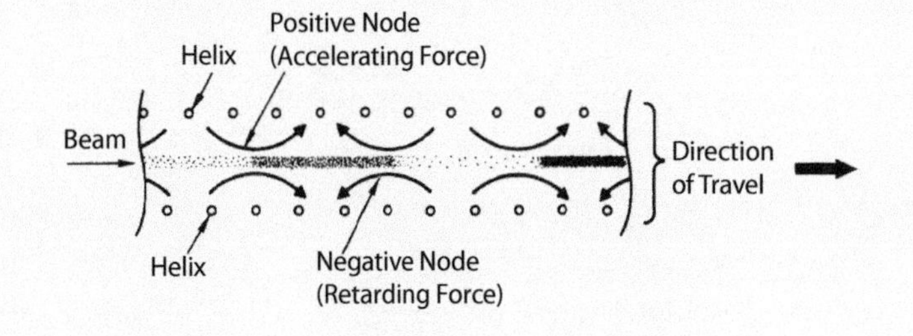

The Remarkable Gridded Traveling Wave Tube *continued*

The traveling bunches in turn produce a strong electromagnetic field. Since it travels slightly faster than the signal, this field transfers energy from the electrons to the signal, thereby amplifying it and slowing the electrons. The longer the helix, the more the signal is amplified. In high gain tubes, attenuators (severs) must be placed at intervals (of 20 to 35 dB gain) along the helix to absorb backward reflections that would cause self-oscillation. They reduce the gain somewhat (about 6 dB each) but have only a small effect on efficiency.

When the signal reaches the end of the helix, it is transferred to a waveguide that is the output port of the tube. The remaining kinetic energy of the electrons (which may amount to as much as 90 percent of the energy originally imparted by the gun) is absorbed as heat in the collector and must be carried away by cooling. Much of the unspent energy, though, can be recovered by making the collector negative enough to decelerate the electrons before they strike it. Kinetic energy is thus converted to potential energy.

High-Power TWTs. Both the average and the peak power of helix TWTs are somewhat limited. As the average power is increased, an increasing number of electrons are intercepted by the helix, and it becomes difficult to remove enough of the resulting heat to avoid damage to the helix. As the required peak power is increased, the beam velocity must be increased, and a point is soon reached where the helix must be made too coarse to provide good interaction with the beam. In high-power tubes, therefore, other slow-wave structures are generally used. The most popular is a series of coupled cavities.

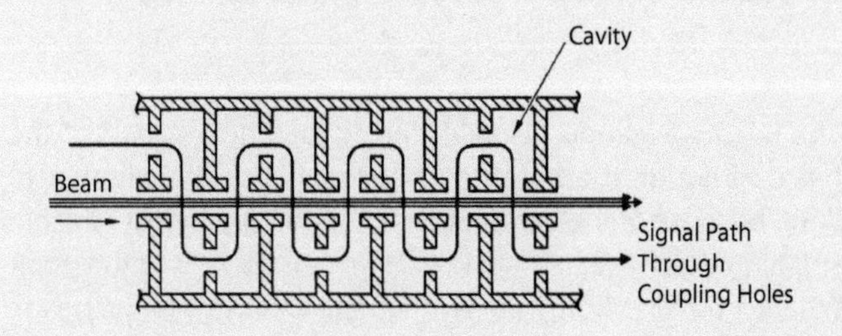

Control Grid. While a pulsed output can be obtained by turning the tube on and off, the pulses can be formed much more conveniently by interposing a grid between the cathode that emits the electrons and the anode whose positive voltage relative to the cathode accelerates them. A low-voltage control signal applied to this grid can turn the beam on and off. To keep the grid from intercepting electrons and being damaged, it is placed in the shadow of a second grid that is electrically tied to the cathode. To eliminate all output between pulses, the low-voltage microwave input signal may be pulsed.

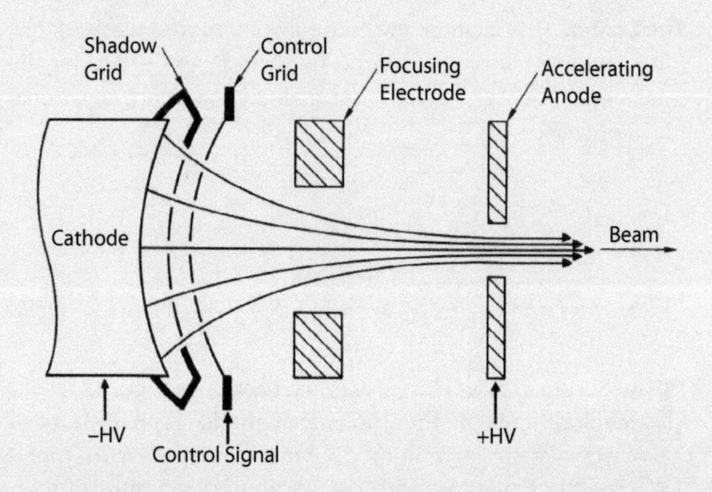

Advantages. Besides the advantages listed earlier, the TWT can provide high-power outputs with gains of up to 10,000,000 or more and efficiencies of up to 50 percent. Low-power helix tubes have the added advantage of providing bandwidths of as much as two octaves (maximum frequency four times the minimum). In high-power tubes, though, where other slow-wave structures must be used, this is generally reduced to between 5 and 20 percent, although some coupled cavity tubes having much greater bandwidths have been built.

Transmitter. The transmitter is typically a high-power TWT amplifier (Fig. 2-18). The TWT is switched on and off to cut coherent pulses from the exciter's signal and then amplifies the pulses to the desired power level for transmission. As explained in the panel above, the tube is turned on and off by a low-power signal applied to a control grid.

By appropriately modifying this signal, the width and repetition frequency of the high-power transmitted pulses can easily be changed. Similarly, by modifying the exciter's low-power signal, the frequency, phase, and power level of the high-power pulses can also be readily be changed or

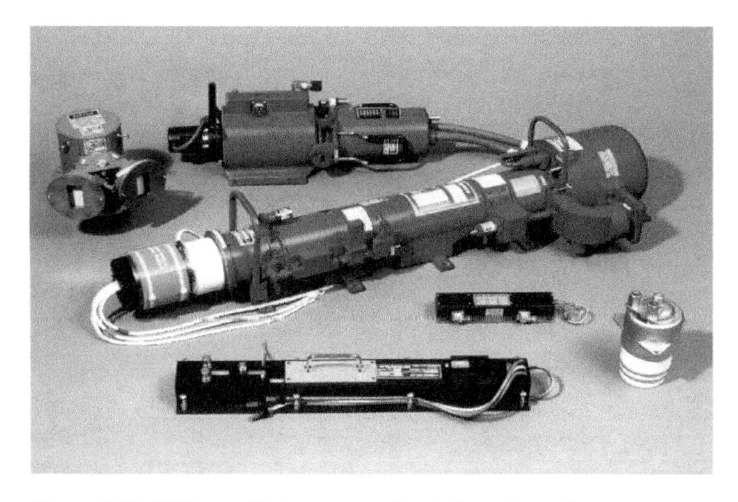

Figure 2-18. TWTs amplify low-power signals from the exciter to the levels required for transmission and can readily be turned on and off with low-power control signal. (Courtesy of DoD.)

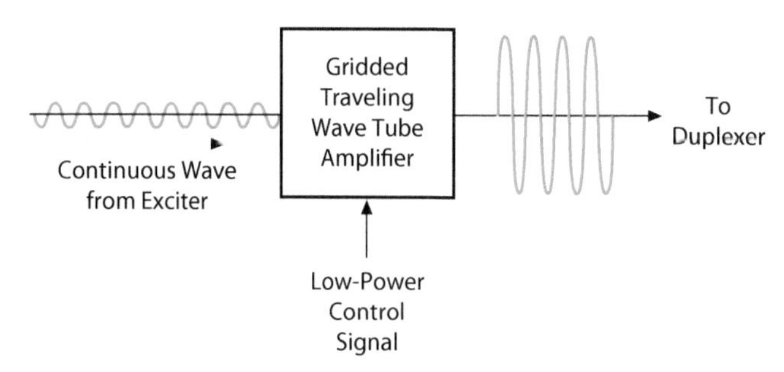

Figure 2-19. By switching the TWT with a low-power control signal, the width and pulse repetition frequency of the high-power pulses can readily be changed. By modifying the low-power input provided by the exciter, the frequency, phase, and power of the transmission waveforms can readily be changed or modulated.

modulated for pulse compression (Fig. 2-19) and diversity of waveform.

Antenna. The antenna could again be a parabolic dish or of a type called a *planar array*, which, instead of employing a central feed that radiates the transmitted wave into a reflector, consists of an array of many individual radiators distributed over a flat surface (Fig. 2-20). The radiators are slots cut in the walls of a complex of waveguides behind the antenna's face.

Though a planar array is more expensive than a dish antenna, its feed can be designed to distribute the radiated power across the array to minimize the radiated sidelobes, which provides enhanced moving target indication performance and clutter rejection. The feed can readily be adapted to enable monopulse measurement of angle-tracking errors.

Receiver. In this example the receiver (Fig. 2-21) differs in a number of respects from earlier descriptions.

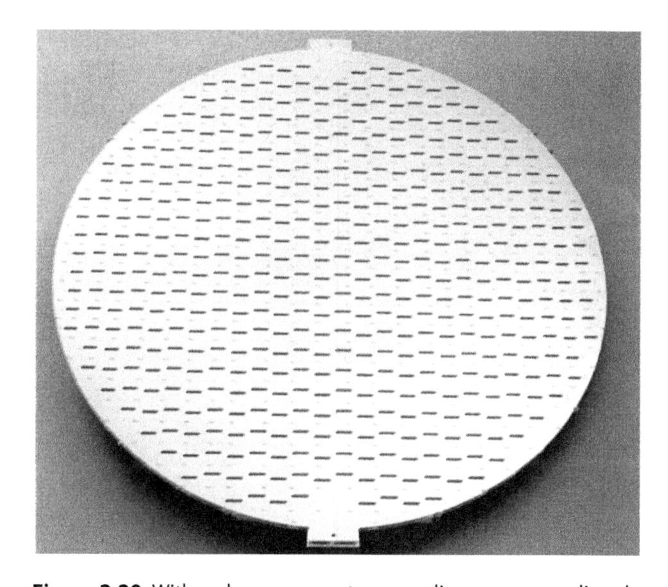

Figure 2-20. With a planar array antenna, radio waves are radiated though slots cut in a complex of waveguides behind the face of the antenna.

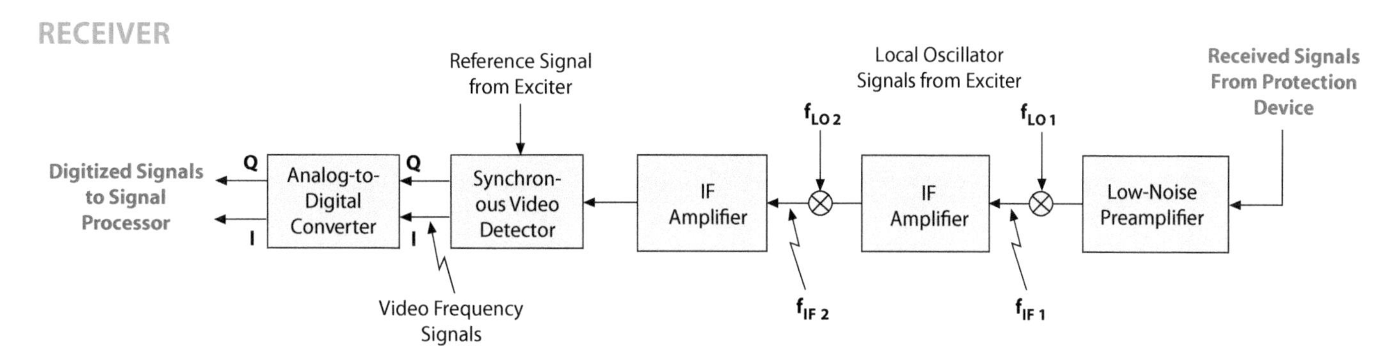

Figure 2-21. In the receiver of a generic coherent pulse-Doppler radar system, to enable digital Doppler filtering, in-phase (I) and quadrature (Q) signals are generated either using video outputs or by direct digital demodulation in a Hilbert transform. To enable monopulse tracking, two receiver channels each with I and Q must be provided.

First, a low-noise amplifier (LNA) ahead of the mixer increases the power of the incoming echoes so that they can better compete with the electrical noise inherently generated in the mixer. The LNA dominates the system noise level and consequently is the ultimate limit of sensitivity of the radar system.

Second, more than one intermediate frequency translation is generally performed to avoid problems with image frequencies (see Chapter 5). This usually results in the signals being "taken down to baseband" (i.e., an IF of zero). However, alternatively in some modern systems, just one IF with digitization may occur at this frequency. Demodulation then follows, digitally, in the signal processor. This not only provides the flexibility to eliminate image frequencies but also allows small errors such as unwanted transmitter modulations to be compensated.

Thus, I and Q samples can be generated at baseband via a 3 dB splitter and a 90° phase retardation (or advancement) of one of the channels, followed by digitization. Alternatively, the output at IF or baseband can be digitized and demodulation accomplished in the digital domain using a Hilbert transform. The vector sum of the I and Q samples is proportional to the energy of the sampled signal, and the inverse tangent of the ratio of Q to I provides the phase measurement.

Signal Processor. This processor is a digital computer specifically designed to efficiently perform the vast number of repetitive additions, subtractions, and multiplications required for real-time signal processing. The data processor loads the program into the signal processor for the currently selected mode of operation.

As required by this program, the signal processor sorts the digitized echo data by time of arrival and hence range (Fig. 2-22), stores the numbers for each range interval in memory locations called *range bins*, and filters out the bulk of the unwanted ground clutter on the basis of its Doppler frequency. By forming a bank of narrowband filters for each range bin via a Fourier transform, the processor then integrates the energy of successive echoes having the same Doppler frequency. Thus, if the target has a significantly different radial velocity from the

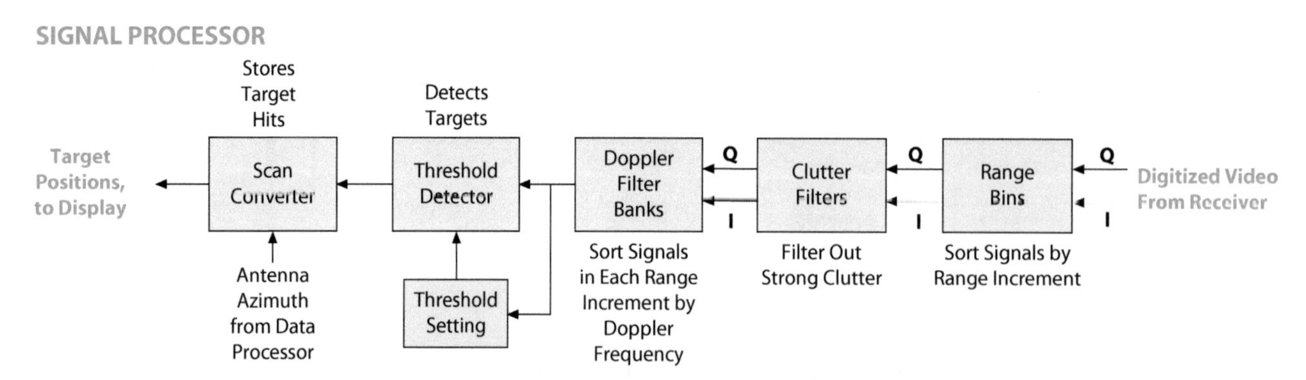

Figure 2-22. The signal processor sorts radar returns by range, storing them in range bins, filters out the clutter, and then sorts the returns in each range bin by Doppler frequency. Targets are detected automatically.

clutter, it may be much more easily detected, often competing only with lower power noise.

By examining the outputs of all the filters, the processor determines the level of the background noise and residual clutter, just as a human operator would do by observing the range trace on an A display. On the basis of increases in amplitude above this level, it automatically detects the target echoes.

Rather than supplying the echoes directly to the display, the processor temporarily stores the targets' positions in its memory. Meanwhile, it continuously scans the memory at a rapid rate and provides the operator with a continuous bright TV-like display of the positions of all targets (Fig. 2-23). This feature, called *digital scan conversion*, gets around the problem of target blips fading from the display during the comparatively long azimuth scan time. Synthetic blips of uniform brightness on a clear background indicate the target positions, and this makes them extremely easy to see. This is a minimal capability, and the flexibility of digital processing and displays makes it possible for much more advanced modes of operation, as will be seen later in this book.

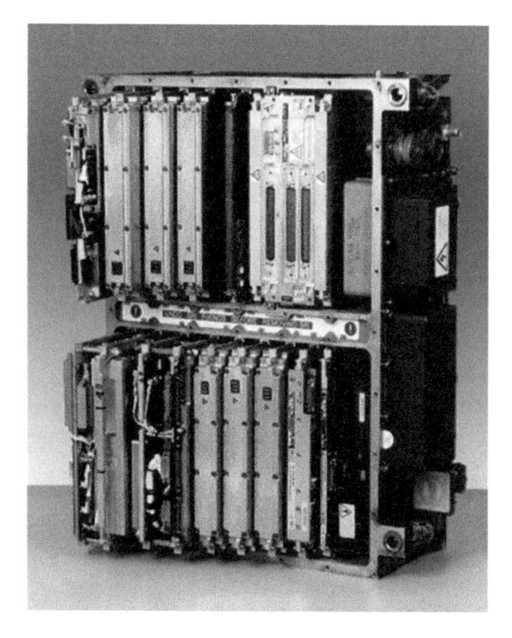

Figure 2-23. In the signal processor, a stored program for the selected mode of operation is automatically entered by the data processor. (Courtesy of Selex ES.)

For example, airborne SAR provides fine-resolution 2D imaging, in range and azimuth. Transmitting wide bandwidth pulses and employing large amounts of pulse compression achieves simultaneously long detection ranges and high range resolutions. To provide fine azimuth resolution, echoes from many pulses in the form of multiple range bins are stored in the processor as the aircraft flies through the sky. These are then integrated for each range bin across successive pulses to form the synthetic aperture and hence the high-resolution image.

Data Processor. This is a general-purpose digital computer; the data processor controls and performs routine computations for all units of the radar. It schedules and carries out the selection of operating modes, such as long-range search, track-while-scan, and SAR mapping. Receiving inputs from the aircraft's inertial navigation system, it stabilizes and controls the antenna during search and track. On the basis of inputs from the signal processor, it controls target acquisition, making it necessary for the operator to only bracket the target to be tracked, with a symbol on the display.

During automatic tracking, the data processor computes the tracking error signals in such a way as to anticipate the effects of all measurable and predictable variables (e.g., the velocity and acceleration of the radar, the bearing of the aircraft, the limits within which the target can reasonably be expected to change its velocity, the signal-to-noise ratio). This process yields extraordinarily smooth and accurate tracking.

Throughout, the data processor monitors all operations of the radar, including its own. In the event of a malfunction, it alerts the operator to the problem and, through built-in tests, isolates the failure to an assembly that can readily be replaced on the flight line.

Monolithic Microwave Integrated Circuits

IN THE 1980S AND EARLY 1990S, DEFENSE ADVANCED RESEARCH PROJECTS Agency (DARPA) Microwave and Millimeter Wave Monolithic Integrated Circuits program made advancements in monolithic microwave integrated circuits (MMICs). These developments made it possible to take full advantage of active electronically scanned array (AESA) radar architectures. MMICs assembled in transmit/ receive modules that are placed at the very front end of the radar result in extremely low loss both in transmit and receive. Modern airborne radars use thousands of these transmit/receive modules in a spatially combined transmitter, and virtually all of them use MMICs.

MMICs are solid-state circuits made with semiconductor wafer processing procedures. They leverage semiconductor processing tool development from the silicon integrated circuit industry; however, for higher frequency circuits, materials other than silicon are used.

This type of circuit is extremely reliable and is predicted to last on the order of a million hours—greater than 100 years. An AESA that uses thousands of MMIC-based modules leads to graceful degradation in performance if some modules fail early. MMICs are smaller and weigh less compared with tube-based transmitters and are therefore highly suitable for use in airborne AESAs. They operate at tens of volts, whereas tube-based amplifiers use thousands of volts.

Gallium Arsenide—A Blessing and a Curse. Gallium arsenide (GaAs) is a compound semiconductor material made from elemental gallium and arsenic. Its electron transport properties are superior to silicon and thus make it well suited for the high-frequency transistors required for MMICs used in X-band and higher-frequency radars. GaAs has a higher saturated electron velocity and higher electron mobility (a measure of how fast electrons can move in a material), which allows transistors that use it to operate at frequencies above 300 GHz. GaAs devices have less internally generated noise, at high frequencies, than silicon devices as a result of higher carrier mobilities and lower device parasitics.

However, the high-frequency performance advantages of GaAs don't come without some drawbacks. The compound is hard to make and is brittle, meaning it breaks easily compared with silicon. GaAs is not abundant in nature like silicon, and large-diameter wafers that are stoichiometrically balanced are expensive. Unlike silicon, GaAs has poor thermal conductivity, so any heat produced in the transistors formed on the GaAs surface must dissipate via conduction through the substrate. In addition, in contrast to silicon, GaAs has no native oxide, so it is difficult to produce mainstream metal oxide semiconductors (MOS) transistors. As a result, metal semiconductor field-effect transistors (MESFETs) are used for GaAs MMIC circuits.

Metal Semiconductor Field-Effect Transistors

The bipolar junction transistor was invented by researchers at Bell Labs in 1949 after attempts to make MESFETs were unsuccessful due to poor material quality. Later on, with improvements in materials, MOS transistors in silicon became available. These structures were not well suited for GaAs devices since there is no native oxide for GaAs.

The first GaAs transistors were made by placing a metal gate directly on the semiconductor and relying on the resultant Schottky barrier junction to control current flow using a gate voltage. Low-resistance ohmic contacts are required to make the drain and source connections. These early transistors were isolated from one another using an etching process that produced mesas where the active transistors were formed. When the source is grounded and the drain is connected to a positive voltage, large numbers of electrons will flow in the active region until a negative gate voltage with respect to the source is applied. At a sufficiently high negative gate voltage, a depletion region will spread across the active region and effectively "pinch off" the channel and stop the electron flow. For well-designed transistors, a small change in gate voltage results in a large change in current flow. This parameter is called the transistor's *transconductance* and is the basis by which amplification is achieved.

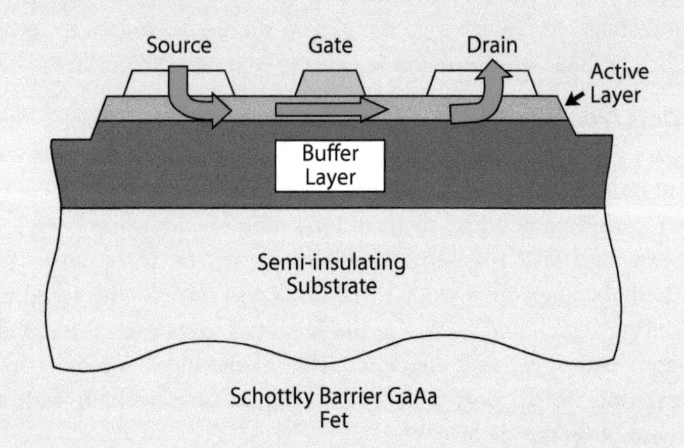

Schottky Barrier GaAa Fet

Pseudomorphic High-Electron-Mobility Transistors. In the basic MESFET structure, efforts to increase transconductance by adding more free electrons to the active region resulted in a decrease in the mobility of those electrons. In the 1980s, advanced growth techniques permitted the inclusion of precise layers of different materials such as InGaAs and AlGaAs. Adding these different materials, called *pseudo-layers* or *band-gap-engineered layers*, resulted in strained layer structures that exhibited enhanced performance. These layers allowed for the introduction of a planar-doped region of silicon atoms and for the presence of free electrons, which are preferred from an

Monolithic Microwave Integrated Circuits *continued*

energy perspective, in the indium gallium arsenide (InGaAs) channel region.

Because of these new layers, a large number of electrons could be added to the materials system. In addition, since they were separated from the parent atoms, they could move at high velocities in the undoped channel region. A high number of electrons could also move at high velocity in a narrow and easily controlled channel. These material enhancements, along with very narrow gate length geometries, gate recess structures, and thick plating, have served as the workhorse transistor technology for MMICs since the mid-1990s. However, there is now a new kid in town!

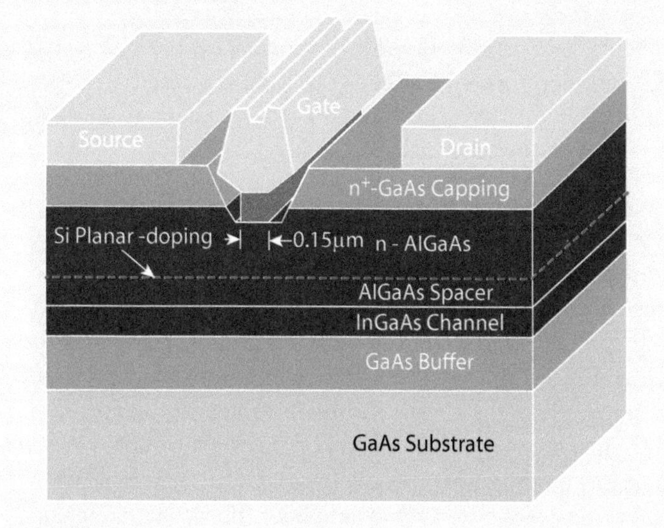

Semiconductor technologies that are available for an AESA application depend primarily on the frequency of operation. At ultra high frequencies, silicon bipolar and silicon laterally diffused metal oxide semiconductors (LDMOSs) are available and are cost-effective. However, in continuous wave or long-pulse-width applications, these technologies may be challenged thermally and at higher frequencies are not suitable. Ground-based, airborne, and shipboard AESAs being developed today are L-band, S-band, or X-band. Technologies appropriate for these frequency bands are silicon carbide (SiC), GaAs, gallium nitride (GaN), and indium phosphide (InP). Silicon germanium (SiGe) operates in these bands but is a low-power technology (<1 W at X-band).

GaN is not without problems. Currently there are no large-diameter bulk GaN wafers. As a result, GaN layers must be grown (deposited) on another substrate. Suitable substrates are sapphire and SiC. For MMICs, SiC substrates are preferred due to their high thermal conductivities. Fortuitously, this material combination provides revolutionary increases in power density for microwave devices. At a given frequency, GaN has at least a factor of five higher power density. This means that GaN-based MMICs can be one-third to one-fourth the size as a similar-power GaAs MMIC or that, for a given area, a GaN-based MMIC can deliver a factor of three or four higher power. As a result of the increase in power per unit area, all advanced AESA development programs are considering GaN in their designs.

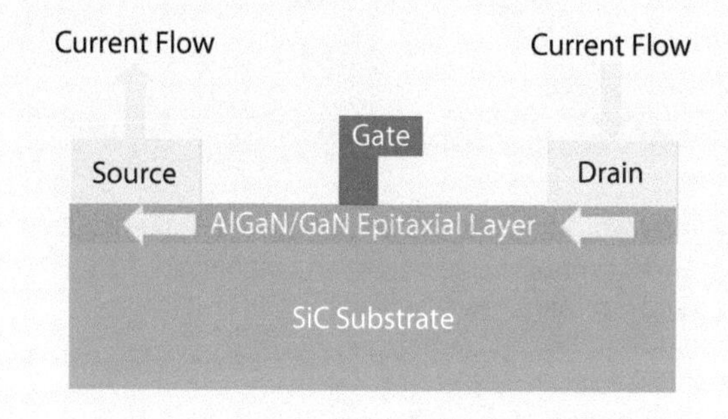

As computing technology continues to advance, the distinction between the signal and data processors is becoming blurred with all functions being carried out within the radar systems computer.

2.3 Exploiting Coherency

Electronically Scanned Radar Systems. Electronically scanned antennas have been a major development of radar systems in the last 15 or more years and are now often the antenna of choice in most new systems. The simplest and most common, the *passive electronically scanned array* (PESA), is a planar array antenna in which a computer-controlled phase shifter is inserted in the feed system immediately behind each radiating

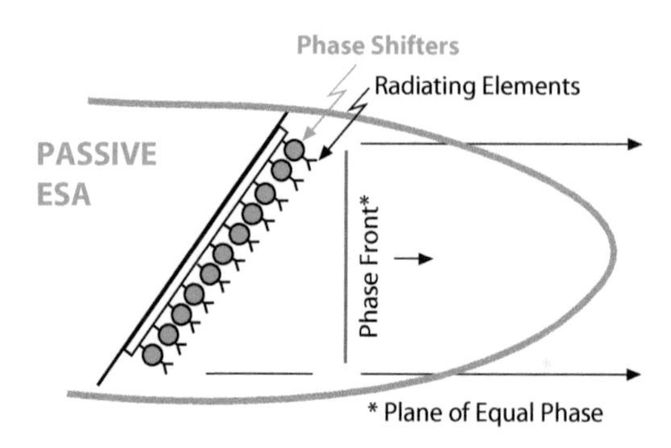

Figure 2-24. By controlling the phase of the signals transmitted and received by each radiating element, the phase shifters in a PESA can steer the radar beam anywhere within the field of regard.

element (Fig. 2-24). By individually controlling the phase shifters, the beam formed by the array can be steered anywhere within a fairly wide field of regard. These antennas are often known as *phased arrays.*

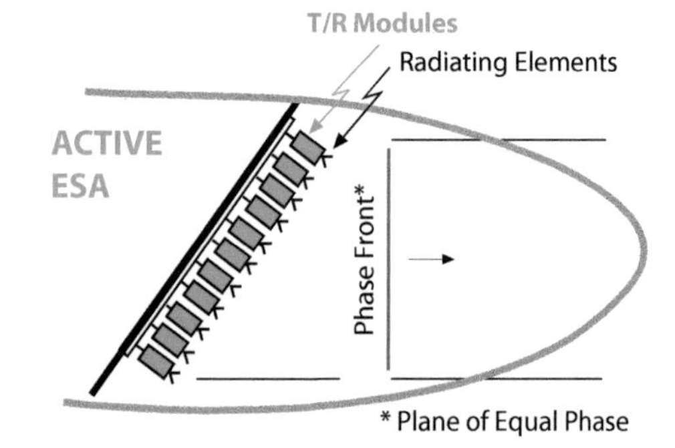

Figure 2-25. In an AESA, transmit/receive modules steer the radar beam by controlling the phase and amplitude of the signals radiated and received by each radiator.

A more versatile, but considerably more expensive, implementation is the *active electronically scanned array* (AESA). It differs from the passive ESA in that it has a tiny solid-state transmitter/receiver (T/R) module inserted behind each radiating element (Fig. 2-25). To steer the beam, provisions are included in each module for controlling both the phase and the amplitude of the signals the module transmits and receives. The transmitter and receiver components are assembled in one single T/R module. This module integrates the electronic phase shifter, the solid-state power amplifier, the LNA, circulators, and a duplexer. It is also common for the module to have self-test features so that the overall performance of the system can be assessed. This is known as a *built-in test environment* (BITE). A limiter is sometimes added between the LNA and antenna. Further details may be found in Chapter 9.

A problem with phase steering arrays is that the large bandwidths required for high range resolution reduce the ability to steer the beam. There are two competing alternatives. One is photonic true-time-delay (TTD) beam steering. Here, the phase of the signals radiated and received by the individual T/R modules of an active ESA is controlled by introducing variable time delays in the elements' feeds in the form of optical fibers. Their lengths, and hence the time the signals take to pass through them, are varied by switching segments of fiber of selectable length into and out of each feed. This greatly broadens the span of frequencies over which the antenna can operate. The approach is to use digital beam steering. This provides the most flexibility and not only is consistent with steering the beam but also enables the beam pattern to be adjusted to avoid ground reflections or to reduce sources of interference.

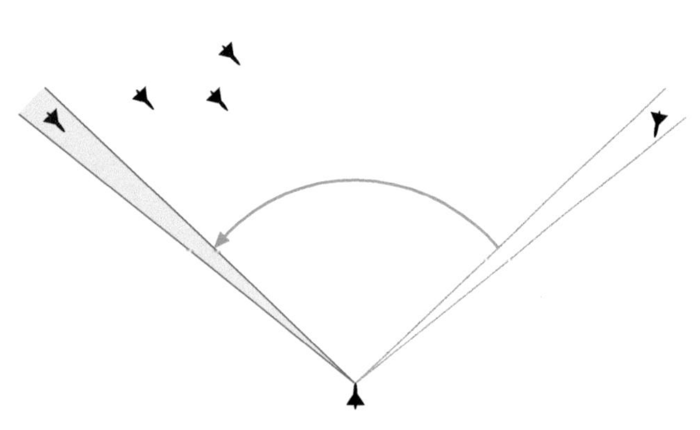

Figure 2-26. Since the radar beam has no inertia, with electronic steering it can be jumped anywhere within the field of regard in less than a millisecond.

ESAs have many advantages, one of the more important of which is extreme beam agility (Fig. 2-26). Because the beam (as opposed to the conventional gimbaled antenna) has no inertia, it can, for example, interactively jump to one or another of several targets at any time. It can dwell on a target for whatever length of time is optimum for tracking without appreciably interrupting the beam's search scan. Among the special advantages of the active ESA is the ability to radiate multiple individually steerable beams on different frequencies (Fig. 2-27).

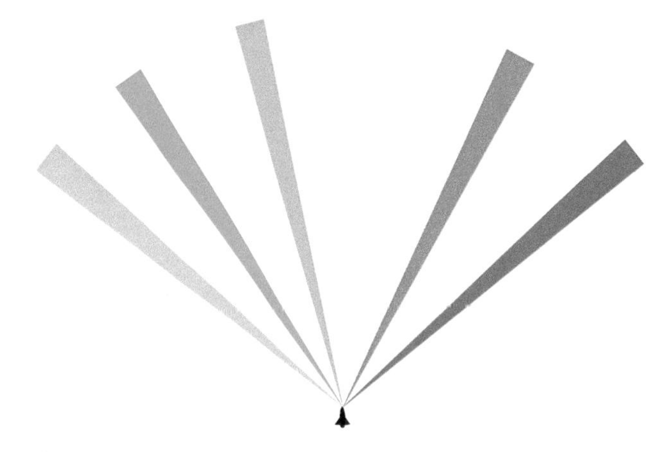

Figure 2-27. With an active ESA, the radar can even simultaneously radiate multiple, independently steerable beams on different frequencies.

Low Probability of Intercept Radar. Keeping a radar's signals from being intercepted and detected by an enemy is especially challenging. As seen in Chapter 1, because of the spreading of the radio waves occurring both on the way out to a target at a range, R, and on the way back to the radar the strength of the target's received echoes decreases as $1/R^4$. The strength of the radar's signals received by the target, however, decreases only as $1/R^2$ (Fig. 2-28).

To get around this huge handicap, an entire family of low probability of intercept (LPI) features has been developed:

- Taking full advantage of the radar's ability to coherently integrate the target echoes
- Interactively reducing the peak transmitted power to the minimum needed at the time for target detection
- Spreading the radar's transmitted power over an immensely broad band of frequencies
- Supplementing the radar data with target data obtained from infrared and other passive sensors and off-board sources
- Turning the radar on only when absolutely necessary

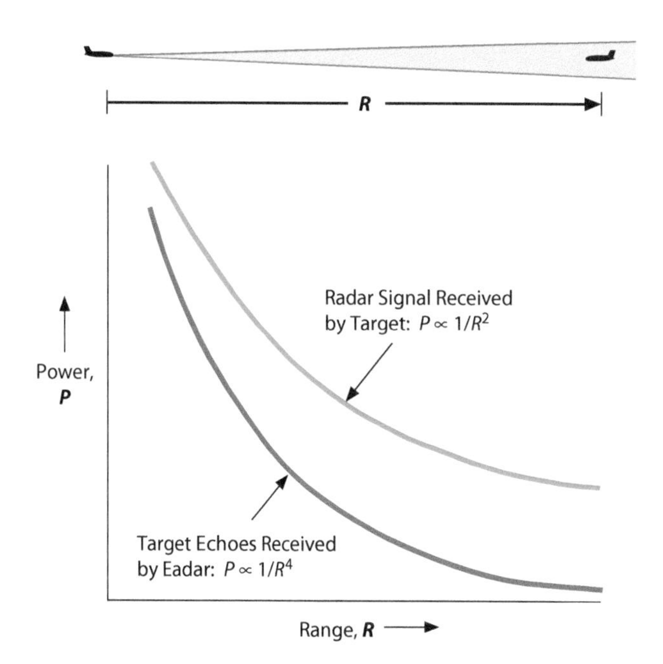

Figure 2-28. This graph shows the handicap surmounted by a radar designed to have a low probability of its signals being usefully intercepted.

Astonishing as it may seem, by combining these and other LPI techniques the radar can detect and track targets without its signals being usefully intercepted by the enemy. Electronically scanned antennas can also play a useful role by transmitting a broad beam and maximizing gain only on receive. In this way the power density that can be intercepted is reduced still further.

Stealthy Radar for Stealthy Aircraft. Viewed broadside, the antenna of a conventional radar alone may have a radar cross section (RCS) many times that of a fighter aircraft. To put such an antenna in the nose of a stealth aircraft would be grossly counterproductive, to say the least. In fact, the first U.S. stealth fighter (Fig. 2-29) didn't even carry a radar.

Figure 2-29. Since no radar at the time had both a low RCS antenna and a low probability of its signals being usefully intercepted by an enemy, the first U.S. stealth fighter, the F-117, was not equipped with a radar.

The first of several measures required to minimize the RCS of a radar antenna is to mount it in a fixed position on the aircraft structure, tilted so that its face will not reflect radio waves in the direction of an illuminating radar. The radar beam cannot then, of course, be easily steered mechanically, but electronic steering offers a solution.

Advanced Processing. Electronic beam steering, null steering, and LPI, along with other advanced techniques, depend critically on immensely high digital processing throughputs.

Orders-of-magnitude increases in throughput have been realized through the use of highly sophisticated technology and the distribution of processing tasks among a great many (hundreds) of individual processing elements, operating in parallel and sharing bulk memories. This trend of increasing processing power is set to continue.

A further enhancement to processing efficiency is *integrated processing*. Rather than providing separate processors for the aircraft's radar, electro-optical, and electronic warfare systems, a single integrated processor serves them all. Size, weight, and cost are thereby reduced.

With dramatically reduced memory costs, it has become possible to perform both signal and data processing in real time with commercial processing elements. Indeed, it is the commercial world that leads the way in this regard, with computer gaming consoles often having more processing power than a fighter radar. Where space and the ability to dissipate heat are at a premium, it is often processor efficiency that limits processing speed. Modern processors require significant prime power and disperse much heat that has to be effectively and efficiently removed.

2.4 Summary

Illustrative of the various approaches to implementation are two generic designs: (1) simple noncoherent pulsed radar; and (2) the more modern coherent pulse-Doppler radar.

The noncoherent pulsed radar employs a magnetron transmitter, a parabolic-reflector antenna, and a superheterodyne receiver. Triggered by timing pulses from a synchronizer, a modulator provides the magnetron with pulses of DC power, which it converts into high-power microwave pulses. These are fed through a duplexer to the antenna. Echoes received by the antenna are fed via the duplexer through a protection device to the receiver, which amplifies and converts them into video signals for display.

The coherent pulse-Doppler radar differs from the simple pulsed radar through measurement of amplitude and phase. The transmitter, such as a traveling wave tube amplifier, cuts pulses of selectable width and PRF from an exciter's low-power continuous wave signal and can be modulated for pulse compression. The antenna may be a planar array having a monopulse feed in both azimuth and elevation. The receiver features a low-noise amplifier and either or both analog and digital demodulation, which results in in-phase and quadrature data channels from which both amplitude and phase can be evaluated. The I and Q data are then further processed in a high-performance computer. It sorts them by range and Doppler frequency, filters out the ground clutter, and automatically detects the target echoes, storing their locations in a memory continuously scanned to provide a TV-like display. All operations of the radar are controlled by a digital computer (radar data processor), which loads the program for the selected mode of operation into the signal processor.

Combined with electronic scanning, the coherent pulse-Doppler radar is able to provide more advanced modes of operation such as low probability of intercept and a reduction in antenna RCS, which contributes to improved aircraft stealth.

Further Reading

S. Kingsley and S. Quegan, *Understanding Radar Systems*, SciTech-IET, 1999.

M. Skolnik, *Introduction to Radar Systems*, 3rd edition, McGraw-Hill, 2002.

P. Lacomme, J. C. Marchais, J. P. Hardange, and E. Normant, *Air and Spaceborne Radar Systems: An introduction*, Elsevier, 2007.

Test your understanding

1. What is the key difference between noncoherent pulsed radar and coherent pulsed radar?
2. In addition to a transmitter, receiver, antenna, and duplexer, what additional components are needed to construct a working radar system?
3. What three further subsystems are required to enable automatic tracking?
4. Name the four elements comprising a TWT, and describe the function of each one.
5. Explain the difference between an active phased array and a passive phased array radar system.

Boeing B-17 Flying Fortress (1938)

The B-17 was developed in response to a 1934 specification for a United States Army Air Corps heavy bomber. It carried a crew of 10, and over 12,000 were produced. The B-17 formed the backbone of the Allied day bombing offensive in Europe and earned a reputation for being able to keep flying in spite of severe battle damage.

3

Representative Applications

F-18 Super Hornet taking off from USS Enterprise

Having introduced basic radar principles and approaches to their implementation, this chapter briefly examines representative practical uses of radar. Some of these, such as air-to-air collision avoidance, ice patrol, and search and rescue are primarily civil applications. Examples of military applications are early warning and missile guidance (Fig. 3-1). Applications such as severe weather detection, storm avoidance, and wind shear warning have both civil and military purposes.

3.1 Weather Phenomena

Weather Detection and Prediction. We are all aware of the havoc severe weather can wreak. Sufficiently early detection can help reduce potentially catastrophic consequences by

Figure 3-1. This Saab Gripen fighter is equipped with a multi-function radar system. (Courtesy of Aviation Explorer.)

Representative Airborne Radar Applications

Weather Phenomenon

- Weather prediction
- Storm avoidance
- Wind shear warning
- Tornado warning

Navigational Aids

- Marking remote facilities
- Facilitating air traffic control
- Avoiding air-to-air collisions
- Altimetry
- Blind low-altitude flight
- Forward range and altitude measurement
- Precision velocity update
- Car collision management

Remote Sensing

- Terrain mapping
- Environmental monitoring
- Law enforcement
- Blind landing guidance
- Change mapping

Reconnaissance and Surveillance

- Search and rescue
- Submarine detection
- Long-range surveillance
- Early warning
- Sea surveillance
- Ground battle management
- Low-altitude surveillance
- Targeting

Fighter/Interceptor Support

- Air-to-air search
- Raid assessment
- Target identification
- Gun/missile fire control
- Missile guidance

Air/Ground Targeting

- Blind tactical bombing
- Strategic bombing
- Defense suppression

Proximity Fuses

- Artillery
- Guided missile

enabling both organizations and individuals to take evasive action. Many countries have networks of ground-based pulse-Doppler radars to measure and predict weather. Indeed, most people are accustomed to interpreting rainfall radar maps when they watch television and Internet weather reports. Weather radars are large-scale systems typically operating within the S- and C-bands. They are highly sensitive, which enables them to see light rainfall as well as intense storms even out to ranges on the order of 150 km. Weather maps for most areas may be generated from several radars whose coverages partially overlap. Additionally, there are three common weather threats to the safety of flight: turbulence; hail; and, particularly at low altitudes, wind shears or microbursts. All three are common products of thunderstorms. One of the most customary uses of airborne weather radar is alerting pilots to these hazards.

Storm Avoidance. In a weather radar system, if the radio frequency of transmitted pulses is appropriately chosen the radar can see through clouds while still receiving echoes from rain within them. The larger the raindrops, the stronger their echoes. By sensing the rate of change of the strength of the echoes range, the radar can detect thunderstorms. By scanning a wide sector, the radar can display those regions in which hazardous weather and turbulence are likely to be encountered. (Fig. 3-2).

Wind Shear Warning. Wind shears are strong downdrafts that can occur unexpectedly in thunderstorms. At low altitudes the outflow of air from the core of the downdraft can cause an aircraft to encounter an increasing headwind when flying into the downdraft and then a strong tailwind when emerging from it (Fig. 3-3). Without warning, this combination of conditions can cause an aircraft that is taking off or landing to crash.

Pulse-Doppler weather radars employed in aircraft are sensitive not only to the intensity of the rainfall but also to its horizontal velocity and therefore to the winds within a storm. By measuring the rate of change of the horizontal winds, these radars can detect a wind shear embedded in rain as much as 8 km ahead, giving the pilot up to around 10 seconds of warning to take avoidance action. Radar has also been used to detect wind shear in clear air by examining the echo response caused by dust particles.

Tornado Warning. A tornado is a violently rotating column of air that is in contact with both the surface of the earth and, most typically, a cumulonimbus cloud. Tornadoes are capable of acquiring huge amounts of energy and hence can cause widespread devastation and loss of life. Coherent pulse-Doppler radar systems have been designed to specifically enable the detection and monitoring of tornadoes. Radar is also used as a tool to investigate the formation of tornadoes so that their threat and movement patterns can be better predicted.

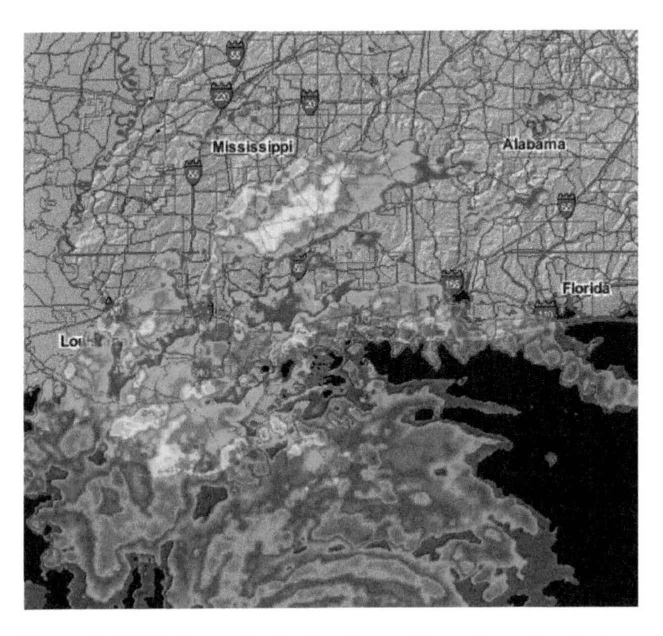

Figure 3-2. This shows the display of a weather radar system's view of Hurricane Katrina, which caused devastation to New Orleans and the surrounding area in 2008. The color-coding indicates intensity of precipitation and turbulence. (Courtesy of National Oceanic and Atmospheric Administration.)

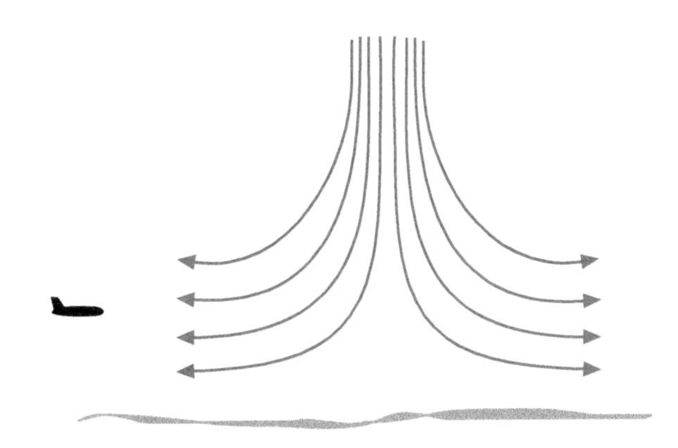

Figure 3-3. In a typical wind shear, as an aircraft approaches the downdraft it encounters increasing headwinds. As it emerges from the downdraft, it encounters strong tailwinds. This can cause instabilities with potentially catastrophic consequences.

3.2 Navigational Aids

Radar has long been used to assist with navigation. It originally helped to guide bombing missions in World War II. Among common navigational uses of airborne radar are marking the locations of remote facilities, assisting air traffic control, preventing air-to-air collisions, measuring absolute altitude, providing guidance for blind low-altitude flight, and measuring the range and altitude of points on the ground ahead. Global Positioning System (GPS) now provides a widely used alternative to radar and in many instances has become the primary navigation tool. Radar systems are also increasingly being employed in ground vehicles. With new cars more and more likely to sport several radar systems, this may become their most common use.

Marking Remote Facilities. For approaching helicopters and airplanes, the locations of offshore drilling platforms, remote airfields, and the like may be marked with radar beacons. The simplest beacon, called a transponder, consists of a receiver, a low-power transmitter, and an omnidirectional

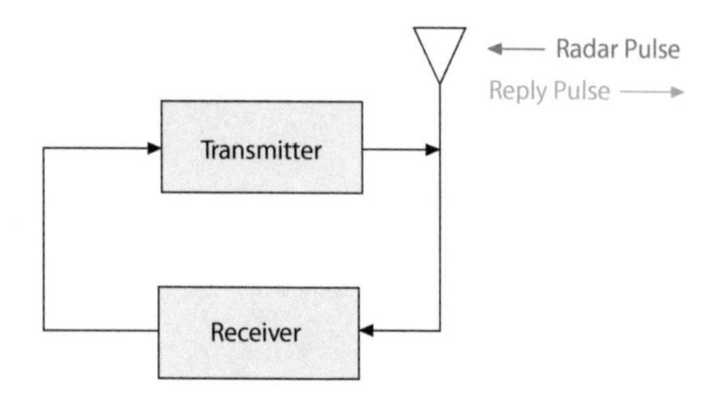

Figure 3-4. Upon receiving a pulse from a radar system, a simple beacon transponder transmits a reply on another frequency. This forms the basis of navigation aids.

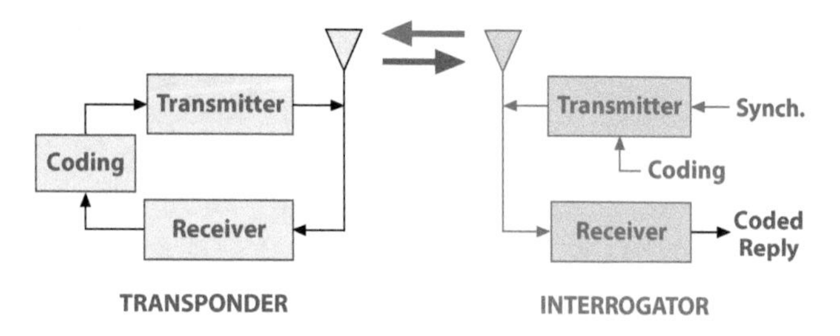

Figure 3-5. In a complete radar beacon system, an interrogator is typically synchronized with a search radar system and the transponder's replies are shown on the display.

Figure 3-6. The antenna of the beacon interrogator is mounted on top of the air traffic control radar antenna. Through coding of the beacon pulses and their replies, the radar is able to identify approaching aircraft and can obtain and display flight information such as altitude and flight number.

antenna (Fig. 3-4). The transponder receives the pulses of any radar whose antenna beam sweeps over it and transmits "reply" pulses on a different frequency. Even though they are low powered, the replies are still designed to be much stronger than the radar's echoes. Since their frequency is different from the radar's, they are not accompanied by clutter and hence stand out clearly on the radar display.

A more capable beacon system (Fig. 3-5) includes an interrogator, which transmits coded pulses to which transponders respond with coded replies. The most common beacons of this sort are those used as part of air traffic control. Until recently these were dominated by the *Air Traffic Control Radar Beacon System* (ATCRBS), which is expected to be phased out. Its replacements are the *Automatic Dependent Surveillance-Broadcast System* (ADS-B) and *Mode Select* (Mode S), which have improved capacity to handle more aircraft and provide more detailed flight information.

Assisting Air Traffic Control. Mode S transponders are carried on all but the smallest private aircraft and have interrogators that operate in conjunction with the air traffic control radar at every major airport. An interrogator's monopulse antenna is mounted atop the radar antenna, so it moves and scans with it (Fig. 3-6). The interrogator's pulses are synchronized with the radar's. The operator can subsequently interrogate an incoming aircraft simply by touching its "blip" on the radar display with a light pen.

Ordinarily the interrogator uses only two of several possible codes. One requests the identification code of the aircraft carrying the transponder, and the other requests the aircraft's altitude. Every beacon-equipped aircraft can thus be positively identified and its position accurately determined in three dimensions.[1] Strictly speaking, this is not a radar system. However, its heritage and use make it a very close cousin, and it is commonly referred to as Mode S radar.

Avoiding Air-to-Air Collisions. The *Traffic Alert and Collision Avoidance System* (TCAS) also uses ATCRBS transponders. Typically integrated with an aircraft's weather radar, TCAS interrogates the air traffic control transponders and whatever aircraft happen to be within the search scan of the radar. From a transponder's replies, TCAS determines their directions, ranges, altitude separations, and closing rates. Based on this information, the system prioritizes threats, interrogates high-priority threats at an increased rate, and, if necessary, gives vertical and horizontal collision avoidance commands.

Altimetry. In many situations, it is desirable to know an aircraft's absolute altitude.[2] Since beneath the aircraft there is usually a large area of ground at very nearly the same range

1. Every aircraft can be assigned a unique identification code because 16 million are available.

2. Distance to the ground.

(Fig. 3-7), a small low-power, broad-beam, downward-looking radar can provide a continuous precise reading of absolute altitude. Interfaced with the aircraft's autopilot, the altimeter can ensure smooth tracking of the glide slope for instrument landings. Altimetry is also used in remote sensing applications to evaluate changes in surface height of the land and sea.

Altimeters may be continuous wave (CW) or pulsed. For military uses, the probability of the altimeter's radiation being detected by an enemy is minimized by transmitting pulses at a very low pulse repetition frequency (PRF) and employing large amounts of pulse compression to spread the pulses' power over a very wide band of frequencies.

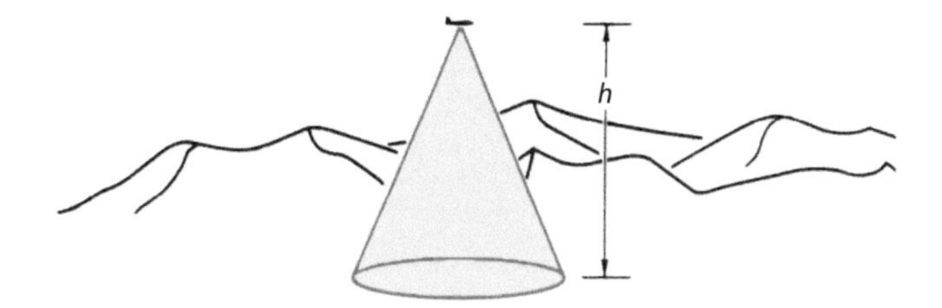

Figure 3-7. An aircraft's absolute altitude can be precisely determined by measuring the range to the ground beneath it with a small low-powered, broad-beam radar.

Blind Low-Altitude Flight. Low-altitude flight is inherently dangerous but can enable a fighter, via a technique termed *hedge-hopping*, to avoid observation and hence enemy attack. To accomplish hedge-hopping, two basic radar modes have been developed: terrain following and terrain avoidance.

In terrain following (Fig. 3-8), an aircraft's forward-looking radar scans the terrain ahead by sweeping a pencil beam vertically to the horizon. From the elevation profile this obtains, vertical steering commands are computed, supplied to the flight control system, and used to automatically fly the aircraft safely at terrain-skimming altitude.

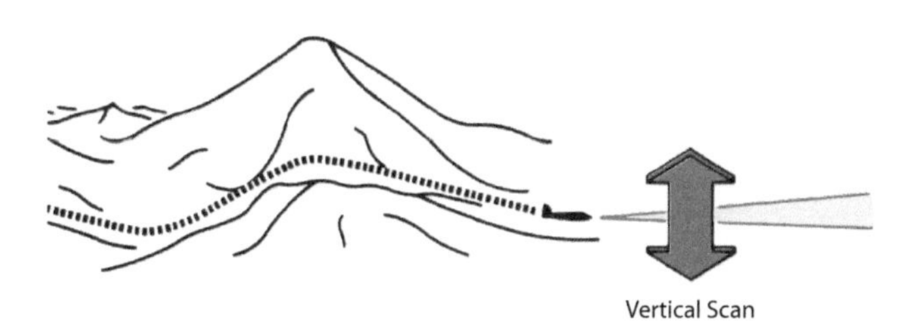

Figure 3-8. For *terrain following*, a radar scans the terrain ahead vertically with a pencil beam.

Terrain avoidance (Fig. 3-9) is similar to terrain following except that periodically the radar also scans horizontally, enabling the aircraft not only to hug the ground but also to fly around obstacles in its path.

Unmanned air vehicles (UAVs) are being used increasingly to replace human-operated systems. UAVs can be placed on a precisely timed, preprogrammed, ground-hugging trajectory along a known contour on a map. Ground clearance is measured with a very low-power radar altimeter. Since it illuminates only the ground beneath the aircraft, the possibility of enemy detection is low. Operating at frequencies for which atmospheric attenuation is high may further reduce this detection probability.

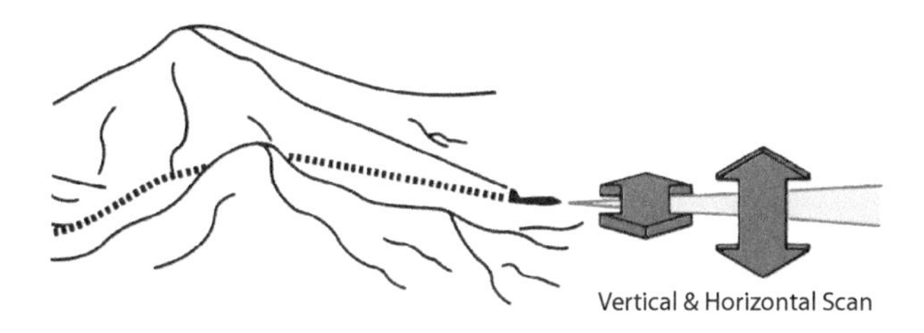

Figure 3-9. For *terrain avoidance*, the radar alternately scans terrain ahead both vertically and horizontally.

Forward Range and Altitude Measurement. On a bombing run over ground that is neither flat nor level, it is often necessary to precisely determine the range and altitude of the aircraft relative to the target. This can be done by training the radar beam on the target and measuring the following two parameters: (1) the antenna depression angle; and (2) the range to the ground at the center of the radar beam (Fig. 3-10).

Monopulse may be used to center the target in the beam (see Chapter 31). Once the target is centered, the range and the depression angle can be measured simultaneously. The simple geometry depicted in Figure 3-10 allows straightforward computation of the altitude of the target relative to the aircraft.

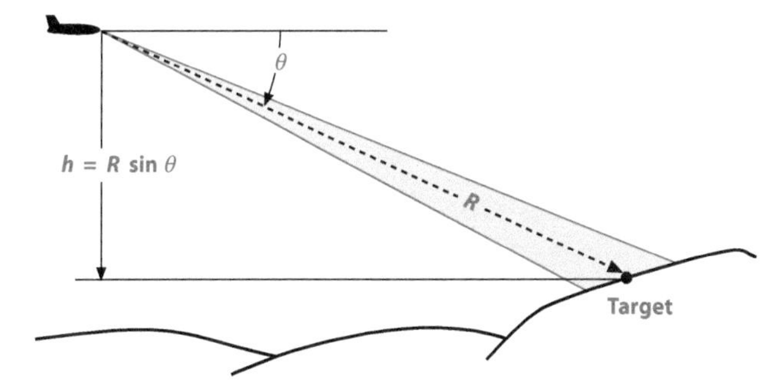

Figure 3-10. The figure depicts measurement of the range and the relative altitude of a point on the ground. The range that the radar measures is the one at which the elevation tracking-error signal is zero.

Figure 3-11. Vehicular radar systems are using low cost but sophisticated technology to improve both traffic safety and traffic flow. (Courtesy of AutoLiv.)

Precision Velocity Update (PVU). As described in Chapter 1, by measuring the Doppler frequency of the returns from three points on the ground ahead forward-looking radar can measure the radar's velocity. Such measurements can be used to update the aircraft's inertial navigation system. If the inertial system fails, the radar can take over. A radar system operating in this way is known as a Doppler navigator.

Car Collision Management. Increasingly, modern vehicles are being equipped with a suite of miniature radar systems that provide for improved collision avoidance and navigation applications. These systems are set to become the most numerous and, consequently, an important form of radar. They typically are designed using a frequency modulated continuous wave (FMCW) and emit average powers of just a few milliwatts at an operating frequency of either 24 GHz or 77 GHz. They can detect a variety of obstacles out to ranges of the order of 100 m, which provides the information necessary for collision avoidance (Fig. 3-11). Although these systems are small and cost just tens of dollars, they offer a sophisticated capability to resolve objects to a few tens of cm. This high range resolution is exploited to determine the precise position of obstacles so that early warning is provided to the driver, or, ultimately in cases of immediate danger, generates command signals that automatically allow safe braking and steering away from the obstacle. The automotive industry hopes this will eventually lead to collisionless and even driverless vehicles.

3.3 Remote Sensing

Radar remote sensing applications have become increasingly important, and they help us to understand the evolution of our planet. These can range from crop monitoring, measurement of deforestation, and assessment of polar ice cap melting to coastal erosion, ice patrol, high-resolution terrain mapping, vehicle speed detection, and autonomous blind landing guidance, to name just a few. This list is ever increasing, as is the number and variety of radar systems designed to meet such applications. Remote sensing of the earth from space continues to grow, fueled by the need to gain a better understanding of how our planet is changing. Indeed, there are so many space-based systems orbiting the earth that parts of it are being mapped several times a day. Here just a few examples are briefly reviewed.

Terrain Mapping. The earliest terrain mapping radars were real-beam, noncoherent systems called *side-looking airborne radars* (SLARs). These have a long linear array antenna that looks out to the side of an aircraft. While the aircraft flies in a straight line, the radar beam is dragged past a scene of interest so that a map-like image can be formed. Most of these systems predate digital processing, so optical scanning processors were employed to record the radar echo data onto photographic film. Figure 3-12 shows an SLAR being used to map ice flows.

SLAR operates at relatively short ranges (a few tens of km) and high frequencies (say, 35 GHz or above) so that adequately high resolutions were possible for applications such as charting passages though ice in waters that freeze over in wintertime. Moreover, because the radar is simple and its antennas are fixed, it is comparatively inexpensive. However, the coherent imaging technique *synthetic aperture radar* (SAR) has become the remote sensing and surveillance tool of choice. High-resolution SAR imaging (a few tens of cm or less) is possible at ranges compatible with operation from a spacecraft.

SAR has proven especially useful for highly accurate, low-cost terrain mapping for both civil and military applications. In addition, high-resolution, three-dimensional maps (Fig. 3-13) can be generated using interferometric techniques. This type of 3D imagery is increasingly being used to map much of the surface of the planet to very high accuracies.

Law Enforcement. Both SLAR and, increasingly, SAR have played important roles in oil spill detection, fishery protection, and the interdiction of smugglers and drug traffickers. Since SAR can provide fine resolution at long ranges, it has the advantage of uncovering illicit activities without alerting the lawbreakers (Fig. 3-14).

Blind-Landing Guidance. The ground directly ahead of an aircraft cannot be mapped with SAR, so other techniques must be employed for landing guidance. One approach is to scan the narrow region ahead with a monopulse antenna. At the short ranges involved, sufficiently fine azimuth resolution may be obtained to enable an aircrew to locate runways and markers

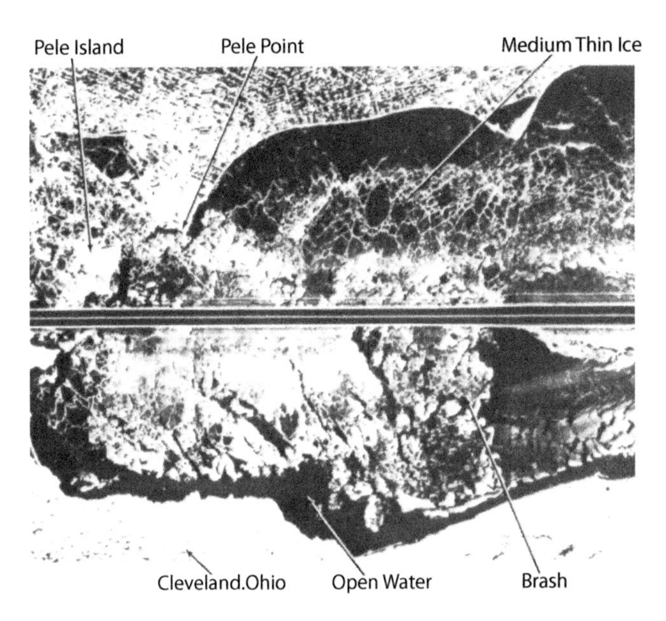

Figure 3-12. An ice flow on Lake Erie is mapped by a real-beam side-looking array radar with a long, fixed array antenna that looks out on either side of the aircraft.

Figure 3-13. This shows a representative interferometric, 3D SAR map. (Crown copyright DERA Malvern.)

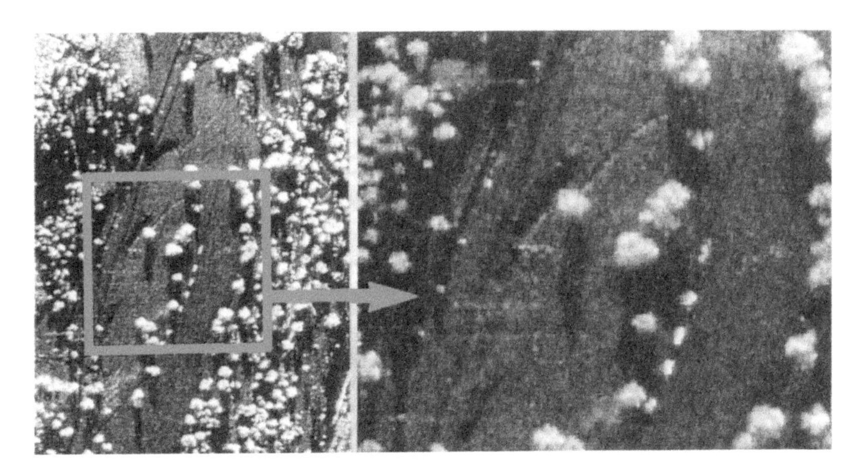

Figure 3-14. A SAR map, such as might be used to interdict smugglers, shows a convoy of trucks on an off-road trail. As indicated by the radar shadows of the trees, the map was made from by a radar system operating at long range some distance above the top of the image.

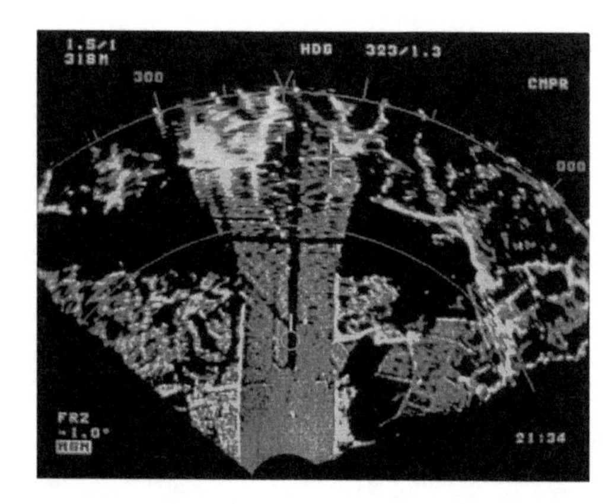

Figure 3-15. A forward-looking radar system, with a monopulse antenna, fills in a gap in a SAR map with real-beam mapping. This provides sufficient resolution to enable blind approaches at landing strips where navigation aids may not be available. (Courtesy of Northrop Grumman Corporation.)

Figure 3-16. Noncoherent change detection showing targets detected in the first image but not the second are indicated in blue. Targets detected in the second image but not the first are shown in red. (Image courtesy of Air Force Research Laboratory, Sensors Directorate, Gotcha Radar Program, Public Release Number: AFRL/WS- 88ABW-2013-5108.)

Figure 3-17. A long-range, long-endurance unmanned reconnaissance aircraft may relay 30 cm resolution SAR maps via satellite directly to users in the field. (Courtesy Northrop Grumman Corp.)

(Fig. 3-15) and therefore to make autonomous approaches to small or unimproved landing strips at night or in bad weather. This technique is known as Doppler beam sharpening and is a close cousin of both SAR and SLAR.

Change Mapping. Detection of changes over time and space provides invaluable information in many ways, such as in coastal erosion, polar ice cap melting, and deforestation. Figure 3-16 offers an example of noncoherent change detection showing targets detected in the first image but not the second in blue. Targets detected in the second image but not the first are shown in red. This allows an image analyst to be alerted to changes on the ground that happen over time.

3.4 Reconnaissance and Surveillance

Reconnaissance and surveillance are military applications of remote sensing, and because of the nature of warfare their radar designs can be very different. In military operations, airborne radar has proven to be vital for its ability to see through smoke, haze, clouds, and rain; to rapidly search vast regions; to detect targets at long ranges; to simultaneously track a great many targets that may be widely dispersed; and, to a degree, to classify targets.

Four representative applications are considered here: long-range air-to-ground reconnaissance; early warning; air-to-ground battle surveillance; and balloon-borne low-altitude surveillance.

Long-Range Air-to-Ground Reconnaissance. Very high-resolution (30 cm) SAR radars provide all-weather surveillance of military targets out to ranges of several hundred km. In fact, as the altitude of the aircraft determines the range to the horizon it also finds the longest operational ranges. The very high resolution provides a level of detail invaluable in pinpointing ground targets for fighters and bombers.

SAR radars have been developed for such missions in small pilotless reconnaissance aircraft capable of long-range endurance flight (Fig. 3-17). These radars also relay radar images of 30 cm resolution via satellite directly to users in the field. The imagery from a SAR system is produced in real time and hence in prodigious quantities. The quantities of data are too great for all of it to be analyzed, even by several humans. However, real-time automatic assessment of imagery using high-speed computers remains a challenge, and much of the detailed interpretation must be done manually.

Early Warning and Sea Surveillance. Airborne radar can detect low-flying aircraft and surface vessels at far greater ranges than can a radar system on the ground or one located on the mast of a ship. Accordingly, to provide early warning for the approach of hostile aircraft and missiles and to maintain surveillance over the seas radars are placed in high-flying loitering aircraft such as the Hawkeye system and airborne warning and control system (AWACS).

Because these aircraft are large and slow, the radars they carry can employ antennas large enough to provide high angular resolution while operating at frequencies low enough that atmospheric attenuation is negligible. They can also transmit very high powers.

They provide 360° coverage, detecting low-flying aircraft out to the radar horizon (which at an altitude of 10,000 m is more than 300 km) and higher altitude targets at substantially greater ranges. In addition, they can simultaneously track hundreds of targets.

Air-to-Ground Surveillance and Battle Management. Very much as AWACS provides surveillance over a vast air space, airborne radar can also provide surveillance over a vast area on the ground. These radars combine high-resolution imaging with an ability to detect ground-moving targets. This is quite an accomplishment in a single system since the requirement to support these two modes results in very different specifications. The moving targets are located more accurately using monopulse, and research has shown that advanced techniques such as *displaced phase center antennas* (DPCA) and *space-time adaptive processing* (STAP) can help to make targets easier to detect against a background of clutter.

Figure 3-18. The passive ESA of Joint STARS radar is housed in an 8 m long radome. The radar performs SAR mapping and ground-moving target detection for tracking and surveillance in battle management. (Courtesy US Air Force.)

Examples of such system include JSTARS, SENTINEL, and AN/AP-12. JSTARS is equipped with a long electronically steered side-looking antenna (Fig. 3-18) and detects and tracks moving targets on the ground with moving target indication (MTI) and detects stationary targets with SAR.

Flying in a racetrack pattern at an altitude of more than 10,000 m and standing off more than 150 km behind a hostile border, the radar can maintain surveillance over a region extending a 150 km or more miles into enemy territory. Through secure communication links, Joint STARS can provide fully processed radar data to an unlimited number of control stations on the ground.

Low-Altitude Air and Sea Surveillance. Maritime or sea surveillance continues to be an important function in which surface targets from submarine periscopes to frigates are detected, tracked and, in the case of ships, classified. High resolution is required to eliminate clutter, to enable the detection of small targets, and to provide the detail necessary for small boat and ship classification.

Figure 3-19. An aerostat carries lightweight solid-state surveillance radar having a large parabolic reflector antenna. Tethered at altitudes of 5000 m, the radar can detect small low altitude aircraft at ranges out to 300 km. An aerostat can stay aloft for 30 days, remain operational in 30 m/s winds, and survive 40 m/s winds. (Courtesy Raytheon Company.)

A novel surveillance application of airborne radar is interdiction of smugglers at sea, especially those carrying drugs. For example, the Customs Service implemented a radar "fence" along the southern border of the United States by placing large-reflector, long-range surveillance radars in tethered balloons (Fig. 3-19), and there are similar projects elsewhere in the world. The relatively stable airborne platform is more suited to the detection of moving targets against a background of sea clutter. Here the craft are often smaller and fast-moving.

Figure 3-20. Equipped with a high-power pulse-Doppler radar, the U.S. Air Force F-16 air superiority fighter can provide surveillance over a huge volume of airspace.

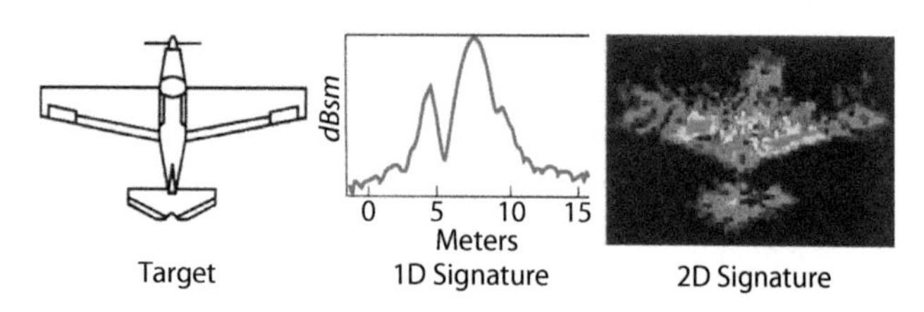

Figure 3-21. These 1D and 2D signatures of aircraft in flight were obtained with an uncooperative target identification system.

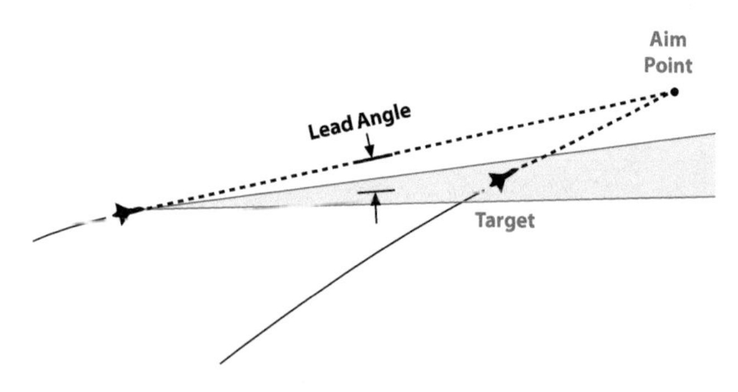

Figure 3-22. In this depiction of a lead-pursuit course for firing guns, the fighter's radar automatically locks onto its target in an air-combat mode and tracks it in a single-target tracking mode.

3.5 Fighter/Interceptor Mission Support

The fighter/interceptor mission is twofold: (1) thwart attacks by aircraft and missiles; and (2) achieve control of the airspace over a given region. In both, the fighter's radar typically plays four vital roles: search; raid assessment; target identification; and fire control.

Air-to-Air Search. The extent to which a fighter's radar must search for targets varies. At one extreme, the fighter may be "vectored" to intercept a target that has already been detected and is being precisely tracked. At the other extreme, the radar may be required to search a huge volume of air space for possible targets (Fig. 3-20).

Raid Assessment. Even if a radar has a narrow pencil beam, at long ranges it may not be able to resolve a close formation of approaching aircraft. Consequently, the fighter's radar is usually provided with a raid assessment mode. This may require the radar to alternate between track-while-scan to maintain situation awareness and single-target tracking of the suspect multiple targets in a mode providing exceptionally fine range and Doppler resolution.

Target Identification. To identify targets that are beyond visual range, some means of radar identification is generally desired. One of these is identification friend or foe (IFF). An IFF interrogator synchronized with the fighter's radar transmits interrogating pulses to which transponders carried in all friendly aircraft respond with coded replies. Despite use of sophisticated codes, the possibility of compromise is always present, so additional means of uncooperative target identification methods have been devised.

These techniques fall into the general category of signature identification. The unique characteristics of the echoes received are used to identify radar targets by type. More typically this approach involves providing sufficiently fine range resolution so that targets may be identified by their 1D range profiles. Going a step further, by employing *inverse synthetic aperture radar* (ISAR), 2D images may be provided. Figure 3-21 illustrates the form of 1D range profiles and 2D ISAR images used for classification.

Fire Control. Depending on a target's range, the pilot may attack it with either the aircraft's guns or its guided missiles.

For firing guns, a selection of close-in combat modes may be provided in which the radar automatically locks onto the target in a single-target tracking mode and continuously supplies its range, range rate, angle, and angular rate to the aircraft's fire-control computer. The latter directs the pilot onto a lead-pursuit course against the target (Fig. 3-22) and, at the appropriate range, gives a firing command. Both steering instructions and firing command are presented on a head-up display so that pilots never need to take their eyes off the target.

Radar-guided missiles, however, are often fired from beyond visual range. These are generally launched while the fighter's radar is operating in a track-while-scan or search-while-track mode. Hence, several missiles may be launched in rapid succession and be in flight simultaneously against different targets.

Initially, the missiles are guided inertially on a lofted trajectory. They then transition to semiactive guidance in which a radar seeker that the missiles carry homes in on the periodic target illumination provided by the fighter's scanning radar (Figure 3-23). At close range, the seeker switches to active guidance in which it provides its own target illumination.

Figure 3-23. An Advanced Medium-Range Air-to-Air Missile (AMRAAM) is test-launched from an F-35 Lightning.

An Advanced Medium-Range Air-to-Air Missile (AMRAAM) (Fig. 3-24) is equipped with a command-inertial guidance system. It steers the missile on a preprogrammed intercept trajectory based on target data obtained by the fighter's radar prior to launch. If the target changes course after launch, update messages are relayed to the missile by coding the radar's normal transmissions. A receiver picks up the messages in the missile and decodes and uses them to correct the course set into the inertial guidance system.[3] For terminal guidance, the missile switches control to a short-range active radar seeker that it carries.

3. If the missile is not in the radar beam at the time, the messages are received via the radar antenna's sidelobes.

3.6 Air-to-Ground Targeting

Radar may play an important role in a wide variety of air-to-ground attacks. To illustrate, hypothetical missions of four different types are examined: (1) tactical-missile targeting; (2) blind tactical bombing; (3) precision strategic bombing; and (4) ground-based defense suppression. In each, the basic strategy is to take advantage of radar's unique capabilities while minimizing radiation from the radar.

Tactical-Missile Targeting. In this hypothetical mission, an attack helicopter lurks behind a hill overlooking a battlefield. Only the antenna pod of a short-range, ultra-high-resolution (millimeter wave) radar situated atop of the rotor mast shows (Fig. 3-25). The radar quickly scans the terrain for potential targets. Automatically prioritizing the targets it detects, the radar

Figure 3-24. An AMRAAM is inertially guided on preprogrammed intercept trajectory; it receives update messages from the radar if the target maneuvers after launch (length 4 m; range 25+ km).

Figure 3-25. A small antenna of high-resolution millimeter-wave radar sitting atop the rotor mast enables an attack helicopter to detect targets for its launch-and-leave missiles while it keeps out of sight from the battlefield.

hands them off to a fire-control system, which aims small, independently guided launch-and-leave missiles.

Blind Tactical Bombing. A strike aircraft is guided by an inertial navigator on a terrain-skimming course to an area where a mobile missile launcher is believed to have been set up (Fig. 3-26). Upon reaching the area, the operator turns on the fighter's radar to update the navigator and then makes a single SAR map. With the map frozen on the radar display, the operator places a cursor over the target's approximate location. Turning the radar on again, the operator makes a detailed SAR map centered on the spot designated by the cursor.

Having identified the target, the operator places the cursor over it. Immediately, the pilot starts receiving steering instructions for the bombing run. At the optimum time, the bomb is automatically released. By briefly breaking radio silence just three times, the radar has provided all the information needed to score a direct hit on the target under conditions of zero visibility.

Precision Strategic Bombing. In this mission, the flight crew of a stealth bomber, flying at some 7000 m altitude, turns the bomber's radar on just long enough to make a high-resolution map of an area where an enemy command center has been activated. This map, too, is frozen, but it is scaled to GPS coordinates. As

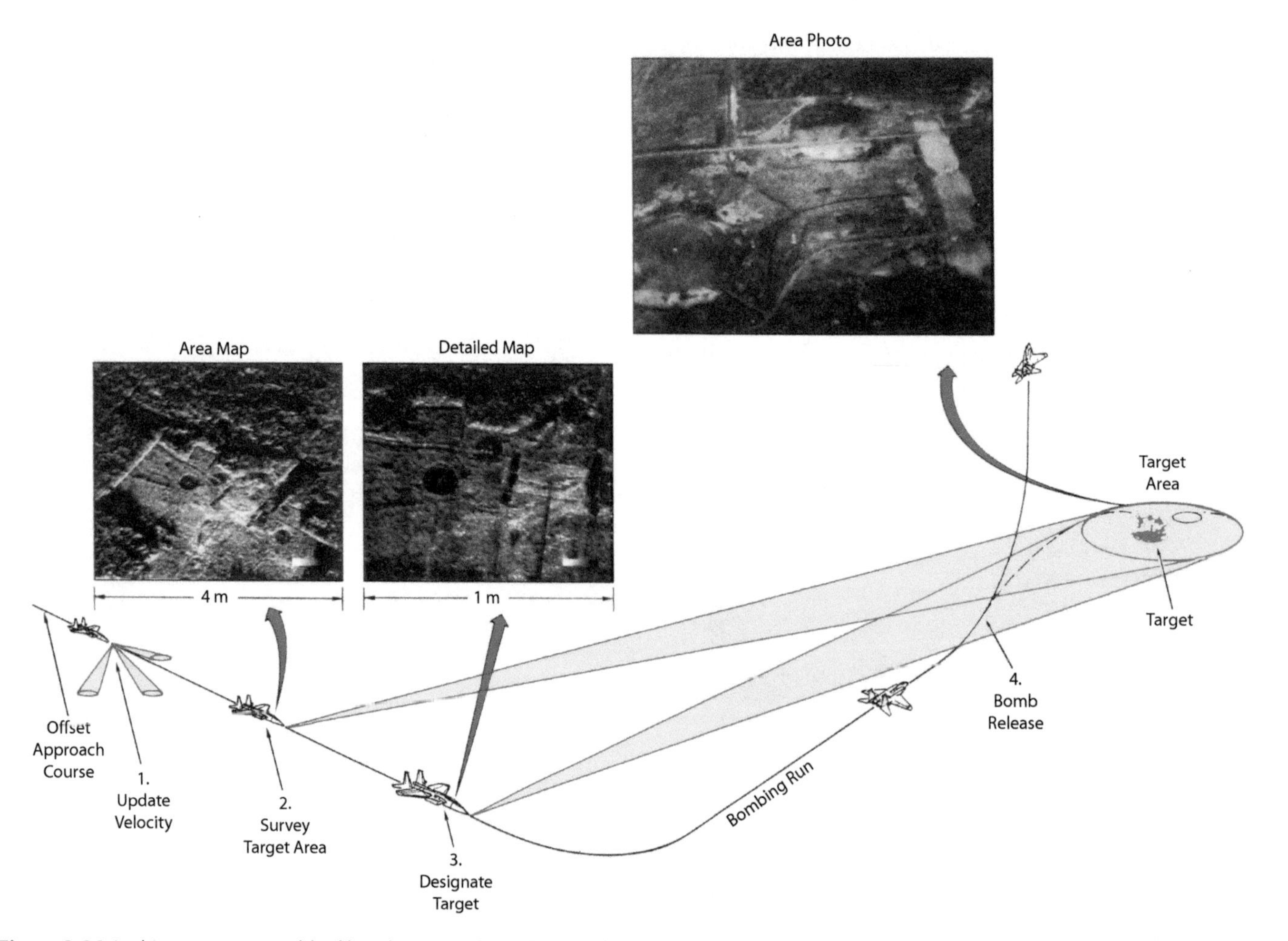

Figure 3-26. In this representative blind bombing run, the strike aircraft scores a direct hit from an offset approach course by turning radar on *just three times.*

soon as the target is identified, the operator places a cursor over it, thereby entering the target's GPS coordinates into the GPS guidance system of a 2000-pound glide bomb.[4] Automatically released at the optimum time, the bomb glides out until it is almost directly over the target (Fig. 3-27) and then dives vertically onto it with an accuracy of better than 1 m.

4. As a hedge against a GPS failure, alternate means of delivery are provided.

Figure 3-27. A GPS-guided bomb glides until it is almost directly over the target designated prior to launch on a SAR map made by the bomber's radar and then dives vertically onto it.

Ground-Based Defense Suppression. Ground-based enemy air-search radars and *surface-to-air missile* (SAM) sites, when radiating, may be put out of action with *high-speed anti-radiation missiles* (HARMs), which home in on emitted radiation from an enemy radar system.

A specially equipped aircraft, lurking at low altitude outside the field of view of an enemy defense radar, determines its direction and range using data received via a data link from other sources. The flight crew preprograms a HARM to search for the radar's signals. Launched in the direction of the radar, the missile soon acquires the radar's signals. It then zooms in on and destroys the radar before the enemy even realizes it is under attack.

3.7 Proximity Fuses

Another important application of airborne radar is proximity fuses (see panel alongside).

Proximity Fuses: Then and Now

An early proximity fuse detonates an artillery shell when the return from the ground reaches a predetermined amplitude or detonates an anti-aircraft shell on the basis of the change in amplitude of the received signal as the shell approaches an aircraft.

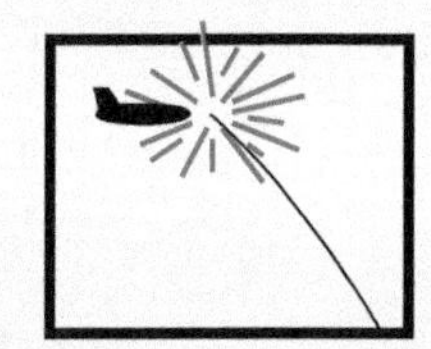

In guided missiles, much more sophisticated fuses are employed. They not only detect the presence of a target but also are able to control the timing of the detonation by measuring the change in Doppler frequency of the radar return as the missile approaches the target.

3.8 Summary

In this chapter it has been possible to look only briefly at just a small range of representative applications. Radar systems come in a wide variety of designs, each tailored to a very specific application.

Further Reading

J. A. Scheer and W. L. Melvin (eds.), *Principles of Modern Radar Volume 3: Applications*, SciTech-IET, 2014.

Test your understanding

1. Name and briefly describe 10 representative applications of pulse Doppler radar.
2. What weather phenomena can be detected and measured using radar?
3. How do systems such as ADS-B differ from air traffic control radar?
4. What is meant by the term "change detection"?
5. What are the advantages of target identification using radar?

De Havilland Mosquito (1941)

The Mosquito was unusual in that its airframe was constructed almost entirely of wood. It served multiple roles including tactical bomber, night high-altitude bomber, day or night fighter, intruder, maritime strike aircraft, and reconnaissance. Due to its wooden construction it had a high maximum speed and was able to outrun almost all enemy aircraft. Its wooden construction also gave it a low radar signature.

PART II

Essential Groundwork

Messerschmitt Me-262 (1944)

The *Schwalbe* (Swallow) was the first jet-powered aircraft to reach operational status. Its advanced design allowed it to be used in a variety of roles such as light bomber, reconnaissance, and night fighter. Its speed and high rate of climb made it extremely difficult to counter. Multiple B-1a trainers were converted into night fighters using the FuG 218 *Neptun* radar, with *Hirschgeweih* antenna array.

4

Radio Waves and Alternating Current Signals

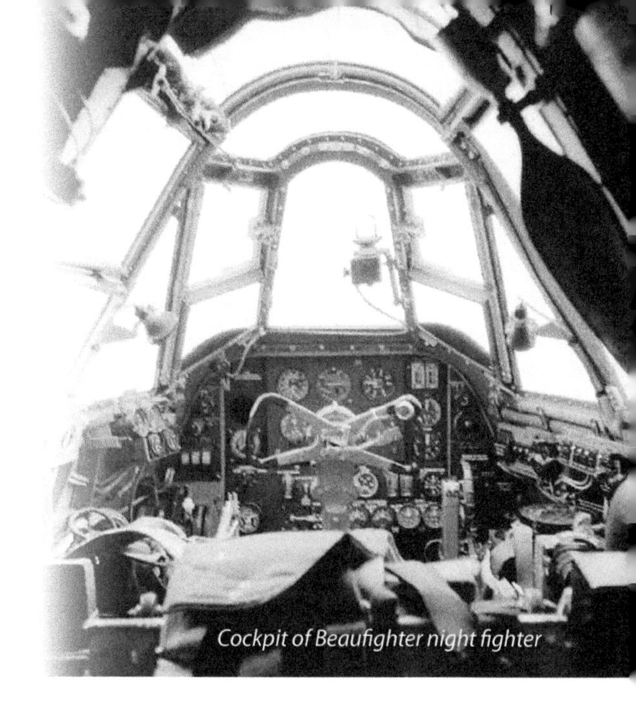
Cockpit of Beaufighter night fighter

Since radio waves and alternating current (AC) signals are vital to all radar functions, any introduction to radar logically begins with them. Indeed, many radar concepts that at first glance may appear quite difficult are, in fact, simple when viewed in the light of radio waves and AC signals.

In this chapter the nature of radio waves is considered together with their fundamental qualities.

4.1 Nature of Radio Waves

Radio waves are perhaps best modeled as energy that has been emitted into space. The energy exists partly in the form of an electric field and partly in the form of a magnetic field. For this reason, the waves are called *electromagnetic*.

Electric and Magnetic Fields. Whenever an electric current flows, a magnetic field is produced. Familiar examples of both of these field types can be identified. Electric fields are created when a charge builds up between a cloud and the ground and produces lightning (Fig. 4-1) or, on a much smaller scale, when a charge builds up on a comb on a particularly dry day, enabling it to attract a scrap of paper. Magnetic fields encircle the earth and cause compasses to react, surround a toy magnet or are produced when current flows through the coil in a telephone earpiece and causes the diaphragm to vibrate and produce sound waves.

Figure 4-1. A common example of an electric field is that which builds up between a cloud and the ground.

The two types of fields are inextricably related. If an electric field varies sinusoidally, so will the magnetic field it produces. If a magnetic field varies sinusoidally, so will the electric

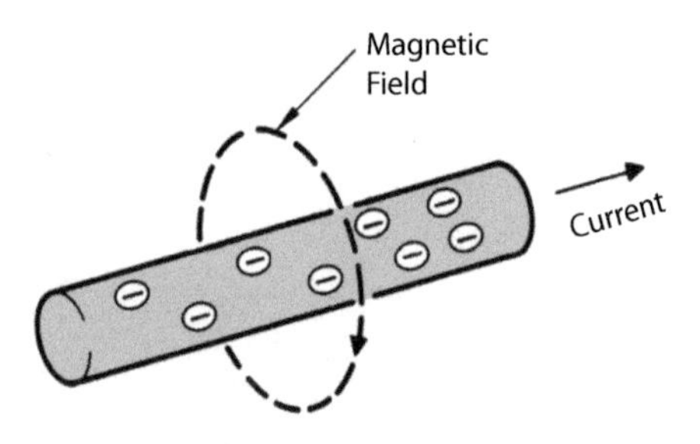

Figure 4-2. Whenever an electric current flows, a magnetic field is produced.

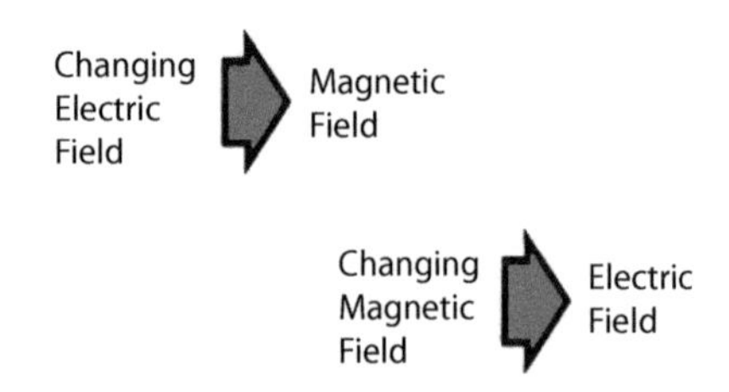

Figure 4-3. Dynamic relationships giving rise to radio waves. If an electric field varies sinusoidally, so will the magnetic field it produces. If a magnetic field varies sinusoidally, so will the electric field it produces.

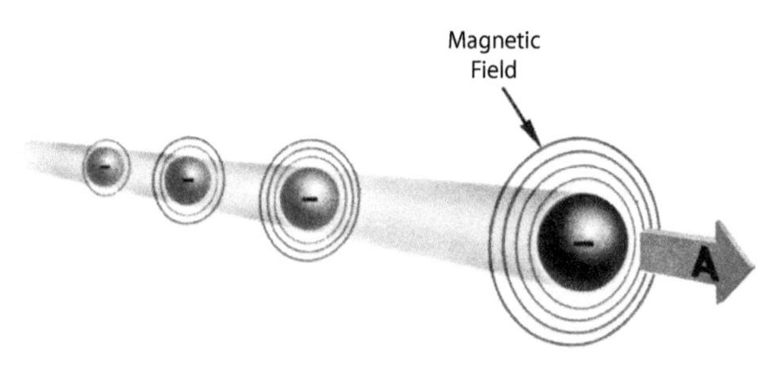

Figure 4-4. Whenever an electric charge accelerates, a changing magnetic field is produced, and electromagnetic energy is radiated.

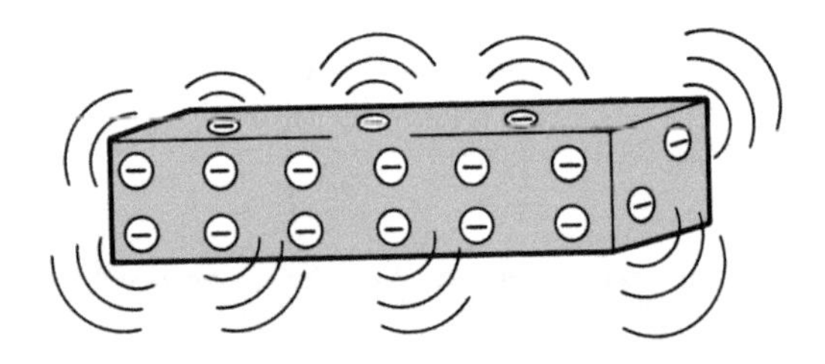

Figure 4-5. Because of thermal agitation, everything around us radiates electromagnetic energy, a tiny portion of which is at radio frequencies.

field it produces. Whenever an electric charge accelerates by changing speed or direction, a changing magnetic field is produced and electromagnetic energy is radiated. Because of thermal agitation of charged particles, everything around us radiates electromagnetic energy, a tiny portion of which is at radio frequencies. For an electric current to flow, whether in a lightning bolt or in a telephone wire, an electric field must exist. And whenever an electric current flows (Fig. 4-2), a magnetic field is produced. The electromagnet is a common example.

If the fields vary with time, the interrelationship extends even further. Any change in a magnetic field, such as an increase or decrease in magnitude produces an electric field. This relationship can be observed in the operation of electric generators and transformers. Similarly, although not so readily apparent, the reverse is true and any change in an electric field produces a magnetic field. A varying electric or magnetic field will cause an electric current to flow in the form of an electromagnetic wave.

In the second half of the nineteenth century, James Clerk Maxwell conceived of the idea that a changing electric field might produce a magnetic field. On the basis of this concept (Fig. 4-3) and the already demonstrated characteristics of electric and magnetic fields, he hypothesized the existence of electromagnetic waves and described their behavior mathematically (Maxwell's equations). Not until some 13 years later was their existence actually demonstrated by Heinrich Hertz.

Electromagnetic Radiation. The dynamic relationship between the electric and magnetic fields gives rise to electromagnetic waves. Because of this, whenever a charge, such as that carried by an electron, accelerates, changes direction or rate of motion, it will change the surrounding fields and electromagnetic energy is radiated (Fig. 4-4). The change in the motion of the charge causes a change in the surrounding magnetic field that is produced by the particle's motion. That change creates a changing electric field a bit further out, which in turn produces a changing magnetic field just beyond it, and on, and on, and on.

It follows that the sources of radiation are countless. As a result of thermal agitation, electrons in all matter are in continual random motion. Consequently, everything around us radiates electromagnetic energy (Fig. 4-5). Most of the energy is in the form of radiant heat (long wavelength infrared). But there is always a tiny fraction in the form of radio waves. Radiant heat, light, and radio waves are, in fact, the same thing: *electromagnetic radiation*. They differ only in frequency.

In contrast to natural radiation, exciting a tuned circuit with a strong electric current produces the waves radiated by a radar system. The waves, therefore, all have substantially the same frequency and contain vastly more energy than the fraction of the natural radiation with the same wavelength.

How Antennas Radiate Energy. For a picture of how radiation takes place, consider a simple elemental antenna in free space. For this purpose, there is no better model than the dipole Hertz used in his original demonstration of radio waves.

This antenna consists of a thin straight conductor with flat plates, like those of a capacitor, at either end (Fig. 4-6). An alternating voltage applied at the center of the conductor causes a current to surge back and forth between the plates. The current produces a continuously changing magnetic field around the conductor. At the same time, the positive and negative charges that alternately build up on the plates as a result of the current flowing in and out of them generate a continuously changing electric field between the plates.

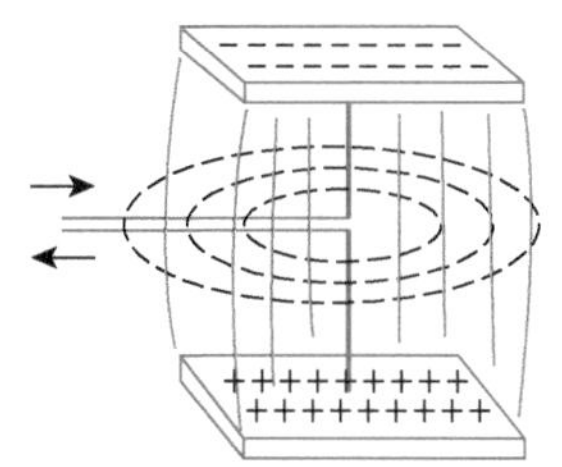

Figure 4-6. This illustration shows a simple dipole antenna such as that used by Hertz to demonstrate radio waves.

The fields are quite strong in the region immediately surrounding the antenna. As with the field of an electromagnet or the field between the plates of a capacitor, most of the energy each field contains returns to the antenna in the course of every oscillation. However, a portion does not. The changing electric field between the plates produces a changing magnetic field just beyond it, which in turn creates a changing electric field just beyond it, and so forth. Similarly, the changing magnetic field surrounding the conductor produces a changing electric field just beyond it, which creates a changing magnetic field just beyond it, and so forth.

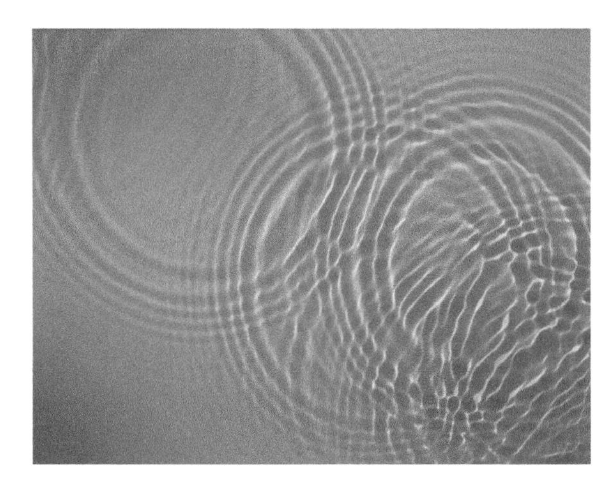

Figure 4-7. Like ripples on a pond, radio waves move outward, long after the disturbance that produced them has ceased.

Within this mutual interchange of energy, the electric and magnetic fields propagate outward from the antenna. Like ripples in a pond around a point where a stone has been thrown in (Fig. 4-7), the fields move outward long after the current that originally produced them has ceased (or in the case of the stone has stopped moving). They and the energy they contain have escaped.

Visualizing a Wave's Field. Although the electric and magnetic fields can't be seen, they can both be visualized quite easily. The electric field may be visualized as the force it would exert on a tiny electrically charged particle suspended in the wave's path. The magnitude of the force corresponds to the field's strength (E) and the direction of the force to the field's direction.[1] As in Figure 4-8, the electric field is commonly portrayed as a series of solid lines whose directions indicate the field's direction and whose density (number of lines per unit of area in a plane normal to the direction) indicates the field strength.

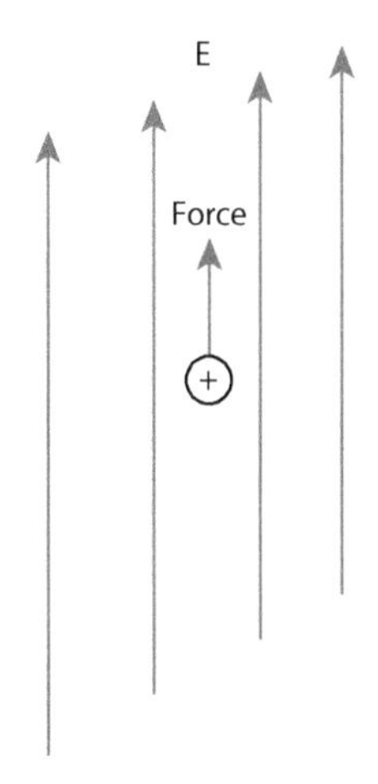

Figure 4-8. The electric field is best visualized as the force it exerts on a charged particle.

The magnetic field may similarly be visualized as the force it would exert on a tiny magnet suspended in the wave's path. The direction of the propagation of the wave is always perpendicular to the directions of both the electric and the magnetic fields. Again, the magnitude of the force corresponds to the field strength (H) and the direction to the field's direction.

1. The direction of travel of the electromagnetic wave is perpendicular to the directions of the electric and magnetic fields that make up the electromagnetic wave.

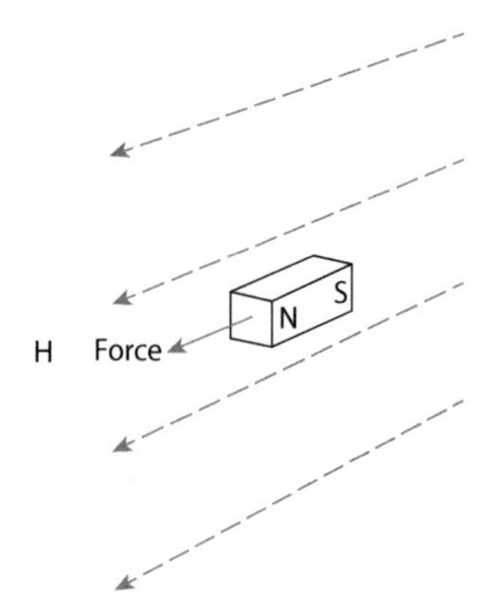

Figure 4-9. The magnetic field is best visualized as the force it would exert on a tiny magnet.

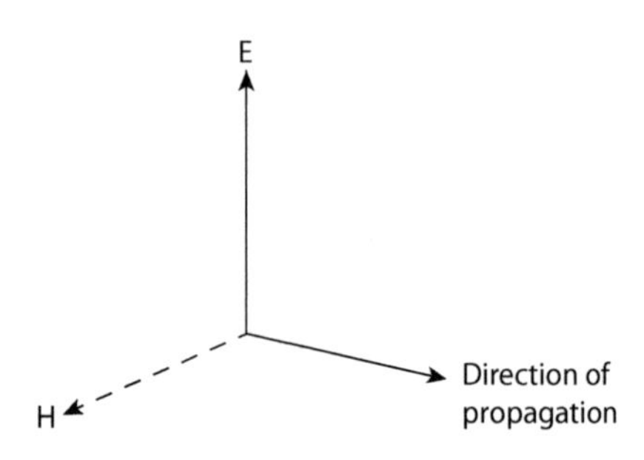

Figure 4-10. The direction of propagation is always perpendicular to the directions of both the electric and the magnetic fields.

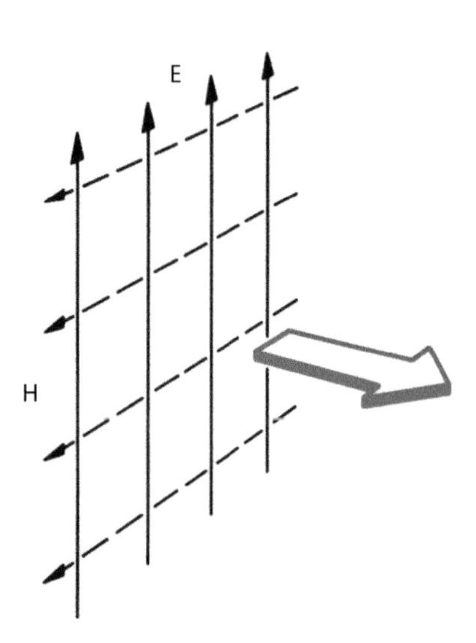

Figure 4-11. In free space, a wave's magnetic field is always perpendicular to its electric field. Direction of travel is perpendicular to both.

This field is portrayed in the same way as the electric field except that the lines are dashed (Fig. 4-9).

4.2 Characteristics of Radio Waves

A radio wave has several fundamental qualities: speed, direction, polarization, intensity, wavelength, frequency, and phase.

Speed. In a vacuum, radio waves travel at constant speed, the speed of light, represented by the letter c. In the earth's atmosphere they travel a tiny bit slower. Moreover, their speed varies slightly not only with the composition of the atmosphere but also with temperature and pressure.

The variation, however, is so extremely small that most practical purposes radio waves can be assumed to travel at a constant speed, the same as that in a vacuum. This speed is very nearly equal to 3×10^8 m/sec. This is the value usually used in radar computations.

Direction. The direction in which a wave travels, that is, the direction of propagation (Fig. 4-10), is always perpendicular to the directions of both the electric and the magnetic fields. These are always such that the direction of propagation is away from the radiator.

When a wave strikes a (retro) reflecting object, the direction of one or the other of the fields is reversed, thereby reversing the direction of propagation. As will be made clear in the blue panel on the next page, which field reverses depends uon the electrical characteristics of the object.

Polarization. This is the term used to express the orientation of the wave's fields. By convention, it is taken as the direction of the electric field (the direction of the force exerted on an electrically charged particle). In free space, outside the immediate vicinity of the radiator the magnetic field is perpendicular to the electrical field (Fig. 4-11) and the direction of propagation is perpendicular to both.

When the electric field is vertical (with respect to the ground), the wave is termed *vertically polarized.* When the electric field is horizontal, the wave is termed *horizontally polarized.*

If the radiating element emitting the wave is a length of thin conductor, the electric field in the direction of maximum radiation will be parallel to the conductor. If the conductor is positioned such that it is vertical, the element is vertically polarized (Fig. 4-12). Conversely, if the conductor

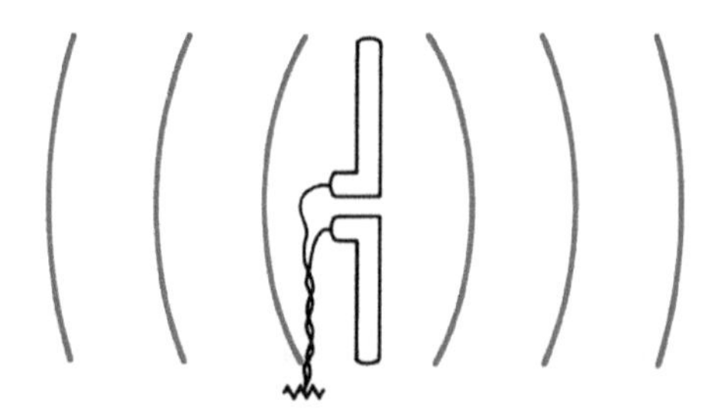

Figure 4-12. If the radiating element is vertical, the element is said to be vertically polarized.

The Speed of Light and Radio Waves

THE SPEED OF LIGHT IN A NONMAGNETIC MEDIUM, SUCH AS THE ATMOSPHERE, IS

$$c = \frac{299.7925 \times 10^6}{(\kappa_e)^{1/2}} \text{ meters/s*}$$

where κ_e is a characteristic, called the dielectric constant, of the medium through which the radiation is propagating. The dielectric constant for air is roughly 1.000536 at sea level. However, this speed is very nearly equal to 3×10^8 m/s and hence is still the value used in the vast majority of radar computations.

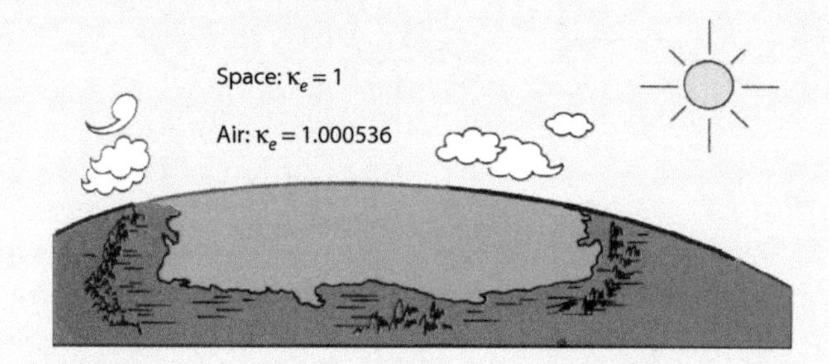

Speed in the Atmosphere. The dielectric constant of the atmosphere varies slightly with the composition, temperature, and pressure of the atmosphere. The variation is such that the speed of light is slightly higher at higher altitudes. The dielectric constant of the atmosphere also varies to some extent with wavelength. As a result, the speeds of light and radio waves are not quite the same, and the speed of radio waves is slightly different in different parts of the radio frequency spectrum.

FREE SPACE: $\kappa_e = 1$

AIR: $\kappa_e = 1.000536$

*From Maxwell's equation, $c = (\mu\varepsilon)^{-1/2}$, where $\mu = \mu_o \mu_m$, and $\varepsilon = \varepsilon_o\kappa_e$. But, $(\mu_0\varepsilon_0)^{-1/2} = 299.7925 \times 10^6$ and, in a nonmagnetic medium, the permeability $\mu_m = 1$.

is positioned so that it is horizontal, the element is horizontally polarized.

A receiving antenna placed in the path of a wave can extract the maximum amount of energy from it if the polarization (orientation) of the antenna and the polarization of the wave are the same. If the polarizations are not the same, the extracted energy is reduced in proportion to the cosine of the angle between them.

When a wave is reflected, the polarization of the reflected wave depends not only on the polarization of the incident wave but also on the structure of the reflecting object. In this way the polarization of radar echoes can, in fact, provide useful information about the object being illuminated.

For the sake of simplicity, the discussion here has been limited to linearly polarized waves (waves whose polarization is the same throughout their length). In some applications, it is desirable to transmit waves whose polarization rotates through 360° in every wavelength (Fig. 4-13). This is called *circular polarization*. It may be achieved by simultaneously transmitting horizontally and vertically polarized waves that are 90° out of phase. In the most general case, polarization is elliptical (i.e., anything other than 90° out of phase). Circular and linear polarizations are special cases of elliptical polarization.

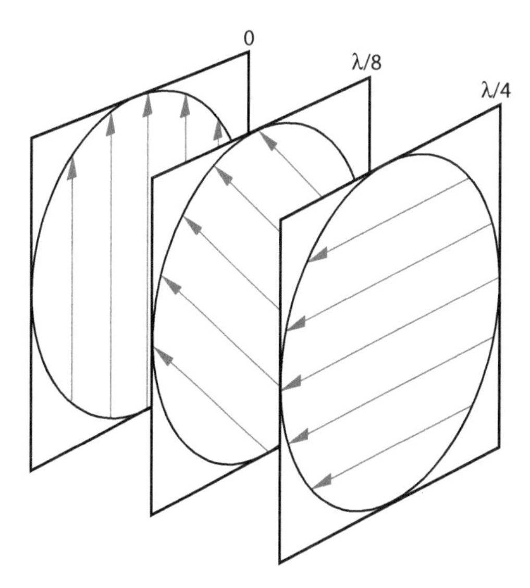

Figure 4-13. Pictured here is polarization of a circularly polarized wave at points separated by 1/8 wavelength. Wave is produced by combining two equal-amplitude waves that are 90° out of phase.

Reflection, Refraction, and Diffraction

ANY OF THREE MECHANISMS MAY CAUSE A RADIO WAVE TO CHANGE DIRECTIONS: reflection (which makes radar possible), refraction, and diffraction or any combination of the three.

Reflection from a Conductive Surface. When a wave strikes a conducting surface, its electric field causes a current to flow that in turn creates a wave whose energy is reradiated, forming the process of *reflection*. From a flat surface (i.e., any irregularities are small compared with a wavelength), reflection is mirror-like and is called *specular*. From an irregular (i.e., any irregularities are of the order of a wavelength or larger) or complex surface (e.g., that of trees or an aircraft), reflection is diffuse, and radiation is scattered in all directions. This is termed *diffuse scattering*.

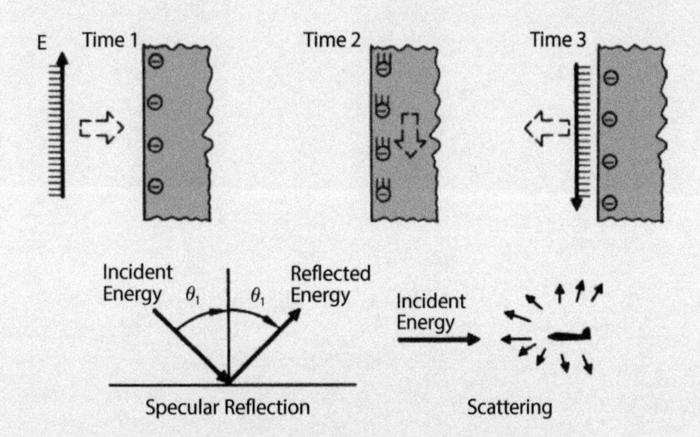

Reflection from a Nonconductive Surface. When a wave enters a nonconducting medium (e.g., Plexiglas) with a different dielectric constant from the medium through which the wave has been propagating (e.g., air), some of the wave's energy is reflected (just as from a conducting surface). The reason is that the dielectric constant, κ_e, of the medium determines the division of energy between the wave's electric and magnetic fields. (In a vacuum, where $\kappa_e = 1$, the energy is divided equally between the two fields.) To adjust the balance to the new dielectric constant, some of the incident energy must be rejected, which occurs through the reflection.

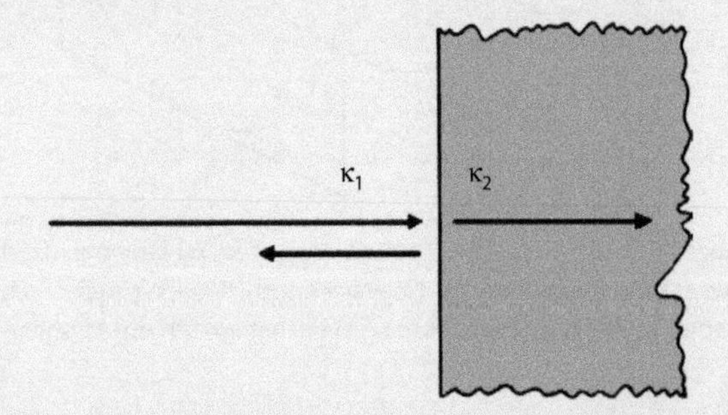

Refraction. If the angle of incidence, θ_1, is greater than zero, when a wave enters a region of different dielectric constant the energy passing through is deflected in a phenomenon called *refraction*. The deflection increases with the angle of incidence and the difference of the two dielectric constants, that is, with the difference between the speeds in the two media.

Assuming that material κ_1 has a higher dielectric constant than κ_2, the wavefront will travels faster in κ_2 than in κ_1. Thus, the portion of the wavefront reaching κ_2 first will travel start to travel at a higher speed in the new material. The portion of the wavefront yet to reach κ_2 continues at its previous speed until it also reaches κ_2. This progressive change in speed of the wavefront causes it to propagate in κ_2 at a larger angle θ_2. The ratio of the velocities in the two media is called the *refractive index*.

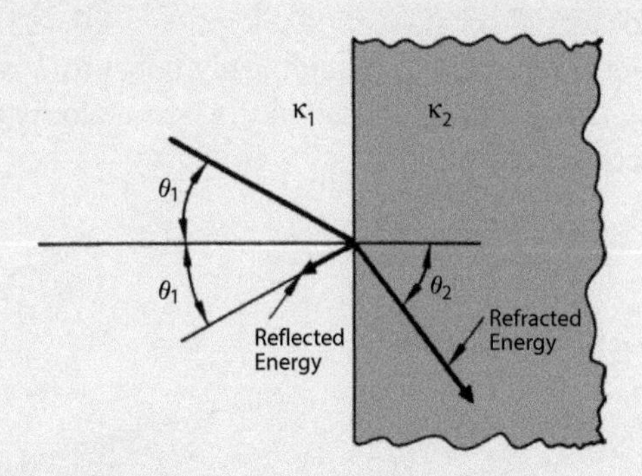

Atmospheric Refraction. A form of refraction occurs in the atmosphere. Because of the increase in the speed of light (decrease in κ_e) with altitude, the path of a horizontally propagating wave gradually bends toward the earth. This phenomenon enables us to see the sun for a short time after it has set. It similarly enables a radar system to see over the horizon.

Diffraction. A wave spreads around objects whose size is comparable to a wavelength and bends around the edges of larger obstructions. For a given size of obstruction, the longer the wavelength, the more significant the effect. That is why radio broadcast stations (operating at wavelengths of a few hundred meters) can be heard in the shadows of buildings and mountains, whereas TV stations operating at wavelengths of only a few meters cannot.

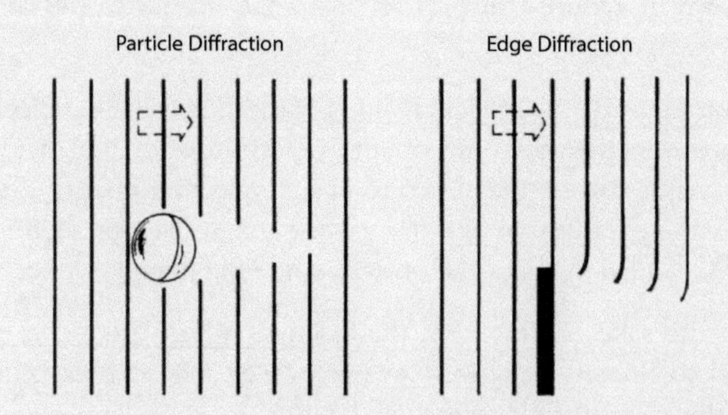

This phenomenon, called *diffraction*, stems from the fact that the energy at each point in a wave is passed on just as if a radiator actually existed at that point. The wave as a whole propagates in a given direction only because the radiation from all points in every wavefront reinforces in that direction and cancels in others. If the wavefronts are broken by an obstruction, cancellation at the edge of the wave is incomplete, which causes the part of the wavefront nearest to the obstruction to propagate differently.

Intensity. The rate at which a radio wave carries energy through space, intensity is defined as the amount of energy flowing per second through a unit of area in a plane normal to the direction of propagation (Fig. 4-14).[2]

2. Other terms for this rate are energy flux and power flow.

The intensity is directly related to the strengths of the electric and magnetic fields. Its instantaneous value equals the product of the strengths of the two fields times the sine of the angle between them. As previously noted, in free space outside the immediate vicinity of the antenna, that angle is 90°; thus, the intensity is simply the product of the two field strengths (EH).

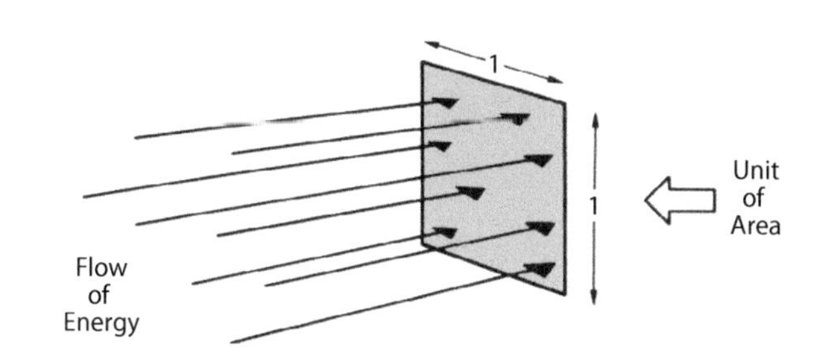

Figure 4-14. Intensity of a wave is the amount of energy flowing per second through a unit of area normal to the direction of propagation.

Generally, what is of interest is not the instantaneous value of the intensity but the average value. If an antenna is interposed at some point in a wave's path, multiplying the wave's average intensity at that point by the area of the antenna gives the amount of energy per second intercepted by the antenna (Fig. 4-15).

In an electrical circuit, the term used for the rate of flow of energy is power. Consequently, in considering the transmission and reception of radio waves, the term power density is often used for the wave's average intensity. The two terms are equivalent. The power of the received signal is the power density of the intercepted wave times the area of the antenna.

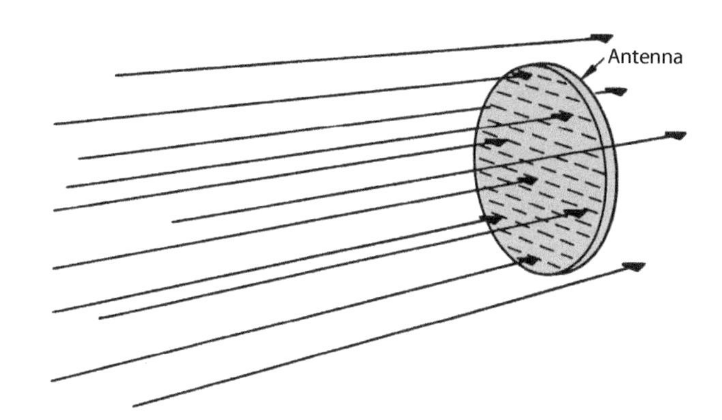

Figure 4-15. Power of received signal equals power density of intercepted wave times area of antenna. (Power density is another term for intensity.)

Wavelength. If a linearly polarized radio wave were frozen in time and its two fields were viewed over some distance in space, two things would be observed. First, the strength of the fields varies cyclically in the direction of the wave's travel. It builds up gradually from zero to its maximum value, returns gradually to zero, builds up to its maximum value again, and so on. The fields in the planes of two successive maxima are shown in Figure 4-16. Second, each time the intensity goes through zero, the directions of both fields are reversed.

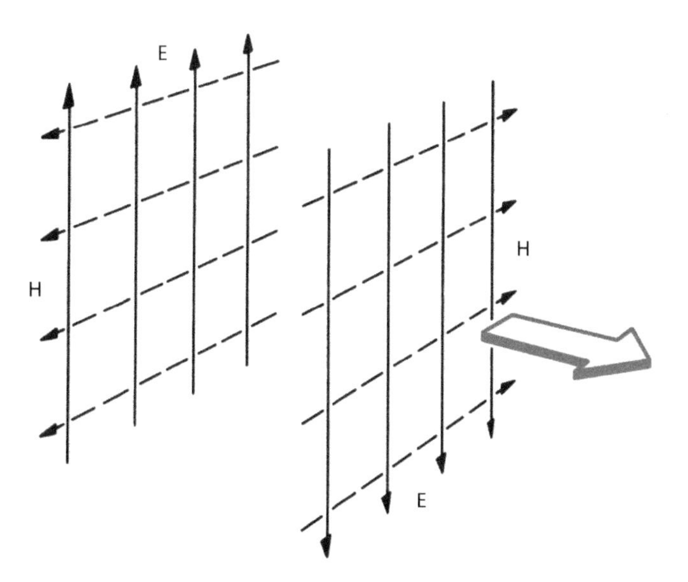

Figure 4-16. The fields of a radio wave, at points of maximum intensity, frozen in space. When intensities go through zero, directions of fields reverse.

The intensity of the fields is plotted versus distance along the direction of travel in Figure 4-17. It is negative when the directions of the forces exerted by the fields are reversed. The shape is the same as a plot of the sine of an angle versus the angle's size (strictly speaking, the wave should be infinitely long). Because of this, radio waves are referred to as sinusoidal, or sine waves.[3]

3. A radio wave will have a pure sinusoidal shape if it is continuous and its peak amplitude, frequency, and phase are constant (i.e., the wave is unmodulated).

Referring again to Figure 4-17, the distance between successive crests, or troughs, is the *wavelength*, which is usually represented by a lowercase Greek lambda, λ, and is expressed in meters, centimeters, or millimeters depending on its length.

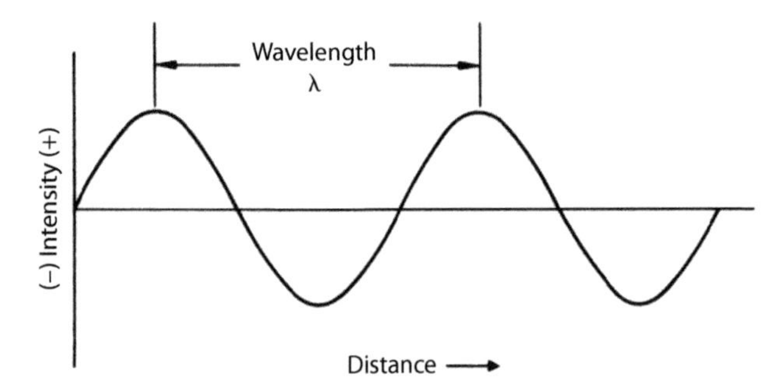

Figure 4-17. This graph shows the variation in intensity of fields in direction of travel. Distance between crests is wavelength.

Frequency. The frequency of a radio wave is directly related to the wavelength. To see the relationship, visualize a radio wave traveling past a fixed point in space. The intensity of the electric and magnetic fields at this point increases and decreases cyclically as the wave goes by.

Placing a receiving antenna in the wave's path and the voltage developed across the antenna terminals is observed on an oscilloscope will reveal that it has the same shape (amplitude versus time) as the earlier plot of the intensity of the fields versus distance along the direction of travel (Fig. 4-17). The number of cycles this signal completes per second is the wave's frequency.

Frequency is usually represented by a lowercase "f" and is expressed in Hertz (Hz) in honor of Heinrich Hertz: 1 Hz is one cycle per second; 1000 Hz is 1 kilohertz (kHz); 1000 000 Hz is 1 megahertz (MHz); 1000 MHz is 1 gigahertz (GHz); and 1000 GHz is a terahertz (THz).

Since a radio wave travels at a constant speed in a given medium, its frequency is inversely proportional to its wavelength. The shorter the wavelength, the more closely spaced the crests and the greater the number of them that will pass a given point in a given period of time—hence the greater the frequency (Fig. 4-18).

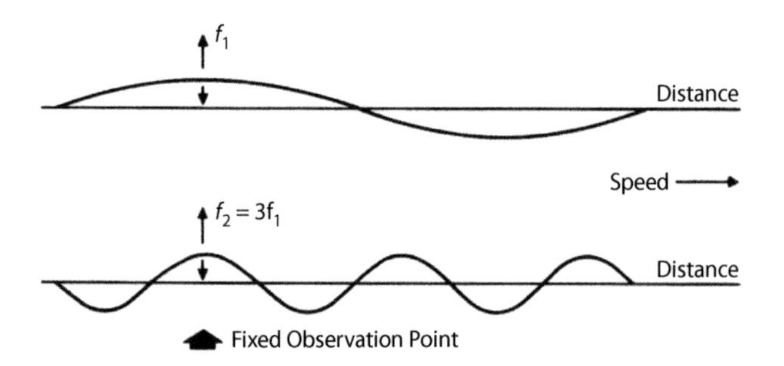

Figure 4-18. Since a radio wave travels at a constant speed, the shorter the wavelength, the higher the frequency.

The constant of proportionality between frequency and wavelength is, of course, the wave's speed. Expressed mathematically,

$$f = \frac{c}{\lambda}$$

where f = frequency, c = speed of the wave (3×10^8 m/sec), and λ = wavelength. With this formula, the frequency corresponding to any wavelength can be quickly found. A wave having a wavelength of 3 cm, for example, has a frequency of 10,000 MHz or 10 GHz.

Knowing the frequency, the wavelength can be found simply by inverting the formula:

$$\lambda = \frac{c}{f}$$

Period. Another measure of frequency is period, T, the length of time a wave or signal takes to complete one cycle (Fig. 4-19). If the frequency is known, the period (in seconds) can be obtained by taking the inverse of the number of cycles per second:

$$\text{Period} = \frac{1}{f}\,(\text{seconds})$$

For example, if the frequency is 1 MHz (i.e., the wave or signal completes one million cycles every second), it will complete one cycle in one-millionth of a second. Its period is one-millionth of a second, or 1 microsecond.

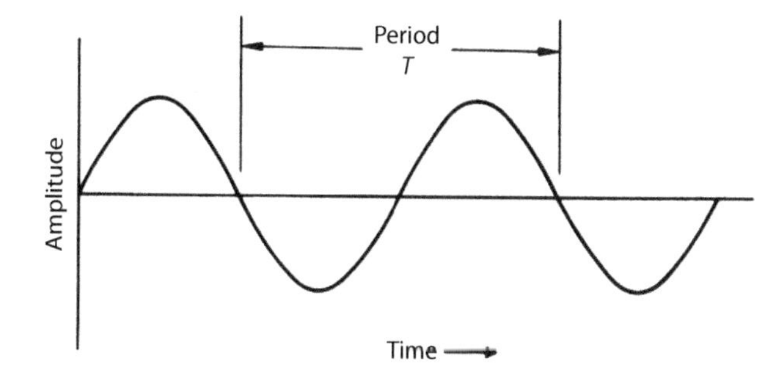

Figure 4-19. Period is length of time a signal takes to complete one cycle.

Phase. This concept is essential to understanding many aspects of radar operation. It is the degree to which the individual cycles of a wave or signal coincide with those of a reference of the same frequency (Fig. 4-20).

Phase is commonly defined in terms of the points in time at which the amplitude of a signal goes through zero in a positive direction. The signal's phase, then, is the amount that these zero crossings lead or lag the corresponding points in the reference signal. This amount can be expressed in several ways. Perhaps the simplest is as a fraction of a wavelength or cycle. However, phase is generally expressed in degrees with 360° corresponding to a complete cycle. If, for instance, a wave is lagging a quarter of a wavelength behind the reference, its phase is $360° \times 1/4 = 90°$. As will be seen later, when the phase of a target echo is repeatedly changing, it indicates the speed of motion of the target or parts of the target.

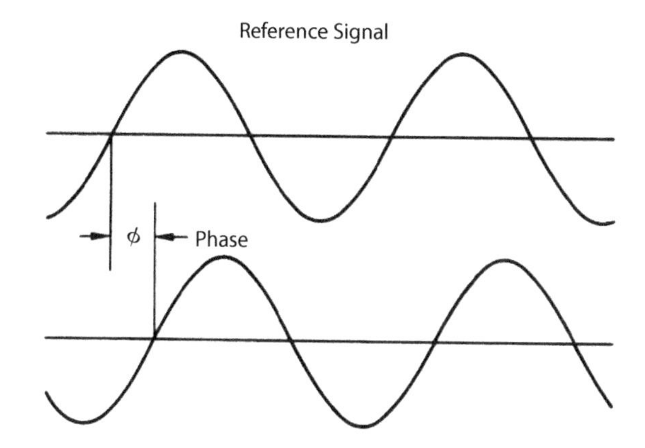

Figure 4-20. Phase is the degree to which the cycles of a wave or signal coincide with those of a reference signal of the same frequency.

4.3 Summary

Radio waves are radiated whenever an electric charge accelerates whether due to thermal agitation in matter or a current surging back and forth through a conductor. Their energy is contained partly in an electric field and partly in a magnetic field. The fields may be visualized in terms of the magnitude and direction of the forces they would exert on an electrically charged particle and a tiny magnet, suspended in the wave's path.

The polarization of the wave is the direction of the electric field. The direction of propagation is always perpendicular to the directions of both fields. In free space at a distance of several wavelengths from the radiator, the magnetic field is perpendicular to the electric field, and the rate of flow of energy equals the product of the magnitudes of the two fields. In an unmodulated signal, the intensity of the fields varies sinusoidally as the wave passes by. The distance between successive crests is the wavelength.

If a receiving antenna is placed in the path of a wave, an AC voltage proportional to the electric field will appear across its terminals. The number of cycles this signal completes per second is the wave's frequency. The length of time the signal takes to complete one cycle is its period. Phase is the fraction of a cycle by which a signal leads or lags a reference signal of the same frequency. It is commonly expressed in degrees. Some relationships to keep in mind are as follows:

- Speed of radio waves $= 3 \times 10^8$ m/s
  $= 300$ m/µs
- Wave length $= \dfrac{300 \times 10^6}{\text{Frequency}}$
- Period $= \dfrac{1}{\text{Frequency}}$

Further Reading

S. E. Schwarz, *Electromagnetics for Engineers*, Oxford University Press, 1995.

K. Lonngren, S. Savov, and R. Jost, *Fundamentals of Electromagnetics with MATLAB®*, 2nd ed., SciTech-IET, 2007.

J. W. Nilsson and S. Reidel, "Sinusoidal Steady-State Analysis," chapter 9 in *Electric Circuits*, Prentice-Hall, 2011.

F. T. Ulaby, E. Michielssen, and U. Ravaioli, *Fundamentals of Applied Electromagnetics*, 6th ed., Pearson, 2014.

Test your understanding

1. How does an antenna radiate energy?
2. What is *diffraction*?
3. Explain what is meant by *polarization*.
4. At what speed do electromagnetic waves travel in free space? How might this differ if the radio waves propagate in the earth's atmosphere?
5. Sketch graphs to show the *wavelength* and *frequency* of a radio wave. Take care to correctly label the axes of the graphs.
6. If a radar system operates at a transmission frequency of 3 GHz, what is the wavelength (the velocity of light can be taken to be 3×10^8 m/s)?
7. For a radar system transmitting at an operating frequency of 3 GHz, what is the period?
8. Describe three mechanisms that cause a radio wave to change direction.
9. Describe what is meant by the phase of a signal.

5

A Nonmathematical Approach to Radar

Modern radar systems are coherent, meaning they measure both amplitude and phase of echo signals. As will shortly be seen, the phase is measured relative to a reference, usually the transmitted signal. Measuring the amplitude and phase provides a powerful basis on which almost all advanced (and some not quite as advanced) techniques are built. A powerful tool often used by the radar engineer to represent the amplitude and phase of a received echo is a graphic device called the *phasor*. Though no more than an arrow, the phasor is key to nonmathematically understanding many seemingly esoteric concepts encountered in radar work such as the spectrum of a pulsed signal, the time-bandwidth product, digital filtering, the formation of real and synthetic antenna beams, and sidelobe reduction.

Unless you are already skilled in the use of phasors, don't yield to the temptation to skip ahead to chapters "about radar." Having mastered the phasor, you will be able to unlock the secrets of many intrinsically simple physical concepts that otherwise you may find yourself struggling to understand. This is because phasors represent the relationships between signals and can be used to combine signals and describe the resultant. As well as being easy to visualize, they have a rigorous mathematical basis so results can be trusted both quantitatively and qualitatively.

This chapter begins by briefly describing the phasor. To demonstrate its application, phasors are then used to explain several basic concepts that are essential to understanding material presented in later chapters. In addition, the decibel (dB) is introduced. It is necessary to become familiar with the dB because it is such a universal measure of many quantities used in radar.

5.1 How a Phasor Represents a Signal

A phasor is nothing more than a rotating arrow (vector), yet it can represent a sinusoidal signal completely (Fig. 5-1). The

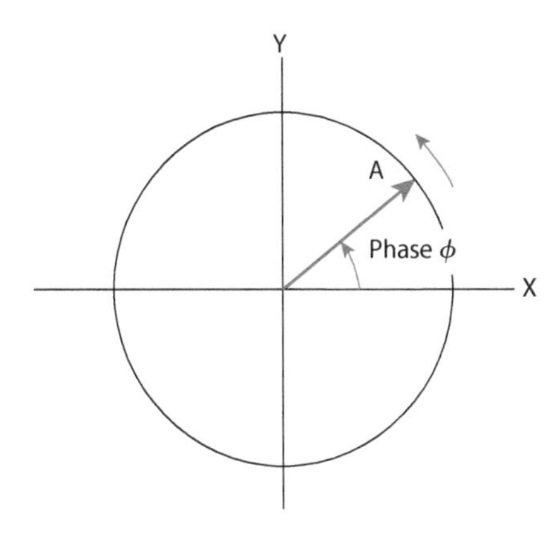

Figure 5-1. A phasor rotates counterclockwise, making one complete revolution for every cycle of the signal it represents.

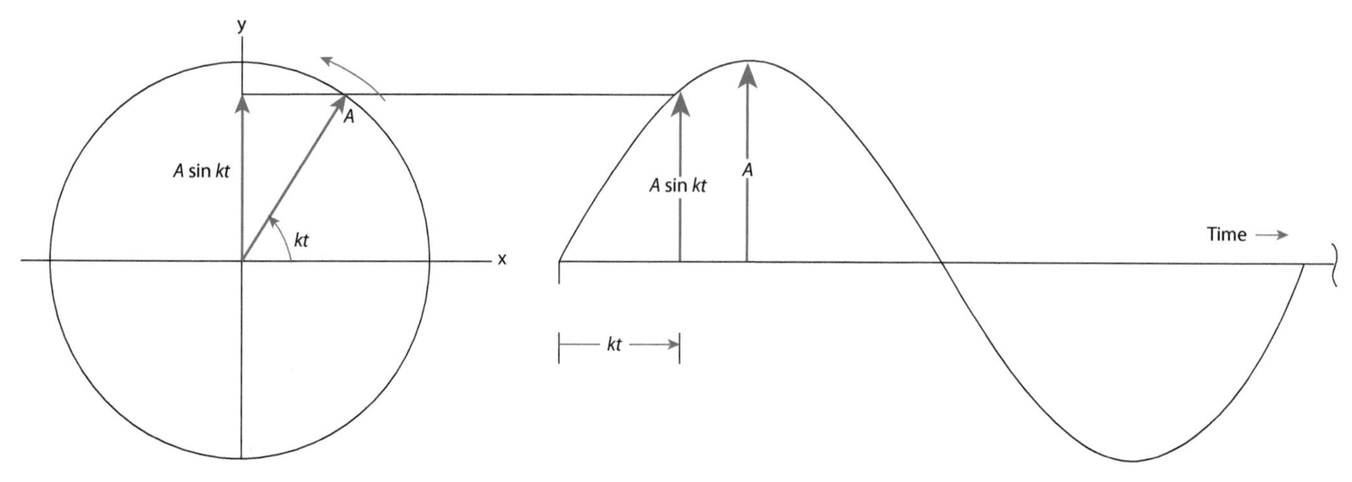

Figure 5-2. For a sine wave, projection of the phasor onto the y axis gives the signal's instantaneous amplitude.

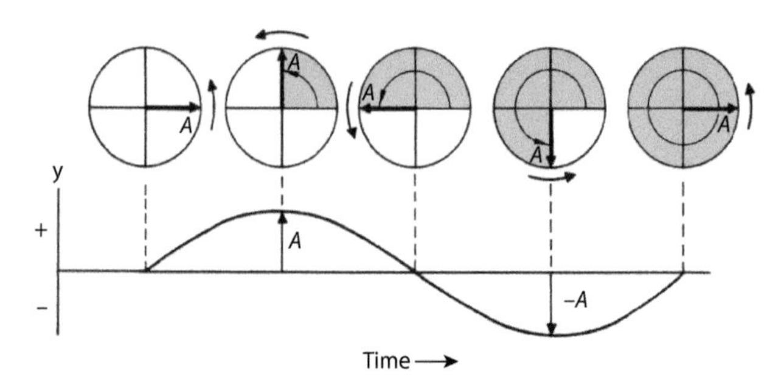

Figure 5-3. As a phasor rotates, projection onto the y axis lengthens to a maximum positive value, returns to zero, lengthens to maximum negative value, and then returns to zero again.

arrow is scaled in length to the signal's peak amplitude. It rotates like the hand of a clock. Phase progression is represented as rotation and is positive in the counterclockwise direction, making one complete revolution for every cycle of the signal. The number of revolutions per second thus equals the signal's frequency.

The length of the projection of the arrow onto a vertical line through the pivot point equals the amplitude times the sine of the angle between the arrow and the horizontal axis (Fig. 5-2). Consequently, if the signal is a sine wave, this projection corresponds to the signal's instantaneous amplitude.

As the arrow rotates (Fig. 5-3), the projection lengthens until it equals the arrow's full length, shrinks to zero, then lengthens in the opposite (negative) direction, and so on, exactly as the instantaneous amplitude of the signal varies with time. If the signal is a cosine wave, the projection on the horizontal axis through the pivot corresponds to the instantaneous amplitude. The 90°-degree angle between the horizontal and vertical axes shows that the cosine wave is a sine wave with a 90°-degree phase shift.

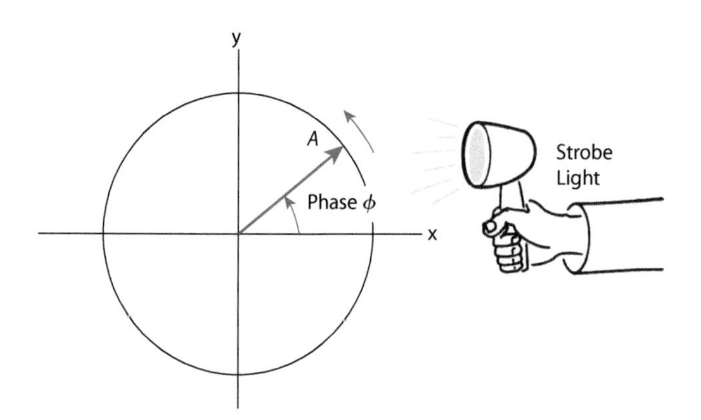

Figure 5-4. A phasor can be thought of as illuminated by a strobe light that flashes on at the same time as a reference phasor would be crossing the x axis. The strobe provides the phase reference.

In the interest of simplicity, the arrow is drawn in a fixed position. It can be thought of as illuminated by a strobe light that flashes on at exactly the same point in every cycle. The strobe point is the instant the arrow would have crossed the x axis had the signal the arrow represents been in phase with a reference signal of the same frequency (Fig. 5-4). In other words, the strobe light is the reference signal or, in radar parlance, the local oscillator (LO) signal.

The angle the arrow makes with the x axis, therefore, corresponds to the signal's phase—and hence the name, phasor. If the signal is in phase with the reference, the phasor will line up with the x axis (Fig. 5-5). If the signal is 90° out of phase (i.e., in quadrature) with the reference, i.e., is in quadrature with it, the phasor will line up with the y axis. For a signal that which leads the reference by 90°, the phasor will point up; for a signal that lags behind the reference by 90°, the phasor will point down.

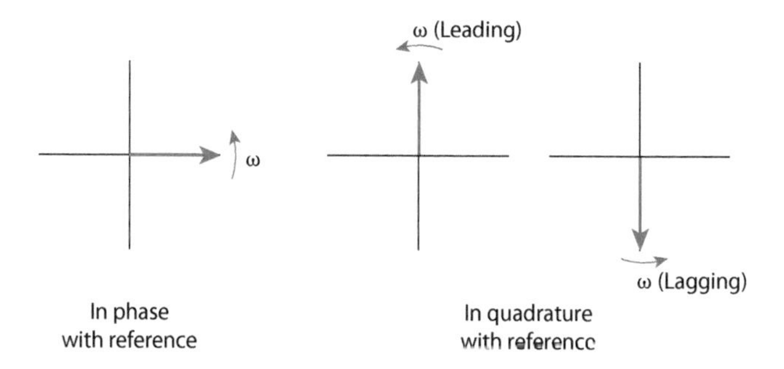

Figure 5-5. If the signal a phasor represents is in phase with the reference (strobe light), the phasor will line up with the x axis. If signal is in quadrature, the phasor will line up with the y axis.

Generally, the rate of rotation of a phasor is represented by the Greek omega, ω. While the value of ω can be expressed in many different units (e.g., in revolutions per second or degrees per second), it is most commonly expressed in radians per second. A radian is an angle that, if drawn from the center of a circle, is subtended by an arc the length of the radius. Since the circumference of a circle is 2π times the radius, the rate of rotation of a phasor in radians per second is 2π times the number of revolutions per second, or the frequency (Fig. 5-6). Thus,

$$\omega = 2\pi f$$

where f is the frequency of the signal, in Hz.

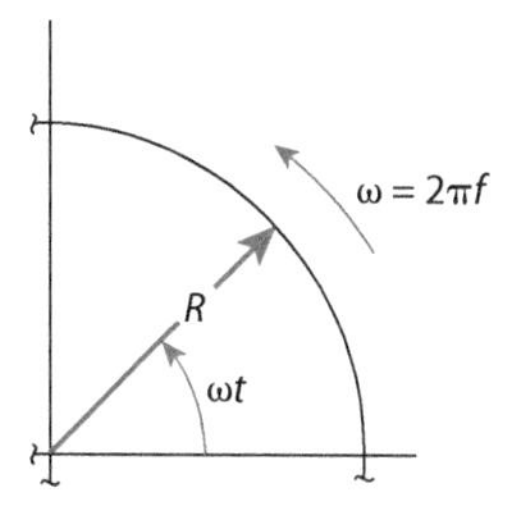

Figure 5-6. Rate of rotation, ω, is generally expressed in radians/second. Since there are 2π radians in a circle, ω = 2πf.

The real power of phasors lies in their ability to represent the relationships between two or more signals clearly and concisely. Phasors may be manipulated to portray the addition of signals of the same frequency but different phases, the addition of signals of different frequencies, and the resolution of signals into in-phase and quadrature components (a key part of modern radar systems). Several common but important aspects of radar operation—including target scintillation, frequency translation, image frequencies, and the creation of sidebands—can illustrate the kind of insights that may be gained from phasors.

5.2 Combining Signals of Different Phase

To see how radio waves of the same frequency but different phases combine, consider drawing two phasors from the same pivot point. Sliding one laterally, one is added to the tip of the other. A third phasor from the pivot point to the tip of the second arrow can then be drawn. This phasor, which rotates counterclockwise in unison with the others, represents their sum (Fig. 5-7).

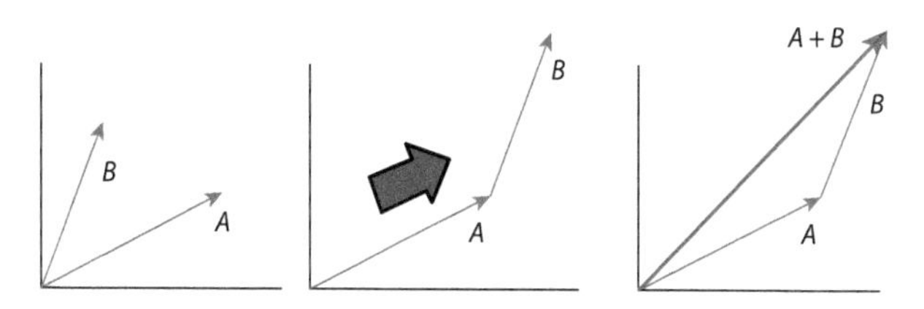

Figure 5-7. To add phasors *A* and *B*, simply slide *B* to the tip of *A*. The sum is a phasor drawn from the origin to the tip of *B*.

The sum can also be obtained without moving the second phasor by constructing a parallelogram with two adjacent sides made up of the phasors to be added. The sum is a phasor drawn from the pivot point to the opposite corner of the parallelogram (Fig. 5-8). The value of such a seemingly simple representation of the sum of two signals can be used to explain target scintillation.

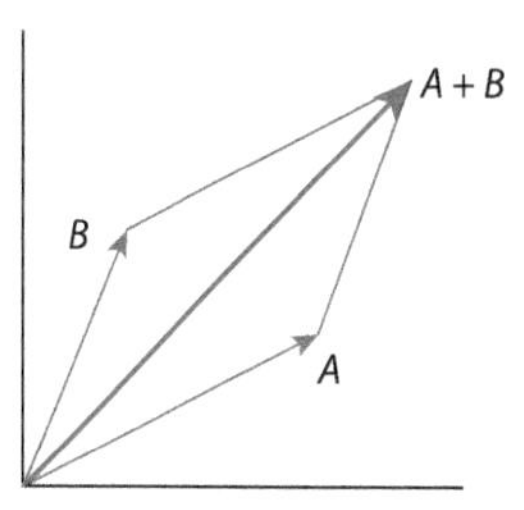

Figure 5-8. Phasors can also be added by constructing a parallelogram and drawing an arrow from the pivot to the opposite corner.

Scintillation. Consider a situation where the reflections of a radar's transmitted waves are received primarily from two

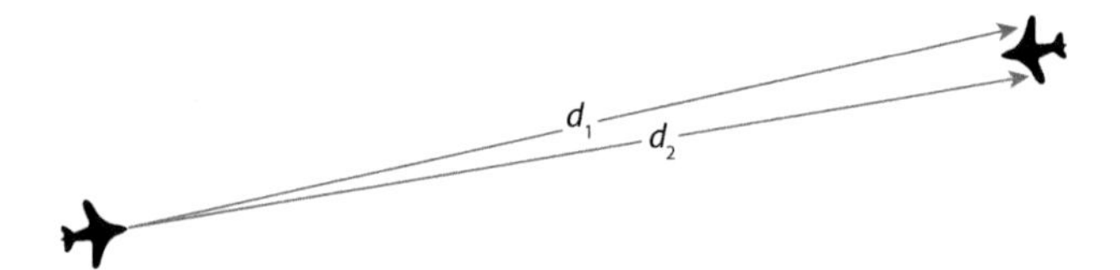

Figure 5-9. In this situation, a radar receives return primarily from two points on a target. Distances to the points are d_1 and d_2.

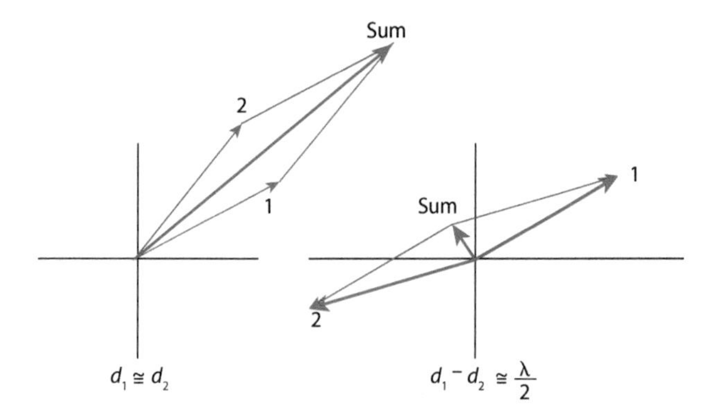

Figure 5-10. If distances d_1 and d_2 to the two points on the target are roughly equal, the combined return will be large, but if the distances differ by roughly half a wavelength, the combined return will be small as they will sum in near anti-phase.

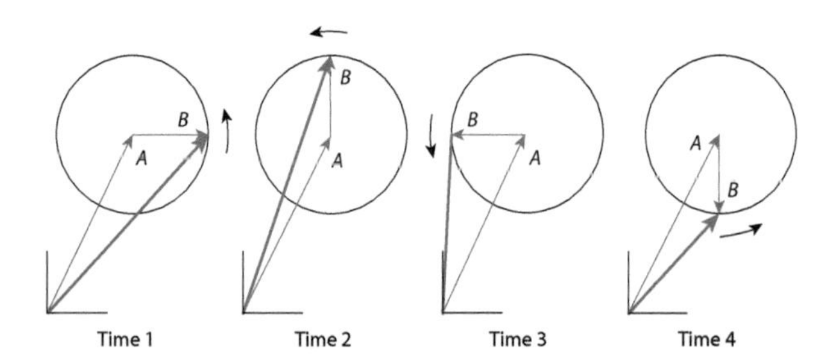

Figure 5-11. How signals of different frequencies combine. If the strobe light is synchronized with the rotation of phasor *A*, it will appear to remain stationary and phasor *B* will rotate relative to it.

parts of a target (Fig. 5-9). The fields of the reflected waves will merge. To see what the resulting wave will be like under various conditions, the waves are represented by phasors.

To begin with, assume that the target's orientation is such that the distances from the radar to the two parts of the target are almost the same (or differ by roughly a whole multiple of a wavelength). Therefore, the two waves are nearly in phase. As illustrated by the left-hand diagram in Figure 5-10, the amplitude of the resulting wave very nearly equals the sum of the amplitudes of the individual waves.

Next, assume that the orientation of the target changes ever so slightly, as it might in normal flight, but enough so that the reflected waves are roughly 180° out of phase. The waves now largely cancel (right-hand diagram in Fig. 5-10).

Clearly, if the phase difference is somewhere between these extremes, the waves neither add nor cancel completely, and their sum has some intermediate value. Thus, the sum may vary wildly from one moment to the next. Recognizing, of course, that appreciable returns may be reflected from many different parts of a target, this wildly varying sum explains why a target's echoes scintillate. This also explains why the maximum detection range of a target is predicted in statistical terms.

What happens to the rest of the reflected energy when the waves don't add up completely? It doesn't disappear. The waves just add up more constructively in directions different to that of the radar receiver.

5.3 Combining Signals of Different Frequency

The application of phasors is not limited to signals of the same frequency. They can also be used to illustrate what happens when two or more signals of different frequency are added together or when the amplitude or phase of a signal of one frequency is varied (i.e., modulated) at a lower frequency.

To see how two signals of slightly different frequency combine, consider drawing a series of phasor diagrams, each showing the relationship between the signals at a progressively later instant in time. If instants are chosen so they are synchronized with the counterclockwise rotation of one of the phasors (i.e., adjusting the frequency of the imaginary strobe light so it is the same as the frequency of one of the phasors), that phasor will occupy the same position in every diagram (phasor *A* in Fig. 5-11).

The second phasor will occupy progressively different positions. The difference from diagram to diagram corresponds to the difference between the two frequencies.

If the difference is positive and the second frequency is higher, the second phasor will rotate counterclockwise relative to the

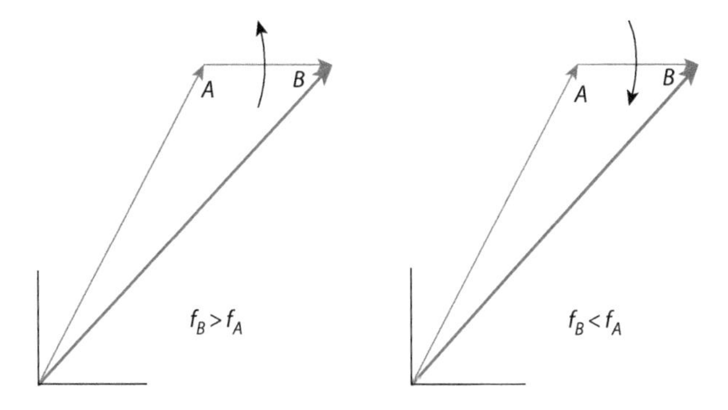

Figure 5-12. If the frequency of *B* is greater than that of *A*, phasor *B* will rotate counterclockwise relative to *A*. Otherwise, it will appear to rotate clockwise.

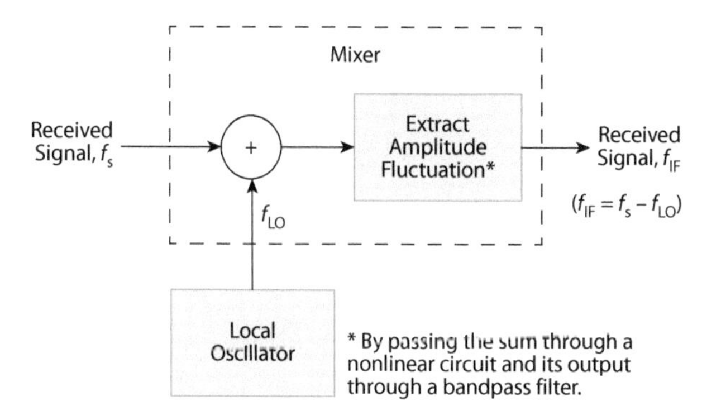

Figure 5-13. A received signal may be translated to a lower frequency f_{IF} by adding it to an LO signal and extracting the amplitude modulation of the sum.

first (Fig. 5-12). If the difference is negative and the second frequency is now lower, the second phasor will rotate clockwise relative to the first.

As the phasors slip into and out of phase, the amplitude of their sum fluctuates (or is modulated) at a rate equal to the difference between the two frequencies. The phase of the sum also is modulated at this rate. It falls behind during one-half of the difference–frequency cycle and slides ahead during the other half. As the phase changes, the rate of rotation of the sum phasor changes: the frequency of the signal is also modulated.

By representing signals of different frequencies in this way, many important aspects of a radar's operation can easily be illustrated graphically using image frequencies or creating sidebands.

Frequency Translation. Since the amplitude of the sum of two phasors fluctuates at a rate equal to the difference between the rates of rotation of the phasors, a signal can be readily shifted down in frequency by any desired amount. Adding one signal to another at a suitably different frequency does this, and then the amplitude fluctuation is extracted. Figure 5-13 shows how this is carried out in a radar receiver. The frequencies of the local oscillator (f_{LO}) and intermediate frequency (f_{IF}) are two very important design parameters for any radar.

In the early stage of virtually every radio or radar receiver, the received signal is translated to a lower intermediate frequency, or IF (Fig. 5-13). Translation is accomplished by mixing the signal with the output of a local oscillator, whose frequency is offset from the signal's frequency by an amount equal to the desired intermediate frequency (f_{IF}).

In one mixing technique, the signal, f_s is simply added to the LO output, as in Figure 5-14, and the fluctuation in the amplitude of the sum is extracted (detected). In another mixing technique, the amplitude of the received signal itself is modulated by the LO output. Amplitude modulation produces image frequencies or sidebands. In this case, the frequency of one of the sidebands is the difference between the frequencies of the received signal and LO signal f_{IF}.[1]

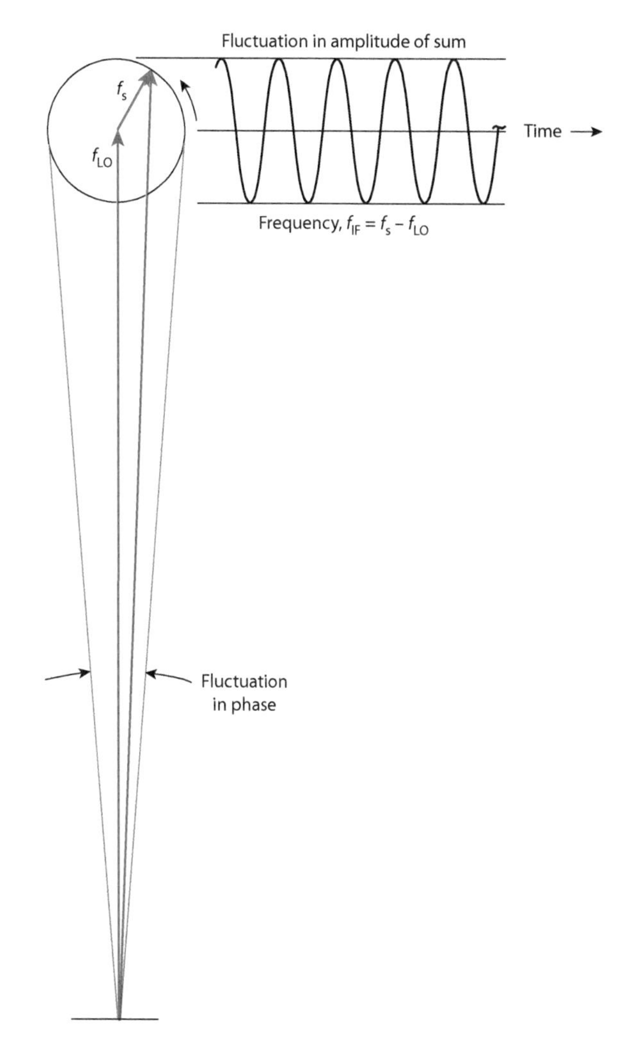

Figure 5-14. If the LO signal is stronger than the received signal, then the fluctuation in amplitude of the sum is virtually identical to the received signal except for being shifted to f_{IF}.

1. For larger frequency differences, these relationships do not necessarily hold. If a phasor's frequency is less than half the reference frequency or is between 1½ and 2, 2½ and 3, 3½ and 4, etc. times the reference frequency, the phasor's apparent rotation will be reversed.

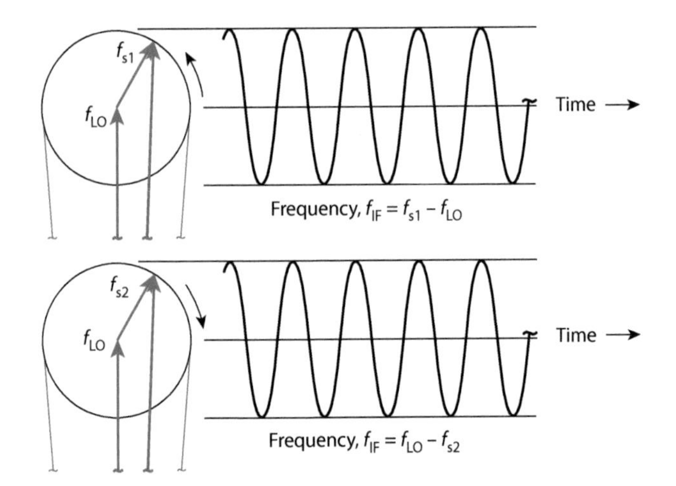

Figure 5-15. Amplitude modulation of sum by signals whose frequencies are above and below f_{LO} by the same amount.

Image Frequencies. The phasor diagram of Figure 5-15 illustrates a subtler aspect of frequency translation. The same amplitude modulation will be produced by a signal whose frequency is above the LO frequency as by one whose frequency is an equal amount below it. The phasors representing the two difference signals rotate in opposite directions, but the effect on the amplitude of the sum is essentially the same. It fluctuates at the difference frequency in either case.

Consequently, if a spurious signal exists whose frequency is the same amount below the LO frequency as the desired signal is above it (or vice versa), both of the signals will be translated to the same intermediate frequency. The spurious signal will thus interfere with the desired signal even though their original frequencies are separated by twice the intermediate frequency. The spurious signal is called an image, and its frequency is called the *image frequency* (Fig. 5-16). Another consequence of images is that noise occurring at the image frequency is added to the noise with which the desired signal must compete. There are solutions to both of these image problems.

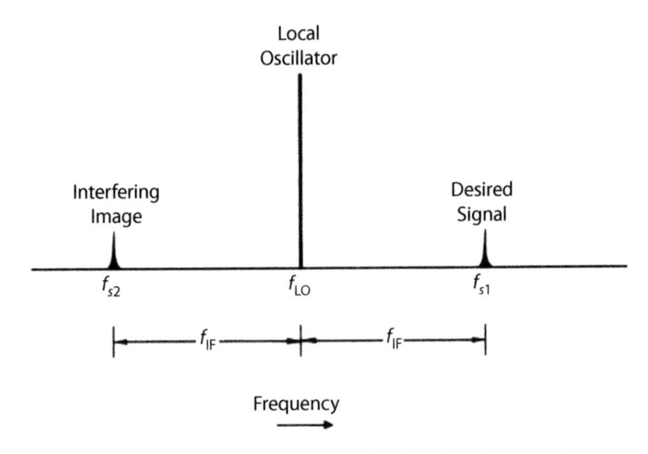

Figure 5-16. If operating frequency is higher than f_{LO}, then the image frequency is $f_{LO} - f_{IF}$, and vice versa.

Creation of Sidebands. When phasors representing two signals of different frequencies are added, the phase modulation of the sum can be eliminated completely by adding a third phasor, which is the same length as the second and rotates at the same rate relative to the first phasor but in the opposite direction (Fig. 5-17). If the counterrotating phasors pass through the axis on the first phasor (vertical axis in Fig. 5-17) simultaneously, the phase modulation will cancel and only the amplitude of the sum will fluctuate. The sum will be a pure amplitude modulation (AM) signal. This is the same sort of signal received from an AM broadcast station.

As in the earlier examples of modulation, the frequency at which the amplitude of the sum is modulated is the difference between the frequency of either one of the counterrotating phasors and the frequency of the fixed phasor. All three phasors rotate in unison with that phasor. But this rotation doesn't show up in the diagram because the imaginary strobe light, which illuminates the phasors, flashes on only once in every cycle of that phasor's rotation.

In some instances AM is actually produced by generating the signals represented by the counterrotating phasors separately and adding them to the signal that is to be modulated. Generally, though, it is the other way around. The signals represented by the counterrotating phasors are the inevitable result of amplitude modulation.

As illustrated by the phasor diagram of Figure 5-18 and readily demonstrated with actual signals, whenever the amplitude of a signal of a given frequency, f_c, is modulated at a

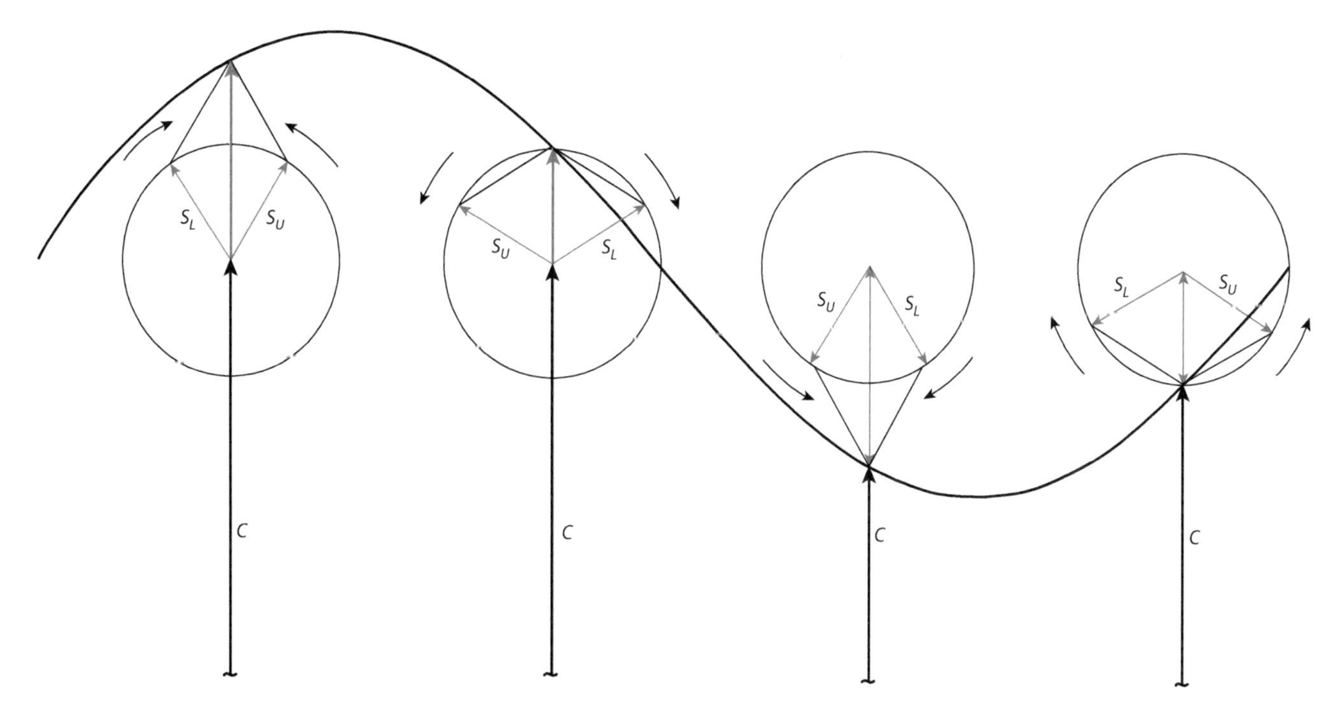

Figure 5-17. If two counterrotating phasors, S_L and S_U, are added to a third phasor, C, and their phases and frequencies are such that all pass through the same axis together, their sum will be a pure amplitude modulated signal.

lower frequency, f_m, two new signals are invariably produced. One of these, represented by the phasor S_U, has a frequency f_m Hz above f_c and another f_m Hz below it, as illustrated in Figure 5-18.

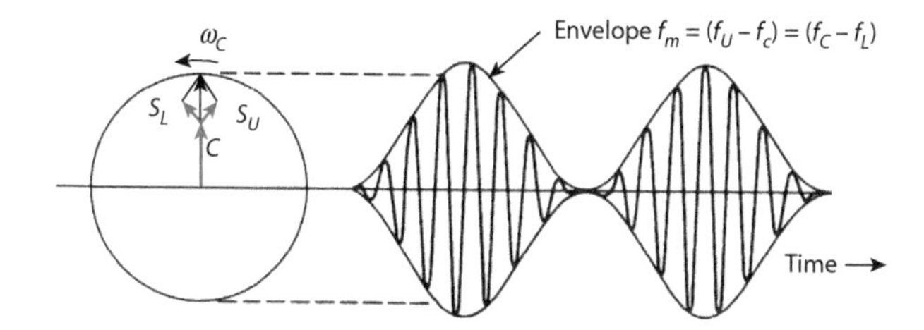

Figure 5-18. If amplitude of a carrier signal C is varied sinusoidally at rate, f_m, two new signals, S_L and S_U, are produced.

Since the frequencies of these signals lie on either side of f_c (Fig. 5-19), the signals are called sideband signals, or simply *sidebands.* Since the signal that is modulated carries the modulation—that is, the modulation is added to and subtracted from the amplitude of this signal—it is called the *carrier.*

The light lines that join the crests of the modulated wave in Figure 5-18 delineate what is called the *modulation envelope.* The frequency of the sidebands is the modulation frequency. The average separation of the sidebands from the baseline is the amplitude of the carrier.

Sidebands are similarly produced when the phase or frequency of a carrier signal is modulated. Only then is the phase

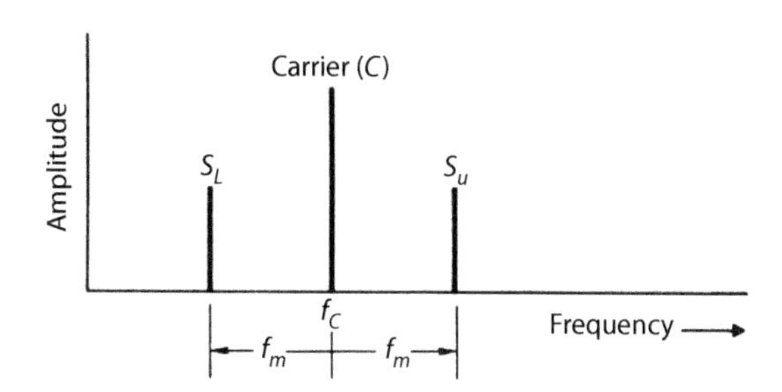

Figure 5-19. Since the frequencies of S_L and S_U are f_m Hz above and below f_c, they are called sidebands.

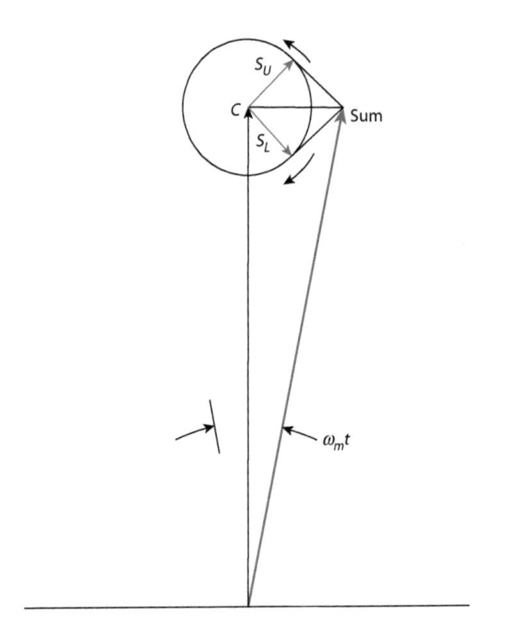

Figure 5-20. Frequency and phase modulation differ from amplitude modulation in that the phase of the sideband signals is shifted by 90°.

relationship of the sidebands to the carrier different (Fig. 5-20). If the percentage by which the phase or frequency varies is large, many sideband pairs separated by multiples of the modulation frequency are created.

The production of sidebands by the transmitter pulsed modulation in some cases causes echoes from a target and a ground patch to be passed by the same Doppler filter even though they have different Doppler frequencies (see Chapter 23 for further details).

5.4 Resolving Signals into In-Phase and Quadrature Components

By resolving a received echo into in-phase (I) and quadrature (Q) components, both phase and amplitude can be recovered. Phase is exploited in techniques such as digital Doppler filtering, synthetic aperture radar (SAR), and electronic beamforming. I and Q components have the same frequency and peak amplitude but differ in phase by 90°. Since a cosine wave reaches its positive peak 90° before a sine wave does, the most convenient way of picturing the two components is as a sine wave ($A \sin \omega\tau$) and a cosine wave ($A \cos \omega\tau$). By convention, the cosine wave is called the I component.[2] Since 90° is one-quarter of a circle, the sine wave is called the Q component.

If the signal is represented by a phasor, the instantaneous amplitude of the I component can be found by projecting the phasor onto the horizontal (x) axis. The instantaneous amplitude of the Q component can be found by projecting the phasor onto the y axis (Fig. 5-21).

For a phasor whose apparent rotation is counterclockwise such that the frequency of the signal (represented by the phasor) is higher than the frequency of the reference signal (strobe light), the I component goes through its positive maximum 90° before the Q component. On the other hand, for a phasor whose apparent rotation is clockwise the frequency of the signal represented by the phasor is lower than that of the reference and the Q component goes through its maximum in a positive direction 90° before the I component.

Distinguishing the Direction of Doppler Shifts. One of the more striking examples of a requirement for resolving signals into I and Q components is found in radars that employ digital Doppler filtering. For digital filtering, the IF output of the receiver must be converted to video frequencies, where the carrier frequency is removed leaving, just the shape or envelope of the signal. Once this conversion has been made, to preserve the sense (positive or negative) of a target's Doppler shift, two video signals must be provided: one corresponding to the cosine of the Doppler frequency (I); and the other to the sine (Q).

A target's Doppler frequency shows up as a progressive shift in the radio frequency phase, ϕ, of successive echoes received from the target, relative to the phase of the pulses transmitted by the radar (see Chapter 15). The phasor diagram in Figure 5-22 illustrates this echo-to-echo phase shift.

2. This convention was adopted because current passing through a resistance is in phase with the voltage across the resistance, whereas a current passing through a reactance either leads or lags behind the voltage by 90°.

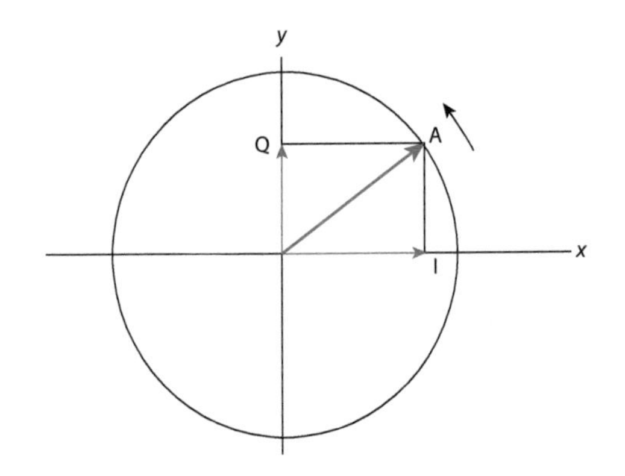

Figure 5-21. Instantaneous values of the I and Q components of a signal are obtained by projecting phasor representation of signal onto both the *x* and *y* axes.

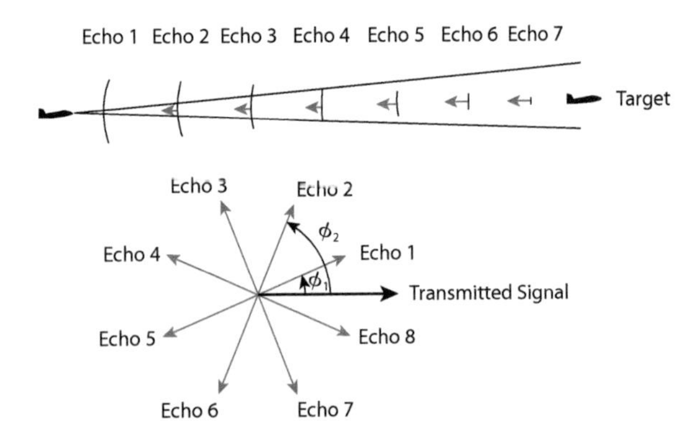

Figure 5-22. A target's Doppler frequency shows up as a pulse-to-pulse shift in phase.

By sensing the progressive phase shift, the radar can produce a video signal whose amplitude fluctuates at the target's Doppler frequency. The signal is illustrated for positive and negative Doppler shifts in Figure 5-23. As the figure clearly shows, however, the fluctuations in the amplitude of this signal are the same for both positive and negative Doppler shifts.

If both the I and Q components of the phase shift are sensed the difference between positive and negative Doppler frequencies may be readily determined. The fluctuation of the Q component will lag behind the fluctuation of the I component if the Doppler shift is positive (Fig. 5-24). Whereas the Q component will lead the fluctuation of the I component if the Doppler shift is negative (Fig. 5-25).

Differentiating between Signals and Images: Image Rejection. Just as it is possible to distinguish between positive and negative Doppler frequencies by resolving the received signals into I and Q components when they are converted from IF to video frequencies, image frequencies can be differentiated from signals when the radar return is translated from the radar's operating frequency to IF. As the phasor diagram of Figure 5-26 illustrates, if a signal's frequency is higher than the LO frequency the Q component of the mixer's output will lag 90° behind the I component. Yet if the signal's frequency is lower than the LO frequency the Q component will lead the I component by 90°. This difference can be exploited in the design of a receiver's mixer stage to reject images.

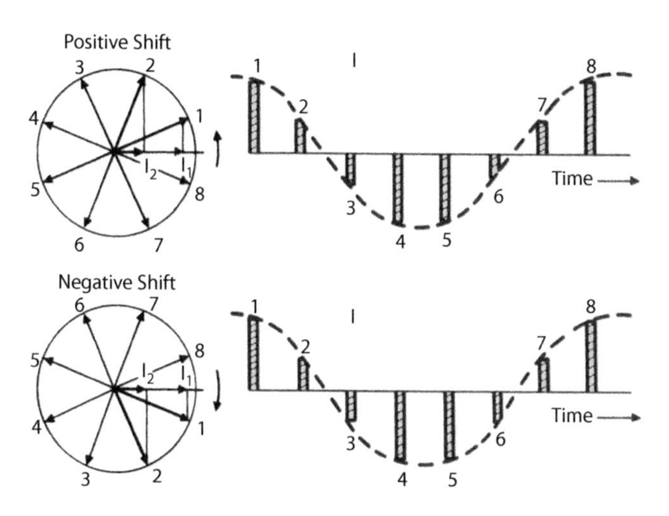

Figure 5-23. Video signal proportional to in-phase component of target echoes fluctuates at the target's Doppler frequency, but fluctuation is the same for both positive and negative Doppler shifts.

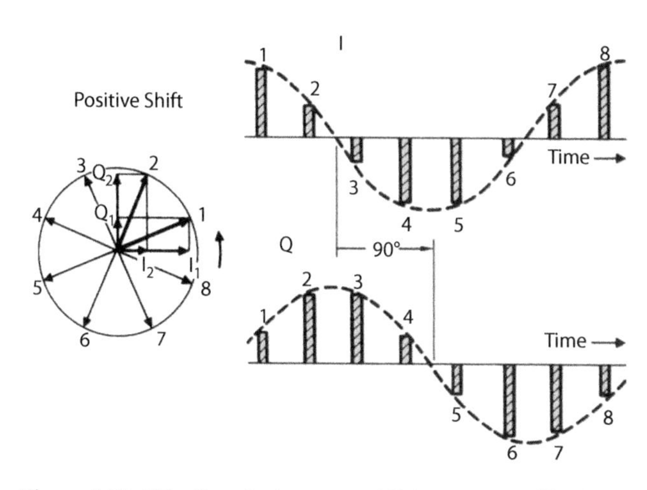

Figure 5-24. If the Doppler frequency shift is positive and both I and Q video signals are provided, Q will lag I by 90°.

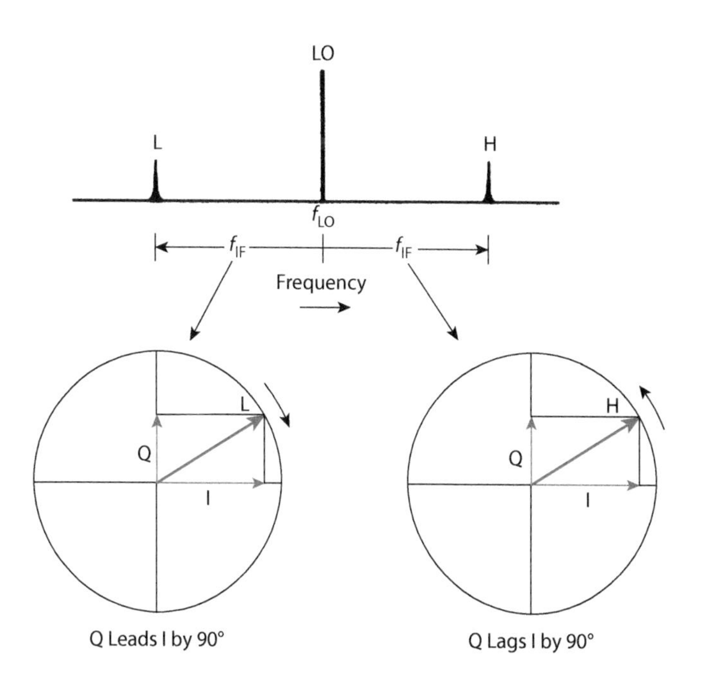

Figure 5-26. The mixer output's Q component will lead the in-phase component if the frequency of the received signal is lower than f_{LO} and will lag behind it if the frequency of the received signal is higher than f_{LO}.

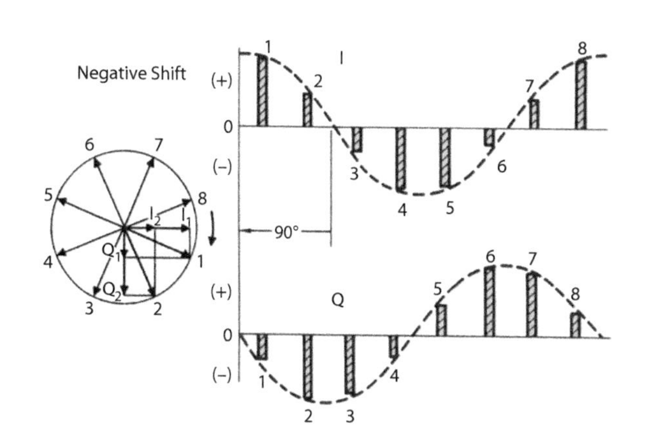

Figure 5-25. If the Doppler frequency shift is negative, Q will lead I by 90°.

The Ubiquitous Decibel

THE DECIBEL (dB) IS ONE OF THE MOST WIDELY USED TOOLS IN THE DESIGN and construction of radar systems. If you are already familiar with decibels, can readily translate to and from them, and feel at ease when the experts start throwing them about, then skip this panel. Otherwise, you will find the few minutes it takes you to read it well worthwhile.

What Decibels Are. The decibel is a logarithmic unit originally devised to express power ratios but is also used today to express a variety of other ratios. Specifically,

$$\text{Power ratio in dB} = 10 \log_{10} \frac{P_2}{P_1}$$

where P_2 and P_1 are the two power levels being compared. For example, if P_2/P_1 is 1000 then the power ratio in decibels is 30.

Origin. Named after Alexander Graham Bell, the unit originated as a measure of attenuation in telephone cable, the ratio of the power of the signal emerging from a cable to the power of the signal fed in at the other end. It so happened that 1 decibel almost exactly equaled the attenuation of 1 mile of standard telephone cable, the unit used until the decibel came along. Also, one decibel relative to the threshold of hearing turned out to be very nearly the smallest ratio of audio-power levels that could be discerned by the human ear, so the dB was soon also adopted in acoustics. From telephone communications, the dB was quite naturally passed on to radio communications and thence to radar.

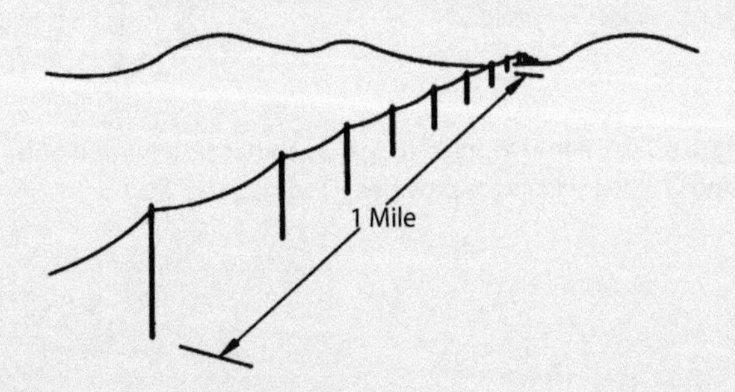

Advantages. Several features of the decibel make it particularly useful to the radar engineer. First, since the decibel is logarithmic, it greatly reduces the size of the numbers required to express large ratios.

A power ratio of 2 to 1 is 3 dB, yet a ratio of 10,000,000 to 1 is only 70 dB. Since the power levels encountered in a radar cover a tremendous range, the compression in the sheer size of numbers that decibels provide is extremely valuable.

In radar, detection performance varies inversely proportional with the fourth power of range. Thus, all other parameters being the same, a change in range from, say, 1 km to 10 km causes a change in detection performance by a factor of 10,000 and such large numbers are typical of radar calculations. In dB 10,000 is just 40, a much smaller number. By tradition, it is usual to express radar parameters in dB.

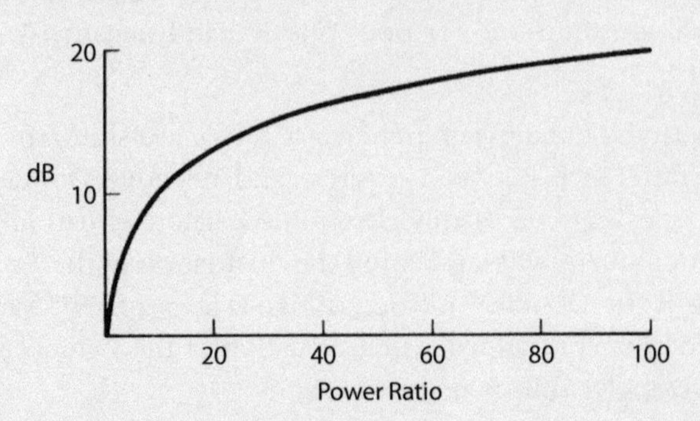

Another advantage also stems from the decibel's logarithmic nature: two numbers expressed as logarithms can be multiplied by simply adding the logarithms. Expressing ratios in decibels therefore makes compound power ratios easier to work with. Multiplying 2500/1 by 63/1 in your head, for example, isn't particularly easy. Yet when these same ratios are expressed in decibels, there is nothing to it: 34 + 18 = 52 dB.

$$\frac{2{,}500}{1} \times \frac{63}{1} = 157{,}500$$

$$34\ \text{dB} + 18\ \text{dB} = 52\ \text{dB}$$

Similarly, with logarithms the reciprocal of a number (one divided by the number) can be obtained by giving the logarithm a negative sign. By merely changing the sign of a ratio expressed in decibels, the ratio can instantly be turned upside down. If 157,500 is 52 dB, then 1/157,500 is –52 dB.

$$52\ \text{dB} = \frac{157{,}500}{1} = 157{,}500$$

$$-52\ \text{dB} = \frac{1}{157{,}500} = 0.000006349$$

When it comes to raising ratios to higher powers or taking roots, these advantages are magnified. If a ratio such as 63 is expressed in decibels, you can square it by multiplying by two: $63^2 = 18\ \text{dB} \times 2 = 36\ \text{dB}$. You can take its fourth root by dividing by four: $\sqrt[4]{63} = 18\ \text{dB} \div 4 = 4\tfrac{1}{2}\ \text{dB}$.

Perhaps the most compelling advantage is that in the world of radar—where detection ranges vary as the one-fourth power of most parameters, target signal powers may vary by factors of trillions, and losses of 20 or 30 percent may be negligible—it is a lot easier to talk and think in terms of decibels than in terms of numbers expressed in scientific notation or ground out of a calculator. Furthermore, by tradition, many radar parameters are commonly expressed in decibels.

To be able to throw decibels about as deftly as a seasoned radar engineer, it is necessary to know only two things: (1) how

The Ubiquitous Decibel *continued*

to convert from power ratios to decibels and vice versa; and (2) how to apply decibels to a few basic characteristics of a radar. If you know the system, both things are surprisingly easy. And the system is really quite simple.

Converting from Power Ratios to dB. You can convert any power ratio (P_2/P_1) to decibels, with any desired degree of accuracy, by dividing P_2 by P_1, finding the logarithm of the result, and multiplying by 10.

$$10 \log_{10} \frac{P_2}{P_1} = \text{dB}$$

Nevertheless, for the accuracy you will normally want, it's not necessary to have a calculator. With the following method, you can do it all in your head—provided you have memorized a few simple numbers.

The first step is to express the ratio as a decimal number, in terms of a power of 10 (scientific notation). A ratio of 10,000/4, for example, is 2500. In scientific notation,

$$2500 = 2.5 \times 10^3$$

When converting to decibels, two portions of this expression are significant: the number 2.5, which we will call the basic power ratio; and the number 3, which is the power of 10.

Now, a ratio expressed in decibels similarly consists of two basic parts: (1) the digit in the ones place (plus any decimal fraction); and (2) the digit or digits to the left of the ones place. The digit in the ones place expresses the basic power ratio: 2.5, in the foregoing example. The digits, if any, to the left of the ones place express the power of 10: in this case, 3.

Incidentally, as you may already have observed, if the power ratio P_2/P_1 is rounded off to the nearest power of 10 (e.g., $2.5 \times 10^3 \approx 10^3$), converting it to decibels is a trivial operation. The basic power ratio then is zero ($\log_{10} 1 = 0$), so the decibel equivalent of P_2/P_1 is simply 10 times the power of 10—in this case 30. Thus,

Power Ratio	Power of 10	dB
1	0	0
10	1	10
100	2	20
1000	3	30
10,000,000	7	70

The basic power ratio, of course, may have any value from 1 to (but not including) 10. So the digit in the ones place can be any number from 0 through 9.999.

The following table gives the basic power ratios for 0 to 9 dB. To simplify the table, all but the ratio for 1 dB have been rounded off to two digits. If you want to become adroit in the use of decibels, you should memorize these ratios.

Power Ratio	dB
1	0
1.26	1
1.6	2
2	3
2.5	4
3.2	5
4	6
5	7
6.3	8
8	9

Returning to our example, if we look up the decibel equivalent of the basic power ratio, 2.5, (or better yet our memory) we find that it is 4 dB. So, expressed in decibels, the complete power ratio, 2.5×10^3, is 34 dB.

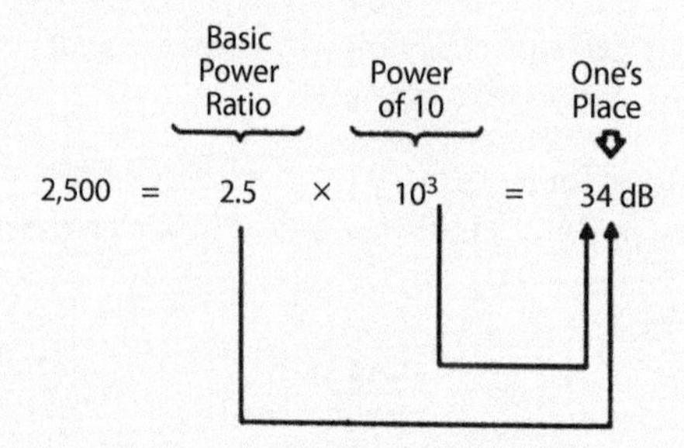

Converting from dB to Power Ratios. To convert from decibels to a power ratio, you can also use a calculator. In this case, you divide the number of decibels by 10 to get the power of 10 and then raise 10 to that power to get the power ratio.

$$\text{Power ratio} = 10^{\text{dB}/10}$$

But you can make the conversion just as easily in your head using the procedure outlined in the preceding paragraphs in reverse.

Suppose, for example, you want to convert 36 dB to the corresponding power ratio. The digit in the ones place, 6, is the dB equivalent of a power ratio of 4. The digit to the left of the ones place, 3, is the power of 10. The power ratio, then, is $4 \times 10^3 = 4{,}000$.

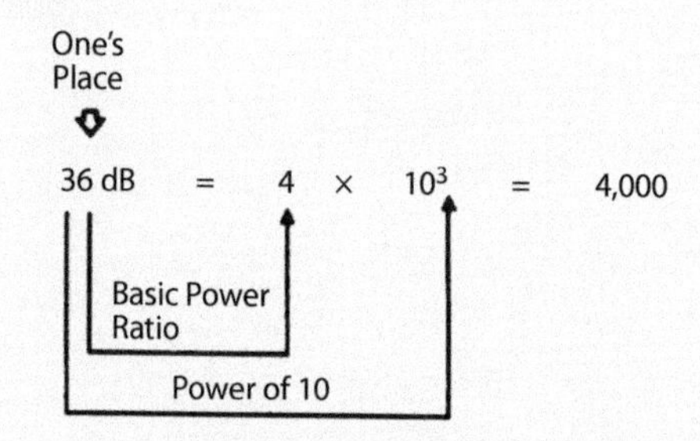

As outlined here, the process may seem a bit laborious, but once you've tried it a few times there is really nothing to it as long as you remember the power ratios corresponding to decibels 1 through 9.

The Ubiquitous Decibel *continued*

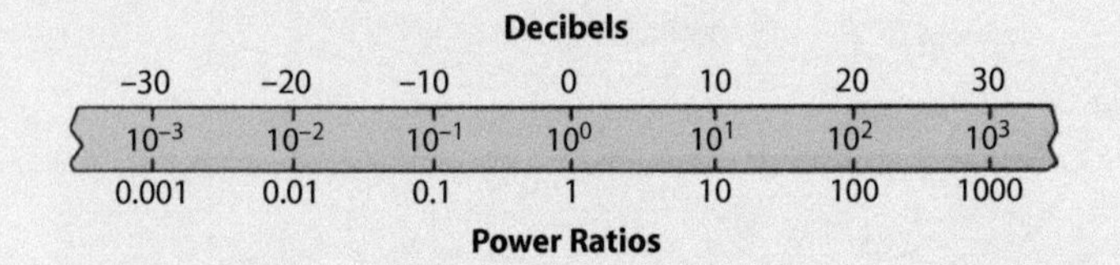

Using Decibels. A common use of decibels in radar work is expressing power gains and power losses.

Gain is the term for an increase in power level. In the case of an amplifier, such as one that might raise a low-power microwave signal to the desired level for radiation by an antenna, gain is the ratio of the power of the signal coming out of the amplifier to the power of the signal going into it.[3]

$$\text{Gain} = \frac{\text{Output power}}{\text{Input power}}$$

If the output power is 250 times the input power, the gain is 250. This ratio (250 to 1) is 24 dB.

Loss is the term for a decrease in power. According to convention, it is the ratio of input power to output power—just the opposite of gain.

$$\text{Loss} = \frac{\text{Input power}}{\text{Output power}}$$

Power Gain in Terms of Voltage. Sometimes it is convenient to express power in terms of voltages. The power dissipated in a resistance equals the voltage, V, applied across the resistance times the current, I, flowing through it: $P = VI$. But the current is equal to the voltage divided by the resistance: $I = V/R$. So the power is equal to (V^2/R).

Accordingly, the power output of a circuit equals $(V_o)^2/R$, and the power input equals $(V_i)^2/R$. If the circuit's input and output impedances are the same, the gain is $(V_o)^2/(V_i)^2$. Expressed in decibels, then, the gain is

$$G = 10\log_{10}\left(\frac{V_o}{V_i}\right)^2 = 20\log_{10}\left(\frac{V_o}{V_i}\right)$$

Decibels as Absolute Units. While decibels were originally used only to express power ratios, they can also be used to express absolute values of power. All that is necessary is to establish some absolute unit of power as a reference. By relating a given value of power to this unit, that value can be expressed with decibels.

A frequently used unit is 1 watt (W). A decibel relative to 1 W is called a dBW. A power of 1 W is 0 dBW; a power of 2 W is 3 dBW; a power of 1 kilowatt (kW) (10^3 W) is 30 dBW.

Another common reference unit is 1 milliwatt (mW). A decibel relative to 1 mW is called a dBm. The dBm is widely used for expressing small signal powers, such as the powers of radar echoes. They vary over a tremendous range. Echoes from a small, distant target may be as weak as −130 dBm or less, while echoes from a short-range target may be as strong as 0 dBm or more. The dynamic range of echo powers is thus at least 130 dB. Considering that −130 dBm is 10^{-13}, or 0.0000000000001 mW the convenience of expressing absolute powers in dBm is striking.

3. Assuming properly matched source and load impedances.

5.5 Summary

This chapter introduced the phasor as a powerful tool for visualizing phase and frequency relationships. Its length corresponds to amplitude; its rate of rotation to frequency; and its angle to phase. The phasor can be drawn in a fixed position by thinking of it as being illuminated by a strobe light that flashes on at the same point in every cycle. If the signal is in phase with the reference, it is drawn horizontally.

If signals of the same frequency are combined, the amplitude of the sum will depend on the relative phases of the signals. Because of this dependence, even a very slight change in target aspect can cause a target's echoes to scintillate.

If signals of different frequency are combined, their sum can be visualized by assuming that the strobe is synchronized with the rotation of one of the phasors, causing it to appear fixed. The other then rotates at the difference frequency.

The amplitude and phase of the sum will be modulated at a rate equal to the difference between the frequencies. The phase modulation can be minimized by making the second signal much stronger than the first. By extracting the amplitude modulation, the first signal can be translated to the difference frequency. At the same time, however, a signal whose frequency is offset from that of the first signal by the same amount in the opposite direction (image) will also be translated to the difference frequency.

Whenever a carrier signal's amplitude is modulated, two sideband signals are produced. Their frequencies are separated from the carrier by the modulation frequency.

Resolution of a signal into in-phase and quadrature components can be visualized by projecting the phasor representing the signal onto the x and y coordinates. Resolving the IF output of a receiver into I and Q components when it is converted to video enables a digital filter to differentiate between positive and negative Doppler frequencies.

The decibel was devised to express power ratios. Being logarithmic, it greatly compresses the numbers needed to express values having a wide dynamic range.

Decibels also make compounding ratios easy. Ratios can be multiplied by adding their decibel equivalents, divided (inverted) by giving them a negative sign, and raised to a power by multiplying them by that power.

A ratio expressed in dB can be thought of as consisting of two parts. The digit in the ones place expresses the basic ratio. The digit to the left of it is the power of 10. To translate from dB to a power ratio in your head, you convert the basic ratio and then place a number of zeros to the right of it equal to the power of ten. To translate to decibels, you do the reverse.

Positive decibels correspond to ratios greater than 1; zero decibels to a ratio of 1; negative decibels to ratios less than 1. There is no decibel equivalent for a ratio of 0.

Decibels are commonly used to express gains and losses. Gain is output divided by input. Loss is input divided by output.

Referenced to absolute units, decibels are also used to express absolute values.

Further Reading

J. W. Nilsson and S. Reidel, "Sinusoidal Steady-State Analysis," chapter 9 in *Electric Circuits*, Prentice-Hall, 2011.

Test your understanding

1. How does a rotating phasor map to a sine wave representation of a radar signal?
2. With the aid of a phasor diagram, show how two signals of the same frequency but different phase combine.
3. How can a phasor diagram represent the combination of signals with different frequencies?
4. A radar system has a transmission frequency of 10 GHz; what is the angular frequency of this signal?
5. Explain the terms *image frequencies* and *sidebands.*
6. How do I and Q components allow both the amplitude and phase of a signal to be recovered?
7. How is a signal resolved into I and Q components?
8. Explain the term *scintillation.*
9. How do I and Q components allow the direction of a Doppler signal to be determined?
10. Two radar echoes differ in amplitude by a factor of 10. What is the power ratio as expressed in dB?

Some Relationships to Keep in Mind

- Power ratio

  $$\text{dB} = 10\log_{10}\frac{P_2}{P_1}$$

- Power ratio in terms of voltages

  $$\text{dB} = 20\log_{10}\frac{V_2}{V_1}$$

- 1 dB = 1¼
- 3 dB = 2
- dBW = dB relative to 1 Watt
- dBm = dB relative to 1 milliwatt
- dBsm = dB relative to 1 square meter of radar cross section
- dBi = dB relative to isotropic radiation

Boeing B-52 Stratofortress (1955)

The B-52 Stratofortress is one of a select number of aircraft that has been in service for more than half of the history of powered flight. It is a long-range, subsonic, jet-powered strategic bomber with a crew of five, designed and built in huge numbers by Boeing, and has been operated by the United States Airforce (USAF) since the 1950s. An upgrade program being conducted through 2015 will take its expected service into the 2040s.

Antenna of Australian Wedgetail AEW radar

6

Preparatory Math for Radar

The previous chapter introduced a nonmathematical way of viewing and manipulating radar signals. Here we provide some of the basic mathematical tools needed to support a more thorough yet still quite simple description of radar systems. These tools are used to synthesize and analyze signals and also to support signal-processing approaches used for target detection. They also relate directly to the phasor description of the last chapter, and both the mathematical and nonmathematical approaches are fully compatible with each other. Further, this chapter shows that phasors are also key to a mathematical understanding of radar enabling both visualization of phase and frequency relationships and their manipulation through complex number representation. First, the different types of signals used within radar systems are formally defined.

However, if you prefer to grasp just the concepts underpinning radar without worrying about the underlying mathematics, feel free to skip this chapter.

6.1 Signal Classification

Signals can be classified in a variety of ways. Perhaps the simplest is the *periodic signal.* (Most radar waveforms are periodic.) A periodic signal, f, repeats itself after a fixed length of time. This can be written as

$$f(t) = f(t + T)$$

where T is the repetition period.

If a signal doesn't repeat itself after a fixed amount of time it is said to be *nonperiodic* or *aperiodic*. Signals are generally classed as being *energy signals* or *power signals.*

An (analog) energy signal is defined as one where the total energy, E, dissipated between the start and end of time (which will be taken to be the start and end of the signal) is nonzero and finite:

$$E = \int_{-\infty}^{\infty} f^2(t)\, dt$$

Radar signals are examples of energy signals as the energy can be obtained simply by integrating the power over the duration of the pulse.

If the average power delivered by a signal from the beginning to the end of time is nonzero and finite, the signal is defined to be a power signal:

$$P = \lim_{T \to \infty} \frac{1}{2T} \int_{-T}^{T} f^2(t)\, dt$$

One example of a power signal is direct current (DC). Power signals are more common in communication systems. For periodic signals the limits of integration can be set by the period leading to an expression for the average power that can be written as

$$P = \frac{1}{T} \int_{0}^{T} f^2(t)\, dt$$

Energy and power signals thus represent two very different classes. If a signal has finite energy its power will be zero and cannot therefore be a power signal. If a signal has finite power its energy will be infinite.

6.2 Complex Numbers

Complex numbers are much less complex than their name suggests and provide a direct relationship to the nonmathematical phasor description of signals. The sine and cosine functions from which phasors were derived can be expressed in exponential as well as trigonometric forms.[1] For example:

$$A \sin \omega t = \frac{A(e^{j\omega t} - e^{-j\omega t})}{2j}$$

$$A \cos \omega t = \frac{A(e^{j\omega t} + e^{-j\omega t})}{2}$$

The letter "j" in the exponential terms has the "value" $\sqrt{-1}$. Because $\sqrt{-1}$ cannot be evaluated, it is said to be an imaginary number. A variable (or number) having an imaginary part and a real part is called a complex variable (or number).

Often, sinusoidal functions are more easily manipulated in exponential form rather than in trigonometric form. At first glance, the exponential terms $e^{j\omega t}$ and $e^{-j\omega t}$ seem to have little physical meaning. However, the functions they represent can be visualized quite easily with phasors. In a phasor diagram, e^{j} is a rotation in a counterclockwise direction and e^{-j} is a rotation in a clockwise direction.

1. The equivalence can be demonstrated by expanding the functions sin x, cos x, and e^{jx} into power series with Maclaurin's theorem.

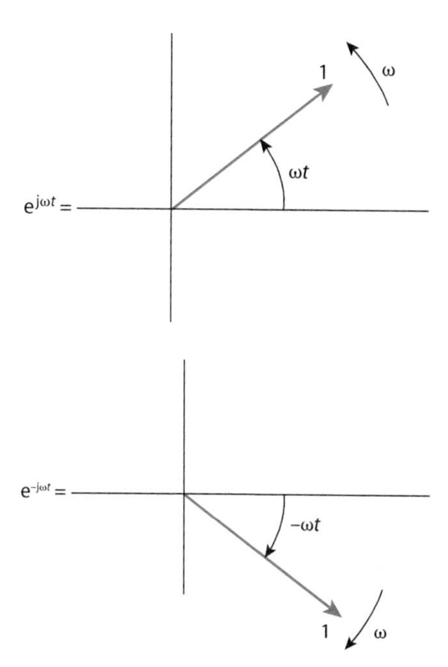

Figure 6-1. Phasors of unit length are pictured here rotating counter-clockwise, $e^{j\omega t}$, and clockwise, $e^{-j\omega t}$, at a rate of ω radians per second.

The term $e^{j\omega t}$ then is represented by a phasor of unit length rotating counterclockwise at a rate of ω radians per second. The term $e^{-j\omega t}$ is a phasor of unit length rotating in a clockwise direction also at a rate of ω radians per second. This is illustrated in Figure 6-1.

The sum $e^{j\omega t} + e^{-j\omega t}$ equals the sum of the projections of the two phasors onto the x axis and is shown in Figure 6-2. This sum is $2\cos\omega t$ (remember to take one of the phasors and start it from the tip of the other to obtain the resultant).

The difference $e^{j\omega t} - e^{-j\omega t}$ equals the projection of the first phasor onto the y axis minus the projection of the second phasor onto the y axis as shown in Figure 6-3. This difference is $2\sin\omega t$ (again take one of the phasors and start it from the tip of the other to obtain the resultant).

Using these basic relationships as building blocks and remembering the values of j raised to various powers,

$$\begin{aligned} j &= \sqrt{-1} \\ j^2 &= \sqrt{-1}\sqrt{-1} = -1 \\ j^3 &= -1\sqrt{-1} = -j \\ j^4 &= (-1)(-1) = +1 \end{aligned}$$

one can easily visualize virtually any relationships involving the complex variable.

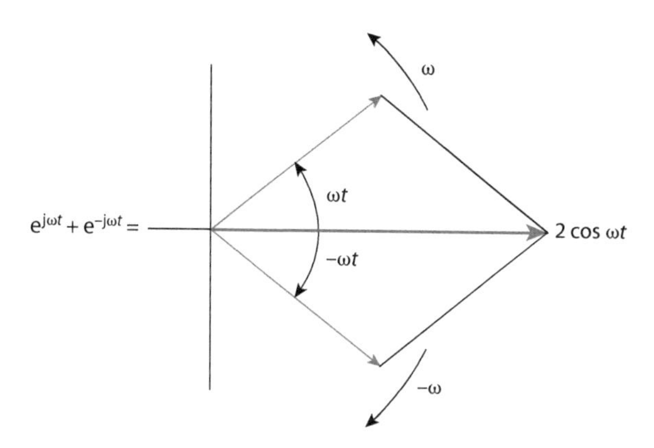

Figure 6-2. This graph illustrates the sum of the projections of two phasors.

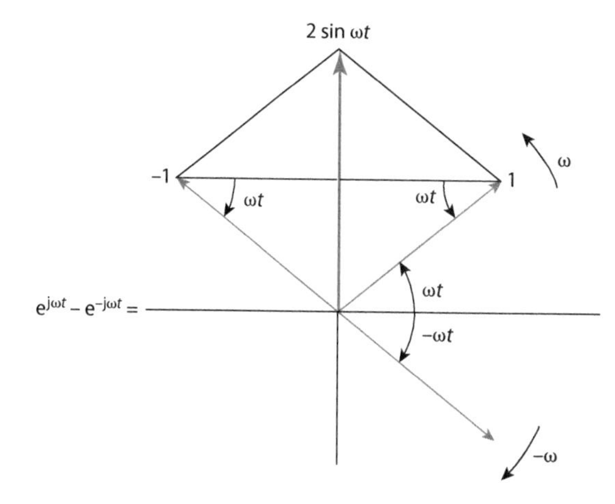

Figure 6-3. This shows the projection of the first phasor onto the y axis minus the projection of the second phasor onto the y axis. This difference is 2 sin ωt.

6.3 Fourier Series

The Fourier series and the Fourier transform (see Section 6.4) are named after Jean Baptiste Joseph Fourier (1768–1830), a French mathematician and physicist. It can be demonstrated both mathematically and graphically that any continuous, periodically repeated waveform, such as a pulsed signal, can be represented by a series of sine waves of specific amplitudes and phases whose frequencies are integer multiples of the repetition frequency of the wave shape. The repetition frequency is called the *fundamental*; the multiples of it are called *harmonics*. The mathematical expression for this collection of waves is the Fourier series.

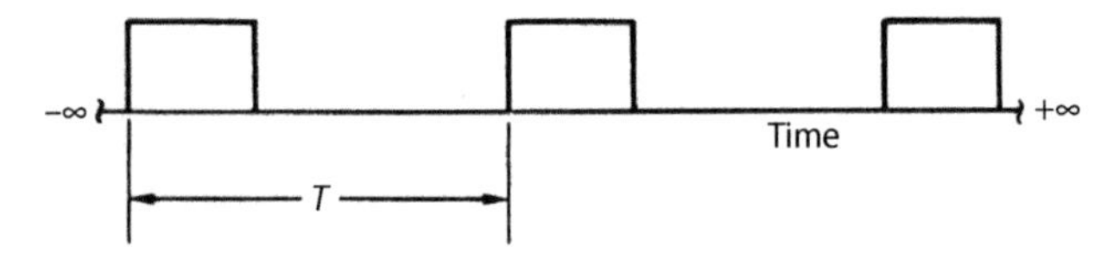

Figure 6-4. This periodic pulse train has period T (=1/f_r).

Mathematical Description of a Fourier Series. The mathematical description begins with any well-behaved periodic function of time, $f(t+T)$, as illustrated in Figure 6-4. T is the period of repetition (and is assumed to continue from the beginning to the end of time).

The function of time, $f(t+T)$, is represented by the sum of a constant, a_0, plus a series of sine terms whose frequencies are integer multiples of the repetition frequency, f_r (= $1/T$). These sine terms comprise the various harmonic components of the periodic signal. For example,

$$f(t) = A_0 + \underbrace{A_1 \sin(\omega_0 t + \phi_1)}_{\text{(First Harmonic)}} + \underbrace{A_2 \sin(2\omega_0 t + \phi_2)}_{\text{(Second Harmonic)}} + \underbrace{A_3 \sin(3\omega_0 t + \phi_3)}_{\text{(Third Harmonic)}} + \underbrace{A_4 \sin(4\omega_0 t + \phi_4)}_{\text{(Fourth Harmonic)}} \cdots$$

where $\omega_0 = 2\pi f_r$, $f_r = 1/T$, and ϕ_1, ϕ_2, ϕ_3, ϕ_4 ... are the phases of the harmonics.

The phase angles can be eliminated by resolving the terms into in-phase and quadrature components as illustrated in Figure 6-5.

$$f(t) = a_0 + \underbrace{a_1 \cos(\omega_0 t) + b_1 (\sin(\omega_0 t)}_{\text{(First Harmonic)}} + \underbrace{a_2 \cos(2\omega_0 t) + b_2 \sin(2\omega_0 t)}_{\text{(Second Harmonic)}} + \underbrace{a_3 \cos(3\omega_0 t) + b_3 \sin(3\omega_0 t)}_{\text{(Third Harmonic)}} \cdots$$

The complete series can be written compactly as the summation of n terms for which n has values of 1, 2, 3...

$$f(t) = a_0 + \sum_{n=1}^{\infty} a_n \cos n\,\omega_0\, t + b_n \sin n\omega_0\, t$$

This is the general form of the Fourier trigonometric series. The coefficients a_n and b_n are the Fourier sine and cosine coefficients representing the amplitudes of the various components comprising the signal. The constant term a_0 represents the

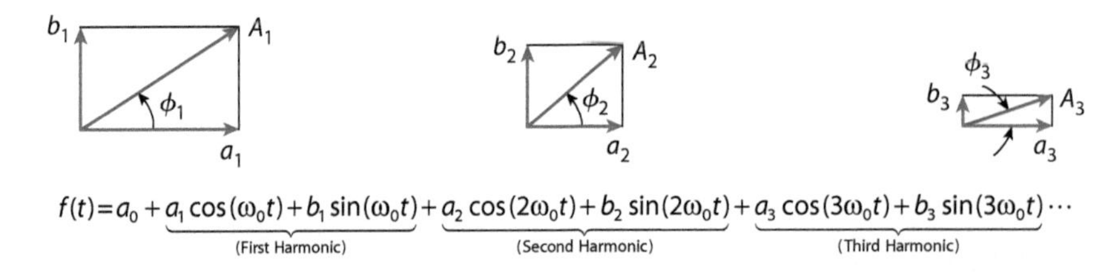

Figure 6-5. This equation resolves the phasors into I and Q components.

constant or DC component of the signal. Although the sum of the series is strictly a periodic variation, it can be used to describe an arbitrary signal over an interval of length T.

Graphical Description of the Fourier Series. The concept underlying the Fourier series is illustrated graphically for a square wave in Figure 6-6.

The sum in Figure 6-6 is obtained by adding the fundamental, third, and fifth harmonics. The reinforcement and cancellation of the different components creates a sum signal that begins to approximate a square wave. By adding more and more harmonics the approximation to the square wave gets better and better. Note that the shape of the composite wave depends as much on the phases of the harmonics as on their amplitudes. To produce a rectangular wave, the phases must be such that all harmonics go through a positive or negative maximum at the same time as the fundamental.

A wave of a more general rectangular shape is illustrated in Figure 6-7. Theoretically, to produce a perfectly rectangular wave an infinite number of harmonics would be required. Actually, the amplitudes of the higher order harmonics are

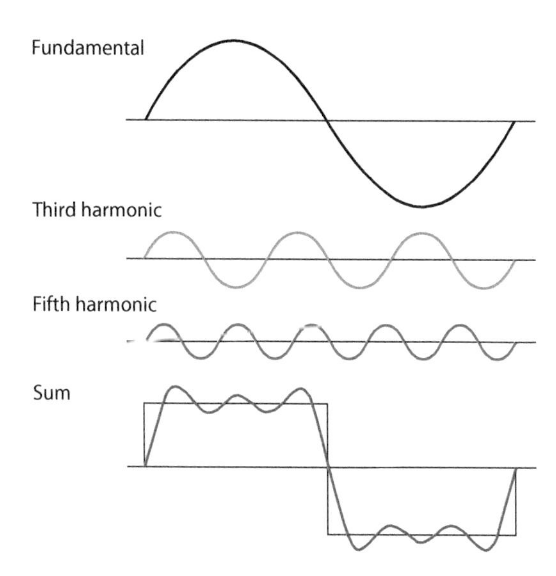

Figure 6-6. This square wave is produced by adding two harmonics to the fundamental. Because positive and negative excursions are of equal duration, amplitude of even harmonics is zero.

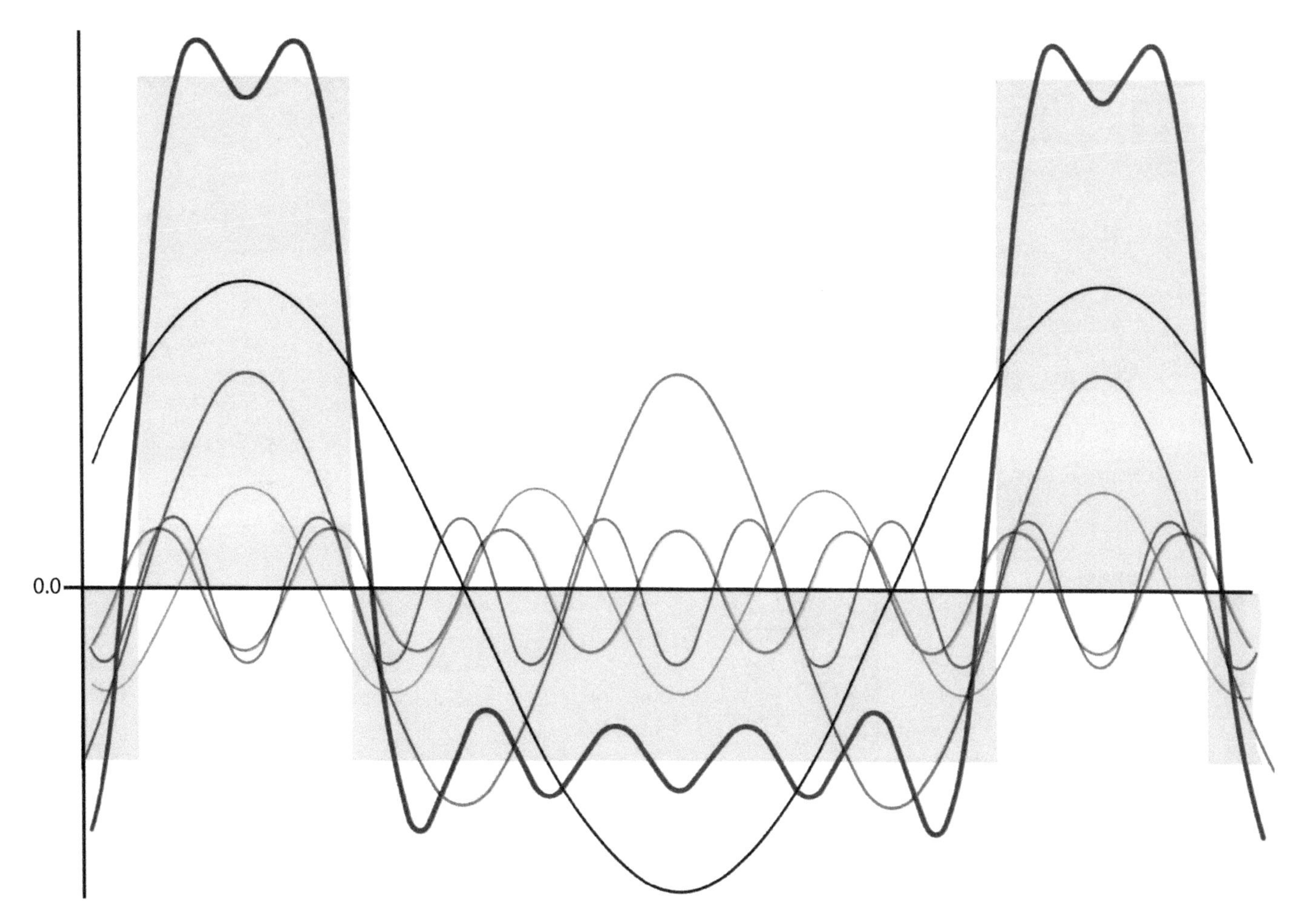

Figure 6-7. This rectangular wave is produced by adding four harmonics to the fundamental. Shape of composite wave is determined by relative amplitudes and phases of harmonics. For shape to be rectangular, all harmonics must go through a positive or negative maximum at the same time as the fundamental.

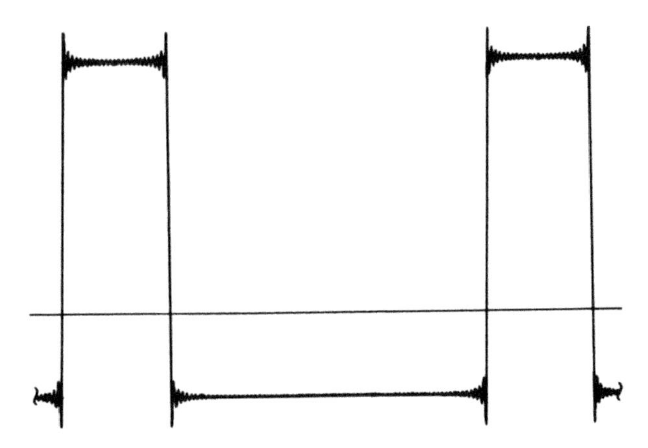

Figure 6-8. This rectangular wave shape is produced by combining 100 harmonics. Note the reduction in ripple.

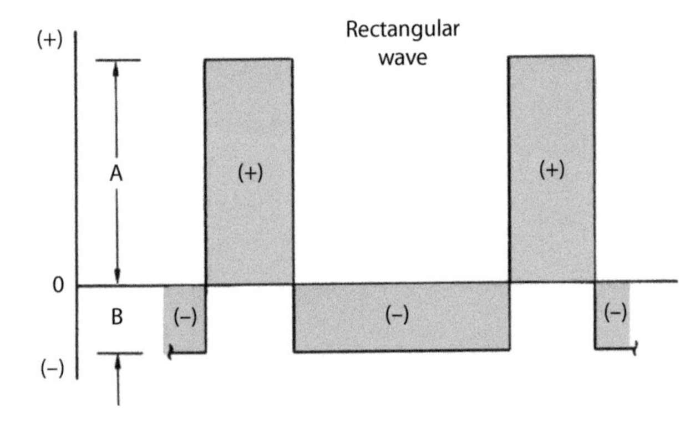

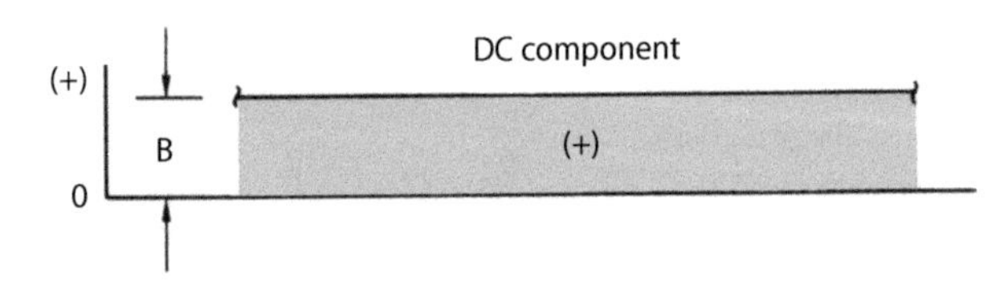

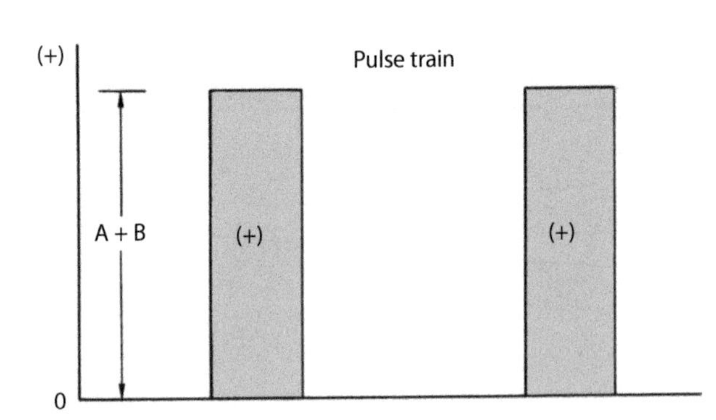

Figure 6-9. To produce a train of rectangular pulses from a rectangular wave, a DC component must be added.

relatively small so reasonably rectangular wave shapes can be produced with a limited number of harmonics. A recognizably rectangular wave was produced in Figure 6-6 by adding only two harmonics to the fundamental. And a still more rectangular wave has been produced in Figure 6-7 by adding only four harmonics.

The more harmonics included, the more rectangular the wave will be and the less pronounced the ripple.

In Figure 6-8, the ripple has been reduced to negligible proportions, except at the sharp corners, by including 100 harmonics.

To create a train of *pulses* (i.e., a waveform whose amplitude alternates between zero and, say, one), such as A + B, with a series of sine waves, a zero-frequency or DC component must be added.

Its value, B, equals the amplitude of the negative loops of the rectangular wave, with sign reversed (Fig. 6-9).

Exponential Form of the Fourier Series. The Fourier series can also be expressed in exponential form using complex numbers. To derive the exponential form of the Fourier series, the sines and cosines are replaced by the complex numbers.

$$a_n \cos(n\omega_0 t) = \frac{a_n}{2}\left[e^{jn\omega_0 t} + e^{-jn\omega_0 t}\right]$$

$$b_n \sin(n\omega_0 t) = \frac{b_n}{2j}\left[e^{jn\omega_0 t} - e^{-jn\omega_0 t}\right]$$

This leads directly to

$$a_n \cos(n\omega_0 t) + b_n \sin(n\omega_0 t) = X_n e^{jn\omega_0 t} + X_{-n} e^{-jn\omega_0 t}$$

where

$$X_n = \frac{1}{2}(a_n - jb_n); \quad X_{-n} = \frac{1}{2}(a_n + jb_n)$$

The general form of this exponential series describing a periodic signal is therefore

$$\begin{aligned} f(t) &= X_0 + X_1 e^{j\omega_0 t} + X_2\, e^{j2\omega_0 t} + \cdots \\ &\cdots + X_{-1} e^{-j\omega_0 t} + X_{-2}\, e^{-j2\omega_0 t} + \cdots \\ &= \sum_{n=-\infty}^{n=\infty} X_n\, e^{jn\omega_0 t} \end{aligned}$$

The coefficients X_n are complex and specify the amplitudes and phases of the harmonic components describing the signal $f(t)$.

6.4 The Fourier Transform

The Fourier transform is a mathematical construct that allows a time domain signal to be represented in the frequency domain and vice versa.

By switching between the time and the frequency domains, signals that can appear very scrambled in one domain may have a very clear structure in the other. The Fourier transform enables signals to be viewed in both the time and frequency domains. Further, in switching from time to frequency the Fourier transform is directly related to Doppler motion and hence is a basic tool in the detection of moving targets. The Fourier transform has a wide variety of other roles and is used in radar (and many other) topics as diverse as antennas (Chapter 8) and synthetic aperture radar (SAR) processing (Section VII).

Time and Frequency Domains. A graph (or equation) relating the amplitude of a signal to time represents the signal in the *time domain* (Fig. 6-10).

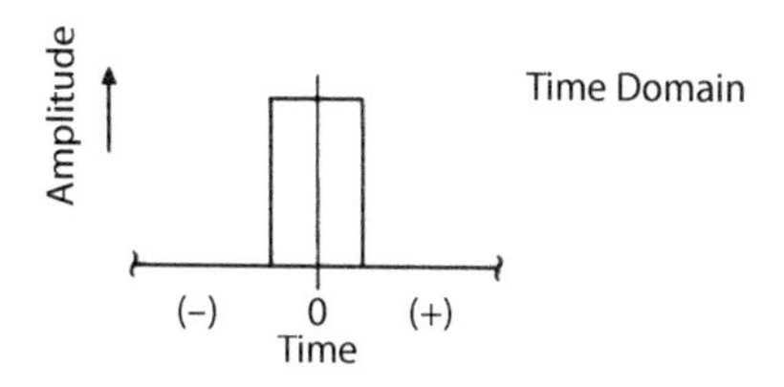

Figure 6-10. This shows a time domain, square wave pulse.

A graph (or equation) relating the amplitude and phase of the signal to frequency (the signal's *spectrum*) represents the signal in the *frequency domain* (Fig. 6-11).

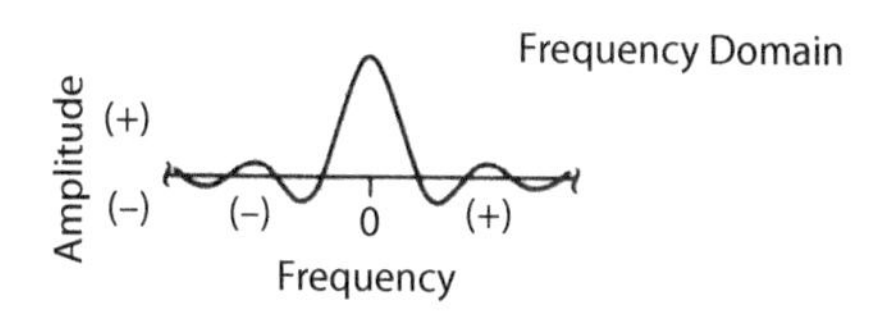

Figure 6-11. This shows a frequency domain (i.e., a Fourier transformed) representation of the time domain signal of Figure 6-10.

A signal can be represented completely in either domain.

Switching between Domains. The representation of a signal in one domain can readily be transformed into the equivalent representation in the other domain through the Fourier transform process (Fig. 6-12).

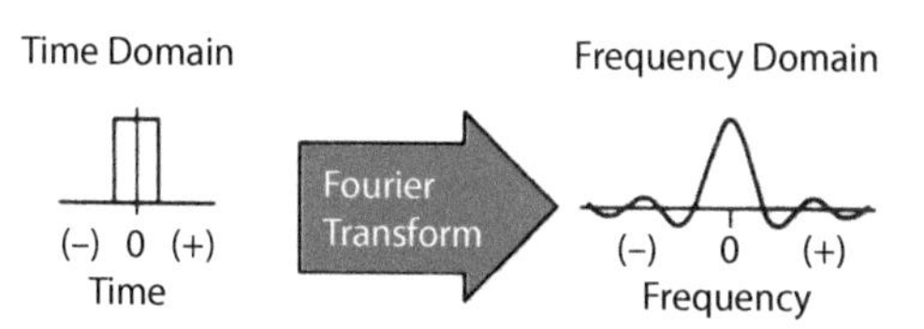

Figure 6-12. This is a Fourier transform of a time domain signal.

The mathematical expression for transforming from the time domain to the frequency domain is called the *Fourier transform.*

The mathematical expression for transforming from the frequency domain to the time domain is called the *inverse Fourier transform* (Figure 6-13). Together, the two transforms are called a *transform pair.* Thus, Figures 6-12 and 6-13 are a Fourier pair. In this way a sine wave in the time domain would transform into a single line (at the frequency of the sine wave) in the Fourier or frequency domain, two very different representations of the same signal.

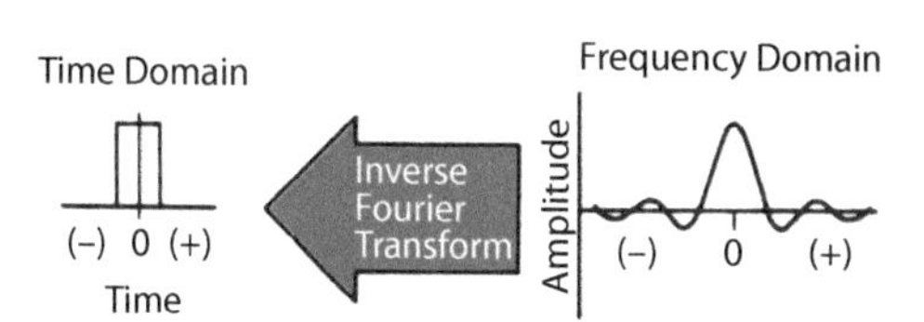

Figure 6-13. This illustrates the inverse Fourier transform of a frequency domain signal.

Calculating the Transforms. To calculate the Fourier transform, an expression for the signal, $f(t)$, is inserted into

$$F(\omega) = \int_{-\infty}^{+\infty} f(t)\, e^{-j\omega t}\, dt$$

and the integration is then performed.

Similarly, to calculate the inverse transform, an expression for the signal, $F(\omega)$, is inserted into

$$f(t) = \int_{-\infty}^{+\infty} \frac{1}{2\pi} F(\omega)\, e^{+j\omega t}\, d\omega$$

Again, the integration is carried out. The variable ω is frequency in radians per second ($\omega = 2\pi f$), and $e^{-j\omega t}$ is the exponential form of the expression, $\cos \omega t - j \sin \omega t$.

However, with the advent of digital signal processing in real radar systems, this is done using a *discrete Fourier transform.*

The continuous time forms of the Fourier transform and inverse Fourier transform become

$$F[k] = \sum_{n=-0}^{N-1} f[n]e^{-j(2\pi/N)nk}, \quad k = 1, 2, 3, 4$$

where $f[n]$ is value of the discrete time domain signal at $t = n$, $X[k]$ is the value of the discrete frequency domain signal at $w = k$, and N is the sampling interval.

Similarly, the discrete inverse Fourier transform is given by

$$f[n] = \frac{1}{N}\sum_{n=0}^{N-1} F[k]e^{j(2\pi/N)kn}, \quad n = 0, 1, 2, 3, 4$$

An important difference between the continuous and discrete time domain variants is due to signals being periodically sampled in time (i.e., digitized). This leads to the spectrum repeating at the inverse of the sampling period (Fig. 6-14). The discrete Fourier transform is usually implemented in a form known as a fast Fourier transform (FFT), whose computations take advantage of some redundancy in the Fourier transform resulting in a very considerable speed up. The good news is that mathematical software typically used to process radar signals will compute the FFT automatically after simply entering the signal vector. However, some care is required to ensure correct results, and the interested reader should delve more deeply than space permits here.

Value of the Concept. The concept of time and frequency domains and the transformation between them is immensely useful. In radar, it is pretty well indispensable. Translating from one representation to the other is crucial to modern signal processing understanding and design. Range resolution and range measurement may be readily generated only in the time domain. Doppler resolution, Doppler range-rate measurement, and certain aspects of high-resolution ground mapping, on the other hand, may be readily generated only in the frequency domain. The Fourier transform is used in almost all radar systems and, in modern systems, usually in discrete form.

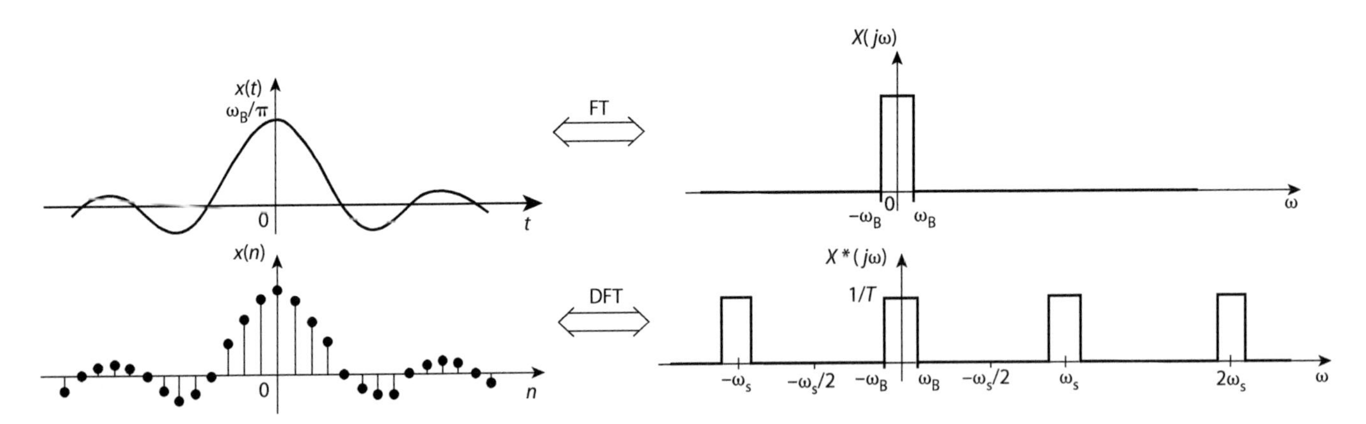

Figure 6-14. This graph demonstrates the Fourier transform of a sampled pulse.

6.5 Statistics and Probability

The world in which we live is statistical in nature, and it should come as no surprise to find that this is also true for radar. Indeed, although radar systems may transmit nicely designed regular signals, the same cannot be said of the echoes that are captured by the receiver. In the simplest terms these either can be random (thermal) noise or can have the same characteristics as noise (e.g., due to scattering from a randomly rough surface). This means that simple detection of targets by radar has to be done against a background of random noise (as might be present in the radar receiver) or by scattering from a rough surface. Hence, probabilistic descriptions characterizing radar performance are routinely used, and this means understanding some basic aspect of statistics. Chapter 12 describes noise and its role in target detection.

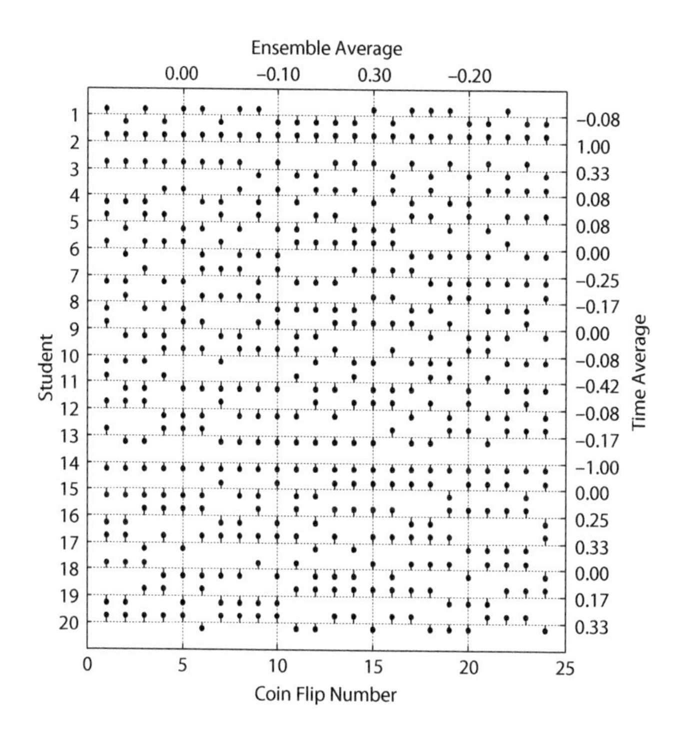

Figure 6-15. These results show 20 students each flipping a coin 24 times.

Realizations of a Random Process. As a practical example of a random process, consider a group of students in a lecture. The students were asked to flip coins and to record the sequence of heads and tails they observed. Figure 6-15 shows an outcome of such an experiment in which 20 students participated, each doing 24 coin flips. An upward stem and dot means the coin flip was a head; a downward dot means the coin flip was a tail.

It is clear that two students had biased coins since student 2 flipped all heads and student 14 flipped all tails. The other students had fair coins with roughly equal numbers of heads and tails. The terminology relating to the experiments is as follows:

The flipping of one coin is a *random event*.

A sequence of 24 coin flips from a student is a *realization*.

The sequences of 24 coin flips from all 20 students constitute an *experiment* that sampled a *random process*.

The collection of all possible realizations in an experiment is called an *ensemble*. Here, 20 sequences or realizations from the ensemble are explored.

Time Average and Ensemble Average. It is important to understand, because of its random nature, the same experiment with the same students and the same coins done on another occasion would give different results. Each student's sequence of coin flips would be different. However, the statistical properties such as the average would be similar. This is also true of radar detection against a background of random noise. Numerical averages can be calculated by assigning a value of +1 for a head and −1 for a tail. Then the averages may be calculated in one of two ways:

1. If the coins are fair, then each student would flip a roughly equal number of heads and tails. Thus, the average of each student's coin flipping sequence should be zero (or close to zero). Taking averages in this way is equivalent to averaging the rows in the experiment shown in the previous figure. It can be seen that, apart from students 2 and 14 who had biased coins the row averages are all quite

close to zero. These averages are called *time averages* or *sequence averages*.

2. The averages may also be done column-wise (i.e., averages across all the students). They are then called *ensemble averages*. There is not room to put all the ensemble averages on the plot, and just a few are shown. For instance, the average across all the students' results for the 10th coin flip was –0.1.

All the ensemble averages are close to zero even though there were some biased coins. That is because the biased coins of student 2 (all heads) and student 14 (all tails) cancel each other out. However, even an unmatched biased coin has little effect because in the ensemble average its results are diluted by those of the other fair coins.

Estimates of Averages and True Averages. It is possible to discover the true averages using only the whole ensemble. The experiment of 20 students doing 24 coin flips is just a small sample of the ensemble (an infinite number f students doing an infinite number of coin flips), and therefore calculation of averages from the experiment gives *estimates* only of the true averages (which is why the estimates of the coin flips are close to zero but not necessarily actually zero).

The estimates get worse as the experiment gets smaller. For instance, the extreme case is an experiment where one student flipped one coin just one time and got a tail, for example. The estimated average and ensemble average based on that one result is –1 even though, for a fair coin, the true average is zero. Thus, in general the estimated averages get closer to the true averages as the experiment gets bigger.

The E-Notation. Ensemble averages are denoted by E, which is known as the *expectation operator*. When dealing with random variables it is important to remember that E is an ensemble average, not a time average.

Let the m-th result from the n-th student be denoted by $f_n[m]$. Using this notation $f_{15}[20] = H$ because student 15 got a head on his or her 20th coin flip. Let the sequence of results from the n-th student be $\{f_n[m]\}$. For example, student 1 had the sequence $\{f_n[m]\} = \{\text{HTHTHH}\ldots\}$ (or $\{f_n[m]\} = \{1,-1,1,-1,1,1\ldots\}$ if heads are assigned a value of 1 and tails a value of –1).

The ensemble average is estimated from a finite number of realizations. For instance, the average value over 20 students of the 20th coin flip (column 10 in Fig. 6-15) is given by the following expression with $N = 20$ and $m = 10$. The index n is the student counter, and m is the coin-flip counter:

$$S_E[m] = \frac{1}{N}\sum_{n=1}^{N} f_n[m]$$

$$S_E[10] = \frac{1}{20}\sum_{n=1}^{20} f_n[10]$$

Note that the numerical calculation given here is an estimate of the true ensemble average for the m-th coin flip

$S_E[m] \approx E(f[m])$. The true ensemble average for the m-th coin flip is found as N becomes very large so that statistical fluctuations are minimized:

$$E(f[m]) = \frac{\lim}{n \to \infty}\left(\frac{1}{N}\sum_{n=1}^{N} f_n[m]\right)$$

In the coin-flipping experiment, the ensemble average gives information about the average properties of all the coins. If all the coins are fair, then the ensemble average is zero for every m. Even if a number of coins are biased toward either heads or tails, the effects of averaging over many coins means that the effects of biased coins tend to cancel each other out.

Calculation of a Time Average. A time or sequence average can be calculated from each student's coin-flip sequence. The time average is taken over M coin flips and the index in the summation is m, the coin-flip number. The time average for the n-th student is given by

$$S_{nT} = \frac{1}{M}\sum_{m=1}^{M} f_n[m]$$

If the number of coin flips is large, then the value of S_{nT} converges toward the true time average. Time averages are usually denoted by a bar or a hat, such as $\bar{f}_n$ or $\hat{f}_n$. Thus,

$$\bar{f}_n = \frac{\lim}{M \to \infty}\left(\frac{1}{M}\sum_{m=1}^{M} f_n[m]\right)$$

The time average contains information about the properties of the n-th coin (or possibly the skill of the n-th student in flipping the coin). If the n-th coin is fair, then $\bar{f}_n = 0$. If the coin is biased toward heads then $\bar{f}_n > 0$, and if it biased toward tails then $\bar{f}_n < 0$.

Variance and Mean Square Deviation. Variance and mean square deviation give a measure of the width about the mean of a distribution. Variance is a term used to describe the variability of an *ensemble* and therefore applies to the columns of an experiment:

$$Var(f) = E((f - E(f))^2)$$

The quantity $f - E(f)$ is termed *mean-centered data*; that is, the ensemble mean or expected value has been subtracted.

Thus, $(f - E(f))^2$ is the square of the deviations or variance from the mean. For instance, if heads are denoted by +1 and tails by –1 and if $E(f) = 0$, then $(f - E(f))^2$ is always +1; therefore, the expected value of the variance is +1. That is, the values are expected to fluctuate by +1 and –1 (which is all that can happen in the coin-flipping experiment).

Thus, for the coin-flipping experiment, $Var(f) = E((f - E(f))^2) = 1$.

The mean square deviation (MSD) is used to estimate variability in a random sequence (i.e., within one realization) and therefore applies to the rows of the coin-flipping experiment. The expression is

$$MSD(f) = \lim_{M \to \infty} \left(\frac{1}{M} \sum_{m=1}^{M} \left(f[m] - \bar{f} \right)^2 \right)$$

The MSD of an unbiased coin is +1. For the biased coin, as used by student number 2, the MSD is zero. All the $f[m]$ were +1 (Heads); thus, the time average is +1 and all the $f[m] - \bar{f}$ values are zero, which leads to a value of zero for the MSD of the biased coin.

Numerical Estimates for Variance and Mean Square Values. The following expression shows how to estimate a variance from experimental data as long as $E(f)$ is known. For example, in the coin-flipping experiment, if the coins are unbiased it is possible to say in advance that $E(f) = 0$. Then the variance is computed using

$$Var(f) \approx \frac{1}{N} \sum_{n=1}^{N} (f_n - E(f))^2$$

If an estimate of $E(f)$ first has to be calculated from the same set of data then the computation of the variance is done using

$$Var(f) \approx \frac{1}{N-1} \sum_{n=1}^{N} \left(f_n - \frac{1}{N} \sum_{n=1}^{N} f_n \right)^2$$

where the unknown $E(f)$ has been approximated using

$$E(f) \approx \frac{1}{N} \sum_{n=1}^{N} f_n$$

In the case where $E(f)$ has been derived approximately from the data, statistical theory shows that the estimate of $Var(f)$ is a little larger than in the case where $E(f)$ is known beforehand, which explains the presence of $N - 1$ rather than N in the denominator of the expression for $Var(f)$. Similar expressions apply to the mean square deviations. If $\bar{f}$ is known beforehand, then

$$MSD(f) \approx \left(\frac{1}{M} \sum_{m=1}^{M} (f[m] - \bar{f})^2 \right)$$

If is $\bar{f}$ estimated from the data, then

$$MSD(f) = \frac{1}{M-1} \sum_{m=1}^{M} \left(f[m] - \frac{1}{M} \sum_{m=1}^{M} f[m] \right)^2$$

Ergodic and Stationary Sequences. An *ergodic* sequence arises from a very special type of random process in which the ensemble and time averages are the same. A coin-flipping experiment using only fair coins would be ergodic. The benefit of ergodicity from a signal processing viewpoint is that properties of the ensemble can be determined from a single realization. In the coin-flipping experiment, it means that an experimenter would be able to assume that 100 coins each flipped once give the same average result as one coin flipped 100 times. This is a reasonable assumption if the coin is fair.

Ergodicity applies to all the statistical properties of the process:

1. The expected value is equal to the time average value.
2. The variance is equal to the mean square deviation.
3. The probability distribution function (PDF) of the random process is the same as the distribution of values in the realization.

In Figure 6-16, the upper panel is a single realization generated from a random noise-like process having a Gaussian distribution (which in this case was known). The bars on the graph of $p(x)$ in the lower panel of were determined from the data in the upper panel, while the smooth curve is the Gaussian distribution for a random process. Because the process is ergodic the statistical distribution of the data from the single realization can be used to estimate the distribution for the process that generated the data.

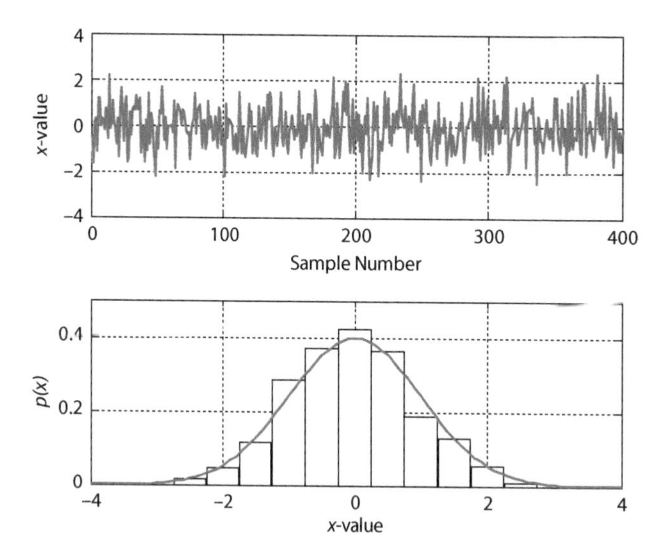

Figure 6-16. A random noise-like sequence has a Gaussian distribution of probabilities. The bars on the $p(x)$ graph are from the data in the upper panel, while the smooth curve is the Gaussian distribution for a random process.

A sequence is *stationary* if its ensemble properties $E(f)$ and $Var(f)$, and more generally its probability distribution function, do not change with time. Otherwise, it is a *nonstationary* sequence or process. A nonstationary process cannot be ergodic. This does occur with radar data and ultimately requires very sophisticated processing, which is outside the scope of this book.

Nonstationary processes also often occur in many forms. One example is the daily temperatures in the Northern Hemisphere for the months January to March. This is because they have a rising trend. The average temperature on January 1 over an ensemble of years such as 2010–2020 will be below the ensemble average for March 31 (averaged over the same years) because the temperature usually rises between January and March. Also note that both these values would be different from the temperature averaged over the period January to March (i.e., the time average) in any given year.

Probability Distributions. Noise in radar systems, such as thermal noise in receivers and in some cases echoes from distributed targets (e.g., the sea or vegetation) can be described by the Gaussian probability distribution function. For this reason it features very centrally when describing the performance of radar systems (as we shall see in a number of the following chapters). Further, it is the statistical nature of noise (and of target echoes) that results in radar detection performance being described in statistical terms. The mathematical expression describing the probability of a quantity that fits a known Gaussian distribution, such as random noise, is

$$p(f) = \frac{1}{\sqrt{2\pi}\sigma} \exp\left(-\frac{(f-\mu)^2}{2\sigma^2} \right)$$

where μ and σ are parameters that define the distribution. σ^2 is the *variance* of the quantity, σ is the *standard deviation,* and μ is the *mean value.*

A second and related PDF is the Rayleigh distribution. Remember that most modern radar systems downconvert and use in-phase and quadrature channels to obtain values of both amplitude and phase. Noise will be present on both the I and the Q channels and will be Gaussian with a zero mean and a variance of σ^2. The resultant amplitude is $\sqrt{(I^2 + Q^2)}$ and has a Rayleigh distribution. The mathematical expression defining a Rayleigh PDF is

$$p(f) = \frac{f}{\sigma^2}\exp\left(\frac{-f^2}{2\sigma^2}\right)$$

where σ is the standard deviation, and σ^2 is the variance. In Chapter 12 the Rayleigh distribution is used to compute the likelihood of noise being observed as a falsely detected target. This is represented as the probability of a false alarm. It is also part of the calculation that determines the likelihood of detecting a true target, the probability of detection.

6.6 Convolution, Cross-Correlation, and Autocorrelation

Convolution and its close cousin *correlation* are invaluable tools in the analysis of signals and are used routinely in many aspects of radar signal processing. Convolution is a basic ingredient in filter design and is used specifically in the matched filter (Chapter 16), a method of maximizing the signal-to-noise ratio. It is also employed in the closely related topic of pulse compression (Chapter 16), a method of generating high range resolution. The cross-correlation function examines the likeness between two signals, and the autocorrelation function evaluates the self-similarity of a given signal to see if and how it contains components that repeat.

Convolution. The discrete form of the convolution of two sampled, time domain signals, $f[n]$ and $g[n]$, can be written as

$$R_{fg}[l] = \sum_{n=-\infty}^{\infty} f[n]g[n-l], \quad l = 0, \pm 1, \pm 2, K$$

where l is the lag and represents a discrete sliding of one signal past the other as the value of l changes, and K is the finite time extent of the signal being *convolved*. Note: this is the convolution of signal $f[n]$ with signal $g[n]$. We can also write similar expression for the convolution of signal $g[n]$ with signal $f[n]$ by simply reversing the order. For example,

$$R_{gf}[l] = \sum_{n=-\infty}^{\infty} g[n]f[n-l], \quad l = 0, \pm 1, \pm 2, K$$

This equation takes one of the signals, "flips" it around, and then slides it across the other as the products of the overlapping parts (determined by a value of l) are integrated (summed). This flipping is a neat way of representing the direction of signals. For example, pulse compression, a technique to achieve high range resolution (Chapter 16), involves *convolving* the echo with a replica of the transmitted signal. Of course the echo travels in the opposite direction and the flipping accounts for this.

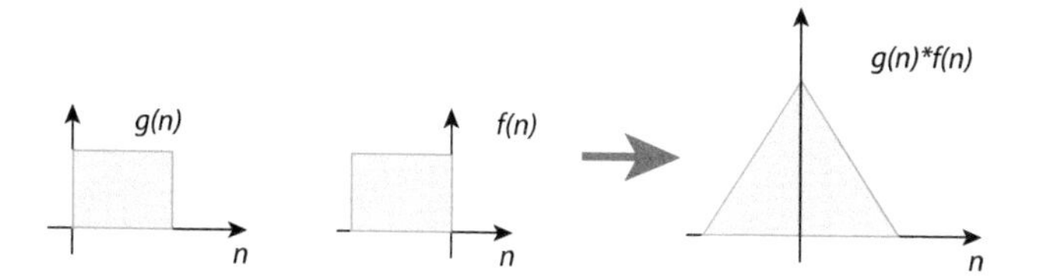

Figure 6-17. This illustration shows the convolution of two rectangular functions.

Figure 6-17 shows an example of convolution. Here two rectangular wave functions of differing magnitudes are convolved with one another, and the slide and integrate mechanism described by the function are indicated. First, the graph of the rectangular function on the right is flipped about the vertical axis and repositioned on the left. It is then slid from left to right across the other rectangular function. The sliding and integration result in a triangular output. Note how its integral is at a maximum when the two rectangular functions exactly overlap (i.e., the lag l is zero). Further, the range of the lag value is dictated by the duration of the signal (for a given sampling rate) and does not have to be extended to plus and minus infinity; that is, there is no point convolving the zero values outside of the signal with more zero values.

If $f[n]$ and $g[n]$ were expressed in the frequency domain, then they could be multiplied together and an inverse Fourier transform could be applied to obtain the same result.

Correlation. Almost identical to convolution, correlation differs only in that there is no flipping of one of the signals. Correlation can be used either to examine how alike two signals, $f[n]$ and $g[n]$, are or to examine especially their periodic properties. The cross-correlation function is written as

$$Rxy[l] = \sum_{n=-\infty}^{\infty} x[n]y[n+l], \quad l = 0, \pm 1, \pm 2, \cdots K$$

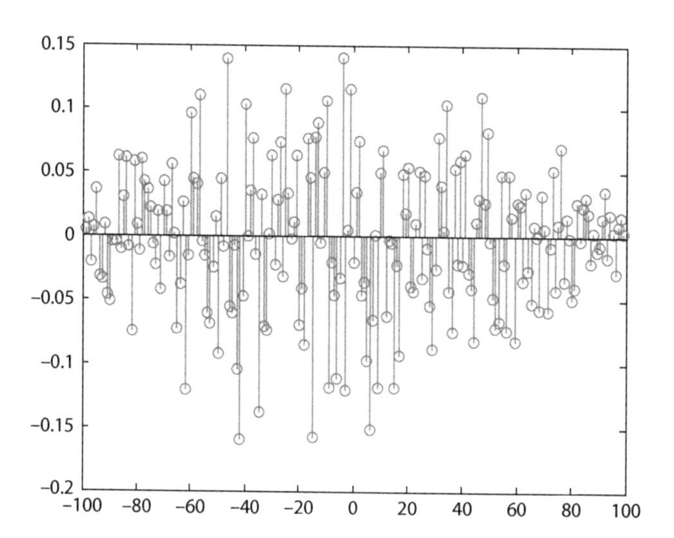

Figure 6-18. This graph shows the cross-correlation of two random noise-like signals.

Figure 6-18 shows the cross-correlation of two random noise-like signals, which indicates that there is no similarity between the two signals (as expected for random noise). If the two signals were periodic with the same period, such as sine waves of the same frequency, then we would get a highly structured and strongly correlated output since the two signals come in and out of phase as a function of the lag value.

Autocorrelation. This is the correlation of a signal $f[n]$ with itself. It is used to reveal properties of a signal such as whether or not it is periodic and also to differentiate a chosen signal from others and from sources of interference such as noise. The autocorrelation of $f[n]$ comes from substituting $f[n]$ for $g[n]$ in the expression for cross-correlation and is thus given by

$$R_{ff}[l] = \sum_{n=-\infty}^{\infty} f[n]f[n+l], \quad l = 0, \pm 1, \pm 2, \ldots K$$

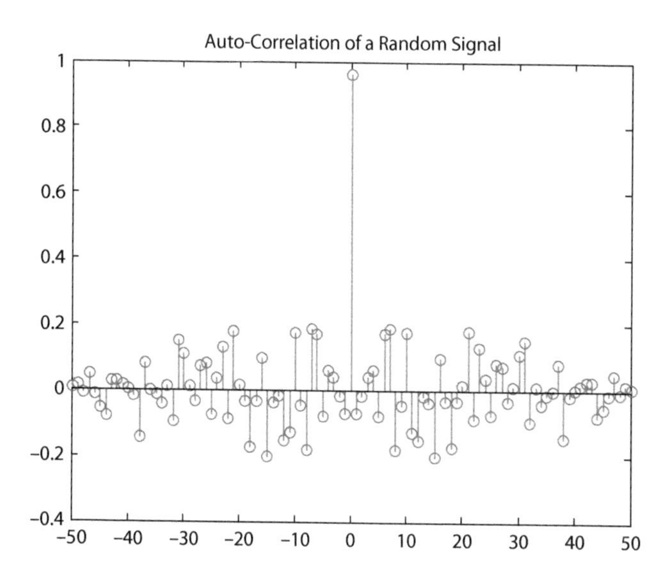

Figure 6-19. This graph shows the autocorrelation of a random signal.

Figure 6-19 shows the autocorrelation of a random noise-like signal. Here we see that it is completely correlated with a lag l of zero. This is true for all signals and is a direct result of the fact that every element of the sum in the autocorrelation function has the same value. However, with only a lag value of one ($l = 1$) the correlation of the signal plunges and hovers, noise-like, around a zero mean. This shows that there are no periodic or repeating components in a random noise-like signal.

Autocovariance. This property of a random process indicates the degree of similarity of a signal in one part of a time series with that taken at a later part in the same time series. It is calculated using the expectation

$$\gamma_{ff}(l) = E(f[n]f[n + l])$$

In this expression $f[n]$ is a mean-centered random variable; that is, the expected value of $E(f) = 0$, or the mean value has been subtracted from f. The equation states that autocovariance γ_{ff} is a function of lag l and is defined as the expected value of f at time n multiplied by f at a later time $n + l$. If $f[n + 1]$ is correlated with $f[n]$ then $\gamma_{ff}(l)$ would be nonzero, whereas if there is no correlation then $\gamma_{ff}(l)$ is zero. The autocovariance can be estimated from a single data sequence of length N. Both $f[n]$ and $f[n + 1]$ must have their means removed before they can be used in the following expression:

$$\gamma_{ff}(l) \approx \frac{1}{N - 1}\sum_{n=1}^{N-1} f[n]f[n + l]$$

where n is the sample number.

Insight into Autocovariance. When $l = 0$, the autocovariance is equal to $Var(f)$, the expression has a maximum value (as $f[n] = f[n]$), and the answer is simply saying that the two signals are identical. As l increases, the value of $Var(f)$ falls; the rate at which it falls as a function of l provides information about the self similarity of the signal (i.e., periodic nature) of the signal. For example, a special case arises when $f[n]$ forms an oscillating signal. Suppose there were exactly 20 samples per cycle of oscillation:

$$f[1] = f[21] = f[41] \ldots$$
$$f[2] = f[22] = f[42] \ldots$$

In other words, apart from random noise present in the signal, $f[n]$ is identical to $f[n + 20]$. Thus, the autocorrelation will also be large when $l = 20$. Moreover, if the peaks of the oscillating signal are at $f[1], f[21], f[41] \ldots$ then there are valleys (minima) at $f[11], f[31], f[41] \ldots$ Thus, the autocovariance will be large and negative when $l = 10$.

This reasoning highlights an important result concerning autocovariance. An oscillating signal's autocovariance is oscillatory with the same period. Moreover, since the calculation for autocovariance includes an average over $N - l$ samples, the autocovariance is much less noisy than the original signal. The autocovariance reveals the degree of periodicity in a signal. This can be completely masked if observed in the time domain only.

Covariance estimation is a technique increasingly used to determine whether or not a range cell contains clutter or a target in advanced concepts such as space-time adaptive processing (STAP).

Cross-Covariance. The cross-covariance of stationary zero-mean signals $\{f[n]\}$ and $\{g[n]\}$ is

$$\gamma_{fg}[l] = E(f[n]g[n + l]) = E(f[n - l]g[n])$$

The notation is γ_{fg}, meaning the cross-covariance between $\{f[n]\}$ and later values $\{g[n]\}$. By contrast, the following is the cross-covariance between $\{g[n]\}$ and later values in $\{f[n]\}$:

$$\gamma_{gf}[l] = E(g[n]f[n + l]) = E(g[n - l]f[n])$$

The lag can also be negative. Here, because the lag is negative $(-l)$, the covariance is between $\{f[n]\}$ and earlier values in $\{g[n]\}$:

$$\gamma_{fg}[-l] = E(f[n]g[n + 1])$$

The cross-covariance function is not symmetrical:

$$\gamma_{fg}[l] \neq \gamma_{fg}[-l]$$

For instance, if $\{f[n]\}$ is an input sequence into a *causal*[2] digital filter whose output is $\{g[n]\}$, then the input sequence $\{f[n]\}$ is expected to correlate with later values in the $\{g[n]\}$ sequence in some way. Thus, $\gamma_{fg}(l)$ would not in general be zero. But if the filter is causal there would be no relationship between $\{f(n)\}$ and earlier values in the $\{g[n]\}$ sequence and so $\gamma_{gf}(-l)$ would be zero. In this way the cross-covariance provides information about the similarity of two signals similar to the way the autocovariance provides information about the self-similarity within a single signal.

2. A system is *causal* if the output at any time depends only on values of the input at the present time or in the past. That is, it cannot have an output until it has an input and cannot anticipate future inputs. A vehicle is a causal system because it cannot anticipate the driver's future actions (at least not yet).

The Wiener-Khinchin Theorem. This theorem shows that the autocovariance function provides a method for estimating a signal's power spectrum (Chapter 19)—directly from the discrete Fourier transform:

$$S_{ff}[k] = \frac{1}{N} \mid F[k]^2 \mid$$

where k is an integer.

The Wiener-Khinchin theorem states that this estimate may also be obtained from the Fourier transform of the autocovariance:

$$S_{ff}[k] = \sum_{l=0}^{N-1} \gamma_{ff}(l) e^{-j2\pi\frac{1}{N}}$$

where l is an integer indicating the number of lags.

Conversely, as might be expected, the autocovariance can be determined from the inverse Fourier transform of an $N - l$ channel power spectrum:

$$\gamma_{ff}[l] = \frac{1}{N} \sum_{k=0}^{N-1} S_{ff}(k) e^{j2\pi l\frac{k}{N}}$$

6.7 Summary

This chapter introduced a number of mathematical concepts and constructs that are widely used in the description of radar performance and in the evaluation and manipulation of radar signals.

Radar signals are examples of energy signals. Complex numbers are a mathematical way of representing phasors and make a useful connection with our nonmathematical way of describing radar signals. Periodic signals are described by the Fourier series that is made up of a number of harmonically related frequencies. The Fourier transforms provides a means for moving between frequency and time domain representations of radar signals.

Noise in a radar system is described by a Gaussian probability distribution function when observed as a zero mean voltage in the I and the Q channels. The overall detected amplitude $\sqrt{(I^2 + Q^2)}$ is described by Rayleigh distribution function.

Convolution, correlation, and autocorrelation provide a powerful means of realizing matched filters, compressing pulses for high range resolution, and forming synthetic aperture, among a host of others, and are also useful tools for determining the properties of radar signals. The Fourier transform of the autocovariance provides an alternative means of estimating the power spectrum of a radar signal.

Further Reading

R. N. Bracewell, *The Fourier Transform and Its Applications*, 3rd ed., McGraw-Hill, 1999.

D. P. Bertsekas and J. N. Tsitsiklis, *Introduction to Probability*, 2nd ed., Athena Scientific, 2008.

M. L. Meade and C. R. Dillon, *Signals and Systems*, Chapman and Hall, 2009.

H. Hsu, *Schaum's Outline of Probability, Random Variables, and Random Processes*, 2nd ed., McGraw-Hill, 2010.

J. F. James, *A Student's Guide to Fourier Transforms; with Applications in Physics and Engineering*, Cambridge University Press, 2011.

H. Hsu, *Schaum's Outline of Signals and Systems*, 3rd ed., McGraw-Hill, 2013.

Test your understanding

1. What is the amplitude and phase represented by the complex number $1 + j$?
2. Sketch the Fourier transform pairs for the following time domain signals: (a) an impulse; (b) a square pulse; and (c) a Gaussian pulse
3. Explain what is meant by the mean and variance of thermal noise.
4. Sketch the autocorrelations functions of (a) a sine wave and (b) thermal noise.
5. Sketch the cross-correlation function of two independent realizations of thermal noise.

PART III

Fundamentals of Radar

English Electric Canberra (1957)

The Canberra was Britain's first-generation jet-powered light bomber to be manufactured in large numbers. It entered service in 1951 and served as a nuclear strike aircraft, tactical bomber, and reconnaissance platform (photographic and electronic). Flying at Mach 0.88, its ability to outpace the jet interceptors of the time and its adaptability made it highly desirable for export. The Canberra set multiple flight records including first nonstop unrefueled transatlantic crossing by a jet aircraft and the altitude record twice (1955 and 1957).

7

Choice of Radio Frequency

Ferranti AI23: the first airborne monopulse radar (Courtesy of Selex ES.)

A primary consideration in the design of virtually every radar is the frequency of the transmitted radio waves—the radar's *operating frequency.* How close a radar comes to satisfying many of the requirements imposed on it—for example, detection range, angular resolution, Doppler performance, size, weight, cost—often hinges on the choice of radio frequency. This choice, in turn, has a major impact on many important aspects of the design and implementation of the radar. In this chapter we will survey the broad span of radio frequencies used by radars and examine the factors that determine the optimum frequency for particular applications.

7.1 Frequencies Used for Radar

Today's radars operate at frequencies ranging from as low as a few megahertz to as high as 300,000,000 MHz (Fig. 7-1).

At the low end are a few highly specialized radars. Sounders measure the height of the ionosphere. Another is *over-the-horizon* (OTH) radars, which take advantage of ionospheric reflection to beyond line of sight and detect targets thousands of kilometers away.

At the high end are laser radars, which operate in the visible and infrared region of the spectrum. Such radars are used to provide the angular resolution needed for such tasks as measuring the ranges of individual targets on the battlefield.

Most radars, however, employ frequencies lying somewhere between a few hundred megahertz and 100,000 MHz. Present airborne radars used for search, surveillance, and multimode operation are predominantly in the 425 MHz to 12 GHz range.

To make such large frequency values more manageable, it is customary to express them in gigahertz. One gigahertz equals 1000 MHz, so a frequency of 100,000 MHz is 100 GHz.

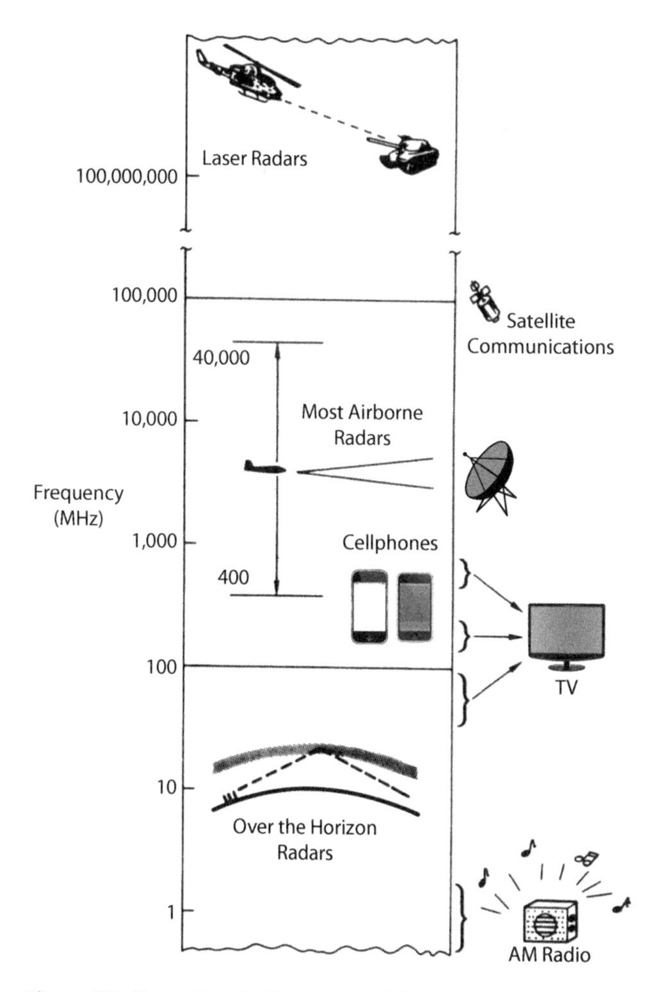

Figure 7-1. Shown here is the portion of the electromagnetic spectrum used for radar.

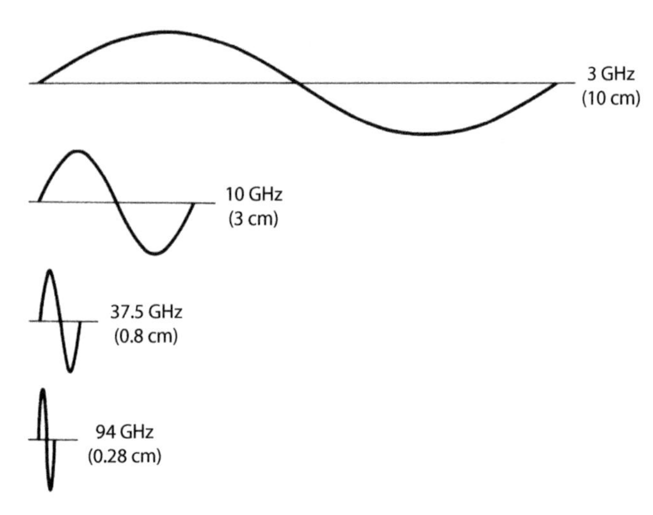

Figure 7-2. These wavelengths used by airborne radars, are shown in their relative size.

Radar operating frequencies are also expressed in terms of wavelength—the speed of light (3×10^8 m/s) divided by the frequency in Hertz (Fig. 7-2).

Incidentally, a convenient rule of thumb for converting from frequency to wavelength is wavelength in centimeters = 30 ÷ frequency in gigahertz. The wavelength of a 10 GHz wave, for example, is 30 ÷ 10 = 3 cm.

To convert from wavelength to frequency you turn the rule around, interchanging wavelength and frequency: frequency in gigahertz = 30 ÷ wavelength in centimeters. The frequency of a 3 cm wave is thus 30 ÷ 3 = 10 GHz.

7.2 Frequency Bands

Besides being identified by discrete values of frequency and/or wavelength, radio waves are also broadly classified as falling within one or another of several arbitrarily established regions of the radio frequency spectrum such as high frequency (HF), very high frequency (VHF), and ultra high frequency (UHF). The frequencies commonly used by radars fall in the VHF, UHF, microwave, and millimeter-wave regions (Fig. 7-3).

During World War II, the microwave region was broken into comparatively narrow bands and assigned letter designations for purposes of military security: L-band, S-band, C-band, X-band, and K-band (military users tend not to like to talk in terms of specific frequencies). To enhance security, the designations were deliberately arranged out of alphabetical sequence. Although long since declassified, these designations have persisted to this day.

The K-band turned out to be very nearly centered on the resonant frequency of water vapor, where absorption of radio waves in the atmosphere is high. Consequently the band was split into three parts. The central portion retained the original designation. The lower portion was designated the Ku-band while the higher portion was designated the Ka-band. An easy way to keep these designations straight is to think of the "u" in Ku as standing for *under* and the "a" in Ka as standing for above the central band. Only a portion of these bands are allocated by the International Telecommunications Union (ITU) for radar use, and radar bands are often further constrained by the bandwidth of radio frequency components.

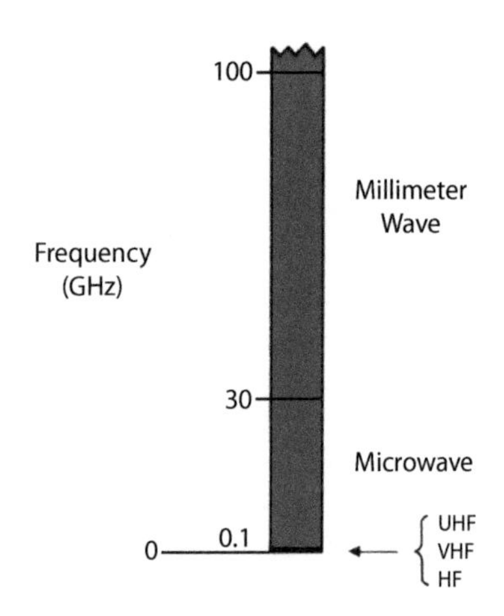

Figure 7-3. Regions of the electromagnetic spectrum commonly used for radar are plotted here on a linear scale. The bandwidth available at millimeter-wave frequencies is much greater than at microwave frequencies.

In the 1970s a completely new sequence of bands—neatly assigned consecutive letter designations from A to M—was devised for electronic countermeasures equipment (Fig. 7-4). Attempts were made to apply these designations to radars as well, but largely because the junctions of the new bands occur at the centers of the traditional bands—about which many radars are clustered—these attempts were unsuccessful. In the United States the "new" band designations are generally used, as originally intended, only for countermeasures.

If you haven't already done so, memorize the center frequencies and wavelengths of these five radar bands:

Band	GHz	cm
Ka (above)	38	0.8
Ku (under)	15	2
X	10	3
C	6	5
S	3	10

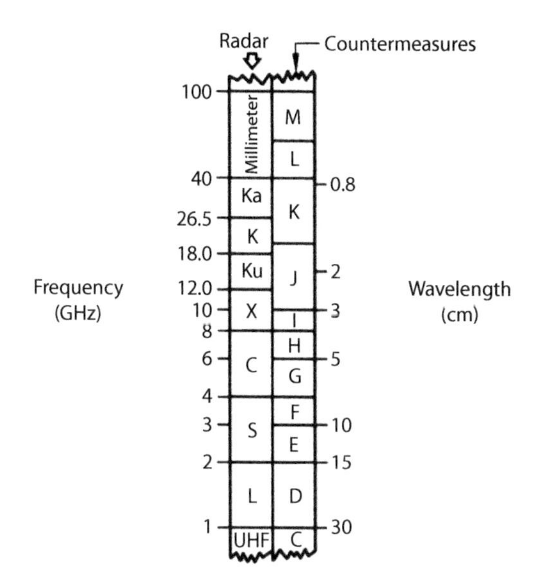

Figure 7-4. Radar and countermeasures band letter designations are shown here with their corresponding wavelengths.

7.3 Influence of Frequency on Radar Performance

The best frequency to use depends on the job the radar is intended to perform. Like most other design decisions, the choice involves trade-offs involving several factors: physical size, transmitted power, antenna beamwidth, and atmospheric attenuation are among the most important.

Physical Size. The dimensions of the hardware used to generate and transmit radio frequency power are in general proportional to wavelength. At the lower frequencies (longer wavelengths), the hardware is usually large and heavy. At the higher frequencies (shorter wavelengths), radars can be put into smaller packages and thus operate in more compact spaces at a lighter weight (Fig. 7-5). The limited space requires more tightly packed electronics, which can present design challenges.

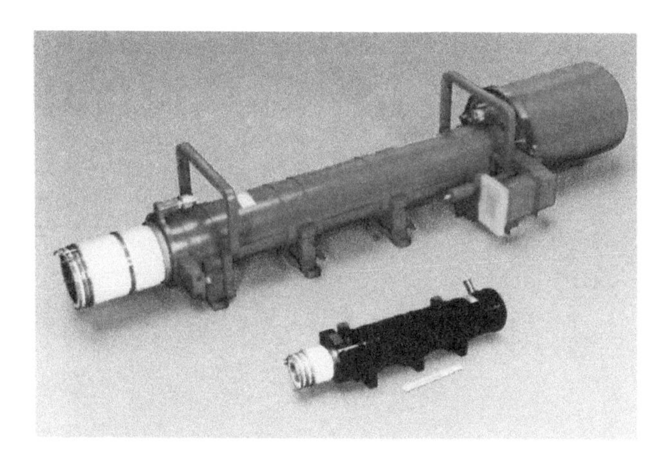

Figure 7-5. The physical size and power-handling capacity of radio frequency components decreases with frequency. A transmitter tube for a 30 cm radar is shown on top while the transmitter tube for a 0.8 cm radar is below it.

Transmitted Power. Because of its impact on hardware size, the choice of wavelength indirectly influences the ability of radar to transmit large amounts of power. The levels of power that can be handled by a radar transmitter are largely limited by voltage gradients (volts per unit of length) and heat dissipation requirements. It is not surprising, therefore, that the larger, heavier radars operating at wavelengths of the order of meters can transmit megawatts of average power, whereas millimeter-wave radars may be limited to only a few hundred Watts of average power.

Most often, though, within the range of available power the amount of power actually used is decided by size, weight, reliability, cost, and detection range considerations.

Beamwidth. As will be explained in Chapter 8, the angular width of a radar's antenna beam is directly proportional to the ratio of the wavelength to the width of the antenna. To achieve a given beamwidth, the longer the wavelength, the wider the antenna must be. At low frequencies, very large antennas must be used to achieve acceptably narrow beams. At high frequencies, small antennas will suffice (Fig. 7-6). The narrower the beam, of course, the greater the power that is concentrated in a particular direction at any one time, and thus the finer the angular resolution.

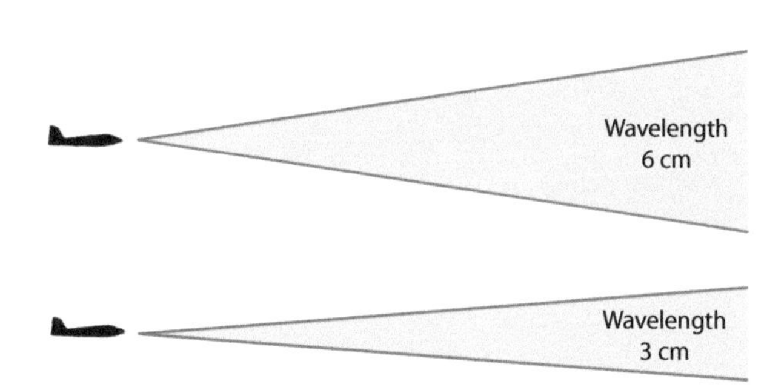

Figure 7-6. For the same size antenna, the angular width of its beam is proportional to wavelength.

Atmospheric Attenuation. In passing through the atmosphere, radio waves are attenuated by two basic mechanisms: absorption and scattering (see blue panel). The absorption is mainly due to oxygen (60 GHz) and water vapor (21 GHz). The scattering is due almost entirely to condensed water vapor (e.g., raindrops). Both absorption and scattering increase with frequency. Below about 0.1 GHz, atmospheric attenuation is negligible; above about 10 GHz it becomes increasingly significant.

Moreover, above about 10 GHz the radar's performance is increasingly degraded by weather clutter competing with desired targets. Even when the attenuation is reasonably low, if enough transmitted energy is scattered back in the direction of the radar, it will be detected. In simple radars that do not employ moving target indication (MTI), this return—called weather clutter—may obscure targets.

While usually not a concern for airborne radars, the effects of the ionosphere on radar signals at UHF and below passing through the ionosphere (attenuation, refraction, dispersion, and Faraday rotation) may also be significant.

Foliage Penetration. In some specialized applications an airborne radar may be required to detect targets hidden under trees. The ability to do this depends on the attenuation properties of the foliage canopy, which are found to increase with frequency. In practice, frequencies of L-band or below are necessary for foliage penetration radars.

Fractional Bandwidth. The fractional bandwidth of a radar is defined as the bandwidth of its signal divided by the center frequency. We'll see later that the bandwidth of a radar signal defines its range resolution, so the greater the bandwidth, the finer the range resolution. However, for a given radar bandwidth, the lower the center frequency, the greater the fractional bandwidth. High fractional bandwidths (greater than about 15%) pose problems for the radar hardware, especially the antenna.

Coexistence with Other Users. The electromagnetic spectrum is used for many other purposes—particularly communications, broadcast, and radionavigation—besides radar. By international agreement the spectrum is allocated among the different users, so some frequency bands are allocated to a particular application on an exclusive basis, while others are shared. All users of the spectrum have requirements for greater and greater bandwidth, yet the electromagnetic spectrum is a strictly finite resource. So even with this regulatory framework, mutual interference can become a problem. Special techniques to improve transmitter spectral purity and to suppress interference, as well as work to understand and quantify the degree of interference that can be tolerated, are active areas of research.

Ambient Noise. Electrical noise from sources outside the radar is high in the HF band. It decreases with frequency (Fig. 7-7), reaching a minimum between about 0.3 and 10 GHz, depending on the level of galactic noise, which varies with solar conditions. From that point, atmospheric noise predominates. It gradually becomes stronger and grows increasingly at K-band and higher frequencies. In many radars, internally generated noise predominates. However, when low-noise receivers are used to meet long-range requirements, external noise can be an important consideration in the selection of frequency.

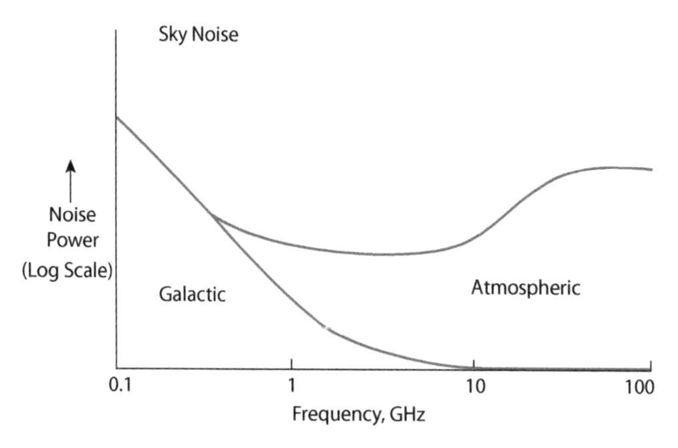

Figure 7-7. Ambient noise reaches a minimum between 0.3 GHz and 10 GHz, depending on the level of galactic noise, which varies with solar conditions.

Doppler Considerations. Doppler shifts are proportional not only to the closing rate of the target, but also to radio frequency. The higher the frequency, the greater the Doppler shift that a given closing rate will produce. As will be made clear in later chapters, excessive Doppler shifts can cause problems. In some cases these tend to limit the frequencies that can be used. On the other hand, Doppler sensitivity to small differences in closing rate can be increased by selecting higher frequencies.

Atmospheric Attenuation

Absorption. Energy is absorbed from radio waves passing through the atmosphere primarily by the gases comprising it. Absorption increases dramatically with frequency.

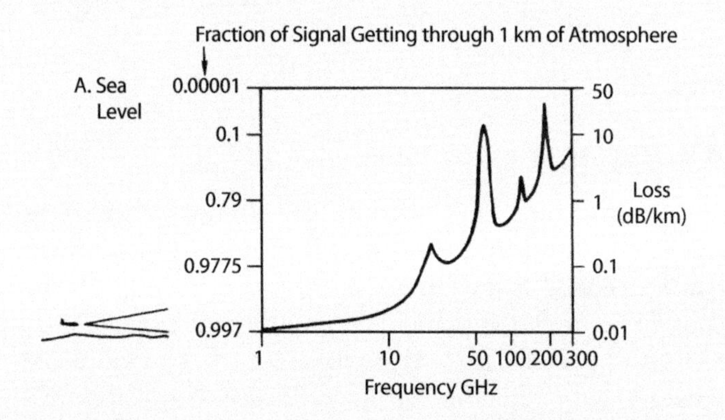

Below about 0.1 GHz, absorption is negligible, while above 5 GHz it becomes increasingly significant. Beyond about 20 GHz, it becomes severe. Typical "windows" are around 35 GHz and 94 GHz.

Most of the absorption is due to oxygen and water vapor. Consequently, it not only decreases at the higher altitudes where the atmosphere is thinner but also with decreasing humidity.

The molecules of oxygen and water vapor have resonant frequencies.

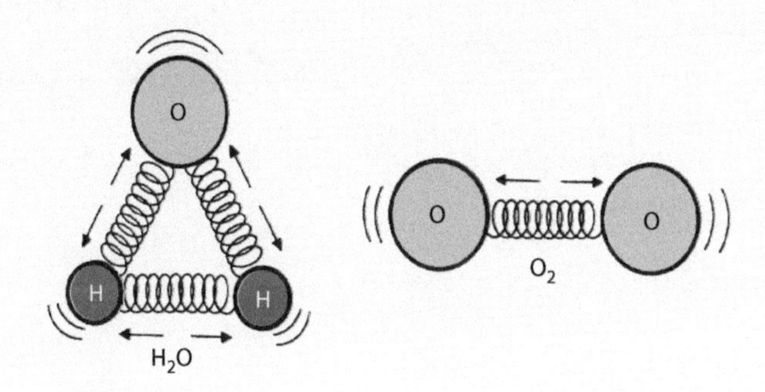

When excited at these frequencies, they absorb more energy—hence the peaks in the absorption curve. The peaks are broadened by molecular collisions and thus are sharper at high altitudes, where the atmosphere is less dense, but their frequencies are the same. (Plot B has the same horizontal axis as A. The vertical axis is shifted down to encompass the lower curve.)

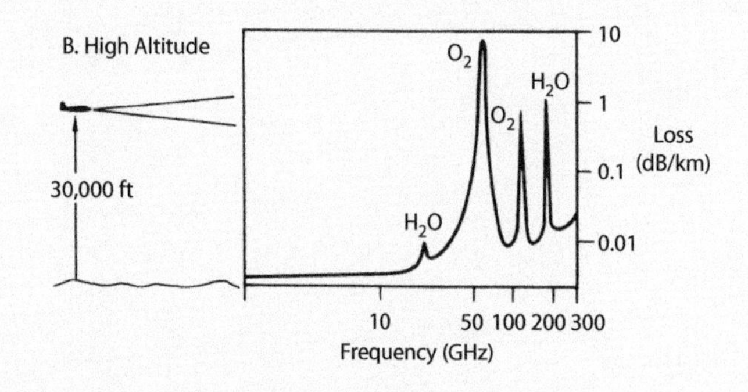

Atmospheric Attenuation *continued*

The peaks at 22 and 185 GHz are due to water vapor; those at 60 and 120 GHz are due to oxygen. The regions between peaks, where the attenuation is lower, are called *windows*.

Energy is also absorbed by particles suspended in the atmosphere, but their principal effect is scattering.

Scattering. Radio waves are scattered by particles suspended in the atmosphere. Scattering increases with the particles' dielectric constant and size relative to wavelength. Scattering becomes severe when the size is comparable to a wavelength.

The principal scatterers are raindrops and, to a lesser extent, hail (because of its much lower dielectric constant). Snowflakes, which contain less water and have slower fall rates, scatter less energy. Clouds, which consist of tiny droplets, scatter even less. Smoke and dust are usually negligible scatterers because of their small particle size and low dielectric constant.

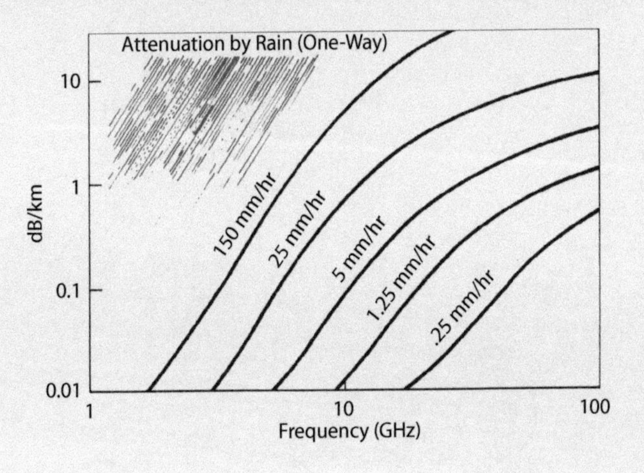

Scattering becomes noticeable in the S-band (3 GHz). At those frequencies and higher, backscattering is sufficient to make rain visible.

Both absorption and scattering by clouds are still negligible in the S-band. Therefore, meteorological radars operating there can measure rainfall rates without being hampered by attenuation or backscatter due to clouds.

Above 10 GHz, scattering and absorption by clouds becomes appreciable. The attenuation is proportional to the amount of water in the clouds.

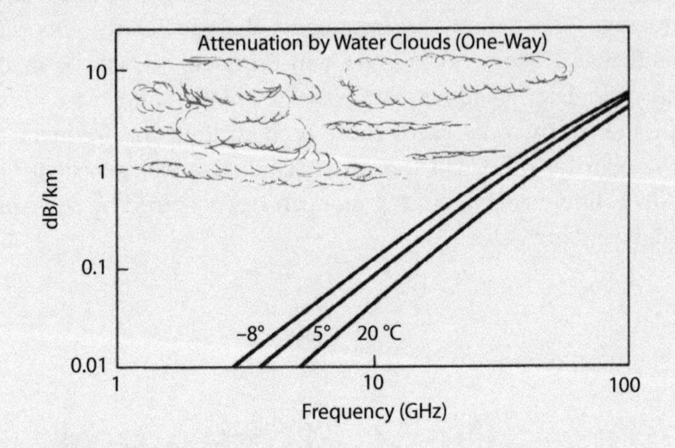

Attenuation increases with decreasing temperature since the dielectric constant of water is inversely proportional to temperature. Ice clouds, however, attenuate less because of the low dielectric constant of ice.

7.4 Selecting the Optimum Frequency

From the preceding, it is evident that selection of the radio frequency is influenced by several factors: the functions the radar is intended to perform, the environment in which the radar will be used, the physical constraints of the platform on which it will operate, and cost. To illustrate, let's consider some representative applications. To put the selection in context we will consider not only airborne applications, but ground and shipboard applications, too.

Ground-Based Applications. These run the gamut of operating frequencies. At one extreme are the long-range multimegawatt surveillance radars. Unfettered by size and weight limitations, they can be made large enough to provide acceptably high angular resolution while operating at relatively low frequencies. OTH radars, as we've seen, operate in the HF band where the ionosphere is suitably reflective. Space surveillance and early warning radars operate in the UHF and VHF bands, where ambient noise is minimal and atmospheric attenuation is negligible.

However, these bands are crowded with communication signals, so their use by radars (whose transmissions generally occupy a comparatively broad band of frequencies) is restricted to special applications and geographic areas. Where such long ranges are not required and some atmospheric attenuation is therefore tolerable, ground radars may be reduced in size by moving up to L-, S-, and C-band frequencies or higher (Fig. 7-8).

Figure 7-8. Ground-based radars commonly operate at lower frequencies where long range is not important. This radar traces the source of mortar fire. X-band and Ku-band may be used for small size and better measurement accuracy. (Courtesy of THALES.)

Shipboard Applications. Aboard ships, physical size becomes a limiting factor in many applications. At the same time, the requirement that ships be able to operate in the most adverse weather puts an upper limit on the frequencies that can be used. This limit is relaxed, however, where extremely long ranges are not required. Furthermore, higher frequencies must be used when operating against surface targets and targets at low elevation angles.

At grazing angles approaching zero, the return received directly from a target is very nearly cancelled by the return from the same target reflected off the water—a phenomenon called *multipath propagation* (Fig. 7-9). Cancellation is due to a 180° phase reversal occurring when the return is reflected. As the grazing angle increases, a difference develops between the lengths of the direct and indirect paths, and cancellation decreases. The shorter the wavelength, the more rapidly the cancellation disappears. For this reason, the shorter wavelength S- and X-band frequencies are widely used for surface search, detection of low-flying targets, and piloting. The same phenomenon is encountered on land when operating over a flat surface.

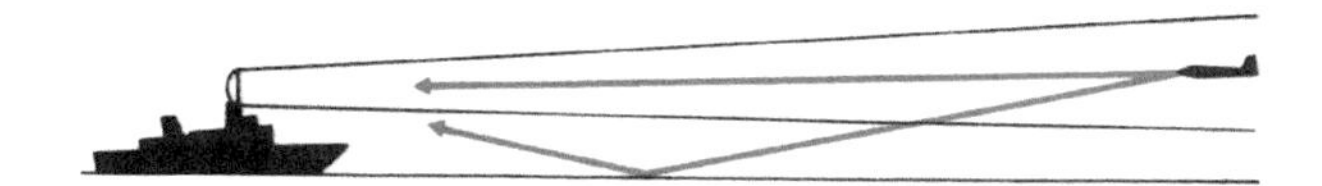

Figure 7-9. At small grazing angles, the return received directly from the target is very nearly cancelled by the return reflected off the water.

Airborne Applications. In aircraft, the limitations on size are considerable. The lowest frequencies generally used here are in the UHF, L, and S-bands, and only for very specific applications. They provide the long detection ranges needed for airborne early warning in the E2 and airborne warning and control system (AWACS) aircraft, respectively (Fig. 7-10). One look at the huge radomes of these aircraft and it is clear why higher frequencies are commonly used when narrow antenna beams are required in smaller aircraft, such as fighters.

The next lowest-frequency applications are in the C-band. Radar altimeters operate here. Interestingly, the band was originally selected because its use made possible light, cheap equipment that could use a triode transmitter tube. These frequencies also enable good cloud penetration. Because altimeters are simple, require only modest amounts of power, and do not need highly directive antennas, they can use these frequencies and still be made conveniently small.

Figure 7-10. Operating in S-band, the AWACS radar provides an early warning, its antenna is very large to provide the desired angular resolution.

Weather radars, which require greater directivity, operate in C-band as well as in X-band. The choice between the two bands reflects a dual trade-off. One is between storm penetration and scattering. If scattering is too severe, the radar will not penetrate deeply enough into a storm to see its full extent. Yet, if too little energy is scattered back to the radar, storms will not be visible at all. The other trade-off is between storm penetration and

Figure 7-11. At X-band, reasonably high angular resolution can be obtained with an antenna small enough to fit in the nose of a fighter.

Figure 7-12. Operating at 94 GHz, this tiny antenna of an air-to-air missile provides the same angular resolution as the much larger antenna pictured in Figure 7-11.

equipment size. C-band radars, providing better penetration and hence longer-range performance, are primarily used by commercial aircraft. X-band radars, providing adequate performance in smaller packages, are widely used by private aircraft.

Most fighter, attack, and reconnaissance radars operate in the X-and Ku-bands with a great many operating in the 3 cm wavelength region of X-band (Fig. 7-11) and the 2 cm wavelength region of Ku-band (Fig. 7-12).

The attractiveness of the 3 cm region is threefold. First, atmospheric attenuation, though appreciable, is still reasonably low—only 0.02 dB/km for two-way transmission at sea level. Second, narrow beamwidths, providing high power densities and excellent angular resolution, can be achieved with antennas small enough to fit in the nose of a small aircraft. Third, because of their wide use, microwave components for 3 cm radars are readily available from a wide range of suppliers.

Where limited range is not a problem and both small size and high angular resolution are desired, higher frequencies can be used. Radars operating in the Ka-band, for example, have been developed to perform ground search and terrain avoidance for some aircraft. But because of the high level of attenuation at these frequencies, to date there has been relatively little utilization of this band.

With the availability of suitable millimeter-wave power-generating components, radar designers are developing extremely small, albeit short-range, radars that take advantage of the atmospheric window at 94 GHz to give small air-to-air missiles high terminal accuracies (Fig. 7-12). At 94 GHz, a 10 cm antenna provides the same angular resolution as a 0.94 m antenna would at 10 GHz (3 cm).

Typical Frequency Selections	
• Early warning radars	UHF, L, and S-bands
• Radar altimeters	C-band
• Weather radars	C- and X-bands
• Fighter/attack	X- and Ku-bands

7.5 Summary

Radio frequencies employed by airborne radars range from a few hundred megahertz to 100,000 MHz (100 GHz), the optimum frequency for any one application being a trade-off among several factors.

In general, the lower the frequency, the greater the physical size of the hardware and the higher the available maximum power. The higher the frequency, the narrower the beam that may be achieved with a given size antenna.

At frequencies above about 0.1 GHz, attenuation due to atmospheric absorption—mainly by water vapor and oxygen—becomes significant. At frequencies of 3 GHz and higher, scattering by condensed water vapor—rain, hail, and to lesser extent, snow—produces weather clutter. It not only increases attenuation, but in radars not equipped with MTI, it can obscure targets. Above about 10 GHz, absorption and scattering become increasingly severe and attenuation due to clouds becomes important.

Noise is minimal between about 0.3 GHz and 10 GHz, but becomes increasingly severe at 20 GHz and higher frequencies.

Doppler shifts increase with frequency, and this may also be a consideration in certain applications.

Further Reading

D. E. Kerr, *Propagation of Short Radio Waves*, IEEE Press, 1986.

M. E. Davis, *Foliage Penetration Radar: Detection and Characterization of Objects under Trees*, SciTech-IET, 2011.

L. W. Barclay, *Propagation of Radiowaves*, 3rd ed., IET, 2012.

Test your understanding

1. Explain the mechanisms by which electromagnetic waves are attenuated when passing through the atmosphere.
2. A particular radar has a center frequency of 15 GHz and a bandwidth of 3 GHz. What is its fractional bandwidth?
3. The Swedish CARABAS FOPEN radar operates over the band 20–90 MHz. What is its fractional bandwidth?
4. An X-band radar signal passes through a severe rainstorm with a horizontal extent of 10 km and an attenuation of 2 dB/km. What is the two-way attenuation of the signal through the rainstorm?

Lockheed U-2 (1957)

As with the SR-71 Blackbird, the U-2 Dragon Lady was designed and manufactured by the Lockheed Skunk Works and is a high altitude reconnaissance aircraft. During the Cold War, its main role was intelligence over the Soviet Union and other countries more recently. It was also flown over Afghanistan and Iran to support North Atlantic Treaty Organization operations.

8

Directivity and the Antenna Beam

"Airborne Cigar" jammer installation in B17 Fortress

The degree to which the antenna concentrates the radiated energy in a desired direction—referred to here as *directivity*—is a key characteristic of virtually every airborne radar. Besides determining the radar's ability to locate targets in angle, directivity can vitally affect the ability to deal with ground clutter and is a major factor governing detection range.

In this chapter, we will learn how the energy radiated by an antenna is distributed in angle and examine the salient characteristics of the radiation pattern: beamwidth, gain, and sidelobes. We will then see how the sidelobes may be reduced; how fast, versatile beam positioning may be accomplished with electronic scanning; and how high angular resolution and angular measurement accuracy may be achieved. Finally, we will learn how the beam may be optimized for ground mapping.

8.1 Distribution of Radiated Energy in Angle

From common simplistic illustrations, it might be supposed that a radar antenna concentrates all of the transmitted energy into a narrow beam—known as a *pencil beam*—within which the power is uniformly distributed. If a pencil beam were trained like a flashlight on an imaginary screen in the sky, it would illuminate a single, round spot with uniform intensity. While this might be desirable, it is even less true of an antenna than of a flashlight.

Like all antennas, a pencil beam antenna radiates some energy in almost every direction. As illustrated in the three-dimensional plot of Figure 8-1, most of the energy is concentrated in a more or less conical region surrounding the central axis, or *boresight line*, of the antenna. This region is called the *mainlobe*. If we slice the plot in two through the central axis of this lobe, we find that it is flanked on either side by a series of

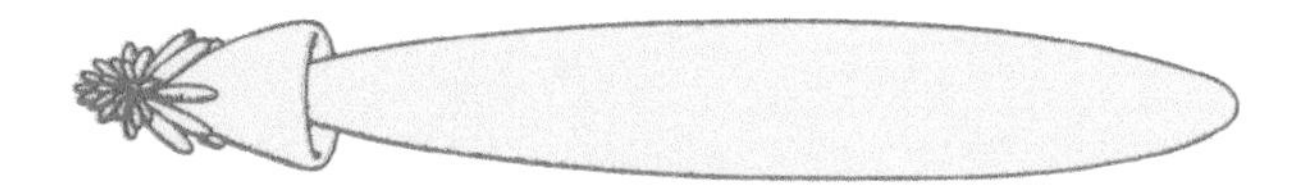
Figure 8-1. This three-dimensional plot shows the strength of the radiation from a pencil beam antenna.

Figure 8-2. In this slice taken through the plot of Figure 8-1, note the series of lesser lobes on either side of the mainlobe.

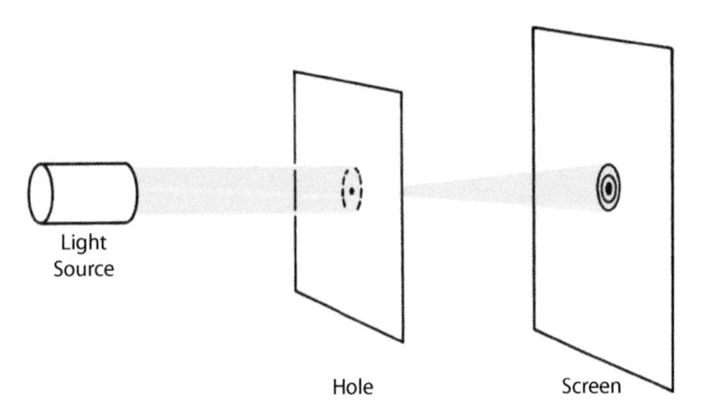

Figure 8-3. The lobular distribution of power shown here is due to diffraction, the process that causes a beam of monochromatic light projected through a tiny hole to spread and become fringed with concentric rings of light.

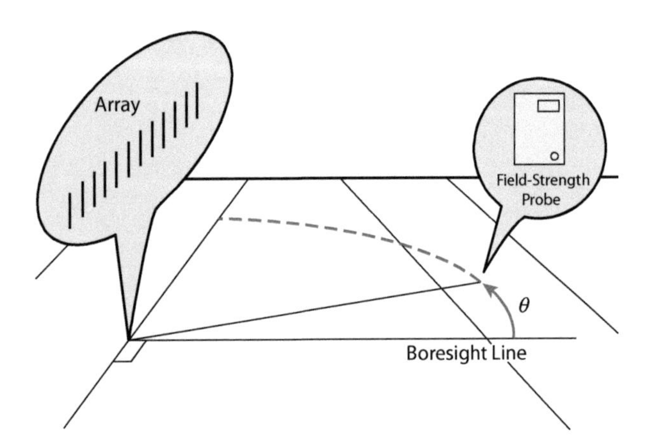

Figure 8-4. To determine the distribution of energy radiated by an array, a field-strength probe is moved along an arc of constant radius. The array consists of a row of closely spaced vertical radiators.

weaker lobes (Fig. 8-2). These are called *sidelobes*; to the rear they are called *backlobes*.

This lobular structure is due to diffraction—the phenomenon observed when a beam of light passes through a small hole (Fig. 8-3). The beam spreads and, if the light is all of one wavelength, becomes fringed with concentric rings of light of progressively decreasing intensity.

The phenomenon is most easily explained if we consider a type of horizontal, one-dimensional antenna called a linear broadside array. It consists of a row of closely spaced radiators, each emitting in all azimuth directions a wave of the same amplitude, phase, and frequency. To measure the combined strength of these waves at various azimuth angles, we place a field strength probe far enough away that the lines of sight from the probe to all radiators are very nearly parallel. Starting at a point on the perpendicular bisector of the array (boresight line), we move the probe along an arc of constant radius from the array center, as shown in Figure 8-4.

At any one point, the field strength (measured in volts per meter) depends on the relative phases of the received waves. The relative phases, in turn, depend on the differences in distance to the individual radiators. These differences can best be visualized if we draw a line from one end of the array, perpendicular to the line of sight to the probe—the line AB in Figure 8-5. The angle this line makes with the array equals the azimuth angle, θ, of the probe.

Now, if θ is zero (i.e., the boresight direction) and the probe is far from the array, the distance from the probe to all of the radiators is essentially the same. (The lines of sight to all radiators, remember, are essentially parallel.) The waves are in phase, and their fields add up to a large sum.

However, if θ is greater than zero, the distance to each successive radiator down the line is progressively greater. As a result, the phases of the received waves are all slightly different, and the sum is not as great.

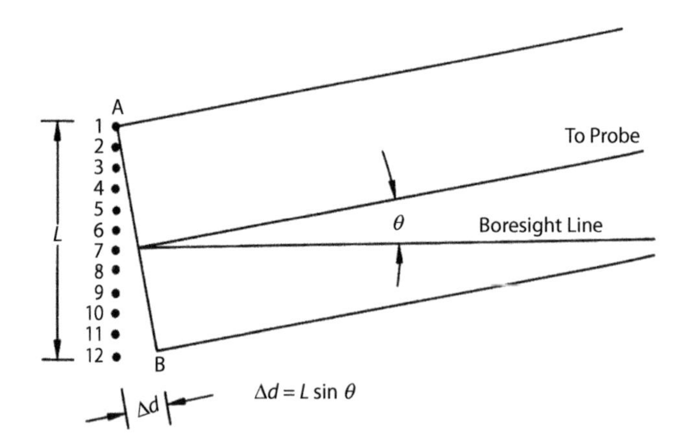

Figure 8-5. Line AB marks off the differences in distance from the individual array elements to the probe. The angle that AB makes with the array equals the azimuth angle, θ, of the probe.

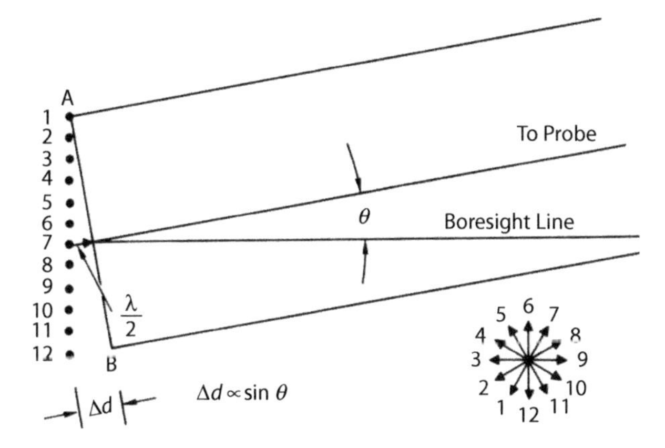

Figure 8-6. When the distance from the probe to radiator No. 7 becomes half a wavelength longer than the distance to radiator No. 1, the signals received from these radiators cancel. So do all the others.

As the azimuth angle increases, the differences in distance increase. A point is ultimately reached (Fig. 8-6) where the distance from the probe to the first radiator beyond the center of the array (No. 7) is a half-wavelength greater than the distance to the radiator at the near end (No. 1). Consequently, the wave received from radiator No. 1 is canceled by the wave received from radiator No. 7. The same is true of the waves received from radiators No. 2 and No. 8 and so on. The sum of the waves received from all of the radiators, therefore, is zero. The probe has reached an azimuth angle where there is a null in the total radiation from the antenna.

If θ is increased further, the waves from the radiators at the ends of the array no longer cancel exactly, and the sum increases. As the difference in distance from the probe to the ends of the array approaches 1.5 wavelengths, another peak is reached (Fig. 8-7). The waves from the radiators in the central portion of the array—Nos. 3 through 10—still cancel. But the waves from the radiators at either end—Nos. 1 and 2 and Nos. 11 and 12—add up to an appreciable sum. The probe is now in the center of the array's first sidelobe.

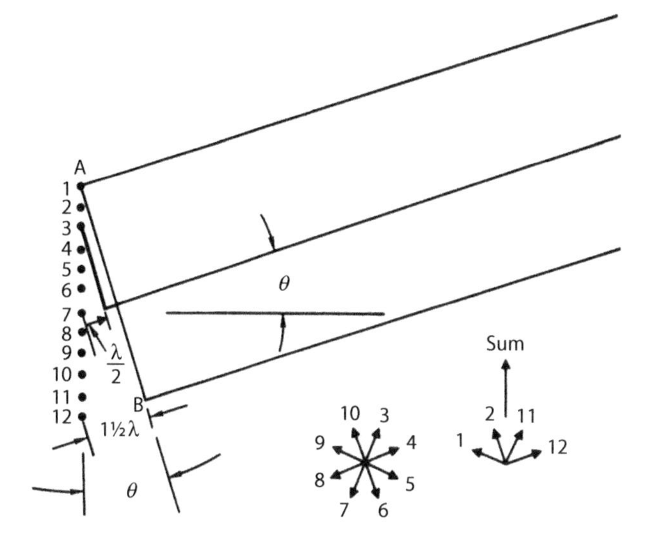

Figure 8-7. As the difference in distance from the probe to the ends of the array approaches 1.5 wavelengths, only those signals from elements 3 through 10 cancel.

If θ is increased still further, the portion of the array for which cancellation occurs increases, and the same general process repeats. The probe thus moves through a succession of nulls and progressively weaker lobes.

The field strength measured in an excursion through several lobes on either side of the mainlobe is plotted versus azimuth angle in Figure 8-8. The shape of this plot is given by the equation

$$E \propto \frac{\sin x}{x}$$

where E is the field strength, and x is proportional to θ. This is called a sine-x-over-x, or *sinc* function shape.

Actually, $x = \pi\ (L/\lambda)\ \sin\ \theta$, where λ is the wavelength. So x is directly proportional to θ only for small values of θ. As θ increases, $\sin\ \theta$ becomes progressively less than θ, with the result that the higher-order sidelobes are spaced progressively farther apart.

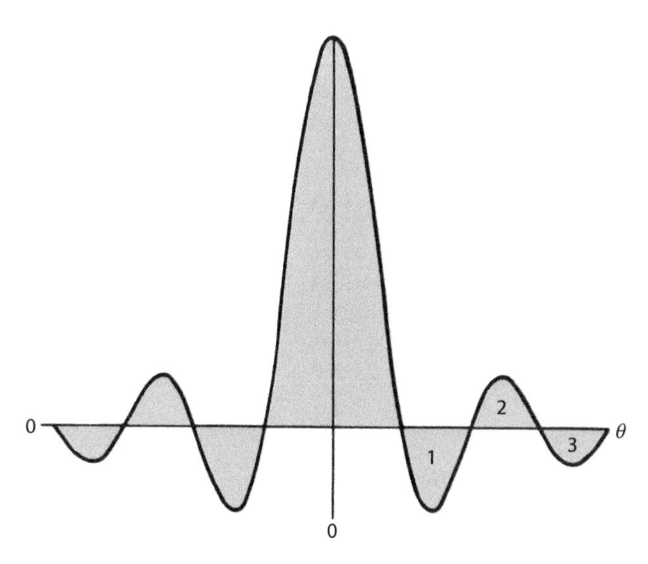

Figure 8-8. This figure represents field strength measured in an excursion through several lobes on either side of the boresight line. The radio frequency phase of odd numbered sidelobes (e.g., 1, 3) is reversed and hence these sidelobes are plotted as negative.

The (sin X)/X Shape

As the angle, Θ, between the line of sight to a distant point and the boresight line of a linear array antenna increases, phasors representing the signals received from the individual radiators fan out, and their sum decreases.

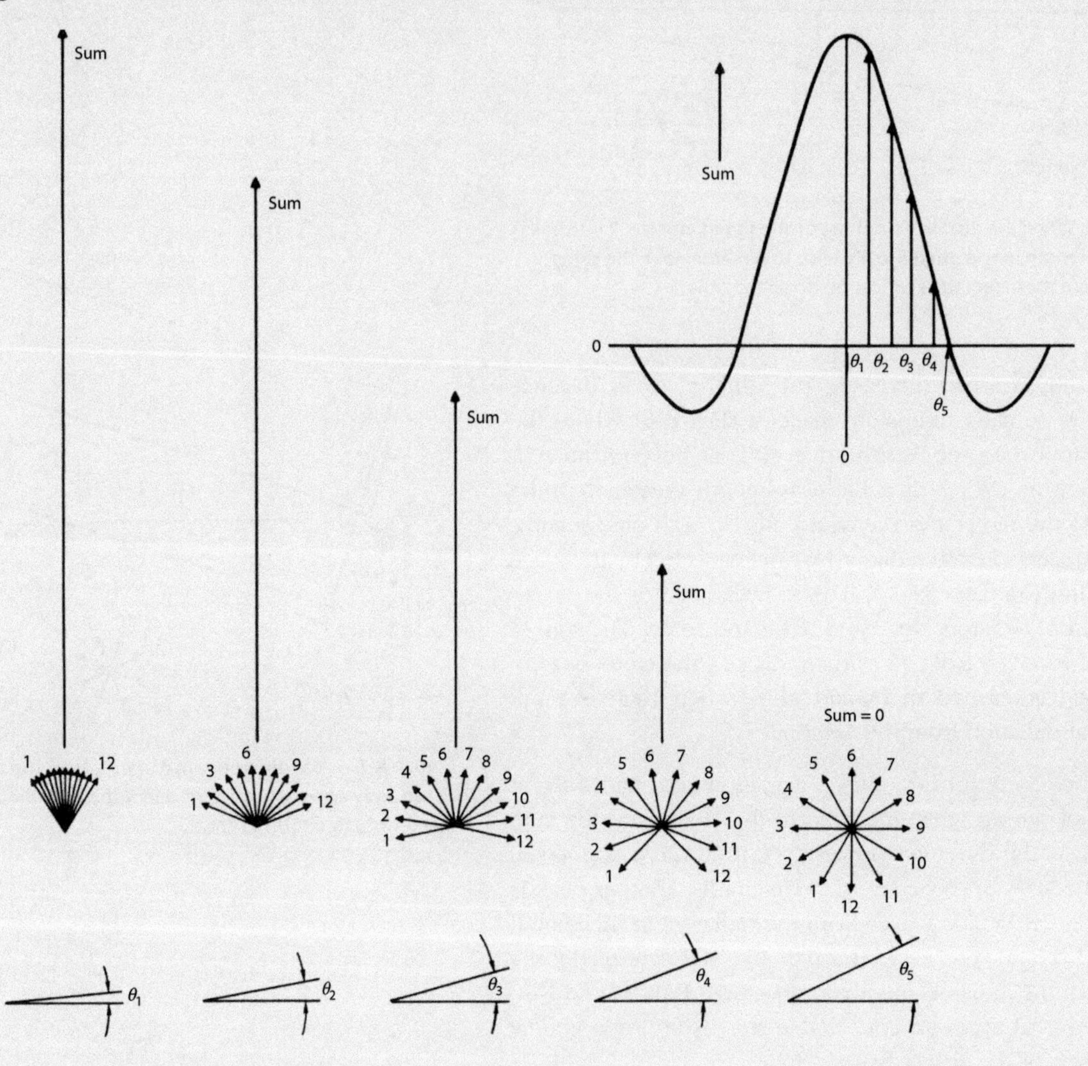

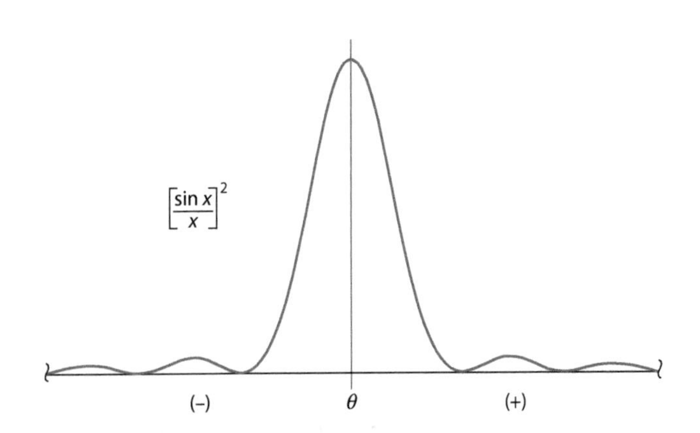

Figure 8-9. The directivity of a linear array is expressed here in terms of power.

The directivity of an array antenna has been explained here in terms of field strength since it is both easily measured and easily visualized.

In a radar, however, what is important is the amount of energy radiated per unit of time: the power of the radiated waves (Fig. 8-9). Power is proportional to field strength squared. Expressed in terms of power, therefore, the equation for the distribution of the radiated energy in angle is

$$\text{Power} \propto \left(\frac{\sin x}{x}\right)^2$$

Two-dimensional planar arrays, such as are commonly used in airborne radars, consist essentially of a number of linear arrays stacked on top of one another.

Two Common Types of Airborne Radar Antennas

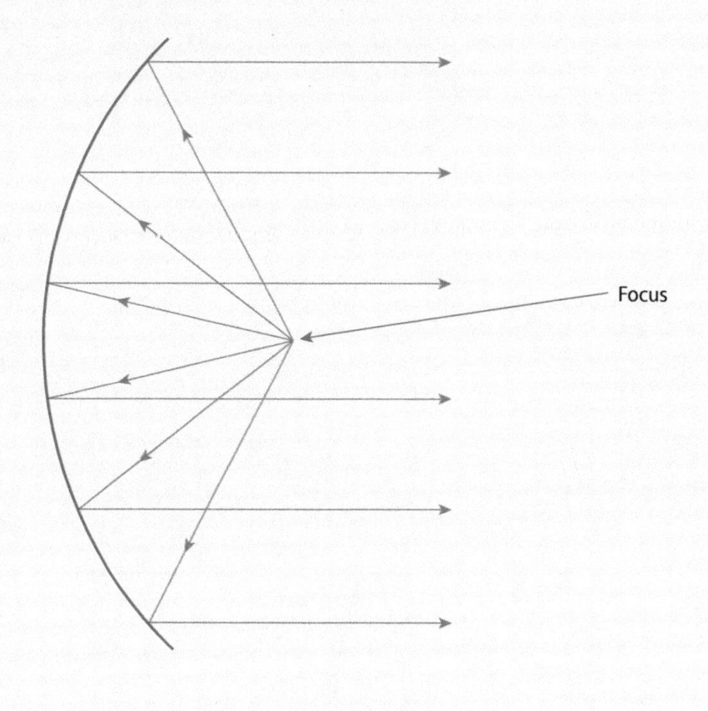

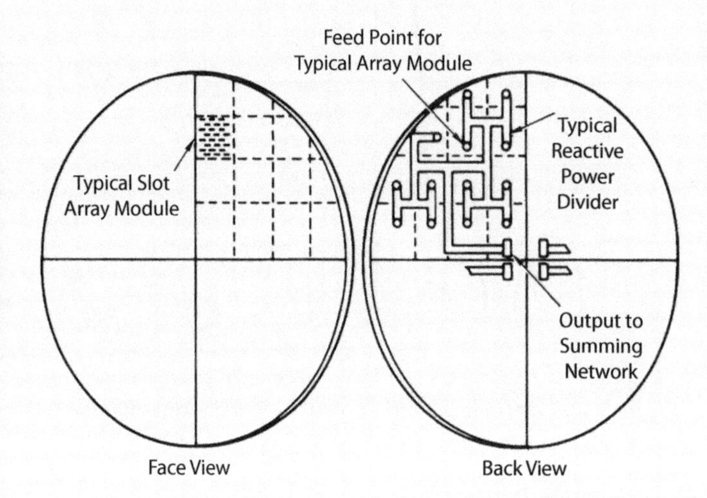

FOR YEARS, *PARABOLIC REFLECTOR ANTENNAS* WERE THE MOST COMMON TYPE used in airborne radars. The feed is located at the focus of a parabola and directs radiation into the dish, which reflects it. The curvature of the parabola is such that the distance from feed to dish to a plane across the mouth (aperture) of dish is the same for every path the radiation can take. Consequently, the phase of the radiation at every point in the plane of the aperture is the same, and a narrow pencil beam is formed. The antenna is simple and relatively inexpensive to fabricate.

In a planar array antenna for an advanced fighter radar, radiation of equal phase is emitted from a two-dimensional array of radiating elements (in this case waveguide slots) in the face. Planar arrays provide relatively high aperture efficiency and low back radiation (spillover). By controlling the excitation of the slots through reactive (nondissipative) power dividers on the back of the antenna, the distribution of energy across the aperture can be shaped to minimize sidelobes. The principal disadvantages are relatively narrow bandwidth (≅ 10 percent) and higher cost. Also, circular polarization, if desired, is more difficult to obtain.

To give an antenna a circular or elliptical shape, the rows of radiators above and below the central ones are progressively shortened. The total radiation from the antenna is the composite of the radiation from the individual radiating elements. Even if the radiation from every element were the same—which it never is—a plot of the total radiation pattern would not have a simple sin *x*/*x* shape. Nevertheless, the general shape of the plot is much the same as the sin *x*/*x* function. (Incidentally, the shape for a uniformly illuminated circular array is exactly the same as the diffraction pattern mentioned earlier for light passing through a small round hole).[1]

1. This pattern has a $J_1(x)$-over-x shape, where J_1 is the Bessel function of the first order.

8.2 Characteristics of the Radiation Pattern

A plot of the power (or field strength) of the radiation from an antenna in any one plane versus angle from the antenna's central axis is called a radiation pattern. In considering directivity, the power at the center of the mainlobe is taken as a reference, and the power radiated in every other direction is taken in ratio to this value. The ratio is normally expressed in decibels and plotted in rectangular coordinates as in Figure 8-10.

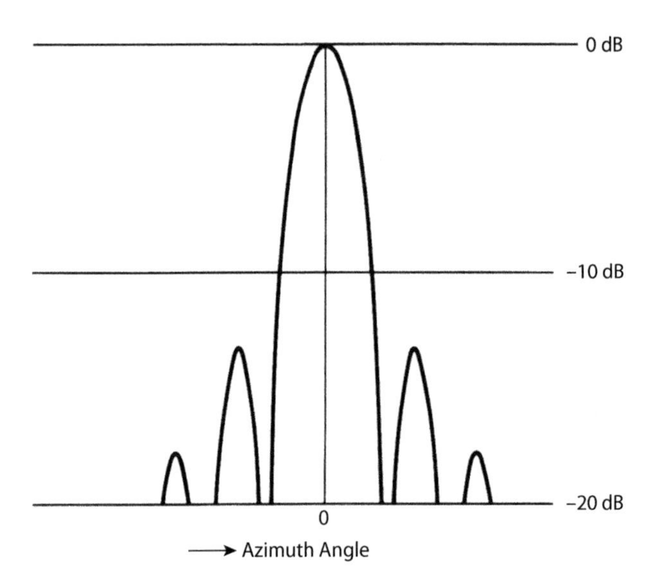

Figure 8-10. A radiation pattern is normally plotted in rectangular coordinates in dB relative to the gain at the center of the mainlobe.

Since the pattern is usually not rotationally symmetric about the center of the mainlobe, cuts must be taken through many different planes to describe an antenna's directivity fully. Also, patterns are generally measured in two polarizations: the polarization for which the antenna was designed; and the polarization orthogonal to this, called the *cross-polarization*.

Generally, three characteristics of a radiation pattern are of interest: the width of the mainlobe; the gain of the mainlobe; and the relative strengths of the sidelobes.

Beamwidth. The width of the mainlobe is called the *beamwidth*. It is the angle between opposite edges of the beam. Since the beam is generally rotationally symmetric, it is common to refer to azimuth beamwidth and elevation beamwidth.

Since the strength of the mainlobe falls off increasingly as the angle from the center of the beam increases, for any value of beamwidth to have meaning, one must specify what the edges of the beam are considered to be.

The beam edges are perhaps most easily defined as the nulls on either side of the mainlobe. In real antennas these nulls are not always distinct, however. From the standpoint of the operation of a radar (Fig. 8-11), it is generally more realistic to define them in terms of the points where the power has dropped to some arbitrarily selected fraction of that at the center of the beam. The fraction most commonly used is ½. Expressed in decibels, a factor of ½ in power is –3 dB. Beamwidth measured between these points, therefore, is called the 3 dB beamwidth.

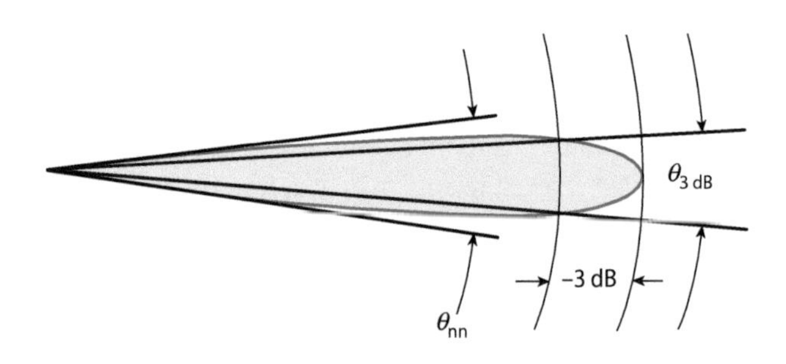

Figure 8-11. Beamwidth is commonly measured between points where power has dropped to one-half of maximum (–3 dB). The 3 dB beamwidth, $\theta_{3\,dB}$, is roughly half of the null-to-null beamwidth, θ_{nn}.

Regardless of how it is defined, beamwidth is determined primarily by the size of the antenna's frontal extent. This area is called the *aperture*. Its dimensions—width, height,

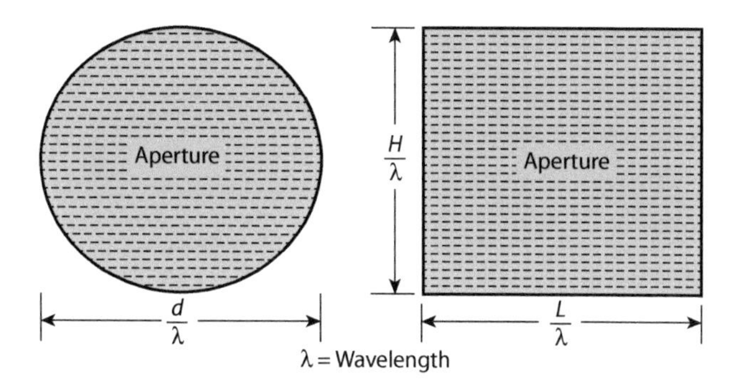

Figure 8-12. Beamwidth is determined primarily by the dimensions of antenna aperture, in wavelengths.

or diameter—are gauged not in inches or centimeters but in wavelengths of the radiated energy (Fig. 8-12).

The larger the appropriate dimension is in relation to the wavelength, the narrower the beam in a plane through that dimension will be. As we saw earlier, the nulls on either side of the mainlobe of a linear array occur at angles for which the distance from the observer to one end of the array is one wavelength longer than to the other end.

Therefore, for either a *linear array* or a *rectangular aperture* over which the illumination is uniformly distributed, the null-to-null beamwidth in radians is twice the ratio of the wavelength to the length of the array (Fig. 8-13).

$$\theta_{\mathrm{nn}} = 2\frac{\lambda}{L}\ \text{radians}$$

where λ is the wavelength of radiated energy, and L is the length of aperture (same units as λ). The 3 dB beamwidth is a little less than half the null-to-null width.

$$\theta_{3\,\mathrm{dB}} = 0.88\frac{\lambda}{L}$$

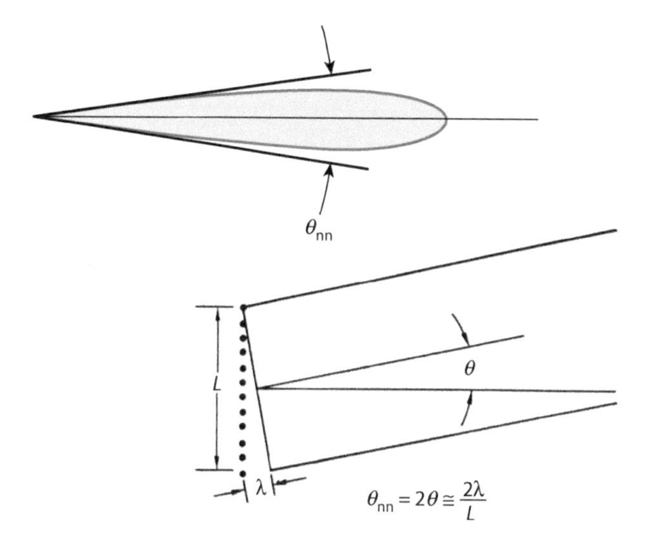

Figure 8-13. For a linear array with uniform illumination, the angle (in radians) from the boresight line to the first null equals the ratio of the wavelength to the length of array. The null-to-null beamwidth is twice this angle.

For a uniformly illuminated *circular aperture* of the diameter, d, the 3 dB beamwidth is a bit greater.

$$\theta_{3\,\mathrm{dB}} = 1.02\frac{\lambda}{d}$$

A circular antenna 60 cm in diameter, radiating energy of 3 cm wavelength, for example, has a beamwidth of $1.02 \times \frac{3}{60} = 0.051$ radian.

One radian equals $360°/2\pi = 57.3°$ (Fig. 8-14). Thus, the beamwidth in degrees is $0.051 \times 57.3 = 2.9°$.

If the antenna has tapered illumination to control the sidelobe levels, such as is typically used in radars for fighter aircraft, the beamwidth will be somewhat greater, typically:

$$\theta_{3\,\mathrm{dB}} \approx 1.25\frac{\lambda}{d}$$

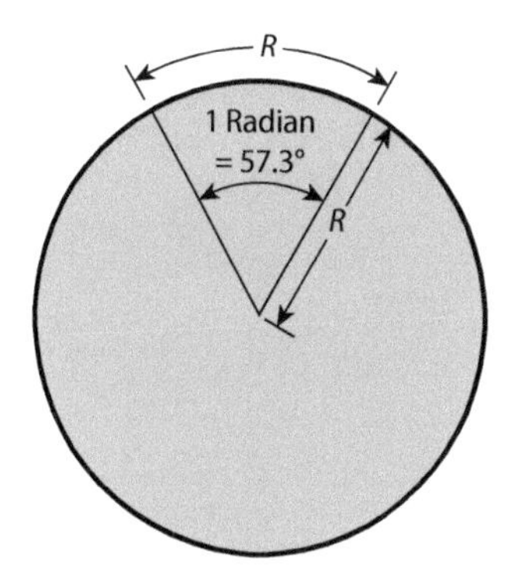

Figure 8-14. A radian is the angle subtended by an arc of length equal to the radius, R. The circumference of a circle equals $2\pi R$. Therefore, 2π radians = 360°, and 1 radian = $360°/2\pi = 57.3°$.

X-Band Beamwidth Rule of Thumb

For tapered illumination:	For untapered illumination:
$\theta_{3\text{dB}} \approx \dfrac{216°}{d}$	$\theta_{3\text{ dB}} \approx \dfrac{178°}{d}$

where d is the diameter in cm.

A 60 cm antenna with tapered illumination would thus have a beamwidth of about 3.6°.

The 3 dB beamwidth of a 60 cm diameter antenna is thus about 216° ÷ 60 = 3.6°. If the illumination is not tapered, 178° should be substituted for 216° in this rule.

Antenna Gain. The gain of an antenna is the ratio of the power per unit of solid angle radiated in a specific direction to the power per unit of solid angle that would have been radiated had the same total power been radiated uniformly in all directions—that is, isotropically (Fig. 8-15).[2] An antenna thus has gain in almost every direction. In most directions, though, the gain is less than one since the gain averaged over all directions is, by the law of conservation of energy, one.

The gain in the center of the mainlobe is thus a measure of the extent to which the radiated energy is concentrated in the direction the antenna is pointing. The narrower the mainlobe, the higher this gain will be.

The maximum gain that can be achieved with a given size antenna is proportional to the area of the antenna aperture in square wavelengths times an illumination efficiency factor. If the aperture were uniformly illuminated and lossless—a practically impossible condition, even if it were desired—the efficiency factor would equal one.

Actually, it ranges somewhere between 0.6 and 0.8 for planar arrays and may be as low as 0.45 for parabolic reflectors. In either case, for a given design, the efficiency factor tends to vary with the width of the band of frequencies the antenna is designed to pass. Typically, the greater the bandwidth, the lower the efficiency.

2. Strictly speaking, the gain referred to here is directivity gain. More commonly, antenna gain connotes directivity gain less whatever power is lost in the antenna.

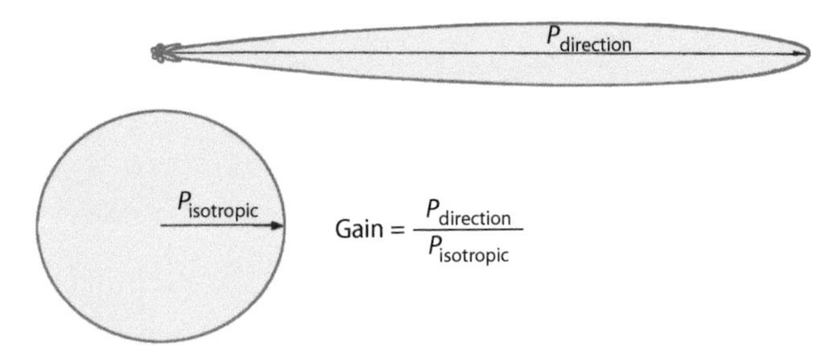

Figure 8-15. Antenna directive gain is the ratio of power radiated in the direction of interest to the power that would be radiated in that direction by an isotropic antenna, that is, one that radiates waves of equal power in all directions.

Relationship Between Antenna Gain and Effective Area

CONSIDER A RECTANGULAR ANTENNA APERTURE OF EFFECTIVE DIMENSIONS $a \times b$. The beamwidths in the respective planes are approximately λ/a and λ/b (radians), which subtend a rectangular area $R\lambda/a \times R\lambda/b$ on the surface of a sphere of radius, R.

The gain of the antenna is the ratio of the power density in the main beam to what would be obtained from an isotropic antenna, that is, the ratio of the total surface area of the sphere to the area $R\lambda/a \times R\lambda/b$:

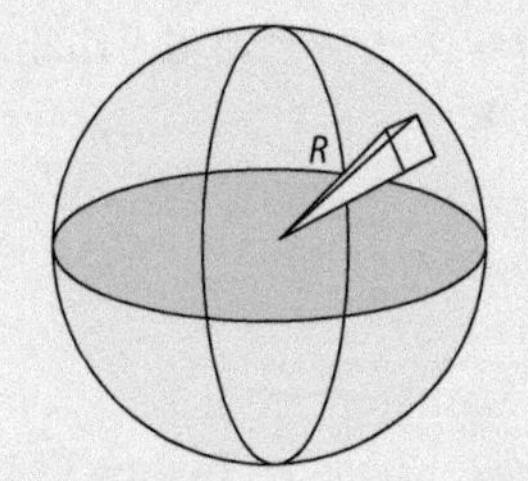

$$G = \frac{4\pi R^2}{R^2\lambda^2/ab}$$

$$= \frac{4\pi A_e}{\lambda^2}$$

where $A_e = a \times b$ is the effective area of the antenna

Thus, antenna gain is often expressed in units of decibels with respect to isotropic, or dBi.

Because of the difficulty of determining the efficiency factor analytically, in practice the gain is determined experimentally and expressed in terms of an effective aperture area,

$$G = 4\pi \frac{A_e}{\lambda^2}$$

where G is the antenna gain at center of mainlobe, λ is the wavelength of radiated energy, and A_e = effective area of aperture (same units as λ^2).

Effective area is equal to physical area times aperture efficiency (which as noted already is virtually always less than 100 percent), so an alternate expression for antenna gain is

$$G = 4\pi \frac{A\eta}{\lambda^2}$$

where A is the physical area of aperture, and η is the aperture efficiency.

Sidelobes. An antenna's sidelobes are not limited to the forward hemisphere. They extend in all directions, even to the rear (backlobes), for a certain amount of radiation invariably "spills over" around the edges of the antenna. Moreover, when the antenna is placed in a radome, the backward radiation is increased. This occurs because the radome scatters some energy from the mainlobe, much as the frosted glass of a light bulb diffuses light from the filament.

The sidelobes are also not neatly defined and have sharp nulls in between. As can be seen from Figure 8-16, the nulls tend to fill in.

For a uniformly illuminated circular aperture, the gain of the strongest (first) sidelobe is only about 1/64 that of the mainlobe. Stated in decibels, the first sidelobe has 18 dB less gain than the mainlobe: it is down 18 dB. The gain of the other sidelobes is substantially lower.

Nevertheless, in aggregate the sidelobes rob the mainlobe of a substantial amount of power. Because of the large solid angle they cover, roughly 25 percent of the total power radiated by a uniformly illuminated antenna occurs outside the mainlobe.

Against most small targets even the strongest sidelobes are sufficiently weak that they can generally be ignored,[3] but against the ground even the weakest sidelobes may produce considerable return. And, as will be explained in Chapter 22, buildings and other structures on the ground form corner reflectors that can return tremendously strong echoes, even when illuminated only by sidelobes.

In military applications, the sidelobes also increase both the radar's susceptibility to detection by an enemy (on transmit) and its vulnerability to jamming (on receive). Interference from a powerful noise jammer, for example, can be much stronger than the echoes of a small or distant target in the mainlobe. Consequently, it is generally desirable for the gain of the sidelobes to be minimized.

Sidelobe Reduction. The degree to which the radiated power is concentrated into the mainlobe is called solid angle efficiency. To make it acceptably high and to minimize problems

Estimating Antenna Gain

X-Band Rule of Thumb:

$G \approx d^2 \eta$

d = diameter in cm

η = aperture efficiency

Example:

Diameter = 60 cm

Aperture efficiency = 0.7

$G \approx 60 \times 60 \times 0.7$
≈ 2520
$\approx 34\,\text{dB}$

General Rule of Thumb:

$G \approx 9\, d^2 \eta$

d = diameter in wavelengths

Example:

Wavelength = 3 cm

Diameter = 60 cm = 20 λ

Aperture efficiency = 0.7

$G \approx 9 \times (20)^2 \times 0.7$
≈ 2520
$\approx 34\,\text{dB}$

3. However, for some targets which in certain aspects reflect a large fraction of the incident energy back in the direction of the radar, sidelobe return can be substantial.

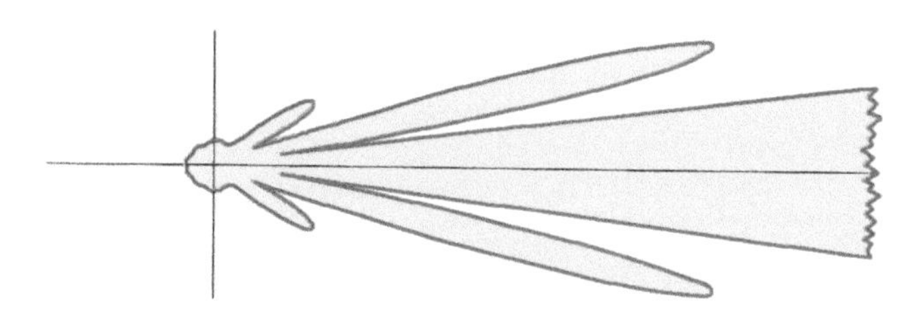

Figure 8-16. An antenna's sidelobes extend in all directions, even to the rear.

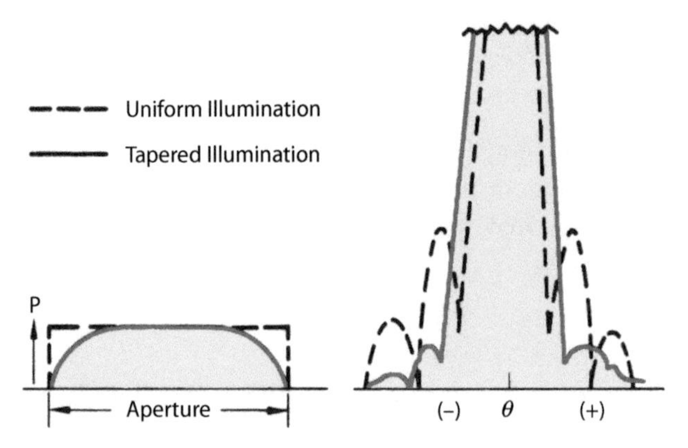

Figure 8-17. Sidelobes may be reduced by tapering illumination at the edges of the aperture.

of ground clutter and jamming, the gain of the sidelobes must generally be reduced. This is done by designing the antenna to radiate more power per unit area through the central portion of the aperture (Fig. 8-17) in a technique called *illumination tapering*. Using this process increases the beamwidth somewhat and hence reduces the peak gain of the mainlobe. But usually this is an acceptable price to pay for reduced sidelobes.

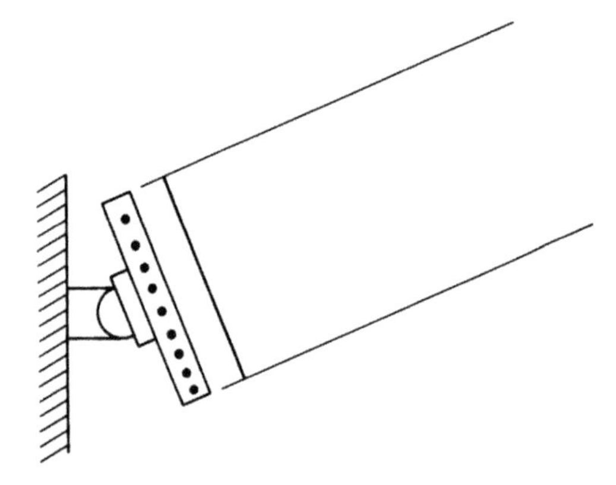

Figure 8-18. A beam is conventionally steered by mechanically deflecting the antenna.

8.3 Electronic Beam Steering

In most airborne radars, the antenna beam is positioned by physically moving the antenna through the desired azimuth and elevation angles (Fig. 8-18). An alternative method made possible with array antennas is to differentially shift the phases of the radio waves emitted by the individual radiators. This technique is called *electronic beam steering* (or *electronic scanning*).

As with the simple linear array described earlier, the direction of maximum radiation from the array (i.e., direction of the mainlobe) is that for which the waves from all of the radiators are in phase. If the phases of the emitted waves are all the same, this direction is perpendicular to the plane of the array. However, if the phases are progressively shifted from one radiator to the next, the direction of maximum radiation will be correspondingly shifted (Fig. 8-19). By appropriately shifting the phases of the inputs to the individual radiators, therefore, the beam can be steered in any desired direction within a large solid angle.

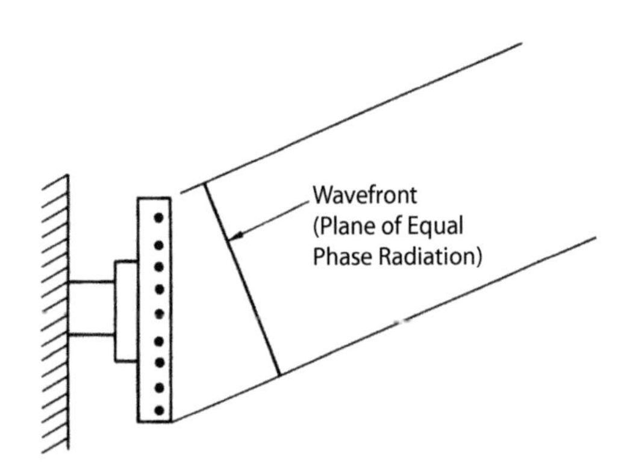

Figure 8-19. With electronic steering, the beam is steered by progressively shifting the phases of the signals radiated by the individual radiators.

Electronic steering has the advantage of being extremely flexible and remarkably fast. The beam can be given any shape, swept in any pattern at a very high rate, or jumped almost instantaneously to any position. It can even be split into two or more beams that radiate simultaneously on different frequencies and can be trained simultaneously on different targets (at the expense of a reduction in detection range).

Figure 8-20. This antenna is for side-looking air-to-ground radar in which a fan-shaped beam is electronically steered in azimuth. The antenna is carried in a pod beneath an aircraft. By rotating it about its longitudinal axis, it can be made to look out on either side of the aircraft.

Depending on the application, electronic steering may be provided in one (Fig. 8-20) or two dimensions. Moreover, it may be combined with either mechanical beam steering or mechanical rotation of the antenna, as in the airborne warning and control system (AWACS) radar.

Naturally, electronic steering also has disadvantages such as increased complexity and degraded performance at large look angles. Performance degradation is caused by aperture foreshortening when viewed from angles off dead center (Fig. 8-21). The length of the foreshortened dimension decreases proportional to the cosine of the angle. The effect is negligible at small scan angles, but it becomes increasingly severe at large angles. The result of the foreshortening (effectively smaller aperture in the direction of illumination) is an increase in beamwidth and more importantly a decrease in gain, which limits the maximum practical look angle to ±60°.

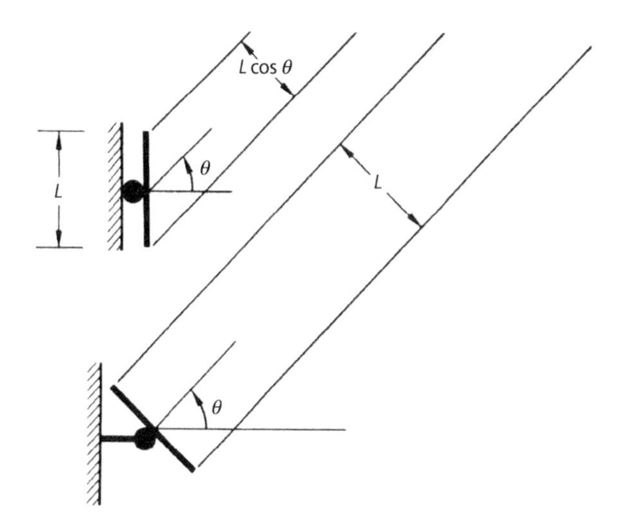

Figure 8-21. With electronic steering, the apparent length of the aperture, L, decreases as cosine of the look angle, θ. With mechanical steering, no such reduction occurs.

With mechanical steering, no such limitation occurs: the plane of the aperture is perpendicular to the direction of the mainlobe for all look angles.

8.4 Angular Resolution

The ability of a radar to resolve targets in azimuth and elevation is determined primarily by the azimuth and elevation beamwidths. This is illustrated simplistically by the two diagrams in Figure 8-22.

In the first diagram, two identical targets, A and B, at nearly the same range are separated by slightly more than the width of the beam. As the beam sweeps across them, the radar receives echoes first from Target A and then from Target B. Consequently, the targets can easily be resolved.

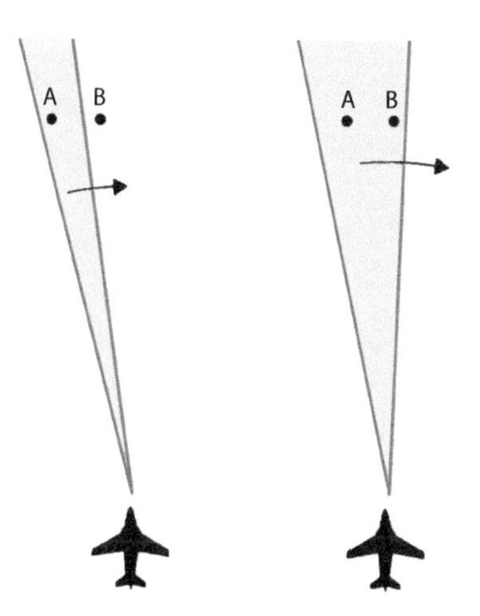

Figure 8-22. The ability to resolve targets in angle is determined primarily by antenna beamwidth. Targets can be resolved if beamwidth is less than their angular separation.

In the second diagram, the same two targets are separated by less than the width of the beam. As the beam sweeps across

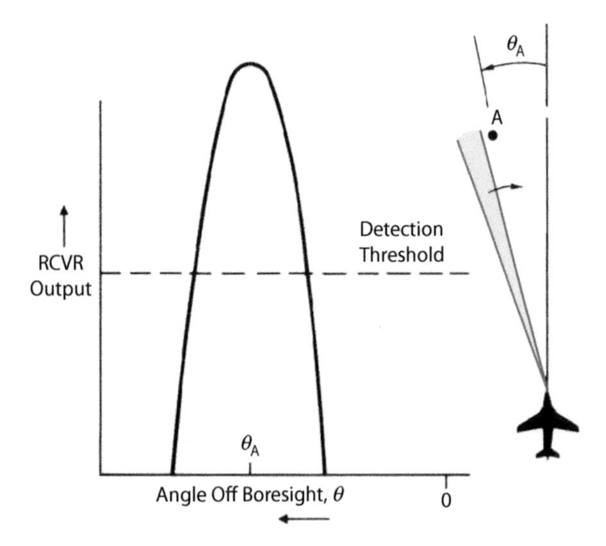

Figure 8-23. Angular accuracy is sharpened by peaking of the receiver output as the beam sweeps across the target. Unless the target echoes are very strong, the azimuth angle over which the return is detected is much less than the null-to-null beamwidth, θ_{nn}.

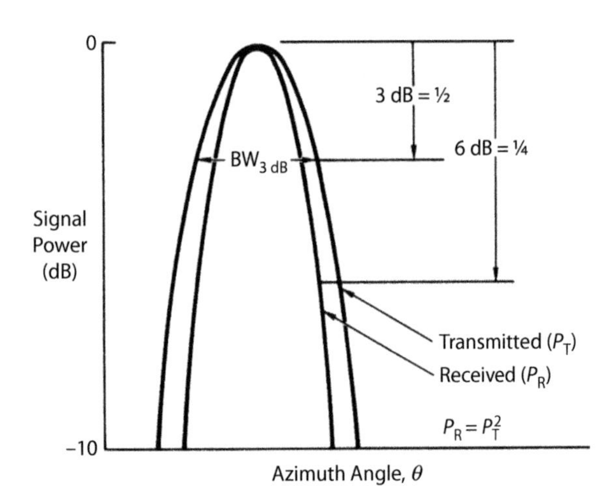

Figure 8-24. Since the antenna's directivity is applied to both transmitted and received waves, the plot of received signal strength versus angle is more sharply peaked.

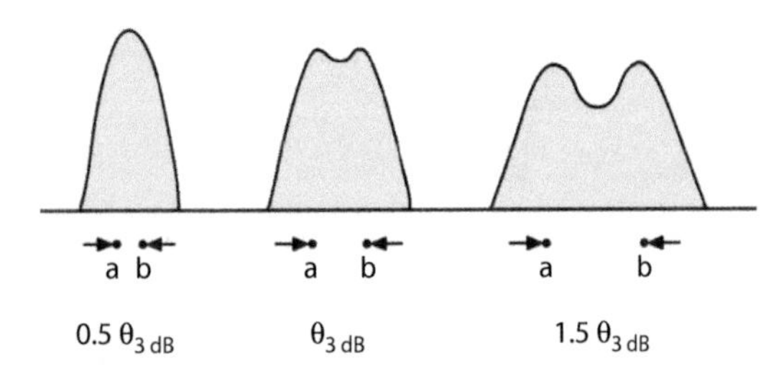

Figure 8-25. As separation between two closely spaced targets is increased, a notch develops in the plot of receiver output versus azimuth angle.

them, the radar again receives echoes first from Target A. However, long before it stops receiving echoes from this target, it starts receiving echoes from Target B. The echoes from the two targets therefore merge together.

Superficially, angular resolution would appear to be limited to the null-to-null width of the mainlobe. But it is actually better than that because the resolution depends not only on the width of the lobe but also on the distribution of power within it.

Figure 8-23 is a plot of strength of the received signal as the mainlobe sweeps across an isolated target. When the leading edge of the lobe passes over the target, the echoes are so weak that they are undetectable. However, their strength increases rapidly and reaches a maximum when the lobe is centered on the target. The strength then drops to an undetectable value again as the trailing edge approaches the target. Note that this curve is not the same shape as the radiation pattern plotted in similar coordinates but instead is more sharply peaked. The reason is that the antenna's directivity applies equally to transmission and reception—a characteristic called *reciprocity*—so the two-way beamwidth is narrower.

To illustrate, suppose the position of a target is such that the power radiated in its direction is half that radiated in the center of the lobe (down 3 dB). When the target echoes are received, their power will again be cut in half. As a result, the received echoes will be only one-quarter as strong (down 6 dB) as when the target is in the center of the lobe (Fig. 8-24).

Because of this compounding, the plot of received signal power is narrower than the radiation pattern. And because the echoes received when the target is near the edges of the lobe are too weak to be detected (unless the target is at short range), the azimuth angle over which the target is detected is narrower than the null-to-null beamwidth.

The net effect of this narrowing on angular resolution is illustrated by the three plots of Figure 8-25. They show a composite of the bell-shaped curves for two equally strong targets, A and B. When the targets are closely spaced, the curves combine to produce a single broad hump. As the spacing increases, a notch develops in the top of this hump. The notch grows until the hump splits in two.

In practice, the notch becomes apparent at a target spacing of 1 to 1.5 times the antenna's 3 dB beamwidth. The 3 dB beamwidth, therefore, has come to be used as the measure of the angular resolution of a radar.

8.5 Angle Measurement

The foregoing should not be taken to imply that the accuracy with which a radar can determine a target's direction is limited to the beamwidth. Since the amplitude of the received echoes varies symmetrically as the beam sweeps across a target, the direction of an isolated target can be determined to within a very small fraction of the beamwidth.

By stopping the antenna's search scan, target angle can be determined with still greater precision. One technique for accomplishing this is *lobing*.

Lobing. During reception, the center of the mainlobe is alternately placed on one side of the target and then the other (Fig. 8-26). If the target is centered between lobes, the received echoes will be the same strength for both lobes. If it is not, the echoes will be stronger for one lobe than for the other.

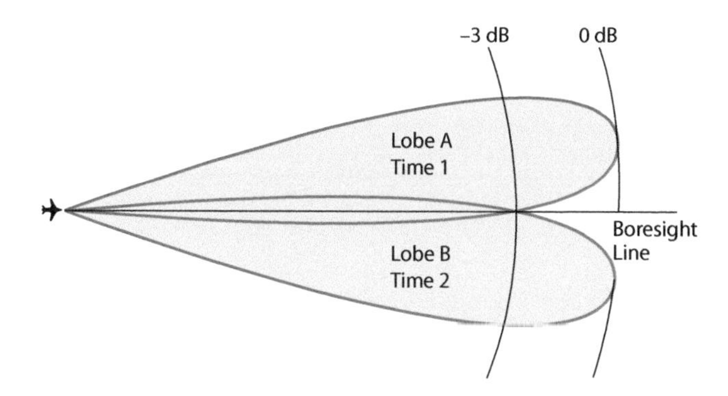

Figure 8-26. With sequential lobing, during reception the angular tracking error is determined by alternately placing the mainlobe on one side and then the other of the antenna boresight line.

Normally, the lobes are separated just enough to intersect at their half-power points. Since the slope of the radiation pattern in this region is relatively steep, a slight displacement of the target from a line through the crossover point results in a large difference in the strength of the echoes received through the two lobes (Fig. 8-27). By positioning the antenna to reduce this difference to zero (i.e., to eliminate the angular error), the antenna can be precisely lined up on the target.

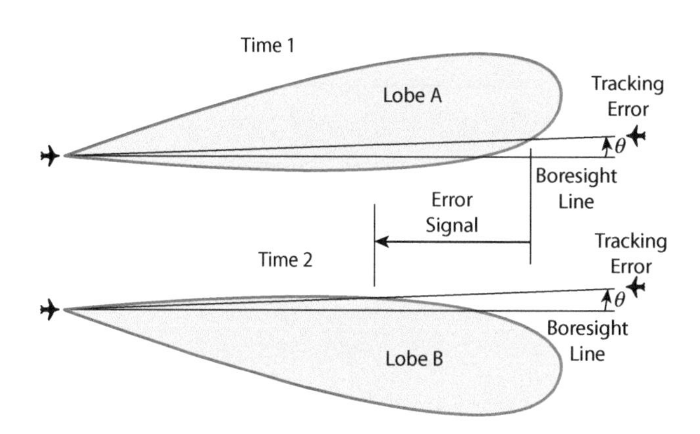

Figure 8-27. If the target is off the boresight line, the return received through one lobe will be stronger than that received through the other. The magnitude of difference corresponds to the magnitude of tracking error, the sign of difference to the direction of error.

Because the lobing is sequential, however, short-term changes in the strength of the target echoes—caused by scintillation or electronic countermeasures—can introduce large, spurious differences in the returns received through the two lobes and thus can degrade tracking accuracy. This problem may be avoided by designing the antenna to produce the lobes simultaneously. Since all the necessary angular tracking information is obtained from one reflected pulse, instead of *simultaneous lobing* it is more commonly called *monopulse operation*.

Monopulse. Monopulse systems are of two general types. They differ in regard to both the direction of the lobes and the way the returns received through opposing lobes are compared.

The first type, called *amplitude comparison monopulse*, essentially duplicates sequential lobing with simultaneously formed lobes (Fig. 8-28). Amplitude comparison monopulse is typically used with reflector antennas.

Because the lobes point in slightly different directions, if a target is not on the boresight line of the antenna the amplitude of the return received through one lobe differs from the amplitude of the return simultaneously received through the other lobe. The difference is proportional to the angular error.

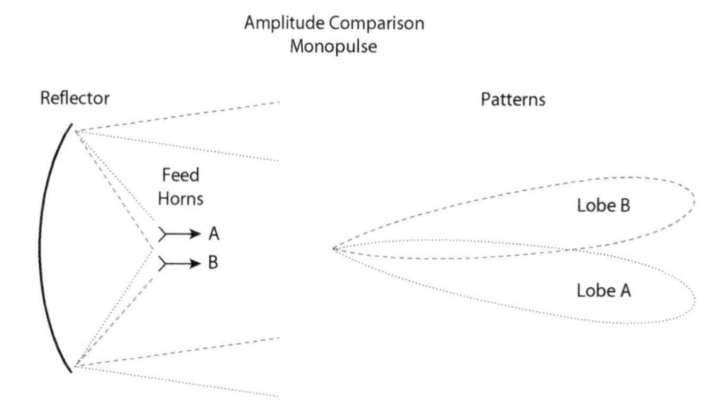

Figure 8-28. In essence, amplitude comparison monopulse duplicates sequential lobing in every respect except that return is received simultaneously through both lobes. The Error signal is the difference between outputs A and B.

By subtracting the output of one feed from the output of the other, an angular tracking error signal, often termed the *difference signal*, is produced. The sum of the two outputs, called the *sum signal*, is used for range tracking.

The second type of monopulse is *phase-comparison monopulse*, which is typically used with planar array antennas. In it, the array is divided into halves, and the resulting lobes

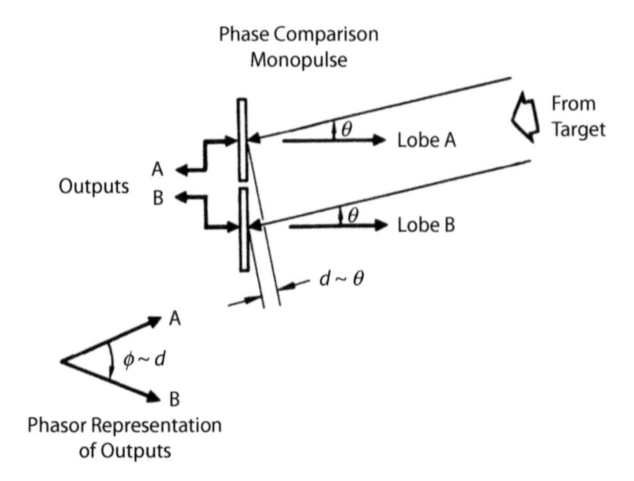

Figure 8-29. In phase-comparison monopulse, since lobes of two antenna halves point in the same direction, amplitudes of outputs A and B are equal. However, their phases differ by angle ϕ, which is proportional to angle error, θ.

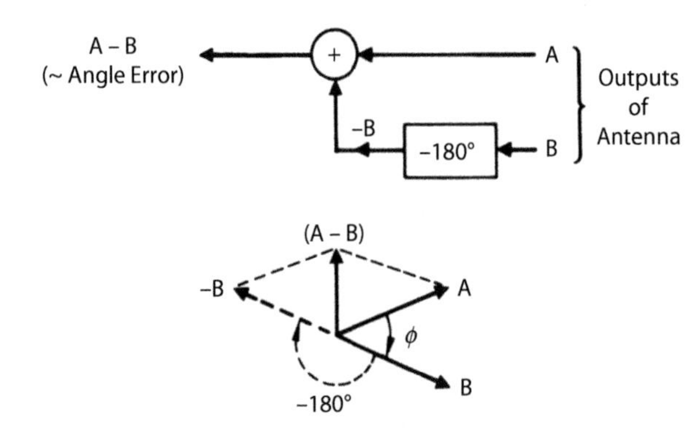

Figure 8-30. Phase difference between outputs of two antenna halves is converted to error signal by introducing a 180° of phase shift in one output and adding the two together.

point in the same direction. Consequently, the return received through one lobe has the same amplitude as that received through the other regardless of the angle of the target relative to the antenna boresight line. However, if an angular error exists, the *phases* of the returns will differ because of the difference in mean distance from the target to each half (Fig. 8-29).

An error signal proportional to the phase difference may be obtained by introducing a 180° phase shift in the output from one half and summing the two outputs. If no tracking error exists, the outputs cancel. If an angular error exists, the resulting phase difference only partially offsets the external phase shift, and a difference output proportional to the tracking error is produced (Fig. 8-30).

By combining the two outputs without the external phase shift, a sum signal is provided for range tracking.

For monopulse tracking in both azimuth and elevation, the antenna is typically divided into quadrants. The *azimuth difference signal* is obtained by separately summing the outputs of the two left quadrants and the two right quadrants and taking the difference between the two sums. The *elevation difference signal* is similarly produced by taking the difference between the sum of the outputs of the two upper quadrants and the sum of the outputs of the two lower quadrants.

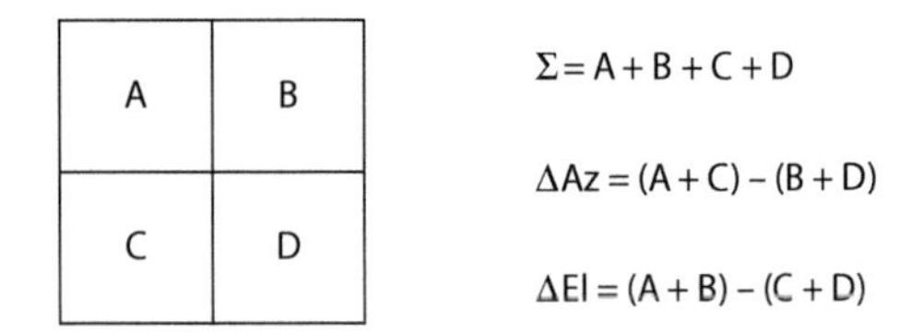

Figure 8-31. Monopulse antenna feed provides sum signal for range tracking; difference signals for angle tracking. Difference signals for azimuth and elevation tracking may be processed on a time-shared basis.

Conventionally, three receiver channels would be provided: one for the azimuth difference signal; a second for the elevation difference signal; and a third for the sum signal. The receiving system can, however, be simplified considerably, by alternately forming the azimuth and elevation difference signals (Fig. 8-31) and feeding them on a time-share basis through a single receiver channel.

How to Calculate the Radiation Pattern for a Linear Array

If you wish, you can readily calculate the radiation pattern for a linear array consisting of any number of radiators having any spacing and any illumination taper.

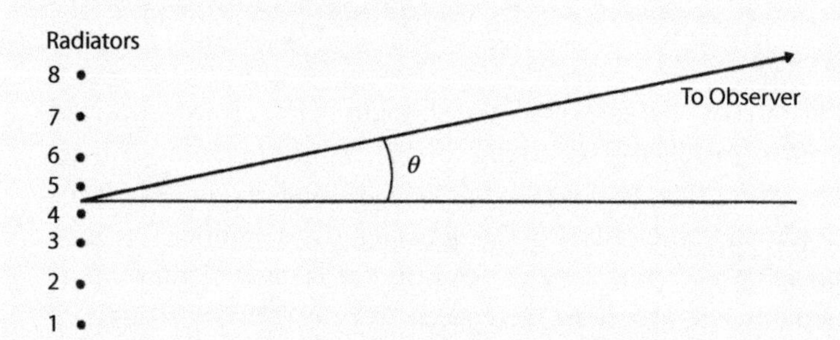

For successive values of the angle, θ, you merely sum the contribution of the individual radiators to the total field strength in the direction, θ. If the array is symmetrical about its central axis, the summation needs to be performed for only half the array, since the other half is a mirror image.

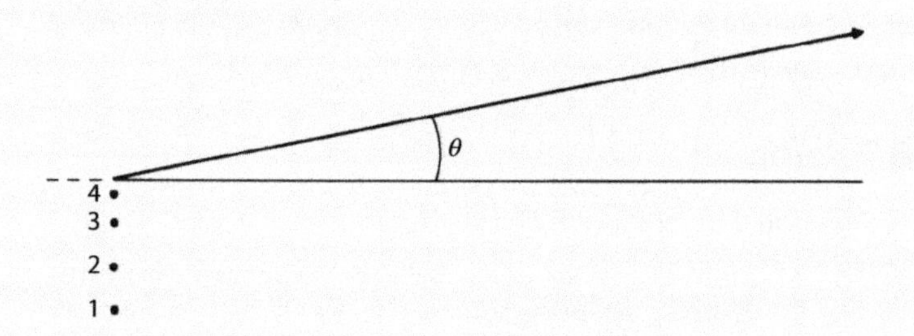

As was illustrated in the blue panel describing the form of the sinx/x function, the contribution of any given radiator, say, No. 2, to the total field strength in a given direction is proportional to the amplitude of the signal, a_2, supplied to the radiator times the complex phasor of the radiation from this radiator relative to the radiation from the radiator at the center of the array (in this example a hypothetical central radiator).

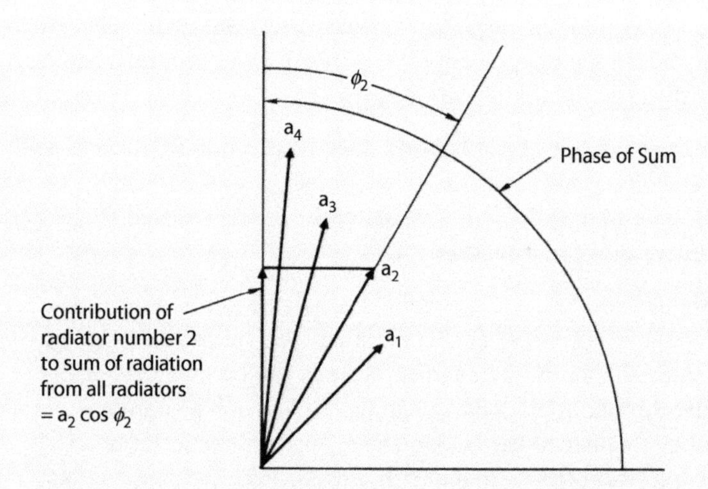

The relative phase, of course, depends on the difference Δd between the distance from radiator No. 2 to an observer (a long way off) in the direction, θ, and the distance from the center of the array to the same observer. That difference equals the distance of the radiator from the array center, d_2, multiplied by $\sin\theta$.

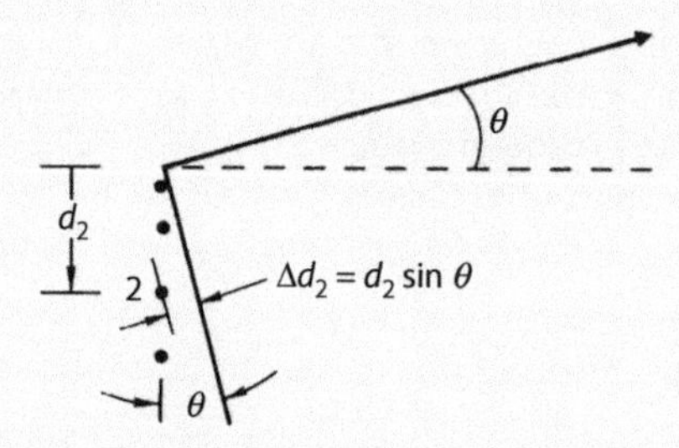

Dividing Δd_2 by the wavelength, λ, and multiplying by 2π yields the phase in radians.

$$\phi_2 = \frac{2\pi \Delta d_2}{\lambda} = \frac{2\pi d_2}{\lambda}\sin\theta$$

Thus, the contribution of radiator No. 2 to the total field strength in the direction, θ, is

$$E_2 \propto a_2 \exp\left(j\frac{2\pi d_2}{\lambda}\sin\theta\right)$$

The total field strength, then, can be found by performing the following summation.

$$E_{total} \propto \sum_{i=1}^{N/2} a_i \exp\left(j\frac{2\pi d_i}{\lambda}\sin\theta\right)$$

By repeating the summation for values of θ from 0° to 90°, the radiation pattern for the array can be obatined.

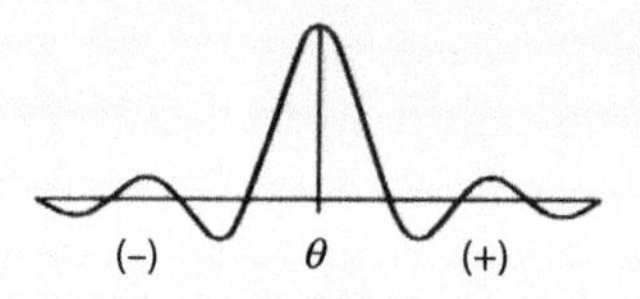

In case you're wondering how this summation is related to the sin *x*/*x* equation given in the text, the relationship is direct. If we assume that the total excitation, *A*, is uniformly distributed over the length of the array, *L*, we can obtain the total field strength simply by integrating this same expression with respect to *d* over the length of the array.

$$E \propto \int_{-L/2}^{L/2} \frac{A}{L}\exp\left(j\frac{2\pi d}{\lambda}\sin\theta\right)dd$$

$$\propto A\frac{\sin\left(\frac{\pi L}{\lambda}\sin\theta\right)}{\left(\frac{\pi L}{\lambda}\sin\theta\right)}$$

$$\propto A\frac{\sin x}{x}, \text{where } x = \frac{\pi L}{\lambda}\sin\theta \cong \frac{\pi L}{\lambda}\theta \quad \text{for small } \theta$$

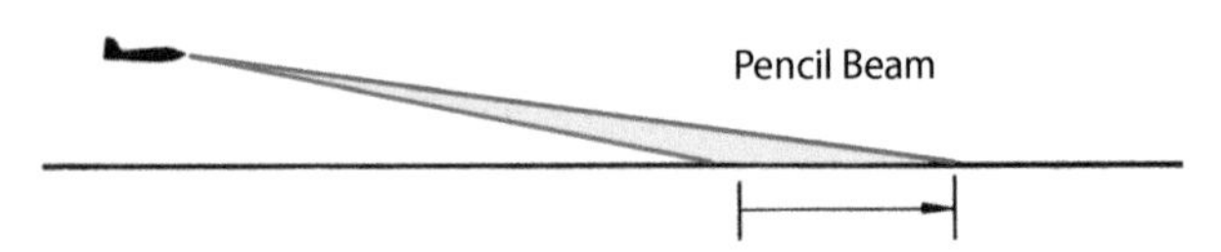

Figure 8-32. For ground mapping, if the radar is at a low altitude, or the range interval being mapped is narrow, a pencil beam can be used. Otherwise, a fan beam is required.

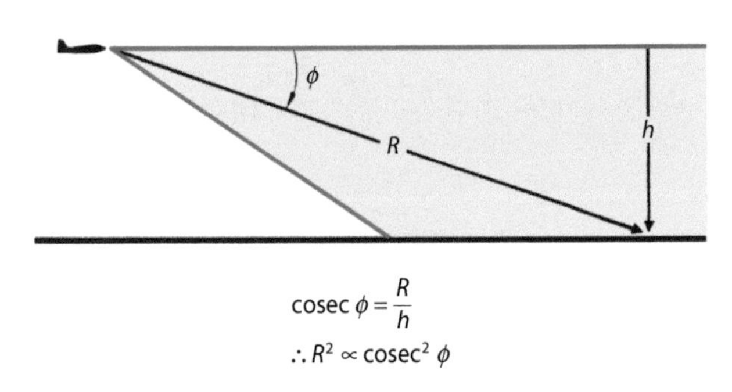

$$\text{cosec}\,\phi = \frac{R}{h}$$
$$\therefore R^2 \propto \text{cosec}^2\,\phi$$

Figure 8-33. To illuminate ground at all ranges uniformly, power radiated at angle ϕ must be proportional to R^2, hence to the cosecant squared of the lookdown angle.

Some Relationships to Keep in Mind

- For a circular uniformly illuminated X-band antenna of diameter, *d*:

$$\theta_{3dB} = \frac{178°}{d} \quad (d \text{ in cm})$$
$$G = d^2\eta$$

(If illumination is tapered, substitute 216° for 178° in expression for beamwidth.)

- For circular, uniformly illuminated antenna and wavelength, λ:

$$\theta_{3dB} = \frac{\lambda}{d}\text{ radians} \quad (d \text{ and } \lambda \text{ in same units})$$
$$G = 9\left(\frac{d}{\lambda}\right)^2 \eta$$

- Angular resolution = $\theta_{3\,dB}$

8.6 Antenna Beams for Ground Mapping

For ground mapping, the entire region being mapped must be illuminated by the antenna's mainlobe (Fig. 8-32). If the radar is operating at low altitudes or if the range interval being mapped is relatively narrow, adequate illumination can be provided by a pencil beam. Otherwise, the antenna must radiate a fan-shaped beam.

Ideally, it is shaped so that the strength of the returns received from equivalent ground targets will be independent of their range. For that, the one-way gain of the antenna must be proportional to the square of the range, *R*, to the ground. This may be achieved by making the gain in the vertical plane proportional to the square of the cosecant of the lookdown angle, ϕ (Fig. 8-33). Hence, the beam is called a *cosecant-squared beam*.

It should be noted that multipurpose antennas, which are not exclusively designed for ground mapping, normally do not have a cosecant-squared beam but a pencil beam. In this case, reduction in strength of the return with range is compensated by increasing the receiver sensitivity with range, a process called *sensitivity time control* (STC) or *automatic gain control* (AGC), described in Chapter 25.

8.7 Summary

A directional antenna radiates a mainlobe surrounded by progressively weaker sidelobes. The width of the mainlobe (beamwidth) is inversely proportional to the width of the antenna aperture in wavelengths.

Antenna directivity is the ratio of the power radiated in a specific direction to the power that would be radiated in that direction if the total power were radiated isotropically (uniformly in all directions). The gain on the axis of the mainlobe is proportional to the area of the aperture in square wavelengths.

Sidelobes rob the mainlobe of substantial power and are a source of undesirable ground clutter. Their gain can be reduced by radiating more power per unit area from the central portion of the aperture than from its edges.

Where extreme versatility and speed are required, the mainlobe of an array antenna may be steered electronically by progressively shifting the phases of the waves radiated by successive radiating elements.

Angular resolution is determined by beamwidth. Angular measurement accuracy much finer than the beamwidth can be achieved and, in single-target tracking, can be made extremely fine through lobing. By designing the antenna to produce the lobes simultaneously (monopulse), angle tracking degradation due to short-term variations in amplitude of the target return can be avoided.

Further Reading

S. Drabowitch, A. Papiernik, H. D. Griffiths, J. Encinas, and B. L. Smith, *Modern Antennas*, 2nd ed., Springer, 2005.

C. A. Balanis, *Antenna Theory: Analysis and Design*, 3rd ed., John Wiley & Sons, Inc., 2005.

L. V. Blake and M. Long, *Antennas: Fundamentals, Design, Measurement*, 3rd ed., SciTech-IET, 2009.

W. L. Stutzman and G. A. Thiele, *Antenna Theory and Design*, 3rd ed., 2012.

Test your understanding

1. A circular antenna, operating at a frequency of 10 GHz, has a diameter of 0.6 m, and has a gain of 34 dBi. What is its aperture efficiency?
2. What is the 3 dB beamwidth of a uniformly illuminated circular antenna of diameter 0.6 m at a frequency of (a) 3 GHz, (b) 10 GHz, and (c) 30 GHz?
3. What are the corresponding 3 dB beamwidths from question 2 if the illumination is tapered?

Northrop Grumman E-8 JOINT STARS (1991)

The E-8 Joint Surveillance Target Attack Radar System (Joint STARS) is developed from the Boeing 707 airframe and is used to track ground targets and transmit imagery and tactical information to ground and air theater command centers. It employs the AN/APY-7 radar with several modes including ground moving target indication (GMTI), fixed target indicator (FTI) target classification, and synthetic aperture radar (SAR).

9

Electronically Scanned Array Antennas

AN/APG-81 Active ESA radar system developed by Northrop Grumman for JSF

Electronically steered array (ESAs) antennas have been employed in surface-based radars since the 1950s.[1] However, because of their greater complexity, weight and cost, they have been slow to replace mechanically steered antennas in airborne applications. With today's technology the ESA has become the dominant antenna used in airborne radar applications. In this chapter we will briefly review the ESA concept, become acquainted with the different types of ESAs, and take stock of the ESA's many compelling advantages as well as a couple of significant limitations.

1. In surface-based radars they were called *phased* arrays, a name that has carried over to airborne applications. They are frequently called electronically *scanned*, as opposed to *steered*, arrays. In light of the versatility of the technique, the more general steered is used here.

9.1 Basic Concepts

An ESA differs from the conventional mechanically steered array antenna in that its beam is steered by individually controlling the phase of the radio waves transmitted and received by each radiating element (Fig. 9-1). The ability to move the beam electronically allows the array to be mounted in a fixed position on the aircraft structure, although the ESA may still be mechanically steered to increase the field of regard.

A digital processor, referred to as the *beam-steering controller* (BSC), translates the desired direction of the beam from broadside (normal to the plane of the antenna) into phase commands for the individual radiating elements.

The incremental phase difference, $\Delta\phi$, that must be applied from one radiating element to the next to move the beam by a desired angle, θ, is proportional to the sine of θ (see blue panel).

$$\Delta\phi = \frac{2\pi d \sin\theta}{\lambda}$$

where d is the element spacing, and λ is the wavelength.

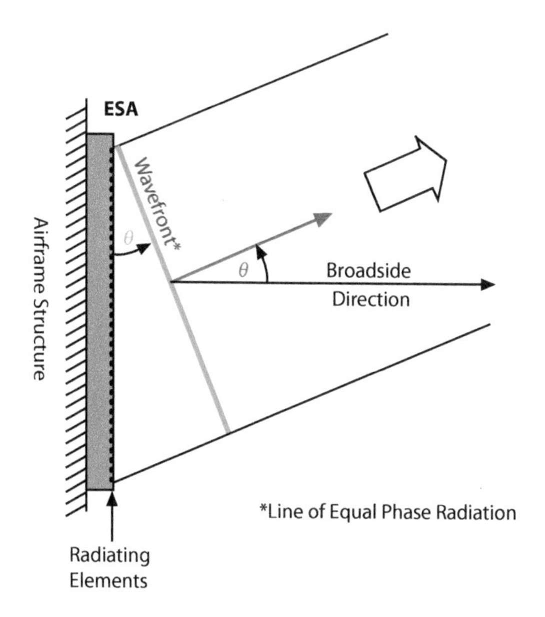

Figure 9-1. The ESA can be mounted in a fixed position on the airframe. Its beam is steered by individually controlling the phase of the waves transmitted and received by each radiating element.

Phase Shift Needed to Steer the Beam

TO STEER THE BEAM θ DEGREES OFF BROADSIDE, THE PHASE OF THE EXCITATION for element **A** must lag that for element **B** by the phase lag, $\Delta\phi$, that is incurred in traveling the distance, ΔR, from radiator **B**.

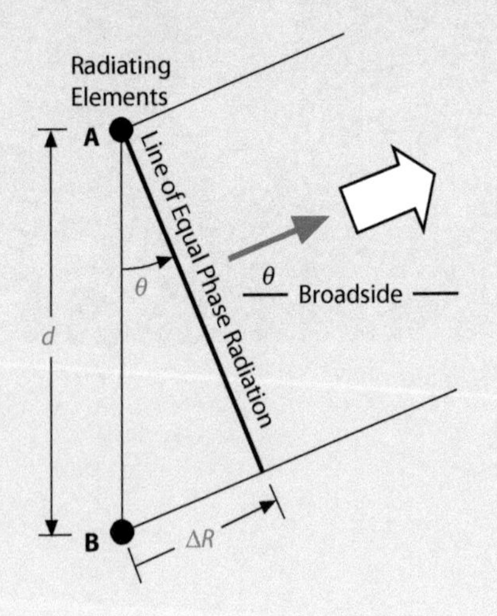

In traveling one wavelength, λ, a wave incurs a phase lag of 2π radians. Therefore, in traveling the distance ΔR, it incurs a phase lag of

$$2\pi\frac{\Delta R}{\lambda}\text{ radians}$$

As can be seen from the diagram,

$$\Delta R = d\sin\theta$$

Hence, the element-to-element phase difference needed to steer the beam θ radians off broadside is

$$\Delta\phi = 2\pi\frac{d\sin\theta}{\lambda}$$

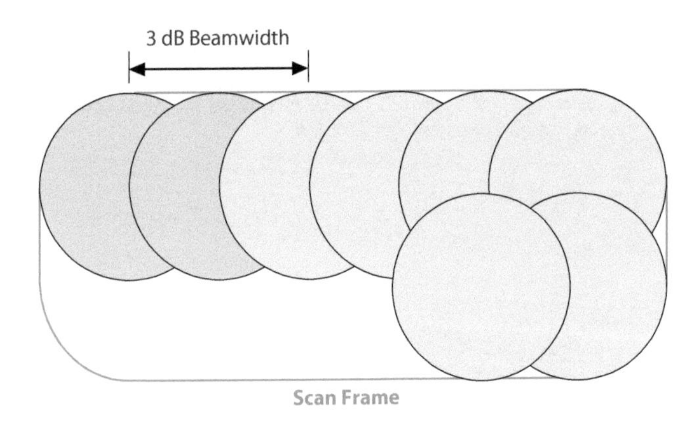

Figure 9-2. For search, the beam steps ahead in increments nominally equal to about 50% of the 3 dB beamwidth, dwelling in each position for a period equal to the desired time-on-target.

For search, the beam is scanned by stepping it in small increments from one position to the next (Fig. 9-2), dwelling in each position for the desired time-on-target, t_{ot}. The size of the steps—typically of the order of half of the 3 dB beamwidth—is optimized by trading off such factors as beam shape loss and scan frame time.

9.2 Types of ESAs

There are two basic types of ESA: passive and active.

Passive ESA. Though considerably more complex than a mechanically steered array (MSA), the passive ESA is far simpler than the active ESA. It operates in conjunction with the same sort of central transmitter and receiver as the MSA. To steer the beam formed by the array, an electronically controlled phase shifter is placed immediately behind each radiating element (Fig. 9-3) or each column of radiating elements in a one-dimensionally steered array. The phase shifter is controlled by either a local processor (the Beam Steering Computer (BSC)) or the central processor.

Active ESA. The active ESA is an order of magnitude more complex than the passive ESA. Distributed within it are both the transmitter power-amplifier function and the receiver front-end functions. Instead of a phase shifter, a tiny dedicated transmit/receive (T/R) module is placed directly behind each radiating element (Fig. 9-4).

The T/R module contains a multistage high power amplifier (HPA), a duplexer (circulator), a protection circuit to block any leakage of the transmitted pulses through the duplexer into the receiving channel, and a low-noise amplifier (LNA) for the

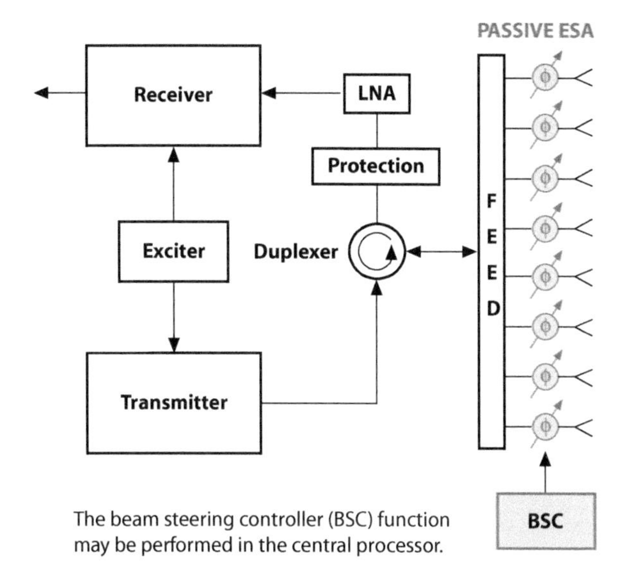

Figure 9-3. The passive ESA uses the same central transmitter and receiver as the MSA. Its beam is steered by placing an electronically controlled phase shifter immediately behind each radiating element.

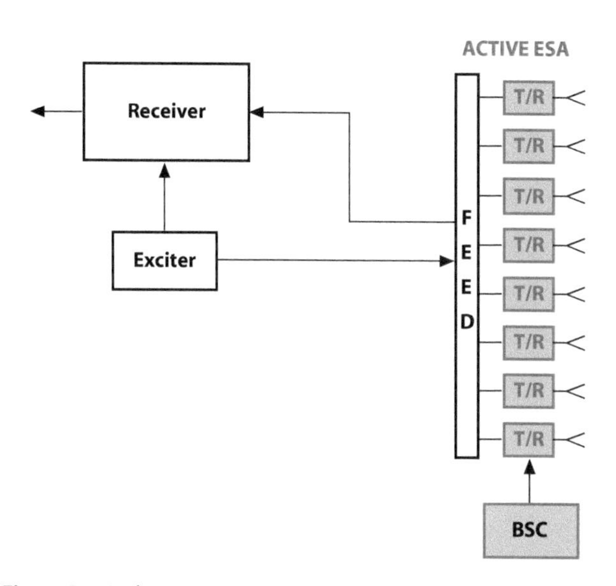

Figure 9-4. In the active ESA, a tiny T/R module is placed immediately behind each radiating element. The centralized transmitter, duplexer, and front-end receiving elements are thereby eliminated.

received signals (Fig. 9-5). The radio frequency (RF) input and output are passed through a variable gain amplifier and a variable phase shifter, which typically are time-shared between transmission and reception. They and the associated switches are controlled by a logic circuit in accordance with commands received from the BSC.

To reduce the cost of the T/R modules and to make them small enough to fit behind the closely spaced radiators, the modules are designed with integrated circuits and are miniaturized (Fig. 9-6). The recent trend is to integrate T/R module chipsets for several radiating elements into a larger module rather than the single-element T/R module package shown in Figure 9-6. These larger T/R modules can feed two or more radiating elements.

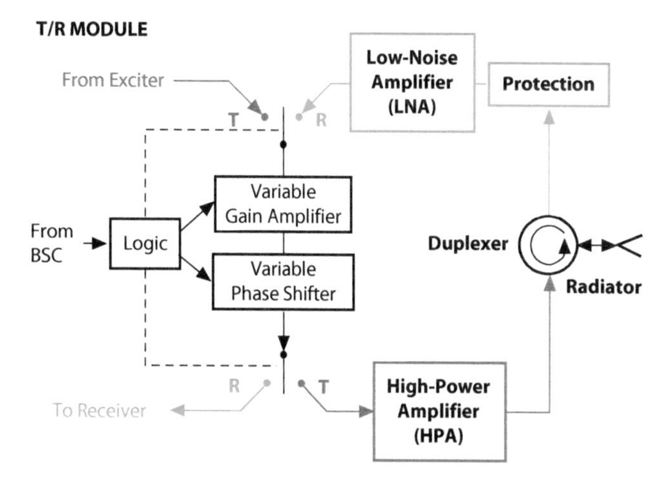

Figure 9-5. Pictured here are basic functional elements of a T/R module. Variable gain amplifier, variable phase shifter, and switches are controlled by the logic element. They may be duplicated for transmit and receive or time shared as shown here.

9.3 Time Delay for Wideband Applications

Most ESAs—both passive and active—use phase shifters to electronically steer the beam. As shown previously, the amount of phase shift required for each element is frequency dependent. For this reason the phase shifters will accurately scan the beam

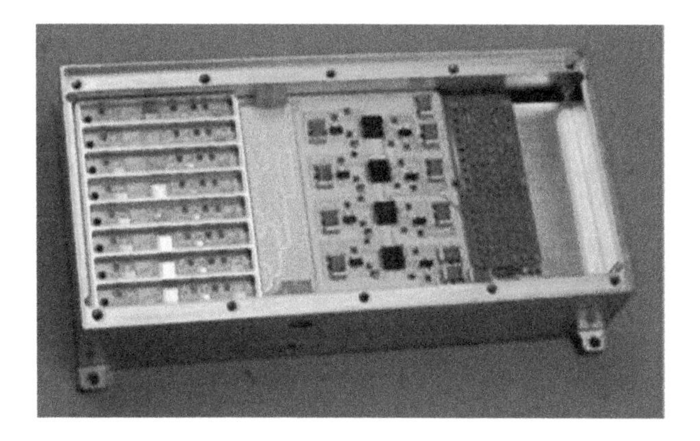

Figure 9-6. With this representative T/R module, even a fairly small ESA could include 2000 to 3000 such modules. (Courtesy of Ball Aerospace and Technologies.)

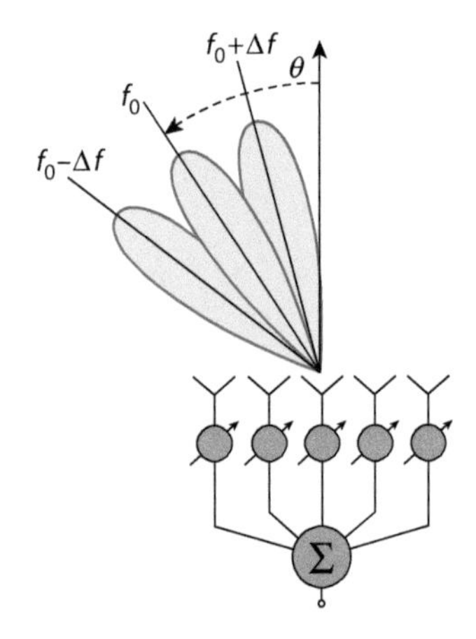

Figure 9-7. Phase shifters will cause the beam to mispoint when the systems operates at frequencies different than the calibration frequency. For this reason time delay is often used for wideband signals.

for only a narrow portion of a wideband signal. Wideband ESAs using phase shifters will exhibit mispointing (or beam squint) as shown in Figure 9-7. The amount of mispointing is given by

$$\Delta\theta = \frac{-\Delta f}{f_0}\tan\theta_s$$

While the phase required to compensate for a free-space path length will change with frequency, the time will not. For this reason wideband arrays use time delay rather than phase shift. The time delay required between two elements in an array is

$$\Delta t = \frac{d\sin\theta}{c}$$

Time delay can be implemented in many ways. For an analog implementation the phase shift is obtained by varying the physical lengths of the feeds for the individual T/R modules. Drawing on the photonic techniques that have proven so valuable in communications systems, a fiber optic feed is provided for each module. The time delay experienced by the signals in passing through the feed is controlled by switching precisely cut lengths of fiber into or out of the feed. By avoiding the limitations on instantaneous bandwidth inherent in electronic phase shifting, the photonic technique makes extremely wide instantaneous bandwidths possible. Time delay can also be implemented using digital beamforming as is discussed in Section 9.6.

9.4 Shared Advantages of Passive and Active ESAs

Both passive and active ESAs have three key advantages that have proven to be increasingly important in military aircraft: (1) they facilitate minimizing the aircraft's radar cross section (RCS); (2) they enable extreme beam agility; and (3) they are highly reliable.

Facilitating RCS Reduction. In any aircraft that must have a low RCS, the installation of a radar antenna is of critical concern. Even a comparatively small planar array can have an RCS of several thousand square meters when illuminated from a direction normal to its face (i.e., broadside). With an MSA, which is in continual motion about its gimbal axes while searching, the contribution of antenna broadside reflections to the aircraft's RCS in the threat window of interest cannot be readily reduced. The issue of MSA RCS can be particularly serious when the radar is in a single-target track mode with its antenna face continuously pointed toward a target. With an ESA, which may be fixed relative to the aircraft structure, the antenna contribution to the aircraft RCS can be minimized. How that is done is explained in Chapter 41.

Extreme Beam Agility. Since no inertia must be overcome in steering the ESA's beam, it is far more agile than the beam of an MSA. To appreciate the difference, consider some typical magnitudes. The maximum rate at which an MSA can be steered, hence the agility of its beam, is limited by the power of the gimbal drive motors to between 100 and 150 deg/s. Moreover, to change the direction of the beam's motion takes roughly 1/10 s.

By contrast, the ESA's beam can be positioned anywhere within a ±60° cone (Fig. 9-8) in less than a few microseconds. This extreme agility has many advantages. It enables:

- Tracking to be established the instant a target is detected
- Single-target tracking accuracies to be obtained against multiple targets
- Targets for missiles controlled by the radar to be illuminated or tracked by the radar even when they are outside its search volume
- Dwell times and waveforms to be individually optimized to meet detection and tracking needs
- Sequential detection techniques to be used, significantly increasing detection range
- Terrain-following capabilities to be greatly improved
- Spoofing to be employed anywhere within the antenna's field of regard

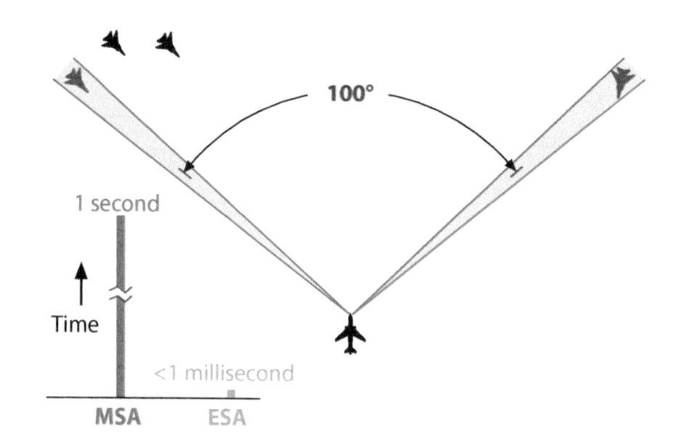

Figure 9-8. To jump the antenna beam from one to another of two targets separated by 100°, an MSA would take roughly a second. An ESA could do it in less than a millisecond.

These capabilities have given rise to a whole new highly versatile and efficient approach to allocating the radar front end and processing resources and to controlling and interleaving the radar's various modes of operation.

High Reliability. ESAs are both reliable and capable of a large measure of graceful degradation. They completely eliminate the need for a gimbal system, drive motors, and rotary joints—all of which are possible sources of failure.

A passive ESA's only active elements are its phase shifters. High-quality phase shifters are remarkably reliable. However, even in random failing, up to 5 percent of them can fail before the antenna's performance degrades enough to warrant replacing them.

The active ESA provides an additional layer of reliability because it replaces the central traveling wave tube (TWT) transmitter with the T/R modules' HPAs, which are inherently dependable. Historically, the former and its high-voltage power supply have accounted for a large percentage of failures in airborne radars. Not only are active ESAs implemented with integrated solid state circuitry, but they also require only low-voltage direct current (dc) power.

In addition, like the phase shifters of the passive ESA, as many as 5 percent of the modules can fail without seriously impairing performance. Even then, the effect of individual failures can be minimized by suitably modifying the radiation from the failed element's nearest neighbors. As a result, the *mean time between critical failures* (MTBCF) of a well-designed active ESA may be comparable to the lifetime of the aircraft!

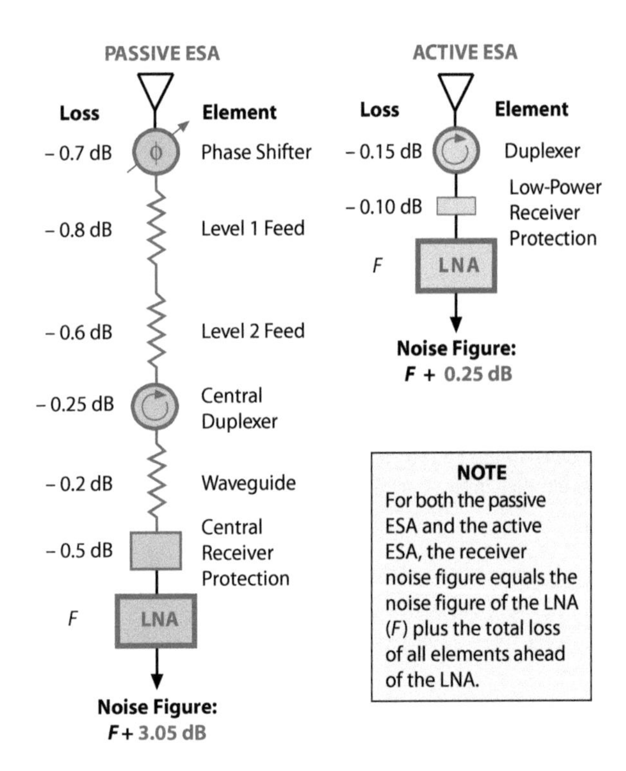

Figure 9-9. By eliminating sources of loss ahead of the LNA, the active ESA achieves a dramatic reduction in receiver noise figure over that obtainable with a comparable passive ESA.

9.5 Additional Advantages of the Active ESA

The active ESA has a number of other advantages over the passive ESA. Several of these accrue from the fact that the T/R module's LNA and HPA are placed almost immediately behind the radiators, thereby essentially eliminating the effect of losses not only in the antenna feed system but also in the phase shifters.

In addition, neglecting the comparatively small loss of signal power in the radiator, the duplexer, and the receiver protection circuit, the LNA establishes the net receiver noise figure (Fig. 9-9). It can be designed to have a very low noise figure. Loss of transmit power is similarly reduced. This improvement, though, may be offset by the difference between the modules' efficiency and the potentially very high efficiency of a TWT.

Amplitude, as well as phase, can be individually controlled for each radiating element on both transmit and receive, thereby providing superior beam-shape agility for such functions as terrain following and short-range synthetic aperture radar (SAR) and inverse SAR (ISAR) imaging. Multiple independently steerable beams may be radiated by dividing the aperture into sub apertures and providing appropriate feeds. Through suitable T/R module design, independently steerable beams

Limitation on Field of Regard

As an ESA's beam is steered off broadside, the apparent width, W', of the effective aperture foreshortens.

$W' = W \cos \theta$

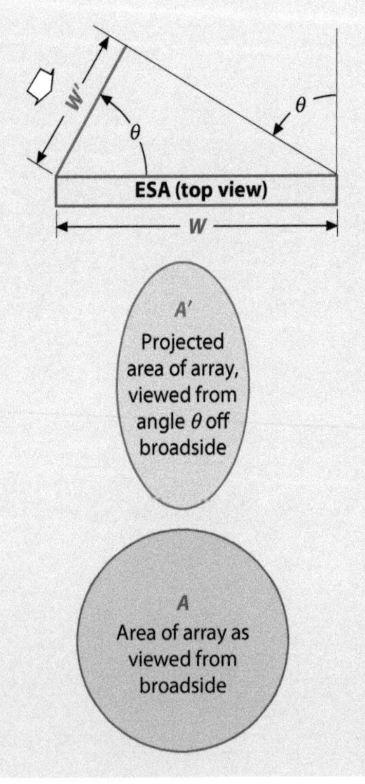

This foreshortening broadens the beam. More importantly, it reduces the projected area, A', of the array, as viewed from angle, θ, off broadside.

$A' = A \cos \theta$

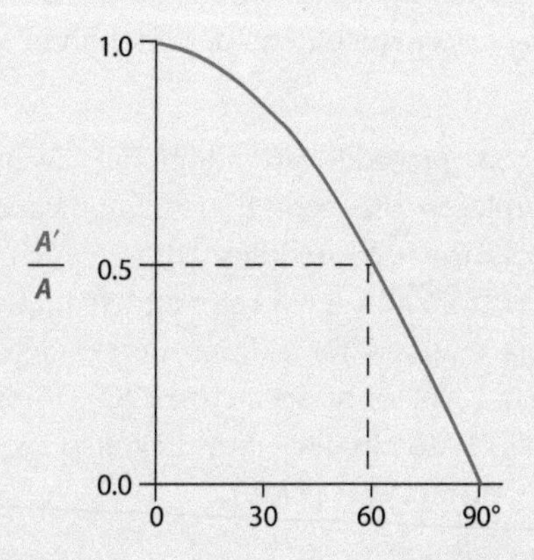

Angle off broadside, θ

Since the gain of the antenna is proportional to the projected area, the maximum practical field of regard for an ESA is limited to about ±60°.

NOTE: This assumes that a typical radiating element with a cos (theta) radiation pattern.

of widely different frequencies may simultaneously share the entire aperture.

9.6 Key Limitations and Their Circumvention

Along with its many advantages, both active and passive ESAs complicate a radar's design in two areas that are handled relatively simply with an MSA: (1) achieving a broad field of regard; and (b) stabilizing the antenna beam in the face of changes in aircraft attitude. These complications and the means for circumventing them are outlined briefly in the following paragraphs.

Achieving a Broad Field of Regard. With an MSA, to whatever extent the radome provides unobstructed visibility, the antenna's field of regard may be increased without in any way impairing the radar's performance. With an ESA, however, as the antenna beam is steered away from the broadside direction, the width of the aperture is foreshortened in proportion to the cosine of the angle off broadside, increasing the azimuth beam width (see left-hand side of blue panel on previous page). More importantly, the projected *area* of the aperture also decreases in proportion to the cosine of the angle, causing the gain to fall off correspondingly.

Depending on the application, the fall-off in gain may be compensated to some extent by increasing the dwell time—at the expense of reduced scan efficiency. Even so, the maximum usable field of regard is generally limited to around ±60°.

While this coverage is adequate for many applications, wider fields of regard may be desired. More than one ESA may then be provided, at considerable additional expense. In one possible configuration, a forward-looking main array is supplemented with two smaller "cheek" arrays, extending the field of regard on either side (Fig. 9-10).

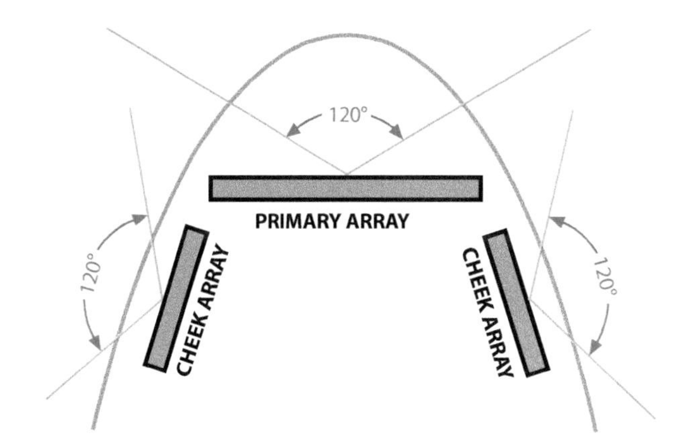

Figure 9-10. Where a broad field of regard is desired, more than one ESA may be used. Here, a central primary array is supplemented with two smaller, "cheek" arrays providing short-range coverage on both sides, for situation awareness.

Beam Stabilization. With an MSA, beam stabilization is not a problem because the antenna is mounted in gimbals and is slaved to the desired beam-pointing direction in spatial coordinates by a fast-acting closed-loop servo system incorporating rate-integrating gyros on the antenna. If the antenna and gimbals are dynamically balanced, this system effectively isolates the antenna from changes in aircraft attitude. Beam steering is required only to trace a search scan pattern or tracking a target, neither of which necessitates particularly high angular rates.

With an ESA, stabilization is not so simple. Since the array is fixed to the airframe, every change in aircraft attitude—be it in roll, pitch, or yaw—must be inertially sensed. Phase commands for steering out the change must be computed for each radiator, and these commands must be transmitted to the antenna's phase shifters or T/R modules and executed. The entire process must be repeated at a high enough rate to keep up with the changes in aircraft attitude.

If the aircraft's maneuvers are at all severe, this rate may be exceptionally high. For a nominal "resteer" rate of 2000 beam positions per second, the phase commands for 2000 to 3000 radiating elements must be calculated, distributed, and executed in less than 500 μs.

Fortunately, with advanced airborne digital processing systems, throughputs of this order can be achieved.

9.7 Trend toward Digital Beamforming

Airborne radar performance was dramatically improved by the move from mechanically to electrically steered arrays. A comparable revolution is now under way with the transition from analog to digital beamforming.

Digital beamforming (DBF) is used to reduce search timelines, enable interference mitigation techniques, and apply time delay for wideband beam steering. The concept of digital beamforming has been discussed for decades, but only recently has digital computing technology progressed to a point where it can be practically implemented.

An array that uses digital beamforming is often referred to as a digital ESA (DESA) and differs from a conventional ESA in a few ways. First, the receive portion of the array is divided into multiple digital channels, and the final stage of beamforming is accomplished in the signal processor. At one extreme, an analog-to-digital (A/D) converter is behind each element. This is called elemental digital beamforming and is shown in Figure 9-11. The other extreme is traditional analog beamforming with a single digital channel. A compromise approach relies on analog subarrays and divides the array into fewer digital channels. The subarray design is shown in Figure 9-12 and is currently the most practical implementation of DBF because it significantly reduces the number of digital channels as well as data throughput and processing requirements. The subarray architecture will be discussed in more detail in Chapter 10.

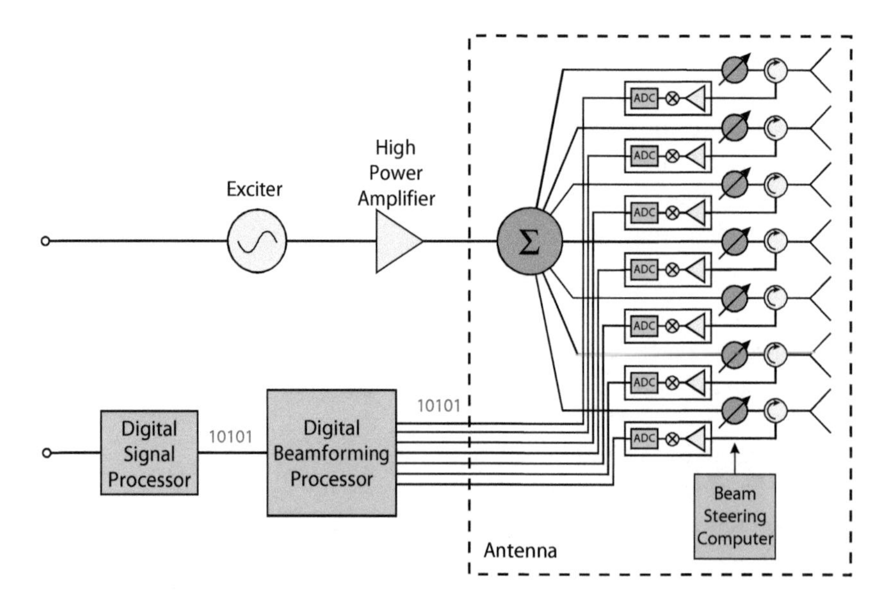

Figure 9-11. With elemental digital beamforming there is a digital receiver behind each element.

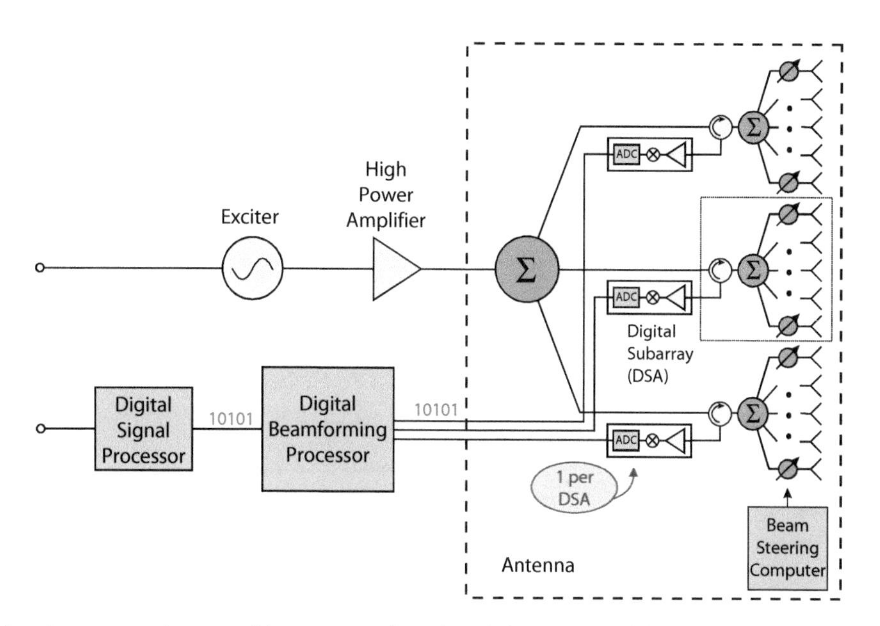

Figure 9-12. Digital beamforming uses a subarray architecture to reduce digital electronics and data processing requirements.

Advantages of Digital Beamforming. There are many benefits to digital beamforming. First, it is possible to form multiple simultaneous receive beams using the same set of digitized signals. The formation of multiple beams allows a given area to be searched more quickly and provides the opportunity for additional processing techniques such as maximum likelihood estimation (MLE) instead of traditional monopulse for more accurate angle estimation.

DBF provides an opportunity to apply time delay digitally to overcome mispointing associated with phase shifters. Also, because amplitude and phase (or time) weights are applied digitally, they have fewer errors than their analog alternatives, which have less precision and are susceptible to temperature fluctuations. For wider bandwidths, true time delays, rather than phase shifts, are necessary, but these are bulkier.

Finally, having digital channels provides an opportunity to cancel, or null, interference sources. As the number of digital channels increases, so does the number of interference sources that can be eliminated. This can be accomplished with prior knowledge of the interferer or with more complicated adaptive array processing techniques. For airborne systems, DBF is the enabling architecture for space-time adaptive processing (STAP), discussed in Chapter 26.

Challenges of Digital Beamforming. The performance benefits of DBF come with many implementation challenges.

Airborne radars typically operate at frequencies that are too high for digital sampling at RF. The A/D converter will instead be located at intermediate frequency (IF) or baseband. For each digital channel there will be one or more mixers as well as a digital receiver, all of which all take up space. Adding these additional components can become a packaging challenge for airborne ESAs that have small interelement spacing and limited volume.

Test your understanding

1. What are three benefits of the ESA?
2. Explain the primary difference between the passive and active ESAs.
3. List a key benefit and key challenge of digital beamforming.
4. Assume that an array has half wavelength spacing between elements. What is the required element-to-element phase shift to electronically steer the beam to 45 degrees?

As the number of digital channels increases, so does the amount of digital data that must be moved from the receiver to the beamforming computer. Once the data have reached the processor, there is another challenge of processing it in real time, especially if adaptive algorithms are to be used. These challenges are magnified as the bandwidth increases.

9.8 Summary

Mounted in a fixed position on the aircraft structure, the ESA produces a beam that is steered by individually controlling the phase of the signals transmitted and received by each radiating element.

A passive ESA operates with a conventional central transmitter and receiver. In contrast, an active ESA's transmitter and receiver front-end functions are distributed within it at the radiator level. The passive ESA is considerably more complex than a MSA; the active ESA is an order of magnitude more complex than the passive ESA but provides additional beam control and increased reliability. DBF is accomplished by using multiple digital receivers and enables enhanced processing techniques. It is complicated and introduces data processing and data throughput challenges.

Both types have three prime advantages: (1) the contribution of their reflectivity to the aircraft's RCS in the threat window of interest can readily be reduced; (2) their beams are extremely agile; and (3) they are highly reliable and capable of graceful degradation. The active ESA also has the advantages of providing an extremely low receiver noise figure, affording beam-shaping versatility, and enabling radiation of independent multiple beams of different frequencies.

The principal limitations of the ESAs are (1) restriction of the maximum field of regard to roughly $\pm 60°$ by the foreshortening of the aperture and consequent reduction in gain at large angles off broadside and (2) the requirement for a substantial amount of processor throughput to stabilize the pointing of the antenna beam in the face of severe aircraft maneuvers.

Further Reading

W. F. Gabriel, "Adaptive Arrays: An Introduction," *Proceedings of the IEEE*, Vol. 64, No. 2, pp. 239–272, February 1976.

M. I. Skolnik (ed.), "Phased Array Radar Antennas," chapter 13 in *Radar Handbook*, 3rd ed., McGraw Hill, 2008.

E. Brookner, "Phased-Array Radars: Past, Astounding Breakthroughs and Future Trends," *Microwave Journal*, Vol. 51, No. 1, January 2008.

M. A. Richards, J. A. Scheer, and W. A. Holm (eds.), "Radar Antennas," chapter 9 in *Principles of Modern Radar: Basic Principles*, SciTech-IET, 2010.

W. L. Melvin and J. A. Scheer (eds.), "Adaptive Digital Beamforming," chapter 9 in *Principles of Modern Radar: Advanced Techniques*, SciTech-IET, 2013.

10

Electronically Scanned Array Design

To fully realize the compelling advantages of the electronically scanned array (ESA), its design and implementation must meet a number of stringent requirements, not the least of which is affordable cost.

This chapter begins first by discussing design considerations that are common to both passive and active ESAs. It then examines design factors that pertain to each of these types of ESAs separately.

10.1 Considerations Common to Passive and Active ESAs

The cost of both passive and active ESAs increases rapidly with the number of phase shifters or transmit/receive (T/R) modules required and hence with the number of radiators in the array.

Consequently, a key design requirement common to both types of ESAs is to space the radiators as widely as possible without creating grating lobes and—if stealth is required—without creating Bragg lobes either. The number of radiators may in some cases be further reduced through judicious selection of radiator lattice.

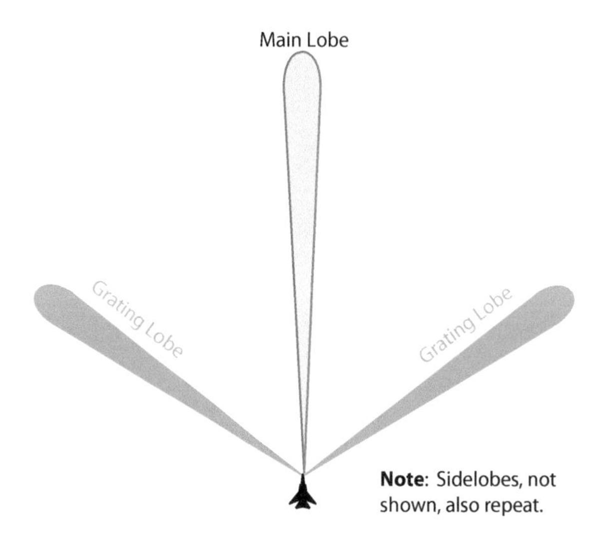

Figure 10-1. Grating lobes are repetitions of the mainlobe. They are produced if the spacing between the radiated elements is too large compared with the wavelength.

Avoiding Grating Lobes. Grating lobes (Fig. 10-1) are repetitions of an antenna's mainlobe[1] that are produced if the spacing of the radiating elements is too large relative to the operating wavelength. They are undesirable because they rob power from the mainlobe, radiate this power in spurious directions, and, from these directions, receive returns that are ambiguous with the returns received through the mainlobe. Also, ground

1. And sidelobes as well.

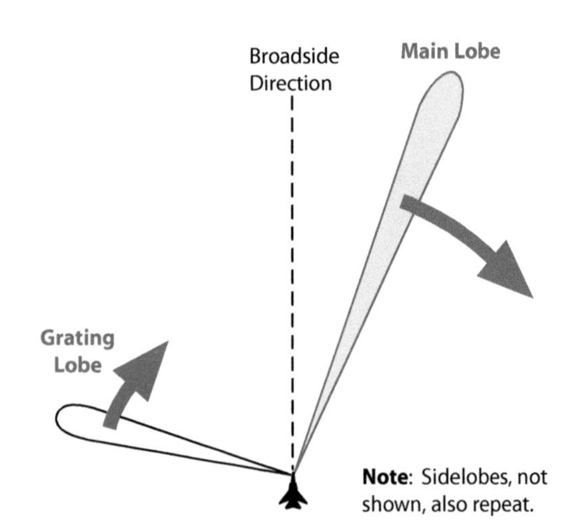

Figure 10-2. With an ESA, if the radiator spacing is not less than one wavelength, as the mainlobe is steered away from broadside, a grating lobe will appear and move into the field of regard.

return or jamming received through the grating lobes may mask targets of interest or desensitize the radar by driving down the automatic gain control (AGC).

Grating lobes are not unique to ESAs. They may be produced by any array antenna if the radiators are too widely spaced. Like the mainlobe, they occur in directions for which the waves received by a distant observer from all of the radiators are in phase. As illustrated by the blue panel below, in the case of a mechanically steered array, where the phases of the waves radiated by all radiators are the same, grating lobes can be avoided even if the radiators are separated by as much as a wavelength.

In an ESA, however, the element spacing cannot be this large. For the angles at which the waves from all radiating elements are in phase depend on both the element spacing and the incremental element-to-element phase shift, $\Delta\phi$, which is applied for beam steering. As the mainlobe is steered away from broadside (i.e., as $\Delta\phi$ is increased from 0), a grating lobe whose existence was precluded by the radiators being no more than a wavelength apart, may materialize on the opposite side of the broadside direction and move into the field of regard (Fig. 10-2).

For an ESA, therefore, the greater the desired maximum scan angle, the closer together the radiating elements must be. The maximum acceptable spacing is

$$d_{max} = \frac{\lambda}{1 + \sin\theta_s}$$

where λ is the wavelength, and θ_s is the maximum desired scan angle. As illustrated in the blue panel on the following page for a maximum scan angle of 60° the radiator spacing is little more than half the operating wavelength. For a wideband array it is important to use the highest operational frequency when calculating the spacing between elements. Grating lobes will occur for the upper edge of the frequency band if the center frequency is used in this calculation.

Incidentally, while the possible locations and movement of grating lobes may be readily visualized for a one-dimensional array, many people find the same task for a two-dimensional array annoyingly difficult. Plotting the lobe positions in so-called *sine space* will ease the task, as explained in the blue panel on page 141.

Radiator Spacing Example

If the maximum scan angle, θ_s, is 30°, what radiator spacing can be used and still avoid grating lobes?

$$d_{max} = \frac{\lambda}{1+\sin 30°} = \frac{\lambda}{1.5} = 0.67\lambda$$

If θ_s is increased to 60°, what must d_{max} be reduced to?

$$d_{max} = \frac{\lambda}{1+\sin 60°} = \frac{\lambda}{1.87} = 0.54\lambda$$

Avoiding Grating Lobes

WHERE GRATING LOBES OCCUR. LIKE THE MAINLOBE, GRATING LOBES OCCUR in those directions, θ_n,

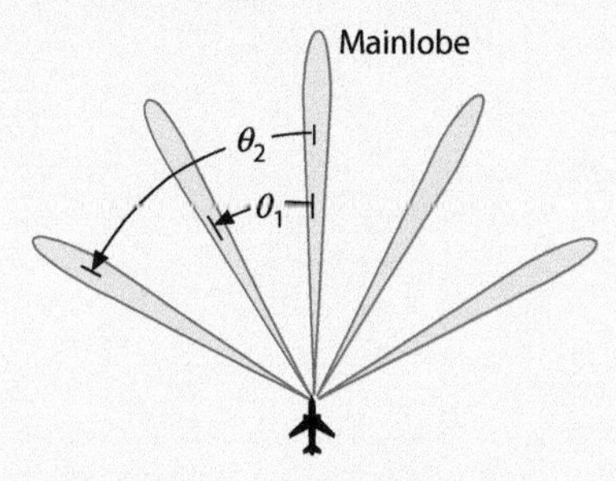

in which the waves received by a distant observer from all of the antenna's radiating elements are in phase.

For a *mechanically steered array* (MSA), where all radiating elements are excited in phase, θ_n is simply the direction in which the incremental difference in range, ΔR_θ, from successive radiating elements to a distant observer is a whole multiple, n, of the operating wavelength, λ.

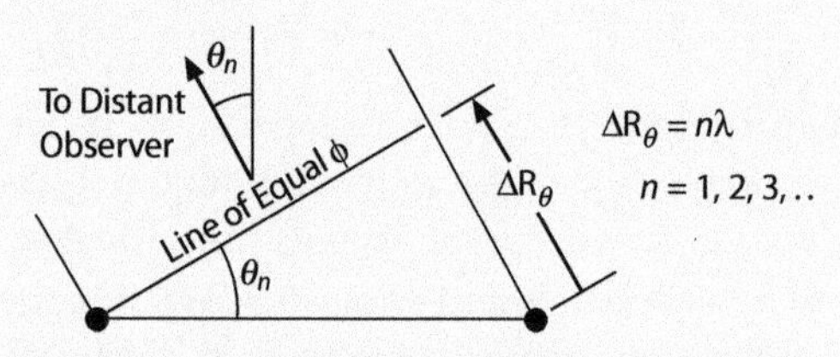

The direction of the n-th pair of grating lobes, θ_n, is thus related to λ and the distance, d, between radiators by the sine function.

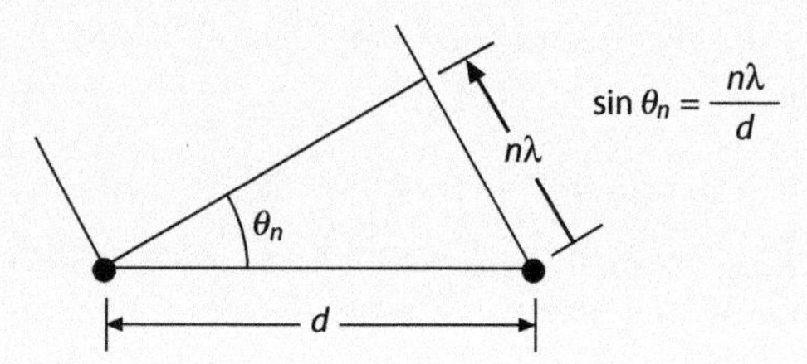

Now, the gain of each radiator usually goes to zero as θ approaches 90°.

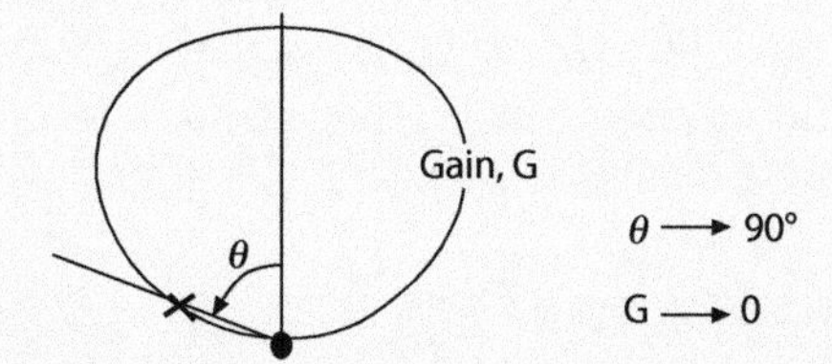

And θ_1, the direction of the first grating lobe, increases to 90° as d is reduced to λ. When d is less than λ, the grating lobe moves into imaginary space and is no longer a problem.

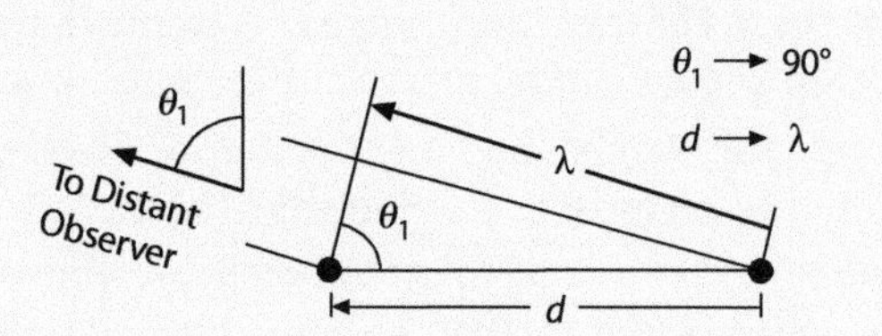

Thus, for an MSA, grating lobes can be avoided by reducing the spacing of the radiators to one wavelength or less:

$$d \leq \lambda$$

For an ESA, avoiding grating lobes is not quite so simple because an incremental phase difference, $\Delta\phi$, is applied to the excitation of successive radiators to steer the mainlobe to the desired scan angle, θ_s.

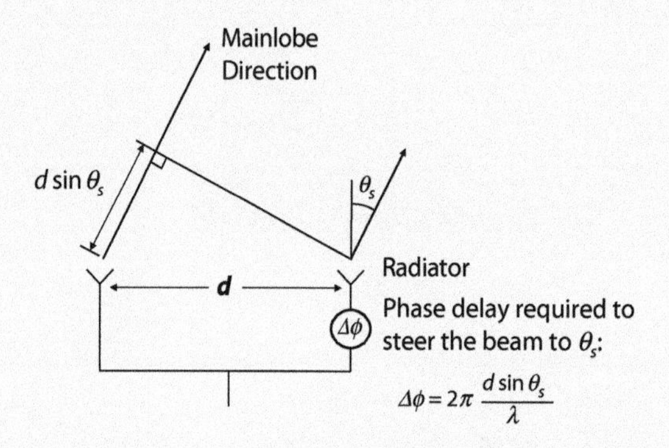

Consequently, for an ESA, grating lobes occur in those directions, θ_n, where the incremental distance, ΔR_θ, from successive radiating elements to a distant observer equals a whole multiple of a wavelength, $n\lambda$, minus the distance, ΔR_θ, corresponding to the phase lag, $\Delta\phi$.

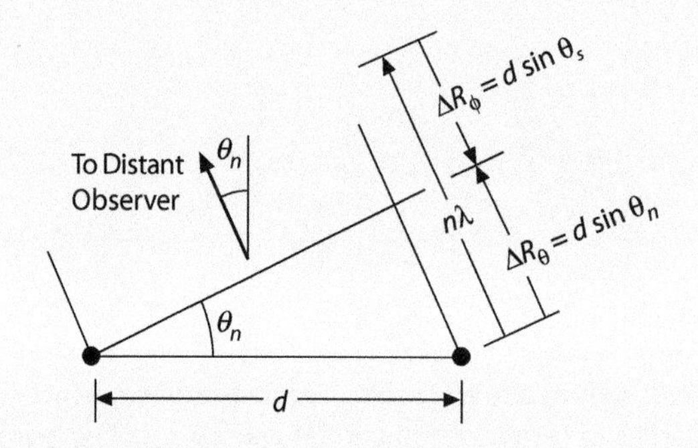

From this simple relationship,

$$d \sin\theta_n = n\lambda - d \sin\theta_s$$

we can obtain the positions of all possible grating lobes. Setting n equal to 1 and θ_s equal to the maximum desired scan angle, yields a "worst-case" equation for the position of the first grating lobe.

$$d \sin\theta_1 = \lambda - d \sin\theta_s$$

As with an MSA, to avoid grating lobes the first grating lobe must be placed at least 90° off broadside. As illustrated in the following diagram, θ_1 approaches 90° as d is reduced to λ minus $d \sin\theta_s$.

Avoiding Grating Lobes *continued*

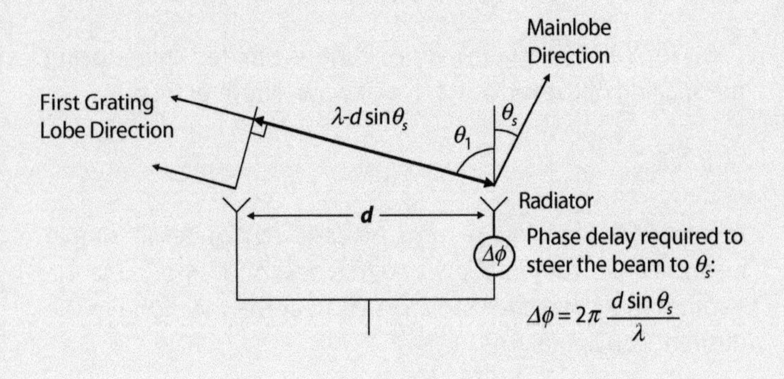

Therefore, since sin 90° = 1, letting sin θ_1 equal 1 and solving the previous equation for d yields the maximum spacing an ESA's radiators may have and avoid grating lobes.

$$d \leq \frac{\lambda}{(1+\sin\theta_s)}$$

Avoiding Bragg Lobes. *Bragg lobes* are retrodirective reflections[2] that may occur if an array is illuminated by another radar from certain angles off broadside. If stealth is required, they must be avoided. As explained in Chapter 41, avoiding Bragg lobes may require a much tighter radiator lattice than is necessary to avoid grating lobes.

2. Energy reflected in the direction it originated.

Choice of Lattice Pattern. The radiator lattice pattern may influence the number of radiators an ESA requires. The most common of these patterns are rectangular and triangular (also referred to as diamond) shaped (Fig. 10-3). With a triangular lattice, the number of radiators may be reduced by up to 14 percent without compromising grating lobe performance. The choice of lattice pattern, though, is also influenced by other considerations, such as radar cross section (RCS) reduction requirements.

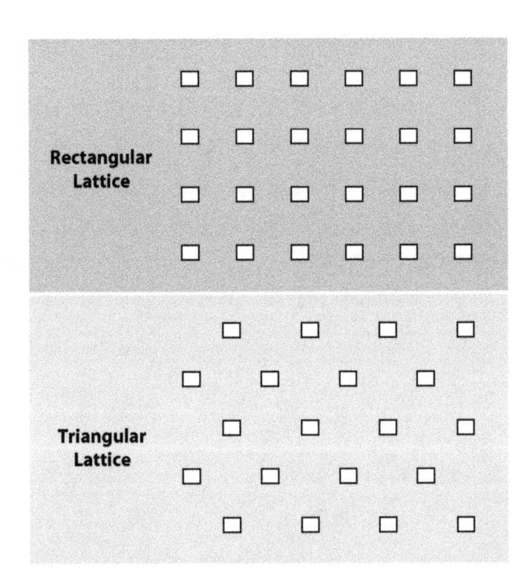

Figure 10-3. Pictured here are common radiator lattice patterns. With the triangular pattern, the number of radiators may be reduced by up to 14 percent without compromising grating lobe performance.

The number of radiators may be reduced still further by selectively thinning the density of elements near the edges of the array in lieu of amplitude tapering for sidelobe control. In assessing thinning schemes, however, their effects on sidelobes and their interaction with edge treatment for RCS reduction must be carefully considered.

In short, no matter what the scheme, some price is always paid for reducing the number of radiators beyond what is achieved by simply limiting their spacing to d_{max}.

10.2 Design of Passive ESAs

Among basic considerations in the design of passive ESAs are the selection of phase shifters, the choice of feed type, and the choice of transmission lines.

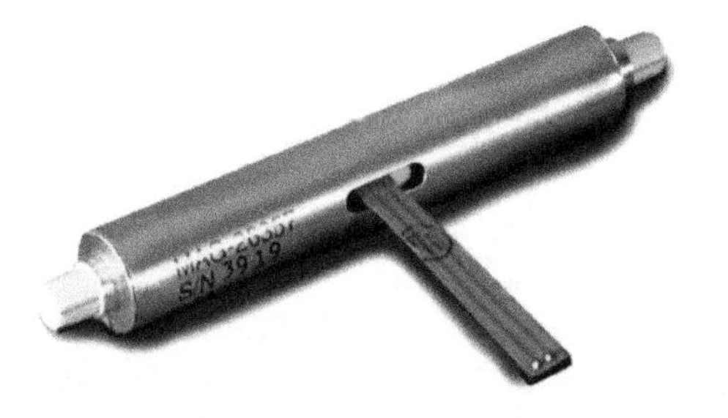

Figure 10-4. This dual-mode ferrite phase shifter of the sort used in passives ESAs is roughly 10 cm long for X-band.

Selection of Phase Shifters. In a passive ESA employing 2000 or more radiators, phase shifters (Fig. 10-4) account for a significant portion of the weight and cost of the array. Consequently, it is critically important that the individual devices be lightweight and low cost. Also, to avoid reducing the radiated power and increasing the receiver noise figure appreciably,

Sin θ Space

FOR EVEN A MECHANICALLY STEERED ARRAY, VISUALIZING THE POSSIBLE POSITIONS of grating lobes is made difficult by the fact that their directions, θ_n, relative to the antenna broadside direction are related to the distance, *d*, between radiators and the wavelength, λ, by the sine function.

$$\sin\theta_n = n\frac{\lambda}{d} \qquad n = 1, 2, 3, \ldots$$

where *n* is the number of the lobe. (The mainlobe is number 0.)

For an ESA, the difficulty is compounded by θ_n being determined not only by the radiator spacing, but also by the steered direction, θ_s, of the mainlobe from broadside.

$$\sin\theta_n = n\frac{\lambda}{d} \pm \sin\theta_s$$

In the case of a 2D ESA, these difficulties are further compounded by the lobes existing in three-dimensional space.

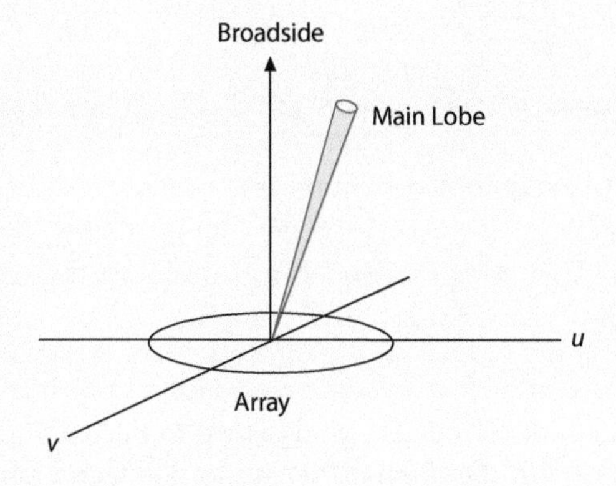

An engineer named Von Aulock elegantly solved all three problems in a single stroke by (1) representing the mainlobe and each grating lobe with a unit vector (arrow one unit long) and (2) projecting the tip of this vector onto the plane of the array.

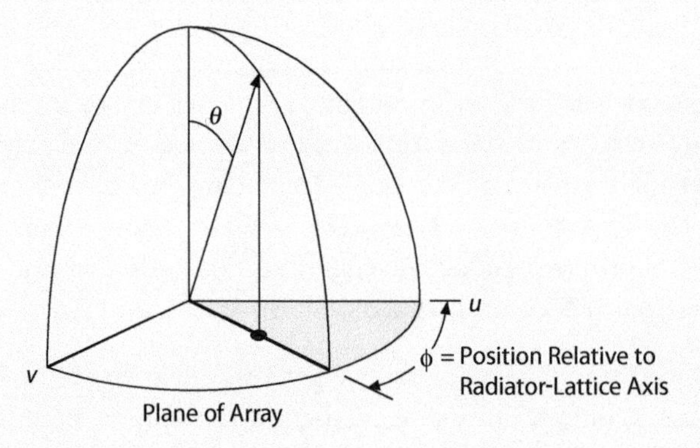

Since the distance from the center of the plane to each point projected onto it is ($1 \times \sin\theta_n$), Von Aulock named the plane *sine space.*

The beauty of sine space is that the position of the mainlobe can be plotted on it simply by scaling off (in the direction ϕ relative to the related lattice axis, *u* or *v*) a distance equal to the sine of the lobe's deflection, θ_s. The positions of any grating lobes can then be predicted by scaling off on either side of the mainlobe distances equal to $n\,\lambda$ divided by the radiator spacings d_u and d_v. Thus:

- Mainlobe distance = $\sin\theta_0$ (at angle ϕ)
- Grating lobe distances = $\pm\frac{\lambda}{d_u}$ and $\pm\frac{\lambda}{d_v}$

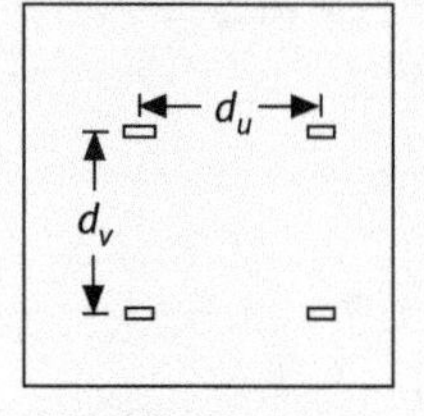

Radiator Lattice

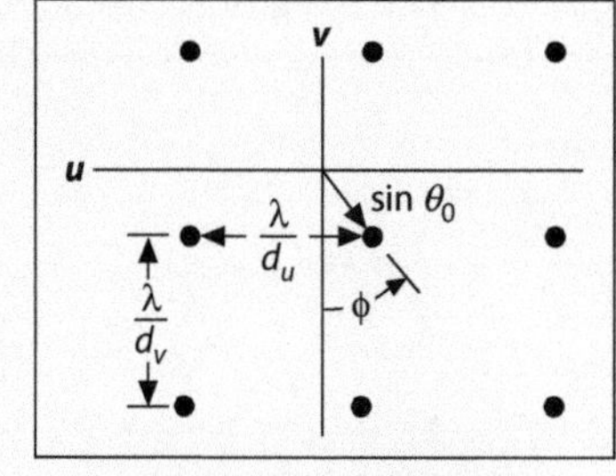

Grating Lobe Diagram Plotted in Sine Theta Space

Since lobes cannot exist at angles greater than 90° off broadside, a circle of radius 1 (the sine of 90°) is drawn around the origin. The area within this circle is termed *real space*, and the area outside it is called *imaginary space.*

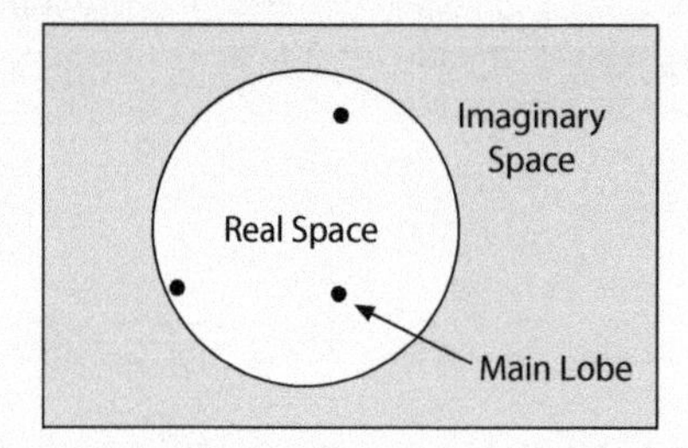

When evaluating radiator lattice patterns and radiator spacing, potential grating lobe positions are often plotted in both real and imaginary space.

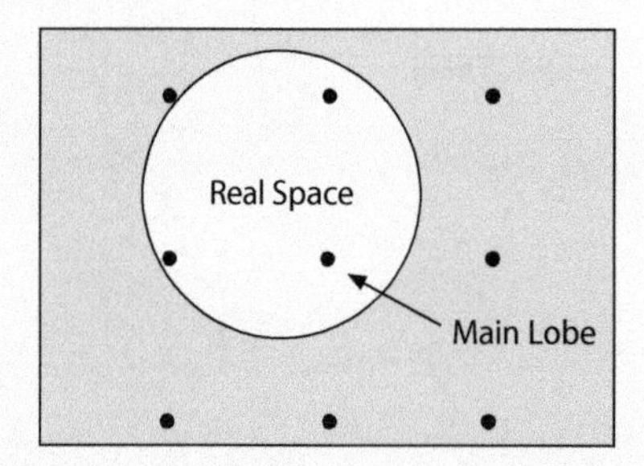

One can then readily see whether any of these lobes will materialize (i.e., move into real space) when the mainlobe is steered to the limits of the desired field of regard.

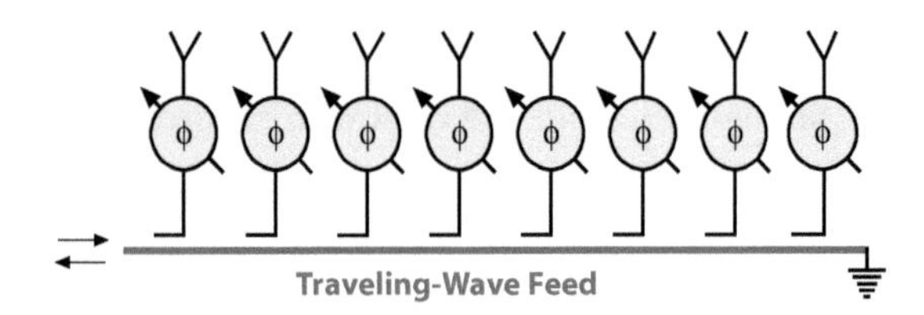

Figure 10-5. A traveling-wave feed is simple and inexpensive. Since the electrical length of the path to each radiator is different, a frequency-dependent phase correction must be made for each element, limiting the instantaneous bandwidth.

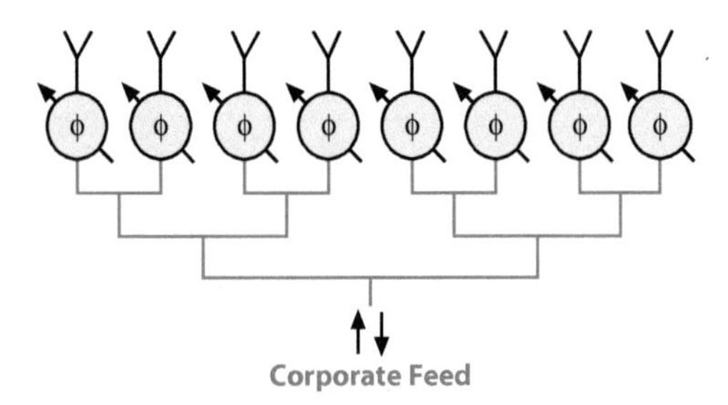

Figure 10-6. Corporate feed makes the electrical length of paths to all radiators the same, eliminating the need for phase corrections and widening the instantaneous bandwidth.

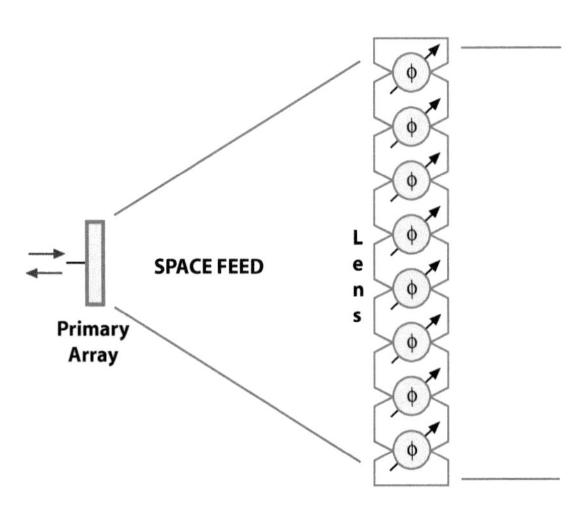

Figure 10-7. The space feed is simple and inexpensive and has an instantaneous bandwidth comparable to a corporate feed. The focal length between the primary array and the lens adds to the depth of the antenna.

the phase shifters' insertion loss must be very low. Other critical electrical characteristics of the phase shifters are accuracy of phase control, switching speed, and voltage standing-wave ratio. While solid-state phase shifters have become the standard for active ESA design, ferrite phase shifters are still used in passive ESAs due to the power handling and low-loss requirements of this array architecture.

Choice of Feed Type. The two basic types of feeds used in passive ESAs are constrained and space. Constrained feeds use transmission lines to distribute the transmitted power to the radiators and to collect the received signal power. Space feeds accomplish this through free space. Constrained feeds may be either the traveling-wave or corporate type.

In a traveling-wave feed, the individual radiating elements, or columns of radiating elements, branch off a common transmission line (Fig. 10-5). The first type of feed is comparatively simple, but it has a limited instantaneous bandwidth. The reason is that the electrical length of the feed path in wavelengths, hence also the phase shift from the common source to each radiator, is different. This causes the beam position to shift if the frequency is changed, so a commanded beam position is usable over only a limited range of frequencies.

The difference may be compensated by adding a suitable correction to the setting of the phase shifter for each radiator. Since the required correction is a function of the wavelength of the signals passing through the feed, any one phase setting provides compensation over only a narrow band of frequencies.

A corporate feed has a pyramidally shaped branching structure (Fig. 10-6). It can readily be designed to make the physical length, hence also the electrical length, of the feed paths to all radiating elements the same, thereby eliminating the need for phase compensation. The instantaneous bandwidth then is limited only by the bandwidths of the radiators and of the phase shifters, transmission lines, and connectors making up the feed system.

Space feeds vary widely in design. Figure 10-7 shows a representative feed. In it, a horn or a small primary array of radiating elements illuminates an electronic lens filling the desired aperture. The lens consists of closely spaced radiating elements, such as short open-ended waveguide sections, each containing an electronically controlled phase shifter.

The space feed is simple, lightweight, and inexpensive. It has low losses and an instantaneous bandwidth approaching that of a corporate feed. The focal length between the primary array and the lens adds considerably to the depth of the antenna.

Also, sidelobe control is difficult to obtain without amplitude tapering at the radiator level.

Choice of Transmission Lines. The two general types of transmission lines commonly used in passive ESA feed systems are stripline and hollow waveguide.

Stripline consists of narrow metal lines (strips) sandwiched between dielectric surfaces (Fig. 10-8). It is lightweight, compact, and low cost. Moreover, it can pass signals having instantaneous bandwidths of up to a full octave. It thus meets the requirements of applications ranging from electronic Protection (EP) and low probability of intercept (LPI) to high-resolution mapping.

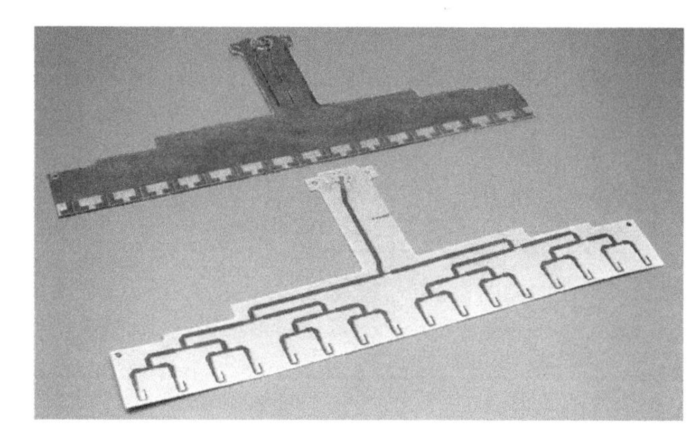

Figure 10-8. Dielectric strip line is commonly used in passive ESAs because it is light weight, compact, low cost and can support large bandwidths.

Hollow metal waveguide (Fig. 10-9) is heavier, more expensive, and has a more limited instantaneous bandwidth but has very low losses. Consequently, it is required for high transmitted powers, weak signal detection, and long runs.

With advances in plastic molding and plating techniques, high-quality, low-cost, metal-coated hollow plastic waveguide has become an attractive option.

10.3 Design of Active ESAs

The key element of an active ESA is the T/R module. Among the many important considerations in its design are the number of different types of integrated circuits required, the power output to be provided, the limits imposed on transmitted noise, and the required precision of phase and amplitude control. Not to be overlooked is the array's crucial physical design. Each of these considerations is discussed briefly in the following sections.

Figure 10-9. This section of hollow metal waveguide is heavier and more expensive than stripline and has a limited instantaneous bandwidth. However, because of its very low losses, it is required in applications with high transmitter powers or weak signal detection.

Chip Set. Ideally, all of a module's circuitry would be integrated on a single wafer. There have been many advances in recent years to achieve this goal; however, because of differences in the requirements of the various functional elements, the necessary technology results in a significant performance compromise. Thus, the circuitry is partitioned by function and placed on more than one chip. The chips are then interconnected in a hybrid microcircuit (Fig. 10-10).

Figure 10-10. This photograph shows a single transmit/receive module. Left: Circulator, power amplifiers, and low-noise amplifier. Right: control electronics. (Courtesy of Selex ES.)

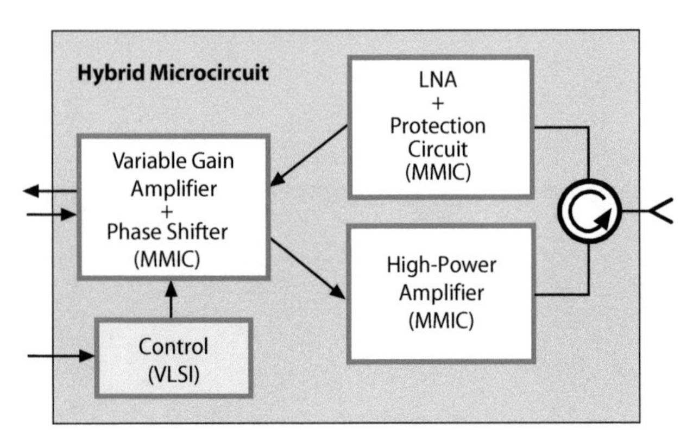

Figure 10-11. Basic chip set for a representative T/R module consists of three monolithic microwave integrated circuits (MMICs) and one digital very large-scale integrated circuit (VLSI).

The basic chip set for a T/R module (Fig. 10-11) includes three *monolithic microwave integrated circuits* (MMICs) plus a digital very large-scale integrated circuit (VLSI):

- High-power amplifier (MMIC)
- Low-noise amplifier (LNA) plus protection circuit (MMIC)
- Variable-gain amplifier and variable phase shifter (MMIC)
- Digital control circuit (VLSI)

Depending on the application, other circuits may be necessary, such as a driver MMIC to amplify the input to the high-power amplifier when high peak powers are required and circuitry for built-in testing.

Most MMICs for X-band and higher frequencies are made of gallium arsenide (GaAs). One limitation of GaAs is its very low thermal conductivity. For GaAs-based circuitry to be adequately cooled, it must be thinned to 2–4 mils thick (mil is 1/1000 of an inch) and be mounted on a thermal spreader. Gallium nitride (GaN) and silicon germanium (SiGe) are emerging as likely materials for future T/R module designs. GaN power amplifiers are capable of providing up to five times more power in the same footprint as GaAs, which enables greater sensitivity and radar-range. SiGe delivers considerably lower power than GaAs, but the material is significantly cheaper, which makes it a likely choice for future low-cost, low-power density radar systems.

Power Output. In general, for a given array size, the array's average power output is dictated by the desired maximum detection range. The realizable average power output, however, is usually constrained by the amounts of primary electrical power and cooling the aircraft designer allocates to the ESA and the module's efficiency. For a given primary power and cooling capacity, the higher the efficiency, the higher the average output power can be.

Regarding module efficiencies, two phrases that often come up are *power-added efficiency* (PAE) and *power overhead.* These are explained in the blue panel on the following page.

In designing the module's high-power amplifier, the required peak power is of greatest concern. It, of course, equals the desired average power per module divided by the minimum anticipated duty factor.

For a given peak power output from the array as a whole, the peak power output required per module is inversely proportional to the number of modules, hence to the area of the array. Thus, to obtain the same peak power from an array having an area of 2 m² as from an array having an area of 4 m², the peak power of each module must be doubled (Fig. 10-12).

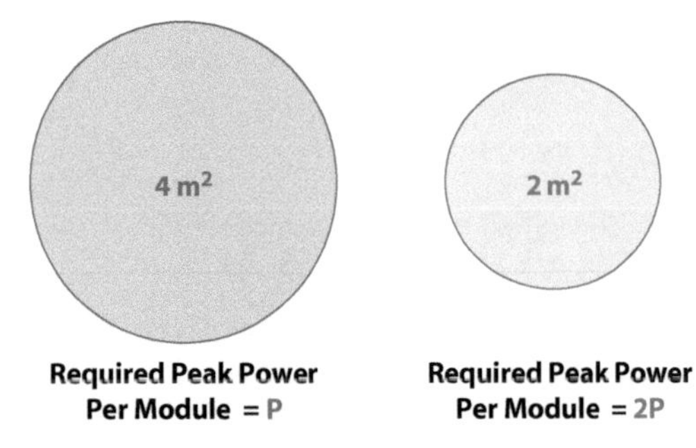

Figure 10-12. The relationship between the peak power per module and the area of an array is represnted in this figure.

Transmitter Noise Limitations. As with a radar employing a central transmitter, noise modulation of the transmitted signal must be minimized. The principal sources of noise modulation in an active ESA are ripple in the direct current (dc) input voltage and fluctuations in the input voltage due to the pulsed nature of the load. Because the voltages are low and the currents are high, adequately filtering the input power is a demanding task. It may require distributing the power conditioning function at an intermediate level within the array or even including a voltage regulator in every T/R module.

Receiver Noise Figure. Since one of the main reasons for going to an active ESA is reduction of receive losses, to fully realize the ESA's potential it is essential that the T/R module have an extremely low receiver noise figure. Typically, the receiver noise figure is quoted for the module as a whole. It equals the noise figure of the LNA plus the losses ahead of the LNA (i.e., in the radiator the duplexer, the protection circuit, and the interconnections) (Fig. 10-13).

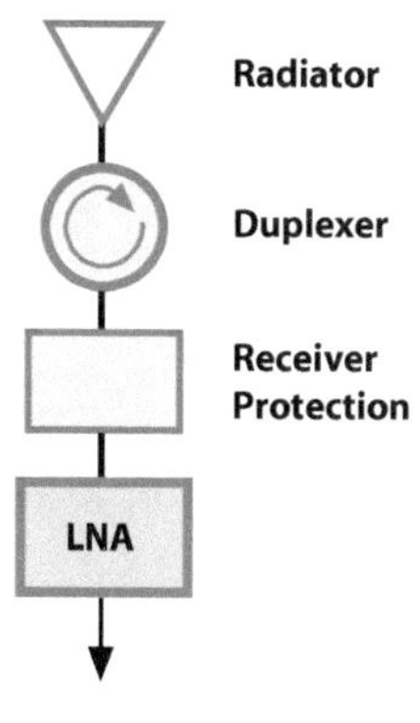

Figure 10-13. The receiver noise figure equals the noise figure for the LNA plus the losses in the elements ahead of the LNA: radiator, duplexer, receiver protection circuit, and interconnections.

Phase and Amplitude Control. The precision with which the phase and amplitude of the transmitted and received signals must be controlled at the radiator level is dictated by the maximum acceptable sidelobe levels of the full array. The lower it is, the smaller the quantization step sizes of the phase and amplitude control circuits must be, wider the amplitude-control range needed to achieve the necessary radiation taper across the array for sidelobe reduction, and the smaller the acceptable phase and amplitude errors.

Array Physical Design. The performance and cost of an active ESA depend not only on the design of the T/R modules but also on the physical design of the assembled array.

In general, the radiators must be precisely positioned and solidly mounted on a rigid back plane. This is essential if the antenna's RCS is to be minimized because any irregularities in the face of the array will result in random scattering that cannot otherwise be reduced.

The modules are typically mounted behind the back plane on cold plates, which carry away the heat they generate. Behind the cold plates are (1) a low-loss feed manifold connecting each module to the exciter and the central receiver; (2) distribution networks providing control signals and dc power to each module; and (3) a distribution system for the coolant that flows through the cold plates.

Just how this general design is implemented may vary widely. One approach, called *stick architecture*, is illustrated in Figure 10-14.

Figure 10-14. This photograph show a small fire control radar AESA (left) and an enlargement of one of its component sticks (right). The antenna uses 16 sticks carrying 20 T/R modules and 8 sticks carrying 16 T/R modules, giving a total of 448 modules.

Measures of Module Efficiency

Power-Added Efficiency. Since a module's high power amplifier (HPA) typically includes more than one stage, the efficiency of the final stage is generally expressed as power added efficiency (PAE).

$$\text{PAE} = \frac{P_o - P_i}{P_{dc}}$$

where

P_o = RF output power

P_i = RF input power

P_{dc} = DC input power

If the gain of the final stage is reasonably high, the PAE is very close to the efficiency of the entire amplifier chain.

Power Overhead. This is the power consumed by the other elements of the module—switching circuitry, LNA, and module control circuit. Because of this overhead, a module's efficiency may be considerably less than the HPA's efficiency, which typically is somewhere between 35 and 45 percent.

Since much of the overhead power is consumed continuously, while the RF output is pulsed, module efficiency may vary appreciably with pulse repetition frequency (PRF).

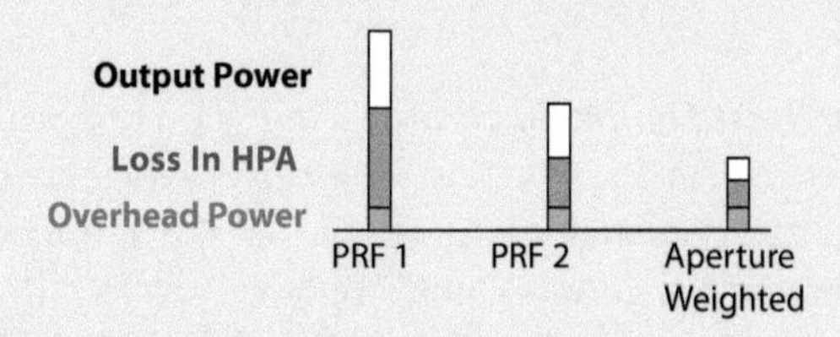

Also, since overhead power is independent of output power, if all modules are identical, as they reasonably would be, aperture weighting can significantly reduce the efficiency of many modules. To minimize this reduction yet to achieve extremely low sidelobes, special weighting algorithms have been developed for active ESAs.

Another approach to the physical design of an active ESA is a so-called *tile architecture.* It employs coin-sized three-dimensional, four-channel modules.

Within each module (Fig. 10-15), successive sections of four T/R circuits are placed on three circuit boards, mounted one on top of the other. Heat generated in the circuits on each board is conducted to the surrounding metal frame.

Figure 10-15. Within the module, successive sections of four T/R circuits are placed on three boards, the heat from which is conducted out to the surrounding metal frame.

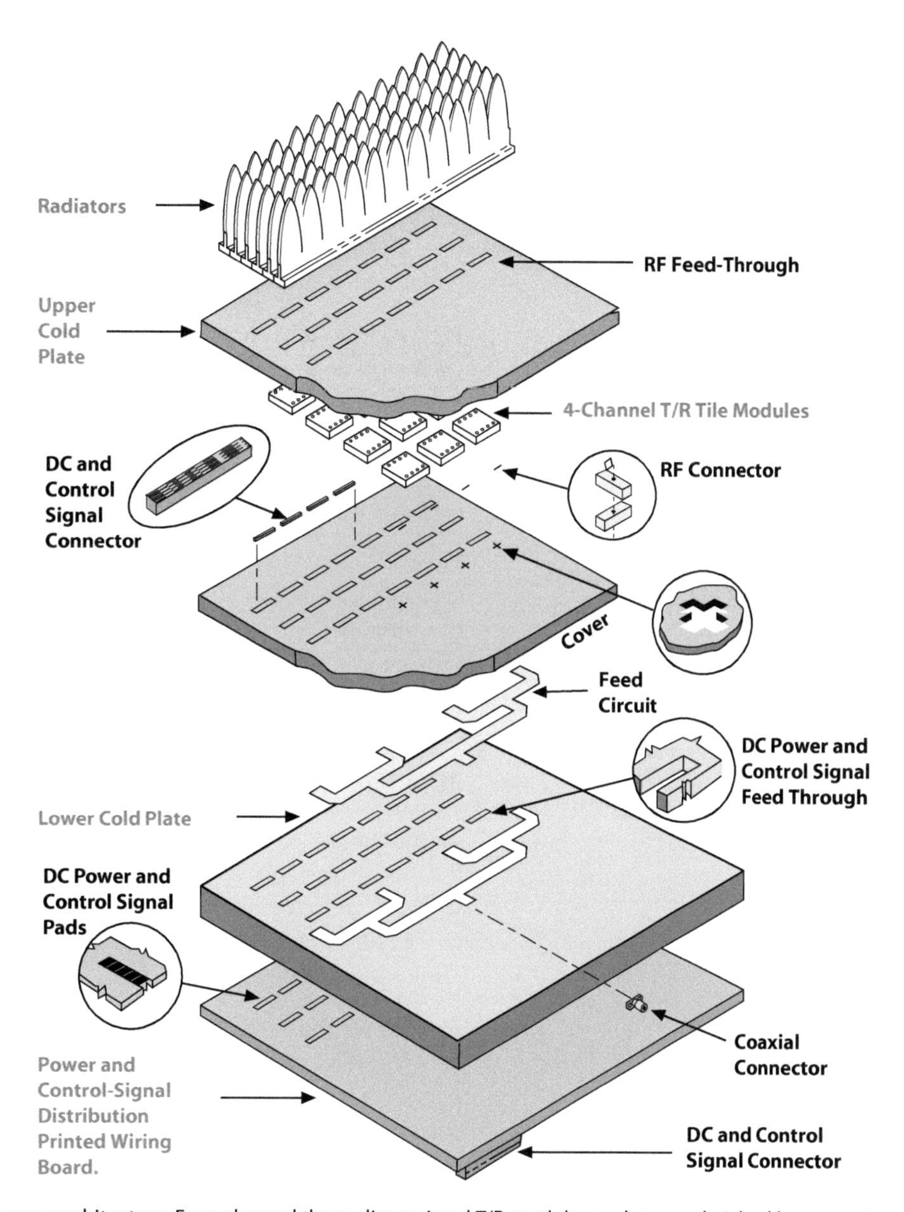

Figure 10-16. In this tile array architecture, Four-channel three-dimensional T/R modules such as sandwiched between two cold plates. RF input and output signals, control signals, and dc power feed through slots in the lower cold plate. RF signals to and from the radiators feed through slots in the upper cold plate.

The modules are sandwiched between cold plates having feed-through slots for the RF signals, dc power, and control signals (Fig. 10-16).

The *subarray architecture* is a third approach that is common with wideband arrays or arrays that implement digital beamforming. A subarray is a subset of the overall antenna that is relatively inexpensive to produce and provides analog beamforming for typically 16 to 64 radiating elements. The subarrays use inexpensive phase shifters to make fine adjustments for beamsteering. As was discussed in Chapter 9, phase shifters will cause mispointing of the beam when operating at frequencies other than the calibration frequency, so each subarray is fed into a time delay unit to provide the bulk of time delay to

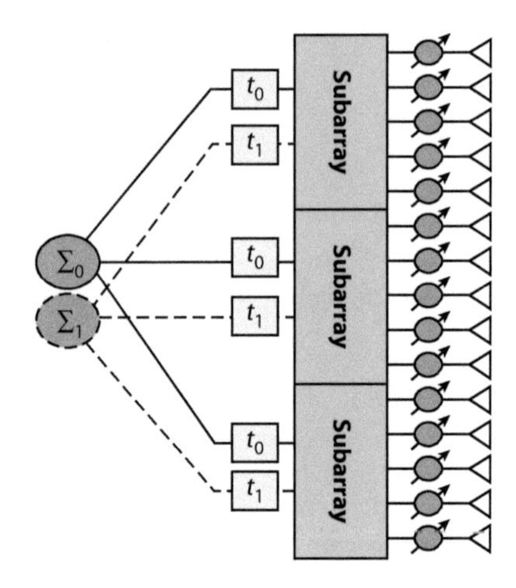

Figure 10-17. The subarray architecture is an economical and practical approach to implement time delay or digital beamforming. The array is divided into smaller subarrays that feed into analog time delay units or digital receivers.

steer the beam (Fig. 10-17). For a digital beamforming application the subarray may feed directly into a digital receiver.

The subarray architecture is a practical and economical way to implement time-delay or digital beamforming with minimal performance degradation. It reduces the number of time-delay units and digital receivers required for the design, which eliminates packaging complexity and reduces cost.

For sidelobe reduction, precise control of phase and gain in each module is essential. Consequently, a comprehensive automatic self-test and calibration capability is provided. To account for manufacturing tolerances, the initial calibration correction for each module is set into a nonvolatile memory in the module's control circuit.

Finally, since more than the maximum acceptable number of modules may malfunction during the operational life of the aircraft, provisions must be included for removing and replacing individual modules—a difficult design task, to say the least.

10.4 Summary

To minimize the cost of an ESA—whether passive or active—the radiating elements must be spaced as far apart as possible without creating grating lobes. The maximum spacing is about half a wavelength. For stealth, still closer spacing may be required to avoid Bragg lobes.

The number of radiators may be reduced by up to 14 percent by using a triangular lattice and further still by thinning the density of elements at the array's edges. However, for such reductions, a price is paid in terms of sidelobe and RCS performance.

The key element of a passive ESA is the phase shifter. They account for more than half the weight and cost of the array and therefore must be lightweight and low cost. Also critical are the transmission lines and feed. For wideband operation, stripline and either a corporate or a space feed must be used. For high power and weak-signal detection, hollow waveguide is required.

The key element of an active ESA is the T/R module. It is implemented with a limited number of monolithic integrated circuits in a hybrid microcircuit. For X-band frequencies and higher, the monolithic circuits are made of gallium arsenide. Critical electrical characteristics are the module's peak power output, precision of phase and amplitude control, receiver noise figure, and noise modulation of the transmitted signal, which must be minimized through filtering of the dc input power.

To minimize the antenna's RCS, the radiators are mounted on an extremely rigid back plane. The T/R modules are mounted on cold plates, immediately behind the back plane. Self-test and self-calibration capabilities are essential.

Further Reading

S. Sabatini and M. Tarrantino, *Multifunction Array Radar—System Design*, Artech House, 1994.

R. Mailloux, *Phased Array Antennas Handbook*, 2nd ed., Artech House, 2005.

S. Drabowitch, A. Papiernik, H. D. Griffiths, J. Encinas, and B. L. Smith, *Modern Antennas*, 2nd ed., Springer, 2005.

T. Jeffrey, *Phased Array Radar Systems*, SciTech-IET, 2010.

W. Wirth, *Radar Techniques Using Array Antennas*, 2nd ed., IET, 2013.

Test your understanding

1. What should be done with element spacing to avoid grating lobes?
2. Why would a designer use a triangular lattice?
3. Why are solid-state phase shifters not used in passive ESAs?
4. You are designing an array that must electronically scan to 45°. What is the maximum spacing (relative to wavelength) between elements to avoid grating lobes?

Avro Vulcan (1956)

The four-engined, delta-winged Vulcan formed the spearhead of the United Kingdom's nuclear deterrent through the 1950s and 1960s. It was equipped with an H2S S-band radar whose design dates back to World War II. It last saw active service in 1982.

11

Pulsed Operation

WWII H2S radar on Halifax aircraft with radome (above) and without (below)

Radars are of two general types: continuous wave (CW) and pulsed. A CW radar transmits continuously and simultaneously listens for the reflected echoes. A pulsed radar, on the other hand, transmits its radio waves intermittently in short pulses and listens for the echoes in the periods between transmissions.

Pulsed radars fall into two categories: those that sense Doppler frequencies and those that do not. The former have come to be called pulse Doppler radars, while the latter are simply called pulsed radars. Nearly all modern radars make use of the Doppler effect in some way, either to measure target velocity or to reject stationary clutter, or both. Here the term pulsed will be used in a general sense to refer to any radar that transmits pulses.

In this chapter we'll consider the advantages of pulsed transmission, characteristics of the pulsed waveform, and the effects of pulsed transmission on transmitted power and energy.

11.1 Advantages of Pulsed Transmission

With the exception of Doppler navigators, altimeters, and proximity fuses, most airborne radars are pulsed. The chief reason is that pulsed operation avoids the problem of the transmitter interfering with reception, commonly referred to as leakage, self-interference, or self-jamming.

There are two principal aspects of the self-jamming problem. One is large-signal interference that can drive the receiver into gain compression or saturation. This can usually be mitigated by using separate transmit and receive antennas with physical isolation. The other issue is the presence of noise sidebands on the transmitted signal that can mask weaker target returns.

Direct signal leakage can usually be managed through a combination of antenna isolation and frequency separation. In Doppler navigators (Fig. 11-1), the Doppler shift provides sufficient frequency separation to keep the transmitted signal from

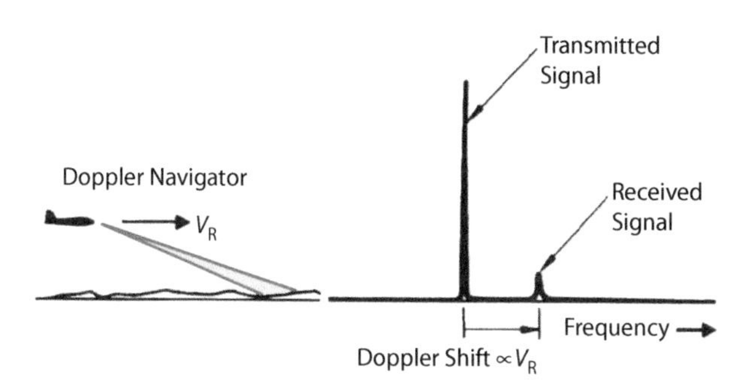

Figure 11-1. In applications where the Doppler frequency of the return is large, the expected Doppler shift prevents the transmitted signal from interfering with reception.

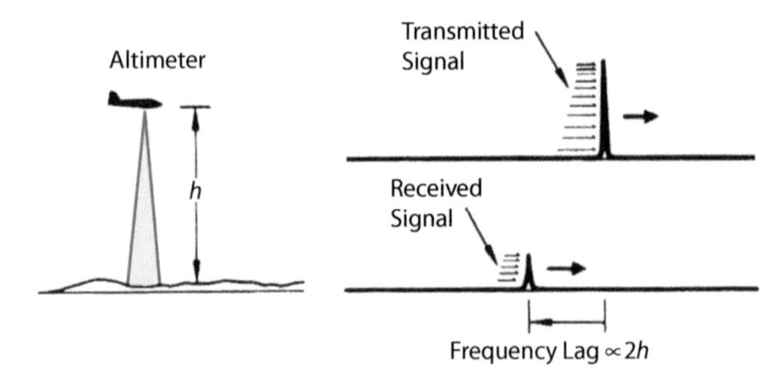

Figure 11-2. When the Doppler shift is negligible, the transmitted signal is prevented from interfering with reception by continuously shifting the transmitter frequency.

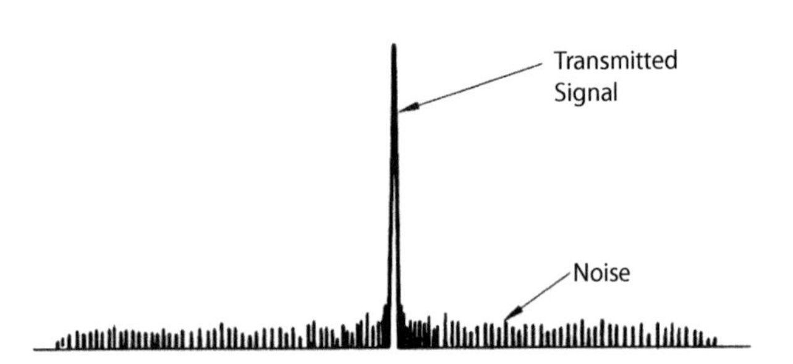

Figure 11-3. Noise sidebands blanket a broad band of frequencies above and below the transmitted signal and are vastly stronger than the echoes of typical airborne targets. Interference from these sidebands can be avoided by transmitting pulses.

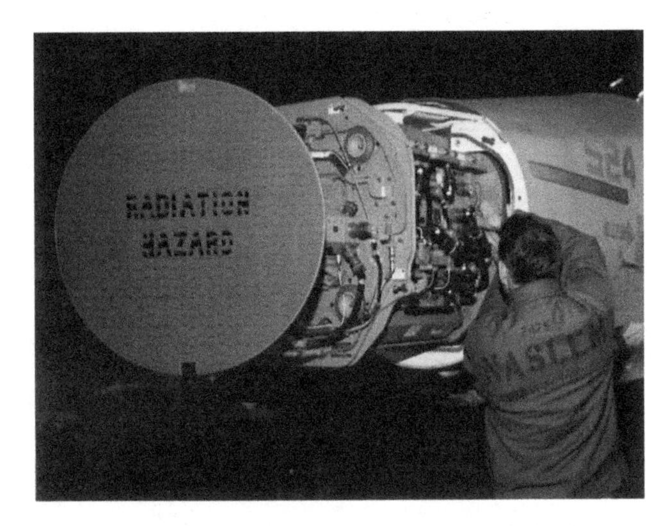

Figure 11-4. When the same antenna must be used for both transmission and reception, pulsed transmission avoids the problem of transmitter noise leaking into the receiver.

interfering with reception. In altimeters (Fig. 11-2), the Doppler shift is usually near zero and interference from the transmitted signal is avoided by continuously shifting the transmitter's frequency. Because of the time the radio waves take to reach the ground and return to the radar receiver, the frequency of the received signal lags behind the frequency of the transmitter, thus the signal is not interfered with.

The problem in most airborne applications is electrical *noise*. Unavoidably generated in every transmitter, this noise modulates the transmitter output. In so doing it creates noise modulation sidebands (see Chapter 5), which blanket a broad band of frequencies above and below the transmitter frequency (Fig. 11-3). Although the power of the noise sidebands may seem infinitesimal (and is negligible compared with that of echoes from the ground at short range), these sidebands are many orders of magnitude stronger than the echoes from the average airborne target.

To keep the noise from interfering with reception, the receiver must be isolated from the transmitter. Adequate isolation can be obtained by physically separating the transmitter and receiver and providing separate antennas for each (as in ground and shipborne CW radars).

In airborne radars, however, because of space limitations, it is usually necessary to use a single antenna for both transmission and reception (Fig. 11-4). When this is done, it is extremely difficult—hence costly—to prevent some of the noise in the transmitter output from leaking through the antenna into the receiver.

11.2 Pulsed Waveforms

If the transmission is pulsed, neither the transmitted signal nor transmitter noise is a problem, as the radar does not transmit and receive at the same time.

Pulsed operation has the further advantage of simplifying range measurement. If the pulses are adequately separated, a target's range can be precisely determined merely by measuring the elapsed time between the transmission of a pulse and reception of the echo of the same pulse.

Overall, the form of the radio waves radiated by a pulsed radar—the transmitted signal—is referred to as the transmitted *waveform* (Fig. 11-5). This waveform has four basic characteristics:

- Carrier frequency
- Pulse width
- Modulation (if any) within each pulse or from pulse to pulse
- Rate at which the pulses are transmitted (pulse repetition frequency)

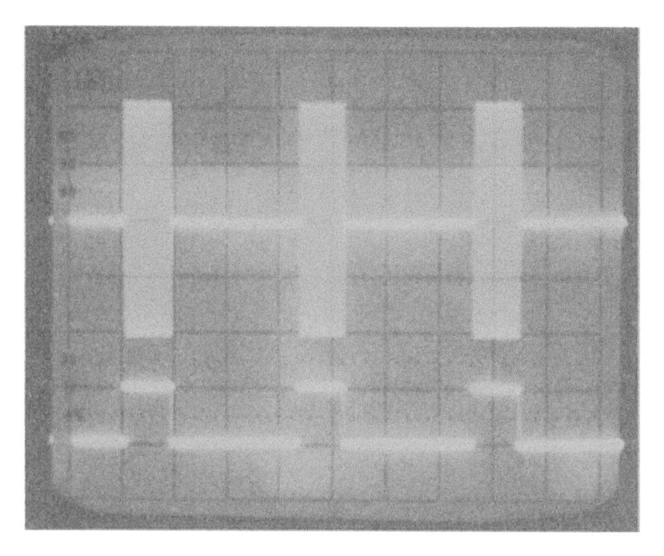

Figure 11-5. Basic characteristics of a transmitter waveform.

Carrier Frequency. This is not always constant, but may be varied in different ways to satisfy specific system or operational requirements. The carrier frequency may be increased or decreased from one pulse to the next. It may be changed at random or in some specified pattern within each pulse, which is referred to as intrapulse modulation.

Pulse Width. The pulse width is the duration of the pulses (Fig. 11-6) and is commonly represented by the lowercase Greek letter τ. Pulse widths can range from a fraction of a microsecond to several tens of milliseconds, depending on the radar application.

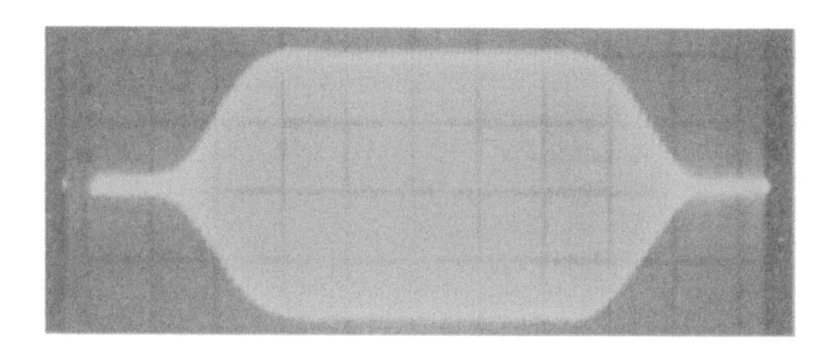

Figure 11-6. A radio frequency pulse as seen on an oscilloscope. The pulse width is the duration of the pulse.

Pulse width may also be expressed in terms of physical length; that is, the distance, at any one instant, between the leading and trailing edges of a pulse as it travels through space. The length, L, of a pulse is equal to the pulse width, τ, times the speed of the propagating waves. That speed is very nearly 3×10^8 m/s. Consequently the physical length of a pulse (Fig. 11-7) is roughly 300 m/μs of pulse width:

Pulse length = 300τ meters,

where τ is the pulse width in microseconds.

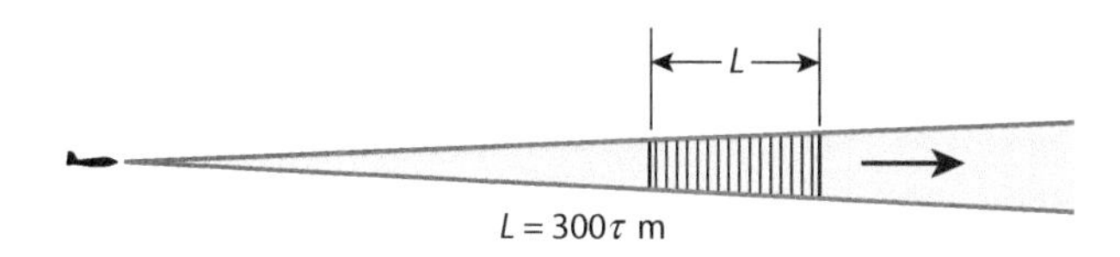

Figure 11-7. Pulse length is the distance from the leading to trailing edge of the pulse as it travels through space.

The pulse length is of great interest. For without some sort of modulation within the pulse, the pulse length determines the ability of a radar to resolve (separate) closely spaced targets in range. The shorter the pulses (if not modulated for compression), the better the range resolution will be.

For a radar to resolve two targets in range with an unmodulated pulse, their range separation must be such that the trailing edge of the transmitted pulse passes the near target before the leading edge of the echo from the far target reaches the near target (Fig. 11-8). To satisfy this condition the range separation must be greater than half the pulse length.

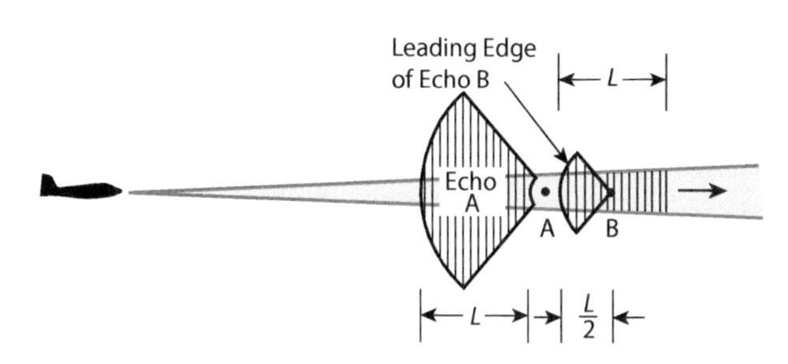

Figure 11-8. To resolve two targets A and B with an unmodulated pulse of length L, their separation (AB) must be greater than $L/2$ so that the echoes do not overlap.

As the pulse length is decreased, the amount of energy contained in the individual pulses decreases. A point is ultimately reached where no further decrease in energy, hence in pulse width, is acceptable. Seemingly this limitation puts a limit on the resolution a radar can achieve. In fact, that is not so, as will be explained in Chapter 16 on pulse compression.

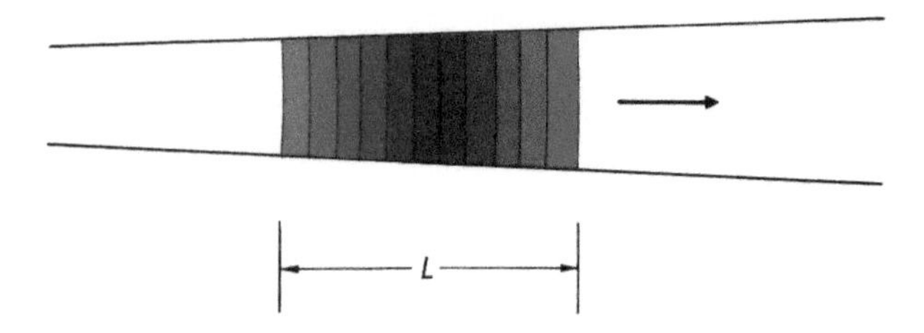

Figure 11-9. If successive segments of transmitted pulse are coded with intrapulse modulation, the same resolution can be obtained as with a pulse the width of a single segment.

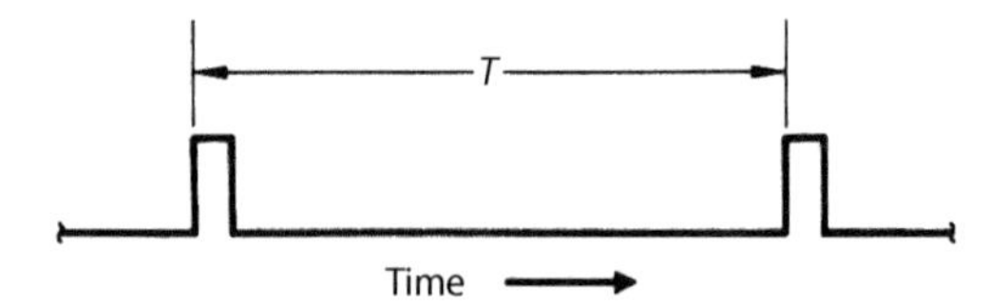

Figure 11-10. The number of pulses transmitted per second is the pulse repetition frequency, PRF. The time between pulses is the pulse repetition interval, *T*.

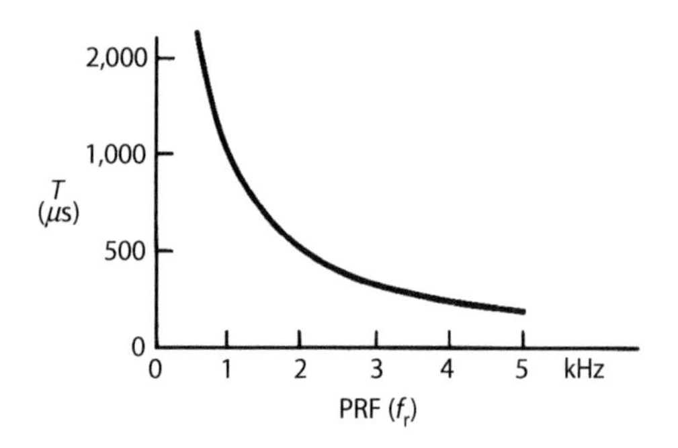

Figure 11-11. The pulse repetition interval, *T*, decreases with an increase in PRF.

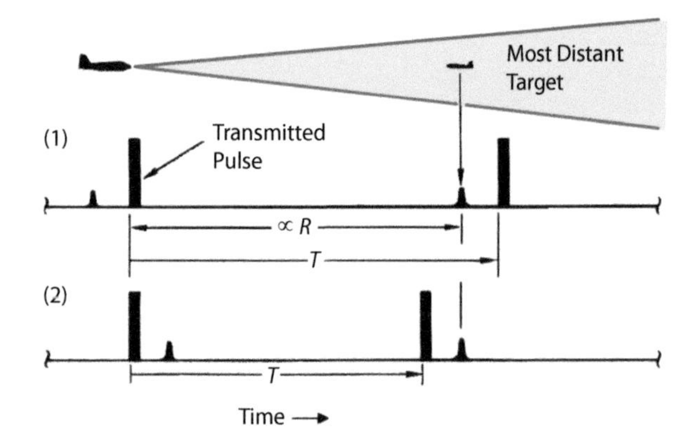

Figure 11-12. (1) If the pulse repetition interval, *T*, is long enough for all echoes from one pulse to be received before the next pulse is transmitted, echoes may be presumed to belong to the preceding pulse, (2) but not if *T* is shorter than this.

Intrapulse Modulation. The limitation that minimum pulse-length requirements impose on range resolution can be circumvented by coding successive increments of the transmitted pulse with phase or frequency modulation (Fig. 11-9). Each target echo will, of course, be similarly coded. By decoding the modulation when the echo is received and progressively delaying successive increments, the radar can, in effect, superimpose one increment on top of another. The resolution thus achieved is the same as if the radar had transmitted a pulse having nearly the same energy as the original pulse but the width of the individual increments. This technique is called *pulse compression*, which will be discussed in detail in Chapter 16.

Pulse Repetition Frequency. Pulse repetition frequency (PRF) is the rate at which a radar's pulses are transmitted, that is, the number of pulses per second (Fig. 11-10), and is commonly represented by f_r. The PRFs of airborne radars range from a few hundred Hz to several hundred kHz. For reasons that will be discussed in subsequent chapters, the PRF can be changed during the course of the radar's operation.

Another measure of pulse rate is its inverse, the period between the start of one pulse and the start of the next pulse. This is called the *pulse repetition interval* (PRI), generally represented by the uppercase letter T. It is sometimes called the *interpulse period*.

The PRI (Fig. 11-11) is equal to 1 s divided by the number of pulses transmitted per second, f_r:

$$T = \frac{1}{f_r}.$$

If the PRF is 100 Hz, for example, the PRI will be 1/100 = 0.01 s, or 10,000 μs.

The choice of PRF is crucial because it determines whether, and to what extent, the ranges and Doppler frequencies observed by the radar will be ambiguous.

Range ambiguities arise as follows. A radar has no direct way of telling to which transmitted pulse a particular echo belongs. If the PRI is long enough for all of the echoes of one pulse to be received before the next pulse is transmitted, this doesn't matter: any echo can be assumed to belong to the immediately preceding pulse (Fig. 11-12). But if the PRI is shorter than this, depending on how much shorter it is, an echo may belong to any one of a number of preceding pulses. Thus the ranges observed by the radar may be ambiguous. This leads to the definition of the *maximum unambiguous range*, R_u, corresponding to a given PRF: as long as the two-way propagation delay to a target at range R_u is less than T, the echo is unambiguously associated with the immediately preceding transmitted pulse, so $R_u = cT/2$, or alternately, $R_u = c/2\text{PRF}$.

Doppler ambiguities arise because the pulsed nature of the signal means that the Doppler shift of the radar echo is sampled from one pulse to the next at a sample rate that is equal to the PRF. The Doppler effect will be explained in greater detail in Chapter 18. Essentially, if the velocity v_r of the target with respect to the radar is such that the two-way range changes by one wavelength (or more generally, by an integer number of wavelengths) during one PRI, then the phase of the echo will be constant from pulse to pulse, just as if the Doppler shift were zero. The values of target velocities for which this condition is met are defined by $v_r \times T = n\lambda/2$, where n is an integer. This means that there are ambiguities at Doppler shifts corresponding to velocities at which this condition is met, that is, at Doppler shifts spaced at velocities of $1/T$.

We can see that the higher the PRF, the greater the spacing of Doppler ambiguities and the closer the spacing of range ambiguities. If the PRF is chosen to be so high that the Doppler ambiguities lie outside the range of practical target velocities, then Doppler can be measured unambiguously.

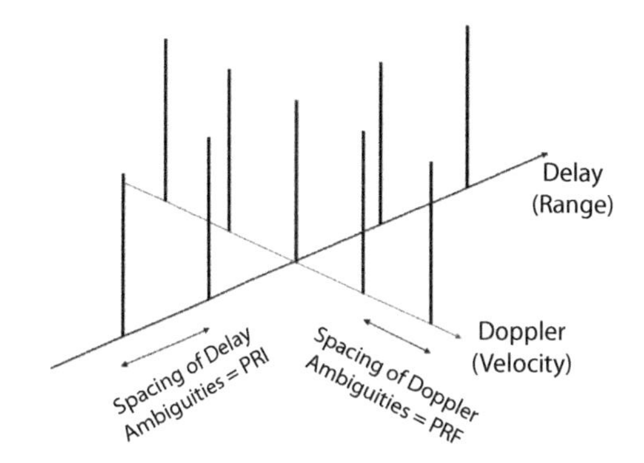

Figure 11-13. This basic form of the ambiguity function shows the ambiguities on the delay and Doppler axes as a function of the radar PRF. Additional off-axis ambiguities (not shown here for clarity) form "bed of nails" in the delay/Doppler space.

11.3 The Ambiguity Diagram

The spacings of the ambiguities in range and in Doppler due to the PRF can be visualized in a particularly elegant way in the form of an ambiguity diagram (Fig. 11-13). This is a two-dimensional plot presented as a function of two-way propagation delay and Doppler shift (or equivalently, range and velocity). The spacing of the ambiguities in delay and Doppler is governed by the choice of PRF (Fig.11-14).

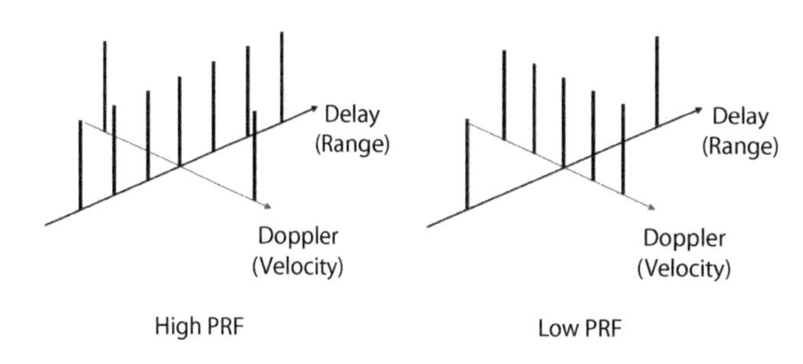

Figure 11-14. The effect of the PRF on the ambiguity spacing in delay and Doppler is depicted here. In the left-hand diagram, a high PRF gives ambiguities that are closely-spaced in delay but widely spaced in Doppler. In the right-hand diagram a low PRF gives ambiguities that are widely spaced in delay but closely spaced in Doppler.

We shall see in Chapters 15 and 16 how the basic idea of the ambiguity diagram can be more formally presented to show the properties of the waveform itself, and how different PRFs can be used to attempt to overcome the constraints imposed by ambiguities.

11.4 Output Power and Transmitted Energy

Before discussing the effect of pulsed transmission on output power and transmitted energy, a review of the relationship between power and energy is in order. As explained at some length in the panel on the next page, power is the rate of flow of energy (Fig. 11-15). Conversely, energy is power integrated over time.

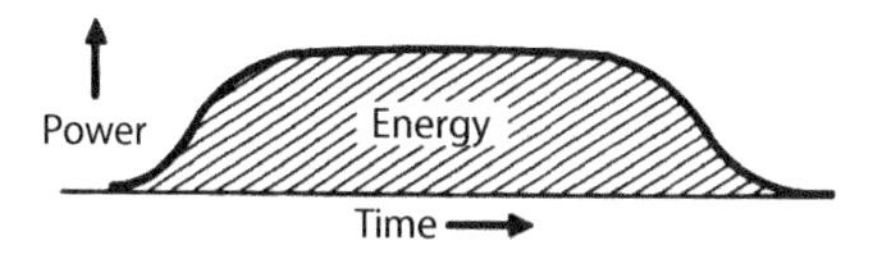

Figure 11-15. Power is the rate of flow of energy. Backscattered energy is what a radar detects.

The amount of energy transmitted by a radar equals the output power times the length of time the radar is transmitting.

Two different measures are commonly used to describe the power of a pulsed radar's output: peak power and average power.

Peak Power. This is the power of the individual pulses. If the pulses are rectangular—that is, if the power level is constant from the beginning to the end of each pulse—peak power is simply the output power when the transmitter is on, or transmitting (Fig. 11-16). Here, peak power is represented by P.

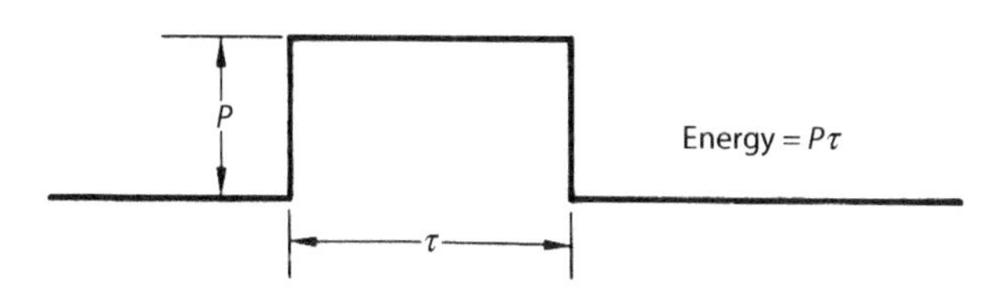

Figure 11-16. Peak power determines both voltage levels and energy per unit of pulse width.

Figure 11-17. Corona is what makes high voltage lines buzz. In a radar it can result in a major loss of power and equipment damage. (Courtesy Electro Power Research Institute.)

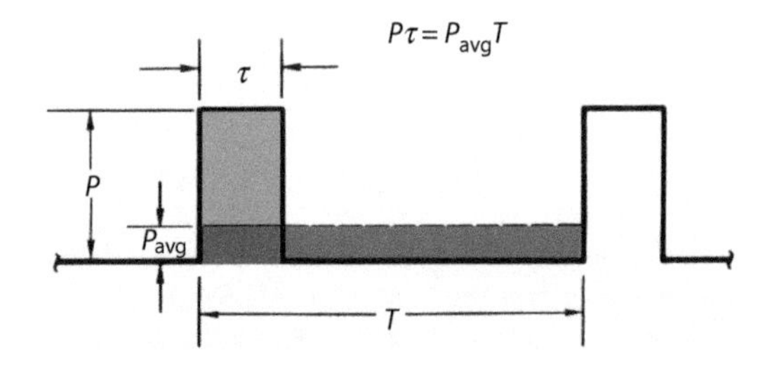

Figure 11-18. Average power is the peak power times the pulse width averaged over the pulse repetition interval.

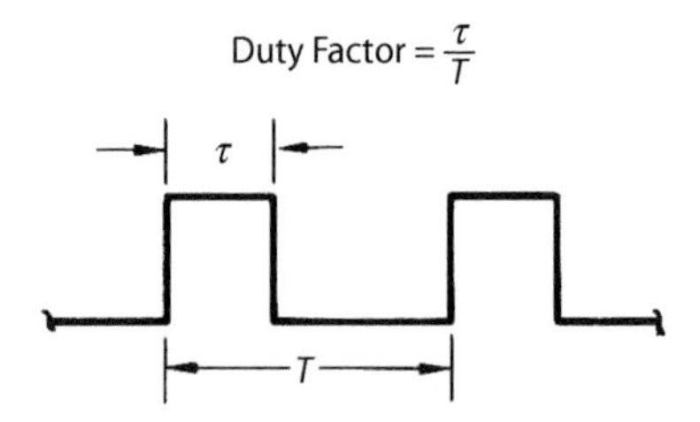

Figure 11-19. The duty factor is the fraction of time the radar is transmitting.

Peak power is important for several reasons. First, it determines the voltages that must be applied to the transmitter. Peak power also determines the intensities of the electromagnetic fields one must contend with: fields across insulators, fields in the waveguides that connect the transmitter to the antenna, etc. If these fields are too intense, problems of corona and arcing will be encountered. Corona is a discharge that occurs when an electric field becomes strong enough to ionize the air; it is what makes high voltage power lines buzz (Fig. 11-17). Arcing occurs when ionization is sufficient for a conductive path to develop through the air. Both effects can result in a major loss of power, as well as equipment damage. Consequently there is an upper limit on the level of peak power deliverable through a single transmission path (waveguide, for example).

Together, peak power and pulse width determine the amount of energy conveyed by the transmitted pulses. If the pulses are rectangular, the energy in each pulse equals the peak power times the pulse width, that is,

Energy per pulse = $P\tau$.

Usually, however, the energy in a train of pulses is what is important. This is related to average power.

Average Power. A radar's average transmitted power is the power of the transmitted pulses averaged over the PRI (Fig. 11-18). Here, average power is represented by P_{avg}.

If a radar's pulses are rectangular, the average power equals the peak power times the ratio of the pulse width, τ, to the PRI, T:

$$P_{avg} = P\frac{\tau}{T}.$$

For example, a radar having a peak power of 100 kW, a pulse width of 1 μs, and a PRI of 2000 μs will have an average power of 100 × 1/2000 = 0.05 kW, or 50 W.

The ratio τ/T is called the *duty factor* of the transmitter (Fig. 11-19). This factor represents the fraction of time the radar is transmitting. If, for example, a radar's pulses are 0.5 μs wide and the PRI is 100 μs, the duty factor is 0.5 ÷ 100 = 0.005. The radar is transmitting 5/1000 of the time it is in operation and is said to have a duty factor of 0.5 percent.

The average output power is important primarily because it is a key factor in determining the radar's potential detection range. The total amount of energy transmitted in a given period equals the average power times the length of the period, T:

$$\text{Transmitted energy} = P_{avg}T.$$

In the interest of maximizing detection range, average power can be increased in any of three ways: by increasing the PRF,

The Distinction Between Energy and Power

Many of us use the terms power and energy loosely and often interchangeably. But if we are to understand the operation of a radar, we must make a clear distinction between the two.

Energy is the capacity to do work. It has many forms: mechanical, electrical, thermal, and so on. Work is accomplished by converting energy from one form to another.

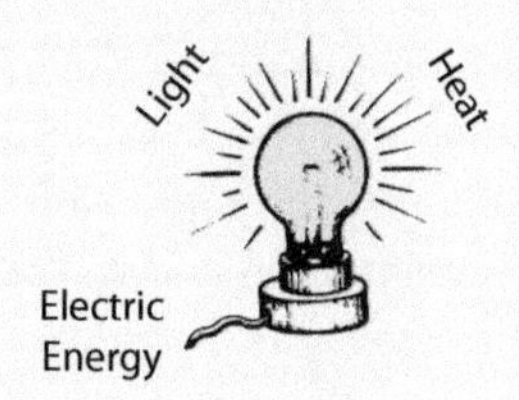

Take an incandescent electric lamp. It converts energy from an electrical form to an electromagnetic form. The result is light. Being inefficient, the lamp also converts a considerable amount of electrical energy to thermal energy. The result is heat, some of which is radiated in electromagnetic form.

Power is the rate at which work is done—the amount of energy converted from one form to another per second.

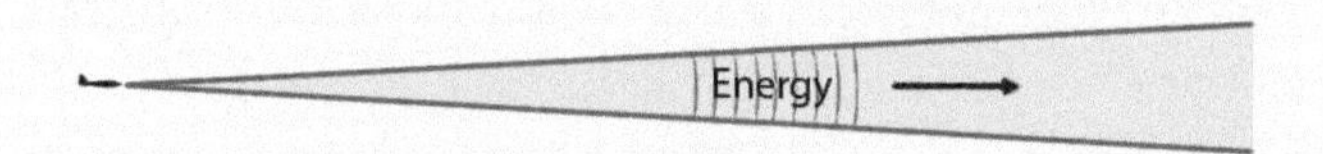

It is also the rate at which energy is transmitted—for example, the amount of energy per second that a radar beams toward a target. The common units of power are the Watt (1 J/s) and the kilowatt (1000 W).

How much energy is converted or transmitted depends on how long the power is present. The common units of energy are the Joule (1 Ws) and the watt-hour (3600 Ws).

A 25 W lamp left on for 4 hours will convert 100 Wh of energy to light and heat—the same as a 100 W lamp left on for only 1 hour.

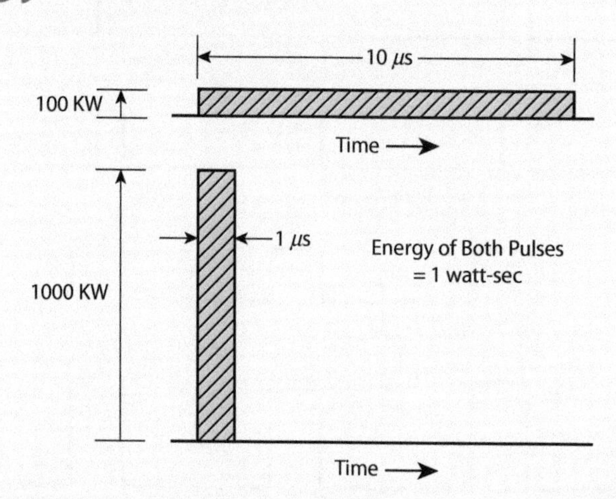

Similarly, a 100 kW radar pulse having a duration of 10 μs will convey as much energy as a 1000 kW pulse having a duration of only 1 μs.

Equipment Rating. Although it is energy that is transmitted and energy that does the work, most electrical equipment is rated in terms of power. Motors are rated in horsepower (746 W = 1 hp) and radio transmitters are rated in Watts or kilowatts.

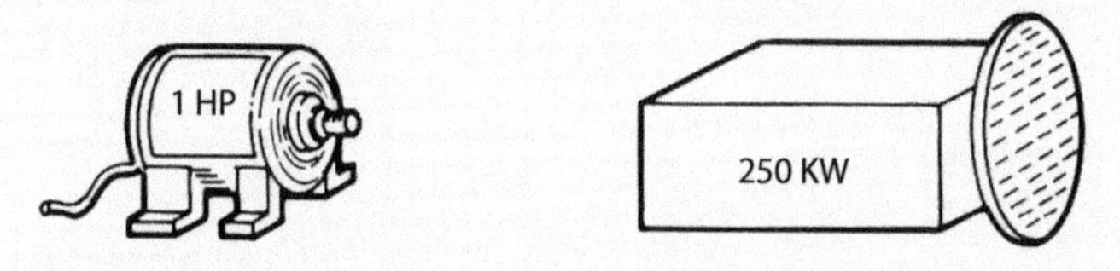

The reason is that the power rating determines the energy handling capacity of the equipment and is a dominant factor in its design.

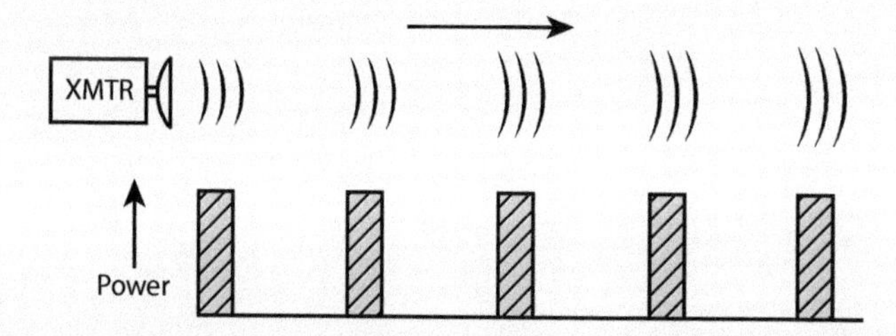

But we must not lose sight of the fact that the amount of radio frequency energy that a radar transmits toward a target equals the power of the transmitted waves times the duration of each pulse times the number of pulses.

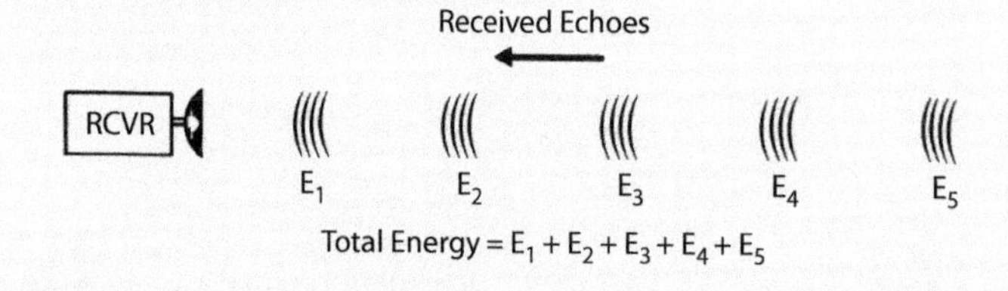

We must also remember that the extent to which the energy of the received echoes can be used to detect the target usually depends on the radar's ability to sum the energy contained in successive echoes.

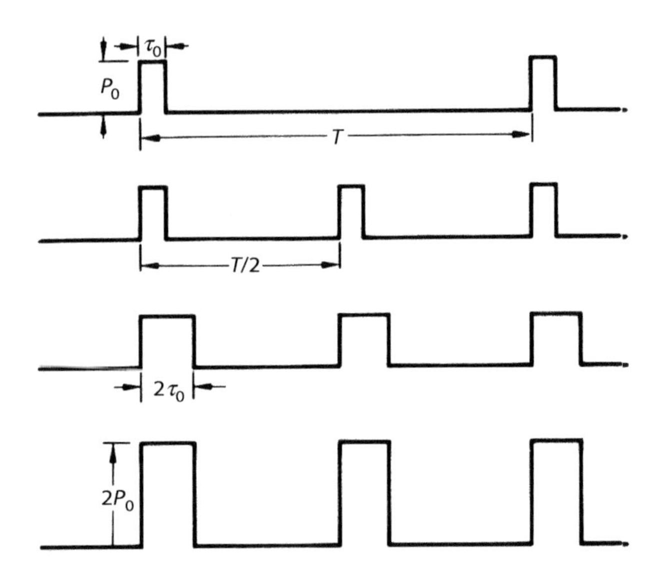

Figure 11-20. There are three independent methods of increasing average power. Combinations of these methods may also be used.

by increasing the pulse width, and by increasing the peak power (Fig. 11-20).

Average power is also of concern for other reasons. Together with the transmitter's efficiency, average power determines the amount of heat due to losses that the transmitter must dissipate. In turn, this determines the amount of cooling required. The average output power plus the losses and efficiency determine the amount of input (prime) power that must be conditioned and supplied to the transmitter. Finally, the higher the average power, the larger and heavier the transmitter tends to be.

11.5 Summary

Because it is difficult to prevent noise sidebands on the transmitted signal from leaking into the receiver, CW transmission is generally practical against small targets only if separate antennas are used for both transmission and reception. Pulsed transmission avoids this problem and provides a simple means of measuring range.

The basic characteristics of the transmitted waveform include radio frequency, pulse width, intrapulse or interpulse modulation, and PRF.

Radio frequency can be varied not only from pulse to pulse, but within the pulses (intrapulse modulation).

Pulse bandwidth determines range resolution. By coding successive increments of each pulse with phase or frequency modulation and decoding the echoes, long pulses can be transmitted to provide greater power output and the received pulses can be compressed (pulse compression) to provide fine resolution.

The PRF determines the extent of range and Doppler ambiguities. The lower the PRF, the less severe the range ambiguities. The higher the PRF, the less severe the Doppler ambiguities. The ambiguities in range and Doppler can be visualized by means of an ambiguity diagram.

Peak power is the power of the individual pulses. The maximum usable peak power is generally limited by problems of arcing and corona.

The average power is the peak power averaged over the PRI. The higher the peak power, pulse width, and PRF, the higher the average power will be.

Energy, not power, is what does the work. The energy in a pulse train equals the average power times the length of the train.

11.6 Some Relationships to Keep in Mind

- Pulse length $\approx 300\tau$ m
- Range resolution $\approx 150\tau$ m

- PRI (interpulse period), $T = \text{PRI} = \dfrac{1}{\text{PRF}}$
- Duty factor $= \dfrac{\tau}{T}$
- Average power, $P_{\text{avg}} = P\dfrac{\tau}{T}$

where

τ = pulse width in microseconds

P = peak power

T = PRI in microseconds

Further Reading

P. M. Woodward, *Probability and Information Theory, with Applications to Radar*, Pergamon Press, 1953 (reprint Artech House, 1980).

C. M. Alabaster, *Pulse Doppler Radar: Principles, Technology, Applications*, SciTech-IET, 2012.

Test your understanding

1. A radar uses a uniform pulse repetition frequency of 500 Hz. At what spacings are its ambiguities in (a) range and (b) Doppler?
2. What is the length, in meters, of a pulse of duration 100 ns?
3. What pulse duration would be needed to give a range resolution of 1 m?
4. A radar has a PRF of 500 Hz, a pulse duration of 10 μs, and a peak transmit power of 10 kW. What are (a) the duty factor, (b) the average power, and (c) the energy per pulse?
5. A radar transmits a pulse of duration 20 μs. Ignoring the effect of receiver recovery time, what is the minimum range at which a target could be detected?

Northrop Grumman E-2 Hawkeye (1964)

The E-2 is an airborne early warning (AEW) twin-turboprop aircraft designed to replace the E-1 Tracer. It is capable of launch from aircraft carriers and is primarily used by the United States Navy. It is one of only two propeller airplanes that operate from carriers (C-2 Greyhound). The current E-2D version carries the APY-9 radar in its radome, which features an active electronically scanned array (AESA).

12

Detection Range

Generally, few things are of more fundamental concern to both designers and users alike than the maximum range at which a radar can detect targets. In this chapter we will discuss what determines that range.

We will begin by tracking down the sources of the electrical background noise against which a target's echoes must ultimately be discerned and discussing what can be done to minimize this noise. We will then trace the factors upon which the strength of the echoes depends and examine the detection process. Finally, we'll see how, by integrating the return from a great many transmitted pulses, a radar can pull the weak echoes of distant targets out of the noise.

12.1 What Determines Detection Range

In principle airborne radars can be designed to detect targets at ranges of hundreds of kilometers. As a rule though, they are designed to operate at much shorter ranges, for at least one compelling reason: obstructions in the line of sight.

Radio waves of the frequencies used by airborne radars behave very much like visible light, except of course that they can penetrate clouds and are not scattered much by aerosols (tiny particles suspended in the atmosphere). They are unable to penetrate liquids or solids very far, and although they bend slightly as a result of the increase in their speed of propagation with altitude (due to the reduction in the refractive index with height) and spread to some extent around obstructions, these effects are slight.

Consequently, no matter how powerful a radar is or how ingenious its design, its range is essentially limited to the maximum unobstructed line of sight. A radar cannot see through mountains, and it cannot see much at low altitudes or on the ground beyond the horizon.

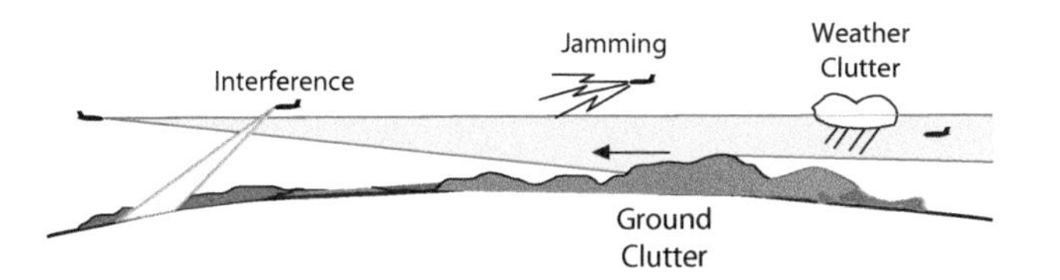

Figure 12-1. Just because a target is within the line of sight does not mean it will be detected. It may be obscured by competing clutter or man-made interference.

Just because a target is within the line of sight, however, does not mean that it will be detected (Fig. 12-1). Depending upon

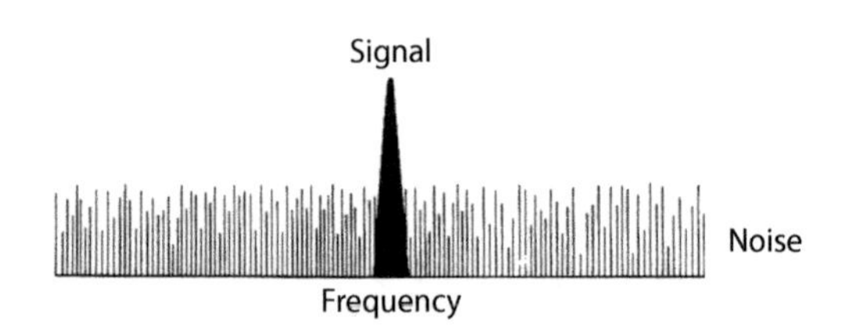

Figure 12-2. In the absence of clutter and interference, whether a target will be detected ultimately depends upon the strength of its echoes relative to the strength of the background noise.

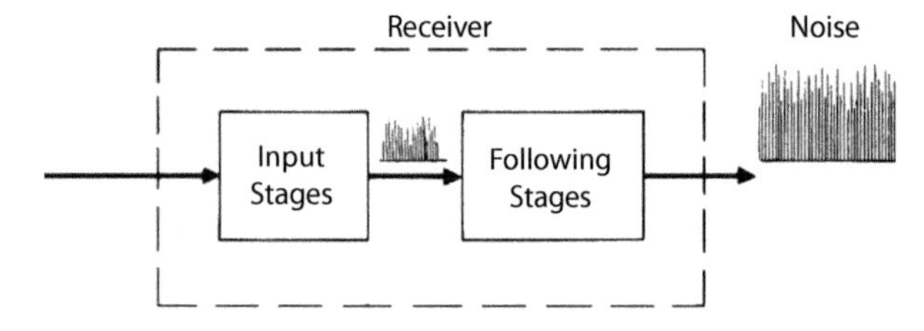

Figure 12-3. Amplified by the full gain of the receiver, noise generated in the input stages swamps the noise generated in the following stages.

the operational situation, a target's echoes may be obscured by clutter returned from the ground or (depending upon the wavelength and the weather) from rain, hail, or snow. A target's echoes may also be obscured by the transmissions of other radars, by jamming, or by other *electromagnetic interference* (EMI).

Clutter can largely be eliminated through Doppler processing (moving target indication [MTI]). And there are ways of dealing with most man-made interference as well.

But depending upon the strength of the transmitted waves, if the target is small or at long range, its echoes may still be obscured by the ever present background of electrical noise.

In a benign environment, then, whether a given target will be detected ultimately depends upon the strength of its echoes relative to the strength of the electrical background noise (Fig. 12-2)—that is, the signal-to-noise ratio (SNR).

12.2 Electrical Background Noise

As the name implies, electrical noise is electrical energy of random amplitude and random frequency. It is present in the output of every radio receiver, and a radar receiver is no exception. At the frequencies used by most radars, the noise is generated primarily within the receiver itself.

Receiver Noise. Most of this noise originates in the input stages of the receiver. The reason is not that these stages are inherently more noisy than others, but rather that, amplified by the receiver's full gain, noise generated in these stages swamps the noise generated farther along the signal path (Fig. 12-3).

Because the noise and the received signals are amplified equally (or nearly so), in computing SNRs, the factor of receiver gain can be eliminated by determining the signal strength at the input to the receiver and dividing the noise output of the receiver by the receiver's gain. Therefore receiver noise is commonly defined as noise per unit of receiver gain:

$$\text{Receiver noise} = \frac{\text{Noise at the receiver output}}{\text{Receiver gain}}$$

This ratio can readily be measured in the laboratory by methods such as are outlined in the following panel.

Since the early days of radio it has been customary to describe the noise performance of a receiver in terms of a figure of merit called the *noise figure*, *F*. It is the ratio of the noise output of the actual receiver to the noise output of a hypothetical, "ideal" minimum-noise receiver of equal gain:

$$F = \frac{\text{Noise output of actual receiver}}{\text{Noise output of ideal receiver}}$$

(Note that since the gains of both receivers are the same, F is independent of receiver gain.)

How the Receiver Noise Figure Is Measured

Although you may never have to measure the noise figure of a radar receiver, you may gain a better feel for its significance if you have a general idea of how it is measured.

Measurement. Basically it is a three-step process known as the *Y-factor method.* This process is based on using a calibrated noise source that delivers a precisely known noise power. With no DC power applied, the noise source behaves as a matched resistor at room temperature ($T_0 = 290$ K). When DC power is applied, it behaves as a noise source whose noise power is defined in terms of a temperature T_H that characterizes its noise power in a bandwidth B, equal to kT_HB. The excess noise ratio (ENR) is simply the ratio of T_H/T_0, and is often expressed in decibels: $10 \log_{10}(T_H/T_0)$. Of course, even though T_H may be several thousand degrees, this value is just a number that characterizes the noise power; the noise source is perfectly cool to touch.

In this way the noise power from the noise source, measured in a bandwidth B, with no DC power applied (the "cold" condition) is kT_0B, and the noise power when DC power is applied (the "hot" condition) is kT_HB. The measurement process is as follows:

First, connect the noise source to the input terminals of the receiver, with no DC power applied, and measure the output power from the receiver.

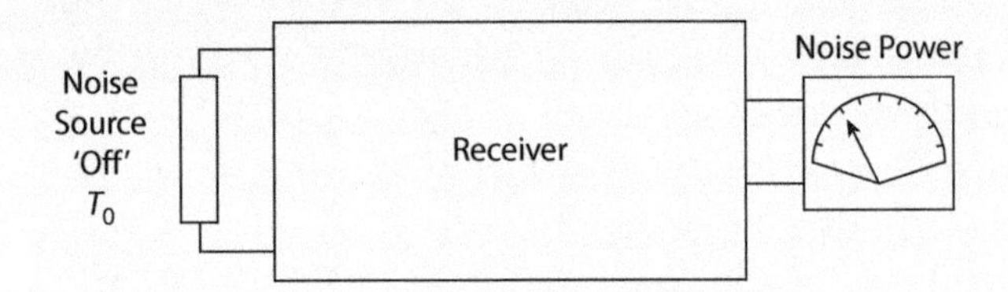

Second, apply DC power (typically 24 V) to the noise generator and measure the increase in noise power.

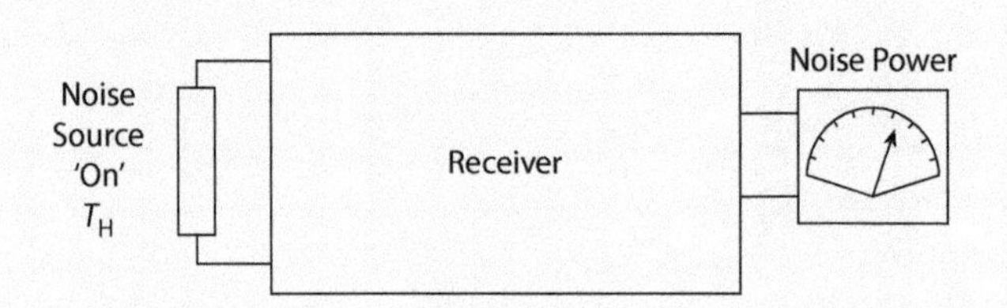

Let's call the ratio of the "off" and "on" noise powers Y, so

$$Y = \frac{kB(T_s + T_H)}{kB(T_s + T_0)},$$

where T_s is the noise temperature of the receiving system. This can be rearranged to give

$$T_s = \frac{T_H - YT_0}{(Y-1)}$$

Third, you insert the measured value for Y and the known values of T_H and T_0 into this equation to give the value of T_s, and hence the noise figure, $F = 1 + T_s/T_0$.

This latter relationship follows directly from the definition of noise figure given above:

$$F = \frac{\text{Noise output of actual receiver}}{\text{Noise output of ideal receiver}} = \frac{k(T_0 + T_s)B}{kT_0B} = 1 + \frac{T_s}{T_0}$$

The noise figure is often expressed in decibels ($10 \log_{10}F$), since it is a ratio of powers.

An ideal receiver, of course, would generate no noise internally whatsoever. The only noise in its output would be noise received from external sources. By and large, that noise has the same spectral characteristics as the noise resulting from thermal agitation in a conductor. Therefore, as a standard for determining F, the sources of external noise for both actual and ideal receivers can reasonably be represented by a resistor connected across the receiver's input terminals (Fig. 12-4). (A resistor is a conductor providing a specified resistance to the flow of current.)

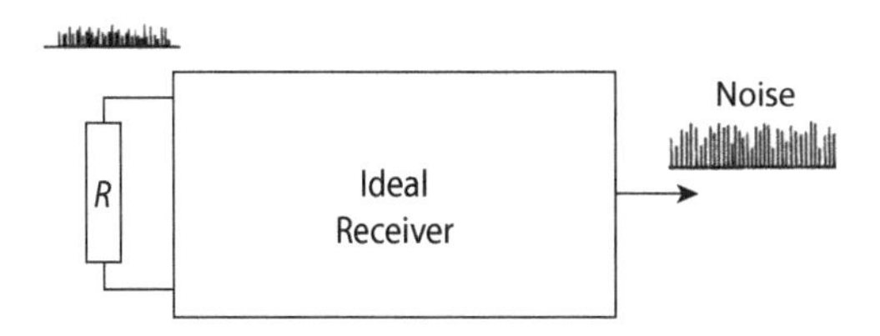

Figure 12-4. The only noise in the output of an ideal receiver would be that received from external sources. This is represented by thermal agitation in a resistor.

Thermal agitation noise is produced by the continuous random motion of free electrons, which are present in every conductor. The amount of motion is proportional to the conductor's temperature above absolute zero (absolute zero is the temperature at which all of this random motion ceases). Quite by chance, at any one instant, more electrons will generally be moving in

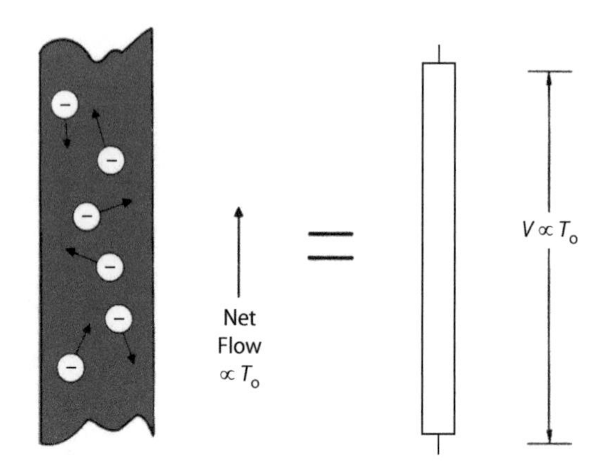

Figure 12-5. Because of thermal agitation, a random voltage proportional to the temperature appears across the electrical resistance of every conductor.

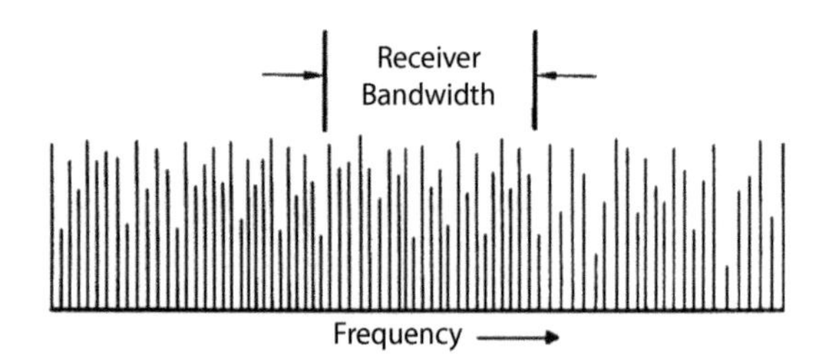

Figure 12-6. Noise in the receiver output is proportional to the bandwidth of the receiver.

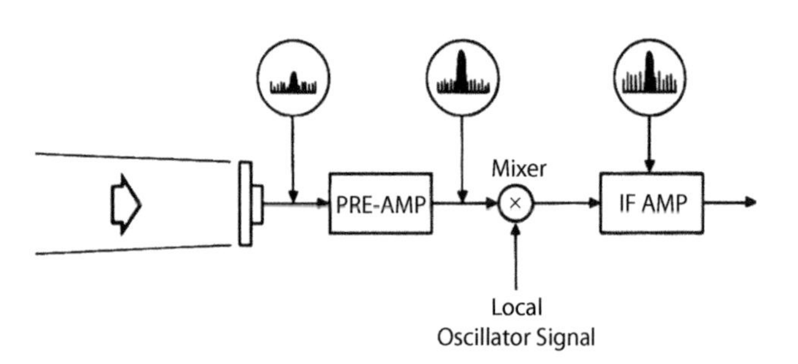

Figure 12-7. Receiver noise may be reduced substantially by providing a low-noise preamplifier ahead of the mixer or using a low-noise mixer.

one direction than in another. This imbalance causes a random voltage proportional to the temperature to appear across the conductor (Fig. 12-5).

Thermal noise is spread more or less uniformly over the entire spectrum. Thus the amount of noise appearing in the output of the ideal receiver is proportional to the absolute temperature of the resistor that is connected across its input terminals times the width of the band of frequencies passed by the receiver—that is, the receiver bandwidth (Fig. 12-6).

The mean power—per unit of receiver gain—of the noise in the output of the hypothetical ideal receiver is thus

$$\text{Mean noise power of an ideal receiver} = kT_0B \text{ W}$$

where

k = Boltzmann's constant, 1.38×10^{-23} W/K/Hz

T_0 = absolute temperature of the resistor representing the external noise (K)

B = receiver bandwidth (Hz)

Since the external noise is the same for both actual and ideal receivers, as long as everyone uses the same value for T_0 in determining the noise figure, the exact value is not critical. By convention, T_0 is taken to be 290 K, which is a good approximation of room temperature and conveniently makes kT_0 a round number (4×10^{-21} W/Hz).

When the internally generated noise is considerably greater than the external noise (as it is in the vast majority of airborne radars in operation today), the noise figure, F, multiplied by the foregoing expression for mean noise power per unit of gain for an ideal receiver is commonly used to represent the level of background noise against which target echoes must be detected:

$$\text{Mean noise power of an actual receiver} = kT_0BF \text{ W}$$

This expression includes both a nominal estimate of the external noise (equivalent of the noise generated in a resistor at room temperature) and the accurately measured internally generated noise.

Although internal noise predominates in many receivers, the noise can be substantially reduced by adding a low-noise preamplifier ahead of the receiver's mixer stage and using a low-noise mixer (Fig. 12-7). The preamplifier increases the signal strength relative to the thermal noise originating in the subsequent stages, while contributing only a minimum amount of noise itself. When a low-noise front end is used, a more accurate estimate may have to be made of the noise received from sources ahead of the receiver.

Noise from Sources Ahead of the Receiver. As explained in Chapter 4, because of thermal agitation, virtually everything around us radiates radio waves. This so-called black-body radiation is extremely weak. Nonetheless, it may be detected by a sensitive receiver and adds to the noise in the receiver output. At the frequencies used by most airborne radars, the principal sources of this natural radiation are the ground, the atmosphere, and the sun (Fig. 12-8).

Radiation from the ground depends not only on the temperature of the ground but also on its "lossiness," or absorption. (The power of the radiated noise is proportional to the absolute temperature times the coefficient of absorption.) Thus, although a body of water may have the same temperature as a land mass, since water is a good conductor and the land usually is not, the water will radiate comparatively little noise. How much of the radiation that is received by the radar varies widely with the gain of the antenna and the direction in which it is looking. For example, far more noise is received when looking down at the warm ground than when looking at a body of water, which reflects the extreme cold of outer space.

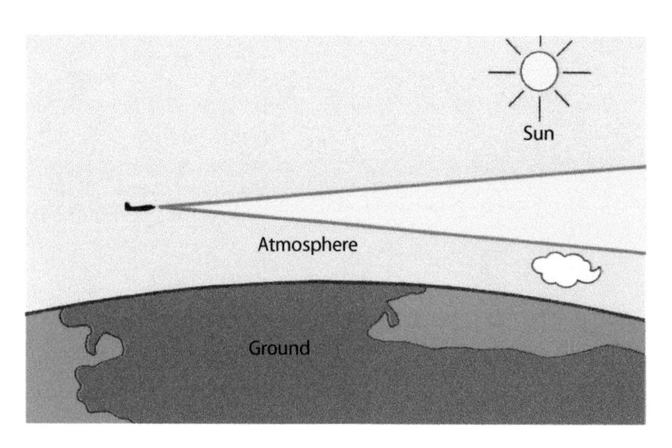

Figure 12-8. The principal sources of noise outside the aircraft vary in amount with antenna gain and direction.

The amount of noise received from the atmosphere depends not only upon the temperature and lossiness of the atmosphere, but upon the amount of atmosphere the antenna is looking through. Since the lossiness varies with frequency, the received noise also depends upon the radar's operating frequency.

Noise received from the sun varies widely with both solar conditions and the radar's operating frequency. Naturally, it is vastly greater if the sun happens to be in the antenna's mainlobe as opposed to its sidelobes.

Within the aircraft carrying the radar, noise is radiated by the radome, the antenna, and the complex of waveguides connecting the antenna to the receiver (Fig. 12-9). Noise from these sources is likewise proportional to their absolute temperature times their loss coefficients.

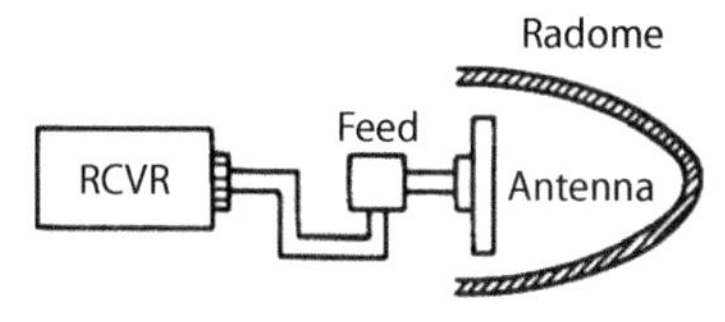

Figure 12-9. There are other sources of noise within an aircraft.

As previously noted, noise from all of these external sources that falls within the receiver passband has essentially the same spectral characteristics as receiver noise. Consequently, when external noise is significant, the noise from each source, as well as the receiver noise, is usually assigned an equivalent noise temperature (Fig. 12-10). These temperatures are combined to produce an equivalent noise temperature for the entire system, T_s. The expression for noise power then becomes

$$\text{Mean noise power of all sources} = kT_sB.$$

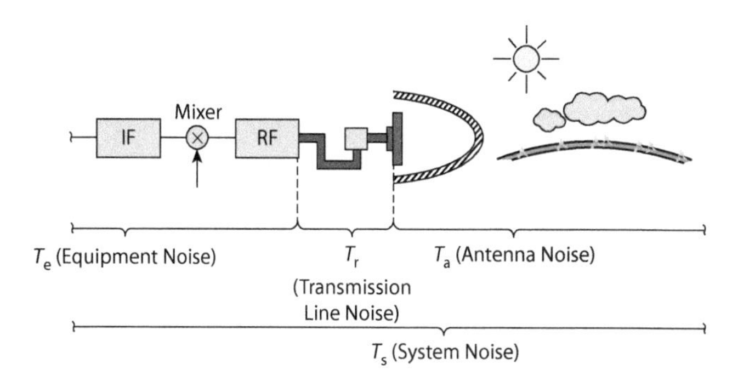

Figure 12-10. When external noise is significant, noise from each source is assigned an equivalent noise temperature.[1]

Competing Noise Energy. Whether noise is expressed in terms of the receiver noise figure, T_0F, or the equivalent noise temperature, T_s, it is noise *energy*, not power, with which a target's echoes must compete. As explained in the last chapter, power is the rate of flow of energy. Noise energy is noise power times the length of time over which the noise energy flows—in this case, the duration of the period in which the return may be received from any one resolvable increment of range. Therefore

$$\text{Mean noise energy} = kT_sBt_n,$$

where t_n is the duration of the noise.

1. Since T_e does not include the noise of the input resistance, as T_0F does, $T_e = T_0(F - 1)$.

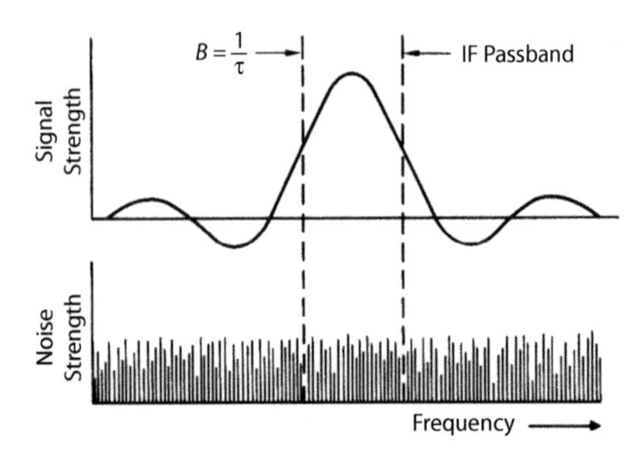

Figure 12-11. The SNR may be maximized by narrowing the passband of the IF amplifier to the point where only the bulk of the signal energy is passed.

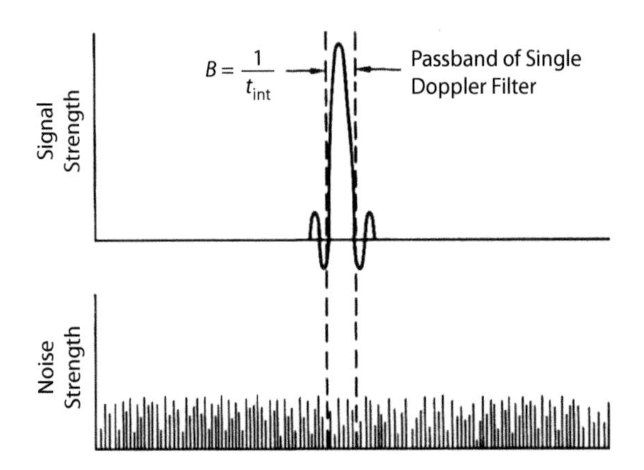

Figure 12-12. In a Doppler radar, the passband is further narrowed by Doppler filtering.

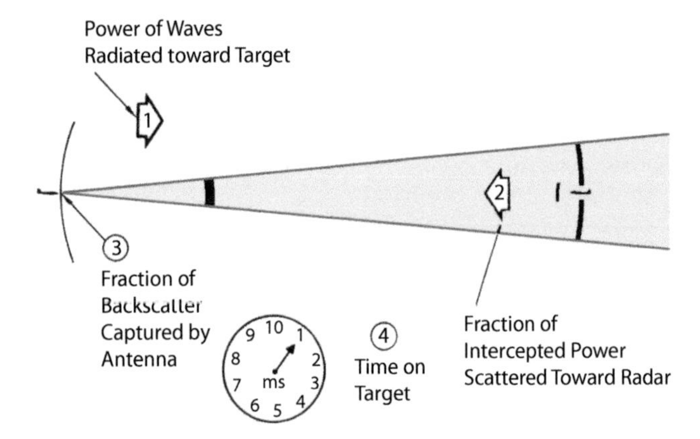

Figure 12-13. Note four factors that determine the energy of the target signal.

For any given noise temperature and duration, the noise can be reduced by minimizing the receiver bandwidth, B. A common practice is to narrow the intermediate frequency (IF) passband until it is just wide enough to pass most of the energy contained in the received echoes. This is called a *matched filter design* (Fig. 12-11).

Another way of looking at this is that the tuned circuits of the receiver IF amplifier integrate the received energy during the width, τ, of each received pulse. They thus accumulate the energy the pulse contains and reject the noise outside the pulse's bandwidth. The optimum bandwidth turns out to be very nearly equal to one divided by the pulse width, τ. When $1/\tau$ is substituted for B in the expression for mean noise, it becomes

$$\text{Mean noise energy of the matched filter design} = \frac{kT_s t_n}{\tau}$$

In Doppler radars, bandwidth is further reduced by Doppler filters, which follow the IF amplifier. (A separate filter is generally provided for every anticipated combination of resolvable range and Doppler frequency.) As will be explained in Chapter 18, the passband of a Doppler filter is approximately equal to $1/t_{int}$, where t_{int} (also known as the coherent processing interval) is the time over which the filter adds up (integrates) the radar returns (Fig. 12-12).

Whereas τ is on the order of microseconds, t_{int} is on the order of milliseconds. Consequently the passband of a Doppler filter is on the order of 1/1000 of the width of the IF passband.

The integration time, t_{int}, is also the length of time over which the noise is received and integrated by the filter. When t_{int} is substituted for t_n and $1/t_{int}$ is substituted for $1/\tau$, the two terms cancel, leaving

$$\text{Mean noise energy of a Doppler radar} = kT_s$$

As noise energy flows into the Doppler filter, the filter's passband (which is inversely proportional to integration time) simultaneously narrows. As a result, the level of the noise energy that accumulates in the filter is more or less independent of the length of the integration period.

Since noise is random, the level of the accumulated energy may vary widely from one integration period to another, but its mean value over a great many integration periods will be kT_s.

On average, therefore, for a target to be detected, enough energy must be received from it to noticeably raise the filter output above this mean level.

Which brings us to the question: what determines how much energy is received from a target; what is the energy of the signal?

12.3 Energy of the Target Echo

Four basic factors (Fig. 12-13) determine the amount of energy a radar will receive from a target during the length of time the antenna is trained on the target, that is, the time on target, t_{ot}. (Note: the time on target is equivalent to an integration period, t_{int}, or multiple integration periods, dependent upon the radar mode):

- Average power—the rate of flow of energy—of the radio waves radiated in the target's direction
- Fraction of the wave's power that is intercepted by the target and scattered back in the radar's direction
- Fraction of that power that is captured by the radar antenna
- Length of time the antenna beam is trained on the target

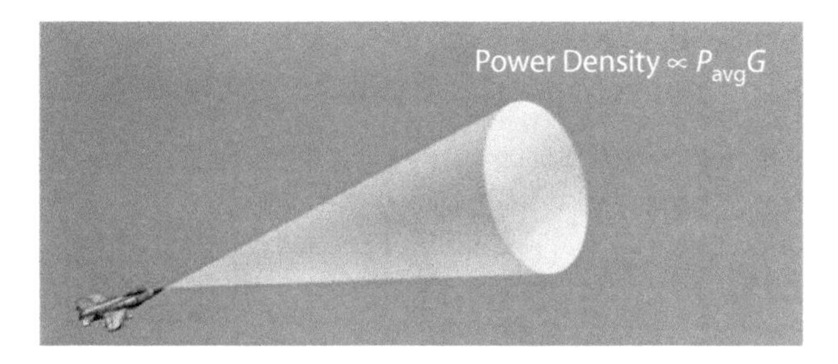

Figure 12-14. The density of power radiated in the target's direction is proportional to the average radiated power times the antenna gain in that direction.

When the antenna is trained on a target, the power density of the radio waves radiated in the target's direction is proportional to the transmitter's average power output, P_{avg}, times the gain, G, of the antenna's mainlobe (Fig. 12-14). (Power density, you will recall, is the rate of flow of energy per unit of area normal to the wave's direction of propagation.)[2]

2. Another term for power density is "power flux."

In transit to the target, the power density is diminished as a result of two things: absorption in the atmosphere and energy spreading. Except at the shorter wavelength, attenuation due to absorption is comparatively small. For the moment, absorption will be neglected, but not the reduction in power density due to spreading.

As the waves propagate toward the target, the energy spreads—like an expanding soap bubble—over an increasingly large area (Fig. 12-15). This area is proportional to the square of the distance from the radar. At the target's range, R, the power density is only $1/R^2$ times what it was at a range of 1 km.

Figure 12-15. As waves travel out to a target, their power is spread over an increasingly large area.

The amount of power intercepted by the target equals the power density at the target's range times the geometric cross-sectional area of the target, as viewed from the radar (the projected area).

What fraction of the intercepted power is scattered back toward the radar depends upon the target's reflectivity and directivity. The reflectivity is simply the ratio of total scattered power to total intercepted power. The directivity—like the gain of an antenna—is the ratio of the power scattered in the direction of the radar to the power that would have been scattered in that direction had the scattering been uniform in all directions.

Customarily, a target's geometric cross-sectional area, reflectivity, and directivity are lumped into a single factor, the *radar cross section* (RCS). This is represented by the Greek letter sigma, σ, and is usually expressed in square meters (see following blue panel.)

Radar Cross Section

A TARGET'S RCS, σ, IS MOST EASILY VISUALIZED AS THE PRODUCT OF THREE factors:

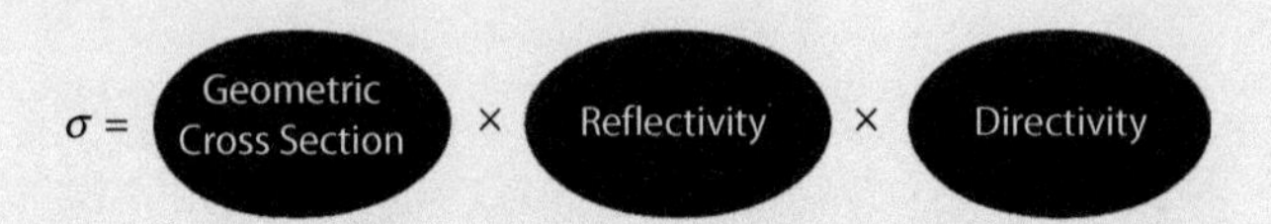

Geometric cross section is the cross sectional silhouette area of the target as viewed from the radar.

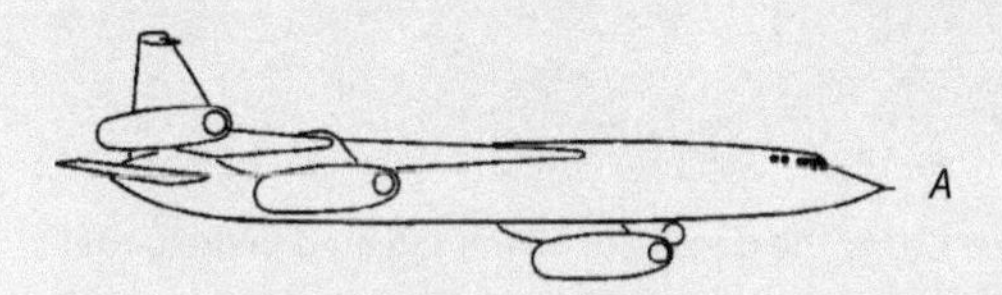

This area determines how much power the target will intercept.

$$P_{\text{intercepted}} = \Phi \times A,$$

where Φ is the power density of the waves incident on the target, in watts per square meter.

Reflectivity is the fraction of the intercepted power that is reradiated (scattered) by the target.

$$\text{Reflectivity} = \frac{P_{\text{scatter}}}{P_{\text{intercepted}}} = \frac{P_{\text{scatter}}}{\Phi A}$$

(The scattered power equals the intercepted power minus whatever power is absorbed by the target.)

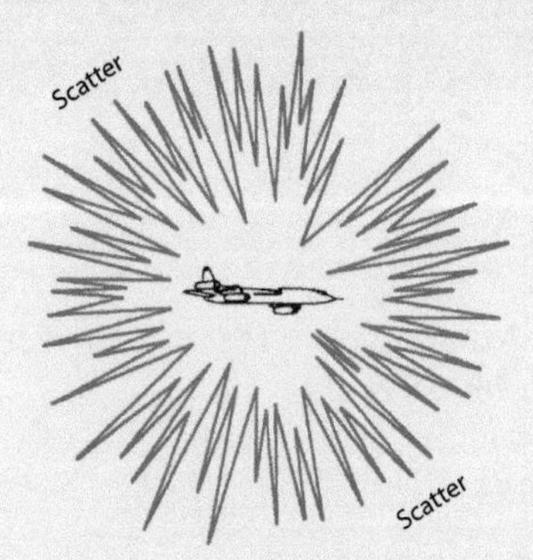

Directivity is the ratio of the power scattered back in the radar's direction to the power that would have been backscattered had the scattering been uniform in all directions, that is, isotropically:

$$\text{Directivity} = \frac{P_{\text{backscatter}}}{P_{\text{isotropic}}}$$

Normally, $P_{\text{backscatter}}$ and $P_{\text{isotropic}}$ are expressed as power per unit of solid angle. $P_{\text{isotropic}}$ thus equals P_{scatter} divided by the number of units of solid angle in a sphere.

The unit of solid angle is the steradian. It is the angle subtended by an area on the surface of a sphere equal to the radius squared.

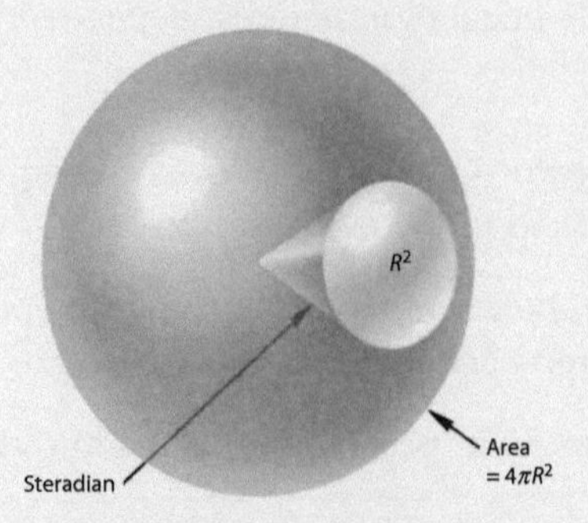

Since the area of a sphere is 4π times the radius squared, a sphere contains 4π steradians. Therefore

$$\text{Directivity} = \frac{P_{\text{backscatter}}}{(1/4\pi)P_{\text{scatter}}}$$

A target may be thought of as consisting of a great many individual reflecting elements (scatterers).

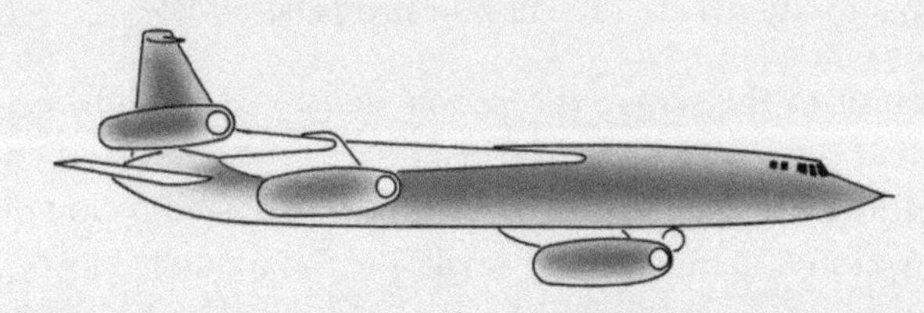

The extent to which the scatter from these combines constructively in the direction of the radar depends upon the relative phases of the backscatter from the individual elements. That in turn depends on the relative distances (in wavelengths) of the elements from the radar. Depending on the configuration and orientation of the target, the directivity may range anywhere from a small fraction to a large number.

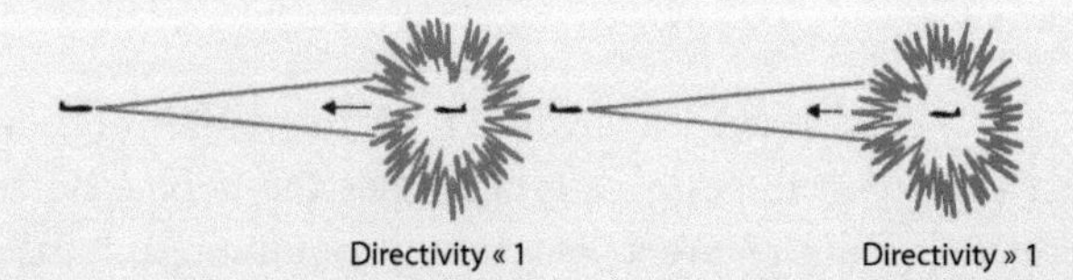

Complete Expression for σ. Expanded in terms of the factors outlined in the preceding paragraphs, the basic expression for the RCS becomes

$$\sigma = A \times \frac{P_{\text{scatter}}}{(A)(P_{\text{incident}})} \times \frac{P_{\text{backscatter}}}{(1/4\pi)(P_{\text{scatter}})}$$

Canceling like terms and spelling out those that remain yields

$$\sigma = 4\pi \frac{\text{Backscatter per steradian}}{\text{Power density of intercepted waves}}$$

This is the common form of the definition of an RCS. It has the advantage of making radar equations easier to write. But expressing σ in terms of geometric cross section, reflectivity, and directivity is more illuminating since the complete expression shows the relationship between σ and the factors that determine its value.

The power density of the waves scattered back in the radar's direction can be found by multiplying the power density of the transmitted waves when they reach the target by the target's RCS (Fig. 12-16). Since the directivity of a target can be quite high, for some target aspects the RCS may be many times the geometric cross sectional area. For other target aspects the RCS may be many times smaller.

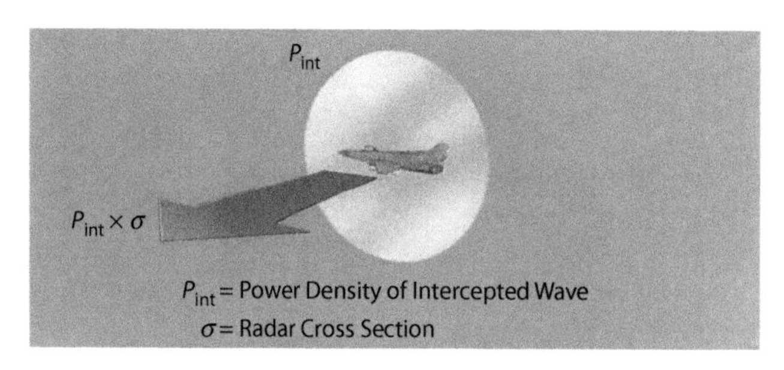

Figure 12-16. The density of power reflected in the radar's direction equals the density of the intercepted wave times the RCS.

As the waves propagate back from the target, they undergo the same geometric spreading as on their way out. Their power density, which has already been reduced by a factor of $1/R^2$, is again reduced by $1/R^2$ (Fig. 12-17). The two factors are compounded, so the power density when the waves reach the radar is only $1/R^2 \times 1/R^2 = 1/R^4$ times what it would be if the target were at a range of only 1 km (or whatever other unit of distance R is measured in).

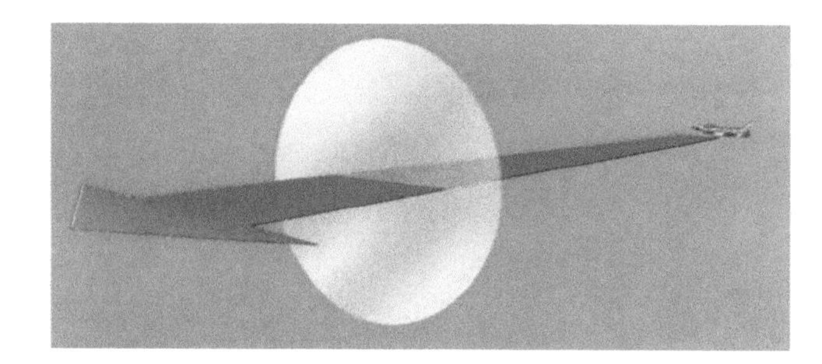

Figure 12-17. Reflected power undergoes an equal amount of spreading in returning to the radar.

To give you a feel for the magnitude of this difference, the relative strengths of the echoes from the same target in the same aspect at ranges of 1 to 50 km are plotted in Figure 12-18. For a range of 1 km, the strength is arbitrarily assumed to equal 1. At 50 km, the relative strength is only 0.00000016, which is too small to be discernible in the figure.

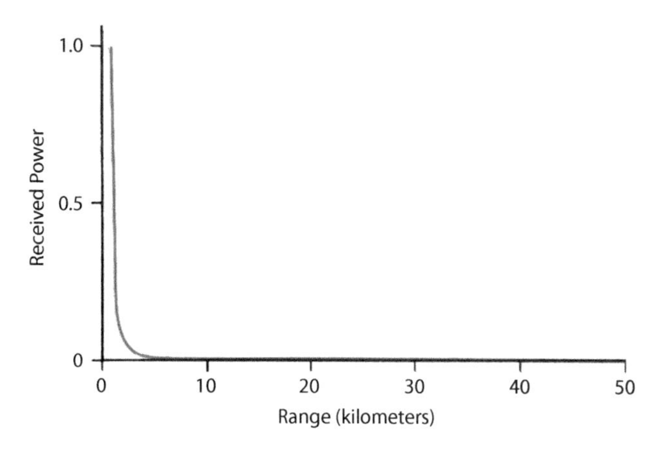

Figure 12-18. Target echoes reduce in strength with range. Echoes from a target at 50 km are only 0.00000016 times as strong as echoes from the same target at 1 km range.

Incidentally, Figure 12-18 dramatically illustrates why the receiver must be able to handle powers of vastly different magnitudes, that is, have a wide dynamic range.

When the backscattered waves reach the antenna, it intercepts a fraction of their power. That fraction equals the power density of the waves times the effective area of the antenna, A_e (Fig. 12-19). The total amount of energy intercepted equals that product times the length of time the antenna is trained on the target, t_{ot}.

As we saw in Chapter 8, the area A_e accounts for the aperture efficiency of the antenna. For all of the intercepted energy to be constructively summed by the antenna feed, the target must be centered in the antenna's mainlobe.

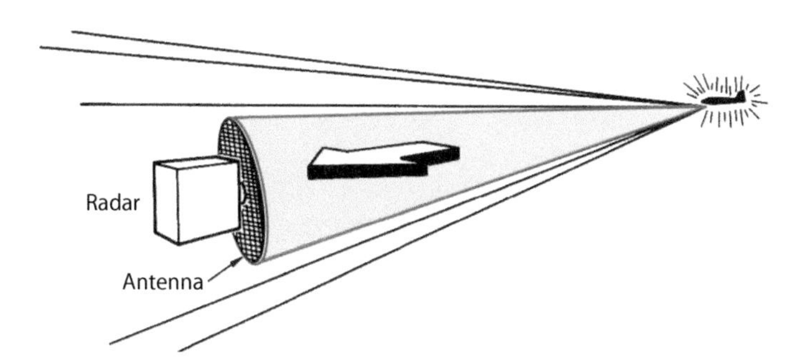

Figure 12-19. Reflected power intercepted by a radar is proportional to the effective area of the antenna.

The Radar Equation

We can combine these concepts to derive the fundamental equation that gives the SNR for a given target at a given range.

Suppose first that a transmit power P_t is fed to an isotropic antenna. The power is radiated equally in all directions, so on the surface of a sphere of radius R the power that passes through an area of 1 m × 1 m is

$\Phi = P_t \times \frac{1}{4\pi R^2}$ W/m², since the surface area of the sphere is $4\pi R^2$.

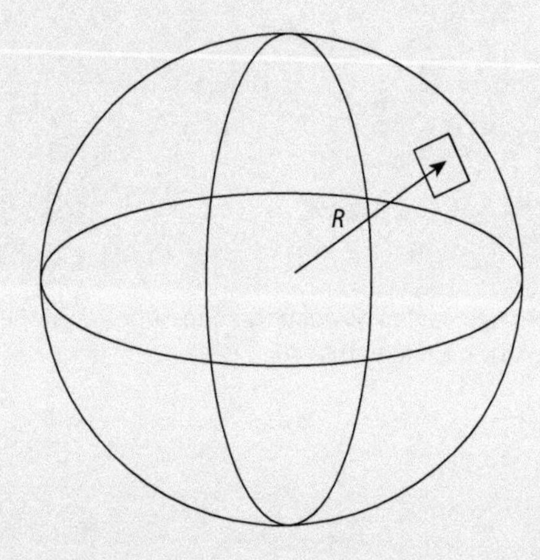

But any practical antenna will be directive, so the power density in the direction of maximum gain is instead

$$\Phi = P_t G \times \frac{1}{4\pi R^2} \quad \text{W/m}^2$$

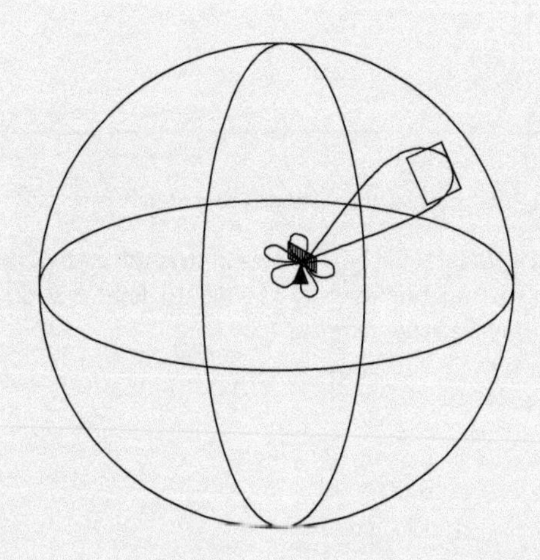

This is incident on the target, whose RCS is σ m², so the power intercepted is

$$P_t G \times \frac{1}{4\pi R^2} \times \sigma \quad \text{W}$$

This is scattered back to the radar, so the power density back at the radar is

$$\Phi' = P_t G \times \frac{1}{4\pi R^2} \times \sigma \times \frac{1}{4\pi R^2} \quad \text{W/m}^2$$

and this is intercepted by the radar antenna, whose effective area is A_e, so the received power P_r is

$$P_r = P_t G \times \frac{1}{4\pi R^2} \times \sigma \times \frac{1}{4\pi R^2} \times A_e \quad \text{W}$$

We saw in Chapter 11 that the relationship between antenna gain and effective area is

$$G = \frac{4\pi A_e}{\lambda^2}$$

so

$$P_r = P_t G \times \frac{1}{4\pi R^2} \times \sigma \times \frac{1}{4\pi R^2} \times \frac{G\lambda^2}{4\pi} = \frac{P_t G^2 \lambda^2 \sigma}{(4\pi)^3 R^4} \quad \text{W}$$

Since the receiver noise power referred to the receiver input is $P_n = kT_0BF$, the SNR is

$$\frac{P_r}{P_n} = \frac{P_t G^2 \lambda^2 \sigma}{(4\pi)^3 R^4 k T_0 B F}$$

which is the basic form of the radar equation. Note that the antenna gain G is squared, since its effect is both on transmit and receive, and the range dependence is $1/R^4$.

The equation is modified by the inclusion of losses, integration gain, the effects of clutter, and the statistical fluctuation of the noise, target, and clutter, so an equation that looks quite simple can ultimately become rather complicated. These ideas are developed in the next chapter.

The received signal can be expressed in terms of energy rather than power:

$$\text{Signal energy} \cong \frac{P_{avg} G^2 \lambda^2 \sigma t_{ot}}{(4\pi)^3 R^4}$$

where

P_{avg} = average transmitted power

G = antenna gain

σ = RCS of the target

t_{ot} = time on target

R = range

This expression *roughly* indicates the total amount of energy that would be received by a radar during the antenna beam's time on target, t_{ot}. Whether all of the energy is actually utilized depends upon the radar's ability to integrate the received waveform.

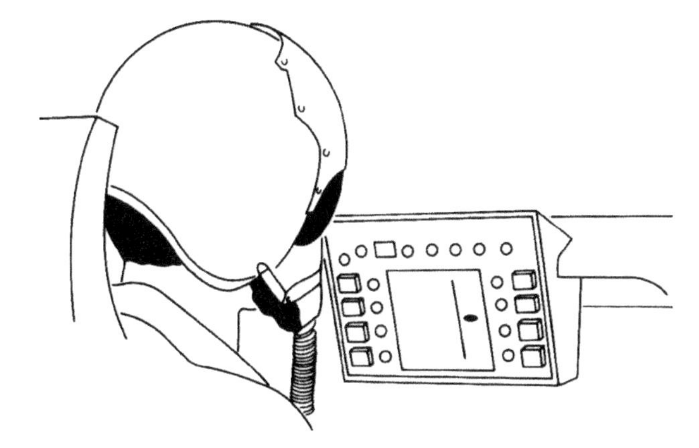

Figure 12-20. In non-Doppler radars, integration takes place on the display and in the eyes and mind of the operator.

In simple, old-fashioned non-Doppler radars, integration was performed by the display (e.g., by the phosphor that caused the image to persist on the face of the cathode ray tube and by the eyes and the mind of the operator (Fig. 12-20)). Because it takes place after detection, this integration is called *postdetection integration* (PDI).

In a Doppler radar, integration is performed primarily by the signal processor's Doppler filters before detection takes place. Provided the integration time, t_{int}, is made equal to t_{ot},[3] the above expression indicates the amplitude of the integrated target signal in the output of a filter at the end of each time on target. Whether the target will be detected depends, of course, upon the ratio of this amplitude to that of the integrated noise, discussed earlier.

3. And provided the target is centered in the passband of one of the filters and the time on target coincides exactly with an integration period.

However, to fully understand the relationship between the SNR and the maximum detection range, we must know a little more about the actual detection process.

12.4 Detection Process

Assume that a small target is approaching a searching Doppler radar from a very great distance. Initially the target echoes are extremely weak—so weak they are completely lost in the background noise.

On first thought, one might suppose the echoes could be pulled out of the noise by increasing the gain of the receiver. But the receiver amplifies noise and echoes equally. Increasing its gain in no way alters the situation.

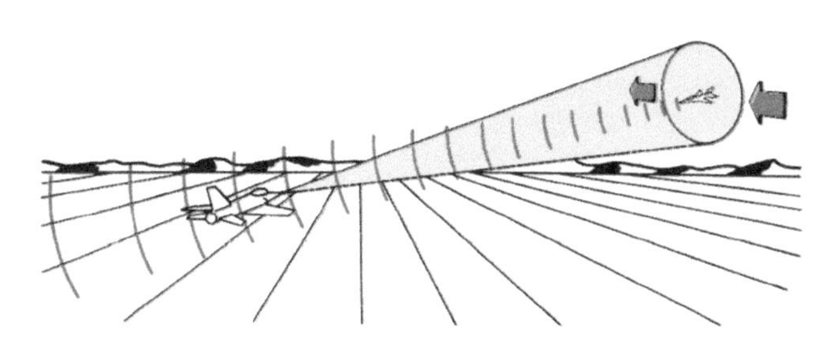

Figure 12-21. Each time the antenna beam sweeps across the target, a stream of echoes is received.

Each time the antenna beam sweeps over the target (Fig. 12-21), a stream of pulses is received. A Doppler filter in the radar's signal processor adds up the energy contained in this stream.

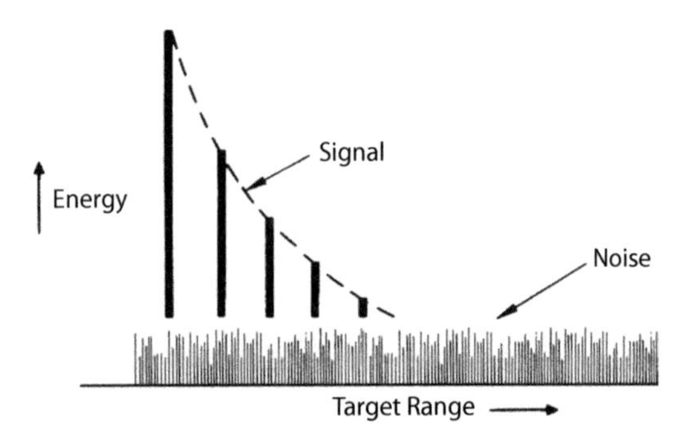

Figure 12-22. Received signal energy for successive times on target. As the range decreases, ratio of signal energy to noise energy increases.

The target signal in the output of the filter thus corresponds fairly closely to the total amount of energy received during the antenna beam's time on target. This energy is indistinguishably combined with the noise energy that has accumulated in the filter during the same period.

As the target's range decreases, the strength of the integrated signal increases. On the other hand, the mean strength of the noise remains about the same. Eventually, the signal becomes strong enough to be detected above the noise (Fig. 12-22).

In Doppler radars, detection is performed automatically. At the end of every integration period, the output of each filter is applied to a separate detector. If the integrated signal plus the accompanying noise exceeds a certain threshold, the detector concludes that a target is present, and a bright, synthetic target blip is presented on the display. Otherwise, the display remains perfectly clear (Fig. 12-23).

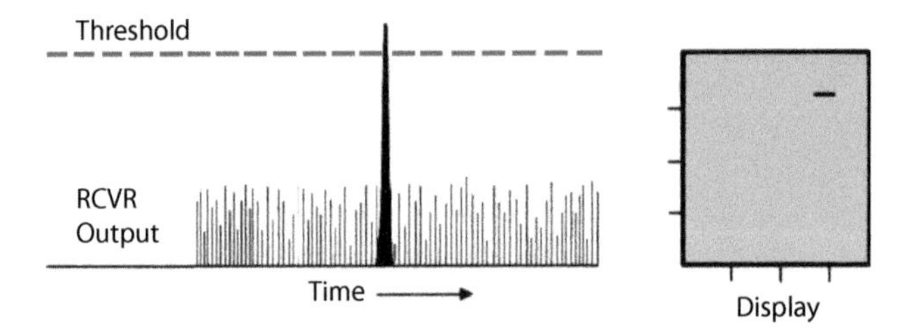

Figure 12-23. If the receiver output exceeds the detection threshold, a bright synthetic blip appears on the display.

The completely random noise alone will occasionally exceed the threshold, and the detector will falsely indicate that a target has been detected (Fig. 12-24). This is called a *false alarm*. The chance that a false alarm will occur is called the *false-alarm probability*. The higher the detection threshold relative to the mean level of the noise energy, the lower the false-alarm probability will be, and vice versa.

Clearly the setting of the threshold is crucial. If it is too high (Fig. 12-25), detectable targets may go undetected. If it is too low, too many false alarms will occur. The optimum setting is slightly higher than the mean level of the noise, just enough to keep the false-alarm probability from exceeding an acceptable value. The mean level of the noise, as well as the system gain, may vary over a wide range. Consequently, the output of the radar's Doppler filters must be continuously monitored to maintain the optimum threshold setting.

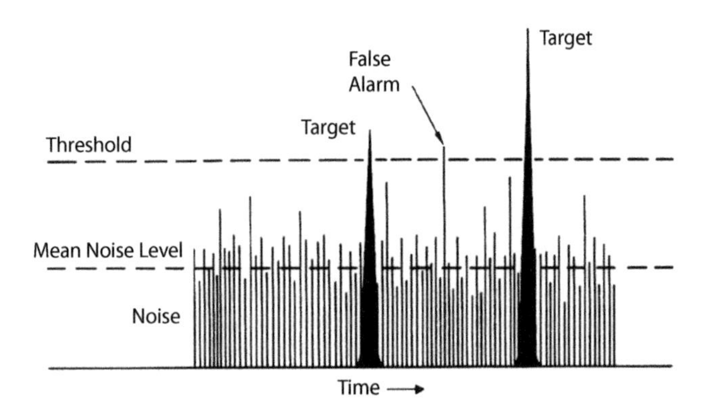

Figure 12-24. The higher the threshold above the mean level of the noise, the lower the probability of a spike of noise crossing it and producing a false alarm.

Generally the threshold for each detector is individually set on the basis of both the probable noise level in the filter whose output is being detected (the "local" noise level) and the average noise level in all of the filters (the "global" noise level). Typically the local level is determined by averaging the outputs

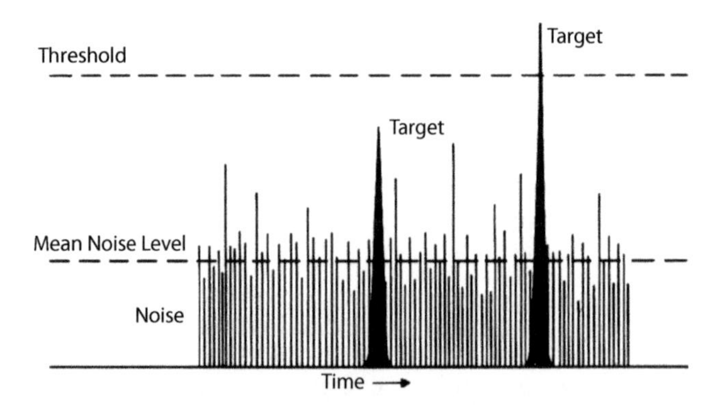

Figure 12-25. If the threshold is too high, some detectable targets may go undetected.

of a group (ensemble) of filters on either side of the one in question. Since most of these outputs will be due to noise, the average can be assumed to approximate the probable noise level in the bracketed filter.

The global noise level is determined by establishing a second noise-detection threshold for every filter. This threshold is set far enough below the target-detection threshold so that in aggregate vastly more threshold crossings are made by noise spikes than by target echoes. By continually counting these crossings and statistically adjusting the count for the difference between the two thresholds, the false-alarm rate for the entire system can be determined.

Exactly how the local ensemble of filters is selected and how the average for the ensemble is weighted in comparison to the system false-alarm rate varies from system to system and mode to mode. As nearly as possible, however, the thresholds are set so as to maintain the false-alarm rate for each detector at the optimum value. If the rate is too high, the thresholds are raised; if it is too low, the thresholds are lowered. For this reason, the automatic detectors are called *constant false-alarm rate* (CFAR) *detectors*.

Regardless of how close to optimum it is, the setting of the target detection threshold, relative to the mean level of the noise, establishes the minimum value of integrated signal energy, s_{det}, that, on average, is required for target detection (Fig. 12-26). Bear in mind, however, that because of the randomness of the noise energy about its mean value, the signal plus the accompanying noise will sometimes exceed the threshold even when the signal energy is less than s_{det}. Likewise, at other times it will fail to reach the threshold, even when the signal energy is greater than s_{det}. Nevertheless, the range at which a given target's integrated signal becomes equal to s_{det} can be considered the maximum detection range (under the existing operating conditions) for that particular target.

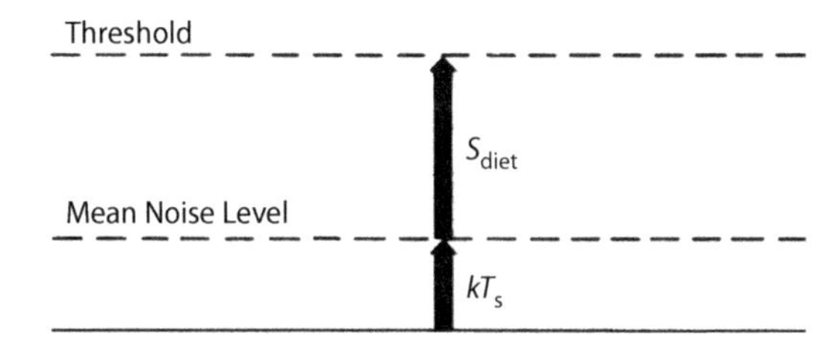

Figure 12-26. Setting the threshold relative to the mean noise level establishes the minimum value of the integrated signal required for detection, s_{det}.

12.5 Integration and Its Effect on Detection Range

Although implicit in the expression for signal energy (page ***), the immense importance of integration in pulling the weak echoes of distant targets out of noise is often overlooked. One can gain valuable insight into this important process by performing a simple experiment.

Experimental Setup. To see how noise energy and signal energy integrate in a narrow bandpass (Doppler) filter we set up a rudimentary radar to look for a test target at a given range and angle. After training the antenna in the expected target's direction, we turn on the receiver for a fixed period of time. Meanwhile, at that point in each interpulse period when return will be received from the expected target's range, we momentarily close a switch (range gate), thus passing a slice of the receiver's IF output—one pulsewidth wide—to a narrowband filter (Fig. 12-27). The filter is tuned to the target's Doppler frequency.

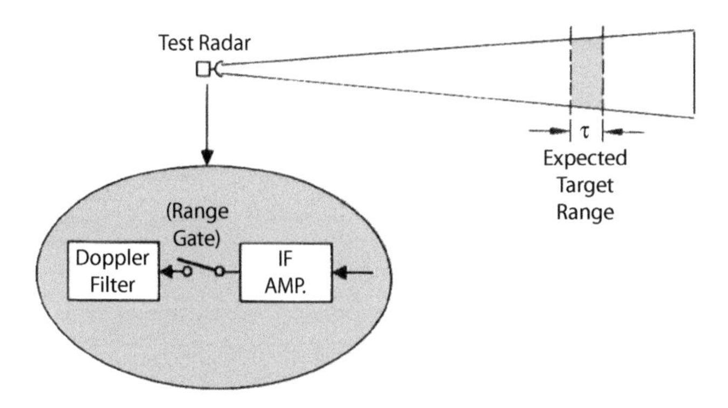

Figure 12-27. A rudimentary radar is set up to look for a target at a given range. A switch closes at the point in the interpulse period when a return is received from the expected target's range.

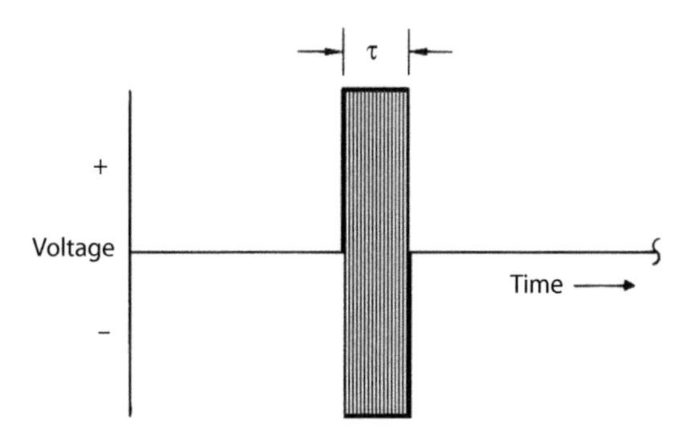

Figure 12-28. After passing through the IF amplifier and being sliced very thin, each noise pulse looks much like a target return.

Noise Alone, Single Integration Period. Initially we perform the experiment with no target present. When the range gate is closed, all the filter receives is a pulse of noise energy.

In this radar, as in most, the passband of the receiver's IF amplifier is just wide enough to pass the bulk of the energy in a target echo (matched filter design). Consequently, after passing through the IF amplifier and being sliced into narrow pulses, the noise looks much like target return (Fig. 12-28). The principal difference as seen by the Doppler filter is this: whereas the phase of the pulses received from a target is constant from pulse to pulse, the phase as well as the amplitude of the noise pulses varies randomly from pulse to pulse. We can see the variation in phase most clearly if we represent the pulses with phasors (Fig. 12-29).

Now, the role of the filter is to further narrow the receiver passband by integrating the energy of successive pulses. What the filter does, in effect, is add up the phasors. In the case of noise, because of the randomness of phase, the pulses largely cancel.[4]

At the end of the integration period, the magnitude of the sum—the integrated noise, $\bar{N}$—is little different than the amplitude of a single noise pulse and only a fraction of the sum of the amplitudes of the individual pulses. The integration period, we will assume, corresponds to a single time on target.

4. Actually, the phase of a target's returns varies from pulse to pulse in proportion to the target's Doppler frequency. But as seen by a filter tuned to this frequency, the phase is very nearly constant.

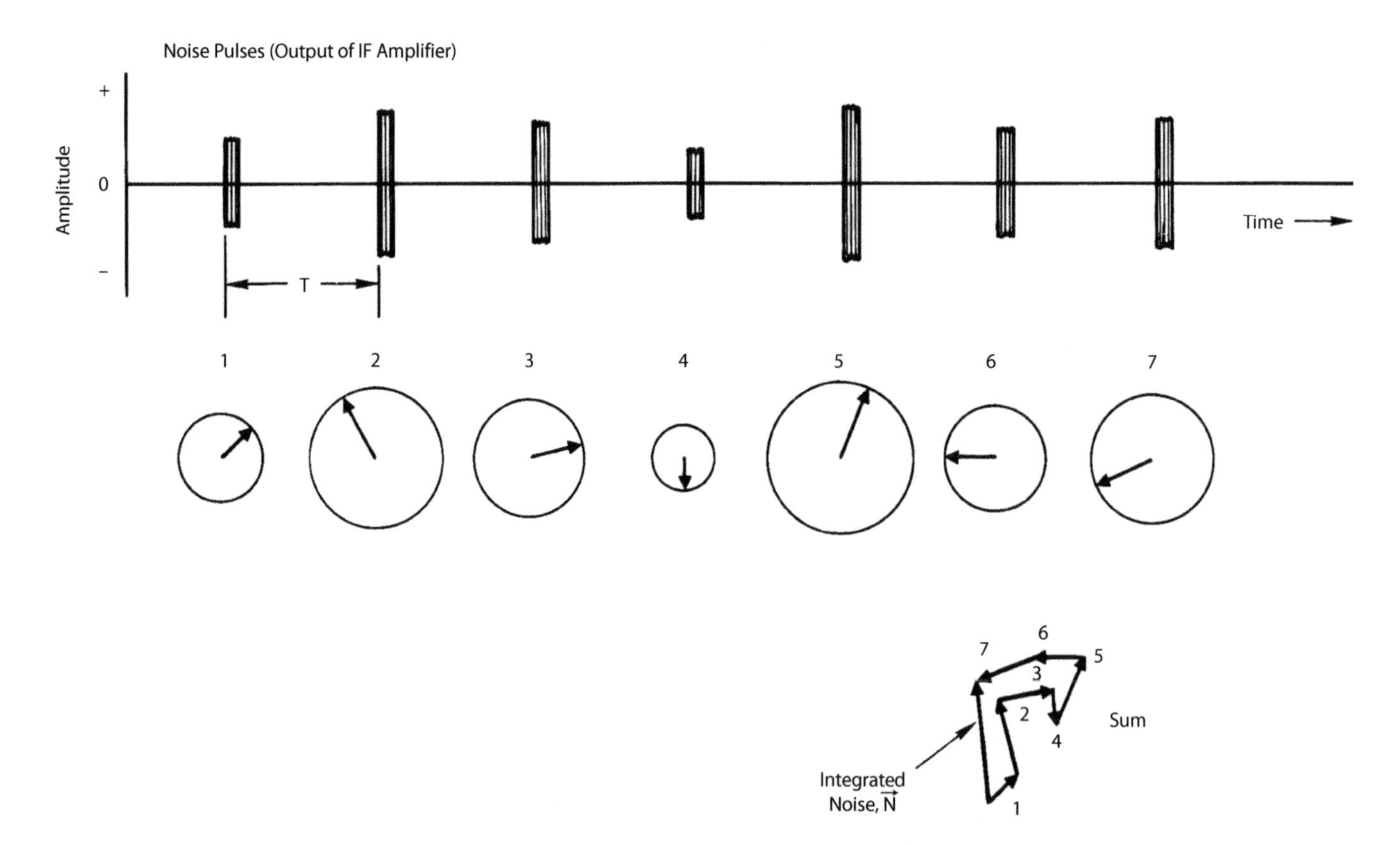

Figure 12-29. This diagram depicts the phasor representation of noise pulses applied to Doppler filter. Because of phase variation, the amplitude of integrated noise is only a fraction of the sum of amplitudes of the individual pulses.

Noise Alone, Successive Times on Target. We repeat the experiment a great many times—each repetition corresponding to a separate time on target. As expected, because of the randomness of the noise, the magnitude and phase of the energy that accumulates in the filter, $\bar{N}$, vary widely from one time on target to the next.

At the end of each time on target, the magnitude of the accumulated energy is "detected" (Fig. 12-30), that is, a voltage (video signal) proportional to the magnitude is produced. Incidentally, since the integration takes place before this detection, the integration is called *predetection integration*.

The video outputs for successive times on target are plotted in Fig. 12-31.

As you can see, over a number of integrated periods, the magnitude of the integrated noise varies randomly about a mean value. Although not illustrated here, the variation in phase is equally random.

Target Signal Only. We repeat the experiment, this time with the target present but (through some magic) with the noise absent. Now each time the range gate is closed, the filter receives a pulse of energy from the target. Unlike the noise pulses, these all have the same phase.[5] When integrated by the filter, they add constructively. At the end of each integration period their sum (Fig. 12-32, top)—the magnitude of the

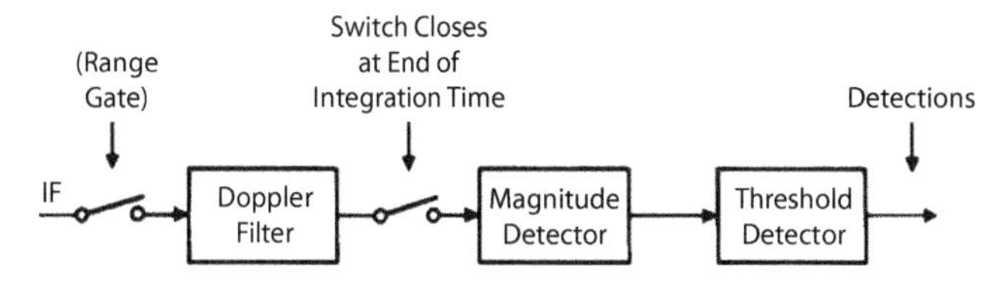

Figure 12-30. At the end of each integration period (time on target), the amplitude of the energy accumulated in the filter is detected and applied to a threshold detector.

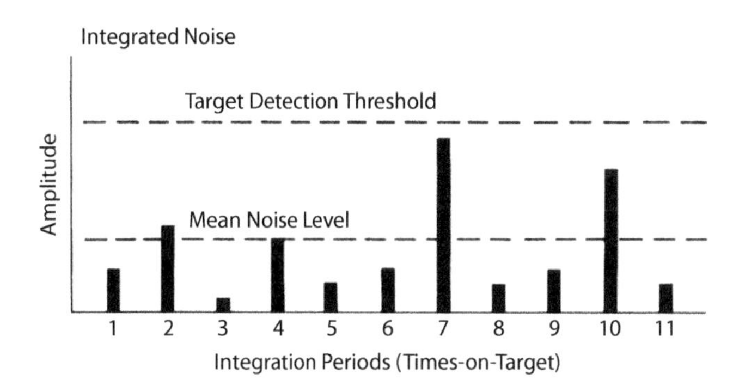

Figure 12-31. Outputs of the amplitude detector of Figure 12-30 are shown at the end of successive times-on-target.

5. During the very short time interval discussed here.

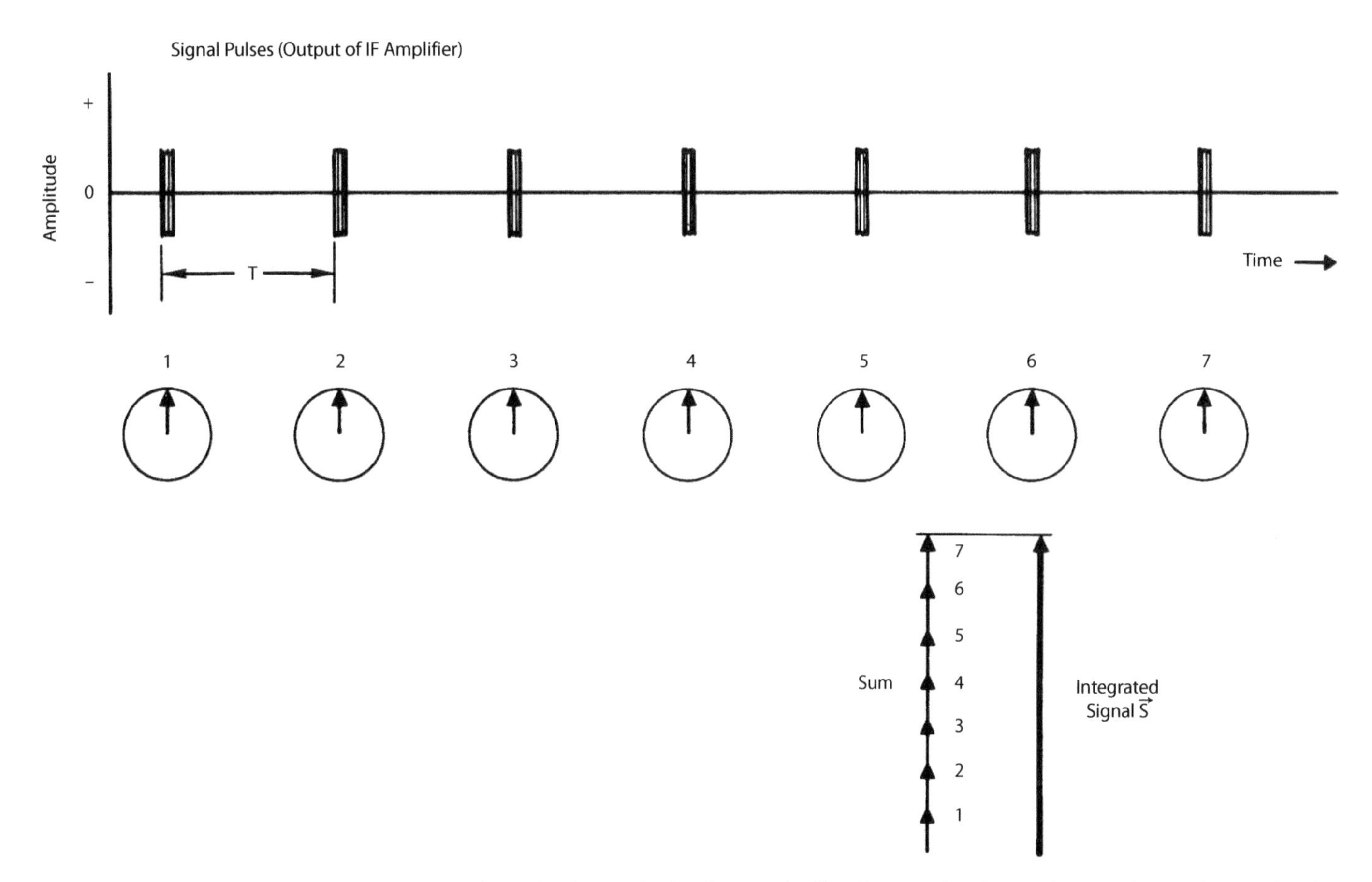

Figure 12-32. Here is the phasor representation of signal pulses applied to the Doppler filter. Because the phase is the same from pulse to pulse, the amplitude of the integrated signal is many times the amplitude of the individual pulses.

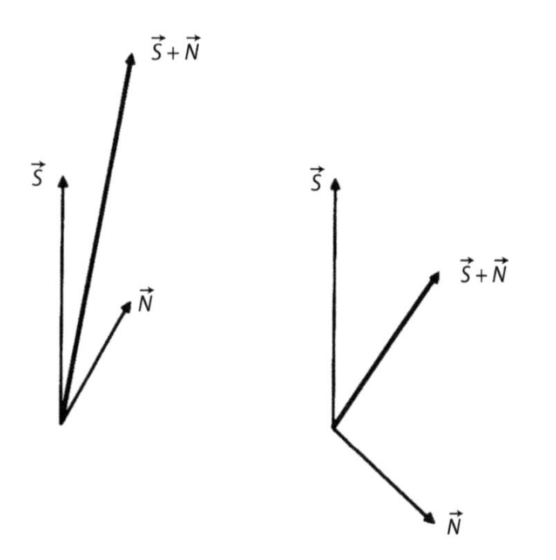

Figure 12-33. The amplitude of an integrated signal plus noise varies widely, depending on the amplitudes and phases of $\bar{S}$ and $\bar{N}$.

integrated signal, $\bar{S}$—very nearly equals the sum of the amplitudes of the individual pulses.

How Signal and Noise Combine. Finally, we repeat the experiment several times with both target signal and noise present. Although they are indistinguishably mixed, and thus integrate simultaneously, we can visualize the result more clearly if we think of the signal and noise as being integrated separately and of their sums, $\bar{S}$ and $\bar{N}$, being vectorially added together at the end of the time on target. Of course, the magnitude of the vector sum depends not only upon the magnitudes of $\bar{S}$ and $\bar{N}$, but upon the phase angle between them (Fig. 12-33). If the noise is in phase with the signal, the two vectors will combine constructively; if the noise is 180° out of phase, they will combine destructively; and there are myriad possible combinations in between. For any one time on target, therefore, the magnitude of the energy that accumulates in the filter equals the magnitude of the integrated signal, $\bar{S}$, plus or minus some fraction of the magnitude of the integrated noise, $\bar{N}$.

Improvement in the SNR. How predetection integration improves the SNR should now be fairly clear. Whereas the noise energy that accumulates in the filter may vary widely from one integration period to another, the mean level of the noise energy is essentially independent of the integration time. The integrated signal energy (target return), on the other hand, increases in direct proportion to the integration time. Thus by increasing the integration time, the SNR can be increased significantly.

An individual target echo, for example, may contain only 1/1000 as much energy as an individual noise pulse, yet after 10,000 pulses have been integrated, the signal may be considerably greater than the noise.

Indeed, the SNR improvement achievable through predetection integration is limited only by (1) the length of the time on target, t_{ot}; (2) the maximum practical length of the integration time, t_{int}, if that is less than t_{ot}; or (3) the length of time over which the target's Doppler frequency remains close enough to the same value for the target echoes to be correlated by the filter (Fig. 12-34). The greater the improvement in the SNR, of course, the weaker the target echoes can be and still be detected, hence the greater the detection range.

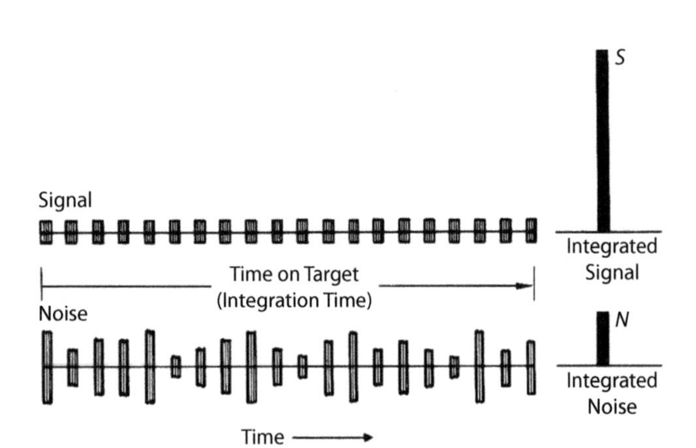

Figure 12-34. The improvement in the SNR is ultimately limited only by the time on target, provided target echoes remain correlated.

12.6 Postdetection Integration

Sometimes, the maximum practical integration time is a good deal less than the time on target. Take, for example, a situation where the Doppler frequencies of expected targets may be subject to rapid change. Since the width of the filter passband is inversely proportional to the integration time (bandwidth ≈ 1/t_{int}), making t_{int} as long as t_{ot} could narrow the passband to the point where the signal may very well move out of it long before the time on target ends (Fig. 12-35).

In such instances, rather than lose any of the signal, the integration time of the Doppler filter is made short enough to provide the required bandwidth, and integration and video detection are repeated throughout the time on target (Fig. 12-36). The video outputs for successive integration periods are then added together (integrated) and their sum is applied to the threshold detector. This second integration process is fundamentally the same as that employed in non-Doppler radars. Since it takes place after video detection, it is called PDI.

Once the output of a Doppler filter (or the output of an IF amplifier in a non-Doppler radar) has been converted to a video signal of single polarity, noise will no longer cancel when integrated. Rather, it will build up throughout the integration time in exactly the same way as the signal. Consequently, with PDI the mean SNR cannot be increased. Nevertheless, an equivalent improvement in detection sensitivity may be achieved. To see why, we must look a little more closely at PDI.

Actually, PDI is nothing more than averaging. It has the same effect as passing the video signal through a low-pass (as opposed to bandpass) filter. You can visualize this most clearly by thinking of the video signal as consisting of a constant (DC) component, the amplitude of which corresponds to the mean level of the signal, plus a fluctuating (AC) component.

The amplitude of the dc component is unaltered by the averaging, but the amplitude of the ac component is reduced. The higher the frequency of the fluctuation and the greater the integration time (i.e., the larger the number of inputs averaged), the greater the reduction will be. Averaging improves detection sensitivity in two important ways.

First, it reduces the average deviation of the integrated noise energy. Consequently, without increasing the false-alarm probability, the target detection threshold can be set closer to the mean noise level (Fig. 12-37). The integrated signal need not be as strong to cross the threshold, and the weaker echoes of more distant targets can be detected.

The second improvement averaging makes is more subtle. As we just saw, when target return is received, the integrated signal is coherently added to the integrated noise. Because of the randomness of the noise, as often as not the noise will be out of phase with the signal and so will combine with the signal destructively.

However, when the integrated signal plus noise is averaged over many integration periods, the fluctuations due to the noise tend to cancel out, leaving only the signal. The possibility of missing an otherwise detectable target because of the destructive combination of it with noise is thus greatly reduced.

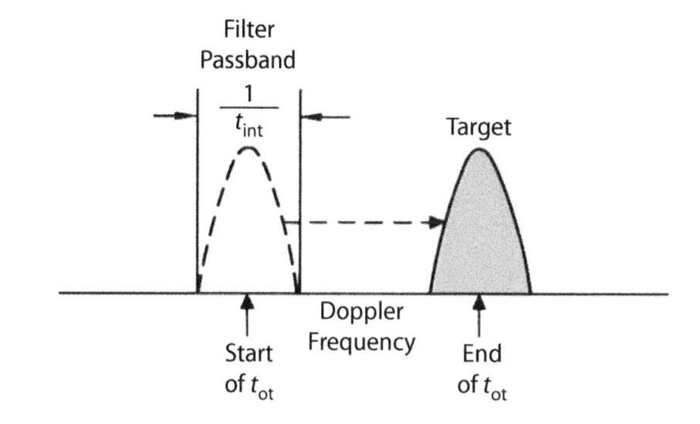

Figure 12-35. Note the situation in which a target's Doppler frequency changes radically during time-on-target, t_{ot}. If the filter integration time, t_{int}, is made equal to t_{ot}, the target will move out of passband before integration is finished.

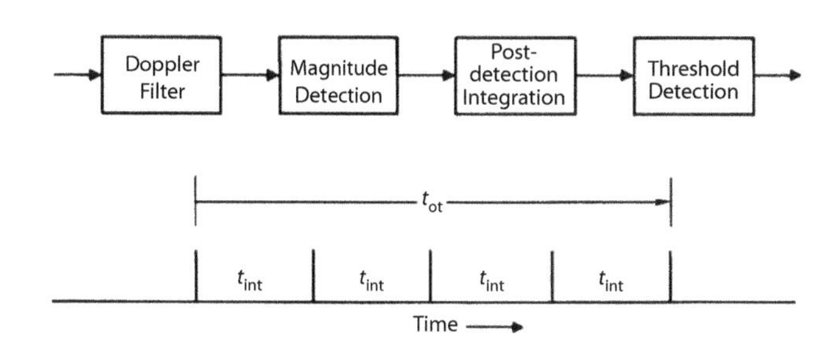

Figure 12-36. The problem is solved by dividing t_{ot} into a number of integration periods short enough to provide adequate Doppler bandwidth and adding up the filter outputs for the entire time-on-target.

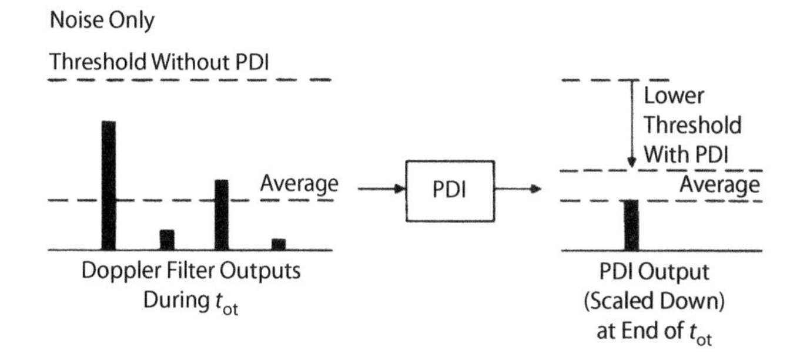

Figure 12-37. By averaging the noise outputs of a Doppler filter during the time on target, PDI enables the target detection threshold to be set much lower, without increasing the false-alarm probability.

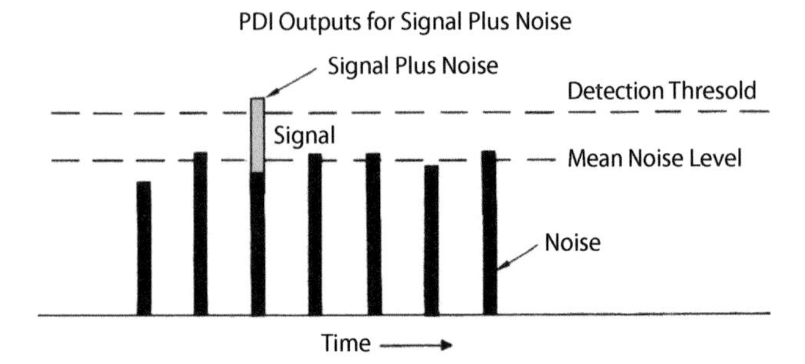

Figure 12-38. The two effects of PDI allow a signal to be detected even when the mean SNR is less than one.

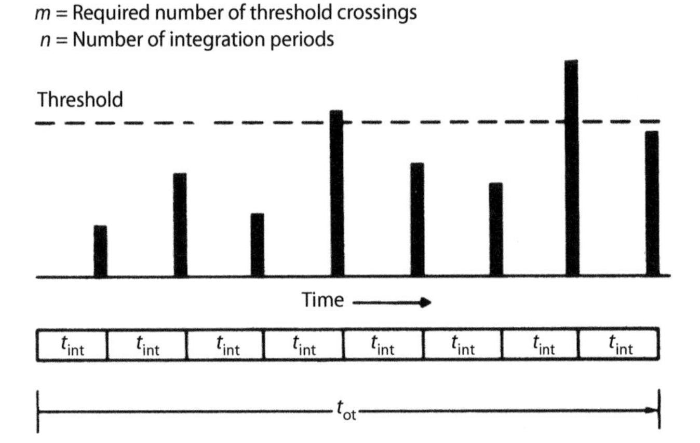

Figure 12-39. Sometimes the equivalent of PDI is obtained by requiring m out of n threshold crossing in a time on target for a detection. Here $m = 2$ and $n = 8$.

Together, these two effects of PDI can substantially reduce the SNR required for detection. As illustrated in Figure 12-38, the fluctuations in the noise and in the signal plus noise can, in the extreme, be reduced to the point where a signal can be detected when the mean SNR is substantially less than one.

Sometimes the equivalent of PDI is approximated by using a so-called m out of n detection criterion. If the time on target spans n predetection integration periods, rather than requiring only one threshold crossing per time on target as a condition for detection, the signal processor requires m crossings (Fig. 12-39). The chance of isolated noise spikes producing false alarms is thereby reduced. The detection threshold can be lowered without increasing the false-alarm probability, and more distant targets can be detected.

12.7 Summary

Since radio waves of the frequencies used by airborne radars travel essentially in straight lines, a target must be within the line of sight to be detected. Range may be further limited by clutter or man-made interference. Ultimately it is determined by the SNR.

The principal source of noise is thermal agitation in the input stages of the receiver. The noise energy is commonly expressed in terms of a figure of merit, F, relating it to an approximation of the external noise provided by thermal agitation in a resistor connected across the receiver's input terminals. In the case of low-noise receivers, external noise sources become more significant, and noise is expressed in terms of an equivalent "system" noise temperature.

How much energy is received from a target depends primarily upon (1) the radar's average transmitted power, antenna gain, and effective antenna area; (2) the time on target; and (3) the target's range, R, and RCS, σ, a factor that accounts for the size, reflectivity, and directivity of the target.

Most radars integrate the return received as the antenna scans across a target. If performed before video detection (predetection integration), the integration increases the SNR in direct proportion to the integration time. If performed after video detection (PDI), the integration accomplishes two things: (1) it averages out the fluctuations in the noise, thereby reducing its peaks, and (2) it averages out the destructive combination of the noise with the signal, thereby reducing the possibility of missing an otherwise detectable target.

For a target to be detected, the integrated signal must exceed a threshold set high enough to keep the probability of noise crossings acceptably low. In Doppler radars, to maintain a constant, optimum false-alarm rate (CFAR), the threshold setting of the magnitude detector for each Doppler filter's output is based on the mean noise level in the outputs of an ensemble of adjacent filters, as well as on measurement of the mean noise level in the outputs of all the filters.

12.8 Some Relationships to Keep in Mind

- Mean noise power = kT_0BF or kT_sB, Watts
- F = Receiver noise figure
- T_0 = Noise temperature (nominally 290 K)
- k = Boltzmann's constant = 1.38×10^{-23} W/K/Hz
- B = Receiver bandwidth, Hz
- $kT_0 = -174$ dBm/Hz
- T_s = System noise temperature (including internal + external noise)
- Mean noise energy = kT_sBt_n, where t_n = duration of the noise
- Mean noise energy of a matched filter = $\frac{k\,T_s\,t_n}{\tau}$
- Mean noise energy of a Doppler radar = kT_s
- Signal energy = $\frac{P_{avg}G^2\lambda^2\sigma}{(4\pi)^3R^4}$,

where

P_{avg} = average transmitted power, watts

G = antenna gain

σ = radar cross section of the target

t_{ot} = time on target

R = range

Further Reading

J. V. DiFranco and W. L. Rubin, *Radar Detection*, SciTech-IET, 2004.

L. V. Blake, *Radar Range Performance Analysis*, Artech House, 1986.

H. L. Van Trees, *Detection, Estimation and Modulation Theory, Part III, Radar-Sonar Signal Processing and Gaussian Signals in Noise*, John Wiley & Sons, Inc., 2001.

D. K. Barton, *Radar Equations for Modern Radar*, Artech House, 2013.

H. L. Van Trees and K. L. Bell, *Detection, Estimation, and Modulation Theory, Second Edition, Part 1—Detection, Estimation and Filtering Theory*, John Wiley & Sons, Inc., 2013.

Test your understanding

1. The Y-factor method is used to measure the noise figure of a receiver. The ENR of the noise source is 15 dB and the ratio of the noise output of the amplifier when the noise source is on compared with that when it is off is measured to be 4 dB. Calculate the noise figure of the amplifier.
2. Show from first principles that the RCS at normal incidence of a square metal plate of side a is given by

$$\sigma = \frac{4\pi a^4}{\lambda^2},$$

where λ is the wavelength.

3. A radar transmits a peak power pulse of 20 kW via an antenna of gain 20 dBi. What is the power density incident on a target at a range of 20 km?
4. If the RCS of the target is 1 m^2 and the time on target is 1 ms, what is the energy incident on the target?
5. An airborne early warning radar uses a transmit power of 1 MW. The radar frequency is 3 GHz and its antenna gain is 38 dBi. The receiver noise figure is 6 dB, total losses are 10 dB, and the effective receiver bandwidth is 50 kHz. What is the SNR, on the basis of a single pulse, of the echo from a target of RCS 1 m^2 at a range of 100 km?

North American Aviation XB-70 Valkyrie (1964)

While never entering production, the XB-70 prototype nuclear deep-penetration bomber is one of the most fascinating projects ever in development. The name "Valkyrie" came from the US Air Force "Name the B-70" contest conducted in 1958. A large six-engined aircraft, it was able to fly above Mach 3 at an altitude of 70,000 ft and avoid interceptors, which were the only effective bomber deterrent at the time. Multiple factors, including the development of high-altitude surface-to-air missiles, budget overruns, the introduction of intercontinental ballistic missiles, and politics, ensured the Valkyrie would never enter service.

13

$$R_0 \propto \sqrt[4]{\frac{P_{avg} A_e^2 \sigma t_{ot}}{k T_s \lambda^2}}$$

The Range Equation for single-search scan: SNR = 1

The Range Equation: What It Does and Doesn't Tell Us

In Chapter 12, we learned that within the line of sight, in the absence of interference and competing ground return, detection range is ultimately determined by the ratio of the energy received from a target—the signal—to the energy of the background noise. We identified the principal factors that define the signal and noise energies and became acquainted with the detection process.

Building on that knowledge, in this chapter we will write a general equation for maximum detection range and analyze it to see how the individual factors we have identified influence the range. We will then narrow down to the special case of volume search. Finally, we will consider the statistical variation in detection range and see how it is accounted for.

13.1 General Range Equation

As we saw in the preceding chapter, when the radar antenna is trained on a target (Fig. 13-1), the energy received from the target during any one integration time is roughly

$$\text{Received signal energy} \cong \frac{P_{avg} G \sigma A_e t_{int}}{(4\pi)^2 R^4}$$

where

P_{avg} = average transmitted power

G = antenna gain

σ = radar cross section of target

A_e = effective antenna area

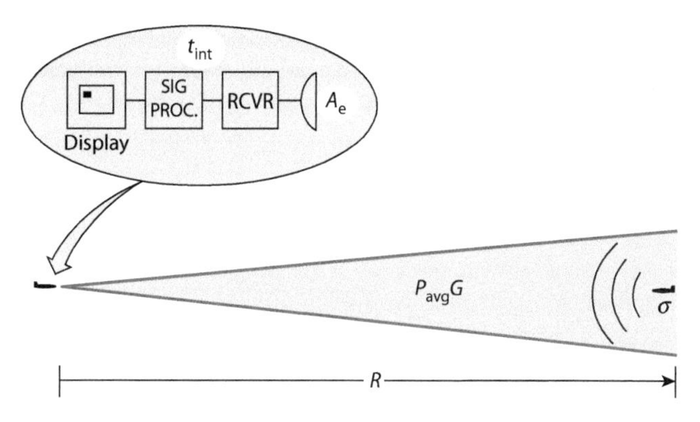

Figure 13-1. The factors shown here determine the received signal energy.

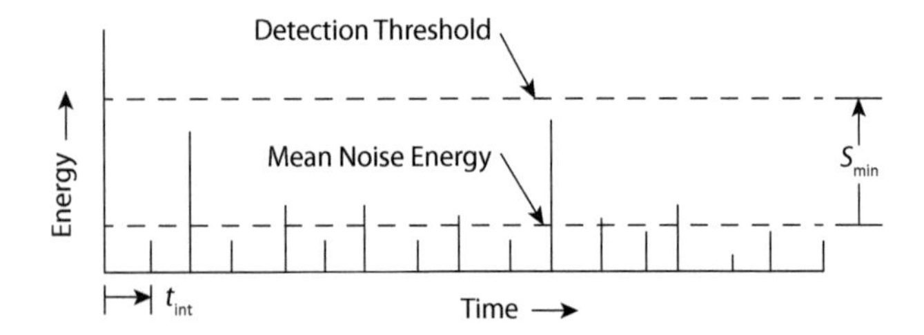

Figure 13-2. In the integrated noise energy at end of successive integration times, t_{int}, shown here, on average, for a target to be detected, the integrated signal energy must be equal to or greater than S_{min}.

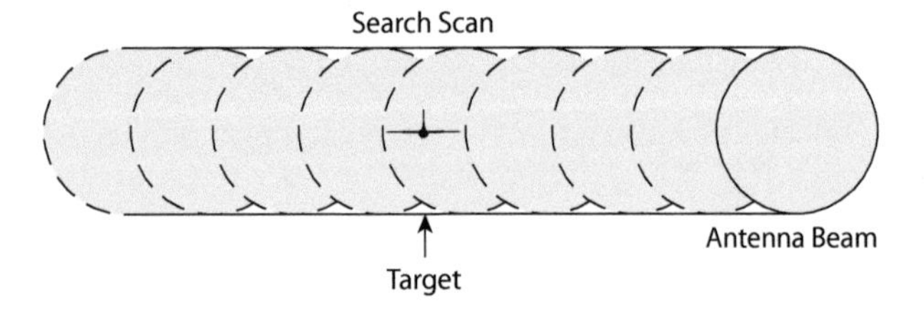

Figure 13-3. The simple range equation can be applied to search by pretending that the transmitted energy is uniformly distributed over the cross section of the antenna beam and the target is centered in the beam's path.

t_{int} = integration time

R = range

For the target to be detected, this energy plus the accompanying noise energy must exceed a certain threshold value. This value is set just high enough above the mean noise level to reduce the probability of noise peaks crossing the threshold—false alarms—to an acceptably low value.

On average, the minimum energy that a signal must have to cross the detection threshold is the difference between the detection threshold and the mean level of the noise, commonly represented by the term S_{min} (Fig. 13-2).

Assuming perfect integration, the maximum range at which a given target will be detected is the range at which the received signal energy becomes equal to S_{min}. Setting the expression for signal energy equal to S_{min} and solving for range, therefore, yields a simple equation for the maximum detection range.

$$R_{max} \cong \sqrt[4]{\frac{P_{avg} G \sigma A_e t_{\mathrm{int}}}{(4\pi)^2 S_{min}}}$$

(Antenna trained on target)

As it stands, the equation applies only when the antenna is continuously trained on the target and the target is in the center of the mainlobe, that is, when a target is being spotlighted. (Bear in mind that though the antenna may be continuously trained on the target, t_{int} is limited to the period of time that the phase of the target signal remains correlated.)

In search, the maximum integration time is limited to the time the antenna takes to sweep across the target: the time-on-target, t_{ot}. Moreover, the beam is actually centered on the target only for an instant, if at all. We can eliminate the first limitation simply by replacing t_{int} with t_{ot}. Temporarily, at least, we can get around the second limitation by pretending that the antenna gain is the same over the entire solid angle encompassed by the mainlobe and that, for the particular scan being used, the target is centered in the beam's path (Fig. 13-3).

Under these conditions, the equation gives the maximum detection range for a single search scan.

$$R_{max} \cong \sqrt[4]{\frac{P_{avg} G \sigma A_e t_{ot}}{(4\pi)^2 S_{min}}}$$

(Single scan of antenna)

Incidentally, if we replace t_{ot} with the pulse width, τ, and P_{avg} with the peak power, P, the equation gives the range for single-pulse detection.

$$R_{max} \cong \sqrt[4]{\frac{P G \sigma A_e t}{(4\pi)^2 S_{min}}}$$

(Single pulse: non-Doppler radar)

Assuming that postdetection integration is accounted for separately, this form of the equation applies to non-Doppler radars (Fig. 13-4).

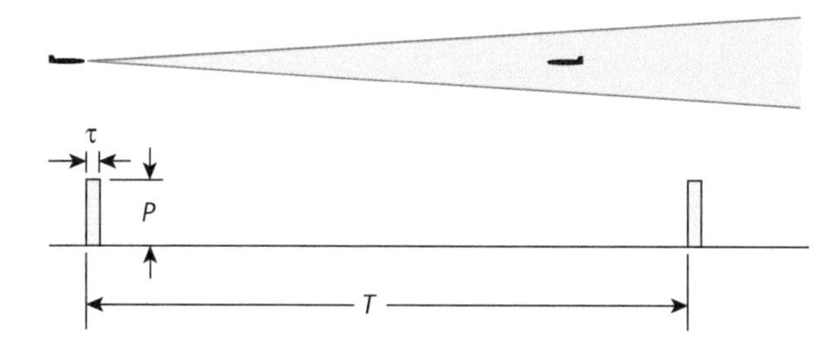

Figure 13-4. The range equation can be applied to non-Doppler radars by substituting the pulse width, τ, for t_{int} and the peak power P, for P_{avg}. Postdetection integration must be accounted for separately.

Omissions. Regardless of which of these forms we use, the equation is incomplete. Among the more obvious omissions are as follows:

- Absorption and scattering in the atmosphere (Fig. 13-5)
- Reduction in signal energy due to the target not necessarily being centered in the path of the scanning antenna beam (called *elevation beamshape loss*)
- The further reduction in signal energy as the beam sweeps across the target (Fig. 13-6) due to the fall-off in two-way antenna gain at angles off beam center (called *azimuth beamshape loss*)
- Losses due to imperfect matching, i.e. some noise being unnecessarily passed or some signal energy being rejected (Fig. 13-7)
- Loss due to the target not necessarily being centered in a Doppler filter
- Degradation of signal-to-noise ratio (SNR) due to imperfect integration of the target return
- Effects of system degradation in the field

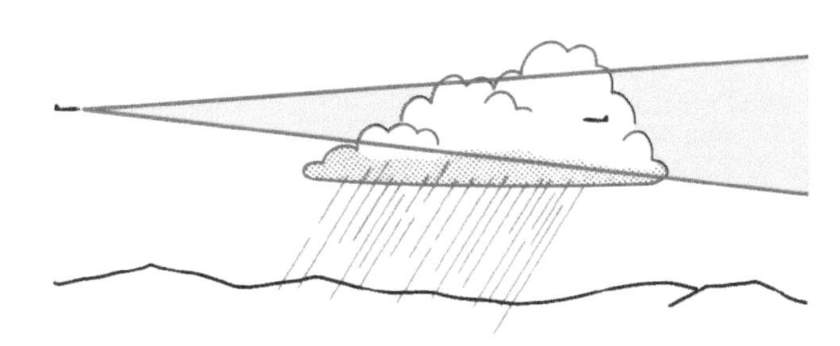

Figure 13-5. One of the many important losses not accounted for by the simple range equation is atmospheric attenuation.

Nevertheless, the equation illustrates the relative contributions of what we have seen to be some of the more fundamental factors.[1]

A More Revealing Form of the Equation. The contribution of a couple of the factors represented by terms in the range equation can be seen more easily if we modify it slightly. First, since S_{min} is related in a fairly complex way to the mean noise energy, kT_s, if we back off from solving for the *maximum* detection range and solve merely for the range at which the integrated SNR is one, we can substitute kT_s, directly for S_{min} (Fig. 13-8). Second, since antenna gain is proportional to effective antenna area divided by wavelength squared ($G \propto A_e/\lambda^2$), we can consolidate terms relating to the antenna.

1. All omitted factors that reduce the SNR are accounted for by including a loss factor, L, in the denominator, where $L \geq 1$.

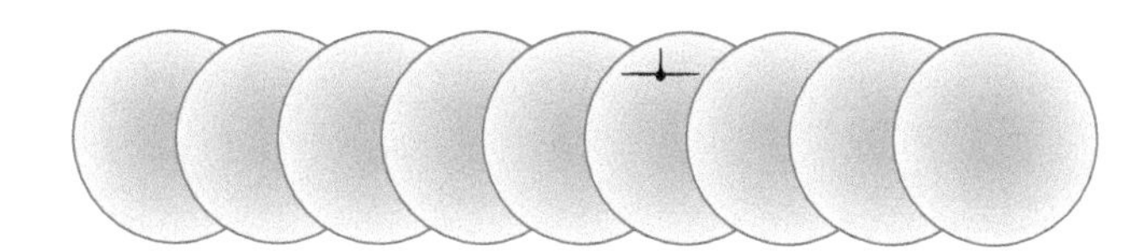

Figure 13-6. Other factors not directly accounted for include the possibility of the target not being centered in the beam's path and the fall off in two-way gain of the antenna at angles off beam center.

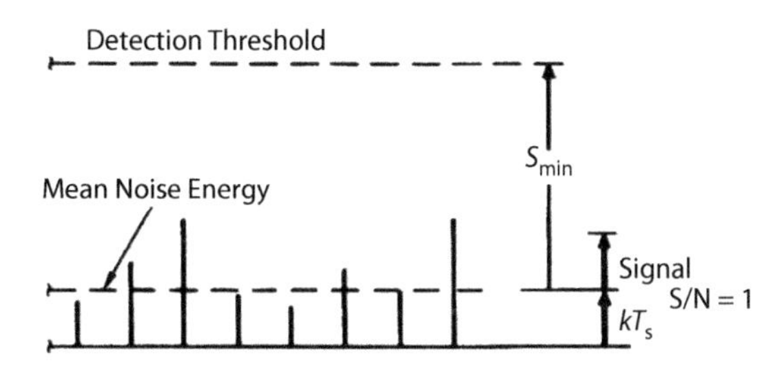

Figure 13-8. Since the detection threshold is related to mean noise level in a complex way, detection range can be expressed in terms of noise energy most simply by solving for the range at which the integrated signal-to-noise ratio is one.

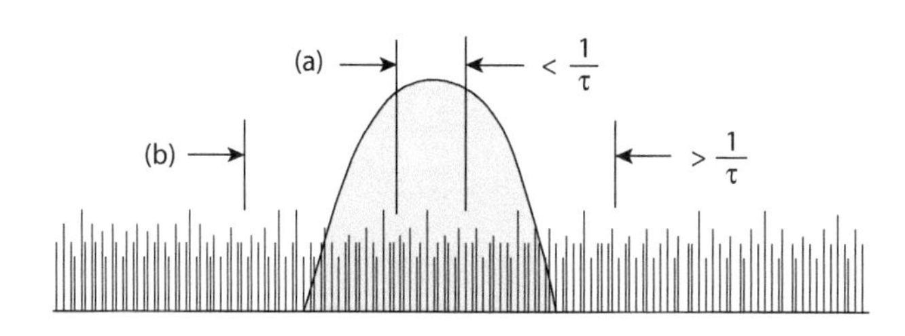

Figure 13-7. Pictured here is the mismatch between intermediate frequency and filter: (a) some signal being rejected that is stronger than accompanying noise; (b) some noise being passed that is stronger than accompanying signal.

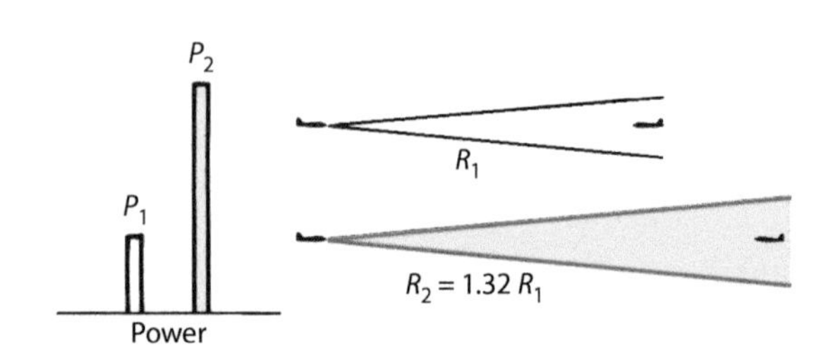

Figure 13-9. Tripling transmitter power would increase detection range by only 32 percent.

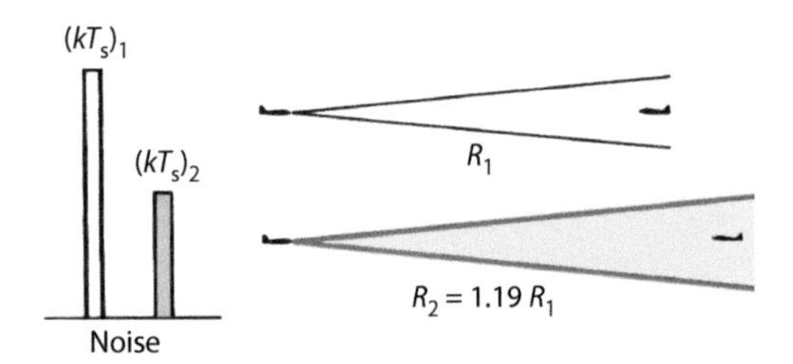

Figure 13-10. Decreasing system noise has the same effect on detection range as increasing power by the same factor.

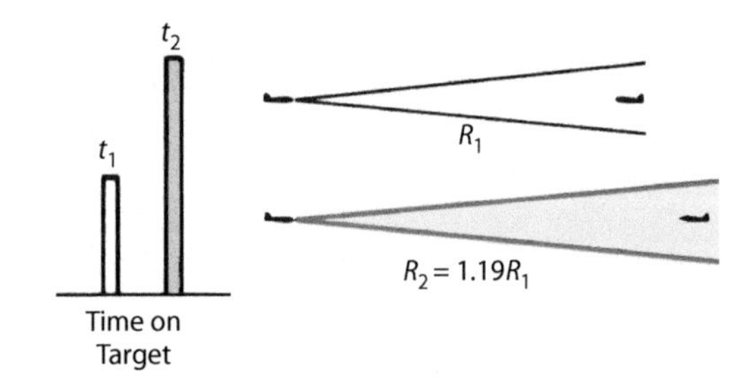

Figure 13-11. Doubling time-on-target would have the same effect as doubling transmitter power.

With these changes, the equation for a single search scan becomes

$$R_0 \propto \sqrt[4]{\frac{P_{avg} A_e^2 \sigma t_{ot}}{k T_s \lambda^2}}$$

(Single search scan: SNR=1)

where R_0 is the range at which the integrated SNR is one, and λ is the wavelength.

Incomplete as it is, the range equation reveals a good deal not only about the effect of changing various parameters but also about some of the trade-offs required in designing a radar.

Average Power. The equation tells us, for example, that increasing the power of the transmitter by a given factor increases the detection range by only about the fourth root of that factor. If we were to increase the power by, say, three times (Fig. 13-9), the detection range would increase by only about 30 percent ($R_2 = R_1\sqrt[4]{3} \cong 1.32$).

Noise. At the same time, the equation tells us that *decreasing* the mean level of the background noise, kT_s, by a given factor has the same effect as *increasing* the average power by the same factor. If we could reduce the noise by 50 percent, for example, the detection range would increase by the same amount (Fig. 13-10) as if we had doubled the power, which is about 20 percent ($R_2 = R_1\sqrt[4]{2} \cong 1.19R_1$).

Time-on-Target. The equation also enables us to predict the effect of changes in time-on-target, or integration time. Suppose that by slowing down the scan we were to double the time-on-target. Provided the target return could still be integrated, this would have the same effect as doubling the power (Fig. 13-11).

Radar Cross Section. The equation further enables us to predict the differences in the ranges at which a given radar can detect targets of different sizes.

Suppose, for example, that the radar detects a target having a certain radar cross section (RCS) at a range of 40 km. Provided that the aspect and directivity remain the same, the radar should be able to detect a target having four times this RCS (Fig. 13-12) at a range of about 66 km ($R_2 = R_1\sqrt[4]{4} \cong 40 \times 1.41 \cong 66$).

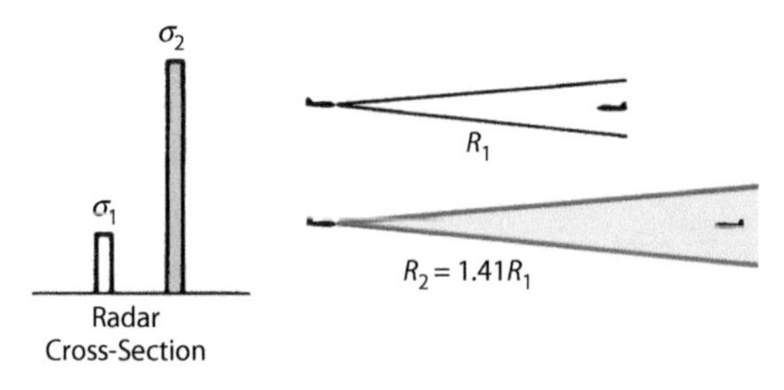

Figure 13-12. An increase in radar cross section has the same effect as a proportional increase in time-on-target.

Antenna Size. Similarly, the equation enables us to predict the effects of changes in size of the antenna. Suppose the antenna is circular and we double its diameter. Assuming that the aperture efficiency remains the same, this increase would increase A_e by 2^2, or $A_e \propto d^2\eta$. The range equation tells us that the increase in A_e (Fig. 13-13) would increase the range at which the radar might detect a given target by a factor of two, $R_2 = R_1\sqrt[4]{(2^2)^2} = 2R_1$, provided we were spotlighting the target.

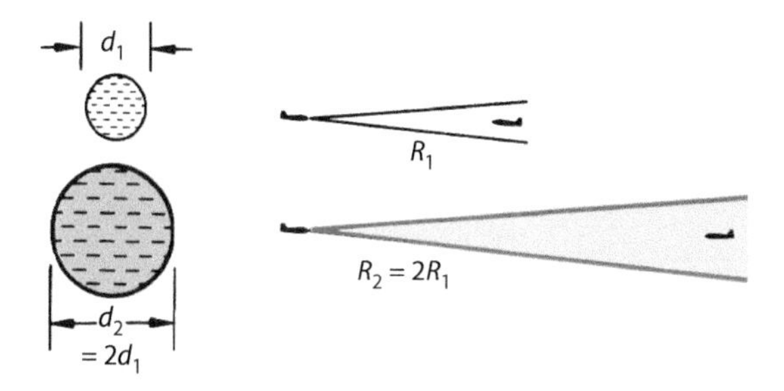

Figure 13-13. Doubling antenna diameter would double detection range, provided scan was slowed to provide the same time-on-target.

Doubling the antenna diameter, however, would cut the beamwidth in half. So if the radar was searching for targets, we would have to slow down the antenna scan to maintain the same time-on-target. If we didn't, t_{ot} would be cut in half, and the range would be increased by a factor of only about 1.68, $R_2 = R_1\sqrt[4]{16 \times 0.5} \cong 1.68R_1$.

Wavelength. Since wavelength squared is in the denominator of the equation, decreasing λ would *appear* to have the same effect on the radar's detection range as increasing the effective area of the antenna, A_e.

But here an important limitation of our simple equation shows up. Depending on what the original wavelength was and how much we decreased it, the first-order effect of decreasing λ might be offset to a considerable extent by such factors as increased atmospheric absorption, one of the factors not accounted for in the equation.

Whereas the range equation indicates that decreasing the wavelength, λ, from 3 cm to 1 cm would increase the radar's detection range by about 70 percent (i.e., $R_2 = R_1\sqrt[4]{1/(1/3)^2} \cong 1.73$), one look at a plot of atmospheric attenuation versus wavelength (Fig. 13-14) tells us that this is simply not so.

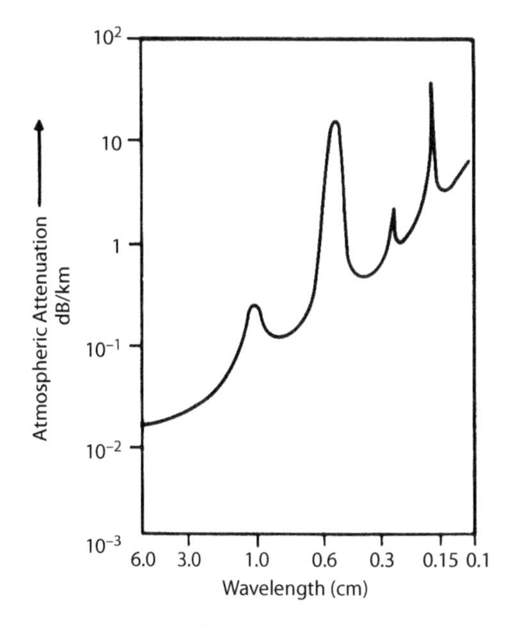

Figure 13-14. First-order effect of decreasing wavelength may be offset by such factors as increased atmospheric attenuation.

As with antenna size, decreasing λ would also decrease the beamwidth, $\theta_{3\,dB} \propto \lambda/d$. In search, therefore, to keep the increase in range from being wiped out by a reduction in t_{ot}, the scan would have to be slowed down (Fig. 13-15).

Because the range equation we have been using doesn't account for the effect of changes in wavelength and antenna size on t_{ot}, it is not as illuminating as it might be for situations where a given volume of target space must be searched for a given period of time. For volume search, therefore, a slightly different form of the equation is commonly used.

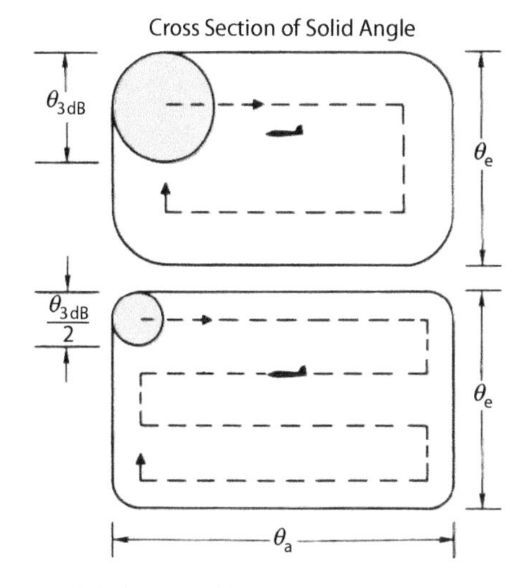

Figure 13-15. If the beamwidth is reduced, the scan must be slowed down to provide the same time-on-target, t_{ot}.

13.2 Equation for Volume Search

To tailor the range equation to volume search, the time-on-target, t_{ot}, must be expressed in terms of (1) the length of time the antenna takes to complete one frame of the search scan, and (2) the size of the solid angle subtended by that frame. The scan frame time is represented by t_f, and the solid angle by the product of the azimuth and elevation

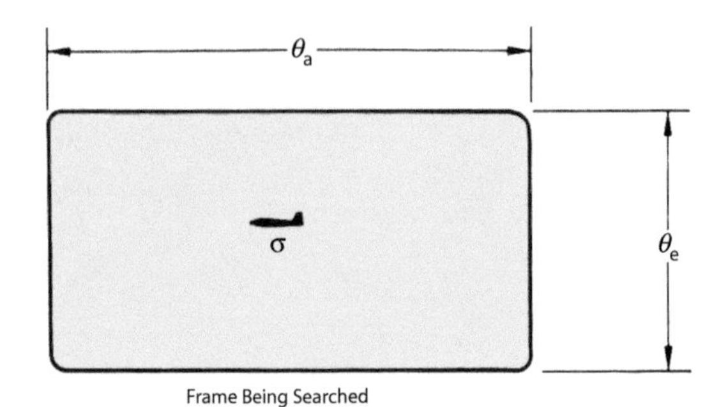

Figure 13-16. During the scan frame time, the total backscattered energy is proportional to the ratio of the radar cross section, σ, of the target to the cross sectional area, $R^2(\theta_a\theta_e)$, of the solid angle scanned at the target's range, $\sigma/(R^2\theta_a\theta_e)$.

angles spanning the frame, θ_a and θ_e, respectively. While this conversion is straightforward (see blue panel), we can get a better physical feel for the fundamental relationships involved in volume search by starting from scratch with a simplified derivation.

Simplified Derivation. The total energy radiated during any one frame time, t_f, equals $P_{avg}t_f$. Assuming that the scan spreads the energy uniformly over the entire solid angle, the fraction of the energy that is intercepted by a target and scattered back toward the radar is proportional to the ratio of (1) the target's RCS to (2) the cross sectional area of the solid angle of the search scan at the target's range (Fig. 13-16). The fraction of the backscattered energy captured by the radar antenna is proportional to A_e (Fig. 13-17).

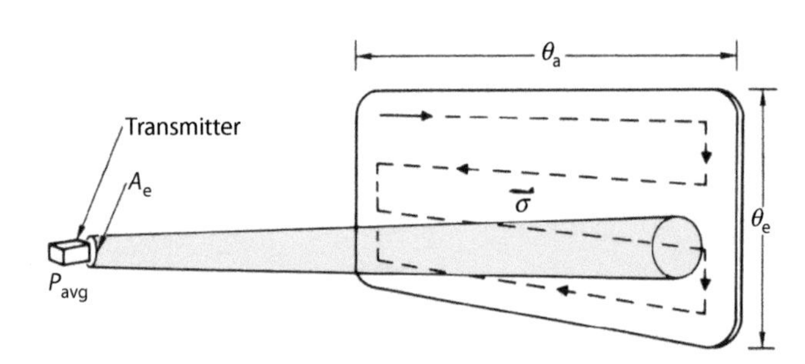

Figure 13-17. The fraction of backscatter intercepted by the radar is proportional to the effective antenna area divided by the range squared, A_e/R^2.

For volume search, therefore, the simplified range equation can be rewritten as

$$R_0 \propto \sqrt[4]{P_{avg}t_f \times \frac{\sigma}{\theta_a\theta_c} \times A_e}$$

where

t_f = frame time

θ_a = azimuth angle scanned

θ_e = elevation angle scanned

Ignoring target RCS, over which we have no control, and rearranging, we get

$$R_0 \propto \sqrt[4]{P_{avg}A_e \times \frac{t_f}{\theta_a\theta_e}}$$

What the Volume Search Equation Tells Us. From this simple equation, we can draw three important conclusions regarding detection range in volume search.

- Only through its secondary influences on atmospheric absorption, available average power, aperture efficiency, ambient noise, target directivity, and so on, does wavelength affect the range.
- For any combination of frame time and solid angle searched, range depends primarily on the product, $P_{avg}A_e$ (Fig. 13-18).
- The greater the ratio of the frame time to the size of the volume searched, the greater the range will be.

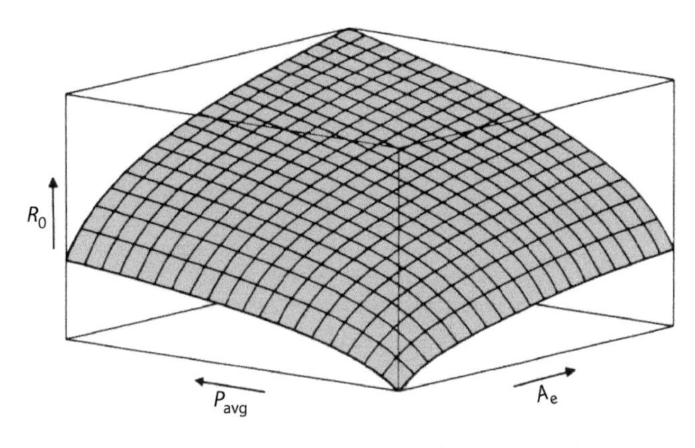

Figure 13-18. For any one combination of frame time and solid angle searched, the detection range can be maximized by using the highest possible power, P_{avg}, and largest possible antenna, A_e.

Frame time, however, may be limited by required system reaction time—which is itself a function of detection range. In addition, the size of the solid angle is dictated by the dispersion of anticipated targets. Therefore, the equation leads to this general conclusion: To maximize detection range for volume search, use the highest possible average power and the largest possible antenna.

Tailoring the Range Equation to Volume Search

In adapting the radar equation to volume search, the antenna beam is conveniently thought of as having a uniform cross section of width $\theta_{3\,dB}$.

This beam is then thought of as jumping a beamwidth at a time through the solid angle that is to be searched.[2]

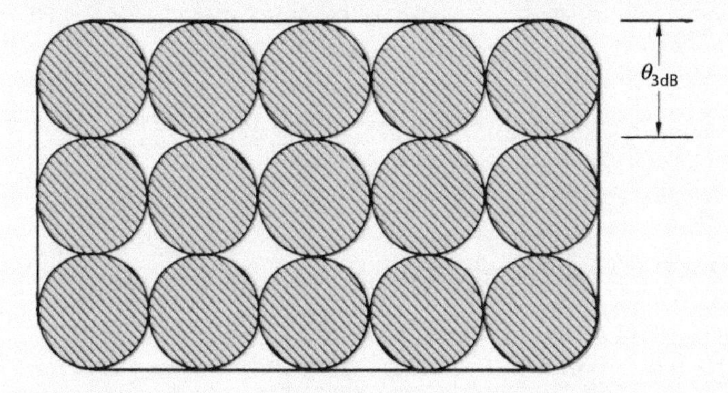

2. Of course, in reality individual beams are not stacked side by side, but the beams are overlapped in both azimuth and elevation to maintain at minimum a complete 3 dB coverage of the search volume. The actual overlap is dependent on the type of search mode being used.

The number of such positions the beam occupies equals the cross-sectional area of the solid angle at unity range divided by the cross section of the beam at the same range

$$\text{Number of beam positions} = \frac{\theta_a \theta_e}{(\theta_{3dB})^2}$$

Since the beam must complete its entire scan in one frame time, t_f, the length of time it dwells on any one target is

$$\text{Dwell time} = t_f \frac{(\theta_{3\,dB})^2}{(\theta_a \theta_e)}$$

Now, the beamwidth is proportional to the ratio of the wavelength, λ, to the diameter of the antenna, d ($\theta_{3\,dB} \propto \lambda/d$). And d, in turn, is proportional to the square root of the effective area of the antenna, $\sqrt{A_e}$. Thus, the dwell time, t_{ot}, is

$$t_{ot} \propto t_f \frac{(\lambda/\sqrt{A_e})^2}{\theta_a \theta_e} = \frac{t_f \lambda^2}{\theta_a \theta_e A_e}$$

Substituting this expression for t_{ot} in the range equation we get

$$R_0 \propto \sqrt[4]{\frac{P_{avg} A_e^2 \sigma t_{ot}}{k T_s \lambda^2}} = \sqrt[4]{\frac{P_{avg} A_e^2 \sigma t_f \lambda^2}{k T_s \lambda^2 \theta_a \theta_e A_e}}$$

Canceling like terms gives

$$R_0 \propto \sqrt[4]{\frac{P_{avg} A_e \sigma t_f}{k T_s \theta_a \theta_e}}$$

Even when all pertinent factors have been included in the radar equation, it cannot tell us with certainty at what range a given target will be detected because background noise and RCS continually fluctuate.

13.3 Fluctuations in Radar Cross Section

Think of a target as consisting of a large number of individual scatterers (Fig. 13-19). The extent to which the scatter from these adds up or cancels in the direction of the radar depends on their relative phases. If the phases are more or less the same, the backscatter will add up to a large sum. If they are not, the sum may be comparatively small.[3]

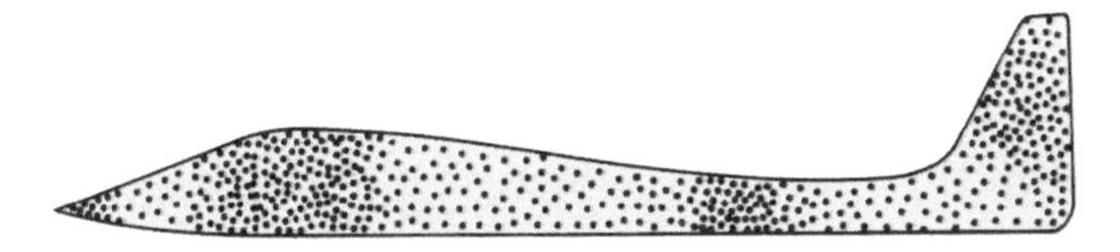

Figure 13-19. A target may be thought of as myriad, tiny reflectors. How their echoes add up depends on their relative phases.

3. Generally, the sums tend to cluster around a median value.

The relative phases count on the instantaneous distances in wavelengths of the reflectors from the radar. Because of the round-trip nature of the transmission, a difference in distance of one-quarter wavelength makes a difference in phase of 180°.

Since the wavelength may be very short, relatively small changes in target aspect, even vibration, can cause the target return to scintillate like the light from a star.[4] And since the configuration of many targets is radically different when viewed from different directions, larger changes in aspect may produce strong peaks or deep fades (Fig. 13-20). Over a period of time these variations will usually average out. However, if the radar-bearing aircraft is approaching on a course that holds a target in the same relative aspect, a peak or fade may persist for some time.

4. More precisely, like the light in a particular spectral line when a star is at a low elevation angle.

Figure 13-20. In this polar plot of the radar cross section, σ, of a typical target, note how widely σ varies with target aspect.

During the early days of the all-weather interceptor, in fact, it was not uncommon to receive complaints from pilots who had picked up a target in a favorable aspect and locked onto it at long range, only to lose lock when the interceptor converted to a constant aspect attack course that happened to place the target in a deep fade. Since they were unfamiliar with the phenomenon, they thought the radar had malfunctioned.

Since the relative phases of the returns from the individual elements vary with wavelength, the target aspects for which the fades occur will generally be slightly different for different wavelengths. One way of getting around the problem of target fading, therefore, is to switch periodically from one to another of several different radio frequencies, thereby providing *frequency diversity*.

13.4 Detection Probability

Because of its randomness, detection performance against targets whose range is limited by thermal background noise is usually stated in terms of probabilities. For search, the most commonly used probability is *blip-scan ratio*, P_d. This ratio is the probability of detecting a given target at a given range any time the antenna beam scans across the target (Fig. 13-21). It is also referred to as *single-scan* or *single-look probability*. The higher the probability specified, the shorter the range will be.

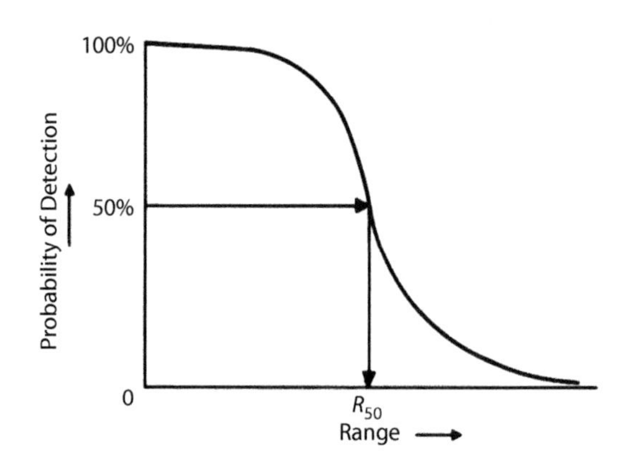

Figure 13-21. Blip-scan ratio is the probability of detecting a given target at a given range on any one scan of the antenna.

The notation used to represent the range is the letter R with a subscript indicating the probability. For instance, R_{50} represents the range for which the probability of detection is 50 percent, and R_{90} is the range for which the probability is 90 percent.

How is the range determined for, say, for a probability of detection of 60 percent? There are five basic steps, explained more fully in the following sections:

1. Decide on an acceptable system false-alarm rate.
2. Calculate the corresponding value of the false-alarm probability for the individual threshold detectors.
3. On the basis of the statistical characteristics of the noise, find the threshold setting that will limit the false-alarm probability to this value.
4. Determine the mean value of the integrated signal-to-noise ratio for which the signal plus the noise will have the specified probability of crossing the threshold (in this case 60 percent).
5. Compute the range at which this signal-to-noise ratio will be obtained.

Deciding on an Acceptable False-Alarm Rate. The average rate at which false alarms appear on the radar display (i.e., the number of false alarms per unit of time) is called the *false-alarm rate* (FAR). The mean time between false alarms is called the *false-alarm time*, t_{fa}, which of course is the reciprocal of the false-alarm rate.

$$t_{\text{fa}} = \frac{1}{FAR}$$

If false alarms occur only once every several hours, they will probably not even be noticed by the radar operator. Yet if the false alarms occur at intervals of the order of a second, they may render the radar useless (Fig. 13-22). What is an acceptable false-alarm time depends on the application. Since raising the detection threshold reduces the maximum detection range, where long range is desired the false-alarm time is usually made no longer than necessary to make the radar easy to operate. In radars for fighter aircraft, for example, a false-alarm time of a minute or so is generally considered acceptable.

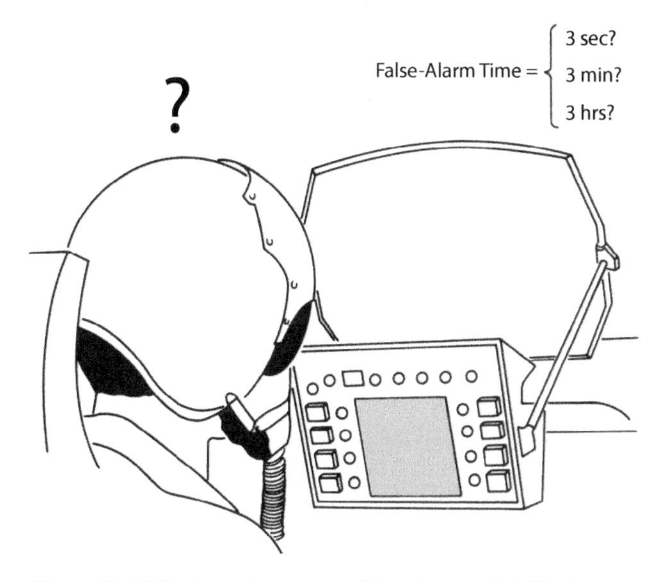

Figure 13-22. To the radar operator false-alarm probability has little direct meaning, but time between false alarms does.

Calculating False-Alarm Probability. The mean time between false alarms is related to the false-alarm probability for the radar's threshold detectors by

$$t_{\text{fa}} = \frac{t_{\text{int}}}{P_{\text{fa}}N}$$

where

t_{fa} = average time between false alarms for the system

t_{int} = integration time of the radar's Doppler filters (plus any PDI)

P_{fa} = false-alarm probability for a single threshold detector

N = number of threshold detectors

If you're puzzled, an analogy using the Lucky 8 Casino may help. Whenever a roulette wheel spins an 8 at this casino, an eight-bell alarm sounds, all bets at that wheel stay put, and everyone who has a bet down is served a free glass of champagne.

Before the casino opened, the question naturally arose: How often will the alarm go off?

Figuring that out was easy. There are 38 compartments in a wheel (Fig. 13-23), so, on average, the ball will land in No. 8 once every 38 spins. If 3 min elapsed between spins, the alarm would sound once every $38 \times 3 = 114$ min for each wheel. The casino would have five wheels, so the alarm would sound five times this often, or once every $114/5 \cong 23$ minutes.

$$\text{1 wheel:}\quad 38\,\frac{\text{spins}}{\text{alarm}} \times 3\,\frac{\text{min}}{\text{spin}} = 114\,\frac{\text{min}}{\text{alarm}}$$

$$\text{5 wheels:}\quad 114\,\frac{\text{min}}{\text{alarm}} \div 5 = 23\,\frac{\text{min}}{\text{alarm}}$$

Figure 13-23. On average an 8 will be spun once every 38 spins.

Since the outcome of each spin is entirely random, the alarm would not, of course, sound at even intervals. There might be two or three alarms in a matter of minutes or none for several hours. But on average the time between alarms would be 23 min.

With the exception of the champagne, the parallel between 8-bell alarms and false alarms on a radar display is direct. The probability of a wheel spinning an 8 corresponds to the false-alarm probability of one of the radar's threshold detectors; the time between spins to the integration time of the radar's Doppler filters; and the number of roulette wheels to the number of threshold detectors.

In general, a bank of Doppler filters is provided for every resolvable increment of range (range gate), and a threshold detector is provided for every filter.

Substituting the product of the number of range gates, N_{RG}, times the number of Doppler filters per filter bank, N_{DF}, for N in the equation above and solving for P_{fa} we get

$$P_{fa} = \frac{t_{int}}{t_{fa} \times N_{RG} \times N_{DF}}$$

Suppose, for example, that the filter integration time, t_{int}, is 0.01 s and the radar has 200 range-gates with 512 Doppler filters per bank (Fig. 13-24). To limit the false-alarm time, t_{fa}, to 90 s, we would have to set the threshold of each detector for a false-alarm probability of about 10^{-9}. Therefore,

$$P_{\mathrm{fa}} = \frac{t_{\mathrm{int}}}{t_{\mathrm{fa}} \times N_{\mathrm{RG}} \times N_{\mathrm{DF}}}$$

$$P_{\mathrm{fa}} = \frac{0.01}{90 \times 200 \times 512} = 1.09 \times 10^{-9}$$

FALSE-ALARM CALCULATION

Problem

Determine the false-alarm probability of a radar's threshold detectors that will limit the system false-alarm time to no more than 90 seconds.

Conditions

Filter integration time...... $t_{int} = 0.01$ second

Number of range gates... $N_{RG} = 200$

Number of filters per bank. $N_{DF} = 512$

Calculation

$$P_{fa} = \frac{t_{int}}{t_{fa} \times N_{RG} \times N_{DF}}$$

$$P_{fa} = \frac{0.01}{90 \times 200 \times 512} \cong 1.09 \times 10^{-9}$$

Figure 13-24. In this calculation of the detector false-alarm probability, the time between system false alarms will be limited to 90 s.

Setting the Detection Threshold. As explained in Chapter 10, the probability of noise crossing the target-detection threshold depends on the setting of the threshold relative to the mean level of the noise. The higher the threshold is, the lower the probability of a crossing.

Just how high the threshold must be to keep P_{fa} from exceeding the specified value depends, of course, upon the statistical nature of the noise. Since the nature of thermal noise is well-known and is essentially the same in all situations, determining the threshold setting that will yield a given false-alarm probability is comparatively simple. The statistical characteristics of the noise are usually represented by what is called a *probability density curve* (Fig. 13-25). This is a plot of the probability that the magnitude of the noise in the output of a narrowband filter will have a given amplitude at any one time.

The probability of the noise exceeding the detection threshold, V_T, equals the ratio of (1) the area under the curve to the right of V_T to (2) the total area under the curve, which is one since by definition the curve encompasses all possible magnitudes. This probability, of course, is the false-alarm probability, P_{fa}.

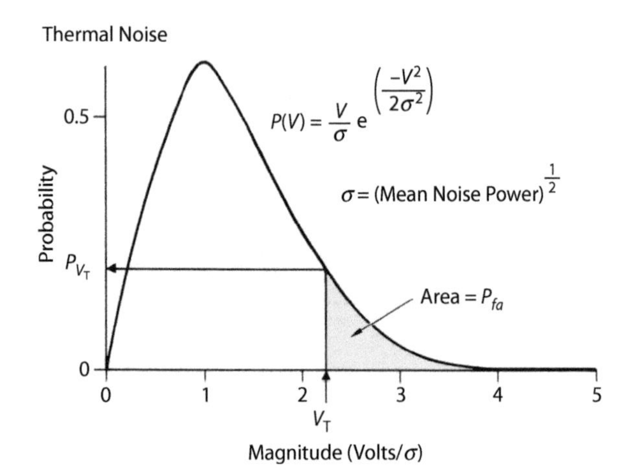

Figure 13-25. Pictured here is the probability density of thermal noise in the output of a narrowband filter. Mean noise power $\sigma^2 = kT_sB$, where kT_s is the integrated noise energy, and B is the filter bandwidth, $1/t_{int}$.

The area under the thermal-noise curve to the right of V_T in Figure 13-25 is plotted versus the value of V_T in Figure 13-26. With a curve like that, one can readily find the required threshold setting for any desired P_{fa}.

$$P_{fa} = \frac{t_{int}}{t_{fa} \times N_{RG} \times N_{DF}}$$

$$P_{fa} = \frac{0.01}{90 \times 200 \times 512} = 1.09 \times 10^{-9}$$

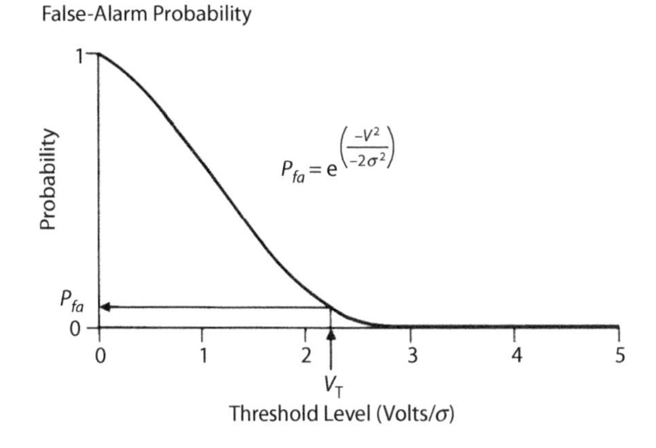

Figure 13-26. This is the area under the thermal-noise probability density curve of Figure 13-25, to the right of the threshold voltage, V_T.

Determining the Required SNR. The probability of a target signal plus the noise exceeding the threshold can similarly be determined. The probability density of the filter output for a representative signal-to-noise ratio is plotted in Figure 13-27, along with a repeat of the probability density curve for the noise alone. As with P_{fa} in the case of noise alone, the area under the curve for signal plus noise to the right of V_T is the probability of detection, P_d. Unlike the fluctuation of the noise, the fluctuation of the signal—which we have seen is due to the variations in RCS—does not have a simple universal characteristic but varies from one target to the next and from one operational situation to another. Nevertheless, statistically it is possible to approximate radar cross sections having common characteristics quite accurately with standard mathematical models.

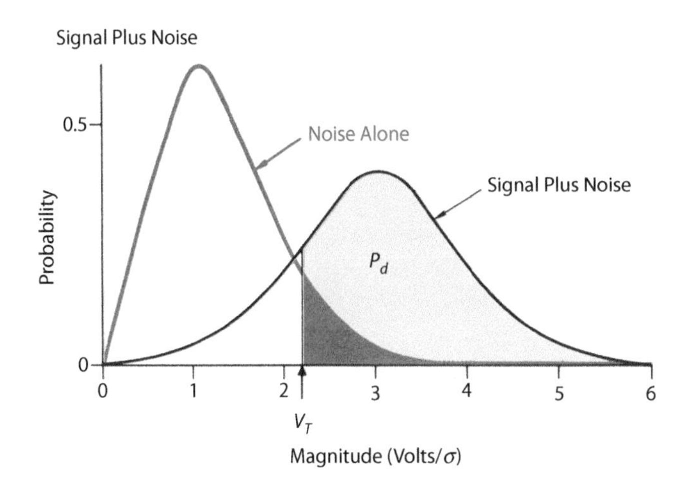

Figure 13-27. This is the probability density of filter output for a representative ratio of signal to noise. The area under the curve to the right of V_T is the probability of detection, P_d. Note that while increasing V_T decreases P_{fa}, it also decreases P_d.

The required signal-to-noise ratio versus detection probability for a wide range of false-alarm probabilities has been calculated for a number of these models. Where specific RCS data are not available or a rigorous calculation is not required, curves based on these results make it easy to find the required SNR.

A commonly used set of curves includes those based on the work of Peter Swerling (Fig. 13-28). They apply to four

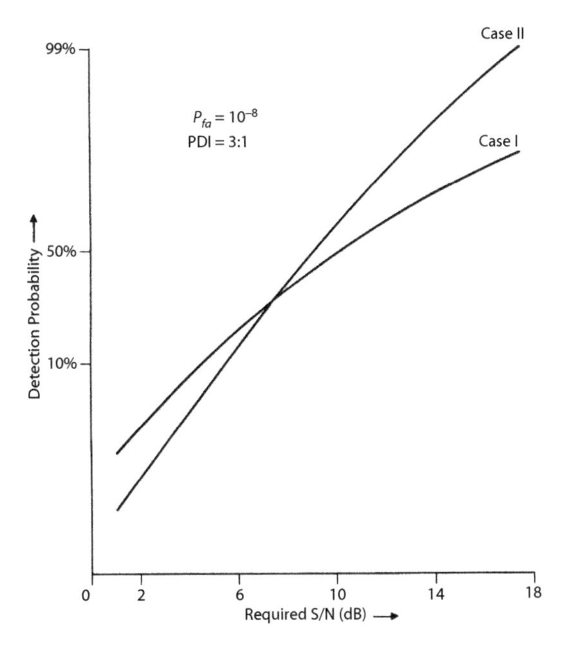

Figure 13-28. Standard curves based on simplified radar cross section models make coarse determination of required signal-to-noise ratio easy (the signal-to-noise ratio is evaluated at the output of the Doppler filter).

Table 13-1. Swerling Model: Four Applicable Cases

CASE	FLUCTUATIONS		SCATTERERS
	Scan-to-Scan	Pulse-to-Pulse	
I	X		Many Independent
II		X	
III	X		One Main
IV		X	

different cases, as shown in Table 13-1. Cases I and II assume a target made up of many independent scattering elements—as in a large (compared with the wavelength) complex target, such as an aircraft. Cases III and IV assume a target made up of one large element plus many small independent elements—as is a small target of simple shape. Cases I and III assume that the RCS fluctuates only from scan to scan; Cases II and IV assume that it also fluctuates from pulse to pulse, such as might be the case where a propeller or helicopter rotor rotates through a significant angle during one pulse repetition interval, changing the phase relationship with which the scattering contributions add. When accounting for fluctuation in detecting aircraft, Case I is used for most radar system designs, since the contributions of the propellers or engine fans are relatively small pulse to pulse whereas the contributions of changing aspect angles scan to scan may be significant.

With curves such as these, for almost any specified false-alarm probability, one can quickly find the integrated signal-to-noise ratio needed to provide any desired detection probability. Except for probabilities that are either very high or very low (and are poorly represented by simplified models) these curves are extremely useful.

Computing the Range. Having found the integrated signal-to-noise ratio needed to provide the desired detection probability, the range at which this ratio will be obtained can be computed with the equation derived earlier for R_0. To adapt the equation for this use, the noise term, kT_s, is multiplied by the required SNR.

$$R_{\mathrm{P_d}} = \left[\frac{P_{\mathrm{avg}} \mathrm{A_e^2} \sigma t_{ot}}{(4\pi)^2 (S/N)_{\mathrm{req}} k T_s \lambda^2}\right]^{1/4}$$

where $R_{\mathrm{P_d}}$ is the range for which the probability of detection is P_d, and $(S/N)_{req}$ is the required signal-to-noise ratio.

13.5 Cumulative Detection Probability

To account for the effects of closing rate, detection range is often expressed in terms of cumulative probability of detection: the probability that a given closing target will have been detected at least once by the time it reaches a certain range.

Cumulative probability of detection, P_c, is related to single-scan probability of detection, P_d, as follows:

$$P_c = 1 - (1 - P_d)^n$$

where n is the number of scans. The term $(1 - P_d)$ is the probability of the target not being detected in a given scan. This term to the n-th power, $(1 - P_d)^n$, is the probability of the target *not* being detected in n successive scans. One minus that probability is the probability of it *being* detected at least once in n scans.

If, for example, $P_d = 0.3$, the probability of the target not being detected in one scan would be $1 - 0.3 = 0.7$. The probability of it not being detected in 10 scans would be $0.7^{10} = 0.03$. The probability of it being detected at least once in 10 scans, therefore, is $1 - 0.03 = 0.97$.

But determining the actual probability is not necessarily as straightforward as that. As the target closes, the value of P_d will increase. Also, a lot relies on how rapidly the target cross section (hence the signal) varies. If the rate is rapid enough for the variation to be essentially random from one scan to the next, over a period of several scans the variation will tend to cancel. If P_d for the range in question has a moderate value, as in the foregoing example, P_c will rapidly approach 100 percent.

Sample Range Computation

Problem: Find the range at which the probability of a given pulse-Doppler radar detecting a given target is 50 percent.

Characteristics of the Radar:

Average power, $P_{avg} = 5$ kW
Effective area of antenna, $A_e = 0.4$ m^2
Wavelength, $\lambda = 0.03$ m
Receiver noise figure, $F = 3$ dB
Total losses, $L = 6$ dB

Target: Fighter, viewed head-on, at constant look angle. Radar cross section, $\sigma = 1$ m^2.

Operating conditions: Radar is searching a solid angle 100° wide in azimuth by 10° wide in elevation. The radar beam's time-on-target, $t_{ot} = 0.03$ s.

Solution: Only two more values are needed to compute the range: the required signal-to-noise ratio, $(S/N)_{req}$, and the noise energy, kT_s.

Since the target is not very large and is viewed head-on from a constant angle, its RCS will probably not fluctuate from pulse to pulse in the 0.03 s time on target. So a rough estimate of $(S/N)_{req}$ can be obtained from Swerling's Case I curve in Figure 13-28, which indicates that for a probability of detection of 50 percent, $(S/N)_{req}$ is about 10 dB.

The value of kT_s can be obtained by multiplying kT_0 by the receiver noise figure, 3 dB (factor of 2). Thus, $kT_s = 8 \times 10^{-21}$.

Plugging these values into the equation for R_{pd} (with the loss term, L, included in the denominator) yields

$$R_{50} = \left[\frac{P_{avg} A_e^2 \sigma t_{ot}}{(4\pi)^2 (S/N)_{req} kT_s \lambda^2 L} \right]^{1/4}$$

$$R_{50} = \left[\frac{(5\times10^3)\times0.4^2\times1\times0.03}{158\times10\times(8\times10^{-21})\times(0.03)^2\times4} \right]^{1/4}$$

$$R_{50} = 1.51\times10^5 \text{ m} = 81.6 \text{ nmi}$$

On the other hand, if there is little change in cross section from scan to scan and the target happens to be in a deep fade, the cumulative probability of detection for the same range may be quite low.

The Many Forms of the Radar-Range Equation

IT IS VERY EASY TO CONFUSE THE MANY DIFFERENT EQUATIONS USED TO CALCULATE SNR. To help you keep them straight in your mind, the constituent expressions for four of them are summarized here.

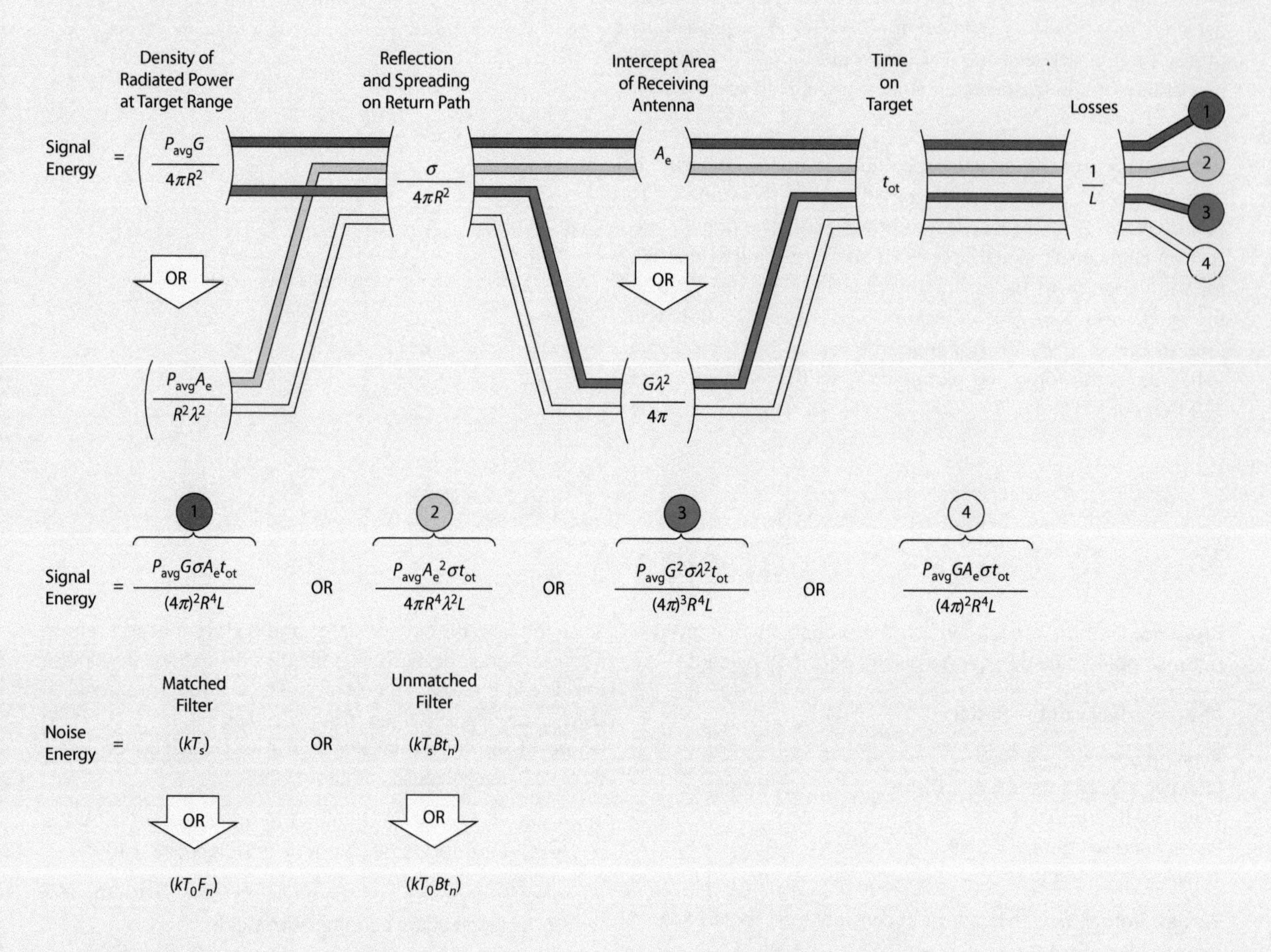

Signal Energy = $\frac{P_{avg}G\sigma A_e t_{ot}}{(4\pi)^2R^4L}$ (1) OR $\frac{P_{avg}A_e^2\sigma t_{ot}}{4\pi R^4\lambda^2 L}$ (2) OR $\frac{P_{avg}G^2\sigma\lambda^2 t_{ot}}{(4\pi)^3R^4L}$ (3) OR $\frac{P_{avg}GA_e\sigma t_{ot}}{(4\pi)^2R^4L}$ (4)

Noise Energy = Matched Filter (kT_s) OR Unmatched Filter (kT_sBt_n)

Matched Filter: (kT_s) OR (kT_0F_n)

Unmatched Filter: (kT_sBt_n) OR (kT_0Bt_n)

$A_e = \frac{G\lambda^2}{4\pi}$ = equivalent area of antenna

B = bandwidth

F = noise figure of receiver

G = antenna gain

k = Boltzmann's constant

L = losses

P_{avg} = average transmitter power

R = range

σ = radar cross section of target

t_n = duration of noise

T_0 = ambient temperature (Kelvin)

t_{ot} = time-on-target (dwell time)

T_s = system noise temperature (Kelvin)

Surface area of sphere = $4\pi R^2$

Antenna gain, $G = \frac{4\pi A_e}{\lambda^2}$

13.6 Summary

From the expressions for signal energy and noise energy a simple equation for detection range may be derived. The expressions tell us the following:

- Range increases as the *one-fourth power* of average transmitted power, target RCS, and integration time
- Range increases as the *square root* of effective antenna area
- A reduction in noise is equivalent to a proportional increase in transmitted power

When adapted to the special case of volume search, the equation tells us that to the first order, range is independent of frequency and can be maximized by using the highest possible average power and the largest possible antenna.

Even when all secondary factors influencing signal-to-noise ratio have been accounted for, the range equation cannot tell us with certainty at what range a given target will be detected, since both noise and RCS fluctuate widely.

Consequently, detection range is usually specified in terms of probabilities. The most common probability for search is blip-scan ratio (also called single-scan or single-look probability). The range at which a given value of this probability may be achieved is determined by (1) establishing an acceptable false-alarm probability, (2) setting the target detection threshold just high enough to realize this probability, and (3) finding the signal-to-noise ratio for this setting that will provide the desired target detection probability, a process which may be simplified through the use of curves based on standard mathematical models of targets. The range at which that ratio will be achieved is then calculated with the range equation.

To account for the effect of closing rate in high-closing rate approaches, detection range may be expressed in terms of cumulative probability of detection—the probability that a given target will be detected at least once before it reaches a given range.

13.7 Some Relationships to Keep in Mind

- Range at which integrated signal-to-noise ratio is 1:

Spotlight $$R_0 \propto \sqrt[4]{\frac{P_{avg} A_e^2 \sigma t_{ot}}{k T_s \lambda^2}}$$

Volume Search $$R_0 \propto \sqrt[4]{\frac{P_{avg} A_e \sigma t_f}{k T_s \theta_a \theta_e}}$$

- False-alarm time:

$$t_{fa} = \frac{1}{\text{False-alarm rate}} = \frac{t_{int}}{P_{fa} N}$$

N = number of threshold detectors

- False-alarm time for a single detector:

$$P_{fa} = \frac{t_{int}}{t_{fa} \times N_{RG} \times N_{df}}$$

N_{RG} = number of range gates

N_{Df} = Doppler filters per bank

- Range for which probability of detection is P_d:

$$R_{P_d} = \left[\frac{P_{avg} A_e^2 \sigma t_{ot}}{(4\pi)(S/N)_{req} k T_s \lambda^2 L}\right]^{1/4}$$

$(S/N)_{req}$ = required signal-to-noise ratio

- Cumulative probability of detection $P_c = 1 - (1 - P_d)^n$

Further Reading

J. V. DiFranco and W. L. Rubin, *Radar Detection*, SciTech-IET, 2004.

L. V. Blake, *Radar Range Performance Analysis*, Artech House, 1986.

P. Lacomme, J.-P. Hardange, J.-C. Marchais, and E. Normant, *Air and Spaceborne Radar Systems: An Introduction*, Elsevier, 2007.

D. K. Barton, *Radar Equations for Modern Radar*, Artech House, 2013.

Test your understanding

1. The single-scan probability of detection of a particular target is 0.25. What is the probability of the target being detected at least once in 16 scans?
2. A radar detects a given target with a given probability at range, *R*. By what factor is *R* increased if (a) the transmit power is doubled (b) the antenna area is doubled; and (c) the target RCS is doubled?
3. A radar is required to scan an angular field of view of 10° (elevation) × 100° (azimuth) with a pencil beam of beamwidth 2°. (a) How many beam positions are required? (b) If the dwell time for each beam position is 0.02 s, what is the frame time?
4. Repeat the sample range calculation on p. 191 to find the range at which the probability of a pulse-Doppler radar detecting a target is 50%, with the following parameters:

Average power	10 kW
Effective antenna area	1 m^2
Wavelength	0.03 m
Receiver noise figure	3 dB
Total losses	10 dB
Target RCS	0.1 m^2
Time on target	0.01 s
Required SNR	13 dB

5. What is the power-aperture product of the radar in question 4?

14

Radar Receivers and Digitization

AESA radiating face
(Courtesy of Selex ES.)

There is no universally accepted definition of what constitutes the *receiver* in a radar system. The most common usage of the term refers to the unit that takes the received radar frequency output from the antenna (usually via a duplexer) and converts the signal into a form that can be used either for display or for subsequent digital signal processing.

In modern radars, this simple segmentation of the radar system has become increasingly blurred. For example, what were formerly receiver front ends are increasingly integrated into phased array antennas, while on the other hand digital processing, which was previously the preserve of a separate processing unit, is increasingly being incorporated into the receiver unit. This chapter deals mostly with receivers in the classical sense, but important aspects of more recent design approaches are also described.

There are two basic types of radar receivers: pulsed and continuous wave (CW). Almost all airborne radars employ pulsed receivers since most of them time-share the same antenna for transmission and reception. Pulsed receivers, which are the focus of this chapter, are more difficult and complex to build.

Receivers have a wide variety of design requirements depending on the system type and the operating environment, but two elements are essential across the board: *sensitivity* (the ability to detect small signals, generally limited by thermal noise); and *selectivity* (the ability to reject unwanted signals, typically by frequency filtering).

This chapter describes the important basic principles underlying radar receiver design and then demonstrates how these work together in several examples of complete receiver systems.

14.1 Basic Principles

The echo received in an airborne radar is extremely small, often well below the level of thermal noise, which is typically only about 10^{-15} W. The job of the radar receiver is to amplify these tiny signals and to filter them out from background noise and clutter—a task that in modern radars employs a great deal of digital signal processing. This chapter deals mainly with the analog parts of the receiver from the incoming radio frequency (RF) signal through to the digitized output, but it does include some aspects of digital filtering, which is an important part of modern designs.

In addition to the basic tasks of amplification, filtering, and digitization, the receiver must avoid any contamination or distortion of the signal since this may result in loss of sensitivity or false detections.

14.2 Low-Noise Amplification

The sensitivity of all receivers is limited by thermal noise, and in radar it is normally thermal noise from the receiver that dominates (as opposed to external noise sources such as the Sun). The first part of a radar receiver is a low-noise amplifier (LNA), which is designed to set the sensitivity of the receiver at the highest realistic level and is connected, usually via some form of duplexing device, to the radar antenna.

The LNA is the key element in determining the basic sensitivity of the receiver, which is characterized by its noise figure, F, defined by

$$F = \frac{\mathrm{SNR_{in}}}{\mathrm{SNR_{out}}}$$

where $\mathrm{SNR_{in}}$ is the signal-to-noise ratio at the receiver input, and $\mathrm{SNR_{out}}$ is the signal-to-noise ratio at the receiver output.

Even the best LNAs degrade the signal-to-noise ratio, but the reason for this initial amplification is to ensure that the contribution of any downstream noise sources is minimized. This can be seen by examining Friis's formula for cascaded noise figure: in a cascade of amplifiers with the first having gain G_1 and noise figure F_1, the second gain G_2 and noise figure F_2, the third gain G_3 and noise figure F_3, and so on, the overall noise figure F is given by

$$F = F_1 + \frac{(F_2 - 1)}{G_1} + \frac{(F_3 - 1)}{G_1 G_2} + \frac{(F_4 - 1)}{G_1 G_2 G_3} + \cdots\cdots$$

If G_1 is high, this makes all contributions other than F_1 negligible, which is the goal in a well-designed system. The system noise figure is largely determined by the first stage in the receiver chain.

In most modern systems, a semiconductor LNA based on gallium arsenide (GaAs) or gallium nitride (GaN) is employed.

These components have revolutionized the design of radar receivers, with common noise figures of 1 dB readily available, some 10 times better than earlier systems.

Of course, nothing is without a price. It is also vital to avoid distortion, so it is crucial that the LNA have *linearity*. A very high gain device (large G_1) will lack linearity. The trade-off between linearity and noise figure is an important aspect of receiver design.

In active electronically scanned array (AESA) radar, an LNA is normally incorporated into each transmit/receive module of the array, which reduces or eliminates the need for an LNA at the input of any subsequent receiver. Any given array has many LNAs—perhaps 1000 or more in a typical airborne AESA radar. Although at first glance this could result in more noise entering the radar receiver, this is not in fact the case. The LNA at a single element will (given sufficient gain) set the noise figure of the complete system, exactly as in a conventional receiver.

14.3 Filtering

Filtering is vital and appears in many aspects of receiver design, the most fundamental of which is to make sure that wanted signals are received with minimal loss while simultaneously minimizing the amount of noise that appears at the receiver output. Filtering also plays a vital role in the design of downconverters and digitizers.

The Matched Filter. Radar receiver designers, unlike designers of an electronic warfare receiver, have the great advantage that they know exactly what signal has been transmitted. Therefore, detecting the echo is immeasurably easier than if the signal is unknown. The key concept here is the *matched filter,* which is designed to match the transmitted signal and maximizes the signal-to-noise ratio at the receiver output.

Matched filter theory is a major branch of radar engineering that originated with D. O. North's 1943 paper. The essence of a matched filter receiver is that it aims to correlate a known signal (i.e., the transmitted signal), with an unknown signal (the received signal) to detect the presence of the known signal in the unknown signal. The ideal matched filter is a time-reversed replica of the transmitted signal.

This works regardless of the characteristics of the transmitted signal: it can be a simple pulse; a frequency-modulated chirp; a binary code sequence; or anything else, provided it is known. The principle is most easily understood just by considering a simple rectangular pulse of width τ. The matched filter for a simple pulse is one whose frequency response is the Fourier transform of the pulse. In simple terms, this equates to a filter with a bandwidth of approximately $1/\tau$.

A typical radar pulse might be 1 μs wide, so the matched filter needs to have a bandwidth of $1/10^{-6}$ Hz, or 1 MHz. This is sufficient to pass the pulse with little degradation but is narrow enough to minimize noise at the output. A narrower bandwidth will reduce the noise but also the signal, whereas a wider bandwidth will admit too much noise.

An ideal radar receiver would implement this filter right at the front end. This would ensure that only desired signals entered the radar receiver and would maximize sensitivity. Unfortunately, it isn't that easy in practice: a 1 MHz filter implemented at the carrier frequency of a typical airborne radar (10 GHz) is extremely difficult to implement because of the very high precision (in relative frequency terms) required. The existing techniques that accomplish this all have very major drawbacks and are rarely used. So in practice this narrowband filtering is usually applied later in the receiver, after the incoming signal has been downconverted to a lower frequency.

14.4 Downconversion

Almost all radar receivers employ frequency downconversion. Although theoretically unnecessary, this is a solution that allows filters and amplifiers to be realized in practice.

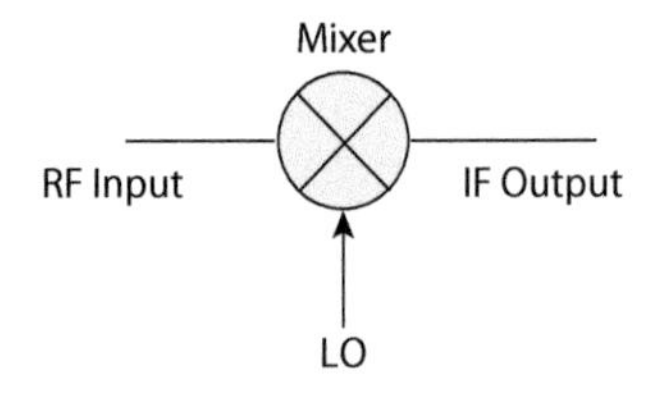

Figure 14-1. A simple mixer stage: the RF input and the Local Oscillator (LO) are mixed together to provide an output at an Intermediate Frequency (IF).

The basic idea is that the incoming RF signal, which carries a modulation with a bandwidth sufficient to represent radar's pulses, is converted to a lower carrier frequency known as the *intermediate frequency* (IF), f_{IF}, but *without* changing the modulation bandwidth. The downconversion process is carried out by a device called a *mixer*.

Mixers (Fig. 14-1) are nonlinear devices that effectively multiply two signals together—in this case the RF signal and a fixed reference signal called the *local oscillator* (LO). We can easily see how this works mathematically.

We can represent the RF signal at its carrier frequency, f_c, by

$$S = \sin(\omega_c t)$$

where $\omega_c = 2\pi f_c$, and t is time. Similarly, the LO signal at f_{LO} is

$$LO = \sin(\omega_{LO} t)$$

Then the output of the mixer, which multiplies the two signals, is

$$\begin{aligned} S \cdot LO &= \sin(\omega_c t) \cdot \sin(\omega_{LO} t) \\ &= \tfrac{1}{2}(\cos((\omega_c - \omega_{LO})t) - \cos((\omega_c + \omega_{LO})t)) \end{aligned}$$

The output of the mixer is now two signals: the *lower sideband* at a frequency below the original RF signal at frequency $f_c - f_{LO}$; and the *upper sideband* above it at frequency $f_c + f_{LO}$. Normally only the lower sideband is desired, so the upper sideband (which is significantly separated in

Mixers

Mixers are a critical element of radar receivers. Mathematically, they can be regarded as a device that multiplies two signals together, which gives the sum and difference frequencies. In practice, such an ideal device does not exist, but good approximations can be built.

The most common type used in radar is the *double balanced mixer*, which uses four switching devices (diodes or transistors). A highly simplified schematic of one type, the *diode ring mixer*, is shown in the following figure.

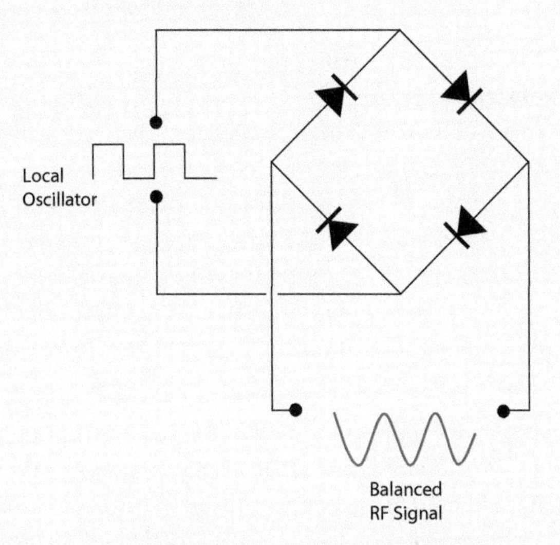

The LO signal is fed to the top and bottom points of the diode ring. In practice, a very high amplitude drive signal (compared with the RF signal) is used, so the LO signal approximates to a square wave (due to saturation effects in the drive amplifiers). The result is that half the time one side of the diode ring is biased "on" and the other side is biased "off." This situation continually reverses at the LO frequency.

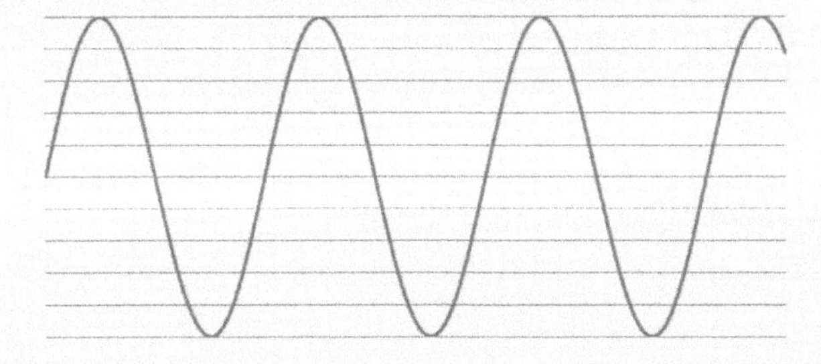

The RF signal is fed to the left and right points of the diode ring in balance; the signal at the left point is 180 degrees out of phase with the signal at the right point. Under the control of the LO, one arm of the balanced RF signal is alternately shorted out by the diodes that are biased on. This gives rise to a complex waveform in the RF circuit, the mixer output, which includes a strong component at the desired difference frequency. This is then filtered out (using other components not shown here) and fed to the next stage of the receiver.

A simple example is shown where the LO is at one-third of the RF signal. The *period* of the LO is three times the period of the RF, so for half the LO period (3/2 × the RF period) we select one arm of the balanced RF signal, and for the next half of the LO period we select the other arm of the balanced RF signal. The resultant waveform repeats every 3/2 × the period of the RF signal; that is, it is at 2/3 the frequency, which is the desired mixer output.

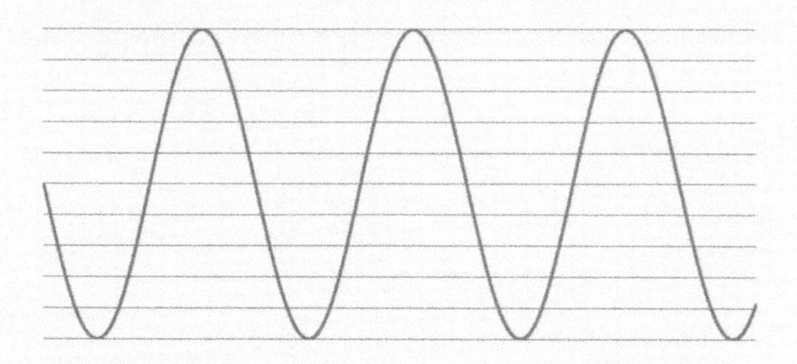

In a practical design, the mixer incorporates *baluns* (short for balanced–unbalanced), which are circuits that can take an unbalanced signal (e.g., RF on a coaxial line) and produce a balanced output to feed the mixer. This is also normally done with the LO.

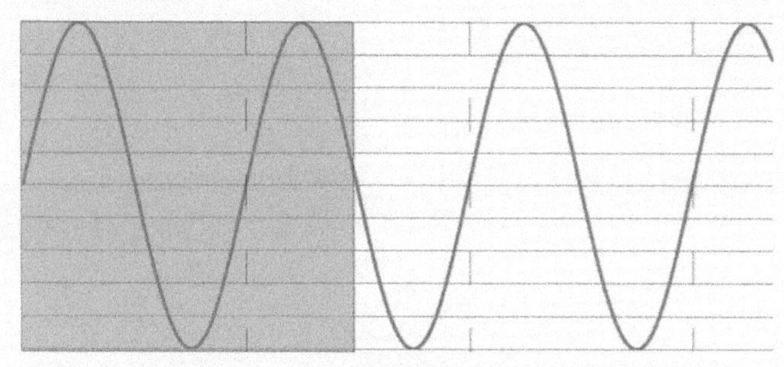

RF signal at the left-hand point of the diode ring

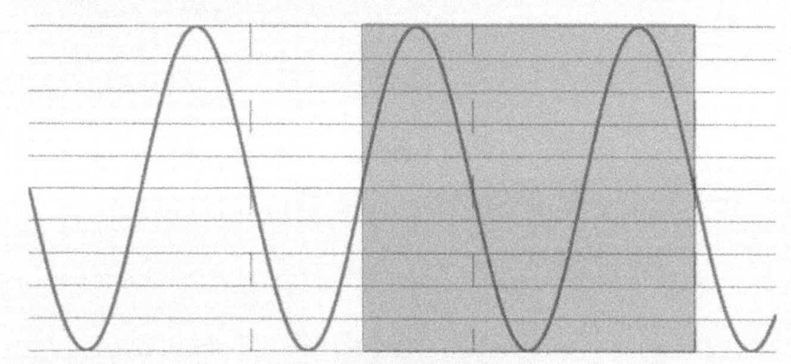

RF signal at the right and point of the diode ring

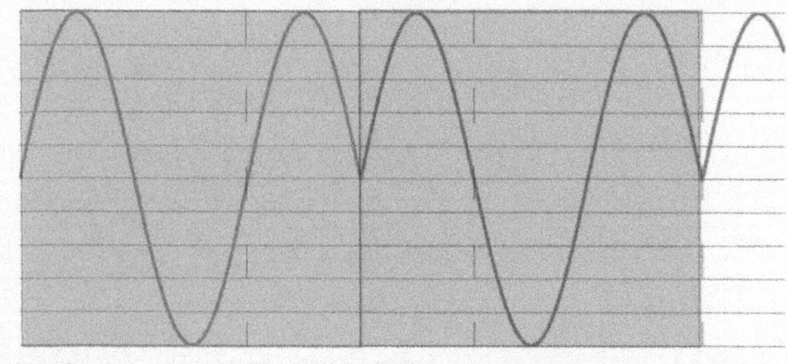

Resultant signal with LO at 1/3 the RF frequency- the signal is now at 2/3 the RF frequency

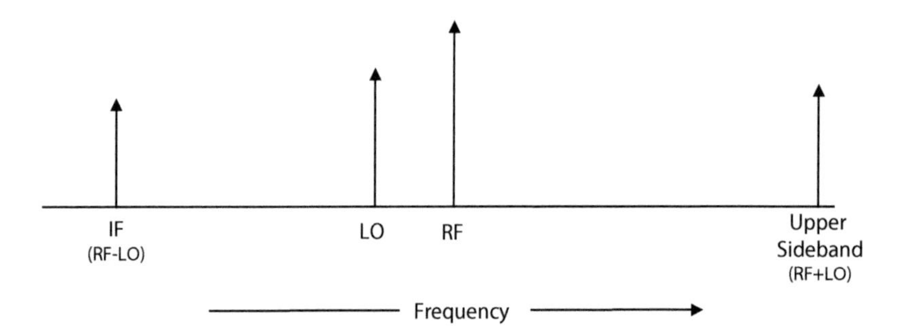

Figure 14-2. These frequencies are present at the input (RF and LO) and the output of the mixer stage.

frequency) is filtered out. The lower sideband is centered on f_{IF} (Fig. 14-2).

Image Rejection. An important issue in the design of this type of receiver is the *image frequency response*. An input at the so-called image frequency, f_i, will also result in an output at f_{IF}, where $f_i = f_c - 2f_{\mathrm{IF}}$ (for the lower sideband case). This is because a signal at the image frequency mixes to a frequency of $f_i - f_{\mathrm{LO}} = -f_{\mathrm{IF}}$, which is indistinguishable from a signal at f_{IF}. Thus, any signal at this image frequency will appear in the IF band at the mixer output, resulting in interference (Fig. 14-3).

To combat this, it is important to filter the RF signal prior to the mixer to minimize any input at the image frequency. Typically the signal is filtered to a bandwidth approximately equal to the IF. The design of such filters is itself a complex matter. It is important to minimize any frequency-dependent amplitude or phase distortion of the signal within the passband; otherwise, high bandwidth signals will be significantly degraded.

Intermodulation. A further issue is intermodulation between different signals at the input. As far as the mixer is concerned, any two signals at different frequencies will be multiplied together to give intermodulation products, regardless of at which port of the mixer the signals are present. Thus, if two frequencies (e.g., the desired echo and a jamming signal) are present at the input and are separated by the IF, they will intermodulate to produce an output at the IF (quite apart from the mixing products produced by intermodulation with the local oscillator). Filtering prior to the mixer can also be used to eliminate this problem, provided

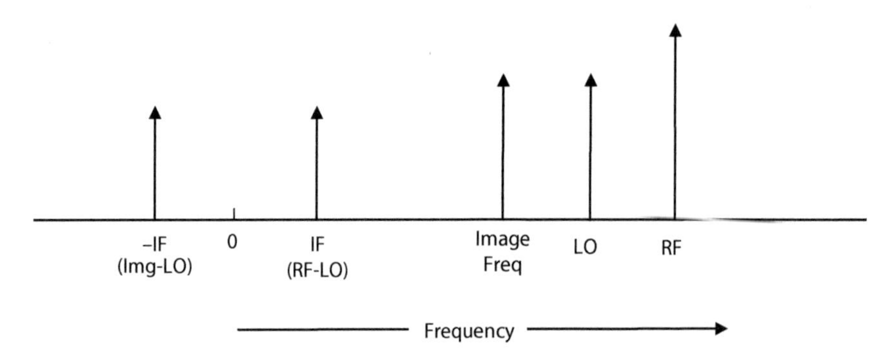

Figure 14-3. Image frequency signals at the image frequency can mix down and be indistinguishable from the desired IF output.

the bandwidth of the input filter is made sufficiently narrow. Use of a high IF makes filtering to reject such signals easier.

A simple rule of thumb in the design of a downconverter is always to filter the input signal to a bandwidth less than the subsequent IF before the mixer. It is easiest to use an example to see how this works. Let's imagine we have a radar operating on a carrier frequency of 10 GHz and we want to downconvert to an IF of 3 GHz, with a total signal bandwidth of 1 GHz. We use a local oscillator of 7 GHz, which produces sidebands centered on 3 GHz and 17 GHz. We then place a 1 GHz bandwidth filter, centered on the IF (2.5 GHz~3.5 GHz), to remove the upper sideband. The image frequency is 4 GHz, and a jamming signal on 7 GHz could produce a damaging output on the IF. However, if we filter the input signal to a bandwidth of 2 GHz (9~11 GHz) before the mixer, this eliminates these signals before they can do any damage: any possible mixing products will lie outside the IF bandwidth (Fig. 14-4).

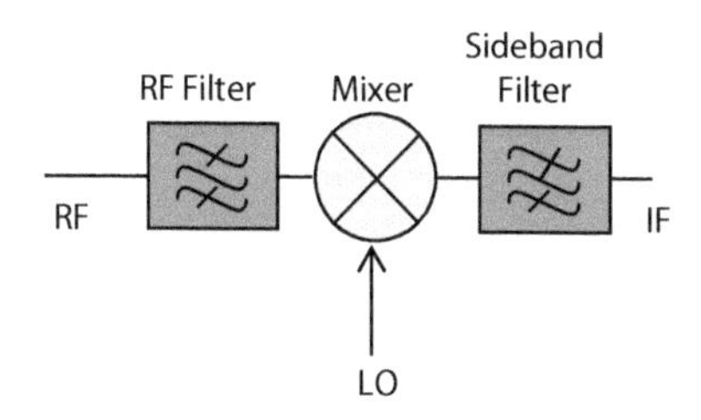

Figure 14-4. The mixer stage is shown here with RF and IF filtering.

Homodyne Receivers. These receivers, also called *direct conversion* or *zero IF* receivers, are the simplest type of downconverting receiver, where f_{LO} is made equal to f_c. This type of receiver converts directly down to baseband (zero IF). Although this is very simple, this type of receiver cannot follow the previously outlined basic principles; it would require a zero bandwidth filter at the input to the mixer. This type of receiver may be found in very simple radars, but it is unsuitable for any system that must be resistant to jamming and interference.

Heterodyne Receivers. Also called *superheterodyne* or *superhet receivers*, in these downconverting receivers f_{LO} is significantly different from f_c. This type of design can incorporate the necessary filtering needed to observe the aforementioned design principles.

Multiple-Stage Downconversion. It is often necessary to downconvert to a low IF to carry out narrow-band filtering or analog-to-digital conversion. If the downconversion is done in a single stage, this makes the receiver vulnerable to jamming and interference as it becomes close to a homodyne receiver. The solution to this is to downconvert in multiple stages, with filtering and amplification at each stage, in which the same principle is observed: before mixing, filter to less-than-subsequent IF. This approach facilitates practical filter designs because it avoids the need for very narrow fractional bandwidths (i.e., filter bandwidth is a very small fraction of the carrier frequency), which are difficult to implement.

A typical X-band radar receiver includes several stages of downconversion: initially to an ultra high frequency (UHF) IF (a few GHz); then to a very ultra high frequency (VHF) IF (a few hundred MHz); then possibly to baseband.

14.5 Dynamic Range

Dynamic range is a simple but often misunderstood concept. The key issue is a receiver's ability to pass large and small signals simultaneously so that the signals are not distorted.

All amplifiers are nonlinear; that is, the output is not simply a multiple of the input at all times. Therefore, if signal is applied to an amplifier, it will not simply be made bigger; it will also be distorted. The bigger the input signal, the greater the distortion; only small signals can be amplified with very little distortion. This is bad news, because a distorted signal can easily give rise to false detections. The trick here is to make sure the amplifier is linear enough to deal with the biggest signal it is likely to encounter, and to do this means increasing the bias currents to the transistors in the amplifier, which in turn increases power consumption. This is the price that must be paid if a high dynamic range is necessary.

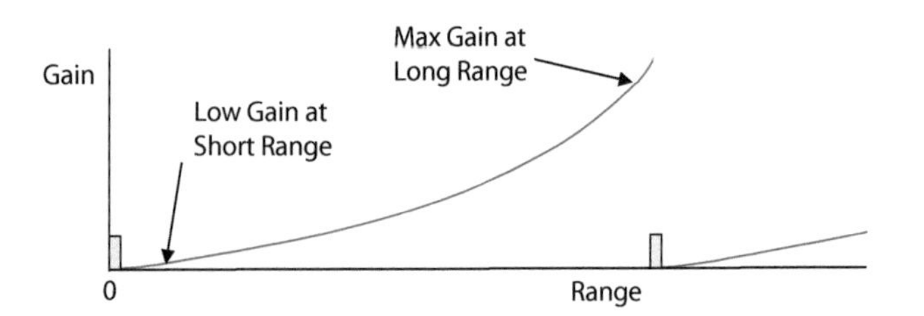

Figure 14-5. Sensitivity time control is a technique where the receiver gain is varied as a function of time (or range).

Other tricks sometimes work. In a radar that measures range without ambiguity, it is possible to vary the gain of the receiver throughout the *pulse repetition interval* (PRI). This is known as *sensitivity time control* (STC), or *swept gain*. It works because strong signals usually originate close to the receiver, whereas weak signals are usually at long range. STC gives just the illusion of high dynamic range because it cannot pass large and small signals simultaneously; they have to be separated in time. Nevertheless, it is a very useful compromise and is used in many systems (Fig. 14-5).

In radars where range is ambiguous, STC cannot be used. Such radars as a rule also employ high levels of Doppler processing. Therefore, the demand for high dynamic range has to be fundamentally addressed; otherwise the small target signals will be lost in spurious signals generated from the large clutter signals. This entails high bias currents and high power consumption in the receiver's amplifiers.

A further negotiation is *automatic gain control* (AGC), which gives only an illusion of greater dynamic range. However, this approach is useful because the level of signals that the radar has to deal with can vary greatly with time, and AGC gives a way of compensating for this.

In absolute terms, a radar would ideally have a dynamic range of perhaps 200 dB (10^{20}:1) to cope with the full range of signals it is likely to encounter in all circumstances. This is not practical, and 100 dB is a more realistic design (although even that is not easy to achieve). A dynamic range of 100 dB, plus an additional 100 dB of AGC, is the sort of compromise typically found in a high-performance radar receiver.

An AGC system measures the radar's operating environment and adjusts the gain accordingly so that the instantaneous dynamic range of the receiver is optimally used. This may be done in a variety of ways, the simplest of which is to provide the operator with a gain control. Most modern systems, however, are automatic and use the measurements of the signal background to adjust the receiver. These control loops need to be carefully designed because they can be exploited by jamming systems to reduce radar sensitivity.

14.6 Spurious Signals and Spectral Purity

Doppler radar uses spectral analysis to separate small targets from large clutter signals. An important issue in this type of radar is spectral purity. The transmitted signal and the local oscillator signals are generally designed to be pure tones at specific frequencies, but in practice they possess components at other frequencies due to practical limitations in the way they are generated. These unwanted components include both discrete spurious frequencies (or spurs) and broadband *phase noise.*

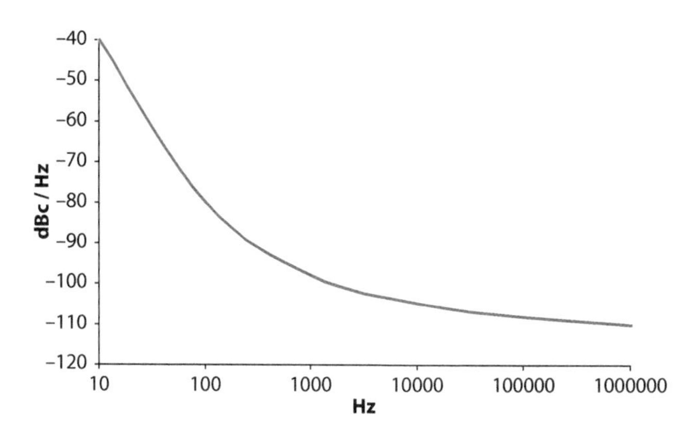

Figure 14-6. This typical phase noise plot shows phase noise spectral density as a function of frequency offset from the carrier.

A typical phase noise plot is shown in Figure 14-6. The plot shows (as a function of frequency on a logarithmic scale) the power spectral density of the noise measured in dB below carrier (dBc) in a 1 Hz bandwidth. An ideal, pure tone would consist simply of a line at zero on the frequency scale, but no practical oscillator is as good as this. In the figure, at a frequency 1000 Hz away from the carrier, the power spectral density is about –98 dBc/Hz. This means that in a typical detection filter bandwidth of 100 Hz the noise from the reference oscillator will be 78 dB below the carrier ($-98 + 10 \log_{10}(100) = -78$). This limits the radar's ability to detect small signals at low Doppler offsets. It's important to note that filtering to narrower bandwidths always helps to reduce the effects of phase noise.

Spurs fall into two classes: signal-related spurs (SRS); and non-signal-related spurs (NRS). The former may increase or decrease in amplitude with the presence of a signal and are normally internally generated, whereas the latter tend to be at a constant level and are caused by nonlinearities in the receiver design, which give rise to intermodulation products.

Spurs differ from phase noise in that they are inherently narrow band, so changing the radar's filtering bandwidth makes little, if any, difference to their level, unlike for phase noise. They are damaging because they can appear as false targets; phase noise results in an increase in the radar's noise level, which reduces its sensitivity.

Control of spurs and phase noise is an issue principally for the radar's signal generation system, but maintaining adequate dynamic range is also critical. It is also vital to eliminate other sources of spurious signals, such as breakthrough from power supplies and interference due to poor screening from other signals within the system. A radar receiver typically uses a great deal of internal screening to isolate sensitive parts of the receiver, so most receivers internally look like a series of separate metal boxes (Fig. 14-7).

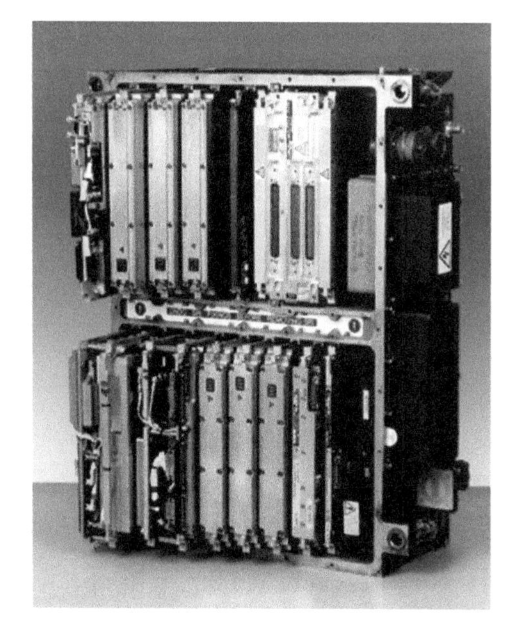

Figure 14-7. This typical airborne radar receiver shows separately screened modules.

14.7 Digitization

Early radar receivers provided an analog output, called the video output, direct to the display, which had been downconverted to the frequency needed to drive the display. This approach is now obsolete, and all radars employ some form of *analog-to-digital* (A/D) conversion at the receiver output. Digitized output greatly facilitates subsequent signal processing and display processing.

An A/D converter provides a sequence of digital values that represent the output voltage of the radar receiver at discrete time intervals. Normally, the output voltage is sampled on a regular basis, and the sample value is held in a *sample-and-hold* (S/H) circuit while it is converted to digital form. An S/H typically consists of a capacitor to hold the voltage and an electronic switch to disconnect the input. The digital representation of the signal, in binary form, can then be processed in a digital signal processor or used to drive a digital display.

The earliest A/D converter designs sampled the radar output after downconversion to a low IF in what was called the video output because in analog radars this was the output that drove the display. The sampling rate chosen was similar to the radar's range gate, so the sequence of samples represented the radar output at various ranges. This approach required all the radar pulse matched filtering to be carried out in analog form prior to A/D conversion. Although this method was effective for radars with a small range of pulse sizes, complex, multifunction radars demanded a range of filters that could easily drive up the complexity of the receiver.

More recent designs, enabled by modern A/D conversion technology, employ much higher sampling rates and carry out digitization at a higher IF. This allows more of the radar's filtering to be performed digitally, which is greatly beneficial since the digital processing is readily reconfigurable and stable.

In theory, this idea could be extended all the way up to the RF, with direct digitization of the incoming signal. The high dynamic range required in radar is what mainly prevents this from happening: A/D conversion technology simply cannot support the required performance.

Nyquist Criterion. The basic principle of sampled-data systems was established by Shannon and Nyquist, who proved that the minimum sampling frequency, f_s, required to represent a signal with maximum frequency, f, is given by $f_s = 2f$. Sampling of higher frequencies leads to *aliasing*, the phenomenon that occurs when the sampled signal appears erroneously as a lower frequency. In practical systems, then, it is important to include a low-pass filter, called the *anti-aliasing filter*, prior to the S/H and A/D conversion.

Effective Number of Bits (ENOB). In radar, dynamic range is all-important. An A/D conversion with n bits can represent 2^n voltage values and in principle can represent a dynamic range of approximately $6n$ dB. However, practical converters are imperfect and achieve a lower dynamic range than the number of bits might suggest. A converter's ENOB is that over which the converter meets a certain specified level of performance in terms of linearity and freedom from spurs. It is difficult to generalize, but the ENOB is usually two to three bits less than the nominal number of bits.

Maximum Input Frequency. Another important characteristic of an A/D conversion, maximum input frequency must be chosen to match the Nyquist criterion for the radar signal. Higher frequency converters always have a lower ENOB. This is for a wide variety of reasons, including the effects of time jitter (which becomes more dominant at higher frequencies) and the difficulty of building converters fast enough to match the sampling rate. In practical radar receiver designs, the downconversion architecture and the A/D converter designs are the subject of a complex trade-off to achieve the required dynamic range and noise performance.

Noise. The digital representation of the signal introduces an additional noise component, known as quantization noise. In power terms, it can be shown that this noise has a power of $Q^2/12$, where Q is the voltage quantization interval. To preserve the noise figure of the receiver, it is essential that the front-end thermal noise of the system is accurately represented at the A/D converter output. If we design a receiver such that front-end thermal noise has a mean voltage of $2Q$, the quantization noise will be about 16 dB below the thermal noise. This is about right to ensure that the back-end noise is negligible compared with the thermal noise floor.

Types of A/D Converters. A/D converters have a variety of designs, two of the most common of which are the *direct conversion*, or *flash*; and the *successive approximation*. The flash is probably the simplest in concept; it includes a large array of voltage sources with each one equal to the value of every possible digital output. The input signal is compared with each of these possible voltages in parallel via a large array of comparators. This allows the digital representation to be obtained very rapidly. This design is used for the fastest A/D converters. The limitation, however, is that the complexity of the comparator network restricts the design to about 8 bits, and an 8-bit system requires 2^8 comparators, where each extra bit doubles the complexity and power consumption. By contrast, successive approximation uses only a single comparator to sequentially compare the input voltage with a range of voltages that is successively narrowed. At each successive step, the converter compares the input voltage to the output of an internal digital to analog converter that has been set to the midpoint of the expected voltage range. The measured error is then used to set a smaller range for the next step. This design exploits the fact that it is easier to build an accurate digital-to-analog (D/A) converter and that by continually narrowing the voltage range higher resolution and accuracy is obtained. The disadvantage is that the time this method takes hampers the maximum input frequency.

Digitization in Noncoherent Radar. Noncoherent radars use just the amplitude of the received radar signal and not the phase information. Thus, the digitizer only must convert the voltage output of the radar receiver for subsequent processing.

Normally, the radar receiver is linear, so the digital output is directly proportional to the RF input signal. However, in some radar designs logarithmic amplifiers are used prior to the A/D conversion to compress the dynamic range on the display. The same function can be achieved using an A/D converter with nonlinear quantization steps, but this is much less common.

Digitization in Coherent Radar. Coherent radars use both the amplitude and phase of the received radar signal in subsequent Doppler processing. To represent this complex quantity digitally, either the phase angle, ϕ, and an amplitude value, A, must be calculated, or, much more commonly, two Cartesian quantities known as the in-phase (I) and quadrature (Q) values are used. These are simply related by

$$\mathrm{I} = A\cos(\phi)$$

$$\mathrm{Q} = A\sin(\phi)$$

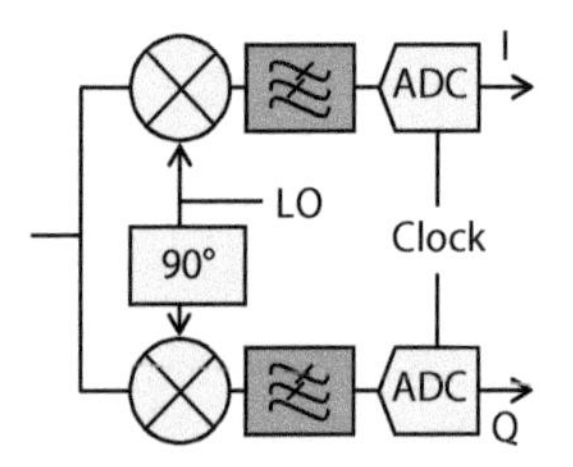

Figure 14-8. A receiver digitization architecture using analog I/Q downconversion and baseband digitization.

Digitizing the radar output to provide I and Q values can be conveniently combined with the final downconversion stage. In this case, the IF signal is split into two equal parts and fed to two parallel mixers. The LO signal is fed to both, but in one case with a 90° phase shift (Fig. 14-8). Two separate A/D converters digitize the output of the two mixers simultaneously, providing the I and Q values. This approach requires only low-frequency A/D converters, which makes it easier to achieve the required dynamic range. However, it is a complex analog design because the two signal paths must be precisely matched in gain, phase, and delay. This is seldom possible in practice, and complex calibration techniques using injected test signals are usually employed to calculate appropriate corrections.

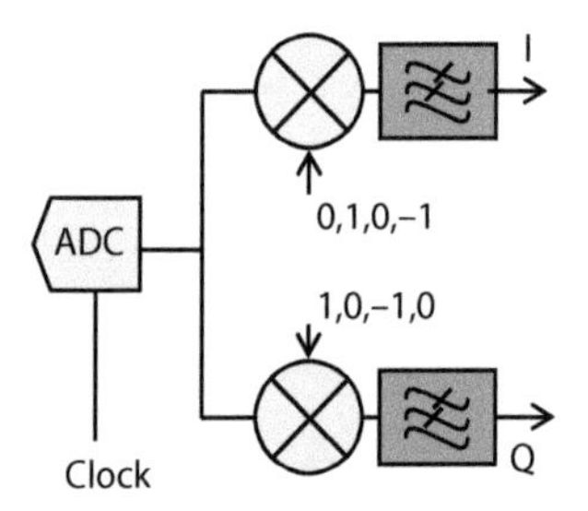

Figure 14-9. A receiver digitization architecture using IF digitization and digital I/Q downconversion achieves the same function as the analog design of Figure 14-8.

An alternative way of achieving the same result is through sampling at the IF and using digital downconversion (Fig. 14-9).

In this architecture a single, much higher frequency A/D converter is employed. Its output is then multiplied by the digital equivalent of the LO, which can be just a series of zeroes and plus or minus ones (which avoids the need for high-speed digital multipliers). This method uses the *Hilbert transform.* The two resultant digital data streams are then digitally low-pass filtered, just as in an analog system, to produce the I and Q outputs.

This design is physically simpler than the analog method, and it avoids the need for precisely balanced circuits or complex calibration. However, it needs a much higher speed A/D converter, which will inevitably have a lower dynamic range. The choice between analog and digital downconversion for coherent systems is a complex trade-off, but today A/D converters have improved to the point that a digital architecture can deliver the levels of performance required.

Figure 14-10. Pictured here is an actual IF digitizer unit with digital downconversion. (Courtesy of Selex ES.)

An example of a modern IF digitizer is shown in Figure 14-10. It consists of a single double-sided printed circuit board with surface-mounted components. The smaller dark components are the A/D converters, and the large silver components are field-programmable gate arrays (FPGAs), which implement the digital downconversion and digital filtering.

14.8 Radar Receiver Architectures

Different receiver architectures have developed as airborne radars have grown in complexity and functionality and as the need to reject jamming and interference have increased. Current designs all use some or all of the fundamental building blocks described in Section 14.1.

Early airborne radars were noncoherent: they relied solely on the *amplitude* of the radar echo, and any *phase* information was discarded. The receivers in these simple radars amplified and downconverted the radar echo to a low video IF, which was then rectified before the amplitude was displayed to the operator. A *detector diode* carried out the rectification step, which discarded the phase information. Although this type of radar was very successful and has remained in use (in one form or another) until today, its major drawback (for an airborne radar) is that it can easily lose a target in the very large ground clutter echo.

To address this drawback, coherent radars were developed, where phase information is not discarded and Doppler filtering can be employed to separate the target from clutter. In early systems this was carried out using a large bank of analog filters, each individually tuned to a different Doppler frequency. Unfortunately, this design was largely incompatible with a pulsed radar design because the analog filters could operate correctly only on continuous wave (CW) signals.

Interrupted CW radars provided a partial solution to this problem by employing a 50/50 transmit/receive duty cycle at the expense of significant signal loss, which allowed for the use of an analog filter bank. This technique was implemented successfully in many airborne radars in the 1960s and 1970s. Still, it suffered from a significant problem with eclipsing: because the receiver was turned off for half the time, target echoes could easily be lost. In addition, the analog filters were very prone to drifting off-tune.

A/C converters provided a major breakthrough in solving the issue of target loss due to large ground clutter echo. They allowed the Doppler filter bank to be mechanized using a digital fast Fourier transform (FFT), with a separate one for each range gate. This avoided all the major drawbacks of the earlier design: the signal losses were directly eliminated, and the digital filter bank was reproducible and stable (unlike its analog predecessor).

Progress since then has been more incremental. Major innovations include higher speed digitization, which allows more filtering to be performed digitally. This is more stable, reproducible, and easily reconfigured. In addition, multiple parallel receiver channels support advanced spatial processing techniques for clutter and jamming rejection.

14.9 Pulsed Noncoherent Receivers

The pulsed noncoherent receiver is the simplest type used in airborne radars. A typical block diagram is shown in Figure 14-11.

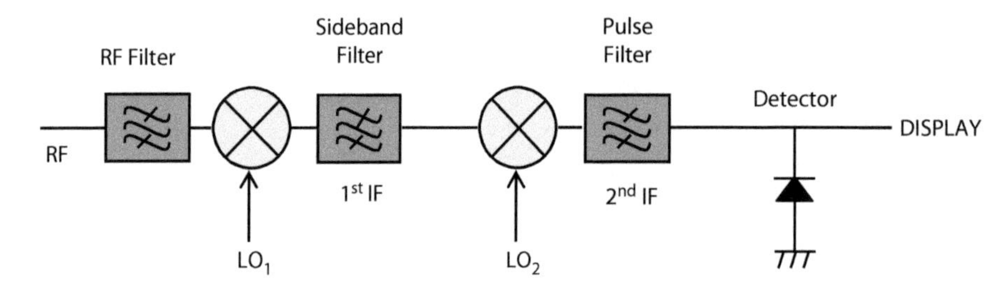

Figure 14-11. This block diagram shows a noncoherent pulse receiver.

It shows the main mixing and filtering steps, but for simplicity the necessary intermediate amplifiers are omitted. The output of the radar antenna is fed to the receiver via an RF filter that limits the band of the input signal to prevent unwanted mixing products and may be preceded by an LNA. Following the initial downconversion to a UHF IF, the unwanted mixing sidebands are filtered out, and a second-stage downconversion to video (typically less than 100 MHz) is carried out. At this point a fixed analog pulse matched filter is used before the detector stage. The pulse matched filter doubles as a filter to remove unwanted sidebands.

This type of radar commonly only has one pulse width, so only one pulse matched filter is required. If the radar has more than one pulse width, different filters are needed with a mechanism for switching between them.

This type of receiver commonly has a fairly poor noise figure, perhaps 10 dB or more. An LNA at the front end will improve matters significantly, and in a well-designed receiver a noise figure of 3 dB is achievable. However, this will require a lot of attention to detail it throughout the signal path, ensuring that losses in filters and mixers are compensated by suitable distributed amplifiers.

Earlier noncoherent receivers would simply feed the detected video to an analog display. This has now been superseded by the insertion of an A/D converter at the receiver output. The digitized video signal can then be conveniently displayed on any suitable digital device as well as put in a form that permits further signal processing.

14.10 Pulsed Coherent Receiver with Baseband Digitization

The majority of modern airborne radar receivers are coherent to support advanced modes of operation including clutter rejection and synthetic aperture radar. A typical block diagram is shown in Figure 14-12.

It shows the main mixing and filtering steps, but for simplicity the necessary intermediate amplifiers are omitted. At the front end, the LNA sets the noise figure of the system by providing sufficient gain that the noise contribution of subsequent stages is minimized. The signal is then downconverted in two stages, with intermediate filtering. As in the noncoherent receiver,

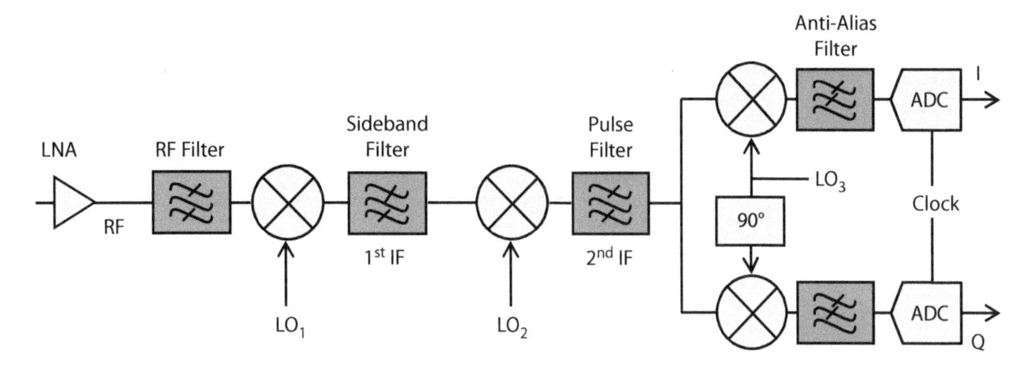

Figure 14-12. This block diagram shows the elements of a coherent pulse receiver with baseband digitization.

pulse matched filtering is usually carried out at a VHF. Here this is the second IF signal, which is split into two, with one replica being downconverted with a 90° phase shifted LO before anti-alias filtering and A/D conversion. The I and Q data streams are then made available to downstream digital signal processing.

Although apparently quite simple and elegant, this architecture is comparatively difficult to implement because it requires several different LO signals, complex matched circuitry, and (in general) active calibration schemes to achieve the desired level of performance.

14.11 Pulsed Coherent Receiver with IF Digitization

More modern coherent receivers employ digital downconversion. A typical block diagram is shown in Figure 14-13.

As before, the LNA at the front end sets the noise figure of the system by providing sufficient gain that the noise contribution of subsequent stages is minimized. The signal is then downconverted to a high IF (typically several GHz) in a single stage. After this the signal is anti-alias filtered before digitization with a high-speed A/D converter. The formation of the I and Q signals and all subsequent filtering is carried out digitally, either in custom digital circuitry, an FPGA, or a programmable digital processor. FPGAs are a preferred approach as they can provide very high throughput digital processing for a relatively low-power budget, but different systems may employ different approaches.

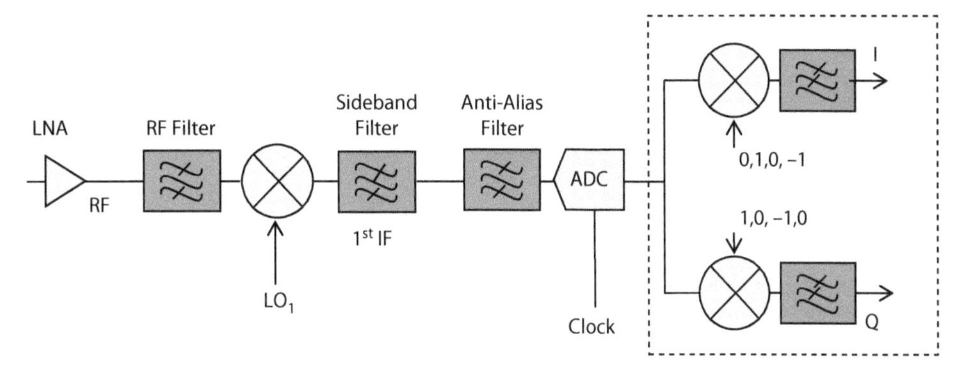

Figure 14-13. This block diagram shows the coherent pulse receiver with IF digitization.

This design greatly simplifies the required analog circuitry and avoids the need for much complex and difficult analog design. Pulse matched filtering is now carried out digitally (e.g., within the dotted boundary shown in Fig. 14-13) and can easily be reconfigured to support a wide range of pulse waveforms, including both unmodulated pulses and coded pulses (for pulse compression).

The problem with this type of design is the need for a very high-speed A/D converter. According to Nyquist's criterion, the sampling rate must exceed twice the maximum input frequency. Thus, with an IF of 2 GHz the A/D converter would appear to run at 4 GHz or more. This is challenging if we need a good dynamic range, but fortunately there is a clever trick that allows the use of slower A/D converters.

Sub-Nyquist Sampling. Nyquist's criterion applies to repetitive signals but makes no particular assumption about their structure other than their frequency content. In the case of a pulsed radar receiver, the signal has certain special characteristics that we can take advantage of so that in fact we can sample slower than Nyquist's criterion would suggest.

The radar signal consists of a signal with finite bandwidth dictated by the radar's range resolution. If the resolution is 150 m, the required signal bandwidth is nominally only 1 MHz. This signal is modulated on the RF carrier and is replicated at IF following downconversion. We can thus imagine the IF signal as being a pure carrier frequency with a limited narrowband modulation superimposed upon it. In principle, it is necessary to sample this signal at only twice the *pulse* bandwidth, not twice the IF. Thus, a lower speed A/D converter can be used.

In practice this is not so simple, and how well it works depends (once again) on the quality of filtering. To use sub-Nyquist sampling, the anti-alias filter becomes a *band-pass* filter, centered on the IF, rather than a *low-pass* filter as is normally envisaged. If this were all that were required, things would be relatively simple, but in fact complex detailed circuit characteristics of the A/D converter are often the limiting factor. Practical A/D converters have a maximum input frequency beyond which the bandwidth of their internal circuitry is exceeded. Therefore, even if a suitably filtered IF signal is put in, the low-pass characteristic of the A/D converter will suppress the wanted signal. As ever, practical designs are a complex trade-off of various factors, but sub-Nyquist sampling is an important and effective technique.

An example of a modern receiver of this type is shown in Figure 14-14.

Figure 14-14. Pictured here is an actual coherent pulse receiver with IF digitization. The downconverter and filtering components can be seen on the upper side, with the digital components on the underside. (Courtesy of Selex ES.)

14.12 Multichannel Receivers

Most airborne radars employ several parallel receiving channels. Early designs employed these for monopulse tracking, where the signal was compared in two precisely matched receivers and the difference was used to calculate the angular tracking error.

Such techniques are still widely used, but it is less common to find precisely matched analog receivers simply due to the sheer difficulty of achieving the required performance. More modern designs use active calibration techniques to measure and correct errors and are therefore much more stable and accurate.

Monopulse Receivers. These are the most common type and use two or sometimes three parallel receivers. One is dedicated to the main antenna sum channel output and the others to the azimuth and elevation difference channel outputs. In a two-channel design the azimuth and elevation difference signals are time-multiplexed. Monopulse receivers are normally just straight replicas of single channel receivers.

Guard Channel Receivers. In some radars a guard channel is incorporated to resolve between antenna mainlobe and sidelobe returns. The requirement here again is for good matching between the main and guard channel receivers in time, amplitude, and frequency so that an accurate comparison can be carried out and unwanted signals rejected. The most sophisticated guard systems carry out cancellation where an amplitude and phase shifted version of the guard channel output is added to the sum channel to cancel out an unwanted signal, such as a jammer.

Multichannel Receivers. These ideas may be extended up to systems with several tens or even hundreds of channels. Adaptive sidelobe cancellation is a generalized form of guard channel cancellation that can deal with many interfering signals simultaneously, and this can be extended to include cancellation of clutter signals. Achieving good balance across a receiver array is important, because any imbalances, while they can be nulled out, reduce the ability of the system to deal with unwanted external signals.

The ultimate in multichannel receivers is to have an individual receiver for each antenna element in an electronically steered array. This can offer enormous flexibility, although at the expense of a very large amount of digital signal processing (and cost). This approach currently remains impractical for most airborne radars, although some low-frequency radars, with a limited number of antenna elements, have already successfully gone this way. It is likely to become more common in the future.

14.13 Specialized Receivers

This section focuses on two specialized types of radar receivers.

Frequency-Modulated CW Receivers. Most of the discussion in this chapter has concerned pulsed receivers since these are by far the most common type in airborne radar, principally because most airborne radars time share the same antenna for transmission and reception. However, some radars, typically very small ones, use separate transmit and receive antennas and thus are able to use CW transmissions. In practice, nearly all such radars use frequency-modulated CW (FMCW) where the modulation frequency is used to measure range.

Designs for this type of receiver differ little from the coherent pulse designs already discussed. The main difference is that the filtering is now defined by the FM bandwidth rather than the pulse bandwidth. In other respects they are very similar.

Stretch Receivers. *Stretch*, or *deramp-on-receive*, receivers are a specialized type commonly used in very high range resolution radars such as synthetic aperture mapping radars. The architecture of the stretch receiver is made so that a relatively narrow band receiver can provide a much wider actual bandwidth.

How is this trick achieved? The idea is to exchange time and frequency. A stretch radar transmits a linear FM pulse, where the carrier frequency is linearly ramped up over a range that defines the total bandwidth of a transmission. So, for example, an X-band radar might ramp the carrier frequency from 8 to 10 GHz over a 100 μs period, providing a total bandwidth of 2 GHz. So far so good, but if our receiver has a bandwidth of only 100 MHz, how do we cope?

The solution is that, at the expected time of arrival of the returned signal, the first LO is then ramped at precisely the same rate as the transmitted signal. The echo and the LO ramp in parallel, and the difference frequency remains constant (for echoes from the same range). Varying the range will thus cause a different frequency to emerge from the first mixer. If this is done correctly, the bandwidth of the signal emerging from the mixer will be kept within the overall bandwidth of the receiver so that very fine range resolution can be achieved by frequency analyzing the signal output. The limitation is that this works over a limited range swath.

This trick of exchanging time for frequency is simple enough to implement in the receiver. It is necessary (in a coherent receiver) only to ensure that the front end has enough bandwidth to pass the full signal and to be able to ramp the first LO appropriately. Most of the complexity falls on the signal generation and timing; the receiver is almost unchanged.

14.14 Summary

Radar receiver designs are a complex trade-off. They must achieve very high dynamic range, low noise figure, and high spectral purity within a compact space and with performance in many cases limited by fundamental physical limits. Simple, elegant designs that digitize at the radar carrier frequency are impractical for all but the simplest radars and are likely to remain so.

Practical receiver designs all employ some form of frequency downconversion, and the filtering design is usually the crucial factor in determining receiver performance. Filtering must be carried out at RF, IF, for anti-aliasing and for matched filtering.

A/D conversion technology is a critical enabler, and modern designs allow more and more filtering to be performed digitally, which is much more stable and repeatable. The penalty is that this places very stringent requirements on the A/D converter.

Further Reading

H. T. Friis, "Noise Figures of Radio Receivers," *Proceedings of the IRE*, pp. 419–422, July 1944.

D. O. North, "An Analysis of the Factors which Determine Signal/Noise Discrimination in Pulsed Carrier Systems," *Proceedings of the IEEE*, Vol. 51, No. 7, pp. 1016–1027, July 1963.

P. P. Vaidyanathan, "Generalizations of the Sampling Theorem: Seven Decades after Nyquist," *IEEE Transactions Circuits Systems I*, Vol. 48, No. 9, September 2001.

H. Nyquist, "Certain Topics in Telegraph Transmission Theory," *Transactions of the AIEE*, Vol. 47, pp. 617–644, January 1928 (reprinted in *Proceedings of the IEEE*, Vol. 90, No. 2, pp. 280–305, February 2002).

J. B. Tsui, *Digital Techniques for Wideband Receivers*, SciTech-IET, 2004.

M. I. Skolnik (ed.), "Radar Receivers," chapter 6 in *Radar Handbook*, 3rd ed., McGraw Hill, 2008.

J. B. Tsui, *Special Design Topics in Digital Wideband Receivers*, Artech House, 2009.

Test your understanding

1. A receiver has a front-end LNA with a gain of 15 dB and a noise figure of 1.5 dB. The second stage of the receiver has a gain of 20 dB and a noise figure of 10 dB, and the third stage has a gain of 30 dB and a noise figure of 15 dB. What is the overall noise figure of the receiver?
2. An airborne radar operating on 10 GHz has a first IF of 2.5 GHz. What is (a) the frequency of the first local oscillator and (b) the image frequency?
3. A coherent airborne radar employs downconversion to baseband (I and Q) and baseband digitization. The two A/D converters each have a sampling rate of 100 MHz. What is the maximum signal bandwidth the receiver can deal with?

Fairchild Republic A-10 Thunderbolt II (1977)

The A-10 is known as "The Warthog" a nickname given by its first pilots for its less than striking appearance. Designed around its main armament, the GAU-8 Avenger, a massive 30 mm rotary cannon, the Thunderbolt is designed solely for close air support of ground forces. Its ability to maneuver at low speeds and altitude allows it to loiter and target more accurately than faster jets, while its strong airframe and triple redundancy systems makes it exceptionally tough.

15

Measuring Range and Resolving in Range

By far the most widely used method of range measurement is pulse-delay ranging. It is simple and can be extremely accurate, because it is easy to make precise timing measurements. However, there is no direct way of telling for sure which transmitted pulse a received echo belongs to, so the measurements are, to varying degrees, ambiguous.

This chapter examines pulse-delay ranging more closely to learn how target ranges are actually measured and to consider the nature of the ambiguities. We will see how ambiguities may be avoided at low pulse repetition frequencies (PRFs) and resolved at higher PRFs. We will then consider ambiguities of a secondary type, called *ghosts*, and see how these may be eliminated. Finally, we will look briefly at how range is measured during single-target tracking and in tracking using electronically scanned antennas.

15.1 Pulse-Delay Ranging

Measuring Range. The vast majority of radars employ pulsed transmissions, where a single antenna is time-shared between transmission and reception. With a pulse radar, the range of a target can be directly determined by measuring the time between the transmission of each pulse and reception of the echo from the target (Fig. 15-1). The round-trip time is divided by two to obtain the time the pulse took to reach the target. This time, multiplied by the speed of propagation, is the target's range. Expressed mathematically,

$$R = \frac{ct}{2}$$

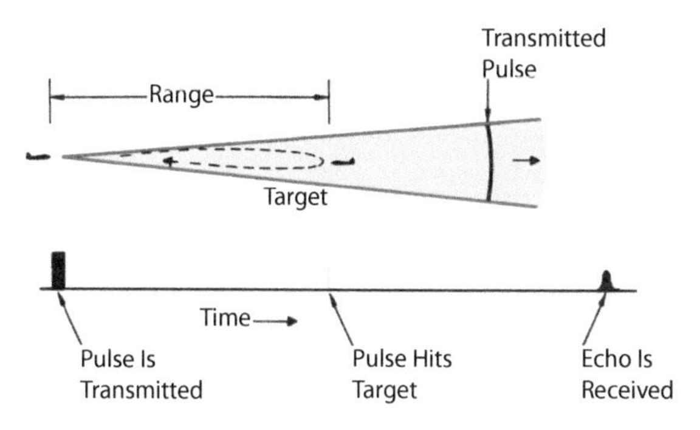

Figure 15-1. Range is determined by measuring the time between transmission of a pulse and reception of the target echo.

where

R = range
c = speed of propagation
t = round-trip transit time

The speed of propagation in air is, for most practical purposes, the same as the speed of light in vacuo (approximately 3×10^8 m/s). Although in this book we generally use metric units, it is common for radars to display ranges in other units, often nautical miles (1 nm = 1852 m).

A useful rule of thumb is this: 10 μs of round-trip transit time equals 1.5 km of range.

APPROXIMATE RANGING TIME

Unit of Distance	μs
1 nm	12.4
1 mi	10.7
1 km	6.67
1.5 km	10.0

If you wish to calculate ranges more accurately, the speed of light in various units of distance is given in Chapter 4. Just how the range is actually measured varies with the type of radar.

Simple Analog Radars. In early radars, and including many radars still in use today, range is measured right on the operator's display. This method is most graphically illustrated by the simple A-scope display of World War II, which used a cathode ray tube in which an electron beam was repeatedly swept across its face (Fig. 15-2). It starts a new sweep each time the radar transmits a pulse, moves at a constant rate throughout the interpulse period, and "flies" back to the starting point again at the end of the period. Each sweep is called a *range sweep*; the line traced by the beam is called the *range trace*. When a target echo is received, it deflects the beam, causing a *pip* to appear on the range trace. The distance from the start of the trace to the pip corresponds to the time between transmission and reception, thus indicating the target's range. (As highly directional radar antennas were adopted, the familiar circular *plan position indicator* type of display was adopted, which could simultaneously display range and angle; in this case the distance of the pip from the center indicated the range.)

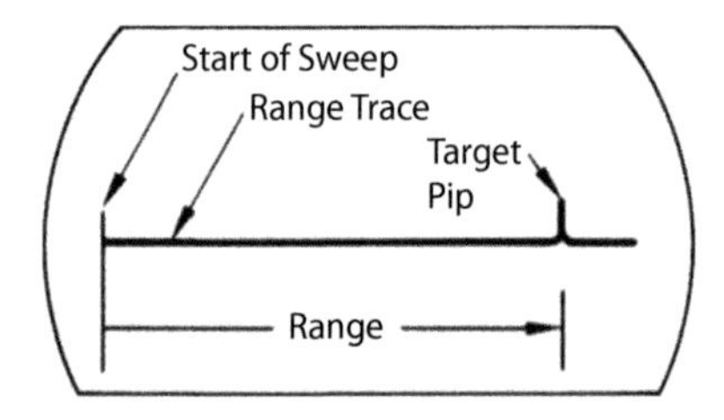

Figure 15-2. In simple analog radars, range is measured on the operator's display. Shown here is an A-scope display of a World War II radar.

Sophisticated Analog Radars. The next stage in radar development introduced the concept of a range gate, in which range is measured by applying the receiver output to a bank of switching circuits, called range gates (Fig. 15-3). The gates are opened sequentially at times corresponding to successive resolvable increments of range: Gate No. 1, then Gate No. 2, and so on. A target's range is determined by noting which gate, or adjacent pair of gates, its echoes pass through.

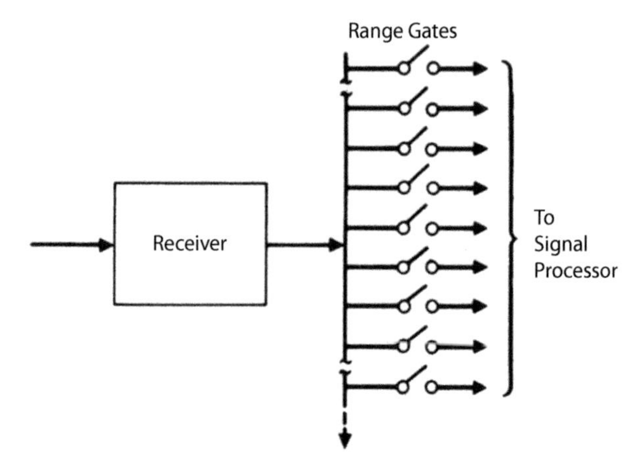

Figure 15-3. In sophisticated analog radars, range gates are sequentially opened (switch closed). Range is determined by noting which gate a target's echoes go through.

Enough range gates are provided to cover either the entire interpulse period or the portion of it corresponding to the range interval of interest. Although this design was a major advance for its time, it has been superseded by the advent of digital systems.

Digital Radars. When digital signal processing is employed, range is essentially measured in the same way as in range-gated analog radars. This is the norm in all modern radars.

The amplitude of the receiver's video output is periodically sampled (Fig. 15-4), and the amplitude is converted to digital form. A *sample-and-hold* circuit is used in conjunction with an analog-to-digital (A/D) converter, which allows the instantaneous level of the receiver's output at each range interval to be obtained. These digital values are then stored in *range bins* prior to subsequent signal processing. A separate bin is provided for each range increment within the interval of interest.

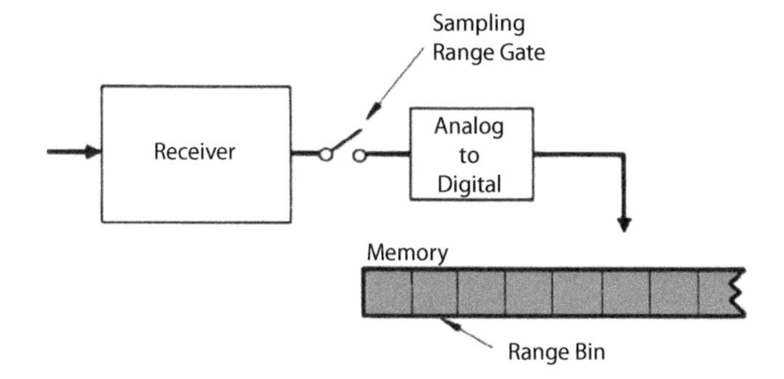

Figure 15-4. In digital radars, receiver output is periodically sampled by a range gate. Converted to a number, each sample is stored in a separate range bin.

As noted in Chapter 2, to enable Doppler filtering after the received signals have been converted to video the receiver must provide both in-phase (I) and quadrature (Q) outputs. Consequently, in digital Doppler radars *two* numbers are stored for each range increment. Together, they correspond to the return passed by a single range gate in an analog system. (More modern systems may employ a faster sampling system and digital downconversion as described in Chapter 14, but for the purposes of this chapter these more advanced techniques are equivalent to the older approach.)

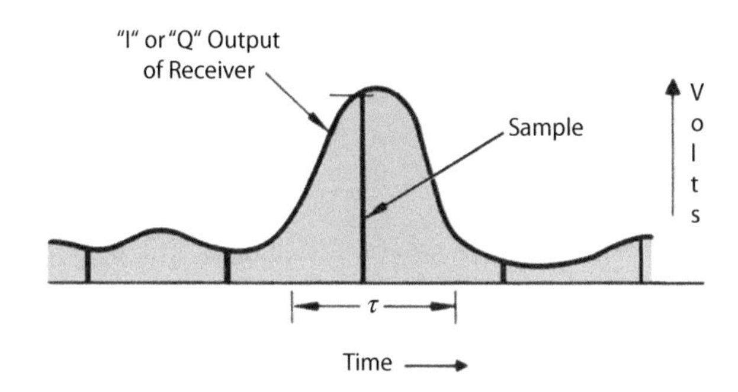

Figure 15-5. A video signal is generally sampled at intervals of the order of a pulse width, τ.

Resolving in Range. The sampling interval is usually chosen to be on the order of the pulse width, τ (Fig. 15-5). Making the sampling interval longer can reduce the complexity of the system, but it results in loss of signal and furthermore degrades the radar's ability to resolve targets in range.

To realize the full range-resolving potential of the pulses as well as to enable more accurate range measurement, samples may be taken at considerably shorter time intervals than the pulse width (Fig. 15-6). Range is then determined by interpolating between the numbers in adjacent range bins. If, for example, the numbers in two adjacent bins are equal, the target is assumed to be halfway between the ranges represented by the two bin positions. Depending on the sampling rate, the signal-to-noise ratio and the pulse width, the measurement can be quite precise.

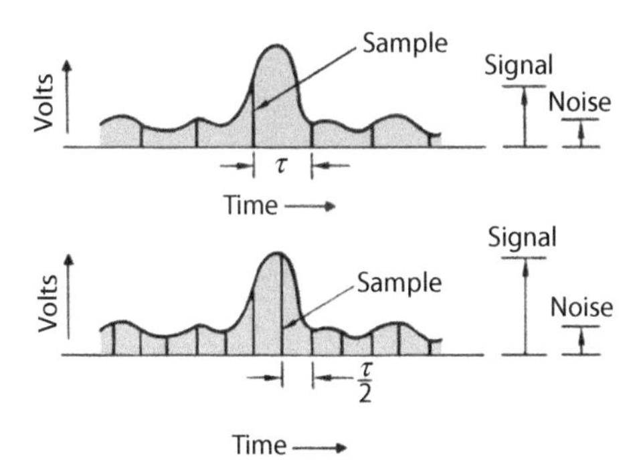

Figure 15-6. To enable more accurate measurement and minimize loss of signal-to-noise ratio, samples may be taken at intervals shorter than a pulse width; range is then computed by interpolating between samples.

Using a comparatively high sampling rate also minimizes the loss in signal-to-noise ratio that occurs when a target's echoes fall partly in one sampling interval and partly in the next. This is called *range-gate straddling loss*. In practice, this loss is determined by both the chosen sampling rate and the characteristics of the radar's pulse filtering (as explained in Chapter 14).

15.2 Range Ambiguities

Pulse-delay ranging works without a hitch as long as the round-trip transit time for the most distant target the radar may detect is shorter than the pulse repetition interval. But if the radar detects a target whose transit time exceeds the pulse repetition interval, the echo of one pulse will be received after the next pulse has been transmitted, and the target will appear, falsely, to be at a much shorter range than it actually is.

Nature of the Ambiguities. To get a more precise feel for the nature of the ambiguities, let us consider a specific example. Suppose the pulse repetition interval, T, corresponds to a range

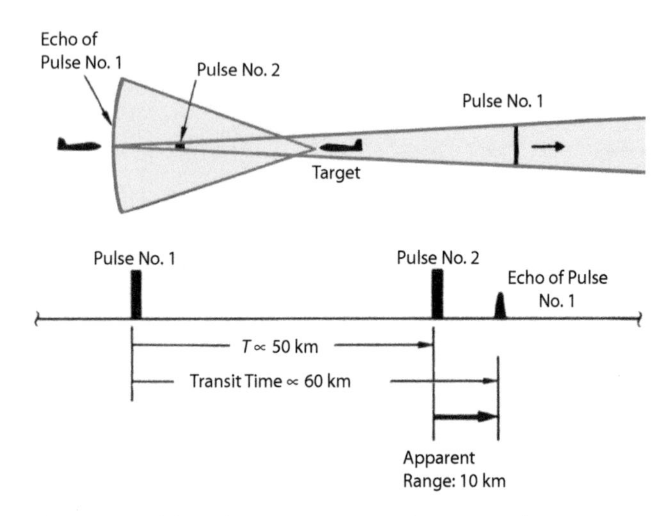

Figure 15-7. If the pulse repetition interval corresponds to 50 km and the transit time to 60 km, the range will appear to be only 10 km.

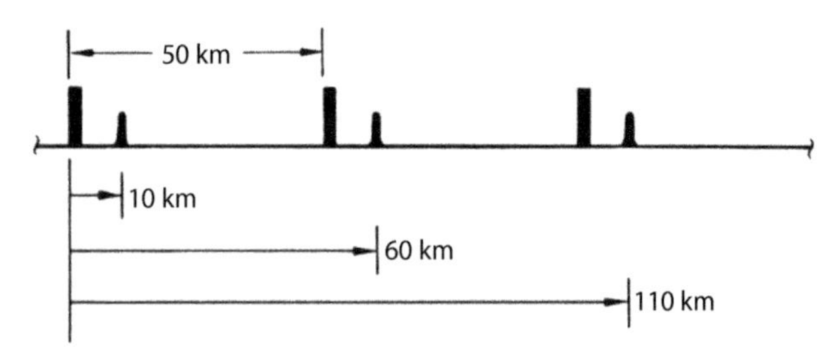

Figure 15-8. There is no direct way of telling whether the true range is really 10 km, 60 km, 110 km, and so forth.

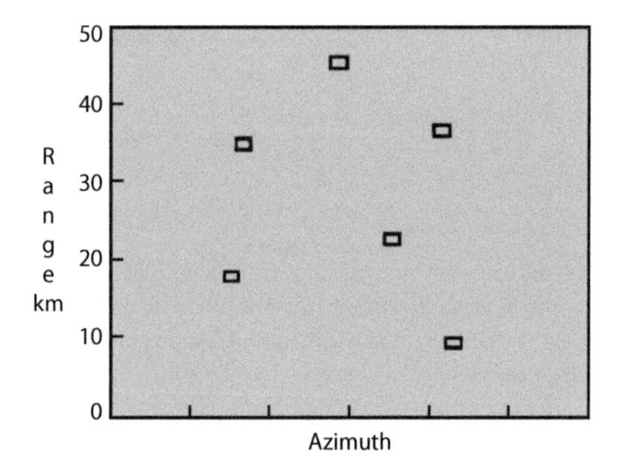

Figure 15-9. The true range of any target appearing on this radar display may be greater than 50 km. Ergo, all ranges are ambiguous.

of 50 km and echoes are received from a target at 60 km (Fig. 15-7). The transit time for this target will be 20 percent greater than the interpulse period (60/50 = 1.2). Consequently, the echo of Pulse No. 1 will not be received until $0.2T$ microsecond after Pulse No. 2 is transmitted. The echo of Pulse No. 2 will not be received until $0.2T$ microsecond after Pulse No. 3 is transmitted, and so on.

If the difference between the time an echo is received and the time the immediately preceding pulse was transmitted is used as the measure of range, the target will appear to be at a range of only 10 km (0.2×50). In fact, there will be no direct way of telling whether the target's true range is 10 km, 60 km, or 110 or 160 km for that matter (Fig. 15-8). This arises simply because to the radar receiver all echo pulses look exactly the same.

Not only that, but as long as there is a possibility of detecting targets at ranges greater than 50 km, the observed ranges of all targets detected by the radar will be ambiguous—even though their true ranges may be less than 50 km. Put another way, if the range indicated by *any* target blip on the radar display can be greater than 50 km, the range indicated by *every* target blip is ambiguous. There is no telling which of the blips represents a target at the greater range (Fig. 15-9). Therefore, range is almost always ambiguous. This point is often overlooked.

This can be a particular problem in a military radar designed to detect small (low radar cross section) aircraft. Such a radar is very sensitive, so it can also pick up echoes from large commercial aircraft at much longer ranges. This can be misinterpreted as smaller targets at close range due to the ambiguity.

The extent of the range ambiguities in the return from a single target is commonly gauged by the number of interpulse periods spanned by the transit time. That is, it is determined by noting during which interpulse period the target's echoes are received (e.g., first, second, third, fourth) following transmission of the pulses that produced them. An echo received during the first interpulse period is called a first-time-around echo. Echoes received during subsequent periods are called multiple-time-around echoes (MTAEs).

Maximum Unambiguous Range. For a given PRF, the longest range from which single-time-around echoes can be received—hence the longest range from which any return may be received without the observed ranges being ambiguous—is called the maximum unambiguous range (or simply the unambiguous range). It is commonly represented by R_u. Since the round-trip transit time for this range equals the interpulse period,

$$R_u = \frac{cT}{2}$$

where

R_u = maximum unambiguous range
c = speed of propagation
T = pulse repetition interval

Since the pulse repetition interval is equal to one divided by the PRF, f_r, an alternative expression is

$$R_u = c / 2f_r$$

A useful rule of thumb is this: R_u in km equals 150 divided by the PRF in kHz (Fig. 15-10). For a PRF of 10 kHz, for example, R_u would be 150/10 = 15 km.

Coping with Ambiguous Ranging. How to deal with range ambiguities depends both on their severity and on the penalty that must be paid for mistaking a distant target for a target at closer range (or vice versa). The severity, in turn, depends on the maximum range at which targets are apt to be detected and on the PRF. Often, the PRF is determined by considerations other than range measurement, such as providing adequate Doppler resolution for clutter rejection. The penalty for not resolving an ambiguity, of course, depends on the operational situation.

Obviously, the possibility of ambiguities could be eliminated altogether by making the PRF low enough to place R_u beyond the maximum range at which any target is apt to be detected (Fig. 15-11). However, since targets of large radar cross section may be detected at very great ranges, it may well be impractical to set the PRF this low, even when a comparatively low PRF is acceptable.

On the other hand, for the expected conditions of use, the probability of detecting such large targets may be slight, and the consequences of sometimes mistaking them for targets at closer range may be of no great importance.

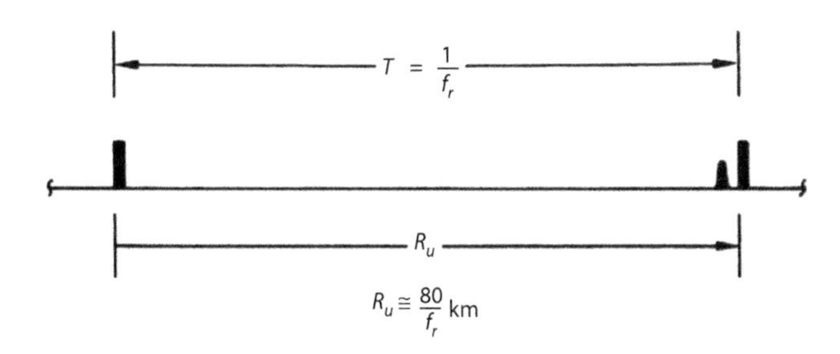

Figure 15-10. The longest range from which an unambiguous return may be received, R_u, corresponds to the interpulse period, T.

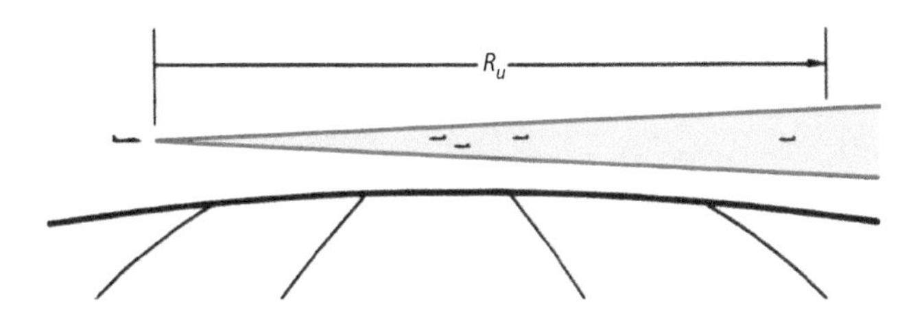

Figure 15-11. Ambiguities can be avoided completely only by making R_u, greater than the range at which any target may be detected.

15.3 Eliminating Ambiguous Returns

If targets at greater ranges than R_u are of no concern to us, we can solve the problem of ambiguities simply by rejecting all returns from beyond R_u (Fig. 15-12). This may sound like a difficult trick, but it can be accomplished quite relatively easily.

One technique is PRF jittering (Fig. 15-13). Any change in PRF results in a change in R_u. Thus, targets at a shorter range than

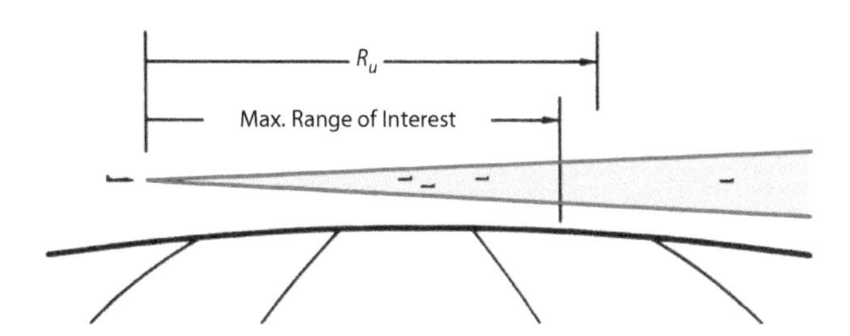

Figure 15-12. If R_u is greater than the maximum range of interest, the problem of ambiguities can be solved by eliminating all return from ranges greater than R_u.

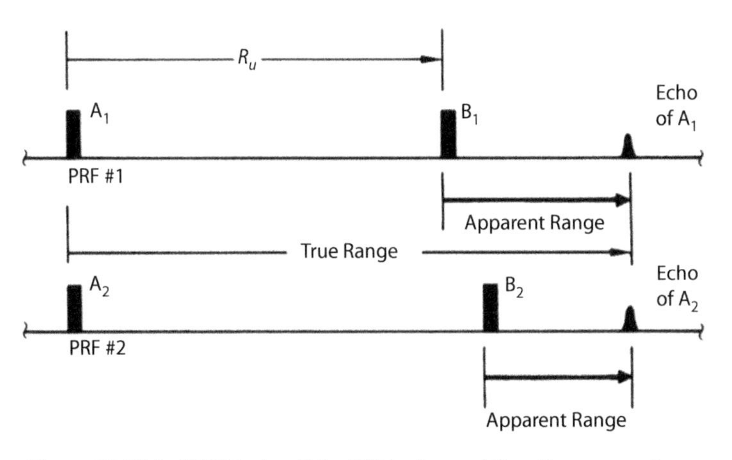

Figure 15-13. In PRF jittering, if the PRF is changed then the apparent range of a target beyond R_u will change—identifying the target as ambiguous.

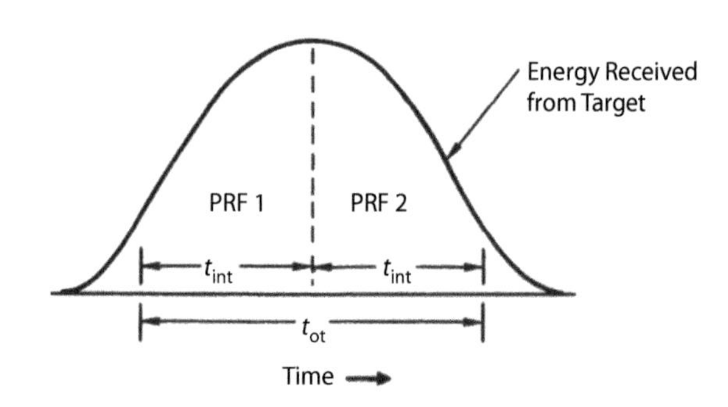

Figure 15-14. The penalty for PRF jittering is that the potential integration time is cut in half, thereby reducing detection sensitivity.

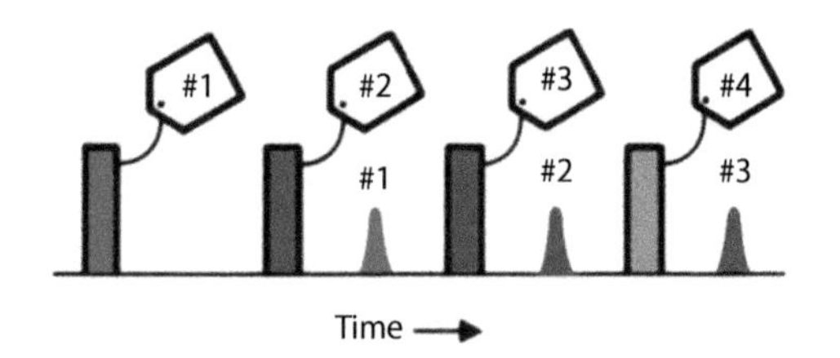

Figure 15-15. By tagging transmitted pulses, we can tell which pulse each belongs to. But except for frequency modulation, tagging has proved impractical.

R_u will always appear at the same time delay from the last transmitted pulse, whereas targets at a longer range than R_u will appear at different time delays from the last transmitted pulse. In other words, shorter range targets will consistently appear in the same range bin, whereas longer range targets will appear in different range bins on each pulse. Since most practical radars employ integration of multiple pulses within the time-on-target, the shorter range targets will integrate up more than the longer range targets so that they can be separated out. Naturally, one pays a price for this improvement. The presence of the MTAEs means that the radar's detection threshold will need to be raised to eliminate them, and this results in some loss in sensitivity. A simple example of this effect is illustrated in Figure 15-14, where two alternating PRFs are employed.

A better technique is to label the pulses in some way that they can be distinguished—the easiest way being to transmit each pulse on a different frequency. In effect, this increases R_u, and the effective PRF is now determined by the interval between pulses of the same frequency. This method, known as frequency agility, is very effective as on each PRI the radar's receiver is tuned only to the first-time-around echoes. Longer range returns will be at a different frequency and are filtered out in the receiver. The approach requires a more complex transmitter and receiver, but it eliminates the losses associated with the PRF jitter technique. This is illustrated in Figure 15-15.

15.4 Resolving Ambiguities

For reasons having nothing to do with ranging, the PRF may have to be made so high that the maximum range of interest is longer than R_u, often many times so. The most common reason for this is that the radar is required to measure Doppler to remove clutter. In this case, the radar is normally designed to resolve range ambiguities, although some systems may employ a *velocity-only* detection mode.

Various methods can be used to resolve range ambiguities, but in essence all of them are essentially techniques to introduce a longer periodicity and hence longer decodable unambiguous range.

Tagging Pulses. Superficially, it might seem that the easiest way to resolve the ambiguities would be to tag successive transmitted pulses—that is, to change (modulate) some characteristic (amplitude, width, or frequency) in some longer cyclical pattern that extends the ambiguity. By looking for corresponding changes in the target echoes, we could then tell which transmitted pulse each echo belongs to and could thereby resolve the ambiguities.

But for one reason or another—problems of mechanization in the case of amplitude modulation and eclipsing and range gate straddling in the case of pulse width modulation—only one of these approaches has as yet proved practical: frequency modulation (Chapter 17). When combined with Doppler processing for air-to-air applications, this approach can be quite complex. Also, earlier analog implementations had serious limitations, principally because of the difficulty of accurately reproducing the desired frequency modulation. The advent of digital waveform generation has largely eliminated this problem in modern systems.

Multiple PRFs. The resolution technique commonly used is an extension of PRF jittering, called PRF switching. It goes a step beyond jittering by taking account of how much a target's apparent range changes when the PRF is changed. By knowing this and the amount the PRF has changed, it is possible to determine the number of whole times, n, that R_u is contained in the target's true range. The principle is essentially exactly the same as that used in a vernier ruler.

The method for determining n is best illustrated by example. We will assume that for reasons other than ranging (Doppler processing) a PRF of 15 kHz has been selected. Consequently, the maximum unambiguous range, R_u, is $150/15 = 10$ km. However, the radar must detect targets out to ranges of at least, say, 75 km, or nearly $8 \times R_u$, and undoubtedly it will detect some targets at ranges beyond that as well.

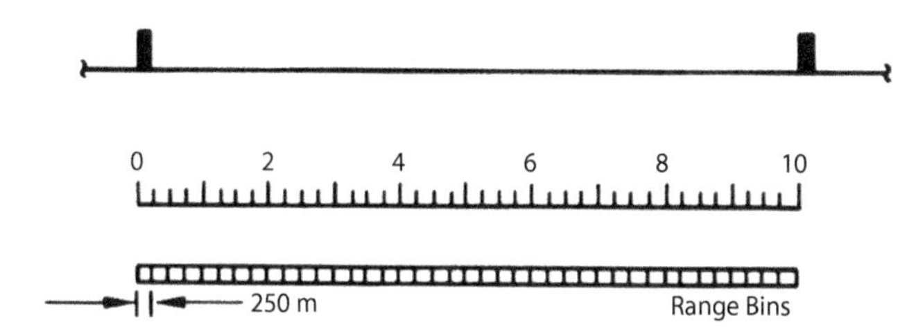

Figure 15-16. To span 10 km ranging interval, a bank of 40 range bins is provided. Each represents a range increment of 0.25 km, or 250 m.

The apparent ranges of all targets will, of course, lie between 0 and 10 km (Fig. 15-16). To span this 10 km interval, a bank of 40 range bins has been provided. Each bin position represents a range interval of 250 m.

A target is detected in bin No. 24. The target's apparent range is $24 \times 0.25 = 6$ km (Fig. 15-17). On the basis of this information alone, we know only that the target could be at any one of the following ranges:

6 km

$10 + 6 = 16$ km

$10 + 10 + 6 = 26$ km

$10 + 10 + 10 + 6 = 36$ km

$10 + 10 + 10 + 10 + 6 = 46$ km

$10 + 10 + 10 + 10 + 10 + 6 = 56$ km

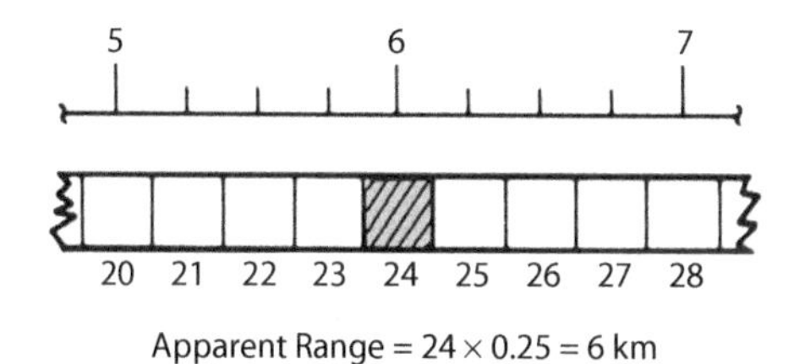

Figure 15-17. A target appears in bin No. 24; its apparent range is 6 km.

To determine which of these is the true range, we switch to a second PRF. To keep the explanation simple, we will assume that this PRF is just enough lower than the first to make R_u 250 m longer than it was before (Fig. 15-18).

What happens to the target's apparent range when the PRF is switched will depend on what the target's true range is. If the true range is 6 km, the switch will not affect the apparent range. The target will remain in bin No. 24.

But if the true range is greater than R_u, for every whole time R_u is contained in the target's true range the apparent range will decrease by 250 m: the target will move one bin position

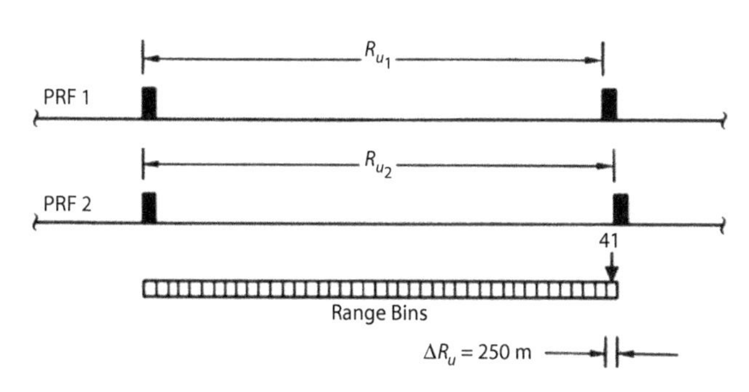

Figure 15-18. PRF is changed to increase R_u by 250 m.

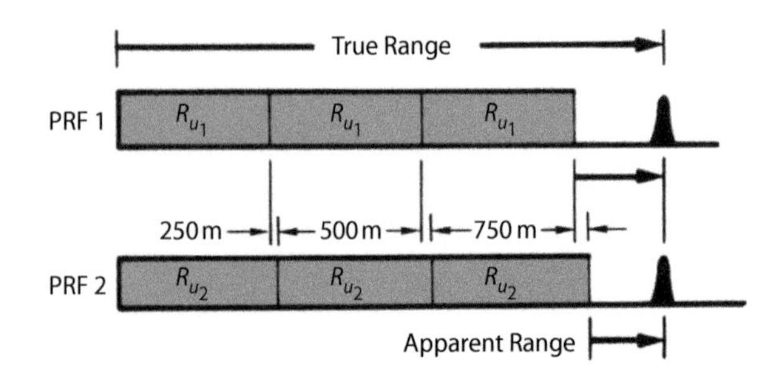

Figure 15-19. For every whole time R_u is contained in true range, apparent range will decrease 250 m when the PRF is switched.

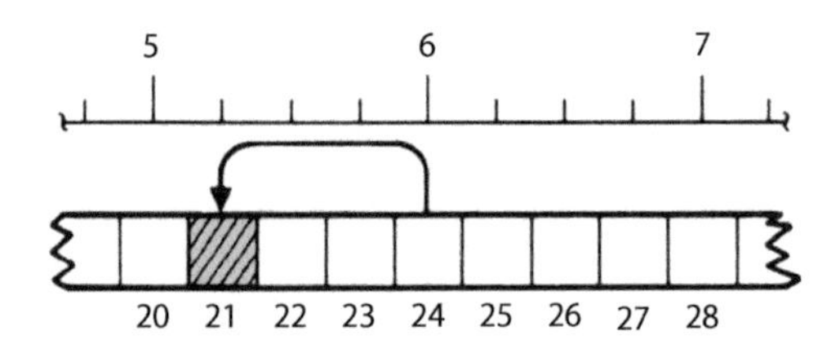

Figure 15-20. If the target jumps three bins, true range is $(3 \times 10) + 6 = 36$ km.

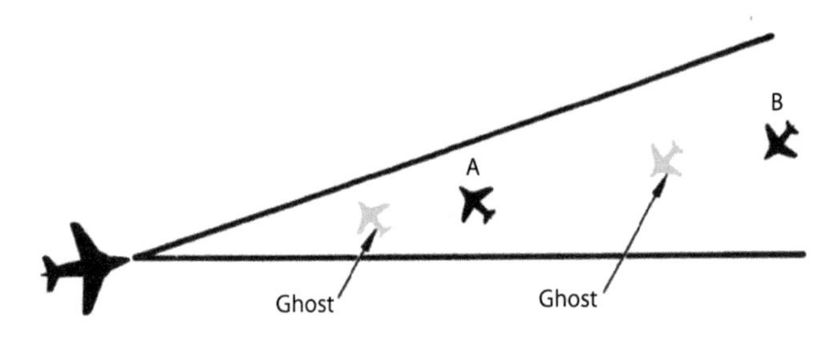

Figure 15-21. If more than one target is detected at the same angle and the targets are not resolvable in Doppler frequency, a problem of ghosts will occur.

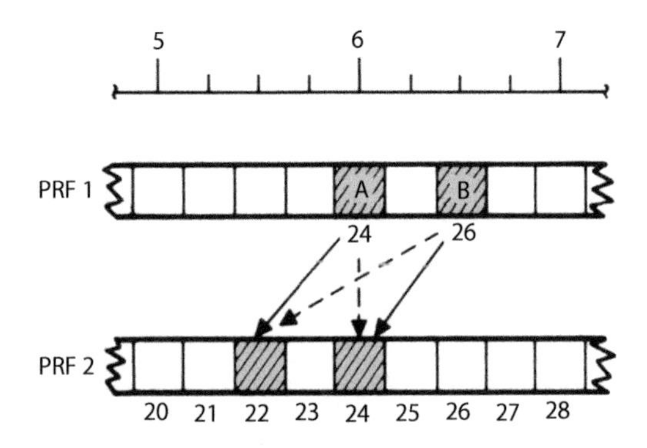

Figure 15-22. When PRF was switched, did Target A move to bin No. 22 and Target B to bin No. 24, or, did Target A stay put?

to the left in Figure 15-19. For the PRFs used here, n equals the number of bins the target shifts.

Computing the Range. We can find the true range, therefore, by (1) counting the number of bin positions the target moves, (2) multiplying this number by R_u, and (3) adding the result to the apparent range.

Suppose the target moves from bin No. 24 (apparent range 6 km) to bin No. 21, a jump of three bins (Fig. 15-20). The target's true range, then, is $(3 \times 10) + 6 = 36$ km.

A practical system would not be mechanized with PRFs so closely spaced. The principle, though, is the same.

From the foregoing, we can draw the following conclusions. The number of whole times, n, that R_u is contained in a target's true range equals the change in apparent range when the PRF is switched, divided by the change in R_u for the two PRFs.

$$n = \Delta R_{\text{apparent}} / \Delta R_u$$

The true range is n times R_u plus the apparent range.

$$R_{\text{true}} = nR_u + R_{\text{apparent}}$$

In a practical system, it is rare to employ only two different PRFs for a variety of reasons, the principal one being avoidance of *blind zones*. These occur at intervals of R_u because the radar receiver has to be blanked during the pulse transmission. To avoid this periodic blind zone, a different PRF can be employed to move the zones around. Thus, a target may be blanked on one PRF but not another. However, this phenomenon can be seen to prevent the ambiguity resolution mechanism from working. The trick is to transmit a *schedule* of n different PRFs and to demand that the target is visible on a smaller number, m, of n. Provided the m different PRFs allow the ambiguities to be resolved, we have a solution. In practice, this is a tricky optimization problem.

Ghosts. When PRF switching is used, a secondary sort of ambiguity, called ghosting, is sometimes encountered. It may occur when two targets are detected simultaneously (i.e., at the same azimuth and elevation angles) and their range rates are so nearly equal that their echoes cannot be separated on the basis of Doppler frequency (Fig. 15-21). Under this condition, when the PRF is switched and one or both targets move to different range bins, we may not be able to tell which target has moved to which bin. Each target will appear to have two possible ranges. One is the true range; the other, in radar jargon, is a ghost.

Example of Ghosts. Figure 15-22 shows two targets, A and B, in the same bank of range bins as used in the preceding example. When the radar is transmitting at the first PRF, the targets are two bins apart: A is in bin No. 24 (apparent range 6 km); B is in bin No. 26 (apparent range 6.5 km). When we switch to the second PRF, the targets appear in bins No. 22 and No. 24. But we have no direct way of telling whether A and B have both moved to the left two bins or whether A has merely stayed put and B has moved four bins to the left and is in bin No. 22.

Each target thus has two possible true ranges (Fig. 15-23). If both A and B have moved two bin positions, the true ranges are

$$\text{Target A: } (2 \times 10) + 6 = 26 \text{ km}$$

$$\text{Target B: } (2 \times 10) + 6.5 = 26.5 \text{ km}$$

On the other hand, if A stayed put and B moved four bin positions, the true ranges are

$$\text{Target A: } (0 \times 10) + 6 = 6 \text{ km}$$

$$\text{Target B: } (4 \times 10) + 6.5 = 46.5 \text{ km}$$

One of the two pairs of ranges are ghosts.

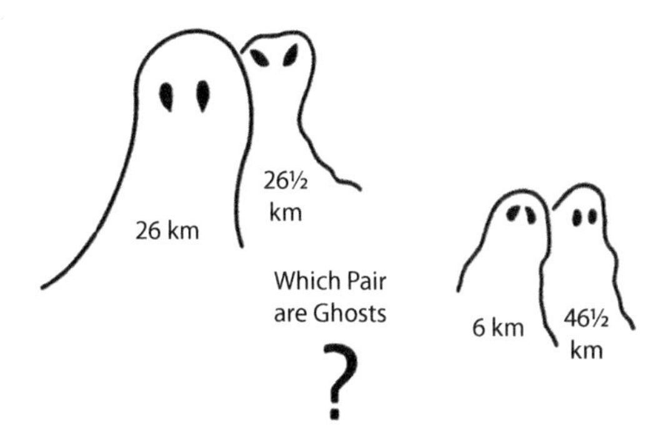

Figure 15-23. Each target shown in Figure 15-22 has two possible true ranges.

Identifying Ghosts. The ghosts may be identified by switching to a third PRF (Fig. 15-24). To simplify the explanation, we'll assume that PRF No. 3 is just enough higher than PRF No. 1 to decrease R_u by 250 m—that is, shorten it by one range bin (from 40 to 39 bins). Accordingly, when PRF No. 3 is used, for every whole time R_u is contained in either target's true range the target will appear one position to the *right* of the bin it occupied when PRF No. 1 was used. This is the same number of positions that appeared to the *left* of that bin when PRF No. 2 was used.

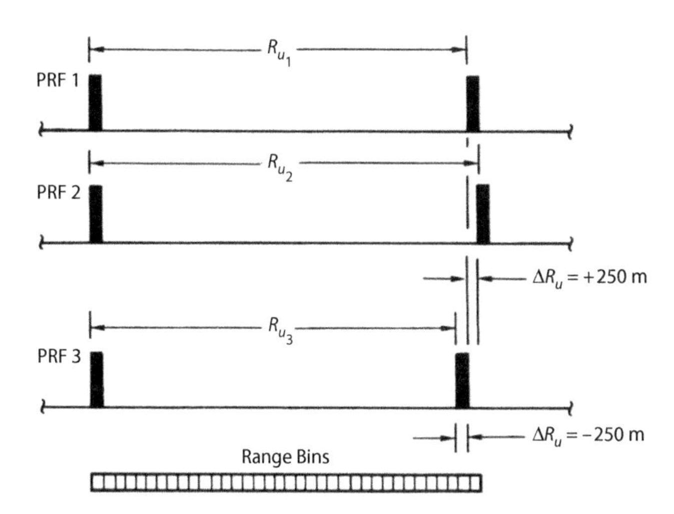

Figure 15-24. To identify the ghosts, a third PRF is added. In this case, it decreases R_u by 250 m.

Let's say that we switch to PRF No. 3 and the targets appear in bins 26 and 28. Which of the two pairs of ranges are ghosts?

As you can see from Figure 15-25, bin 26 is two positions to the right of bin A, which it originally occupied. Likewise, bin 28 is two positions to the right of bin B, which it originally occupied. Since, when we switched earlier to PRF No. 2, one target appeared two positions to the left of the bin A originally occupied and the other target appeared two positions to the left of the bin B originally occupied, we conclude that $n = 2$ for both targets. Their true ranges are 26 km and 26.5 km. The other pair of ranges are ghosts.

It may be instructive to consider where the targets would have appeared when we switched to PRF No. 3 if the first

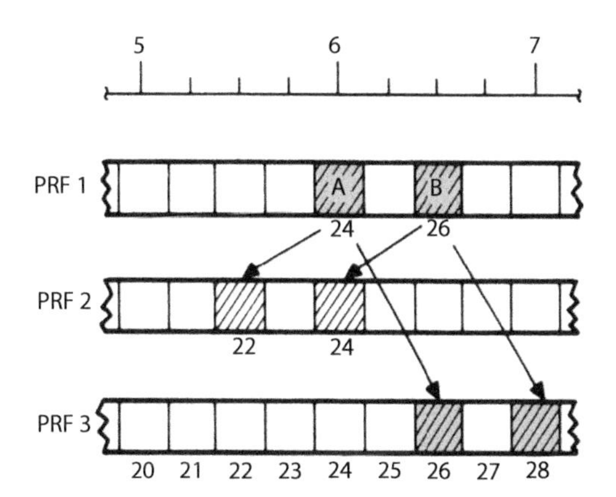

Figure 15-25. When radar is switched to PRF 3, targets jump to bins 26 and 28. The value of n for both targets must be 2.

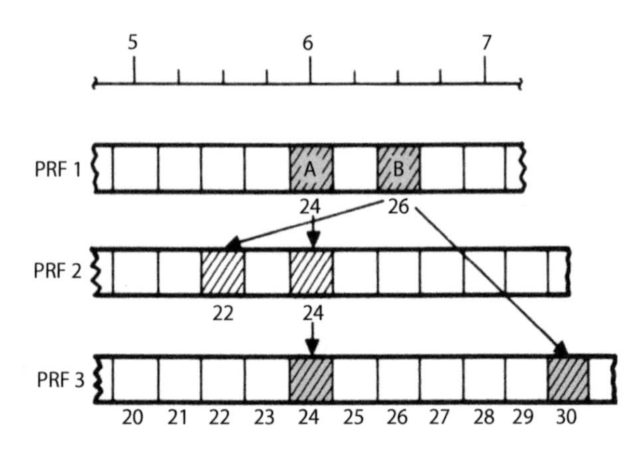

Figure 15-26. If A's true range had been 6 km, it would have stayed put when the radar was switched to PRF 3, and B would have jumped four positions to the right.

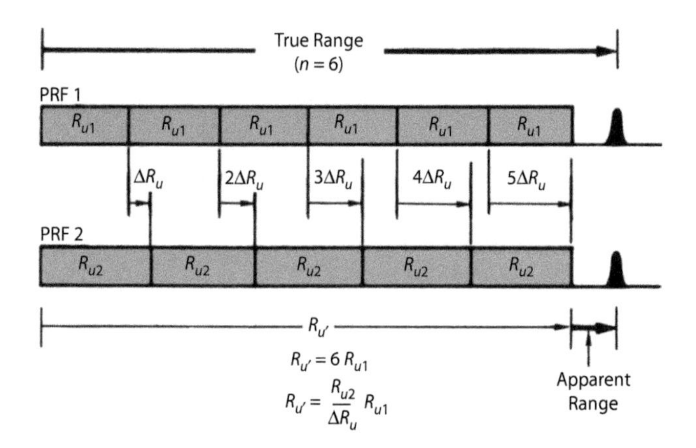

Figure 15-27. Depicted here is the range for which ambiguities can no longer be resolved by switching between two PRFs. Since $5R_{u2} = 6\,R_{u1}$, apparent range does not change when PRF is switched. R_u', is maximum unambiguous range for this combination of PRFs.

pair of ranges had been ghosts and the second pair (6 km and 46.5 km) had been the true ranges. In that case (Fig. 15-26), since $n = 0$ for 6 km target A would have stayed put. Since $n = 4$ for 40 km, target B would have moved four positions to the *right*—the same distance (for these particular PRFs) that it must have moved to the *left* when earlier we switched to PRF No. 2.

15.5 How Many PRFs?

From what has been said so far, it might appear that no more than three PRFs would ever be required: one for measuring range; another for resolving range ambiguities; a third for deghosting simultaneously detected targets. This is not so, however.

Number of PRFs for Resolving Ambiguities. Depending on how great the detection ranges are and how high and widely spaced the PRFs are, more than one PRF (besides the first) may be required to resolve ambiguities. Figure 15-27 illustrates why.

The true range in that example includes six whole multiples of the unambiguous range for PRF No. 1 ($n = 6$). This is clear. But the difference in the unambiguous ranges for the two PRFs (ΔR_u) is such that five times the unambiguous range for PRF No. 2 exactly equals six times the unambiguous range for PRF No. 1. Consequently, for the target range assumed here (Fig. 15-28), when the PRF is switched the apparent range remains the same just as though $n = 0$.

If the true range were long enough to make $n = 7$ or more, the apparent range would again change when the PRF was switched, but the change then would indicate only how much n exceeds 6. This particular combination of PRFs extends the maximum unambiguous range to six times the unambiguous range for PRF No. 1 but no farther (Fig. 15-29).

In fact, a more general expression for the true range than that given earlier is

$$\text{True range} = n'R_u' + nR_u + R_{\text{apparent}}$$

where R_u' is the unambiguous range for the combination of the two PRFs, and n' is the number of whole times R_u' is contained

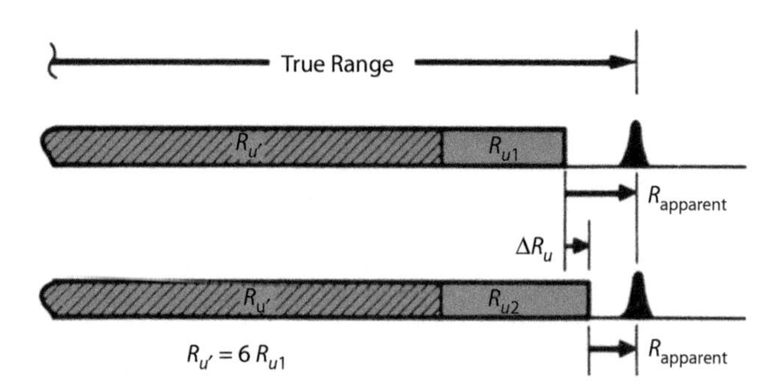

Figure 15-28. If the true range is increased beyond R_u', the apparent range will change when PRF is switched but (in this case) only by an amount corresponding to $(n - 6)$.

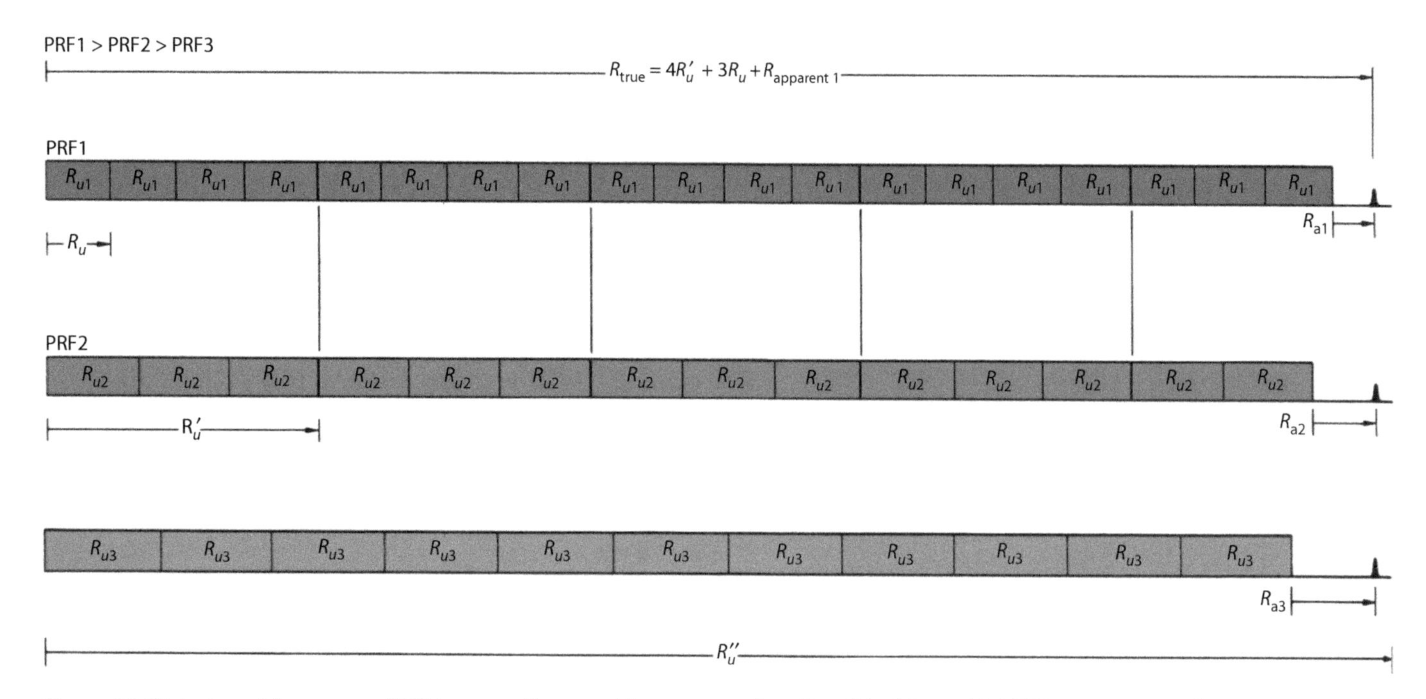

Figure 15-29. Just as adding a second PRF increases the unambiguous range from R_u to R_u', adding a third PRF increases it to R_u''. For any one combination of R_{u1}, R_{u2}, R_{u3}, and $R_{apparent}$, there is only one possible value of the true range. It is uniquely indicated by the values of the three apparent ranges, R_{a1}, R_{a2}, and R_{a3}.

in the true range. To find the value of n' we must switch to a third PRF.

With the aid of a diagram like Figure 15-29, it can be shown that for every additional PRF the unambiguous range for the combination increases by the ratio of (1) R_u for the added PRF to (2) the difference between that value of R_u and the value for the preceding PRF (Fig. 15-30). Thus, if the unambiguous ranges for three PRFs taken individually are 3, 4, and 5 km, the unambiguous range for the combination is $3/1 \times 4/1 \times 5/1 = 60$ km. How many PRFs are required for resolving range ambiguities, then, depends on the desired maximum unambiguous range and the values of R_u for the individual PRFs.

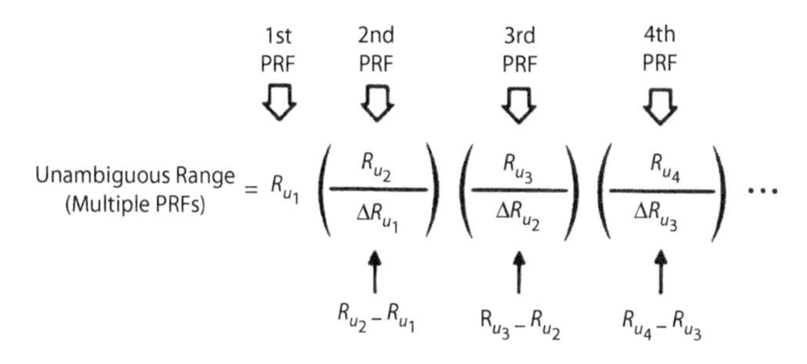

Figure 15-30. For each additional PRF, the unambiguous range for the combination is increased by the ratio of the unambiguous range, R_u, for the added PRF to the difference between R_u for that PRF and R_u for the preceding PRF.

Number of PRFs for Deghosting. More PRFs may also be required for deghosting. To perform this on all possible combinations of the observed ranges of more than two simultaneously detected targets, an additional PRF must be provided for each additional target. Thus, if a single PRF suffices to resolve range ambiguities, a radar employing N PRFs can uniquely measure the ranges of $(N - 1)$ simultaneously detected targets.

The Trade-Off. As with PRF jittering, a price is paid for PRF switching. Each additional PRF not only reduces the integration time—hence reducing detection range—but also increases the complexity of mechanization. The number of PRFs actually used, therefore, is a compromise between these costs and the cost of occasionally having to contend with ambiguous ranges and unresolved ghosts (Fig. 15-31).

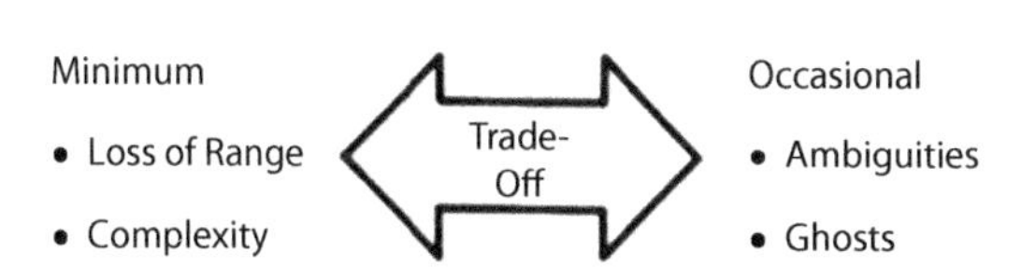

Figure 15-31. The number of PRFs actually used is always a compromise.

The optimum number of PRFs naturally varies with the application. For most of the fighter radar applications in which the required PRFs are low enough to make PRF switching practical, one additional PRF as a rule suffices for resolving ambiguities and another for deghosting, making a total of three. However, other factors—chiefly blind zone elimination and clutter rejection—often require significantly more PRFs, and it is not uncommon to transmit a schedule of typically eight or nine different PRFs, of which any three are used to resolve ambiguities and eliminate ghosts.

Measurement Accuracy. In PRF switching designs, the size of the range bins is of crucial importance. The more accurately range is measured in a single PRF, the easier it is to resolve range and eliminate ghosts. But as ever, there is a trade-off: a target will typically move during the radar dwell, and if the range gates are too small the target echo will be smeared across multiple gates, reducing sensitivity and further complicating range measurement. In addition, signal-to-noise affects the accuracy of range measurement. These factors must be carefully considered in designing a PRF switching system. In general, practical radars will incorporate additional PRFs to ensure that their ambiguity-resolving mechanizations are robust and reliable.

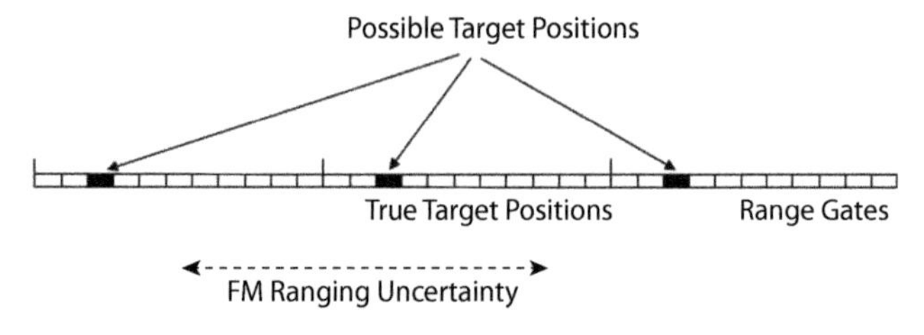

Figure 15-32. This range-gated HPRF is a way to measure range accurately with an FM ranging system.

15.6 Enhanced Pulse Tagging (Range-Gated High PRF)

A very effective technique is to use a combination of frequency-modulated (FM) ranging and conventional range gating. A limitation of FM designs is that they have comparatively poor range resolution. In a high PRF (HPRF) design (where the PRF is chosen to allow unambiguous Doppler processing), the resolution is typically of the same order as R_u, perhaps only 2 or 3 km. It is very desirable to measure range more accurately than this and to resolve any closely spaced targets.

In range-gated HPRF designs, this is achieved by carrying out FM ranging on multiple range gates in parallel. Provided the FM ranging can achieve an accuracy better than R_u, this resolves the ambiguity, and the accurate range can then be determined as the number of multiples of R_u plus the fine range gate measurement that lies closest to the FM range measurement (Fig. 15-32).

15.7 Single-Target Tracking

During single-target tracking, range measurement is simplified in two respects.

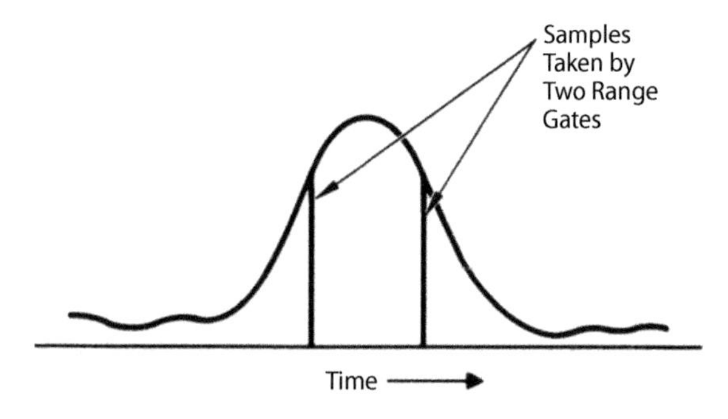

Figure 15-33. For single-target tracking, only two range gates are needed. By positioning them to equalize their outputs, they are centered on a target.

First, only two adjacent range gates must be provided (Fig. 15-33). The time delay between the transmission of a pulse and the opening of these gates is automatically adjusted to equalize the output of the two gates, thereby centering them on the target. By measuring this delay, the target's apparent range may be precisely determined. An alternative technique, more commonly employed in digital radars, is to interpolate the outputs of adjacent gates to achieve a similar result.

Second, once the ambiguities in the target's range have been resolved, no further resolution of ambiguities is necessary. Accurate track can be kept of the true range simply by keeping continuous track of the changes in apparent range.

It is worth noting, however, that these techniques can be open to deception jamming that attempts to steal the gate with a false jamming return. This false echo can then be delayed, fooling the victim radar into following it; it can then be turned off, and the victim radar will lose its track. Practical systems are typically more complex than described here so that they are not vulnerable to such countermeasures.

15.8 Electronically Scanned Radars

Electronically scanned radars are increasingly becoming the norm for airborne radars. The ability of such systems to rapidly steer their beam brings many advantages, including in the area of range measurement and ambiguity resolution.

When searching for new targets, an electronically scanned radar may employ any classical waveform with any of the range measurement techniques described earlier. Nonetheless, it may choose to use a different waveform, such as an unmodulated HPRF waveform, which is not capable of decoding range. The advantage of this approach is that the simpler waveform design can be made more sensitive.

If an initial detection is made with a simple unmodulated waveform, the electronically scanned radar then looks back and concentrates more time in the direction of the initial detection, changing its waveform to allow range to be measured. This may use any suitable waveform and ranging method. This allows a track to be rapidly established at a longer range than would otherwise be possible.

Thereafter, track updates can operate on a similar principle to single-target tracks. The true range is known, and updates need to measure only changes, not resolve ambiguities. Track update dwells can be adaptively optimized to measure only the essential information. This allows multiple targets to be tracked with minimal expenditure of time.

15.9 Summary

With pulse-delay ranging, range is determined by measuring the time between transmission of a pulse and reception of an echo. In rudimentary radars, the measurement is made on the range trace of the display. In sophisticated analog radars, it is made by opening a succession of range gates. Digital radars accomplish the equivalent by periodically sampling the receiver output, converting the samples to numbers, and storing them in a bank of range bins.

The range for which the round-trip transit time equals the interpulse period is called the maximum unambiguous range, R_u. A target at greater range will appear to have a range equal

Test your understanding

1. What is the speed of electromagnetic radiation traveling through air?
2. What is the maximum unambiguous range for a radar with a PRF of 400 Hz?
3. A radar has a PRF of 2 kHz, and a target return is observed at an apparent range of 17 km. At what other possible ranges might the target be?
4. A radar uses two PRFs to measure range. Both have an integer number of 150 m range cells; the first has a PRF of 14,706 Hz and the second 13,158 Hz. A target echo is observed in cell 20 of the first PRF and cell 56 of the second PRF. (a) What are the unambiguous ranges for the first and second PRFs, and (b) what is the true range of the radar echo?

to its true range minus some multiple of R_u. So long as there is a possibility of detecting any targets at ranges greater than R_u, all observed ranges are ambiguous.

What is done about range ambiguities depends on their severity and the penalty for ambiguous measurements.

If the PRF can be set low enough to make R_u greater than the maximum range of interest, ambiguities can be avoided by discarding the return from those targets beyond R_u. These can be identified by jittering the PRF and looking for a corresponding jitter in the apparent target ranges or using frequency agility to eliminate multiple-time-around returns.

If higher PRFs are required, ambiguities must be resolved. This can be done by switching between two or more PRFs and measuring the changes, if any, in the apparent ranges.

If two or more targets are detected simultaneously, each target may appear to have two possible ranges, one of which is a ghost. Ghosts can be eliminated by switching to additional PRFs.

Besides increasing complexity, using more than one PRF decreases detection range. The optimum number of PRFs is a compromise between these costs and the cost of occasionally having to contend with unresolved ambiguities and ghosts.

A combination of pulse-delay ranging and FM ranging can be very effective provided the FM ranging can be made sufficiently accurate to allow it to resolve the range ambiguity of the pulse-delay ranging.

Some relationships to remember are as follows: (1) ranging time: 10 μs = 1.5 km of range, 12.4 μs = 1 nmi of range; (2) maximum unambiguous range (km): R_u= 150/PRF (kHz); (3) when PRF switching is used to resolve range ambiguities: $R_{true} = nR_u + R_{apparent}$, $n = \Delta R_{apparent}/\Delta R_u$

Further Reading

E. Aronoff and N. M. Greenblatt, "Medium PRF Radar Design and Performance," in D. K. Barton, *CW and Doppler Radars*, Vol. 7, Artech House, pp. 261–276, 1978.

S. A. Hovanessian, *Radar System Design and Analysis*, Artech House, 1984.

P. E. Holbourn and A. M. Kinghorn, "Performance Analysis of Airborne Pulse Doppler Radar," in *Proceedings of the IEEE International Conference RADAR '85*, Washington DC, pp. 12–16, May 1985.

A. M. Kinghorn and N. K. Williams, "The Decodability of Multiple-PRF Radar Waveforms," in *Proceedings of the IEEE International Conference RADAR '97*, pp. 544–547, October 1997.

P. Z. Peebles, "Range Measurement and Tracking in Radar," chapter 11 in *Radar Principles*, John Wiley & Sons, Inc., 1998.

16

Pulse Compression and High-Resolution Radar

Philip Woodward, originator of the ambiguity function

Ideally, to obtain both long detection range and fine range resolution, extremely narrow pulses (for fine resolution) of exceptionally high peak power (for long range) should be transmitted. However, there is a practical limit on the amount of peak power that subsequently limits the detection range. This peak power limit forces the use of long pulses at the expense of range resolution.

The solution to this dilemma is pulse compression, in which coding is modulated onto long, peak power–constrained pulses during transmit, followed by "compression" of the received echoes by decoding their modulation. This provides the necessary average power for an achievable level of peak power. This chapter introduces the fundamental principles of pulse compression and the various classes of modulation coding, otherwise known as the radar *waveform*.

16.1 Pulse Compression: A Beneficial Complication

Pulse compression might appear to be an unnecessary complication to the notion of how radar operates. Narrow pulses can easily provide the desired range resolution by setting the pulse width. For relatively short-range operation this arrangement is acceptable. However, if one wishes to have long-range detection capability, it becomes clear from the radar-range equation (see Chapter 13) that increasingly high peak powers are necessary. However, there are practical limits to what can be made available from a realistic radar transmitter. The necessary extension to longer pulses subsequently establishes a set of trade-offs to design the appropriate transmitted signal and

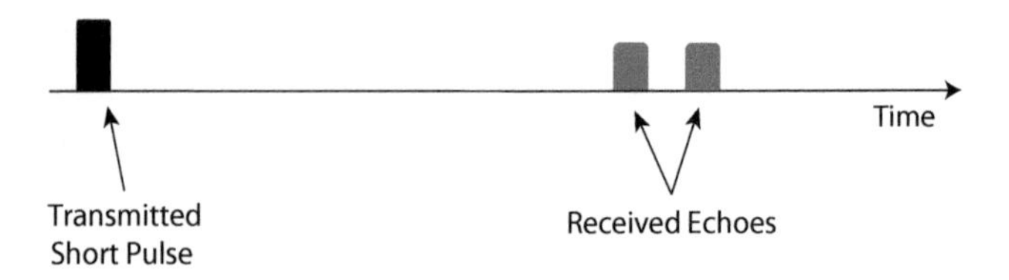

Figure 16-1. With a short pulse, closely spaced targets can be resolved. However, the limit on peak power likewise limits the maximum detectable range.

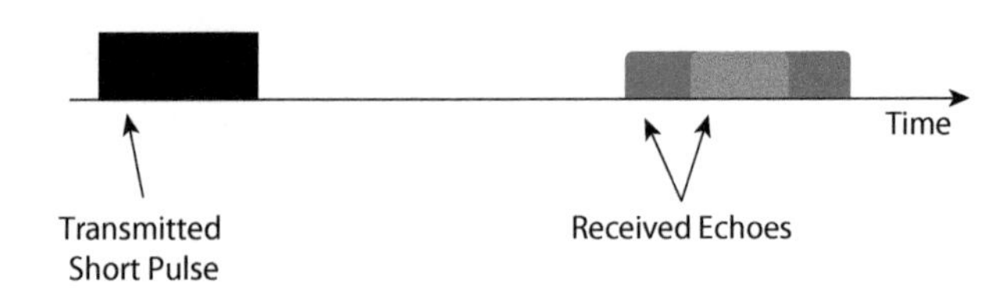

Figure 16-2. With the increased energy from a long pulse, the maximum detection range can be extended. Pulse compression is now required to separate the closely spaced targets.

receive filtering to perform the desired radar sensing function. It is also worth noting that echo-locating mammals seemed to have developed this capability long before radar engineers ever thought of it.

The Pulse Width Dilemma. Figure 16-1 illustrates an example of the transmission of a short pulse (at peak power) and the resulting echoes from two targets that are closely spaced in range. As long as these targets are separated by more than the pulse width it is possible to distinguish one from the other. However, because there is a limit to the amount of peak power the transmitter can achieve, this short pulse approach severely limits the maximum range from which targets can be reliably detected.

To extend the maximum detection range, more energy is required to be "put on the target." Since the peak power is bounded, the pulse width must be increased (in time). Figure 16-2 shows an example of what occurs when the short pulse from Figure 16-1 is extended in time (pulse width) by a factor of 5. The energy that is incident onto, and thereby reflected from, a target also increases by a factor of 5, thus extending the maximum detection range. However, now there is overlap between the two closely spaced targets such that they cannot be distinguished from one another. The solution to this problem is pulse compression.

Waveforms. In radar, the waveform is simply the transmitted signal. This may be a continuous signal or it may be a pulse. The notion of a radar waveform is extended here to include a modulation imparted upon a pulse. In principle, this modulation could be in terms of frequency/phase, amplitude, or polarization, though the former is by far the most common. Taken as a whole, pulse compression involves the transmission of a modulated, pulsed waveform followed by filtering of the received echoes, where the filter is coherently matched to the waveform.

There are often numerous objectives to be considered when designing a waveform, including

- Total energy of the modulated pulse (this relates to the SNR of received echoes)
- Discrimination between delay-shifted versions of the waveform (for both range resolution and sensitivity)
- Impact of Doppler shift
- Low probability of intercept (by a potential adversary)

The pulse energy is maximized when the amplitude envelope of the pulse is constant. The delay and Doppler characteristics of a waveform are collectively referred to as the waveform *ambiguity function* (see Chapter 11). The intercept probability of a waveform is dependent upon whether it appears to be man-made or naturally occurring (noise radar is an example of the latter).

Most commonly, a waveform can be ascribed to one of the following classes: frequency modulated chirp (linear or nonlinear) or phase-coded waveform (biphase or polyphase).

Linear frequency modulation (LFM) chirp is the most widely used of all waveforms due to its simplicity of implementation on transmit, its robustness to Doppler shift, and the existence of a useful wideband receiver filtering structure known as *stretch processing*. However, due to relatively high time-delay (range) sidelobes resulting from LFM matched filtering, nonlinear frequency modulation (NLFM) and phase-coded waveforms have been devised as possible alternatives.

The Matched Filter. What actually happens when an echo passes through a filter that is matched to the transmitted waveform can be visualized if the echo is thought of as consisting of a sequence of subpulses, or *chips*, each with a distinct phase. As depicted in Figure 16-3 the matched filter is likewise a sequence of chips, though each possesses the conjugate phase (i.e., reflected about the real, or horizontal, axis). If the aligned sets of chips are piecewise multiplied, they all produce the same value (here, set to an arbitrary phase of $e^{j0} = 1$ for simplicity) such that they add constructively in phase.

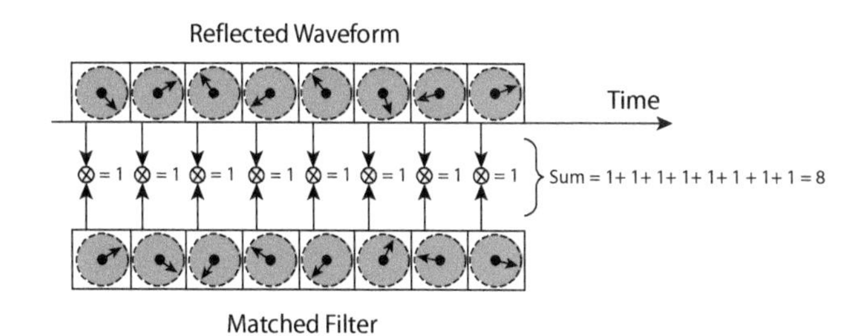

Figure 16-3. For a waveform represented as a sequence of 8 chips, the matched filter constructively combines the segments to yield a processing gain, also known as the *pulse compression ratio*, of 8.

Figure 16-3 depicts the precise point in time when an echo aligns with the matched filter, thus producing a gain on the echo. At other delay shifts a different phenomenon is observed. For example, Figure 16-4 illustrates what occurs when the echo is shifted in time by just one chip interval compared with the matched case of Figure 16-3. This time, when the aligned chips are piecewise multiplied, a set of phase values is produced that are out of phase with each other and thus combine destructively when added. The resulting summation will typically be much smaller than the matched case of Figure 16-3. For other delays, different sets of phases are produced by the matched filter, which subsequently yields different destructive combinations that vary as a function of delay in a way that is characteristic to each individual waveform.

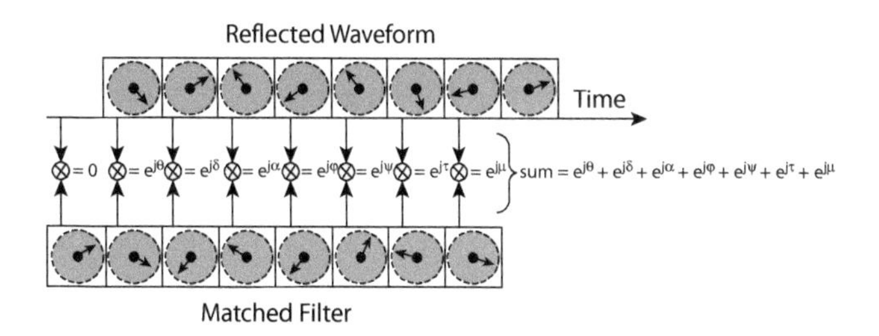

Figure 16-4. For delays different from the match point, the segments of the echo do not match the phase sequence of the matched filter, thereby combining destructively to produce a smaller value (here, much less than 8).

In reality, a physical waveform must be continuous. For the discrete illustrations in Figures 16-3 and 16-4 the chips can be thought of as representing basic phase shapes that enable the adjacent chips to connect in a continuous manner over the extent of the waveform (and likewise the matched filter). When considered in this way, the matched filter concept extends to all types of frequency modulated and phase-coded waveforms.

It is becoming increasingly common to perform matched filtering digitally, thus requiring sampling of the received echoes and a digital representation of the filter. The determination of the sampling rate involves a trade-off between higher computational complexity and the acceptable degree of loss from *range straddling* (also known as *range cusping*) that occurs when an echo is not sampled precisely at its matched position.

The continuum of delay shifts comprising the matched filter response to a single echo (with no Doppler) is actually the

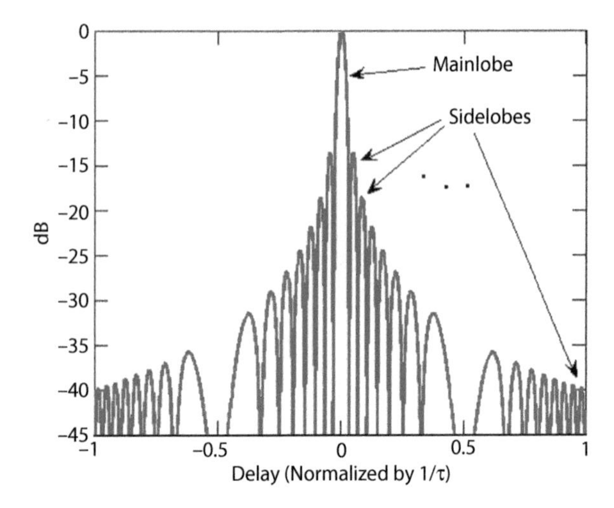

Figure 16-5. The matched filter response (waveform autocorrelation) for an LFM chirp with uncompressed pulse width τ illustrates the mainlobe and sidelobes in delay that would result from a single target echo.

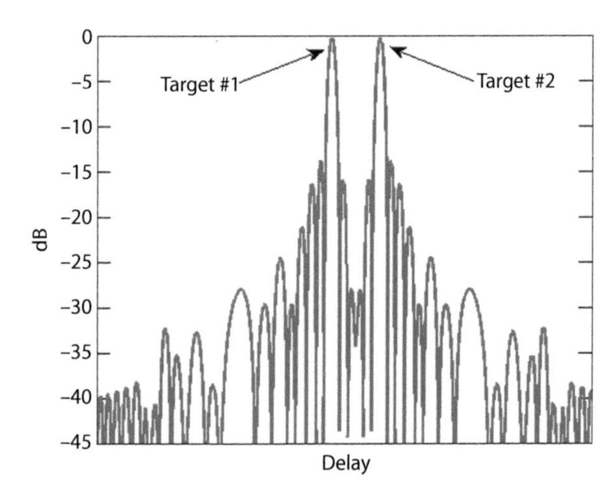

Figure 16-6. The echoes from two closely spaced targets may be resolved if they have similar receive powers and are not too close together.

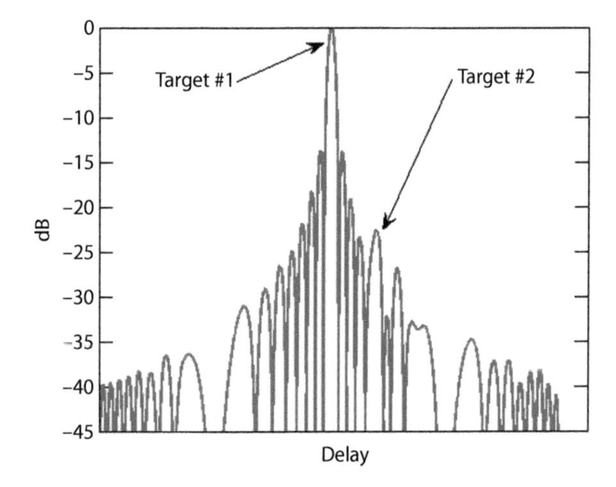

Figure 16-7. If the receive echoes from two closely spaced targets have sufficiently disparate receive powers, then the smaller target may be lost among the range sidelobes of the larger target.

autocorrelation of the transmitted waveform. For example, Figure 16-5 illustrates the autocorrelation for an LFM chirp that is normalized so the match point is at 0 dB. Figure 16-5 shows the ideal pulse compressed response having −13 dB peak sidelobes (these may be reduced using weighting; see the section "Amplitude Weighting").

Resolution and Range Sidelobes. Similar to an antenna radiation pattern, the matched filter mainlobe is the delay region immediately surrounding the matched position. Using the LFM matched filter response in Figure 16-5 as an example, it is predominantly the width of the mainlobe that determines if two closely spaced targets in range can be resolved. Therefore, if the matched filter is applied to the echoes generated by the two targets of Figure 16-2 (assuming the pulse was modulated with an LFM waveform), the pulse compressed output would look like the result shown in Figure 16-6.

As it is much shorter than the pulse width, the width of the mainlobe enables improved range resolution. The range resolution is now inversely proportional to the bandwidth of the waveform. A convenient point of reference is that a range resolution of 30 cm corresponds to a waveform bandwidth of approximately 500 MHz.

Referring again to Figure 16-5, the smaller peaks surrounding the mainlobe are known as *range sidelobes*. For the LFM chirp, the largest sidelobe is approximately 13 dB lower than the value at the matched position and defines the *peak sidelobe level* (PSL). Range sidelobes are one of the performance trade-offs of pulse compression, as they limit the sensitivity of the radar. For example, if the received power of the two target echoes depicted in Figure 16-6 were very different, the matched filter response would instead look like the result in Figure 16-7, in which the range sidelobes induced by the higher-power target can actually mask the mainlobe of the lower-power target.

Doppler Effects and the Ambiguity Function. The discussion thus far has been limited to the case where no Doppler effects are present. Doppler is a shift in frequency that is induced by radial motion between the radar and the subject of the radar illumination (see Chapter 18 for a detailed discussion). For example, a police radar measures the amount of frequency shift of the echo from a moving vehicle to measure its speed relative to the position of the radar. Relative motion towards the radar causes a positive frequency shift (i.e., a higher frequency echo), while relative motion away from the radar causes a negative shift (i.e., a lower frequency echo).

With regard to pulse compression, the impact of motion-induced Doppler frequency shift is an altering of the phase progression of the waveform echo. As a result, the gain from constructive combining at the matched position (see Figure 16-3) can be degraded or even completely lost depending on the degree of Doppler shift and the nature of the waveform.

A plot of the matched filter response versus Doppler frequency shift is shown in Figure 16-8. This is defined as the *ambiguity function* (see Chapter 11).

The matched position is located where both delay and Doppler are zero. The zero Doppler cut (horizontally across Doppler = 0 Hz) reveals the waveform autocorrelation (and is the same result shown in Fig. 16-5). In the Doppler dimension the mainlobe width is inversely proportional to the pulse width. Away from the mainlobe, range-Doppler sidelobes can be observed.

In current fielded radar systems, the two most commonly employed waveforms are the LFM chirp and the biphase (or binary phase)-coded waveform. The following sections outline the benefits and deficiencies of each.

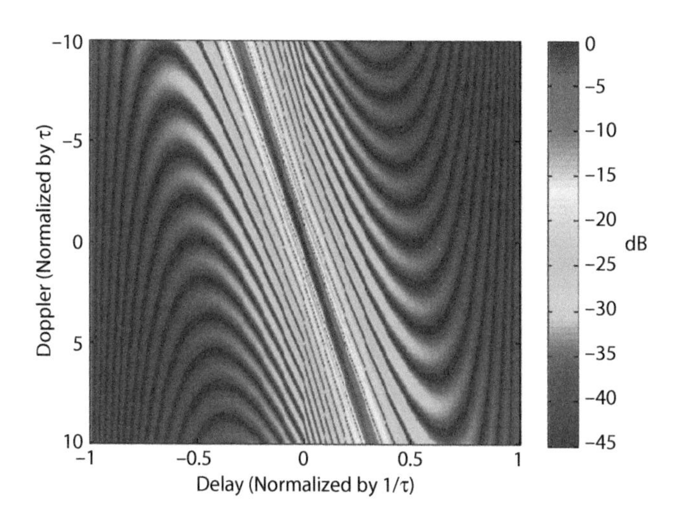

Figure 16-8. This graph shows the delay/Doppler ambiguity function for the LFM chirp (brightness scale in decibels).

16.2 Linear Frequency Modulation (Chirp)

Because of its similarity to the chirping of a bird, its inventors called this form of modulation a "chirp." Since it was the first pulse compression technique, the term chirp is still in common usage and is synonymous with pulse compression.

For LFM chirp coding, the frequency of the transmitted pulse is increased (an "up-chirp") or decreased (a "down-chirp") at a constant rate throughout its length (see Figure 16-9), thus every echo has the same linear increase/decrease in frequency.

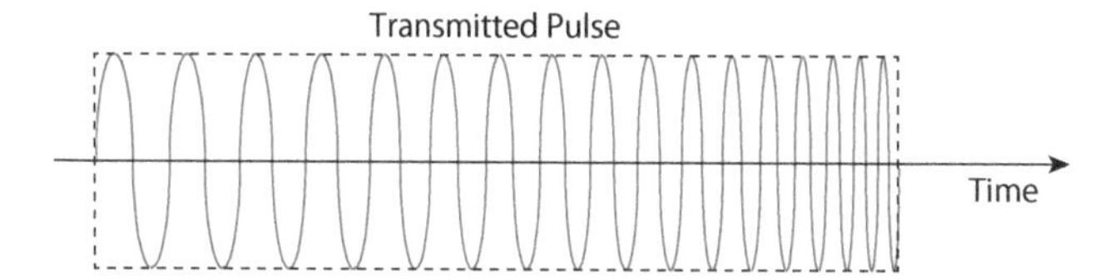

Figure 16-9. Shown here is an LFM up-chirp.

LFM Implementation. A major benefit of LFM chirp is the ease with which it can be implemented. The transmitter needs only to sweep linearly from some starting frequency at the beginning of the pulse to some ending frequency at the tail of the pulse, which can be accomplished in many different ways in both analog and digital hardware.

Filtering may be done with an analog device—such as an acoustical delay line—or, more common in modern systems, digitally. For a narrow range swath the LFM chirp can be decoded using a technique called *stretch processing*, which can accommodate a very large waveform bandwidth, thus enabling very fine range resolution.

For stretch processing (described in detail in the accompanying panel, the echo delay time (range) is converted to frequency. As a result, the return from any one range corresponds to a constant frequency, and the returns from different ranges may be separated with a bank of narrowband filters implemented with the efficient fast Fourier transform (see Chapter 21). Range is determined by measuring the instantaneous difference between the frequencies of the transmitted and received signals.

Incidentally, stretch processing is similar to the FM ranging technique used by continuous wave (CW) radars (see Chapter 17). The principal differences are that instead of transmitting pulses, the CW radar transmits continuously, and the period over which the transmitter's frequency changes in any one direction is many times the round-trip ranging time.

Stretch Processing of LFM Chirp

For a narrow range swath, such as is mapped by a synthetic aperture radar (see Chapter 33), LFM chirp modulation is commonly decoded by a technique called *stretch processing* or *deramping*.

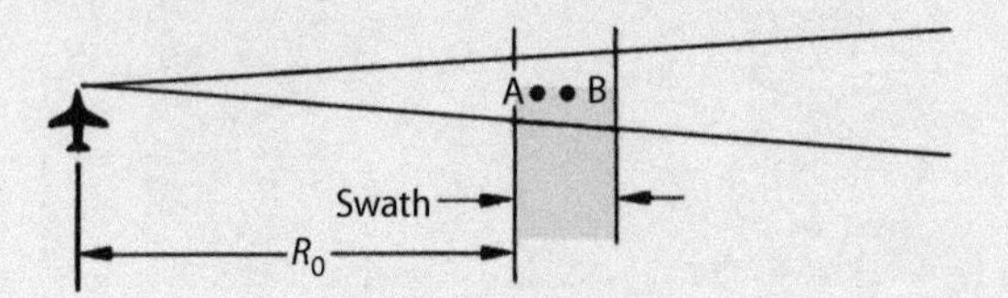

For the up-chirp example, as the return from the swath is received, its frequency is subtracted from a reference frequency that increases at the same rate as the transmitter frequency. However, the reference frequency increases continuously throughout the entire interval over which echoes are received.

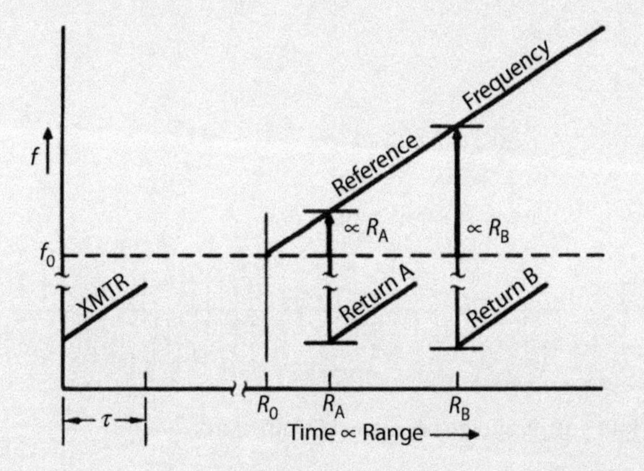

Consequently, the difference between the reference frequency and the frequency of the return from any particular point on the ground is constant. Moreover, as can be seen from the above figure, if we subtract the reference frequency's initial offset, f_0, from the difference already obtained, the result is proportional to the range of the point from the near edge of the swath, R_0. Range is thus converted to frequency.

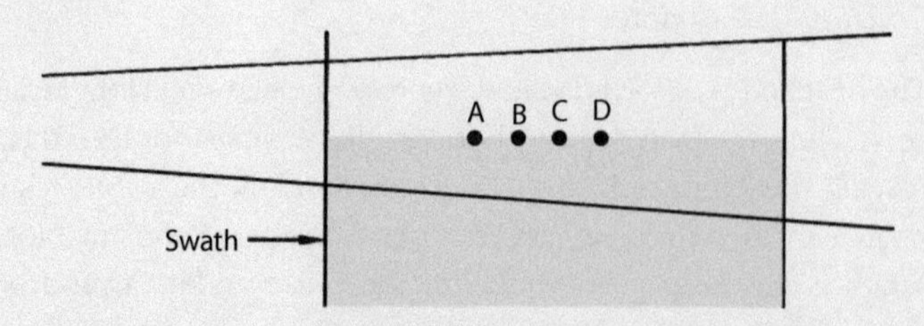

To see how fine resolution is achieved, consider the returns from four closely spaced points after the subtraction has been performed.

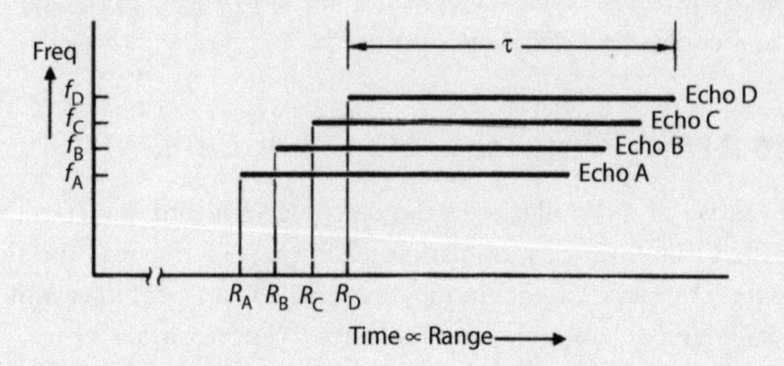

Although the returns were received such that their pulse echoes almost completely overlap, the slight stagger in their arrival times results in clearly discernible differences in frequency.

As indicated in the figure below, the continuously changing reference frequency may be subtracted at one of three points in the receiving system. One is the mixer, which converts the radar returns to the receiver's intermediate frequency (IF). The second point is the synchronous detector, which converts the output of the IF amplifier to video frequencies. And the third point is in the signal processor, after the video has been digitized.

To sort the difference frequencies, the video output of the synchronous detector is applied to a bank of narrowband filters, implemented with the fast Fourier transform.

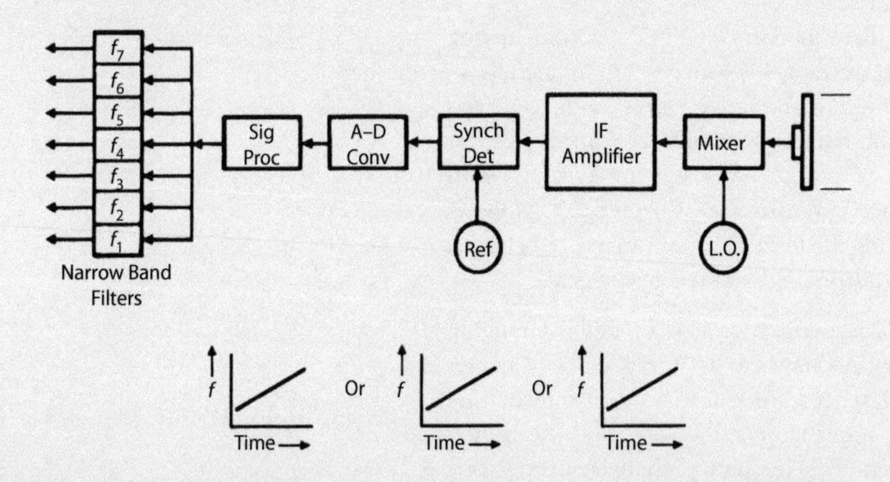

Pulse Compression Ratio. The straightforward nature of the LFM provides a convenient framework with which to better explain the processing gain and range resolution enhancement provided by pulse compression. The pulse compression ratio, represented by the factor 8 for the example in Figure 16-3, is the ratio of the uncompressed pulse width τ to the compressed width τ_{comp}. Whereas the previous example explained the phenomenon in terms of phase-modulated subpulses, LFM chirp allows us to consider it in terms of frequency sensitivity.

If returns received simultaneously from two slightly different ranges are to be separated on the basis of the difference in their frequencies, besides providing a delay proportional to frequency (refer to the panel on stretch processing), a second requirement must also be satisfied. The frequency difference must be large enough for the signals to be resolved by the filter.

As will be made clear in Chapter 20, the frequency resolution of the matched filter response increases (becomes narrower) as the uncompressed pulse width increases (see Figure 16-10). Specifically, the frequency resolution Δf is related to the uncompressed pulse width as

$$\Delta f = \frac{1}{\tau}.$$

In other words, as illustrated by Figure 16-11, for the LFM matched filter to resolve two closely spaced echoes, the instantaneous difference in their delay-shifted frequencies must meet or exceed the inverse of the uncompressed pulse width τ.

Furthermore, the compressed pulse width τ_{comp} is the period of time over which the frequency of the uncompressed LFM pulse changes by Δf (see Figure 16-12). By extension, if the frequency of the uncompressed LFM pulse changes at a rate of $\Delta f/\tau_{comp}$ (in hertz per second), then the total change in frequency, ΔF, over

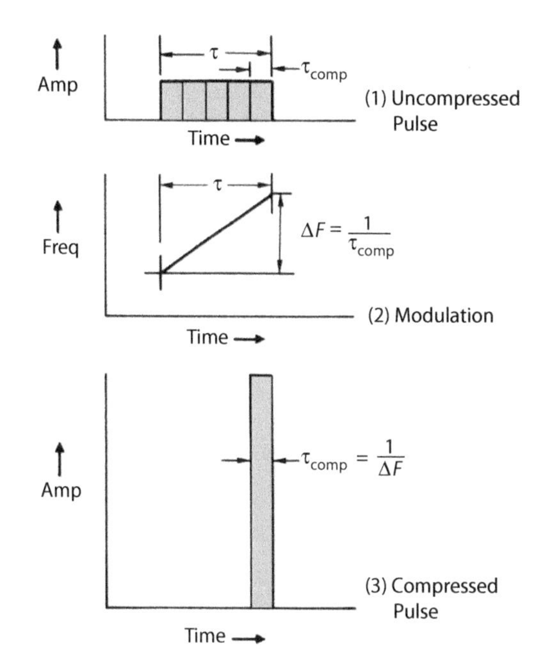

Figure 16-10. Conceptual relationship between uncompressed pulse width, chirp modulation bandwidth ΔF, and compressed pulse width for an LFM waveform. The compressed pulse width corresponds to the mainlobe from Figure 16-5.

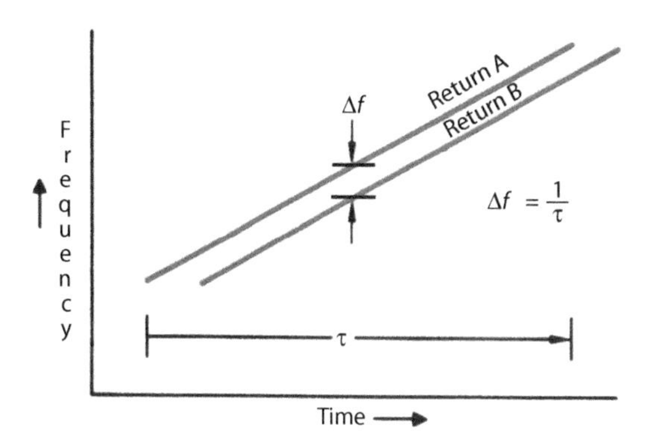

Figure 16-11. For a filter to resolve two concurrently received LFM returns, the instantaneous difference in their frequencies (Δf) must equal at least $1/\tau$.

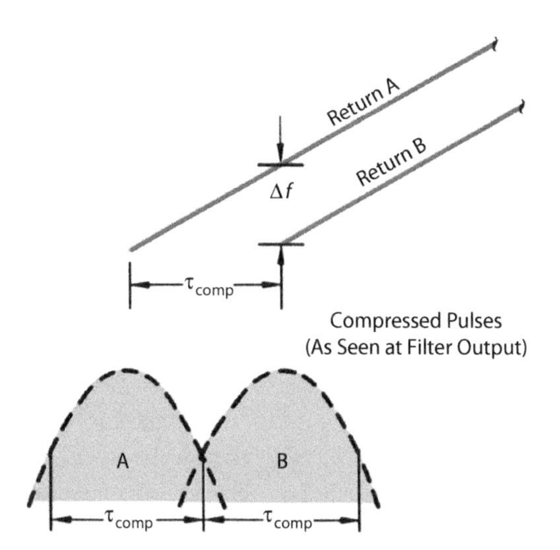

Figure 16-12. If the minimum resolvable frequency difference is Δf, the time in which the frequency of the uncompressed LFM pulse changes by Δf is the width of the compressed pulse, τ_{comp}.

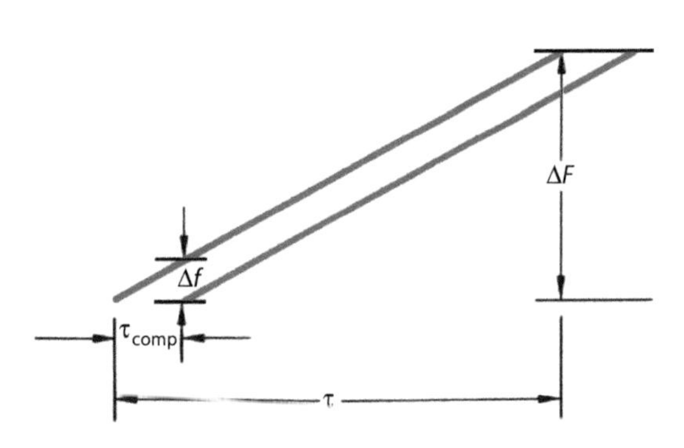

Figure 16-13. The ratio of uncompressed pulse width, τ, to compressed pulse width, τ_{comp}, equals the ratio of the total change in frequency over the pulse width, ΔF, to minimum resolvable frequency difference, Δf.

the duration of the uncompressed pulse will be this rate times the uncompressed pulse width, τ.

The rate of frequency change $\Delta f/\tau_{comp}$ is known as the *chirp rate*:

$$\text{chirp rate} = \frac{\Delta f}{\tau_{comp}} \text{ (in Hertz per second).}$$

The total change in frequency, ΔF, is the *bandwidth* for the LFM chirp:

$$\text{bandwidth} = \Delta F = \left(\frac{\Delta f}{\tau_{comp}}\right)\tau = \text{chirp rate} \times \tau.$$

As is apparent from the geometry of Figure 16-13, the pulse compression ratio, τ/τ_{comp}, equals the ratio of ΔF to Δf:

$$\text{Pulse compression ratio} = \frac{\tau}{\tau_{comp}} = \frac{\Delta F}{\Delta f}.$$

Substituting $1/\tau$ for Δf, the pulse compression ratio equals the uncompressed pulse width times ΔF:

$$\text{Pulse compression ratio} = \tau\Delta F.$$

The quantity $\tau\Delta F$ is also called the *time–bandwidth product.*

This simple relationship—pulse compression ratio equals time–bandwidth product—tells us a lot about the LFM chirp. To begin with, for a given uncompressed pulse width τ, the compression ratio increases directly with an increase in bandwidth ΔF. Conversely, for a given bandwidth ΔF, the compression ratio increases directly with an increase in the uncompressed pulse width τ.

If the time–bandwidth product is set equal to τ/τ_{comp} as

$$\tau\Delta F = \frac{\tau}{\tau_{comp}},$$

τ cancels out so that

$$\tau_{comp} = \frac{1}{\Delta F}.$$

In other words, the width of the compressed pulse is determined entirely by the bandwidth ΔF of the transmitted pulse; that is, the greater the frequency change, the narrower the compressed pulse width. Rearranging this last equation tells us that the total change in transmitter frequency (the LFM bandwidth) must be

$$\Delta F = \frac{1}{\tau_{comp}}.$$

This relationship provides a useful benchmark for the transmitter bandwidth necessary to achieve a desired bandwidth (and therefore range resolution) for arbitrary waveforms. It should be noted, however, that the equality only holds for the LFM chirp, which spends an equal amount of time (and thus power) in each of the frequencies due to its linear frequency sweep. Different waveforms that occupy some frequencies longer than others or employ a weighting across frequencies may require a higher bandwidth to achieve the same compressed pulse width as the LFM.

To get a feel for the relative values involved for LFM chirp, consider a couple of representative examples.

- Using LFM to provide the same compression as in the 8-chip matched filter discussed earlier, $\tau/\tau_{comp} = 8$. If the original pulse is 1 μs, the range resolution has now improved to 18.75 m. This would separate aircraft targets except for very tight formations of small planes.
- It is assumed that in order to provide adequate "energy on target," the width of a radar's transmitted pulse must be $\tau = 10$ μs. To provide the desired range resolution of 1.5 m, a compressed pulse width of $\tau_{comp} = 0.01$ μs is required. Therefore the pulse compression ratio must be

$$\frac{\tau}{\tau_{comp}} = \frac{10}{0.01} = 1000$$

To achieve a compressed pulse width of 0.01 μs (10^{-8} s), the change in transmitter frequency, ΔF, over the duration of each transmitted pulse must be $1/10^{-8} = 10^8$ Hz, or a bandwidth of 100 MHz.

Since the duration of the uncompressed pulse is 10 μs (10^{-5} s), the rate of change of the transmitter frequency (the chirp rate) will be $10^8/10^{-5} = 10^{13}$ Hz/s, or 10,000 GHz/s. This arrangement equates to a total linear frequency modulation excursion of 100 MHz over the duration of the pulsed waveform.

Incidentally, these values explain why stretch processing is practical only for relatively narrow range intervals. The ranging time for an interval of 100 km, for instance, is $13.3 \times 50 = 665$ μs. If the receiver local-oscillator frequency is shifted at a rate of 10 GHz/s throughout that time (see Figure 16-14), the total frequency shift would be $10{,}000 \times 620 \times 10^{-6} = 6.65$ GHz. Such a large shift was deemed impractical in previous editions of this book and is only now beginning to enter the realm of possibility.

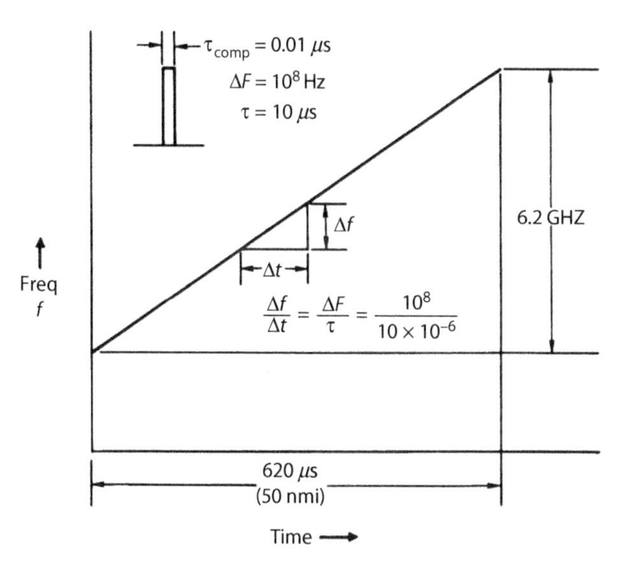

Figure 16-14. If stretch processing is used over a 100 km range interval to decode a 10 μs pulse modulated for a 1000:1 compression ratio, the receiver local oscillator would have to be swept over 6.65 GHz.

LFM Ambiguity Function. LFM allows very large compression ratios to be achieved with a relatively simple implementation. To assess the performance capability of LFM, consider its delay and Doppler characteristics that were illustrated via the ambiguity function in Figures 16-5 and 16-8.

One disadvantage of LFM is the high sidelobes that occur in the range dimension. These high sidelobes have driven the development of alternative waveforms and filtering strategies.

Another possible disadvantage is the ambiguity that occurs between range and Doppler (the range–Doppler ridge), shown in Figure 16-8. If an echo possesses a sufficient Doppler shift, it will also appear to be shifted in range, thereby limiting the ultimate accuracy for which true range may be determined. However, this *Doppler tolerance* also allows for simpler receiver processing by precluding the need for a bank of matched filters tuned to different possible Doppler frequencies.

Amplitude Weighting. A well-known approach to reduce range sidelobes for LFM is to apply an amplitude weighting to the waveform that reduces the power in the regions nearer the ends of the pulse. Due to the frequency-swept nature of LFM, this weighting results in a deemphasizing of the frequencies near the extremities of the bandwidth, which results in lower sidelobes in the time domain due to the relationship between time and frequency (from Chapter 6).

The trade-off for this significant reduction in range sidelobes is reduced transmit power, which directly impacts detection sensitivity. This weighting can also cause a degradation in range resolution due to reduced power in the outer frequencies, which is essentially a reduction in bandwidth. A typical compromise is to allow the resolution size to increase by approximately 50%, which enables sidelobe levels to be around −35 to −40 dB or less.

From an implementation standpoint, weighting the transmitted pulse may also be prohibitive if high-efficiency, nonlinear amplification is required.

A common compromise is to transmit the standard LFM waveform while applying a receive filter that is weighted. This form of *mismatched filtering* has the advantage of still enabling the maximum power on transmit as well as power-efficient nonlinear amplification. The trade-off is a small mismatch loss between the waveform and filter that is acceptable for many radar applications.

16.3 Phase Modulation

In this type of coding the waveform is represented as a discrete sequence of increments, with each increment corresponding to one from a set of phase values modulated onto a *subpulse* (or *chip*). The set of possible phase values is often referred to as the phase *constellation*. For practical reasons, it is often desirable for the nature of the subpulse shape to provide continuous transitions between adjacent subpulses.

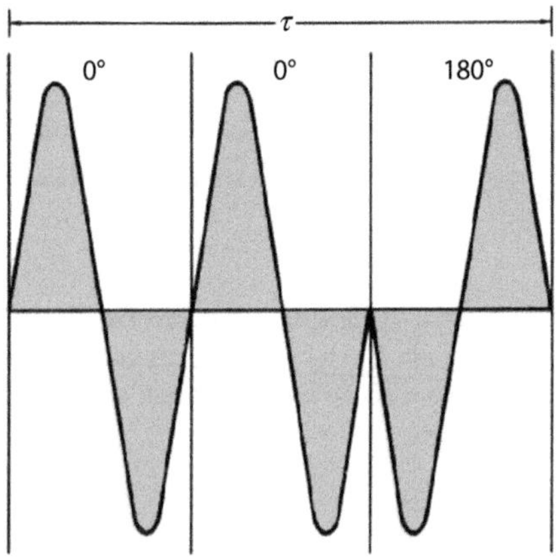

Binary Phase Modulation. The simplest form of phase modulation employs a constellation of two opposite phase values (usually 0° and 180°) that are modulated onto a subpulse. The radio frequency phase of certain subpulse segments is shifted by 180° (or −1), according to a predetermined binary code. The subpulse is comprised of a multiple number of wavelengths of the carrier frequency.

Figure 16-15 illustrates an exemplary three-segment code. (So you can readily discern the phases, the wavelength has been arbitrarily increased to the point where each segment contains only one cycle.)

Figure 16-15. Binary phase coding of a transmitted pulse. The pulse is marked off into segments and the phases of certain segments (here, the third) are reversed.

A common shorthand method of indicating the coding is to represent the segments with + and – signs. An unshifted segment (0°) is represented by a + sign and a shifted segment (180°) by a – sign. The signs making up the code are referred to as digits. The number of digits indicates the pulse compression ratio of the code.

The received echoes are passed through a tapped delay line (Figure 16-16) that provides a time delay exactly equal to the duration of the uncompressed pulse, τ. The delay line may be implemented either with an analog device or digitally. Clearly the tapped delay line for the binary-coded waveform is an implementation of the matched filter previously shown in Figures 16-3 and 16-4.

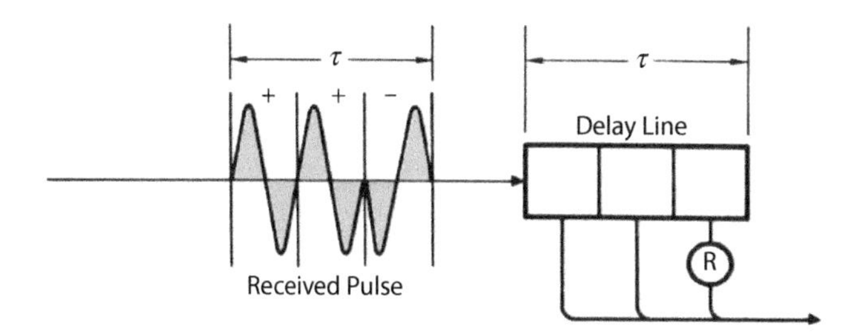

Figure 16-16. Received pulse echoes are passed through a tapped delay line filter. A separate tap is provided for each segment of the pulse. Here, the third tap is reversed R to represent a 180° phase shift.

Like the transmitted pulse, the delay line is divided into segments. An output tap is provided for each segment. The taps are all tied to a single output terminal. At any instant, the signal at the output terminal corresponds to the sum of whatever segments of a received pulse currently occupy the individual segments of the line.

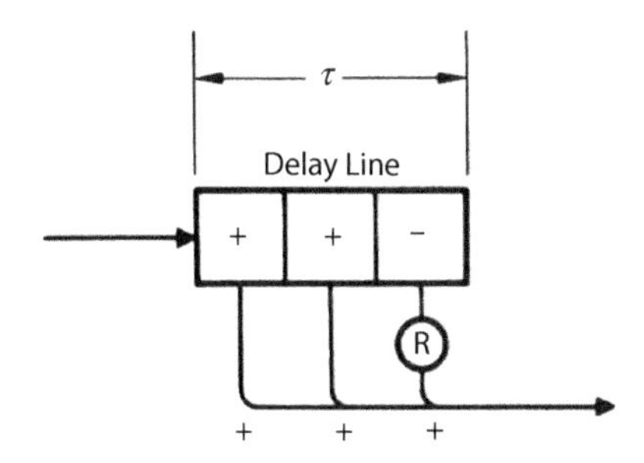

Figure 16-17. The phase reversal ® is placed so that when a pulse completely fills the delay line, outputs from all taps will be in phase.

Now, in certain taps, 180° phase reversals are inserted. Their positions correspond to the positions of the phase-shifted segments in the transmitted pulse. Thus when a received echo has progressed to the point where it completely fills the line, the outputs from all of the taps will be in phase (Figure 16-17). Their sum will then equal the amplitude of the pulse times the number of segments it contains.

To see step by step how the binary-coded pulse is compressed, consider a simple three-segment delay line and the three-digit code, illustrated in Figure 16-15.

Suppose the echo from a single-point target is received. Initially the output from the delay line is zero. When segment 1 of the echo has entered the line, the signal at the output terminal corresponds to the amplitude of this segment (Figure 16-18). Since its phase is 180°, the output is negative: –1.

An instant later, segment 2 has entered the line. Now the output signal equals the sum of segments 1 and 2. Since the segments are 180° out of phase, however, they cancel: the output is 0.

When segment 3 has entered the line, the output signal is the sum of all three segments. At this point segment 1 has reached the tap containing the phase reversal. The output from this tap is, therefore, in phase with the unshifted segments 2 and 3 such that the combined output of the three taps is three times the amplitude of the individual segments: +3.

As segments 2 and 3 pass through the line, this process continues. The output drops to 0, then becomes –1, and finally returns to 0 again.

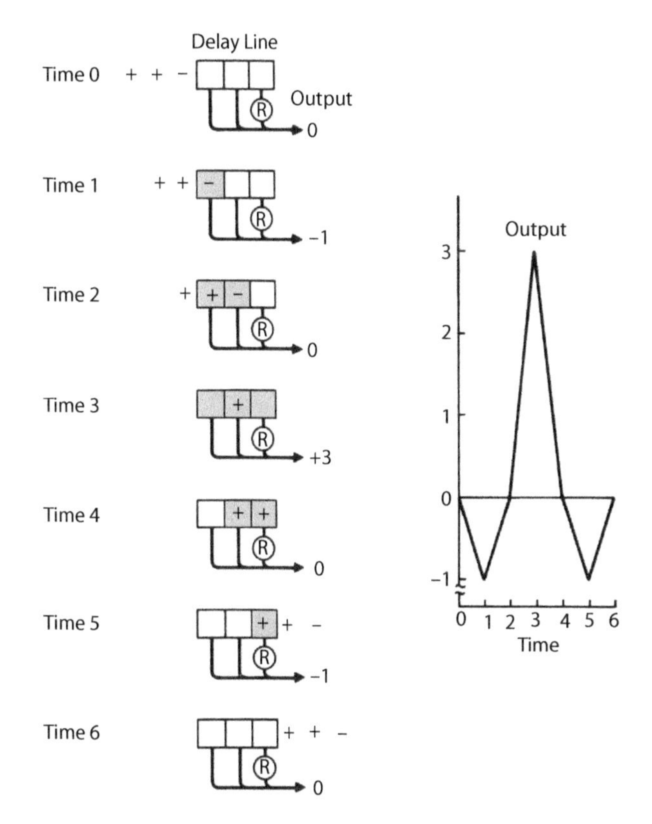

Figure 16-18. Step-by-step progress of a three-digit binary phase modulated pulse through a tapped delay line.

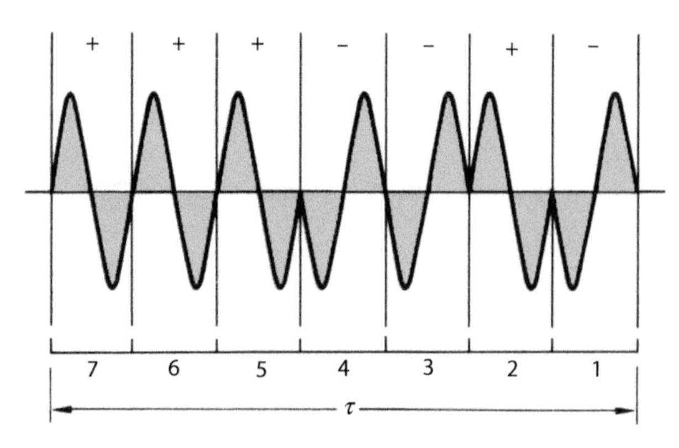

Figure 16-19. A seven-digit binary phase code.

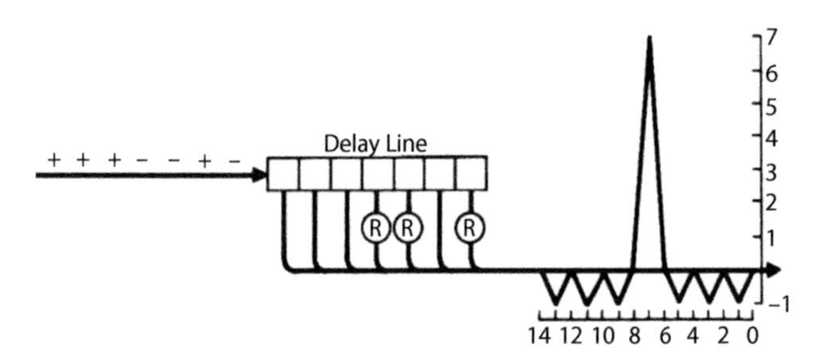

Figure 16-20. This is the output produced when a seven-digit binary phase code is passed through a tapped delay line with phase reversals in the appropriate taps.

N	Barker Codes
2	+ – Or (+ +)
3	+ + –
4	+ – + – Or (+ – – –)
5	+ + + – +
7	+ + + – – + –
11	+ + + – – – + – – + –
13	+ + + + + – – + – + – +

Note: Plus and minus signs may be interchanged.
(+ + – changed to – – +); order of digits may be reversed.
(+ + – changed to – + +); codes in parentheses are complementary codes.

Figure 16-21. Barker codes come very close to the goal of producing no sidelobes. However, the largest Barker code contains only 13 digits.

A somewhat more practical example is shown in Figure 16-19. This code has seven digits. Assuming no losses, the peak amplitude of the compressed pulse is seven times that of the uncompressed pulse, and the compressed pulse is only one-seventh as wide.

To see why the code produces the output it does, transfer the code to a sheet of paper and slide it across the delay line plotted in Figure 16-20, digit by digit, noting the sum of the outputs for each position. (A minus sign, –, over a tap with a reversal ® in it becomes a +, while conversely the reversal of a + becomes a –.) You should obtain the output shown in the figure.

Barker Codes. Ideally, for all positions of the echo in the line—except the central one—the collection of 0° or 180° outputs would cancel and there would be no range sidelobes.

One set of codes, called the Barker codes, comes very close to meeting this goal (Figure 16-21). Two of these codes have been used in the above examples. As has been seen, they produce sidelobes whose amplitudes are no greater than the amplitude of the individual code segments. Consequently the ratio of mainlobe amplitude to sidelobe amplitude, as well as the pulse compression ratio, increases with the number of segments into which the pulses are divided—that is, the number of digits in the binary code.

Unfortunately, the longest Barker code contains only 13 digits. Arbitrary binary codes can be made practically any length, but their sidelobe characteristics, though reasonably good, do not possess this desirable property of the Barker codes. Such codes require an exhaustive computer search and are called *minimum peak sidelobe* codes.

Complementary Codes. It turns out that the four-digit Barker code has a special feature that enables us to build codes of greater length and even eliminate the sidelobes altogether (under certain conditions). This feature arises because the four-digit code, as well as the two-digit code, has a complementary form. The sidelobe structures produced by the complementary forms have opposite phases (Figure 16-22). Therefore, if successive transmitted pulses are alternately modulated with the two forms of the code and filter each with their corresponding delay line, the returns from successive pulses can be added such that the sidelobes cancel.

Furthermore, by chaining the complementary forms together according to a certain pattern, codes of much greater length can be built. As illustrated in Figure 16-23, the two forms of the four-digit code are just such combinations of the two forms of the two-digit code, and these are just such combinations of the two fundamental binary digits, + and –.

Unlike the unchained Barker codes, the chained codes (also called *nested codes*) produce sidelobes having amplitudes greater than one. However, since the chains are complementary, these larger sidelobes—like the others—cancel when successive pulses are added (at least in the absence of Doppler).

Doppler Sensitivity. Compared with LFM chirp, coded modulations can be much more sensitive to Doppler frequency shift. If all segments of a phase-coded pulse are to add constructively when the pulse is centered in the delay line, while cancelling when it is not, very little additional shift in phase over the length of the pulse can be tolerated.

A Doppler shift of 10 kHz amounts to a phase shift of 10,000 × 360°/s, or 3.6°/μs. If the uncompressed pulse width is as much as 50 μs (Figure 16-24), this shift will itself equal 180° over the length of the pulse, and performance will deteriorate. For the scheme to be effective, either the Doppler shifts must be comparatively small or the uncompressed pulses must be reasonably short.

One way to contend with phase-coding sensitivity to Doppler is through "Doppler tuning," in which a bank of Doppler-shifted versions of the delay line matched filter outputs are applied. While this approach increases the overall hardware (analog filtering) or computational (digital filtering) requirements, it does have the benefit of avoiding the range–Doppler ambiguity problem of the LFM chirp.

The sidelobe cancellation property of complementary code sets is very sensitive to pulse-to-pulse Doppler shift. The sidelobes do not perfectly subtract when Doppler is present, so residual sidelobes emerge.

Polyphase Codes. Phase coding is not limited to just two increments (0° and 180°). Codes with phase constellations comprised of more than two possible values are collectively referred to as *polyphase codes*. Here a particular example is considered and is taken from a family called Frank codes.

The fundamental phase increment ϕ for a Frank code is established by dividing 360° by the number of different phases in the constellation, P. The coded pulse is then built by chaining together P groups of P segments each. The total number of segments in a pulse, therefore, equals P^2.

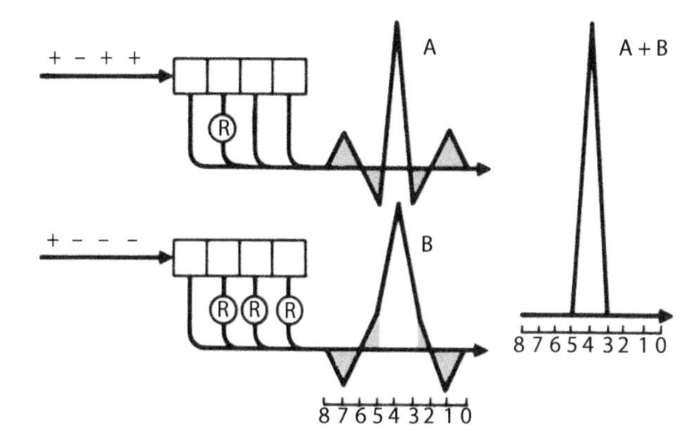

Figure 16-22. Echoes from complementary phase coding received from the same target in alternating pulses. When echoes are added, the time sidelobes cancel.

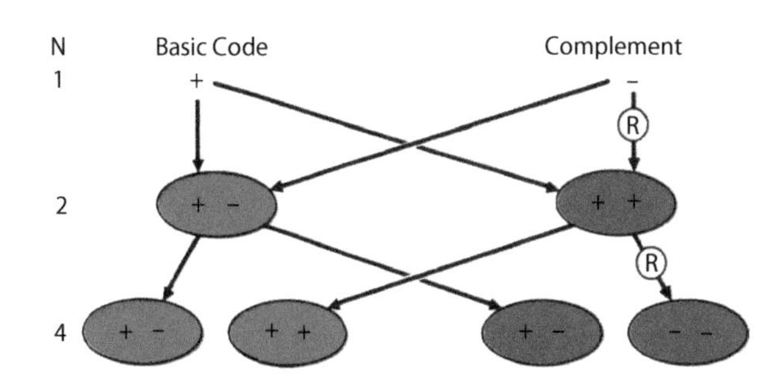

Figure 16-23. How complementary codes are formed. The basic two-digit code is formed by chaining a basic binary digit (+) to its complement (–). A complementary two-digit code is formed by chaining a basic binary digit (+) to its complement with the sign reversed (+). The basic four-digit code is formed by chaining a basic two-digit code (+ –) to its complementary two-digit code (+ +). A complementary four-digit code is formed by chaining a basic two-digit code (+ –) to its complementary two-digit code with the sign reversed (– –), and so on.

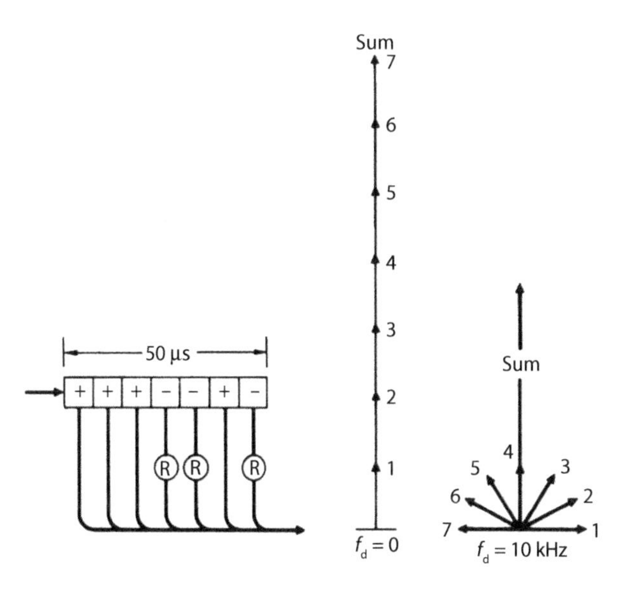

Figure 16-24. The reduction in peak output of a tapped delay line for a 50 μs, phase-coded pulse resulting from a Doppler shift of 10 kHz.

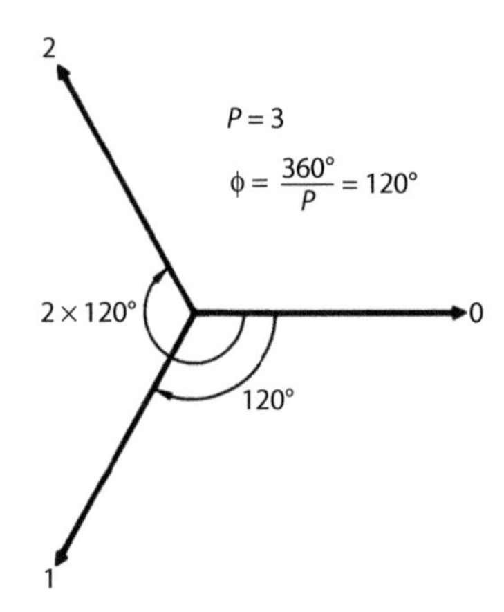

Figure 16-25. Phase increments for a Frank code in which the number of phases P is three.

In a three-phase code (Figure 16-25), for example, the fundamental phase increment is $360° \div 3 = 120°$, making the phases 0°, 120°, and 240°. The coded pulse consists of three groups of three segments—a total of nine segments.

Group 1			Group 2			Group 3		

Phases are assigned to the individual segments according to two simple rules: (1) the phase of the first segment of every group is 0°, that is, 0° ___ ___, 0° ___ ___, 0° ___,___; and (2) the phases of the remaining segments in each group increase in increments of

$$\Delta\Phi = (G - 1) \times (P - 1) \times \phi°,$$

where

G = group number

P = number of phases

ϕ = basic phase increment.

For a three-phase code ($P = 3$, $\phi = 120°$, $P - 1 = 2$), then $\Delta\Phi = (G - 1) \times 2\phi$. So the phase increment in Group 1 is 0°, the phase increment for Group 2 is 2ϕ, and the phase increment for Group 3 is 4ϕ.

Written in terms of ϕ, the nine digits of the code for $P = 3$ are

Group 1	Group 2	Group 3
0, 0, 0	0, 2ϕ, 4ϕ	0, 4ϕ, 8ϕ

Substituting 120° for ϕ and dropping multiples of 360°, the code becomes

Group 1	Group 2	Group 3
0°, 0°, 0°	0°, 240°, 120°	0°, 120°, 240°

Echoes are decoded by passing them through a tapped delay line (or the digital equivalent) in the same way as binary phase-coded echoes (Figure 16-26). The only difference is, the phase shifts in the taps have more than one value.

For a given number of segments, a Frank code provides the same pulse compression ratio as a binary phase code and the same ratio of peak amplitude to sidelobe amplitude as a Barker code. Yet, by using more phases (increasing P), the codes can be made of the greater length, P^2. As P is increased, however, the size of the fundamental phase increment decreases, making performance more sensitive to externally introduced phase shifts (e.g., transmitter distortion) and imposing more severe restrictions on uncompressed pulse width and maximum Doppler shift.

Frank codes are an example of a class of codes in which the discrete phase sequence can be viewed as a sampled version of the LFM chirp. Other such codes are the Zadoff-Chu code, the "P" codes, and Golomb codes. Like the minimum peak

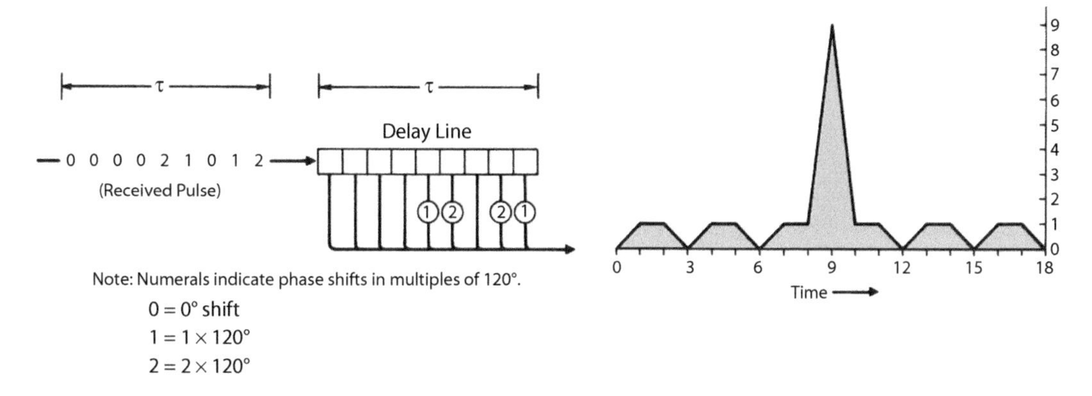

Figure 16-26. Processing of Frank codes is similar to that of binary codes. Phase shifts introduced in taps complement shifts in corresponding segments of the coded pulse. If the phase of a segment is shifted by 1 × 120°, the corresponding tap adds a shift of 2 × 120°, making the total shift when the pulse fills the line equal 3 × 120° = 360°. This phase relationship is identical to the matched filter.

sidelobe codes for the binary phase constellation, it is also possible to perform an exhaustive computer search for polyphase codes of arbitrary length and phase constellation.

While binary codes are in widespread use, the implementation of polyphase codes is more limited. The reason is that binary codes can be implemented in a phase-continuous manner in the transmitter while, until very recently, polyphase codes could not. These phase discontinuities at the chip transitions produce spectral spreading and can also limit the fidelity with which a polyphase-coded waveform can be generated by a practical transmitter. However, the design freedom provided by polyphase codes serves as the basis for new emerging radar capabilities. This topic is discussed further in Chapter 45.

16.4 Summary

Since radar transmitters are peak power limited, pulse compression provides the means to achieve sufficient energy on target for detection while enabling the requisite range resolution. Pulse compression comprises transmission of a modulated waveform and receiver filtering to compress the resulting echoes in range.

The most commonly used pulse compression techniques are the LFM chirp and binary phase coding.

With LFM, the frequency of each transmitted pulse is continuously increased or decreased. Applying the receive filter that is matched to the waveform results in a compressed pulse width of $1/\Delta F$, where ΔF is the total change in frequency (i.e., the bandwidth) of the waveform. The LFM range sidelobes may be reduced by amplitude weighting the receiver matched filter at a cost of reduced range resolution and mismatch loss.

When only a narrow range swath is of interest, the LFM chirp can be decoded using stretch processing, whereby range is converted to frequency in the receiver. Differences in frequency are resolved by a bank of tuned filters implemented with the efficient fast Fourier transform. With the LFM waveform and stretch processing, very large compression ratios and fine range resolution can be achieved. The LFM is rather insensitive to Doppler frequency shift, though such a shift produces an ambiguity in range.

In binary phase modulation, each pulse is marked off into segments, with the phase of certain segments reversed. Received echoes are passed through a tapped delay line having phase reversals in taps corresponding to those in the code. Binary codes are more sensitive to Doppler frequency shift than the LFM chirp.

Barker codes represent a form of binary phase modulation in which the mainlobe-to-sidelobe ratio equals the pulse–compression ratio, though the longest Barker code is only 13 digits.

Sidelobes may be eliminated by alternately transmitting complementary codes that are obtained from chained Barker codes. However, this property requires little to no Doppler shift.

Polyphase (e.g., Frank) codes can also be used, but these are more sensitive to Doppler shift than binary codes, due to smaller phase increments. Polyphase codes tend to produce phase discontinuities, which results in spectral spreading and sensitivity to transmitter distortion.

Further Reading

N. Levanon and E. Mozeson, *Radar Signals*, John Wiley & Sons, 2004.

M. I. Skolnik (ed.), "Pulse Compression Radar," chapter 8 in *Radar Handbook*, 3rd ed., McGraw Hill, 2008.

G. Brooker, "High Range-Resolution Techniques," chapter 11 in *Sensors for Ranging and Imaging*, SciTech-IET, 2009.

M. A. Richards, J. A. Scheer, and W. A. Holm (eds.), "Fundamentals of Pulse Compression Waveforms," chapter 20 in *Principles of Modern Radar: Basic Principles*, SciTech-IET, 2010.

Test your understanding

1. To achieve a range resolution of 0.5 m with a 20 μs pulse, determine the required chirp rate and the associated pulse compression ratio.
2. Using the process described in Figure 16-18, determine the tapped delay line output for the following binary codes:
   a. Length 11 Barker code as defined in Figure 16-21
   b. [+++−−−+−−++] (Barker 11, but with the last digit flipped)
3. Using the tapped delay line outputs from problem 2, determine the largest sidelobe value relative to the mainlobe. This ratio is known as the PSL.
4. For $P = 4$ phase values, determine the length 16 Frank code and compute its output from the tapped delay line.

17

Frequency-Modulated Continuous Wave Ranging

Thales SCOUT FMCW radar
(Courtesy of Thales.)

This chapter briefly describes the principle of frequency modulation (FM) ranging. It explains how Doppler frequency shifts, which would otherwise introduce gross measurement errors, are taken into account and how a problem of ghosting similar to that encountered in pulse repetition frequency (PRF) switching is handled. Finally, it briefly considers the resolution and accuracy that may be obtained with FM ranging.

17.1 Basic Principle

With FM ranging, the time lag between transmission and reception is converted to a frequency difference. By measuring it, the time lag—hence the range—is determined.

In simplest form, the process is as follows. The radio frequency of the transmitter is increased at a constant rate. Each successive transmitted pulse thus has a slightly higher radio frequency. The linear modulation is continued for a period at least several times as long as the two-way propagation delay for the most distant target of significance (Fig. 17-1). Over the course of this period, the instantaneous difference between the frequency of the received echoes and the frequency of the transmitter is measured. The transmitter is then returned to the starting frequency, and the cycle is repeated.

Just how the measured frequency difference is related to a target's range is illustrated in Figure 17-2 for a static situation, such as a tail chase, where the range rate is zero.

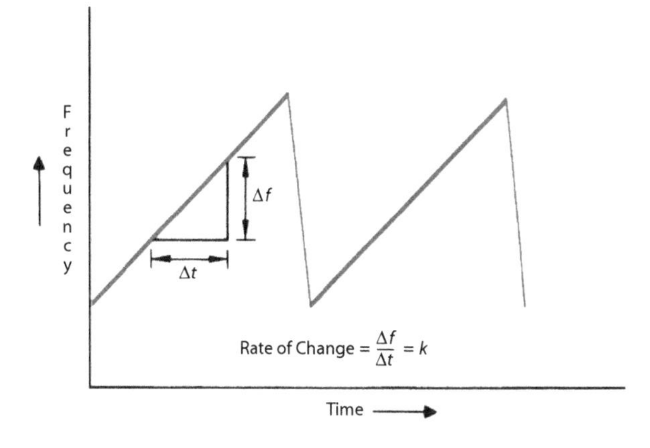

Figure 17-1. In simplest form, FM ranging involves changing the transmitter frequency at a constant rate. The length of slope is generally many times the maximum two-way propagation delay.

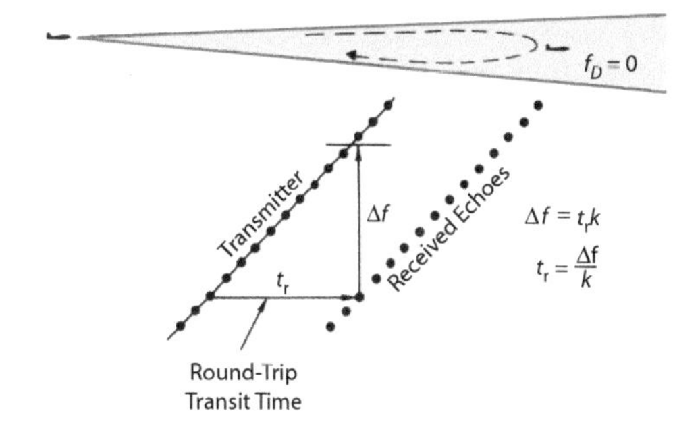

Figure 17-2. The difference between the frequency of an echo and the frequency of the transmitter at the time an echo is received, Δf, is proportional to two-way propagation delay, t_r.

In this figure, the radio frequency of both the transmitter and the echoes received from a target are plotted versus time. The frequency sweep may be (1) realized as a sequence of pulses, in which case the dots on the plot of transmitter frequency represent individual transmitted pulses, or (2) continuous, in which case the technique is called frequency-modulated continuous wave (FMCW). The horizontal distance between each of these dots and the dot representing the received target echo is the two-way propagation delay. The vertical distance between the echo dot and the line representing the transmitter frequency is the difference, Δf, between the frequency of the echo and the frequency of the transmitter when the echo is received.

As you can see, this difference equals the rate of change of the transmitter frequency—hertz per microsecond—times the two-way propagation delay. By measuring the frequency difference and dividing it by the rate (which we already know), we can find the two-way propagation delay and hence the range.

Suppose, for example, that the measured frequency difference is 10,000 Hz and the transmitter frequency has been increasing at a rate of 10 Hz per μs. The two-way propagation delay is

$$t_r = \frac{10{,}000\,\text{Hz}}{10\,\text{Hz}/\mu\text{s}} = 1{,}000\,\mu\text{s}$$

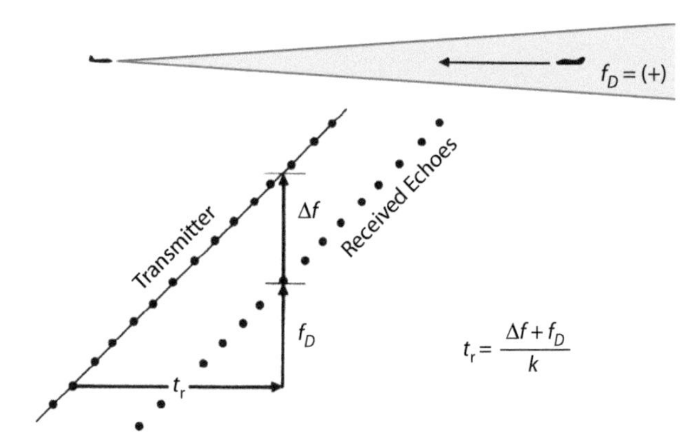

Figure 17-3. The frequency difference, Δf, between transmitter and received echoes is reduced by the target's Doppler frequency, f_D. To find the two-way propagation delay, f_D must be added to Δf.

Since 6.67 μs of two-way propagation delay correspond to 1 km of range, the target's range is equal to 1,000/6.67 = 150 km.

17.2 Accounting for the Doppler Shift

Actually, the process is more complicated than what was just described because the range rate is rarely zero. The frequency of a target echo is not equal solely to the frequency of the transmitted pulse that produced it but to that frequency plus the target's Doppler frequency. To find the two-way propagation delay, we must add the Doppler frequency, f_D, to the measured frequency difference (Fig. 17-3).

Including a Constant-Frequency Segment. As you may have surmised, the Doppler frequency can be found by interrupting the frequency modulation at the end of each cycle and transmitting at a constant frequency for a brief period. During this period, the difference between the echo frequency and the transmitter frequency will be due solely to the target's Doppler frequency. By measuring that difference (Fig. 17-4) and adding it to the difference measured during the sloping segment, we can find the two-way propagation delay and hence the target range.

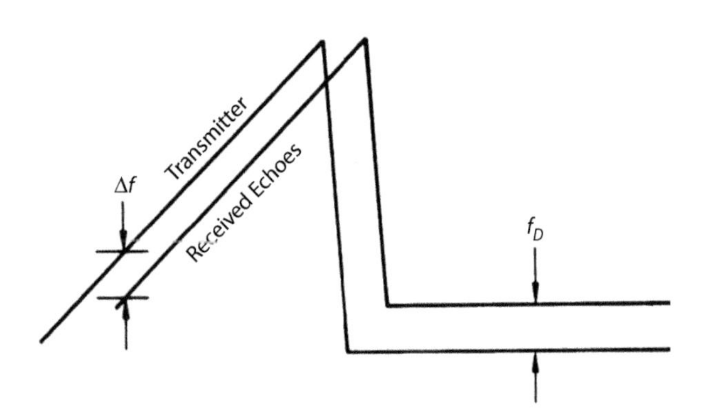

Figure 17-4. The target's Doppler frequency, f_D, may be measured by adding a constant frequency segment to the modulation cycle.

Alternate, Two-Slope Cycle. It turns out that the Doppler frequency can be added just as easily by employing a two-slope modulation cycle. The first slope is the same as the rising-frequency slope just described. Once it has been traversed, the frequency is decreased at the same rate until the

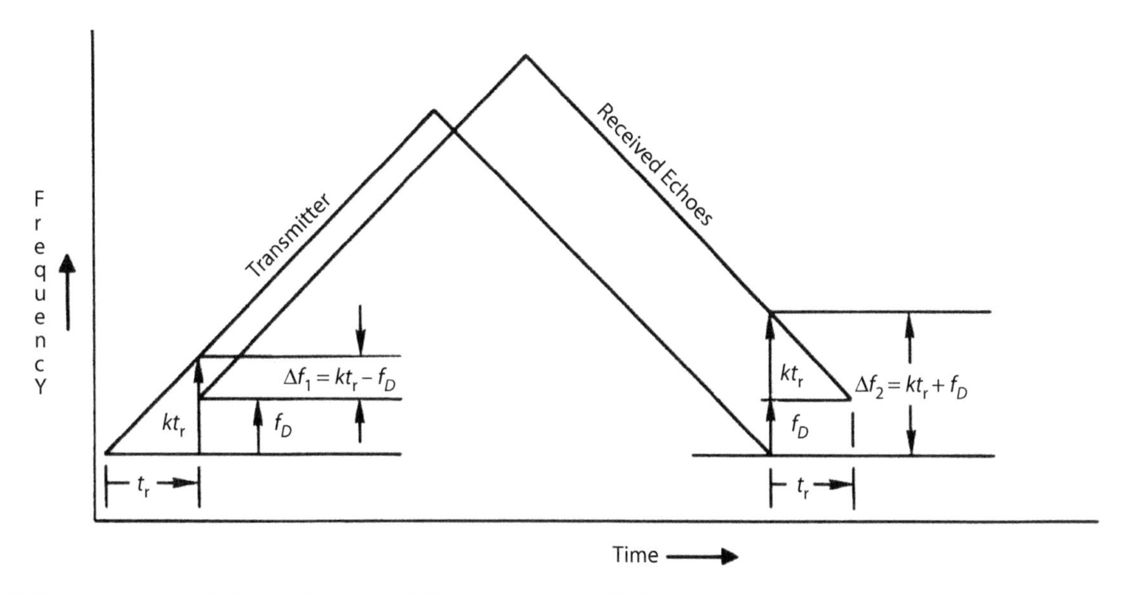

Figure 17-5. With two-slope modulation, the frequency difference is decreased by f_D during the rising slope and increased by f_D during the falling slope.

starting frequency is again reached (Fig. 17-5). The cycle is then repeated.

If the target is closing—that is, has a positive Doppler frequency, f_D—the difference between the frequency of the transmitter and the frequency of the received echoes will be decreased by f_D during the rising-frequency segment and increased by f_D during the falling-frequency segment. (The reverse will be true if the target is receding). Consequently, if the frequency differences for the two segments are added, the Doppler frequency will cancel out. Of course, if the target is receding, the sign of the shift of f_D will be reversed.

$$\begin{aligned} \Delta f_1 &= kt_r - f_D \\ \Delta f_2 &= kt_r + f_D \\ \hline \Delta f_1 + \Delta f_2 &= 2kt_r + 0 \end{aligned}$$

The sum, then, will be twice the frequency difference, kt_r, due to the two-way propagation delay. The latter can be found by dividing the sum by twice the rate of change of the transmitter frequency.

$$t_r = \frac{\Delta f_1 + \Delta f_2}{2k}$$

where

t_r = round-trip transit time
Δf_1 = difference between transmitter and echo frequencies during rising-frequency segment
Δf_2 = difference between transmitter and echo frequencies during falling-frequency segment
k = rate of change of transmitter frequency

$$t_r = \frac{7\text{ kHz} + 13\text{ kHz}}{2 \times .01\text{ kHz}/\mu\text{s}}$$
$$= \frac{20\text{ kHz}}{.02\text{ kHz}/\mu\text{s}}$$
$$= 1{,}000\ \mu\text{s}$$

Again, knowing the propagation delay, we can readily calculate the target range.

Suppose the target used in the previous example ($k = 10$ Hz/µs, $kt_r = 10$ kHz) had a Doppler frequency of 3 kHz. During the rising-frequency segment, the measured frequency difference would have been 10 – 3 = 7 kHz. During the falling-frequency segment, it would have been 10 + 3 = 13 kHz. Adding the two differences and dividing by $2k$ (20 Hz/µs) gives the same propagation delay, 1,000 µs, as when the Doppler frequency was zero.

Although in both this example and the illustrations the Doppler frequency is positive, the equation works just as well for negative Doppler frequencies.

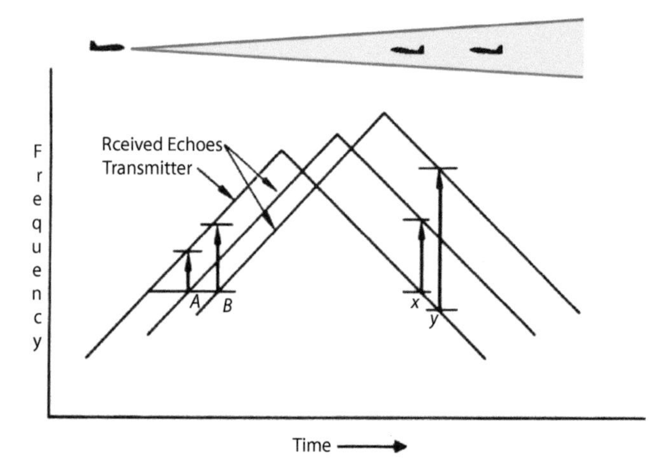

Figure 17-6. If two targets are detected simultaneously, two frequency differences will be measured during each segment of the cycle.

17.3 Eliminating Ghosts

If the antenna beam encompasses two targets at the same time, a problem of ghosting may be encountered, as with PRF switching. There will be two frequency differences during the first segment of the modulation cycle and two during the second (Fig. 17-6). This is true, of course, regardless of whether both segments are sloped or one is sloped and the other is not.

Why Ghosts? Although we can tell from the continuity in the plots of frequency versus time which frequency differences belong to the same target, the continuity is not visible to the radar. As will be explained in detail in Chapter 24, for a radar to discern small frequency differences such as are normally encountered in FM ranging, it must receive echoes from a target for an appreciable period of time. In essence, all the radar observes are two frequency differences at the end of the first segment and two (probably different) frequency differences at the end of the second segment.

This is illustrated in Figure 17-7. There, the first two differences are referred to as A and B and the second two as x and y. Without some additional information it is impossible to tell for sure how these differences should be paired—whether A and x pertain to the same target or A and y.

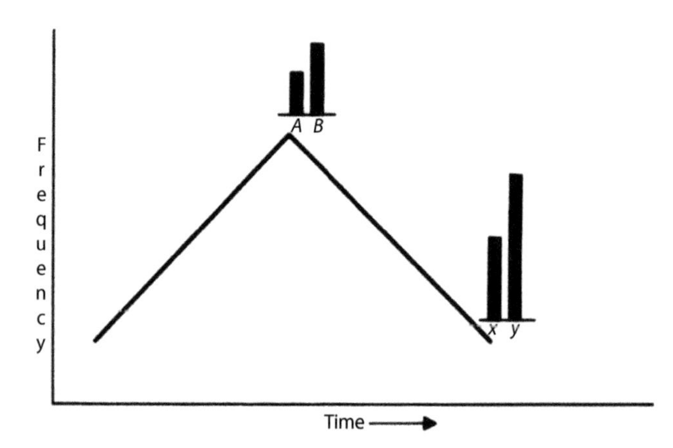

Figure 17-7. All the radar sees are two frequency differences at the end of each segment. Radar has no way of telling whether A should be paired with x or y.

Identifying Ghosts. In applications where ghosts are apt to be encountered, they may be eliminated by adding another segment to the modulation cycle—much as they are by adding another PRF when PRF switching is used.

A representative three-slope cycle consists of equal increasing and decreasing frequency segments—such as we just considered—plus a constant-frequency segment (Fig. 17-8). The latter, of course, provides a direct measure of the targets' Doppler frequencies.

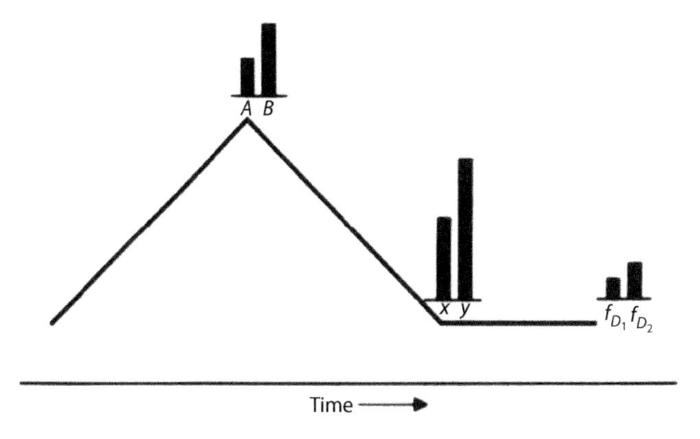

Figure 17-8. The problem is solved by adding a third segment in which Doppler frequencies are separately measured.

There is, however, no direct way of pairing the measured Doppler frequencies with *A* and *B* or *x* and *y* either. But the correct pairing of *A* and *B* with *x* and *y* can quickly be found by knowing the Doppler frequencies. Just as Doppler frequency cancels out when we *add* the frequency differences for positively and negatively sloping segments, so propagation delay cancels out when we *subtract* the differences. The result then is twice the Doppler frequency.

$$\begin{array}{r} \Delta f_2 = kt_r + f_D \\ -(\Delta f_1 = kt_r - f_D) \\ \hline \Delta f_2 - \Delta f_1 = 0 + 2f_D \end{array}$$

Therefore, by subtracting *A* (or *B*) from *x* (or *y*) and comparing the results with the measured Doppler frequencies, we can tell which of the two possible pairings is correct (Fig. 17-9). If we say that $(x - A)$ is twice one of the measured Doppler frequencies, then the pairing should be as follows:

x with *A*
y with *B*

Otherwise, *y* should be paired with *A* and *x* with *B*.

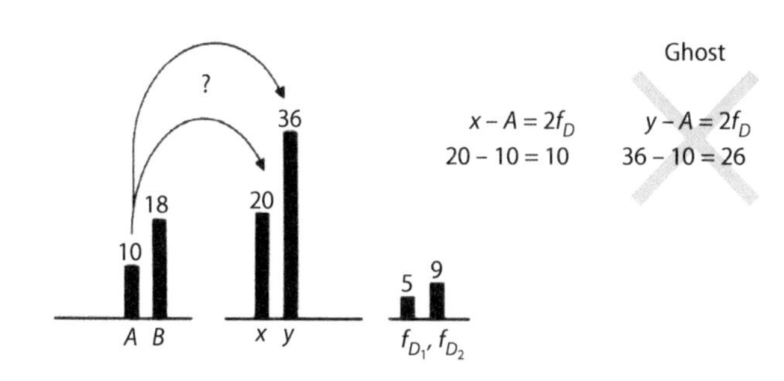

Figure 17-9. Knowing the two Doppler frequencies, the radar can readily tell whether *x* and *A* or *y* and *A* should be paired.

Doppler Frequency Greater than kt_r. In the illustrations shown so far, the Doppler frequency has been less than the frequency difference due to the ranging time, kt_r. While this is true of applications such as altimetry, it is not true of air-to-air applications. For these, the rate of change of the transmitter frequency is typically made low enough so that the maximum value of kt_r will be only a small fraction of the highest Doppler frequency normally encountered. In that case, a plot of the frequency of the echoes from a closing target during the rising-frequency portion of the modulation cycle appears as in Figure 17-10.

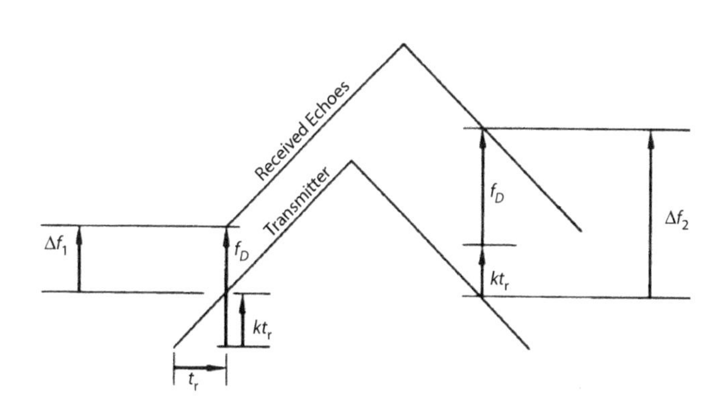

Figure 17-10. When Doppler frequency, f_D, is greater than frequency difference due to ranging time, kt_r, echo frequencies are higher than transmitter frequency during the rising frequency segment.

The relationships between the measured frequency differences for two or more simultaneously detected targets can then be seen more clearly if the differences for each segment of the cycle are plotted on separate horizontal scales—one above the other—as in Figure 17-11. The differences for the rising-frequency segment (*A* and *B* in the figure) appear on the negative half of the frequency scale and the differences for the falling-frequency segment (*x* and *y*) on the positive half. The differences can be paired by drawing horizontal arrows, between them, of lengths corresponding to the Doppler frequencies measured in the third segment of the cycle.

In this case, we find that *y* is separated from *A* by two lengths of the arrow, f_{D_2}. These frequency differences therefore belong to the same target. The point where they abut corresponds to the frequency difference due to the two-way propagation delay for the target, kt_{r_A}.

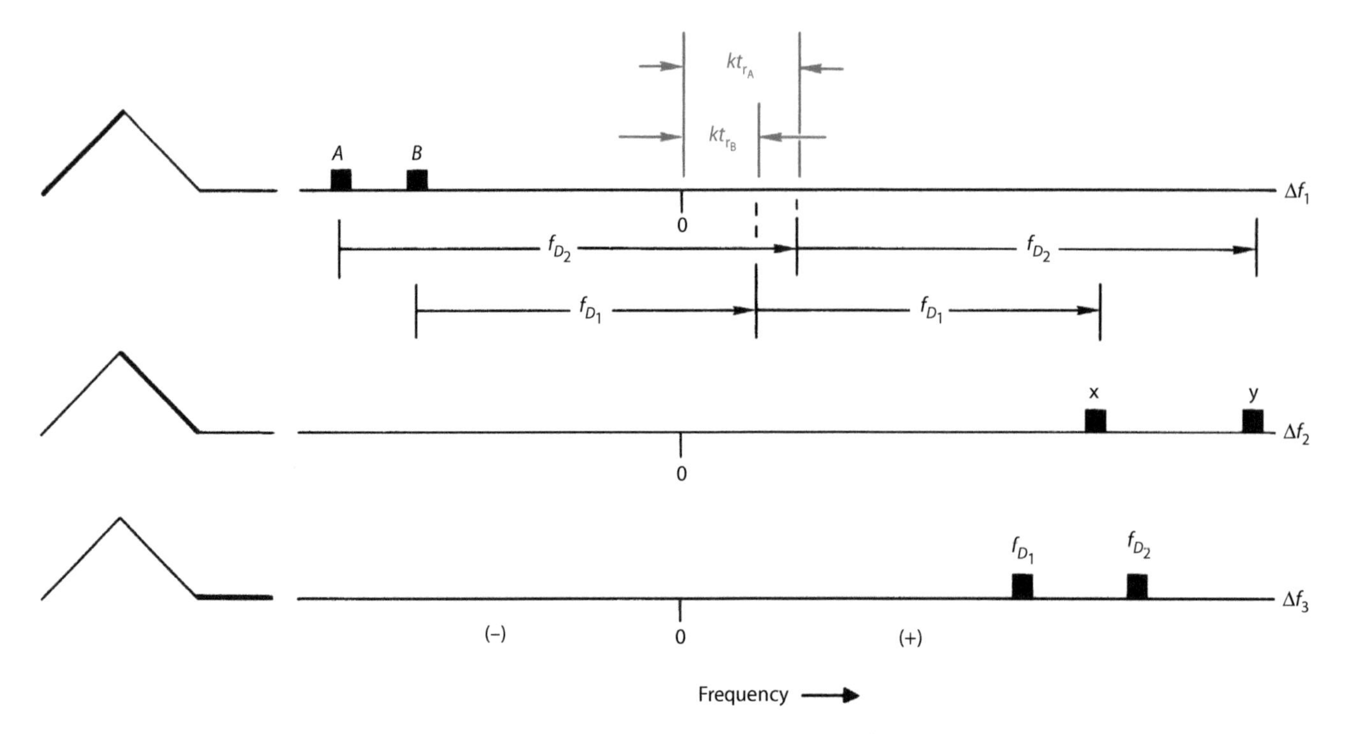

Figure 17-11. The relationships between frequency differences measured during the rising and falling slopes can be seen more clearly if plotted on separate line for each slope.

$$A + f_{D_2} = kt_{r_A}$$

Similarly, x is separated from B by two lengths of the arrow, f_{D_1}, and the point where they abut corresponds to the frequency difference due to the propagation delay for the target, kt_{r_B}.

Three Targets Detected Simultaneously. Figure 17-12 plots the measured frequency differences for three targets. They, too, can be paired easily. Comparing C with x, y, and z, we find that it is separated from z by $2f_{D_3}$. There are still two possible combinations of A and B with x and y: A with x and B with y, or A with y and B with x. But with C out of the way we can

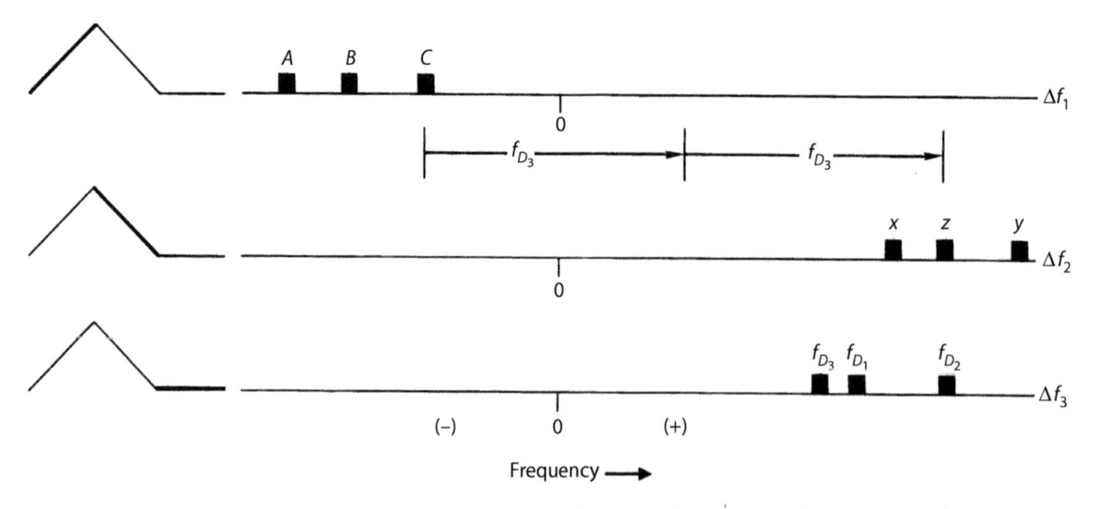

Figure 17-12. When three targets are detected simultaneously, once one combination of frequency differences has been paired the others may be paired in the same way as when only two targets are detected.

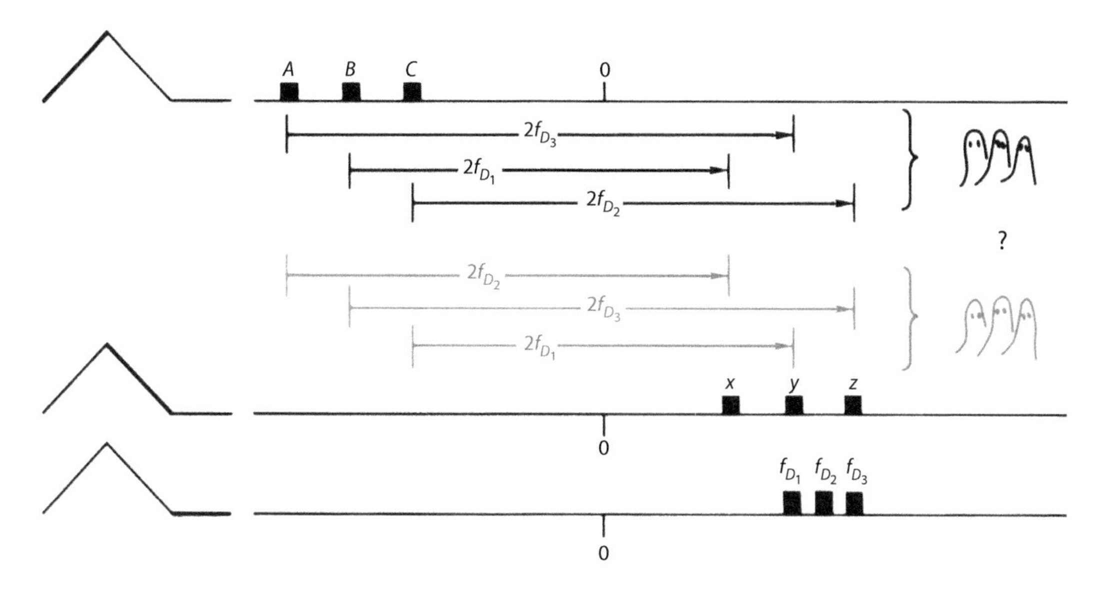

Figure 17-13. With only three slopes, certain combinations of three targets will leave unresolved ghosts, but these combinations are rare.

readily tell which of these are ghosts—just as we did when only two targets were detected to begin with.

Certain combinations of ranges and Doppler frequencies may occur, however, for which more than one pairing of A, B, and C with x, y, and z are possible. One of these is illustrated in Figure 17-13. The frequency differences shown there can readily be paired as follows:

$$A + 2f_{D_3} = y$$
$$B + 2f_{D_1} = x$$
$$C + 2f_{D_2} = z$$

But a second pairing is also possible.

$$A + 2f_{D_2} = x$$
$$B + 2f_{D_3} = z$$
$$C + 2f_{D_1} = y$$

The ranges indicated by one or the other of these pairings are ghosts, and with only three PRFs we cannot tell which. As the number of simultaneously detected targets increases, the number of these potential ghost-producing combinations, though small, goes up.

They can be eliminated by adding more slopes to the cycle. As with PRFs in pulse-delay ranging, if N is the number of slopes, all possible combinations of $(N - 1)$ simultaneously detected targets can be deghosted. But the problem of ghosts is usually much less severe with FM ranging than with pulse-delay ranging because in situations where it is normally used neither range nor Doppler frequency is ambiguous.

Recognizing that there will always be some possibility of encountering unresolved ghosts, a three-slope modulation cycle usually suffices.

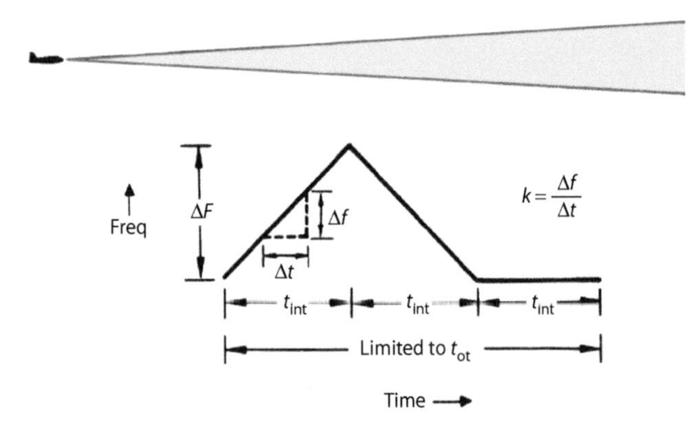

Figure 17-14. Frequency measurement accuracy is limited by time-on-target, t_{ot}.

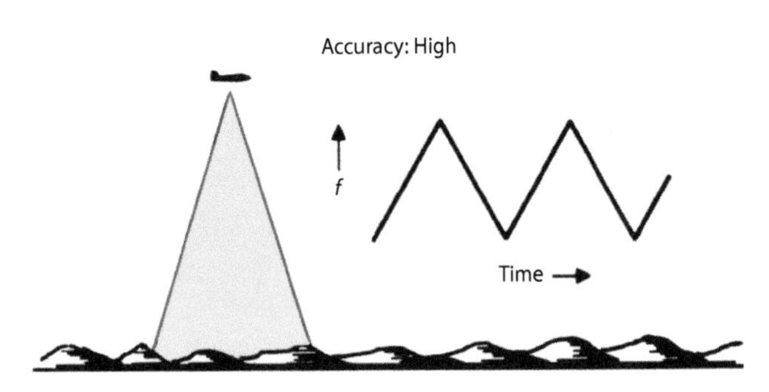

Figure 17-15. In altimeters, modulation sweep bandwidth can be made broad enough to provide highly accurate range measurement since the time-on-target is not constrained.

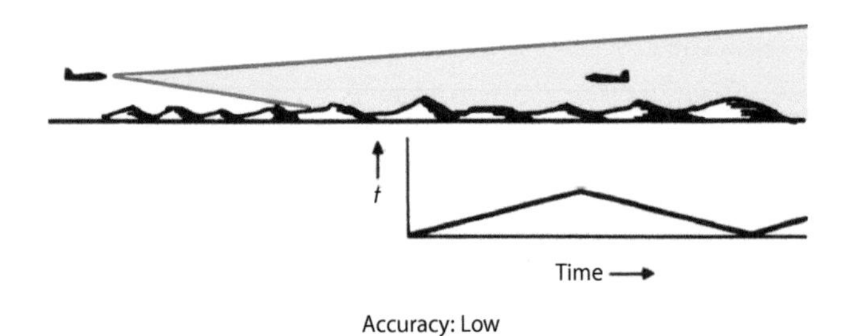

Figure 17-16. For air-to-air applications, slopes must be made shallow to avoid smearing the spectrum of the ground return. The result is low accuracy.

17.4 Performance

The accuracy of FM ranging depends on two basic factors: (1) the rate k at which the transmitter frequency is changed; and (2) the accuracy with which the frequency differences are measured.

The greater the value of k, the greater the frequency difference that a given two-way propagation delay will produce. The greater this difference and the greater the accuracy with which frequency can be measured, the more accurately the range will be determined and finer resolution of separated targets will be possible.

Frequency Measurement Accuracy. This increases with the length of time, t_{int}, over which the measurement is made—the length of the segments of the modulation cycle (Fig. 17-14).

In search operation, the length of the segments is limited by the length of time the antenna beam takes to scan across a target: time on target, t_{ot}.

Since the time-on-target is typically fixed by other considerations, the steepness of the sloping segments of the cycle—the rate, k—becomes the controlling factor for range measurement resolution.

Steepness of Slope, k. In applications such as low-altitude altimeters, k can be made sufficiently high to provide extremely precise range measurements (Fig. 17-15).

However, as will be explained in detail in Chapter 27, in air-to-air applications the value of k is severely limited. As k is increased, ground return—which may be received from ranges out to hundreds of kilometers—is smeared over an increasingly broad band of frequencies. A point is quickly reached where the clutter blankets the targets, even though the Doppler frequencies of targets and clutter may be quite different (Fig. 17-16).

Because of the limitations on k, in these applications FM ranging is fairly imprecise. Whereas pulse-delay ranging yields accuracies on the order of tens of cm, FM ranging yields accuracies on the order of km. However, in applications where k is not constrained, very high resolution and accuracy can be obtained. The radar altimeter shown in Figure 17-15 is one example. Another example is the use of mm-wave radar for vehicle collision avoidance or cruise control.

FMCW ranging can also be seen as an alternative way of realizing pulse compression processing of a linear FM signal. Waveforms of high time-bandwidth product can be used, giving high processing gain. This also allows low peak transmit powers, giving the radar low probability of intercept (LPI) properties.

Radar Equation. The same basic radar equation as was developed in Chapter 12 is relevant to FM radar:

$$\frac{P_r}{P_n} = \frac{P_t G^2 \lambda^2 \sigma}{(4\pi)^3 R^4 \; kT_0 BF}$$

An important difference from a conventional pulsed radar, though, is that the appropriate value of receiver noise bandwidth, B, in this equation is the inverse of the sweep duration rather than the sweep bandwidth. Since the ratio of these bandwidths may be several orders of magnitude, there may be a substantial processing gain associated with FM radar—in other words, for the same signal-to-noise ratio the peak transmit power may be substantially lower.

To illustrate this point, suppose that a particular FMCW radar uses a sweep with a bandwidth of 50 MHz and a sweep duration of 100 μs. The receiver noise bandwidth is not 50 MHz, but 1/100 μs = 10 kHz, which is a difference of 2×10^4, or 37 dB.

Linear FM Sweep Generation. Linear FM swept signals can be generated by various means. Digital generation is increasingly practical, even for very broad bandwidths. Voltage-controlled oscillators (VCOs), which are fed with a linear voltage tuning ramp, are usable, though the frequency-versus-tuning voltage characteristic is unlikely to be perfectly linear and may vary with temperature and with the load impedance presented to the VCO. Sweep linearization techniques are therefore attractive.

17.5 Summary

With FM ranging, the time lag between transmission and reception is converted to a frequency shift. By measuring this shift, the range is determined. Typically, the transmitter frequency is changed at a constant rate. The change is continued over a considerable period of time so the frequency difference can be accurately measured.

To cancel the contribution of the target's Doppler frequency to the measured frequency difference, a second measurement is made. This is done either while transmitting at a constant frequency or while changing the transmitter frequency in the opposite direction. The second measurement is then subtracted from the first.

To resolve ambiguities occurring when two targets are detected simultaneously, a third measurement may be made. In general, N measurements are needed to resolve $(N - 1)$ simultaneously detected targets.

For long-range applications, FM ranging is more complicated and as a rule less is accurate than pulse-delay ranging and reduces the radar's detection range.

However, FM radar provides a processing gain that can be substantial, so the peak transmit power requirement may be much lower than an equivalent pulsed radar. This can give an FM radar useful LPI properties.

Further Reading

H. D. Griffiths, "New Ideas in FM Radar," *IEE Electronics and Communication Engineering Journal*, Vol. 2, No. 5, pp. 185–194, October 1990.

A. G. Stove, "Linear FMCW Radar Techniques," *IEE Proceedings Part F*, Vol. 139, No. 5, 343–350, October 1992.

M. Jankiraman, *Design of Multi-Frequency CW Radars*, SciTech-IET, 2008.

Test your understanding

1. An FMCW radar altimeter has a waveform bandwidth of 300 MHz. What is its range resolution?
2. A vehicle collision avoidance radar uses the FMCW principle, and operates at a frequency of 77 GHz. The sweep bandwidth is 100 MHz, and the sweep duration is 1 ms. What is the two-way propagation delay associated with a target at a range of 60 m, and what is the resulting echo beat frequency?
3. The radar is modified to use successive up-chirp and down-chirp waveforms, each of 100 MHz bandwidth and 1 ms duration. The echo beat frequencies observed from a particular target are 55.13 kHz (down-chirp) and 44.87 kHz (up-chirp). Calculate the range and velocity of the target.
4. Tarsier® is an FMCW radar developed by the British company QinetiQ to detect debris on airport runways. Some of its parameters are as follows:

Carrier frequency	94.5 GHz
FM sweep bandwidth	600 MHz
FM sweep period	2.6 ms
Transmit power	150 mW
Antenna width	0.88 m
Antenna height	0.088 m
Antenna rotation rate	3°/sec
Receiver noise figure	6.5 dB

   How many FM sweeps per azimuth scan illuminate a given target?

5. It is claimed that this radar can detect a metal bolt of radar cross section 0.01 m^2 on the runway at a range of 1 km. Justify this claim, stating any assumptions that you make.

Tupolev Tu-160 Blackjack (1987)

The Tu-160 is a supersonic strategic bomber developed in response to the US Air Force B-1 bomber project. It is capable of Mach 2 speeds and is only surpassed by the failed XB-70 as the fastest bomber every produced. However, it is currently the largest supersonic combat aircraft in the world and is also the variable-sweep wing aircraft ever built. Here it is seen escorted by a Royal Air Force Tornado F3.

PART IV

Pulse Doppler Radar

Panavia Tornado GR-4 (1979)

The Tornado was developed by Panavia Aircraft GmbH, a consortium of contractors from Britain, West Germany, and Italy. It is a variable-sweep wing multirole aircraft designed to penetrate enemy defenses at low altitude but serves multiple roles including interceptor, electronic combat, and fighter/bomber. The Tornado ADV variant carries the AI.24 Foxhunter radar for air defense operations and capable of tracking 20 targets at ranges of up to 160 km (100 miles).

18

The Doppler Effect

Relativistic Doppler Effect: A source of light waves moving to the right, relative to observers, with velocity 0.7c

By sensing Doppler frequencies, a radar system not only can measure range-rates but also can separate moving target echoes from stationary clutter or produce high-resolution ground maps. Since these are important functions of many of today's radars, it is important to understand the Doppler effect.

This chapter examines the Doppler shift more closely, first, via the compression or expansion of wavelength and, second, via the continuous shift of phase. In this way the factors that determine the Doppler frequencies of the echo return from both moving targets and the ground are pinpointed. Finally, the chapter considers the special case of the Doppler shift of a target's echoes as observed by a semi-active missile.

18.1 The Doppler Effect and Its Causes

The Doppler effect is a shift in the frequency of a wave radiated, reflected, or received by an object in motion. As illustrated in Figure 18-1, a wave radiated from a point source is compressed in the direction of motion and is spread out in the opposite direction. In both cases, the greater the object's speed, the greater the effect will be. Only at right angles to the motion is the wave unaffected. Since frequency is inversely proportional to wavelength, the more compressed the wave, the higher its frequency will be and vice versa. Therefore, the frequency of the wave is shifted in direct proportion to the object's velocity.

In the case of radar, it is the relative motion between the radar and detected objects that produces Doppler shifts (Fig. 18-2). If the distance between the radar and a reflecting object is decreasing, the waves are compressed. Their wavelength is shortened, and their frequency is increased. If the distance is increasing, the effect is just the opposite.

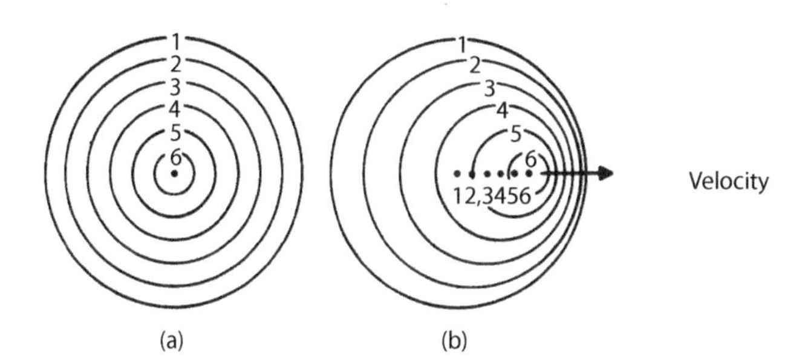

Figure 18-1. In a wave radiated from a point source when stationary (a) and when moving (b), it is compressed in the direction of motion, spreads out in the opposite direction, and is unaffected in the direction normal to motion.

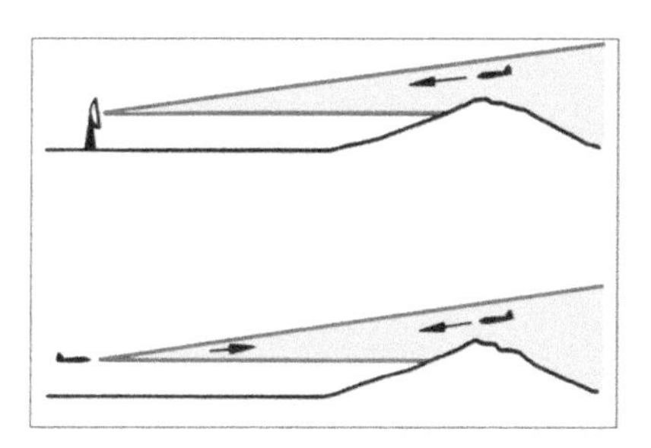

Figure 18-2. With a ground-based radar, relative motion is due entirely to the target's motion. With airborne radar, it is due to the relative motion of both the radar and the target.

With ground-based radars, any relative motion is essentially caused by target movement. Echo returns from the ground have little or no Doppler shift. (Objects on the ground, such as wind-blown crops or vehicles, may be moving, but this is ignored for now.) Therefore, differentiating between ground clutter and the echoes of moving targets such as aircraft is comparatively straightforward.

With airborne or moving radars, on the other hand, the relative motion may be due to the motion of either the radar or the targets or both. Even in aircraft such as hovering helicopters, the radar is always in motion. Consequently, both target echoes and ground return have Doppler shifts. This complicates the task of separating target echoes from ground clutter, especially if the radar and the targets move such that they have the same effective Doppler as clutter. This is because pulse Doppler radar can differentiate between the targets and clutter only on the basis of differences in the magnitudes of their Doppler shifts. Before discussing that, however, we consider how the Doppler shift actually occurs.

18.2 Where and How the Doppler Shift Takes Place

If both radar and target are moving, the radio waves may be compressed (or stretched) at three points in their travel: transmission; reflection; and reception. The compression in wavelength occurring in the simple case of a radar closing on a target, head-on, is illustrated in Figure 18-3.

In these simplified diagrams, the slightly curved vertical lines represent plane waves (viewed edge-on), called *wavefronts,* at

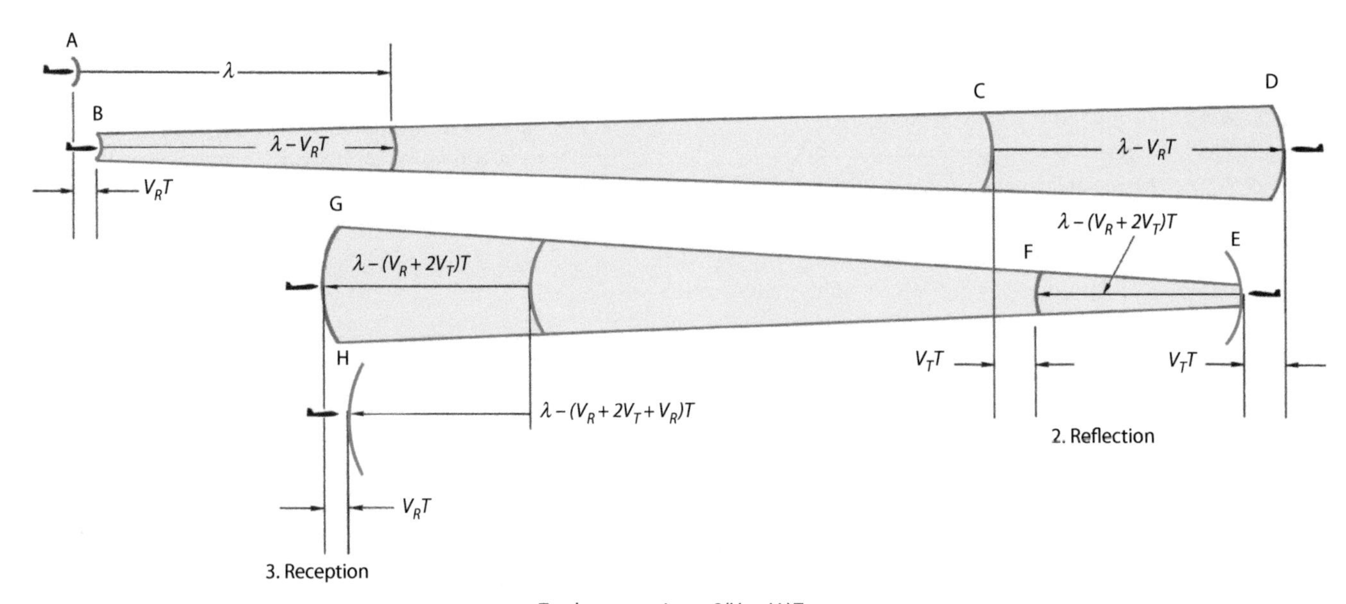

Figure 18-3. Compression in wavelength occurs during transmission, reflection, and reception. The first wavefront is noted in red, and the second wavefront is noted in blue.

every point on which the phase of the wave's field is the same. Figure 18-3 shows wavefronts on which the fields have their maximum intensity in a positive direction. In other words, they represent wave *crests*. Two successive wavefronts (wavefront 1 in red, wavefront 2 in blue) are shown at each of the points in question.

For the sake of readability, the diagrams have not been drawn to scale. Note that the wavelength (the spacing between successive wavefronts of the same phase) is only a small fraction of the length of most targets. For example, at X-band the wavelength is approximately 3 cm, and a target such as a vehicle might be several meters in length. Also, since the speed of light is 3×10^8 m/s, in a given period of time an aircraft travels only a minuscule fraction—just a few microns—of the distance the waves travel.

The radar is at point A when it transmits the first wavefront (red). By the time it transmits the second wavefront (blue), it has advanced to point B, decreasing the wavelength by a distance equal to the velocity of the radar, V_R, times the time between transmissions of the two wavefronts. That time, of course, is the period of the wave, T, or the inverse of the radar frequency, f. The space between wavefronts as the wave travels out to the target, therefore, is $\lambda - V_R T$.

The top part of Fig. 18-3 illustrates the compression occurring when the wave is reflected by the target. When the first wavefront (red) is reflected, the target is at point D and the second wavefront (blue) is at point C. By the time the second wavefront (blue) is reflected, the target has advanced to point E, shortening the distance the wavefront has had to travel from point C to reach the target by an amount equal to the velocity of the target, V_T, times the period, T. Meanwhile, the reflection of the first wavefront (red) has traveled an equal distance (D to F). But the target's advance has reduced the separation between this reflected wavefront and the reflection of the second wavefront (blue), which is just now leaving the target, by $V_T T$.

The space between wavefronts of the reflected wave as it travels back to the radar, therefore, is $\lambda - (V_R + 2V_T)T$.

The bottom part of Fig. 18-3 illustrates the reception of the two wavefronts by the radar. The radar is at point G when it receives the first wavefront (red). The second wavefront (blue) is one compressed wavelength away, but by the time it is received the radar has advanced to point H. Thus, during reception, the wavelength is compressed additionally by a distance $V_R T$, that is, the same amount as during transmission.

In all, the wavelength is compressed by twice the sum of the two velocities times the period of the transmitted wave, T.

$$\text{Total compression} = 2(V_R + V_T)T$$

Since T is very short, the compression is extremely slight. For an X-band radio wave and values of V_R and V_T of 300 m/s, the compression is only about 12 μm. Nevertheless, since the radio frequency of an X-band wave is very high (10 GHz), the resulting frequency shift, at X-band, is 40 kHz.

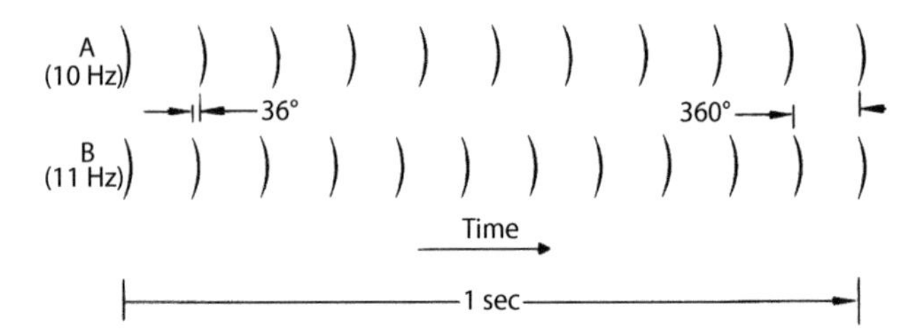

Figure 18-4. These two waves show wavefronts of slightly different frequency. The difference is tantamount to a continuous shift in phase, here 36° per cycle.

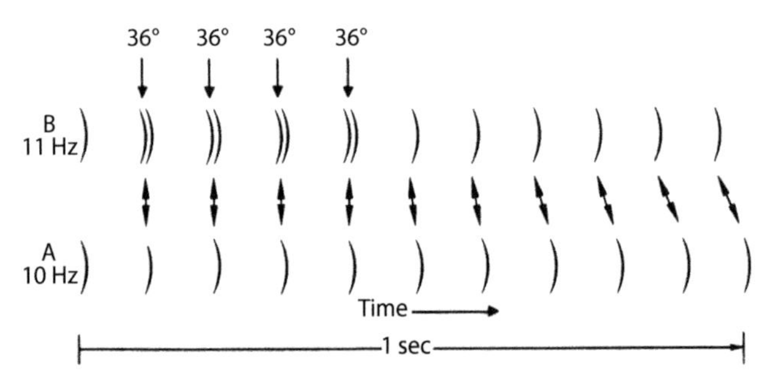

Figure 18-5. Frequency can be decreased an increment of 1 Hz (from 11 to 10 Hz) by inserting a 36° phase shift between wavefronts. When insertion is discontinued, the wave reverts to its original frequency.

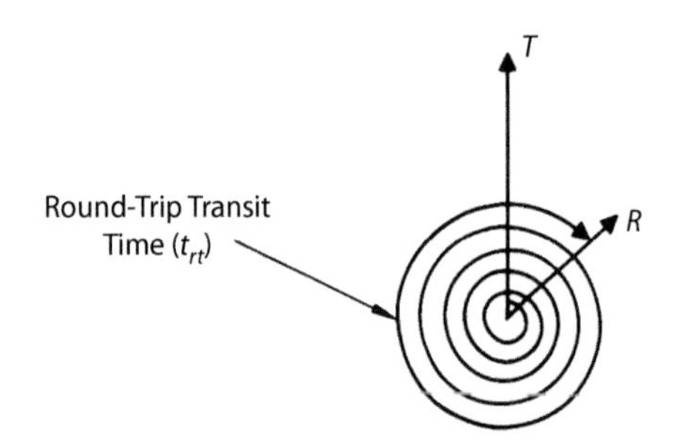

Figure 18-6. The phase of the received wave lags that of the transmitted wave by the round-trip transit time. Every time the round-trip component crosses the *T* vector, the wavefront cycles through 360° and the last little bit is ϕ, the residual phase difference between the radar and the target.

18.3 Magnitude of the Doppler Frequency

Although we can sense the Doppler effect by observing the compression in wavelength due to the relative motion of a radar system and a target, the Doppler frequency is calculated much more simply on the basis of the shift in phase of the received wave.

Frequency, a Continuous Phase Shift. You may not think a change in the frequency of a wave is tantamount to a continuous shift in phase. Figure 18-4 shows a 1 s sample of two waves, A and B. Their frequencies are 10 Hz and 11 Hz, respectively. At 11 Hz, B completes one more cycle per second than A. In other words, at every second, the phase of B relative to A advances 360° (or 2π radians). Since A completes 10 cycles every second, the gain in phase per cycle of A is $360°/10 = 36°$ (or 0.2π radians).

Shifting the frequency of B down to 10 Hz can be accomplished simply by inserting a time delay equivalent to 36° of phase between successive wavefronts (Fig. 18-5).

As long as the wave's phase is continually shifted, the frequency shift will persist. But if is stopped, B will revert to its original frequency. By shifting phase in the opposite direction, that is, decreasing the time between wavefronts, a wave's frequency can be similarly increased.

This is also the case with the Doppler shift in the signal received from a target. However, in this case phase is not shifted through the arbitrary insertion or removal of increments of time between wavefronts but instead as the result of the continuous change in the time that the radio waves take to travel from the radar to the target and back (i.e., the change in the round-trip transit time).

Phasor Representation of Doppler Frequency. Doppler frequency shifts in radar can be visualized using the phasor concept introduced in Chapter 5. The simple phasor diagram illustrated in Figure 18-6 portrays the phase of the received wave relative to that of the transmitted wave. Phasor *T* represents the transmitted wave; phasor *R* represents the received wave. (To make the relationship between the two phasors easier to visualize, assume that the radar transmits continuously, though that is not necessary.) At any one instant, the phase of the received wave, *R*, lags that of the transmitted wave, *T*, by the round-trip transit time, t_{rt}. If t_{rt} were a whole number of wavelengths, the two phasors would coincide. If t_{rt} were a whole number of wavelengths minus half a wavelength, *R* would lag half a revolution behind *T*.

More generally, let us suppose that t_{rt} is 100,000 times the period of the transmitted wave plus some fraction, ϕ (Fig. 18-7). The rotation of R, though it is 100,000 complete revolutions behind the rotation of T, will be out of phase with it by only the fraction of a complete revolution (cycle), ϕ.

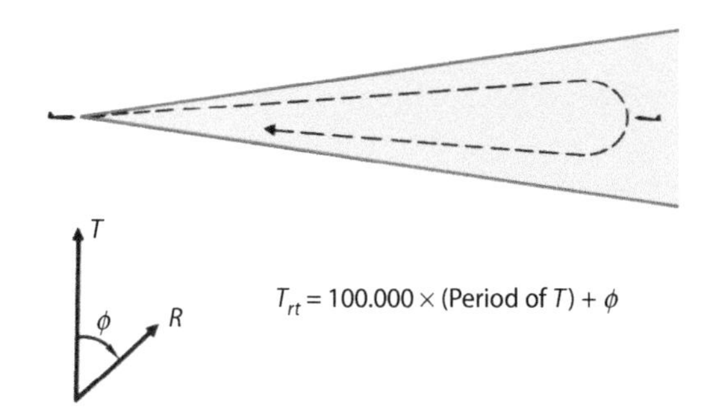

Figure 18-7. If round-trip transit time is 100,000 times the period of the transmitted wave plus a fraction ϕ, R will be out of phase with T by only the fraction ϕ.

Now, if the transit time is constant (range-rate = 0), the phase lag, too, will be constant, and the angle, ϕ, will remain the same. The two phasors, therefore, will rotate at the same rate. The frequencies of the transmitted and received signals will be the same. In other words, both radar and target are not moving; therefore, nothing changes, and hence any phase difference between the target and the radar remains constant.

However, if the transit time decreases slightly, the total phase lag will decrease, reducing the angle, ϕ. If the decrease continues (decreasing range), R will rotate counterclockwise relative to T (Fig. 18-8). The frequency of the received wave will be greater than that of the transmitted wave. The effect of a changing distance between the radar and the target is to create a corresponding change in the phase and hence in the Doppler frequency.

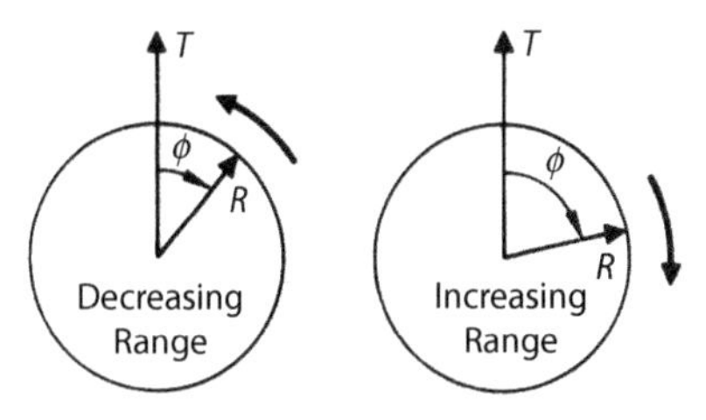

Figure 18-8. If range decreases, ϕ will decrease, causing R to rotate counterclockwise relative to T and thus to have a higher frequency.

Essentially, the same thing happens if the transit time increases (positive range-rate). The only difference is that the phase lag increases, and R, though still rotating counterclockwise in absolute terms, rotates clockwise relative to T. The frequency of the received wave is less than that of the transmitted wave (Fig. 18-8).

The difference in frequency between the transmitted and received waves, known as the target's Doppler frequency, f_D, is proportional to the rate of change of ϕ. That is, by measuring the phase change, the relative velocity between a target and the radar can be measured.

Equation for f_d Derived. If the rate of change of the phase angle, $\dot{\phi}$, is measured in whole revolutions per second (1 revolution = 2π radians = 360° = 1 whole cycle per second), the Doppler frequency in Hz equals $\dot{\phi}$. Since the phasor R makes one revolution relative to phasor T every time the round-trip distance, d, to the target changes by one wavelength, λ, the Doppler frequency equals the rate of change of d in wavelengths.

$$f_D = -\frac{\dot{d}}{\lambda}$$

The minus sign accounts for the fact that if $\dot{d}$ is negative (target moving towards the radar), the Doppler frequency is positive.

Since d is twice the target's range ($d = 2R$), the rate of change of d (Fig. 18-9) is twice the range-rate ($\dot{d} = 2\dot{R}$). The target's Doppler frequency, therefore, is twice the range-rate divided by the wavelength.

$$f_D = -2\frac{\dot{R}}{\lambda}$$

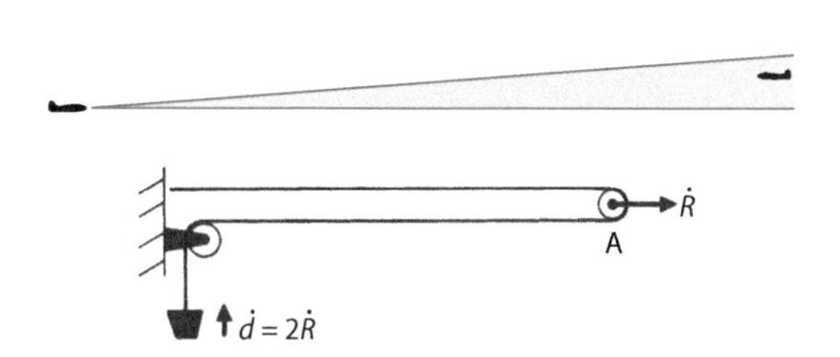

Figure 18-9. As illustrated by this simple mechanical analogy, the round-trip distance from radar to target changes at twice the range-rate. If pulley A moves to right at rate $\dot{R}$, the weight moves up at rate $\dot{d}$, which is twice $\dot{R}$.

where

f_D = Doppler frequency, Hz

$\dot{R}$ = range-rate, m/s

λ = transmitted wavelength, same units as R

Since wavelength equals the speed of light divided by the frequency of the wave, an alternative expression for Doppler frequency is

$$f_D = -2\frac{\dot{R} f}{c}$$

where f is the frequency of the transmitted wave, and c is the speed of light.

Doppler Shift in a Nutshell

FOR EVERY HALF-WAVELENGTH PER SECOND THAT A TARGET'S RANGE DECREASES, the radio frequency phase of the received echo advances by the equivalent of one whole cycle per second.

$$\therefore f_D = \frac{-\dot{R}}{\lambda/2} = \frac{-2\dot{R}}{\lambda}$$

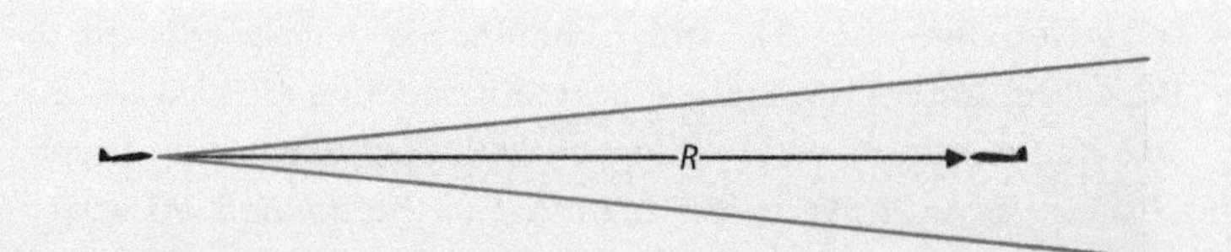

where

f_D = Doppler shift (positive for decreasing R)

$\dot{R}$ = radial component of relative velocity

λ = wavelength

18.4 Doppler Frequency of an Aircraft

Either of these expressions for the Doppler frequency enables quick and accurate calculation of the Doppler frequency of any target for any radar. Take an X-band radar with a wavelength of 3 cm (3×10^{-2} m). Suppose the radar is closing on a target at 300 m/s ($\dot{R} = -300$ m/s). The target's Doppler frequency is $(-2 \times -300)/0.03 = 20{,}000$ Hz, or 20 kHz.

If the wavelength were only half as long—1.5 cm instead of 3 cm—the same closing rate would produce twice the Doppler shift (i.e., 40 kHz instead of 20 kHz).

The equations apply equally to targets whose range is increasing. In this case, f_D has a negative sign, signifying that the radio frequency of the echoes is f_D Hz less than the transmitter frequency.

A simple rule of thumb for estimating Doppler frequencies for X-band radars is: 1 m/s of range-rate produces 70 Hz of Doppler

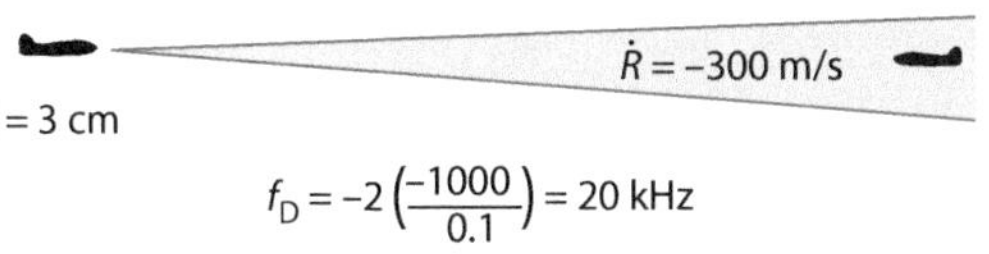

$\dot{R} = -300$ m/s

$\lambda = 3$ cm

$$f_D = -2\left(\frac{-1000}{0.1}\right) = 20 \text{ kHz}$$

Figure 18-10. With this expression the Doppler frequency of any target can easily be calculated.

shift. By this rule, a target whose closing rate is 300 m/s would have a Doppler frequency of $300 \times 70 = 21$ kHz (Fig. 18-10). Turning the rule around, a target whose Doppler frequency is 7 kHz would have a range-rate of $7000/70 = 100$ m/s.

For other wavelengths, the constants can simply be scaled to the wavelength: for example, 21 Hz per m/s for S-band ($\lambda = 10$ cm); 42 Hz for C-band ($\lambda = 5$ cm).

A target's range-rate, of course, depends on the velocities of both the radar and the target. For a radar approaching a target head-on (Fig. 18-11a), the range-rate is simply the numerical sum of the magnitudes of the two velocities.

$$\dot{R} = -(V_R + V_T)$$

Consequently,

$$f_D = -2\frac{\dot{R}}{\lambda} = 2\frac{V_R + V_T}{\lambda}$$

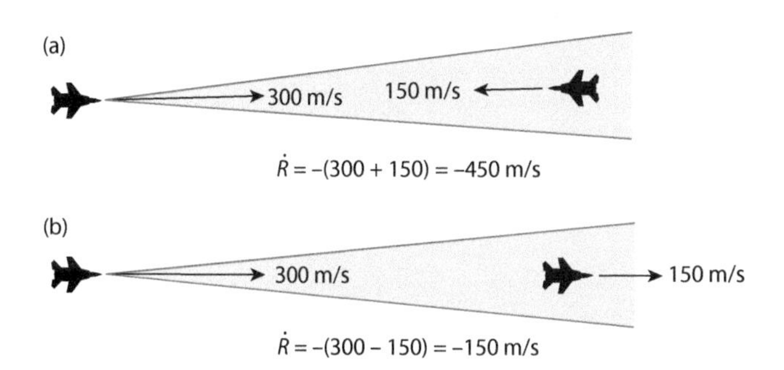

Figure 18-11. (a) For a target approaching nose-on, range-rate is sum of the magnitudes of aircraft velocities. (b) For tail-on approach, range-rate is the difference between them.

For a target tail-on (Fig. 18-11b), the rate is the difference between them. If the radar's velocity is greater than the target's, the range rate will be negative (decreasing range). If the radar's velocity is less than the target's, the range-rate will be positive (increasing range). If the two velocities are equal, the rate will be zero.

For the more general case where the velocities are not collinear, the range rate is the sum of the projections of the radar velocity and the target velocity along the line of sight to the target. This is usually known as the relative radial velocity. As illustrated in Figure 18-12, if the projection of the target velocity is toward the radar, the range will be decreasing. But if it is not, whether the range is decreasing or increasing depends on the relative magnitudes of the two projections (as in the collinear tail-on case).

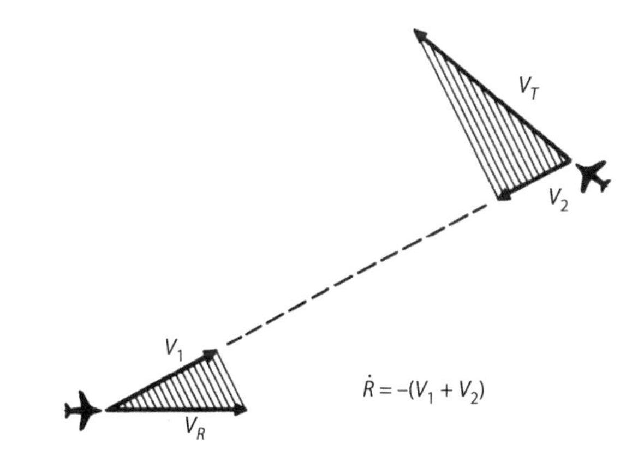

Figure 18-12. In general, range-rate of a target is the sum of the magnitudes of the projections of radar velocity and target velocity on line of sight to target.

A target's Doppler frequency, therefore, can vary widely depending on the operational situation. In nose-on approaches, it is always high. In tail-on approaches, it is generally low. In between, its value depends on the look angle and the direction the target is flying.

18.5 Doppler Frequency of Ground Return

The Doppler frequency of the return from a patch of ground is also proportional to the range-rate divided by the wavelength.

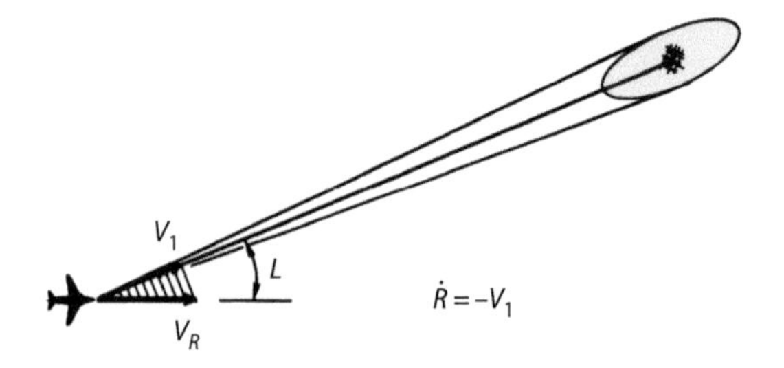

Figure 18-13. The range-rate of a ground patch is the magnitude of the projection of radar velocity on the line of sight to the patch.

The only difference is that the range-rate of a patch of ground is due entirely to the radar's own velocity (Fig. 18-13).

Therefore, the projection of the radar's velocity on the line of sight to the patch can be substituted for $-\dot{R}$. For a ground patch dead ahead, this projection equals the radar's full velocity, V_R. For a ground patch directly to the side or directly below, the projection is zero. In between, it equals V_R times the cosine of the angle, L, between V_R and the line of sight to the patch.

The Doppler frequency of the return from a patch of ground, therefore, is

$$f_D = 2\frac{V_R \cos L}{\lambda}$$

where

f_D = Doppler frequency of echo from ground patch, Hz

V_R = velocity of radar, m/s

L = angle between V_R and line of sight to patch

λ = transmitted wavelength, (m)

Suppose, for example, the velocity and wavelength are such that $2V_R/\lambda = 10{,}000$, and return is received from a patch at an angle of 60°. Since the cosine of 60° is 0.5, the Doppler frequency is $10{,}000 \times 0.5 = 5$ kHz.

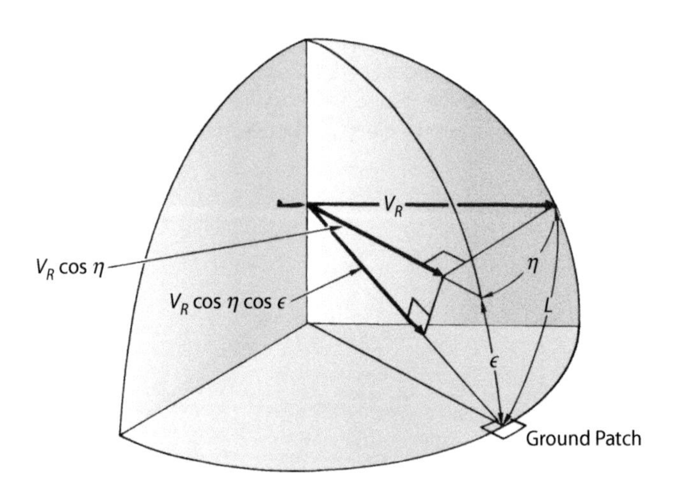

Figure 18-14. Radar velocity, V_R, is projected onto line of sight to ground patch in terms of azimuth angle, η, and depression angle, ε. Projection of V_R onto line of sight then equals $V_R \cos \eta \cos \varepsilon$.

If the angle, L, is resolved into its azimuth and elevation components, the term cosine L in the previous equation must be replaced by the product of the cosines of the azimuth and elevation angles of the patch (Fig. 18-14)

$$f_D = 2\frac{V_R \cos \eta \cos \varepsilon}{\lambda}$$

where

η = azimuth angle of patch

ε = lookdown angle of patch

As a rule, ground return is received not from a single small patch but from a great many patches at a great many different angles. The return therefore covers a broad spectrum of frequencies.

Finally, the relative radial velocity between the radar and the target can be zero or near to zero, as in the case of a side-looking radar. In this instance, target and clutter will have the same Doppler and will therefore be more difficult to discriminate. This continues to drive radar researchers to increase Doppler sensitivity so that discrimination can be maximized under as wide a range of conditions as possible.

18.6 Doppler Frequency Seen by a Semiactive Missile

A semiactive missile homes in on the scatter from a target that is illuminated by a radar carried in the launch aircraft.

Therefore, the Doppler frequency of the target as seen by the missile may be quite different from that seen by the illuminating radar.

This is illustrated for a simple collinear case in Figure 18-15. The distance, d, from radar to target to missile changes at a rate equal to the radar velocity plus two times the target velocity plus the missile velocity, V_M.

$$\dot{d} = -(V_R + 2V_T + V_M)$$

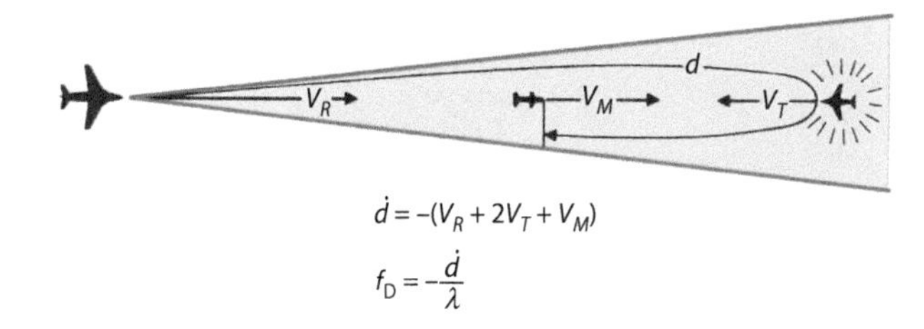

Figure 18-15. This target's Doppler frequency as seen by a semiactive missile is proportional to rate of change of distance, $\dot{d}$, from radar to target to missile.

The missile velocity equals the radar velocity plus the incremental velocity of the missile relative to the radar, $V_M = V_R + \Delta V_M$. With this substitution,

$$\dot{d} = -(2V_R + 2V_T + \Delta V_M)$$

The range-rate of the target relative to the radar is $\dot{R} = -(V_R + V_T)$, and the Doppler frequency is $-\dot{d}/\lambda$. Therefore, expressed in terms of relative velocities, the target's Doppler frequency as seen by the missile is

$$f_{D_M} = \frac{-2\dot{R} + \Delta V_M}{\lambda}$$

$$\dot{d} = -(V_R + 2V_T + V_M)$$

$$f_D = -\frac{\dot{d}}{\lambda}$$

$$\dot{d} = \dot{R}_{R-T} + \dot{R}_{M-T}$$

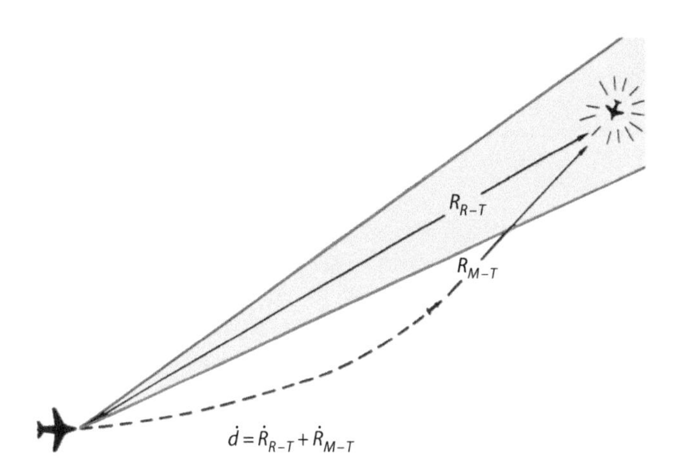

Figure 18-16. The rate of change of the distance to the missile is sum of the range-rate of the target relative to the radar plus the range-rate of the target relative to the missile.

The foregoing equation applies only if the velocities are all collinear and the missile is on the line of sight from the radar to the target. More generally, the rate at which the distance from the radar to the target to the missile changes equals the range-rate of the target relative to the radar plus the range-rate of the target relative to the missile (Fig. 18-16). The latter is the sum of the projections of V_M and $-V_T$ on the line of sight from the missile to the target. That is, we see again that the relative radial velocities are those that apply.

Initially, f_{D_M} may be comparatively high. However, as the attack progresses, f_{D_M} may fall off considerably, particularly if the missile is drawn into a tail chase, as shown in Figure 18-17.

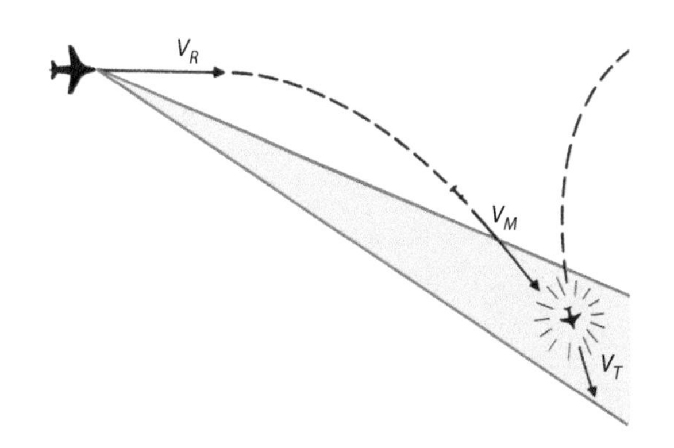

Figure 18-17. The target's Doppler frequency as seen by the missile may decrease as the attack progresses, particularly if the missile is drawn into a tail chase.

18.7 Summary

In the case of radar echoes, the Doppler effect can be visualized as the crowding (or spreading) of wavefronts caused by motion of the reflecting object relative to the radar. Since frequency is tantamount to a continuous phase shift, the resulting shift in frequency is equal to the rate (wavelengths per second) at which the round-trip distance traveled by the radio

Test your understanding

1. You walk directly toward a sound source emitting a constant frequency tone. How does the frequency you hear differ from that being emitted by the sound source?
2. Two aircraft carrying X-band radar systems are approaching each other at a relative velocity of 500 m/s. What is the apparent change in wavelength observed by both radar systems?
3. An aircraft traveling at a velocity of 300 m/s and carrying an X-band (10 GHz) radar observes a stationary object on the ground. What is the amount of Doppler shift observed by the radar system?
4. In 3, the object begins to move at a constant velocity of 1.5 m/s toward the radar. What is the new Doppler shift as observed by the radar system?
5. In 4, what is the necessary integration time required to observe the difference between the moving object and other stationary objects of the same echo strength?

waves is changing, that is, twice the range-rate divided by the wavelength.

The range-rate of a moving target is determined by the velocities of the radar and target and by the angle of the line of sight to the target relative to the direction of the radar's velocity. In nose-on approaches, the range-rate is usually greater than the radar's velocity; in tail-on approaches, it is less.

The range-rate of a patch of ground is determined solely by the radar's velocity and the angle to the patch. Since return may be received from ground patches in many directions, the ground return generally covers a broad band of frequencies.

18.8 Some Important Relationships to Keep in Mind

- Doppler frequency of a target:

$$f_D = -2\frac{\dot{R}}{\lambda}$$

where $\dot{R}$ = range rate

λ = wavelength

- Doppler frequency of a ground patch:

$$f_D = -2\frac{V_R \cos L}{\lambda}$$

where V_R = radar's velocity

L = look angle to the patch

- Doppler shifts at X-band for common velocities:

1 m/s = 70 Hz of Doppler shift

300 m/s = 21 kHz of Doppler shift

Further Reading

P. Z. Peebles, "Frequency (Doppler) Measurement," chapter 12 in *Radar Principles*, John Wiley & Sons, Inc., 1998.

V. N. Bringi and V. Chandrasekar, *Polarimetric Doppler Weather Radar Principles and Applications*, Cambridge University Press, 2007.

G. Brooker, "Doppler Measurement," chapter 10 in *Sensors for Ranging and Imaging*, SciTech Publishing-IET, 2009.

D. C. Schleher, *MTI and Pulse Doppler Radar with MATLAB®*, Artech House, 2009.

M. A. Richards, J. A. Scheer, and W. A. Holm (eds.), "Doppler Processing," chapter 17 in *Principles of Modern Radar: Basic Principles*, SciTech-IET, 2010.

19

The Spectrum of a Pulsed Signal

The (sin x)/x spectrum of a simple pulsed signal

In the previous chapter we saw that any relative movement between a radar sensor and a target will result in a Doppler shifted echo. In this way Doppler provides a means of distinguishing between stationary objects, such as the ground, and moving targets, such as motor vehicles, aircraft, etc. In order to exploit this effect, it is important to have a good understanding of how spectra are generated, especially in relation to coherent pulsed Doppler.

By performing a few simple experiments, it can be seen how the spectrum of a pulsed signal is a function of pulse width, pulse repetition frequency (PRF), and the duration of a signal. Along the way we will come to understand what coherence is and why it is so vital in a Doppler radar.

To get a feel for the relationships in question, consider performing a series of simple conceptual experiments. All that is required is just two pieces of equipment (Fig. 19-1). First you will need a microwave transmitter, which for the initial experiments consists simply of an oscillator. Its output signal is assumed to have constant amplitude and a constant, highly stable wavelength. A switch is provided with which the transmitter can be turned on or off at any desired instant. The other piece of equipment you will need is a microwave receiver that detects the transmitted signal. This receiver is highly selective at any given frequency[1] but can select over a very broad band of frequencies. A meter indicates the amplitude of the receiver's output.

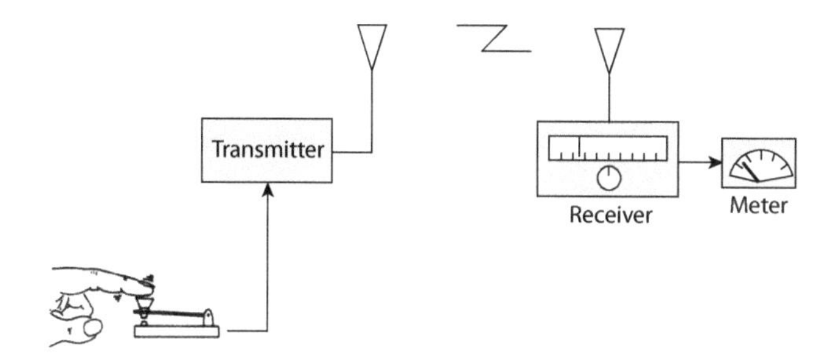

Figure 19-1. To determine the effect of pulse modulation on radio frequency, a series of simple experiments are performed with a microwave transmitter and receiver.

1. The receiver's passband is only 1 Hz wide; outside this band, its sensitivity is negligible.

19.1 Bandwidth

To find what determines the bandwidth of a pulsed signal, we perform two experiments.

Experiment No. 1: CW Signal. In this experiment, a continuous wave (CW) is transmitted at a frequency, f_0, and slowly tuned,

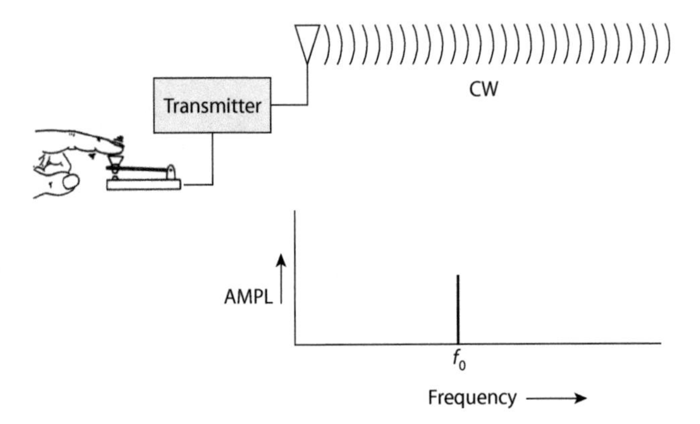

Figure 19-2. A continuous-wave signal produces an output from the receiver only when it is tuned to a single frequency.

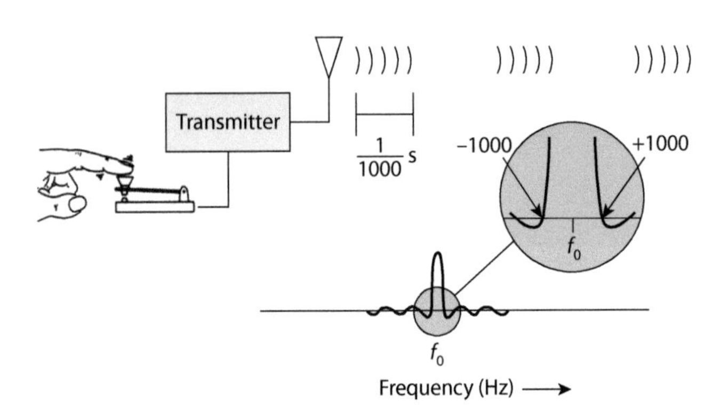

Figure 19-3. A train of independent pulses having a pulse width of 10 ms and a constant PRF produces a receiver output that is continuous over a band of frequencies 2 kHz wide.

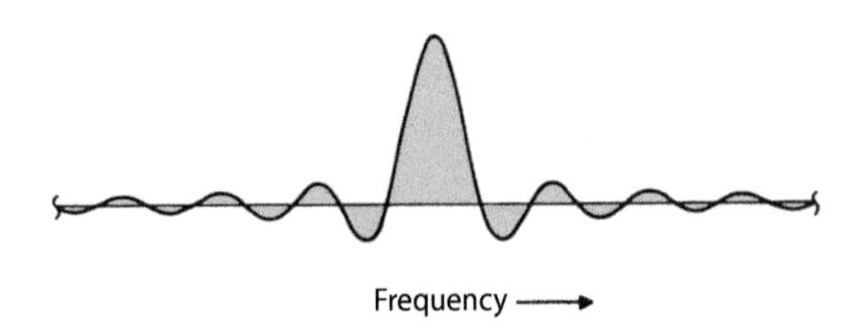

Figure 19-4. Plot of receiver output versus frequency has a (sin *x*)/*x* shape. Sidelobes half the width of the central lobe and continuously diminishing in amplitude extend above and below the mainlobe.

1 Hz at a time, through the receiver's frequency range in search of the transmitted signal (Fig. 19-2). As might have been predicted, the signal produces a strong output from the receiver at a single frequency, f_0. Although the entire tuning range is searched, there is no trace of the signal at any other frequency. If the amplitude of the receiver output is plotted versus frequency, it appears as a narrow vertical line (i.e., 1 Hz wide).

Experiment No. 2: Stream of Independent Pulse. In the second experiment, the transmitter is periodically switched on and off so that it transmits a continuous stream of pulses having a constant PRF (Fig. 19-3). It should be noted that although the switching is precise, the radio frequency phases of successive pulses are not the same, but vary randomly from pulse to pulse. This is equivalent to a "noncoherent" radar.

Each pulse is exactly 1 ms long. While 1 ms is a very short time, bear in mind that it is on the order of 1000 times longer than the pulses of a great many radars. Typically radar pulse durations are in the range of 1 to 10 μs.

Because of the signal's lower average power (the transmitter is "on" only a fraction of the time, whereas before it was on all the time), the receiver output is not as strong as before, but it still occurs at the same point, f_0, on the dial. However, the plot of receiver output versus frequency is not quite as sharp as before. In fact, when expanded, it is continuous over a band of frequencies extending from 1 kHz below f_0 to 1 kHz above it. In other words, the null-to-null bandwidth is 2 kHz.

The signal also produces an output in a succession of contiguous bands above and below this band. Within these bands, which are half as wide as the central one, the output is very much weaker, becoming more so the farther the bands are removed from f_0. The plot of receiver output versus frequency (Fig. 19-4) has the same (sin *x*)/*x* shape as the radiation pattern of a uniformly illuminated linear array antenna. It is the uniform illumination or constant amplitude that causes this. Although these spectral sidelobes are important, for the time being only the central band is considered.

The width of this central band can be determined by either the PRF or the pulse width, or both. To see if it is the PRF, the experiment is repeated at several progressively lower PRFs. But, except for a reduction in receiver output due to the lower duty factor, the receiver output is unchanged. For a signal of the type emitted by a simple transmitter, the PRF does not affect the spectrum.

Carrying this finding to its logical extreme in which the interpulse period is stretched to days, it can be further concluded that the spectrum of a single pulse is exactly the same as that of a stream of independent pulses. Therefore, the PRF does not determine bandwidth.

What about pulse width? To find the relationship between bandwidth and pulse width, the experiment is repeated several times using progressively narrower pulses. The final pulse width is 1 μs.

The result of narrowing the pulses is striking. As the pulse width decreases, the bandwidth increases tremendously (Fig. 19-5). For the final pulse width of 1 μs, the band extends from 1 MHz below f_0 to 1 MHz above it. The total bandwidth, from null to null, is 2 MHz.

A frequency of 2 MHz is 2 divided by 1 μs. Similarly, a frequency of 2 kHz is 2 divided by 1 ms. Consequently, we conclude that the null-to-null width of the spectral lobe of a stream of independent pulses is

$$BW_{nn} = \frac{2}{\tau},$$

where

BW_{nn} = null-to-null bandwidth

τ = pulse width

The null-to-null bandwidth of a 0.5 μs pulse, for example, is 2 ÷ 0.5 μs = 4 MHz (Fig. 19-6).

But this raises a serious question. If at X-band the Doppler shift is only 34 kHz per 500 m/s of closing rate, the Doppler frequencies encountered by most airborne radars will be no more than a few hundred kHz.

If a pulsed signal has a null-to-null bandwidth on the order of a few megahertz, roughly 10 times the highest Doppler shift, then the Doppler shift will be too tiny to distinguish. How then can pulsed radar ever detect Doppler frequencies? The answer is that it can't, unless the received pulses are in some way coherent.

19.2 Coherence

In radar, the term coherence is used to mean a consistency, or continuity, in the phase of a signal from one pulse to the next. One way to think of coherence is to consider a CW signal that is chopped into pulses, as shown in Figure 19-9. The points at which the chopping of the pulse starts and stops are always the same. In this way each pulse is phase coherent with the others. This is illustrated in Figure 19-7, where the first wavefront in each pulse is separated from the last wavefront in the preceding pulse by some integral number of wavelengths. For example, if the wavelength is exactly 3 cm, the separation may be 3 000 000 or 3 000 003 or 3 000 006 cm, etc., but not, say, 3 000 001 or 3 000 0033.15 cm.

In experiment 2, switching the transmitter on and off formed pulses. Although the switching was precise, the radio frequency phases of the individual pulses, that is, their "starting" phases, varied at random from pulse to pulse. In other words, the transmitted signal was incoherent. This is not surprising when you consider that the period of, say, an X-band signal is

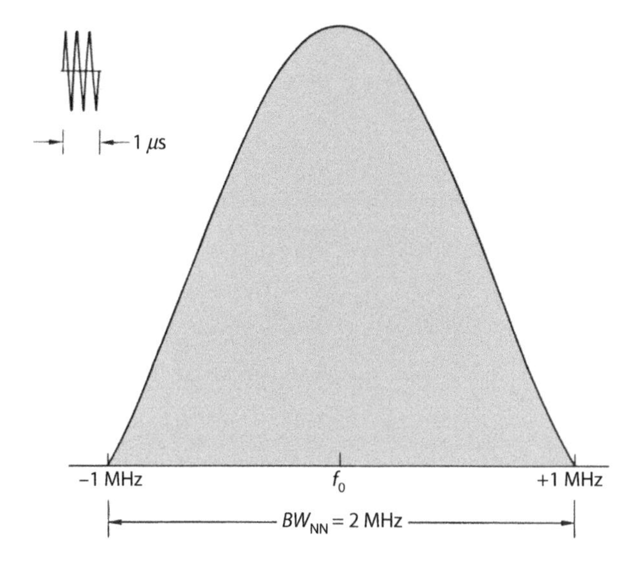

Figure 19-5. For a 1 μs pulse width, the null-to-null bandwidth of the central lobe is 2 MHz.

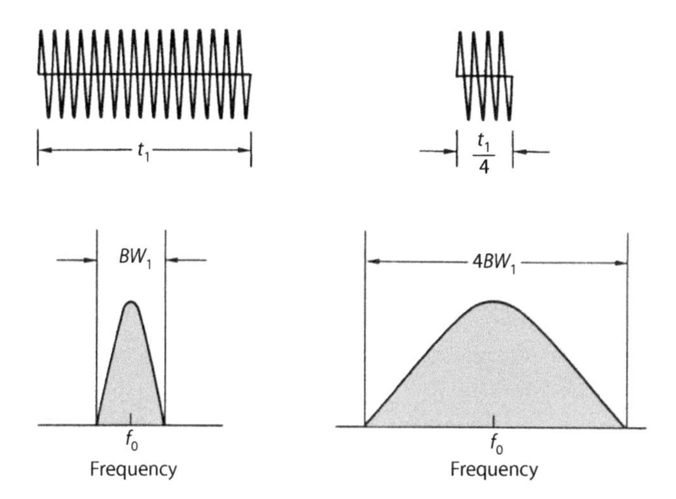

Figure 19-6. The narrower the pulses, the wider the central spectral lobe.

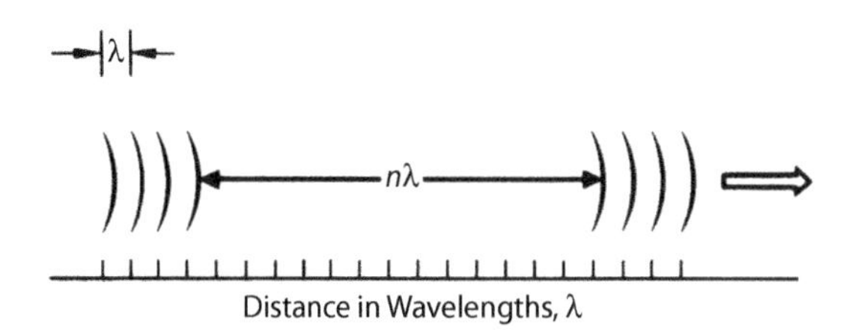

Figure 19-7. The common form of coherence is when the first wavefront in the second pulse is separated from the last wavefront of the same phase in the first pulse by a whole number of wavelengths.

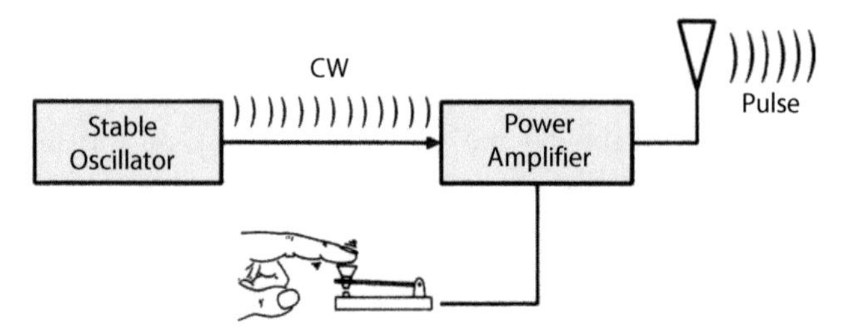

Figure 19-8. A coherent pulse train can be produced with a master oscillator–power amplifier. The oscillator runs continuously; the amplifier is keyed "on" to produce pulses.

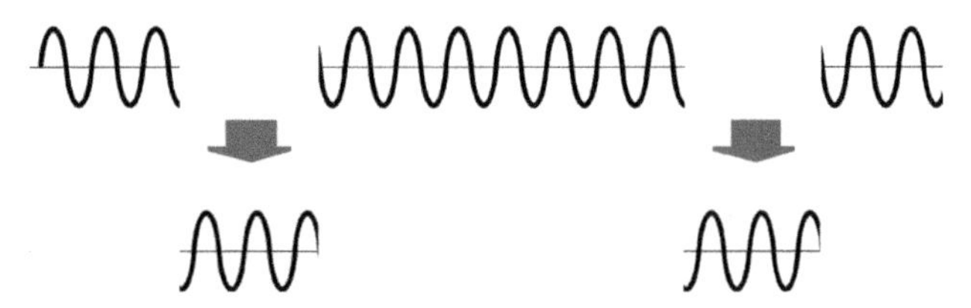

Figure 19-9. Pulses of a master oscillator–power amplifier are, in effect, cut from a CW, hence they are coherent.

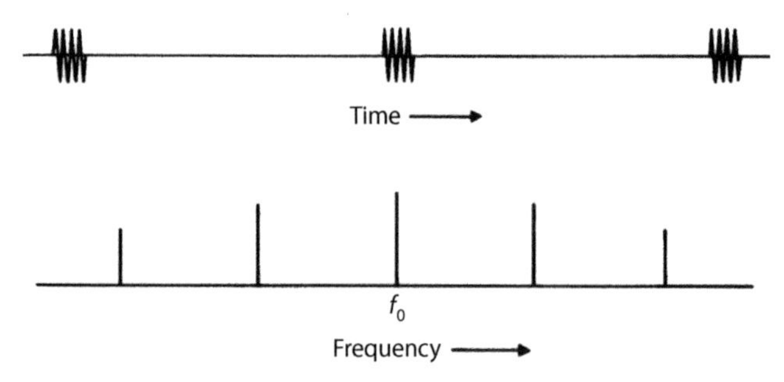

Figure 19-10. Coherent pulses produce output from a receiver at evenly spaced intervals of frequency.

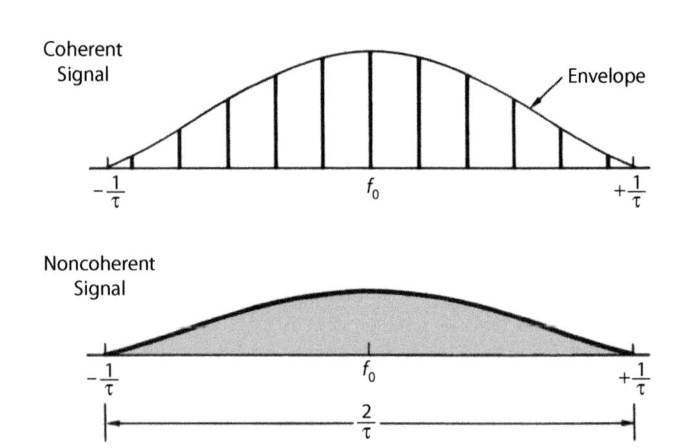

Figure 19-11. Spectral lines of a coherent signal fit within an envelope having the same shape (sin x/x) as the spectrum of a noncoherent pulse train having an equal pulse width, τ.

only 1/10,000 μs, and a hence a single degree of phase is only 1/3,600,000 μs.

Achieving Coherence. With a somewhat more elaborate transmitter, coherence can be achieved. The type of transmitter most commonly used in Doppler radars is called a *master oscillator–power amplifier* (Fig. 19-8). In essence, it consists of an oscillator that produces a low-power signal of highly stable wavelength and an amplifier that amplifies the signal to the power level needed for transmission. The oscillator runs continuously; the power amplifier is switched on and off to produce the pulses. Although the switching is no more precise than in the simple, noncoherent transmitter, the radio frequency phases of successive pulses are exactly the same as if the pulses had been cut from a CW (Fig. 19-9). The separation between the last wavefront in one pulse and the first wavefront in the next pulse is thus always exactly equal to a whole number of wavelengths. The pulses are phase coherent.

Experiment No. 3: Effect of Coherence. To see what effect coherence has on the bandwidth of a pulsed signal, experiment no. 2 is performed again, but this time the master oscillator–power amplifier combination is used.

The effect of changing to coherent transmission is remarkable (Fig. 19-10). With noncoherent transmission the signal's central spectral lobe is spread over a broad band of frequencies, whereas with coherent transmission it peaks up almost as sharply as the CW did. However, there is one important difference. Instead of appearing at only one frequency, the coherent pulsed signal appears at many different frequency positions. In fact, its spectrum consists of a series of evenly spaced lines.

Comparing this spectrum with the corresponding spectrum for the noncoherent signal (the same PRF and same pulse width), two things are observed. First, at those frequencies where the coherent signal produces an output, it is a great deal stronger than the output produced by the noncoherent signal, evidently because the energy has been concentrated into narrow lines (it has to go somewhere). Second, the "envelope" within which these lines fit (Fig. 19-11) has the same shape $(\sin x)/x$ and the same null-to-null width $(2/\tau)$ as the spectrum of the noncoherent signal.

Suspecting that the spacing of the lines is related to the PRF, the experiment is repeated several times, at progressively higher PRFs. As the PRF is increased, the lines move farther apart. In every case the spacing exactly equals the PRF (Fig. 19-12).

Incidentally, it is should be noted that since a constant pulse width was maintained, as the PRF was increased, the number of lines decreased. Had the PRF been progressively increased, a point would ultimately have been reached where all of the power was concentrated into a single line. This point would be equivalent to transmitting a CW signal as in experiment no. 1, where the result was also a single line.

The important conclusion to be drawn from this experiment is that the spectrum of a coherent pulsed signal consists of a series of lines that (1) occur at intervals equal to the PRF on either side of f_0 and (2) fit within an envelope having a $(\sin x)/x$ shape with nulls at multiples of $1/\tau$ above and below f_0.

As will be seen, unless a pulse train is infinitely long (which no pulse train could possibly be), the spectral lines have a finite width. This width is a function of the duration of the pulse train.

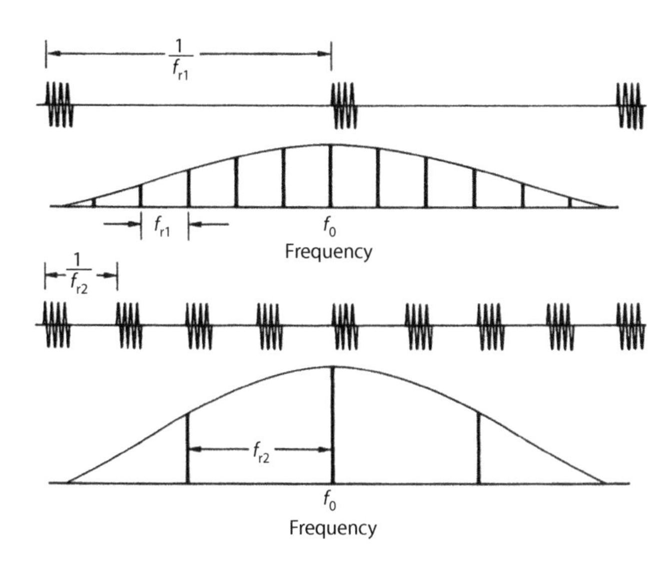

Figure 19-12. Spacing of spectral lines for a coherent pulse train equals PRF, f_r.

Earlier Methods of Achieving Coherence

Largely because the master oscillator–power amplifier was expensive to implement with the components then available, in early airborne Doppler radars various other techniques were used to achieve coherence. Some of these are still in use today.

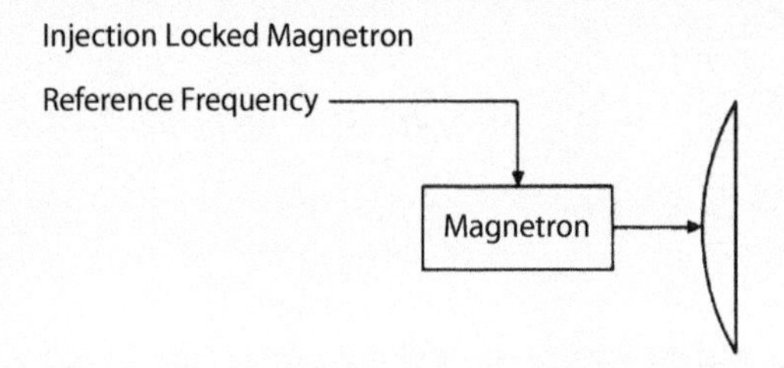

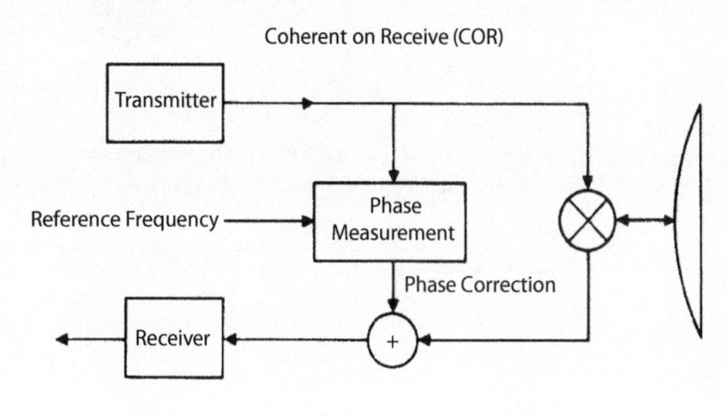

In one, called *injection locking*, the starting phase of a simple noncoherent transmitter such as a magnetron is "locked" to the phase of a highly stable, continuously generated, low-power signal that is injected into the magnetron cavity. Unfortunately, with injection locking, the degree of coherence generally less than desirable.

In another, called *coherent on receive* (COR), the phase of each transmitted pulse is measured relative to a continuously generated reference signal. An appropriate phase correction is then applied to the return received during the immediately following interpulse period.

With COR, only the first-time return is coherent, since the phase correction is only valid for a return from the immediately preceding transmitted pulse.

In still another approach, called noncoherent or clutter-referenced moving target indication, the equivalent of coherence is achieved by detecting the "beat" between the target echoes and the simultaneously received ground return. But this technique has serious limitations.

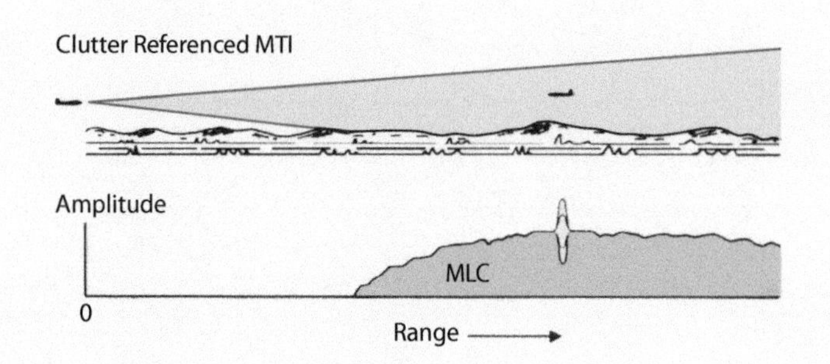

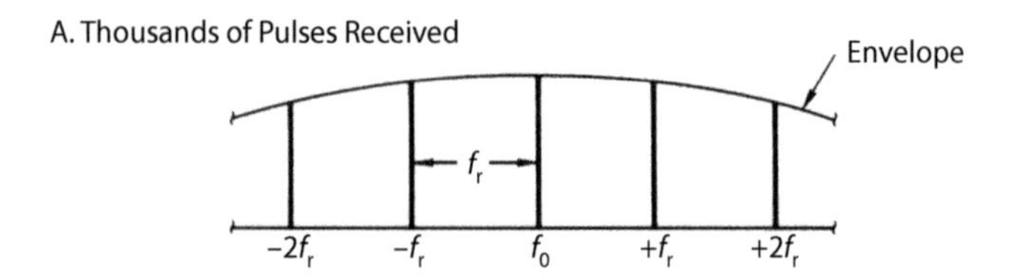

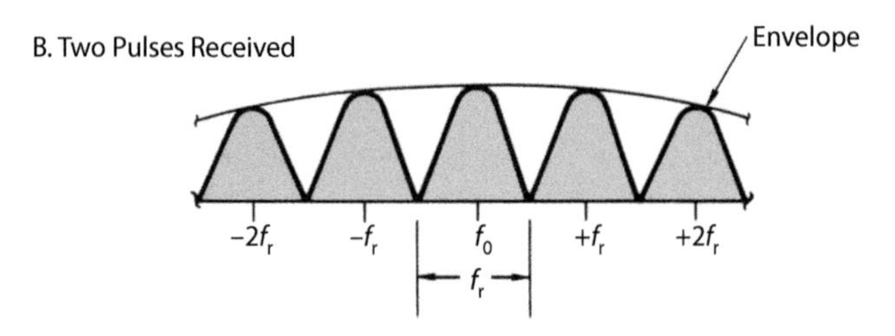

Figure 19-13. When thousands of pulses are received, spectral lines are narrow and sharply defined, whereas when only two pulses are received, spectral lines broaden until they are contiguous.

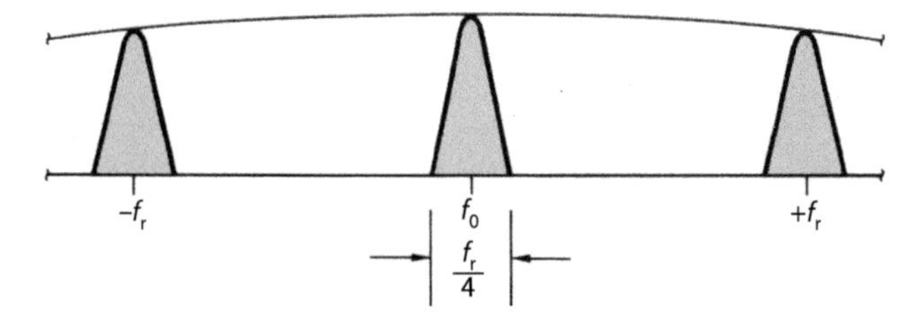

Figure 19-14. When eight pulses (instead of two) are received, the null-to-null width of spectral lines is only one-fourth as great.

19.3 Line Width versus the Duration of the Pulse Train

To find the relationship between line width and the length of a pulse train, two more experiments are performed.

Experiment No. 4: Two-Pulse Train. For this experiment, the same receiver and coherent transmitter are used as before, but holding the PRF constant, only two pulses are transmitted. The results are shown in Figure 19-13, along with a repeat of the results of experiment no. 3 for the same PRF and pulse width.

When the pulse train was 1000 or more pulses long, the receiver output peaked sharply at each multiple of the PRF. But when it is only two pulses long, the plot of receiver output versus frequency is almost continuous. The output still reaches its maximum values at multiples of the PRF and falls off on either side of each peak, but it only reaches zero halfway between peaks. The null-to-null "line width" is given by the PRF, f_r Hz.

Experiment No. 5: Eight-Pulse Train. The experiment is repeated using the same PRF and pulse width, but this time we transmit four times as many pulses for each dial setting: eight as opposed to two. Although the signal still produces an output over a fairly broad band around each multiple of the PRF (Fig. 19-14), the spectral lines are now only one-fourth as wide, or $f_r/4$ Hz as opposed to f_r Hz.

General Relationships. From the results of these two experiments, it is concluded that the width of the spectral lines is inversely proportional to the number of pulses in the pulse train. Since for two pulses the line width equals the PRF, it can be further concluded that for N pulses the line width equals $2/N$ times the PRF;

$$LW_{nn} = \left(\frac{2}{N}\right) f_r,$$

where

LW_{nn} = null-to-null width

f_r = pulse repetition frequency

N = number of pulses in the train

$$LW_{nn} = \frac{2}{N} f_r$$

$$f_r = \frac{1}{T}$$

$$\therefore LW_{nn} = \frac{2}{NT}$$

If, for example, a pulse train contains 32 pulses, the line width is 2/32, or 1/16 of the PRF.

The primary factor determining the width of the spectral line is not the number of pulses but rather is the duration of the pulse train. This becomes clear if we replace f_r in the expression for LW_{nn} with $1/T$, where T is the interpulse period. The expression then becomes $LW_{nn} = 2/(NT)$. Since N is the number of interpulse periods in the train, NT is the train's total length. Accordingly,

$$LW_{nn} = \frac{2}{\text{Length of train (seconds)}} \text{ Hz.}$$

Thus it is the length of the time period over which echoes are recorded that determines the width of the frequency lines (the frequency resolution), not the duration of a single pulse. The longer the time period, the better the ability of the radar to discern different Doppler frequency shifts. Also, the higher the PRF, the wider the frequency gap between repeating lines. This gap is normally set such that the highest anticipated Doppler frequency shift will be less than the frequency of the first repeat. This avoids any ambiguities between targets moving at different velocities. As we shall see in Chapters 28–30, on low, medium, and high PRF modes of operation, this is not always possible and other techniques have to be used to resolve any ambiguities.

Equivalence of Pulse Train to Long Pulse. Interestingly, the results of this experiment are consistent with those of experiment no. 2 in which the null-to-null bandwidth for a single pulse is 2 divided by the length of the pulse in seconds: $BW_{nn} = 2/\tau$. If a single pulse the length of a train of N pulses is transmitted, its null-to-null bandwidth would be exactly the same as the null-to-null line width of the pulse train (Fig. 19-15). Thus there is only one difference between the spectrum of a train of coherent pulses and the spectrum of a single pulse the same length as the train: the spectrum of the pulse train is repeated at intervals equal to the PRF.

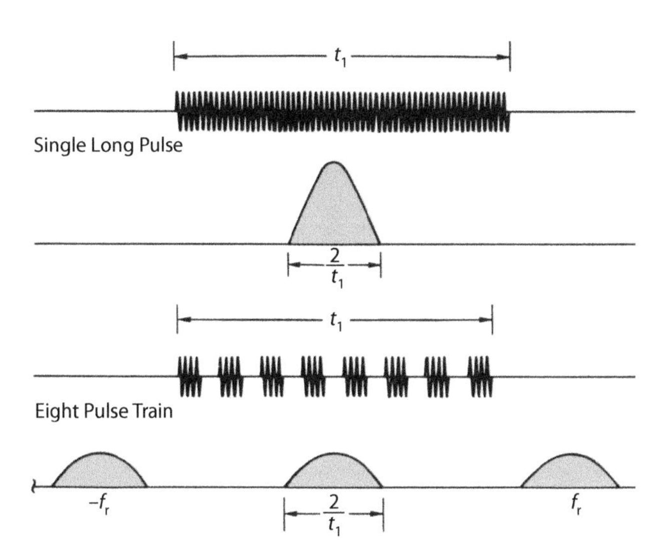

Figure 19-15. Individual spectral lines for a coherent pulse train differ from the spectrum of a single pulse of the same length only in being repeated at intervals equal to the PRF (f_r).

The parallel between the spectra of a coherent pulse train and a single long pulse is noted here for two reasons. First, it makes remembering the spectrum of a pulsed signal a bit easier, and second, it will prove illuminating when explaining the pulsed spectrum in the next chapter.

19.4 Spectral Sidelobes

What about spectral sidelobes? Just as they flank the mainlobe of the spectrum of a single pulse, sidelobes of half the null-to-null line width flank each "line" of the spectrum of a pulse train (Fig. 19-16). The line itself has a $(\sin x)/x$ shape.

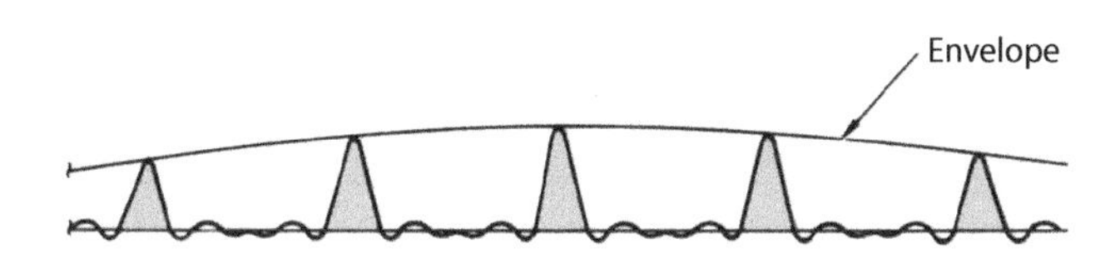

Figure 19-16. Just as sidelobes flank the central spectral lobe of a single pulse, they also flank each line in the spectrum of a coherent pulse train.

Since the length of the train of pulses received from a target during any one scan of the radar antenna is invariably limited, the sidelobes are an important concern to the radar designer. Sidelobes tend to fill in the gaps between the spectral lines. Fortunately, by suitably designing the Doppler filters of the radar's signal processor, the sidelobes can generally be reduced to an acceptable level.

Conclusions Drawn from the Experiments. The conclusions drawn from the five simple experiments are summarized graphically in the blue panel. By way of underscoring their significance, consider the question raised earlier in this chapter:

If the spectral width of a radar pulse can be many times the highest Doppler frequency, how can a pulsed radar discern small Doppler shifts in what may be an extremely weak target return buried in strong ground clutter?

In light of the illustrations in the panel, the answer becomes abundantly clear. A pulsed radar can readily discern these shifts if the following conditions are satisfied:

- The radar is coherent.[2]
- The PRF is high enough to spread the lines of the spectrum reasonably far apart.
- The duration of the pulse train is long enough to make the lines reasonably narrow.
- The Doppler filters are suitably designed to reduce the spectral sidelobes.

2. If ranges are extremely long, it is possible to transmit perfectly coherent pulses and have some loss of coherence in the medium through which the waves propagate.

Results of the Experiments

Continuous wave infinite length

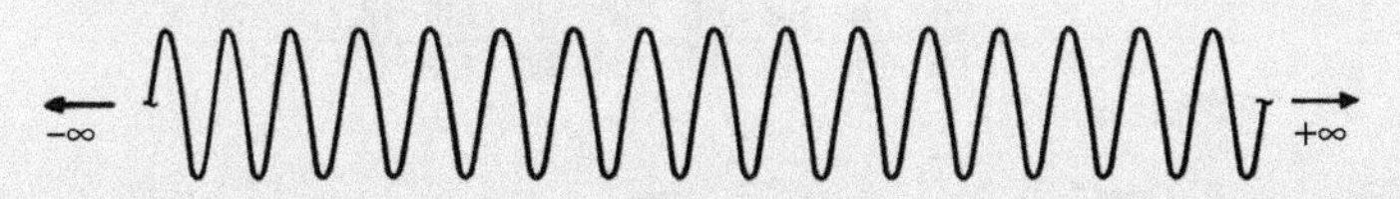

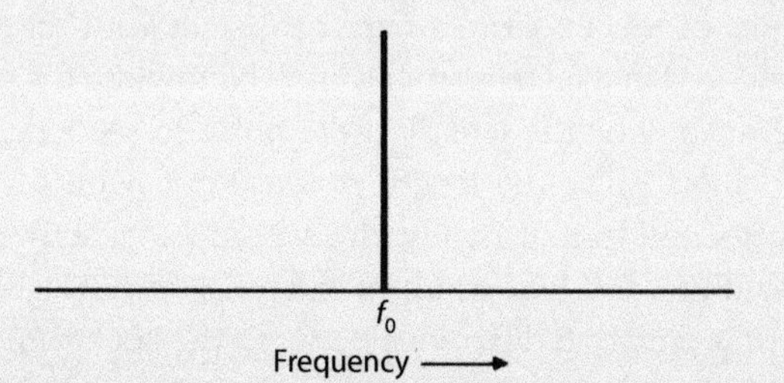

Single pulse

Train of noncoherent pulses (random starting phases)

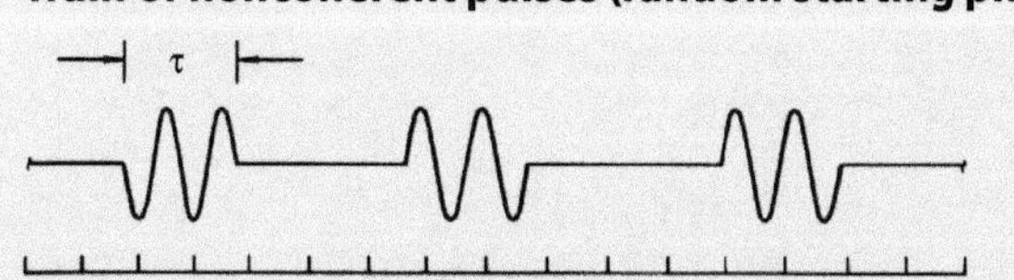

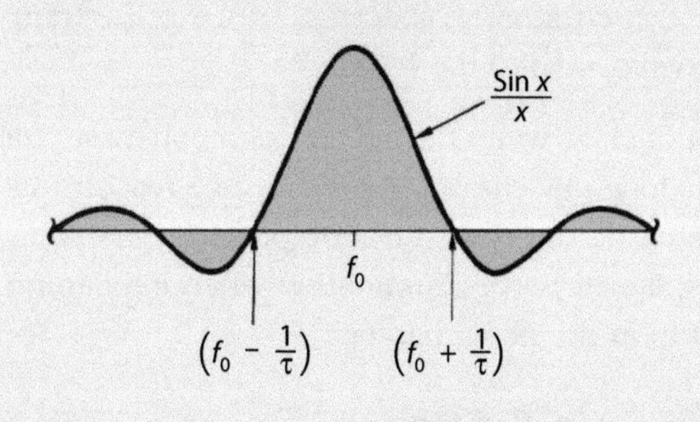

Train of coherent pulses infinite length

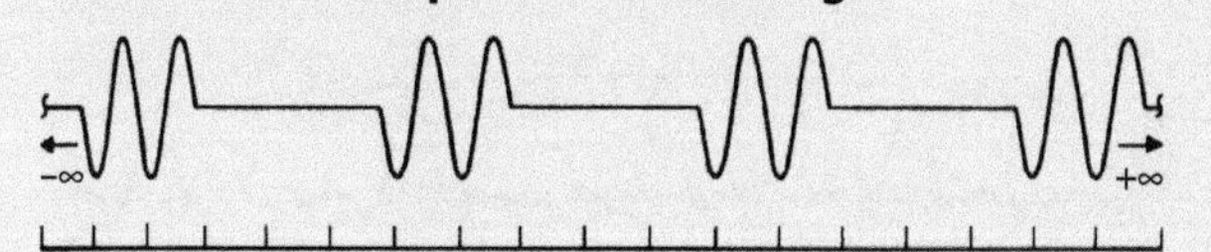

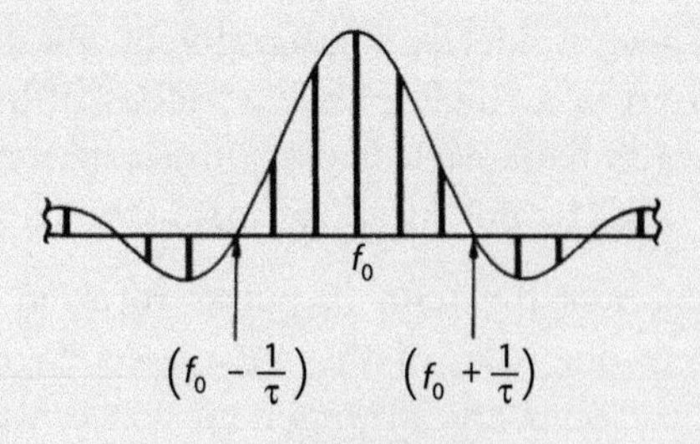

Train of coherent pulses limited length

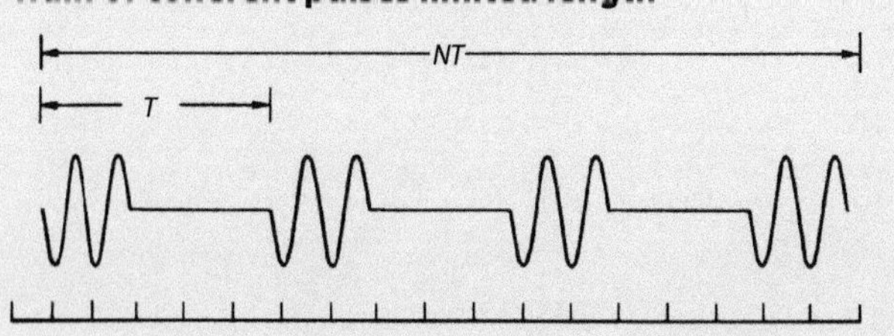

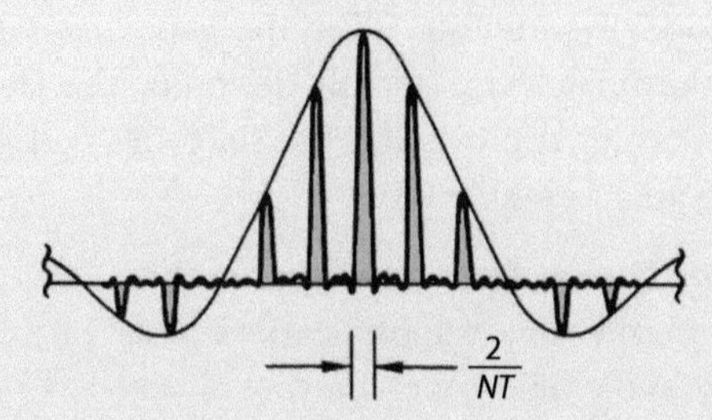

19.5 Summary

Transmitting a radio frequency signal in pulses markedly changes in the signal's spectrum. Whereas the spectrum of a CW of constant wavelength consists of a single line, the spectrum of a single pulse of the same wavelength covers a band of frequencies and has a $(\sin x)/x$ shape. The width of the central lobe of this spectrum varies inversely with pulse width. If the pulses are as narrow as those used in many radars, the central lobe may be several megahertz wide.

A train of pulses of random starting phase is said to be noncoherent. Its spectrum has the same shape as that of a single pulse.

Coherence is a consistency or continuity in the phases of successive pulses. The pulses are essentially cut out of a CW.

The spectrum of a coherent pulse train of infinite length consists of lines at intervals equal to the PRF within an envelope having the same shape as the spectrum of a single pulse. If the coherent pulse train is not infinitely long, the individual lines have a finite width and the same shape as the spectrum of a single pulse the length of the train. Line width is thus inversely proportional to the length of the train.

Test your understanding

1. What is meant by the term "coherence"?
2. A pulsed radar transmits a rectangular pulse of duration 500 ns. What is the null-to-null bandwidth of the transmitted pulse?
3. A pulsed radar processes echoes over a total time duration of 0.5 s. What is the width of the resulting spectral lines?
4. If the PRF of a pulsed radar is increased, what happens to the spacing between the spectral lines?

19.6 Some Relationships to Keep in Mind

- For a single pulse:

$$\text{Null-to-null bandwidth} = 2/\tau,$$

where τ = pulse width.

- For a coherent pulse train:

$$\text{Line spacing} = f_r$$

$$\text{Null-to-null line width} = \frac{2}{N} f_r = \frac{2}{NT},$$

where

f_r = pulse repetition frequency

N = number of pulses in the train

T = interpulse period.

Further Reading

R. N. Bracewell, *The Fourier Transform and Its Applications*, 3rd ed., McGraw-Hill, 1999.

W. L. Melvin and J. A. Scheer (eds.), "Doppler Phenomenology and Data Acquisition," chapter 8 in *Principles of Modern Radar: Basic Principles*, vol. 1, SciTech-IET, 2010.

AgustaWestland AW101 Merlin

The Merlin was developed by Westland Helicopters in the UK and Agusta in Italy to serve as a naval utility helicopter and anti-submarine warfare replacement for the Westland Sea Kings. The AW101 carries the Blue Kestrel search and detection radar which has 360 degree scanning capability and can detect targets up to 25 nautical miles away. Some variants are also equipped with anti-submarine systems from processing sonographic data from sonobuoys for detection and targeting.

20

The Pulsed Spectrum Unveiled

The Royal Navy's next-generation helicopter Wildcat, lands onboard HMS Iron Duke in Portsmouth, prior to three days of intensive tests.

The preceding chapter showed the dramatic effect that a pulsed coherent transmission has on the spectrum of a radio wave. While merely to memorize the relationships would suffice, a deeper insight into the operation of radar systems will be gained if the reasons for them are understood.

This chapter provides those reasons. It begins by raising the fundamental question of exactly what is meant by the spectrum of a signal. This, as will be seen, is the crux of the matter. The chapter goes on to explain the spectrum of a pulsed signal in two quite different ways:

1. In terms of the Fourier series, a conceptually simple but powerful analytical tool
2. In terms of what physically takes place when a radio frequency signal passes through a lossless narrowband filter

The essence of both explanations is presented in more precise, mathematical terms via the Fourier transform at the end of the chapter. An introduction to the mathematics of both Fourier series and the Fourier transforms can be found in chapter 6.

20.1 Spectra

Spectrum Defined. Broadly speaking, the spectrum of a signal is the distribution of the signal's energy over the range of possible frequencies. It is commonly portrayed as a plot of amplitude versus frequency (Fig. 20-1).

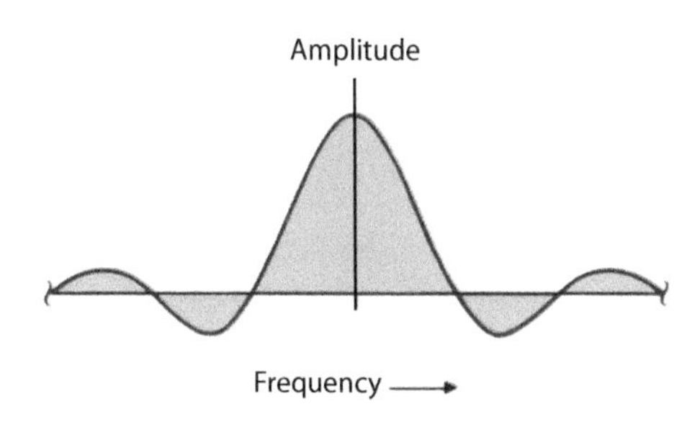

Figure 20-1. The spectrum of a signal is commonly portrayed as plot of amplitude versus frequency.

A rough physical feel for the spectrum of a pulsed signal was gained in the last chapter by measuring the output that a signal produced when it was applied to a highly selective receiver. The receiver was tuned, one Hz at a time, through a broad

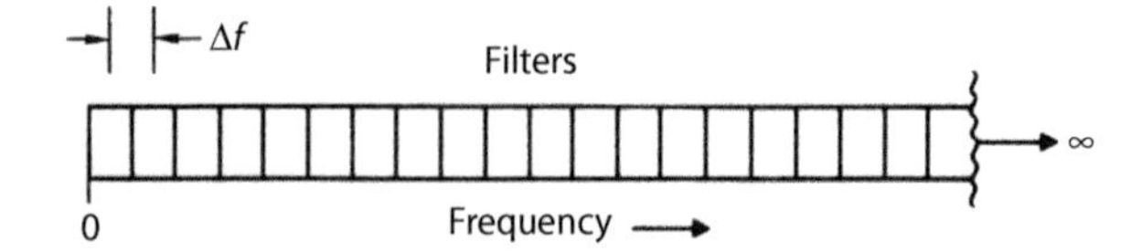

Figure 20-2. To explain its spectrum, a signal may be envisioned as being applied simultaneously to myriad lossless narrowband filters whose frequencies are infinitesimally closely spaced.

band of frequencies. Actually, the spectrum can be defined quite rigorously in these terms, provided they are refined them as follows. Instead of envisioning the signal as being applied to a receiver whose frequency is periodically changed, think of it as being applied simultaneously to myriad lossless narrowband filters whose frequencies are infinitesimally closely spaced and cover the entire range from zero to infinity (Fig. 20-2). A signal's spectrum, then, is a plot of the amplitudes of the filter outputs versus the frequencies of the filters and would look like that of Figure 20-1.[1]

What a Lossless Narrowband Filter Does. The most easily visualized mechanical analogy to a lossless narrowband filter is a pendulum suspended from a frictionless pivot in a vacuum (Fig. 20-3).

The frequency of the filter is the pendulum's natural frequency, i.e. the number of cycles per second that it would complete if deflected and allowed to swing freely.

The input signal is applied to the pendulum by a tiny electric motor at the center of the pendulum mass. On the shaft of this motor is an eccentric flywheel. The speed of the motor is such that for every cycle of the input signal the flywheel makes one complete revolution. Because of the flywheel's imbalance, a sinusoidally varying reactive force is exerted on the pendulum[2]. This force tends to make the pendulum swing alternately right and left. The effect is similar to that of a child "pumping" a swing (Fig. 20-4).

The filter's output is the amplitude to which the swing builds up over the duration of the input signal. That is, we consider a pendulum of a given natural frequency as a loss-less filter.

By means of this analogy, it's not too difficult to explain why even the simplest ac signal has a broad frequency spectrum. Consider a signal that turns the flywheel at a rate of 1000 revolutions per second and has duration of 1/10th of a second. To see what

1. To be completely rigorous, besides plotting the amplitude of each filter's output, one must also indicate its phase.

2. A more exact analogy is a mass suspended on a spring, since its restoring force is *directly* proportional to the displacement. But for small displacements, the analogy to a pendulum is very close.

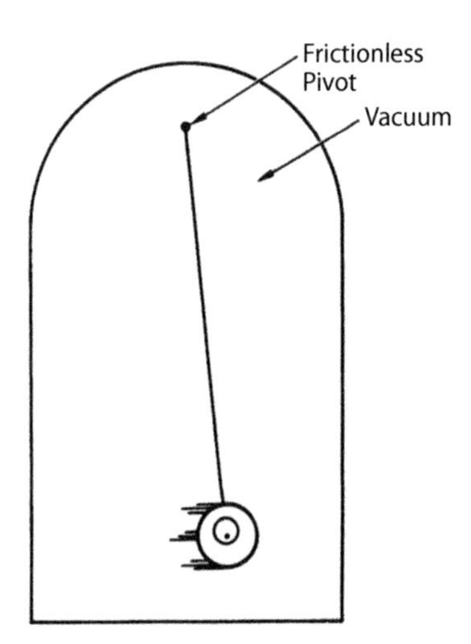

Figure 20-3. A lossless narrowband filter is analogous to a pendulum suspended from a frictionless pivot in a vacuum. The amplitude of the swing corresponds to filter's output.

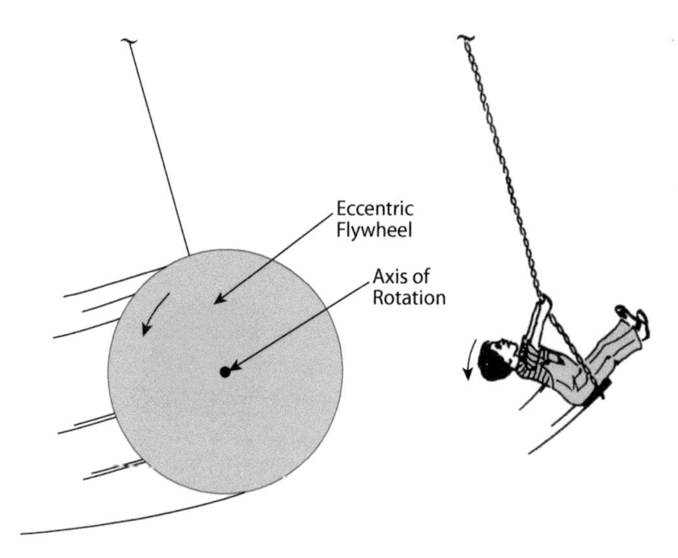

Figure 20-4. Motor-driven eccentric flywheel makes one revolution for each cycle of the input signal. Reactive force is similar to that produced by a child "pumping" a swing.

its spectrum is like, the signal is applies simultaneously to the myriad filters, where each filter is a pendulum with a different natural frequency. As the flywheels start turning, all of the pendulums begin to swing. The extent to which each pendulum's swing builds up depends upon the pendulum's natural frequency.

In the case of the pendulum whose frequency is exactly 1000 Hz (Fig. 20-5), with every turn of the flywheel, the amplitude of the swing increases by the same amount. The swing remains in phase with the sinusoidally varying forces exerted by the flywheel. After 1/10th of a second has elapsed and the flywheel has made 100 turns, the pendulum is swinging with an amplitude 100 times as great as when the input completed its first cycle.

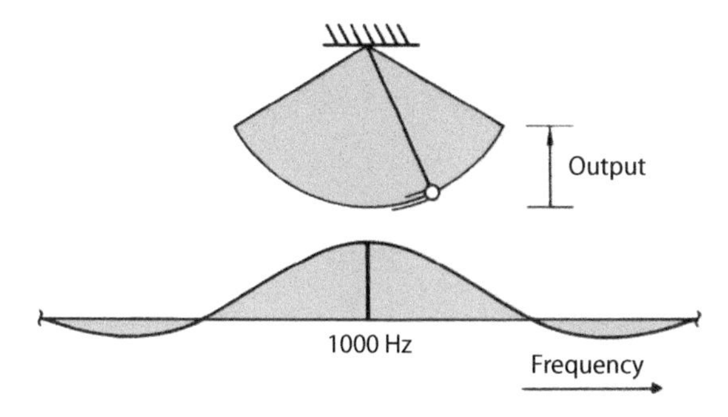

Figure 20-5. Swing of pendulum whose frequency is 1000 Hz stays in phase with reactive force of flywheel and builds up.

In the case of a pendulum whose frequency is, say, 995 Hz, i.e. 5 Hz less than 1000 Hz (Fig. 20-6), the swing starts building up in the same way. But because of the pendulum's lower natural frequency, the phase of the swing gradually falls behind that of the flywheel's rotation. Consequently, the momentum of the pendulum and the reactive forces of the flywheel work against each other over a correspondingly increasing fraction of each cycle. When the input stops, the amplitude of this pendulum's swing is considerably less than that of the pendulum whose frequency is 1000 Hz. Nevertheless, the swing is substantial.

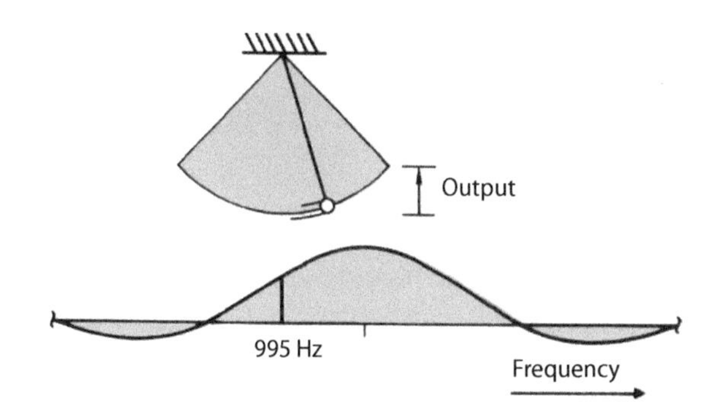

Figure 20-6. Momentum of pendulum whose frequency is 995 Hz works against flywheel part of the time, so buildup is not as great.

But in the case of the pendulum whose frequency is 990 Hz, i.e. 10 Hz less than 1000 Hz (the frequency of the input signal's first spectral null), the phase of the swing falls behind at a high enough rate that the swing is completely damped out by the time the input ends (Fig. 20-7).

For the pendulum whose frequency is 985 Hz, i.e. 15 Hz less than 1000 Hz (in the middle of the first sidelobe), the phase of the swing falls behind at a sufficiently high rate that the swing builds up and damps out and builds up once again before the input ends. Though the final amplitude of the swing is only a fraction of that of the pendulum whose frequency is 1000 Hz, this fraction is considerable and is roughly 21 percent (Fig. 20-8).

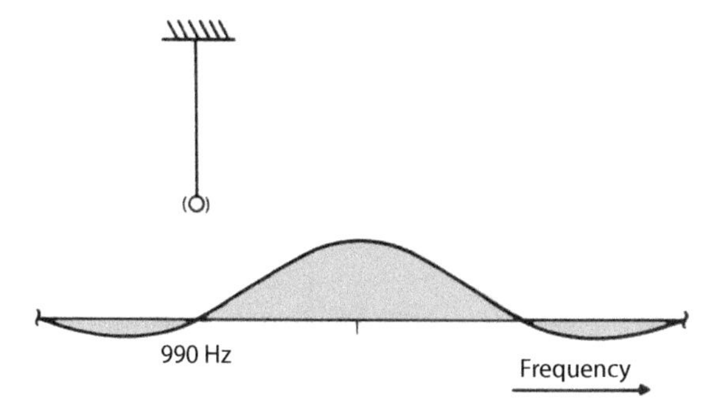

Figure 20-7. The swing of this pendulum whose frequency is 990 Hz builds up initially but is completely damped out when signal ends.

For pendulums whose frequencies are farther and farther below 1000 Hz, the familiar pattern of lobes and nulls is observed. At corresponding points within successive lobes, the farther the lobe is from 1000 Hz, the more nearly the total

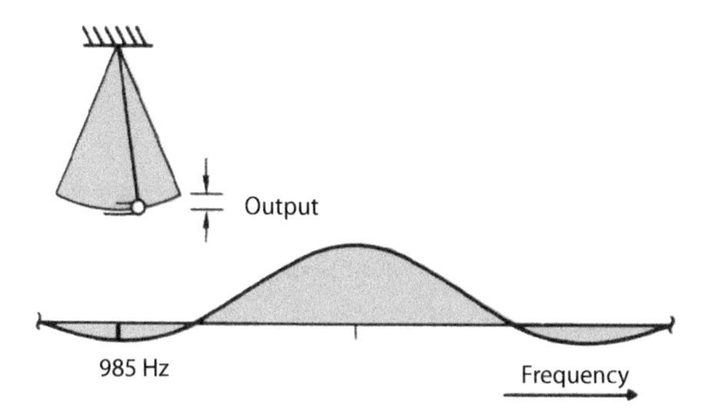

Figure 20-8. The swing of this pendulum whose frequency is 985 Hz falls behind sufficiently fast that it builds up again before input ends.

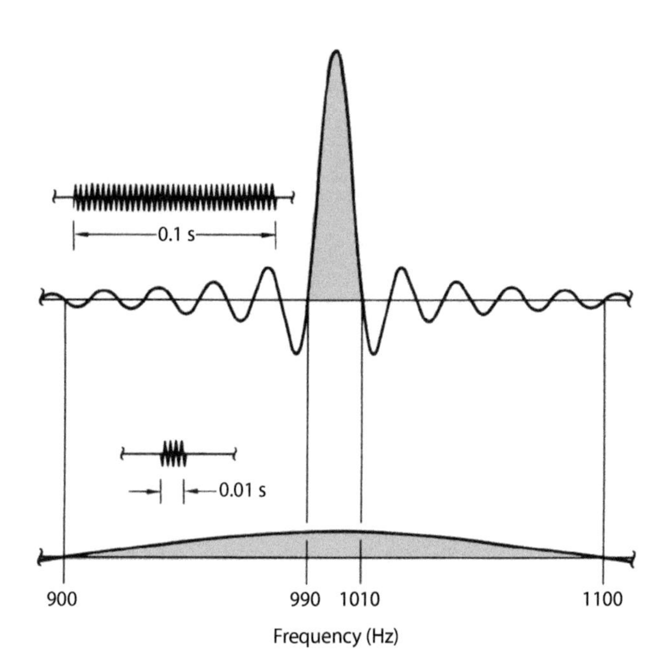

Figure 20-9. Complete spectrum of 1/10 second pulse discussed in the text (top). Although most of the energy is centered on the carrier frequency (1000 Hz), if the pulse's duration is decreased to 1/100 second (bottom), the central spectral lobe alone spreads over a band 200 Hz wide.

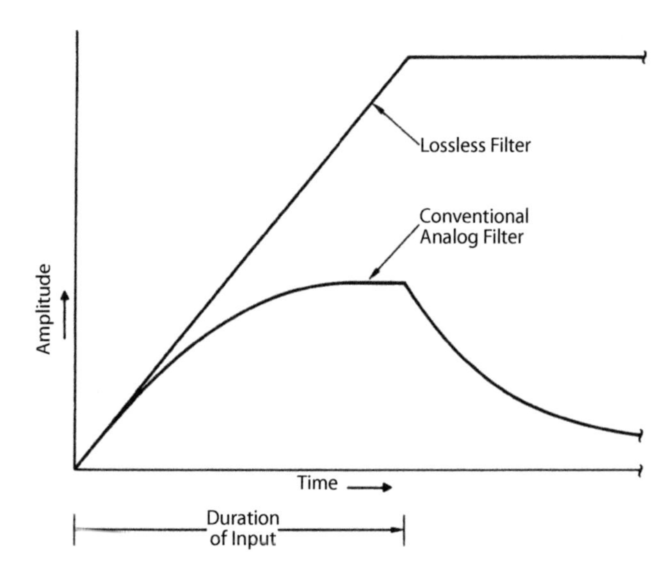

Figure 20-10. The difference between the outputs of a lossless filter and a conventional analog filter is shown that the lossless filter is a perfect integrator.

time during which pendulum and flywheel work against each other equals the total time during which they work together. Hence, the less the final amplitude of the pendulum's swing is. But no matter how far down in frequency the pendulums go, *only* at those frequencies for which the periods of buildup and damping are *exactly* equal is a pendulum completely at rest when the input ends.

For the pendulums whose frequencies are greater than 1000 Hz, the responses are similar.

Thus, the spectrum of every signal encountered covers an immensely broad band of frequencies. It is true that the energy at most of these frequencies is minuscule. But the shorter the signal, the more widely its energy is spread. For example, if the duration of the signal above is reduced by a factor of 10, when the input ends, the shift in phase of each pendulum's swing will be only 1/10th as great as before.

This means the nulls on either side of the signal's central spectral lobe will be 10 times farther apart than they were (Fig. 20-9) and the distribution of the signal's energy will be proportionately broader.

Following this general line of reasoning, a far less cumbersome graphic model of a lossless filter may be used. At this point, though, one thing more should be said about narrowband filters. A lossless filter differs from most of the filters with which we are familiar in two important respects.

First, whereas the output of a *conventional* filter builds up fairly quickly to a "steady-state" value when a constant-amplitude input of the filter's frequency is applied, the output of a *lossless* filter continues to build up as long as the input continues, after all, there is no loss (Fig. 20-10)!

Second, whereas the output of a *conventional* filter decays after the input stops, the output of a lossless filter retains its last value for an unlimited time, unless the output is dumped in some way.

In short, a lossless filter can be thought of as perfectly integrating the energy of that component of the input signal that has the same frequency as the filter.

That's all very well, but how can a purely sinusoidal signal that completes a given number of cycles per second *really* have a component of energy at any other frequency? The fact is, it does. To see why, though, the definition of frequency must be examined a little more closely.

Definition of "Frequency." The frequency of a sinusoidal signal is the number of cycles the signal completes per second. But recall, that definition was qualified as applying strictly to a continuous, unmodulated signal.

Although not generally thought of in this way, a pulsed radio wave transmitted by a radar system, is actually a continuous wave (the carrier) whose amplitude is modulated by a pulsed video signal. In other words it is a continuous wave signal that is periodically switched on and off to form a train of pulses.

The pulses have an amplitude of unity and the amplitude of the interval between pulses is zero (Fig. 20-11).

As introduced in Chapter 5, any wave whose amplitude is modulated invariably has sidebands and a portion of the wave's energy is contained in each of these.

One way of explaining how the energy of a pulse is distributed in frequency is to visualize the spectrum in terms of the sidebands produced by the pulse. This can be determined from the nature of the sidebands with the help of Fourier series (see Chapter 6).

Spectrum of a Train of Pulses. A portion of an infinitely long train of rectangular pulses is plotted in Figure 20-12. Beneath it is a plot of amplitude versus frequency for the individual waves that would have to be added together to produce the waveform, that is, the wave's spectrum. The relationship between time and frequency representations of a signal is given by the *Fourier Transform*.

Each line of this spectrum, except the zero-frequency line, represents a sine wave that goes through a maximum at the same time as the fundamental frequency. The phases of the waves are thus implicit in the plot. In alternate lobes of the envelope, the phase of each of the harmonics is shifted by 180°, indicated by plotting the amplitudes of these harmonics as negative.

Spectrum of a Pulse Modulated Radio Wave. Chapter 5 outlined that when the amplitude of a carrier wave of frequency, f_c, is modulated by a single sine wave of frequency f_m, two sidebands are produced. One is a frequency f_m above f_c, the other is f_m below f_c.

Therefore, when the carrier of a coherent transmitter is modulated to form a train of pulses, such as that illustrated in Figure 20-13, the sine wave represented by each line

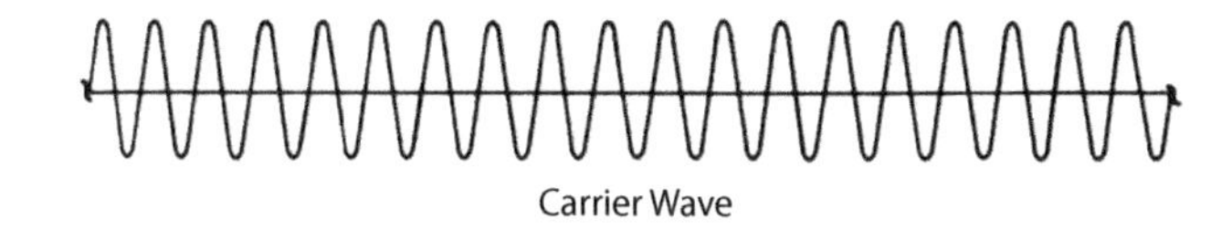

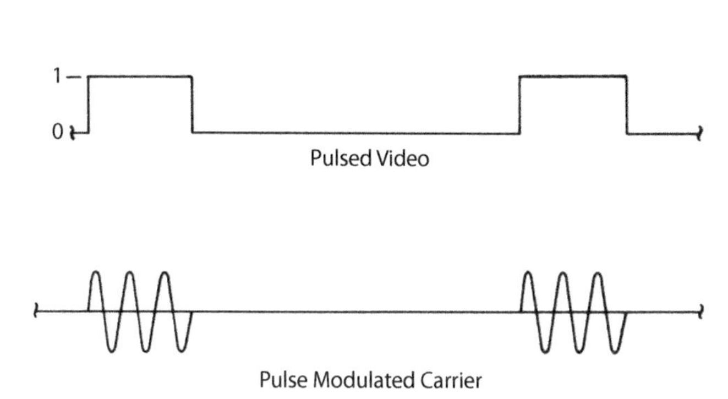

Figure 20-11. A coherent pulsed radio-frequency signal is actually a continuous wave (carrier) whose amplitude is modulated by a pulsed video signal.

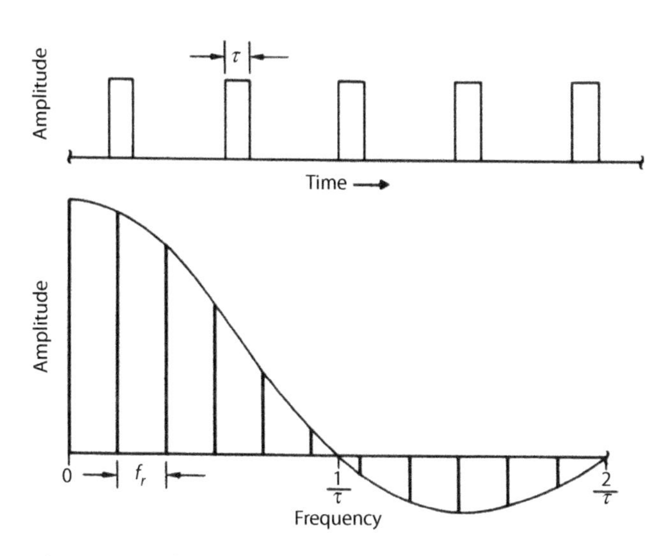

Figure 20-12. These graphs show a portion of an infinitely long rectangular pulse train and the spectrum of the train.

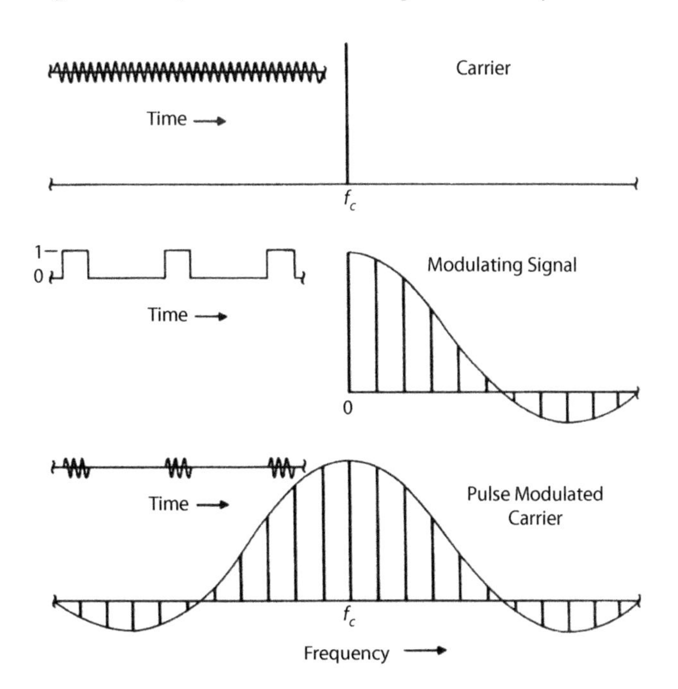

Figure 20-13. When a continuous carrier wave is modulated to form an infinitely long train of pulses, each harmonic of the video signal produces a sideband above and below the carrier frequency.

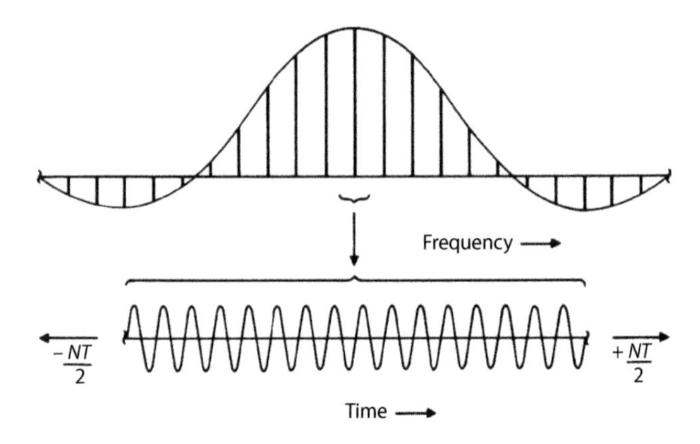

Figure 20-14. Each line in the spectrum of a pulse modulated carrier represents a single sine wave the length of the pulse train.

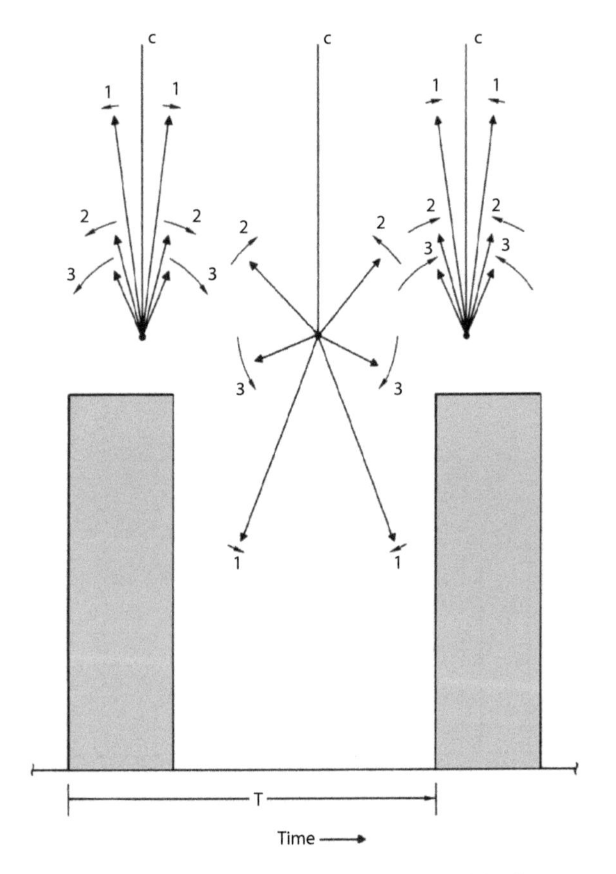

Figure 20-15. Phasors here represent the carrier and the first several sidebands of an infinitely long pulse train. During the pulses they combine constructively. Between pulses, they combine destructively and cancel.

in the spectrum of the modulating signal produces two sidebands.

The fundamental produces sidebands f_r Hz above and below the carrier. The second harmonic produces sidebands $2f_r$ above and below the carrier, and so on. The zero frequency line produces an output at the carrier frequency. The spectrum of the envelope is thus mirrored above and below the carrier frequency. The resulting radio frequency spectrum is exactly the same as the spectrum obtained for a continuous train of coherent pulses in Experiment 3 of the preceding chapter.

What the Spectral Lines Represent. One aspect of the spectrum of a pulsed carrier wave that seems difficult is that each of the individual spectral lines represents a continuous wave. i.e. each line is a wave of constant amplitude and constant frequency that continues uninterrupted in time from the beginning to the end of the pulse train (Fig. 20-14). How can this be, when the transmitter is "on" for only a fraction of each interpulse repetition time?

The amplitudes and phases of the fundamental and its harmonics are such that they completely cancel the carrier, as well as each other, during the periods between pulses. Yet they combine to produce a signal having the carrier's wavelength and the full power of the transmitter, during the brief period of each pulse.

This is illustrated in a cursory way for a pulse train by the phasor diagrams of Figure 20-15.

Figure 20-15 shows how the carrier and the first three sidebands above and below the carrier combine to produce the transmitted pulses. The phasor representing the carrier is synchronized with the strobe that provides the phase reference for the phasors (as described in Chapter 5).

Therefore, the phasors representing the upper sidebands rotate counterclockwise and the phasors representing the lower sidebands rotate clockwise. The higher the order of the individual sidebands, the more rapidly the phasors rotate.

Since the harmonics are all integer multiples of the fundamental frequency (which equals the pulse repetition frequency) once every repetition period all of the phasors line up and add constuctively.[3] Thereafter, the counter-rotating phasors rapidly fan out, pointing essentially in opposite directions and canceling the carrier and each other for the balance of the period. They come together once again at the beginning of the next period.

A pulsed signal has a true line spectrum only if the pulse train is infinitely long. Otherwise, the spectral lines have a finite width. How does the Fourier series tell us what the width is? This question is answered most simply in terms of the spectrum

3. Except the phasors representing harmonics in the odd numbered sidelobes, not shown in the figure. They are 180° out of phase with the others.

of a single pulse. So let us first examine what the Fourier series tells us about the spectrum of a single pulse.

Spectrum of a Single Pulse. Strictly speaking, the Fourier series applies to a signal only if the signal has a repetitive waveform that can be assumed to continue uninterrupted from the beginning to the end of time. In some cases, though, this assumption is sufficiently valid, even though the waveform may not be repetitive at all.

For example, this is even true in the case of a single rectangular pulse. Let's start with a continuously repetitive form of the pulse shown in Figure 20-16a.

Keeping the pulse width constant, the repetition frequency is gradually decreased and the lines of the pulsed signal's spectrum move closer and closer together (Fig. 20-16b). The envelope within which they fit retains its original shape, since that is determined solely by the pulse width.

If this process is continued, where the time between pulses is stretched to weeks, years, eons, and eventually to an infinite number of eons, the separation between spectral lines ultimately disappears and they merge into a continuum.

This is now equivalent to having a single pulse, which therefore has a continuous spectrum with exactly the same shape as the envelope of the line spectrum of the continuously repetitive waveform (Fig. 20-16c). This is what the spectrum of a single pulse was found to be in Experiment 2 of the previous chapter.

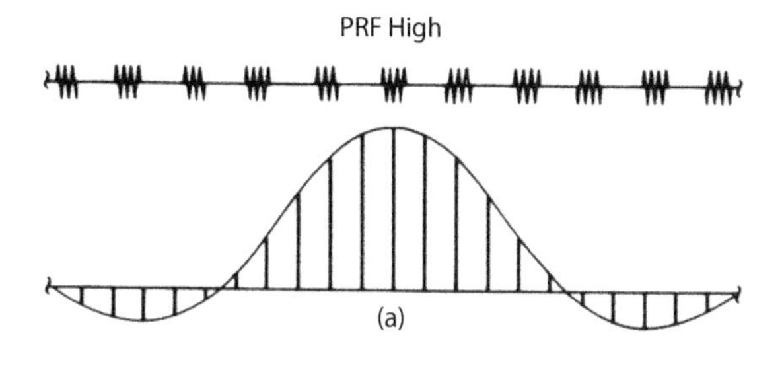

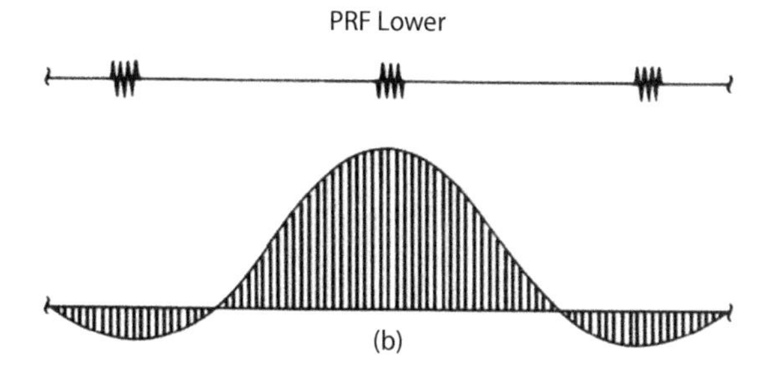

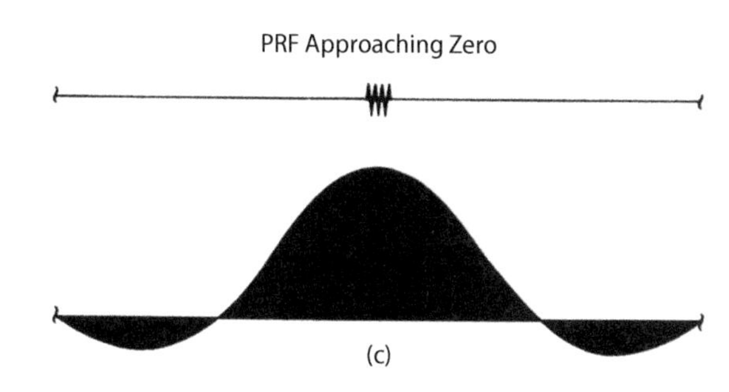

Figure 20-16. Consider a continuous pulse train of infinite length and its spectrum. As PRF is reduced, spectral lines move closer together. As PRF approaches zero, spectrum becomes continuous.

Incidentally, if the above logic is pursued a step further, it leads to an interesting conclusion. Since the pulse train, of which this single pulse is a part, is infinitely long, every point in the pulse's spectrum represents a continuous wave of infinite duration. How can that be?

Of course, no wave extends to the end of time. However, the spectra of the signals being considered here are exactly the same as if the signals were comprised of infinitely long waves.

So, in modeling spectral characteristics, it matters little whether such long waves actually exist. With that point settled, let us return to the question of what the Fourier series tells us about spectral line width.

Line Width. Knowing the spectrum of a single pulse, the spectrum of a pulse train of limited length, such as a radar system would receive from a target can be easily found. Let's consider the spectrum of a train of N pulses, having an interpulse repetition time T, hence a total length NT.

Start by imagining an infinitely long pulse train. Each line of the spectrum of this train represents a continuous wave having a single frequency and an infinite duration (i.e., a true CW

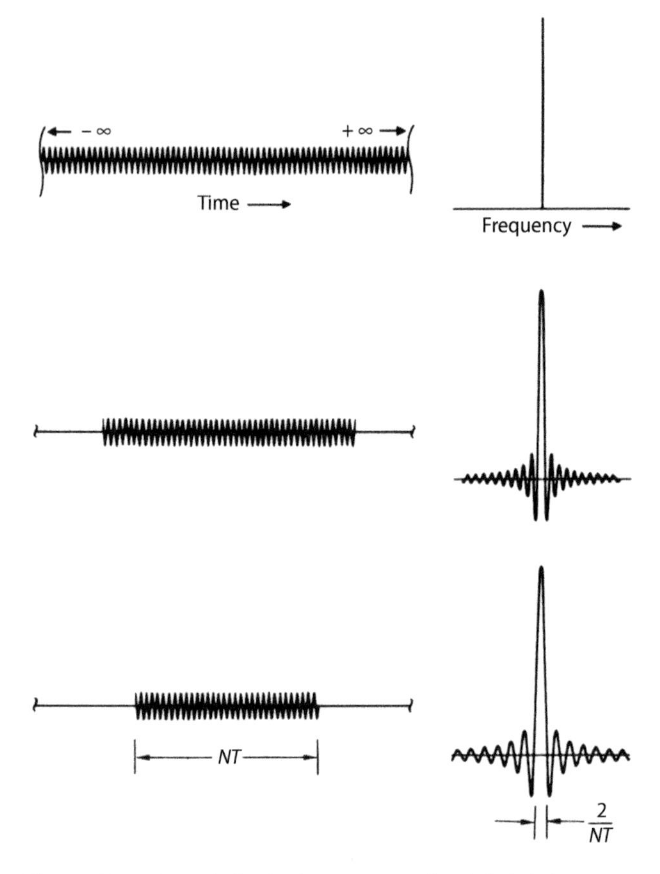

Figure 20-17. A single line in the spectrum of an infinitely long pulse train represents a CW wave. As the length of the train is reduced, this wave becomes a single pulse, and its spectrum broadens into a sin x/x shape.

signal). Holding the PRF and pulse width constant, the length of the train is gradually reduced (Fig. 20-17).

Since the constituent CW signals are the same length as the pulse train, each of them now becomes a single long pulse. As the length of this pulse decreases, the spectral line representing it gradually broadens into a sin x/x shape. When we finally reach the length, NT, of the pulse train in question, the null-to-null width of the central lobe of this "line" equals $2/(NT)$.

Thus, the Fourier series indirectly shows that the spectrum of a pulse train of limited length differs from the spectrum of a train of infinite length, only in that each spectral line has a sin x/x shape. The null-to-null width of the line is inversely proportional to the length of the pulse train.

$$\text{Line width} = \frac{2}{\text{Length of pulse train}}$$

This is exactly what the line width was found to be in Experiments 4 and 5 of the preceding chapter.

20.2 Spectrum Explained from a Filter's Point of View

As explained in section 20.1, a lossless narrowband filter integrates the energy of a signal wave in such a way that the filter's output builds up to a large amplitude only if the frequency of the signal is the same as that to which the filter is tuned.

In essence, the filter determines how close the frequencies are to being the same by sensing the shift, if any, in the phases of successive cycles of the input signal, relative to a signal whose frequency is that of the filter.

Analogy of a Filter to a Ruler. If the wave crests of the input signal are represented graphically by a series of vertical lines spaced at intervals equal to a wavelength, the filter can be thought of as measuring the spacing between wave crests with an imaginary ruler.[4] On this ruler, marks are inscribed at intervals of one wavelength for the frequency to which the filter is tuned.

If the filter is tuned to the exact frequency of the wave, when the first mark is lined up with a wave crest, all subsequent marks will similarly line up (Fig. 20-18).

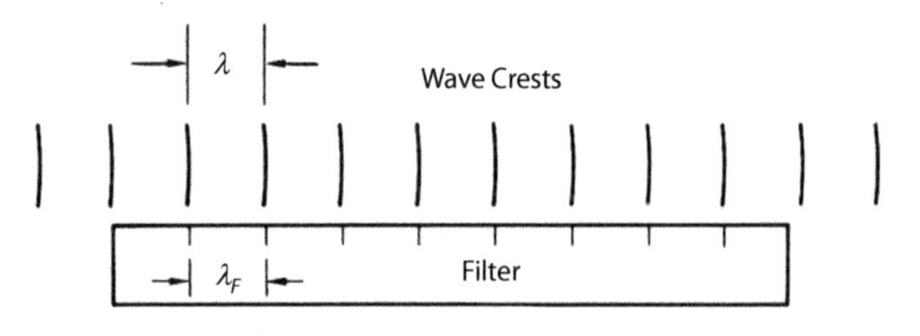

Figure 20-18. A lossless narrowband filter can be thought of as measuring the spacing of a signal's wave crests with an imaginary ruler.

4. As $f = c/\lambda$, the closer the marks on the ruler, the higher the frequency. This relationship allows the analogy of a filter to a ruler

If the filter is tuned to a slightly different frequency, the first mark beyond the initial one will be displaced slightly from the next wave crest; the second mark will be displaced twice as much from the following wave crest; the third mark, three time as much and so on (Fig. 20-19). The displacements correspond to the phases of the individual cycles of the signal as seen by the filter; the progressive increase in displacement corresponds to the progressive shift in phase from once cycle to the next.

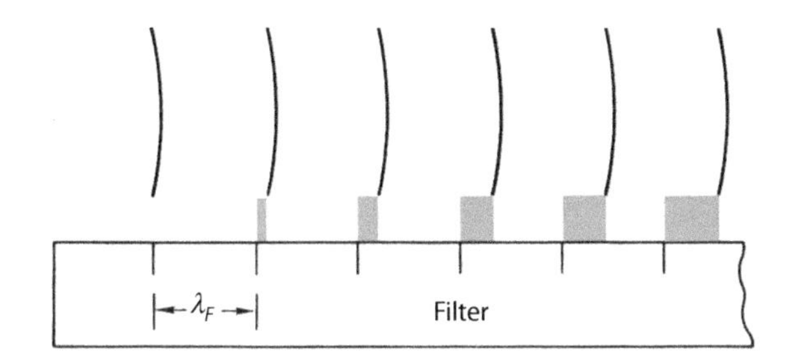

Figure 20-19. If the filter's frequency is higher than the signal's, the phase shift between the wave crests and the marks on the ruler builds up.

Now, the amplitude of each cycle of the wave, as well as the phase of that cycle relative to the corresponding mark on the ruler, can be represented by a phasor (Fig. 20-20). What the narrowband filter does is integrate the phasors for successive cycles (Fig. 20-21). If n phasors point in the same direction such that the cycles they represent have the same phase, then the sum will be n times the length of the phasors. If they point in slightly different directions, the sum will be less. And if they point in opposite directions, where the cycles are 180° out of phase, they will cancel.

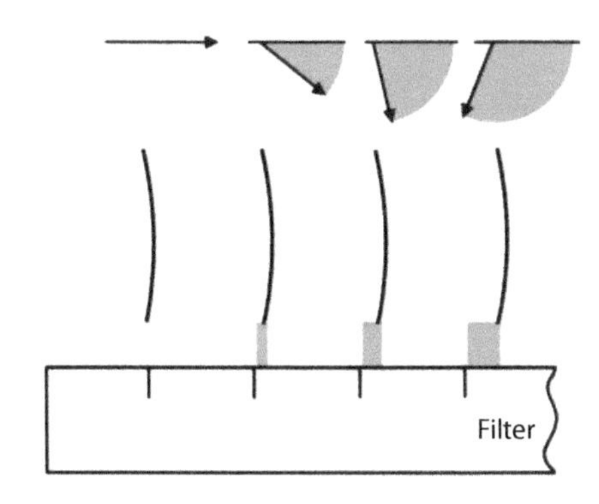

Figure 20-20. The amplitude and phase of each cycle of the wave can be represented by the phasor shown at the top of the Figure.

With this simple analogy in mind, the results of some of the experiments performed in the preceding chapter can be analyzed.

Spectrum of a Single Pulse. Earlier it was stated that the spectrum of a single pulse is continuous over a band of frequencies $2/\tau$ Hz wide where τ is the length of the pulse. To see why this is so, a pulse, τ seconds long, is measured with four different rulers. Each ruler represents a narrowband filter tuned to a different frequency and thus each has marks at a different spacing.

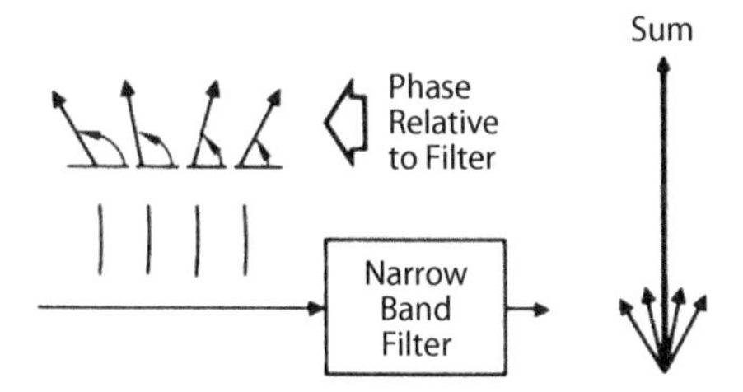

Figure 20-21. In essence, the filter adds up the phasors for successive cycles.

In Figure 20-22(a), the filter has the same frequency as the pulse's carrier (f_c). Consequently, the phases of the wave crests relative to the marks on the ruler are all the same. The phasors representing the individual cycles of the wave all point in the same direction. The pulse is eight cycles long. Assuming that the length of each phasor is one, their sum is eight.

In Figure 20-22 the same pulse is applied to a filter having a higher frequency ($f_c + \Delta f$); the wavelength marks are closer together. As a result, there is a progressive shift in the phases of the wave crests relative to the marks. Over the length of the pulse, the shift builds up to a quarter of a wavelength. The phasors, therefore, fan out over 90°. Even so, their sum is nearly seven, in other words the component Δf has caused a small reduction in the integration gain.

In Figure 20-22(c), the filter has a considerably higher frequency ($f_c + 2\Delta f$). The total accumulated phase shift over the length of the pulse now is half a wavelength (180°). Still, the sum is nearly half what it was for the filter tuned to f_c, i.e., the integration gain has reduced further to a value of 4.4 as the frequency deviates more from f_c.

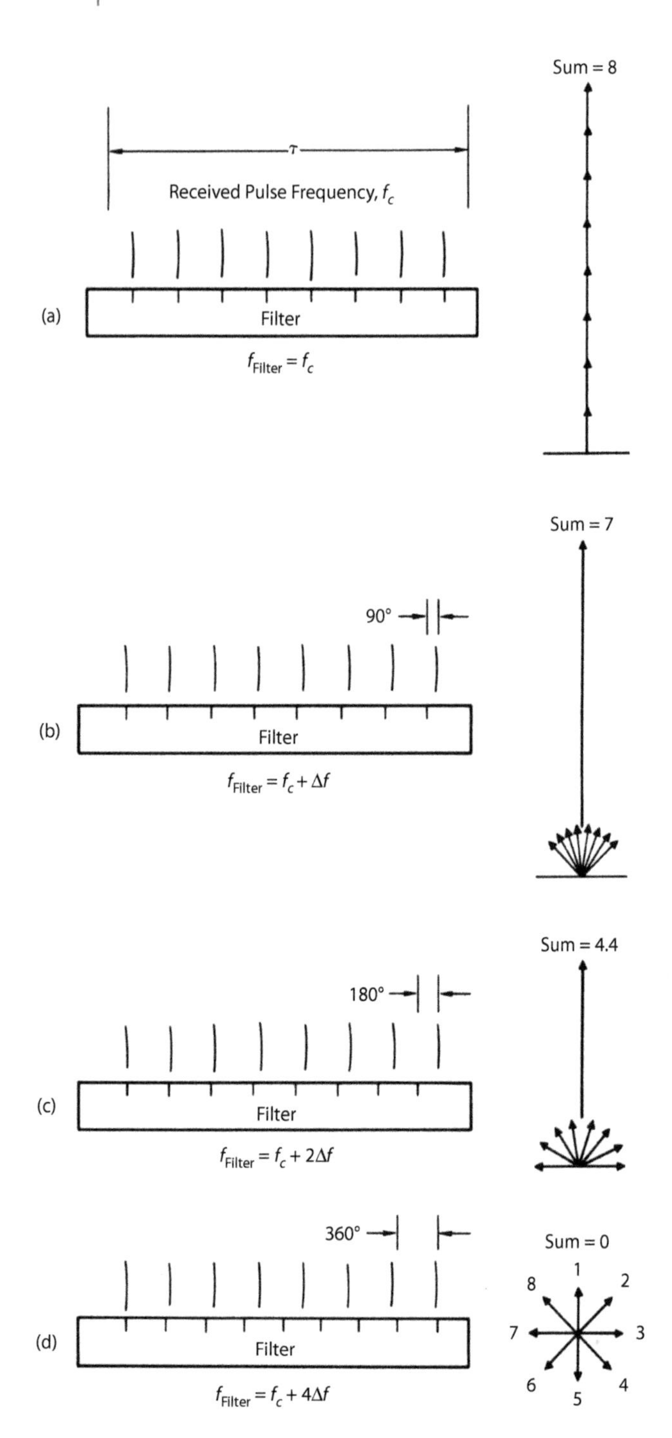

Figure 20-22. If a filter is tuned to progressively higher frequencies, the cumulative phase shift over the length of a pulse increases.

In Figure 20-22(d), the filter has a sufficiently high frequency ($f_c + 4\Delta f$) that the phase shift over the length of the pulse is one whole wavelength. As a result, the phasors are uniformly spread over 360°. Pointing in opposite directions, the phasors for cycles No. 1 and No. 5 cancel. So do the phasors for cycles No. 2 and No. 6, No. 3 and No. 7, and No. 4 and No. 8. The pulse produces no output from the filter; in other words a frequency has been reached where there is a null in the pulse's spectrum.

What is this frequency? Over the duration of the pulse the oscillation of the filter that was tuned to the null frequency completed one more cycle than the pulse's carrier (Fig. 20-23). The duration of the pulse was τ seconds. So, the filter's frequency is the inverse of this duration, i.e., $1/\tau$ cycles per second (Hz) higher than the carrier frequency, f_c. The null frequency, therefore, is ($f_c + 1/\tau$).

Following the same line of reasoning, a null will be found $1/\tau$ cycles per second below f_c (Fig. 20-24). The null-to-null bandwidth of a single pulse, therefore, is $2/\tau$ Hz, exactly as was observed in Experiment No. 2 in Chapter 18.

The reason for the filter's response is that a difference in frequency is actually a continuous linear shift in phase. When a pulsed signal is applied to a filter, the rate of this shift in cycles per second equals the difference between the signal's carrier frequency and the frequency of the filter. In the case of a single pulse, only when this difference is large enough to make the total phase shift over the duration of the pulse equal one whole wavelength, do the individual cycles of the received wave entirely cancel. The shorter the pulse (Fig. 20-25), the greater the frequency difference must be to satisfy this condition. Conversely, the longer the pulse, the less the frequency difference must be. Thus, for a pulse whose duration is 1 microsecond, the null-to-null width of the central spectral lobe is $2 \div 10^{-6} = 2$ MHz. For a pulse whose duration is 1 second, the spectral line width is 2 Hz. And for a pulse whose duration is 1 hour, the spectral line width is only $2 \div (60 \times 60) \cong 0.00056$ Hz.

Spectrum of a Coherent Pulse Train. To see why the null-to-null width of the central spectral lobe is drastically reduced when

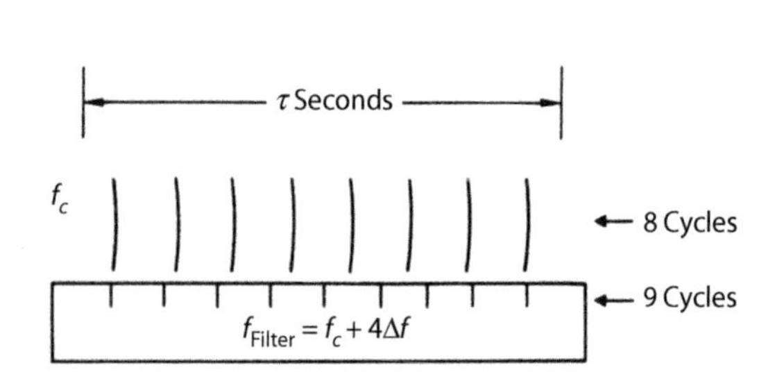

Figure 20-23. At null, in τ seconds filter completes one more cycle than signal.

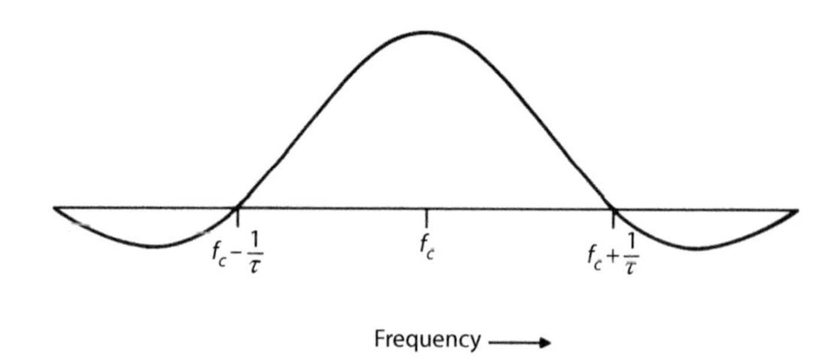

Figure 20-24. A plot of the phasor sums has a sin x/x shape with nulls $1/\tau$ Hz above and below the carrier frequency, f_c.

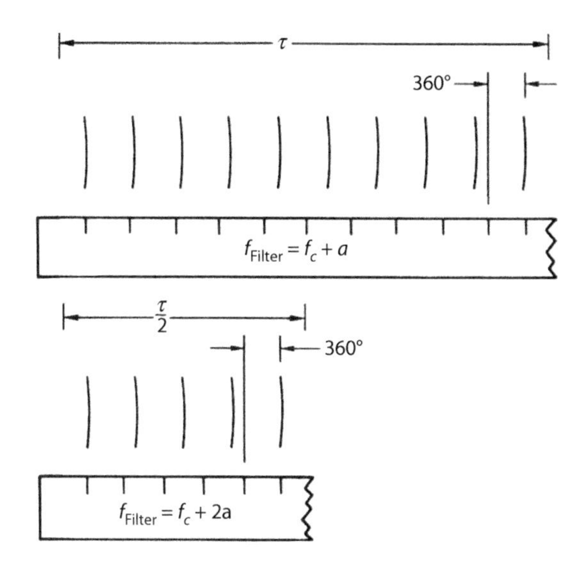

Figure 20-25. The shorter a pulse is, the greater the frequency difference must be to produce a 360° phase shift over the length of the pulse.

the filter integrates a train of coherent pulses, consider each pulse represented by a single phasor (Fig. 20-26).

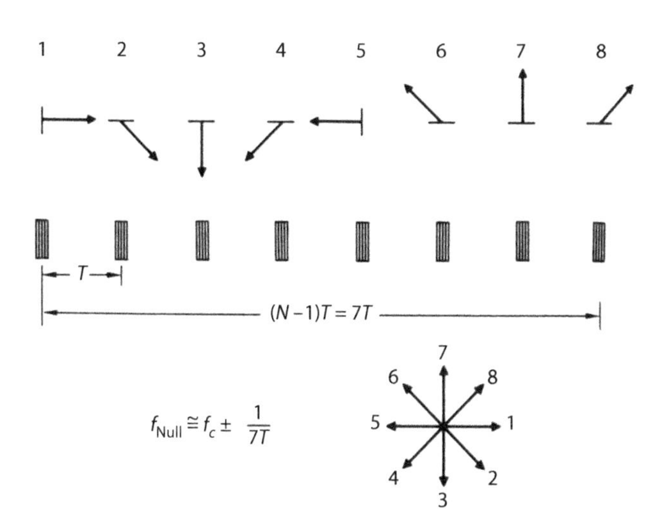

Figure 20-26. The amplitude and phase of each pulse in a train may be represented by a single phasor. Spectral nulls occur when phase shift over length of train is 360°.

The nulls now occur when the total phase shift over the length of the pulse train is one wavelength. Since the train is many times longer than a single pulse, the frequency difference that produces a phase shift of one wavelength is many times smaller for the train than for a single pulse.

For example, consider a train of 32 pulses (Fig. 20-27). Suppose repetition time is 100 times the pulse width. The duration of the train will be roughly 31 × 100 = 3100 times the duration of a single pulse, making the null-to-null bandwidth of the line only 1/3100 that of a single pulse.[5]

Now consider what happens if every other pulse in this train is deleted. Because the length of the pulse train is essentially the same, the phasors for the remaining 16 pulses still cancel at almost the same frequency; so the null-to-null bandwidth is about the same.

However, since there are only half as many pulses, the amplitude of the filter output will be only half as great.

In addition, since the PRF is only half as great, the pulses will produce an output from the filter at twice as many points within the envelope established by the pulse width. In other words the width (or resolution) is reduced by a factor of two.

Why does the pulse train produce an output at intervals equal to the PRF in the first place?

Repetition of Spectral Lines. Recalling that when a train of pulses of a given carrier frequency is applied to a filter the output of the filter falls off as the filter is tuned away from the carrier frequency. This is due to the pulse-to-pulse difference in the phase of the carrier, as seen by the filter. Since phase repeats every 360°, there is no way of telling whether

5. Although the train contains 32 pulses, it is only one pulse width longer than 31 inter-pulse periods.

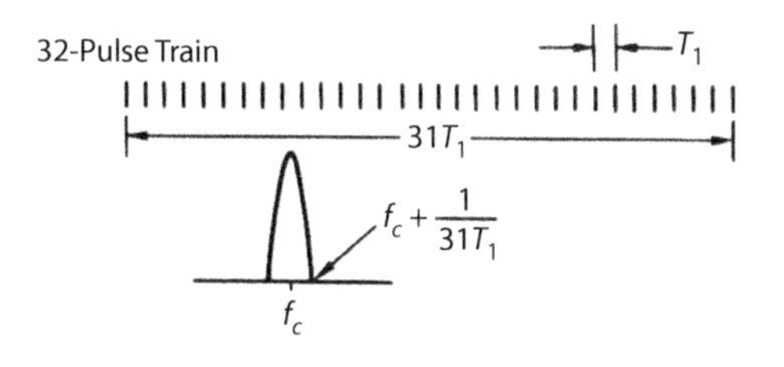

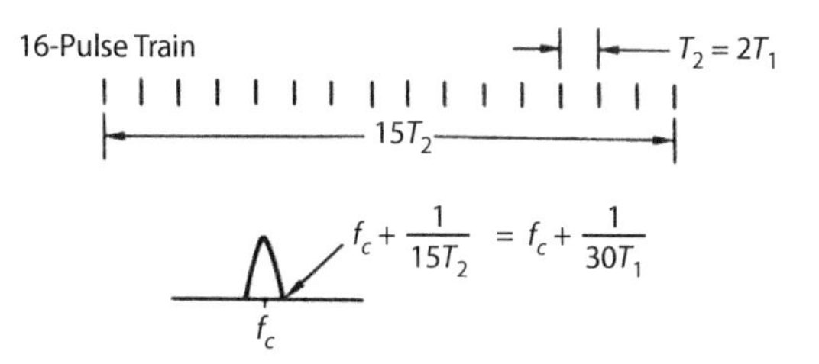

Figure 20-27. If every other pulse in a train is deleted, the output is reduced, but the bandwidth remains essentially unchanged.

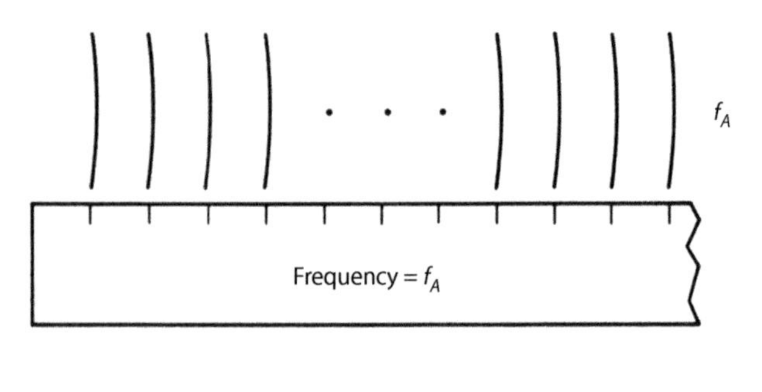

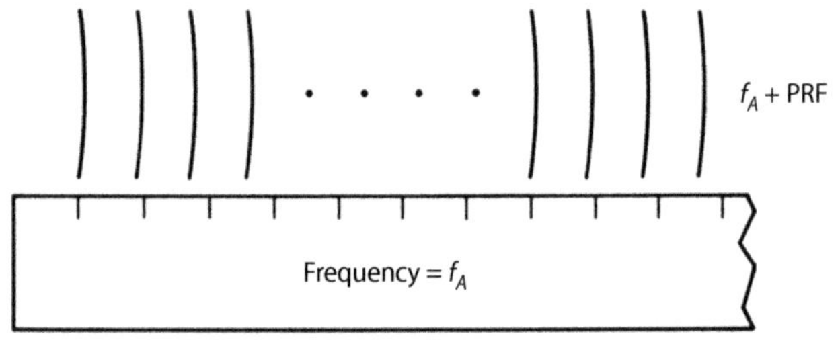

Figure 20-28. Consider pulse trains whose carrier frequencies equal the filter frequency, f_A, (top) and f_A plus the PRF (bottom). Only difference in the outputs produced by the two trains is that due to the cycle-to-cycle phase shift within each pulse of the second train.

the phase of any one pulse is the same as that of the preceding pulse or has been shifted by 360° or some multiple of 360°.

A pulse-to-pulse shift of 360° is equivalent to one cycle per interpulse repetition time. This corresponds to an increment of frequency equal to the PRF (Fig. 20-28). Consequently, there may be very little difference between a filter's response to a pulse train whose carrier frequency is the same as the filter frequency and its response to a pulse train whose carrier frequency is some integer multiple of the PRF (above or below the filter frequency). In fact, the only difference is that due to the phase shift occurring from cycle to cycle over the duration of each pulse. Unless the multiple of the PRF is very high or the pulse width is a fairly large fraction of the inter-pulse period the difference is slight. In other words, unless the carrier frequency is near one end or the other of the envelope established by the pulse width.

20.3 Mathematical Explanation of the Pulsed Spectrum

The spectrum of a pulsed signal is derived mathematically in the following panel, building on the explanations of Fourier series and the Fourier transform introduced in Chapter 6. If your interest is not so mathematical, then skip ahead to "Results."

Mathematical Explanation of the Pulsed Spectrum

IN THE TEXT, THE SPECTRUM OF A PULSED SIGNAL IS EXPLAINED IN SEVERAL quite different nonmathematical ways. While hopefully these explanations have provided some helpful insights, the spectrum can be explained much more rigorously *and* succinctly in purely mathematical terms.

Accordingly, a mathematical derivation of the spectrum of a simple, perfectly rectangular pulsed signal is presented on the third and fourth pages of this panel.

A brief preliminary explanation of the derivation is given on the first two pages.

General Approach. Basically, the panel shows two things: first, the derivation of a mathematical expression for a pulse modulated carrier signal as a function of time, $f(t)$, and second the transformation of this expression from the time domain to the frequency domain, in other words, the Fourier transform of the signal.

The expression for the pulse-modulated signal is derived by writing separate expressions for each of the following.

1. An infinitely long pulsed video signal, $f_1(t)$, having an amplitude of 1, a pulse width τ, an interpulse repetition time T, and a pulse repetition frequency (expressed in radians per second) of ω_0 where $\omega_0 = 2\pi/T = 2\pi f_r$.

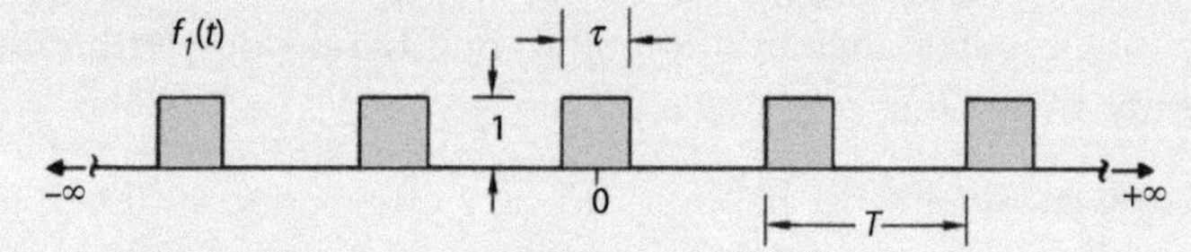

2. A signal, $f_2(t)$, having an amplitude of 1 and a duration equal to the length of a train of N pulses whose interpulse repetition time is T.

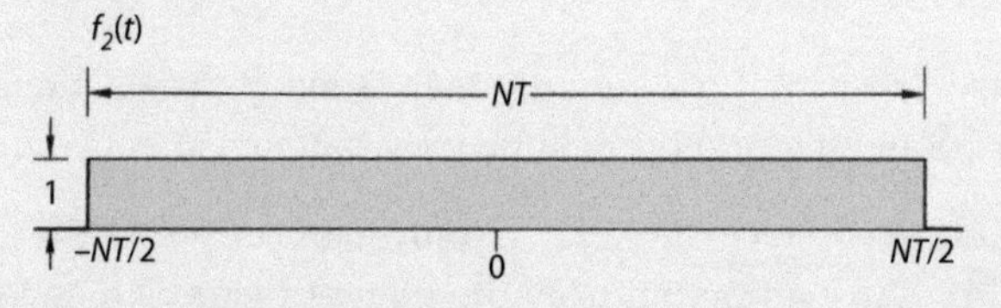

- An infinitely long carrier wave, $f_3(t)$, having an amplitude A, and a frequency expressed in radians per second of ω_c, where $\omega_c = 2\pi f_c$.

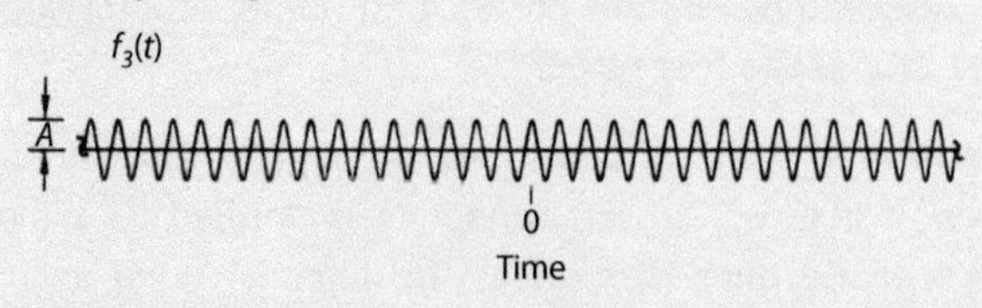

* By positioning zero on the time axis in the center of one of the pulses, the coefficients of the sine terms are reduced to zero. (The signal has even symmetry.)

Mathematical Explanation of the Pulsed Spectrum *continued*

The expression for the pulsed video signal, $f_1(t)$, is obtained by evaluating the coefficients (a_0, a_2 a_4. . .)* of the Fourier series in terms of the pulse width, τ, and interpulse repetition time, T, of the signal and substituting these values into the series.

Multiplying the first two functions, $f_1(t)$ and $f_2(t)$, together gives an equation for the pulsed modulating signal. Multiplying the function for the carrier wave, $f_3(t)$, by this product yields the desired equation for the pulse modulated carrier, $f_4(t)$.

The Fourier transform of this function is then derived, yielding the spectrum of the pulse-modulated signal. The essence of both derivations is briefly outlined on the next page.

Essence of the Derivations. In evaluating the coefficients of the Fourier series for the pulsed video signal, the key operation is multiplying the equation for the signal by cos ωt for the frequency of the harmonic whose coefficient we wish to obtain. As illustrated in the figure below, when the instantaneous amplitudes of two sine waves of the same phase and frequency are multiplied together, their product is positive for both halves of every cycle. (The same is true of cosine waves.)

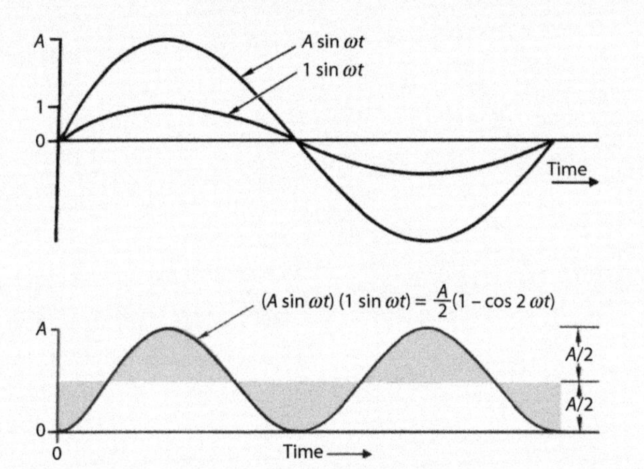

Consequently, if the product is integrated over a complete cycle, the result divided by the period of the cycle is one-half the product of the peak amplitudes of the two waves. If the amplitude of one of the waves is one, then the product is one-half the peak amplitude of the other wave.

Yet if the frequencies of the two waves are not the same, the sign of the product will alternate between (+) and (–). If the frequency of one wave is an integer multiple of the frequency of the other then, when the product is integrated over the period of the lower frequency wave, the result will be zero.

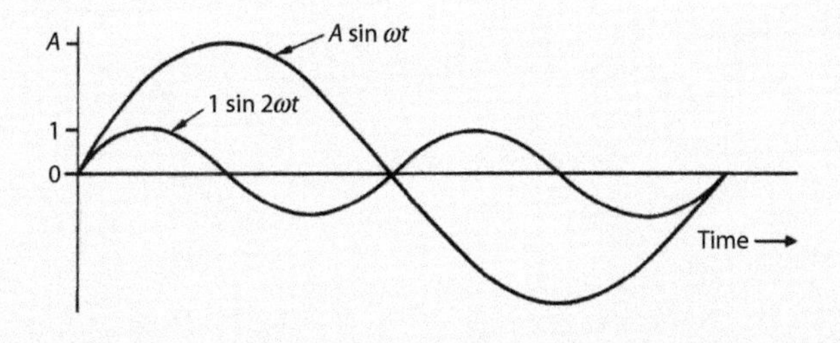

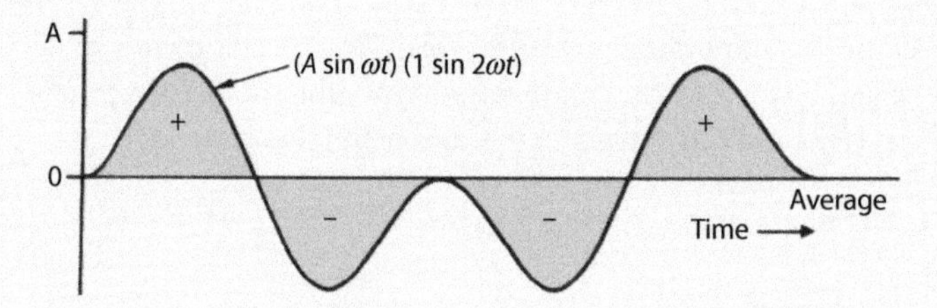

Thus, the coefficients of the Fourier series for a continuously repeating pulsed video signal can be found by multiplying the mathematical expression for the waveform by the cosines of $\omega_0 t$, $2\omega_0 t$, $3\omega_0 t$, . . . $n\omega_0 t$, in turn, integrating each product over the waveform's repetition time, T, and dividing by $T/2$. The dc coefficient (i.e., average amplitude) is found by integrating the expression for the wave over the period T and then dividing by T.

Similarly, in deriving the Fourier transform, that component of the pulse modulated wave having a particular frequency, ω, can be found by multiplying the equation for the wave (as a function of time) by cos ωt – j sin ωt and integrating the product. In this case, since ω is not necessarily an integer multiple of the fundamental of the modulated wave, the product must be integrated over the entire duration of the pulse train, i.e., from $-NT/2$ to $+NT/2$. As with the Fourier series, the dc component is found by integrating the expression for the wave alone over the same period and dividing by its duration.

In deriving the Fourier transform, the sinusoidal functions are expressed most conveniently in exponential form. The relationships between the two forms were explained with phasors in Chapter 5 and are summarized in the following diagram.

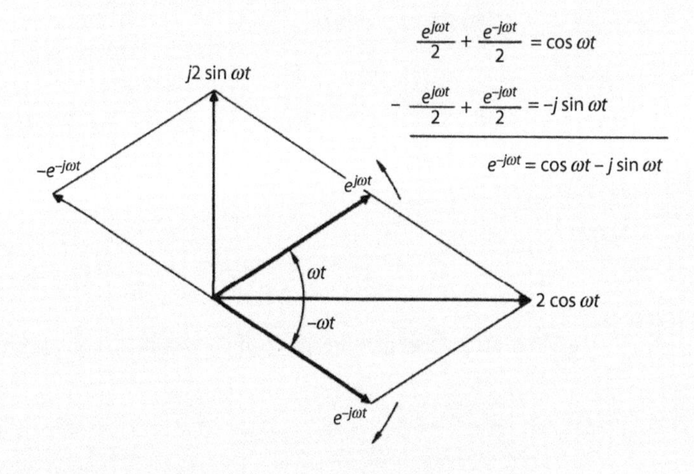

The only calculus you need to know to follow the derivation is that the integral of the cosine of ωt is $1/\omega$ times the sine of ωt and the integral of $e^{-j\omega t}$ is $-1/j\omega$ times $e^{-j\omega t}$.

$$\int \cos(\omega t)dt = \frac{1}{\omega}\sin\omega t$$

$$\int e^{-j\omega t}dt = \frac{1}{-j\omega}e^{-j\omega t}$$

Mathematical Explanation of the Pulsed Spectrum *continued*

One other reminder. If a quantity raised to a given power is multiplied by the same quantity raised to another power, the product is the quantity raised to the sum of the two powers. Thus,

$$e^{j\omega t} \times e^{j\omega_0 t} = e^{(j\omega t + j\omega_0 t)} = e^{j(\omega+\omega_0)t}$$

With the above relationships in mind, let us proceed with the derivations. The expression for the pulse modulated carrier is derived on the next page and the Fourier transform of this expression on the facing page.

1. Continuous Pulsed Modulation Signal (Expressed as a Fourier Series)

$$f_1(t) = a_0 + \sum_{n=1}^{\infty} a_n \cos n\,\omega_0 t \qquad \omega_0 = 2\pi\frac{1}{T} = 2\pi f_r$$

$$a_0 = \frac{1}{T}\int_{-\tau/2}^{\tau/2} dt = \left.\frac{t}{T}\right|_{-\tau/2}^{\tau/2} = \frac{\tau}{T}$$

$$a_n = \frac{2}{T}\int_{-\tau/2}^{\tau/2} \cos(n\,\omega_0 t)\,dt = \left.\frac{2}{T\,n\,\omega_0}\sin n\,\omega_0 t\right|_{-\tau/2}^{\tau/2} = \frac{2}{T\,n\,\omega_0}\Big[\underbrace{\sin n\,\omega_0\frac{\tau}{2} - \sin n\,\omega_0\frac{-\tau}{2}}_{-\sin(-\alpha)=\sin\alpha}\Big] = \underbrace{2\frac{\tau}{T}\frac{\sin n\,\omega_0\frac{-\tau}{2}}{n\,\omega_0\frac{\tau}{2}}}_{\text{Numerator and denominator multiplied by }\frac{\tau}{2}}$$

$$f_1(t) = \frac{\tau}{T}\left[1 + 2\sum_{n=1}^{\infty}\frac{\sin n\,\omega_0\frac{\tau}{2}}{n\,\omega_0\frac{\tau}{2}}\cos n\,\omega_0 t\right]$$

2. Duration Pulse Modulation

$$f_2(t) = 1 \qquad \frac{-NT}{2} \le t \le \frac{NT}{2}$$

$$= 0 \qquad t < \frac{-NT}{2} \text{ and } t > \frac{NT}{2}$$

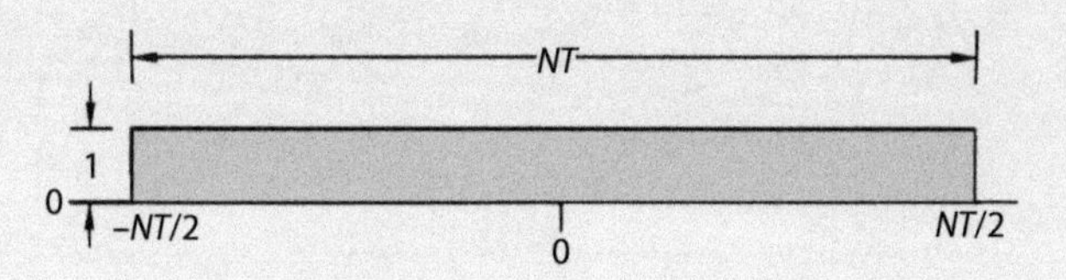

3. Unmodulated Carrier

$$f_3(t) = A\cos\omega_c t$$

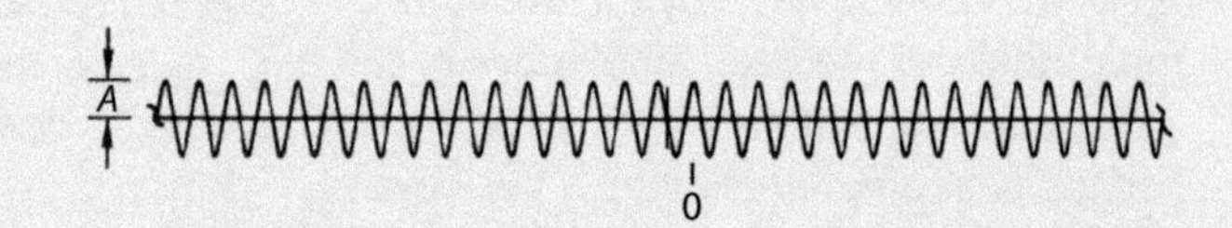

4. Pulse Modulated Carrier (Product of expressions 1, 2, and 3)

$$f_4(t) = f_1(t)\cdot f_2(t)\cdot f_3(t)$$

$$= \frac{A\tau}{T}\left\{1 + 2\sum_{n=1}^{\infty}\frac{\sin n\,\omega_0\frac{\tau}{2}}{n\,\omega_0\frac{\tau}{2}}(\cos n\,\omega_0 t)\right\}\cos\omega_c t \qquad \frac{-NT}{2} \le t \le \frac{NT}{2}$$

$$= \frac{A\tau}{T}\left\{\cos\omega_c t + \sum_{n=1}^{\infty}\frac{\sin n\,\omega_0\frac{\tau}{2}}{n\,\omega_0\frac{\tau}{2}}\left[\cos(\omega_c + n\,\omega_0)t + \cos(\omega_c - n\,\omega_0)t\right]\right\}$$

Mathematical Explanation of the Pulsed Spectrum *continued*

5. Fourier Transform of Pulse Modulated Carrier

$$F(j\omega)=\int_{-\infty}^{\infty} e^{-j\omega t} f_4(t)\,dt$$

$$F(j\omega)=\frac{A\tau}{T}\left[\overbrace{\int_{-NT/2}^{NT/2} e^{-j\omega t}\cos\omega_c t\,dt}^{❶}+\sum_{n=1}^{\infty}\frac{\sin n\omega_0\frac{\tau}{2}}{n\omega_0\frac{\tau}{2}}\left\{\overbrace{\int_{-NT/2}^{NT/2} e^{-j\omega t}\cos(\omega_c+n\omega_0)t\,dt}^{❷}+\overbrace{\int_{-NT/2}^{NT/2} e^{-j\omega t}\cos(\omega_c-n\omega_0)t\,dt}^{❸}\right\}\right]$$

$$❶=\frac{1}{2}\int_{-NT/2}^{NT/2} e^{-j\omega t}(e^{j\omega_c t}+e^{-j\omega_c t})\,dt=\frac{1}{2}\int_{-NT/2}^{NT/2} e^{-j(\omega+\omega_c)t}\,dt+\frac{1}{2}\int_{-NT/2}^{NT/2} e^{-j(\omega-\omega_c)t}\,dt$$

$$=\left.\frac{e^{-j(\omega+\omega_c)t}}{-2j(\omega+\omega_c)}\right|_{-NT/2}^{NT/2}+\left.\frac{e^{-j(\omega-\omega_c)t}}{-2j(\omega-\omega_c)}\right|_{-NT/2}^{NT/2}$$

$$=\frac{e^{-j(\omega+\omega_c)NT/2}-e^{j(\omega+\omega_c)NT/2}}{-2j(\omega+\omega_c)}+\frac{e^{-j(\omega-\omega_c)NT/2}-e^{j(\omega-\omega_c)NT/2}}{-2j(\omega-\omega_c)}$$

$$=\frac{NT}{2}\left\{\frac{\sin\left[(\omega+\omega_c)\frac{NT}{2}\right]}{(\omega+\omega_c)\frac{NT}{2}}+\frac{\sin\left[(\omega-\omega_c)\frac{NT}{2}\right]}{(\omega-\omega_c)\frac{NT}{2}}\right\}$$

$$❷=\frac{NT}{2}\left\{\frac{\sin\left[(\omega+\omega_c+n\omega_0)\frac{NT}{2}\right]}{(\omega+\omega_c+n\omega_0)\frac{NT}{2}}+\frac{\sin\left[(\omega-\omega_c-n\omega_0)\frac{NT}{2}\right]}{(\omega-\omega_c-n\omega_0)\frac{NT}{2}}\right\}$$

$$❸=\frac{NT}{2}\left\{\frac{\sin\left[(\omega+\omega_c-n\omega_0)\frac{NT}{2}\right]}{(\omega+\omega_c-n\omega_0)\frac{NT}{2}}+\frac{\sin\left[(\omega-\omega_c+n\omega_0)\frac{NT}{2}\right]}{(\omega-\omega_c+n\omega_0)\frac{NT}{2}}\right\}$$

$$F(j\omega)=\frac{A\tau N}{2}\left[\underbrace{\frac{\sin\left[(\omega+\omega_c)\frac{NT}{2}\right]}{(\omega+\omega_c)\frac{NT}{2}}}_{\text{Carrier}}+\sum_{n=1}^{\infty}\underbrace{\frac{\sin n\omega_0\frac{\tau}{2}}{n\omega_0\frac{\tau}{2}}}_{\text{Envelope}}\left\{\underbrace{\frac{\sin\left[(\omega+\omega_c+n\omega_0)\frac{NT}{2}\right]}{(\omega+\omega_c+n\omega_0)\frac{NT}{2}}}_{\text{Lower Sidebands}}+\underbrace{\frac{\sin\left[(\omega+\omega_c-n\omega_0)\frac{NT}{2}\right]}{(\omega+\omega_c-n\omega_0)\frac{NT}{2}}}_{\text{Upper Sidebands}}\right\}\right.$$

$$\left.+\frac{\sin\left[(\omega-\omega_c)\frac{NT}{2}\right]}{(\omega-\omega_c)\frac{NT}{2}}+\sum_{n=1}^{\infty}\frac{\sin n\omega_0\frac{\tau}{2}}{n\omega_0\frac{\tau}{2}}\left\{\frac{\sin\left[(\omega-\omega_c+n\omega_0)\frac{NT}{2}\right]}{(\omega-\omega_c+n\omega_0)\frac{NT}{2}}+\frac{\sin\left[(\omega-\omega_c-n\omega_0)\frac{NT}{2}\right]}{(\omega-\omega_c-n\omega_0)\frac{NT}{2}}\right\}\right]$$

Results. The final equation obtained in the blue panel is the Fourier transform for a train of N perfectly rectangular pulses having the following characteristics:

- Carrier frequency, $\omega_c = 2\pi f_c$
- Pulse width, τ
- PRF (angular frequency), $\omega_0 = 2\pi f_r$
- Interpulse repetition time, T
- Duration, NT

The transform consists of two similar sets of terms. The first set applies to frequencies having negative values; the second set, to frequencies having positive values.

The positive-frequency portion of the transform is repeated in Figure 20-29. The first term inside the brackets represents the spectrum of the central spectral line (the carrier). Immediately following the summation sign is the sin x/x term giving the envelope within which the other spectral lines fit. The remaining terms represent the lines above and below the carrier.

By substituting appropriate values for N (the number of pulses), the same equation can be applied to pulse trains of virtually any length.

Beneath the equation is a plot of the spectrum given by its amplitude versus frequency in radians per second. It is obtained by evaluating the equation for values of ω covering a wide enough range of positive frequencies to include the entire central lobe of the envelope. The first pair of nulls in the envelope occurs $2\pi/\tau$ radians per second above and below the carrier frequency ω_c. Within the envelope, spectral lines occur above and below the carrier frequency, at intervals equal to the PRF, ω_0. Each line has a sin x/x shape, with nulls $2\pi/NT$ above and below the line's central frequency.[6]

6. 2π radians equals 360°

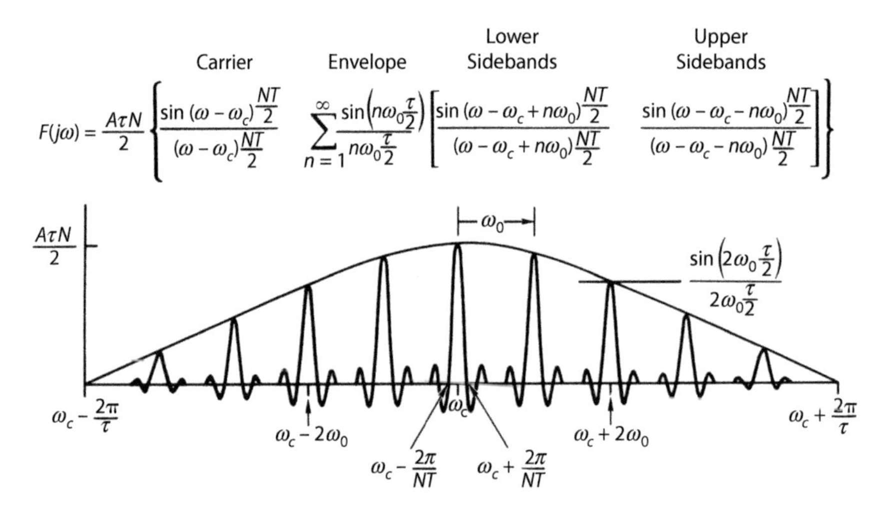

Figure 20-29. Positive-frequency portion of the Fourier transform for a rectangular train of N pulses. The pulses have a width τ, a carrier frequency of ω_c, a PRF of ω_0, and an interpulse repetition time of T.

Significance of the Negative-Frequency Terms. Many people are puzzled by the negative-frequency components of the Fourier transform. It does seem odd but has a perfectly rational logic.

It so happens that the negative components reflect the difference between the transform for a signal whose carrier is a cosine wave and the transform for a signal whose carrier is a sine wave. In the case of the transform for a cosine wave (Fig. 20-30A), such as the one just derived, the algebraic signs of the negative-frequency terms are the same as the signs of the corresponding positive-frequency terms. Whereas, in the case of the transform for a sine wave, the signs of the negative-frequency terms (Fig. 20-30B) are the opposites of the signs of the corresponding positive-frequency terms.

If the signal is considered alone, the negative-frequency terms have no significance, i.e., they contribute no additional information regarding what frequencies are present. Since $\cos(-\omega t) = \cos \omega t$, the energy represented by the negative-frequency terms of the transform for a cosine wave merely adds to the energy represented by the corresponding positive-frequency terms. Further, since $\sin(-\omega t) = -\sin \omega t$, the energy represented by the negative-frequency terms of the transform for a sine wave likewise merely adds to the energy represented by the positive frequency terms.

However, in the case of a signal that has been resolved into I and Q components, the negative-frequency terms do contribute additional information. When the signal is translated to the video range, the signal's Fourier transform will have only negative-frequency terms if the frequency of the original signal was lower than that of the reference signal used in the frequency translation. The transform will have only positive-frequency terms if the frequency of the original signal was higher.

What the Amplitudes Represent. One important question remains to be answered. All of the spectra shown thus far have been plots of amplitude versus frequency. But nothing has been said about how this amplitude relates to the amplitude of the wave in the time domain or even what units it is expressed in.

Examining the first term of the Fourier transform, the term that establishes the peak amplitude of the envelope can clear this up.

$$\frac{A\tau N}{2}$$

The factor, A, was defined in the panel as the peak amplitude of the carrier wave. Assuming that A is a voltage, then the spectrum is a plot of voltage versus frequency.

Going a step further, since power is proportional to voltage squared, by squaring the values of amplitude given by the

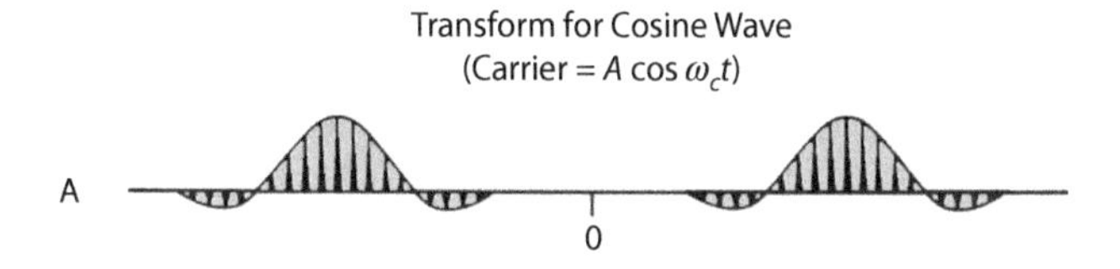

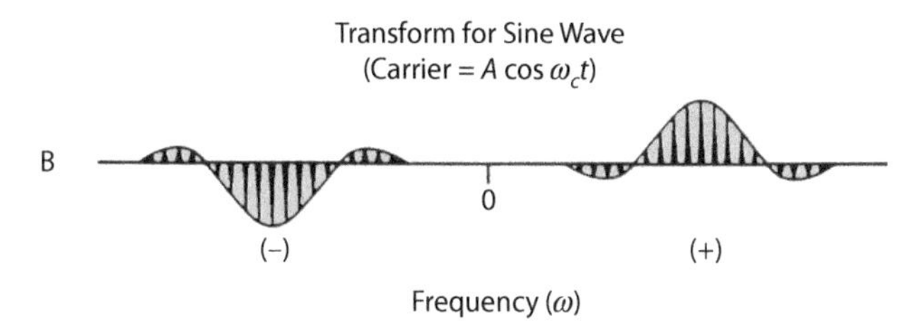

Figure 20-30. Shown is a comparison of frequency spectra for a signal whose carrier is a cosine wave and a signal whose carrier is a sine wave. Note that the algebraic signs of the negative frequency terms are reversed for the sine wave.

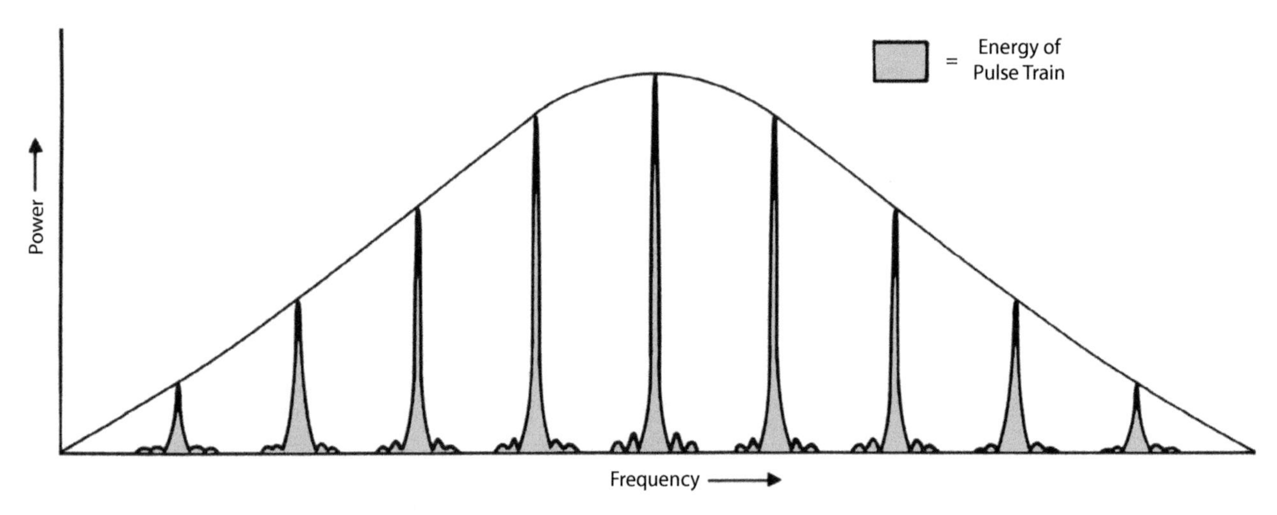

Figure 20-31. If the amplitude represented by a signal's Fourier transform is a voltage, then a plot of the amplitude squared versus frequency is the signal's power spectrum, and the area under this plot corresponds to the signal's energy.

Fourier transform, the *power* spectrum of the pulsed signal can be obtained (Fig. 20-31).

Energy, of course, is power times time (see chapter 6). It can be shown mathematically[7] that the total area under the power spectrum equals the total *energy* of the pulsed signal. The power spectrum thus illustrates how the energy of the signal is distributed in frequency. For example, by measuring the area encompassed by the central line of the power spectrum and dividing it by the total area encompassed by the spectrum, we can tell what fraction of the signal's energy is contained in that line.

7. It is Parseval's theorem that shows the total area under the power spectrum equals the total energy of a pulse signal.

20.4 Summary

The spectrum of a signal is the distribution of the signal's energy over the range of possible frequencies. One way of explaining this distribution is to envision the signal as being applied simultaneously to myriad lossless narrowband filters whose frequencies are infinitesimally closely spaced and cover the complete range of frequencies. Each filter can be envisioned as a pendulum suspended from a frictionless pivot in a vacuum and driven by the reactive force of an eccentric flywheel rotating at the frequency of the input signal.

A pulsed radio frequency (RF) signal, such as that transmitted by a radar system, is actually a continuous wave (carrier) whose amplitude is modulated by to a value of one during each pulse and zero between pulses. Another way of explaining how the energy of a pulse is distributed in frequency is in terms of the sidebands produced by the video modulating signal.

A continuous train rectangular pulses can be constructed by adding together a series of sine waves of appropriate amplitudes and phases, whose frequencies are multiples of the wave's repetition frequency, plus a DC signal of appropriate amplitude (the Fourier series). When the amplitude of the carrier wave is modulated by the pulsed wave, each of these sine waves produces sidebands on either side of the carrier frequency.

The spectrum of a single pulse may be found by starting with a pulse modulated wave that is endlessly repetitive and decreasing the repetition frequency to zero (i.e., the time between pulses is infinite). The spectrum of a pulse train of limited length can then be found by treating each of the sine waves comprising the pulse modulated wave as a single pulse that is the length of the train.

The spectrum of a pulsed carrier may also be explained in terms of the progressive phase shift of the carrier relative to the frequency to which a narrowband filter is tuned. For a single pulse, nulls in the filter output occur at frequencies for which the phase shift over the length of the pulse is 360° or a multiple thereof.

Further Reading

R. N. Bracewell, *The Fourier Transform and Its Applications*, 3rd ed., McGraw-Hill, 1999.

V. N. Bringi and V. Chandrasekar, *Polarimetric Doppler Weather Radar Principles and Applications*, Cambridge University Press, 2007.

D. C. Schleher, *MTI and Pulse Doppler Radar with MATLAB®*, Artech House, 2009.

Test your understanding

1. What is a "pulsed spectrum"?
2. What is the relationship between frequency and wavelength?
3. What is the wavelength of an S-band (3 GHz) radar pulse?
4. How does the spectrum of a chain of pulses differ from that of a single pulse?
5. What is the relationship between pulse length and spectral width?
6. What is the spectral width of a pulse of duration 2 ns?

McDonnell Douglas F-15 Eagle (1976)

The Eagle is a twin-engined air superiority fighter initially manufactured by McDonnell Douglas (now Boeing). While it has been criticized as too large to be a dedicated dogfighter, it is among the most successful modern fighters with over 100 aerial combat victories and no losses. Starting in 2007, nearly 200 US Air Force F-15Cs were retrofitted with the AN/APG-63(V)3 active electronically scanned array radar and another upgrade to infrared search and track (IRST) system is currently in development.

21

Doppler Sensing and Digital Filtering

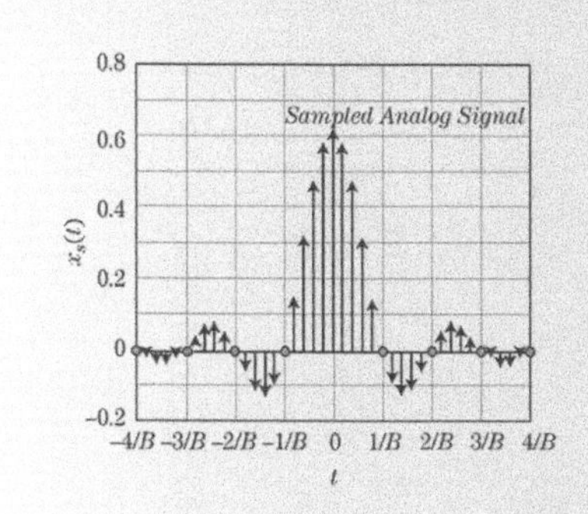

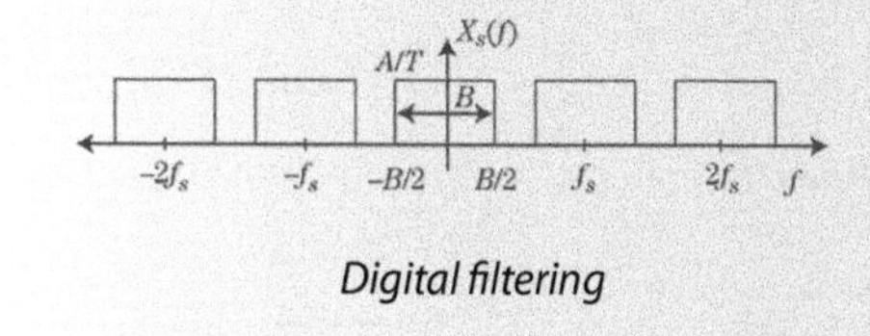

Digital filtering

The three main reasons for sensing Doppler frequencies are (1) to separate or resolve returns received simultaneously from different objects; (2) to determine range rates; and (3) to separate or resolve multiple frequencies from a single object (e.g., separating the rotation of an aircraft engines turbine blades from the bulk velocity of the fuselage).

This chapter describes sensing Doppler frequencies and detecting differences between them using a bank of Doppler filters. The key filtering methods are examined, concentrating most attention on digital filters. Finally, some of the more practical aspects of implementing Doppler filters are reviewed.

21.1 Doppler Filter Bank

How can a radar detect echoes from many different sources simultaneously and then sort them out on the basis of differences in Doppler frequency? Conceptually, it is quite simple. The received signals are applied to a bank of filters, commonly referred to as Doppler filters (Fig. 21-1).

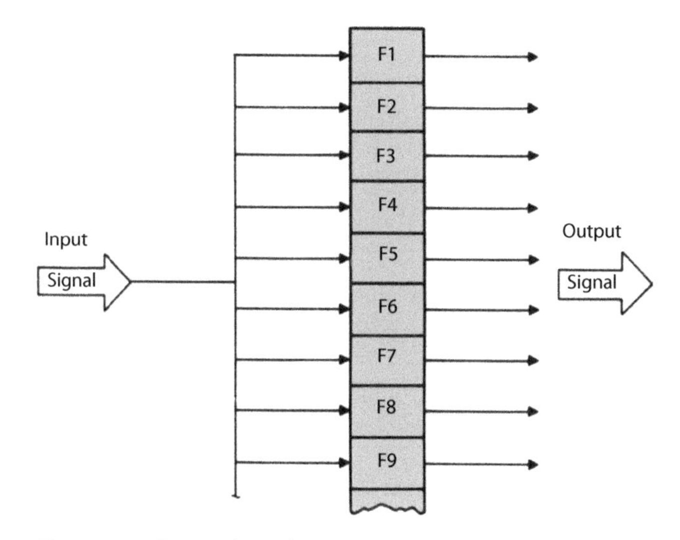

Figure 21-1. Received signals are separated in frequency in a parallel bank of filters.

Each filter is designed to pass a narrow band of frequencies (Fig. 21-2). Ideally each filter produces an output only if the frequency of a received signal falls within its band. Actually, because of filter sidelobes, it may produce some output for signals whose carrier frequencies lie outside the band. If the return is to be sorted by range as well as Doppler frequency, a separate filter bank is provided for each range increment (range cell or bin).

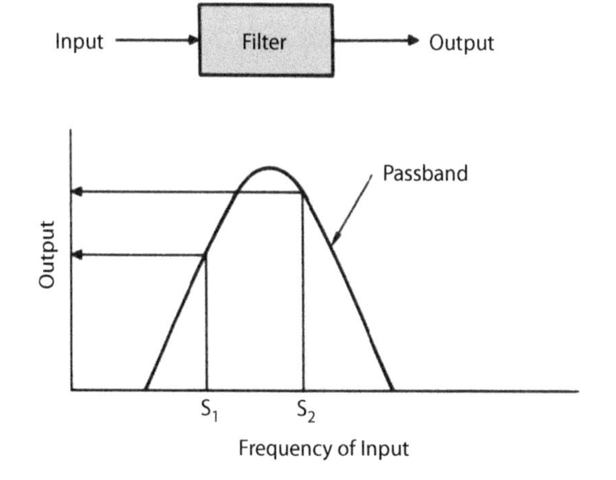

Figure 21-2. Each filter passes only a narrow band of frequencies (neglecting sidelobes). The closer a signal is to the center frequency, the greater the output.

Moving up the bank from the lower end, each filter is tuned to a progressively higher frequency. To minimize the loss in signal-to-noise ratio occurring when adjacent filters straddle a target's frequency, the center frequencies of the filters are

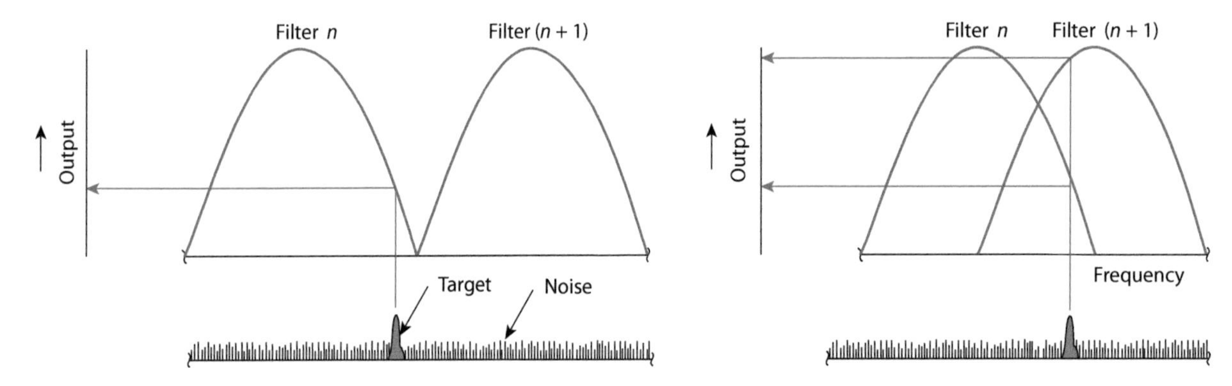

Figure 21-3. To minimize the loss of output when a signal lies between the center frequencies of two filters, the passbands overlap.

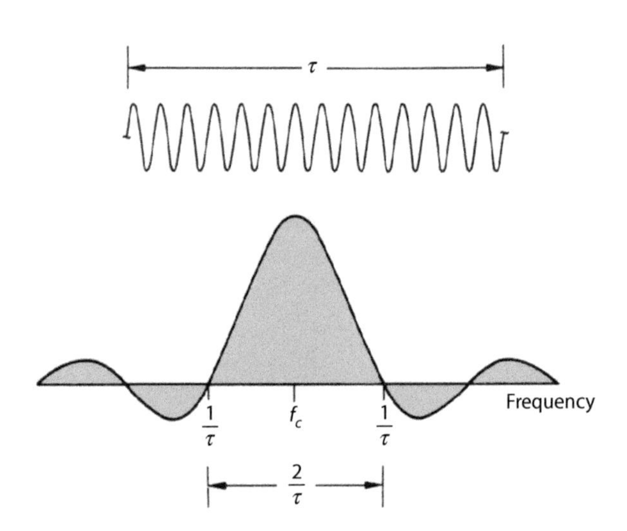

Figure 21-4. Shown here is the spectrum of sinusoidal signal of duration, τ.

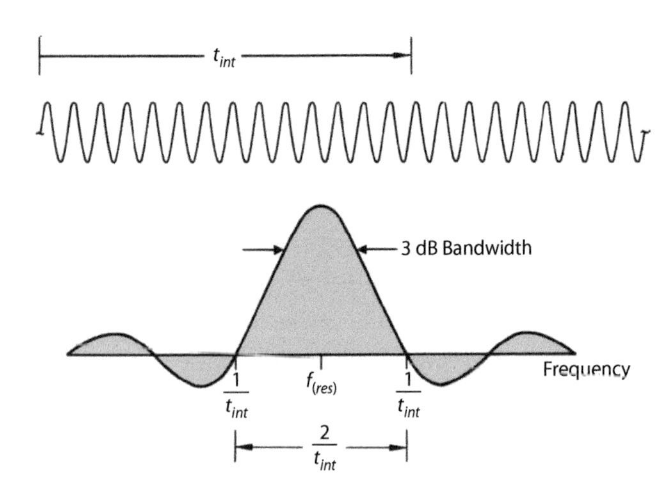

Figure 21-5. This is an output characterstic of a narrowband filter to which a signal at least as long as the filter integration, t_{int}, is applied.

spaced so that the passbands overlap (Fig. 21-3). Thus, if a target's Doppler frequency gradually increases, an output is produced, first, primarily from one filter; next, more or less equally from that filter and the next filter up the line; then, primarily from the second filter, and so on.

Bandwidth of the Filters. As we saw in the previous chapter, a narrowband filter achieves its selectivity by integrating the signals applied to it over a period of time. The width of the band of frequencies passed by the filter depends primarily on the length of the integration time, t_{int}.

The previous chapter also discussed how the spectrum of a sinusoidal signal of duration, τ, i.e., a (single pulse), has a sin x/x shape such as that shown in Figure 21-4. Each point on this plot corresponds to the output the signal would produce from a narrowband filter that integrates the signal throughout its entire duration. The plot was obtained using a method that progressively tunes the filter to each of a great many different frequencies.

We can find the relationship between the filter's bandwidth and t_{int} simply by using the aforementioned method in a slightly modified form, as follows:

- Hold the tuning of the filter constant and progressively change the frequency of the applied signal.
- Limit the filter's integration time, t_{int}, and make the signal at least as long as t_{int}.

Now, instead of representing the spectrum of the applied signal, the plot represents the output characteristic of the narrowband filter.

The central lobe of this characteristic is the filter's passband, and the center frequency of the central lobe is the filters resonant frequency, $f_{(res)}$ frequency. Since t_{int} in this last case corresponds directly to τ in the earlier one, the filter's null-to-null bandwidth is $2/t_{int}$ (Fig. 21-5). For ease of comparison, the horizontal scale factor used in this figure was adjusted to make the positions of the nulls the same as in Figure 21-4. Bear in mind, though, that integration times are generally on the order of milliseconds, whereas pulse widths are typically a thousand times less, on the order of microseconds.

As with the mainlobe of an antenna radiation pattern, a more useful measure of filter bandwidth than the null-to-null width is the width of the central lobe at the points where the power of the output is reduced to half its maximum value, the 3 dB bandwidth. Similar to a uniformly illuminated antenna, that width is approximately half the null-to-null width.

$$BW_{3\,\text{dB}} \cong \frac{1}{t_{int}}$$

To realize this bandwidth, the duration of the applied signal must at least equal t_{int}. In fact, the filter bandwidth is usually selected on the basis of the maximum available integration time.

If the radar is pulsed, the number of pulse that must be integrated to achieve a given bandwidth is equal to t_{int} times the pulse repetition frequency (PRF). A useful rule of thumb derived from this relationship is that the 3 dB bandwidth of a filter equals the PRF divided by the number of pulses integrated.

The bandwidth given by the previous equation is the *minimum* achievable bandwidth. Depending on application, a practical filter may have a substantially broader passband as a result of losses or, in digital designs, deliberately introduced *weighting* (see Section 21.5).

Filter Bank Passband. Enough filters must be included in the bank to bracket the anticipated range of Doppler frequencies. This is to cover the range of expected target velocities (remembering that radars only measure radial velocities). For example, if the maximum anticipated positive Doppler frequency is 100 kHz and the maximum anticipated negative Doppler frequency is –30 kHz (Fig. 21-6), then the passband of the filter bank, f_r, would have to be at least 100 + 30 = 130 kHz wide to pass the return from all targets and the radar PRF would have to exceed 130 kHz as shown in Figure 21-6.

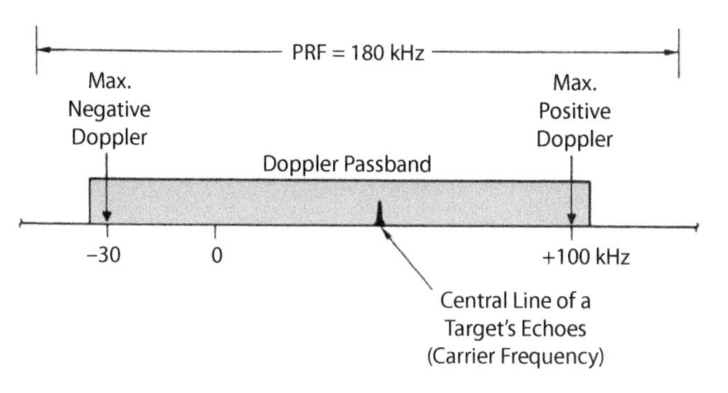

Figure 21-6. When the PRF is greater than the spread between the maximum positive and negative Doppler frequencies, the Doppler passband should be made wide enough to encompass these frequencies.

On the other hand, if the PRF is less than the anticipated spread in Doppler frequencies (as it often must be to reduce range ambiguities), the passband of the bank should be made no greater than the PRF. The reason is that the spectral lines of a pulsed signal occur at intervals equal to the PRF, and it is desirable that any one target appears at only one point in the filter bank's passband. This is another way of stating the Nyquist theorem: any signal must be sampled at least at twice its highest frequency to avoid ambiguities.

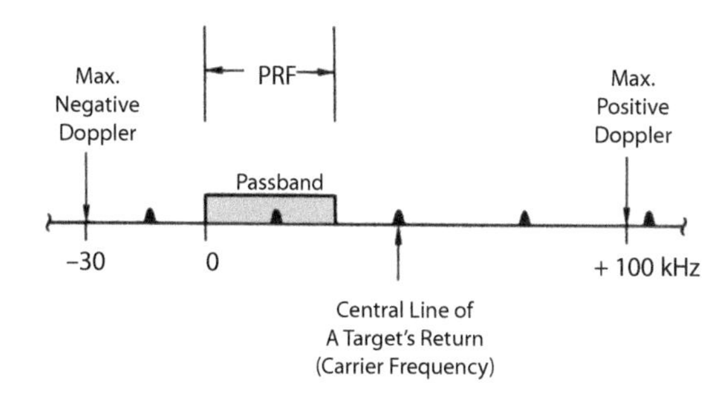

Figure 21-7. When the PRF is less than the spread of Doppler frequencies, the passband should be made no wider than the PRF so that a target will appear at only one point within the band.

Depending on the target's Doppler frequency, the spectral line falling within the passband in this case may not be the target's central one (carrier frequency). It may be one of the lines (e.g., sideband frequencies) above or below it (Fig. 21-7). But since the lines are harmonically related, which one it is doesn't matter. What is important for each target is that one and only one line falls within the passband.

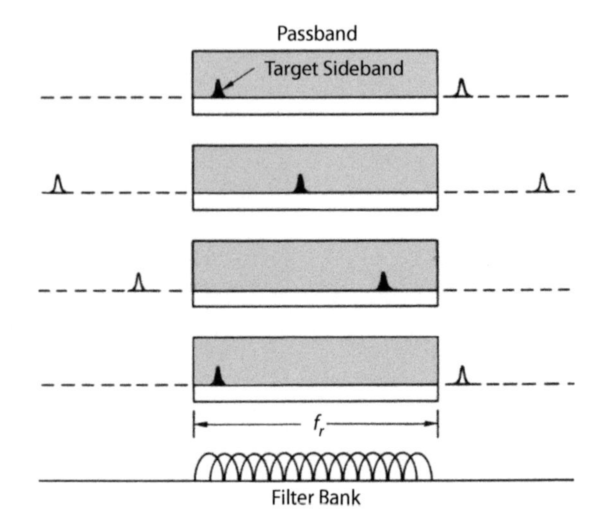

Figure 21-8. If the width of the filter bank's passband equals f_r or less, only one line of the target's spectrum will fall within it, regardless of the target's Doppler frequency.

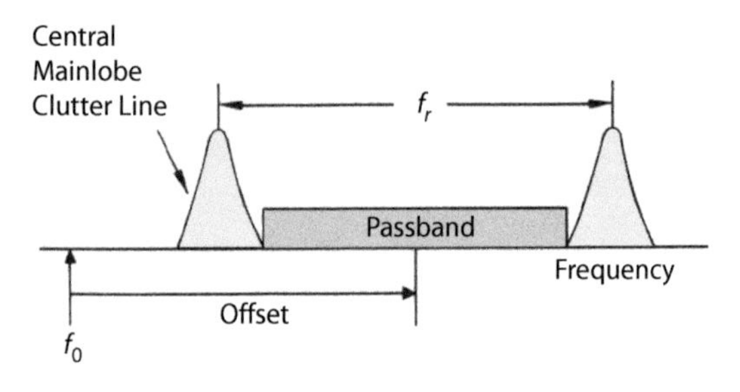

Figure 21-9. The passband may be offset from f_0 to avoid mainlobe clutter.

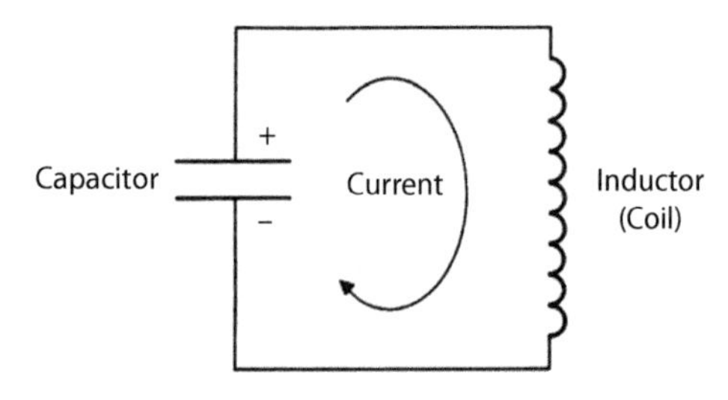

Figure 21-10. An analog filter is a tuned electrical circuit. In its simplest form it consists of a capacitor and an inductor.

Figure 21-8 illustrates how this requirement is satisfied when the width of the passband equals the PRF or f_r. It shows a portion of the spectrum of a target's echoes for each of several progressively higher Doppler frequencies, which all happen to be such that the target's central line (carrier frequency) lies outside the figure. Superimposed over the spectrum is a mask with a window in it, representing the passband of a filter bank f_r Hz wide.

In the first plot of Figure 21-8, one of the target's spectral lines falls in the lower end of the passband. With the progressive increase in Doppler frequency, in subsequent plots this same line appears farther and farther up in the band. In the last plot, the Doppler frequency is sufficiently high that the line is actually above the passband. However, the next lower frequency line now appears in the lower end of the passband.

It can similarly be shown that the target will always appear somewhere within the passband regardless of where we position it. Without causing any problems, therefore, the passband can be shifted up or down relative to the transmitter frequency, f_0. In low and medium PRF radars, for example, the passband is generally made somewhat less than f_r Hz wide and shifted up in frequency so it conveniently lies between the central and next higher lines of the ground return received through the antenna's mainlobe (Fig. 21-9). In this way the stationary clutter is separated or resolved from moving targets. (Actually, to simplify the design, the frequencies of the Doppler filters are not changed. Instead the spectrum of the radar return is shifted relative to the filter bank. The net result, however, is the same.)

Filters' Basic Function. Filters are actually sensitive to phase shifts rather frequency because a Doppler frequency is, in fact, a continuous phase shift.

In its simplest form an analog filter is a tuned electrical circuit consisting of a capacitor and an inductor (Fig. 21-10). If a charge is placed on the capacitor, a current surges back and forth between the plates of the capacitor through the inductor, alternately discharging the capacitor and charging it back up again with the opposite polarity. The number of these cycles completed per second is contingent on the capacitance of the capacitor, and the inductance of the inductor and is called the *resonant frequency of the circuit.*[1] The inductor and capacitor naturally have some losses (resistance). Consequently, the passband is invariably wider than $1/t_{int}$. The lower the losses, the closer the passband approaches this limit.

Here we concentrate on digital filtering, which is the norm in modern radar systems.

21.2 Digital Filtering

While an analog filter is implemented with circuit elements whose electrical characteristics are analogous to mathematical operations, a digital filter is implemented with the logic

1. The resonant frequency is $1/2\pi\sqrt{LC}$, where L is the inductance, and C is the capacitance.

of a digital computer, which performs these same operations numerically. Why perform the filtering this way?

There are several reasons, but accuracy, reliability, and flexibility are the main ones. Once the radar return has been accurately converted to digital numbers, all subsequent signal processing is essentially error-free. (There are quantization and round-off errors, but these can be kept within acceptable bounds through proper system design.)

Also, all results are repeatable, no adjustments are required, and performance doesn't degrade with the passage of time. Where a great many Doppler filters are required and a variety of operating modes is desired, the size and weight of the equipment needed to implement the radar can be substantially reduced through digital filtering. In fact, it is only through digital filtering that many of today's advanced multimode airborne radars are even feasible.

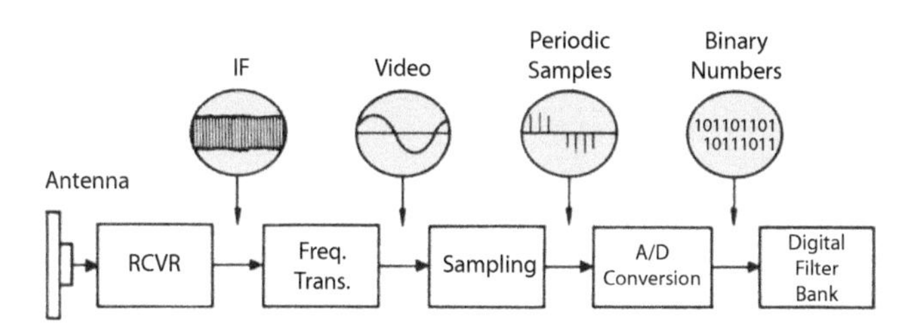

Figure 21-11. For digital filtering, the intermediate frequency output of the receiver must be translated to video frequencies, sampled, and converted to binary numbers.

Converting the radar return into digital form for processing requires some additional operations. Generally, a radar receiver's intermediate frequency is too high to make analog-to-digital conversion convenient, so at the outset (Fig. 21-11) the receiver output is translated downward to an offset frequency in the video frequency range that might be anywhere from zero (direct current, or DC) to several megahertz. Since this signal is continuously varying and the numbers into which it will be converted are discrete,[2] the signal must be sampled at regular short time intervals. The sampling rate is again determined by the Nyquist rate and is usually set by the modulation bandwidth of the transmitted signal (which must therefore be sampled at twice the highest modulation frequency). Finally, each sample must be converted to an equivalent binary digital number. The numbers are applied as inputs to the computer that forms the filters.

2. Discontinuous in time, that is, the value of each number is separate and distinct from that of the preceding number.

Translation to Video Frequencies. The radar receiver's intermediate frequency (IF) output signal is translated to video frequencies by comparing it with a reference signal whose frequency corresponds to the transmitter frequency, f_0, translated to the receiver's IF. In some cases, an offset is added to the reference frequency, but we will assume no offset here (i.e., the IF becomes zero Hz). It is worth noting, though, that as the performance of analog-to-digital (A/D) converters improves, it is becoming increasingly more common for digitization to take place at IF. In fact, for the lower radar bands this can occur at radio frequency (RF), which leads to the concept of all-digital radar. This improvement typically comes with little in the way of A/D cost penalties.

The relationship between the reference signal and the IF output produced by a target is illustrated for three representative situations by the phasor diagrams in Figure 21-12, in which the imaginary strobe light that illuminates the phasors is synchronized with the reference signal so that the phasor representing it remains fixed.

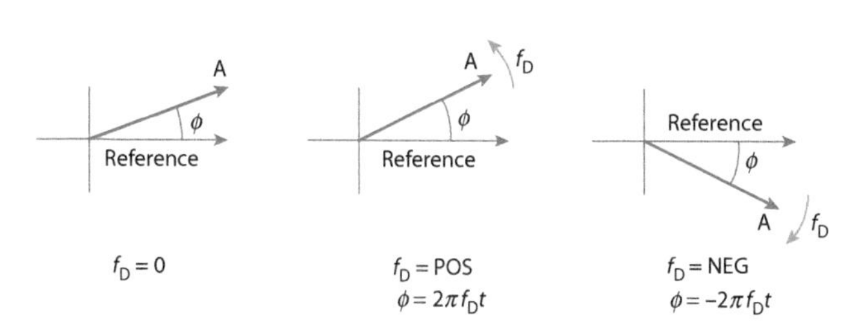

Figure 21-12. Three possible relationships between the reference signal supplied to the synchronous detector and the IF output, A, produced by the return from a target.

How the Synchronous Detector Works

Basic Function. The synchronous detector compares a Doppler-shifted input signal with an unshifted reference signal and produces an output whose amplitude is proportional to the amplitude, A, of the input signal times the cosine of the phase, ϕ, of the input signal relative to the reference signal.

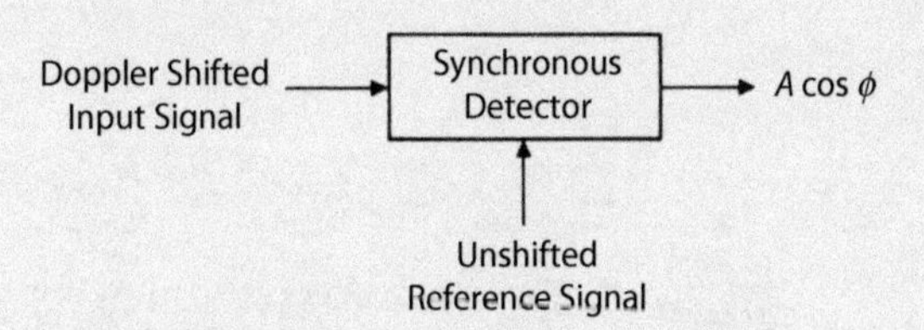

For purposes of explanation, it is assumed here that the reference signal has an amplitude of k and a frequency of ω_0 rad/s.

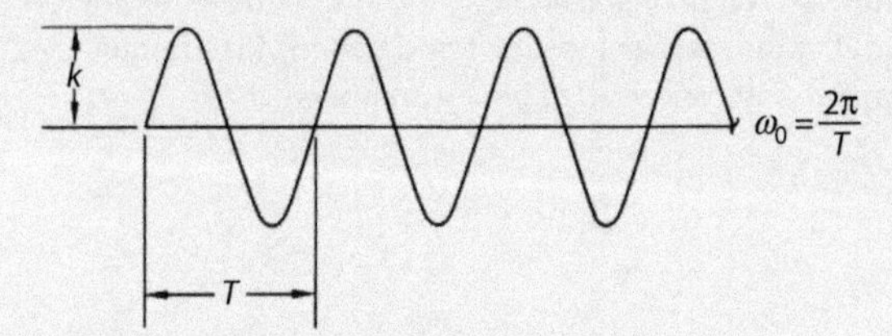

Reference Signal = $k \sin \omega_0 t$.

Since a Doppler frequency shift is a continuous phase shift, at any instant the Doppler-shifted input signal can be thought of as having a frequency equal to the reference frequency, ω_0, but being shifted in phase relative to the reference signal by ϕ radians.

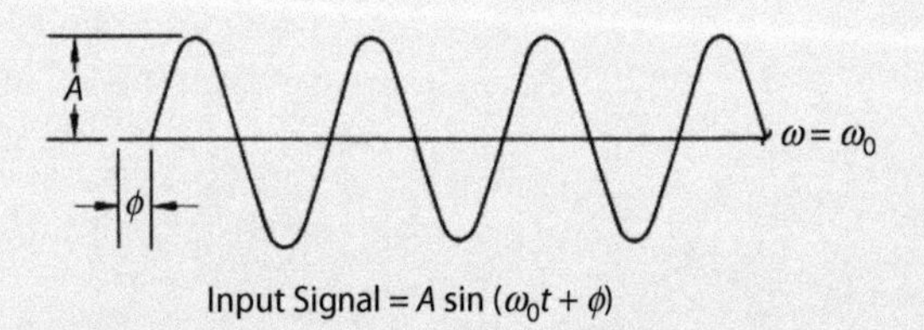

Input Signal = $A \sin(\omega_0 t + \phi)$

Detector's Function. In essence, the detector does two things: (1) multiplies the instantaneous value of the input signal by the instantaneous value of the reference signal; and (2) applies the resulting signal to a low-pass filter.

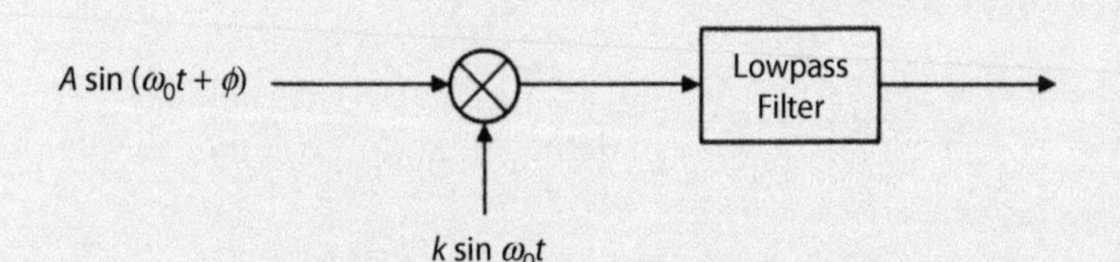

The filter's passband is wide enough to pass the highest Doppler frequency that may be encountered but narrow enough to reject completely any signal whose frequency is as high as or higher than ω_0.

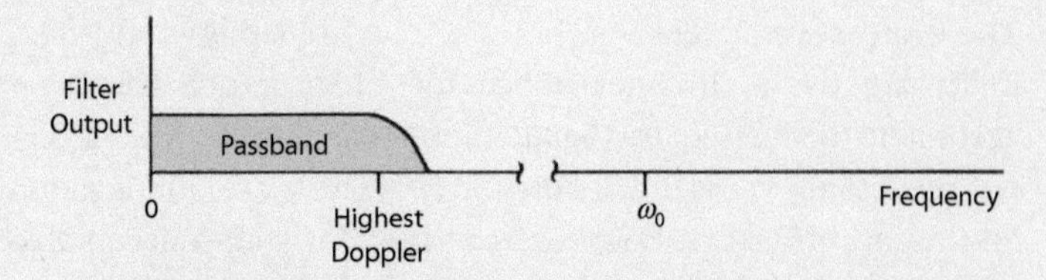

Multiplication Products. By means of a simple trigonometric identity, the input signal can be shown to consist of two components:

$$A \sin(\omega_0 t + \phi) = \underbrace{A(\sin\phi)(\cos\omega_0 t)}_{(1)} + \underbrace{A(\cos\phi)(\sin\omega_0 t)}_{(2)}$$

When term 1 is multiplied by the expression for the reference signal ($k \sin \omega_0 t$), the product is

$$= kA(\sin\phi)(\cos\omega_0 t)(\sin\omega_0 t)$$
$$= kA(\sin\phi)(\sin 2\omega_0 t)$$

Because of its high frequency, $2\omega_0$, the signal represented by this product is rejected by the low-pass filter.

However, when term 2 is multiplied by $k \sin \omega_0 t$, the product expands mathematically into two terms:

$$= kA(\cos\phi)(\sin^2\omega_0 t)$$
$$= kA(\cos\phi)\left[\frac{1}{2} + \frac{1}{2}\cos 2\omega_0 t\right]$$

Because of its high frequency, $2\omega_0$, the signal represented by the second of these terms is also rejected by the low-pass filter. The sole output of the filter, then, is

$$= \frac{kA}{2}(\cos\phi)$$

If k is taken as being equal to two, then

$$V_{output} = A \cos\phi$$

where A is proportional to the amplitude of the input signal, and ϕ is the signal's phase relative to the reference signal. Since this is a cosine function, it is called the in-phase (I) output.

Reference Shifted 90°. If we shift the phase of the reference signal—that is, insert a delay that makes the signal applied to the detector equal $k \sin(\omega_0 t - 90°)$—the same input signal will produce an output equal to $A \cos(\phi - 90°)$. Since the cosine of any angle minus 90° equals the sine of the angle, the output voltage is proportional to the sine of ϕ.

$$V_{output} = A \sin\phi$$

Again, A is proportional to the amplitude of the input signal, and ϕ is the signal's phase relative to the unshifted reference. Since this is a sine function, it is called the quadrature (Q) output.

In the first diagram, the frequency of the target signal equals f_0, or no Doppler shift. Consequently, the phasor representing the target return also remains fixed. The angle, ϕ, corresponds to the phase of the target signal relative to the reference signal.

In the second diagram, the target has a positive Doppler frequency. The target phasor therefore rotates counterclockwise, with ϕ increasing at a rate proportional to the Doppler frequency, f_D.

$$\dot{\phi} = 2\pi f_D \text{ radians per second}$$

In the third diagram, the target's Doppler frequency is negative, so the target's phasor rotates clockwise. Again, the phase angle, ϕ, changes at a rate proportional to the Doppler frequency.

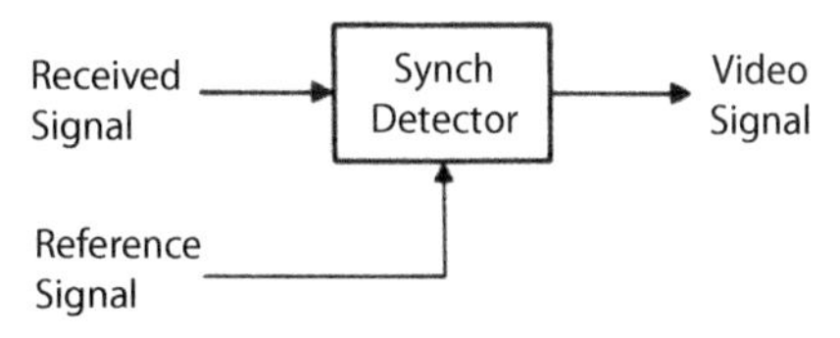

Figure 21-13. For digital filtering, a synchronous detector translates the received signal to the video frequency range.

The IF output signal is then compared with the reference signal by a circuit called a *synchronous detector* (Fig. 21-13). It produces an output voltage proportional to the amplitude of the received signal times the cosine of the phase angle, ϕ, relative to the reference signal.

$$V_{output} = A \cos \phi$$

where A is proportional to the amplitude of the received signal, and ϕ is its phase (Fig. 21-14).

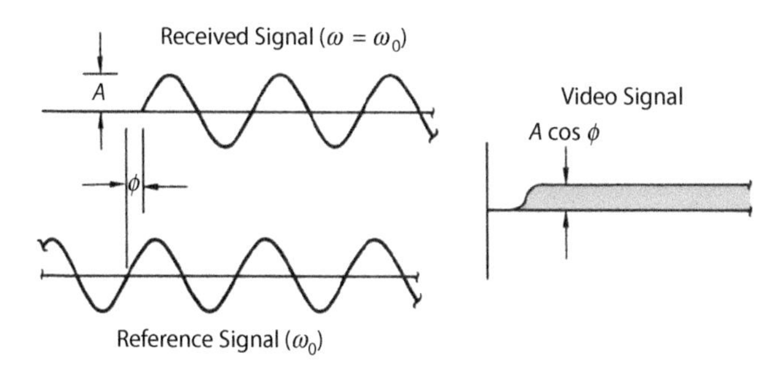

Figure 21-14. Amplitude of output pulse is proportional to cosine of received pulse's phase relative to the reference signal.

The detector's output can be conveniently visualized as the projection of the phasor representation of the received signal on the x axis (Fig. 21-15). If the target's Doppler frequency is zero, the output voltage, x, will be constant.

Its exact value may lie anywhere between zero and A, depending on the signal's phase. If the target's Doppler frequency is not zero, the output, x, will be a cosine wave with amplitude A and a frequency equal to the target's Doppler frequency.

If the radar is pulsed, unless the duty factor is very high the output pulse produced by each target echo will represent only a fraction of a cycle of the target's apparent Doppler frequency.

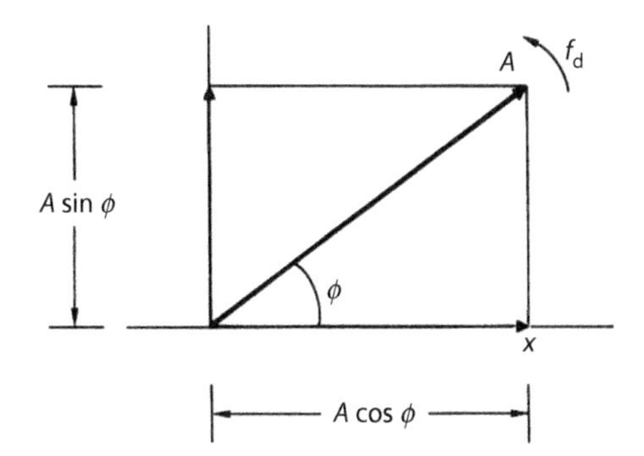

Figure 21-15. The output that a received signal produces from a single synchronous detector may be visualized as the projection of the phasor representation of the received signal on the x axis.

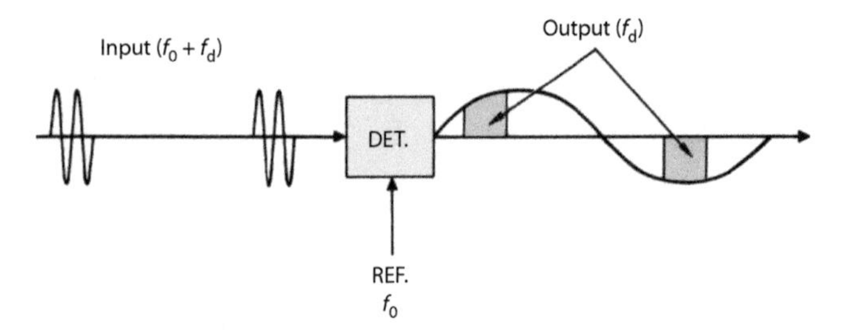

Figure 21-16. The output of the synchronous detector for a pulsed input signal having a duty factor of 25 percent and an apparent Doppler frequency equal to half the PRF.

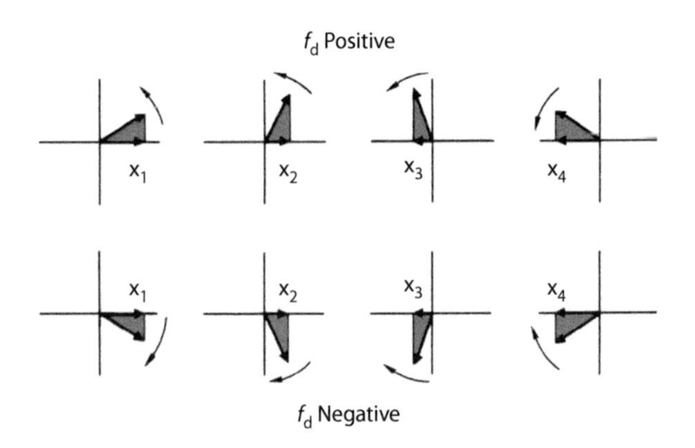

Figure 21-17. The detector output for successive echoes. The in-phase or quadrature component alone is the same for both positive and negative Doppler frequencies.

3. To avoid imbalances, a single detector and A/D converter may instead be used for I and Q channels. The sampling rate must then be doubled.

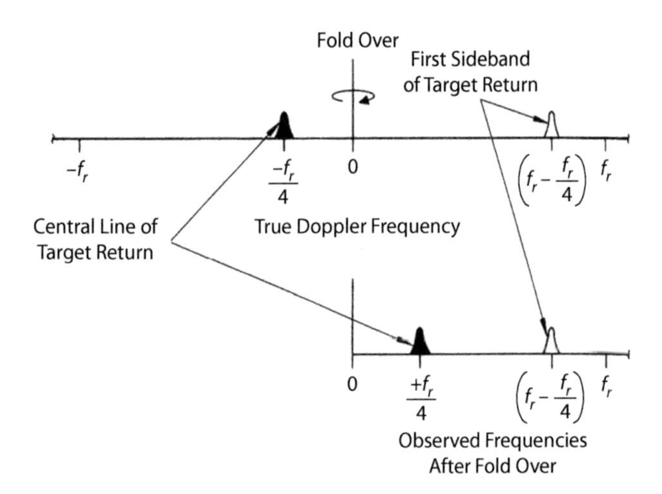

Figure 21-18. If only one component of the return is processed, the negative portion of the Doppler spectrum will be folded over onto the positive portion.

Nevertheless, by observing successive pulses the amplitude of the target return is slowly revealed and Doppler frequency can be determined (Fig. 21-16).

However, since x varies cyclically as the phasor rotates, on average half the received energy (the component $A \sin \phi$ in Fig. 21-15) is thrown away. Also, in some applications where the time-on-target is short compared with the period of the Doppler frequency, the echoes may all be received when $\cos \phi$ is so small that they cannot be detected.

More importantly, it won't be possible to tell in which direction the phasor is rotating. For a given rate of rotation, the projections of the phasor on the x axis are the same, whether the phasor rotates clockwise or counterclockwise (Fig. 21-17). On the basis of these projections alone, there is no way of telling whether the target's Doppler frequency is positive or negative (i.e., whether a target is approaching or receding from the radar). Indeed, in simple moving target indication (MTI) radars, which process only one component of the return, all Doppler frequencies are indicated as positive. The negative half of the Doppler spectrum is said to be *folded over* onto the positive half. Thus, a target whose Doppler frequency is $-0.25f_r$ will also appear to have a Doppler frequency of $0.25f_r$ (Fig. 21-18).

These limitations can be eliminated by simultaneously applying the IF output to a second synchronous detector to which the same reference signal is applied but with a 90° phase lag.[3] Since the cosine of (ϕ – 90°) equals the sine of ϕ, the output voltage of this detector is proportional to the amplitude of the target return times the sine of the phase angle, ϕ, relative to the unshifted reference. This is known as *I/Q detection*, enabling both amplitude and phase to be measured.

$$V_{output\ 2} = A \sin \phi$$

It is convenient to visualize I/Q detection as the projection of the phasor representation of the target return onto the x and y axes, as in Figure 21-19. If the Q detector's output, y, lags behind the I detector's output, x, the phasor is rotating counterclockwise and the target's Doppler frequency is positive. On the other hand, if y leads x, the phasor is rotating clockwise and the target's Doppler frequency is negative.

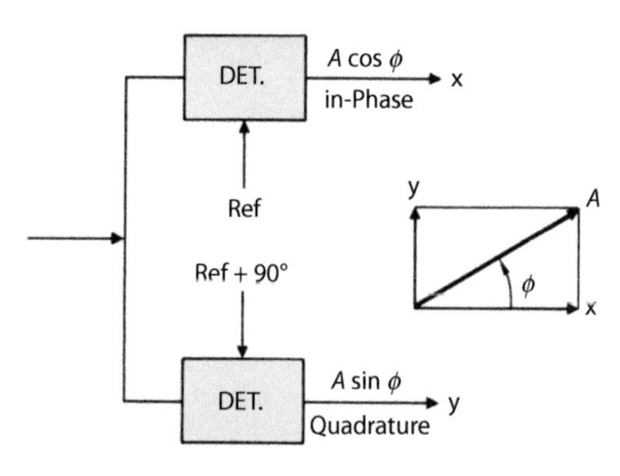

Figure 21-19. In a two-channel detector system, the reference frequency for the quadrature channel has a 90° phase lag.

Together, the I and Q projections describe the phasor completely. Their vector sum equals the length of the phasor, A.

Their ratio, together with their algebraic signs, unambiguously indicates the phase angle, ϕ, including both the rate and the direction of the phasors rotation.

Sampling the Video Signals. To convert the continuously varying outputs of the I and Q detector to digital numbers, they must be sampled at the Nyquist rate. Because the outputs may vary rapidly, the sampling rate must be precisely controlled to avoid introducing errors. Depending on the design of the radar, the rate may fluctuate anywhere from a few hundred thousand to over one billion samples per second. This variation is largely determined by the modulation applied to the transmitted signal to enable simultaneous high range resolution and long-range detection (see Chapter 16). Typically, from a single transmitted pulse, the radar will form a number of adjoining samples (*range gates* or *range cells*) in the form of a *range profile*. The formation of the range profile is repeated over successive pulses for a chosen integration time, t_{int}, creating a time series of adjoining resolution or range cells. More details can be found in Chapter 14. An example of a successive series of range profiles is shown in Figure 21-20.

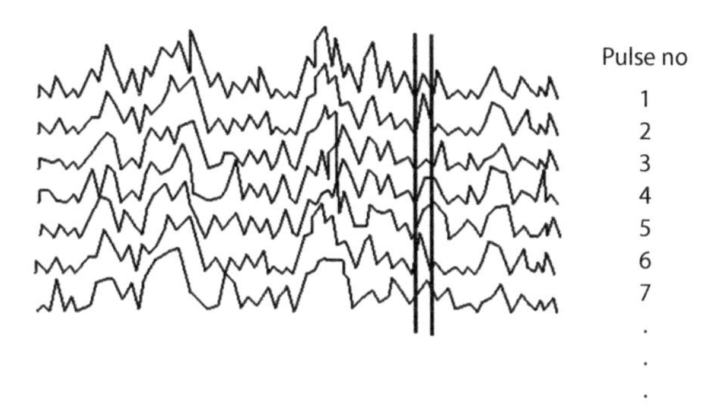

Figure 21-20. In this time series of range profiles, each pulse forms a range profile consisting of a number of range gates or cells. The two vertical bars show the time history echo values of a single range gate.

Forming the Filter. During each successive integration time, t_{int}, the radar processor or computer forms a bank of Doppler filters for every range gate in the range profile.

Each filter in the bank for a given range gate receives as inputs the same set of complex numbers (x_n, y_n) from the A/D converter (Fig. 21-21). If an echo is being received from a target, each pair of numbers are the I and Q components of one sample of a signal. From these the amplitude and phase can be determined, where the former corresponds to the power of the target return and the frequency (derived from the latter) is the target's Doppler frequency. The job of the filter is to integrate these numbers in such that if the Doppler frequency is the same as the filter's frequency, the sum will be large; otherwise, it is small.

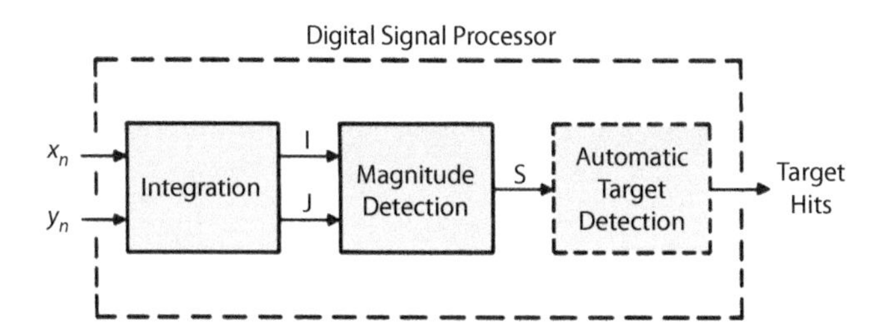

Figure 21-21. In these functions performed by a digital filter, during each successive integration time, t_{int}, S is the magnitude of the echo signal.

21.3 Inputs to the Filter

It is helpful to have a tangible idea of what is represented by the samples that form the inputs to the filter. This can be done quite easily by viewing the output of one of a pulsed radar's two synchronous detectors on a range trace.

Detector Output Displayed on a Range Trace. Suppose that the output of the I detector is supplied to the vertical deflection circuit of an oscilloscope on which is displayed a horizontal range trace. It has already been established that the I output equals A cos ϕ, where A is the amplitude, and ϕ is the RF phase of the target return relative to the reference signal supplied to the detector. The radar's PRF is, say, 8 kHz. Echoes are being received from four targets. Their Doppler frequencies are 0, 1, 6, and 8 kHz. So that the echoes of each target can be isolated on the range trace, the targets have been positioned at progressively greater ranges (Fig. 21-22). To isolate the effects of the

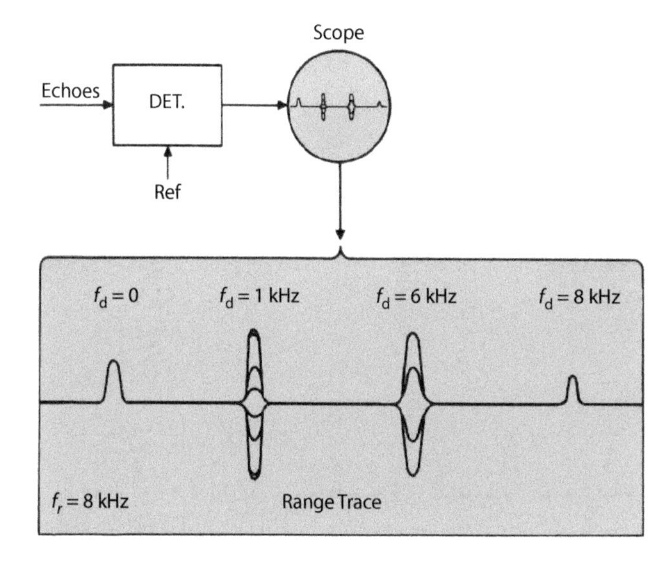

Figure 21-22. In this output of a single-channel synchronous detector displayed on a range trace, echoes of equal amplitude are being received from four targets all having different Doppler frequencies.

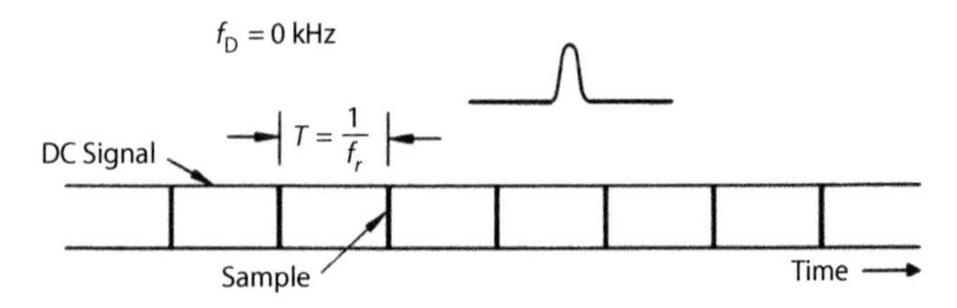

Figure 21-23. This detector output for a target with zero Doppler frequency has a constant amplitude.

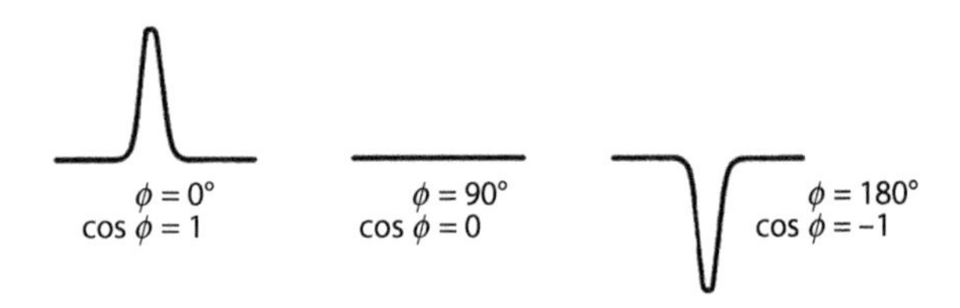

Figure 21-24. Depending on the echo's radio frequency phase, the amplitude of the detector output for a zero Doppler frequency may be anywhere between +1 and –1 times the echo's amplitude.

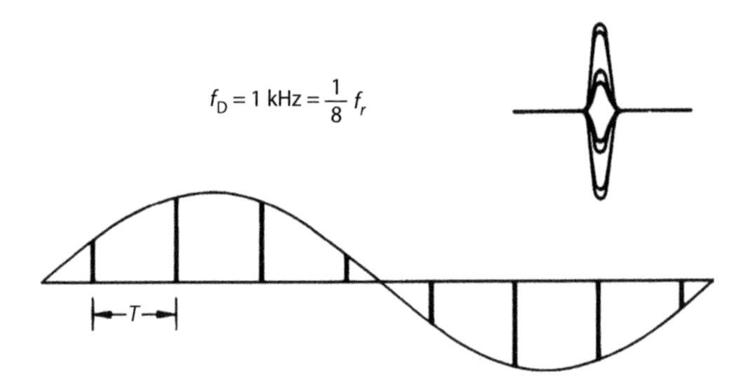

Figure 21-25. The detector output varies sinusoidally from pulse to pulse.

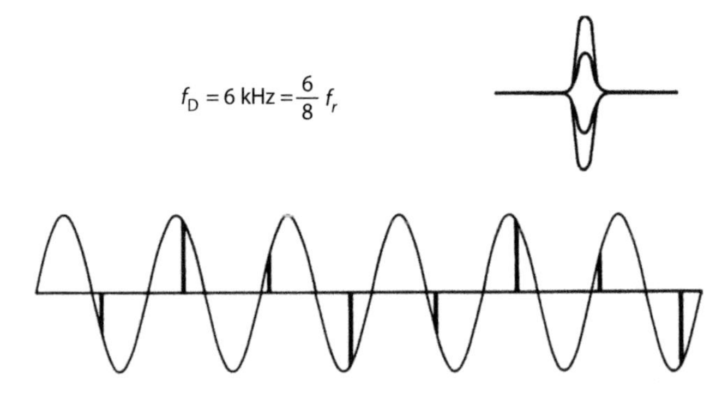

Figure 21-26. The detector output alternates between positive and negative values.

frequency differences, it is assumed that the amplitudes of the received echoes are all the same.

Despite this similarity, the height among the *pips* produced by the four targets is quite different. Some pips' height fluctuates because the height of a target pip drawn on the oscilloscope during any single range sweep (interpulse period) corresponds to the detector output produced by a single target echo. Since, in this case, the period of even the highest Doppler frequency is a great deal longer than the width of a radar pulse, each pip is essentially a sample taken at a single point in a cycle of the target's Doppler frequency.

For the target having zero Doppler frequency, the height of successive pips is constant (Fig. 21-23). Since the target echoes have the same frequency as the reference signal, their phase, relative to it, does not change from one echo to the next. The detector output for the range increment in which this target resides is a pulsed DC voltage. As previously explained, the amplitude of this voltage may lie anywhere between zero and $+A$ or $-A$ depending on the phase, ϕ, of the target echoes (Fig. 21-24).

One of the reasons for providing both I and Q channels is to eliminate this variability. Since the output of the I detector equals $A \cos \phi$ and the output of the Q detector equals $A \sin \phi$, the magnitude of the vector sum of the two outputs for all values of ϕ equals A.

From the standpoint of filtering, though, the important characteristic of I and Q samples when the Doppler shift is zero is that their individual amplitudes do not fluctuate because ϕ doesn't change.

Moving on to the target whose Doppler frequency is 1 kHz (Fig. 21-25), the amplitude of its pips fluctuates widely from pulse to pulse. Because the echoes do not have the same radio frequency as the reference signal, their phase relative to it changes from pulse to pulse. The amount of change is 360° times the ratio of the target's Doppler frequency to the PRF. In this case (Doppler frequency 1 kHz, PRF 8 kHz), the ratio is 1/8. The Doppler frequency wave is, in effect, being sampled at intervals of 360° × 1/8 = 45°. (The magnitude of the vector sum of the I and Q samples equals A, but the phase of the sum cycles through 360° at a rate equal to the Doppler frequency.)

For the target whose Doppler frequency is 6 kHz (Fig. 21-26), the story is the same. The only difference, in this case, is that the samples are taken at intervals of 360° × 6/8 = 270°.

As a result, not only does the amplitude of the detector's output fluctuate widely, but also, as the target return slides into and out of phase with the reference frequency, the samples alternate between positive and negative signs.

For the target whose Doppler frequency is 8 kHz, though, the pips once again have a constant amplitude (Fig. 21-27). The reason, of course, is that the Doppler frequency and the PRF are equal. The samples are all taken at the same point in the Doppler frequency cycle. There is, in fact, no way of telling whether the Doppler frequency is zero, or f_r, or some integer multiple of f_r.

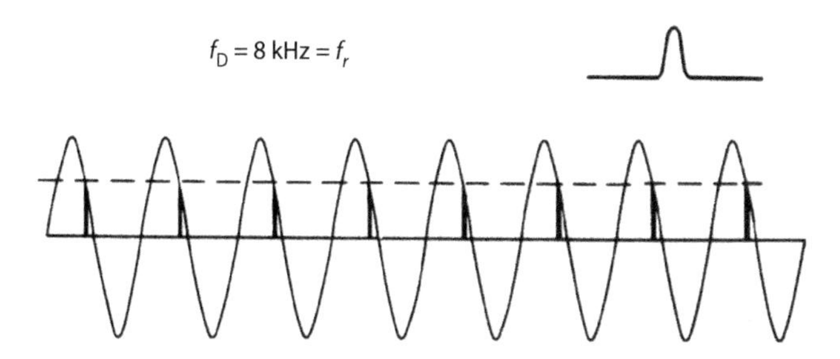

Figure 21-27. The detector output is similar to that for a target with zero Doppler frequency.

Similarly, if echoes are received from a target having a Doppler frequency of 9 kHz (Fig. 21-28), the pips it produces will fluctuate at exactly the same rate as those produced by the target having a Doppler frequency of 1 kHz. The observed frequency is ambiguous.

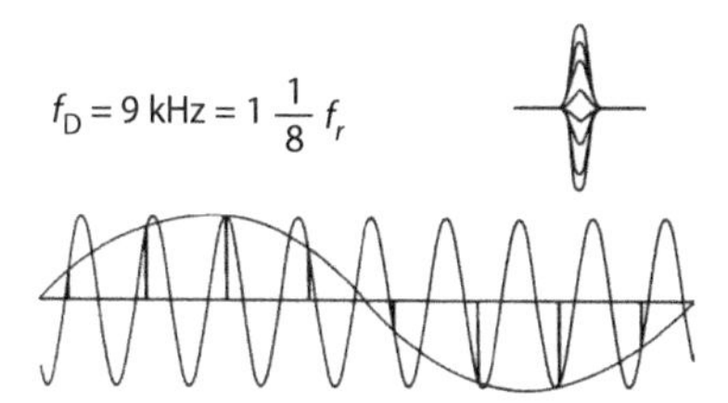

Figure 21-28. The detector output fluctuates at exactly the same rate as for a target whose Doppler frequency is 1 kHz (9 kHz – 8 kHz).

Phasor Representation of the Samples. The detector output at a point on a range trace corresponding to a particular target's range can be presented neatly in a phasor diagram showing the outputs of both I and Q detectors simultaneously (Fig. 21-29). The length of the phasor corresponds to the amplitude, A, of the target's echoes. The angle, ϕ, that the phasor makes with the X axis corresponds to the RF phase of the echoes relative to the reference signal. The length, x, of the projection of the phasor on the X axis corresponds to the output of the I detector; the length, y, of the projection on the Y axis corresponds to the output of the Q detector.

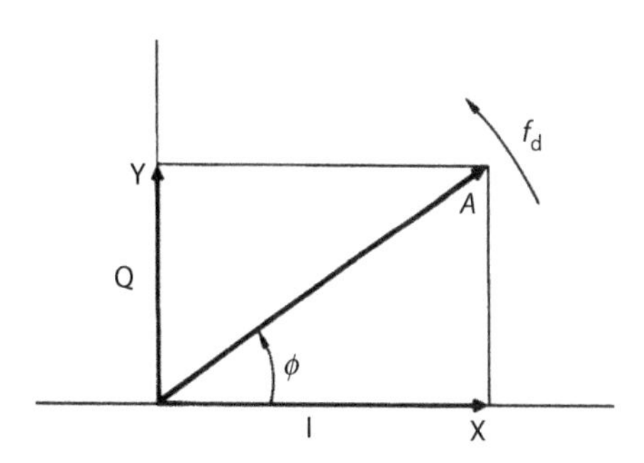

Figure 21-29. If the sine wave is represented by a phasor, A, the I component is the projection of the phasor on the X axis, and the Q component is the projection of the phasor on the Y axis.

The phasor rotates at the target's apparent Doppler frequency, that is, its true Doppler frequency or true Doppler frequency plus or minus an integer multiple of the sampling rate. If this frequency is positive (i.e., greater than the reference frequency) then rotation is counterclockwise (Fig. 21-30), and if it is negative then rotation is clockwise. The amount that the phasor steps ahead from sample to sample, $\Delta\phi$, is 2π radians (360°) times the Doppler frequency times the length of the sampling interval:

$$\Delta\phi = 2\pi f_D T_s$$

where f_D is the apparent Doppler frequency, and T_s is the sampling interval. If the sampling rate is the PRF (as it generally is in all-digital signal processors), T_s is the interpulse period, T.

21.4 What the Digital Filter Does

Digital filtering is simply a clever way of adding up or integrating successive samples of a continuous wave so that they produce an appreciable sum *only* if the wave's frequency lies within a given narrow band (i.e., a sum equivalent to the output that would be produced if the continuous wave were applied to a narrowband analog filter). If the variation in amplitude from sample to sample corresponds closely to the resonant frequency of the equivalent analog filter, the sum builds up; otherwise, it does not.

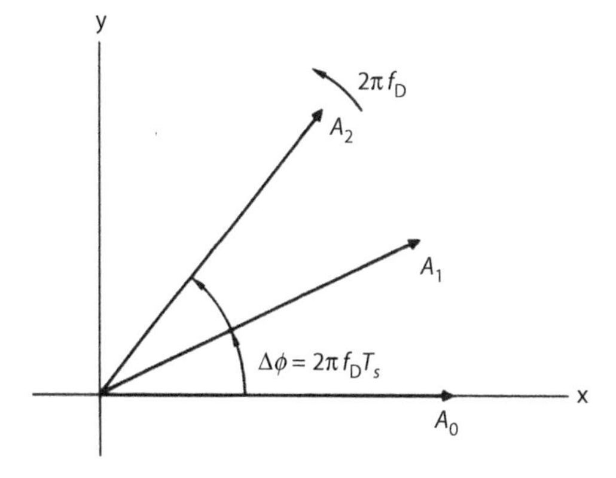

Figure 21-30. The amount that the phasor steps ahead from sample to sample, $\Delta\phi$, is proportional to the target's Doppler frequency.

What the filter does, in effect, is project the x and y components of the phasor representation of the samples onto a

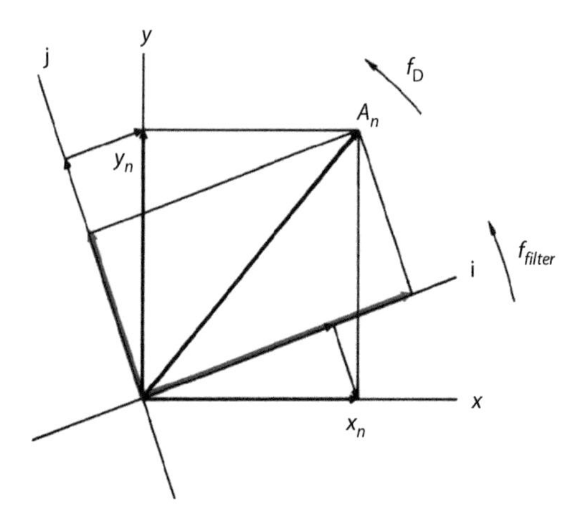

Figure 21-31. The filter projects the *x* and *y* components of the phasor, A_n, onto a coordinate system (i, j) that rotates at the frequency to which the filter is tuned.

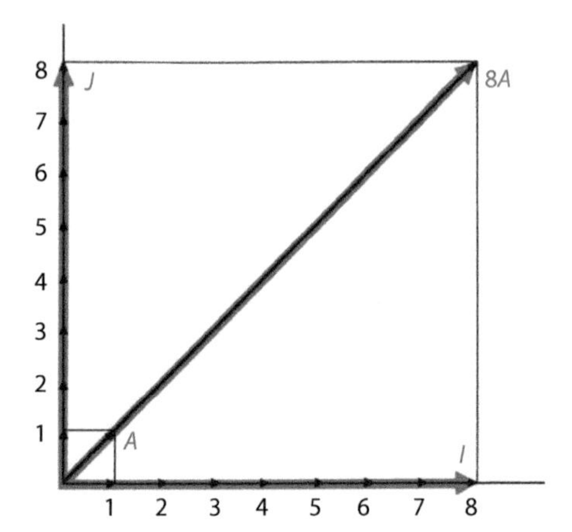

Figure 21-32. If the target's Doppler frequency and the filter frequency are the same, after eight pulses have been integrated, the magnitude of the vector sum of the phasor's projections on *I* and *J* will be eight times the phasor's amplitude.

rotating coordinate system (i, j; Fig. 21-31). The rate at which the coordinates rotate (number of revolutions per second) is made equal to the center frequency of the band the filter is intended to pass. This rate, f_f, can be thought of as the filter's resonant frequency, or the frequency to which it is tuned.

If the frequency of the sampled wave is the same as that of the filter, the angle between the phasor, *A*, and the rotating coordinate system will be the same for every sample (Fig. 21-32).

Consequently, after *N* samples have been received, the sum of the projections of the *x* and *y* components of *A* on the i axis will be *N* times as great as after a single sample has been received. The same will be true of the sum of the projections on the j axis.

On the other hand, if the frequencies of the sampled wave and the filter differ sufficiently, the angle between the phasor and the rotating coordinate system (i, j) will vary cyclically and the projections will tend to cancel.

At the end of the integration time, the sum of the projections on the i axis, *I*, is added vectorially to the sum of the projections on the j axis, *J*. The magnitude of the overall vector sum is the output of the filter. The quantities *I* and *J* are then dumped, and the integration is repeated for the next *N* samples.

We can visualize the process most easily if we imagine that we are riding on the rotating coordinate system (Fig. 21-33). We then see the true phasor rotation relative only to the i and j axes—that is, the rotation, $\dot{\Phi}$ due to the difference, Δf, between the frequency of the sampled wave and the frequency of the filter. As we just saw, if Δf is zero then the phasor will be in the same relative position each time a sample is taken, but if the frequencies are different then the phasor will be in progressively different positions (Fig. 21-34).

The phase difference, $\Delta\Phi$, between successive positions is directly proportional to the frequency difference. In radians

$$\Delta\Phi = (2\pi\ T_s)\ \Delta f$$

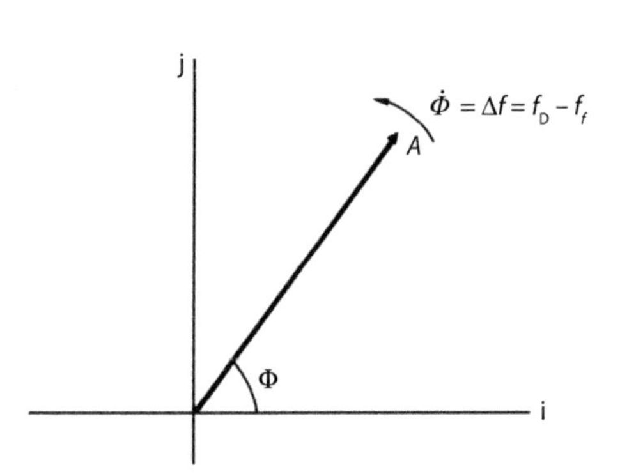

Figure 21-33. If we were to ride on the rotating coordinate system (i, j), we would see only the rotation, $\dot{\Phi}$ of the phasor, *A*, due to the difference between the frequencies of the sampled wave and the filter.

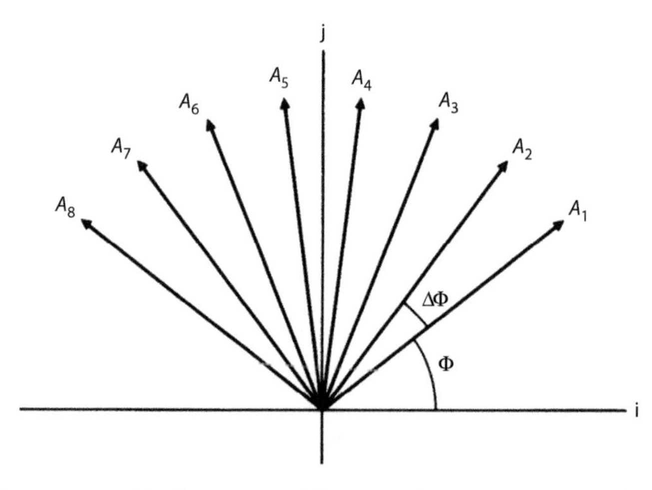

Figure 21-34. If the frequencies of the sampled wave and the filter differ, the phases of successive samples will differ by $\Delta\Phi$.

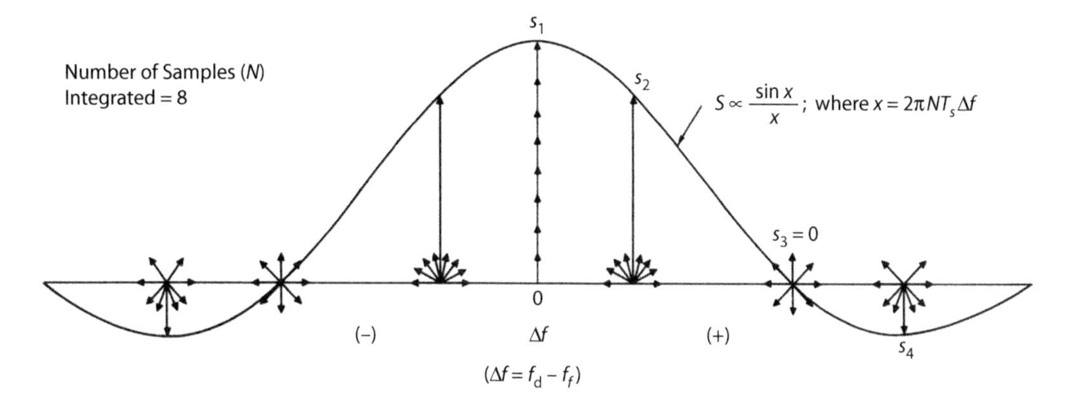

Figure 21-35. If the difference, Δf, between the frequencies of a sampled signal and the filter is increased, the phasors representing successive samples will fan out increasingly and the magnitude of their sum, S, will vary as $\sin x/x$.

where 2π is the number of radians in one revolution, T_s is the sampling interval, and Δf is the difference between the frequencies of the sampled wave and the filter.

As can be see from the phasor diagrams in Fig. 21-35, if the frequency difference Δf (hence also $\Delta\Phi$) is gradually increased, the phasor positions fan out increasingly, and the extent to which the samples cancel correspondingly increases.

A point is soon reached where the phasor positions are fanned out over a full 360°. The sum of the samples is then zero; a null in the filter characteristic has been reached. Beyond this null, the filter output goes through a succession of sidelobes.

For a given amplitude of the sampled signal, a plot of the amplitude of the filter output versus Δf has a sin x/x shape. The band of frequencies between the first pair of nulls is the filter's passband.

Since 360° is 2π radians, after N samples have been integrated the value of $\Delta\Phi$ for which the first nulls occur is 2π divided by N.

$$\Delta\Phi_N = \frac{2\pi}{N}$$

By substituting the expression we derived earlier for $\Delta\Phi N$ ($2\pi T_s \Delta f$), we can find the difference, Δf, between the frequencies of the wave and the filter at the nulls.

$$2\pi T_s \Delta f = \frac{2\pi}{N}$$

$$\Delta f = \frac{1}{NT_s}$$

The number of samples integrated, N, times the sampling interval, T_s, is the filter integration time, t_{int}. Therefore, the null-to-null bandwidth is

$$BW_{nn} = \frac{2}{NT_s} = \frac{2}{t_{\text{int}}}$$

The 3 dB bandwidth of a sin x/x curve is roughly half the null-to-null bandwidth, so

$$BW_{3\,\text{dB}} \cong \frac{1}{t_{\text{int}}}$$

Thus, the more samples integrated and the longer the sampling interval, the greater t_{int} and the narrower the passband.

As noted previously, if the samples received as inputs by the filter are due to the return from a target, the amplitude, A, of the phasor representation of the sampled signal will be proportional to the power of the target echoes. The filter output then will be proportional to the power of the echoes times the integration time and thus will be proportional to the total *energy* received from the target during t_{int}. The constant of proportionality will have its maximum value if the target's Doppler frequency is centered in the filter passband. The constant will be zero if the Doppler frequency is the same as one of the null frequencies. Otherwise, the constant will have some intermediate value determined by the filter's sin x/x output characteristic.

Processing Returns from Successive Ranges. If returns from more than one range gate are to be processed, instead of receiving an input of only one number in every interpulse period the computer receives a continuous stream of numbers (e.g., corresponding to all range gates in a range profile). During the first interpulse period, the computer successively multiplies the number, x_1, for each range gate by cos $\Delta\theta$, storing the individual products in separate registers. During the next interpulse period, the computer multiplies the number, x_2, for each range gate by cos $2\Delta\theta$, adds this product to the previously stored product, and so on.

Thus, over the same integration period, the computer forms a separate filter, tuned to the same frequency, for the returns from every range gate. If, for example, returns from 100 range gates are processed, the computer forms 100 filters during every integration time for just this one Doppler frequency.

In simple processors, the products may be stored in a *shift register* (Fig. 21-36). It has as many storage positions as there are range gates. Each time a new product is produced, the stored sums are all shifted one position (to the right in the figure). The new product is added to the sum that has spilled out of the last

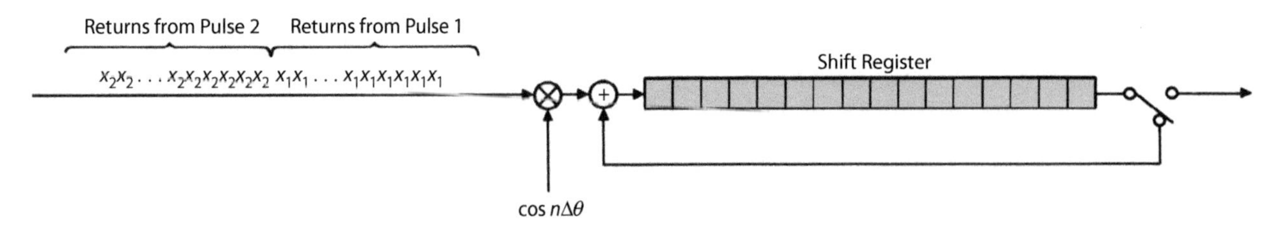

Figure 21-36. When the returns from successive range increments are processed, the sums for the individual increments may be stored in a shift register. As each new number is received, the sums shift one position to the right.

position, and the resulting sum is stored in the memory position that has just been vacated (at the left end of the register).

Since the simple filters considered so far process only one component of the return, they cannot discriminate between positive and negative Doppler frequencies. If the frequency to which a filter is tuned is 10 kHz, for example, the filter will pass return whose frequency is f_0 – 10 kHz just as well as return whose frequency is f_0 + 10 kHz. To differentiate between positive and negative Doppler frequencies, both in-phase and quadrature components must be processed. The computing then is done essentially in two parallel channels.

Algorithm for Approximating $\sqrt{I^2 + J^2}$

ALTHOUGH IT LOOKS SIMPLE ENOUGH, TAKING THE VECTOR SUM $\sqrt{I^2 + J^2}$ is a comparatively long process because the square root can be found only through an iterative series of trials.

To save computing time, therefore, the value of $\sqrt{I^2 + J^2}$ is commonly approximated. The simplest of several possible algorithms for making this approximation is the following:

1. Subtract I from J (or vice versa) to find which is smaller.
2. Divide the smaller quantity by two. (Doing this in binary arithmetic is easy: you just shift the number right one binary place.)
3. Add the result to the larger quantity. The sum is $\sim\sqrt{I^2 + J^2}$

The error of approximation varies with the value of the phase Φ. But at most it is only a fraction of a decibel.

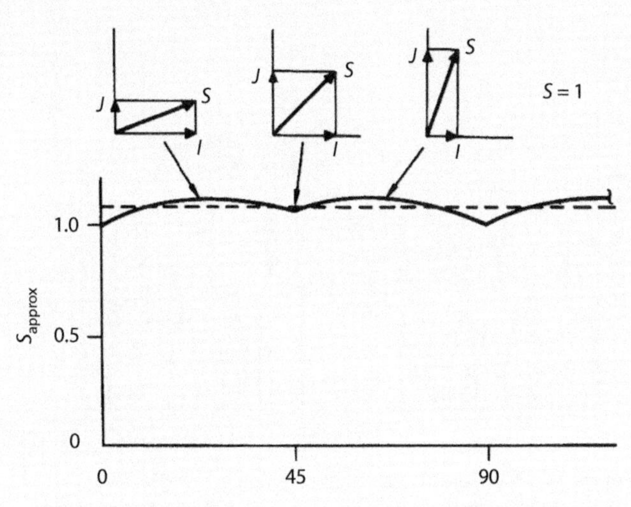

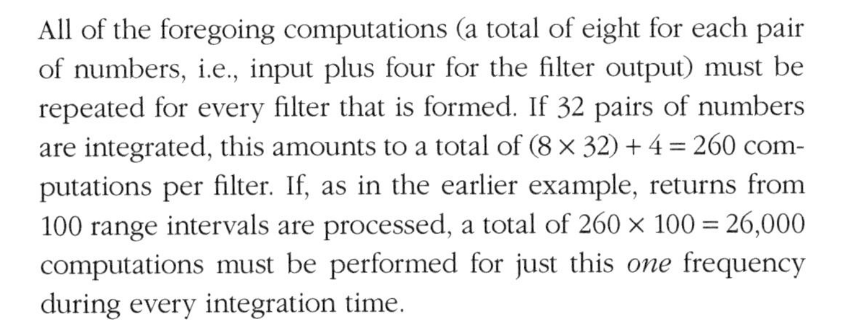
All of the foregoing computations (a total of eight for each pair of numbers, i.e., input plus four for the filter output) must be repeated for every filter that is formed. If 32 pairs of numbers are integrated, this amounts to a total of (8 × 32) + 4 = 260 computations per filter. If, as in the earlier example, returns from 100 range intervals are processed, a total of 260 × 100 = 26,000 computations must be performed for just this *one* frequency during every integration time.

21.5 Sidelobe Reduction

As mentioned earlier, the passband of a digital filter has sidelobes similar to those of a linear array antenna. Unless something is done to reduce these, particularly strong echoes from a target may be detected in the outputs of several adjacent Doppler filters or, if they are extremely strong, in the outputs of a considerable portion of the filter bank.

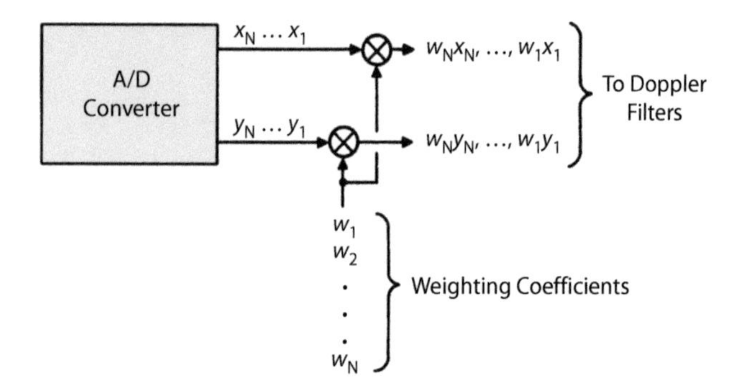

Figure 21-37. The outputs of this A/D converter are multiplied by weighting coefficients before being supplied to the Doppler filters.

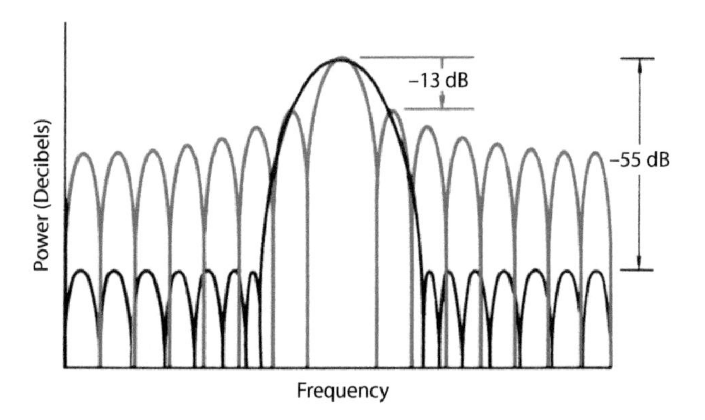

Figure 21-38. Pictured here is the reduction in sidelobe level achieved by weighting the inputs to a representative Doppler filter. Note the broadening of the passband to nearly double the original filter bandwidth.

Fortunately, filter sidelobes yield to the same reduction technique as antenna sidelobes. Just as antenna sidelobes are due to the radiation from the radiators at the ends of the array, filter sidelobes are caused by the pulses at the beginning and end of the pulse train. By progressively reducing the amplitudes of these pulses, the spectral sidelobes can be substantially reduced.

This process, called *amplitude weighting*, is carried out before the digitized video is supplied to the Doppler filters (Fig. 21-37). Following every transmission, the numbers representing the I and Q components of the return from each range gate are multiplied by a weighting coefficient. The coefficient is changed from one pulse to the next according to a prescribed pattern, which is repeated for each train of pulses that is to be integrated. If this pattern has been suitably selected, the sidelobes can be reduced to an acceptable level. In the process, the passband is widened somewhat just as the mainlobe of an antenna is widened by illumination tapering. But this is generally a small price to pay for the reduction achieved in the sidelobes.

What is an acceptable sidelobe level? Naturally this depends on the application. The characteristic of a weighted filter for a representative fighter application is shown in Figure 21-38, where the sidelobes have been reduced from –13 dBs to –55 dBs.

Incidentally, even when a Doppler filter's sidelobes have been acceptably reduced, some return will invariably get through the sidelobes. If strong enough, therefore, return outside a filter's passband can still be detected in the filter's output. Therefore, it is essential that strong ground return are filtered out *before* the radar return is applied to a bank of Doppler filters.

21.6 Filtering Actual Signals

In the foregoing discussion, a somewhat artificial situation was considered in which the numbers supplied to the filter represented a continuous train of echoes from a single target and nothing more. How does the filter respond in the real world, where echoes may be received simultaneously from more than one target, where target echoes may be accompanied by strong ground return, and where sometimes there are no echoes at all, only noise?

Many mathematical software packages are available to process signals on a general-purpose digital computer. Once the essentials of the Doppler filter have been mastered, a simple approach is to "call" a fast Fourier transform (FFT) routine that will take the time-domain data and convert it into the frequency domain.

Dynamic Range. The input to the filter is the algebraic sum of the instantaneous values of all of the simultaneously received signals, multiplied by the system gain up to that point. This approach assumes that the radar receiver and signal processor are reasonably linear and that saturation has been avoided in all stages of receiving and signal processing up to the filter. The output of the filters will be the same as if each of these signals had been integrated individually and the individual outputs had then been superimposed.

Suppose, for example, that the Doppler frequency of a given signal, S_1, lies in the center of the filter's passband and the Doppler frequency of a stronger signal, S_2, lies outside (Fig. 21-39). The ratio of the outputs produced by the two signals will equal the ratio of the powers of the two signals times the ratio of the filter's gain at its center frequency to its gain at the frequency of S_2. If S_2 is 30 dB stronger than S_1 but the filter's gain at S_2's frequency is 55 dB less than the gain at the center of the passband, the output produced by S_1 will be 55 dB − 30 dB, that is, 25 dB stronger than the output produced by S_2.

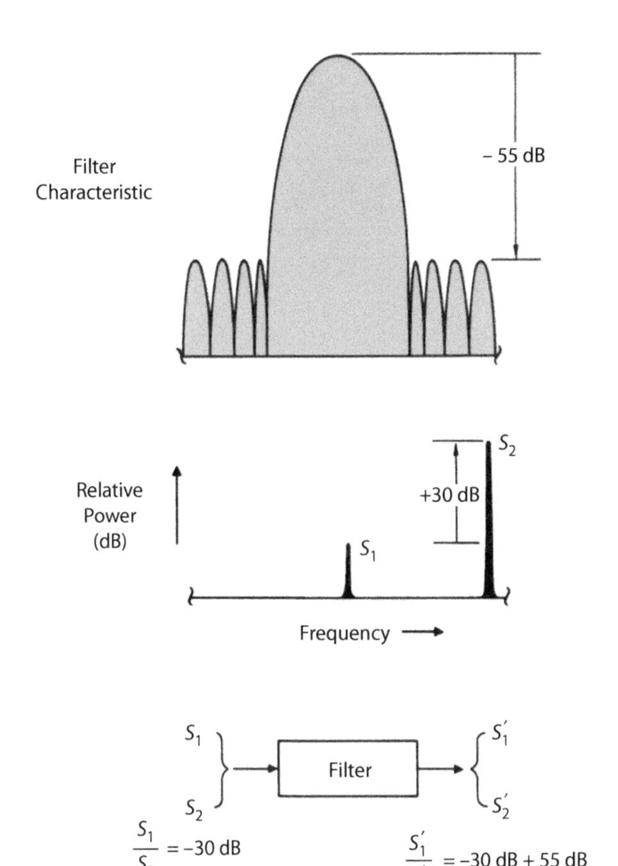

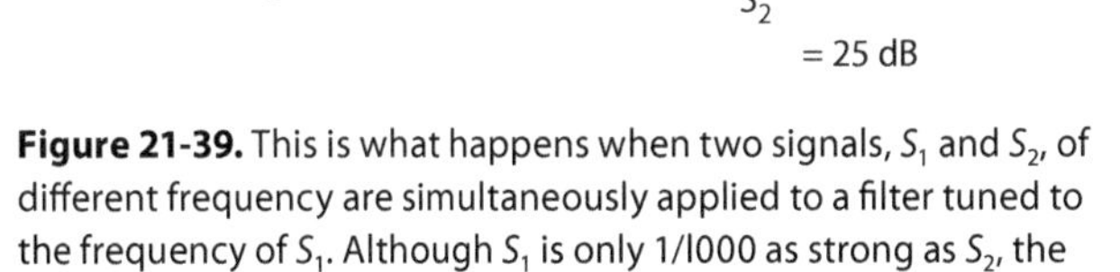

Figure 21-39. This is what happens when two signals, S_1 and S_2, of different frequency are simultaneously applied to a filter tuned to the frequency of S_1. Although S_1 is only 1/1000 as strong as S_2, the output produced by S_1 will be 25 dB stronger.

The system must be able not only to handle signals of maximum strength but also provide to a wide enough range of output levels at any one time to detect small differences in output due to simultaneously received echoes from small distant targets. The solution is to provide adequate dynamic range.

Dynamic range is the spread between (1) the minimum incremental change in the amplitude of the input to a circuit or system that will produce a discernible change in output and (2) the maximum peak-to-peak amplitude the input can have without saturating the output (i.e., without reaching a point where the output no longer responds to a further increase in input). Beyond this point, the output becomes a distorted representation of the input. This limit is often determined by the specification of the A/D converters because there is a limit on the number of bits an A/D converter can assign between the smallest and largest echo voltages.

Providing adequate dynamic range is an important consideration in the design of the receiving and signal processing system of any radar, but it is crucial in radar systems that must sense Doppler frequencies. If the dynamic range is not adequate, weak signals may be masked by strong signals, and spurious signals will also be created. These signals, whose frequencies may be quite different from those of the received signals, may appear falsely as apparent target echoes or may interfere with the detection of true targets.

Noise may also be present. Depending on its relative phase, noise falling in the filter passband may combine with a target signal either destructively or constructively (or somewhere in between). As a result, the filter output produced by an otherwise detectable target may sometimes fail to cross the detection threshold and vice versa. At times, too, the integrated noise alone may exceed the threshold. Most radar systems are set to toggle the lowest bit (or two bits) of the A/D converter with noise.

Spurious Signals. There are two types of spurious signals: *harmonics* and *cross-modulation products*.

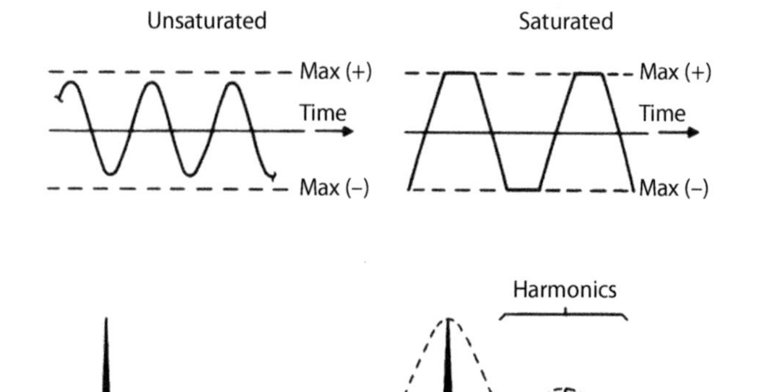

Figure 21-40. Harmonics are created when a signal's peak amplitude is limited by saturation.

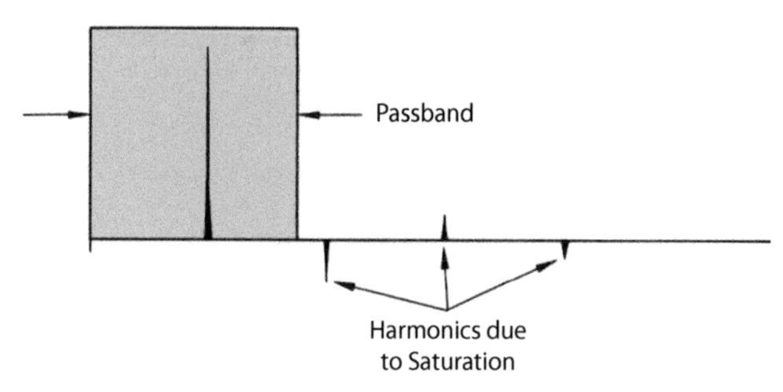

Figure 21-41. If the passband is narrow enough, harmonics due to saturation may be eliminated.

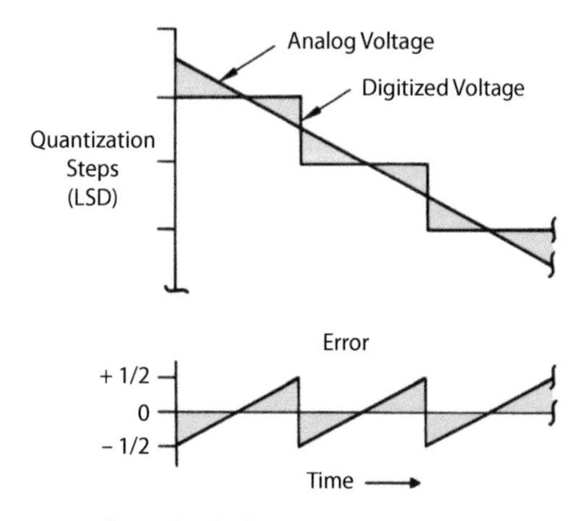

Figure 21-42. If a gradually changing voltage is represented by digital numbers, the error due to quantization has a triangular shape and a peak amplitude equal to half the value of the least significant digit (LSD).

Harmonics are signals whose frequencies are multiples of another signal's frequency. Simply lopping off the top and bottom of a sine wave demonstrates how harmonics are created when a system's output is limited by saturation (Fig. 21-40).

The result is a nearly square wave, which is made up of a series of sine waves whose frequencies are multiples of the frequency of the square wave. If a system's passband is narrow enough, the harmonics may lie outside the passband and thus may be rejected (Fig. 21-41). Otherwise, they may cause problems.

Cross-modulation is the modulation of one signal by another. It is produced if the sum of two or more signals of different frequency is limited by saturation. The products of cross-modulation are sidebands. They occur both above and below the frequencies of the modulated signals.

Consequently, if the frequencies of the saturating signals are closely spaced, a great many cross-modulation products will be passed by a system regardless of the width of its passband.

Avoiding Saturation. The creation of harmonics and cross-modulation products may be avoided by avoiding saturation.

Toward this end, in designing a signal processing system, the average signal level is usually kept as low as possible without risking the loss of weak signals in locally generated noise. Enough dynamic range is then provided to prevent strong signals from saturating the system. Generally, this approach leads to a trade-off between saturation on one hand and low-level noise on the other.

Dealing with Quantization Noise. The problem of low-level noise is exacerbated by the presence of so-called quantization noise resulting from A/D conversion. It is the inevitable result of representing signal amplitudes that are continuously variable with digital numbers that are graduated in finite steps (quanta).

Figure 21-42 illustrates this effect for a linearly changing signal. After being digitized, the signal actually consists of the sum of two signals: (1) a quantized replica of the original analog signal; and (2) a triangular error wave having a peak amplitude equal to half the value of the least significant digit (LSD).

If the original signal is composed of periodic samples of the return from a given range gate, a simple triangular error wave is generally not produced. Successive samples are about as likely to fall at one point as another between the steps of the A/D converter's reference voltage. Consequently, this undesirable by-product of digitization is more or less random and thus is customarily categorized as noise.

Quantization noise in both the A/D converter and the processor puts a lower limit on the signal levels that can be handled by a system. A common figure of merit for an A/D's dynamic range is the ratio of (1) the maximum peak signal voltage the A/D can handle to (2) the root mean square (RMS) value of the quantization error voltage.[4]

4. For a triangular wave shape, the RMS value is approximately $(1 + \sqrt{12}) \times$ LSD.

To avoid degrading signal-to-noise ratios, the quantization noise should contribute only negligibly to the overall system noise. For that, the level of the incoming signals must be set high enough that the level of the noise accompanying the signals is substantially higher than the quantization noise—ideally, on the order of 10 times higher (Fig. 21-43).

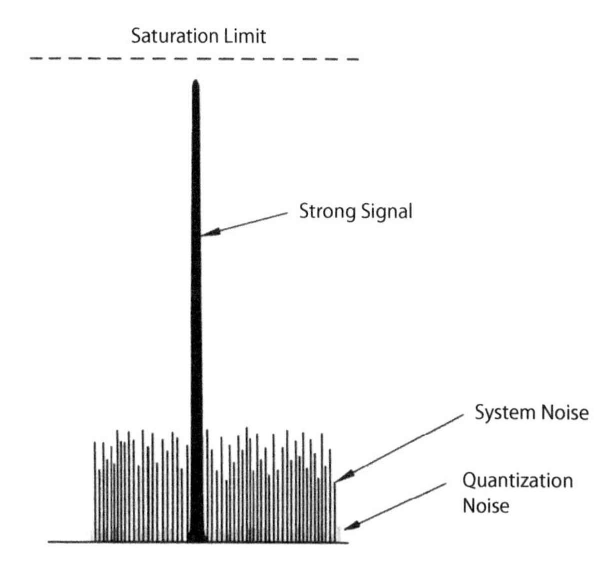

Figure 21-43. Ideally the quantization noise would be 1/10 or less of the system noise, and the saturation limit would be sufficiently far above the system noise to accommodate the strong signals.

To prevent saturation by strong signals, the dynamic range must be correspondingly increased. This may require increasing the number of digits in the numbers used to represent the signals or handling the processing in a cleverer way (e.g., attenuating closer ranges where targets have larger echo signals) or both.

What if the reception of a train of target echoes is not synchronized with the filter's integration period? Suppose the first echo of the train is received halfway through t_{int}. Synchronization between the radar antenna's time-on-target and the integration time of the Doppler filters is entirely random. The first pulse in a train of target echoes is as likely to arrive in the middle of the integration period as at the beginning (Fig. 21-44). Consequently, on an average, the integrated signal in the output of the filter generally falls short of the maximum possible value. In calculating detection probabilities, this difference is normally accounted for by including a loss term in the range equation (see Chapter 13).

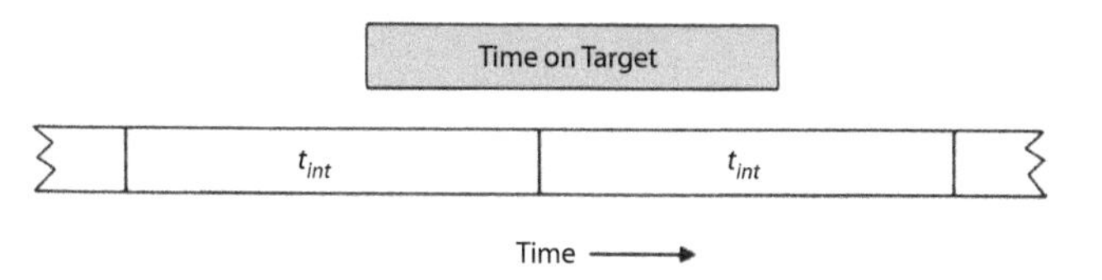

Figure 21-44. The synchronization of the antenna's time on any one target and the filter's integration time, t_{int}, is completely random.

All told, though, performance of a digital filter in a real-life situation is very much as has been described here. If saturation has been avoided and strong ground return has largely been rejected in advance, a well-designed digital filter will separate target echoes from clutter and noise on the basis of their differences in Doppler frequency just as effectively as a well-designed analog filter.

21.7 Summary

To sort out the radar return from various objects according to Doppler frequency, the receiver output is applied to a bank of narrowband filters. If sorting by range is also required, a separate bank is provided for each range increment. The width of the passband of a narrowband filter is primarily determined by the filter's integration time but is increased by losses. So that return will not be lost when a target straddles two filters, the passbands are made to overlap. So that only one line of a target's spectrum will fall within the band of frequencies bracketed by the bank, the passband of the bank is made no greater than the PRF.

For digital filtering, the IF output of the receiver is translated to video frequencies by applying it to a pair of synchronous

Some Relationships to Keep in Mind

- Filter passbands:

  Null-to-null $= \frac{2}{t_{int}}$

  Between half power points $= \frac{1}{t_{int}}$

  where t_{int} = filter integration time)

- Operations per sample required to form a filter with the DFT

  Multiplications = 4

  Additions = 4

- Operations required to approximate $\sqrt{I^2 + J^2}$

  Subtractions = 1

  Division by 2 = 1

  Additions = 1

Test your understanding

1. What are the benefits of Doppler sensing and Doppler filtering?
2. A pulsed radar has a pulse repetition frequency of 1000 Hz. What will be the 3 dB Doppler filter bandwidth after integrating echo pulses for a duration of 1 second?
3. A signal with a bandwidth of 1 MHz is to be digitally sampled and fully reconstructed. At what frequency should it be sampled?
4. What is the (approximate) relationship between the null-to-null spectral width and the 3 dB spectral width of a Doppler filter passband?
5. Explain how amplitude weighting reduces sidelobes.
6. What is meant by the term *dynamic range*, and why is it so important in the design of pulsed radar systems?

detectors, along with a reference signal whose frequency corresponds to that of the transmitter. The outputs of the detectors represent the I and Q components of the return the latter of which are needed to preserve the sense of the Doppler frequencies.

In pulsed radar, sampling corresponds to the range gating of an analog processor. Each sample is converted to a binary number by comparing its voltage with a succession of progressively higher voltages of precisely known value. The numbers are then supplied to a special purpose computer or digital processor, which implements the filters.

As inputs, a digital Doppler filter receives a succession of pairs of digital numbers. If echoes from a target are being received, each pair constitutes the x and y components of a phasor representing one sample of a signal whose amplitude corresponds to the power of the target echoes and whose frequency is the target's Doppler frequency. The job of the filter is to integrate these numbers in such a way that if the Doppler frequency is the same as the filter's frequency, the sum will be large but otherwise will not be.

In essence, the filter projects successive x and y components onto a coordinated system that rotates at the frequency the filter is tuned to and sums the components separately. At the end of the integration period, the magnitude of the integrated signal is computed by vectorially adding the two sums. The simple algorithm, which must be repeatedly computed, to perform the integration and obtain the magnitude of the vector sum is called the *discrete Fourier transform* (DFT) (see Chapter 6).

A plot of the filter output versus Doppler frequency for a pulse train of given length and power has a sin x/x shape. Its peak value is proportional to the total energy of the pulse train, and its nulls occur at intervals equal to $1/t_{int}$ on either side of the central frequency. To reduce the sidelobes of this pattern, the numbers representing the pulses at the beginning and end of the train are progressively scaled down—a process called *amplitude weighting*.

Barring nonlinearities and saturation, when several signals are received simultaneously the filter output is the same as if the signals had been integrated separately and the results had been superimposed. Since receipt of a train of pulses cannot be synchronized with the filter's integration time, the filter output is on an average less than the potential maximum value.

Further Reading

V. N. Bringi and V. Chandrasekar, *Polarimetric Doppler Weather Radar Principles and Applications*, Cambridge University Press, 2007.

D. C. Schleher, *MTI and Pulse Doppler Radar with MATLAB®*, Artech House, 2009.

22

Measuring Range-Rate

In many radar applications, knowing a target's present position (angle and range) relative to the radar is not enough. Often one must be able to predict the target's position at some future time. For that, the target's angular rate and its range-rate also need to be known.

Range-rate may be determined by one of two general methods. In the first, called *range-differentiation*, the rate is computed on the basis of the change in the measured range with time. In the second method, which is generally superior, the radar measures the target's Doppler frequency, which is directly proportional to the range-rate.

In this chapter, both methods are briefly examined.

22.1 Range Differentiation

If target range is plotted versus time, the slope of the plot is the range-rate (Fig. 22-1). A downward slope corresponds to a negative rate, an upward slope, to a positive rate.

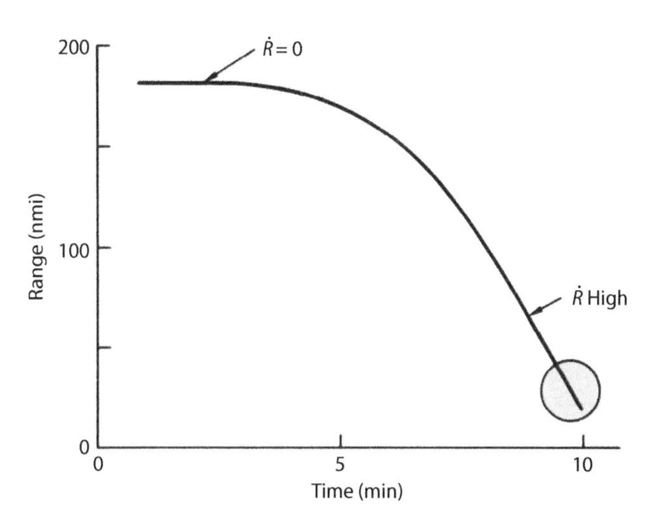

Figure 22-1. The range-rate, $\dot{R}$, corresponds to the slope of a plot of range versus time.

Determining the slope and hence the range-rate, is easy. Two points are selected on the plot separated by a small difference in time and the difference in their ranges is measured. Dividing the range difference by the time difference yields the range-rate, such as

$$\dot{R} = \frac{\Delta R}{\Delta t}$$

where

$\dot{R}$ = range-rate

ΔR = difference in range

Δt = difference in time

If ΔR is taken as the difference between the current range and the range Δt seconds earlier, $\dot{R}$ corresponds to the current range-rate. This process approximates differentiation (Fig. 22-2).[1]

1. With differentiation, the time difference (Δt) is infinitesimally small.

In essence, range-rate is measured in this way both by non-Doppler radars and by Doppler radars when operating under conditions where Doppler ambiguities are too severe for range-rate to be measured directly by sensing Doppler frequency. In short, radars measure range and range measured over successive time intervals leads to measurement of range-rate.

If the range-rate is changing, the shorter the Δt is made, the more closely the measured rate, $\dot{R}$, will follow the changes in the actual rate; i.e., the less the measured rate will lag behind the actual rate (Fig. 22-3)—a quality referred to *as good dynamic response.*

Unfortunately, a certain amount of random error, or "noise" is invariably present in the measured range. Though small in comparison to the range itself, the noise can be appreciable in comparison to ΔR. In fact, the shorter Δt is made, the smaller ΔR will be and therefore the greater the extent to which the noise will degrade the rate measurement (Fig. 22-4).

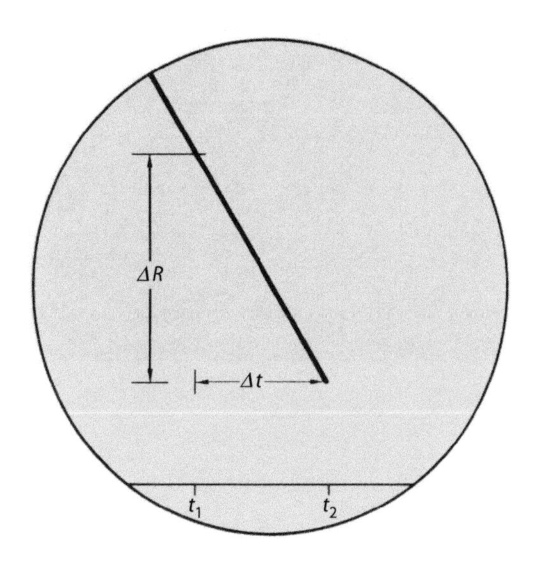

Figure 22-2. Slope of range plot can be found by taking difference in range (ΔR) at points separated by a short increment of time (Δt).

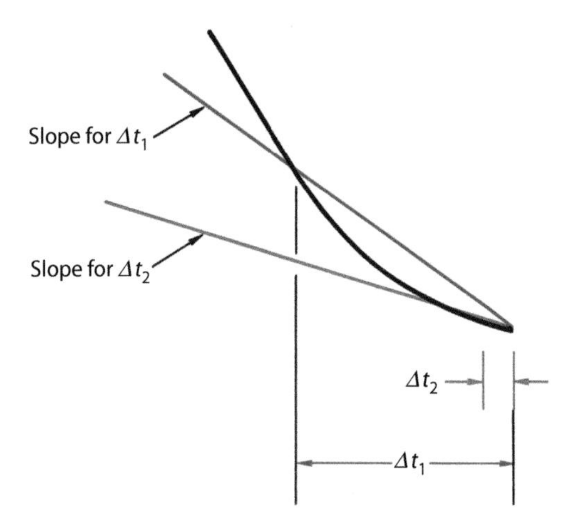

Figure 22-3. The shorter Δt is made, the closer the measured slope will follow changes in the actual range-rate.

Smoothing the measured range-rate can reduce measurement noise, but the effect of smoothing is essentially the same as that of increasing Δt. The performance achieved with this method of range-rate measurement is thus a compromise between smooth tracking and good dynamic response.

When a target is tracked without stopping the antenna's search scan (while in track-while-scan mode), the dynamic response is still further limited by the fact that Δt is stretched to the time for a full scan. If the radar's range measurement sensitivity is sufficiently high, a useful estimate of the range-rate can be provided during the dwell time by extrapolation.

22.2 Doppler Method

The Doppler method of measuring range-rate is much more commonly used as it enables a more accurate determination and combined with digital signal processing, offers greater flexibility. A Doppler radar not only measures range-rates

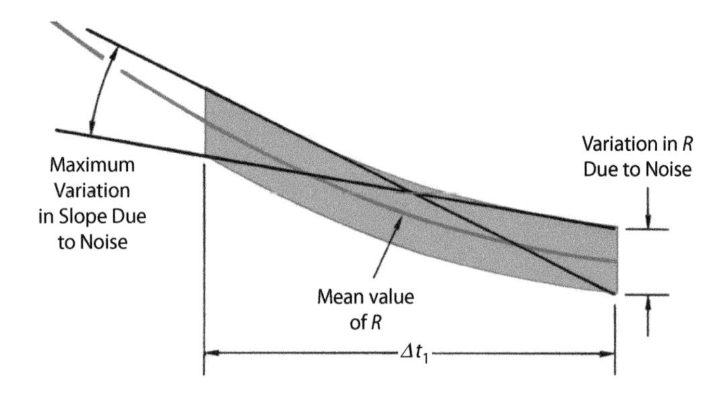

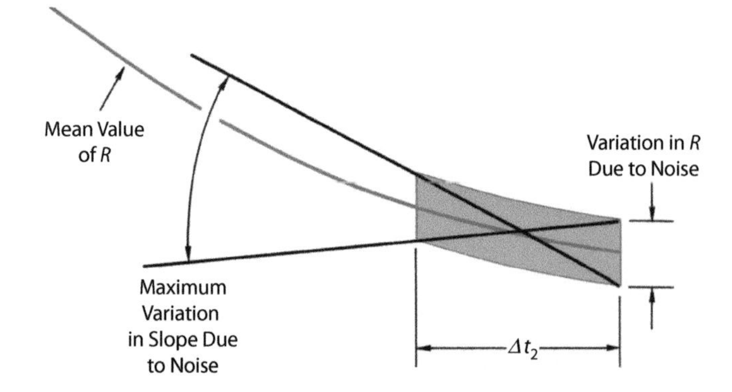

Figure 22-4. The shorter Δt is, the more the measured range-rate may differ from the actual rate as a result of noise in the measured range.

with greater precision but also makes the measurement directly. It does so using phase rather than changes in target position. Measurement of phase is essentially a fine measurement of range and it is this that leads to improved accuracy.

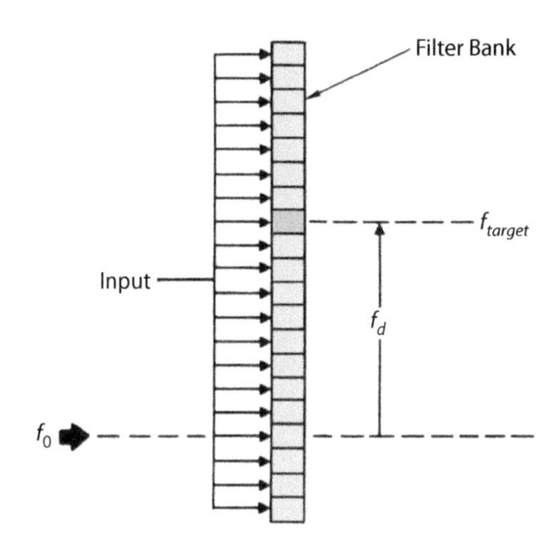

Figure 22-5. A target's Doppler frequency may be determined simply by noting the target's position in the filter bank.

In the absence of Doppler ambiguities, a target's Doppler frequency may be determined simply by noting in which filter of the Doppler filter bank the target appears (Fig. 22-5). However, if it straddles two adjacent filters, by interpolating between the center frequencies of the filters on the basis of the difference in their outputs its Doppler frequency may be determined. In translating the Doppler spectrum to the frequency of the filter bank an accurate track of the relative position of the transmitter frequency, f_0, must be kept. To measure the Doppler frequency, one need only count down the bank from the target's position to the frequency corresponding to f_0. Alternatively, if the Doppler spectrum has been offset from f_0, the mainlobe clutter is placed at zero frequency and then you count to the bottom of the filter and add the offset.

The Doppler frequency is determined with greater precision during single-target tracking (see Chapter 31). In this mode, the receiver output is usually applied in parallel to two adjacent Doppler filters, whose passbands overlap near their –3 dB points (Fig. 22-6).

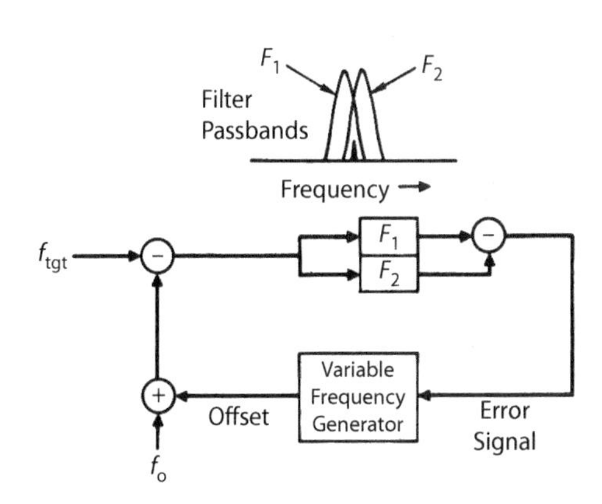

Figure 22-6. In single-target tracking, an offset is added to the target's frequency to center it exactly between two filters. The Doppler shift is then determined by precisely measuring the offset.

An automatic tracking circuit shifts the Doppler spectrum just enough to cause the target to produce equal outputs from the two filters. This shift is maintained equal to the target's Doppler frequency and the Doppler frequency is then given by measuring the shift.

The range-rate or velocity is computed from the Doppler frequency with the inverse of the expression for Doppler frequency derived in Chap. 18:

$$\dot{R} = -\frac{f_D \lambda}{2}$$

where

$\dot{R}$ = range-rate

f_D = Doppler frequency

λ = wavelength

The rules of thumb learned in chapter 18 can similarly be inverted. For example, an X-band radar system, with a wavelength of 3 cm, has a range-rate, in meters per second, nearly equal to the Doppler frequency in hertz divided by 70 (or Doppler frequency = range-rate × 70).

But how can it be known that it is the target echoes' carrier frequency we have observed and not a sideband or aliased frequency, some multiple of f_r above or below the carrier? Isn't any measured Doppler frequency and hence the computed range-rate, inevitably ambiguous?

Seemingly unfortunately, the answer is yes. However, whether the ambiguity is significant depends on the PRF and the magnitudes of the closing rates between the radar system and the target that are encountered.

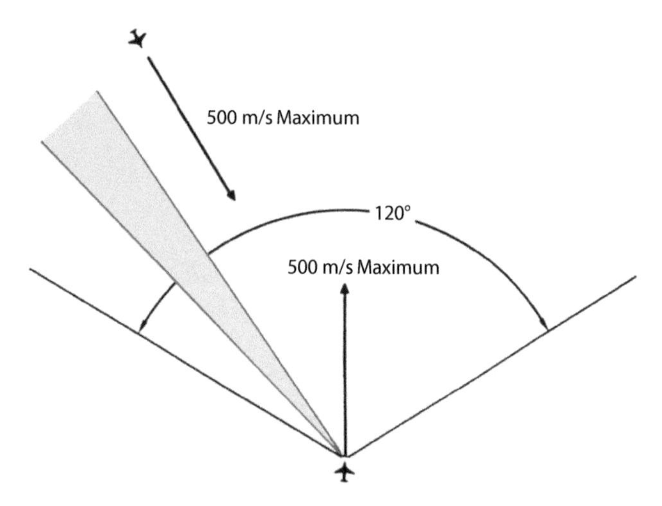

Figure 22-7. This hypothetical situation illustrates conditions under which Doppler measurements may be significant ambiguous.

22.3 Potential Doppler Ambiguities

To get a feel for the significance of Doppler ambiguities at different PRFs, consider a hypothetical operational situation.

A hypothetical Situation. Assume that a radar is operating against targets detected anywhere within a 120°-wide sector located dead ahead (Fig. 22-7). The targets may be flying in any direction.

Their speeds may vary but are not expected to exceed 500 m/s. The maximum speed of the aircraft carrying the radar is, say, also 500 m/s.

Under these conditions, the maximum closing rate that the radar might encounter (i.e., the rate when the radar-bearing aircraft and the target are flying at maximum relative speed) is nose-on (Fig. 22-8). This is –500 m/s – 500 m/s = –1000 m/s. At X-band, this maximum rate would produce a Doppler shift of roughly 1000 × 70 = 70 kHz.[2] Note the negative sign convention for the relative velocities between the radar and the target.

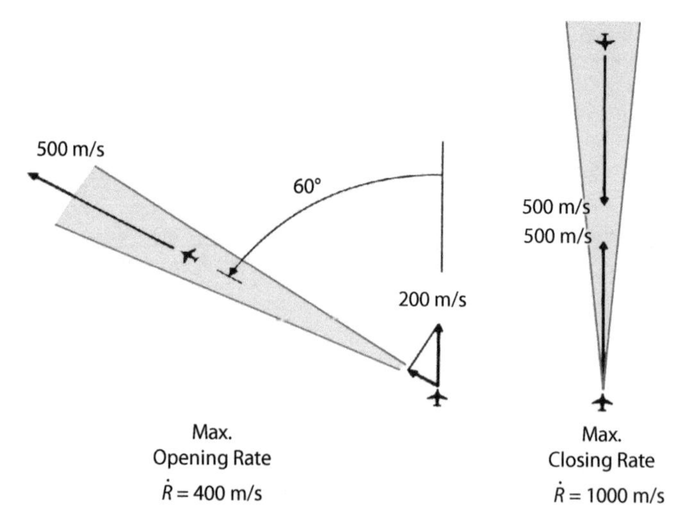

Figure 22-8. Flight geometries that produce maximum negative Doppler frequency (left) and maximum positive Doppler frequency (right).

2. Remember that negative range-rates (range decreasing) result in positive Doppler frequencies and vice versa according to:

$$f_D = -\frac{2\dot{R}}{\lambda}$$

Thus the maximum opening rate occurs if a target were at the largest azimuth angle (60°) and flying at maximum speed away from the radar and the radar-bearing aircraft was flying at its minimum speed. If the radar-bearing aircraft is travelling at a speed of 200 m/s this maximum opening rate will be +500 m/s – (0.5 × 200 m/s) = 400 m/s. This produces a Doppler shift of approximately –400 × 70 = –28 Hz.

Thus, provided the radar does not encounter a significant target whose speed exceeds 500 m/s or whose azimuth exceeds 60°, the spread between maximum positive and negative Doppler frequencies would be 70 – (–28) = 98 kHz (Fig. 22-9).

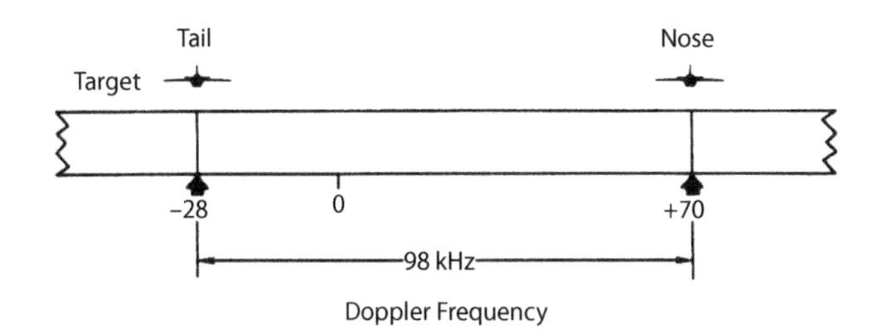

Figure 22-9. The spread between the maximum positive and the maximum negative Doppler frequencies for the hypothetical situation.

PRF Greater Than Spread of Doppler Frequencies. Suppose now, that in the situation described above, the radar's PRF is 120 kHz. To cover the band of anticipated Doppler frequencies (–28 kHz to +70 kHz) with a little room to spare, let's say a Doppler filter bank is used that has a bandwidth extending from a little below –28 kHz to a little above +70 kHz (e.g. the shaded area in Fig. 22-10).

If a target has the maximum anticipated closing rate (a Doppler frequency of +70 kHz) the carrier frequency of echoes will fall just inside the high frequency end of the passband. Since the first pair of sidebands are separated from the carrier frequency by the PRF (120 kHz), the sideband nearest the passband will have a frequency of 70–120 = –50 kHz. This is well below the lower end of the passband (just below –28 kHz).

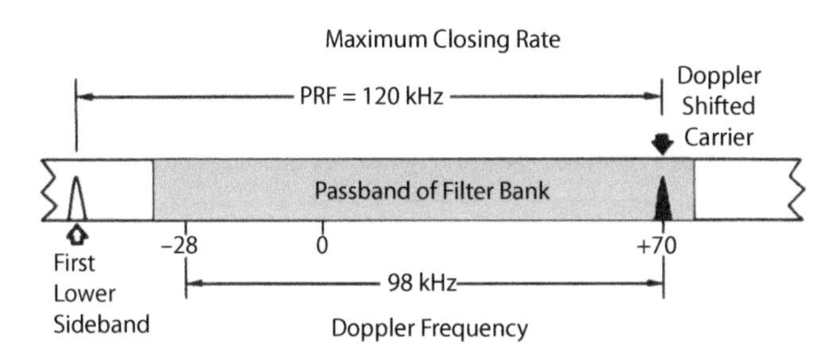

Figure 22-10. If the PRF *exceeds* the spread between the maximum positive and maximum negative Doppler frequencies, the carrier of the most rapidly closing target will fall in the passband and nearest sideband will lie below it.

Similarly, if we encounter a target having the maximum anticipated negative Doppler frequency (–28 kHz), the carrier frequency of its echoes will fall just inside the lower end of the passband (Fig. 22-11). The nearest sideband in this case will have a frequency of –28 + 120 = 92 kHz, well above the upper end of the passband (just above 70 kHz).

Thus, if the PRF is greater than the spread between the maximum anticipated positive and negative Doppler frequencies, the only spectral line producing an output from the filter bank is due to the carrier of the target echoes. The difference between this frequency and the transmitter's carrier frequency is the target's true Doppler frequency. Hence, it may be concluded that if the PRF is greater than the Doppler spread then no significant ambiguities will exist.

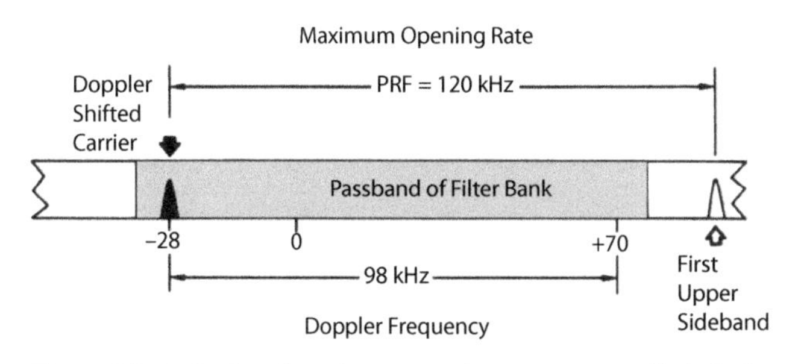

Figure 22-11. Carrier of maximum opening-rate target will similarly fall in passband and nearest sideband will lie above it.

This would *not* be the case, however, if the PRF were less than the spread between the maximum positive and maximum negative Doppler frequencies. In fact this is often the situation, as the PRF must be set to satisfy other operational requirements such as avoiding range ambiguities.

PRF Less Than Spread of Doppler Frequencies. Suppose that in this same hypothetical situation where the difference between maximum anticipated positive and maximum negative

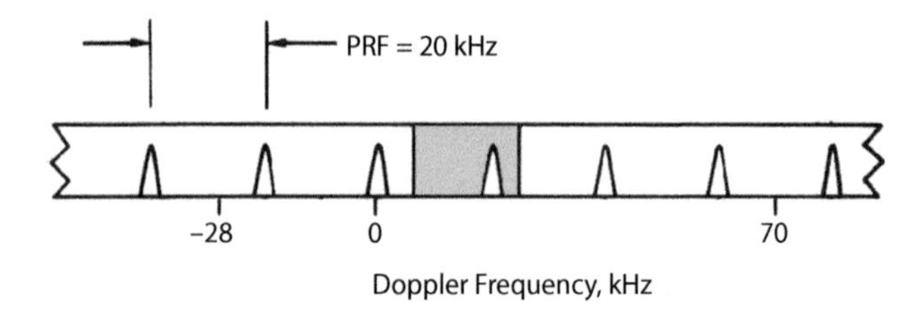

Figure 22-12. If the PRF is less than the spread of the maximum closing rate, the radar has no direct way of telling which repetition of carrier frequency it is observing and the Doppler frequency measurement is ambiguous.

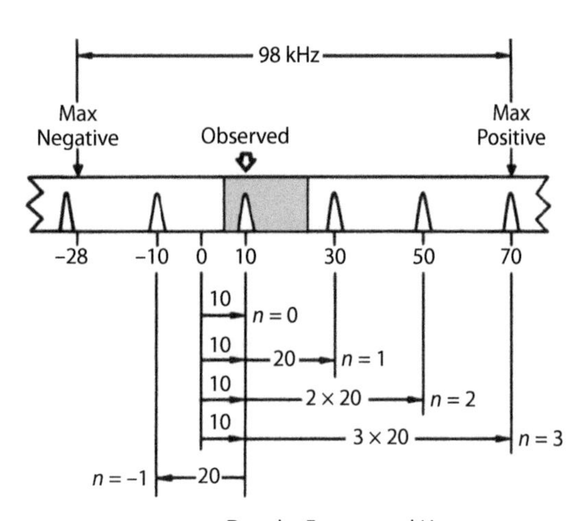

Figure 22-13. If the PRF is 20 kHz and the observed Doppler frequency is 10 kHz, the true Doppler could have any of the following values: –10, 10, 30, 50, and 70 kHz.

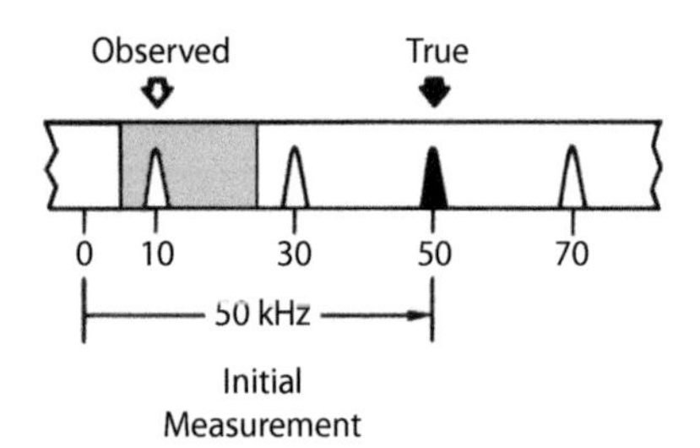

Figure 22-14. By making an initial measurement using the differentiation method the true Doppler frequency and hence the value, *n*, can be determined.

Doppler frequencies is 98 kHz, the PRF is reduced to 20 kHz (Fig. 22-12). The separation between a target echoes' carrier frequency and first pair of sidebands, as well as between successive sidebands above and below, is now only one-sixth of what is was before.

So that the return from any one target will appear at only one point within the passband, it must be made less than 20 kHz wide. However, if the sidebands are only 20 kHz apart, no matter where the passband is positioned, there is no direct way of telling whether the target return that appears in the filter bank output is the echoes' carrier or corresponds to a sideband, or even which sideband it might be. Thus the true Doppler frequency is ambiguous. To determine the target's true Doppler frequency and hence its range rate, the ambiguity has to be resolved.

22.4 Resolving Doppler Ambiguities

To resolve Doppler ambiguities, there has to be some way of telling what whole multiple of the PRF (if any) separates the observed frequency of the target echoes from the carrier frequency. If not too great, the multiple, n, may readily be determined. There are two common ways to do this: *range differentiation* and *PRF switching*.

Range Differentiation. In general, the simplest way to determine n is to make an approximate initial measurement of the range-rate by the differentiation method. From this rate, the approximate value of the true Doppler frequency can be computed. Subtracting the observed frequency from the computed value of the true frequency and dividing by the PRF yields a value for n.

For example, suppose the PRF is 20 kHz and the observed Doppler frequency is 10 kHz (Fig. 22-13). The true Doppler frequency then could be –10 kHz plus any whole multiple of the PRF (20 kHz) up to 70 kHz. The approximate value of the true Doppler frequency computed from the initial range-rate measurement, let's say, turns out to be 50 kHz. The difference between this frequency and the observed Doppler frequency is $50 - 10 = 40$ kHz. Dividing the difference by the PRF, we get $n = 2$ ($40 \div 20$). Thus, the carrier frequency of the target echo is separated from the observed Doppler frequency by two times the PRF (Fig. 22-14).

Although in this simple example it is assumed that the initial range-rate measurement was fairly precise, in practice it may not be particularly accurate. As long as any error in the Doppler frequency computed from the initial rate measurement is less than half the PRF, it is still possible to tell in which PRF interval the carrier lies and hence to tell what the value of n is. The initially computed "true" Doppler frequency, for example, might have been only 42 kHz, almost half way between the two nearest possible exact values (30 and 50 kHz) (Fig. 22-15).

Nevertheless, this rough initially computed value of 42 kHz would still be accurate enough to find the correct value of n. The difference between the initially computed value of the Doppler frequency and the observed value is 42 – 10 = 32 kHz. Dividing the difference by the PRF, gives a value for n of 1.6 (32 ÷20). If this is rounded off to the nearest whole number then $n = 2$.

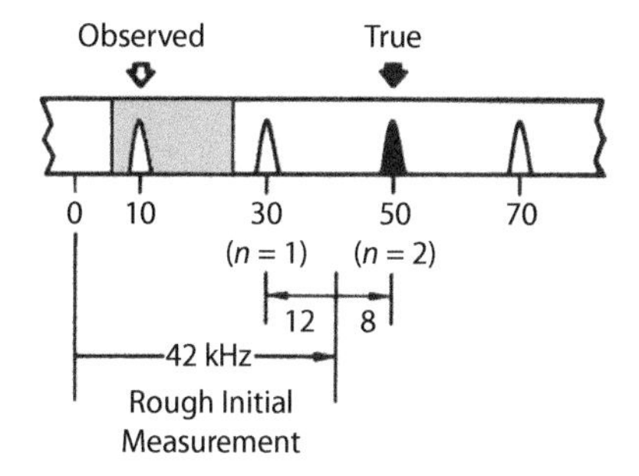

Figure 22-15. Initial measurement of true Doppler frequency need not be particularly accurate. If error is less than half the PRF, value of n can still be found.

Having determined the value of n just once, by tracking the target continuously the true Doppler frequency is continuously computed, with high precision, solely on the basis of the observed frequency.

PRF Switching. The value of n can also be determined with a PRF switching technique similar to that used to resolve range ambiguities (see Chapter 15). In essence, this technique alternately switches the PRF between two relatively closely spaced values and the change, if any, in the target's observed frequency is noted.

Switching the PRF has no effect on the target echoes' carrier frequency, f_c, which equals the carrier frequency of the transmitted pulses plus the target's Doppler frequency and is completely independent of the PRF. However, this is not the case for the sideband frequencies above and below f_c. Because these frequencies are separated from f_c by multiples of the PRF, when the PRF is changed, the sideband frequencies also correspondingly change (Fig. 22-16).

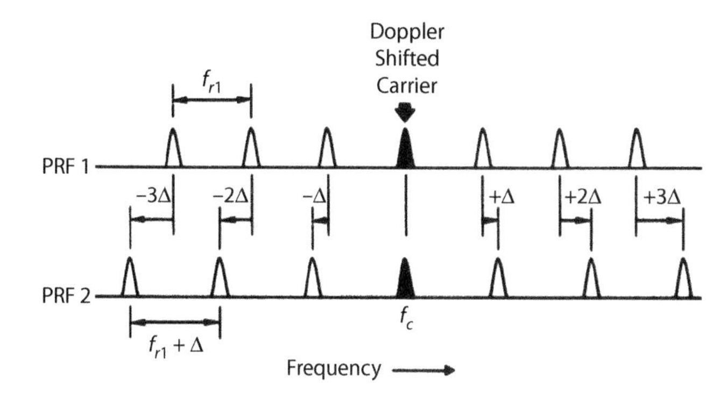

Figure 22-16. If the PRF is changed, each sideband frequency shifts by amount, $n\Delta$, proportional to multiples of f_r separating it from carrier.

Which direction a particular sideband frequency moves, up or down, depends upon two things:

1. Whether the sideband frequency is above or below f_c
2. Whether the PRF has been increased or decreased

An upper sideband will move up, if the PRF is increased and will move down if it is decreased. A lower sideband, on the other hand, will move down if the PRF is increased and up if it is decreased.

How much the observed Doppler frequency moves also depends on two things:

1. By how much the PRF has been changed
2. What multiple of the PRF separates the observed frequency from f_c.

If the PRF is changed by 1 kHz, the first set of sidebands on either side of f_c will move 1 kHz; the second, 2 kHz; the third, 3 kHz, and so on. If the PRF is changed by 2 kHz, each set of sidebands will move twice as far, and so on.

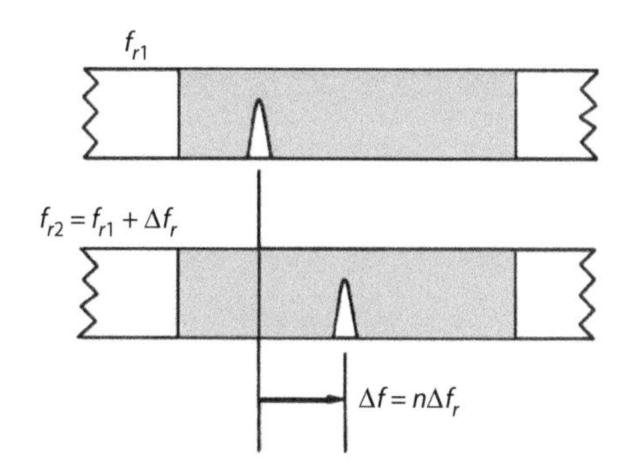

Figure 22-17. By noting the change in observed frequency when PRF is switched, multiple (n) of f_r contained in true frequency can be determined.

By noting the change in the target's observed Doppler frequency (if any), where f_c is relative to the observed frequency (Fig. 22-17) can easily be found. If the observed frequency does not change, it is at f_c. If it does change, the direction of the change determines whether f_c is above or below the observed frequency. The amount of the change tells us what multiple of the PRF that f_c is away from the observed frequency.

Thus, the factor n by which the PRF must be multiplied to obtain the difference between the echoes' carrier frequency f_c and the observed frequency is

$$n = \frac{\Delta f_{obs}}{\Delta f_r}$$

where

Δf_{obs} = change in target's observed frequency when PRF is switched

Δf_r = amount PRF is changed

For example, if an increase in PRF (Δf_r) of 2 kHz causes a target's observed Doppler frequency to increase by 4 kHz, the value of n would be $4 \div 2 = 2$.

In order to avoid the possibility of "ghosts" when echoes are simultaneously received from more than one target, the PRF must be switched between three values instead of two, just as when resolving range ambiguities. Switching the PRF has the disadvantage of reducing the maximum detection range. In practice, PRFs may even be switched between four or five values, depending on the role of the radar system and numbers of targets to be detected.

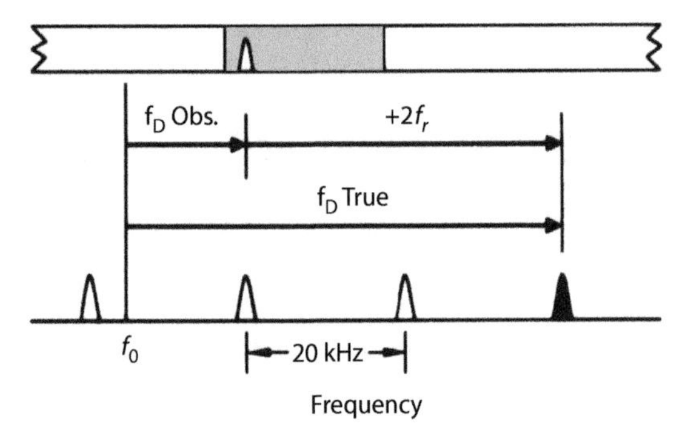

Figure 22-18. True Doppler frequency is computed by adding n times f_r to observed Doppler frequencies. (Here $n = 2$.)

Calculating the Doppler Frequency. Having determined the value of n by either of the methods just outlined, the target's true Doppler frequency, $f_d f_D$, can be computed by simply multiplying the PRF by n and adding the product to the observed frequency (Fig. 22-18)

$$f_d f_D = nf_r + f_{obs}$$

where

f_r = PRF before the switch

f_{obs} = target's observed Doppler frequency

22.5 Summary

A target's range-rate may be determined either by continuously measuring its range and calculating the rate at which the range changes. This is a process that approximates differentiation. Alternatively, range-rate can be determined by measuring the target's Doppler frequency. Because of inevitable random errors in the measured range, the differentiation method tends to be less accurate and provides poorer dynamic response.

The Doppler method can be extremely precise and can be nearly instantaneous. The observed Doppler frequencies, however, are inherently ambiguous. Unless the spread between the maximum anticipated positive and negative Doppler frequencies is less than the PRF (and the consequence of occasionally mistaking a very high-speed target for a lower speed one is negligible) the ambiguities must be resolved.

To resolve ambiguities, the number of times, n, that the PRF is contained in the difference between the observed frequency and the true frequency must be determined. If n is not too large, it can readily be found either by measuring the range-rate initially with the differentiation method or by switching the PRF and observing the direction and amount that the observed Doppler frequency changes.

Further Reading

P. Z. Peebles, "Frequency (Doppler) Measurement," chapter 12 in *Radar Principles*, John Wiley & Sons, Inc., 1998.

V. N. Bringi and V. Chandrasekar, *Polarimetric Doppler Weather Radar Principles and Applications*, Cambridge University Press, 2007.

D. C. Schleher, *MTI and Pulse Doppler Radar with MATLAB®*, Artech House, 2009.

Test your understanding

1. Two aircraft, carrying X-band (10 GHz) pulsed radar systems are approaching each other at a relative velocity of 500 m/s. What is the Doppler shift measured by each radar system?
2. For the case described in 1 above, what is the PRF required to avoid aliasing?
3. What is the maximum unambiguous range for a stationary pulsed Doppler radar that has a PRF of 50 kHz?
4. How could the maximum unambiguous range of a pulsed Doppler radar be increased?

General Atomics MQ-9 Reaper (2007)

The MQ-9 Reaper (formerly named Predator B) is an unmanned aerial vehicle (UAV) capable of remote controlled or autonomous flight operations and developed by General Atomics Aeronautical Systems (GA-ASI) primarily for the United States Air Force. The MQ-9 and other UAVs are referred to as Remotely Piloted Vehicles/Aircraft (RPV/RPA) by the U.S. Air Force to indicate their human ground controllers.

PART V

Clutter

Lockheed SR-71 Blackbird (1966)

While other aircraft in the 1960s were falling by the wayside because of the increased accuracy and speed of surface-to-air missiles, the SR-71 simply outran any threat. Developed as a black project by Lockheed's Skunk Works division, it was an advanced, long-range, Mach 3+ strategic reconnaissance aircraft. It was the first interceptor design to be equipped with a pulse Doppler radar, the Hughes ASG–18. Concepts proven with the ASG–18 were later refined and incorporated into subsequent Hughes radars for the F-14, F-15, and F-18 fighters.

23

Sources and Spectra of Ground Return

Thresholding noise

Airborne radar nearly always has to contend with unwanted returns (clutter) from the earth's surface while trying to detect targets of interest. Often these clutter returns have a much larger magnitude than those from targets of interest, such as other aircraft. The radar designer must take steps to minimize clutter returns through suitable radar design and must devise special detection techniques to distinguish between targets and clutter.

In this chapter we look at clutter due to reflections from the ground, although many of the general principles apply equally to sea clutter. We identify how the clutter returns are affected by the antenna beamshape and transmitted waveform, together with the radar platform height and speed.

Most of the clutter return is usually received through the mainlobe of the radar antenna, if it illuminates the ground. However, we also need to understand the nature of the clutter returns received through the antenna sidelobes, which can also be significant when compared with the returns from a small target. A particular problem for airborne radars is the altitude return, which is due to the sidelobe returns for clutter immediately below the aircraft. These regions of clutter return are illustrated in Figure 23-1.

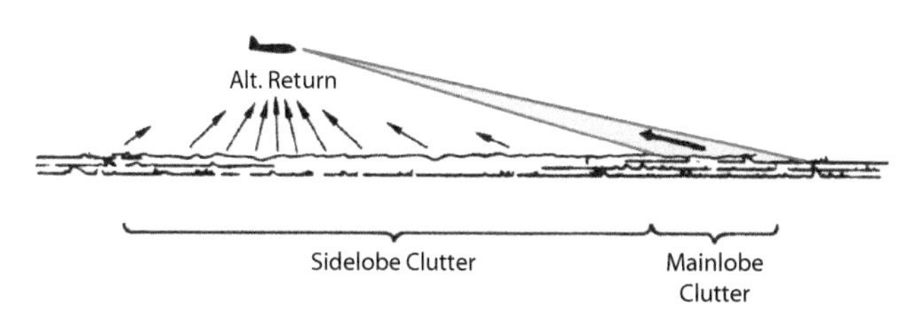

Figure 23-1. This illustration shows ground clutter received by a radar.

The principal means of discerning target echoes from ground clutter is Doppler resolution, combined with their relative amplitudes. In ground-based applications, separating targets from clutter is relatively straightforward. Since the radar is stationary, all of the clutter has essentially one Doppler frequency—zero. In airborne applications, however, this is far from true. Consequently, the way in which the clutter is distributed over the band of possible frequencies—its Doppler spectrum—and the relationship of this spectrum to the Doppler frequencies of anticipated targets critically influence the radar's design. We need to understand the amplitude and Doppler spectra of the clutter returns under different conditions, and how they compare with typical target returns in varying situations. This will help to provide a better understanding of the performance of airborne radars so we can develop methods for discriminating between clutter and targets.

For simplicity, we assume that the radar is transmitting at a sufficiently high pulse repetition frequency (PRF) that Doppler ambiguities are avoided. The effects of ambiguities, which can make ground clutter much more difficult to counter, are covered in the next chapter.

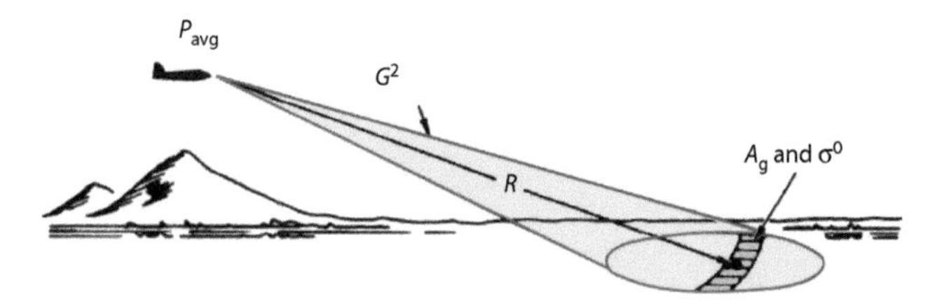

Figure 23-2. Various factors determine the power of the return from a patch of ground: two-way gain of the radar antenna, range to the patch, area of the patch, and backscattering coefficient, σ^0.

23.1 The Amplitude of the Ground Return

In general, ground return is governed by the same basic factors as return from an aircraft. For a given transmitter frequency, the power of the return received from a small patch of ground (Fig. 23-2) is

$$P_r \propto \frac{P_{avg} G^2 \sigma^0 A_g}{R^4}$$

where

P_{avg} = average transmitted power

G = gain of the radar antenna in the direction of the patch (G^2 is the two-way gain)

σ^0 = clutter reflectivity or normalized radar cross section (NRCS)

A_g = resolved area of ground (clutter patch)

R = range to clutter patch

The clutter reflectivity, σ^0, is the radar cross section per unit area of the clutter patch.

The ground surface illuminated by the radar is modeled as an area containing very many individual scatterers, approximately uniformly spatially distributed over the illuminated patch. Consider, for example, a plowed field, woodland or vegetation. We expect the backscattered signal to be proportional to the illuminated area. Clearly, different types of ground have different reflectivity, as will be discussed below. In Chapter 25 we will see how the returns from clutter fluctuate from one patch to the next. However, we observe that the average power of the return per unit area from a given type of terrain can be modeled as a constant value, with local values fluctuating about this mean level. Thus we characterize different types of clutter by their normalized radar cross section or backscatter coefficient, σ^0.

When the appropriate value of σ^0 is multiplied by the area of a particular patch of ground illuminated by the radar, the product is the radar cross section (σ) of the patch. We usually want to know the power of the clutter in a particular range cell of the radar return. The area of the clutter patch is then the area resolved by the combination of the radar azimuth beamwidth and the range resolution of the radar (defined by its compressed pulse length, τ_{comp}) (Fig. 23-3). At large grazing angles, the area on the ground may be defined by the azimuth and elevation beamwidths of the antenna (Fig. 23-3a). At lower grazing angles, the area of the patch will be defined by the azimuth beamwidth and the compressed pulse length (Fig. 23-3b). As we shall see later, the angular resolution of the radar may also be a function of its Doppler resolution, so this needs to be considered when assessing the effective area of the clutter patch, A_g.

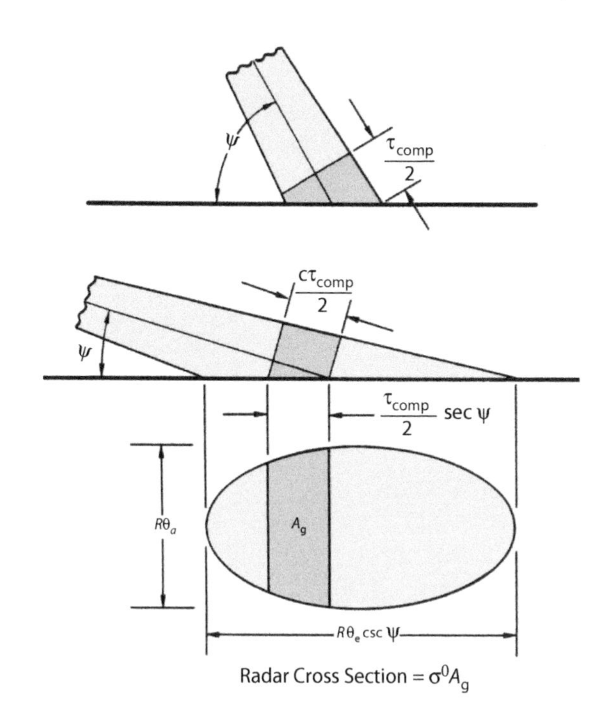

Figure 23-3. The radar cross section of a patch of ground equals the backscattering coefficient, σ^0, times the resolvable ground area, A_g. At a steep grazing angles (ψ), A_g may be determined solely by the radar's Doppler and angular resolution and ψ. Generally, at shallow angles A_g also is limited by the compressed pulse width, τ_{comp}.

The approach to estimating the effective size of the clutter patch (Fig. 23-3) assumes that the antenna has a "square" beamshape with constant gain over the beam and no sidelobes. We shall see later that sidelobes can make a significant contribution to the Doppler spectrum of the clutter, but the bulk of the backscattered energy is received through the mainbeam. The method of estimating performance illustrated in Figure 23-3 gives a sufficiently accurate estimate of the total clutter energy for most applications.

To be able to predict the radar performance in clutter under different conditions, we need to be able to assign appropriate values to σ^0. These values have generally been estimated by analysis of very large numbers of measurements under a wide range of different conditions. Many researchers have contributed to these data over many years. σ^0 is dependent on the type of terrain, the grazing angle, the radar frequency, and the radar polarization. A perfectly smooth conducting surface has no backscatter. It behaves like a perfect mirror with all the incident energy reflected forward. But as the surface roughens, the backscatter reflectivity increases and the amount of forward-scattered energy decreases. For a given roughness, σ^0 will vary with the grazing angle. This is illustrated in Figure 23-4. A rough surface viewed at low grazing angles appears increasingly smooth as the grazing angle decreases and its backscatter decreases. Individual scatterers (e.g. rocks, lumps of earth, etc.) may also shadow each other, which also decreases the backscatter. This range of grazing angles, where there is a marked reduction of σ^0, is known as the interference region, as illustrated in Figure 23-4. As the grazing angle, ψ, increases, the value of σ^0 increases and over a large range of grazing angles varies approximately in proportion to sin(ψ). This is known as the plateau region. Then at very high grazing angles (approaching normal incidence) the ground again also appears almost mirrorlike, now giving a very large quasi-specular backscatter. This is why we observe a particularly large response through the sidelobes at these grazing angles, known as the altitude return (Fig. 23-1).

The value of σ^0 for horizontally polarized signals may be less than for vertically polarized signals, as illustrated in Figure 23-4. This is particularly observed for relatively smooth surfaces (such as water or flat fields). However, for many types of ground clutter, dependence on polarization may be relatively small.

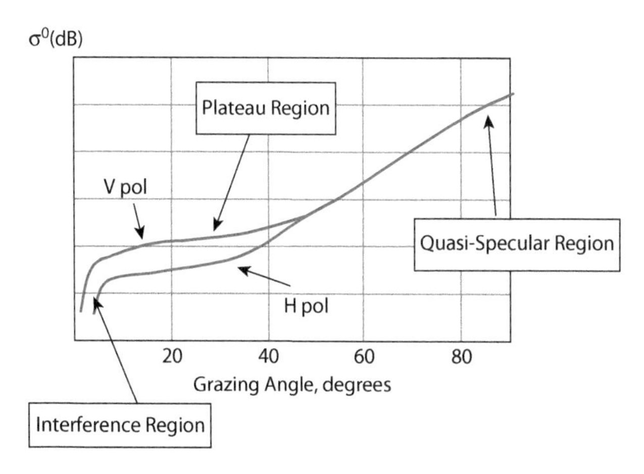

Figure 23-4. Shown here is a variation of σ^0 with grazing angle. (Adapted from Long, as in POMR Figure 5. [9]).

There is a strong dependence on radar frequency for the values of σ^0 for some types of terrain. The apparent roughness of a surface is dependent on the radar wavelength. As the wavelength decreases, scatterers appear proportionately larger and the surface then appears rougher, giving larger values for σ^0. However, such dependence is not always observed. For some types of terrain, such as forest or farmland, there may be little dependence on radar frequency, and the reflectivity of mountainous areas may be higher at VHF than at X-band.

The values of σ^0 appropriate to different conditions have been widely studied and a summary can be found in POMR, Chapter 5. As mentioned before, in the plateau region, σ^0 is proportional to $\sin(\psi)$, and a useful guide to values of σ^0 in different types of terrain can be obtained from what is known as the constant γ model, where

$$\gamma = \frac{\sigma^0}{\sin(\psi)}$$

Table 23-1 shows some typical values for σ^0 for different types of terrain, at a frequency of 10 GHz and a grazing angle of 10°. Equivalent values of γ are also presented. These backscatter coefficients are expressed in units of dBm²/m², signifying the ratio in decibels (dB) of the radar cross section (m²) to the clutter patch area (m²). From Table 23-1, small values of reflectivity result in large negative values of backscatter coefficient, expressed in decibels, while very large values of reflectivity can result in positive values in decibels, as observed for backscatter from cities.

Table 23-1 Typical backscatter coefficients.

	σ^0 dBm²/m²*	γ dBm²/m²
Smooth water	−53	−45.4
Desert	−20	−12.4
Wooded area	−15	−7.4
Cities	−7	0.6

*Values for a 10° grazing angle and 10 GHz frequency.

23.2 Doppler Spectra of Ground Clutter Returns

Mainlobe Return. If the radar antenna mainlobe is illuminating the ground, as when looking down from high altitude or flying at low altitude and not looking up, the radar will receive returns from the ground. Even when flying at high altitude and looking straight ahead, the lower part of the main beam may intersect the ground at long ranges. The average magnitude of the returns for a given range and bearing can be calculated, as discussed above, from a knowledge of the backscatter coefficient and applying the radar-range equation. In addition to the clutter amplitude, we are also interested in any Doppler shift it may have. Because the radar is on a moving platform, the returns from the ground will indeed have a Doppler shift, which will vary according to the viewing geometry.

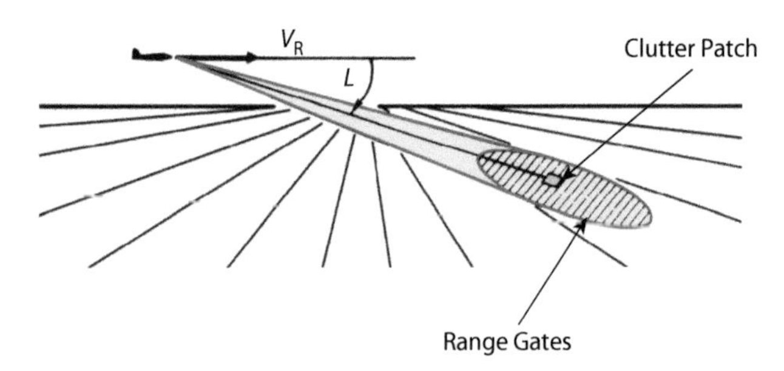

Figure 23-5. The area illuminated by the mainlobe may be thought of as consisting of many small ground patches, each at a different look angle.

Doppler Frequency versus Angle. The spectral characteristics of mainlobe return are best understood by visualizing the ground area illuminated by the mainlobe as consisting of a large number of small, individual patches (Fig. 23-5). The Doppler frequency, f_D, of each small patch is proportional to the cosine of the angle, L, between the radar velocity vector, V_R, and the line of sight to the ground patch:

$$f_D = \frac{2V_R \cos L}{\lambda}$$

where

V_R = velocity of the radar

L = angle between V_R and the line of sight to the ground patch

λ = wavelength

The angle L is not the same for every patch. As a result, the collective return occupies a band of frequencies.

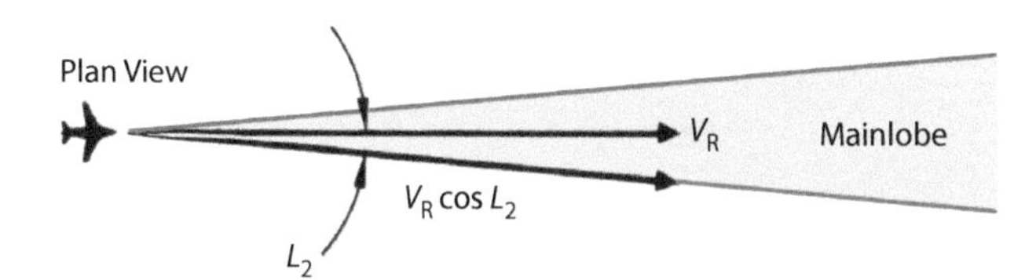

Figure 23-6. When looking straight ahead, the relative velocity (closing rate) for all look angles within the beam is about the same, so that $\cos L_2 \approx 1$ and $f_D \approx 2V_R/\lambda$.

When the antenna is looking straight ahead (Fig. 23-6), the Doppler frequency of the return from patches near the center of the illuminated area ($L \approx 0$) very nearly equals its maximum possible value: $f_{d_{max}} = 2V_R/\lambda$.

The Doppler frequencies of patches farther from the center are lower. But since the angles to these patches are small and the cosine of a small angle is very nearly one, the band of frequencies covered by the mainlobe return when looking straight ahead is quite narrow.

From the perspective of the radar, the antenna will be steered in azimuth, measured in the horizontal plane, and in elevation or depression, measured in the vertical plane. Generally, an antenna pointing down below the horizon in level flight will be described as having a positive depression angle. As the azimuth and depression angle of the antenna increase (Fig. 23-7), the angle L increases and thus the cosine of L for patches at the center of the illuminated ground area decreases. Consequently, the Doppler frequency of these patches decreases. At the same time, the spread between the values of cos L for patches at the two edges of the area increases, causing the band of Doppler frequencies covered by the mainlobe clutter to become wider.

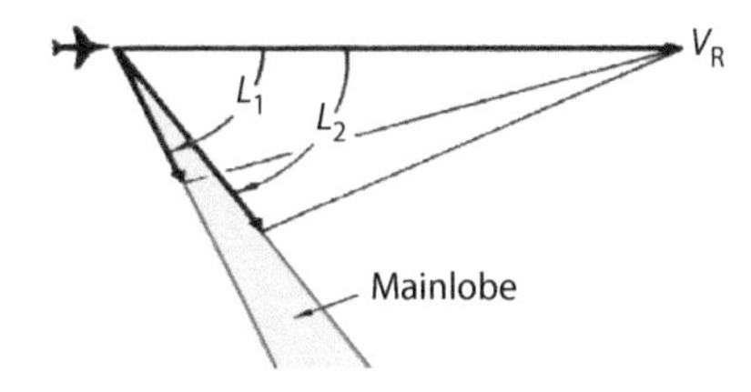

Figure 23-7. As angle L increases, the value of cos L for patches at the center of the beam decreases and the spread between values for patches at the edges of the beam increases.

To give you a quantitative feel for these relationships, the cosine of the angle L off the center of the illuminated area is plotted in Figure 23-8 for values of L between +90° and −90°. The vertical scale gives the corresponding Doppler frequencies for a radar velocity of 250 m/s and a wavelength of 3 cm.

Superimposed over the graph are two vertical bands. Each brackets those angles encompassed by a mainlobe having a beamwidth of 4°. The band in the center is for an antenna azimuth angle of zero. The other band is for an antenna azimuth angle of 60°. (In both cases, the antenna depression angle is zero and the aircraft is assumed to be at very low altitude.)

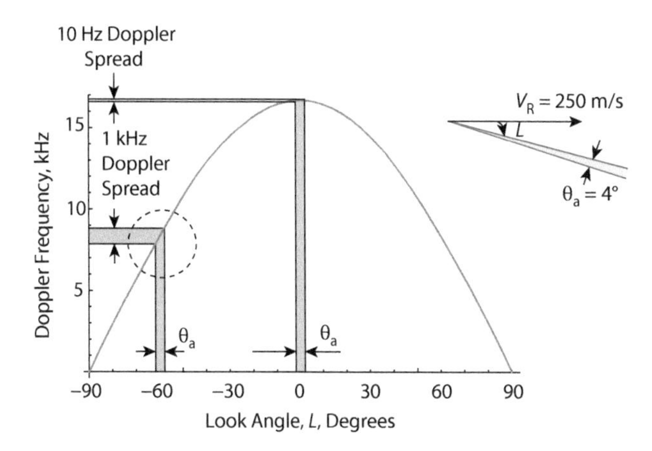

Figure 23-8. This figure shows the variation in the Doppler frequency of mainlobe clutter versus the look angle for λ = 3 cm and V_R = 250 m/s. Vertical bands represent the Doppler spread across the antenna mainlobe for θ_a = 4°.

When the azimuth is 0°, the central Doppler frequency of the return is 16.66 kHz. Yet, when the azimuth increases to 60°, this frequency is only 8.33 kHz—a decrease of 50 percent (cos 60° = 0.5). On the other hand, the width of the band of frequencies spanned by the return (the Doppler spread) is much greater at the larger antenna angle. When the azimuth is 0°, the

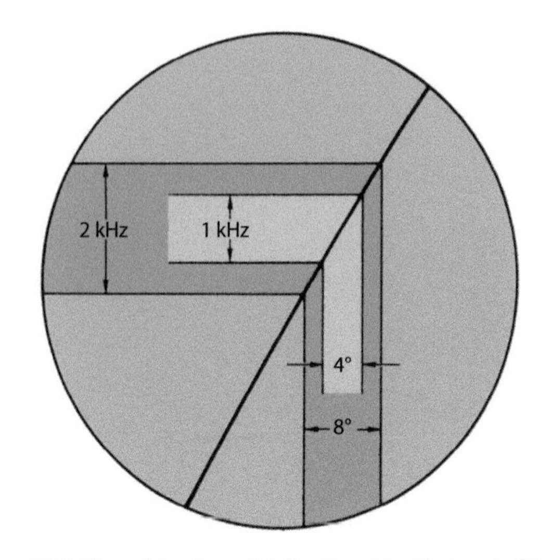

Figure 23-9. The wider the mainlobe, the wider the band of the mainlobe clutter frequencies.

Doppler frequency of a patch at the edge of the illuminated ground area ($f_{D_{max}}$ cos 2°) is so close to that of a patch at the center ($f_{D_{max}}$) that the difference cannot be read from the graph. Actually, it is about 10 Hz. Yet, since the cosine changes much more rapidly at large angles, when the azimuth is 60°, the return spans a band of frequencies just over 1 kHz wide—$f_{D_{max}}$ (cos 58° – cos 62°) = 16.66 (0.53 – 0.47) ≈ 1 kHz.

Influence of Beamwidth, Speed, and Wavelength. For any one antenna azimuth (and/or depression) angle, the wider the mainlobe, the wider the band of mainlobe frequencies. Figure 23-9 shows the area within the dashed circle in Figure 23-8, for different beamwidths. If the beamwidth is increased from 4° to 8°, the width of the band for an antenna angle of 60° would be 2 kHz—twice that for the 4° beam.

Both the center frequency and the width of the band vary directly with the speed of the radar ($f_{D_{max}} \propto V_R$): if it decreases, they decrease; if it increases, they increase. Suppose the center frequency is 8 kHz. If the speed is doubled, this frequency as well as the frequencies at the edges of the band will double. Not only will the entire band shift up by 8 kHz, but its width will double (Fig. 23-10).

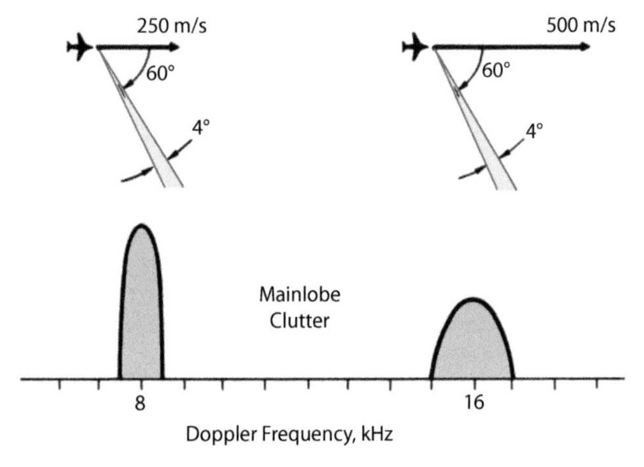

Figure 23-10. If the speed of the radar is doubled, both the center frequency and the width of the spectrum will double.

Also, the width and center frequency vary inversely with wavelength ($f_{D_{max}} \propto 1/\lambda$): the longer the wavelength, the narrower the band will be, and vice versa. Other conditions being the same, at S-band wavelengths (10 cm), the band is only 3/10 as wide as at X-band wavelengths (3 cm) (Fig. 23-11).

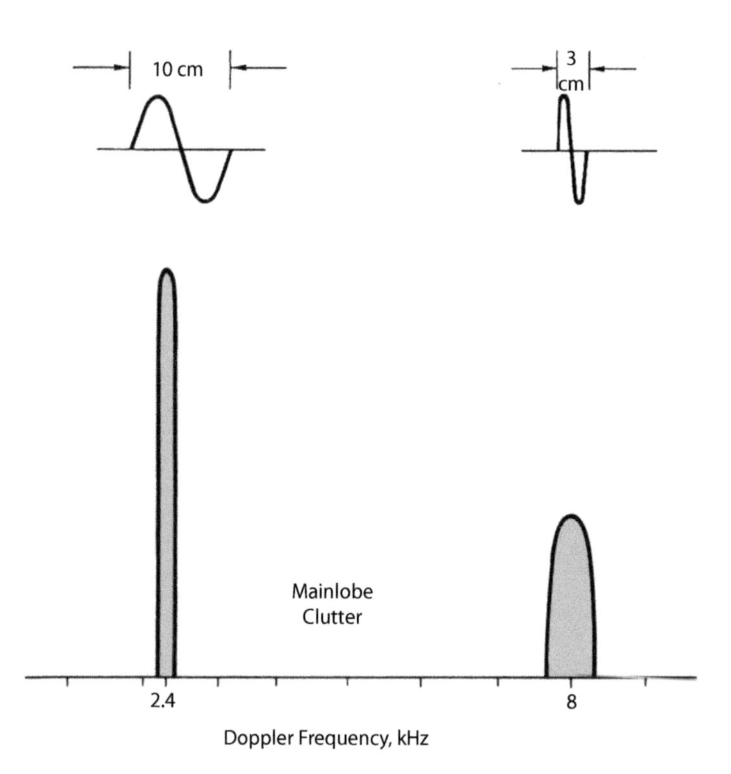

Figure 23-11. If the wavelength of the radar transmitter is decreased, both the center frequency and the spectral width of the mainlobe clutter will increase proportionately.

Effect of Antenna Scan. In a typical radar search mode, the antenna scans back and forth through an azimuth angle that may be ±70° or more. As it sweeps from one extreme to straight ahead (Fig. 23-12), the mainlobe clutter band moves up in frequency and simultaneously squeezes into a narrow line. As the sweep continues to the other extreme, the clutter moves down in frequency and spreads to its original width. Thus the band appears to "breathe."

Significance. Because of its strength, spectral width, and variability, the mainlobe return can be difficult to counter when searching for aircraft. On the other hand, the strength and spectral width are advantageous when ground mapping. In the latter case, the stronger the mainlobe return, the better, and the wider the band of frequencies it occupies, the higher the angular resolution that can be obtained through Doppler processing.

Sidelobe Clutter. The radar return received through the antenna's sidelobes is always undesirable and is called *sidelobe clutter*. Excluding the altitude return, sidelobe clutter is not nearly as concentrated (less power per unit of Doppler frequency) as mainlobe clutter, but it covers a much wider band of frequencies.

Frequency and Power. Antenna sidelobes extend in all directions, even to the rear, although they may be quite low in some directions. This is discussed further in Chapter 8 (Directivity and the Antenna Beam), where it is noted that the sidelobes close to the mainbeam direction may have a

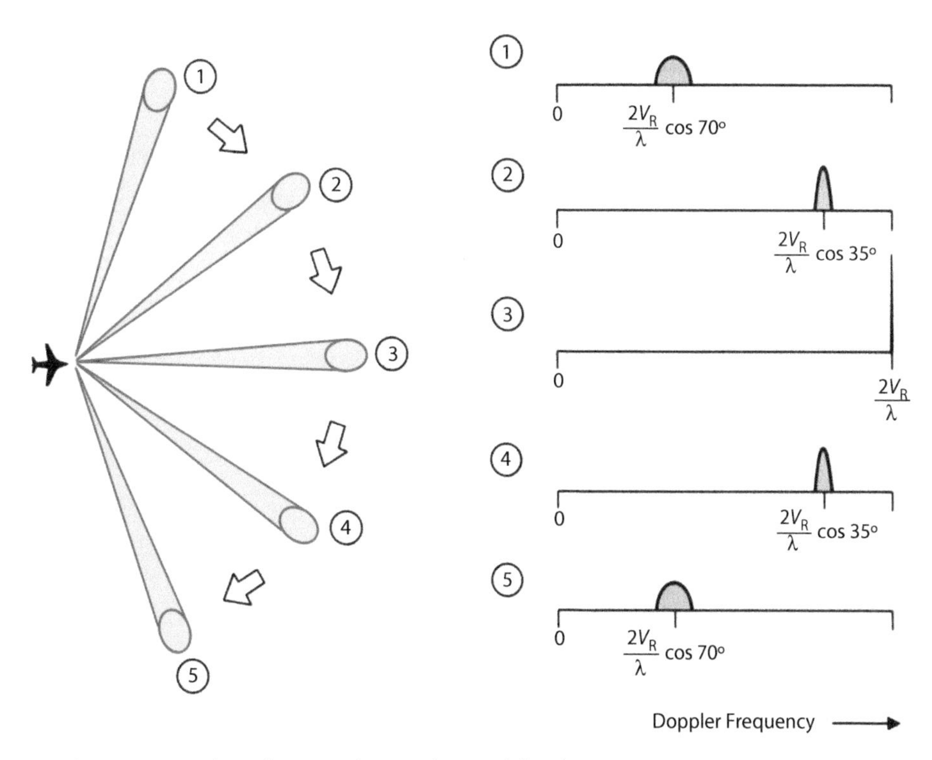

Figure 23-12. As the antenna beam sweeps through its search scan, the mainlobe clutter spectrum moves out to its maximum frequency and squeezes into a narrow line, then returns again.

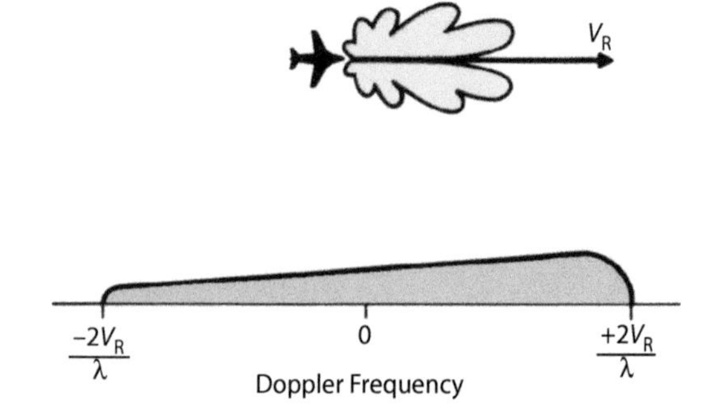

Figure 23-13. Because sidelobes are radiated in all directions, sidelobe clutter extends from a positive frequency corresponding to the radar's velocity to an equal negative frequency.

peak gain of perhaps 20 dB less than the mainbeam, while the far sidelobes may have peak gains 30 dB less than the mainbeam. Although the sidelobe gain may be much less than the mainbeam, the power received from the ground through the sidelobes can still be significant compared with the return from a small wanted target. So, regardless of the antenna look angle, we must be concerned with the signals received from sidelobes pointing ahead, behind, and at every angle in between. When the range between the radar and the clutter is decreasing, the sidelobe clutter will have positive Doppler frequencies and when the range is increasing it will have negative Doppler frequencies. The band of frequencies covered by the sidelobe clutter extends from a positive frequency corresponding to the radar's velocity ($f_D = 2V_R/\lambda$) to an equal negative (less than the transmitter's) frequency (Fig. 23-13).

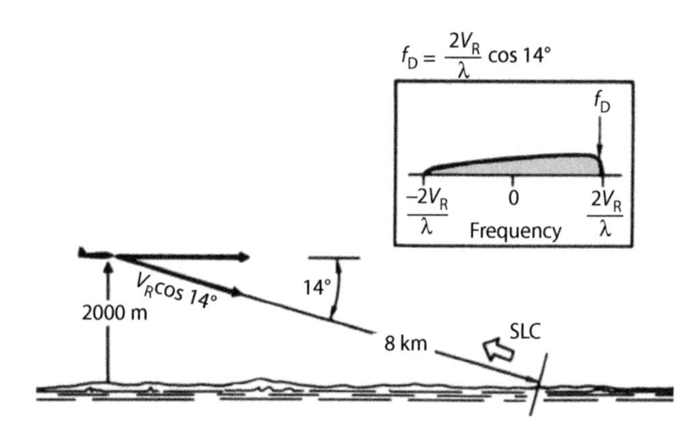

Figure 23-14. At an altitude of 2000 m, the sidelobe return from a range of only 8 km will have a Doppler frequency almost equal to the maximum sidelobe clutter frequency.

While the power radiated in any one direction through the sidelobes is relatively small, the area illuminated by the sidelobes is large. As we shall see in Chapter 24, the returns from a target at long range may also have to contend with clutter from short ranges, due to range ambiguities in the radar waveform. Even if the mainbeam is only illuminating the ground at long range, the sidelobe returns will include strong clutter from relatively short ranges. This is so, even out to the ends of the sidelobe clutter spectrum. As illustrated in Figure 23-14, since the cosine of a small angle is nearly equal to one, if a radar is at an altitude of 2000 m, return from a range of only 8 km (depression angle = 14°) will have a Doppler frequency only 3 percent less than the maximum ($2V_R/\lambda$).

In summary, not only can sidelobe clutter power be substantial, but the clutter may be spread over a broad band of Doppler frequencies.

Impact on Target Detection. The extent to which the clutter interferes with target detection depends on the frequency discrimination the radar provides. Figure 23-15 shows the geometry that determines a radar's ability to discriminate on the basis of differences in range and Doppler frequency. Returns from equal range lie on the surface of a sphere, centered on the radar. For returns from the ground, we are interested in the intersection of the ground plane with this sphere. Similarly, returns from the ground with equal Doppler shift are determined by the intersection of the ground plane with a cone around the radar's velocity vector. Since the angle between this vector and every point on the cone is the same, the return from every point along the contour has the same Doppler frequency.

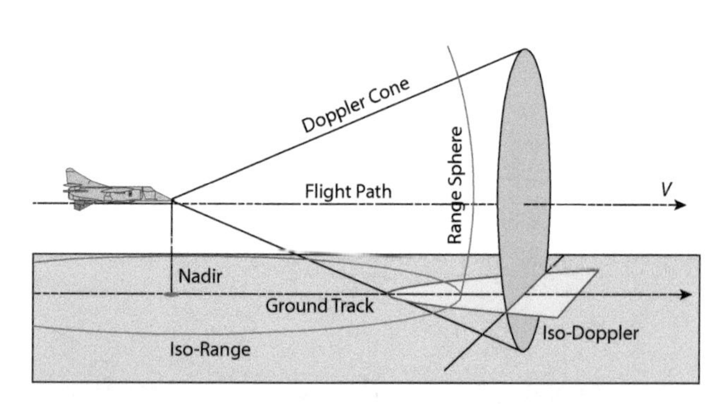

Figure 23-15. Lines of iso-range and iso-Doppler for reflections from the ground surface. (From J. Ender, Aerospace Radar tutorial slides.)

Lines of constant Doppler frequency, called iso-Doppler contours are illustrated in Figure 23-16. Just as the distances between contour lines on a relief map correspond to a fixed interval of elevation, so the distances between iso-Doppler lines correspond to a fixed interval of Doppler frequency.

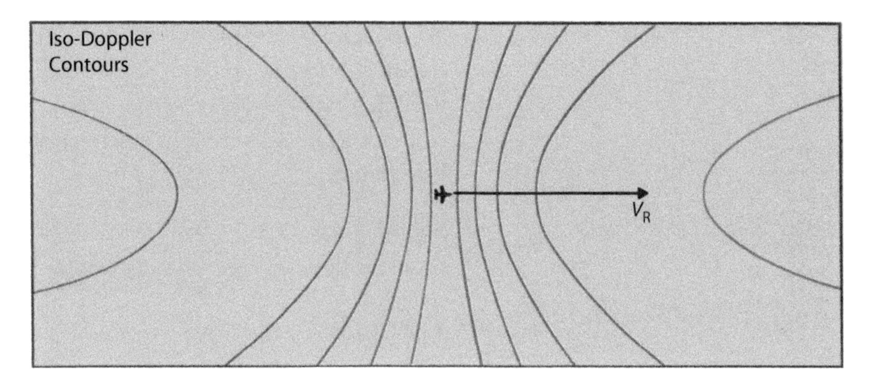

Figure 23-16. Note the lines of constant Doppler frequency—iso-Doppler contours—plotted on a plan view of the ground. Each contour corresponds to the intersection of a cone about the radar's velocity vector with the ground.

Let us suppose now that the Doppler interval corresponds to the minimum difference in Doppler frequency that can be discerned by a particular radar—that is, its Doppler resolution. If the radar differentiates between targets and clutter solely on the basis of Doppler frequency, a target falling amid the sidelobe clutter must compete with the return from the entire strip of ground between the contours bracketing the target's Doppler frequency.

And, as is made clear in Figure 23-17, depending upon the target's range rate, much of this ground may be at substantially closer range than the target.

To appreciate this point, remember that the strength of the radar return is inversely proportional to the fourth power of the range from which it is received. For given values of antenna gain and backscattering coefficient, the power of the return from a ground patch at a range of, say, 1 km is $(10/1)^4 = 10{,}000$ times (40 dB greater than) that of the return from a ground patch of the same size at a range of 10 km.

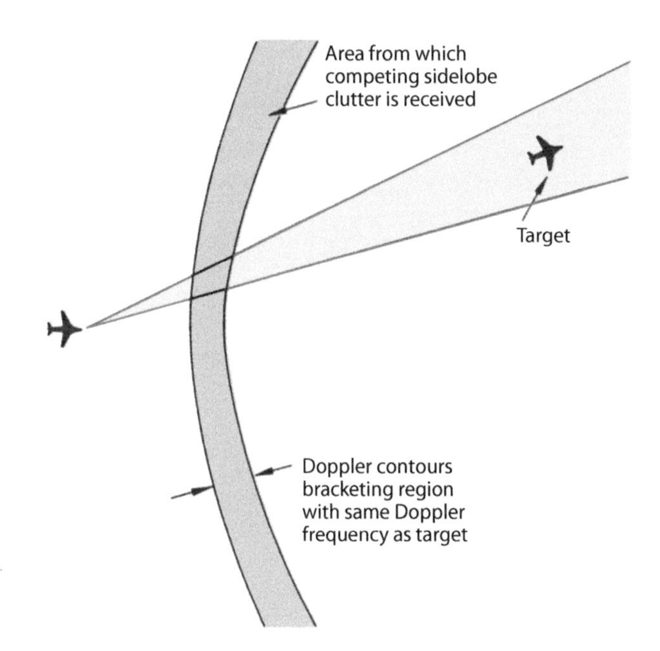

Figure 23-17. If a radar differentiates between target echoes and ground return solely on the basis of Doppler frequency, sidelobe clutter can present a serious problem.

If the radar also provides range resolution (as by range gating), the target must compete only with that portion of the clutter passed by the same range gate as the target's echoes. The range contours will be circles on a plot such as Figure 23-18, which illustrates the combined effect of range and Doppler processing, where now the contours are shown for selected range gates and Doppler resolution intervals. The spectrum at a particular range is found from the intersection of a given range gate with the iso-Doppler contours.

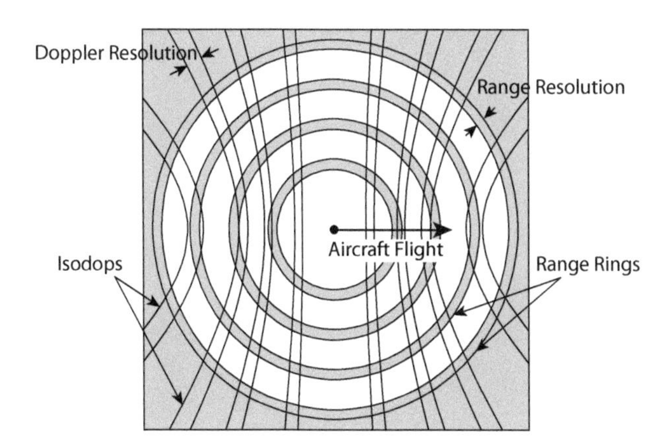

Figure 23-18. Shown in this figure are selected range rings and iso-Doppler contours (isodops) for a radar using range-Doppler processing.

Other Factors Governing Sidelobe Clutter Strength. The strength of the sidelobe return from any one patch of ground depends upon several factors besides range. One is the gain of the particular sidelobe within which the patch lies. In a representative fighter radar, the two-way gain of the first sidelobe beyond the mainlobe is on the order of 100 times (20 dB) stronger than that of the weaker sidelobes. The strength of the return also varies widely with the nature of the terrain included in the patch—its scattering coefficient.

As explained previously, as the grazing angle increases, the backscattering coefficient also increases. Consequently, even though a radar is closer to the ground at low altitudes, sidelobe clutter may be most severe when flying at moderate rather than low altitudes.

Significance. Clearly, the extent to which sidelobe clutter is a problem depends upon many things.

- Frequency resolution provided by the radar
- Range resolution provided by the radar
- Gain of the sidelobes
- Altitude of the radar
- Scattering coefficient and grazing angle

Also, as already noted, certain man-made objects can be immensely important sources of sidelobe clutter (see the separate discussion at the end of this chapter).

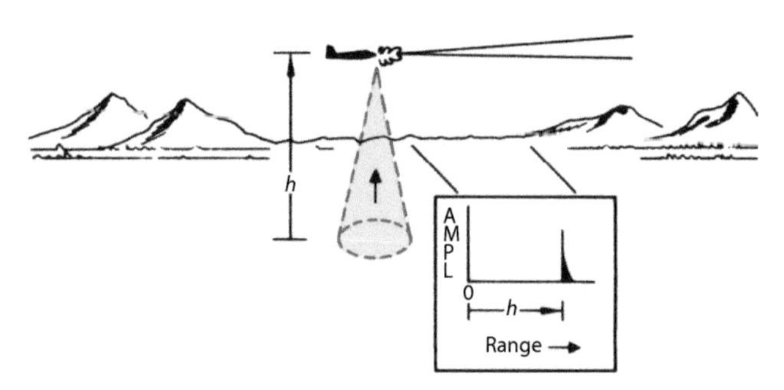

Figure 23-19. Altitude return comes from a large area, often at very close range.

Altitude Return. Beneath an aircraft there is usually a large region within which the ground is so close to being at a single range that the sidelobe return from it appears as a spike on a plot of amplitude versus range (Fig. 23-19).

The range of this *altitude return* equals the radar's absolute altitude.

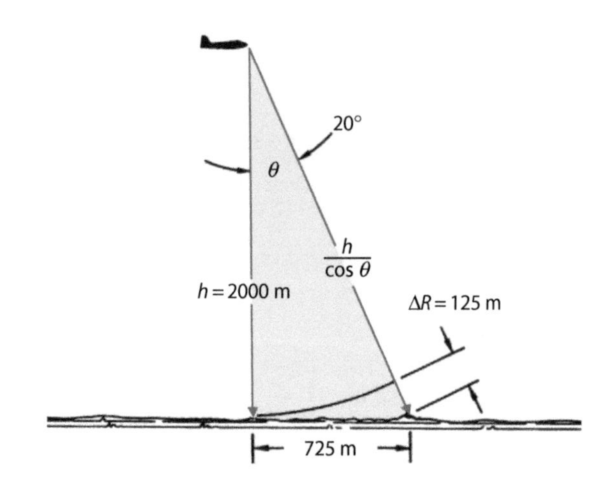

Figure 23-20. At an altitude of 2000 m, even at an angle of incidence, θ, of 20°, the slant range to the ground is only 125 m greater than the altitude.

Relative Strength. The altitude return is not only much stronger than the surrounding sidelobe clutter but may be as strong or stronger than the mainlobe clutter. The area from which it comes not only may be very large, but it is often at extremely close range. In addition, as discussed earlier, the backscatter coefficient, σ^0, may be very high close to normal incidence.

This point is illustrated in Figure 23-20. A radar is at an altitude (h) of 2000 m over flat terrain. The slant range to the ground at an angle of incidence θ equals $h/\cos\theta$. Even when θ is as much as 20°, its cosine is only slightly less than one and therefore the slant range to the ground is only 125 m greater than the altitude (vertical range). Unless the radar's range resolution is better than 125 m, one range cell will receive all the return from this area below the aircraft.

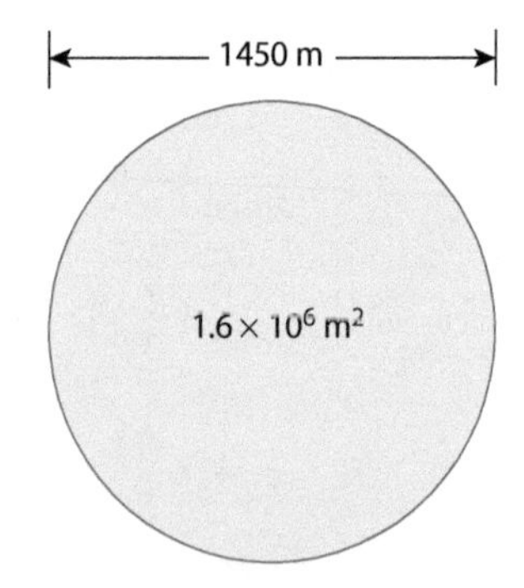

Figure 23-21. At an altitude of 2000 m, an angle of incidence of 20° encompasses a circle with area of about 1.6 million m².

If the slant range at an angle of incidence of 20° is rotated about the vertical axis, it traces a circle on the ground having a diameter of roughly 1450 m (Fig. 23-21).

A circle this size has an area of about 1.6×10^6 m². Thus the radar receives all of the backscatter from this area, at a range of just 2 km, in the round-trip transit time for a range increment of 125 m. That time is only about 1 μs.

Furthermore, as mentioned above, at near-vertical incidence, the backscattering coefficient tends to be very large.

Over water, the coefficient is enormous. Thus the altitude return appears as a sharp spike in a plot of amplitude versus range.

Doppler Frequency. The altitude return also peaks in a plot of amplitude versus frequency, but not sharply. The reason can be seen in Figure 23-22.

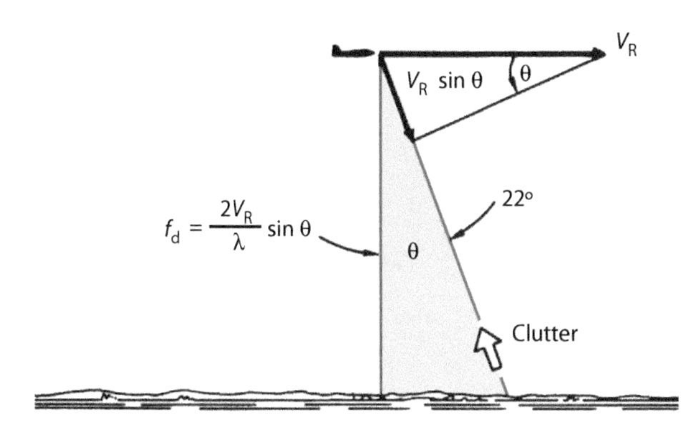

Figure 23-22. Doppler frequency of the return from a given angle of incidence, θ, is proportional to the projection of radar velocity onto the slant range at that angle, hence sin θ.

The projection of the radar velocity, V_R, on the slant range to the ground equals V_R sin θ. Unlike the cosine, the sine of an angle changes most rapidly as the angle goes through zero. While the Doppler frequency of the clutter is zero when θ is zero, it increases to nearly 40 percent of its maximum value ($2V_R/\lambda$) at an angle θ of only 22°. The return from a circle of ground that produces a sharp spike in a plot of amplitude versus range thus produces a broad hump in a plot of amplitude versus Doppler frequency (Fig. 23-23).

Normally the Doppler frequency of the altitude return is centered at zero. However, if the altitude of the radar is changing—as when the aircraft is climbing, diving, or flying over sloping terrain—this will not be the case. For a dive, the Doppler frequency will be positive (Fig. 23-24); for a climb, it will be negative. Even though the frequency is generally fairly low, it can be considerable. In a 30° dive, for instance, the altitude changes at a rate equal to half the radar velocity.

Significance. Despite its strength, the altitude return is usually less difficult to deal with than the other ground returns. Not only does it come from a single range, but its range is predictable; and, as we have seen, its frequency is generally close to zero—although that unfortunately is the Doppler frequency of a target pursued at constant range (e.g., a tail chase with zero closing rate, $\dot{R} = 0$).

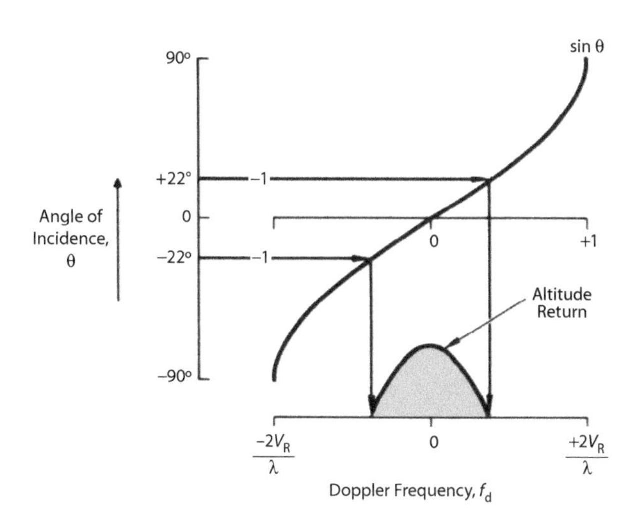

Figure 23-23. Since sin θ changes most rapidly as θ goes through zero, the altitude return is spread over a comparatively broad band of Doppler frequencies.

23.3 Relation of Clutter Spectrum to Target Frequencies

Having become familiar with the characteristics of mainlobe clutter, sidelobe clutter, and altitude return individually, let us look briefly at the composite clutter spectrum and its relationship to the frequencies of the echoes from representative airborne

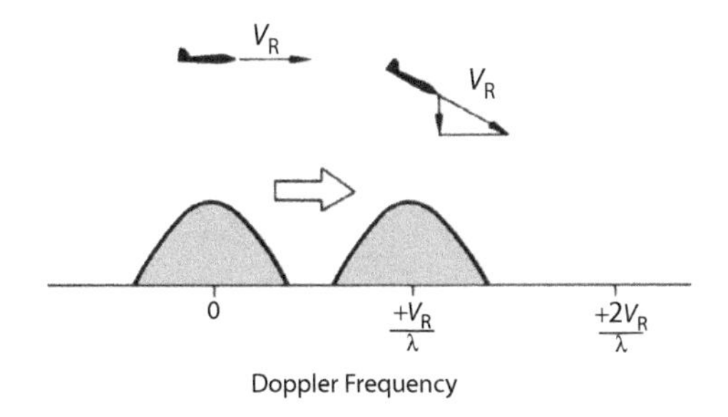

Figure 23-24. Doppler frequency of altitude return is normally low but may be quite high in a dive.

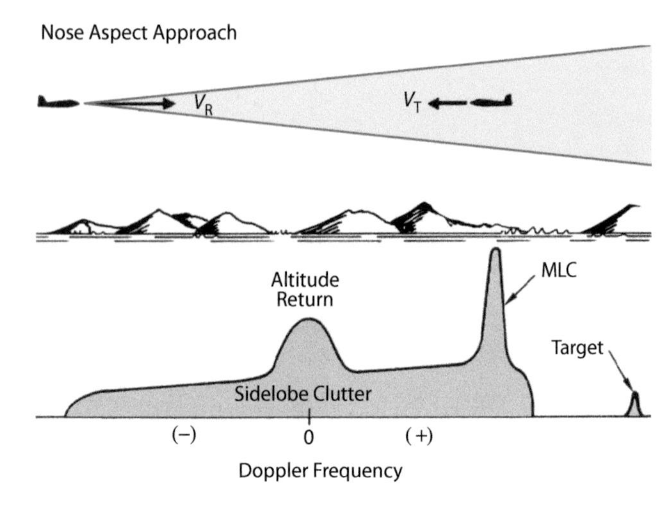

Figure 23-25. For a nose-aspect approach, the Doppler frequency of the target is greater than that of any ground return. MLC = mainlobe clutter.

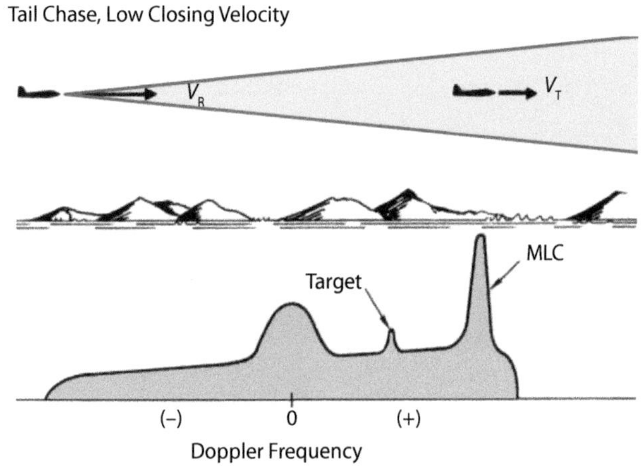

Figure 23-26. For a tail chase, the Doppler frequency of the target falls within the sidelobe clutter. MLC = mainlobe clutter.

targets in typical operational situations. Again, we assume that the PRF is high enough to avoid Doppler ambiguities.

Figure 23-25 illustrates the relationship between target and clutter frequencies for a nose-aspect approach. Because the relative velocity between the radar and the target is greater than the radar's velocity, the target's Doppler frequency is greater than that of any of the ground return.

Figure 23-26 shows the relationship for a tail chase. Because the target's range rate as measured by the radar is less than the radar's velocity, the target's Doppler frequency falls within the band of sidelobe clutter, at a Doppler frequency dependent upon the range rate.

In Figure 23-27, the target's velocity is perpendicular to the line of sight from the radar. Now, the target has the same Doppler frequency as the mainlobe clutter. Fortunately, a target will attain such a relationship only occasionally and usually will remain in it fleetingly.

In Figure 23-28, the target's closing rate is zero. Now the target has the same Doppler frequency as the altitude return.

Figure 23-29 shows two targets that are moving away from the radar. Target A has a rate of change of range relative to the radar that is greater than the radar's ground speed (V_R), so this target appears in the clear beyond the negative-frequency end of the sidelobe clutter spectrum. On the other hand, target B has a relative rate of change of range less than V_R. So this target appears within the negative-frequency portion of the sidelobe clutter spectrum.

With these situations as a guide, the relationship of the Doppler frequencies of target return and ground return for virtually any situation can easily be pictured (Fig. 23-30). Bear in mind, however, that at lower PRFs, Doppler ambiguities can occur that may cause a target and a ground patch having quite different range rates to appear to have the same Doppler frequency.

The consequences of such ambiguities will be discussed in the next chapter. The signal processing commonly performed

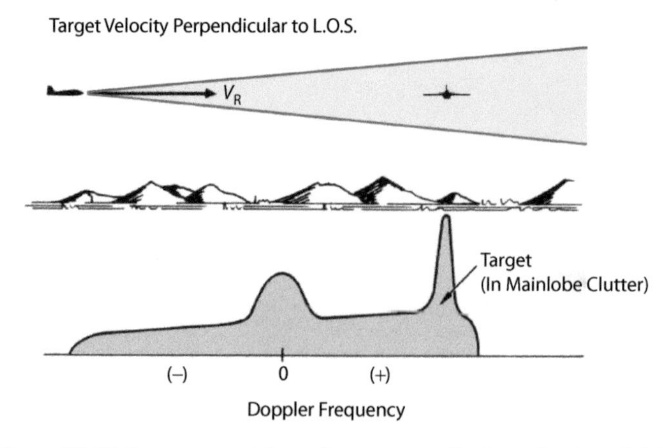

Figure 23-27. For a target with a velocity perpendicular to the line of sight (LOS), the target return is obscured by mainlobe clutter.

to separate target echoes from ground clutter in the various operational situations presented here, as well as when operating at PRFs that make Doppler frequencies ambiguous, will be described in Chapters 27–30.

23.4 Return from Objects on the Terrain

The return from certain man-made structures can be very strong. For example, viewed straight on by an X-band radar, a smooth, flat metal sign only 1 m square has a radar cross section on the order of 14,000 m^2 (Fig. 23-31) compared with 1 m^2 or less for a small aircraft in some aspects.

This may sound odd, but not if you think about it. Most of the power intercepted by the sign when viewed straight on is reflected back in the direction of the radar. The sign is a specular (mirrorlike) reflector. It acts like an antenna that is trained on the radar and reradiates all of the transmitted power it intercepts. For example, an antenna has a directional gain, G, given by

$$G = \frac{4\pi A_e}{\lambda^2},$$

where A_e is the area of the effective antenna aperture, and λ is the radar wavelength. At X-band frequencies ($\lambda = 3$ cm), an antenna with $A_e = 1$ m^2 has a gain of around 14,000. Multiply the area of the sign by this gain ($1 \times 14{,}000$) and you get a radar cross section of 14,000 m^2.

In principle, the radar return from a flat reflecting surface, such as a sign, is directly comparable to the intense reflections one frequently gets from the windshield of a car or the window of a hillside house when it is struck from just the right angle by the early morning or late afternoon sun.

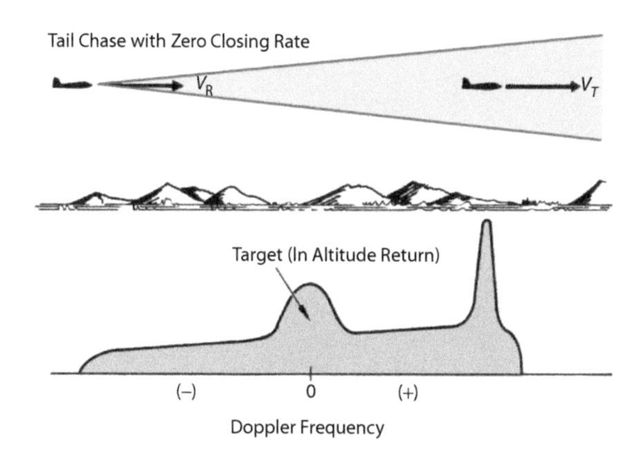

Figure 23-28. For a tail chase with zero closing rate, the target is buried in the altitude return.

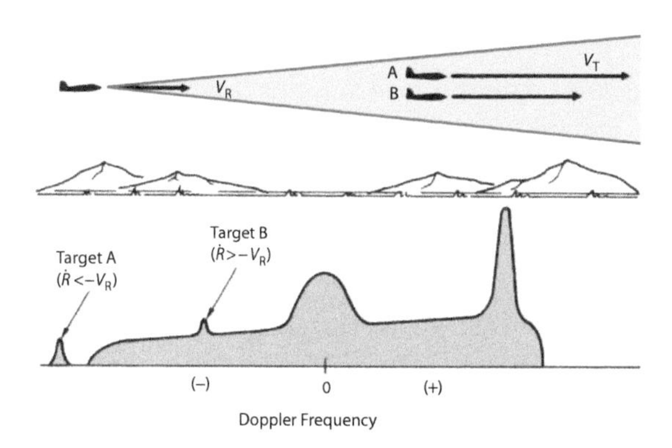

Figure 23-29. If the relative range rate is more negative than $-V_R$, the target appears in the clear (A) below the sidelobe clutter spectrum; otherwise, it appears in the negative-frequency half of the spectrum (B).

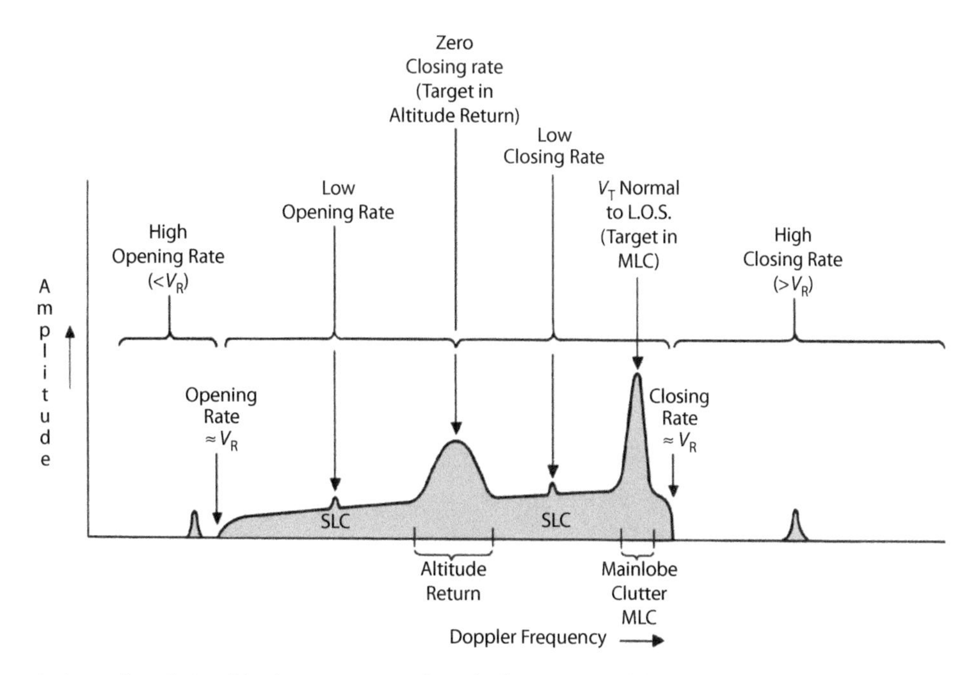

Figure 23-30. This graph shows the relationships between target Doppler frequency and the ground clutter spectrum for various target closing rates. (Doppler frequencies are assumed to be unambiguous.)

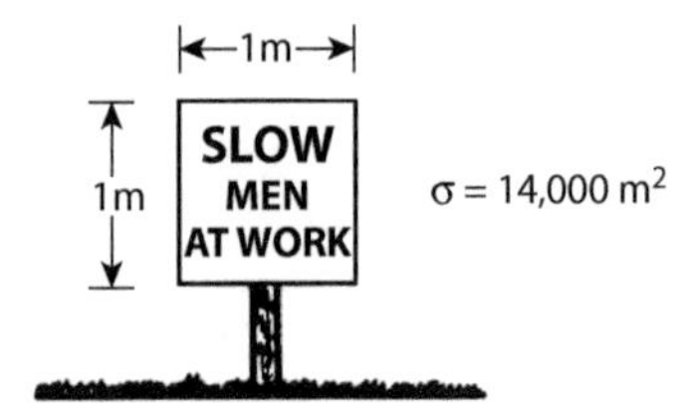

Figure 23-31. Viewed straight on, a smooth, flat plate has a very large radar cross section.

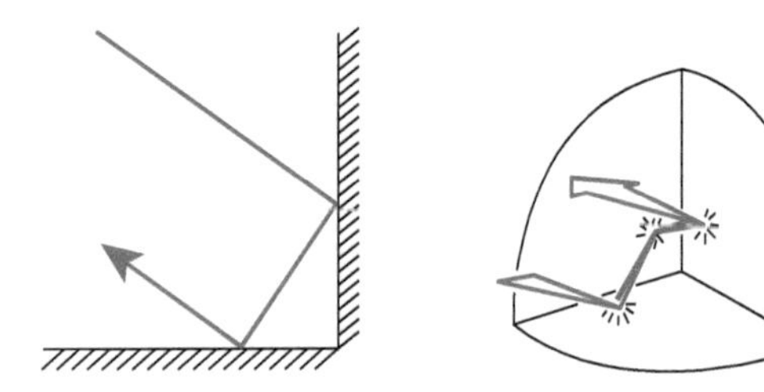

Figure 23-32. A corner formed by two flat surfaces is retroreflective over a wide range of angles, and that formed by three surfaces over an even wider range.

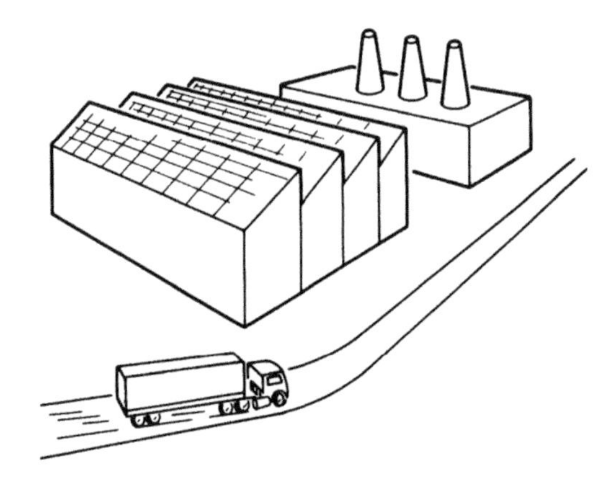

Figure 23-33. Portions of a building may act like corner reflectors, while a truck may act like several corner reflectors.

While a single flat surface such as a sign must be viewed from nearly straight on to reflect the incident energy back to the radar, two surfaces forming a 90° corner will do so over a wide range of angles in a plane normal to the intersection of the surfaces. These surfaces are called *retroreflective*. If a third surface is added at right angles to the other two (forming a corner reflector), the range of angles over which the surfaces will be retroreflective is increased to nearly a quarter of a hemisphere (Fig. 23-32). Incidentally, this is how bicycle reflectors work. Portions of a clutter source such as a large building may act like corner reflectors, and a vehicle such as a truck may look like a group of corner reflectors (Fig. 23-33).

Because of their very large radar cross sections, retroreflective objects on the ground can produce sidelobe returns as strong as or stronger than the echoes from distant aircraft received through the mainlobe. Furthermore, because the objects are of limited geographic extent—they are discrete as opposed to distributed reflectors—all of the return from one of them has very nearly the same Doppler frequency and comes from very nearly the same range. Thus the return may appear to the radar exactly as if it came from an aircraft in the mainlobe.

Naturally, since these objects are virtually all man-made, they are much more numerous in urban rather than rural areas. Nevertheless, they may be encountered almost anywhere. In rural areas, farm buildings such as large barns or silos may have very large radar cross sections (see Figure 23-34). Electricity towers or wind turbines may also give large returns.

Depending on the use of the radar, special measures may be required to reduce or eliminate sidelobe returns of this sort. Mainlobe returns from such objects are usually not a problem, since their Doppler frequencies are generally different from that of targets of interest. But if the objects are moving (e.g. a wind turbine) or have extraordinarily large radar cross-sections, the mainlobe return, too, can be a problem.

Figure 23-34. Even in rural areas there may be numerous structures with large radar cross sections.

23.5 Summary

Backscattering from the ground is modeled in terms of the backscatter coefficient, σ^0, which is multiplied by the area of a ground patch to obtain its radar cross section, σ. The coefficient σ^0 varies with the grazing angle, radar frequency, polarization, electrical characteristics of the ground, roughness of the terrain, and nature of the objects on it.

The most important ground return—and the only return of interest for ground mapping—is that received through the antenna's mainlobe. When the antenna is looking straight ahead, the Doppler frequency of this return corresponds to the radar's full ground speed. As the look angle increases, the Doppler frequency decreases and spreads over an increasingly broad band. Both the center Doppler frequency and the width of this band increase directly with the radar's velocity and are inversely proportional to wavelength.

The ground return received through the sidelobes, although comparatively weak at any one frequency, extends from a positive frequency corresponding to the radar's full velocity ($2V_R/\lambda$) to an equal negative frequency. The portion received from directly below—the altitude return—is especially strong, particularly over water. It appears as a spike on a plot of amplitude versus range and as a broad hump on the Doppler spectrum. Its center Doppler frequency is normally zero.

Man-made objects on the ground may be highly retroreflective and can produce sidelobe return as strong as the target echoes received through the mainlobe.

If the PRF is high enough to eliminate Doppler ambiguities, some targets may have Doppler frequencies that allow the radar to distinguish them from the ground return. This will depend upon the target's velocity relative to the radar. As long as is the target's opening or closing rates relative to the radar are greater than the radar's velocity, the echoes will lie outside the ground return. Otherwise, they must compete with sidelobe return. Only if the target is flying at right angles to the line of sight from the radar will its echoes have the same Doppler frequency as the mainlobe return, and only if the closing rate is zero will target echoes have the same Doppler frequency as the altitude return.

However, as we shall see in the next chapter, Doppler ambiguities can cause the target and unwanted ground patch returns, with quite different range rates, to appear to have the same Doppler frequency, thereby greatly compounding the problem of separating target echoes from clutter.

Further Reading

F. T. Ulaby and M. C. Dobson, *Handbook of Radar Scattering Statistics for Terrain*, Artech House, 1989.

W. C. Morchin, *Airborne Early Warning Radar*, Artech House, 1990.

F. E. Nathanson, J. P. Reilly, and M. N. Cohen, "Sea and Land Backscatter," chapter 7 in *Radar Design Principles: Signal Processing and the Environment*, 2nd ed., SciTech-IET, 1991.

G. Morris and L. Harkness, *Airborne Pulse Doppler Radar*, 2nd ed., Artech House 1996.

M. W. Long, *Radar Reflectivity of Land and Sea*, 3rd ed., Artech House, 2001.

M. I. Skolnik, "Radar Clutter," chapter 7 in *Introduction to Radar Systems*, 3rd ed., McGraw Hill, 2001.

J. B. Billingsley, *Low-Angle Radar Land Clutter: Measurements and Empirical Models*, William Andrew Publishing, 2002.

M. I. Skolnik (ed.), "Ground Echo," chapter 16 in *Radar Handbook*, 3rd ed., McGraw Hill, 2008.

M. A. Richards, J. A. Scheer, and W. A. Holm (eds.), "Characteristics of Clutter," chapter 5 in *Principles of Modern Radar: Basic Principles*, SciTech-IET, 2010.

Test your understanding

1. What is the radar cross section of a patch of ground with an area of 100 m^2 and a reflectivity of –20 dBm^2/m^2?
2. How does clutter reflectivity vary with grazing angle within the plateau region?
3. What are typical values in X-band for a 10° grazing angle of normalized radar cross sections, σ^0, of (a) deserts and (b) cities?
4. For an airborne radar antenna with a fixed beamwidth, for what azimuth look direction relative to the aircraft's velocity vector will the spread in Doppler frequency of the mainbeam ground clutter return be the greatest?
5. If the speed of the radar is halved and the transmitted frequency is doubled, how will the width of the Doppler spectrum of the ground clutter be affected?
6. When will the Doppler shift of the return from an air target be the same as that of the altitude return in the radar?

24

Effect of Range and Doppler Ambiguities on Ground Clutter

E-3D operator position

In the last chapter we surveyed the sources of ground returns and became acquainted with their Doppler spectrum. However, we did not consider the significant effects of range and Doppler ambiguities on the ground return. Although we discussed both types of ambiguities in detail in earlier chapters, the discussions there involved only target returns. If a radar is searching for or tracking a target in the presence of ground clutter, however, the consequences of ambiguities in the clutter are quite different from the consequences of the same ambiguities in the target return.

In the case of a target, we are interested in the target itself and in the *value* of its range or Doppler frequency. Since a target such as an aircraft is essentially a point source, ambiguities simply give the observed range or Doppler frequency more than one possible value. If the ambiguities are not too severe, we can resolve them through techniques such as pulse repetition frequency (PRF) switching.

In the case of ground clutter, on the other hand, we are interested in *differences* in range and Doppler frequency, which will enable us to separate the clutter from the target echoes. Since the sources of the clutter generally are widely dispersed, ambiguities tend to reduce these differences.

In this chapter, after briefly considering the dispersed nature of ground clutter, we examine the effects of ambiguities on the range and Doppler profiles for a representative flight situation

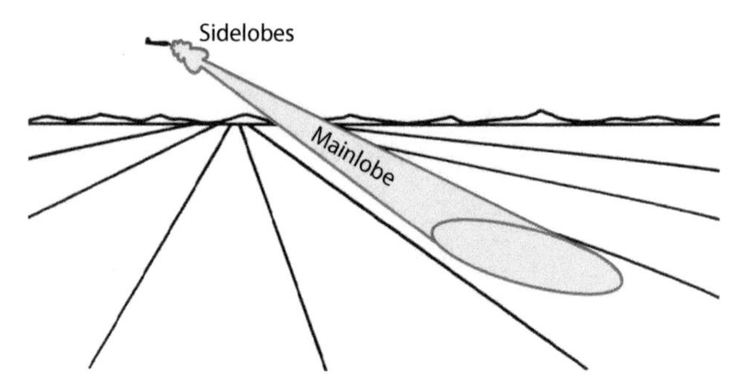

Figure 24-1. When the antenna beam strikes the ground, it illuminates an extensive area in both azimuth and elevation. Additionally, sidelobes illuminate the ground in all directions.

and see how they compound the problem of separating target echoes from clutter.

24.1 Dispersed Nature of the Clutter

As we saw in Chapter 23, when the antenna beam strikes the ground, it usually illuminates an area that is extensive in both range and angle. Furthermore, the antenna invariably has sidelobes through which it radiates an appreciable amount of energy (Fig. 24-1).

Ground return of various amplitudes is thus received from many different ranges and directions. Since the direction to a point on the ground in large measure determines the point's range rate (i.e. the relative rate of change of the range between the radar and the point on the ground), the return also covers a broad band of Doppler frequencies (the Doppler shift, f_D, from a scatterer is given by $f_D = 2v_r/\lambda$ Hz, where v_r is the range rate and λ is the radar wavelength).

Naturally, any spread in range and Doppler frequency of the ground return makes the problem of separating target echoes from it more difficult. Ambiguities in range and Doppler frequency compound the problem by causing clutter from more than one block of ranges and more than one portion of the Doppler spectrum to be superimposed. The effect of this can be visualized most clearly by examining separately the range and Doppler frequency profiles for a representative flight situation. We assume that a radar-equipped aircraft is flying at low altitude over terrain from which a considerable amount of ground return is received. The radar antenna is looking down at a slightly negative elevation angle and at an angle of about 30° to the direction of flight. The Doppler shift imposed on a target return to the radar is determined by the relative velocity between the target and the radar. If the range to the target is reducing with time, the target is described as having a closing rate, which implies a positive Doppler shift, while if the range is increasing it is described as having an opening rate (or sometimes a negative closing rate), giving a negative Doppler shift.

Two air targets, A and B, are in the antenna's mainlobe. Target A is being overtaken from the rear and has a low closing rate. Target B is approaching the radar head-on and has a high closing rate (Fig. 24-2).

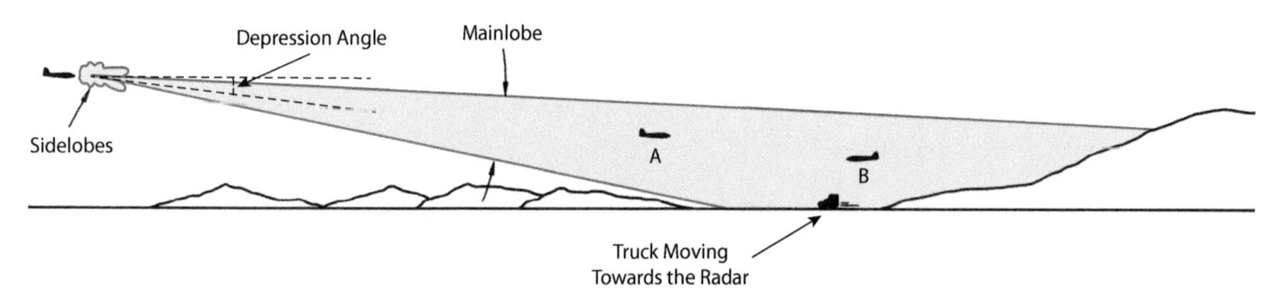

Figure 24-2. In this side view representative flight situation, targets include both low and high closing rate aircraft plus a truck.

For purposes of illustration, target A has been placed at a range from which ground clutter is only being received through the sidelobes; target B is at a range from which both mainlobe and sidelobe ground returns are being received (Fig. 24-3).

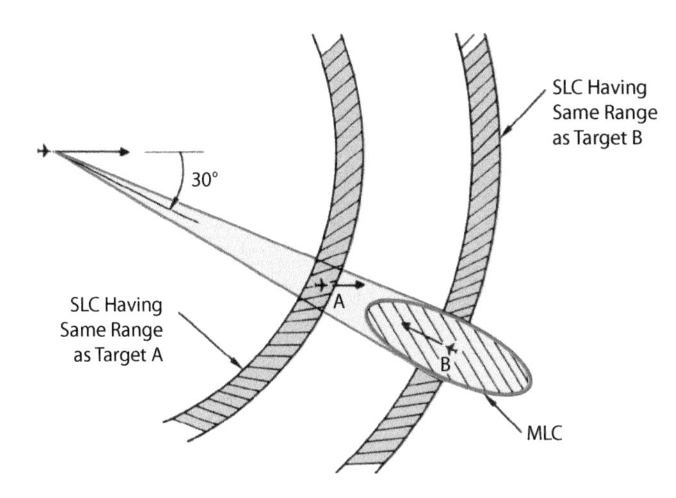

Figure 24-3. Representative flight situation (plan view). Target A is at a range from which only sidelobe ground clutter return is received; target B is at a range from which both mainlobe and sidelobe ground returns are received. SLC = sidelobe clutter; MLC = mainlobe clutter.

Within the ground patch illuminated by the mainlobe is a truck. It is heading toward the radar and thus has a slightly higher closing rate than the ground over which it is traveling.

The flight situation diagram is repeated in Figure 24-4, with the corresponding "true" range profile beneath it. This profile is a plot of the amplitude of the radar return versus the range of its sources relative to the radar, that is the slant range, as opposed to the horizontal range on the ground. Sidelobe clutter, you will notice, extends outward from a range equal to the radar's altitude. Notice how rapidly it decreases in amplitude as the range increases.

The echoes from target A stand out clearly above the sidelobe clutter. In contrast, the echoes from target B and the truck are completely obscured by the much stronger mainlobe clutter. Even though we know exactly where to look for these echoes, we cannot distinguish them from the clutter on the basis of amplitude.

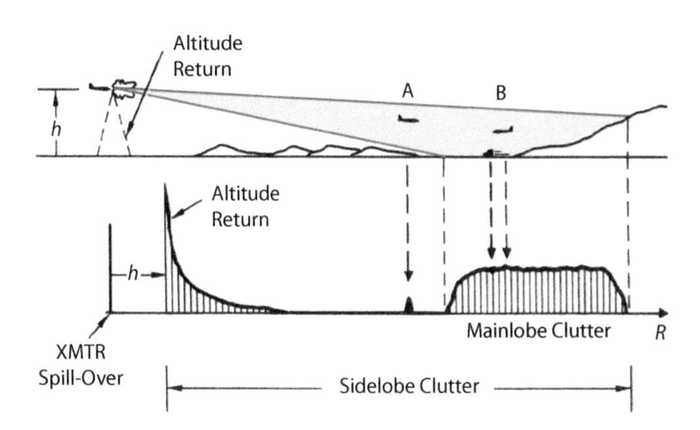

Figure 24-4. In this true range profile of the representative flight situation, air target A can be seen clearly above the sidelobe clutter, but air target B and the truck are obscured by the mainlobe clutter.

Toward the left end of the range profile there is a strong spike. This is the altitude return, received at a slant range h, equal to the aircraft's altitude.

It should be noted that during the pulse transmission the receiver will be blanked, so no signals are seen until after the pulse transmission is completed. Thus the radar pulse length determines the minimum range at which the radar can detect a target.

24.2 Range Ambiguities

Range ambiguities arise when all of the echoes from one pulse are not received before the next pulse is transmitted. As explained in detail in Chapter 11, when a return echo is received from beyond the unambiguous range, R_u, it is impossible to tell which transmitted pulse generated the particular echo (Fig. 24-5). But far more important from the standpoint of clutter rejection is that the returns from ranges separated by R_u are received simultaneously. Therefore the echoes from a target must compete with ground return not only from the target's own range, but from every range that is separated from it by a whole multiple of R_u.

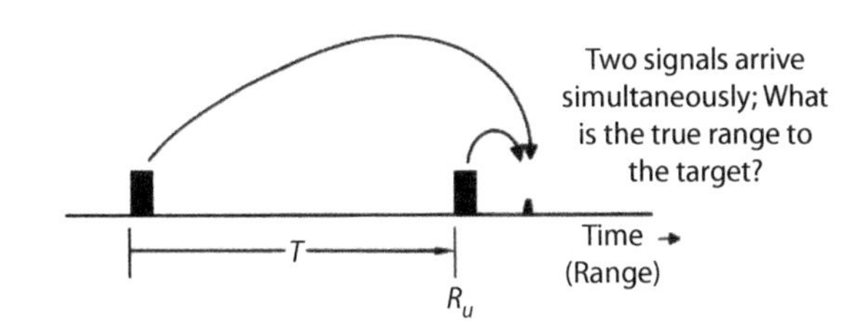

Figure 24-5. Range ambiguities occur when the return from one pulse is being received after the next pulse is transmitted. The range corresponding to the interpulse period, T, is R_u.

Returns from Ranges Separated by R_u. To illustrate the effect of range ambiguities on the range profile observed at the output

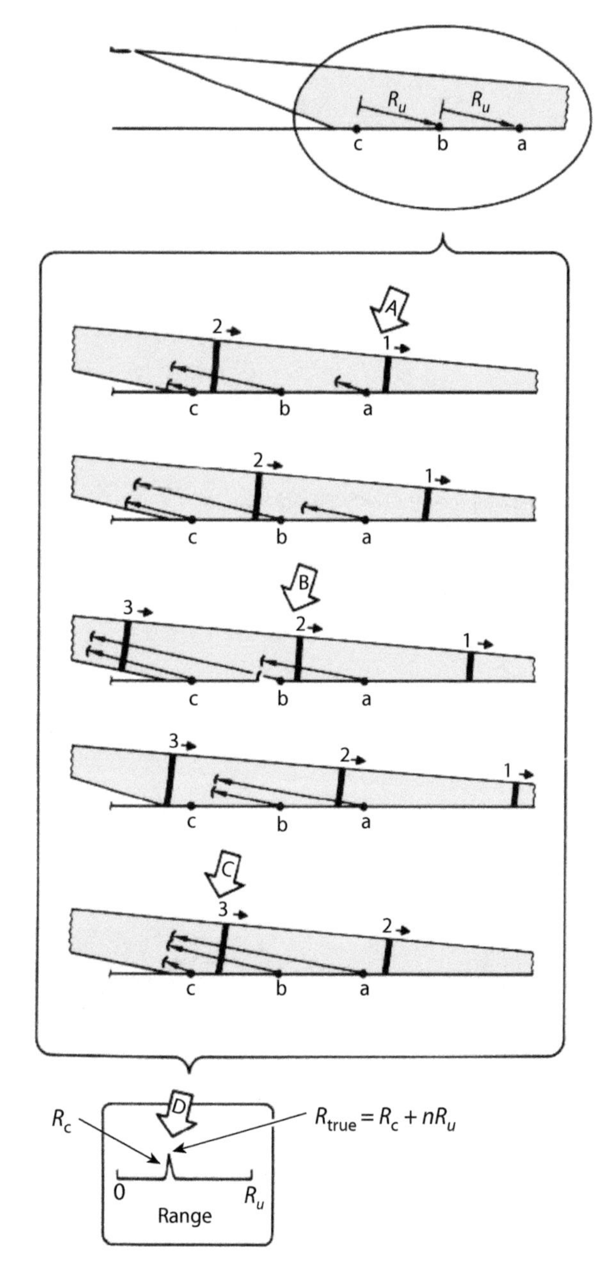

Figure 24-6. This figure shows the return of echoes of three successive transmitted pulses (labeled 1, 2, and 3) from points on the ground (a, b, and c) that are separated in range by the unambiguous range, R_u.

of the receiver, Figure 24-6 traces the paths of the echoes of three successive transmitted pulses from three points on the ground—a, b, and c—which are separated in slant range by R_u. The key points in the figure (called out by the large lettered arrows) are the following.

A. An echo of pulse 1 is reflected from the most distant point, a.

B. This echo reaches the next most distant point, b, just as it is reflecting an echo of pulse 2.

C. Traveling together, these two echoes similarly reach the near point, c, just as it is reflecting an echo of pulse 3. The three echoes travel the remaining distance to the radar together.

D. They arrive simultaneously and appear on the radar display as though received from a single range. All three points have the same apparent range as point c, R_c. The true range is unknown but will be given by $R_{true} = R_c + nR_u$, for integer $n \geq 0$.

An instant later, the echoes of the same pulses from points just beyond a, b, and c arrive simultaneously. An instant after that, so do the echoes from points just beyond these, and so on.

Thus the range profile is, in effect, broken into segments, R_u wide, that are superimposed, one over the other (Fig. 24-7). The actual range, R_{true}, to a given point in this interval is given by $R_{true} = R_c + nR_u$, where, in general, n is unknown.

Range Zones. The range of point a in Figure 24-6 was selected more or less at random. It could have been any range within the region from which the return is received.

Let us assume now that the range of a is such that it places point c at zero range (right at the radar antenna). Then the echoes of all ranges between c and b would be first-time-around echoes; that is, they would be echoes of the immediately preceding (last) transmitted pulse. The echoes from all ranges between b and a would be second-time-around echoes; they would be echoes of the pulse before the immediately preceding one. Likewise, the echoes from all points between a and a point R_u beyond a would be third-time-around echoes. The particular

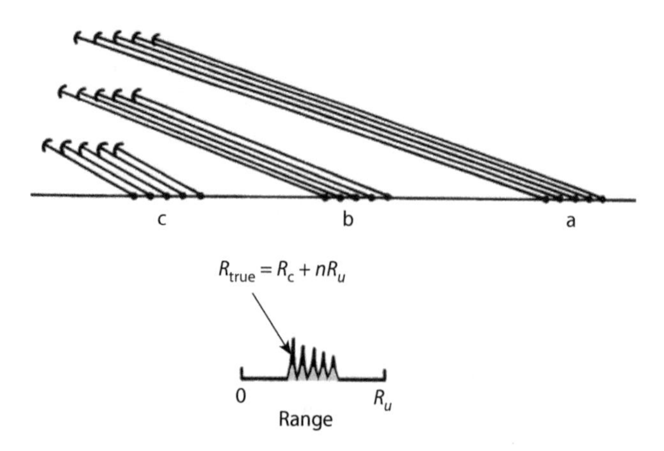

Figure 24-7. Echoes from points just beyond c, b, and a are received simultaneously. So are echoes from points just beyond these, and on and on.

segments into which the range profile would in this case be divided (Fig. 24-8) are called *range zones* (or *ambiguity zones*).

Although the true range profile could similarly be divided into any number of different sets of contiguous zones R_u wide, this particular set was chosen for two reasons. First, it conveniently starts at zero range. Second, the true range of every point within any one zone is the point's apparent range plus the same whole multiple of R_u.

Now, the higher the PRF, the shorter R_u will be, hence the narrower the range zones. The narrower the zones, the greater the number of segments into which the true range profile will be divided and from which returns will be received simultaneously.

As shown in Chapter 11, $R_u = 150{,}000/f_r$ km, where f_r is the PRF in Hertz. If, for example, the PRF is 3 kHz, the range zones will be 50 km wide. If returns are received from ranges out to 150 km, the true range profile will be broken into three zones.

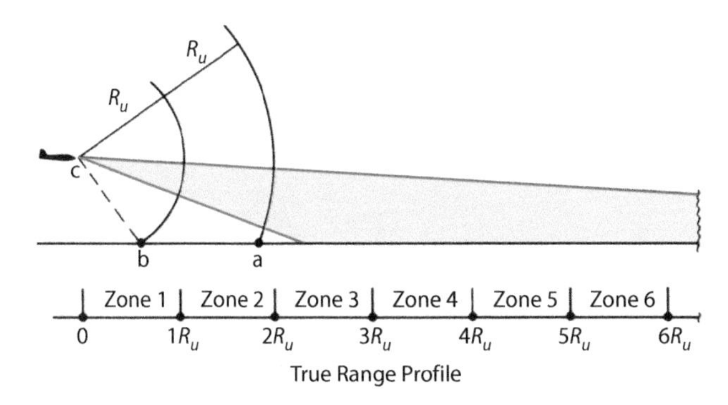

Figure 24-8. If point c in the preceding example had been moved to zero range, and points b and a had been moved equally, the true range profile would have been broken into segments called range zones.

Range Zones Superimposed. The effect of breaking the true range profile for our representative flight situation into three range zones is illustrated in Figure 24-9. There the returns from zones 2 and 3 are placed beneath the return from zone 1 and the corresponding ranges within the zones are lined up. Beneath these plots is the composite profile that would appear at the input to the receiver.

As you can see, superimposed over the echoes from target A are not only the sidelobe clutter from the target's own range, but the much stronger close-in sidelobe clutter from the corresponding range in zone 1 and the still stronger mainlobe clutter from the corresponding range in zone 3. Similarly, superimposed over the echoes from target B and the truck are not only the mainlobe clutter from their own ranges, but the sidelobe clutter from the corresponding ranges in zones 1 and 2.

As the PRF is increased and the range zones narrow, the amount of clutter that may be superimposed over any one target's echoes increases (Fig. 24-10). If the PRF is increased

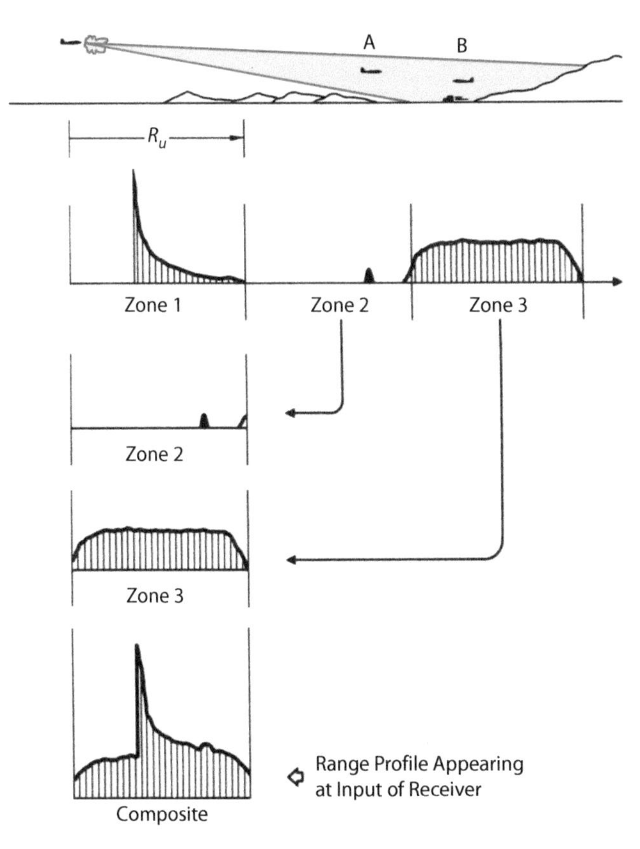

Figure 24-9. The result of R_u being one-third of the maximum range from which signals are received. The range profile for a representative flight situation is divided into three range zones. The mainlobe clutter dominates the composite profile seen by the radar.

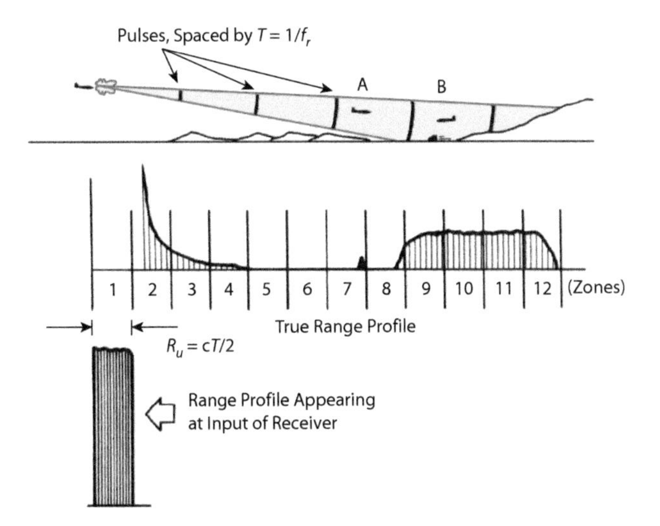

Figure 24-10. The higher the PRF, the narrower the range zones and the more deeply the clutter is piled up.

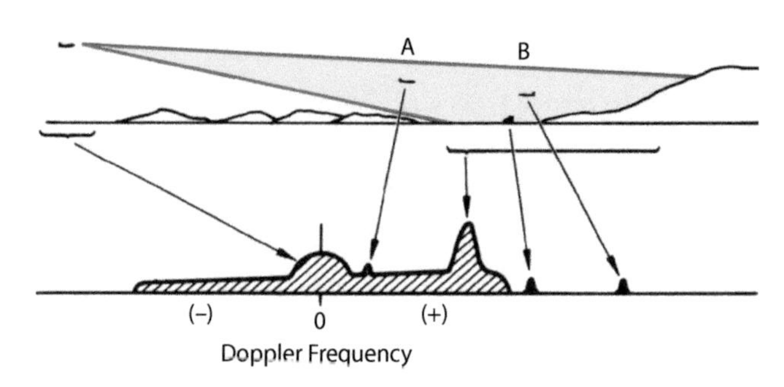

Figure 24-11. In this true Doppler frequency profile of a representative flight situation, air target B and the truck, having higher closing rates than the ground, appear in the clear.

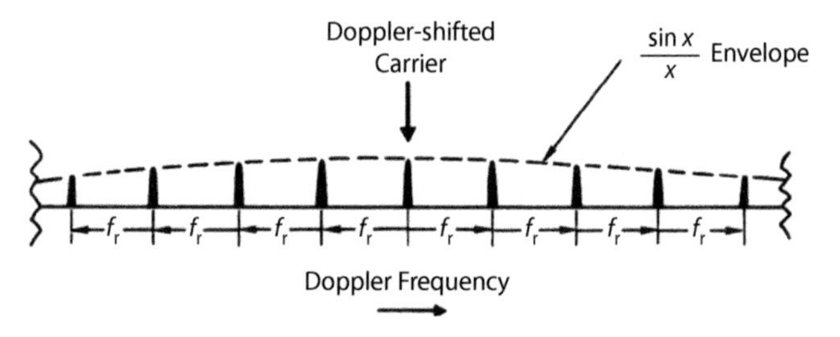

Figure 24-12. Each element of radar return has sideband frequencies separated from the Doppler-shifted carrier by multiples of the PRF, f_r.

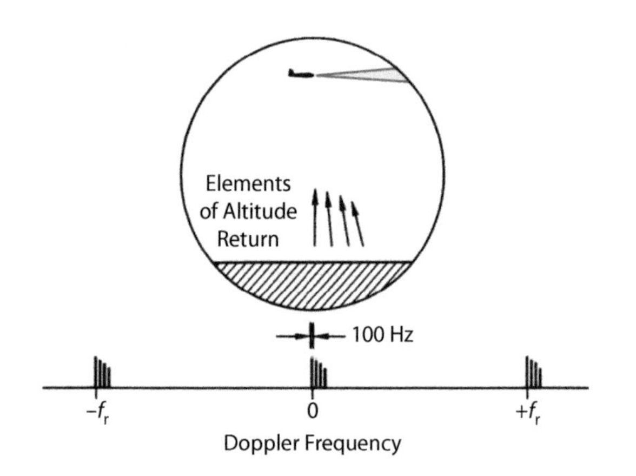

Figure 24-13. The elements shown in this spectrum of a portion of the altitude return are separated in Doppler frequency by 100 Hz. Each element has sidelobes separated from the Doppler-shifted carrier frequency by multiples of the PRF.

without limit, a point is ultimately reached where the radar transmits continuously. A target's echoes must then compete with the ground clutter received from all ranges.

Clearly, the shorter the unambiguous range interval, the less able the radar will be to isolate target echoes from the clutter on the basis of differences in range, and the greater the extent to which the radar must depend on other means, such as differences in Doppler frequency, for clutter rejection.

24.3 Doppler Profile

The flight profile is shown with the true Doppler frequency profile beneath it in Figure 24-11. This profile is a graph of the amplitude of the radar return versus Doppler frequency. In plotting it, no attempt was made to differentiate the return received at one point in the interpulse period from the return received at another. The profile represents the return from all ranges.

Sidelobe clutter extends from zero Doppler frequency out in both positive and negative directions to frequencies corresponding to the radar's full velocity ($f_D = \pm 2V_R/\lambda$). The spike at zero frequency is the transmitter spillover. The broad hump under it is the altitude return. The narrower hump near the maximum positive sidelobe clutter frequency is mainlobe clutter.

Target A is being overtaken, so its Doppler frequency falls below that of the mainlobe clutter, in the band of frequencies occupied by sidelobe clutter. Because a good deal of this clutter comes from shorter ranges than the target range, target echoes may be difficult to detect above the clutter. If the target were smaller or at much greater range, its echoes would be very much smaller than the clutter and it would not be detectable.

Since target B and the truck are approaching the radar and are nearly straight ahead, they have higher Doppler frequencies than any of the clutter. They will then only compete with receiver noise for detection, as discussed in Chapter 12.

Doppler Ambiguities. As we learned in Chapter 20, when the radar transmission is pulsed, each element of the radar return has sideband Doppler frequencies separated from the Doppler-shifted carrier frequency by multiples of the pulse repetition frequency, f_r (Fig. 24-12). Thus, the portion of the altitude return that has zero Doppler frequency also appears to have Doppler frequencies of $\pm f_r$, $\pm 2f_r$, $\pm 3f_r$, $\pm 4f_r$, etc.

Similarly, the portion of the return having a true Doppler frequency of, say, +100 Hz also appears to have Doppler frequencies of $100 \pm f_r$, $100 \pm 2f_r$, $100 \pm 3f_r$, $100 \pm 4f_r$, etc.

The same is true of the return received from every other point in the true Doppler frequency profile (Fig. 24-13). The entire profile is therefore repeated at intervals equal to f_r above and below the carrier frequency of the transmitted pulses. Because it is made up of Doppler-shifted return from the central spectral line of the transmitted signal, the true profile is commonly

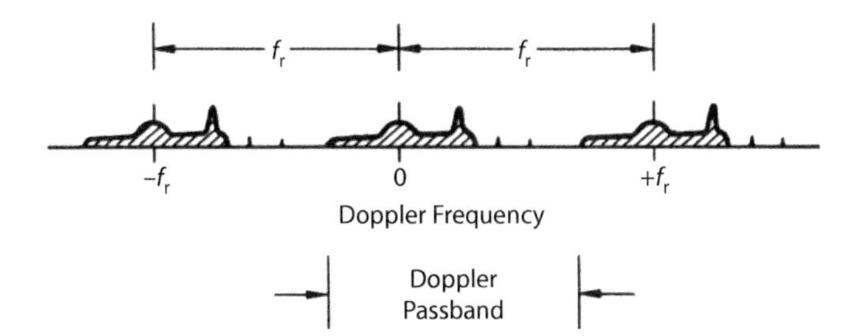

Figure 24-14. If f_r is high enough, the nearest sidebands will be entirely outside the passband.

referred to as the *central line return*. Repetitions of the spectrum, or of portions of it such as the mainlobe clutter, are then referred to as *PRF lines*.

If f_r is sufficiently high, there will be gaps between the repetitions of the Doppler spectra, as shown in Figure 24-14. Targets that have Doppler shifts (which will also be ambiguous around multiples of f_r) that fall into this region will only have to compete with the thermal noise of the receiver for detection.

However, if f_r is less than the width of the true Doppler profile of the ground return (as it often must be made to reduce or eliminate range ambiguities), the repetitions will overlap and the observed Doppler frequencies of the ground clutter spectrum will be ambiguous. This condition is illustrated in Figure 24-15 for a value of f_r that is only one-half the width of the unambiguous ground clutter spectrum. In this case, the unambiguous spectrum is overlapped by the repetitions immediately above and below it. For clarity, each repetition is plotted on a separate baseline. In actuality, they will all merge into a single composite spectrum, which is shown at the bottom of the figure.

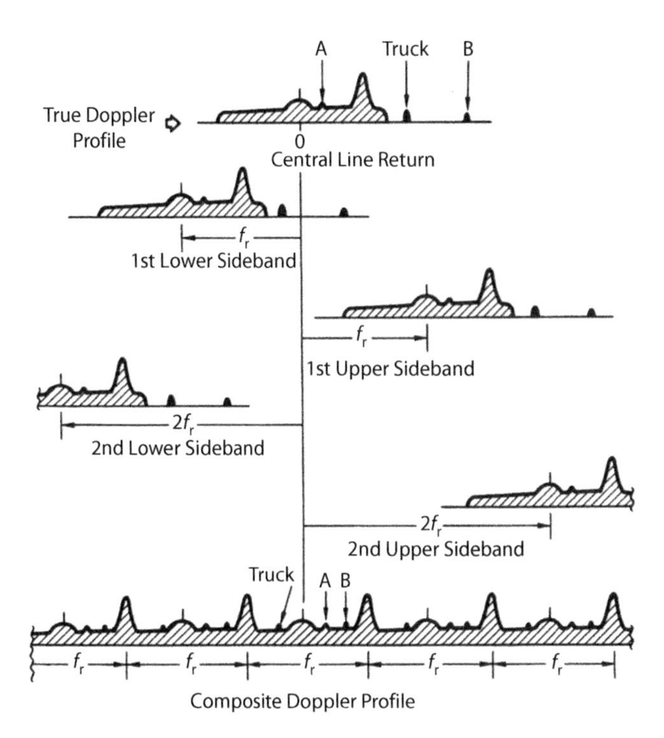

Figure 24-15. If f_r is less than the width of the true Doppler spectrum, repetitions of the spectrum due to sideband frequencies will overlap and actually merge to form the single composite profile shown at the bottom of the figure.

Any overlapping of the repetitions of the profile, such as that just illustrated, will result in a target's echoes and ground clutter passing through the same Doppler filter(s), regardless of the true Doppler shift of the target. Examples of this are air target B and the truck, in Figure 24-15. Whereas the true Doppler frequencies of these targets are actually higher than those of any of the clutter, in the composite profile, both targets are competing with sidelobe clutter for detection.

Because the sideband frequencies are separated from the central line frequencies by multiples of the PRF, any one segment of the composite spectrum is identical to every other segment of the same width on either side. As noted in previous chapters, it is for this reason that the passband of the Doppler filter bank does not need to be any more than f_r Hz wide.

The repetitions of the true Doppler profile overlap increasingly as the PRF is reduced (Fig. 24-16). From the standpoint of clutter rejection, reducing the PRF has two main effects. First, more and more sidelobe clutter piles up in the space between successive mainlobe clutter lines. Second, and more important, the mainlobe clutter lines move closer together. Since the width of

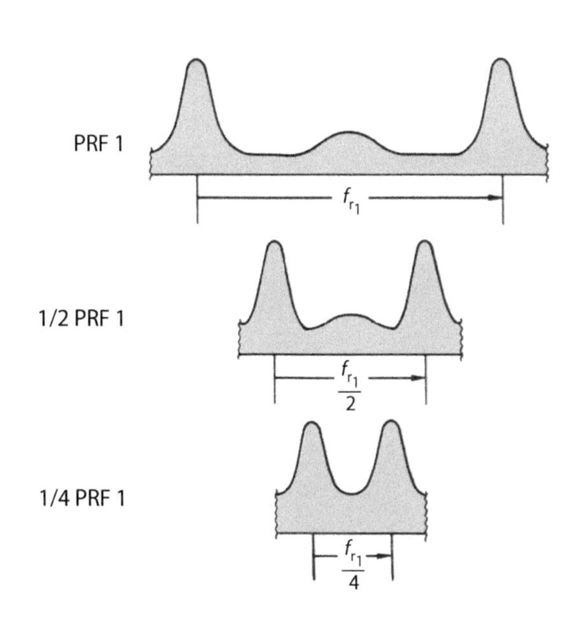

Figure 24-16. As PRF is decreased, repetitions of the mainlobe clutter spectrum move closer together, leaving less room in which to detect targets.

these lines is independent of the PRF, reducing the PRF causes the mainlobe clutter to occupy an increasingly larger percentage of the receiver passband and causes the altitude return and other close-in sidelobe clutter to pile up increasingly in the space between.

As the percentage of the passband occupied by mainlobe clutter increases, it becomes increasingly difficult to reject even the mainlobe clutter on the basis of its Doppler frequency without at the same time rejecting a large percentage of the target echoes. If carried to its extreme, the overlap will ultimately reach a point where the mainlobe clutter completely fills the receiver passband.

Clearly, the lower the PRF, the more severe the effect of Doppler ambiguities on ground clutter.

Test your understanding

1. If a radar has a PRF of 1 kHz, what are
   a. the maximum unambiguous range?
   b. the maximum range of Doppler frequencies that can be measured unambiguously in a coherent radar?
2. If a radar has a PRF of 1 kHz, and it detects a target that is at a range of 200 km, from what other ranges will the radar simultaneously receive returns from other targets or the ground?
3. An airborne radar has a wavelength of 3 cm and is on an aircraft flying at a low altitude at a velocity of 90 m/s; what is the minimum PRF that will allow any targets to be detected with Doppler shifts greater or less than the Doppler shifts of the returns from ground clutter through the antenna mainlobe and sidelobes? What will be the Doppler shifts of such targets?
4. If a radar has a PRF of 1 kHz, what is the bandwidth of a Doppler processor that will be able to process the Doppler shifts associated with targets of any speed?

24.4 Summary

Since ground clutter is widely spread in range and Doppler shift, range and Doppler ambiguities greatly compound the problem of isolating target echoes from the clutter. In effect, range ambiguities break the range profile into zones, which are superimposed on one another. Because of this superposition, a target's echoes may be received simultaneously with clutter not only from the target's own range, but from the corresponding range in every other range zone. Increasing the PRF narrows the range zones and increases the number of zones that are superimposed, thereby making it increasingly difficult to isolate the target echoes.

Doppler ambiguities cause successive repetitions of the Doppler profile to overlap. Because of this, a target's echoes may have to compete with clutter whose true Doppler frequency is quite different from that of the target. Increasing the PRF moves successive repetitions of the mainlobe clutter spectrum farther apart, thereby making it easier to isolate the target echoes. Thus the relationship between PRF and Doppler ambiguities is just the opposite of that between PRF and range ambiguities. The lower the PRF, the more severe the effect of Doppler ambiguities on ground clutter; and the higher the PRF, the more severe the effect of range ambiguities.

Further Reading

W. C. Morchin, *Airborne Early Warning Radar*, Artech House, 1990.

G. Morris and L. Harkness, *Airborne Pulse Doppler Radar*, 2nd ed., Artech House, 1996.

C. M. Alabaster, *Pulse Doppler Radar: Principles, Technology, Applications*, SciTech-IET, 2012.

25

Representing Clutter

North Pacific storm waves, winter 1989 (Courtesy of NOAA.)

In chapters 23 and 24, we showed how returns from the ground, or clutter, affect the performance of airborne radar. Clutter returns can be very large, masking the returns from smaller targets of interest. In some circumstances the relative Doppler shifts of clutter and targets can be used to help separate wanted target returns from clutter. In other cases we may only be able to detect a target if the amplitude of its return is significantly larger than that of the clutter.

Returns classed as clutter can include unwanted reflections from the sea surface, precipitation (rain, snow), clouds, birds and insects, as well as from the ground. Here we investigate a bit further the characteristics of ground clutter, but many of the principles we discuss can be applied to other types of clutter.

We saw in Chapter 23 that the magnitude of the ground clutter return was determined by the size of the resolvable ground area, A_g, and the normalized radar cross section (NRCS), σ^0. The value of the NRCS can vary greatly, depending on the terrain type and factors such as the grazing angle, radar wavelength, and radar polarization. Values of NRCS have been widely measured and some typical examples are shown in Chapter 23, using the constant gamma model, where for a given terrain type and a grazing angle ψ it is found that

$$\gamma = \frac{\sigma^0}{\sin\psi}$$

In a typical scene, the ground returns are very complex. For example, in an agricultural area there may be large areas of fields and woods, but there will also be hedges, roads, electricity transmission pylons, rocks, vehicles, and farm buildings. An urban scene will be even more complex. In trying to model the returns from such scenes, we usually try to distinguish between discrete scatterers, such as buildings, vehicles, and towers, and distributed clutter, that is, fields or woods.

Farm Near Hayfield
(Image courtesy of Tom Curtis / FreeDigitalPhotos.net)

Wind Power In The Mountains
(Image courtesy of xedos4 / FreeDigitalPhotos.net)

London Tower Blocks
(Image courtesy of Tom Curtis / FreeDigitalPhotos.net)

Figures: Ground clutter scenes: undulating fields with hedges, trees, buildings, etc.; wooded area with wind turbines; London tower blocks.

In order to understand the performance of radar detection systems, and indeed to design such systems, we need to understand in more detail the characteristics of these ground clutter returns. Large discrete targets are modeled separately, as will be discussed later. For distributed clutter we need to take a more statistical approach, to account for the random fluctuations of the signals we see as we look over areas of similar composition, such as fields or woods. It should be noted that for many types of ground clutter, the signals seen in the radar may be similar to those caused by thermal noise. Thermal noise and its effects on radar detection performance are described in Chapter 12. In this chapter we look at the equivalent characteristics of clutter signals and show how they are similar and how they differ.

Much of the understanding of distributed clutter signals is captured in different model types. We saw in Chapter 23 how we can use the NRCS model to predict the average magnitude of the clutter return. The actual clutter signals fluctuate considerably about their average values as the radar looks from one clutter patch to the next. The returns for a single clutter patch may also fluctuate in time, due to internal motion, or will change if the radar changes frequency. We will model these fluctuations of the clutter returns about their mean values.

25.1 Clutter as Noise

We saw in Figure 23-3 how the return from the ground at a given range and bearing from the radar is determined by the resolvable ground area, A_g. A_g is defined, at low grazing angles, by the compressed radar pulse length, τ_{comp} (see Chapter 14), the range to the patch from the radar, R, the azimuth beamwidth of the antenna, θ_a, and the grazing angle, ψ. A good approximation to the effective value is given by

$$A_g = R\theta_a c(\tau_{comp}/2)\sec\psi$$

Here the range extent of the clutter patch projected onto the ground is $c(\tau_{comp}/2)\sec\psi$ and the azimuth extent of the patch is $R\theta_a$. A representative set of values for a specific situation might be

$\theta_a = 2°$

$\tau_{comp} = 100$ ns

$\psi = 5°$

$R = 25$ km

Now $\sec 5° \approx 1$ and so the range extent of the patch size is about $c(\tau_{comp}/2)\sec\psi = 3\times10^8$ m/s $\times (100\times10^{-9}$ s$)/2 = 15$ m. The azimuth extent of the patch is $R\theta_a = (25\times10^3$ m$)\times(2°\times\pi/180$ radians$) = 872$ m (note that the azimuth beamwidth is required in radians here).

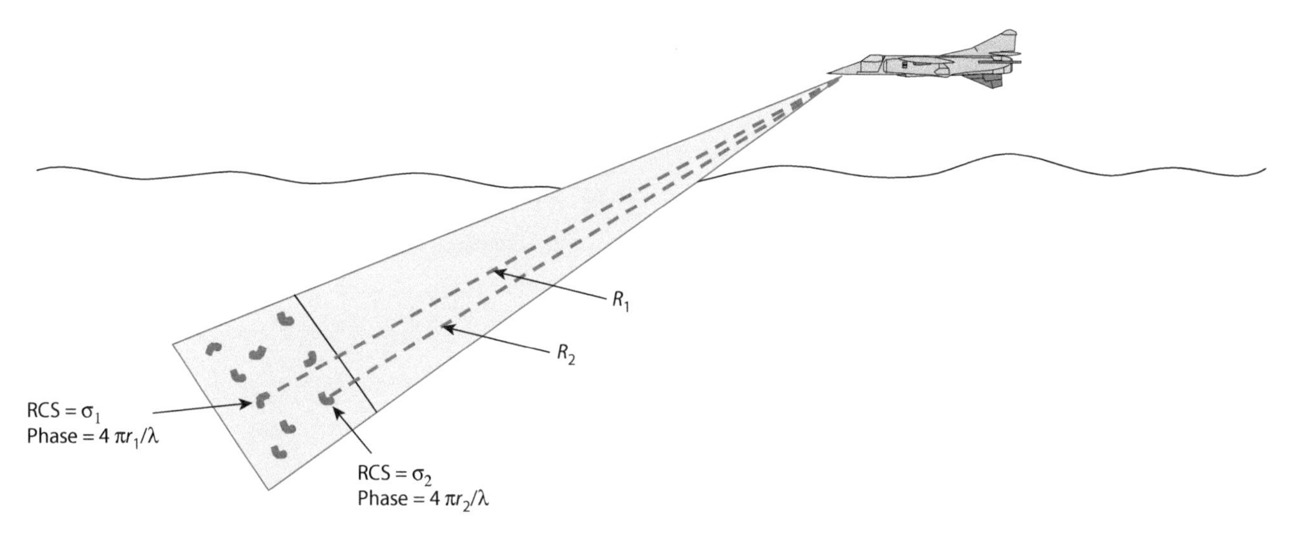

Figure 25-1. Multiple scatterers in a clutter patch add their returns in the radar beam. Each has it own RCS and a phase, θ, determined by its range from the radar.

Over typical terrains, such as rocky desert, fields, or woods, such a patch will include numerous features that will reflect the radar signals. Such objects may range from grains of sand, to individual stones, to plants and trees. The way that all these reflected signals combine in the radar receiver depends on their radar cross sections and their individual ranges from the radar. This is because the reflected signal from each individual scatterer has its own amplitude and phase angle, as illustrated in Figure 25-1, so that the scattered signals within an illuminated patch add as random vectors. This vector addition of the voltages received from several scatterers is illustrated in Figure 25-2 (see also Chapter 6).

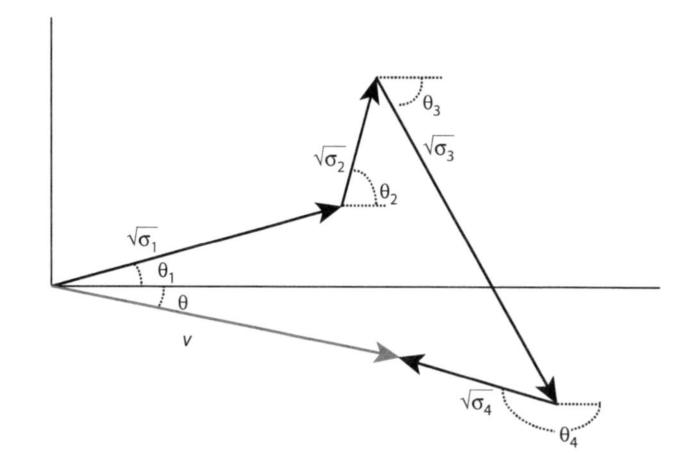

Figure 25-2. Vector addition of voltages from four individual scatterers, with amplitude $\sqrt{\sigma_i}$ and phase θ_i, giving resultant voltage $v\angle\theta = \sqrt{\sigma_1}\angle\theta_1 + \sqrt{\sigma_2}\angle\theta_2 + \sqrt{\sigma_3}\angle\theta_3 + \sqrt{\sigma_4}\angle\theta_4$.

The amplitude of each signal is effectively a voltage value in the radar receiver and so is proportional to $\sqrt{\sigma}$, where σ is the radar cross section (RCS) of the scatterer, which is a measure of the power in the return. The associated phase angle, θ, is determined by the range between the radar and the scatterer and the radar wavelength. If the range from a scatterer to the radar is R, then the number of wavelengths in the path from the scatterer to the radar is R/λ. A change of range of one wavelength is equivalent to a phase change of 2π radians. The total phase change over a two-way path (from the radar to the scatterer and return) is then given by $\theta = 2 \times R/\lambda \times 2\pi = 4\pi R/\lambda$ radians. Remembering that the wavelength, λ, is, for example, 3 cm in X-band, then a very small change in range can greatly affect the phase of the signal for a given scatterer. If we now consider all the scatterers in the resolved clutter patch, these may number in the hundreds or thousands. The resultant signal seen by the radar receiver is the vector sum of all these signals. Figure 25-3 illustrates these signals and their sum for just 200 random-size scatterers.

If the radar looks at the same scene from a different viewing angle or a similar scene in an adjacent clutter patch, the resultant

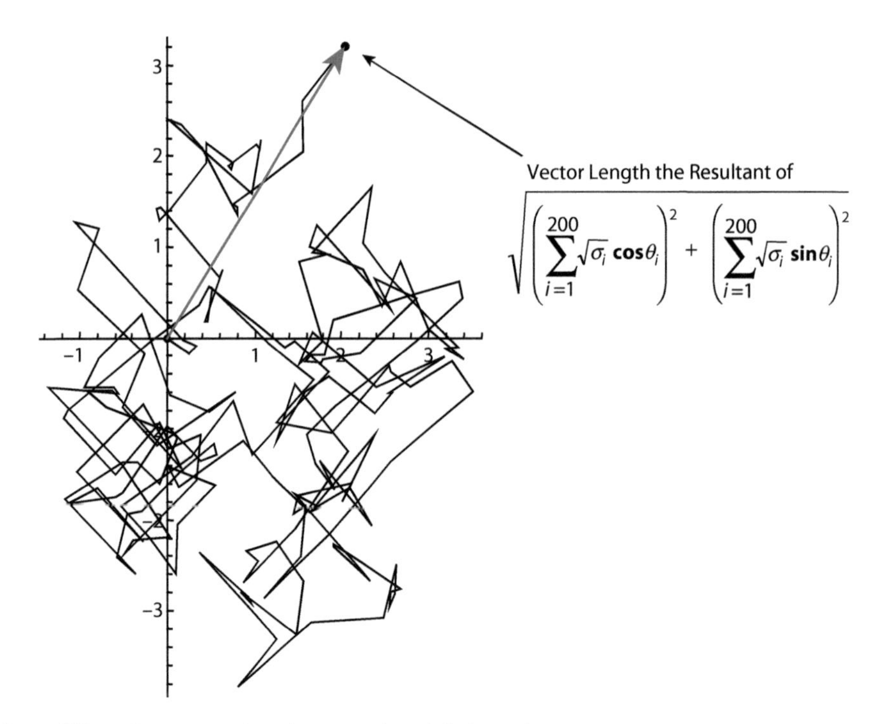

Figure 25-3. This graph shows 200 random vectors in a clutter patch and their resultant sum.

vector sum may be quite different. The average power of the returns, measured over all similar clutter patches or all viewing angles, is the sum of all the values of the radar cross section, σ_i, of the scatterers. However, at any instant the signal may fluctuate significantly above or below this mean level. Provided that the number of scatterers in the clutter patch is high (say, greater than 10 to 20), then the phase angle of the various scatterers will appear to be random and the central limit theorem applies to the sum of these individual vectors. The resulting vector appears to be very similar to thermal noise. This is illustrated in Figure 25-4, which shows the power (i.e., the squared amplitude of the resultant vectors, such as that shown in Figure 25-3) of the signal from 500 successive clutter patches that have the same area and NRCS. The average power of these signals is unity, but it can be seen that in this example, some values are as high as five or six, while some are close to zero.

Thus the ground clutter return in our simple model appears to be similar to noise. As discussed in Chapter 12, we model

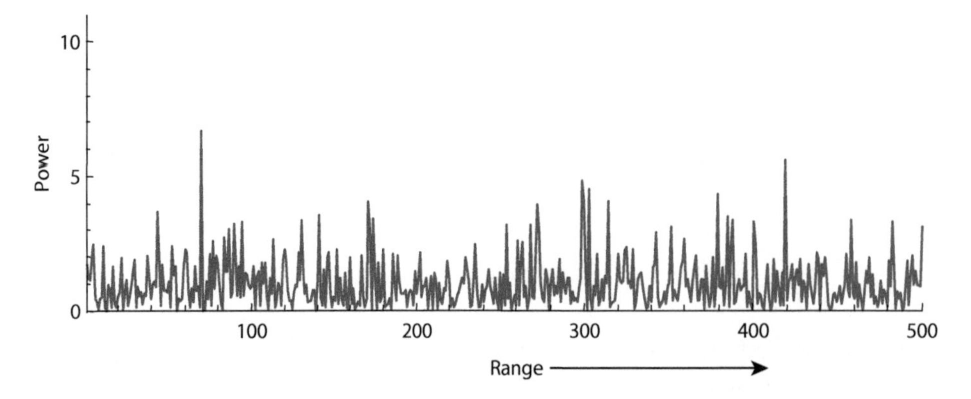

Figure 25-4. This graph shows a noise-like clutter signal.

fluctuations of this sort by using a probability density function, PDF, which is a plot of the probability that the noise (or clutter) will have a given amplitude at any instance. For the amplitude of noise signals it is known as the Rayleigh PDF. The area under the curve, as shown in Figure 25-5, gives the probability that any clutter sample will be larger than the threshold v_T. If v_T is a detection threshold in a radar receiver, the probability of the noise or clutter exceeding this level is known as the probability of false alarm, P_{fa}. The total area under the curve (i.e., when $v_T = 0$) must equal a probability of 1.

We will see later how this model helps us predict performance when detecting targets against a background of clutter, in the same way as this was described in Chapter 12 for targets in noise.

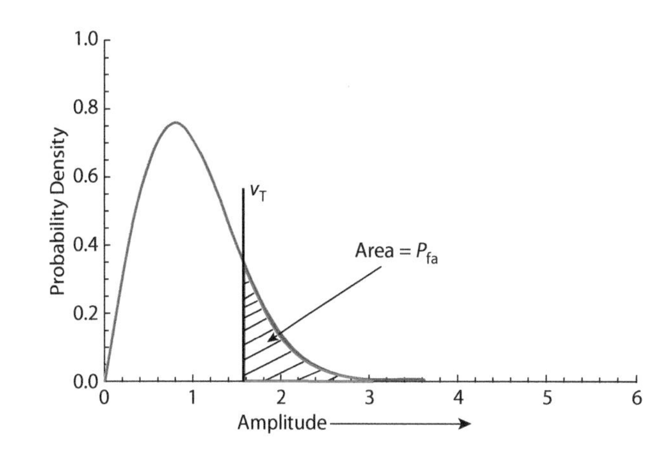

Figure 25-5. This graph shows the Rayleigh probability density function for noise, showing the probability of false alarm, P_{fa}, for a threshold v_T.

25.2 Limitations of the Noise Model for Ground Clutter

Our simple model for clutter predicts that it will have noise-like amplitude statistics (Gaussian statistics, following from the central limit theorem). This is a good model for the backscatter from large areas of uniform terrain, such flat grasslands or dense woodland. Incidentally, this is also a very good model for the amplitude statistics of rain clutter, due to scattering from very large numbers of spherical raindrops. However, it is often observed that the clutter signals from the ground fluctuate more widely in amplitude than would be predicted from a simple noise model. This could be due to, for example, the NRCS itself fluctuating from one clutter patch to the next. The returns from a single patch are still the result of scattering from very many individual scatterers, but now the mean power (the sum of the radar cross sections of the scatterers in a patch) may vary from one patch to the next. When observed over many patches, the backscattered signals become what is sometimes called "spiky." In such a case, the NRCS predicted by σ^0 will be the reflectivity averaged over all these local fluctuations.

In Figure 25-6 we illustrate the comparison between the amplitude (voltage) of signals from noiselike clutter and the returns from very spiky clutter. The mean of these signals, averaged over all clutter patches, is one. It can be seen that the spiky signals frequently have amplitudes up to six times the mean level, while the noisy signals have their maxima mainly in the range of two to three times the mean.

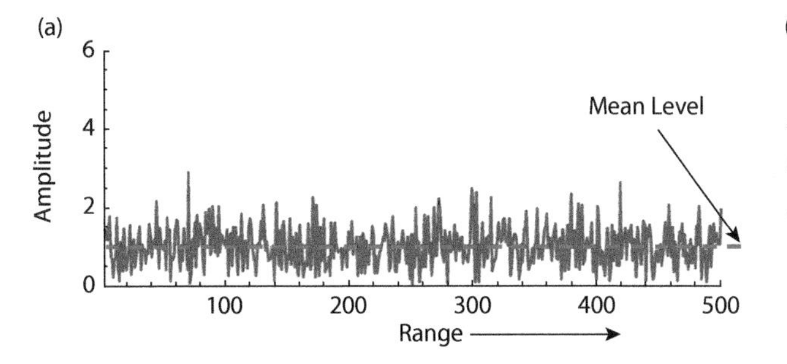

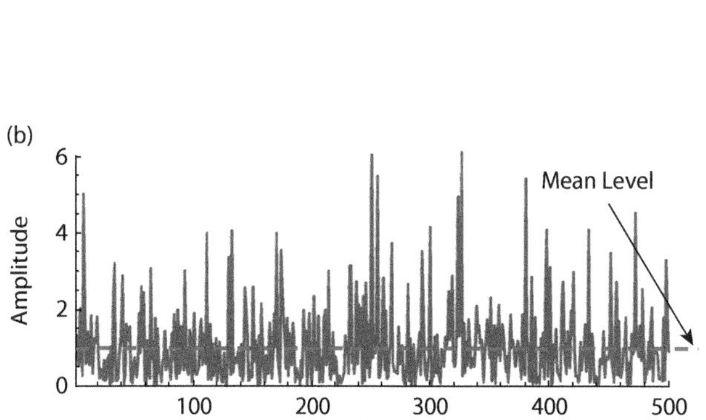

Figure 25-6. These plots give the amplitude against range (in arbitrary units) for (a) noise and (b) spiky clutter; in each case the mean level of the signals is 1.

This type of spiky behavior is often observed in real ground clutter data. It may be due to local variations in reflectivity from one patch to the next, say, due to changes in the ground slope, or the presence of features such as roads or hedges that are included in the overall scene being modeled.

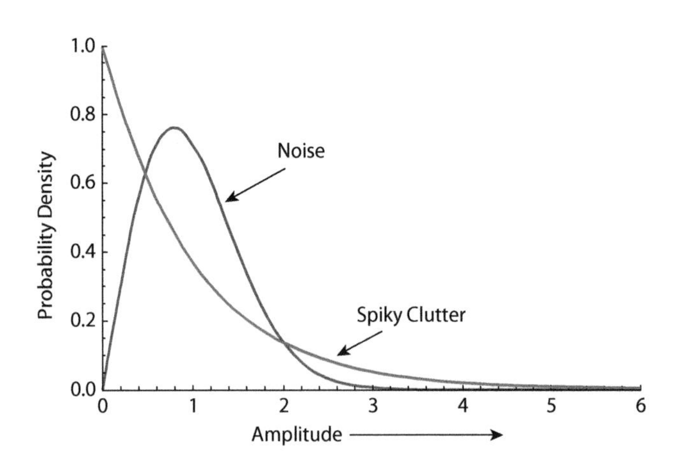

Figure 25-7. PDFs of noise and spiky clutter are graphed here.

25.3 Improved Clutter Models

We model different types of amplitude fluctuation by using different PDFs to represent the amplitudes. Figure 25-7 shows the density functions for the amplitudes of the envelopes of noise and the spiky data represented in Figure 25-6. In general, the statistics of clutter amplitude may vary from noise-like to very spiky and the full range of cases is often modeled by a family of density functions that can represent the varying degrees of "spikiness" encountered. Figure 25-8 shows the range of probability density functions from a family known as the *K* distribution. The spikiness in this case is characterized by a shape parameter ν, which varies in practice from about 0.1 (very spiky) to ∞ (noise). Other families of distribution that are used include the Weibull and lognormal distributions. They are all characterized by having a shape parameter that can be used to adjust the spikiness to fit the observed data.

It must be remembered that all these different representations are models that we use to improve our understanding of the radar's performance and to design better detection signal processing. Real life rarely conforms to a chosen model for long; nevertheless, these models are useful for understanding the likely performance of our radars under different conditions.

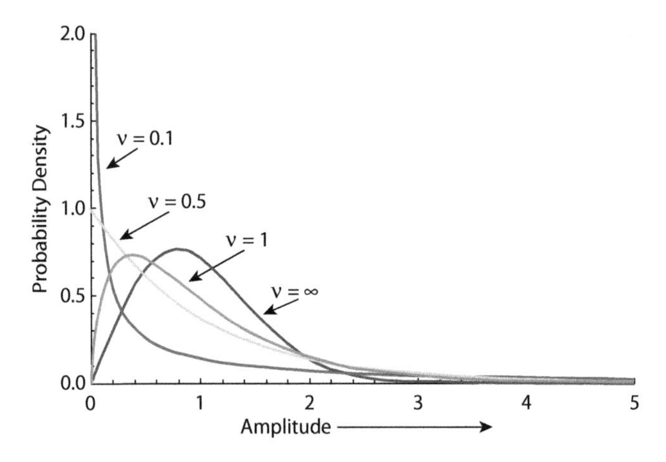

Figure 25-8. Graphed here is the family of PDFs (*K* distribution).

25.4 Other Characteristics of Ground Clutter

Thus far we have discussed clutter signals in terms of their mean power level, determined by the NRCS and the resolved area of the clutter patch, and the fluctuations of the amplitude about that mean level, modeled by families of PDFs. Other features of ground clutter also distinguish them from thermal noise signals.

Doppler Spectrum. One important characteristic of clutter is its Doppler spectrum. This was discussed in Chapters 23 and 24, where it was observed that the shape of the Doppler spectrum of the returns from the ground for an airborne radar was mainly determined by the antenna mainlobe and sidelobe structure and the aircraft velocity. In one example of ground clutter, the returns were due to backscatter from rocks or sand distributed in a spatially uniform way over the clutter patch. If the radar is stationary, the returns from such a patch will not vary with time and the clutter patch will have no inherent Doppler spectrum. The platform motion introduces differential Doppler shifts across the patch, giving a Doppler spectrum. In addition, sometimes the clutter itself may move over time. A good example is trees or vegetation blowing in the wind. In this case, the clutter return will have an inherent

Doppler spectrum, albeit with a zero average velocity for a stationary radar. A typical spread of velocities might give a standard deviation of the velocity spectrum width, σ_v, of about 1 m/s. The equivalent standard deviation of the frequency spectrum is given by $\sigma_f = 2\sigma_v/\lambda$ Hz. In X-band, $\lambda = 3$ cm, so $\sigma_f = 2 \times 1$ m/s/0.03 m ≈ 66 Hz. When observed from an airborne platform, this inherent Doppler spectrum will be subject to an additional spreading of the mainlobe spectrum, described in Chapter 23. Note that the Doppler spectrum of thermal noise is uniform over the whole unambiguous Doppler space, defined by the pulse repetition frequency, f_r (see Chapter 24). An example of the power spectrum of clutter and thermal noise is shown in Figure 25-9. This clutter has a Gaussian-shaped spectrum, with a standard deviation of 66 Hz, equivalent to a 1 m/s standard deviation of velocity, as discussed above. The mean Doppler shift is 0 Hz and the clutter-to-noise ratio is 20 dB. The PRF in this example is 2 kHz and the spectrum is plotted over the unambiguous interval of ±1 kHz. The figure shows an instantaneous example of the spectrum, which appears noisy. The long-term average that is obtained by averaging over many spectra is shown in Figure 25-10. Also shown in this figure is the effect of platform motion on the spectrum. The plot shows the effect of observing the clutter from an airborne platform having a velocity, V, of 100 m/s. The antenna has an azimuth beamwidth of 2° and is pointing normal to the direction of travel of the aircraft. The spectrum due to the antenna sidelobes extends over $\pm 2V/\lambda = 2 \times 100/0.03 = 6.6$ kHz, but it is assumed that the energy coming through the sidelobes is below the noise level in this example.

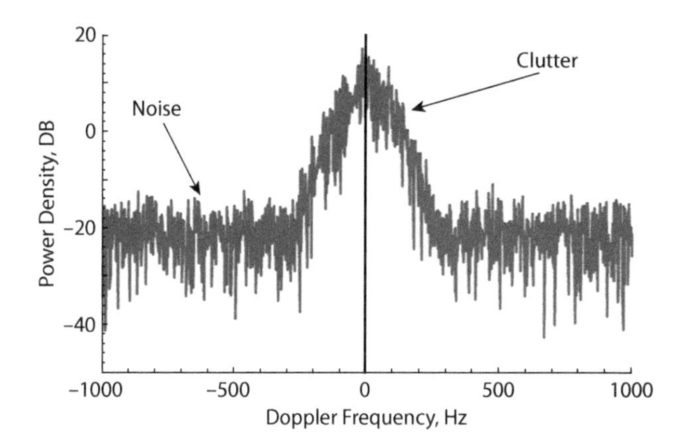

Figure 25-9. Power spectral density of clutter in noise.

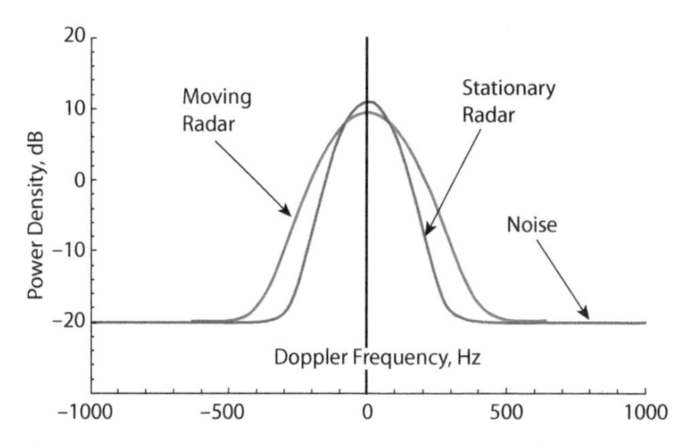

Figure 25-10. Average power spectral densities seen from a stationary radar and from a moving radar, showing the additional spectrum spread due to platform motion.

Spatial Correlation. Another characteristic of clutter that is different from a simple thermal noise model is the spatial variation of the signals. For the standard noise model, the mean clutter power is constant from one clutter patch to the next. In spiky clutter, the average mean level is still determined by the NRCS, but now the local mean level may fluctuate about the overall average level. A classic example is the returns from the sea surface, where the local mean level fluctuates as it follows the patterns of waves or swell. In ground clutter, the equivalent might be local undulating ground, such as the sand dunes shown in Figure 25-11.

A typical example of spiky clutter with some spatial structure is shown in Figure 25-12. Compare this representation with the data shown in Figure 25-6b, which have the same amplitude statistics and same mean level but are fluctuating randomly from one clutter patch (range cell) to the next. Such spatial structure may have an effect on the detection signal processing described in the following section.

Figure 25-11. The Mesquite Sand Dunes at Sunrise in Death Valley National Park illustrate a scene where the clutter returns may exhibit periodic spatial structure. (Image courtesy of dexchao/ FreeDigitalPhotos.net.)

25.5 Discrete Scatterers

In addition to the distributed clutter previously discussed most ground scenes will also include discrete objects, which are also

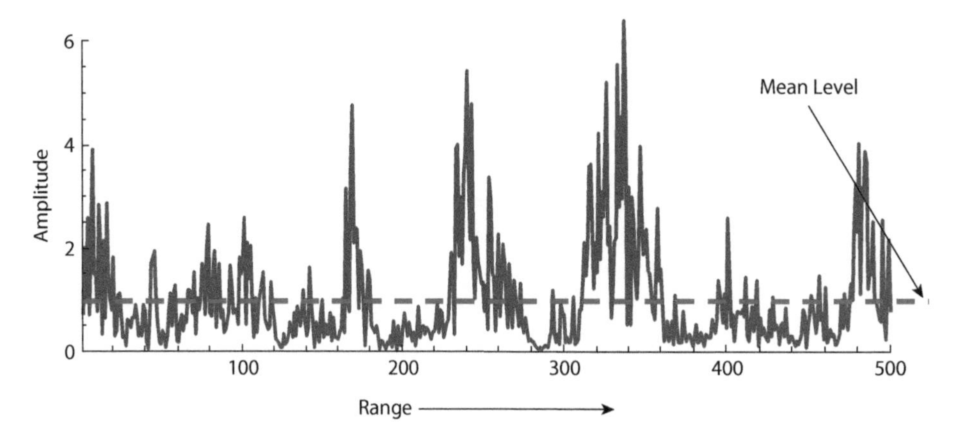

Figure 25-12. This graph plots amplitude against range (in arbitrary units) for spatially correlated clutter.

classified as clutter (as discussed in Chapter 23). An example was given of a small flat sheet with area 1 m^2 having an RCS of about 14,000 m^2 at X-band. So it can easily be understood that reflections with an effective RCS of up 10^6 m^2 may be seen from time to time. A small aircraft may have an RCS of 1 m^2, and scattering from such large discrete scatterers entering the radar through its sidelobes may have a comparable response. We cannot easily model the returns from these isolated discrete targets using our probability models, and their characteristics depend on the specific scene being observed. We must treat them as if they are individual stationary targets in any analysis of performance. Following the analysis of results in the literature, Barton (2005) has suggested that a useful guide to the expected number of such scatterers in a rural scene might be to assume typical average numbers per unit area of 0.2/km^2 for scatterers with an RCS of 10^4 m^2, 0.5/km^2 for scatterers with an RCS of 10^3 m^2, and 2/km^2 for scatterers with an RCS of 10^2 m^2. To put these RCS values in context, a farm tractor might have an RCS of 100 m^2, while a large building, such as a barn, may have an RCS greater than 10^4 m^2. Clearly, much higher densities of scatterers may be encountered in specific situations. An individual wind turbine may have an RCS of more than 10^3 m^2 and there may be, say, four or five per square kilometer. Urban scenes will be dominated by large numbers of very big scatterers.

25.6 Predicting Detection Performance

We saw in Chapter 12 how we predict detection range in noise. Using the radar-range equation we can predict the power received from a target, given the target RCS, the range to the target, and the radar parameters. Similarly, if we know the receiver noise figure, we can calculate the noise energy in the receiver. From these values we can estimate the signal-to-noise (S/N) ratio that will be achieved for a target at a given range. To know whether the target can be detected, we must discern the minimum required S/N ratio, $(S/N)_{req}$, to achieve a particular probability of detection at a specified probability of false alarm.

We also saw in Chapter 12 that $(S/N)_{req}$ depends on the fluctuation characteristics of the target, described by the various Swerling models. Another important parameter was the time on target, t_{ot}, of the radar beam. This latter parameter determines how much signal energy is used to detect the target. For a pulse radar, it is a measure of how many pulses, N, are used in a beam dwell (the period over which the antenna illuminates the target), with $N = f_r t_{ot}$, where f_r is the pulse repetition frequency.

With this information, we saw in Chapter 12 that the detection range, R_{P_d}, for targets in thermal noise and assuming coherent processing can be expressed as

$$R_{P_d} = \left(\frac{P_{avg} A_e^2 \sigma\, t_{ot}}{(4\pi)(S/N)_{req} k T_s \lambda^2 L} \right)^{1/4}$$

where for the range in meters

P_{avg} = average transmitter power (W)

A_e = antenna aperture (m^2)

σ = RCS of the target (m^2)

t_{ot} = time on target (s)

$(S/N)_{req}$ = signal-to-noise ratio required to achieve the desired probability of detection, P_d

k = Boltzmann's constant (1.38×10^{-23} J/K)

T_s = receiver noise temperature in Kelvin (K)

λ = radar wavelength (m)

L = system losses

We can take a similar approach to the detections of targets in clutter. The RCS of a clutter patch is given by $\sigma = \sigma^0 A_g$. If we substitute this value in the previous range equation we get the range at which the clutter itself can be detected against the receiver noise. However, we want to know about detecting a target against a clutter background, which may be much larger than the noise level. Now, although we have described clutter as being in some respects noiselike, detection performance must be calculated a bit differently.

We now have three components of our signal: target return, clutter return, and receiver noise. As we discuss we can no longer simply incorporate a time on target, t_{ot}, into our range equation, so usually we work with a baseline of the values achieved by a single pulse. We calculate the detection performance for a single pulse and then extend our analysis to incorporate the effects of signal processing using multiple pulses to illuminate a target. First, we write down the S/N power ratio, the clutter-to-noise (C/N) power ratio, and the signal-to-clutter (S/C) power ratio of our various signal components. The single-pulse S/N ratio is given by

$$S/N = \frac{P G^2 \sigma \lambda^2 \tau}{(4\pi)^3 R^4 k T_s L_s}$$

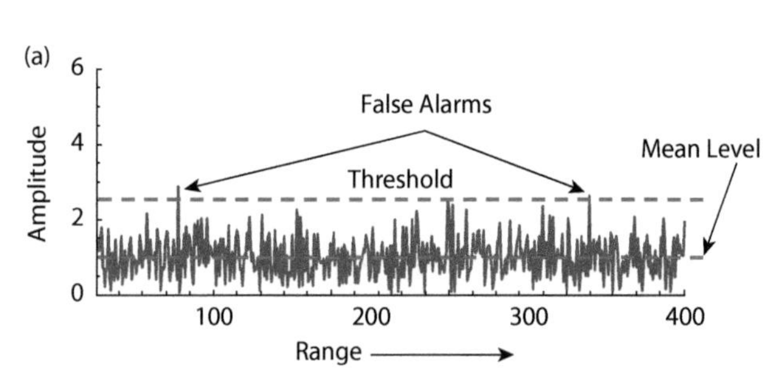

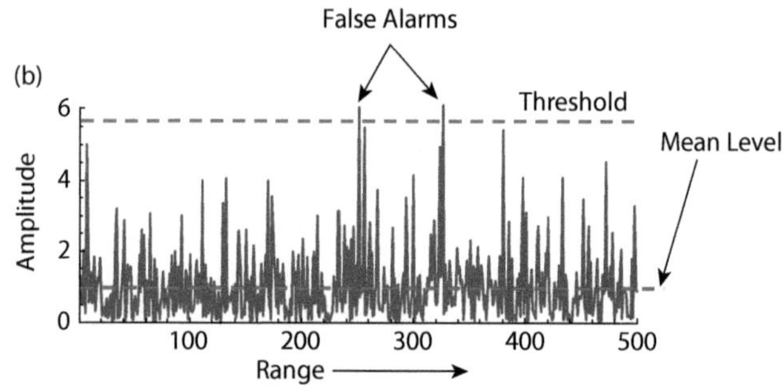

Figure 25-13. These two graphs show the following: (a) Threshold in noise. (b) Threshold in spiky clutter.

where P is the peak radar power ($P_{avg} = Pf_r\tau$), τ is the uncompressed pulse length, and L_s is all the radar losses associated with the target signal.

The C/N ratio is

$$C/N = \frac{PG^2\sigma^0 A_g \lambda^2 \tau}{(4\pi)^3 R^4 k T_s L_c}$$

where $A_g = R\theta_a c(\tau_{comp}/2)\sec\psi$ and L_c is all the radar losses associated with the clutter signal (usually, $L_c \approx L_s$).

The S/C ratio is

$$S/C = \frac{\sigma}{\sigma^0 A_g}\frac{L_c}{L_s} \approx \frac{\sigma}{\sigma^0 A_g}$$

From these results we can also calculate a signal-to-interference ratio (SIR), which is the ratio of the signal to the clutter-plus-noise powers: $SIR = \left(\frac{S}{C+N}\right) = \frac{(S/N)}{1+(C/N)}$

Single Pulse Detection. With these equations we can calculate the total clutter, target, and noise powers in the return from a single pulse. For detection using the returns from a single pulse, the calculation is similar to that for noise. In the case where $(C/N) \gg 1$, we must calculate the threshold that will give the required P_{fa}, just as we did for thermal noise, using the PDF for the clutter amplitude (see, e.g., Fig. 25-5). If we use our noise model for clutter, the calculation is the same as that for noise. However, if clutter is spiky, then we do the equivalent calculations for its representative PDF. For spiky clutter we must set a higher threshold relative to the mean level of the clutter than we would if the simple noise model was usable. In Figure 25-13, it can be seen that the spiky clutter needs a higher threshold to control the false alarm rate. If the K distribution family of PDFs represents the whole range of clutter characteristics we might encounter, we can calculate the thresholds to set as the clutter becomes more spiky. Figure 25-14 shows a family of curves representing the variation of P_{fa} as a function of the threshold setting for various members of the K distribution family, with different values of the shape parameter, ν. These curves show that for a smaller number of false alarms (i.e., a more negative value of $\log_{10}P_{fa}$), a higher threshold is

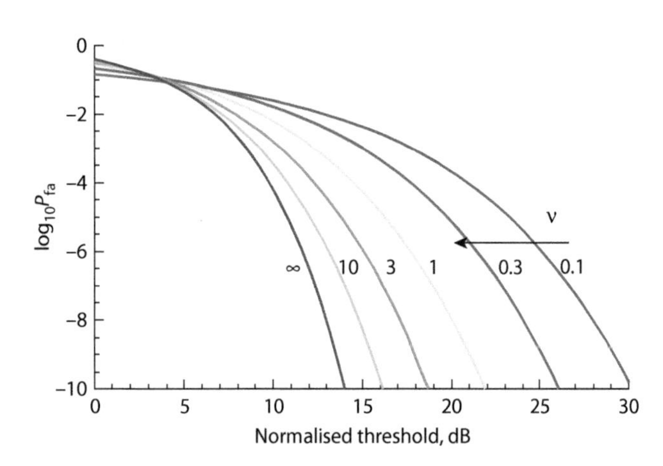

Figure 25-14. This graph plots the $\log_{10}P_{fa}$ versus the threshold for clutter intensity (threshold normalized to the mean clutter power) for the K distribution family of clutter PDFs ($\nu = \infty$ is equivalent to thermal noise, while $\nu = 0.1$ is very spiky clutter).

required. Also, for a given value of P_{fa}, the smaller the value of the shape parameter, the larger the threshold required. For example, it can be seen that for $P_{fa} = 10^{-6}$ (i.e., $\log_{10}P_{fa} = -6$ in the figure), the threshold required in noise ($\nu = \infty$) is about 11 dB, while in very spiky clutter ($\nu = 0.1$) a threshold of about 25 dB must be used. This is a very large range of threshold setting to be used as different clutter conditions are encountered by the radar. Now for every decibel that the threshold is raised to control false alarms, the target must be equivalently larger to be detected above the threshold with the same P_d. A useful rule of thumb is that the S/C required to achieve $P_d = 0.5$ is approximately given by the value of the threshold relative to the mean level required to set the desired P_{fa}. The exact calculations are quite complex, requiring numerical computing methods to provide the required answers. As the clutter power reduces, so does C/N. In the limit of no clutter we are back to detecting in noise. In intermediate cases we can still calculate performance, but we must know the PDF of clutter plus noise.

Multiple Pulse Detection. When multiple pulses are used in a dwell (the period of time for which a target is illuminated), we calculate the effect of processing over the full t_{ot}. For detection against thermal noise, coherent processing over t_{ot} (i.e., a dwell of N pulses) has the effect of dividing the noise equally among the N Doppler bins of a fast Fourier transform (FFT) filter. This means that after coherent processing, in a given Doppler bin, $S/N \propto t_{ot}$, as reflected in the equation for R_{P_d}, above. However, for clutter returns we saw in Chapters 23 and 24 that the spectrum is mainly confined to the Doppler frequencies associated with mainbeam clutter, with smaller levels in the sidelobe spectrum regions. If our target is moving fast enough, with a high enough opening or closing rate (i.e., the target is moving fast enough away from or toward the radar platform that its Doppler shift is greater than the Doppler shift of any ground clutter returns; see Figure 23-30), then detection will only be against noise. Then the equation for R_{P_d} given above can be used. For targets with a low or zero closing rate, we must work out how much clutter is falling into the same Doppler bin as the target. Then we calculate $\left(\frac{S}{C+N}\right)$ for that bin and then calculate P_d, given P_{fa}, in the same way as for the single-pulse case.

We can also try to use the energy from multiple pulses to improve detection with postdetection (noncoherent) processing, which uses only the amplitudes of the pulse returns. This is quite effective for targets in thermal noise, as discussed in Chapter 10, since the sampled noise is random. However, clutter returns from one pulse to the next may be effectively constant. In that case, there is little improvement in the effective S/C following noncoherent pulse-to-pulse integration. Some improvement may be achieved from pulse-to-pulse integration if the radar uses pulse-to-pulse frequency agility, because changes in radar frequency may cause the clutter to fluctuate about its mean level.

25.7 Summary

Detailed modeling of clutter is a major area of study in the field of radar. In this chapter we only scratched the surface of the problem and its potential solutions. There are some relatively simple models for clutter that may be quite representative of real life in some circumstances and that have the advantage of being quite amenable for use in calculating detection performance. However, real clutter often behaves quite differently from these simple models and performance prediction using the simple models may be greatly overoptimistic. In particular, real clutter can exhibit spiky behavior, which forces the radar to set higher

Clutter Models

Figure 25-8 illustrates the family of PDFs known as the K distribution, often used to model the amplitude statistics of clutter. This family of distributions, together with the Weibull distribution includes the Rayleigh distribution as one of its members. Another PDF family often used to characterize ground clutter is the lognormal distribution. There are no "correct" distributions to represent clutter. All these PDF families are models used by researchers in an attempt to represent in a single parameterized form the wide range of data characteristics they have observed. The choice of a suitable model when designing a radar may depend on which one is believed to best match the conditions that the radar is expected to encounter. However, it must always be remembered that these models are intended to be representative of the spread of typical conditions and real returns on a particular trial may often be quite different from modeled predictions.

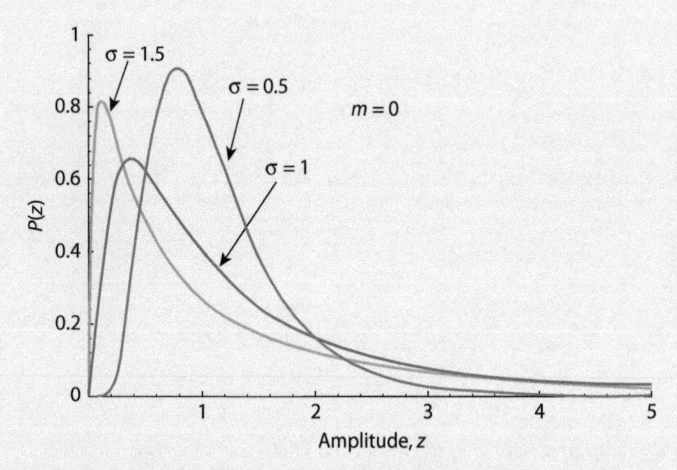

The lognormal family of PDFs, for $m = 0$ and selected values of σ; larger values of σ can represent extremely spiky clutter.

Lognormal Distribution. The lognormal distribution has a PDF of intensity z given by

$$P(z)=\frac{1}{z\sqrt{2\pi\sigma^2}}\exp\left(-\frac{\left(\log_e[z]-m\right)^2}{2\sigma^2}\right);\quad z\geq 0$$

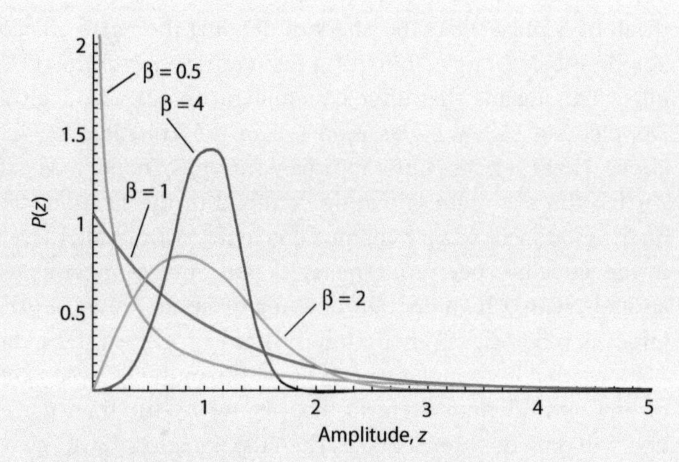

Weibull PDF family plotted for $a = 1$ ($\beta = 2$ is a Rayleigh distribution, $\beta = 1$ is an exponential distribution and $\beta < 1$ is very spiky clutter).

where m and σ are the scale and shape parameters, respectively. The distribution arises from a normal or Gaussian distribution, with mean m and variance σ^2, of $\log_e z$. Some typical examples of this family of PDFs are shown in Figure 25-15.

Weibull Distribution. The Weibull model has a PDF of

$$P(z)=\beta\frac{z^{\beta-1}}{a^{\beta}}\exp(-(z/a)^{\beta});\quad z\geq 0$$

with scale parameter a and shape parameter β. Some examples are illustrated in Figure 25-16.

thresholds to maintain acceptable false alarm rates. Advanced models of PDFs have been developed to model performance in such cases. Coherent processing with multiple pulses can bring considerable improvements in performance depending on the relative velocities of the targets and the clutter.

This chapter has identified some important points to note when designing radars for target detection in clutter:

1. **Radar power:** Increasing radar peak power will increase S/N and C/N, but will not affect S/C. This means that if $C \gg N$, then increasing peak power will have no effect on detection performance, which is determined by S/C. If $C \ll N$, then performance is determined by S/N, which is improved by increasing the transmitter power.
2. **Time on target, t_{ot}:** For coherent detection, increasing t_{ot} will increase S/N, but its effect on C/N and S/C will depend on the clutter spectrum and the target Doppler shift.
3. **Target velocity:** Target velocity has no effect on detection against noise (provided that the target does not move out of the range gate during the illumination period) but will have a big effect on detection against clutter, depending on the value of the target Doppler shift relative to the clutter.
4. **Postdetection integration:** If the single-pulse target detection is dominated by clutter, then postdetection integration may have a limited effect. This is because clutter returns, unlike thermal noise, may be substantially unchanged from one pulse to the next or from one coherent dwell to the next.

Further Reading

F. E. Nathanson, with J. P. Reilly and M. N. Cohen, "Sea and Land Backscatter," chapter 7 in *Radar Design Principles*, 2nd ed., SciTech-IET, 1999.

M. W. Long, *Radar Reflectivity of Land and Sea*, 3rd ed., Artech House, 2001.

J. B. Billingsley, *Low-Angle Radar Land Clutter: Measurements and Empirical Models*, William Andrew Publishing, 2002.

D. K. Barton, *Radar System Analysis and Modeling*, Artech House, 2005.

M. I. Skolnik (ed.), "Sea Clutter," chapter 15 in *Radar Handbook*, 3rd ed., McGraw Hill, 2008.

M. I. Skolnik (ed.), "Ground Echo," chapter 16 in *Radar Handbook*, 3rd ed., McGraw Hill, 2008.

M. A. Richards, J. A. Scheer, and W. A. Holmes (eds.), "Characteristics of Clutter," chapter 5 in *Principles of Modern Radar: Basic Principles*, SciTech-IET, 2010.

K. D. Ward, R. J. A. Tough, and S. Watts, *Sea Clutter: Scattering, the K Distribution and Radar Performance*, 2nd ed., IET, 2013.

Test your understanding

1. Why do clutter returns from a patch of ground change if the radar changes viewing angle or frequency?
2. What are the limitations of modeling ground clutter in the same way as thermal noise?
3. Name some probability density function (PDF) models that are commonly used to model the amplitude fluctuations of ground clutter.
4. What effect does "spiky" clutter have on the ability of a radar to detect targets in clutter?
5. Will increased transmitter power improve the detectability of a target in the presence of strong clutter from the same range?
6. How can pulse-to-pulse integration help to improve the detection of a target seen against a background of strong clutter at the same range?

Mikoyan MiG-35 (2013)

The "Fulcrum-F" is a Mach 2+ air-to-air and air-to-ground fighter under development by Mikoyan. Developed from the MiG-29 Fulcrum-E, there are 10 current prototypes in testing of what Mikoyan has classified as a 4++ generation jet fighter. It includes the Phazotron Zhuk-AE active electronically scanned array radar which is thought to have a detection range of 160 km for air targets and 300 km for ships.

26

Separating Ground Moving Targets from Clutter

Tank convoy

26.1 Introduction

In earlier chapters very little was said about the detection of moving targets on the ground, such as cars, trucks, and tanks. Except that low PRFs are generally required for air-to-ground operation, it was tacitly assumed that separating *ground moving targets* (GMTs) from competing ground return on the basis of differences in Doppler frequency is a similar problem to that of separating airborne moving targets from ground return. However, this assumption is only partially true. The radial component of the velocity of many ground moving targets is so low that the returns from them are embedded in the mainlobe clutter seen from a moving platform and cannot be separated from it by conventional moving target indication (MTI) techniques.

In this chapter, after briefly examining the problem, we introduce some techniques for detecting such targets. These techniques are all aimed at reducing the effects of platform motion, which spreads the clutter spectrum, to improve detection of slow moving targets that would otherwise be embedded in the mainlobe clutter. One of the earliest methods applied to this problem is *displaced phase center antenna* (DPCA). Another technique is *notching* or *clutter nulling*. In their standard forms, these two techniques require knowledge of platform velocity and antenna direction and make certain assumptions about the nature of the clutter spectra. The third technique that is discussed is *space-time adaptive processing* (STAP). As its name suggests, this method continuously adapts to the observed clutter spectra as a function of space (look direction and range) and time. Simply detecting a moving target is only part of the problem for a practical system; we must

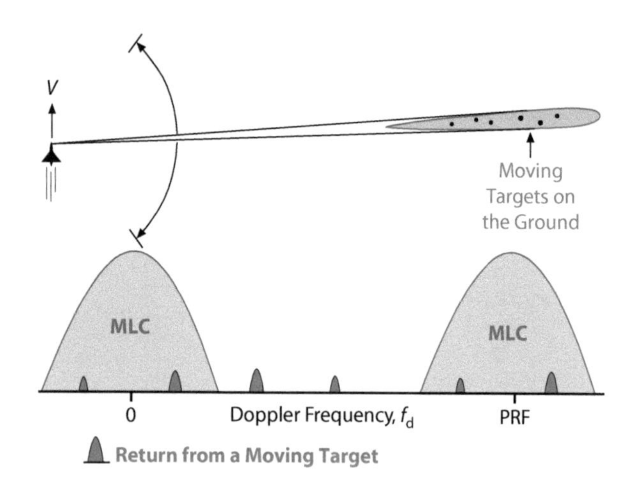

Figure 26-1. Pictured here is the Doppler spectrum of ground moving targets. Special techniques are used to detect targets whose *true* Doppler frequencies fall within mainlobe clutter (MLC).

also accurately measure the target's position on the ground. We describe how these moving target detection techniques are often combined with precise angle measurements. When combined with measurements of range, these can provide the coordinates of a target on the ground.

26.2 Problem of Detecting "Slow" Moving Targets

The chief problem in detecting moving targets in low-PRF modes—whether on the ground or in flight—is separating the target returns from the mainlobe clutter. As explained in Chapter 23, by employing a reasonably long antenna (i.e., one with a narrow beamwidth) and flying at comparatively low speeds, we can reduce the width of the mainlobe clutter spectrum and spread its repetitions far enough apart to provide a fairly wide clutter-free region in which to detect moving targets (Fig. 26-1). The true Doppler frequency may be ambiguous, dependent on the PRF, but any target whose *apparent* (ambiguous) Doppler frequency falls within the mainlobe clutter might be periodically moved into the clutter-free region by switching the PRF among several different widely separated values.

The Antenna Phase Center

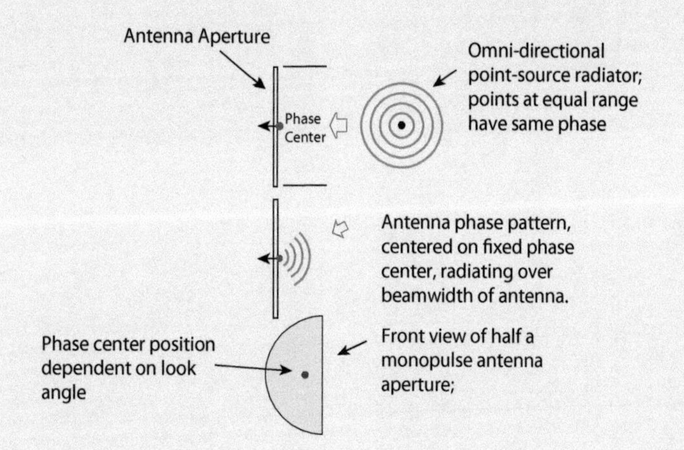

EVERY ANTENNA HAS A PHASE CENTER. IT IS THAT POINT IN SPACE WHERE, if a hypothetical omnidirectional point-source radiator is placed there, the signals received by it from any source within the field of the real antenna would have the same radio frequency phase as signals from the same source received by the real antenna.

If weighting of the antenna (for sidelobe reduction) is symmetrical, the position of the phase center will be the same for all look angles. But if the weighting is nonsymmetrical—as in a half aperture of a monopulse antenna—the position of the phase center will be a function of the look angle.

However, if the radial component of a target's velocity is so low that its true Doppler frequency lies within the mainlobe clutter, no amount of PRF switching will move the target's returns out of the clutter. Consequently, in many applications a special "slow moving target" indication capability is needed. Conceptually, the simplest is DPCA.

DPCA. This technique takes advantage of the fact that the Doppler shift in the frequency of the returns received from the ground is due mainly to the aircraft's velocity (there will also be some Doppler spread due to the "internal motion" of the clutter scatterers, such as with wind-blown trees, which may exhibit a spread of velocities of about ±1 or 2 m/s). Specifically, this shift—which is seen as a progressive pulse-to-pulse shift in the phase of the returns from a scatterer at any one range—is the result of the forward displacement of the radar antenna's phase center (defined in the panel on the preceding page) from one interpulse period to the next.

Therefore, for any two successive pulses, the shift can be eliminated by displacing the antenna phase center by an equal amount in the opposite direction before the second pulse of the pair is transmitted. The second pulse will then be transmitted from the same point in space as the first.

So, how can an antenna's phase center be displaced? One method is to provide the radar with a two-segment side-looking antenna. The aircraft's velocity and the radar's PRF are adjusted so that during each interpulse period the aircraft will advance a distance precisely equal to that between the phase centers of the two antenna segments (Fig. 26-2).

Successive pulses are then alternately transmitted by the two segments:

- Pulse n by the forward segment, pulse $n+1$ by the aft segment;
- Pulse $n+2$ by the forward segment, pulse $n+3$ by the aft segment, and so on.

As a result, the pulses of every pair—e.g., n and $n+1$—are transmitted from exactly the same point in space.

The returns of each pulse are received by the antenna segment that transmitted the pulse. When the return from any one range, R, is received, the phase center of that segment will have advanced a distance equal to the aircraft velocity, V, times the round-trip transit time, t_R, for the range R. But if V is constant, this advance will be the same for both pulses. Therefore the round-trip distance traveled by the pulses to any one point on the ground will be the same, making the phases of the returns the same (Fig. 26.3).

So for each resolvable range interval, the radar returns received from the ground may be canceled simply by passing the digitized video outputs of the radar receiver through a single-pulse-delay clutter canceler. As illustrated in Figure 26-4, it

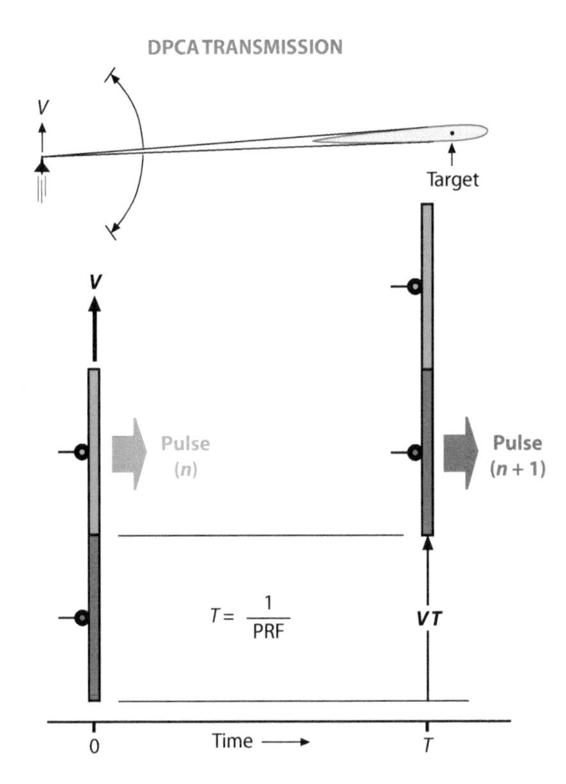

Figure 26-2. In DPCA, the radar transmits alternate pulses with forward and aft segments of an antenna with two phase centers. The velocity, V, and PRF are adjusted so the radar advances between pulses a distance VT that equals the distance between the phase centers of the segments; pulses n and $n+1$ are then transmitted from the same point in space.

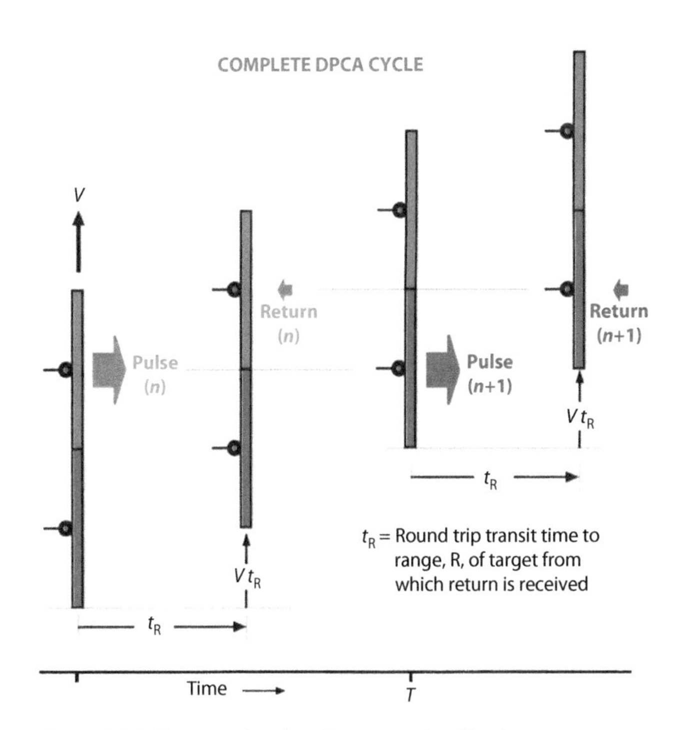

Figure 26.3. Returns of each pulse are received by the same segment that transmitted the pulse. Consequently the round-trip distances traveled by both pulses n and $n+1$ are equal.

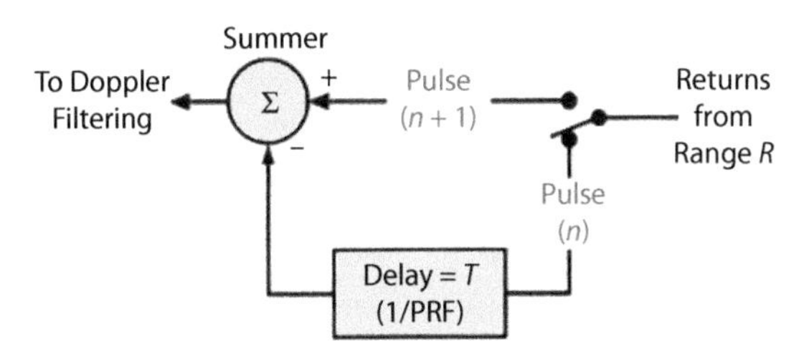

Figure 26-4. Pictured here is a clutter canceler.

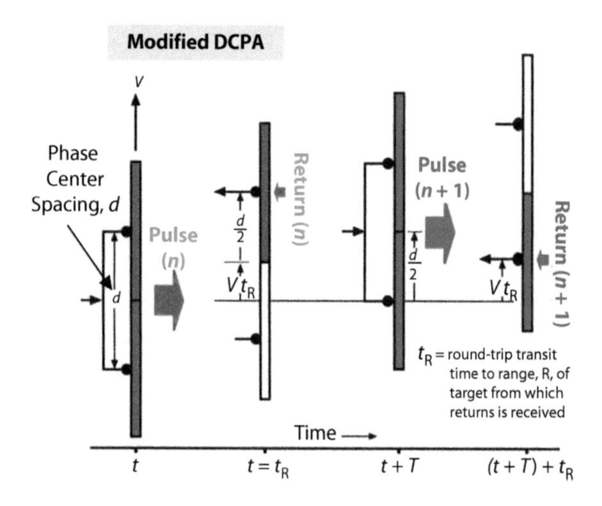

Figure 26-5. To transmit with full aperture, the velocity, V, and PRF are adjusted so the radar travels only half the distance between the phase centers during the interpulse period. Returns of pulse n are received by the forward antenna segment; returns of pulse $n+1$ are received by the aft segment. Thus the round-trip distance traveled by both pulses is the same.

Table 26-1. DPCA using the full aperture for transmitting

Pulse	Displacement of Phase Centers		
	Transmit	Receive	Total
(n)	0	$V t_R + \frac{d}{2}$	$V t_R + \frac{d}{2}$
(N + 1)	$\frac{d}{2}$	$V t_R$	$V t_R + \frac{d}{2}$

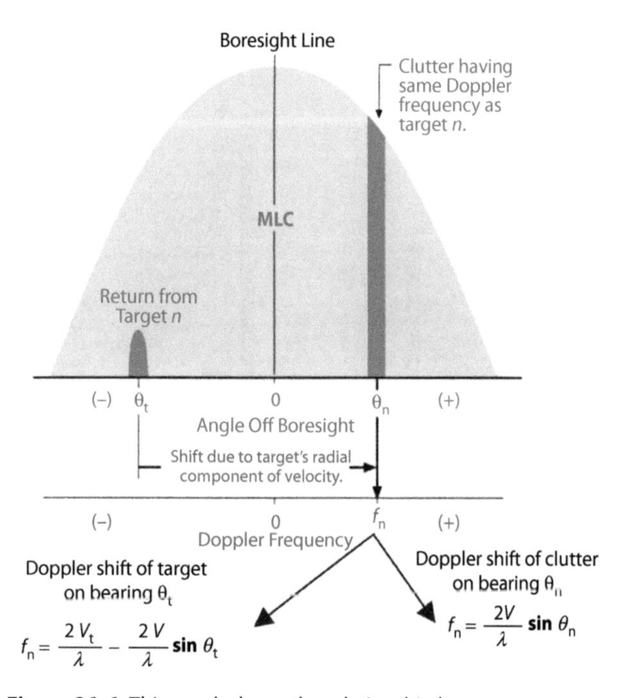

Figure 26-6. This graph shows the relationship between return received from a slowly moving target on the ground and mainlobe clutter. Because of the target's radial component of velocity, clutter having the same Doppler frequency as the target, f_n, is received from a different angle off the boresight line.

delays the return of pulse n by the interpulse period, T, and subtracts it from the return of pulse $n + 1$.

Successive returns from a *moving* target, however, will differ in phase as a result of the radial component of the target's velocity. Consequently they will not cancel, but will produce a useful output.

Effective as this technique is, it has four limitations:

1. The PRF is tied to the aircraft's velocity.
2. Very tight constraints are placed on aircraft and antenna motion.
3. The phase and amplitude characteristics of the two antenna segments and of the receive channels for both segments must be precisely matched.
4. Only half of the aperture is used at any one time.

The fourth limitation may be partially removed by adjusting the velocity and PRF so that during the interpulse period the phase centers advance by only half the distance between them (Fig. 26-5). The entire aperture may then be used for transmission, but still only half the aperture may be used for reception. Returns of pulse n must be received by the forward segment, while returns of pulse $n + 1$ must be received by the aft segment.

Although both pulses are not transmitted from the same point in space and returns from the same ranges are not received at the same points in space, the result is the same as if they were. As indicated in Table 26.1, the phase centers' total displacement for transmission and reception is the same for both pulses. Therefore, the round-trip distance traveled to any one point on the ground is the same for both pulses—just as when the pulses are alternately transmitted by the fore and aft antenna segments.

If the antenna has three receive apertures, a similar approach can be used, but now providing the opportunity to use a two-delay canceler, which is effectively two of the single-delay cancelers of Figure 26-4 placed in series. As we will see later when we discuss STAP, these ideas can be generalized to any number of antenna phase centers.

Notching Technique. Notching has the advantage over classical DPCA in that it does not require the PRF be tied to aircraft velocity and it relaxes the constraints on aircraft and antenna motion. Mainlobe clutter is rejected without rejecting target returns by taking advantage of the Doppler shift due to the radial component of a target's velocity, small as it may be. Because of this shift, clutter having the same Doppler frequency as target n comes from a slightly different angle, θ_n, off the boresight line (Fig. 26-6). Here, the clutter return received in a side-looking antenna has a Doppler shift that is determined by the angle off the boresight line of the clutter scatterers and the radar platform's velocity. For a platform velocity V, clutter from an angle θ_n off the boresight line will have a Doppler shift of $\frac{2V}{\lambda}\sin\theta_n$. The target is on a bearing of $-\theta_t$, but because of its own radial velocity toward the radar of V_t, the total Doppler shift will be $\frac{2V_t}{\lambda} - \frac{2V}{\lambda}\sin\theta_t$. There will be a value of θ_n such that $\frac{2V}{\lambda}\sin\theta_n = \frac{2V_t}{\lambda} - \frac{2V}{\lambda}\sin\theta_t$. Therefore, by placing a notch in the antenna receive pattern at θ_n, the clutter can be rejected without rejecting the return from target n (Fig. 26-7).

Moreover, the return from target n is isolated from clutter received from other directions—and therefore having other Doppler frequencies—by Doppler filtering.

Because a target's angular position and radial component of velocity generally are not known in advance, and because returns from targets in different directions may be received simultaneously, a separate notch must be formed for each of the N resolvable Doppler frequencies. To avoid rejecting target returns along with the clutter, notching must be performed after, not before, Doppler filtering.

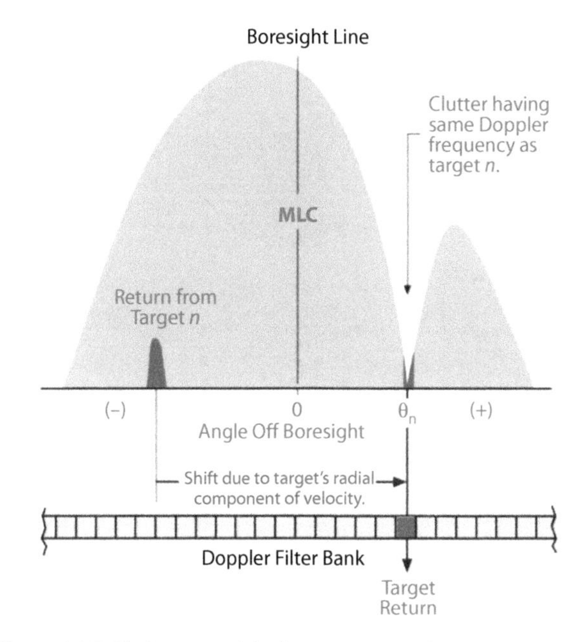

Figure 26-7. Placing a notch in the antenna receive pattern at angle, θ_n, from which the radar receives mainlobe clutter having the same Doppler frequency, f_n, as target n prevents clutter from interfering with the target's detection. Doppler filtering isolates the target return from other clutter.

How a Notch is Made

In a Two-Segment Antenna's Receive Pattern at an Angle θ_n Off the Boresight Line

1. Calculate the difference in distance, Δ_d, traveled to the two phase centers, A and B, by returns from a distant point at the angle, θ_n, off the boresight line.

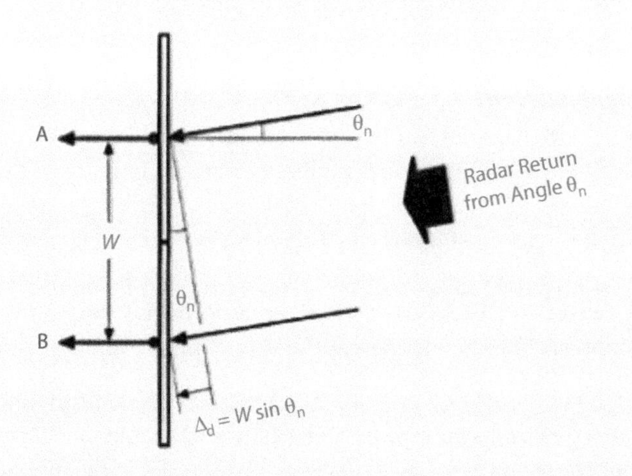

2. Convert Δ_d to phase ϕ:

$$\phi = \frac{2\pi}{\lambda}\Delta_d = \frac{2\pi}{\lambda}W\sin\theta_n$$

3. Subtract ϕ from π radians (180°). In the example here we wish to apply equal and opposite phase corrections to A and B to place them in antiphase. We therefore divide $\phi - \pi$ by 2 and the result is the phase rotation, $\Delta\phi$, which—when made in opposite directions to antenna outputs A and B—will increase the phase difference, ϕ, between returns received from θ_n to 180°, as shown in the following figure.

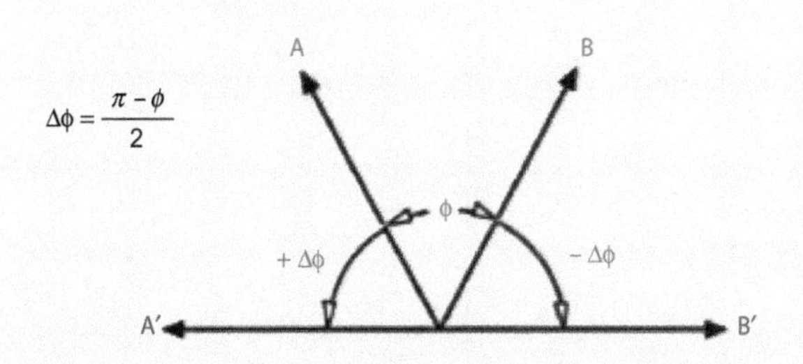

4. Sum the phase-rotated outputs, A′ and B′. The returns received from θ_n will then cancel, producing the equivalent of a notch in the antenna receive pattern at angle θ_n.

5. To produce a notch on the opposite side of the boresight line, reverse the directions of the two phase rotations.

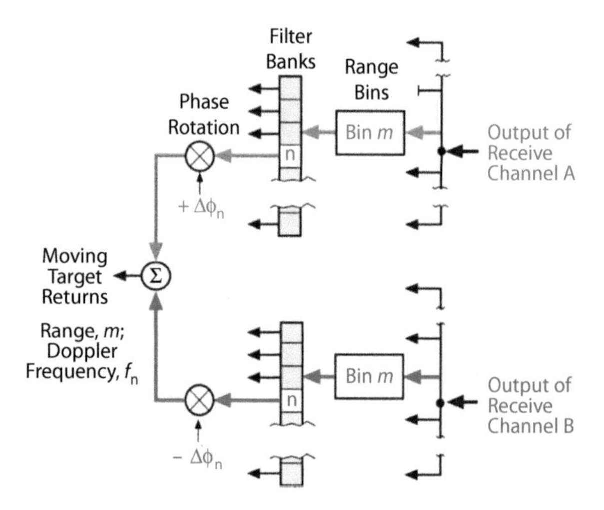

Figure 26-8. This illustration shows the implementation of the notching technique. Video outputs of receive channels A and B are collected in range bins. For each range bin, *m*, a Doppler filter bank is formed. Output of each filter, n, is rotated in phase in channel A by $+\Delta\phi_n$ and in channel B by $-\Delta\phi_n$. Rotated outputs are then summed, creating the equivalent of a notch in the antenna receive pattern at θ_n, while passing returns from targets at other angles whose Doppler frequency is f_n.

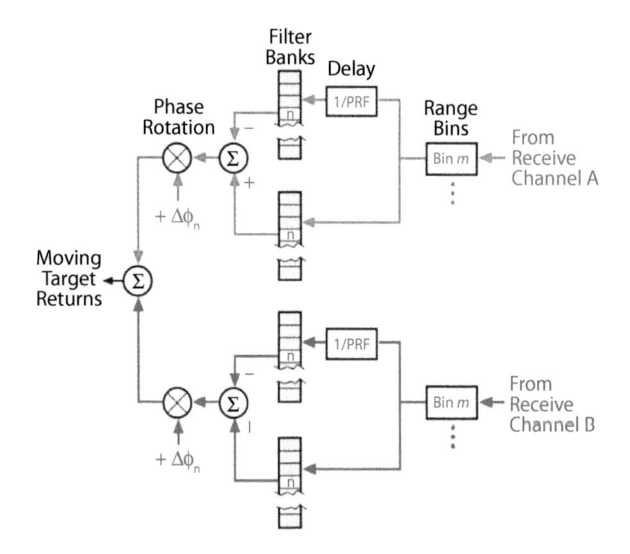

Figure 26-9. Combination of classical DPCA and notching. The technique eases the constraints DPCA places on aircraft and antenna motion and/or improves the clutter rejection performance of notching in applications that limit dwell time.

The notches are produced with an interferometric technique similar to that used in phase-comparison monopulse angle tracking (see panel, left). As with classical DPCA, a two-segment electronically steered antenna is typically used. So that very small differences in Doppler frequency may be resolved, dwell times are increased to allow predetection integration over long periods, t_{int}. So that the notching can be done in the signal processor, a separate receive and signal processing channel is generally provided for each antenna segment.

Implementation of the notching process is illustrated in abbreviated form in Figure 26-8. For every range bin in both channel A and channel B, a separate Doppler filter bank is formed. The outputs of each pair of filters, *n*, passing returns of the same Doppler frequency, f_n, from the same range, *m*, are then rotated in opposite directions through the angle $\Delta\theta_n$. This rotation causes the returns received by the two antenna segments from the ground at an angle θ_n off the boresight line to be 180° out of phase.

The phase-rotated returns are summed, with the result that the ground returns from θ_n cancel, while returns from targets at any other angle off the boresight line whose Doppler frequency is f_n do not.

For radars in which monopulse sum and difference signals for angle tracking are produced ahead of the receiver (i.e., at microwave frequencies), notching is performed similarly with the outputs of the sum and difference channels. In that case though, rather than being phase rotated and summed, the outputs of the corresponding Doppler filters, f_n, are weighted and summed to shift the null of the difference output to the angle θ_n off boresight.

Many targets on the ground will have high enough radial velocities to fall in the clutter-free portion of the Doppler spectrum, and therefore notching is generally time shared with conventional MTI processing.

Combined Notching and DPCA. Generally, notching provides very good mainlobe clutter cancellation, but under some conditions—such as when frame-time requirements limit dwell times and hence achievable Doppler resolution—clutter rejection performance can be substantially improved by combining notching and DPCA. This improvement may be traded at any time for an easing of DPCAs strict constraints on aircraft and antenna motion and/or for uncoupling PRF from aircraft velocity.

Implementation differs from that just described, primarily in that for each range bin, two Doppler filter banks are formed from the outputs of each receive channel and the inputs to one of these banks are delayed by the interpulse period, *T* (Fig. 26-9).

Although further improvements in clutter rejection performance can be expected through further algorithm development, remember that since the cancellation techniques rely on clutter scatterers being stationary, cancellation ultimately will be limited by the "internal motion" of the clutter.

STAP. As described above, the creation of a different directional null in each Doppler bin requires a knowledge of the platform velocity and assumes a known fixed relationship between the Doppler shift of a point on the ground and its bearing relative to the aircraft's direction of travel. The techniques of DPCA and notching are open-loop filtering systems. As mentioned in the discussion on DPCA, these techniques are not restricted to antennas with two phase centers. In general, we can form nulls at different Doppler frequencies and in different directions using an array of N antenna elements, such as in an active electronically scanned array (AESA) system.

In a practical system, the aircraft velocity continuously changes. Also the Doppler shift of returns from the ground may have a more complex relationship to the look direction than assumed above due, for example, to the effects of varying terrain slope and height over the scene. It is very desirable to have a system that automatically adapts to the observed spectrum and adjusts the filter response at each Doppler frequency accordingly. The simple DPCA approach typically uses the returns from just two pulses in a simple canceler filter. The notching filter steers a null in space, again using just two elements. If we have a phased array antenna with N elements and we process the returns from M pulses, we can design much more complex filter responses in space (i.e., directional nulls) and time (i.e., Doppler filters). The filter responses can now adapt to the spectrum of the clutter as a function of look direction, as we shall see below.

In a typical AESA application, each of N antenna array elements is used for transmission and reception. Returns at each element are stored for M successive pulses as the platform moves along its track. Usually the antenna will be looking broadside to the aircraft track, as in DPCA processing. The array of received signals is illustrated in Figure 26-10, which shows an array of

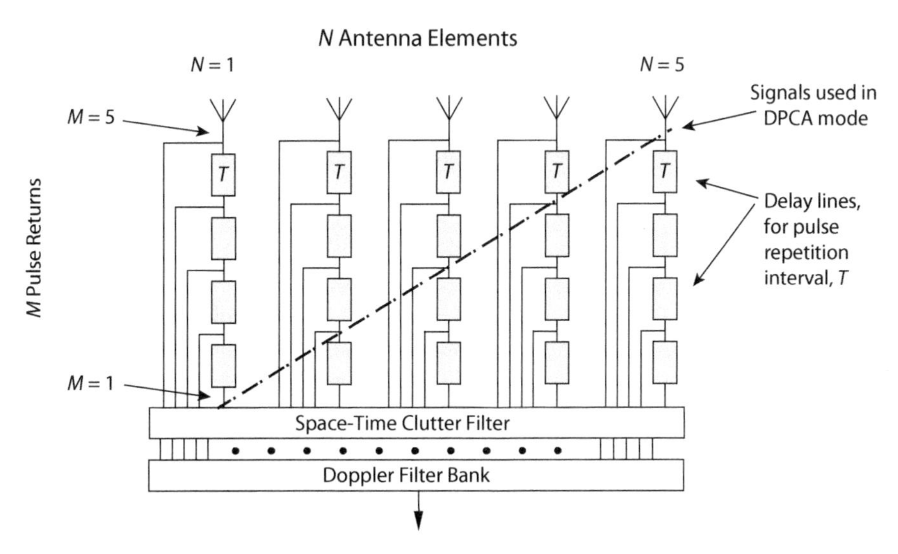

Figure 26-10. This illustration shows AESA pulse storage for adaptive space and time filtering.

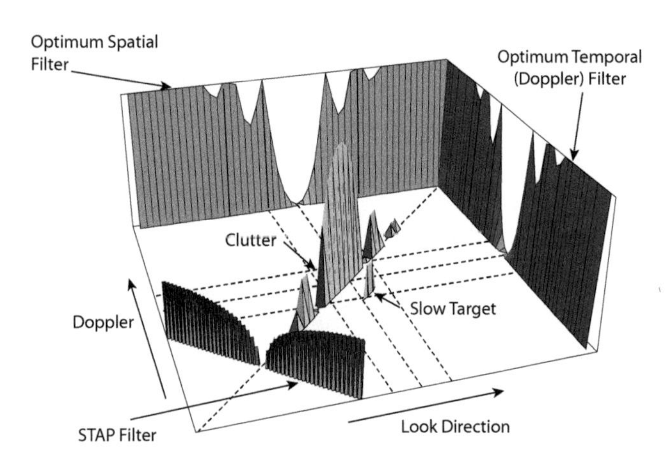

Figure 26-11. The space-time filter response shows the advantage of STAP over individual space or time filters. (After R. Klemm, "Introduction to Space-time Adaptive Processing," *Electronics & Communication Journal, IEE*, Vol 11, No. 1, February 1999, pp. 5–12.)

five antenna elements and a filter using the returns from five pulses. The returns from each pulse, spaced in time by a pulse repetition interval, *T*, at each antenna element are weighted and summed in the space–time clutter filter. This processing is undertaken for each range cell of the radar. After the space–time filter, the returns are then Doppler filtered, to detect targets moving at different speeds.

We suitably weight the returns at each element to create a filter that uses the information in space (*N* antenna elements) and time (*M* pulses). By analyzing these data, we can create a filter that minimizes the clutter residue in an adaptive manner (i.e., closed-loop as opposed to the previously open-loop techniques). The resulting filter response (Fig. 26-11) is effectively an extension of the result achieved by clutter notching using just two pulses, but now we use many more pulses, supporting a much better filter shape and permitting targets to be detected at much lower velocities relative to the clutter spectrum. If the radar antenna moves by exactly the spacing between adjacent antenna elements during one pulse repetition interval, we will have the same outcome as an *N* pulse DPCA system. See Figure 26-10, where the dotted line shows the data samples that would be used in a five phase-centre DPCA system. The filter response in Figure 26-11 is the same as that with a notching system, but now the filters have automatically adapted to the clutter rather than being simple notches directed to a predetermined angle at each Doppler frequency. Figure 26-11 also shows the filter responses if we filtered only in Doppler or only in the look direction (space). It can be seen that in neither of these cases could our slow moving target be detected.

To adapt the STAP filters in this way, the radar collects samples of the clutter over a number of range gates and over time. We assume the clutter characteristics are the same over all these data samples. If the clutter is very nonhomogeneous, the radar signal processing may have difficulty in estimating the correct filter response, even in the closed-loop system.

26.3 Precise Angle Measurement

While the DPCA and notching techniques detect a target that would otherwise be hopelessly embedded in mainlobe clutter, they don't tell at what angle within the antenna beam the target is located. For although the direction of the interfering clutter can be determined from the frequency of the Doppler filter that passes the target return, without knowing the target's radial velocity, and thus its Doppler frequency, it is impossible to tell directly how far removed from that angle the target actually is.

In conventional operation of a two-segment antenna and a two-channel receiving system, a target's precise direction can be obtained by comparing the phases, or amplitudes, of the outputs the target produces from the two channels. In contrast, the slow moving target detection techniques fully utilize the outputs of both channels for clutter rejection and target detection.

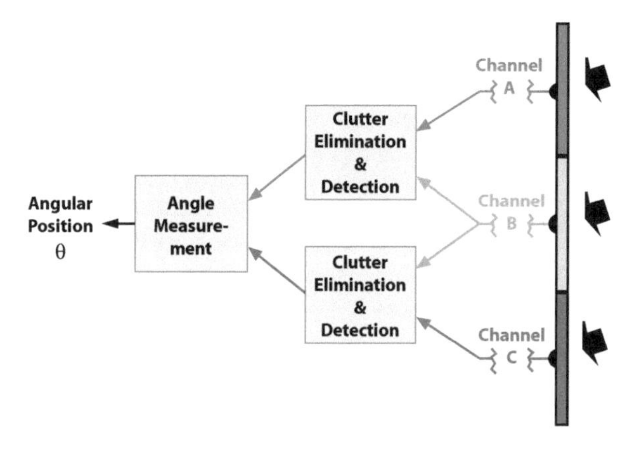

Figure 26-12. An approach to precise angle measurement. Since returns received by two antenna segments are required for clutter elimination and target detection, a third segment must be provided to determine the target's angle, θ, within the radar's beam.

Accordingly, where precise angle measurement is required, a three-segment antenna and three receive channels are used. As illustrated in Figure 26-12, the outputs of receive channels A and B are used to provide clutter rejection and target detection for an effective phase center halfway between the phase centers of antenna segments A and B. The outputs of channels B and C are similarly used to provide clutter rejection and target detection for an equivalent phase center halfway between those of antenna segments B and C. The target's precise direction is then estimated from the distance between the two effective phase centers and the difference in phase of the two output signals.

For a STAP system, the N-elements array can be processed to create sum and difference channels, as in a monopulse radar (see Chapter 8), as well as adaptively filtering the clutter.

A practical ground moving target indication (GMTI) radar must not only detect slow moving ground targets, but, to be useful to a user, it must then be able to accurately position them on a map. The previously described techniques show how the slow targets may be detected and their range and bearing from the radar measured. These measurements must then be transformed from the measurement frame of the radar into a ground coordinate system, such as height, latitude, and longitude. This requires a very accurate knowledge of the radar's own position, usually achieved using a satellite navigation system such as GPS or Galileo, and an accurate terrain elevation map or digital terrain elevation data (DTED). With this additional information, the radar can then accurately position moving targets on the ground. Figure 26-13 is an example of a GMTI radar display, showing moving motor vehicles on the ground. This example shows trails for the complete history for an area 40 km × 35 km gathered from several full rotations of the radar antenna over a period of 9 minutes. The trails are colored yellow and cyan to indicate approaching and receding targets in the line of sight of the radar, respectively. The platform heading is northeast and a wide area strip of plots is built up until the coastline is reached, which is apparent from the lack of plots in the northeast corner of the image. Although this is not what is normally seen by an operator, who would usually be interested in monitoring individual movements over a shorter time period, the large number of targets allows traffic densities on the road network to be easily identified, with major highways distinguished from smaller, less travelled routes.

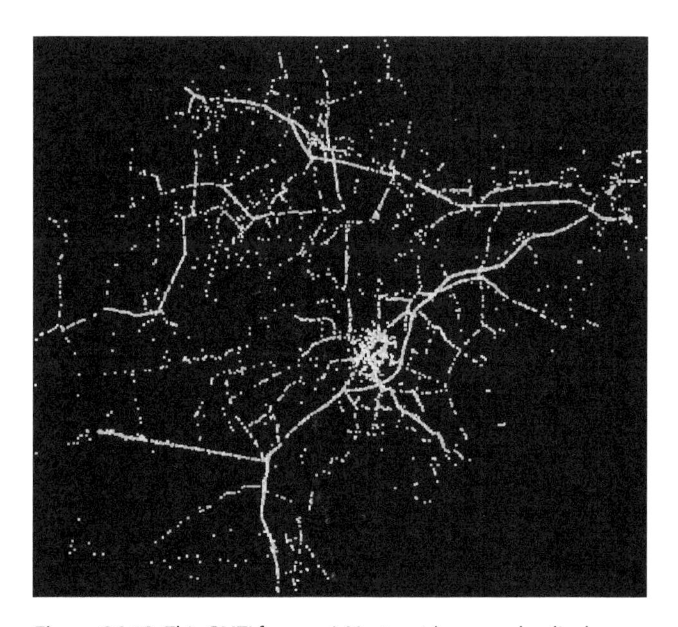

Figure 26-13. This GMTI from an I-Master airborne radar display shows moving vehicles on roads. (Courtesy of Thales UK.)

26.4 Summary

Targets whose true Doppler frequencies fall in mainlobe clutter may be separated from the clutter with either DPCA or

Test your understanding

1. Why are slow moving ground targets often more difficult to detect than air targets?
2. A side-looking airborne antenna has two phase centers, separated by 0.5 m in the along-track direction. If the aircraft's velocity is 200 m/s, what value of PRF should I use to employ displaced phase center antenna (DPCA) processing on successive pulses?
3. What are the disadvantages of the basic DPCA technique?
4. If an airborne radar antenna with a beamwidth of $\theta°$ is pointing normal to the direction of flight, with an aircraft velocity v m/s and a radar wavelength λ, what are the Doppler shifts of the ground returns seen at the edges of the beam (assume zero depression angle)?
5. Why does space–time adaptive processing (STAP) offer the potential for better performance than DPCA or notching?

notching, both of which employ a two-segment side-looking antenna.

For DPCA, aircraft velocity and PRF are adjusted so the radar advances the distance between the phase centers of the antenna segments during the interpulse period. By transmitting and receiving alternate pulses with fore and aft segments, both pulses travel the same round-trip distance to any one point on the ground, so mainlobe clutter can be eliminated by a clutter canceler.

For notching, radar returns are sorted with a Doppler filter bank and a notch is placed in the antenna receive pattern for each filter. Because of the Doppler shift due to a target's velocity, mainlobe clutter having the same Doppler frequency as the target will come from a different direction than the target return. Thus the clutter is "notched out" without rejecting target returns.

By combining DPCA with notching, greater flexibility is obtained than with either technique alone. However, notching assumes an a priori known relationship between the clutter Doppler shift and the look direction. In general this is not the case, given unknown ground height variations and other anomalies. Hence this open-lope approach is suboptimal. An advanced closed-loop approach to this problem is STAP. In STAP the radar uses an array of antenna receive elements to estimate the clutter characteristics and produce a filter response that adapts in space and time to reject the observed clutter returns.

Finally, a practical radar must not only detect a slow moving target but must also estimate its bearing within the antenna beam. When using simple two-channel DPCA or notching techniques, a third antenna element must be provided to achieve this additional measurement. Similar approaches are used with STAP, where the processing creates sum and difference patterns as well as adapting to the clutter characteristics.

Further Reading

R. Klemm, "Introduction to Space-Time Adaptive Processing," *IEE Electronics & Communication Engineering Journal*, Vol. 11, No. 1, 1999, pp. 5–12.

M. I. Skolnik, "MTI and Pulse Doppler Radar," chapter 3 in *Introduction to Radar Systems*, 3rd ed., McGraw Hill, 2001.

R. Klemm, *Principles of Space-Time Adaptive Processing*, IEE, 2002.

PART VI

Air-to-Air Operation

Lockheed C-130K Hercules and Vickers VC10

The four-engined C-130 Hercules first saw service in 1957 and has been widely used ever since in numerous different roles. Here a C-130K and VC10 from 1312 Flight, Royal Air Force, practice air-to-air refueling.

27

PRF and Ambiguities

A Pilot's View: RAF Aerobatic Team, The Red Arrows, practicing their maneuvers over RAF Scampton, Lincoln.

The choice of pulse repetition frequency (PRF) in pulsed radar, especially in pulsed Doppler radar, is a key design consideration. Other conditions remaining the same, the PRF determines to what extent the observed ranges and Doppler frequencies will be ambiguous. In turn, it also defines the ability of the radar to measure range and Doppler velocity directly and to reject ground clutter. In situations where substantial amounts of clutter are encountered, the radar's ability to reject clutter and pass legitimate targets crucially affects its overall performance.

This chapter surveys the wide range of pulse repetition frequencies employed by airborne radars to reveal the regions in which significant range and Doppler ambiguities may occur. Three basic categories of pulsed operation—low, medium, and high—are identified, and their relative merits are discussed.

27.1 Primary Consideration: Ambiguities

Pulse repetition frequencies can vary from a few hundred Hz to several hundred kHz (Fig. 27-1). Exactly where, within this broad spectrum, a radar system will perform best under a given set of conditions depends on a number of considerations, the most important of which are range and Doppler ambiguities.

Range Ambiguities. For range to be unambiguous, the echoes from the most distant detectible targets must be

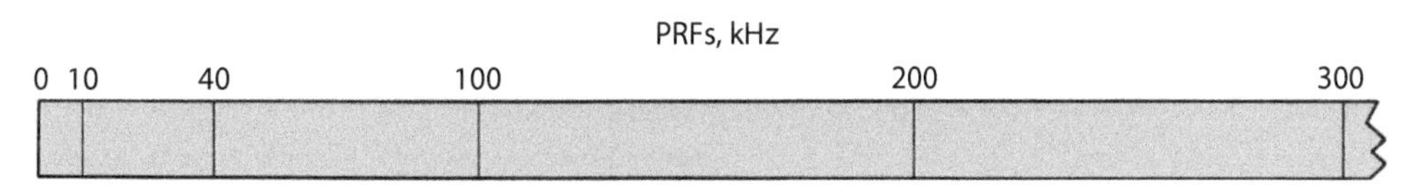

Figure 27-1. PRFs can range all the way from a few hundred Hz to several hundred kHz.

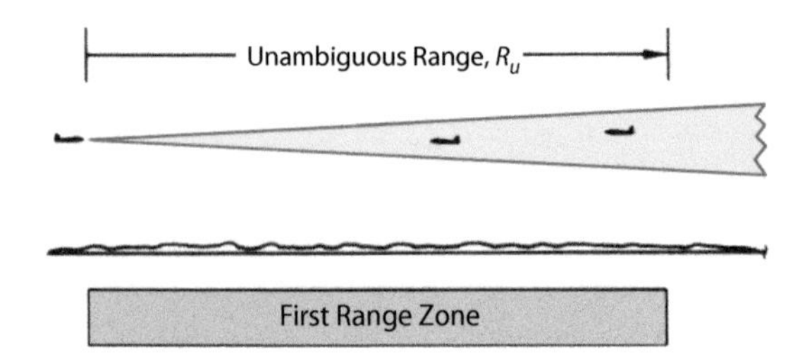

Figure 27-2. Assuming all echo returns from beyond the first range zone are rejected, this zone is a region of unambiguous range.

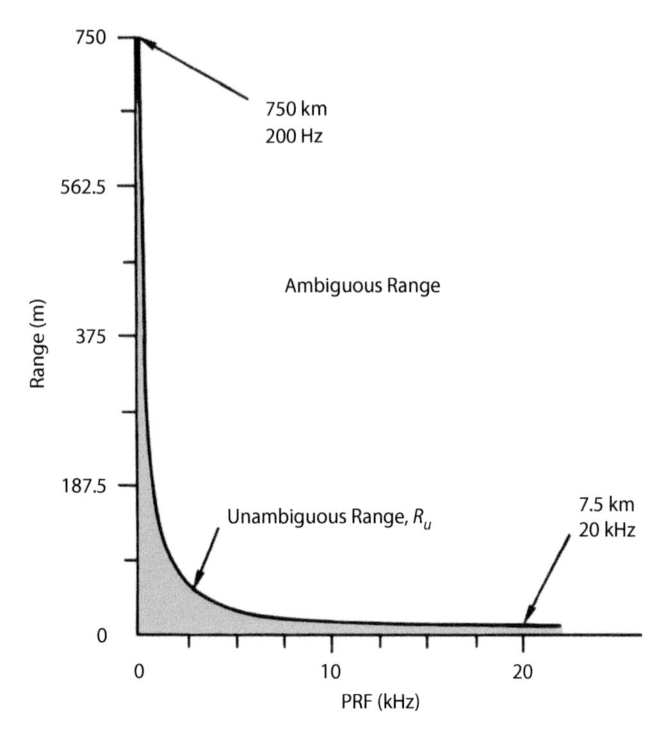

Figure 27-3. The area under the curve encompasses every combination of range and PRF for which a target's observed range will be unambiguous. This assumes that all echo returns from beyond the first range zone are either negligible or are rejected.

first-time-around echoes. In other words, all echoes caused by a transmission pulse must be received before the next pulse is emitted. The range over which this occurs is termed *the first range zone*. Thus, the first range zone is the unambiguous range, R_u (Fig. 27-2). However, it is possible that echoes from large cross section targets or clutter located beyond the first range zone are still detected (as the radar pulse continues to travel on and on). Thus, in effect range is almost always ambiguous.

Indeed, if the PRF is sufficiently low so that the maximum required operating range falls within or is equal to the first range zone, longer range ambiguities can still be eliminated. Techniques for such rejection include PRF jitter (Chapter 15). For our purposes it is assumed that the first range zone is a region of unambiguous range.

The first range zone extends to a range approximately equal to 150 km divided by the PRF in kHz. In Figure 27-3, this range is plotted versus PRF. The area under the curve encompasses every combination of PRF and range for which range is *unambiguous*. The area above the curve encompasses every combination of PRF and range for which range is *ambiguous*.

Notice how rapidly the curve plunges as the PRF is increased. From a range of 750 km at a PRF of 200 Hz, it drops to 18.75 km at a PRF of 8 kHz and to only 7.5 km at a PRF of 20 kHz.

Doppler Ambiguities. Like range, Doppler frequency is inherently ambiguous. Whether the ambiguities are significant, however, depends on the PRF, the wavelength, and the difference between the maximum opening and closing rates. The maximum closing rate is the velocity of the most rapidly approaching target. The maximum opening rate may be either that of a target moving relatively slowly compared with the radar or the relative velocity of the ground from which sidelobe clutter is received behind the radar (Fig. 27-4). In fighter applications it is generally the latter. This rate is very nearly equal to the maximum velocity, V_r, of the aircraft carrying the radar system.

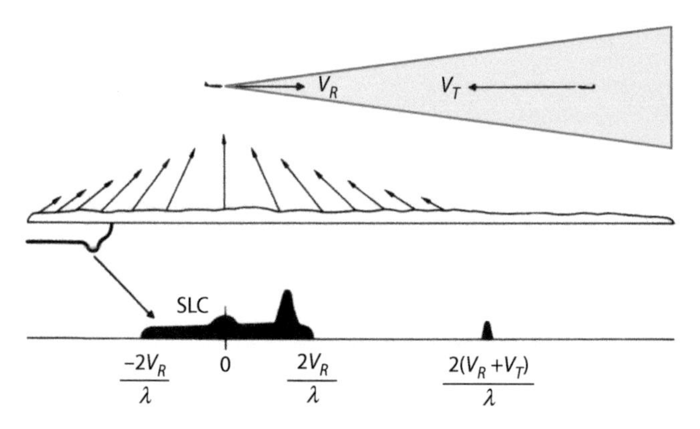

Figure 27-4. The maximum opening rate is usually that of the ground from which sidelobe clutter is received behind the radar. The maximum closing rate is caused from the fastest approaching target.

The relationship between the PRF and the Doppler frequency at which ambiguities arise in a clutter environment is illustrated in Figure 27-5.

Figure 27-5 shows the Doppler profile for a typical flight situation, including both the true profile (central line frequencies) and its next higher frequency repetition (first upper sideband image). Corresponding points in the two profiles are separated by the pulse repetition frequency, f_r. A high closing rate target (B) appears in the clear region above the highest actual clutter frequency. If this target's closing rate were progressively increased, the target would move up the Doppler frequency scale (to the right in the figure). Eventually it moves into the repeated sidelobe clutter image spectrum but will also appear in the true sidelobe clutter spectrum, which is separated by f_r. On the basis of Doppler frequency alone, the radar would have no way of separating the target echoes from the sidelobe clutter even though their true Doppler frequencies are quite different.

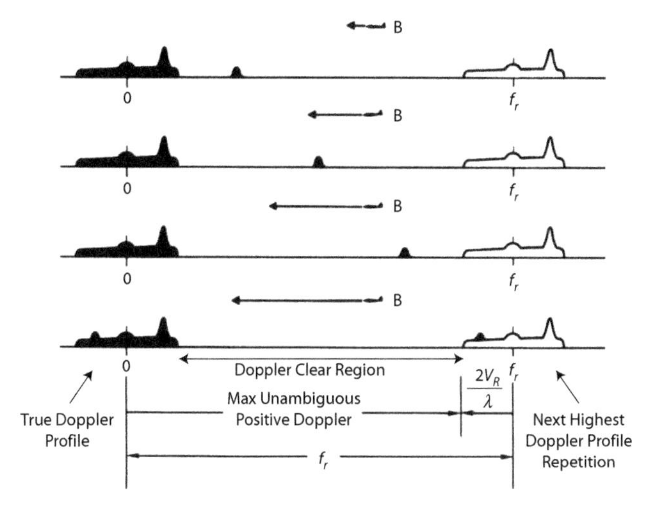

Figure 27-5. This Doppler profile represents a typical flight situation and shows the true Doppler profile and its next higher repetition. As the target velocity increases, the target moves through the Doppler clear region and ultimately enters the repetition of the negative frequency of sidelobe clutter. In the first image it is centered at a frequency of one PRF, f_r, away from zero Doppler. The target then appears ambiguously in the same relative position in the true Doppler sidelobe clutter and cannot be distinguished on the basis of Doppler alone. This sets an upper target velocity limit for targets that can be detected without ambiguity.

Therefore, from the standpoint of clutter rejection, the highest unambiguous Doppler frequency that a target can have (i.e., the highest frequency at which the target will not have to compete with clutter whose true Doppler frequency is different from the target's) is equal to the pulse repetition frequency minus the maximum sidelobe clutter frequency. The latter frequency, as just noted, corresponds to the radar's velocity.

$$\text{Maximum unambiguous Doppler} = \text{PRF} - \left(\frac{2V_R}{\lambda}\right)$$

The maximum closing rate for which the Doppler frequency will be unambiguous in a clutter environment is plotted versus PRF in Figure 27-6. A wavelength of 3 cm and a radar velocity of 500 m/s are assumed. The plot decreases linearly from a closing rate of about 4000 m/s at a PRF of 300 kHz to a closing rate of 500 m/s (radar's ground speed) at a PRF of 70 kHz, at which point it is terminated since at lower PRFs the maximum positive and negative sidelobe clutter frequencies overlap.

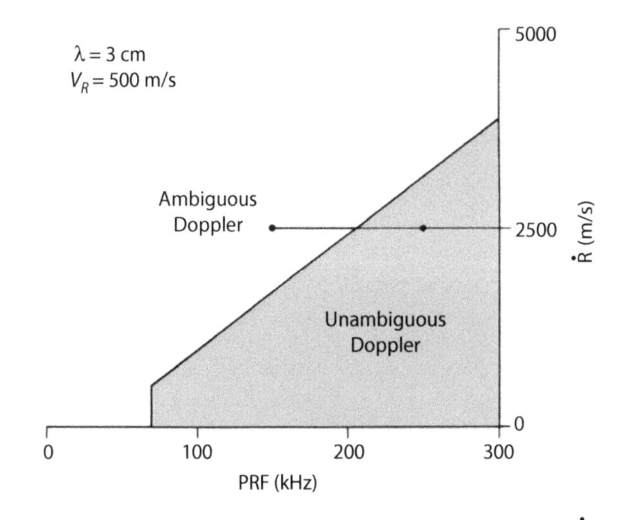

Figure 27-6. The combinations of PRF and target closing rate, $\dot{R}$, are graphed with the observed unambiguous Doppler frequencies shown in green.

In Figure 27-6 the area beneath the curve encompasses every combination of PRF and closing rate for which the observed Doppler frequencies are unambiguous. Conversely, the area above the curve encompasses every combination for which the Doppler frequencies are ambiguous. For example, at a PRF of 250 kHz and a closing rate of 2500 m/s the Doppler frequency is unambiguous, whereas at a PRF of 150 kHz and the same closing rate the Doppler frequency is ambiguous.

If the aircraft's ground speed is greater than 500 m/s, the area beneath the curve will be correspondingly reduced and vice versa.

Since Doppler frequency is inversely related to wavelength, the shorter the wavelength, the more limited the region of unambiguous Doppler frequencies. To illustrate the profound effect

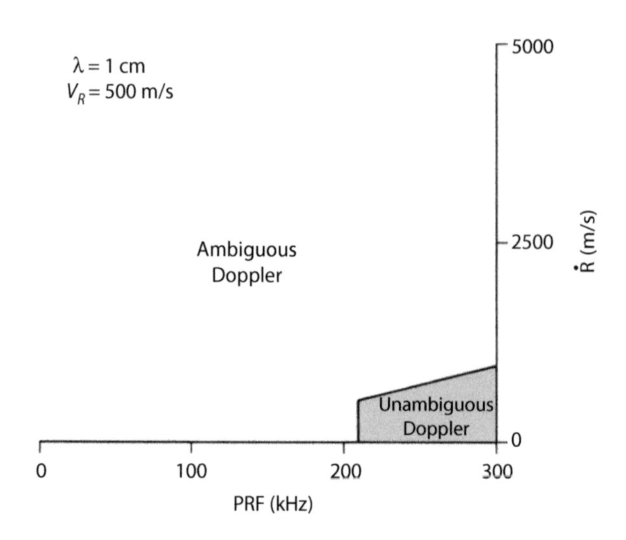

Figure 27-7. The dramatic reduction in the region of unambiguous Doppler frequencies result from a decrease in wavelength λ from 3 to 1 cm.

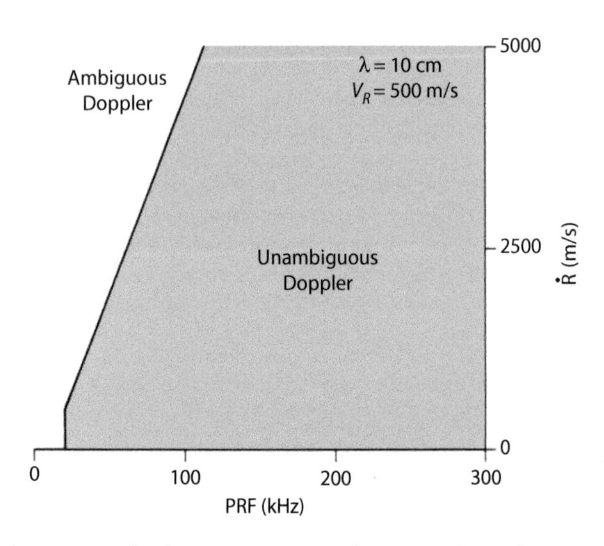

Figure 27-8. The dramatic increase in the region of unambiguous Doppler frequency result from an increase in wavelength to 10 cm.

of wavelength on Doppler ambiguities, Figure 27-7 plots the maximum target closing rate at which the Doppler frequency will be unambiguous for a wavelength of 1 cm (i.e., a transmission frequency of approximately 35 GHz). Not only is the area under the curve comparatively small, but even at a PRF of 300 kHz the maximum closing rate at which the PRF is unambiguous is less than 1000 m/s.

Yet at a wavelength of 10 cm (a frequency of 3 GHz), the area under the curve extends from 500 m/s at 21 kHz to about 13,500 m/s (off the scale) at 300 kHz (Fig. 27-8).

Putting the Plots in Perspective. Unambiguous measurement of detection range and Doppler velocity depends on the selected value of the PRF. Indeed, PRF is a key radar design parameter and must be chosen with care, especially for radar systems operating from fast moving jet aircraft.

The regions of unambiguous range and unambiguous Doppler frequency (for $\lambda = 3$ cm and $V_R = 500$ m/s) are shown together in Figure 27-9. Drawn to the scale of this figure, the region of unambiguous range is quite narrow, yet the region of unambiguous Doppler frequencies is comparatively broad. In between is a region of considerable extent within which range and Doppler frequency are both ambiguous.

Thus, the choice of PRF is a compromise. If the PRF is increased beyond a relatively small value, the observed ranges will be ambiguous. However, unless the PRF is raised to a sufficiently high value, the observed Doppler frequencies will be ambiguous.

While both range and Doppler ambiguities make clutter rejection difficult, their effects on a radar systems operation are quite different. It turns out that by designing a radar system to operate over a wide range of PRFs and by judiciously selecting the PRF to suit the operational requirements at the time, the difficulties can be almost completely obviated.

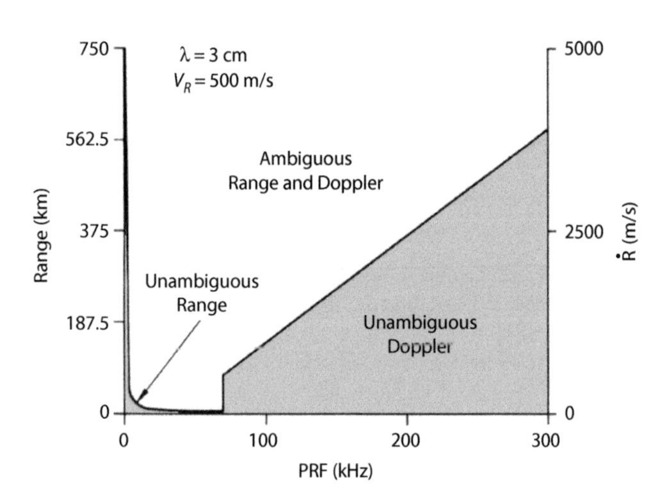

Figure 27-9. When the region of unambiguous range is plotted to same scale as the PRF, it becomes apparent that the choice of PRF is, at best, a compromise.

27.2 The Three Basic Categories of PRF

It is common to classify airborne radars in terms of their PRFs because of how tremendously they impact performance. Recognizing that the regions of unambiguous range and unambiguous Doppler frequency are very nearly mutually exclusive, three basic categories of PRF have been established: *low*, *medium*, and *high*.

These are defined in terms, not of the numerical value of the PRF per se, but of whether the PRF is such that the observed ranges or Doppler frequencies are ambiguous. Although exact definitions vary, all are similar. The following descriptions are widely used but should be considered as guides rather than as formal definitions:

- *A low PRF* is one for which the range of all detected target is unambiguous (i.e., the target lies in the first range zone).
- *A high PRF* is one for which the observed Doppler frequencies of all detected targets are unambiguous.
- *A medium PRF* is one for which neither of these conditions is satisfied. Both *range* and *Doppler* frequency are ambiguous.

CATEGORIES OF PRF

PRF	RANGE	DOPPLER
HIGH	Ambiguous	Unambiguous
MED	Ambiguous	Ambiguous
LOW	Unambiguous	Ambiguous

The low, high, or medium category is assigned depending considerably on operating conditions. For example, a PRF of 4 kHz implies an unambiguous range of approximately 37.5 km (c/2.PRF) (Fig. 27-10). Thus, a target detected at a range of 30 km would be unambiguous so it would be placed in the low PRF category. Yet the same PRF of 4 kHz would be deemed medium both if the maximum range were greater than 37.5 km and if the spread between maximum positive and negative Doppler frequencies exceeded 4 kHz (as would be the case for a fighter aircraft traveling at a velocity of 100 m/s carrying a radar operating at a wavelength of 3 cm).

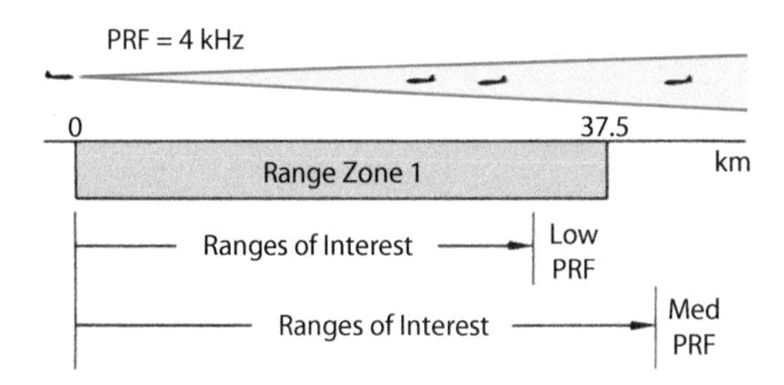

Figure 27-10. A PRF of 4 kHz would be low if the maximum range of interest was less than 37 km but medium if maximum range of interest was greater than 37 km.

Similarly, a PRF of 20 kHz might also be considered medium for X-band radar (3 cm wavelength) yet high for S-band radar (10 cm wavelength) if the radar's velocity were 100 m/s and the velocity of the fastest target was 500 m/s (thus giving a maximum closing rate of 600 m/s; Fig. 27-11).

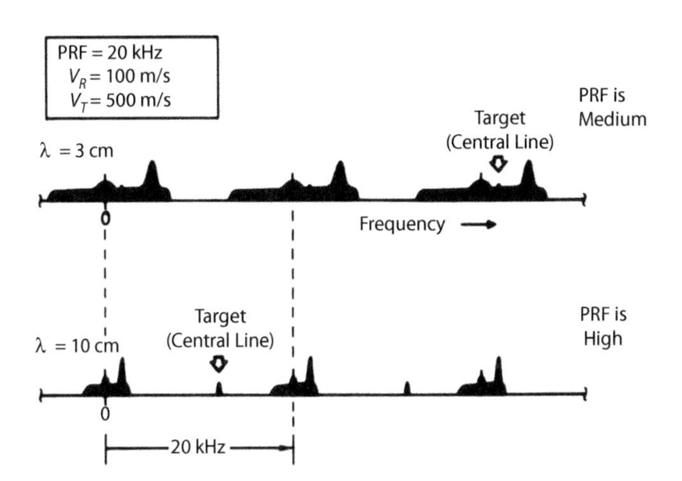

Figure 27-11. A PRF of 20 kHz might be medium at a wavelength of 3 cm yet high at a wavelength of 10 cm.

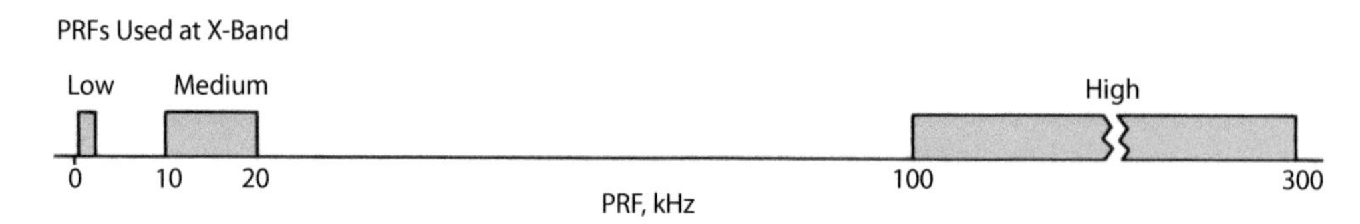

Figure 27-12. In practice, not all of the PRFs within each category are used.

In practice, not all of the possible PRFs within each category are used for any one radar band (Fig. 27-12). At X-band, for example, PRFs in the low category typically run from 250 to 4000 Hz; PRFs in the medium category are of the order of 10 to 20 kHz; and PRFs in the high category may range anywhere from 100 to 300 kHz.

All of this does not mean that the PRF categories are technicalities of little practical importance. To the contrary, in the everyday world of radar development and application, they have proved to be immensely useful. Whereas changing the PRF within any one category does not alter the radar's design in any fundamental way, changing it from one category to another quite radically affects both the radar's signal processing requirements and its performance.

LOW PRFs

ADVANTAGES	LIMITATIONS
1. Good for air-to-air lookup and ground mapping.	1. Poor for air-to-air lookdown (much target return may be rejected along with mainlobe clutter).
2. Good for precise range measurement and fine range resolution.	2. Ground moving targets can be a problem.
3. Simple pulse delay ranging possible.	3. Doppler ambiguities generally too severe to be resolved.
4. Normal sidelobe return can be rejected through range resolution.	

Low PRF Operation. Because range is unambiguous at low PRFs, this mode of operation has two important advantages. First, range may be measured directly by simple precise pulse-delay ranging. Second, as will be explained in the next chapter, virtually all sidelobe returns can be rejected through range resolution (except for echoes from point targets of exceptionally large radar cross section).

However, unless the mainlobe clutter is separated in range from targets, it can be rejected only on the basis of differences in Doppler frequency. Because of the overlapping of successive repetitions of the Doppler spectrum at low PRFs, the clutter cannot be rejected without also rejecting the returns from a considerable portion of the spectrum in which targets may appear. If the wavelength is long enough, the ground speed is low enough ($f_d \propto V_R/\lambda$), and the antenna is large enough ($\theta_{3dB} \propto \lambda/d$), the mainlobe clutter spectrum will be sufficiently narrow so that the possible loss of target return is quite tolerable.

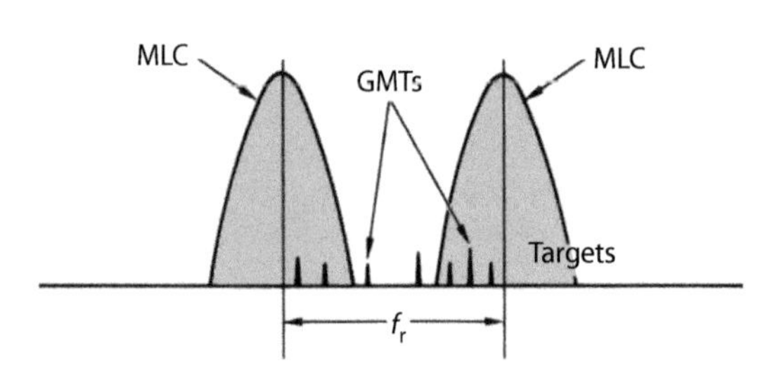

Figure 27-13. If the PRF is made low enough to provide reasonably long unambiguous ranges, most of target echoes will be rejected along with mainlobe clutter. In addition, ground moving targets cannot be directly discerned from airborne targets.

However, the parameters and conditions under which most fighter radars must operate are (1) a short wavelength, (2) a small antenna, and (3) a potentially high ground speed. Thus, if the PRF is low enough to extend the first range zone out to reasonably long ranges (e.g., 55 to 75 km), the unambiguous Doppler spectrum occupies most of the Doppler passband (Fig. 27-13). Consequently, when the clutter is rejected, the echoes from most of the target region will also be rejected. Further, since target echoes of widely different true Doppler frequencies are intermixed, not only is it impossible to resolve

Doppler ambiguities, but the radar is also susceptible to interference from ground moving targets.

Because of the severity of the mainlobe clutter problem, the use of low PRFs for air-to-air operation in fighters (which employ short wavelengths and comparatively small antennas) is restricted largely to situations where mainlobe clutter can be avoided:

- When flying over water, which has a relatively low backscattering coefficient at moderate to low grazing angles (because of its nearly mirror-like surface)
- When looking up in search of targets at higher altitudes
- When the mainlobe strikes the ground beyond the maximum range of interest (Fig. 27-14) and clutter from beyond the first range zone is rejected through other means than Doppler resolution

For ground mapping and synthetic aperture radar (SAR) imaging, low PRFs are ideal. This is because the mainlobe ground return is now the only return of interest and its overwhelming strength is an asset, not a liability. Moreover, the unambiguous observation of range, provided by a low PRF, is essential.

In SAR imaging (Fig. 27-15), the unambiguous observation of Doppler frequencies is also essential. Fortunately, the PRF can generally be made high enough to prevent the repetitions of the mainlobe clutter spectrum from overlapping while it provides an adequately long maximum unambiguous range.

High PRF Operation. The problem of mainlobe clutter can be solved by operating at high PRFs. The width of the mainlobe clutter spectrum is designed to occupy only a small fraction of the width of the band of true target Doppler frequencies. Thus, at high PRFs mainlobe clutter does not appreciably encroach on the region of the spectrum in which targets are expected to appear. Moreover, since all significant Doppler ambiguities are eliminated at high PRFs, mainlobe clutter can be rejected on the basis of Doppler frequency without at the same time rejecting echoes from targets. Only if a target is flying nearly at right angles to the line of sight from the radar will its echoes have the same Doppler frequency as the clutter and thus be rejected (i.e., the measured radial velocity of the target is very low even if its actual velocity is high).

Operation at high PRFs has other important advantages. First, between the central-line sidelobe clutter frequencies and their first repetition, a region opens up in which there is no clutter (Fig. 27-16). It is here that the Doppler frequencies of approaching targets lie. Second, closing rates can be measured directly by sensing Doppler frequencies. Third, for a given peak power, average transmitted power can be maximized simply by increasing the PRF until a duty factor of 50 percent is reached. High duty factors can also be obtained at low PRFs, but this requires increasing the pulse width and employing

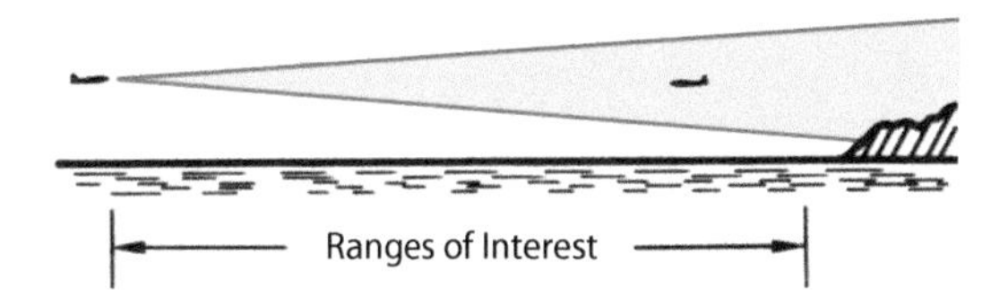

Figure 27-14. Use of low PRFs for air-to-air operations in fighter-type aircraft is limited to situations where mainlobe clutter is not a problem. Examples include over water or where the mainlobe does not strike the ground within ranges of interest.

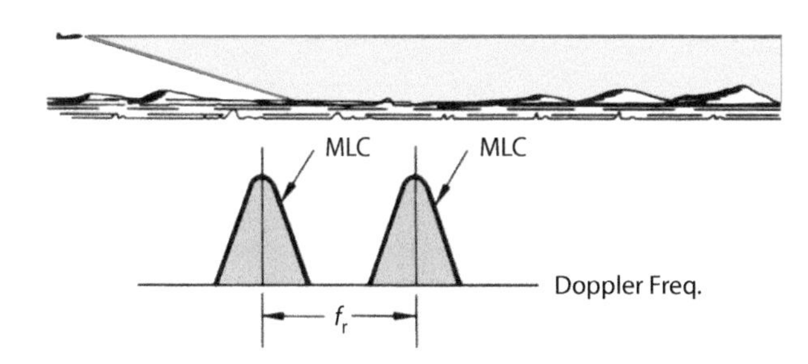

Figure 27-15. For ground mapping, the unambiguous range provided by low PRFs is essential. However, for SAR mapping, the PRF must also be high enough that repetitions of mainlobe clutter do not overlap.

HIGH PRFs

ADVANTAGES	LIMITATIONS
1. Good nose-aspect capability (high closing-rate targets appear in clutter-free region of spectrum).	1. Detection range against low closing-rate targets may be degraded by sidelobe clutter.
2. A high average power can be provided by increasing the PRF. (Only moderate amounts of pulse compression, if any, are needed to maximize average power.)	2. Precludes use of simple, accurate pulse delay ranging.
3. Mainlobe clutter can be rejected without also rejecting target echoes.	3. Zero closing-rate targets may be rejected with altitude return and transmitter spillover.

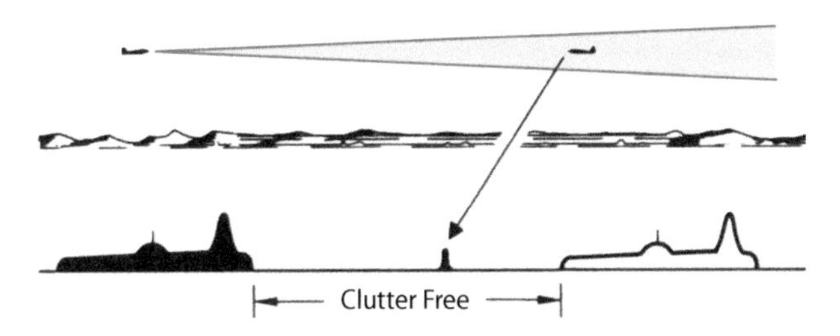

Figure 27-16. High PRFs provide a clutter-free region in which to detect high closing-rate targets.

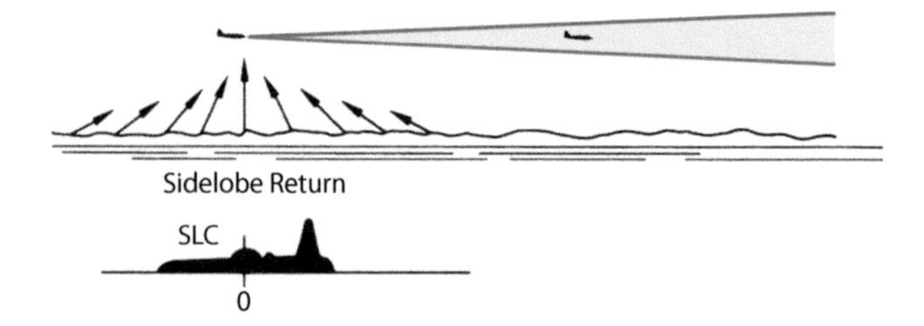

Figure 27-17. Echoes from low closing-rate targets must compete with sidelobe clutter, much of it from short range.

large amounts of pulse compression to provide the degree of range resolution that is essential in low PRF operation.

The principal limitation of high PRF operation is that detection performance may be degraded by sidelobe clutter when operating against low closing-rate, tail-aspect targets. Much of the sidelobe clutter falling within the same resolvable frequency increment as a target's echoes will have been reflected from very much shorter ranges and so will be quite strong (Fig. 27-17).

When flying at moderate to low altitude over terrain that has a high backscattering coefficient, unless the target has a large radar cross section or is at short range its echoes may be lost in the clutter. Further, if little or no range discrimination is provided, zero closing-rate targets will be rejected along with the altitude return.

Another disadvantage of very high PRFs is that they make pulse-delay ranging more difficult. As the PRF is increased, range ambiguities become more severe. To resolve them, the radar must switch among many PRFs.

Overall, when mainlobe clutter is a problem and long detection ranges are desired against approaching targets, the advantages of high PRFs far outweigh the disadvantages.

MEDIUM PRFs

ADVANTAGES	LIMITATIONS
1. Good all-aspect capability—copes satisfactorily with both mainlobe and sidelobe clutter.	1. Detection range against both low and high closing-rate targets can be limited by a sidelobe clutter.
2. Ground moving targets readily eliminated.	2. Must resolve both range and Doppler ambiguities.
3. Pulse-delay ranging possible.	3. Special measures needed to reject sidelobe return from strong ground targets.

Medium PRF Operation. Medium PRFs were conceived as a solution to the problems of detecting tail-aspect targets in the presence of both mainlobe and strong sidelobe clutter, thereby providing good all-aspect coverage. If the maximum required operating range is not exceptionally long, the PRF can be set high enough to provide adequate separation between the periodic repetitions of the mainlobe clutter spectrum without incurring particularly severe range ambiguities.

Mainlobe clutter can then be isolated from the bulk of the target return on the basis of its Doppler frequency. Individual targets can be isolated from the bulk of the sidelobe clutter through a combination of range and Doppler resolution.

Further, ground moving targets (being close to the frequency of the mainlobe clutter) can be rejected without also rejecting an unacceptable additional amount of possible valid targets (Fig. 27-18). However, at medium PRFs there will be both range and Doppler ambiguities.

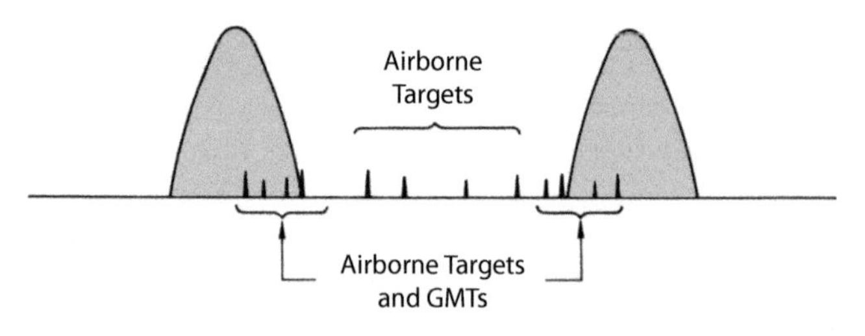

Figure 27-18. With medium PRFs, repetitions of mainlobe clutter are separated widely enough so that both it and return from ground moving targets can be rejected without also rejecting an undue amount of airborne targets.

Range ambiguities are more easily resolved than at high PRFs so that pulse-delay ranging is possible. The ambiguities that occur in Doppler frequencies are more moderate and thus easier to resolve.

However, because there are both range and Doppler ambiguities, nose- and tail-aspect targets may have to compete with close-in sidelobe clutter (Fig. 27-19). This problem can be avoided by switching among several different PRFs.

The resulting reduction in the integration time for each PRF limits the maximum detection range. Where extremely long detection range is not required (as when operating at moderate to low altitudes, in lookdown situations, or in tail chases), adequate detection range can generally be achieved.

Another consequence of the range and Doppler ambiguities encountered at medium PRFs is that sidelobe echoes from large radar cross section ground targets can be a serious problem. Special measures must be taken to eliminate these echoes, or they can be confused with echoes from legitimate airborne targets.

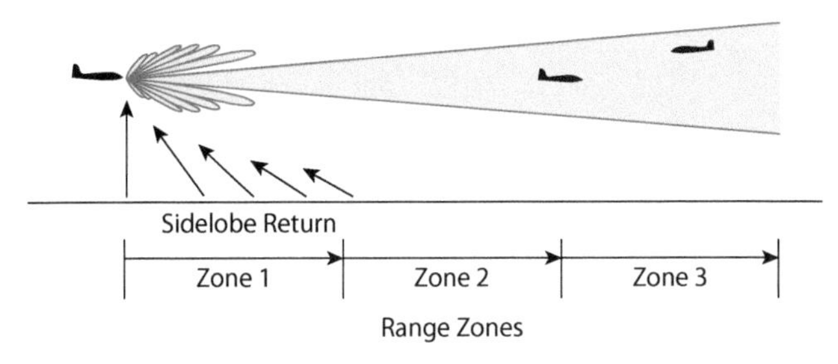

Figure 27-19. While targets may still have to compete with close-in sidelobe clutter, this can be avoided by switching among several different PRFs.

27.3 Summary

Pulse repetition frequencies used by airborne radars range from a few hundred Hz to several hundred kHz at X-band. Generally, only at low PRFs is range unambiguous (and then only if all echoes from beyond the first range zone are excluded or are of negligible strength). Conversely, only at considerably higher PRFs are Doppler frequencies largely unambiguous. Thus, the choice of PRF is generally a compromise.

Three categories of PRF have been established: low, high, and medium. The category into which a particular PRF falls will depend on the operational situation.

CATEGORIES OF PRF

PRF	RANGE	DOPPLER
HIGH	Ambiguous	Unambiguous
MED	Ambiguous	Ambiguous
LOW	Unambiguous	Ambiguous

A low PRF is one for which the maximum required operating range falls within the first range zone. Simple pulse-delay ranging can be used, and sidelobe clutter can be almost entirely removed through adequate range resolution. But in fighter radars mounted on high-speed aircraft, Doppler ambiguities are generally so severe that mainlobe clutter cannot be rejected without rejecting many possible targets. Ground moving targets may also be a problem.

A high PRF is one for which Doppler frequencies of all significant targets are unambiguous. Mainlobe clutter can be rejected without rejecting targets, as there is a clutter-free region in which approaching targets can be detected. High average power can be obtained by increasing the PRF. While this mode is excellent for nose-aspect targets, because of range ambiguities sidelobe clutter may severely limit performance against tail-aspect targets. Range rates can be measured directly, but pulse-delay ranging may be difficult or impractical because of severe range ambiguities.

A medium PRF is one for which both range and Doppler frequencies are ambiguous. But if the value of the PRF is judiciously selected, the ambiguities are comparatively easy to resolve. Consequently, good all-aspect performance can be provided despite mainlobe and sidelobe clutter as well as ground moving targets. Maximum detection range, however, is limited by close-in sidelobe clutter, and sidelobe return from objects on the ground with large radar cross sections may be a problem.

There are applications in which the resolution of range and Doppler ambiguities still remains a challenge, and this is a topic of active research.

Further Reading

G. V. Morris and L. Harkness, *Airborne Pulse Doppler Radar*, Artech House, 1996.

W. L. Melvin, J. A. Scheer, and W. A. Holm (eds.), "Doppler Phenomenology and Data Acquisition," chapter 8 in *Principles of Modern Radar: Basic Principles*, Vol. 1, SciTech-IET, 2010.

C. M. Alabaster, *Pulse Doppler Radar: Principles, Technology, and Applications*, SciTech-IET, 2012.

M. A. Richards, "Doppler Processing," chapter 5 in *Fundamentals of Radar Signal Processing*, 2nd ed., McGraw-Hill, 2014.

Test your understanding

1. Why do range ambiguities arise in pulse-Doppler radar?
2. Why do Doppler ambiguities arise in pulse-Doppler radar?
3. An X-band (10 GHz) pulse-Doppler radar operating from an aircraft traveling at a velocity of 150 m/s has a PRF of 500 kHz. What is the maximum unambiguous Doppler against a stationary target?
4. How does the answer to question 3 change if the radar has an operating frequency of 100 GHz?
5. What is meant by *low, medium,* and *high PRF regions*?

28

Low PRF Operation

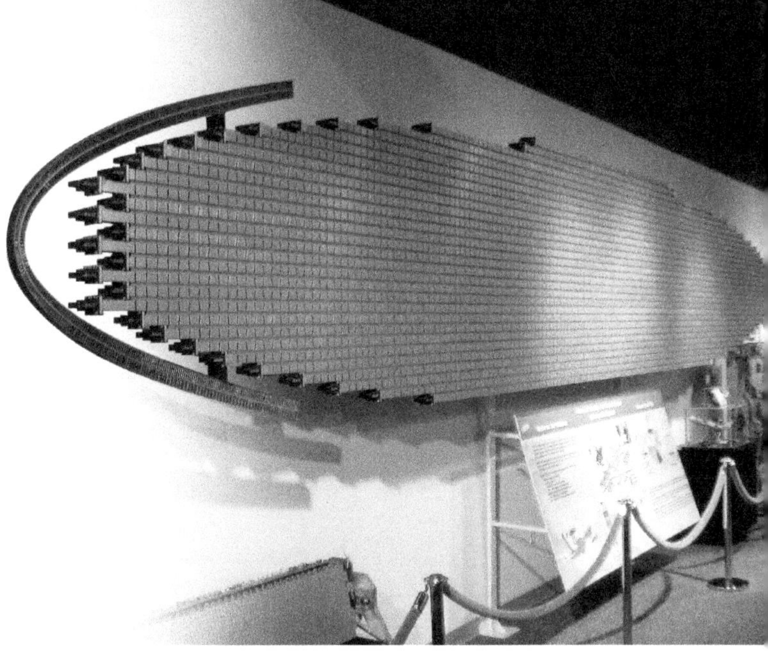

Westinghouse AN/APY-1 radar antenna from an E-3A Sentry AWACS (National Electronics Museum, Baltimore)

A low PRF is, by definition, one for which the first range zone (the zone from which first-time-around echoes are received) extends at least to the maximum range the radar is designed to handle (Fig. 28-1). In the absence of echoes from beyond this zone, the observed ranges are unambiguous. The first range zone extends out to the so-called maximum unambiguous range, R_u.

Typically, low pulse repetition frequencies (PRFs) range from around 250 Hz ($R_u = 600$ km) to 4000 Hz ($R_u = 37.5$ km). Unfortunately, at such PRFs, unless the wavelength is relatively long or the target velocities are relatively low, the observed Doppler frequencies are highly ambiguous.

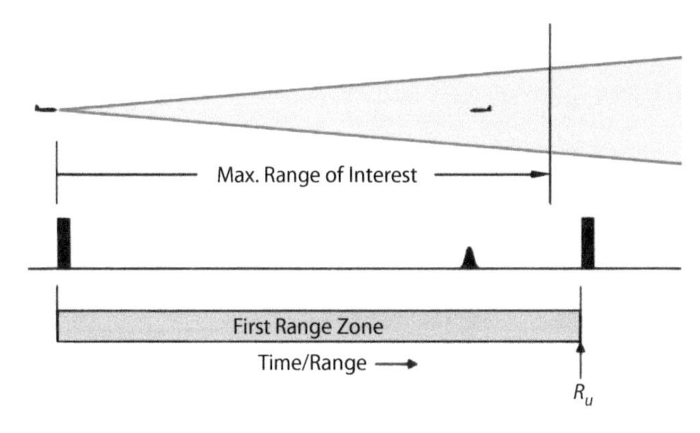

Figure 28-1. A low PRF is one for which the first range zone extends at least to the maximum range the radar is designed to handle.

Low PRFs are essential for most air-to-ground uses such as ground moving target (GMT) detection and synthetic aperture radar (SAR) and are superior to both medium and high PRFs for certain air-to-air applications, such as early warning. But for use in fighter aircraft, where target echoes generally compete with mainlobe clutter, they have serious limitations.

This chapter more closely examines low PRF operation and discusses how target echoes may be separated from ground clutter and how the signal processing may be performed. Subsequently, low PRF operation advantages and disadvantages are outlined to see how these limitations may be alleviated.

28.1 Differentiating between Targets and Clutter

To see what must be done to separate target echoes from ground clutter, we consider the range and Doppler profiles that would be observed by a low PRF radar for a fast-flying aircraft.

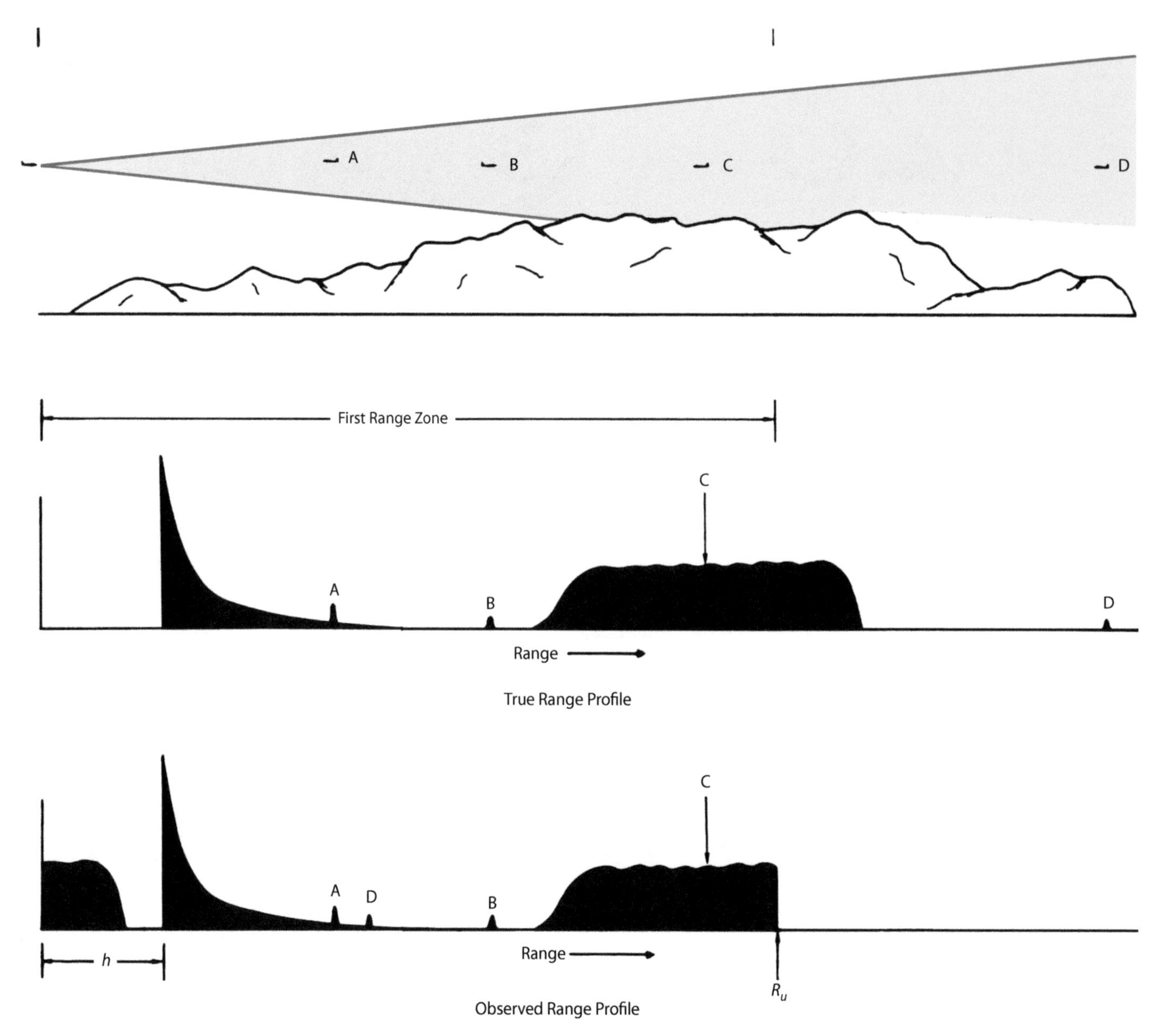

Figure 28-2. Ranges observed by the radar in this range profile for a low PRF radar in a typical operational situation correspond directly to true ranges all the way out to the maximum range of interest.

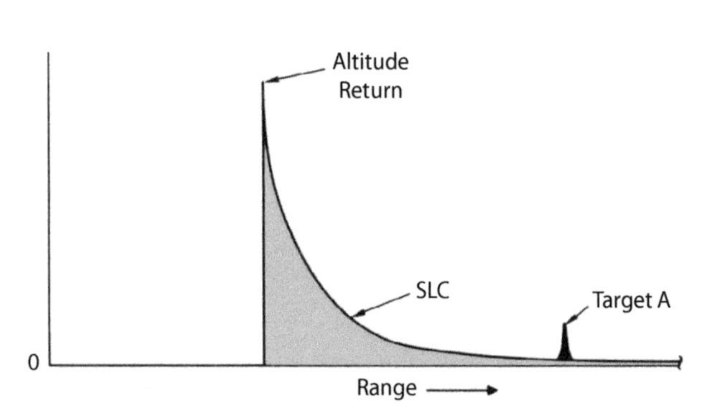

Figure 28-3. A target's echoes are generally stronger than the sidelobe clutter at the range of the target.

Range Profile. As illustrated in Figure 28-2, within the first range zone the observed ranges are true ranges. The altitude return, sidelobe clutter, and mainlobe clutter are all clearly identifiable, as is the return from targets A and B, which are outside the mainlobe clutter. Target C, however, is completely obscured by mainlobe clutter. Target D, which is beyond the first range zone, appears falsely at a much closer range (as shown in the bottom panel of the figure).

From even a cursory inspection of the portion of the profile in which sidelobe clutter alone appears, one thing is immediately clear (Fig. 28-3): the echoes from a target will generally be stronger than the sidelobe clutter received from the target's own range.

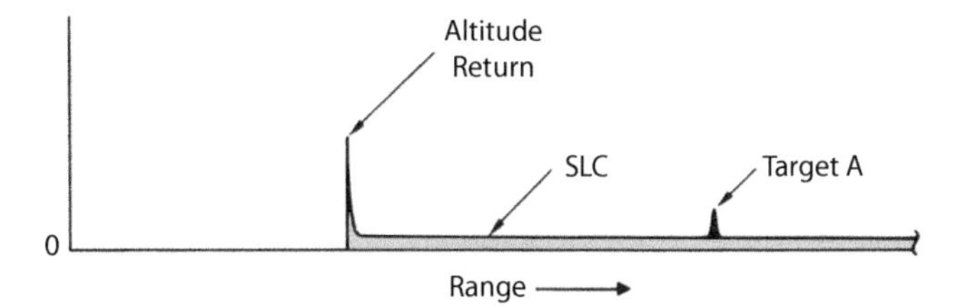

Figure 28-4. In a range profile at the receiver output, the sensitivity time control (STC) makes the amplitude of the output independent of range, preventing saturation by close-in echo returns.

This is perhaps more clearly illustrated by the range profile observed at the output of the receiver (Fig. 28-4). Figure 28-4 shows the amplitudes of both the target echoes and the sidelobe clutter to be more or less independent of range over the portion of the profile in which strong sidelobe clutter is received. This characteristic is due to a feature called *sensitivity time control* (STC), which is generally employed when operating at low PRFs.

Now, consider slicing the range into increments (range bins), where each increment matches the width of the received pulses.[1] The energy contained in each range bin can be isolated. This enables the target to be distinguished from sidelobe echoes by the differences in the amplitude from one range bin to the next (Fig. 28-5). The amplitudes in each range bin are collectively termed a *range profile*. The time duration (i.e., the bandwidth of the transmitted pulse) determines the range extent of the range bin and equates to the range resolution of the radar system (see Chapter 16).

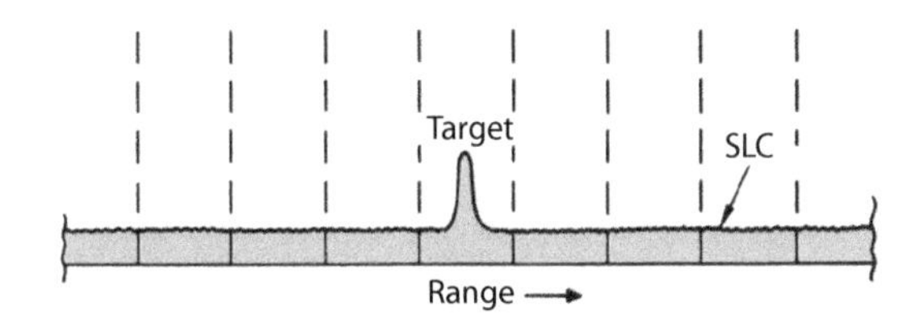

Figure 28-5. If the range is sliced into narrow increments (range bins) matching the transmission pulse width, a target's echoes can be detected in sidelobe clutter on the basis of their different amplitudes in the range profile.

1. Compressed width, if pulse compression is used.

But this is not necessarily the case with mainlobe clutter. Isolating the returns using a range profile only moderates the problem. The mainlobe clutter at the same range as a target is generally much stronger than the target echoes. Moreover, even when mainlobe clutter is received from a range that is some multiple of R_u beyond the target's range (multiple-time-around return), it may well be stronger than the target echoes. To differentiate between target echoes and simultaneously received mainlobe clutter, differences in Doppler frequency must be exploited.

Doppler Profile. Since range is largely *unambiguous* when low PRFs are used, the appearance of the Doppler profile varies considerably with the point in the interpulse repetition period at which the profile is observed. In other words, the returns from different range bins (in the range profile) may have quite different Doppler profiles.

The Doppler profile for the range increment in which target C resides is illustrated in Figure 28-6. The most prominent features of this profile are the periodic repetitions of the mainlobe clutter spectrum. These occur at intervals equal to the pulse repetition frequency, f_r. Though they may be quite wide, they are commonly called lines, or PRF lines. (The central line is the carrier frequency; the others are sideband frequencies.)

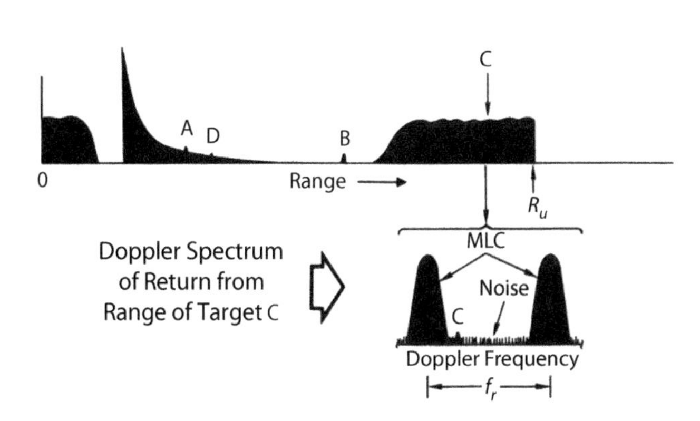

Figure 28-6. A target at a range from which mainlobe clutter is also received can be detected only if its Doppler frequency is different from that of the clutter.

Sensitivity Time Control

AT LOW PRFs, SATURATION OF THE RECEIVING SYSTEM BY STRONG ECHO returns from short ranges is commonly avoided, without loss of detection sensitivity at greater ranges, through a feature called *sensitivity time control*.

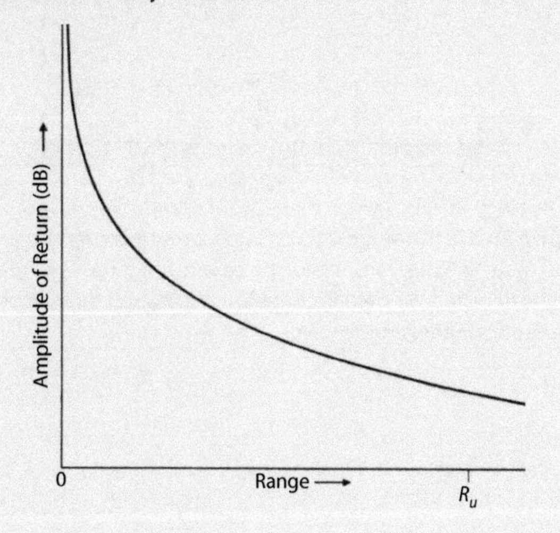

After each pulse has been transmitted, the system gain, which initially is greatly reduced, is increased with time to match the decrease in amplitude of the radar return with range. Maximum gain is usually reached well before the end of the interpulse period.

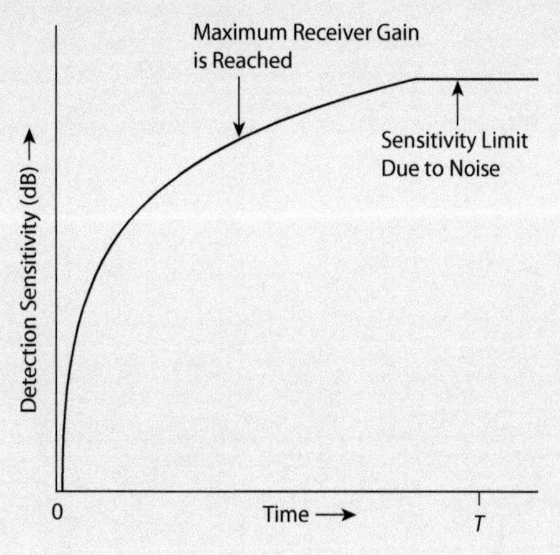

Thereafter, lowering the detection threshold until the noise limit is reached continues the sensitivity increase. This is the point where the threshold is just far enough above the mean noise level to limit the false alarm probability to an acceptable value.

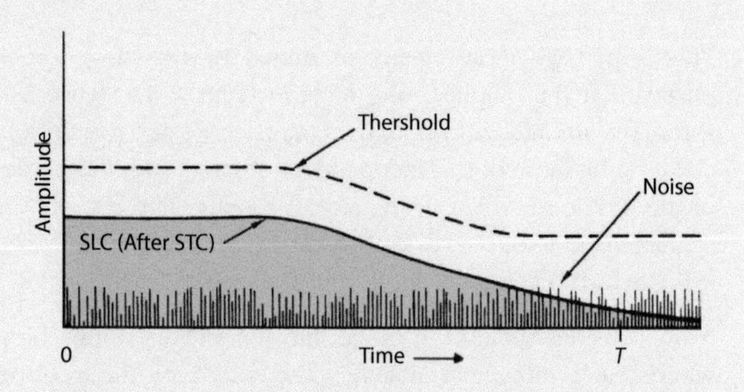

Thus, maximum sensitivity is provided at long ranges, where it is needed to detect the weak echoes of distant targets, while at the same time the strong return from short ranges is prevented from saturating the system.

STC may be applied at various points in a system, from where it helps prevent saturation in all following stages.

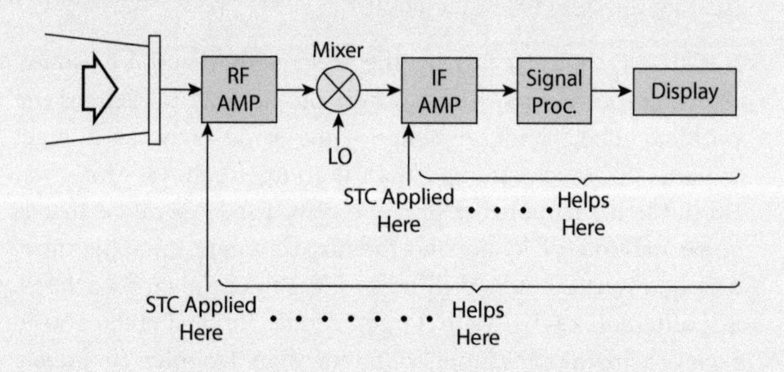

Thermal background noise and target C can now are both seen between successive PRF lines. Sidelobe clutter from this range happens to be so weak that it is below the noise level. Note that if target C's Doppler frequency had been a little lower the mainlobe clutter might have obscured it.

Target B is also at a sufficiently long range that the accompanying sidelobe clutter is below the noise level. Since, in this particular situation, mainlobe clutter is not received from target B's range, the target will appear in the clear regardless of its Doppler frequency. It must compete only with background noise.

At shorter ranges, such as that of target A, sidelobe clutter is much stronger than noise. Nevertheless, because the accompanying sidelobe clutter comes from the same range as the target's echoes, the target appears above the clutter (Fig. 28-7).

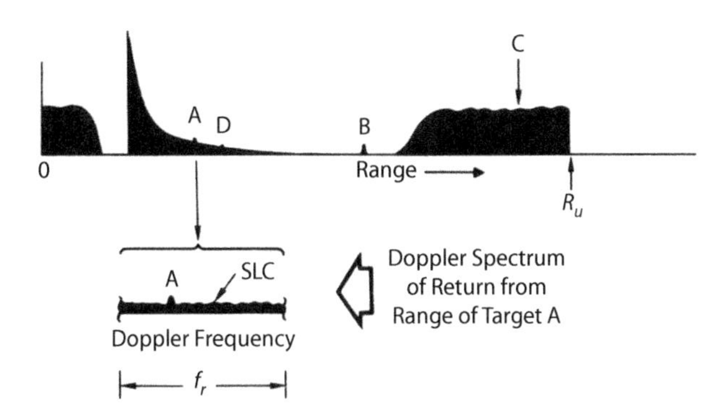

Figure 28-7. Target A, at short range, must compete with sidelobe clutter. But since the target is at the same range as the clutter, the target echoes are stronger than the clutter.

Target D is a second-time-around target and thus is not wanted. Although it happens to be stronger than the accompanying first-time-around sidelobe clutter, it can be prevented from reaching the display by using multiple PRFs or *PRF jittering* [changing the PRF in an unpredictable way to remove ambiguities.

Figure 28-8 shows the Doppler profile at the range of the altitude return, which is generally spread over a band of frequencies whose width exceeds most low PRFs. Consequently, in the Doppler profile, for a range increment from which altitude return is received, the altitude of return is indistinguishable from the other sidelobe clutter.

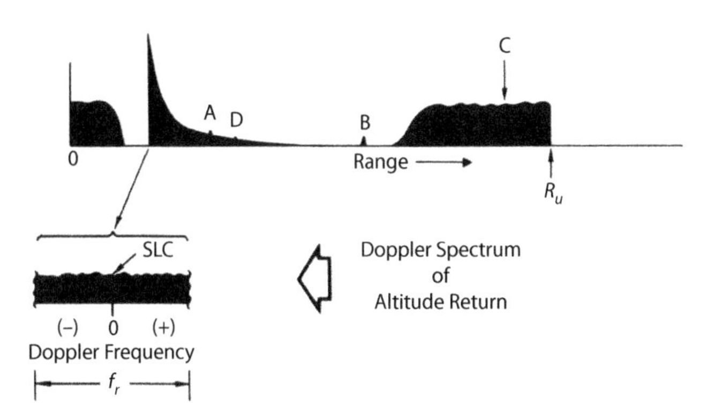

Figure 28-8. The altitude return is spread over such a broad band of frequencies it is indistinguishable from other sidelobe clutter.

Because the altitude return is spread over such a broad band of Doppler frequencies, if Doppler filtering is employed a target may be detected above the altitude return, provided the target's echoes are very strong (as they may well be at very short ranges). For the altitude return to be at a short range, of course, the radar must be at low altitude.

Figure 28-9 shows a fairly broad portion of the Doppler spectrum for a range from which mainlobe clutter is received. To eliminate the clutter, the band of frequencies in which the central line lies must be rejected. In addition, bands of equal width at intervals equal to f_r throughout the receiver's intermediate frequency (IF) passband must also be rejected. Along with the clutter, some target echoes may also be rejected. Only sidelobe clutter and noise plus the (unrejected) target echoes will remain. The target echoes may then be separated from the sidelobe clutter and noise on the basis of differences in amplitude and Doppler frequency.

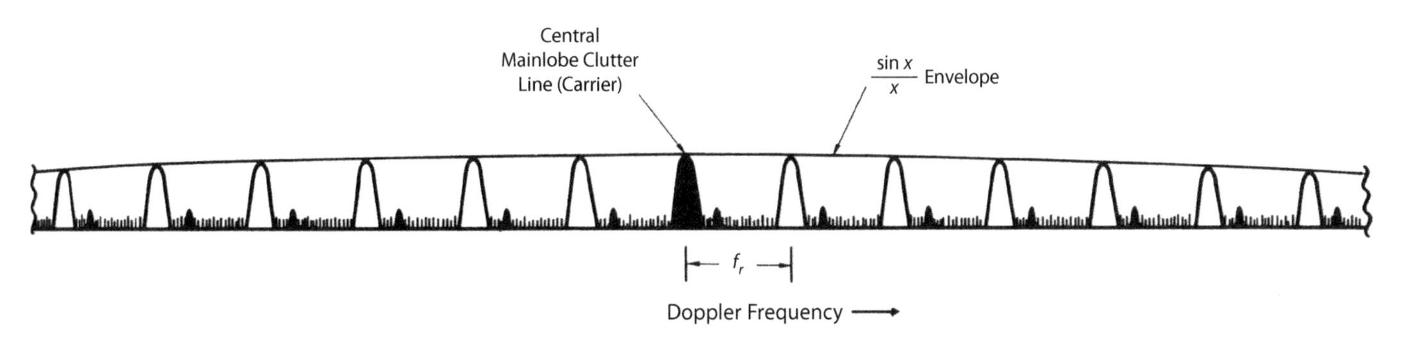

Figure 28-9. The Doppler profile for the range from which mainlobe clutter is received consists of periodic repetitions of the mainlobe clutter spectrum, with the sidelobe clutter and echoes from the target in between.

28.2 Signal Processing

One approach to processing the echoes is illustrated by the block diagram in Figure 28-10.

The IF output of the receiver is fed to a synchronous detector that converts it to in-phase (I) and quadrature (Q) video signals. The frequency of the reference signal supplied to the detector is designed to place the central line of mainlobe clutter at zero frequency (direct current). The central line is picked since its frequency doesn't change when the PRF is changed, whereas the frequencies of the other lines do.

An analog-to-digital (A/D) converter samples the video signals at intervals matching the width of the transmitted pulses.[2] The output of the converter therefore is a stream of numbers representing the I and Q components of the returns from successive range increments. The numbers are sorted by range increment into separate range bins.

2. Compressed width, if pulse compression is used.

To reduce the amount of mainlobe clutter, the numbers for each range bin are passed through a separate clutter canceler. As with the A/D converter, each of these has both I and Q channels.

To reduce the mainlobe clutter *residue* (from the clutter canceler) and to minimize the amount of noise (and simultaneously received sidelobe clutter) with which a target must compete, a bank of Doppler filters processes the output of each clutter canceler. This bank is implemented with a fast Fourier transform (FFT). The passband of the filter bank is made equal to the PRF. For each range bin, a number of echoes corresponding to a sequence of pulses are Fourier transformed from the time domain to the frequency domain so that moving targets are separated from stationary (or near-stationary) clutter. The frequency bin width is determined by the total duration over

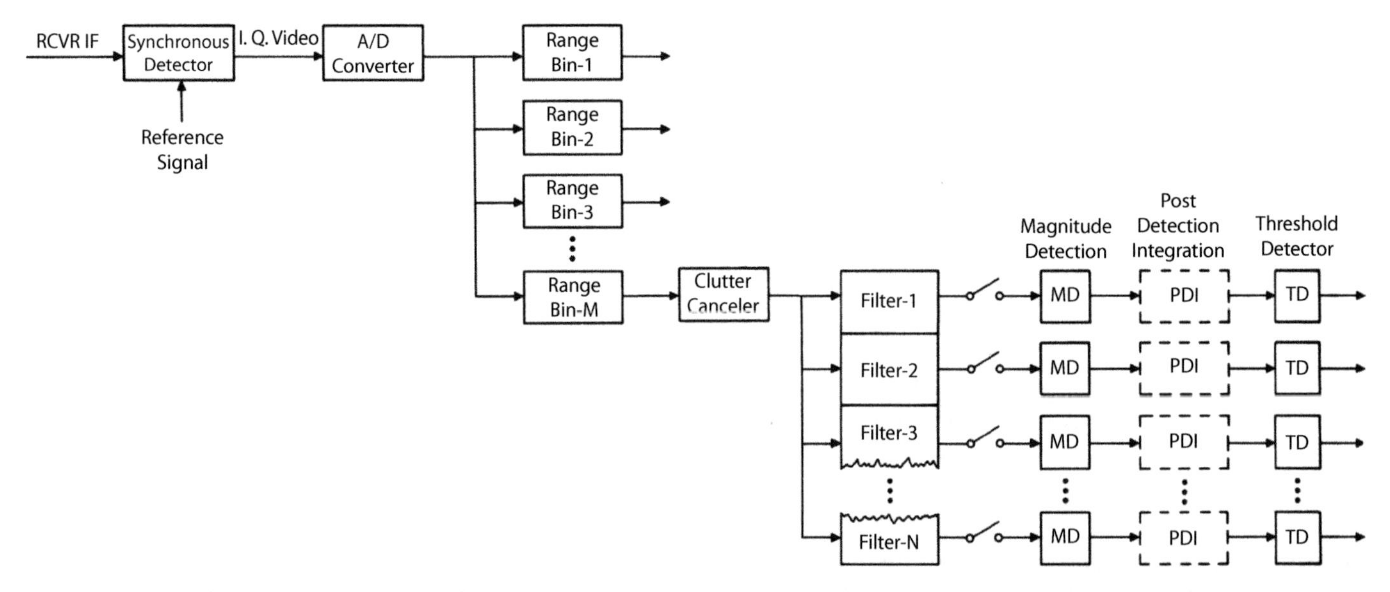

Figure 28-10. This shows the signal processing functions in a low PRF radar employing Doppler filtering. In applications where the mainlobe clutter is avoided, clutter cancelers may be eliminated.

which pulses are collected. The number of bins is equal to the number of pulses.

The output of each Doppler filter is applied to a threshold detector. Values exceeding the threshold are declared as a target detection.

Three aspects of this approach warrant elaboration: (1) how the clutter canceler works; (2) how the detection threshold is set; and (3) how the central mainlobe clutter line is maintained at dc.

Clutter Canceler. In simplest form, each channel of a digital clutter canceler consists of a short-term memory and a summer (Fig. 28-11). The memory holds each of the numbers received from the A/D converter for one interpulse period ($1/f_r$). The summer then subtracts the stored number from the currently received number and outputs the difference. Thus, each number output by the canceler corresponds to the *change* in amplitude of the return from a particular range during the preceding interpulse period. Thus, if the stationary clutter is unchanged it will be canceled, and if there is a moving target it will be observed as an amplitude fluctuation. This is also known as a *delay-line canceler.*

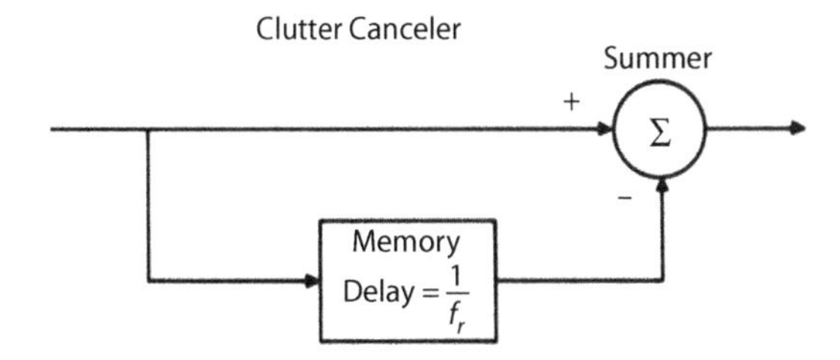

Figure 28-11. In its simplest form, a clutter canceler consists of short-term memory and a summer. The memory holds the signal for one interpulse period, and the summer subtracts the delayed signal from the undelayed signal.

The measured echoes for successive returns at any given range bin are instantaneous samples of a signal whose amplitude corresponds to the amplitude of the returns and whose frequency is the Doppler frequency of the returns. Illustrated in Figures 28-12, 28-13, and 28-14 are successive samples of three such signals. The frequencies of the signals are 0, f_r, and $f_r/2$. All of the samples are taken at intervals equal to the interpulse period, $1/f_r$.

The Classic Delay-Line Clutter Canceler

THE ORIGINAL APPLICATION OF THE CLUTTER CANCELER WAS TO PROVIDE FOR moving target indication (MTI) in ground-based radars. Their initial mechanization was analog.

Bipolar video from a phase-sensitive detector in the receiver is passed through a delay line (e.g., a quartz crystal), which introduces a delay equal to the interpulse repetition period. The delayed signal is then subtracted from the undelayed signal. Since ground return has no Doppler shift, the video signal produced by the return from the ground at any one range is essentially constant from one interpulse repetition period to the next. The video signal produced by a moving target, however, fluctuates at the target's Doppler frequency. The clutter, therefore, cancels whereas the target signal does not.

Analog cancelers were used to provide MTI in airborne early warning radars and the first types of fighter radars. Digital cancelers, however, have the compelling advantages of avoiding delay instability and being adjustment free. Consequently, although most delay-line cancelers are still analog, the cancelers used in all modern airborne radars are digital.

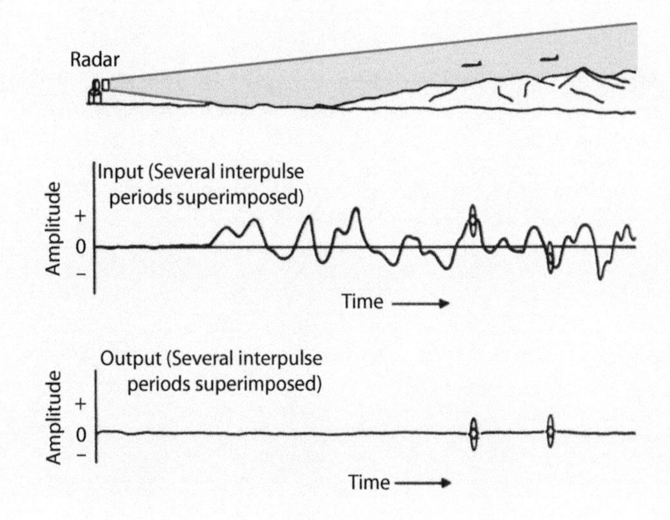

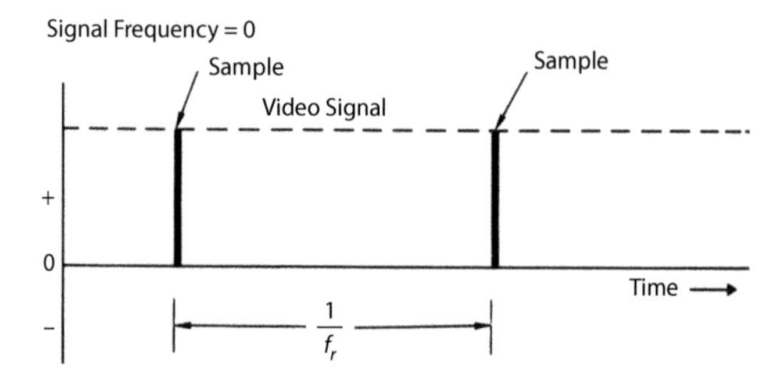

Figure 28-12. The periodic samples of this video signal have a frequency of zero (the dc signal) and the same amplitude and algebraic sign.

Successive samples of a video signal having zero frequency, such as the central mainlobe clutter line, have the same magnitude and the same algebraic sign (Fig. 28-12). Therefore, when one sample is subtracted from the other, they cancel.

The same is true for a signal whose frequency is f_r, such as the first mainlobe clutter line above the central one. Since the sampling interval is equal to the period of the wave, the samples are all taken at the same point in every cycle (Fig. 28-13).

But for a frequency of $f_r/2$, the result is just the opposite. Because the sampling interval is only half the period of the wave, the samples are alternately positive and negative (Fig. 28-14). When one is subtracted from the other, the difference is twice the magnitude of the individual samples.

For frequencies above and below $f_r/2$, the differences become progressively smaller. As a result, a plot of the canceler's output versus frequency for a constant-amplitude input has an inverted U shape (Fig. 28-15). The inverted U effectively suppresses low Doppler clutter and allows a fast-moving target to be amplified.

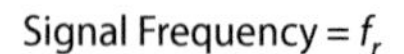

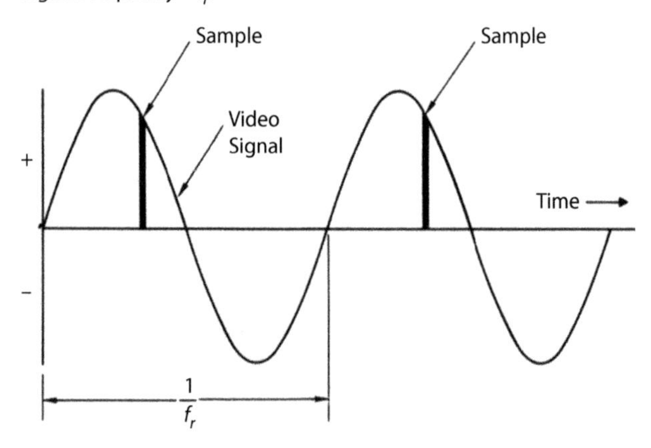

Figure 28-13. The samples of this video signal have a frequency equal to the PRF, f_r, and the same amplitude and algebraic sign.

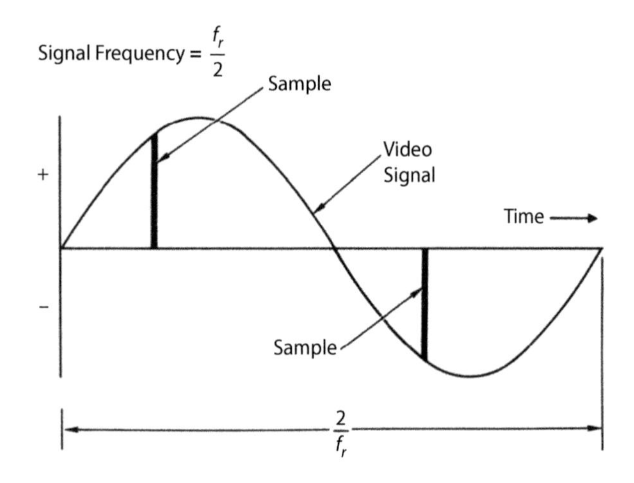

Figure 28-14. The samples of this video signal have a Doppler frequency equal to $f_r/2$, the same amplitude, and alternating algebraic signs.

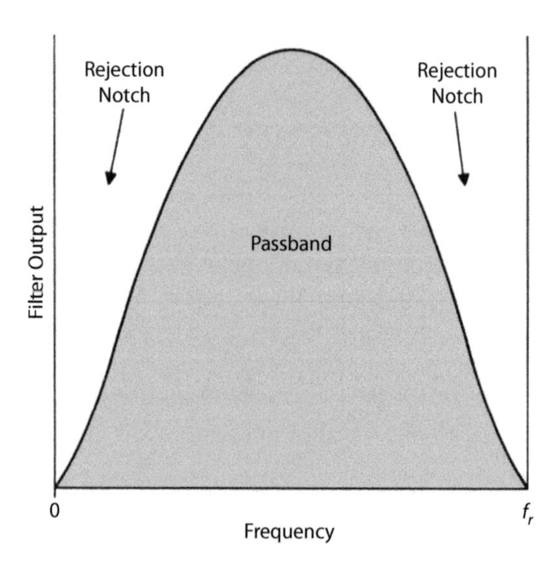

Figure 28-15. The graph shows the output produced by a simple single-delay clutter canceler for constant amplitude input.

What about frequencies higher than f_r? This is illustrated with the simple phasor diagrams shown in Figure 28-16. When a signal is sampled at a given rate (in this case f_r), the samples will be exactly the same if the signal's frequency is $f_D + f_r$ as they would be if their frequency was f_D.

The same is true if the signal's frequency is f_D plus any multiple of f_r.

The canceler's output characteristic therefore repeats identically at intervals of f_r from 0 (dc) and so on up in frequency (Fig. 28-17).

The regions in which the output approaches zero (i.e., at dc and multiples of f_r) are called *rejection notches*. If any one of the mainlobe clutter lines is placed at dc (normally this is the central line), every line will fall in a rejection notch and the clutter will tend to cancel (hence the name *clutter canceler*).

The notches of the simple canceler are generally much narrower than the clutter lines. This means that they allow some clutter to get through, as the cancellation is not perfect. The clutter that is not canceled is called the *clutter residue*. The clutter notches can readily be widened. The simplest way is to connect more than one canceler together in series, known as *cascading*.

If the rejection notches are made sufficiently wide and the mainlobe clutter is centered in them, it will largely cancel (Fig. 28-18). The output then will represent target echoes, sidelobe

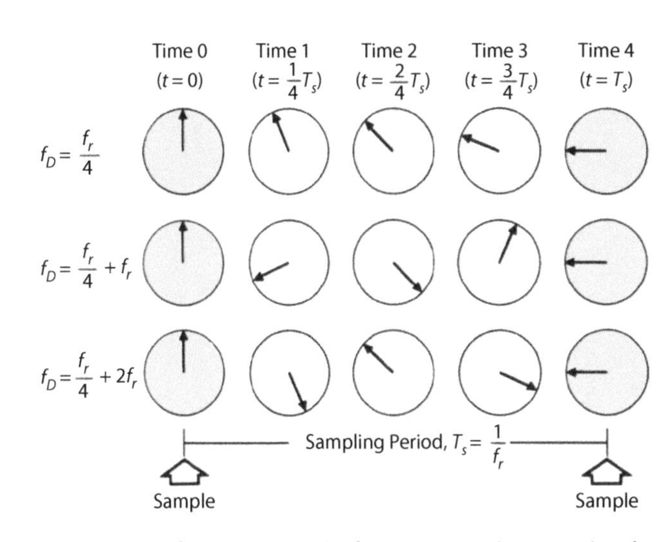

Figure 28-16. If a target's Doppler frequency equals some value, f_D, plus a whole multiple of the sampling rate, f_r, the output from the canceler will be the same as if the Doppler frequency was f_D.

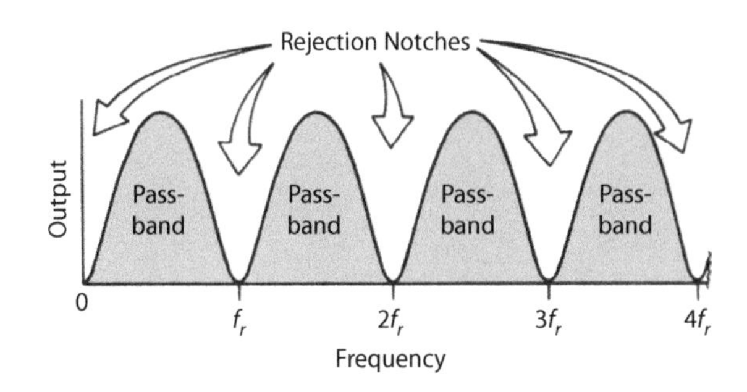

Figure 28-17. The clutter canceler's output characteristic repeats at intervals of f_r from dc and up.

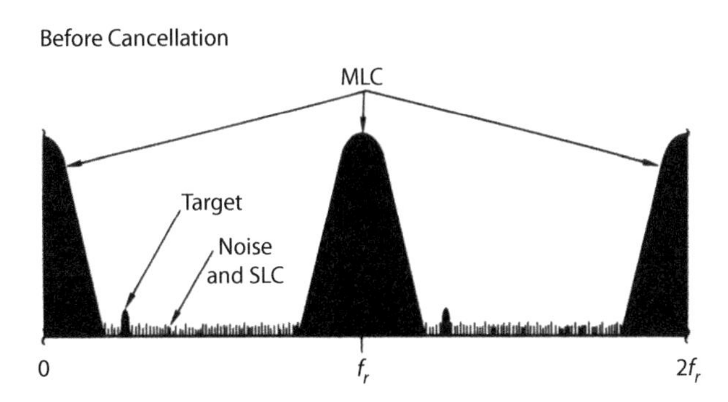

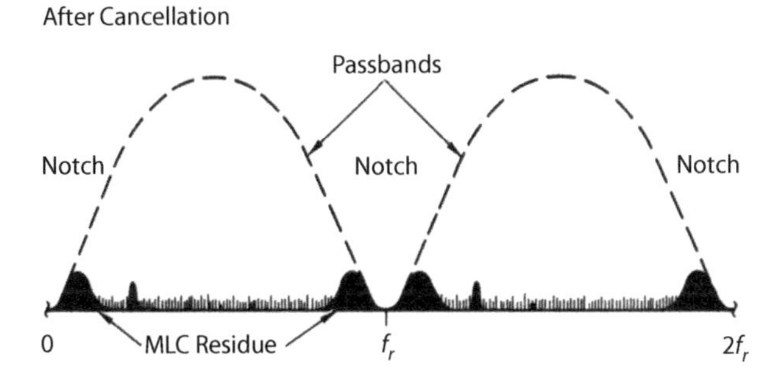

Figure 28.18. If the clutter canceler's rejection notches are made wide enough, the mainlobe clutter will largely cancel. The output will consist of mainlobe clutter residue, target echoes, sidelobe clutter, and noise.

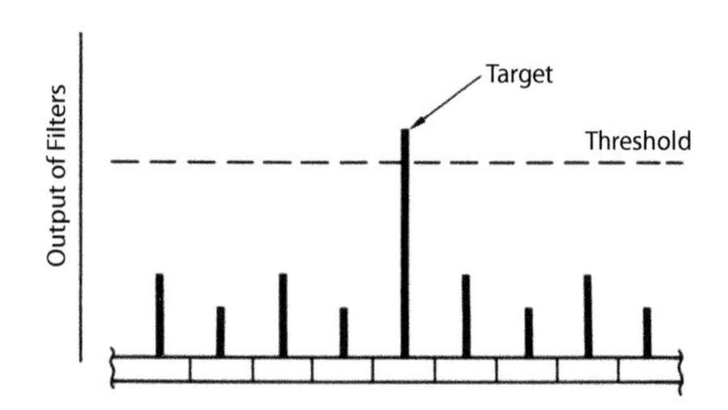

Figure 28-19. The threshold for target detection in the output of each filter bin is set far enough above the average of the outputs of the adjacent filter bins so that the probability of clutter crossing the threshold can be reduced to an acceptable value.

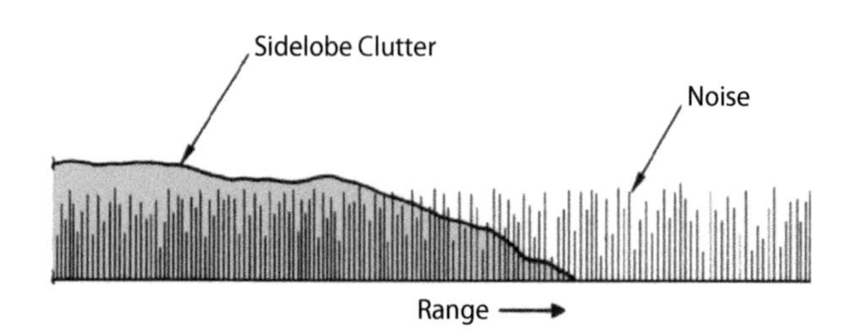

Figure 28-20. This shows the "long-range end" of a range profile after clutter removal. The sidelobe clutter ultimately becomes submerged in receiver noise.

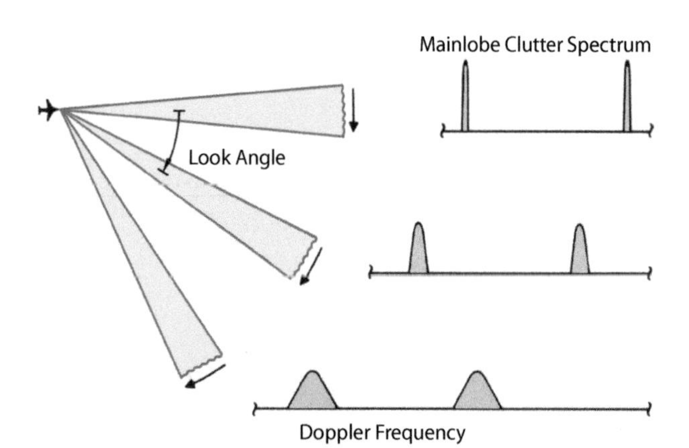

Figure 28-21. As the look angle increases, the mainlobe clutter spectrum broadens and shifts down in frequency.

clutter, and background noise (plus, of course, any mainlobe clutter residue).

The Doppler filters following each clutter canceler not only eliminate most of the mainlobe clutter residue but also substantially reduce both the amplitude of the competing sidelobe clutter and the mean level of the noise. By suitably setting the target detection threshold, we can still further reduce the possibility of clutter and noise producing false alarms.

However, it should also be noted that as the notch width is increased there is also the possibility of removing legitimate moving targets. In other words, the minimum *detectable velocity* is increased. This can be important in detecting crossing targets and ground moving targets.

Detection Threshold. For each Doppler filter, a threshold is set to a predetermined level that is higher than the average of the outputs of the corresponding filters for several (Doppler) bins on either side of the one being examined. If the value of the "test" bin is higher, then a target is declared (Fig. 28-19). Provided the threshold offset has been correctly chosen and the averaging has been properly done, the probability of clutter crossing the threshold may be reduced to an acceptably low value while it provides adequate sensitivity for the detection of target echoes.

Consider the range profile of the receiver output after mainlobe clutter removal. As the range increases, a point is eventually reached where the sidelobe clutter is submerged beneath the background noise (Fig. 28-20).

Beyond this range, noise determines the detection threshold. Thus, when low PRFs are being used, detection range is usually limited not by sidelobe clutter but only by background noise.

Tracking the Mainlobe Clutter. The mainlobe clutter spectrum varies continually. As the antenna look angle increases, the center frequency of the spectrum decreases and the width increases from nearly a line to a broad hump (Fig. 28-21). For a given target and aircraft velocity, the center frequency decreases because the radial component of the target velocity decreases (as it becomes a crossing target). The spectrum broadens as the look angle increases because the range of radial velocities within the beam increases. As the speed of the radar increases, both the frequency and the width increase.

Consequently, to keep the mainlobe clutter lines in the clutter canceler's rejection band, the frequency offset provided by the synchronous detector must track these changes in clutter frequency. From knowledge of the antenna look angle and the ground speed, the frequency of the clutter lines can readily be predicted. Changes in frequency can then be tracked by appropriately adjusting the reference frequency supplied to the synchronous detector (Fig. 28-22).

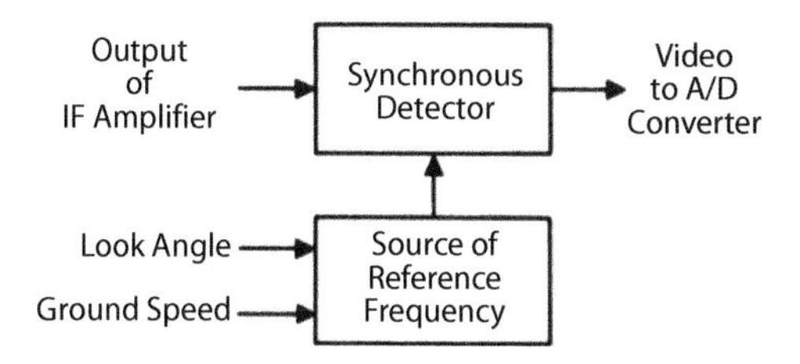

Figure 28-22. By continuously adjusting the reference frequency supplied to the synchronous detector to account for changes in look angle and ground speed, mainlobe clutter can be kept in rejection notches of clutter canceler.

28.3 Advantages and Limitations of Low PRF Operation

Low PRF operation has both great advantages and limitations.

Advantages

- Target ranges can be measured directly by the simple, highly precise pulse-delay method.
- Sidelobe clutter can largely be rejected through range resolution.
- STC can be used to provide wide dynamic range.
- Signal processing requirements can be met quite simply.
- Detection range is usually limited only by background noise.

Limitations

- If a target's Doppler frequency is such that the target's echoes fall within one of the clutter filter's rejection notches, the radar will be blind to the target.
- First-time-around echoes are received from all ranges out to the maximum range the radar is designed to handle. However, there is little other than STC and obstructions in the line of sight to prevent multiple-time-around echoes of strong targets beyond R_u appearing falsely within the radar's range. Note that mainlobe clutter from beyond R_u is, of course, rejected on the basis of Doppler frequency, just as is the mainlobe clutter from ranges out to R_u.
- In fighters, limitations on antenna size require use of wavelengths so short that Doppler ambiguities are severe. Not only is direct measurement of closing rates impractical, but also airborne targets are difficult to distinguish from moving targets on the ground.

28.4 Getting around the Limitations

Certain steps can be taken to help avoid low PRF limitations such as Doppler blind zones, multiple-time-around echoes, low duty factor, and moving targets on the ground.

Doppler Blind Zones. Perhaps the most significant limitation of low PRF operation is the so-called *Doppler blind zone.* This is referred to as such because received echoes with a Doppler frequency that fall within the rejection band of the clutter canceler will also be rejected, and this also occurs at multiples of the PRF (Fig. 28-23).

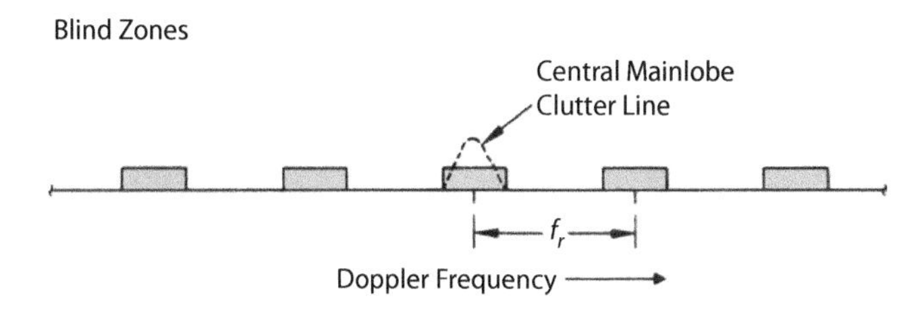

Figure 28-23. These frequency bands fall within rejection notches of the clutter canceler and the Doppler filter bank. The radar is blind to any target whose true Doppler frequency lies within one of these bands.

At low PRFs a target's Doppler frequency may be many times the PRF, so the target is about as likely to appear at any one point within a span of frequencies equal to the PRF as at any other. Therefore, the probability of a target being in the blind zones at any one time is roughly equal to the ratio of the width of the rejection notches to the PRF.

This probability can be reduced in several ways. One is simply to increase the PRF, thereby spreading the blind zones farther apart. The desired maximum detection range, however, limits the extent to which the PRF can be raised. For example, a desired maximum detection range of 37.5 km results in a PRF upper limit of 4 kHz.

Another way of reducing the severity of the blind zones is to reduce their width. The extent to which that can be done is limited by the width of the mainlobe clutter lines. These can be narrowed, for example, by (1) increasing the size of the antenna, hence reducing the beamwidth or allowing use of longer wavelengths; (2) limiting the speed of the radar-bearing aircraft, hence reducing the spread of the mainlobe clutter frequencies; or (3) limiting the maximum antenna look angle, hence further reducing the spread of the mainlobe clutter frequencies.

In radar systems used for applications such as early warning and surveillance, where the speed of the radar-bearing aircraft is low, blind zones can be tapered to the point that they are not a serious problem by employing large antennas. As illustrated in the upper half of Figure 28-24, for a radar having a 3 m long antenna and a velocity of only 150 m/s, even in the worst case (azimuth angle of 90°) the Doppler-clear region will at least be as wide as the blind region at a PRF as low as 200 Hz.

However, in radars for fighter applications, where antenna size is limited and radar speeds can be high, blind zones can

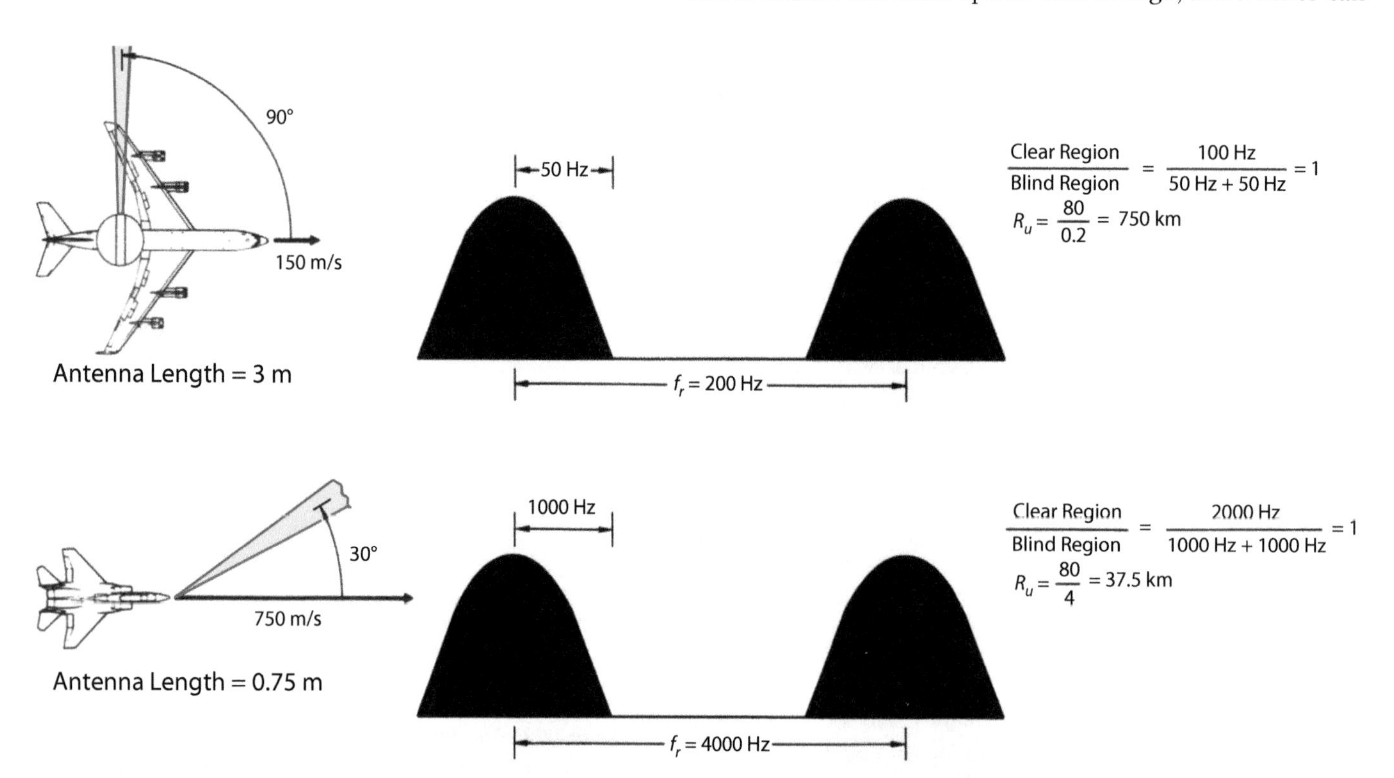

Figure 28-24. Where long antennas are practical and radar velocities are low, the ratio of the Doppler clear to the blind region is sufficiently large such that blind zones are not a serious problem. However, in fighters the blind zones force the use of higher PRFs or limited look angles or both.

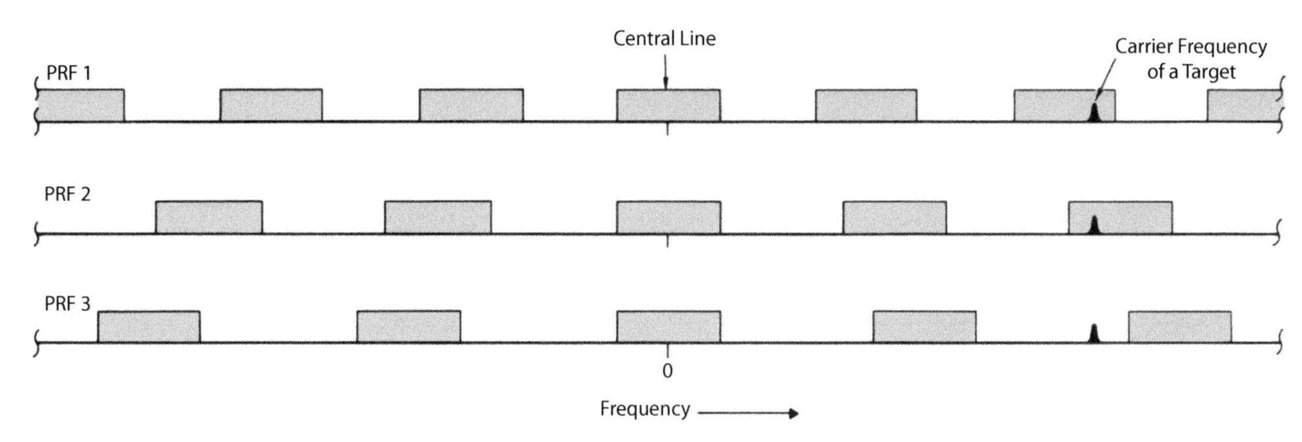

Figure 28-25. By periodically changing the PRF, blind zones can be shifted. This reduces the possibility that any one target will remain in a blind zone for the entire time-on-target.

occupy an excessive portion of the Doppler spectrum. About the only recourse (beyond increasing the PRF) is to limit the maximum look angle.

As illustrated in the lower half of Figure 28-25, the radar for a fighter, whose antenna is generally of the order of 0.75 m in diameter and whose maximum velocity may well be around 750 m/s or more, a one-to-one ratio of clear-to-blind regions can be achieved only by limiting the azimuth angle to a maximum of no more than 30° and raising the PRF to 4 kHz. Usually neither such a severe restriction of azimuth angle nor such a high PRF is attractive. Consequently, in those fighter applications where mainlobe clutter is a problem, medium or high PRFs are commonly used.

The probability of a target remaining in a blind zone can be substantially reduced throughout an entire time-on-target. Since the blind zones are all separated from the zone at zero frequency by multiples of the PRF, changing the PRF can move them about. The central line will remain at dc, since it is the carrier frequency. In principle, enough different, widely separated PRFs are used; it is possible to periodically uncover every part of the spectrum (Fig. 28-25). However, since the time-on-target is divided among the different PRFs, PRF switching reduces detection sensitivity. The more PRFs that are used, the more the sensitivity will be reduced.

A common alternative is to *jitter*, or sweep, the PRF between two values. If these are suitably chosen, targets whose closing rates fall within a limited span of interest, such as rates corresponding to aircraft velocities around Mach 1 (around 340 m/s), can be kept continuously in the clear.

If a target's Doppler frequency is already known, the target may be kept out of the blind zones by adaptively selecting the PRF. That is, the PRF may be selected so the zones will straddle the target's frequency (Fig. 28-26). The necessary a priori Doppler information may be obtained by detecting the target

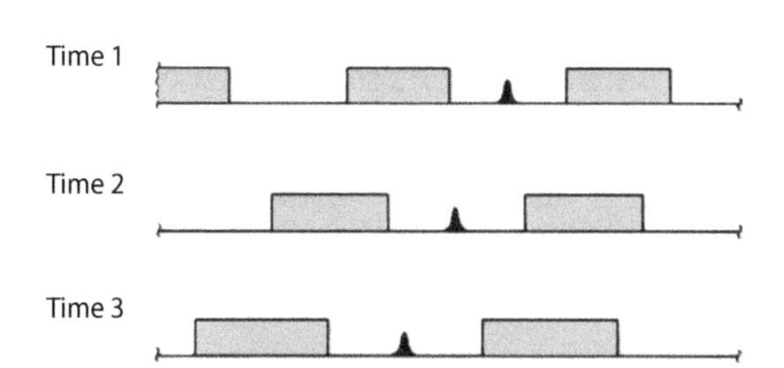

Figure 28-26. If a target's Doppler frequency is known, it can be kept continually in the clear by adaptively changing the PRF.

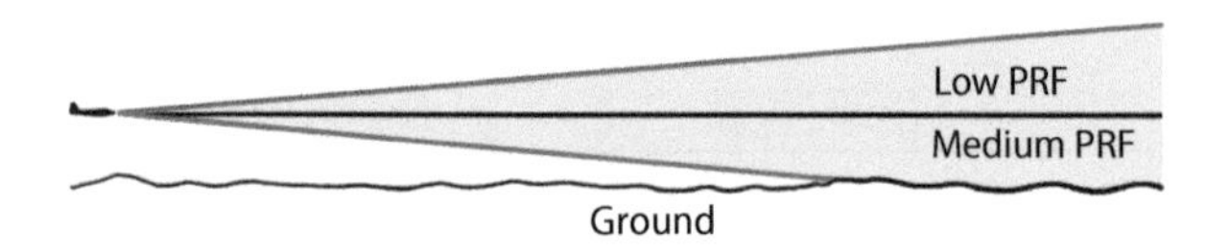

Figure 28-27. A possible mode in low-altitude applications employs low PRFs on the upper bar of the search scan, where the beam does not strike the ground, and medium PRFs on lower bar.

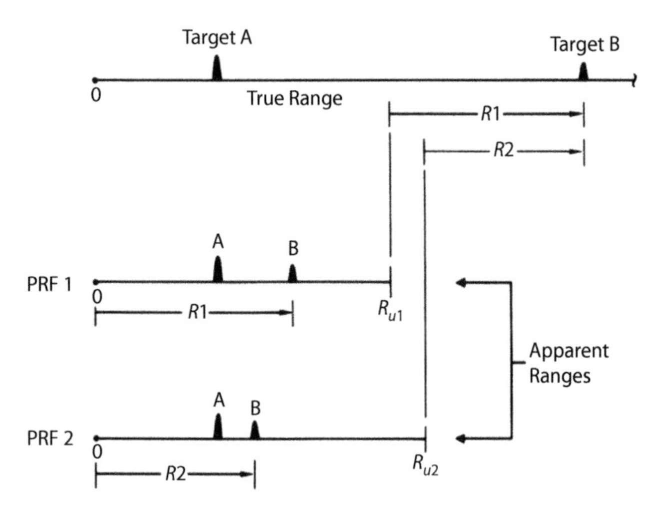

Figure 28-28. If the PRF is changed by a small amount, the observed range of a target beyond the unambiguous range, R_u, will change but the observed range of a target in the first range zone will not.

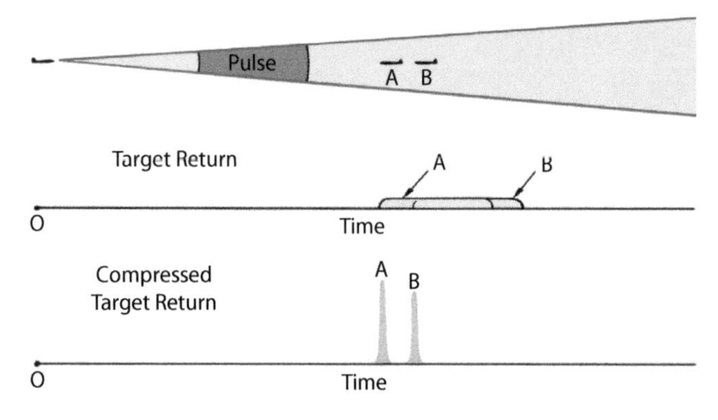

Figure 28-29. The duty factor can be increased by transmitting very long pulses and using large amounts of pulse compression to obtain the desired range resolution.

in a high PRF search mode. Alternatively, it may be available as a result of tracking the target in range.

If mainlobe clutter is not a problem (e.g., if the mainlobe intercepts the ground only at shorter or longer ranges than those of interest or if it does not intercept the ground at all), blind zones can be avoided simply by not discarding any return—that is, by eliminating the clutter cancelers and processing the outputs of all the Doppler filters.

In low-altitude applications (Fig. 28-27), a possible mode is one in which the radar employs low PRFs (for long-range detection) on the upper bar of the antenna search scan, where mainlobe clutter is not encountered, and medium or high PRFs on the lower bars (for good performance in mainlobe clutter).

Multiple-Time-Around Echoes. The problems of multiple-time-around target echoes may be moderated to some extent by STC. To illustrate, let us assume that the unambiguous range is 37 km. If return is received from a target at 39 km, it will appear to have a range of 2 km. However, its echoes will be $(2/39)^4$ times weaker (i.e., reduced by 53 dB) than echoes from a target of the same radar cross section and aspect that is truly at a range of 2 km. With STC, because detection sensitivity is greatly reduced during the initial portion of the interpulse period, it is likely that this unwanted target would not be detected.

On the other hand, if the target were at a range of, say, 72 km, this would not necessarily be so. The target would then have an apparent range of 35 km. Its echoes would be $(35/72)^4$ times weaker (i.e., reduced by 12.5 dB) as those of an equivalent target at 35 km, and hence strong targets might be detected.

If the PRF is changed by a small amount, the observed ranges of multiple-time-around targets will correspondingly change, whereas the observed ranges of the first-time-around targets will not. Therefore, by periodically changing the PRF and looking for changes in the observed target ranges the multiple-time-around targets can be spotted and prevented from reaching the display (Fig. 28-28).

Low Duty Factor. Within the capabilities of the transmitter, reasonably high duty factors can be achieved at low PRFs by transmitting very long pulses and employing large amounts of pulse compression to achieve the desired range resolution (Fig. 28-29). If this is done in the absence of mainlobe clutter, low PRF radar can actually obtain greater search detection ranges than high PRF radar employing the same average power.

This difference is due to the losses incurred by the high PRF radar due to eclipsing, that is, the echo being received while the radar is transmitting and the receiver is blanked out. Even at low PRFs, a considerable amount of return may be lost as a result of eclipsing, but this is much less of a problem at low than at medium and high PRFs.

For as long as a returned pulse is not received at exactly the same time as the transmitter is transmitting, some of the pulse will get through to the receiver. As the range increases, the portion of the return getting through also increases. For targets at ranges greater than one pulse length, none of the return is lost.

This is true only up to the point where the trailing edge of the echo from a target at the maximum range of interest is received as the leading edge of the next pulse is being transmitted. In other words, the interpulse repetition period must be at least one pulse width longer than the round-trip transit time for the most distant target of interest (Fig. 28-30).

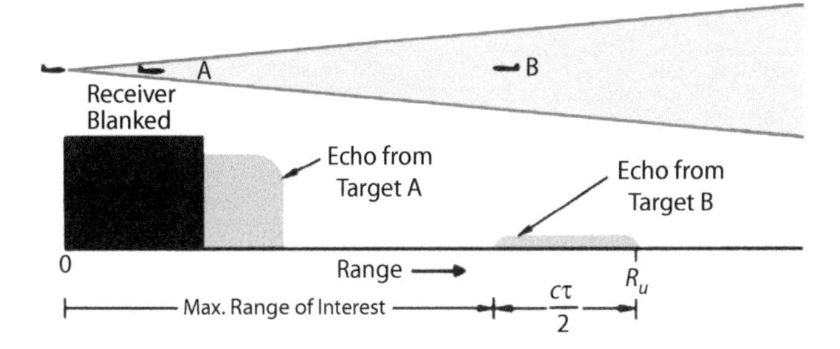

Figure 28-30. To avoid eclipsing of long-range targets, the interpulse repetition period must be at least one pulsewidth longer than the transit time for the most distant target of interest.

Provided this requirement is met, duty factors of up to 20 percent may be used by a radar system operating at low PRFs without incurring a significant eclipsing loss. In contrast, at medium and high PRFs, because of range ambiguities, the severity of eclipsing is independent of the range from which the return is received. The eclipsing loss increases directly with the duty factor.

Moving Targets on the Ground. In air-to-ground radar, the detection of ground moving targets may be a primary objective. However, in air-to-air operations, *rejecting* GMTs may be essential. When a radar is operating over a region where there are hundreds of moving vehicles on the ground (e.g., cars, trucks, trains; Fig. 28-31), it may detect a great many more GMTs than airborne targets. The former may clutter up the display so much that the operator cannot discern the real targets of interest even though the latter have been solidly detected and are clearly displayed.

GMTs may be identified by observing the effect that switching the PRF has on their apparent Doppler frequencies. Because of the greater ground speeds of airborne targets, at such low PRFs an aircraft's apparent Doppler frequency will generally be its true Doppler frequency minus some multiple of the PRF. Consequently, the apparent frequencies of these targets will

Figure 28-31. When searching for aircraft over areas where hundreds of vehicles may be moving on the ground, means must be provided to eliminate these targets from the radar display.

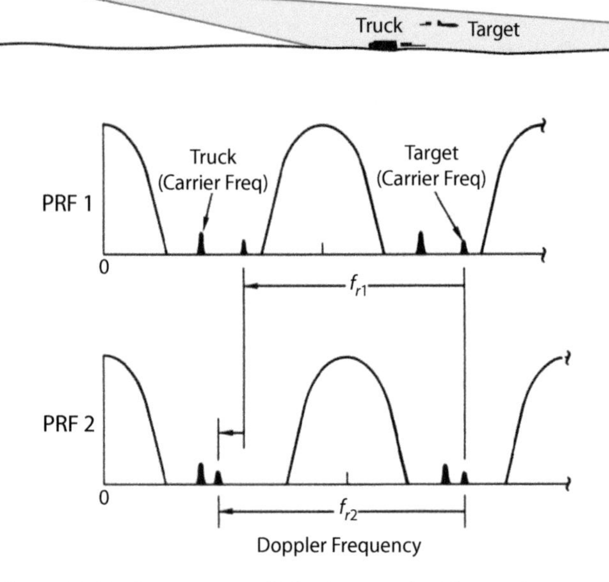

Figure 28-32. GMTs may usually be distinguished from airborne targets because their apparent Doppler frequencies do not change if the PRF is changed slightly.

LOW PRFs

ADVANTAGES	LIMITATIONS
Good for air-to-air look-up and ground mapping.	Poor for air-to-air look-down—much target return may be rejected along with mainlobe clutter.
Good for precise range measurement and fine range resolution.	Ground moving targets can be a problem.
Simple pulse delay ranging possible.	Doppler ambiguities generally too severe to be resolved.
Normal sidelobe return can be rejected through range resolution.	Higher peak powers or larger amounts of pulse compression generally required.

usually change when the PRF is switched. On the other hand, the observed Doppler frequencies of GMTs, whose speeds are much lower, will generally be true frequencies and so will not change (Fig. 28-32). By disregarding the threshold crossings that occur in the same Doppler filter after the PRF has been switched, GMTs may largely be prevented from appearing on the display.

In air-to-ground applications where GMTs rather than airborne moving targets are of interest, the previously described logic is inverted. Thus, airborne moving targets are prevented from appearing on the display by discarding those threshold crossings that do not occur in the same Doppler filter after the PRF has been switched.

28.5 Summary

At low PRFs, cancellation and Doppler filtering may largely eliminate mainlobe clutter. To keep the clutter in the canceler's rejection notches, the offset must be varied with radar speed and antenna look angle.

Sidelobe clutter, mainlobe clutter residue, and background noise are minimized through a combination of range gating and Doppler filtering. The maximum detection range is often limited only by receiver noise.

The principal limitation of low PRFs is Doppler blind zones. These are regions in the Doppler spectrum for which a target's observed Doppler frequency is the same as that of the mainlobe clutter. The zones are the same width as the mainlobe clutter lines and are spaced at intervals equal to the PRF. Although this is not a serious problem where large antennas and low radar speeds are practical, for radars in fast-flying aircraft, blind zones can be acceptably reduced only by employing such a high PRF that R_u, the maximum operating range, is severely reduced or by limiting the maximum look angle.

The possibility of a target remaining in a blind zone for an entire time-on-target may be minimized by switching among widely separated PRFs. Alternatively, a limited span of Doppler frequencies may be kept clear by jittering or sweeping the PRF. A third alternative is to keep the Doppler frequency of a given target clear by adaptively selecting the PRF.

Jittering the PRF may identify multiple-time-around target echoes, which can then be discarded and not displayed. Both PRF switching and PRF jittering, however, reduce detection sensitivity.

In fighter applications, severe Doppler ambiguities make it difficult to discriminate between airborne and ground moving targets. The problem can be alleviated at the cost of wider blind zones by discarding the outputs of a larger number of filters at the ends of the Doppler filter bank or by noting whether a target appears in the same Doppler filter after the PRF has been changed slightly.

Because of the blind zone problem, low PRFs are generally used only where mainlobe clutter can be avoided or where large antennas and low radar speeds are practical.

Further Reading

G. V. Morris and L. Harkness, *Airborne Pulse Doppler Radar*, Artech House, 1996.

C. M. Alabaster, *Pulse Doppler Radar: Principles, Technology, and Applications*, SciTech-IET, 2012.

Test your understanding

1. Describe how clutter can be canceled in a low PRF radar system.
2. How do Doppler ambiguities arise in a low PRF radar system?
3. What methods can be used to eliminate Doppler ambiguities in a low PRF radar system? Use diagrams to illustrate your answer.
4. If a low PRF radar system is to have a maximum unambiguous range of 30 km, what should the maximum PRF be?
5. In question 4, a target is detected ambiguously at a range of 32 km as being at 2 km. By how much is it attenuated compared with a target of equal echoing area genuinely located at a range of 2 km?

Mitsubishi F-2 (2000)

The JASDF's Mitsubishi F-2 fighter was developed with Lockheed Martin as the major subcontractor under the management of the Japanese Technical Research and Development Institute. It is equipped with a J/APG-1 active, phased-array radar, which was the first AESA radar in the world to enter service on a fighter.

29

Medium PRF Operation

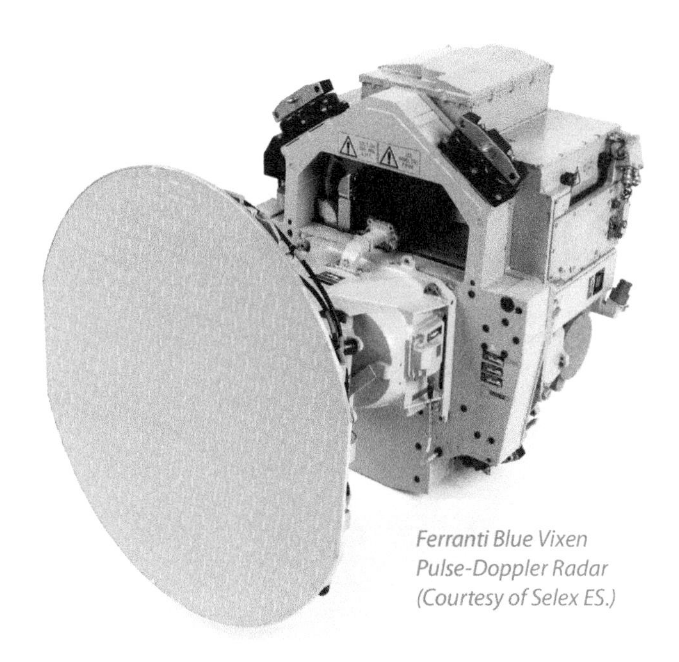

As discussed in Chapter 27, a medium pulse repetition frequency (PRF) is one for which range and Doppler frequency are both ambiguous (Fig. 29-1). In practice, only the lower reaches of the relatively wide band of PRFs satisfying this definition are actually used. Moreover, the optimum value increases with the radar's radio frequency. For X-band, medium PRFs typically range from about 8 to 16 kHz, slightly higher than the top of the low PRF range, which falls somewhere between 2 and 4 kHz.

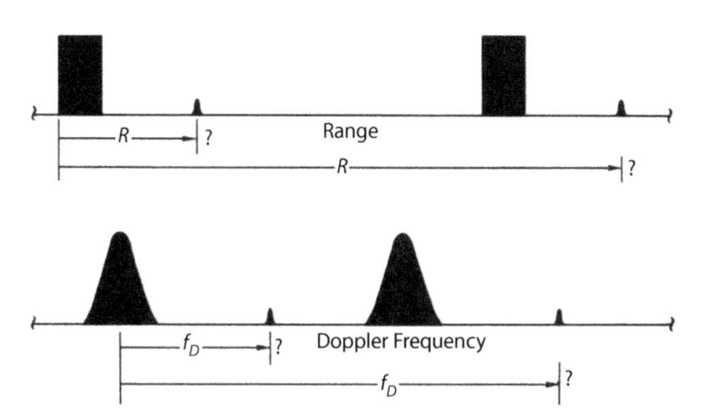

Figure 29-1. A medium PRF is one for which both range and Doppler frequencies are ambiguous.

Medium PRF operation was conceived as a means of getting around some of the limitations of low and high PRFs in fighter applications. The primary reason for operating above the low PRF region is to improve the radar's ability to contend with mainlobe clutter and ground moving targets (GMTs). Conversely, the primary reason for operating below the high PRF region is to improve the radar's ability to contend with sidelobe clutter in the tail hemisphere approaches (low closing rates).

This chapter takes a closer look at medium PRF operation—specifically how to separate targets from clutter and how the signal processing is performed. Problems of rejecting GMTs, eliminating blind zones, minimizing sidelobe clutter, and rejecting sidelobe return from those targets on the ground that have exceptionally large radar cross sections are examined.

29.1 Differentiating between Targets and Clutter

To get a clear picture of the problem of rejecting ground clutter, consider the range and Doppler profiles for a representative flight situation where the radar has a PRF of 10 kHz and that the maximum range of interest is 45 km.

Range Profiling. The top part of Figure 29-2 shows a radar illuminating an area of terrain to a maximum range of 45 km.

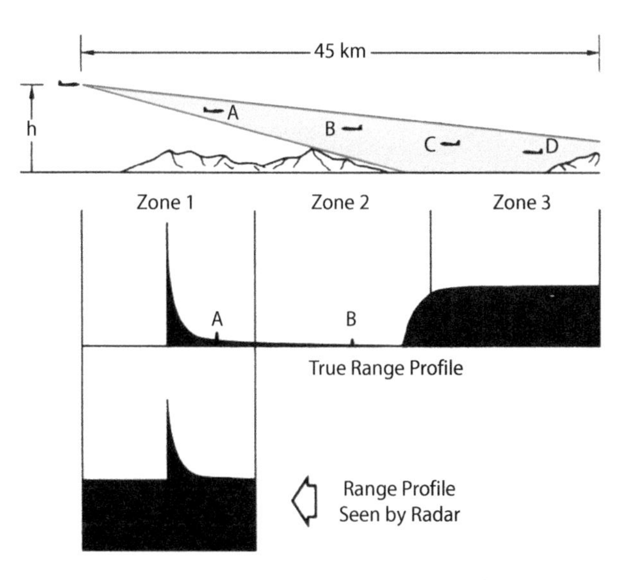

Figure 29-2. This range profile is representative of a flight situation. The PRF is such that the maximum range of interest is broken into three range zones; however, the radar sees these three zones superimposed together into a single zone.

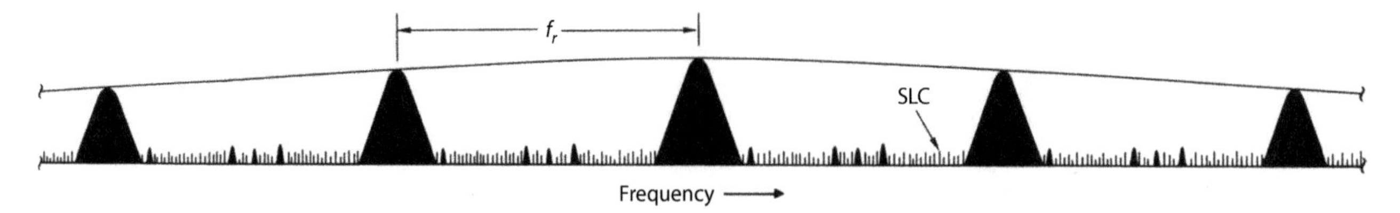

Figure 29-3. In this Doppler profile of a representative flight situation, the mainlobe clutter lines are much more widely spaced than for low PRF operation (other conditions remaining the same).

The bottom part of Figure 29-2 is the true range profile as seen by the radar.

The width of the first unambiguous range zone is determined by the PRF of 10 kHz and equates to 15 km. The 45 km maximum range of interest therefore can be considered to consist of three segments, each 15 km long.

However, as seen by the radar, these are indistinguishably superimposed. That is, the three 15 km segments are all merged into a single segment extending 15 km from the radar. In this segment ground clutter from zone 3 completely blankets the observed range interval. Mainlobe clutter extends from one end to the other. Strong sidelobe clutter received from short ranges in zone 1 also covers a substantial portion of the profile seen by the radar. This amalgam of clutter makes it impossible to detect any of the targets.

Except in the case of very large targets in comparatively light clutter, no amount of range discrimination on its own is going to enable the radar to isolate the target echoes from the clutter. To reject mainlobe and sidelobe clutter, Doppler frequency discrimination must be used.

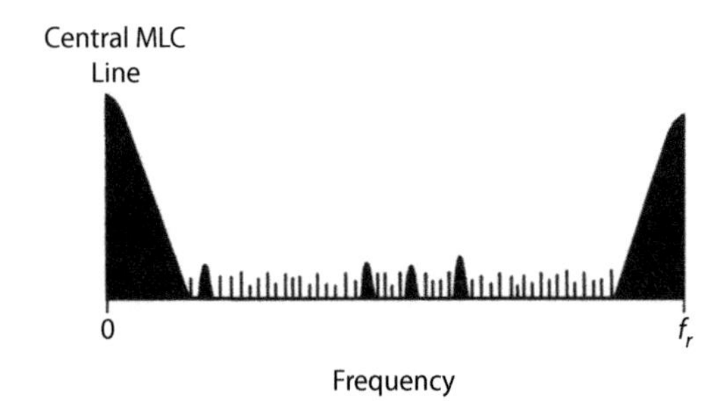

Figure 29-4. In this portion of the Doppler profile processed by the radar, its spectrum is normally shifted to place the central line of the mainlobe clutter at zero frequency (direct current).

Doppler Profiling. As in the case of low PRFs, this profile consists of a series of mainlobe clutter lines separated by the pulse repetition frequency, f_r (Fig. 29-3). Between any two consecutive lines appears most, but not all, of the sidelobe clutter and the return from most, but not necessarily all, of the targets (Fig. 29-4). The rest of the sidelobe clutter and target return is indistinguishably intermixed with the mainlobe clutter.

Rejecting Mainlobe Clutter. While the Doppler profiles for low and medium PRF operation are similar, there is one important difference. At medium PRFs, the mainlobe clutter lines are spread farther apart (other conditions remaining the same). Since the width of the line is independent of the PRF, there is considerably more clear room between them in which to detect targets. Even if the mainlobe clutter is reasonably broad, it can be rejected on the basis of its Doppler frequency without, at the same time, rejecting the return from an inordinately large fraction of the radar's true targets.

Rejecting Sidelobe Clutter. Because of the more severe range ambiguities, this is not as simple as at low PRFs. This is illustrated in Figure 29-5, where the range profile, as seen by the radar, is repeated with the mainlobe clutter removed. There are two things to notice in this plot: (1) the sidelobe clutter has a sawtooth shape; and (2) only the short-range target, A, can be discerned above the clutter—targets B and C are still obscured.

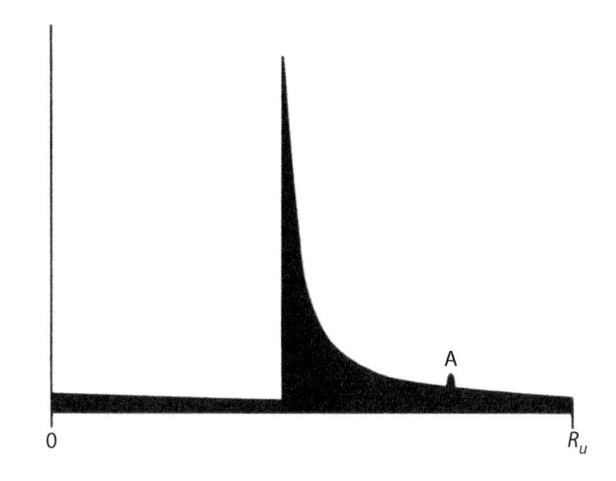

Figure 29-5. This is the range profile as seen by the signal processor with mainlobe clutter removed. Only the short-range target can be discerned above the clutter.

The sawtooth shape is caused by the strong sidelobe return from the first range zone being superimposed over the weaker return from subsequent range zones (Fig. 29-6). As for the obscured targets, target B, in the second range zone, must compete not only with sidelobe clutter from its own range but also with the far stronger clutter from the corresponding range in the first range zone. Targets C and D, in the third range zone, must compete not only with sidelobe clutter from their own range but also with the much stronger return from the corresponding ranges in the first and second zones.

The clutter can be substantially reduced. It comes not only from different ranges but also from different angles. Since returns from different angles have different Doppler frequencies, they can be differentiated from the target echoes and a great deal of the competing sidelobe clutter if they are sorted by both range and Doppler frequency.

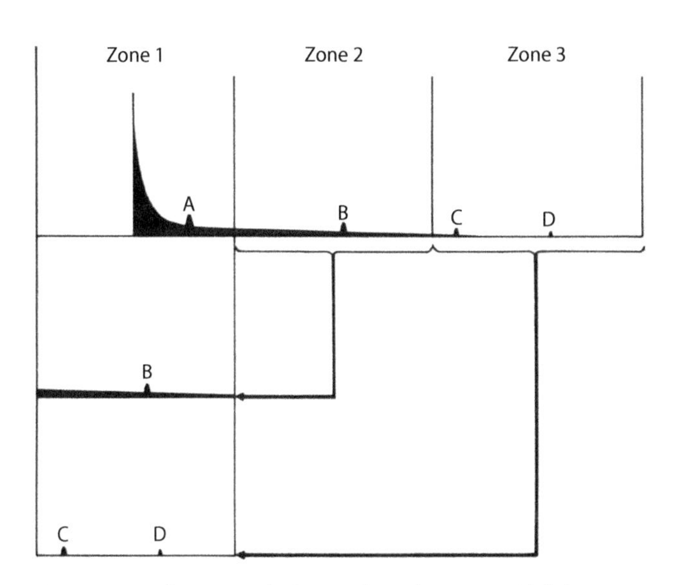

Figure 29-6. The sawtooth shape is from the strong sidelobe clutter echoes from the first range zone being superimposed over the weaker ones from the second and third zones.

Sorting by range may be done by range gating (sampling), just as in low PRF operation. The range gates will isolate the returns received from relatively narrow strips of ground at constant range. Because of range ambiguities, though, the return passed by each gate will come from not just one strip but several. In addition, as already noted, one or more of these strips may lie at relatively short range.

Still, the reduction in clutter obtained through range gating will be substantial (Fig. 29-7).

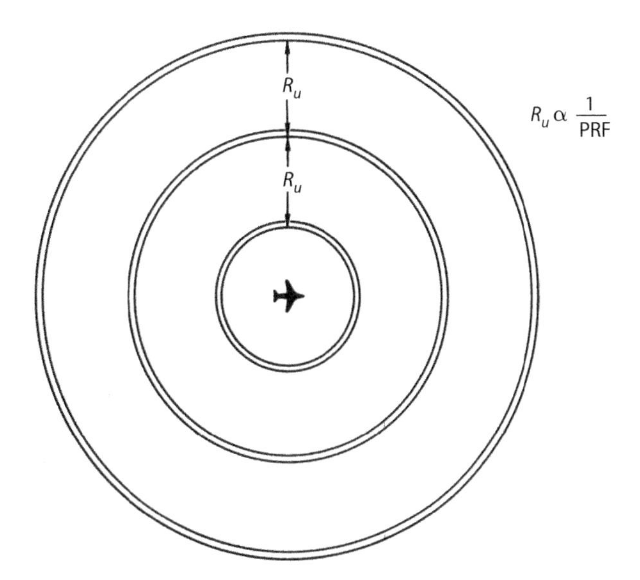

Figure 29-7. Each range gate passes the echo from a series of circular strips (only three of which are shown here) each separated by R_u.

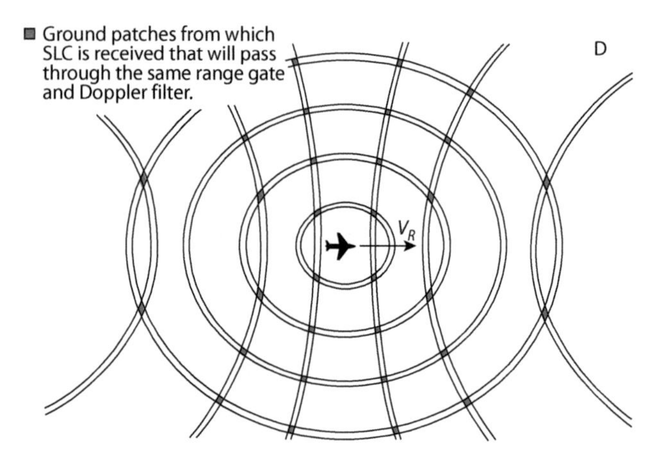

Figure 29-8. Each Doppler filter receives only the portion of the total sidelobe echo that is passed by a single range gate. The filter passes only that fraction of this echo return that comes from strips of ground whose angles relative to the radar's velocity are such that the return falls in the filter's passband.

Sorting by Doppler frequency may be accomplished by applying the output of each range gate to a bank of Doppler filters. These will isolate the echo returns received from strips of ground lying between lines of constant angle relative to the radar's velocity (Fig. 29-8).

Because of Doppler ambiguities any one filter will pass the return from not just one strip but several. Nevertheless, the amount of clutter with which a target's echoes must now compete will be only a fraction of that passed by the range gate.

29.2 Signal Processing

As illustrated in Figure 29-9, for medium PRFs the signal processing is quite similar to that for low PRFs, with three main differences:

1. Because range ambiguities preclude the use of sensitivity time control (STC), additional automatic gain control is needed to avoid saturation of the analog-to-digital (A/D) converter.
2. To further attenuate sidelobe clutter (range ambiguities pile up more deeply at medium PRFs), the passbands of the Doppler filters may have to be made considerably narrower.
3. Additional processing is required to resolve range and Doppler ambiguities.

As at low PRFs, the first step in processing the intermediate frequency (IF) output of the radar receiver is to shift the Doppler spectrum to place the central mainlobe clutter line at direct current (dc). Again, the shift must be dynamically controlled to account for changes in radar velocity and antenna look angle. The in-phase (I) and quadrature (Q) outputs of the synchronous detector that performs this shift are likewise sampled at intervals on the order of the transmitted pulse width[1] and are then digitized.

1. Compressed pulse width when pulse compression is used.

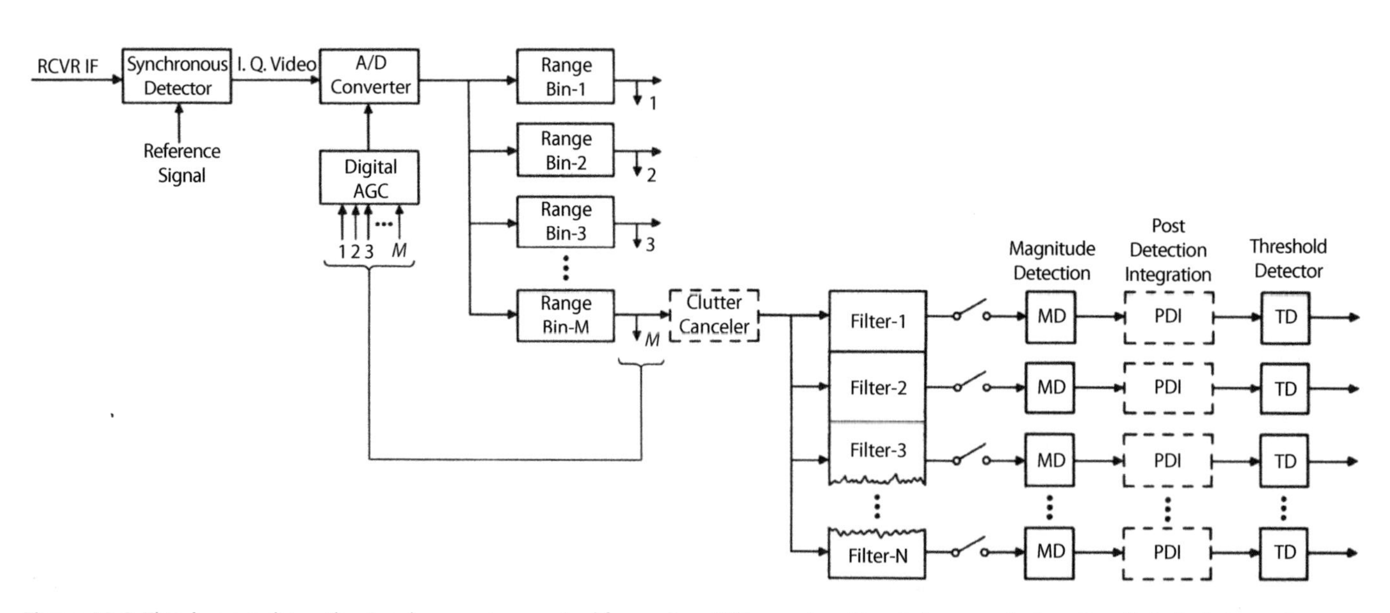

Figure 29-9. This diagram shows the signal processing required for medium PRF operation. The clutter canceler is optionally included to reduce the dynamic range required of the Doppler filters. Post-detection integration (PDI) may be provided if the filter integration time is less than the time-on-target.

However, to reduce the dynamic range required of the A/D converter, automatic gain control is provided ahead of the converter. For this, the converter's output is monitored, and a continuously updated profile of the output over the course of the interpulse repetition period is stored. On the basis of this profile, a gain control signal is produced and applied to the amplifiers ahead of the A/D converter.

By reducing the gain when the mainlobe clutter and strong close-in sidelobe clutter are being received, the control signal keeps the converter from being saturated. At the same time it maintains the input to the converter well above the local noise level when weaker echoes are coming through. Because the control signal is derived after the return is digitized, this is called *digital automatic gain control* (DAGC).

To reduce the dynamic range required in the subsequent processing, once the output of the A/D converter has been sorted into range bins an optional next step is to get rid of the bulk of the mainlobe clutter in each bin. This may be accomplished with a clutter canceler (Fig. 29-10).

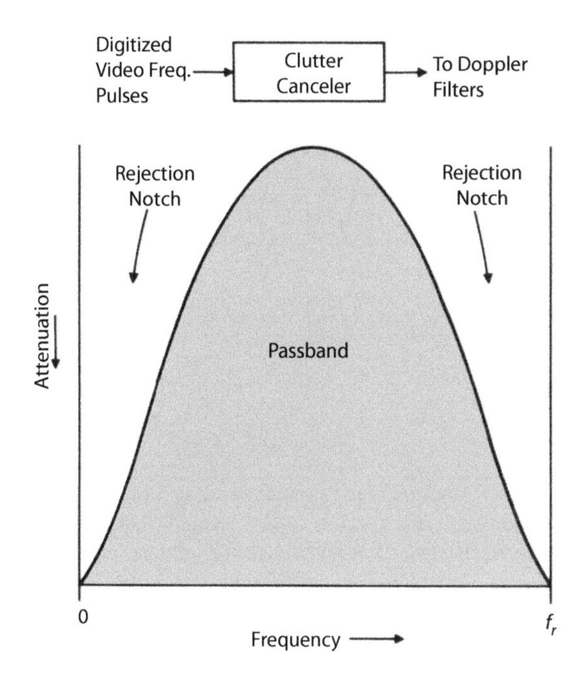

Figure 29-10. Mainlobe clutter is reduced by passing the digitized video frequency output of the receiver through a simple clutter canceler with a passband characteristic of this form.

The return in each bin is next applied to a bank of Doppler filters. At the end of every integration period, the magnitude of each filter's output is detected.

Subsequently, the integrated return passed by each Doppler filter during every time-on-target is applied to a separate threshold detector, which is adaptively set to keep the probability of clutter producing false alarms at an acceptably low level. The setting of this threshold may be based on the average level of the clutter for (1) several range increments on either side of the one in question, (2) several Doppler frequencies on either side of the one in question, (3) several integration periods before and after the one in question, or (4) some combination of these. In general, the optimum averaging scheme for a clutter background is different from that for a noise background.

When a target is detected, its apparent range can be determined by observing in which bin (or adjacent bins) it was detected. Similarly, its apparent Doppler frequency and hence range rate can be determined by observing in which Doppler filter (or adjacent filters) the detection occurred.

The observed range and Doppler frequency will be ambiguous. Range ambiguities are resolved by PRF switching. Doppler ambiguities may be resolved by the methods described in Chapter 22.

29.3 Rejecting Ground Moving Targets

GMTs in medium PRFs are not nearly the problem they are at low PRFs. In the former, the mainlobe clutter lines are spread sufficiently far apart that GMTs appear only near the ends of the region between the clutter lines. Targets with positive closing rates appear at the lower end; targets with negative closing rates appear at the upper end ($f_r - f_D$). GMTs therefore can

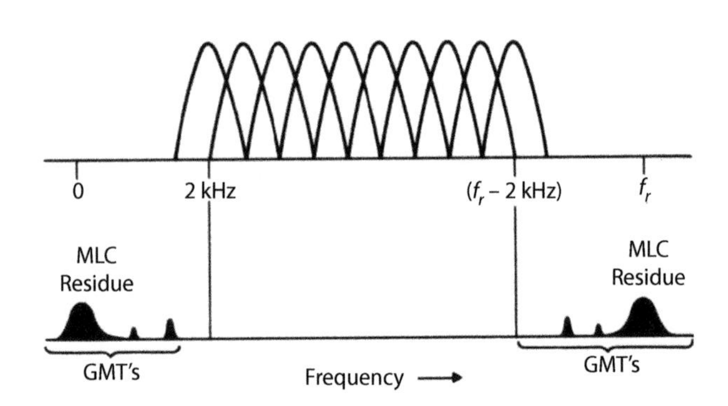

Figure 29-11. Most GMTs and mainlobe clutter (MLC) residue that is passed by the clutter canceler can be rejected by discarding the return between 0 and 2 kHz and between (f_r – 2 kHz) and f_r.

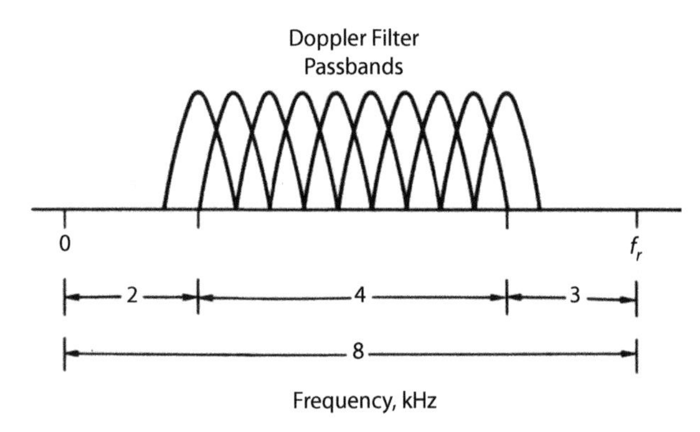

Figure 29-12. If 4 kHz of Doppler spectrum is discarded to eliminate GMTs, the PRF must be at least 8 kHz to meet the criterion that 50 percent of the Doppler spectrum is clear.

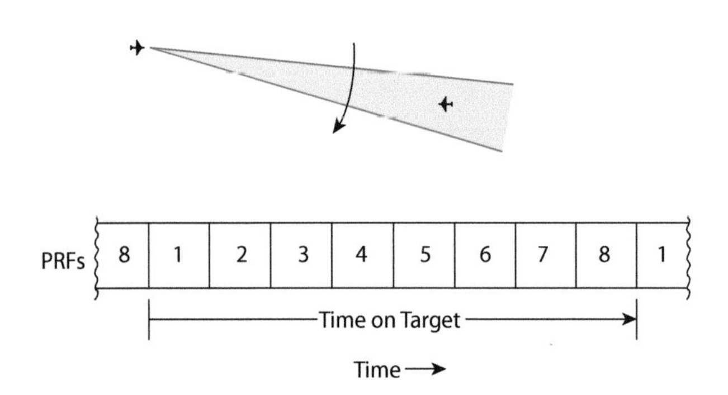

Figure 29-13. To eliminate blind zones and resolve range ambiguities, a radar system may cycle through a number of widely spaced PRFs.

be eliminated without losing an unreasonable fraction of the target return simply by discarding any return in the frequency bands where GMTs may appear.

In general, the maximum measured Doppler velocities for GMTs are known based on highway speed limits, the frequency of illumination, and the viewing angle of the radar system. These are used as a means of setting the rejection criteria. For example, most surface vehicles travel at less than 110 km/hour. At X-band we have 18.75 Hz per km per hour of closing rate. Thus, the maximum Doppler shift of GMTs relative to the center frequencies of the mainlobe clutter lines would be about 2 kHz ($110 \times 18.75 = 2062.5$ Hz).

GMTs with a component of velocity toward the radar (positive Doppler shift) or away from the radar (negative Doppler shift) can be eliminated by discarding all echoes whose frequencies are less than 2 kHz above the center of each mainlobe clutter line (Fig. 29-11) or 2 kHz below the center of each line, respectively. In the latter, the mainlobe clutter residue would also simultaneously be eliminated.

The anticipated maximum Doppler frequency of the GMTs relative to the Doppler frequency of the mainlobe clutter usually puts a lower limit on the selection of the PRF. Suppose we establish a design criterion that at least 50 percent of the Doppler spectrum must be clear for the detection of targets (i.e., is not covered by blind zones). If GMTs are eliminated by discarding all echoes whose frequencies are within 2 kHz of the center of each clutter line (Fig. 29-12), the filter bank's passband must be at least 4 kHz wide. To accomplish this, the PRF must be at least $2 + 4 + 2 = 8$ kHz.

Since the Doppler shift is inversely proportional to wavelength ($f_D = 2V/\lambda$), the shorter the wavelength, the higher the minimum PRF and vice versa. Take a wavelength of 1 cm, for example. At this wavelength, the maximum relative Doppler shift for a 110 km/hour vehicle is 6 kHz as opposed to 2 kHz. Consequently, if the previous design criteria are applied to a 1 cm wavelength radar system, the minimum PRF is $6 + 12 + 6 = 24$ kHz.

29.4 Eliminating Blind Zones

Blind zones are still a problem for medium PRF radars. In fact, because of range ambiguities the radar must contend with blind zones not only in the Doppler spectrum but also in the range interval being searched.

Doppler Blind Zones. Because mainlobe clutter covers a much smaller portion of the Doppler frequency spectrum, Doppler blind zones are far less severe at medium PRFs than at low PRFs and so can be eliminated by switching among fewer PRFs. However, additional PRFs are required to resolve range ambiguities and eliminate ghosts.

Typically, the radar is cycled through a fixed number of fairly widely spaced PRFs (Fig. 29-13). If a target is in the

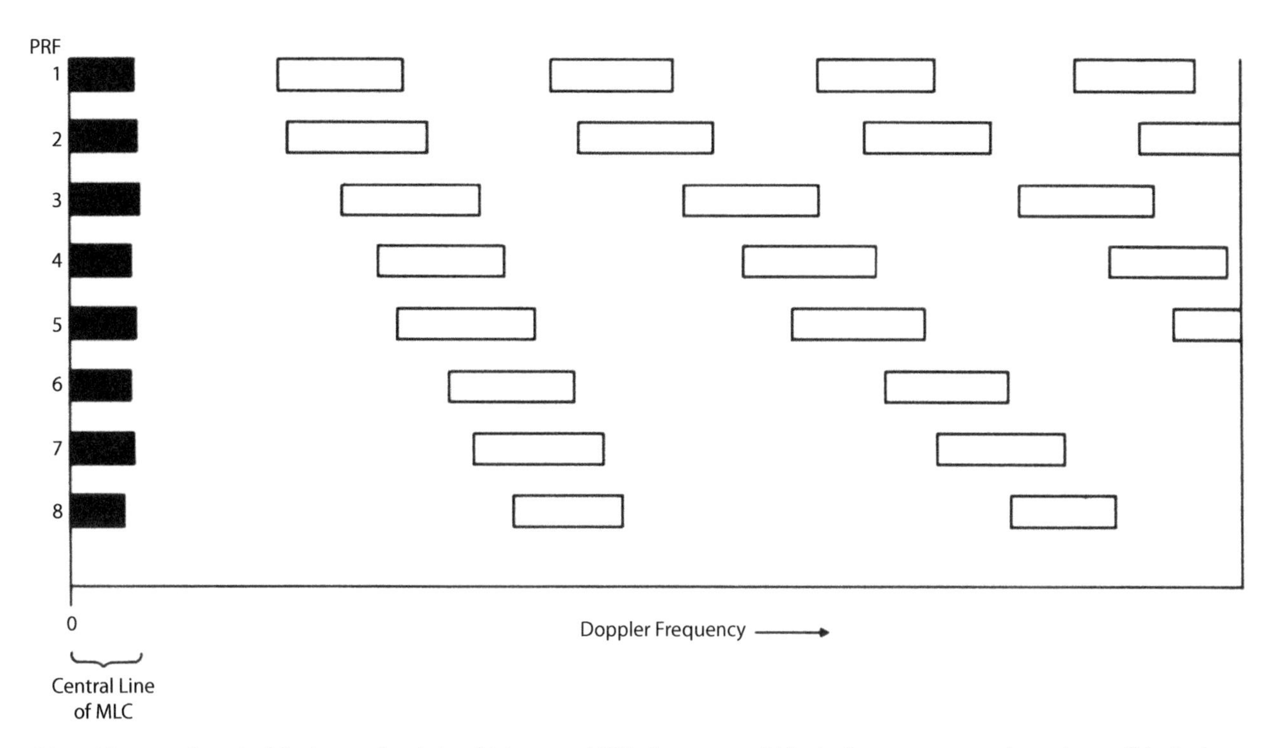

Figure 29-14. There are Doppler blind zones for eight widely spaced PRFs. Any target within the frequency range shown here will be in the clear for at least three PRFs.

clear on any three of these and its echoes exceed the detection threshold on all three, the target will be declared as having been detected. The range ambiguities will then be resolved, which is called *deghosting*. The optimum number of PRFs varies with the operational situation. A typical waveform might be called 3:8. This means it cycles through 8 PRFs, any 3 of which must be clear for detection. This can be seen in Figure 29-14, where a vertical line drawn anywhere on the Doppler frequency axis will show at least three PRF clear zones.

Range Blind Zones. These are ranges at which targets will generally not be detected because their echoes are drowned out by sidelobe clutter simultaneously received from shorter ranges or the targets are eclipsed by the transmitted. Eclipsing means that the receiver is turned off while transmitting, so it follows that echoes are not received during this time.

Just how these blind zones come about is best illustrated graphically. Figure 29-15 plots the strength of a target's echoes as range increases. It goes from a relatively few km out to the maximum range of interest (i.e., well beyond the unambiguous range of zone 1). Superimposed over this plot is a periodically repeated graph of the strength of the sidelobe clutter received over the course of the interpulse repetition period. Each repetition of the sidelobe clutter plot represents the clutter background against which target echoes must be detected when a target is in a different ambiguous range zone.

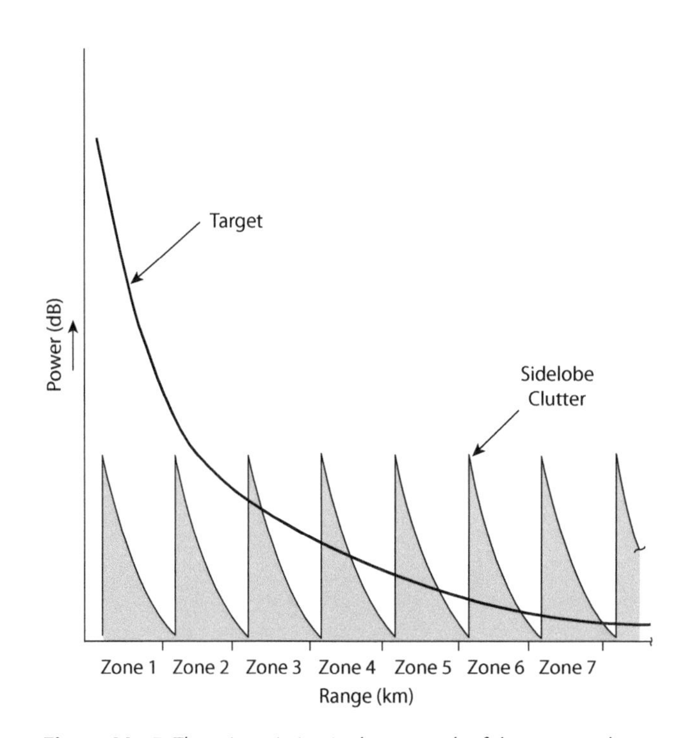

Figure 29-15. There is variation in the strength of the target echo return with range. The strength of the sidelobe clutter with which the target must compete is superimposed. The blind zones occur where the plots overlap.

If a target is in the first or second range zone, it will be substantially stronger than any of the sidelobe clutter. If it is in the

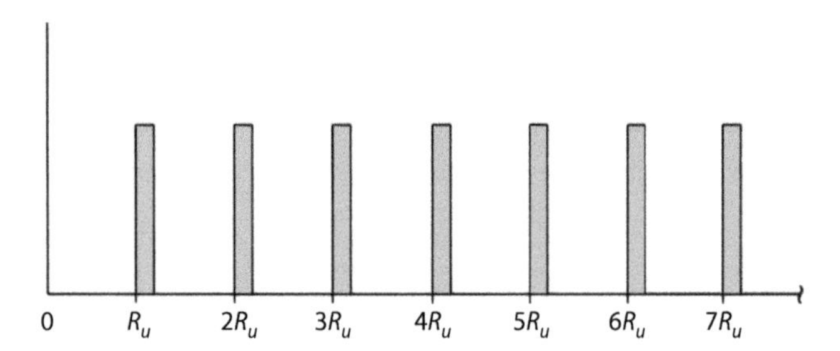

Figure 29-16. The diagram shows the range blind zones due to the eclipsing caused by the transmitted pulses. As the width of the transmitted pulses is increased, these zones become appreciable.

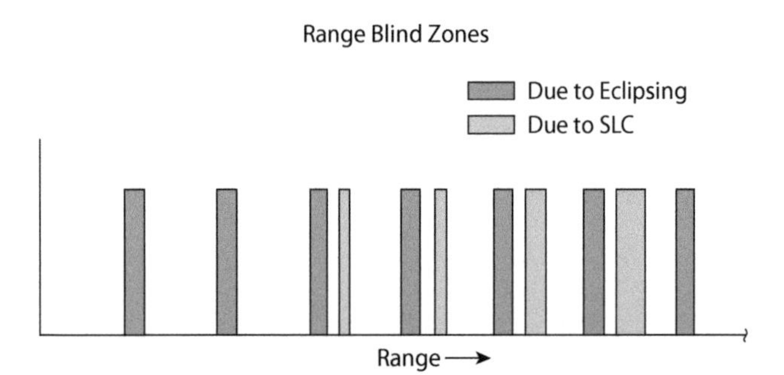

Figure 29-17. This diagram shows the combined range blind zones due to eclipsing and due to sidelobe clutter for a representative medium PRF radar system.

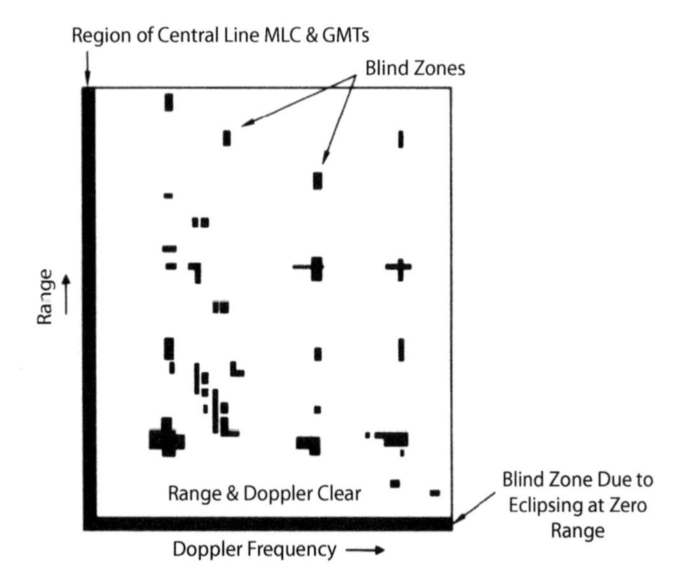

Figure 29-18. This is the region in which a representative radar system is both range clear and Doppler clear for at least three of eight widely spaced PRFs.

third range zone, it will still be stronger than most of the clutter but not as strong as the peak produced by the altitude return and the sidelobe return immediately following it. If the target is in the fourth range zone, it will be stronger than the clutter over a smaller portion of the zone, and so on.

At ranges where the clutter is as strong or stronger than the target echoes, the target will go undetected, just as it would if it were masked by receiver noise. The radar is thus blind to the target. The extent of the range blind zones increases with the strength of the clutter. The stronger the clutter, the wider the blind zone will be. The strength of the clutter, in turn, depends on several things, including the sidelobe gain, terrain type, and radar altitude.

Added to the range blind zones, because of strong sidelobe clutter, are blind zones caused by eclipsing (Fig. 29-16). While the radar is transmitting (and for a very short recovery time immediately thereafter), the receiver is blanked. Consequently, if a target's echoes are received at such times that they overlap these periods (as they invariably will be if the target's range is a multiple of R_u), not all of the target echo returns will get through the receiver; thus, the target may not be seen. The resulting blind zones may be narrow enough to be inconsequential. However, they can become significant if the pulses are long, as is the case in some medium PRF radars.

The combined range blind zones due to sidelobe clutter and eclipsing for a representative radar system are illustrated in Figure 29-17.

As with Doppler blind zones, the positions of the range blind zones shift with changes in the PRF. Fortunately, the shift is such that the same PRF switching used to reduce Doppler blind zones will also largely reduce the range blind zones (Fig. 29-18).

Bear in mind that it is not enough for a target to be in a Doppler clear region for one set of PRFs and in a range clear region for another. For a target to be detected, it must be in a region that is clear of both Doppler and range ambiguities for the same set of PRFs.

If the target's Doppler frequency falls in a Doppler blind zone, its echoes will not get through a Doppler filter and be detected even though it is in a range-clear region. Conversely, if the target is in a range blind zone, even though its echoes may get through a filter they will be buried in the accompanying sidelobe clutter. The effect is to drive the detection threshold up to a point where the target will not be detected.

Note that the foregoing discussion of blind zones all pertains to search. Applications demanding the use of medium PRF operation are increasing, and hence there is a continuing quest to seek PRF combinations that allow ambiguities to be removed. This is an ongoing area of research.

29.5 Minimizing Sidelobe Clutter

At medium PRFs, sidelobe clutter must be kept to a minimum. Not only does it determine the extent of the range blind zones, but it also limits the maximum detection range.

Since most of the sidelobe clutter comes from relatively short ranges, the background of clutter against which targets must be detected is generally stronger than the background noise falling in the passband of a Doppler filter. Therefore, no matter how powerful the radar or how large a target's radar cross section, if the target's range is continuously increased (Fig. 29-19) a point will ultimately be reached where its echoes become lost in the clutter. The stronger the clutter, the shorter this range will be.

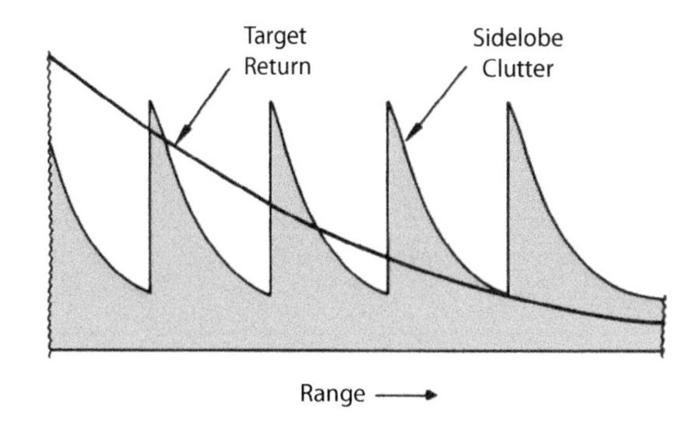

Figure 29-19. As a target's range is increased, its echoes may eventually be engulfed in sidelobe clutter unless special measures are taken to minimize the sidelobe clutter.

Several things can be done to minimize sidelobe clutter (Fig. 29-20). The most important is to design the radar antenna so that the gain of its sidelobes is as low as possible. Antenna sidelobes can be reduced by tapering the distribution of radiated power across the antenna (as discussed in Chapters 8, 9, and 10).

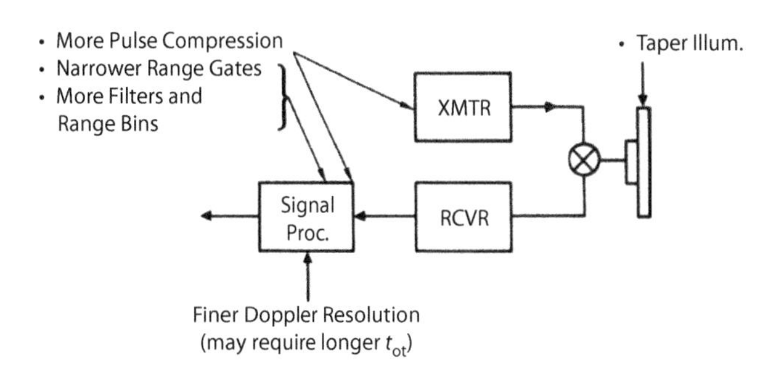

Figure 29-20. Some measures can be taken to reduce sidelobe clutter.

Narrowing the radar's pulses and correspondingly narrowing the range bins can also reduce the amount of clutter with which a target's echoes must compete. If, for example, the pulse width is reduced by a factor of 10, the sidelobe clutter will be reduced by roughly the same amount. Narrowing the pulses, of course, requires adding more range bins and forming more Doppler filters (a separate bank of filters being required for every range bin). Improvements in signal processing power has helped facilitate this and has resulted in higher levels of target detection performance. However, narrower pulses reduce detection range since there is an overall loss in transmit signal (if all else remains the same).

To overcome this sensitivity loss, pulse compression can be employed (see Chapter 16). Pulse compression allows high range resolution (narrow range bins) without reducing the average transmitted power (Fig. 29-21). Enough pulse compression is then provided to achieve the desired range resolution.

The narrowing of the range bins effectively removes clutter with which targets must compete; hence, detection performance improves. There is a limit as to how narrow the range bins can be made, largely determined by the physical size of the target. If the range bins are made too narrow, then the target is over-resolved such that its echo strength in any one range bin will be diminished, leading to degradation of detection performance.

Narrowing the frequency resolution of the Doppler filters can further reduce the sidelobe clutter with which a target must compete. To achieve this, the total duration over which pulses are collected for Doppler processing must be extended (improving Doppler resolution). Again, there is a notional limit dictated by the range of Doppler frequencies that a target may possess over the pulse collection duration. Usually, the filters must be made a little wider than this to minimize filter straddle loss. Changes in Doppler frequency may be due to acceleration

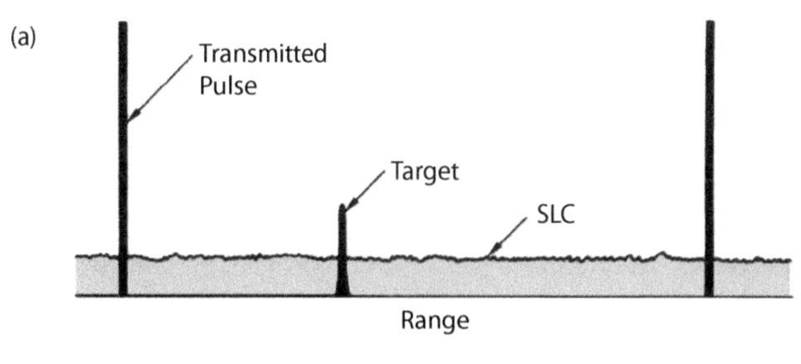

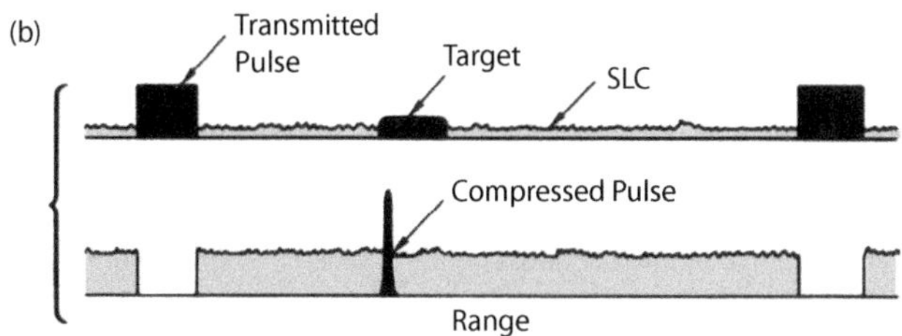

Figure 29-21. Two ways of reducing sidelobe clutter through range resolution: (a) transmit very narrow pulses of high enough peak power to provide adequate detection range; (b) transmit wider pulses of the same average power and use pulse compression to provide the desired range resolution.

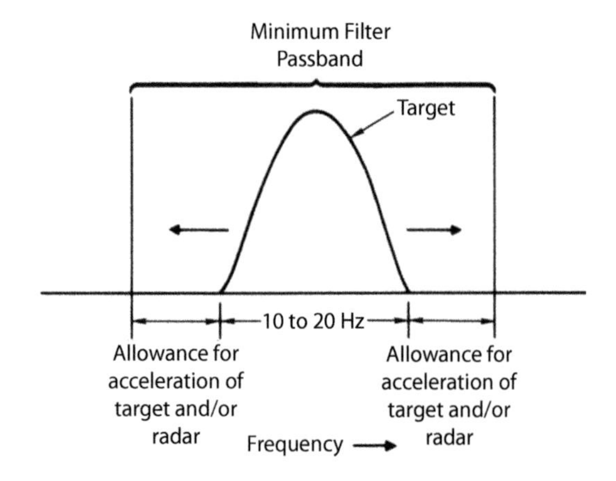

Figure 29-22. The passband of a Doppler filter must be wide enough to accommodate the target return and allow for changes in the Doppler frequency during the integration period.

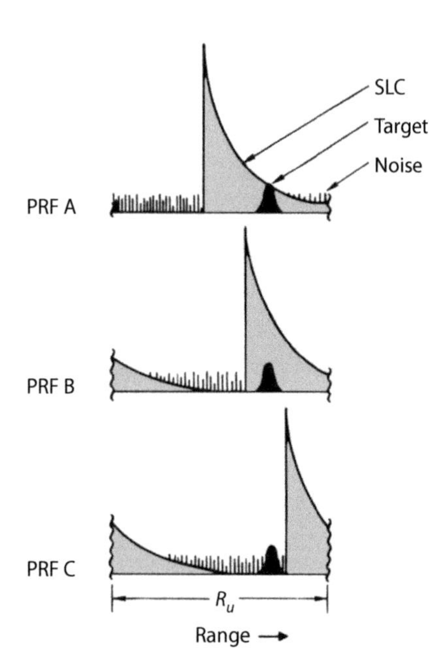

Figure 29-23. By switching among enough PRFs, the sawtooth pattern of sidelobe clutter can be shifted so that virtually every target may be detected against a background only of noise.

of either the aircraft carrying the radar or the target during the filter formation time (Fig. 29-22). This approach also increases the signal-processing requirement, which must be accounted for in the radar processor design and specification.

In the ways previously discussed, sidelobe clutter may be reduced to a point where it falls below the noise level at the ends of the interpulse period (Fig. 29-23). The detection ranges of targets will then be limited only by noise.

By switching among a large enough selection of PRFs, these clutter-free regions can be shifted about so that the detection range of virtually all targets will be limited only by noise. As more PRFs are added, the available integration time for each PRF decreases, which will limit the detection range.

Detection ranges achievable with medium PRFs are thus invariably somewhat less than those achievable under similar conditions with high PRFs against nose-aspect targets or with low PRFs in situations where mainlobe clutter is not a problem.

29.6 Sidelobe Return from Targets of Large RCS

One important form of sidelobe return not yet considered is that from structures on the ground such as buildings and trucks (Fig. 29-24), which can have exceptionally large radar cross sections. Ground clutter can be immensely varied and generally requires careful treatment. Even when large radar cross section objects are in the sidelobes, they can return clutter echoes every bit as strong as those received from the mainlobe. If a GMT's Doppler frequency falls in the filter bank's passband (as it most often will when the GMT is in a sidelobe), it will be detected no different from if it were an aircraft in the mainlobe.

Since these unwanted targets are point-like reflectors, no amount of range or Doppler resolution will make them less

Figure 29-24. Because of range and Doppler ambiguities at medium PRFs, a radar system is vulnerable to unwanted ground targets. Wind farms present clutter that has both static and moving components.

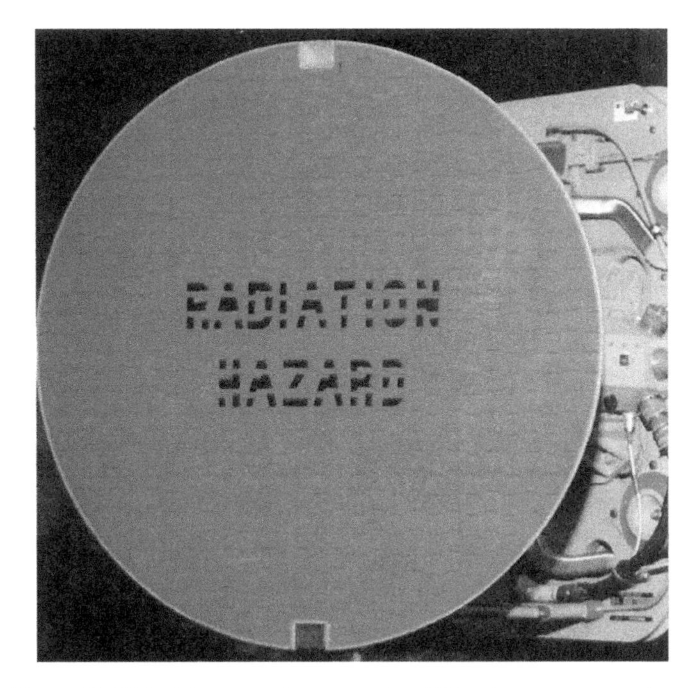

Figure 29-25. In this antenna from a medium PRF radar system, note the horn antenna for the guard receiver located at the top of the main antenna.

likely to be detected. On the contrary, the greater the resolution provided, the greater the extent to which the surrounding sidelobe clutter will be attenuated and the more prominently the point-like targets will protrude above it.

Although a radar is vulnerable to such targets when operating at low PRFs, it is much more so when operating at medium PRFs because of the more severe range ambiguities. Some special means, therefore, must be provided to keep these targets from reaching the radar display.

One way of dealing with these unwanted targets is to provide the radar with a guard channel. This can consist of a separate receiver whose input is supplied by a small horn antenna mounted on the main radar antenna (Fig. 29-25).

The width of the horn's mainlobe is sufficient to encompass the entire region illuminated by the radar antenna's principal sidelobes. The gain of the horn's mainlobe is greater than that of any of the sidelobes (Fig. 29-26).

Any detectable target in the radar antenna's sidelobes therefore will produce a stronger output from the guard receiver than from the main receiver.

On the other hand, because the gain of the radar antenna's mainlobe is much greater than that of the horn, any target in the radar antenna's mainlobe will produce a much stronger output from the main receiver than from the guard receiver.

Consequently, by comparing the outputs of the two receivers and inhibiting the output of the main receiver when the output of the guard receiver is stronger, any targets that are in the sidelobes can be prevented from appearing on the radar display (Fig. 29-27).[2]

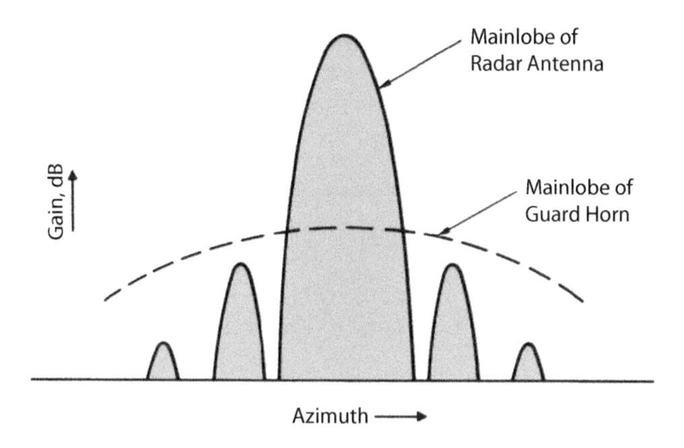

Figure 29-26. The gain of the horn's mainlobe is greater than that of the radar antenna's sidelobes but less than that of the radar antenna's mainlobe.

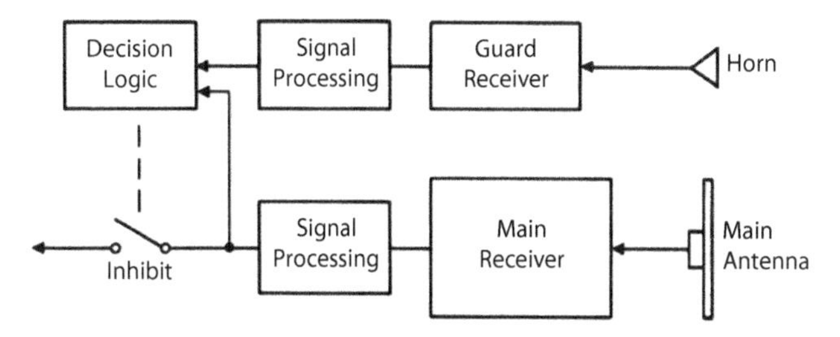

Figure 29-27. The output of the main receiver is inhibited when a target is detected simultaneously through the guard and main receiver channels.

2. Mainlobe return from the unwanted ground target will, of course, fall in the rejection notch of the clutter canceler. If the return is extremely strong, however, a detectable fraction of it may get through.

Advantages and limitations of medium PRF operation

Advantages	Limitations
Good all-aspect capability—that is, copes satisfactorily with both mainlobe and sidelobe clutter.	Detection range against both low and high closing-rate targets potentially limited by sidelobe clutter.
Ground moving targets readily eliminated.	Must resolve both range and Doppler ambiguities.
Pulse-delay ranging possible.	Special measures needed to reject sidelobe return from strong ground targets.

29.7 Summary

In medium PRF operation, the PRF is usually set just high enough to spread the mainlobe clutter lines such that mainlobe clutter and any GMTs can be rejected without also rejecting the return from an unreasonably high percentage of targets.

Through a combination of range and Doppler discrimination, the background of sidelobe clutter against which the target echoes must be detected can be reduced to an acceptable level.

Because of the increased separation of the mainlobe clutter lines, switching among a small number of fairly widely spaced PRFs largely eliminates Doppler blind zones. Because distant targets must compete with close-in sidelobe clutter, the peaks of this clutter result in range blind zones.

If the clutter is not too strong, the same PRF switching used to eliminate Doppler blind zones will also largely eliminate range blind zones together with those due to eclipsing. However, even in the Doppler clear regions sidelobe clutter usually limits detection range. It is essential that a low sidelobe antenna be used.

Increasing the range and Doppler resolution can further reduce sidelobe return. To eliminate sidelobe return from ground targets of exceptionally large radar cross section, a guard channel may also be provided.

Further Reading

G. V. Morris and L. Harkness, *Airborne Pulse Doppler Radar*, Artech House, 1996.

C. M. Alabaster, *Pulse Doppler Radar: Principles, Technology, and Applications*, SciTech-IET, 2012.

Test your understanding

1. A medium PRF radar system has a maximum range of interest at 10 km and a PRF of 10 kHz. What is the maximum unambiguous range? What is the maximum unambiguous velocity for an illuminating radar frequency of 10 GHz?
2. Describe how the dynamic range requirements of an analog-to-digital converter can be reduced.
3. Describe the differences between range blind zones caused by eclipsing and in those caused by sidelobe clutter.
4. What steps can be taken to minimize sidelobe clutter in a medium PRF radar system?

30

High PRF Operation

A high pulse repetition frequency (PRF) is one for which the observed Doppler frequencies of all significant targets are unambiguous. The observed ranges, however, are generally ambiguous.

High PRF operation has three principal advantages:

1. Since Doppler frequencies are unambiguous, mainlobe clutter can be rejected without rejecting any target echoes whose Doppler frequencies are different from that of the clutter.
2. By employing a high enough PRF, the lines (more realistically bands) of the clutter spectrum can be spread far enough apart to open up an entirely clutter-free region between them, where high relative velocity[1] (closing rate) targets will appear (Fig. 30-1).
3. Increasing the PRF rather than the pulsewidth can be used to increase transmitter duty factors. This enables high average powers without the need for pulse compression or very high peak powers.

Detection range increases with the ratio of the signal energy to the energy of the background noise and clutter. By employing a high duty factor-high PRF waveform, long detection ranges can be obtained even in a clutter environment. However, when strong sidelobe clutter is encountered, detection ranges against low relative velocity targets may be impaired because of range ambiguities.

This chapter examines a high duty factor, high PRF waveform to see what must be done to separate targets from ground returns. It addresses the problems of range measurement and eclipsing loss as well as the steps necessary for improved performance against low relative velocity targets.

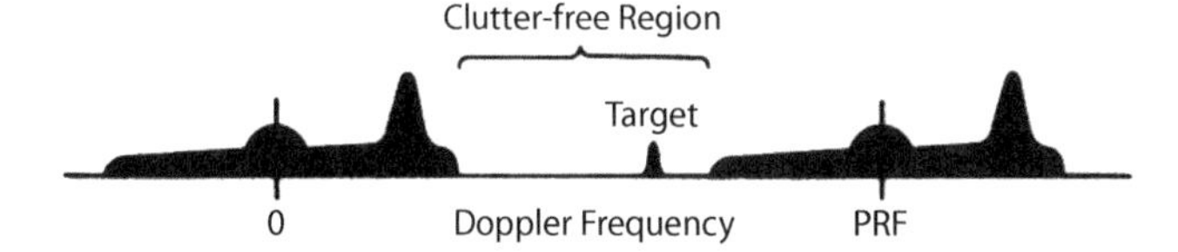

Figure 30-1. High PRF operation spreads the clutter bands far enough apart to open up a large clutter-free region in which high relative velocity targets will appear.

1. Targets whose relative velocities are greater than the radar's ground speed.

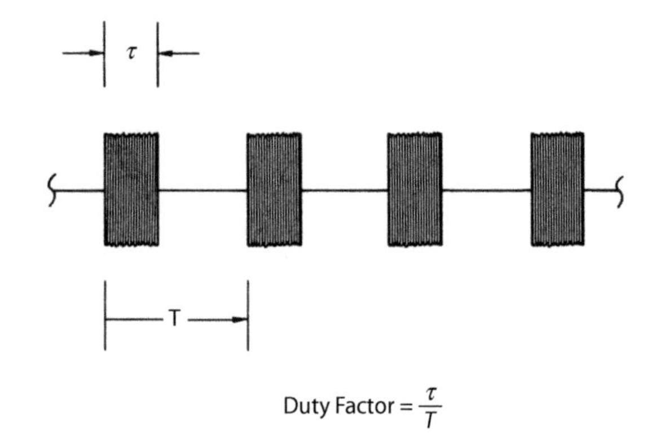

Figure 30-2. This high duty factor, high PRF waveform is typical of those used in radars for fighter aircraft. The duty factor is generally somewhat less than 50 percent.

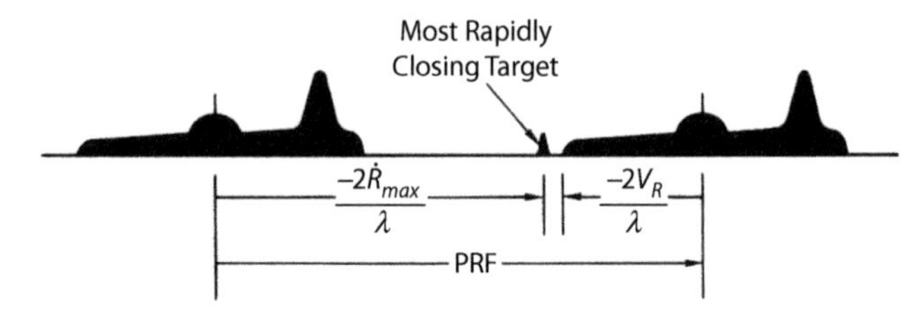

Figure 30-3. For the Doppler clear region to encompass all high closing-rate targets, the PRF must exceed the Doppler frequency of the most rapid target plus the maximum sidelobe clutter frequency.

30.1 High PRF Waveform

A representative high duty factor, high PRF waveform is shown in Figure 30-2. Because the radar receiver must be *blanked* (i.e., switched out) during transmission and because the duplexer has a finite recovery time, the maximum useful transmitter duty factor is generally somewhat less than 50 percent.

As for the PRF, if the clutter-free (i.e., Doppler-clear) region is to encompass all significant high closing-rate targets, the PRF must be greater than the sum of (1) the Doppler frequency of the most rapidly closing target; and (2) the maximum sidelobe clutter frequency (determined by the radar's velocity). The maximum sidelobe clutter frequency is twice the radar velocity divided by the wavelength. The target Doppler frequency is twice the target velocity divided by the wavelength (Fig. 30-3).

The shorter the wavelength, the higher the Doppler frequencies of both clutter and target and hence the higher the required PRF. Typically, in fighter applications at X-band frequencies the PRF is on the order of 100 to 300 kHz.

30.2 Isolating the Target Returns

To get a clear picture of the problems of differentiating between target echoes and ground clutter and of isolating target echoes from clutter and as much of the background noise as possible, we look at the range and Doppler profiles for the representative flight situation shown in Figure 30-4.

Assume that the radar is operating at a PRF of 200 kHz with a duty factor of 45 percent; that the aircraft carrying the radar has a ground speed of 750 m/s; and that echoes are being received from three targets. Targets A and B are flying in the same direction as the radar. Target A has a relative velocity of 300 m/s. Target B has zero relative velocity. Target C is approaching the radar from long range and has a relative velocity of 1500 m/s.

Range Profile. The range profile is illustrated in Figure 30-5. The width of the range profile, as observed by the radar, R_u, is less than 0.75 km (150 ÷ 200 = 0.75 km). This is because returns from every 0.75 km increment of range out to the maximum

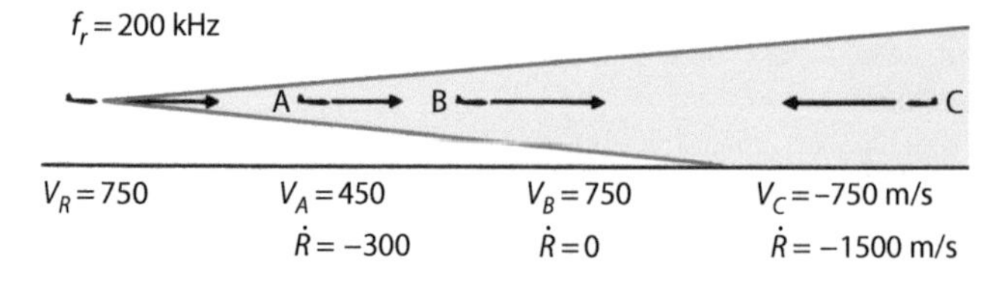

Figure 30-4. In this representative flight situation, target A has a low relative velocity, target B has a zero closing rate (zero relative velocity) and target C has a high closing rate (high relative velocity).

range from which returns are received are collapsed into this narrow interval. This includes all of the mainlobe clutter, the altitude return, all of the remaining sidelobe clutter, plus the transmitter spillover and the background noise. Buried in the midst of all this are echoes from the three targets. Virtually the only way they can be separated from the clutter and noise is to sort out the return by Doppler frequency.

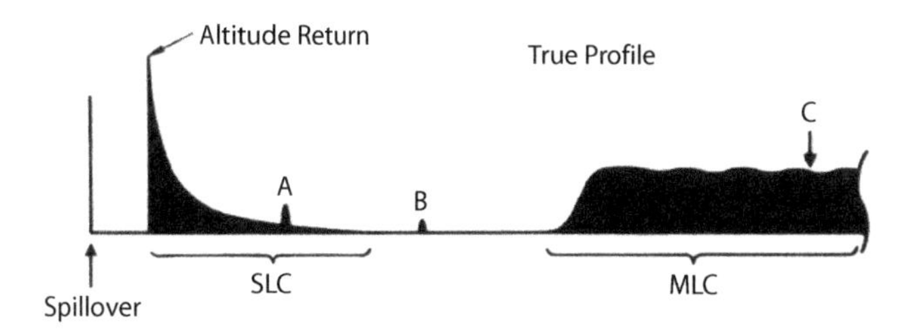

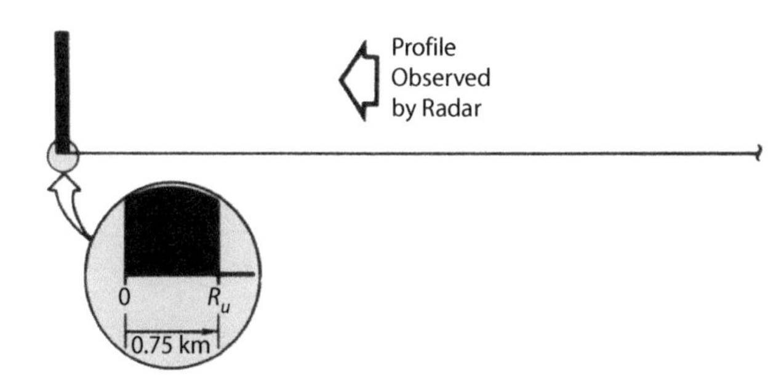

Figure 30-5. In this range profile for a representative flight situation, returns from virtually all ranges are collapsed into a band of observed ranges that is less than 1 km wide.

Doppler Profile. The Doppler profile is shown in Figure 30-6. As with low and medium PRF operation, the profile is a composite of the entire true Doppler spectrum of the radar return and is repeated at the PRF. However, there is one important difference. Since the width of the spectrum is less than the PRF, the repetitions (bands) do not overlap; that is, they are unambiguous.

The central band is shown in more detail in Figure 30-7. Here, the following features can be identified: transmitter spillover; altitude return; sidelobe clutter; and mainlobe clutter. The width of the sidelobe clutter region varies with the radar velocity. The width and frequency of the mainlobe clutter line varies continuously with the antenna look angle as well as with the radar velocity.

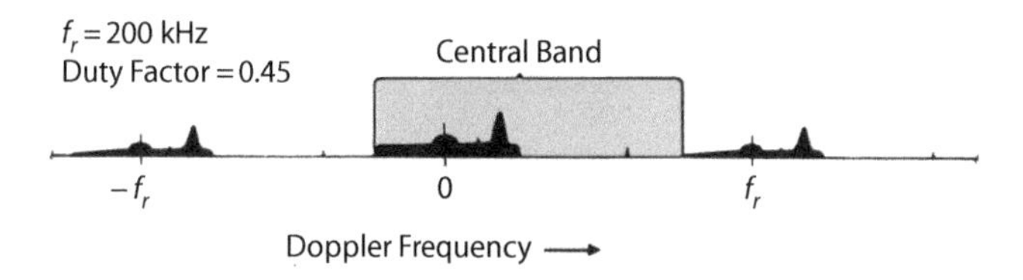

Figure 30-6. In this Doppler profile for a representative flight situation, repetitions of the true profile do not overlap.

Barely poking up above the sidelobe clutter are the echoes from target A, the low relative velocity target. In the clear between the high-frequency end of the sidelobe clutter region and the low-frequency end of the next higher band are the echoes from target C, the high relative velocity target. Provided the other clutter is removed, this target need compete only with thermal noise to be detected.

The echoes from the zero closing-rate target (B) are nowhere to be seen. However, they are there—they have merged with the combined altitude return and transmitter spillover, which has zero Doppler frequency.

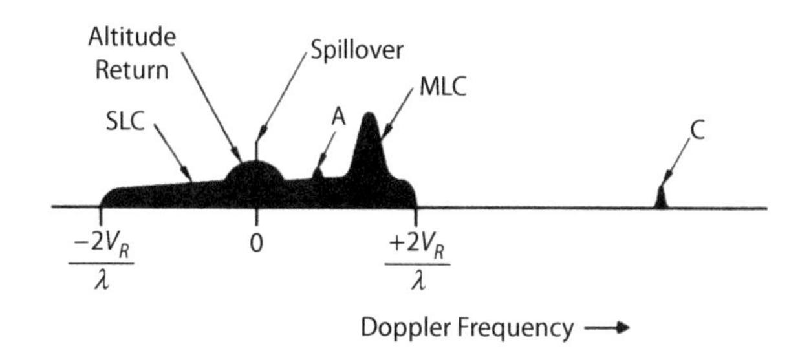

Figure 30-7. This depicts the central band of the Doppler profile.

Rejecting the Strong Clutter. Figure 30-7 suggests that a logical first step in isolating the target return is to reject the spillover and the strong ground return, the mainlobe clutter (MLC), and the altitude return.

Where little or no range discrimination is provided, this return may be as much as 60 dB stronger than a target's echoes (Fig. 30-8) and must therefore be removed. A power ratio of

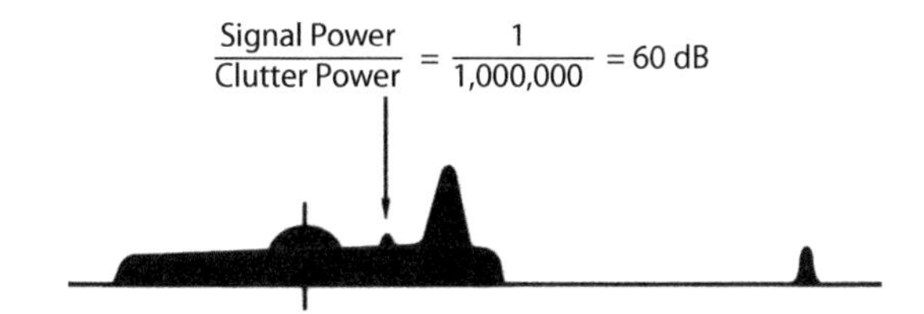

Figure 30-8. The power of the mainlobe clutter and the altitude return may be 60 dB stronger than that of a target's echoes.

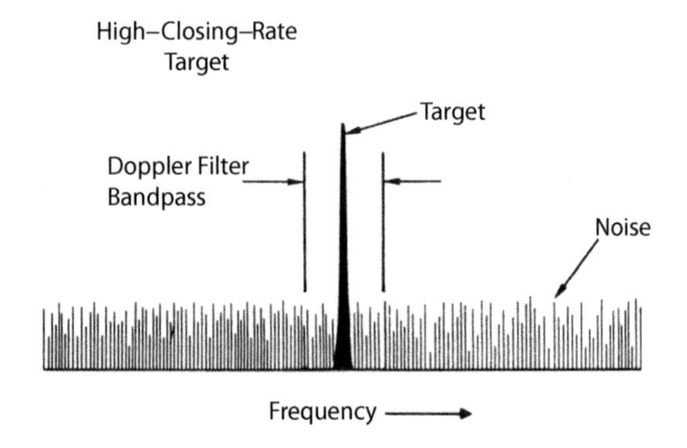

Figure 30-9. The Doppler filter isolates the high closing-rate target from all other returns and all but the immediately surrounding noise.

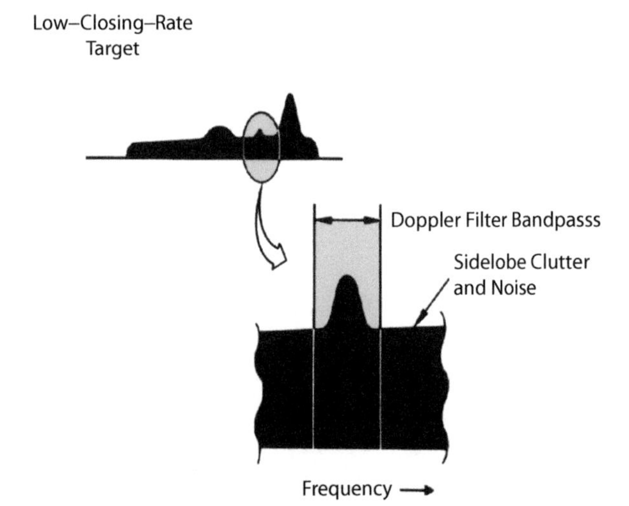

Figure 30-10. The low closing-rate target must compete with immediately surrounding sidelobe clutter, much of which may come from a far closer angle than that of the target.

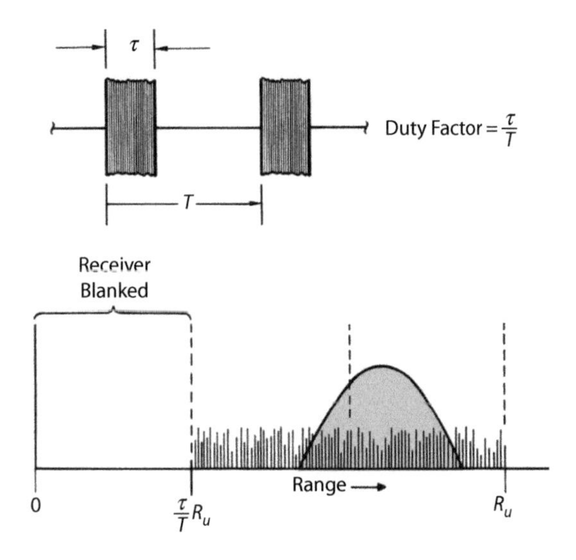

Figure 30-11. Providing more than one range gate may reduce the amount of noise or clutter with which a target must compete. This requires the duty factor to be less than 50 percent.

1 million to 1 is 60 dB. A Doppler filter bank alone simply cannot cope with such strong clutter. This is because, even though the clutter may be widely separated from a target's frequency, the attenuation that a Doppler filter provides outside its passband can be insufficient to keep the clutter from drowning out target echoes (that also fall outside the Doppler filter passband).

If targets are to be detected in both sidelobe clutter and Doppler clear regions, the spillover and altitude return (which have essentially zero Doppler frequency) must be rejected separately from the mainlobe clutter (which has a widely varying frequency).

Doppler Resolution. Once the strong clutter has been removed, the target echoes can be isolated through Doppler filtering. The role of the Doppler filters is slightly different for high relative velocity targets than for low.

In the case of high relative velocity targets (those having closing rates greater than the radar's ground speed), the Doppler filters serve three basic functions: (1) separating the target echoes from all of the remaining sidelobe clutter, including the residual mainlobe clutter; (2) reducing the amount of noise with which the echoes must compete by reducing the spectral width of the background noise accompanying the echoes of any one target (Fig. 30-9); and (3) isolating the echoes received from different targets (provided they have sufficiently different Doppler frequencies).

It is this noise that ultimately limits the maximum range at which high relative velocity targets can be detected. The more the noise is reduced, the greater the detection range will be.

In the case of low relative velocity targets, the Doppler filters perform the same target isolation function. However, they cannot completely separate a target's echoes from the sidelobe clutter (Fig. 30-10) because some of this clutter has the same Doppler frequency as the target and therefore limits the range at which low radial velocity targets may be detected.

Because of the more severe range ambiguities, however, the competing clutter is much stronger than that encountered under the same conditions in medium PRF operation. For this reason, when high PRFs are used in situations where appreciable sidelobe clutter is received, detection ranges against low relative velocity targets are degraded.

Range Gating. With duty factors approaching 50 percent, there is little or no possibility of isolating the return from different ranges with range gates.

However, if the duty factor is much less than 50 percent (i.e., if the interpulse period is much more than twice the pulse width), the opportunity for employing additional range gating arises (Fig. 30-11). By providing more than one range gate, the amount of noise or sidelobe clutter with which a target must compete may be reduced and the loss in signal-to-noise or signal-to-clutter ratio due to targets not being centered in the gate may be cut. The lower the duty factor, the greater the improvement that may be realized by adding range gates. If the average transmitted power is held constant by increasing the peak power, reducing the duty factor and gating in range allows the maximum detection range to be significantly increased. The reduction in duty factor cuts the eclipsing loss and the range gates reduce the competing noise or clutter.

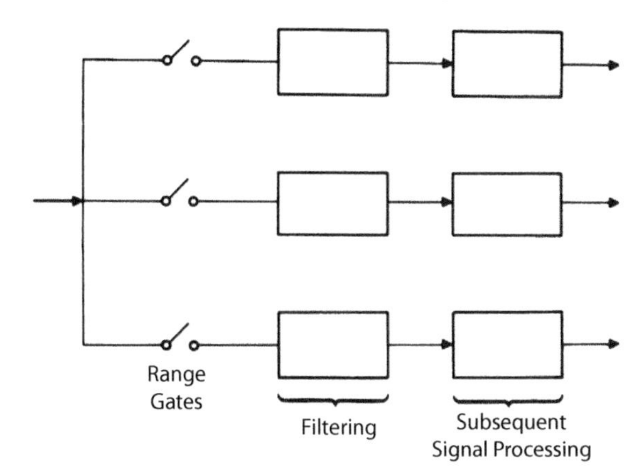

Figure 30-12. For every range gate that is provided, all filtering and subsequent signal processing must be duplicated.

Since range gating must precede Doppler filtering, the entire Doppler filter bank and all subsequent signal processing must be duplicated for every range gate provided (Fig. 30-12). Increasing the number of range gates from one to just two entails forming twice as many Doppler filters, detecting the magnitudes of twice as many filter outputs, setting twice as many detection thresholds, and so on. Moreover, where the value of the PRF is very high, a great many more Doppler filters are required to provide the same Doppler resolution than is the case for medium PRF. The cost of range gating, therefore, can be much higher in a high than in a medium PRF radar system.

30.3 Signal Processing

Just how these signal-processing functions are actually performed varies widely from one radar system to another. Digital implementation is now the norm. However, many systems still use analog processing, and sometimes a combination of analog and digital is advantageous. For example, to reduce the dynamic range required of the analog-to-digital (A/D) converter, the initial filtering may be analog and only the final Doppler filtering digital. In some cases the processor may be designed to look for targets in both sidelobe clutter and Doppler clear regions. In others, targets are searched for only in the Doppler-clear region.

Analog Implementation. In analog processors, the intermediate frequency (IF) output of the receiver is applied at the outset to a band-pass filter (Fig. 30-13) that passes only the central band of Doppler frequencies.[2]

2. The central band is selected because it contains the most power. Loss of signal power in the other bands is no problem since the noise and clutter they contain are rejected too.

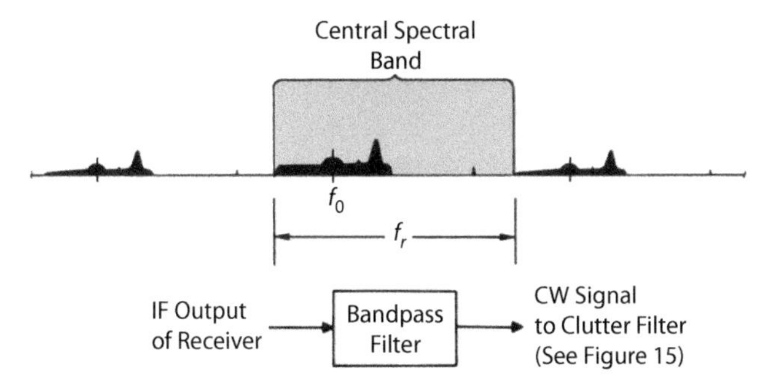

Figure 30-13. For analog processing, the IF output of the receiver is fed to a filter that passes only the central spectral band.

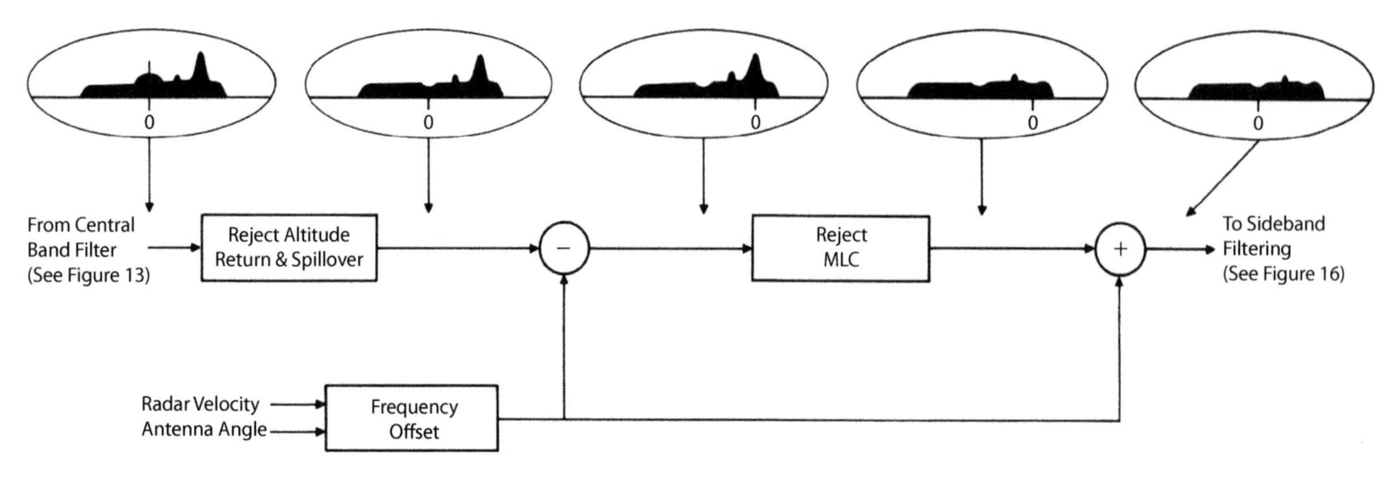

Figure 30-14. The first steps in signal processing are as follows: (1) reject the altitude return and the transmitter spillover; (2) offset the spectrum to track mainlobe clutter; (3) reject mainlobe clutter; (4) remove the frequency offset.

The output of the central band filter (Fig. 30-14) is passed through a second filter to remove the clutter with zero Doppler frequency (transmitter spillover and altitude return). Along with these, the echoes of any zero closing-rate targets will unavoidably be rejected.

The entire Doppler spectrum is then shifted in frequency to center the mainlobe clutter in the rejection notch of a second rejection filter. As with low and medium PRF operation, this shift must be dynamically controlled to match the changes in the clutter frequency due to changes in the radar velocity and antenna look angle. Once the mainlobe clutter has been removed, the same frequency offset may be applied in reverse to center the Doppler spectrum, again at a fixed frequency.

To keep the stronger return in the target regions from saturating subsequent stages, its relative amplitude is reduced through automatic gain control (AGC). For this, the Doppler spectrum is commonly broken into contiguous subbands (Fig. 30-15), one or more of which may span the sidelobe clutter region and the Doppler clear region. By applying the AGC separately in each subband, the strong return in one band is prevented from desensitizing the other bands.

Once signal levels are equalized, the subbands are recombined and applied to a single, long bank of Doppler filters (Fig. 30-16).

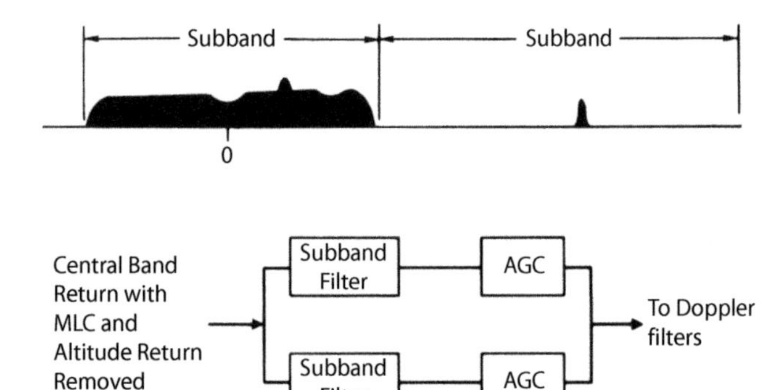

Figure 30-15. To avoid saturation by strong signals, the echo return is segregated into subbands prior and subjected to automatic gain control (AGC).

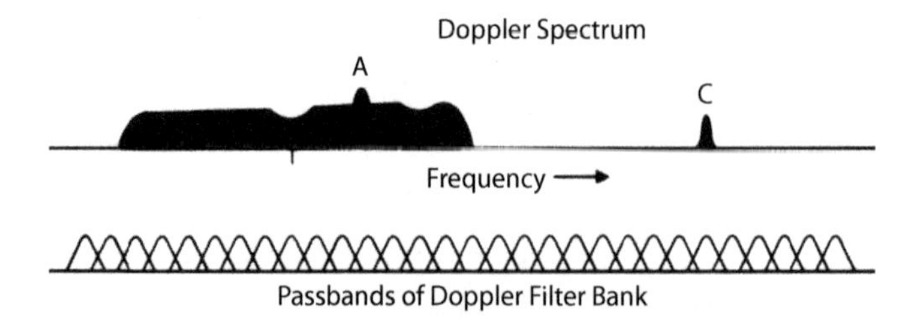

Figure 30-16. After the altitude return, transmitter spillover, and mainlobe clutter are filtered out, the radar return is applied to a bank of Doppler filters.

At the very end of every filter integration time, t_{int}, the amplitude of the signal that has built up in each filter is detected. If t_{int} is less than the time-on-target for the radar antenna, the detector outputs for the complete time-on-target are added up in post-detection integrators (PDIs). Finally, the integrated output of each Doppler filter is supplied to a separate threshold detector.

Digital Implementation. The IF output of the receiver is applied to an in-phase and quadrature (I/Q) detector (Fig. 30-17). Its outputs, which are video, are then sampled at intervals equal to the compressed pulse width by an A/D converter or is digitally demodulated using a Hilbert transform. To prevent saturation, digital automatic gain control (DAGC) is applied to amplifiers ahead of the converter.

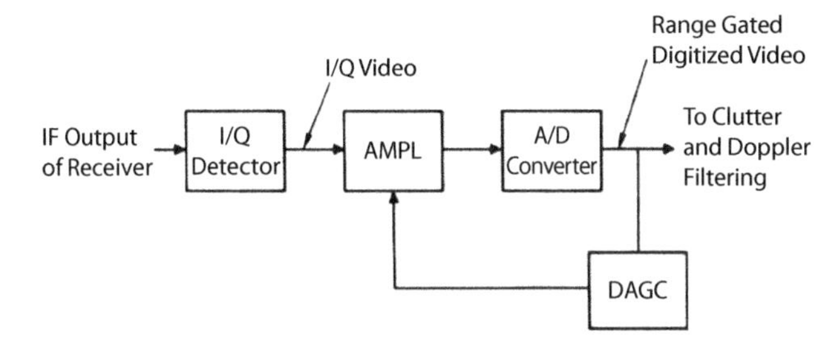

Figure 30-17. For digital processing, the output of the I and Q detector is sampled at intervals equal to the (compressed) pulse width and digitized. Digital AGC prevents saturation of the A/D converter.

In contrast to the output of an analog central band filter, the samples taken by the A/D converter represent the power in *all* of the Doppler bands passed by the receiver's IF amplifier. However, since the power includes signal as well as noise plus clutter, the signal-to-noise or signal-to-clutter ratio is essentially the same as when central band processing is employed.

Following A/D conversion, all of the same steps may be performed as in analog processing. This provides for greater design flexibility such as adaptive filtering.

30.4 Ranging

Because of the difficulty of pulse-delay ranging at high PRFs, frequency modulation (FM) ranging is usually employed. Its accuracy is proportional to the ratio of the frequency resolution of the Doppler filter bank and the rate of change of the transmitter frequency, $\dot{f}$ (Fig. 30-18). The finer the frequency resolution and the greater $\dot{f}$, the more accurately the ranging time can be measured. Frequency resolution roughly equals the 3 dB bandwidth of the Doppler filters, so

$$\text{Range accuracy} \approx \frac{BW_{3\text{dB}}}{\dot{f}}$$

To illustrate, let's say that the 3 dB bandwidth of the Doppler filters is 100 Hz and the rate of change of the transmitter frequency $\dot{f} = 3\,\text{MHz}$ per second. The accuracy with which the ranging time, t_r, can be measured then is $100 \div (3 \times 10^6) = 33\ \mu\text{s}$. At 12.4 μs per nautical mile of range, this corresponds to an accuracy of approximately 5 km.

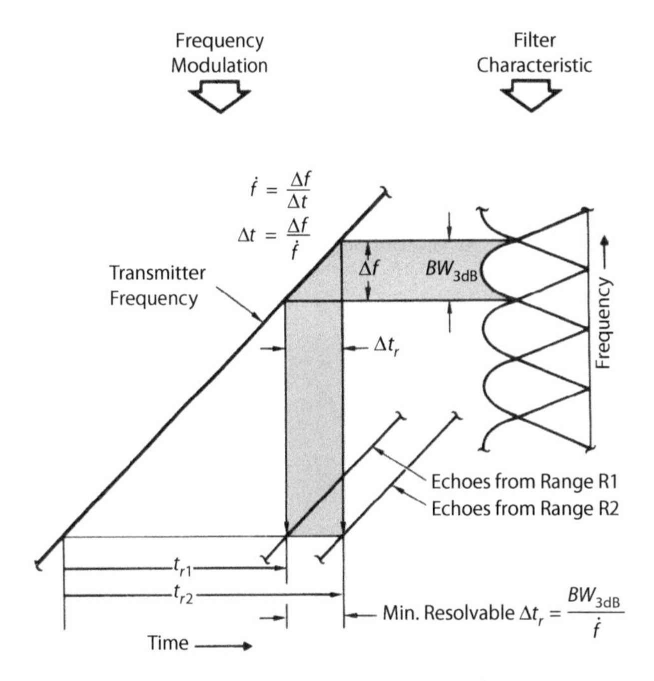

Figure 30-18. The slope of the modulation curve, $\dot{f}$, and the bandwidth of the filters, $BW_{3\text{dB}}$, determine the minimum resolvable difference in ranging time, Δt_r.

One might suppose that by simply narrowing the filters or increasing $\dot{f}$ virtually any degree of range accuracy is possible. However, there are practical limits on both.

The filter bandwidth is limited by the integration time ($BW_{3\text{dB}} \cong 1/t_{int}$). Furthermore, with three-slope modulation, the

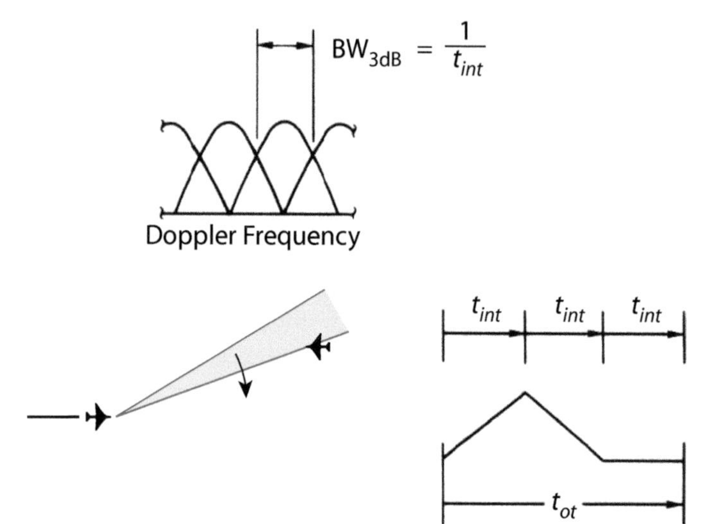

Figure 30-19. The filter bandwidth is inversely proportional to the integration time. With three-slope ranging, the integration time is less than 1/3 of the time-on-target.

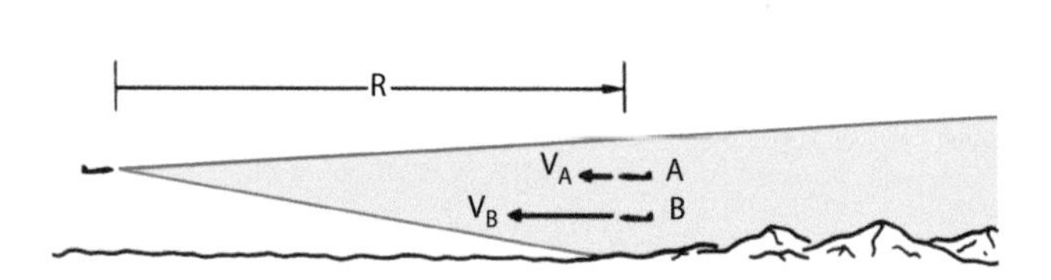

a) Two targets appproaching from long range. One has much higher closing rate than the other.

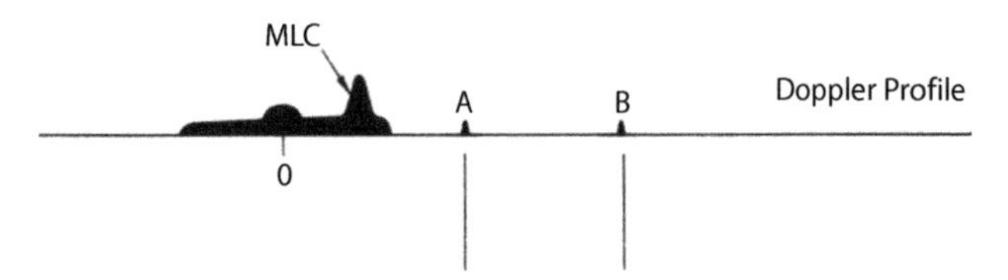

b) Doppler profile, without FM ranging–no clutter spreading.

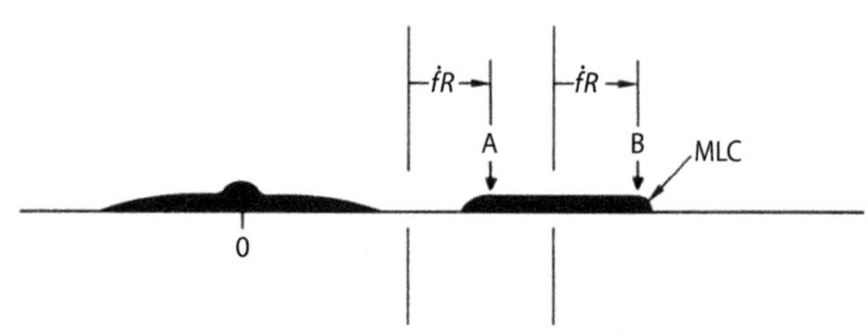

c) Doppler profile, with FM ranging and *large* value ḟ. Mainlobe clutter spread shifts out in frequency and spreads over Doppler clear region, obscuring targets.

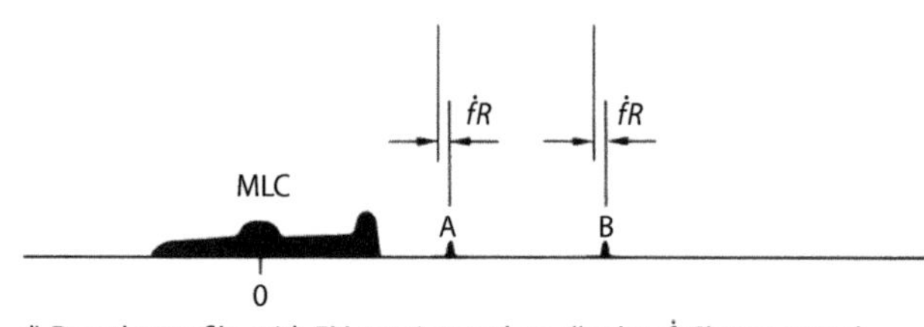

d) Doppler profile, with FM ranging and *small* value ḟ. Clutter spread only moderately, leaving targets in the clear.

Figure 30-20. Clutter spreading limits the rate, $\dot{f}$, at which the transmitter frequency may be changed for FM ranging. (a) Two targets approaching from long range. One has much higher closing rate than the other. (b) Doppler profile, without FM ranging, no clutter spreading. (c) Doppler profile, with FM ranging and *large* value $\dot{f}$. Mainlobe clutter spread shifts out in frequency and spreads over Doppler clear region, obscuring targets. (d) Doppler profile, with FM ranging and *small* value $\dot{f}$. Clutter spreads only moderately, leaving targets in the clear.

maximum integration time is less than 1/3 of the time-on-target (Fig. 30-19). The minimum possible filter bandwidth, therefore, is roughly $3/t_{ot}$ Hz.

Also, $\dot{f}$ is limited by the spreading of the clutter spectrum, which is due to the clutter being received from a wide span of ranges. To keep spreading within acceptable bounds, the maximum shift in the frequency of the radar return, caused by range, must be no more than a small fraction of the maximum Doppler shift of the clutter.

The reason is illustrated (for the descending slope of the modulation cycle) in Figure 30-20. It shows a radar system detecting two long-range targets: one has a high Doppler frequency; and the other has a Doppler frequency only slightly higher than the highest clutter frequency. The antenna's mainlobe illuminates the ground for a very great distance.

Beneath the situation diagram (a) are three frequency profiles. The first is a plot of the Doppler frequencies of the targets and the ground return (b).

What happens if the frequency shift corresponding to the targets' range is comparable to the Doppler shift of the higher closing rate target? As can be seen in (c), the mainlobe clutter not only shifts into the normally clutter-free region but also spreads to the point where it blankets both targets.

In (d), the frequency shift corresponding to range is a small fraction of the Doppler shift. Although the clutter still spreads, it does not spread enough to interfere with target detection.

For a typical fighter application the constraints on minimum filter bandwidth and maximum rate of change of transmitter frequency are such that the range accuracy is on the order of a few km. This is substantially poorer than can be obtained with pulse-delay ranging.

While range information is always highly desirable, it is not essential when searching for targets at extremely long ranges. What is most important, however, is to detect targets and to know their direction. Determining whether a target in a given direction is 200 or 300 km away can come later.

Because a target must be detected on all three slopes of the modulation cycle and the integration time per slope is only one-third of what it would be without FM ranging, the price paid for range measurement is a reduction in detection range.

Accordingly, for situations where extremely long detection ranges are desired, a special mode may be provided in which range is simply not measured. In this mode, called *velocity search* or *pulse-Doppler search*, targets are displayed in range rate versus azimuth (Fig. 30-21). Once a target is detected, the operator can switch to the more standard range-while-search mode, in which range is measured by FM ranging and the targets are presented on a range-versus-azimuth display.

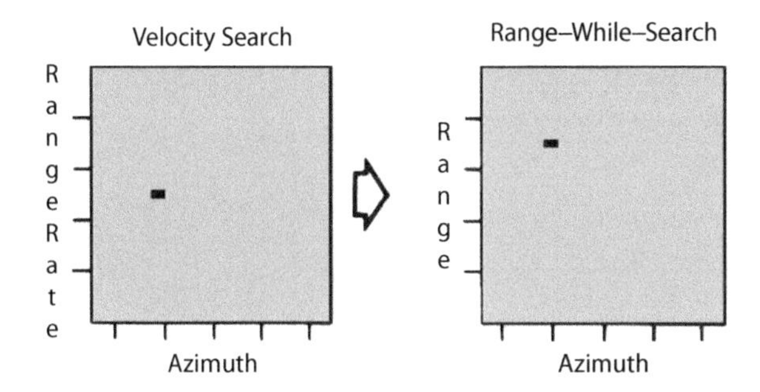

Figure 30-21. For the maximum detection range, a velocity-search mode may be provided. Once the target is detected, the operator may switch to the range-while-search mode.

30.5 Problem of Eclipsing

When operating at very high duty factors, a considerable amount of target return is lost as a result of eclipsing, that is, when echoes are received in part or in whole when the radar is transmitting and the receiver is blanked or switched off.

Eclipsing, however, is not always as severe a problem as it might at first seem. A target is totally eclipsed only when its range is such that its echoes are received during the period when the receiver is blanked. Otherwise, at least a portion of the return gets through (Fig. 30-22). As the degree of coincidence is reduced, so is the eclipsing loss.

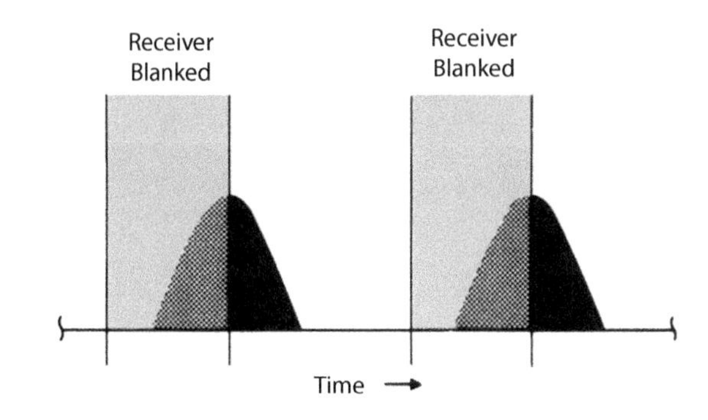

Figure 30-22. As long as the received echoes and periods of receiver "blanking" do not coincide exactly, a portion of the return will get through.

Even so, eclipsing reduces the signal-to-noise ratio sufficiently to leave periodic holes of appreciable size in the radar's range coverage. Fortunately, when searching for targets approaching from very long ranges we are concerned mainly with the cumulative probability of detection (the probability that a target will be seen at least once before it has approached to within a given range). Moreover, once a target has been detected, it generally does not need to be detected continuously because a rapidly approaching target will not remain at an eclipsed range for very long. As the range decreases and the signal strength increases, the gaps in range coverage tend to disappear (Fig. 30-23).

In applications where relative velocities may be comparatively low, nearly continuous detection is required. Switching the PRF among different values may reduce the length of time that any one range remains eclipsed. In single-target tracking, by periodically changing the PRF at appropriate times a target can

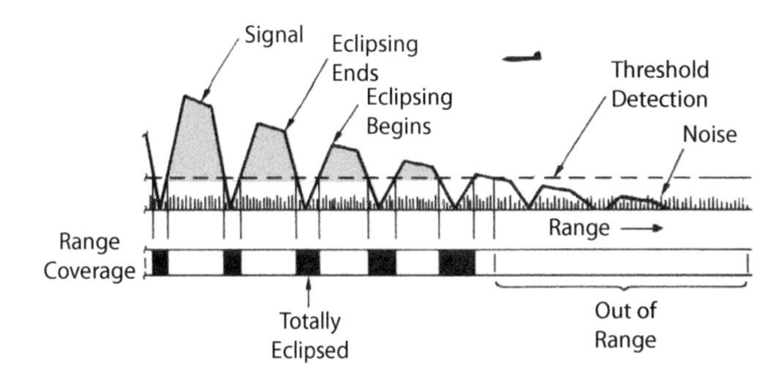

Figure 30-23. A reduction in signal-to-noise ratio occurs due to eclipsing for an approaching target. As range decreases, the gaps in range coverage grow narrower.

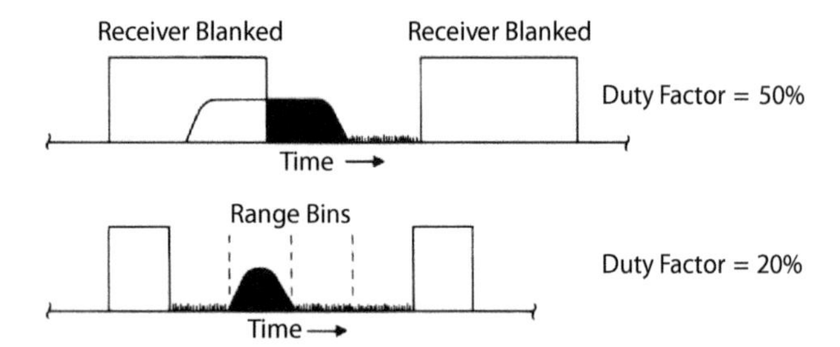

Figure 30-24. Reducing the duty factor and employing multiple range gates may reduce eclipsing loss.

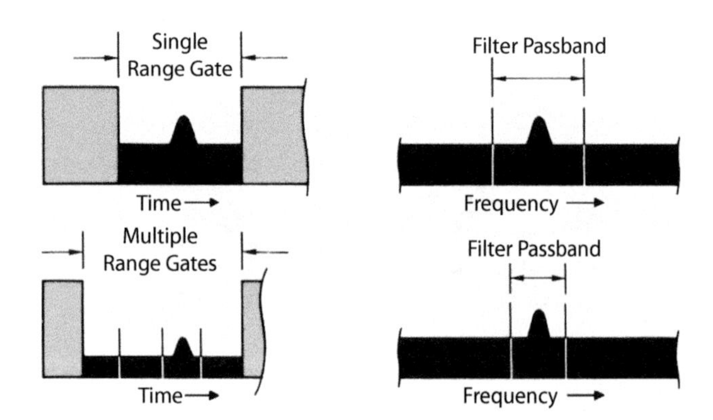

Figure 30-25. The signal-to-clutter ratio—and hence performance against low closing-rate targets—may be improved by reducing the duty factor and providing multiple range gates or by narrowing the passband of the Doppler filters.

be kept largely in the clear. At high duty factors, though, the holes in range coverage are not easily eliminated, particularly at short ranges. Furthermore, PRF switching introduces losses that reduce the maximum detection range in the Doppler clear region.

Lowering the duty factor may also reduce eclipsing, but this will also reduce the average transmitted power. However, using multiple range gates can compensate for this reduction. Suppose, for example, that the duty factor is reduced from 50 to 20 percent and that four range gates are provided (Fig. 30-24). If the peak transmitted power remained the same, the average power and hence the total received energy would be decreased by a factor of $0.2 \div 0.5 = 0.4$. But, as can be seen from the figure, with four range gates the noise energy with which the signal would have to compete at any one time would be reduced by the same factor; thus, these two effects would cancel. For a continuously closing target the signal-to-noise ratio would increase in direct proportion to the increase in the fraction of the time the receiver is not blanked. In this case, the increase would be on the order of $0.5 \div 0.2 = 2.5$.

Thus, by reducing the duty factor and providing multiple range gates, the detection range will be increased, and also the holes in range coverage due to eclipsing may be correspondingly narrowed. As noted earlier, providing multiple range gates substantially increases the complexity and therefore also the cost of implementation.

30.6 Improving Tail Aspect Performance

Several approaches may be taken to improve performance against low relative velocity tail aspect targets in severe clutter. Since the root of the problem is sidelobe clutter, a logical first step is to minimize the antenna sidelobes.

For a given sidelobe level, the amount of sidelobe return with which a low relative velocity target must compete may be further reduced by narrowing the Doppler filter passbands (Fig. 30-25). This, of course, entails adding more filters, and there are practical limits on how much the passbands can be narrowed.

At the expense of still greater complexity and a lower duty factor, the competing return may be further reduced by narrowing the pulses and employing more range gates.[3] Even then, because of the transmitter spillover and altitude return the radar will be blind to zero closing-rate targets (those being pursued at constant range).

A particularly attractive solution to the problem is to employ high PRFs when long detection range against forward-located targets is essential and to interleave high and medium PRFs when long detection ranges are required against both forward- and tail-located targets.

An effective way of accomplishing this is illustrated in Figure 30-26. High and medium PRF modes are employed on alternate

3. For a given duty factor, and hence degree of eclipsing, the number of range gates may be increased still further, without loss of signal, through pulse compression.

bars of the antenna scan pattern. The bars assigned to the high PRF mode in one frame are assigned to the medium PRF mode in the next frame and vice versa. Since adjacent bars overlap, virtually complete solid-angle coverage is achieved in both modes. Rapidly approaching targets, beyond the reach of the medium PRF mode, are detected in the high PRF mode. Low relative velocity targets, as well as any shorter-range targets that may be eclipsed in the high PRF mode, are detected in the medium PRF mode.

In the high PRF mode, when PRF interleaving is used the complexity of the signal processor can be substantially reduced by processing only the return that falls in the Doppler clear region. This return, of course, must first be isolated from the clutter: mainlobe, sidelobe, and altitude return. This can readily be accomplished by passing the receiver output through one or more broad band-pass filters. After performing automatic gain control, their outputs are supplied to a suitably long bank of Doppler filters (Fig. 30-27).

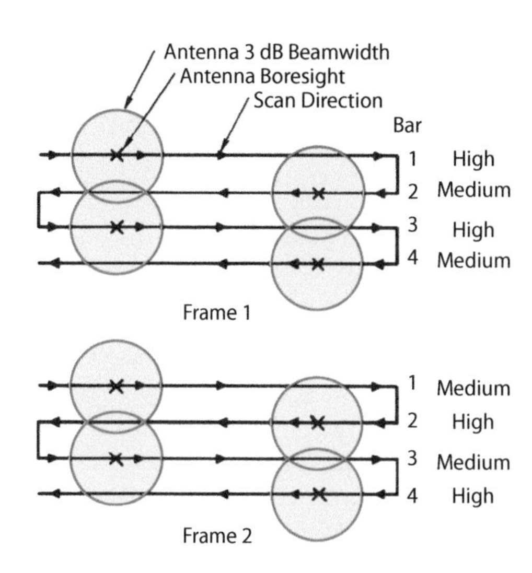

Figure 30-26. Interleaving of high and medium PRFs on alternate bars of the search scan helps to achieve maximum detection range in both forward and tail aspects.

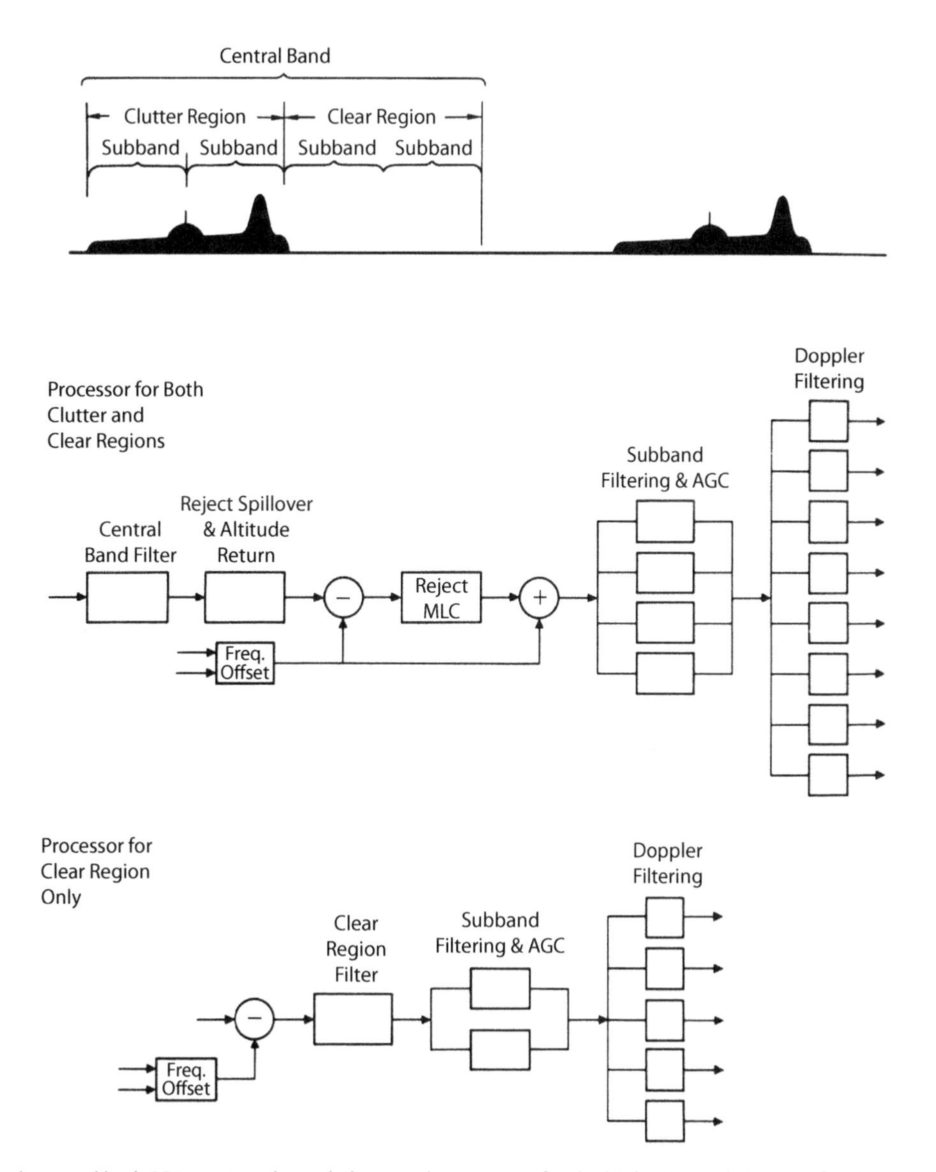

Figure 30-27. When medium and high PRFs are interleaved, the signal processing for the high PRF mode is simplified by processing only the return in the clutter-free region.

Illuminating Targets for Semiactive Missile Guidance

If the pulsed transmission of a high PRF radar is used to illuminate targets for semiactive missiles to home in on, both the PRF and the duty factor may be made somewhat higher than they would be for normal searching and tracking.

Increased PRF. Because of the high velocity of the missile relative to the launch aircraft, a target's Doppler frequency is generally much higher as seen by the missile than as seen by the radar in the launch aircraft. To ensure that the minimum acceptable PRF results in unambiguous velocity data obtained by the missile, it is necessary to add half the velocity of the missile relative to the radar to the maximum target relative velocity.

Increased Duty Factor. A semiactive missile must be launched at long ranges. Since for a given peak transmitted power detection range increases with duty factor, it is desirable to make the duty factor as high as possible. However, the maximum useful duty factor that a radar system may employ for detecting and tracking targets is limited by the eclipsing loss to somewhat less than 50 percent. When the radar's pulsed transmission is used to illuminate a target for a missile, this limitation does not necessarily hold. Because the missile is remote from the radar, blanking is not needed to keep transmitter noise from leaking directly into the missile receiver when the radar is transmitting. Consequently, if the missile's seeker must be capable of long detection range, the duty factor may be made considerably higher than 50 percent. The maximum acceptable duty factor is, of course, limited by eclipsing losses in the radar (this is a feature of bistatic operation; see Chapter 43).

What about the directly received signal from the radar? Because of the Doppler shift, the frequency of the radar's transmitted signal is sufficiently different from the frequency of the target echoes that the seeker in the missile can separate the two. However, care must be taken in the design of the radar to minimize the radiation of transmitter noise, some of which invariably has the same frequency as a target's echoes.

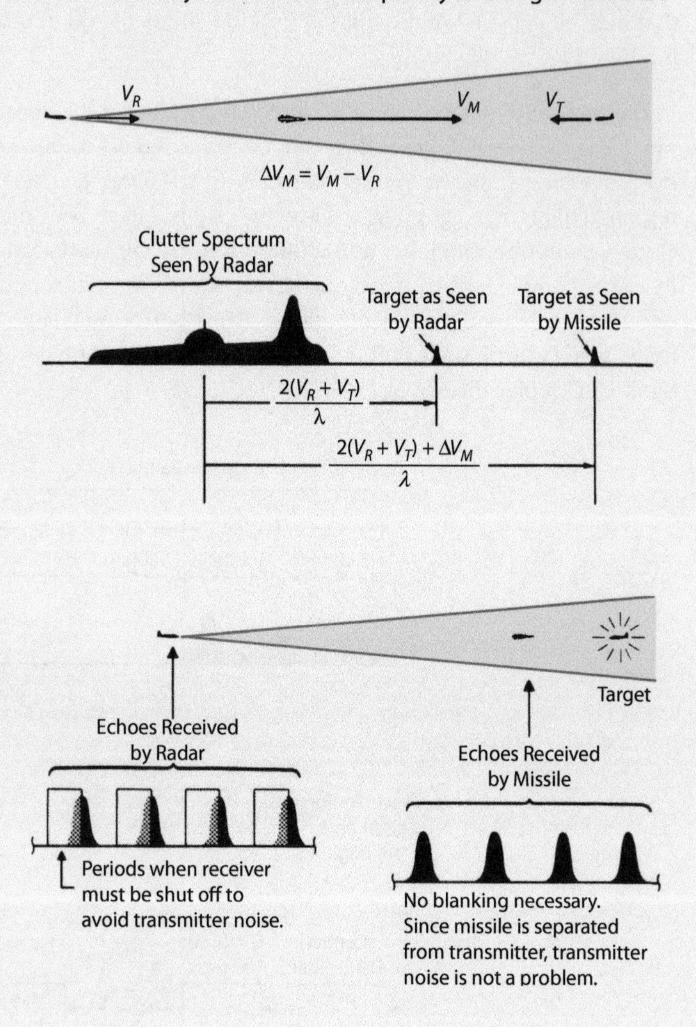

30.7 Summary

To maximize detection range, high PRF waveforms for fighter applications usually have duty factors approaching 50 percent. Even higher duty factors may be used when illuminating targets for long range semiactive missiles.

To provide an adequate Doppler clear region, the PRF must at least equal the maximum sidelobe clutter frequency plus the Doppler frequency of the highest closing rate target. For semiactive missile guidance an allowance must be added for the velocity of the missile relative to the radar.

Successive filters are used to reject the combined spillover, as well as altitude return, and mainlobe clutter. The remaining return may then be divided into subbands for the application of AGC. The return is then applied to a bank of Doppler filters. If a target's relative velocity is greater than the radar's ground speed, the target echoes compete only with the noise passed by the same Doppler filter that passes the echoes. But if the relative velocity is less than this, they must compete with sidelobe clutter passed by the filter, much of which may come from comparatively close range.

In many high duty factor radars, the only range gating is that provided by receiver blanking. At the cost of increased complexity, using multiple range gates may reduce noise and sidelobe clutter.

Range must generally be measured with FM ranging. Because the rate at which the transmitter frequency may be changed is limited by the spreading of the clutter spectrum, range accuracy is poor. Since range measurement reduces detection range, when maximum detection range is desired, a special velocity search mode may be provided in which range is not measured.

When operating at high duty factors, eclipsing losses are significant. Switching PRFs or providing multiple range gates may minimize them. PRF switching, though, reduces detection range in the Doppler clear region, and employing multiple range gates increases the cost of implementation.

Performance against low closing-rate targets may be improved by providing a low sidelobe antenna, greater Doppler resolution, and multiple range gates. One attractive approach is to interleave high and medium PRF operation on alternate bars of the antenna scan.

Further Reading

G. V. Morris and L. Harkness, *Airborne Pulse Doppler Radar*, Artech House, 1996.

C. M. Alabaster, *Pulse Doppler Radar: Principles, Technology, and Applications*, SciTech-IET, 2012.

Test your understanding

1. What is the effect of a high duty cycle in a high PRF radar system?
2. In a high PRF radar system, what factors govern the choice of the PRF?
3. In a high PRF radar system, what determines Doppler resolution?
4. How is ranging accomplished in a high PRF radar system?
5. In a semiactive missile seeker how is the signal received directly from the illuminating radar separated from that of echoes from genuine targets?

British Aerospace Sea Harrier (1993)

The Harrier was developed from the radical P-1127 Kestral VSTOL (Vertical/Short Takeoff and Landing) prototype in the late 1950s and early 1960s and saw service with the Royal Air Force and the Royal Navy. The USMC version, built by McDonnell Douglas, was designated AV-8B and was equipped with the APG-73 radar.

31

Automatic Tracking

Conical scanning antenna feed on SCR-720 radar, in nose of P-61 aircraft

This chapter examines techniques for tracking detected targets. Tracking is achieved using both the radar hardware and the radar signal processing, which results in a closed-loop system. *Single-target tracking* (STT) and *track-while-scan* (TWS) modes (introduced in Chapter 2) are examined. Before we look at tracking measurements and methods, we need to define some terminology.

Estimate, accuracy, and *precision* are routinely used to describe different aspects of tracking. *Estimate* is applied to the value of any parameter that is (1) measurable only in combination with corrupting interference, such as thermal noise (Fig. 31-1); and (2) not directly measurable, such as range-rate based on a sequence of range measurements.

According to this definition, every parameter measured or computed by a radar system, no matter how precisely, is an estimate.

Next, two important parameters are distinguished: *accuracy* and *precision*. In general, both refer to the measurement of a quantity, which in tracking includes the target parameters such as true range, velocity, and bearing. Thus, the measurements represent an estimate of a target's true parameters as made by the radar system.

Accuracy indicates how close a measurement is to the true value, whereas precision describes how much variability there is in a number of measurements of the same parameter. Together they form the basis of the estimate, as made by a radar system, of true target parameters. Figure 31-2 shows an example where accuracy and precision can be seen as quite different and (sometimes) independent of one another. The goal in tracking radar is to have both high accuracy and high precision.

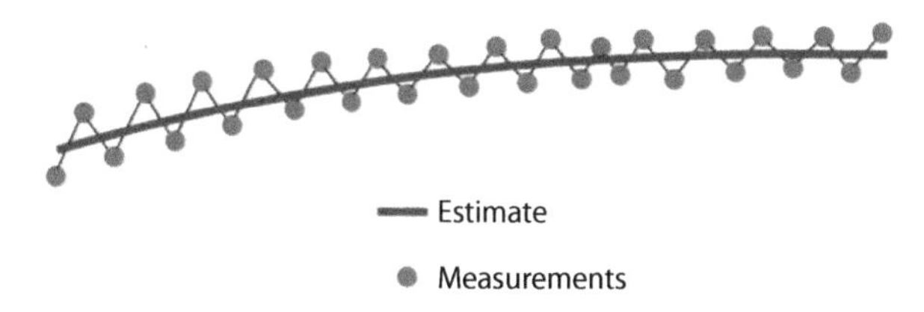

Figure 31-1. The value of any measured parameter is termed an estimate.

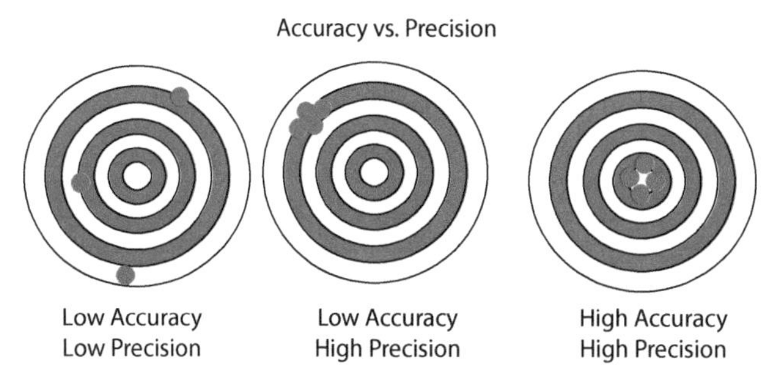

Figure 31-2. Accuracy and precision describe two aspects of a set of measurements.

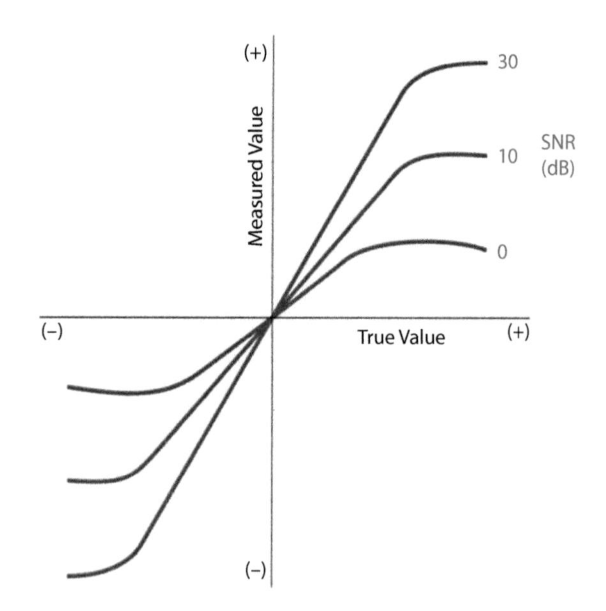

Figure 31-3. A tracking discriminant may be represented by a normalized plot of the measured value of the tracking error versus the true value. The steeper the linear portion of the discriminant, the more sensitive the measurement.

Another term used in tracking is *discriminant,* which quantifies calibration of the measurement function. It is commonly represented by a plot of the output of the hardware or software that performs the measurement versus the true value of the tracking error (Fig. 31-3). The slope of the linear portion of the plot is the discriminant and determines the sensitivity of the measurement. Typically, the slope increases as the signal-to-noise ratio increases.

An important feature of discriminants is that they are typically normalized and therefore dimensionless. Consequently, precise measurement of voltage or power levels is not necessarily required. Moreover, except for the influence of signal-to-noise ratio, the measured values of the tracking error do not vary with signal strength. They are independent of the target's size, range, maneuvers, and radar cross section (RCS) fluctuations. If desired, a discriminant can be given a dimension by multiplying it by a precomputed constant. Discriminants are used throughout tracking and have the aim of improving the estimate of a targets' measured parameters, such as range, Doppler, elevation, and azimuth angle.

31.1 Single-Target Tracking

Single-target tracking provides continuous and accurate current data about a target's position, velocity, and acceleration, all of which may be continuously changing. To achieve this, separate semi-independent tracking loops are typically established for range, range-rate (Doppler frequency), and angle.

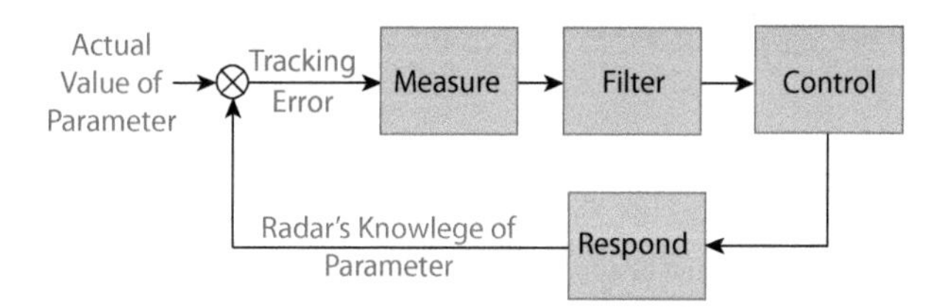

Figure 31-4. Basic functions performed by a single-target tracking loop provide feedback.

Functions Included in a Tracking Loop. The tracking loop can be divided into four basic functions: *measurement, filtering, control,* and *response* (Fig. 31-4).

Measurement is the determination of the difference between the latest value of the parameter (e.g., the target's range) and the radar's current knowledge of the parameter. This is the tracking error.

Filtering processes successive measurements to minimize random variations (noise) due to target scintillation, thermal agitation, and other corrupting sources of interference. Tracking accuracy depends critically on how effectively filtering is done. A tracking filter may be thought of as a low-pass filter whose key parameters are cut-off frequency and gain. These constraints are constantly adjusted in light of the signal to noise ratio, the target's potential maneuvers, and the radar-bearing aircraft's actual maneuvers to eliminate as much noise as possible without introducing excessive time lag (especially during a maneuver).

Control is generation of a command calculated from the filter's outputs to reduce the tracking error (as close as possible to zero).

Response is the action of the hardware or software to which the command is given. The difference between the response and the current actual value of the parameter feeds back to the input, closing the loop so that the entire process repeats. Through successive iterations, the parameter may be tracked with very high precision.

Improving Range Estimation. The radar estimate of range to a single target can be improved by using a technique known as *early-gate, late-gate*. A range bin is divided into two parts (or gates) where one is shifted by half a range bin with respect to the other. A target may therefore appear in the two gates simultaneously, as shown in Figure 31-5.

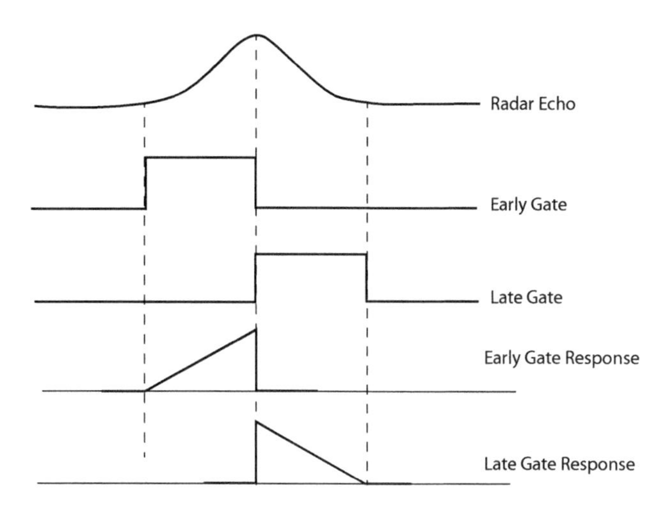

Figure 31-5. The early-gate, late-gate technique improves the estimation of target range. In this example the target is at the center of the range resolution bin as indicated by the equal values in the early and late gates.

In this example, the target is located in the center of a range bin, and hence its response is equally divided between the early and late gates. If more of the target echo were located in the early gate than in the late gate, the voltage measured in the early gate would be greater. This is termed the *range discriminant* (Fig. 31-6). Thus, by measuring the voltage difference between the responses in the early and late gates, the target position is more accurately determined with the accuracy better than implied by range resolution.

Evaluation of the range discriminant is performed in the range-tracking loop (Fig. 31-4).

Range-Tracking Loop. The range-tracking loop measures the target's current range and keeps a range bin centered on the target's echoes (isolating the target for Doppler and angle tracking). The tracking error, e, is proportional to the difference between the early and late samples, R_E and R_L. To keep the range bin centered on the target echoes, the range discriminant is formed by measuring the difference between the amplitudes of the two range gates, $R_L - R_E$. The measurement is normalized by dividing it by the sum of the amplitudes. Subsequently, the sampling times must be shifted to center the range gate on the target's echoes as a function of the difference between R_L and R_E.

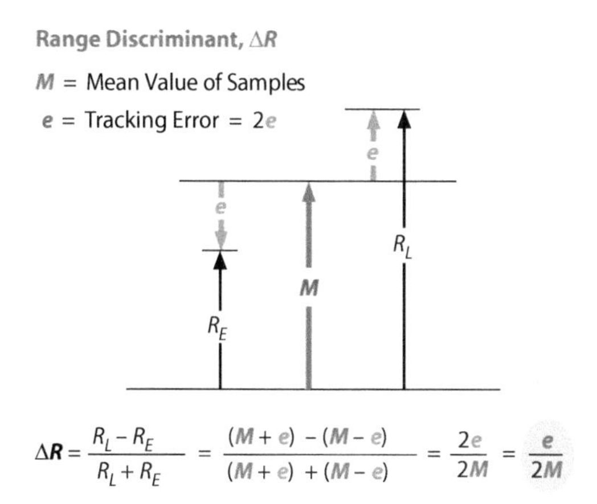

Figure 31-6. The range-tracking error is proportional to the difference between the magnitudes of the samples stored in the early and late range bins. Dividing by their sum yields a nondimensional ratio of the error, the range discriminant.

On the basis of the range discriminant and the previous range-bin command, the range filter produces best estimates of the target's range and range-rate, a measure of the range acceleration, and a new range-gate command (Fig. 31-7).

The range-gate command is a prediction of what the target's range will be when the next target echo is sampled. This is computed by taking the filter's latest estimates of the target's range and range-rate, and extrapolating them to calculate the new range.

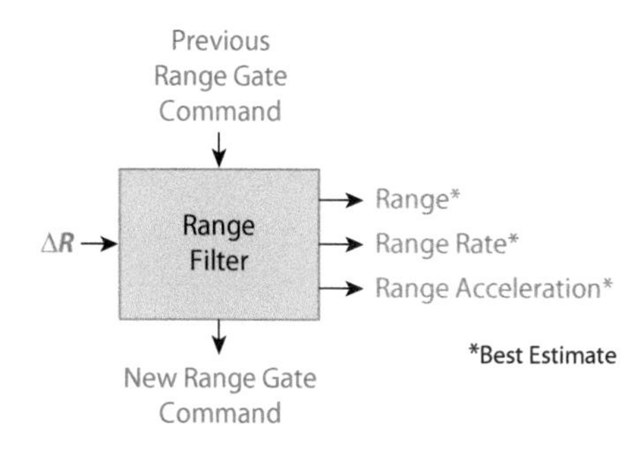

Figure 31-7. In these inputs and outputs of the range filter, ΔR is the range discriminant.

To carry out the range-gate command, the predicted target range is first corrected for radar peculiarities (e.g., sampling-time granularity) and distortion of the pulse shape in going through the receiver and the pulse-stretching low-pass filter. The prediction is then converted into units of time measured from the trailing (or leading) edge of the immediately

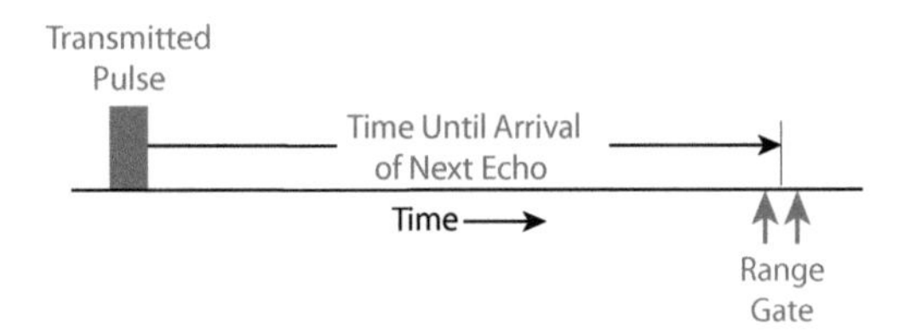

Figure 31-8. To position the range-gate in response to the range-gate command, the predicted range is converted to time.

1. Two separate banks of filters are formed by integrating the samples collected in the early and late range bins. The velocity gate may be formed in either or both of them.

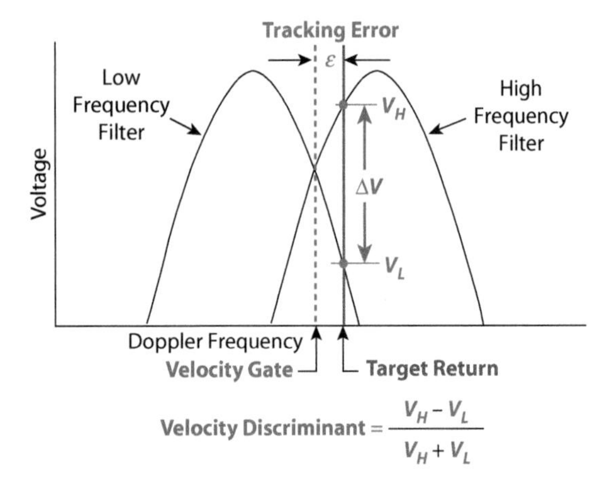

Figure 31-9. The simplest velocity gate is the intersection of two adjacent Doppler filters. The velocity discriminant is the difference between the output voltage the target return produces from the two filters divided by the sum of the two voltages.

2. If the pulse repetition frequency (PRF) is less than the target's Doppler frequency, some multiple, *n*, of the PRF must be added to this sum.

preceding transmitted pulse and hence to the estimated arrival time of the next echo (Fig. 31-8).

Improving the Doppler Estimation. The estimate of a target's Doppler is improved in a manner conceptually very similar to improving range location as just described. Two velocity (Doppler) gates are used in place of the two range gates to generate the improved Doppler estimate.

The simplest approach examines the crossover point of two adjacent Doppler filters,[1] called the *low-frequency* and *high-frequency* filters (the Doppler equivalent to early and late range-gates). Any error in the alignment of the velocity gate shows up as a difference between the outputs of these filters. The Doppler or velocity discriminant is formed by taking the difference between the magnitudes of the outputs, $V_H - V_L$, and normalizing it by dividing by their sum (Fig. 31-9). The result is supplied to the velocity filter.

The functions of the Doppler or velocity filter almost exactly parallel those of the range filter. The velocity filter's outputs are simply more accurate estimates of the target's range-rate and range acceleration.

Doppler (Range-Rate) Tracking Loop. This loop isolates the target's returns for angle tracking by keeping a "velocity gate" centered on the target's Doppler frequency.

Based on the velocity filter's most current range-rate and range acceleration estimates, a velocity-gate command is produced. It predicts what the target's Doppler frequency will be when the next set of Doppler filters is formed.

The command is applied to a variable radio frequency (RF) oscillator. Its output is mixed with the received signal, thereby shifting the received signal's frequency such that the target's predicted Doppler frequency will be centered in the velocity gate. The sum of the oscillator's frequency and the velocity gate's fixed frequency is the target's predicted Doppler frequency (Fig. 31-10).[2]

Improving the Angle Estimation. Chapter 1 introduced three techniques for improving the bearing angle estimate of target position: sequential lobing; amplitude-comparison monopulse; and phase-comparison monopulse. Only amplitude-comparison monopulse will be considered here in

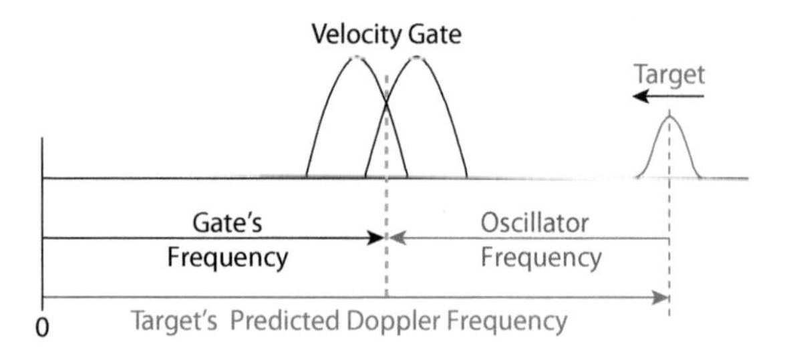

Figure 31-10. When the oscillator has moved the target into the gate, the sum of the oscillator's frequency and the velocity gate's fixed frequency is the target's predicted Doppler frequency.

Common Coordinate Systems

MEASUREMENTS OF DISTANCES AND ANGLES MAKE SENSE ONLY IF REFERENCED to a coordinate system.

Several common systems are shown here.

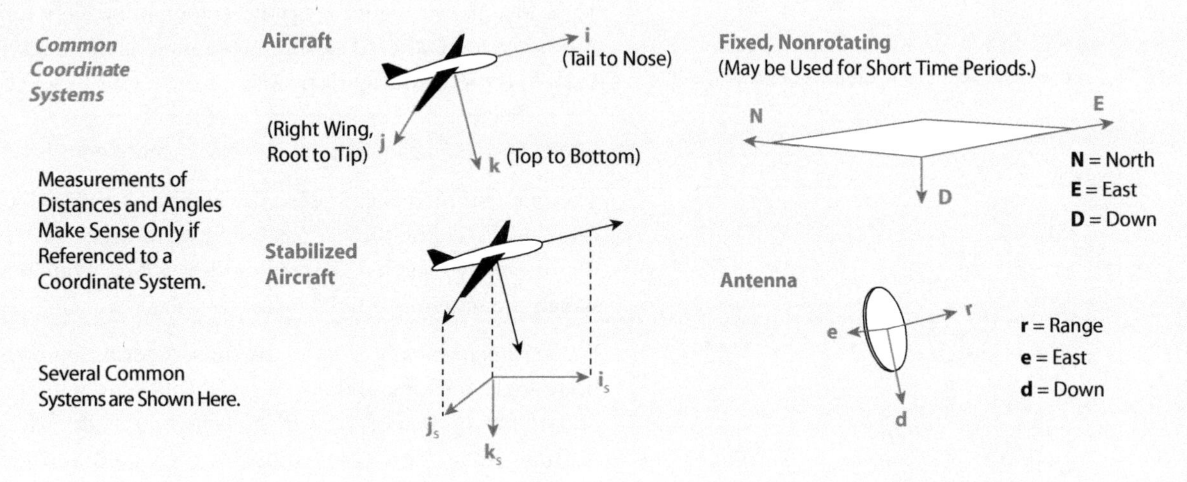

relation to tracking. In this technique, during reception the antenna's radiation pattern is split into two lobes that cross at their half power points, as illustrated in Figure 31-11. Again, this is similar in concept to that of early–late range-gate tracking and Doppler filtering.

In Figure 31-12, the difference between the amplitude of the target's echoes as received through the left and right lobes, $V_L - V_R$, is roughly proportional to the angular difference between the antennas line of boresight (i.e., its pointing angle) and the angle at which the target lies. Dividing this difference by the sum of the two amplitudes yields a dimensionless discriminant for the azimuth component whose value is directly related to this angular difference. A discriminant for the elevation component is similarly formed.

Angle-Tracking Loop. This loop keeps the antenna boresight precisely trained on the target. Commonly used coordinate systems are defined in the blue panel.

The angle-tracking loop measures the angle between the antenna boresight and the line of sight to the target. This angle, ε, is called the *angle off boresight* (AOB) and is generally resolved into azimuth and elevation coordinates (Fig. 31-12).

The measured components of the target angle with respect to the pointing angle of the antenna are supplied to the angle-tracking filter along with the following environmental information:

- Signal-to-noise ratio
- Radar-bearing aircraft's velocity

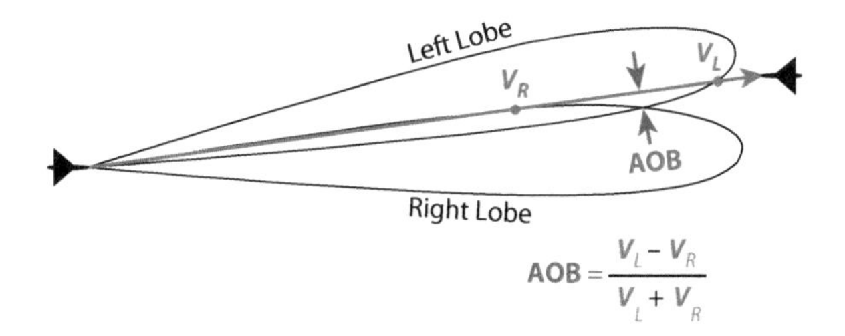

Figure 31-11. Pictured here is the angle-tracking discriminant for amplitude-comparison monopulse. The antenna lobes cross on the boresight line; so the angle off boresight (AOB) is roughly proportional to the difference between the voltage of returns received through the two lobes.

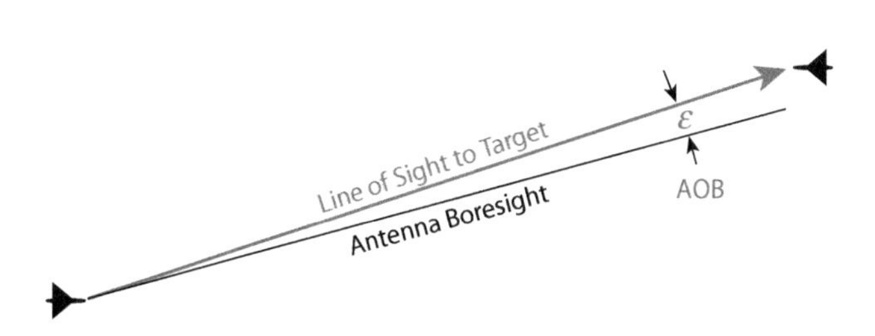

Figure 31-12. The angle-tracking loop measures the angle between the line of sight to the target and the antenna boresight line.

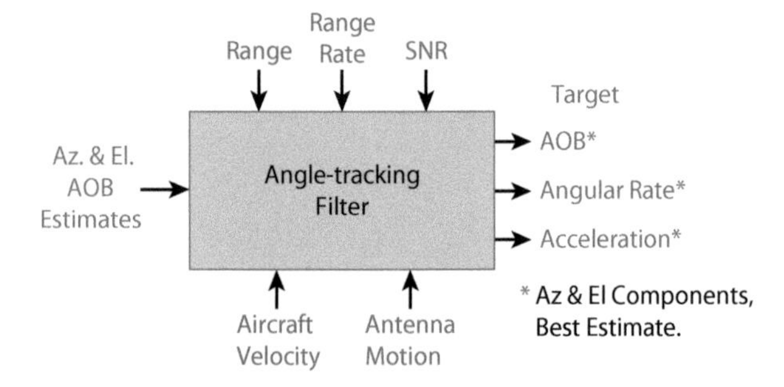

Figure 31-13. Pictured here are inputs and outputs of the angle-tracking filter.

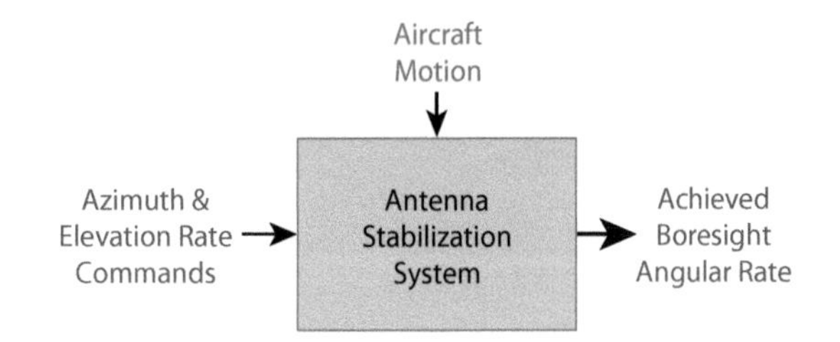

Figure 31-14. The antenna is stabilized against changes in aircraft attitude by slaving it to azimuth and elevation axes established by rate-integrating gyros mounted on it. The rate commands precess the gyros.

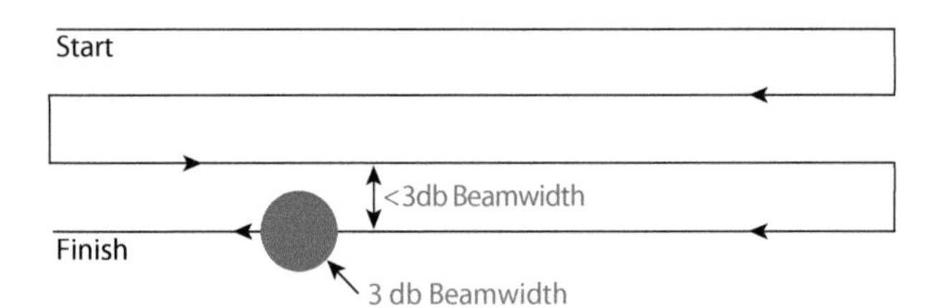

Figure 31-15. In this representative four-bar raster scan, to prevent targets from being missed, the spacing of bars is less than the 3 dB beamwidth. Consequently, the same target may often be detected on more than one bar—one of several conflicts TWS resolves.

- Target range and range-rate
- Antenna's current angle rate

From these inputs, the filter produces best estimates of the azimuth and elevation components of the angular difference, the angle rate of the line of sight to the target, and the target's acceleration (Fig. 31-13).

To reduce the AOB and keep the antenna boresight trained on the target, azimuth and elevation rate commands are generated. Each of these is the algebraic sum of the filter's best estimate of the respective line of sight rate and a rate proportional to the filter's best estimate of the respective component of the AOB.

The rate commands are fed to the antenna stabilization system (Fig. 31-14). There, they control the rate of precession of gyros that inertially establish azimuth and elevation axes in space to which the antenna is tightly slaved.

For an electronically steered antenna, steering commands for both angle tracking and space stabilization must be provided. To correct continuously for changes in aircraft attitude, no matter how small, new commands are computed and fed to the antenna at a very high rate.

31.2 Track-While-Scan

Track-while-scan is an elegant combination of searching and tracking. To search for targets, the radar repeatedly scans a raster of one or more bars (Fig. 31-15). Each scan is independent of all the others. Whenever a target is detected, the radar typically provides both the operator and the TWS function with estimates of the target's range, range-rate (Doppler), azimuth angle, and elevation angle. For any single detection, the estimates are referred to collectively as an *observation*.

In pure search, the operator must decide whether targets detected on the current scan are the same as those detected on a previous scan or scans. With TWS, however, this decision is made automatically.

In the course of successive scans, the TWS maintains an accurate track of the relative flight path of each valid target. This process is carried out iteratively via five steps: *preprocessing*; *correlation*; *track initiation and deletion*; *filtering*; and *gate formation* (Fig. 31-16).

Preprocessing. In this step, two important operations are performed on each new observation. First, if a target having the same range, range-rate, and angular position has been detected

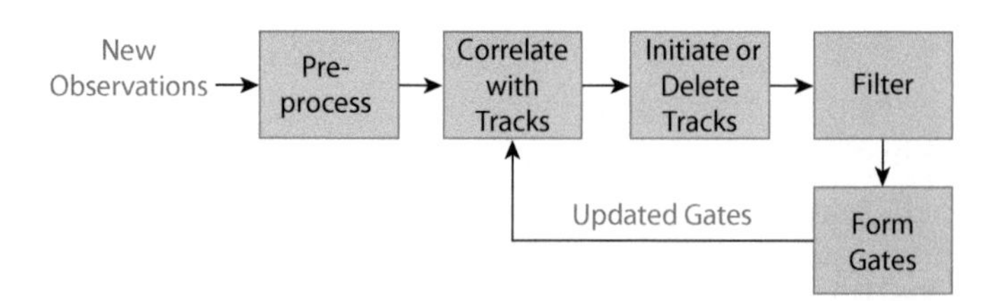

Figure 31-16. Track-while-scan processing has five steps.

on a preceding scan, the observations are combined. Second, if not already so referenced, each observation is translated to a fixed coordinate system, such as the NED described in the blue panel. The angle estimates are conveniently formulated as *direction cosines.* In this case, these are cosines of the angles between the direction of the target and the N, E, and D axes. Range and range-rate may be projected onto the N, E, and D axes simply by multiplying them by the respective direction cosines.

Correlation. This step determines whether a new observation should be assigned to an existing track. On the basis of the observations assigned to the track thus far, tracking filters accurately extend the values of the N, E, and D components of each parameter of the track to the time of the current observation. The filters then predict the values of these components for the time of the next observation.

On the basis of accuracy statistics derived by the filters, a gate scaled to the maximum error in measurement and prediction is placed around each component of the prediction for the track, as illustrated in Figure 31-17. If the next observation falls within all of the gates for the track, the observation is assigned to the track.

Naturally, when closely spaced observations are received, conflicts in assignments are likely to occur. To resolve these conflicts, a statistical distance of each observation from the track (or tracks) is computed by normalizing and combining the differences between measurements and predictions for all components of the observation. Each track is centered in a gate (Fig. 31-18), the radius of which corresponds to the maximum possible statistical distance between measurement and prediction.

A representative conflict is illustrated in Figure 31-19. Observation O_1 falls within the gates of two different tracks: T_1 and T_2. Observations O_2 and O_3 both fall within the gate of track T_2. Conflicts such as this are typically resolved as follows.

- Observation O_1 is assigned to track T_1 because it is the only observation within the gate of T_1, while T_2 has other observations, O_2 and O_3, within its gate.
- Observation O_2 is assigned to track T_2 because its distance, $d_{2,2}$, from the center of the gate is less than that of O_3.[3]

Track Creation or Deletion. When a new observation, such as O_4 in Figure 31-19, does not fit in the gate of an existing track, a tentative new track is established. If, on the next scan (or possibly the next scan after that), a second observation correlates with this track, the track is confirmed. If not, the observation is assumed to have been a false alarm and is dropped. Similarly, if for a given number of scans no new observation correlates with an existing track, the track is removed.

Filtering. This is similar to the filtering performed in single-target tracking. On the basis of the differences between the predictions and new measurements for each track, the track is updated, new predictions are made, and accuracy statistics for both observations and predictions are derived.

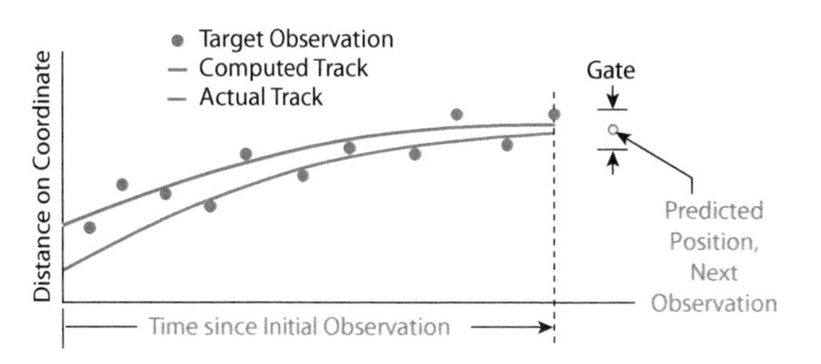

Figure 31-17. This diagram shows a representative track of one component (N, E, or D) of one of a target's parameters. Its predicted value at the time of the next observation and the gate for correlating the observation with the track are illustrated.

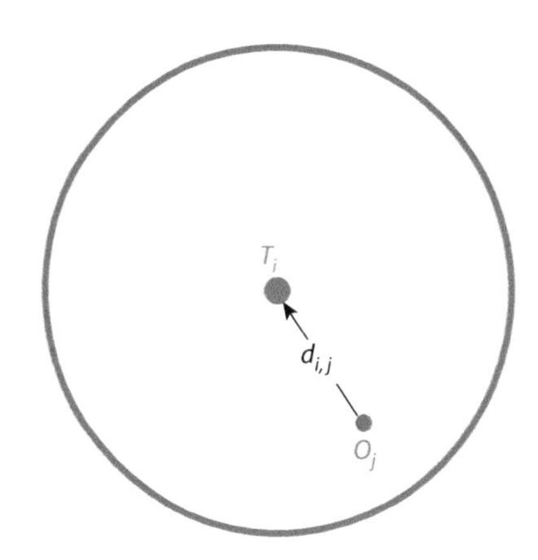

Figure 31-18. The gate for correlating an observation, O_j, with a track, T_i. The size of the gate corresponds to the maximum possible statistical distance, d, a valid observation may be from the track.

3. A restriction applied in this case is that a tentative track cannot be initiated for an observation that falls within the gate of an existing track. Accordingly, because a competing observation is assigned to the track in which O_3 falls, this observation is discarded.

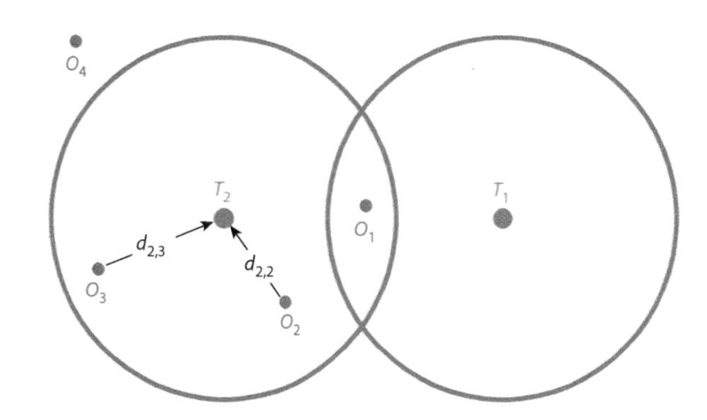

Figure 31-19. In this typical conflicts arising when targets are closely spaced, gates for tracks T_2 and T_1 overlap. Observation O_1 falls in both gates, and observations O_2 and O_3 both fall in the gate for track T_2.

Gate Formation. From the prediction and accuracy statistics derived by the filter, new gates are formed and supplied to the correlation function.

As a result of the filtering, the longer a target is observed, the more accurately the new gates are positioned and the closer the computed track comes to the actual track.

31.3 Track Filtering

Track filtering estimates the trajectory of a target track from radar measurements assigned to that track. Since it is an estimate, it is subject to measurement and process errors, but it aims to determine the target trajectory as accurately as possible. Track filtering is typically accomplished using a combination of prediction and correction. Figure 31-20 illustrates the components that comprise track filtering.

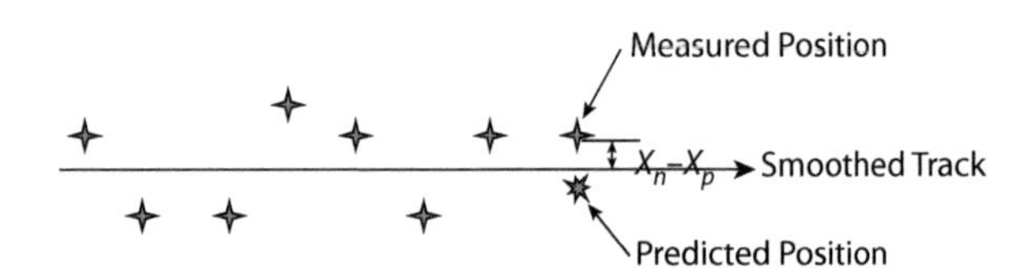

Figure 31-20. Track filtering closes the gap between measurements and predictions over extended measurements.

The stars in Figure 31-20 represent a series of measurements of a target moving at a constant velocity. From the measurements the filter makes a prediction of where the target is likely to be by the time of the next measurement. By comparing the predicted position with the measured position, a new smoothed position can be selected that represents a compromise between measurement noise and process noise (from the prediction process). This filtering process can be written as

$$x_s(k) = x_p(k) + \alpha(x_m(k) - x_p(k))$$

where $x_s(k)$ is the new *filtered* position at time $t = k$, on the *smoothed* track. $x_p(k)$ is the predicted position at time $t = k$, (i.e., the output of the previous application of the filter), and $x_m(k)$ is the measured position at time $t = k$. The parameter α controls the filtered position by placing it either nearer to the prediction or nearer to the measurement. $x_s(k)$ then becomes the prediction in the next cycle of the tracking filter. Figure 31-21 shows how selection of α results in very different behaviors.

When $\alpha = 1$ then $x_s(k) = x_m(k)$, the filtered positions are the same as the measured positions and consequently the track is formed by simply joining the measurements together. Conversely, when $\alpha = 0$ then $x_s(k) = x_p(k)$, the measurements are ignored and the track is formed arbitrarily if at all. For values between 0 and 1, the filter uses the position measurements only.

A more complex but better performing filter uses both position and velocity measurements in the form of the $\alpha - \beta$ tracking filter.

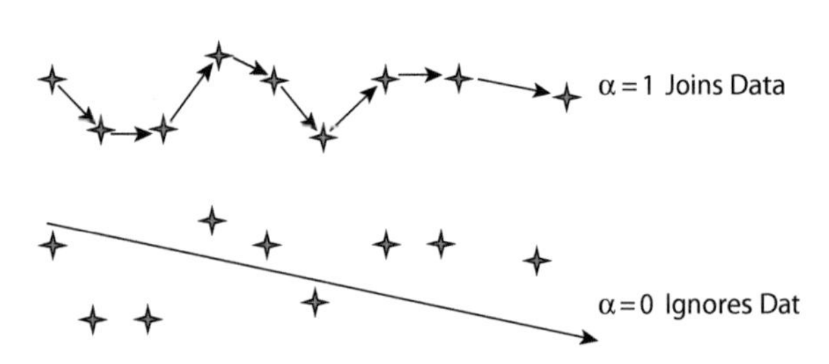

Figure 31-21. Track filtering when $\alpha = 1$ (joins data) or 0 (ignores the data) shows extremes of track behavior.

The $\alpha - \beta$ Track Filter. The previous equation representing the simple filter process will always have an output that lags the data as predictions are based on prior position measurements only. The $\alpha - \beta$ filter overcomes this by additionally using Doppler or velocity information. For velocity we have

$$v_s(k) = v_s(k-1) + \beta(x_m(k) - x_p(k)) / T$$

where $v_s(k)$ is the *filtered* velocity, $v_s(k-1)$ is the previous value of the filtered velocity, and T is the time between updates. The equations for $x_s(k)$ and $v_s(k)$ are known as the equations of the $\alpha-\beta$ tracking filter and enable the new *corrected* position to be computed through

$$x_p(k+1) = x_s(k) + v_s(k)T$$

where $x_p(k+1)$ is the new corrected position which is derived from the previous smoothed position and velocity values. Together these equations make up the *predictor-corrector* construct at the heart of most common tracking filters.

The choice of α and β is a trade-off between reducing sensitivity to noise and reducing sensitivity to target motion changes (maneuvers). Large values of α and β results in higher tracking noise but greater responsiveness to target motion changes. Typically α and β will lie between 0.1 and 0.9, and their choice is determined by user requirements and subsequent system design constraints.

For many tracking problems the $\alpha-\beta$ tracking filter provides an adequate solution. However, for more complex cases the constant values of α and β represent a limitation. The Kalman filter is an extension of the $\alpha-\beta$ tracking filter in which α and β are variables that are set so that variance in the measurement and process noise is minimized. The Kalman tracking filter is also of the predictor–corrector form.

31.4 Summary

For single-target tracking, semi-independent tracking loops are generally provided for range, Doppler frequency, azimuth, and elevation. Each loop includes four basic functions: measurement, filtering, control, and system response.

The range-tracking error is measured by taking the difference between early and late samples of the target echoes; the Doppler-tracking error, by taking the difference between the outputs of two adjacent Doppler filters; and the angle-tracking errors, by taking the difference between the returns received through two antenna lobes.

The scale factor of each measurement, commonly represented by a plot of the measured value of the tracking error versus the true value, is called a discriminant. It is normalized so that the measurement will be largely independent of signal strength and precise measurement of voltages or powers will not be required.

Successive measurements are passed through a low-pass filter, whose gain and cut-off frequency are constantly adjusted based on the signal-to-noise ratio, potential target maneuvers, and the aircraft's own maneuvers to eliminate as much noise as possible without introducing excessive lag.

From the filter outputs, a command calculated to reduce the tracking error to zero is produced. For range tracking, the command adjusts the radar's sampling times; for Doppler tracking, it shifts the frequency of the received echoes; for angle tracking it precesses the rate gyros of the antenna stabilization system.

In track-while-scan, targets detected in successive search scans are tracked by filtering their parameters, much as in single-target tracking. For each track, gates based on the filtered parameters are used to determine whether new detections should be assigned to existing tracks or new tentative tracks should be established for them and whether any existing tracks should be dropped.

Tracking filters are required for producing smoothed, accurate estimates of a target's true position and trajectory. $\alpha-\beta$ and Kalman filters are examples of predictor–corrector approaches to tracking and are commonplace in many tracking radar systems.

Further Reading

E. Brookner, *Tracking and Kalman Filtering Made Easy*, Wiley, 1998.

S. Blackman and R. Popoli, *Design and Analysis of Modern Tracking Systems*, Artech House, 1999.

M. I. Skolnik (ed.), "Automatic Detection, Tracking, and Sensor Integration," chapter 7 in *Radar Handbook*, 3rd ed., McGraw Hill, 2008.

M. I. Skolnik (ed.), "Tracking Radar," chapter 9 in *Radar Handbook*, 3rd ed., McGraw Hill, 2008.

G. Brooker, "Tracking Moving Targets," chapter 14 in *Sensors for Ranging and Imaging*, SciTech-IET, 2009.

M. A. Richards, J. A. Scheer, and W. A. Holm (eds.), "Radar Measurements," chapter 18 in *Principles of Modern Radar: Vol. 1, Basic Principles*, SciTech-IET, 2010.

M. A. Richards, J. A. Scheer, and W. A. Holm (eds.), "Radar Tracking Algorithms," chapter 19 in *Principles of Modern Radar: Vol. 1, Basic Principles*, SciTech-IET, 2010.

W. L. Melvin and J. A. Scheer (eds.), "Multitarget, Multisensor Tracking," chapter 15 in *Principles of Modern Radar: Vol. 2 Advanced Techniques*, SciTech-IET, 2012.

Test your understanding

1. Explain the terms *accuracy* and *precision.*
2. In an automatic tracking radar system, what is the role of a discriminant?
3. Describe a method for improving angle, range, or Doppler estimation in an automatic tracking radar system.
4. What is the role of the "correlator" in track-while–scan radar systems?
5. Describe how the predictor–corrector process works in an alpha–beta tracking algorithm.

PART VII

Imaging Radar

CARABAS II VHF SAR (1996)

CARABAS is an experimental low-frequency SAR operating over the 20–90 MHz band, for foliage penetration (FOPEN), to detect targets under forest canopies. It was developed by the Swedish Defence Research Agency FOI. The radar is carried by the Rockwell TP86 Sabreliner aircraft and flown by the Swedish Air Force. Conspicuously visible in the image are the two ultra-wideband antennas.

32

Radar and Resolution

SAR image of Pentagon.(Courtesy of Sandia Labs.)

High-resolution imaging radar has become an essential tool for remote sensing and military surveillance. High-resolution imaging allows finer details to be discerned than at low resolutions, and this can be exploited with great effectiveness. A radar sensor of moderate resolution allows only detection of objects, such as an aircraft, together with an estimate of location and relative velocity. As both *along-track resolution* (in the direction of the aircraft trajectory) and *across-track resolution* (perpendicular to the aircraft trajectory) is improved, more and more parts of the object are separately resolved, resulting in a greater level of detail being revealed. Indeed, very fine resolutions can permit identification or classification of an object. This chapter defines radar image resolution and introduces methods for achieving fine resolution in along-track. Together, these can result in the generation of high-resolution 2D imagery.

A radar system whose resolution is significantly finer than the size of an object can be used to create a scattering map or "radar image" of the object and its structure. Because of the phase-coherent nature of a radar transmitter, the imagery consists of complex values having both magnitude and phase. Generally, the phase value is discarded after image formation, and the resultant magnitude is displayed as a gray-scale image. This is similar to an optical photograph, although there are important differences such as frequency and direction of illumination. However, in radar images, if the resolution is sufficiently high, the size, shape, and orientation of an object as well as more detailed features can be inferred.

The methods for creating a radar image are quite different from those used in optical cameras. Fine resolution in a radar image is created using a combination of fine across-track (range) resolution (achieved using a wide-bandwidth radar transmission) coupled with fine along-track resolution via aperture synthesis.

As we shall see, synthetic aperture radar (SAR) has become an indispensible tool in both civil and military remote sensing.

32.1 How Resolution is Defined

The quality of imagery produced by a radar sensor is gauged primarily by the ability to resolve between closely spaced objects. This ability may be defined in terms of *resolution distance* and *cell size*. The resolution distance is the minimum distance in the radar image by which two scatterers of equal echoing area may be separated and still be discerned as individual scatterers. The scatterers might be individual objects, such as two separate aircraft, or could be component parts of a single object, such as the nose, cockpit, engines, wings, and tail of a single aircraft. This separation of scatterers is usually expressed in terms of a across-track component, d_r, and an azimuth or along-track component, d_a (the component at right angles to the radar's radial line of sight).[1]

1. The across-track (i.e., perpendicular to the aircraft trajectory) resolution is sometimes expressed as the radial resolution and is equivalent to range resolution in conventional radar. Similarly along-track resolution is sometimes expressed as the cross-range or azimuth resolution.

A resolution cell is a rectangle whose sides, d_r and d_a, define the size of the cell (Fig. 32-1). Because an object may be oriented in any arbitrary direction, d_r and d_a are often made equal, which causes the cell's shape to be square. However, the resolution is not necessarily square. In real-beam radar, for instance, where fine azimuth resolution is difficult to obtain, d_r is typically a small fraction of d_a. For example, at X-band (3 cm wavelength), a 1 m antenna results in an azimuth beamwidth of 30 mrads, and this determines d_a. At a range of 100 km, d_a is 3 km; compare this with d_r, which may be 30 cm, as determined by a bandwidth of 500 MHz.

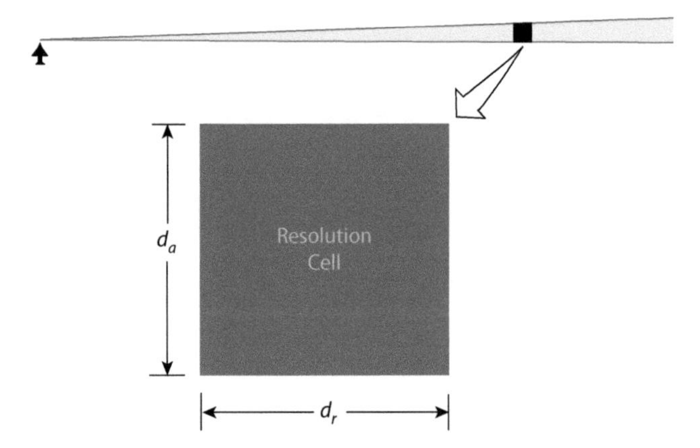

Figure 32-1. Resolution distance is the minimum distance by which two scatterers can be separated and still be discerned separately. A resolution cell is a rectangle whose sides are the range and azimuth resolution distances.

Strictly speaking, the resolution cell is not necessarily a sharply delineated rectangle, as shown in Figure 32-1. Rather, it is usually a rounded rectangular "blob" whose brightness falls off at the edges. Nevertheless, for most radar imagery square pixels with sharp delineation are assumed, and we will follow this convention.

32.2 Factors Influencing Choice of Resolution Cell Size

Among the more important considerations influencing the choice of cell size are (1) the sizes of the objects or scatterers to be resolved; (2) the type and quantity of signal processing required to produce the imagery; (3) cost; and (4) the task of interpreting the imagery (Table 32-1).

Table 32-1. Resolution Required for Various Mapping Applications

Features To Be Resolved	Cell Size
Coastlines, large cities, and the outlines of mountains	150 m
Major highways, variations in fields	20–30 m
Road map details: city streets, large buildings, small airfields	10–15 m
Vehicles, houses, small buildings	1–3 m

Size of Objects to be Resolved. The size of the resolution cells and the generation of useful radar imagery depends on the application. To discern gross features of terrain, such as coastlines separating land and sea or the overall outlines of cities and mountains, a relatively coarse resolution is acceptable (~150 m). To recognize major highways while distinguishing between crops of fields and wooded areas, an improved resolution is needed (~20–30 m). To recognize individual city streets, large buildings, and small airfields and the sort of details commonly included in a road map, a still finer resolution might be required (~10–15 m). To recognize the shapes of objects on the ground, such as vehicles, houses, and small buildings, the resolution must be even finer (~1–3 m). Exactly how fine varies with the sizes and the shapes of the objects. As a rough rule, the required resolution distance is somewhere between 1/5 and 1/20 of the major dimension of the smallest object to be recognized.

This relationship between resolution and the ability to recognize features and objects is indicated in Figure 32-2, where two

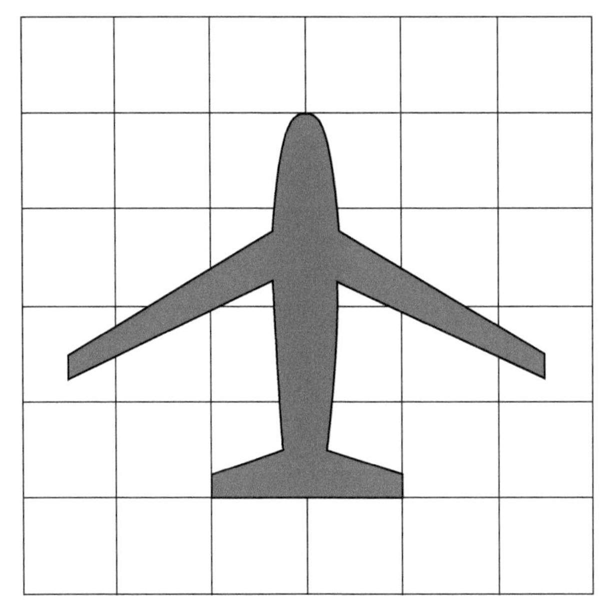

Cell Size: 1/5 Major Dimension

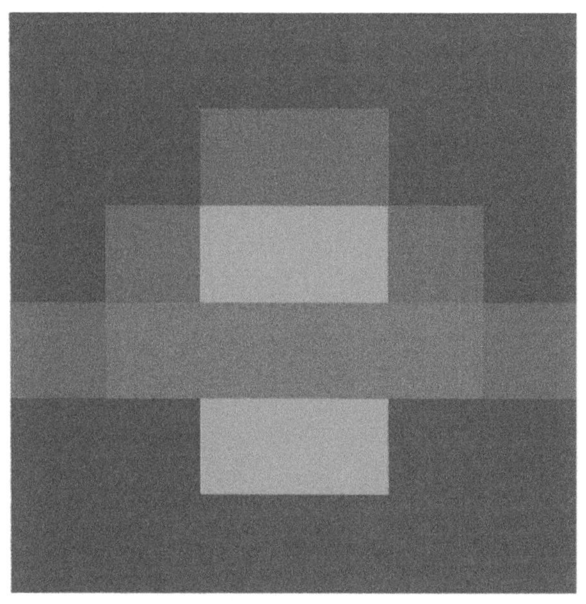

Corresponding Map

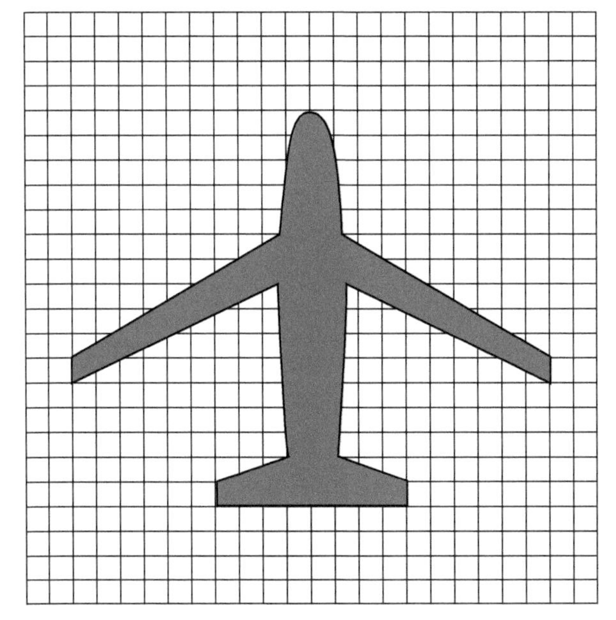

Cell Size: 1/20 Major Dimension

Corresponding Map

Figure 32-2. These boxes illustrate the effect of cell size on shape. The silhouettes (left) are identical. The radar images (right) are simplified representations. Assuming that all elements of the plane reflect equally in the direction of the radar, a cell size of 1/20 of the silhouette's major dimension enables good representation of shape.

silhouettes of the same airplane are shown. A grid of resolution cells is superimposed over one silhouette whose sides are 1/5 of the wingspan. Over the other, a grid of cells is superimposed whose sides are 1/20 of the wingspan. Alongside each outline is a simplified representation of the image corresponding to the indicated resolution cell size. In these images, cells that are completely filled by the silhouette are shown as yellow; cells that are partly filled are shown in shades of green and correspond to the percentage filled. Cells that do not include the airplane at all are shown as dark green. For this particular shape, a resolution of 1/5 of the major dimension enables some degree of shape recognition, while a resolution of 1/20 of the major dimension clearly shows the shape to be that of an aircraft.

However, in forming Figure 32-2, all elements of the aircraft are assumed to reflect the incident radio waves in the direction of the radar's receiver equally. In fact, for any one combination of look angle, radio frequency, and polarization, only a few bright scattering centers might be observed in the image. Thus, even though the cell size might be as small as 1/20 of the major dimension, the airplane's shape might still prove difficult to recognize. This is because at radar wavelengths aircraft appear very smooth, and hence most of the reflected energy is in the forward direction, away from the radar.

However, as later chapters will show, by repeatedly mapping the same area from different directions and with different radio

Figure 32-3. This SAR image of part of Karlsruhe, Germany, was made by PAMIR, an X-band airborne radar system designed and operated by the Fraunhofer Institute for High Frequency Physics and Radar Techniques. The imaged area is too large to show the full resolution.

Figure 32-4. This high-resolution image of Karlsruhe, Germany, was made by PAMIR, the same airborne radar system that created the image in the previous figure. The imaged area is still too large to show the full resolution.

frequencies and polarizations, the fraction of an object's surface from which reflections are received can be substantially increased. Via such techniques an image can be formed similar to that indicated in Figure 32-2, allowing the shapes of individual objects to be discerned. In many practical situations this is not possible, but the less complete form of imagery generated is still immensely useful.

Signal Processing Requirements. A major potential constraint on resolution fineness is the amount of signal processing required. Figure 32-3 shows a SAR image of part of Karlsruhe, Germany,[2] with much fine detail. However, to display such a large imaged area, it is not possible to simultaneously observe the full resolution. We get a sense of this in Figure 32-4, a zoomed-in image of the area outlined in red in Figure 32-3. Even here the imaged area is still too large to allow the full resolution to be displayed. Figure 32-5 shows an even closer zoom of the area outlined in red in Figure 32-4. Now the full resolution of 10 cm is displayed and reveals fine details such as small structural features on the roof of the building. Contrast this with Figure 32-3, where even the building is difficult to pick out.

In general, to map an area of a given size, such as that shown in Figure 32-3, the amount of processing increases in proportion to the number of resolution cells per unit area. If the cells are *square*, the number of cells is proportional to the square of the resolution. Thus, improving resolution by a factor of

2. For more details of PAMIR and high-resolution radar imaging see A. R. Brenner and L. Roessing, "Radar Imaging of Urban Areas by Means of Very High-Resolution SAR and Interferometric SAR," *IEEE Transactions on Geoscience and Remote Sensing* 46(10), pp. 2971–2982, 2008.

Figure 32-5. This high-resolution image of Karlsruhe, Germany, shows the resolution of 10 cm. Image created by PAMIR, an airborne radar system designed and operated by the Fraunhofer Institute for High Frequency Physics and Radar Techniques FHR.

two quadruples the number of cells. However, the impact on processing load may be even higher as the requirements for error correction become more severe at higher resolutions.

Costs and Data Management. It is not easy to generalize regarding this important parameter. However, as the resolution is made finer, the complexity and quantity of both the radar hardware and the signal processing increase. As a consequence, the cost also increases, as does processing time. Depending on the particular case, a further increase in resolution may become prohibitively expensive. Yet as digital technology continues to advance, the cost of providing a given resolution tends to decrease. As might be expected, there is a balance between cost and achievable resolution. A larger number of resolution cells in a given area also results in an increase in data volume, and this must be managed with appropriate storage, data transfer, and computational resources.

Interpreting the Images. Superficially, image interpretation hardly seems an important consideration when determining the required resolution. In fact, it is of the highest importance and drives the choice of resolution, which in turn drives many of the radar parameters. Regardless of resolution, interpreting radar imagery has to be done with care.

A great many features of objects and terrain appear quite differently in a radar image than they do, for example, in an optical image. The longer wavelengths of radar scatter differently than those at optical frequencies. Since radar provides its own illumination, shadows in imagery are always in a direction away from the radar. As we have seen, radar is a coherent sensor. This results in imagery that has a grainy appearance due to the constructive and destructive interference between all the scatterers that contribute to each pixel in a radar image. This phenomenon is called *speckle*.

For these and other reasons, radar imagery is somewhat unique and requires training before imagery can be interpreted with relative ease. Consider the example of a single-seat attack aircraft streaking across the countryside at a speed of 1440 km/h (400 m/s or Mach 1.2). Rapid interpretation of radar imagery is critical, and a pilot must analyze an image in just a matter of seconds in addition to performing other duties (like flying the aircraft). To make this job manageable, only small patches of ground are mapped. If the resolution is increased to enable positive identification of objects, the area covered by the individual images is correspondingly reduced.

Radar imagery is typically produced in prodigious quantities. Satellite systems map most of the surface of the earth every few days. Such huge quantities of data must be carefully managed, and interpretation of imagery necessitates automatic approaches. However, the complexities of using automatic computational techniques are considerable, and as a consequence image interpretation is still the subject of intense research. Both the resolution requirements and the sizes of the areas mapped vary widely according to application. It follows that the approaches to implementation also reflect this variation.

32.3 Achieving Fine Resolution

Since fine resolution in range (across-track) is generally more readily obtained than it is in the along-track direction, we consider it first.

Across-Track (Range) Resolution. As explained in Chapter 11, the resolution obtained in range amounts to about 150 m for every μs of pulse width. Fine range resolution can, in principle, be obtained simply by reducing the duration of pulses. For example, a 0.1 μs pulse yields a resolution of about 15 m, and a 0.01 μs pulse a resolution of about 1.5 m, and so on.

A first limitation on how short the duration of pulses may be made is the width of the band of frequencies that can be passed by the transmitter and receiver. To pass the bulk of the power contained in the pulse, the 3 dB bandwidth must be of the order of $1/\tau$ Hz, where τ is the pulse width (Fig. 32-6). This means that for a 0.01 μs pulse width, the bandwidth must be of the order of 100 MHz.

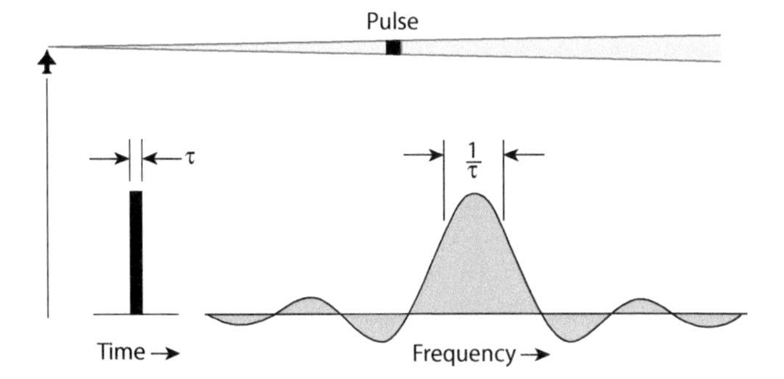

Figure 32-6. The range resolution distance decreases with pulse width. As pulse width is shortened, the required bandwidth increases.

The ease with which wide bandwidths may be obtained depends primarily on the radar's operating frequency. For a given transmitter frequency, as the required bandwidth is increased, a point is ultimately reached beyond which the

Example: Azimuth Resolution

For a real antenna (from Chapter 8):

$$\theta_{3dB} \cong \frac{\lambda}{L} \text{ radians}$$

$$d_a \cong \theta_{3dBR}\, R = \frac{\lambda}{L}$$

with wavelength, $\lambda = 3$ cm; length of antenna, $L = 3$ m; and range, $R = 100$ km, the calculation is

$$d_a \cong \frac{0.03 \times 100\,000}{3} = 1 \text{ km}$$

hardware becomes increasingly difficult to design and build and too costly. As a crude rule of thumb, depending on the particular case, this point lies somewhere between 3 and 10 percent of the operating frequency. Thus for X-band (10 GHz), a bandwidth of 100 MHz would be only about 1 percent. At 1 GHz (L-band), the same bandwidth would be 10 percent. Among hardware items for which bandwidth is more critical are the antenna and some of the radio frequency components.

A second limitation arises if the peak power and the pulse repetition frequency (PRF) are kept constant. Transmission of extremely narrow high range resolution pulses greatly reduces the average transmitted power (the peak power multiplied by the pulse duration multiplied by the PRF). This problem can be avoided by employing pulse compression (Chapter 16). With a pulse compression ratio of 1000:1, a radar system can transmit pulses of 10 μs duration and after compression still achieve a range resolution of 1.5 m (10/1000 = 0.01 μs or a 100 MHz bandwidth). The required bandwidth is determined by the compressed pulse width and is therefore 100 MHz, in this example.

Along-Track (Cross-Range) Resolution. As we saw earlier, the resolution associated with the azimuthal beam width of an antenna can be defined as the 3 dB extent of the antenna beamwidth (in radians) multiplied by range. The 3 dB azimuth beamwidth of an antenna approximately equals the wavelength divided by the length of the antenna. Thus, for a given range, fine resolution can be obtained by operating at a very short wavelength or by employing a very long antenna, or some combination of both. Because of severe attenuation by the atmosphere at shorter wavelengths, the minimum practical wavelength for long-range mapping is around 3 cm. In airborne applications, the length of the radar antenna is usually limited by the carrying abilities of the aircraft. Thus, a long antenna, of, say, 3 m and a wavelength of 3 cm results in a relatively poor cross-range resolution of 1 km at a range of 100 km. However, if the maximum range of interest is reasonably short and the resolution requirements are not too demanding, an antenna of practical size can provide a narrow "real" beam to yield adequate results. At ranges of up to 18 to 20 km, a side-looking radar system, having a 5 m long antenna operating at X-band, can provide resolution suitable for identifying features such as oil slicks and also resolving small craft. At a range of 20 km the resolution is 120 m. Real-beam airborne radars of this type are called sideways-looking airborne radars (SLARs). Figure 32-7 shows an SLAR image in which small craft and an oil slick can be identified. Alternatively, shorter-range applications can make use of higher frequency (shorter wavelength) radar systems to enable finer resolution.

By and large, though, resolutions fine enough for recognizing the shapes of even fairly large objects at long ranges imply either an impractically long antenna or the use of wavelengths so short

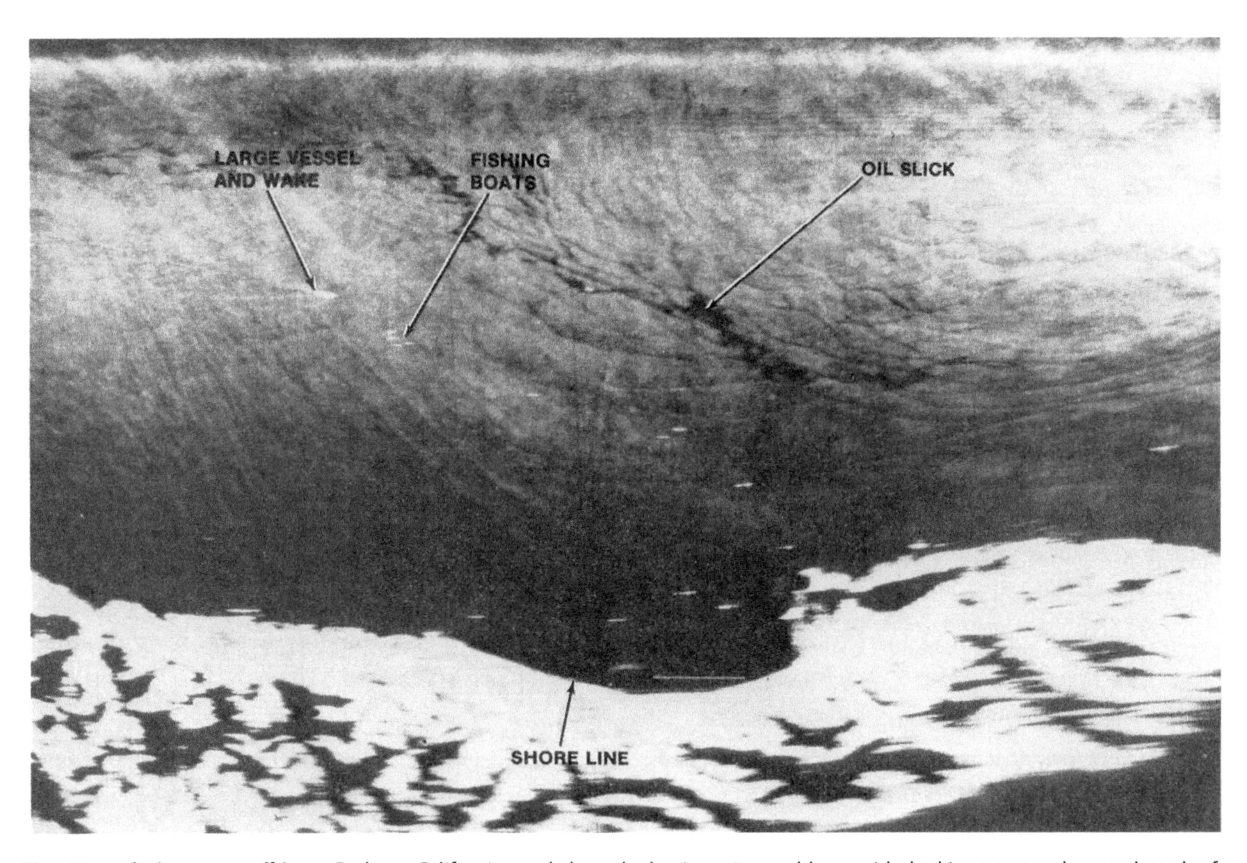

Figure 32-7. Map of oil seepage off Santa Barbara, California, made by radar having a 5 m real-beam side-looking array and a wavelength of 3 cm. The radar's flight path is along the top of the map. The range to the coast is about 28 km. At 9 km, azimuth resolution is approximately 54 m. (Courtesy Motorola Inc.)

that the radar must contend with severe attenuation in the atmosphere. The answer to this dilemma is to use a coherent radar system and measure the phase change caused by movement of a radar system with respect to an illuminated scene.

By implementing phase-based processing, finer resolution can be achieved in the along-track dimension via the synthesis of very large virtual apertures. Consider an aircraft flying a straight and level trajectory carrying a sideways-looking coherent radar. As the aircraft progresses, a sequence of pulses is transmitted and received. These pulses can be combined together to form a long virtual or synthetic aperture (hence Synthetic Aperture Radar or SAR). The aperture is created by exploiting the phase shifts due to continually changing position of the radar with respect to any point on the ground. The total angle over which the illuminating radar views a point on the ground determines the length of the synthetic aperture and therefore the fineness of the along-track resolution. Because a change in phase over time equates to a Doppler frequency, SAR imagery can be formed using Doppler processing techniques.

32.4 Summary

The quality of a radar image is gauged by the size of the resolution cell. Adequate resolution is determined primarily by the size of the smallest objects that must be recognized.

Test your understanding

1. Why is high resolution important in radar imaging?
2. How is fine resolution in across-track generated?
3. At a range of 150 km, what is the along-track resolution of a real beam radar system that has an antenna of length 3 m and an illumination frequency of 3 cm?
4. What are the differences between optical images and radar images?

Resolution requirements are tempered by the amount of signal processing that must be done, the task of interpreting the details of the maps, data management, and cost. Fine across-track (range) resolution may be found with reasonable levels of peak power by using pulse compression. Fine along-track (cross-range) resolution may be obtained by using short wavelengths and long antennas, or with synthetic aperture processing leading to synthetic aperture radar or SAR imaging.

Further Reading

D. R. Wehner, *High-Resolution Radar*, Artech House, 1995.

W. G. Carrara, R. S. Goodman, and R. M. Majewski, *Spotlight Synthetic Aperture Radar: Signal Processing Algorithms*, Artech House, 1995.

P. Z. Peebles, *Radar Principles*, John Wiley & Sons, Inc., 1998

M. Soumekh, *Synthetic Aperture Radar Signal Processing with MATLAB® Algorithms*, Wiley, 1999.

C. J. Oliver and S. Quegan, *Understanding Synthetic Aperture Images*, SciTech-IET, 2004.

G. Brooker, "High Range-Resolution Techniques," chapter 11 in *Sensors for Ranging and Imaging*, SciTech-IET, 2009.

M. A. Richards, J. A. Scheer, and W. A. Holm (eds.), "An Overview of Radar Imaging," chapter 21 in *Principles of Modern Radar: Basic Principles: Volume 1, Basic Concepts*, SciTech-IET, 2010.

C. V. Jakowatz, D. E. Wahl, P. H. Eichel, D. C. Ghiglia, and P. A. Thompson, *Spotlight-Mode Synthetic Aperture Radar: A Signal Processing Approach*, Springer, 2011.

J. J. van Zyl, *Synthetic Aperture Radar Polarimetry* (e-book), Wiley, 2011.

W. L. Melvin and J. A. Scheer (eds.), "Spotlight Syntehtic Aperture Radar," chapter 6 in *Principles of Modern Radar: Volume 2, Advanced Techniques*, SciTech-IET, 2012.

W. L. Melvin and J. A. Scheer (eds.), "Stripmap SAR," chapter 7 in *Principles of Modern Radar: Volume 2, Advanced Techniques*, SciTech-IET, 2012.

W. L. Melvin and J. A. Scheer (eds.), "Interferometric SAR and Coherent Exploitation," chapter 8 in *Principles of Modern Radar: Volume 2, Advanced Techniques*, SciTech-IET, 2012.

33

Imaging Methods

There are many different types of radar imaging methods. Each results in differing capabilities, usually driven by user requirements. In this chapter we are introduced to the most common and useful, beginning with the most widely used of all: *synthetic aperture radar* (SAR).

33.1 SAR

As an aircraft or satellite flies over an area of ground, an onboard SAR can provide fine-resolution images of the terrain below. High-performance optical sensors (cameras) provide an alternate method for imaging terrain and in many ways are superior to SAR systems. With less processing, optical systems can produce fine-resolution images across the entire visual and infrared spectrum. Optical images are generally more easily interpreted because they operate in the same part of the electromagnetic spectrum used in human vision. Further, because optical imagery is noncoherent, it is free of a phenomenon known as *speckle*.[1] Optical systems are also *passive* because the sun provides the illuminating source; in contrast, radar provides its own illumination and is therefore known as *active*.

1. Speckle arises because radar is a coherent sensor. It is a manifestation of scattering from multiple sources that leads to constructive and destructive interference. This interference "scintillates" with small changes in viewing angle. It gives SAR imagery the "grainy" appearance we saw in the previous chapter.

However, in many other ways a SAR imaging system is superior to optical imaging. Since radar is an active sensor operating in the radio frequency (RF) part of the electromagnetic spectrum, it can image during both day and night. Moreover, at the much longer wavelengths of the microwave spectrum, radar imaging is unaffected by cloud cover and can create imagery in almost all weather conditions.

Although SAR images can be more difficult to interpret, they can provide additional information regarding an illuminated area. For example, a low-frequency SAR image allows detection of large objects under camouflage or under a vegetation canopy that are otherwise opaque at optical frequencies.

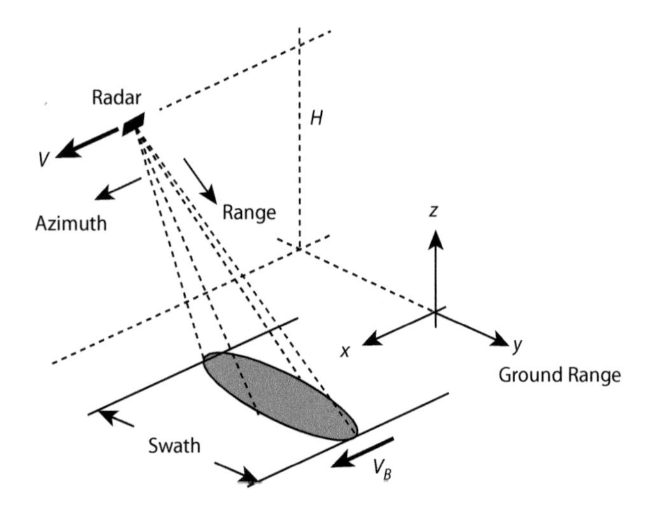

Figure 33-1. As the radar flies a straight and constant altitude flight trajectory, it maps out a continuous swath of imagery in the along-track direction.

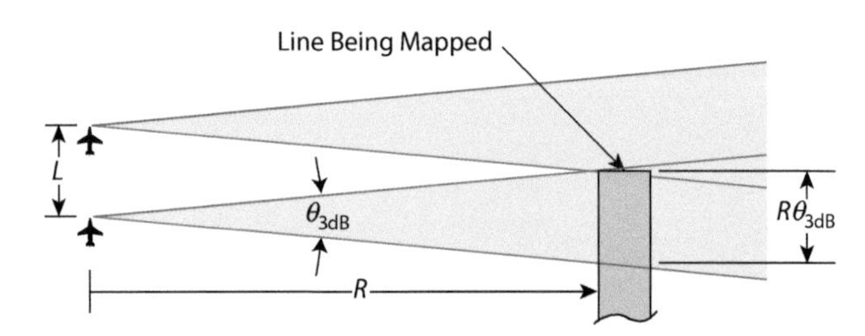

Figure 33-2. For the lower aircraft position, the line being mapped enters the real beam. At the higher aircraft position, the line being mapped exits the real beam. In this way each line being mapped will first enter and then exit the real beam as the radar traverses the entire length of the synthetic array. This determines the maximum synthetic aperture length, *L*.

SAR systems also allows for long-range imaging. Aircraft altitude ultimately determines the maximum imaging range of an airborne SAR, since this is what decides the local horizon. This can be of the order of 300 km for an altitude of 15 000 m. The much higher orbits of space-borne SAR leads to even greater imaging ranges of the order of 7500 km.

SAR Swath (or Strip) Map Imaging. Here the antenna beam is pointed in a direction perpendicular to that of the travel of the aircraft (Fig. 33-1). Assume the radar to fly along a perfectly straight path at a constant altitude. Pulses are transmitted and echoes are received from a patch or swath of ground illuminated by the real beam. As the real beam is "dragged" along the ground, a continuous image is formed as the swath moves in unison with the aircraft in the along-track direction.

The positions of the aircraft at which a particular point on the ground will enter and exit this moving beam mark the maximum length of the synthetic aperture (Fig. 33-2). The different complex values measured for each received echo are combined to create the synthetic aperture and form an image. As the aircraft moves, new, slightly displaced apertures are continuously synthesized, allowing continuous image formation in the along-track direction.

To achieve fine resolution in the across-track dimension, a wide transmit bandwidth is needed. To accomplish this in the along-track resolution, a sequence of received echoes must be combined to synthesize a long antenna.

Since the radar is moving, collecting scattering responses at different points in time is equivalent to collecting scattering responses at different points in space. Thus, integrating the SAR's collected echoes over the radar's traveled distance forms the equivalent of a very long linear antenna array. Summing the echoes at each point forms the array. This is called a *synthetic array*, or a *synthetic aperture*. Since this synthetic aperture can be very long, its effective azimuthal beam width will be very narrow. Thus, it results in very fine along-track resolution. Using aperture synthesis to generate high resolution in cross-range and combining with high resolution in cross-track (using wideband waveforms, Chapter 16), two-dimensional imagery can be formed.

Aperture synthesis can also be expressed in terms of Doppler. The finite beamwidth of the real antenna, coupled with the movement of the aircraft, results in a changing measured Doppler frequency as points on the ground move into and out of the real beam. More specifically, as a point on the ground enters the real beam, the range from the radar is a maximum. When the radar is directly adjacent to the object, this range is a minimum. As the point subsequently leaves the beam, the range again becomes a maximum. Thus, the range to a fixed point on the ground is continuously changing. In turn, this causes a continuous change in measured phase over the time for which an object is illuminated by the real radar beam. This governs the time for which an aperture can be synthesized and therefore the length of the synthetic aperture.

From Chapter 18 we recall that a time-varying phase is equivalent to Doppler. As a point on the ground enters the beam, the Doppler is a maximum since the phase is changing at its most rapid rate (Panel 33-1). When the aircraft has moved so that the point is now in the middle of the beam, there will be no change in range or phase, and the Doppler is zero. When the point exits the real beam, there is again a maximum component of Doppler equal and opposite to that when the point entered the real beam.

SAR Swath Mapping and Along-Track Resolution. In swath-mapping SAR, the along-track resolution is established by the physical size of the real antenna. In fact, smaller real antennas allow longer synthetic arrays to be generated. The reason for this is simple enough. For the echo from any one line on the ground to be received by all elements of the synthetic array, the line must lie within the real beam for the synthetic array to be formed (Fig. 33-2). As a consequence, the width of the real beam, at the range where the line is located, will be equal to the maximum length of the synthetic array. Therefore, because smaller antennas have wider real beams, a longer array can be synthesized and finer resolution achieved.

For a given real-antenna size, how fine can the resolution be?

Before answering this question, an important difference between the beamwidth of a synthetic array and the beamwidth of a real array must be considered. A real array has both a one-way and a two-way radiation pattern. The one-way pattern is formed upon transmission as a result of the progressive difference in the distances from successive array elements to any point off the boresight line (Fig. 33-3). This pattern has a sin x/x shape. The two-way pattern is formed upon reception through the same mechanism. Since the phase shifts are the same for both transmission and reception, the two-way pattern is essentially a compounding of the one-way pattern and thus has a $\left(\frac{\sin x}{x}\right)^2$ shape.

A synthetic array, on the other hand, has only a one-way pattern. The array is synthesized only from the echoes received by the real antenna (i.e., not on transmit). The received echoes that combine to form the synthetic aperture assume the role of successive array elements. However, because each element receives echoes only from its own transmission, the element-to-element phase shifts from a given point on the ground correspond to the differences in the *round-trip* distances from the individual elements to that point and back. This is equivalent to saying that the two-way pattern of the synthetic array has

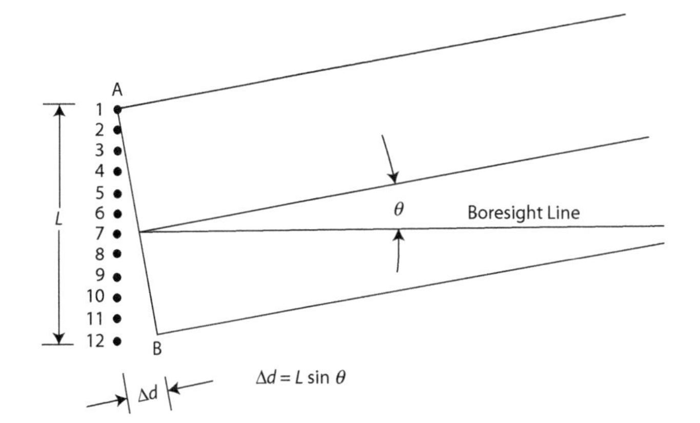

Figure 33-3. The one-way radiation pattern of a real array is formed during transmission as a result of progressive differences in distance from successive array elements to the observation point.

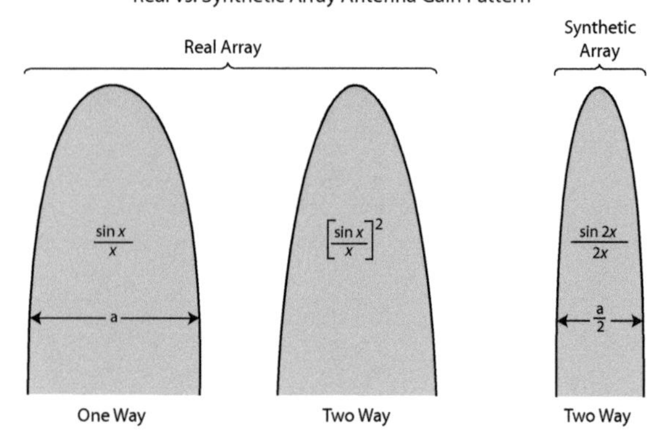

Figure 33-4. In comparing the mainlobes of real and synthetic arrays of the same length, the synthetic array has a one-way radiation pattern since the array is synthesized only from the radar return.

2. The minus 4 dB condition is commonly used, but you may see it stated as the 3 dB case. Either way, the simpler 4 dB expressions for beamwidth and azimuth resolution can be adopted.

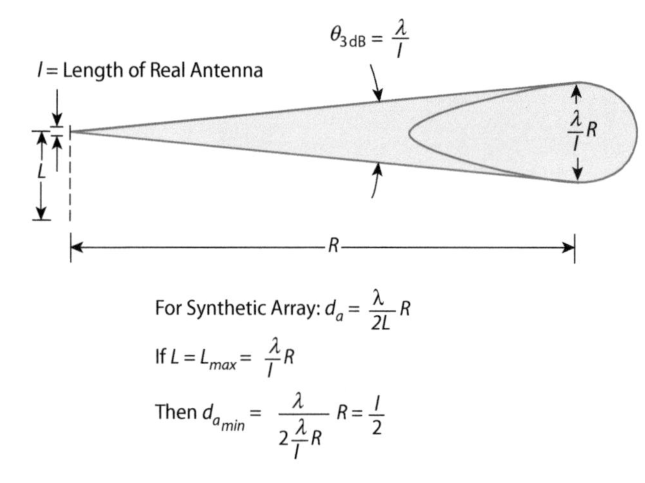

Figure 33-5. The maximum synthetic aperture length, L_{max}, is limited to the size of the real beamwidth at range, R, to $\frac{\lambda}{l}R$. For this condition the azimuth resolution of the synthetic array will equal one-half of the length of the real antenna.

the same shape as the one-way pattern of a real array of *twice* the length, $\left(\frac{\sin 2x}{2x}\right)$ (Fig. 33-4).

For a uniformly illuminated *real array*, the one-way 3 dB beamwidth is 0.88 times the ratio of the wavelength to the array length. Consequently, for a uniformly illuminated *synthetic* array, the two-way 3 dB beamwidth is

$$\theta_{3\,dB} = 0.44\frac{\lambda}{L} \text{ radians}$$

The point on the radiation pattern where the beamwidth is defined is fairly arbitrary. It turns out that by measuring the beamwidth at a point 1 dB lower down the factor 0.44 can be increased to approximately 0.5. To simplify the beamwidth equation, therefore, the minus 4 dB point is commonly used,[2] for example,

$$\theta_{4\,dB} \approx \frac{\lambda}{2L} \textit{radians}$$

and similarly the azimuth resolution is approximated by

$$d_a \approx \frac{\lambda}{2L}R$$

Thus, returning to our question, if the length of the synthetic array is limited by the width of the beam of the real antenna, how fine can the resolution of the synthetic array be?

If the real antenna is a linear array and its length is l, then its *one-way* 4 dB beamwidth is λ/l (Fig. 33-5). Multiplying this expression by the range, R, to the swath being mapped gives the maximum length, L_{max}, of the real beam on the ground at a range, R.

$$L_{max} = \frac{\lambda}{l}R$$

Substituting L_{max} for L in this expression for azimuth resolution, d_a, we find that the best resolution distance $d_{a_{min}}$ is half the length of the real antenna.

$$d_{a_{min}} = \frac{l}{2}$$

This, then, is the finest resolution of a swath-mapping SAR whose beam is positioned at a fixed angle relative to the flight path. It's quite a remarkable result because the best resolution is independent of parameters as fundamental as wavelength. Note how a smaller real antenna results in finer resolution. However, making an antenna smaller also reduces the sensitivity and hence reduces the maximum imaging range of the radar system.

Swath Mapping Along-Track Resolution: The Doppler Method

THE RESOLUTION LIMIT OF SWATH-MAPPING SAR CAN ALSO BE DERIVED IN terms of Doppler frequencies. Swath-mapping SAR uses a real beam that is pointed in a fixed direction with respect to the aircraft track. Here we assume this to be perpendicular to the direction of travel of the aircraft (or spacecraft) as shown in the following figure. By dragging this beam along the ground, targets, located at position *P*, will enter and then exit. For the duration that they are illuminated by the real beam, the synthetic aperture can be formed.

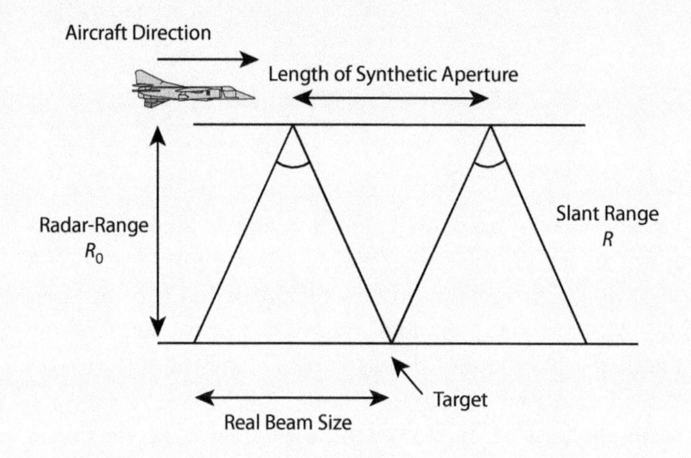

For a swath-mapping geometry the phase change across the synthetic aperture can be derived from the change in range. The range to a point on the ground changes from R to R_0 and back to R as the target progresses through the real beam:

$$R=(R_0^2+x^2)^2$$

where x is the distance traveled by the aircraft in mapping out the synthetic aperture. This expression can be expanded as a power series:

$$R=R_0^2\left(1+\frac{x^2}{R_0^2}\right)^{1/2}$$

$$R=R_0\left(1+\frac{x^2}{2R_0^2}+\frac{x^4}{8R_0^4}+\cdots\right)$$

Taking just the first two terms gives

$$R=R_0+\frac{x^2}{2R_0}$$

and the phase can be written as

$$\phi(x)=-\frac{2\pi}{\lambda}2R$$

The negative sign is adopted for a phase that is decreasing as the aperture is mapped out. Inserting the expression for the range gives

$$\phi(x)=\phi_0-\frac{2\pi x^2}{R_0\lambda}$$

where ϕ_0 is the phase at the minimum range, R_0. This tells us that the phase change across the synthetic aperture is a quadratic function with respect to x. It has the form shown in the following graph.

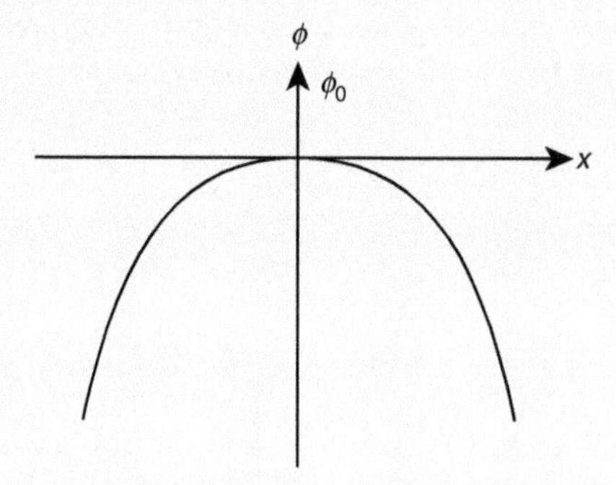

As described in Chapter 18, a rate of change of phase as a function of time equates to a change in Doppler across the synthetic aperture.

$$f_d=\frac{1}{2\pi}\frac{d\phi}{dt}$$

Therefore

$$f_d=-\frac{2vx}{R_0\lambda}$$

Thus, the Doppler changes linearly as a function of x in synthesizing the aperture. This is shown schematically in the following figure and is very reminiscent of the linear frequency modulated (FM) waveforms used in pulse compression described in Chapter 16.

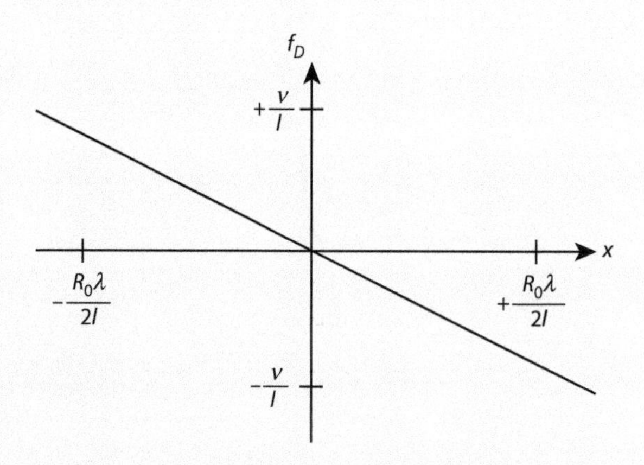

Expressing aperture synthesis in terms of Doppler also enables us to derive the finest along-track resolution.

Swath Mapping Along-Track Resolution: The Doppler Method *continued*

The limit on the Doppler resolution is inversely proportional to the maximum time taken to form the synthetic aperture. It is equivalent to the Doppler bandwidth occupied by a patch on the ground of finite extent $d_{a_{min}}$. This allows the minimum resolvable distance to be expressed as

$$d_{a_{min}} = \frac{R_0\lambda}{2vT}$$

where v is the aircraft velocity, T is the aperture synthesis time, and vT is the synthetic aperture length. By simple geometry the aperture synthesis time is the length of the real beam at range, R_0, divided by the aircraft velocity:

$$T = \frac{R_0\lambda}{vl}$$

where l is the length of the real antenna

Substituting for T in the equation for $d_{a_{min}}$ leads to

$$d_{a_{min}} = \frac{R_0\lambda}{2v}\frac{vl}{R_0\lambda} = \frac{l}{2}$$

This shows, again, that the best resolution achieved by swath-mapping SAR is equal to half the length of the real aperture.

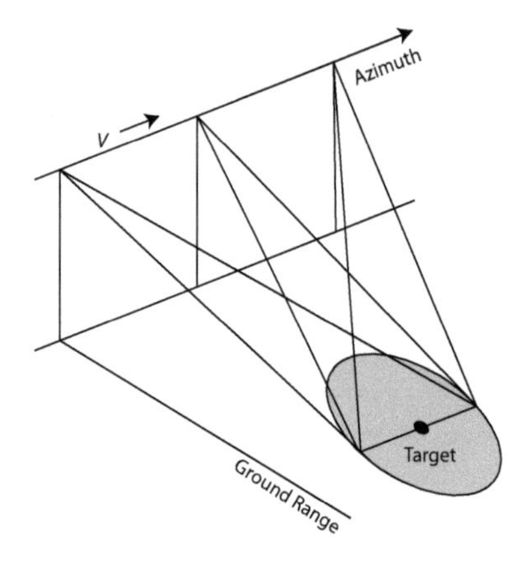

Figure 33-6. In spotlight SAR the real beam is steered continuously to a single point on the ground, thus enabling longer apertures to be synthesized.

33.2 Spotlight SAR

Spotlight SAR is used to generate imagery at finer resolutions than is possible using the swath-mapping method. It does this by imaging a limited region of interest. In spotlight SAR the radar continually steers the beam, illuminating a fixed area of ground. Consequently, it can fly a much longer synthetic aperture, thereby obtaining much finer along-track resolutions (Fig. 33-6). However, imaging takes places over a reduced area, limited to the size of the real beam that intercepts the ground.

In summary, spotlight SAR differs from swath-mapping SAR in three distinct ways:

- Since the beam is continuously trained on the area being mapped, the length of the synthetic array (or Doppler processing time) is not limited by the azimuthal beam width of the real antenna.
- Since the size of the real antenna does not limit the length of the synthetic array, the real antenna size can be increased without reducing the array length. By using a larger antenna, the mainlobe beamwidth is decreased, the gain is increased, and the signal-to-noise ratio (SNR) correspondingly improved but with a smaller "spot size."
- Spotlighting improves the quality of an image by filling in gaps in the backscatter from points on the ground or on an object. When swath-mapping SAR illuminates an object, such as a parked airplane, echo returns may be received from only a few main scattering centers. This results in the airplane's shape not necessarily being as easily recognized as might be expected, given the ratio of the aircraft's principal dimensions to the size of the radar's resolution cells.

By using spotlight SAR, the number of angles over which an image is formed is increased. This leads to more scatterers

Figure 33-7. A composite spotlight SAR image of a car park was formed from an aircraft flying a circular trajectory around the target over 360°. Subsequently the imagery has been rendered in 3D. The images were taken at slightly different times and hence some differences can be observed. (Courtesy of Air Force Research Laboratory, Sensors Directorate, Gotcha Radar Program, Public Release Number: AFRL/88 ABW-11-575).

contributing to the final image and thus results in a more complete representation of the target. An example is shown in Figure 33-7, where the imagery has been formed over all 360° available (and eight passes of the spotlight SAR). Additionally, the image has been rendered to provide a 3D effect. Comparison with the optical photograph below it in the figure (taken at a slightly different time) indicates the fine detail available from such images.

33.3 Inverse SAR Imaging

Both swath mapping and spotlight forms of SAR use a moving platform and rely on the imaged target or scene being stationary. As a consequence, they are not well suited for imaging targets in motion, such as vehicles, ships, and aircraft. Target motions induce phase shifts that, if not accounted for, will defocus the synthetic aperture and degrade the imagery.

However, let's consider a stationary radar system with a real beam staring in a fixed direction, observing a moving target. The SAR method can be inverted to allow an image to be formed. Suppose an aircraft target enters and exits the fixed beam following a path perpendicular to the direction of look of the radar. At the point of entry there is a maximum Doppler in the direction of the radar. At the center of the beam, there is no radial component of motion, and the Doppler takes a value of zero. As the target exits the beam, there is an equal but opposite Doppler to that of entering.

This imaging technique is the exact inverse of SAR swath mapping. Unsurprisingly, it is called *inverse SAR* (ISAR). It can also work with targets that have other types of motion, such as the pitching, rolling, and yaw of a ship at sea. Objects high on the ship will rotate at a faster rate than those at a lower height and thus will have higher rates of change of phase and higher Doppler. Again, in ISAR, Doppler processing can be used to form the images. As in SAR, across-track resolution is obtained using wide bandwidths. Resolution in the orthogonal direction[3] is determined by the angle and time over which the target is measured.

3. There is no equivalent sense of along track in ISAR. Instead, we use the term cross-range to denote resolution in the orthogonal direction to radial range.

Consider the example of a ship undergoing a pitching motion with respect to an illuminating radar system, as shown in Figure 33-8. The rate of phase change measured for points higher on the ship's superstructure such as P_1 will be higher than for points lower down, P_2. In other words, in a given period of time, P_1 will "pitch" farther toward the radar than P_2 resulting in a larger rate of change of phase. Consequently, the Doppler of the echoes from P_1 and all points between it and P_2 will differ in their individual values. Consequently, a map of Doppler versus range will yield an image where the larger values of Doppler frequency are due to scattering from higher parts of the ship resulting in the formation of a silhouette (Fig. 33-8).

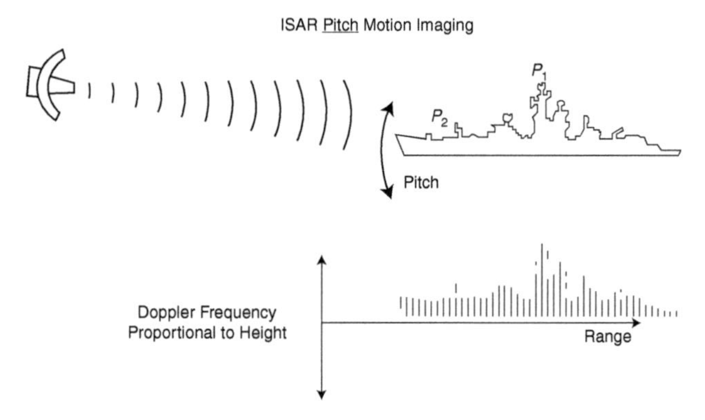

Figure 33-8. In this ISAR image of a ship undergoing a "pitching" motion, the differences in height are due to the angular rotation of the target as seen by the radar and enable a "silhouette" mage to be formed.

ISAR image cross-range resolution, d_n, is proportional to the ratio of the Doppler filters' 3 dB bandwidth, $BW_{3\,dB}$, to the target's rate of rotation, $\dot{\theta}$.

$$d_n = \frac{BW_{3\text{dB}}}{2\dot{\theta}}\lambda$$

The cross-range dimension, though, is not necessarily horizontal as with conventional SAR but is perpendicular to the axis about which the target happens to be rotating. No image is formed, of course, if that axis is colinear with the radar's line of sight (i.e., if the target has no rotational motion as viewed from the radar).

Besides imaging targets that have rotational motion, ISAR has another important difference compared to conventional SAR. This is illustrated in Figure 33-9.

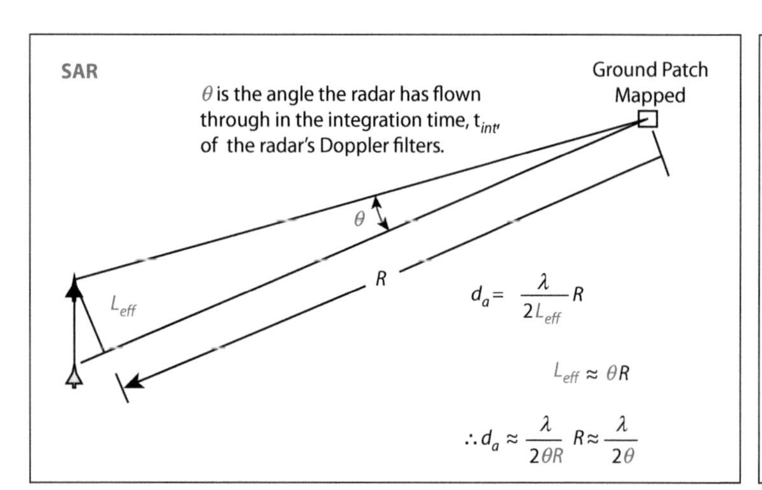

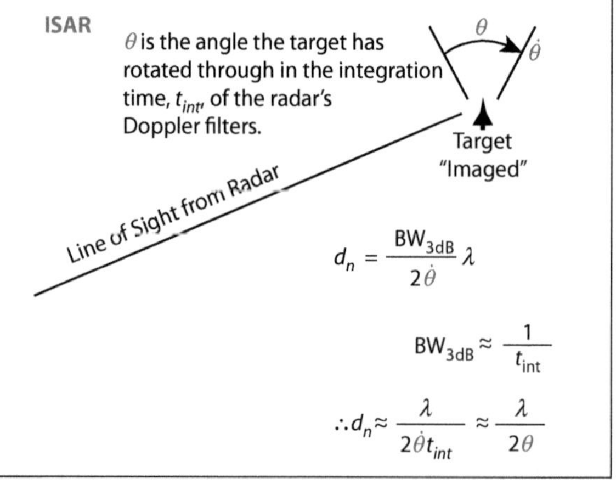

Figure 33-9. With conventional SAR, the cross-range (along-track) resolution, d_a, is inversely proportional to the angle, θ, the radar flies through with respect to a point on the ground during the Doppler filters' integration time, t_{int}. With ISAR, the cross-range resolution distance, d_n, is inversely proportional to the angle, θ, that the target rotates through in the aperture synthesis time, t_{int}.

Cross-Range Resolution of ISAR Images

The cross-range resolution in ISAR is a function of the change in viewing angle between the radar and target. This causes a differential rate of phase change that results in different Doppler frequencies. The different Doppler frequencies correspond to different cross-range positions on the target. Doppler resolution will determine the image resolution in the cross-range direction. The following diagram indicates the more general case of a rotating target illuminated by a stationary radar system. The farther away a scatterer is from the point of rotation, the greater the rate of change of phase as measured by the stationary radar system. Here we consider a line of scatterers distributed along a radius emanating from the center of rotation.

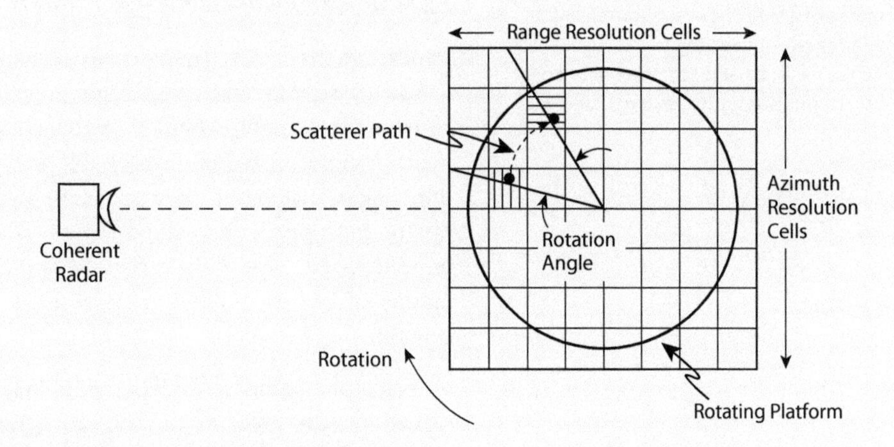

For a rotating target consisting of multiple scatterers (e.g., distributed along the radius of the circle in the previous diagram), the line of scatterers rotates in an anticlockwise fashion, and We can define the following:

$\dot{\theta}$ = the rate of rotation of the scatterers

d_n = the desired cross-range resolution

$\Delta\dot{R}$ = the incremental increase in range-rate of scatterers separated by the distance, d_n, normal to the line of site of the radar

$$\Delta\dot{R} = d_n\dot{\theta}$$

The resulting difference in the Doppler of radar echoes from scatters separated by a distance, d_n, is:

$$\Delta f_d = \frac{2\Delta\dot{R}}{\lambda} = \frac{2d_n\dot{\theta}}{\lambda}$$

The minimum difference in Doppler frequency that the radar can resolve approximately equals the 3 dB bandwidth of its Doppler filters. Therefore

$$\frac{2d_n\dot{\theta}}{\lambda} = BW_{3dB}$$

rearranging gives an expression for the cross-range resolution distance

$$d_n = \frac{BW_{3dB}}{2\dot{\theta}}\lambda$$

In airborne ISAR, the aircraft carrying the radar system will have its own forward motion. The aircraft velocity must be corrected in applying the ISAR technique. If the radar is imaging a rotating object with forward motion (e.g., a ship traveling at sea), additional phase corrections must be made. These must be done over the time for which echoes are collected for image production. An ISAR image can be formed in a relatively straightforward way. A sequence of echoes received for each range cell is applied to a bank of Doppler filters, and an image is produced directly from their outputs. The bank of Doppler filters can be formed via a Fourier transform. This is typically referred to as *range-Doppler mapping*.

With SAR, the cross-range (along-track) resolution, d_a, is inversely proportional to the angle, θ, through which the radar flies with respect to a point on the ground during the integration time, t_{int}, of the Doppler filters. With ISAR, the cross-range resolution, d_n is inversely proportional to the angle, θ, through which the target rotates during t_{int}.

Figure 33-10. In this ISAR image of a ship target superimposed on the actual ship, the correspondence between the two is clear (the ships superstructure is shown as vertical height in the ISAR image). The color-coding indicates scattering strength: red, high; blue, low.

For ISAR, the radar system need not be moving. An image can be formed if there is a viewing angle change between the radar and target.

Figure 33-10 shows an example of an ISAR image of a ship target superimposed on an optical image of the ship. In the image the ships superstructure is shown as vertical height in the ISAR image. The color-coding indicates scattering strength. The correspondence between the two is clear.

33.4 Interferometric SAR

Interferometric SAR (InSAR) can measure the variations in height of terrain with very high precision. Combined with conventional high-resolution SAR mapping, the interferometric height measurements enable the production of three-dimensional topographic maps. Employed by satellite-borne radars, InSAR provides accurate, high-resolution, global topographic maps for a wide variety of applications. Height resolutions of between a few meters and a few centimeters are possible.

In principle, multiple independent antennas could be displaced on the SAR aircraft or spacecraft, but the relatively small physical size of these structures severely restricts the height resolution of the resulting array. Another option is to construct a large *synthetic* vertical array by flying repeat parallel passes of the same SAR, each at an increasing altitude. Thus, all the measurements enable a two-dimensional array to be synthesized. However, this requires many flight paths at the different altitudes, all of which have to be combined to form a two-dimensional synthetic aperture. This isn't really feasible in practice.

Consider instead a SAR image constructed using one pass of the radar and another constructed using a second pass along a slightly different (e.g., elevated) altitude. Assuming nothing has changed on the ground, the two images will appear to be nearly identical. However, they have one significant difference, and this does not appear when comparing the magnitude of the scattering between pixels of each image but instead is exhibited as a difference in the phase values between the pixels. The estimated scattering value for each pixel is a complex number, with both a magnitude and a relative phase. The phase is typically discarded when displaying SAR images, but for estimating surface height phase plays a key role.

The relative phase of each pixel of a complex-valued SAR image depends on the range distance between the SAR and the scattering objects observed. This distance, in turn, relies on the height of the object. Objects at higher elevations will be closer to the SAR than those at lower elevations. Thus, the phase of a SAR pixel depends on the height of an object at that location. For two SAR images constructed from repeat, parallel passes, the phase at corresponding pixels will be different because of the difference in range for each pass. If the ground is flat and level, this phase difference can be predicted geometrically. However, for terrain of v arying elevation, the phase difference is a function of the height of the ground. Since the two flight tracks are still relatively close, it can be assumed that this difference in phase is due *only* to this difference in range and thus depends only on the height of the surface. In this way, the pixel phase difference between two repeat-parallel-pass images can be used to directly estimate the height of the surface at each and every pixel location. Figure 33-11 shows an example of a 3D SAR image of Death Valley, California, that includes the topography evaluated from the interferometric SAR processing.

The process by which such topographic images are formed is known as *phase interferometry*. This technique provides the very precise estimates of surface height. The measured phase difference is the difference in distance between the SAR system and the surface of the ground. A difference of a half-wavelength causes a phase shift of 2π rad. The phase difference can easily exceed this number, though, which causes ambiguities in measuring height. These ambiguities are reduced using a technique called *phase unwrapping*. If the phase difference can be accurately measured to within a few degrees, the range difference, and thus the surface height, can be estimated to accuracy within centimeters, although more usually this is tens of centimeters.

Figure 33-11. Shown here is an interferometric SAR image of Death Valley, California.

InSAR

InSAR RADAR OBTAINS THE ELEVATION DATA NEEDED FOR THREE-DIMENSIONAL mapping by determining the elevation angle of the line of sight to the center of each resolution cell in the swath or patch that is imaged. From this angle, θ_e, the height, H, of the radar, and the slant range, R, to the cell, the cell's height and horizontal distance from the radar are computed.

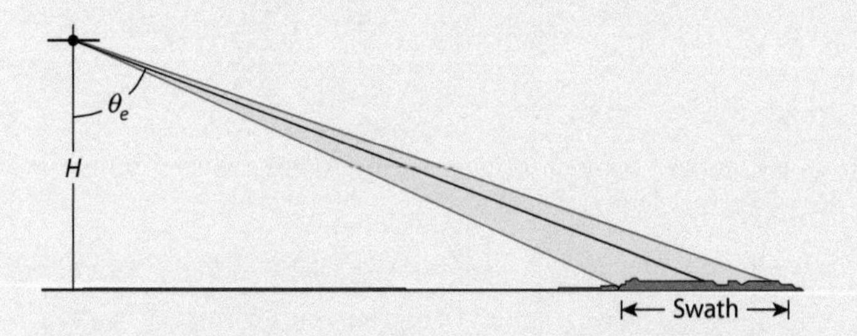

To obtain the values of elevation needed for three-dimensional mapping, the SAR measures the elevation angle, θ_e, of the line of sight to the center of each resolution cell in the swath being image.

The radar determines the elevation angle, θ_e, of the line of sight to a resolution cell in much the same way as a phase-comparison monopulse system determines a tracking error. As illustrated below, two antennas separated by a relatively short distance, B, on a cross-track baseline, receive radar echoes from a point p, in the center of the cell. The baseline is tilted a prescribed amount to illuminate the area being mapped. Ranges R_1 and R_2 from the two antennas to p differ by an amount roughly equal to B times the sine of the angle, θ_L, between the line of sight to p and a line normal to the baseline.

The phases of the coherent radar returns received by the two antennas differ in proportion to the difference in the two ranges:

$$\Delta\phi = \frac{2\pi}{\lambda}(R_1 - R_2)\,\text{radians}$$

By measuring $\Delta\phi$, the elevation angle, θ_L, between the line of sight to p and the normal B may be computed.

InSAR maps can be created using either repeat passes of a single transmitter-receiver SAR system or a single transmitter and two spatially separated receivers on a single pass. An equation for ϕ, in terms of R_1, λ, and the length and tilt of B, can be derived directly from the geometry illustrated in the following figure for the single-pass case ($\kappa = 1$) and the repeat-pass case ($\kappa = 2$). Based upon that equation, an exact equation for θ_L can similarly be derived.

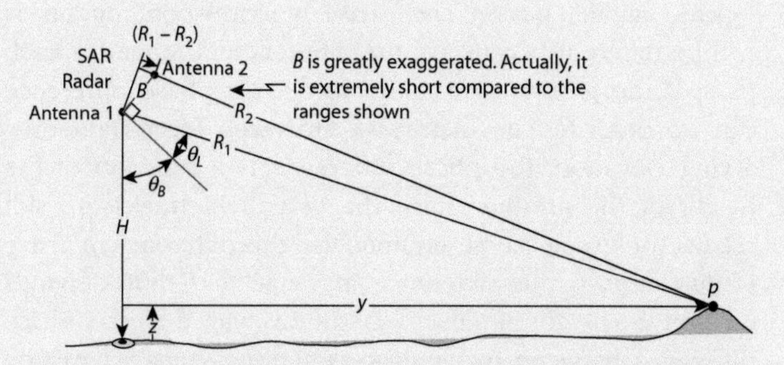

The figure shows the parameters that an InSAR measures to determine the elevation, z, and horizontal range, y, of p in the center of a resolution cell. Angle, θ_L, between the line of sight to p and the line normal to the baseline, B, is determined by sensing the phase difference, $\Delta\phi$, between the returns received by the two antennas as a result of the difference in slant range $(R_1 - R_2)$ from p.

$$\phi\Delta = \frac{2\kappa}{\lambda}[R_1 - (R_1^2 + B^2 + 2R_1B\sin\theta_L)^{1/2}]$$

$$\theta_L = \sin^{-1}\left[\frac{\lambda^2\phi^2}{8(k\pi)^2R_1B} - \frac{\lambda\phi}{2\kappa\pi B} - \frac{B}{2R_1}\right]$$

$K = 1$ for one pass mapping

$K = 2$ for two pass mapping

Having computed the value of θ_L, all that must be done to obtain the elevation angle, θ_e, is to add to θ_L the angle, θ_B, between the normal to the baseline and the vertical axis.

$$\theta_e = (\theta_L + \theta_B)$$

From this sum and the range R_1, the horizontal position, y, and elevation, z, of p may readily be computed.

$y = R_1 \sin\theta_e$

$z = H - R_1 \cos\theta_e$

Ambiguities and their resolution. In collecting returns from successive range increments across the full width of the swath being mapped, the range difference, $R_1 - R_2$, and hence ϕ, increase continuously. Since the wavelength is comparatively short, the value of ϕ may cycle repeatedly through 2π radians (360°) and so will be ambiguous.

The two InSAR images are coregistered, and the pixel to pixel phase difference is computed to form an interferogram. This is illustrated in the following three images of the Brecon Beacons area in Wales. The upper image shows the basic SAR image made from echoes received by one of the radar's two antennas. The center image shows an interferogram produced by coregistering and computing the pixel-to-pixel phase difference for the

InSAR *continued*

images produced with the echoes received by the two antennas. Finally, in the lower image the original SAR map has been "draped" over the 3D rendition to form a topographic map. The images were obtained with DERA Malvern's C-band InSAR radar.

a

b

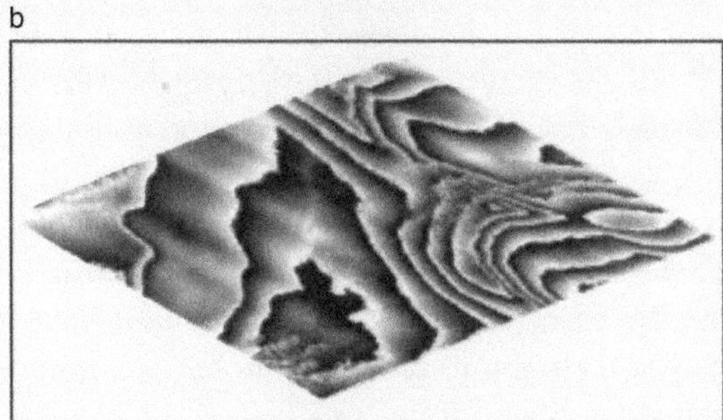

c

By adding 2π to the value of ϕ each time a fringe in the interferogram is crossed (i.e., phase unwrapping), the ambiguities are removed. The horizontal position, *y*, and elevation, *z*, of each cell are then accurately computed, and the map can be topographically reconstructed.

The topographic accuracy depends critically on the accuracy with which the phase unwrapping is performed. Provided the SNR is reasonably high and the fringes are not too close together, this is a straightforward process. It can become complicated, however, if steep slopes or shadows are encountered. Steep slopes may result in some points at which fringes are crossed being overlaid by others, and the shadows may cause some points to be missed.

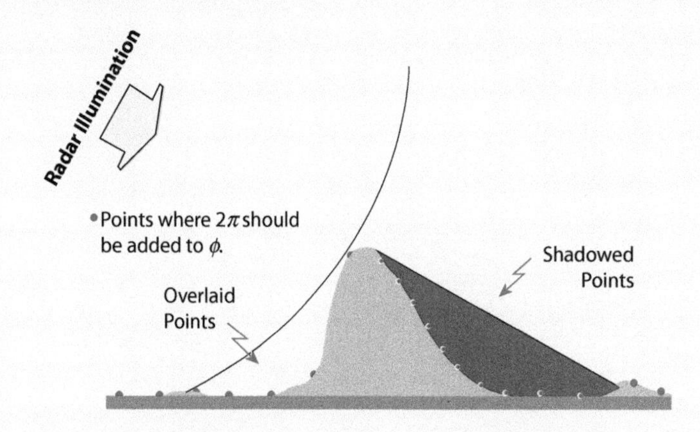

Possible sources of error. Steep slopes may result in echoes from some points being overlaid with echoes from points nearer in range. In the figure the top of the slope appears at the same slant range as the small bump a little closer to the radar. Other points may be missed as they lie in shadows and no echo is received and thus no height can be produced.

As a result, InSAR imagery is used to provide accurate and detailed surface topography measurements over wide areas—in fact, most of the earth's surface. In addition to generating 3D surface maps of the earth, InSAR has found many other applications. For example, because the surface topographic maps are so accurate, they are used to detect slight changes in topography over time, including those due to geophysical phenomena such as seismic events. This is done by examining the difference in two InSAR topographic maps taken before and after a seismic event. Consequently, InSAR has been used to monitor and measure the effects of earthquakes, the subtle movement of glaciers, coastal erosion, and the reduction of polar ice sheets.

Although InSAR can theoretically achieve centimeter elevation estimates, a number of issues degrade accuracy. First, the flight track of each repeat-pass SAR data acquisition must be known with high precision. Second, the sensor phase-measurement must also be precise (to a few degrees or better). Third, if repeat parallel passes are used, *everything* associated with the measurement must remain constant during the time between passes. The SAR hardware, the illuminated surface, and even the atmospheric conditions should ideally be identical for each pass. Of course, this is not realistic since even the wind blowing through surface vegetation can alter scattering response between passes. Each of these issues will have some negative effect on InSAR accuracy. In other words, InSAR requires the scene from which the two images are formed to be unchanged from pass to pass. If this condition is not met, due to aircraft turbulence or scene changes, then the quality of the height estimates will be degraded.

With careful calibration and measurement, the accuracies associated with an InSAR system can be very high. Alternatively, the aircraft can carry, in effect, two radars and form the InSAR image on a single pass. This eliminates the problems associated with changes caused by the environment. Two separate radars can be mounted vertically on the aircraft fuselage or laterally on the each wing of the aircraft. In fact, only one of the radars has to transmit, providing the echo is received at two separated locations.

However, there is one limit on the accuracy of surface height that cannot be overcome with more measurement precision. InSAR purports to measure the elevation of the terrain for each resolution cell (i.e., at each pixel), but the surface over the extent of the cell is almost never flat and smooth. Instead, the surface is rough and sloped and often covered in vegetation. Just what is the surface height for this case? How is it defined?

Consider a forest. Is the elevation defined at the top of the tree canopy, at the bottom, or somewhere in the middle? Centimeter measurement accuracy has little meaning when the definition of the measurement is so murky. A SAR image actually represents the summation of all scattering from all the individual objects (e.g., leaves and branches) within the pixel. In other words, the elements that make up a single pixel in a SAR image are distributed across different heights. Thus, the height provided by an InSAR estimate is an averaged value that depends on the scattered magnitude and position of each of the individual elements. Nevertheless, InSAR provides a satisfactory means of generating 3D SAR imagery, and it is a technique that has found widespread application.

33.5 Polarimetric SAR

A standard SAR image provides a scattering map of illuminated terrain, where the scattering magnitude is typically indicated as a gray-scale value. Black is generally used to indicate an absence of scattering, and the largest scattering magnitudes are white. This scattering magnitude depends on the physical content within a resolution cell, size, structure, and material properties. Given its dependence on this diverse set of physical attributes, it is difficult to infer from the scattering magnitude anything specific about the physical makeup of the elements within a single SAR pixel.

However, the scattering strength from an object depends also on the characteristics of the illuminating electromagnetic wave, including frequency and polarization. Terrain with a large scattering magnitude at one frequency or polarization may have a completely different scattering magnitude at another. The scattering response of an object across frequency and polarization thus provides important information about an object or about terrain type. From subsequent analysis, much can be inferred about the physical attributes of the imaged area.

Consequently, SAR systems are sometimes designed to have different polarizations on both transmit and receive. This allows the polarimetric scattering properties of a scene to be measured. For example, the transmitter could produce a vertically polarized incident wave and then receive both vertically polarized (i.e., copolarized) and horizontally polarized (i.e., cross-polarized) echoes. The next pulse can then be horizontally polarized, and again both vertical (cross-polarized) and horizontal (copolarized) echoes are received. Consequently, a total of four polarization states are measured (i.e., *VV*, *HV*, *HH*, and *VH*). These *polarimetric SAR* systems provide four separate SAR images of the same terrain (at the same time), with each map corresponding to a different transmit/receive polarization pair.

It is common that the scattering magnitudes of *VV*, *HH*, and *HV* (or *VH*: the cross-pol measurements are theoretically identical due to reciprocity) are used to form a false-color image. By assigning each of their magnitudes to one of the colors red, blue, and green, a composite polarimteric SAR image can be formed. The contrast in the resulting colors allows for easier identification and classification of different terrain types.

This is illustrated in the image of a rural scene shown in Figure 33-12, which was created using a C-band fully polarimetric radar system. The image shows the difference between the agricultural fields expressed in the different colors. The colors are derived by the combination of different polarizations. In this example red = *HH*, blue = *VV* and green = *HV*. Thus, the purple color in Figure 33-12 is a combination of strong *VV* and *HH* scattering. The green shows significant depolarization, suggesting there is a change in polarization after reflection, perhaps indicating whether or not a field contains crops or bare soil.

33.6 Tomographic SAR

Tomographic SAR is essentially an extension of SAR. Tomography is less common than the other SAR methods, but it is also capable of generating 2D and even 3D imagery. Tomography exploits multiple passes of the SAR sensor

Figure 33-12. This polarimetric color-coded composite SAR image was produced by DLR's dual pol E-SAR system.

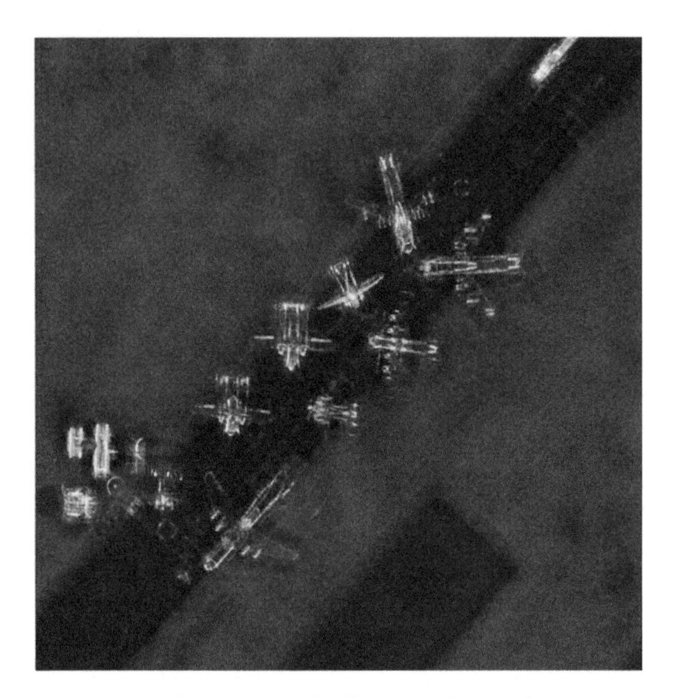

Figure 33-13. This is an example of tomographic SAR showing aircraft grouped on a runway apron. (Courtesy of Air Force Research Laboratory, Sensors Directorate, Gotcha Radar Program, Public Release Number: AFRL/88 ABW-11-5753.)

where each is at a different geometry (or direction of look). Simplistically, if two synthetic apertures of the same scene are formed perpendicularly to one another, then summing the complex values for each "pixel" can form a 2D image where the resolution cells of the two synthetic apertures intersect.

Many synthetic apertures can be generated at different angles to one another. Then, by using signal processing similar to that used in tomographic imaging (e.g., a CT scan), these multiple apertures can be used to form a complete SAR image.

Tomographic SAR can also be used to estimate the vertical distribution of scattering elements within each pixel. This process can be used to determine the thickness of a vegetation canopy or the vertical height and structure of tall buildings or structures in the same way as InSAR. The quality of tomographic SAR imagery is dependent on the number of apertures generated. More passes will generally increase the quality of the image. However, the time taken to gather all the apertures means that any changes caused by the environment (e.g., wind blowing the leaves of a tree) will degrade the resulting imagery. An example of a high-resolution tomographic SAR image is shown in Figure 33-13 in which the shapes of aircrafts grouped on a runway apron can be clearly seen.

33.7 Summary

Radar imaging can be accomplished in numerous ways. At short ranges, azimuth resolution adequate for many high-resolution applications can be obtained with real-beam antennas. But to recognize the shapes of even fairly large objects at long ranges, sufficient resolution can be obtained only by synthesizing the output of a long array antenna from the returns received over a period of time by the real antenna, SAR. The equivalent of an antenna thousands of feet long may be realized.

Operationally, SAR has many compelling advantages.

- It affords excellent resolution with a small antenna even at very long ranges.

- Where desired, it provides resolution as fine as a few tens of centimeters or so.
- It enables resolution to be made independent of range, and
- It is exceptionally versatile.

Aperture synthesis is achieved by viewing an object over a range of angles and processing the phase shifted echoes.

SAR, spotlight SAR, ISAR, InSAR, tomographic SAR, and polarimetric SAR are the most common forms of radar imaging.

Spotlight SAR allows higher image resolution but at the expense of coverage area. ISAR uses target motion to form images. The imaging radar can be stationary. across-track resolution is achieved using wideband waveforms and pulse compression. Cross-range resolution exploits a change in viewing angle between the radar system and the target or scene being imaged. Polarimetric SAR provides an additional means of classifying different areas and objects in SAR imagery. Tomographic SAR enables high image resolutions without for wide-band transmit signals.

Some Relationships to Keep in Mind

Minimum resolution requirements:

Road map details: 10 to 15 m

Shapes: 1/5 to 1/20 of major dimension

Achievable resolution

$d_r = 150\,\tau$ m (where τ is in μs)

τ = compressed pulse width

required bandwidth = $1/\tau$

$d_a \cong \frac{\lambda}{l} R$ (for a real antenna)

$d_a \cong \frac{\lambda}{2L} R$ (for a synthetic array)

l = real antenna length, m

L = synthetic array length, m

Further Reading

D. R. Wehner, *High-Resolution Radar*, Artech House, 1995.

C. V. J. Jakowatz, D. E. Wahl, P. H. Eichel, D. C. Ghiglia, and P. A. Thompson, *Spotlight-Mode Synthetic Aperture Radar: A Signal Processing Approach*, Springer, 1996.

M. Soumekh, *Synthetic Aperture Radar Signal Processing with MATLAB® Algorithms*, Wiley, 1999.

I. G. Cumming and F. H. Wong, *Digital Processing of Synthetic Aperture Radar Data: Algorithms and Implementation*, Artech House, 2005.

R. J. Sullivan, *Radar Foundations for Imaging and Advanced Concepts*, SciTech-IET, 2005.

M. A. Richards, J. A. Scheer, and W. A. Holm (eds.), "An Overview of Radar Imaging," chapter 21 in *Principles of Modern Radar: Basic Principles: Volume 1, Basic Concepts*, SciTech-IET, 2010.

J. J. van Zyl, *Synthetic Aperture Radar Polarimetry* (e-book), Wiley, 2011.

W. L. Melvin and J. A. Scheer (eds.), "Spotlight Synthetic Aperture Radar," chapter 6 in *Principles of Modern Radar: Volume 2, Advanced Techniques*, SciTech-IET, 2012.

W. L. Melvin and J. A. Scheer (eds.), "Stripmap SAR," chapter 7 in *Principles of Modern Radar: Volume 2, Advanced Techniques*, SciTech-IET, 2012.

W. L. Melvin and J. A. Scheer (eds.), "Interferometric SAR and Coherent Exploitation," chapter 8 in *Principles of Modern Radar: Volume 2, Advanced Techniques*, SciTech-IET, 2012.

Test your understanding

1. Explain why, in swath-mapping SAR, the finest achievable resolution is limited to approximately one-half of the length of the real beam antenna used to collect the image data?
2. How does spotlight SAR enable finer resolution imaging than is possible with swath-mapping SAR?
3. Consider the ISAR imaging of an object on a rotating turntable illuminated by a radar located at a fixed position. Show, by means of a diagram, how the inverse synthetic aperture is formed.
4. What are the advantages and disadvantages of using the repeat, parallel pass form of InSAR imaging?
5. What can cause the scattering response from an object to be different for different illuminating polarizations?

Cosmo-SkyMed

COSMO-SkyMed (COnstellation of small Satellites for the Mediterranean basin Observation) is an Earth observation satellite system funded by the Italian Ministry of Research and Ministry of Defense, and conducted by the Italian Space Agency (ASI). It is intended for both military and civilian use including seismic hazard analysis, environmental disaster monitoring, and agricultural mapping. It comprises four identical medium-sized satellites each equipped with and X-band synthetic aperture radar, in sun-synchronous polar orbits that allow observations of an area of interest to be repeated several times a day.

34

SAR Image Formation and Processing

SIR-C polarimetric C-band and L-band images of Great Wall of China (Courtesy of NASA.)

In Chapter 33 we saw how aperture synthesis solves the problem of providing fine along-track resolution, even at very long ranges. The principles of radar imaging are founded primarily on a combination of antenna theory and signal processing concepts. Synthetic aperture radar (SAR) is the most widely used of the radar imaging techniques. Therefore, in this chapter we examine further the principles of SAR, especially the fundamentals of digital processing.

SAR takes advantage of the forward motion of the radar to produce the equivalent of a long antenna, as we have already seen. Each time a pulse is transmitted, the radar occupies a position a little farther along on the flight path. By pointing a relatively small antenna out to one side and summing the returns from successive pulses, it is possible to synthesize a very long side-looking linear aperture.

34.1 Unfocused SAR

The synthesis of an aperture is most easily visualized by studying a simplified unfocused SAR system. Consider an aircraft, carrying an X-band radar system, flying in a straight line at constant speed and altitude. The radar antenna is pointed downward slightly (i.e., at the ground) and aligned at a fixed angle of 90° relative to the flight path (Fig. 34-1).

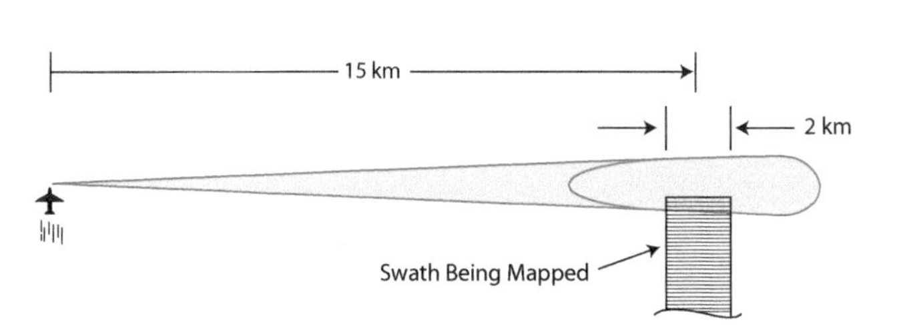

Figure 34-1. In the operation of a SAR radar system with the real antenna trained at a fixed angle 90° to the flight path, the radar maps a 2 km wide swath at a range of 15 km.

As the aircraft progresses, the beam is swept across a broad swath of ground parallel to the flight path. Only a relatively narrow portion of this swath is considered. A portion of the overall swath is a strip 2 km wide, offset from the flight path by about 15 km.

The aircraft's mission requires the ground within this strip be mapped with *across-track* and *along-track* resolutions of 15 m. As will be explained shortly, to provide an along-track resolution of 15 m at a range of 16 km, the SAR radar must synthesize an array roughly 15 m long.

1. An aircraft's ground speed is the horizontal speed of the aircraft relative to the ground.

For an aircraft ground speed[1] of 300 m/s and a pulse repetition frequency (PRF) of 1 kHz, every time the radar transmits a pulse the center of the radar antenna is 30 cm farther along the flight path. The synthetic array can thus be thought of as consisting of a line of elemental radiators 30 cm apart (Fig. 34-2). To synthesize the required 15 m long aperture, 50 such elements are required. In other words, the returns from 50 consecutive transmitted pulses must be combined.

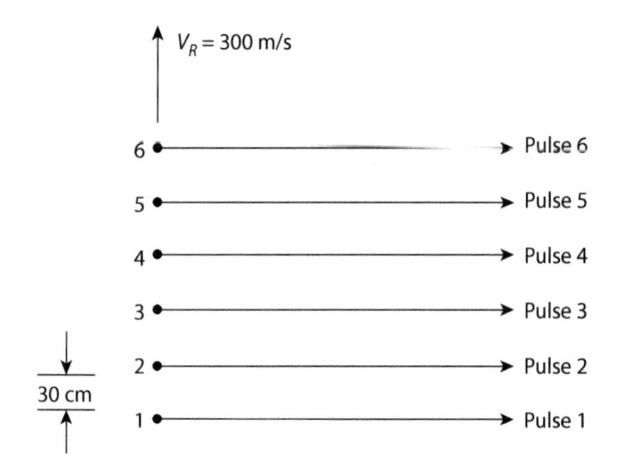

Figure 34-2. These points represent the positions of the center of the antenna when successive pulses are transmitted. Each point constitutes one "element" of synthetic array.

The combining is done after the receiver's output has been digitized. A provided bank of range bins spans the 2 km range swath being mapped (Fig. 34-3). Following every transmission, the return from each resolvable range bin within this interval is added to the contents of the previous equivalent range bin.

The echo return from the first pulse is received entirely by element number one, the return from the second pulse is received entirely by element number two, and so on. The beam-forming operation of the synthetic array can be thought of as being similar to a real array receiving radiation emanating from a particular point of the ground.

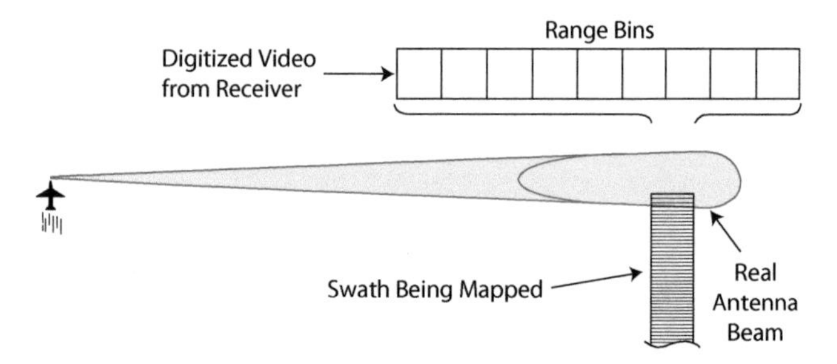

Figure 34-3. Returns received by successive elements of the synthetic array are summed in bank of range bins spanning the range interval being mapped.

Provided that the range is long compared with the array length, the distance from a patch of ground on the boresight line (perpendicular to the flight path) to each array element is approximately equal. Thus, echoes received from the patch by all elements also have a radio frequency (RF) phase that is approximately equal. When added up in the boresight direction, the phases produce a maximal coherent sum (Fig. 34-4).

On the other hand, for a ground patch that is *not* quite on the boresight line, the distance (or phase) from the patch to successive array elements is progressively different. In this case

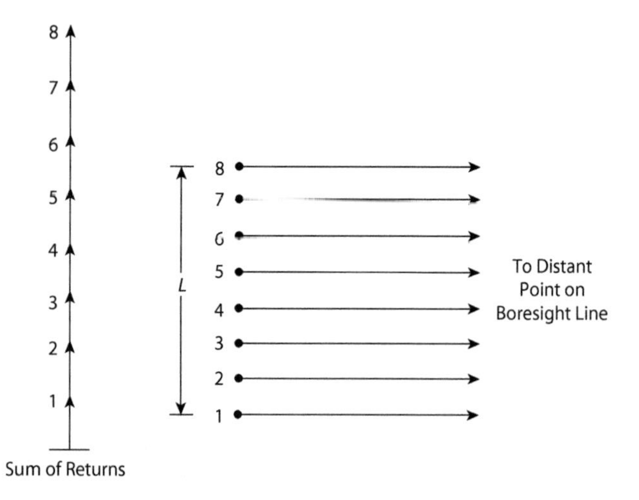

Figure 34-4. Distances from successive array elements to a distant point on the boresight line add up coherently in phase.

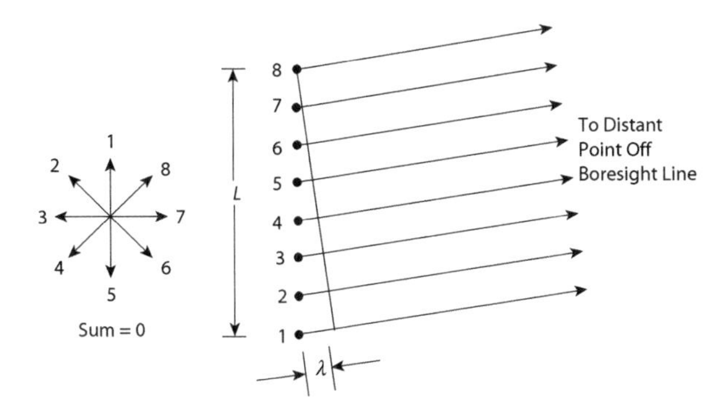

Figure 34-5. Distances from successive array elements to a point off the boresight line are progressively different, so returns from the point tend to cancel. The null condition is shown here.

the echoes received from the off-boresight patch by successive elements have progressively different phases and consequently tend to cancel. This constructive and destructive summation of the echoes leads to the production of a very narrow antenna beam (Fig. 34-5) much as in a real beam array antenna (Chapter 9).

When the returns from the 50 pulses required to form the array have been combined, the result that has built up in each range bin comes quite close to representing the total echo from a single range/azimuth resolution cell (Fig. 34-6). The contents of the bank of bins (pixels), therefore, represent the echoes from a single row of resolution cells spanning the 2 km wide range swath being mapped.

At this point, the contents of the individual range bins can be transferred to corresponding locations in computer memory ready for display (Fig. 34-7). The signal processor then begins

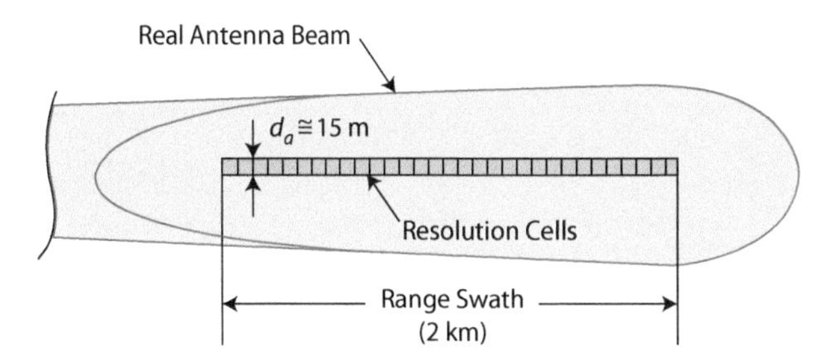

Figure 34-6. When returns from 50 pulses have been integrated, the contents of the bins (or pixels) represent the echoes from a single row of range/azimuth resolution cells.

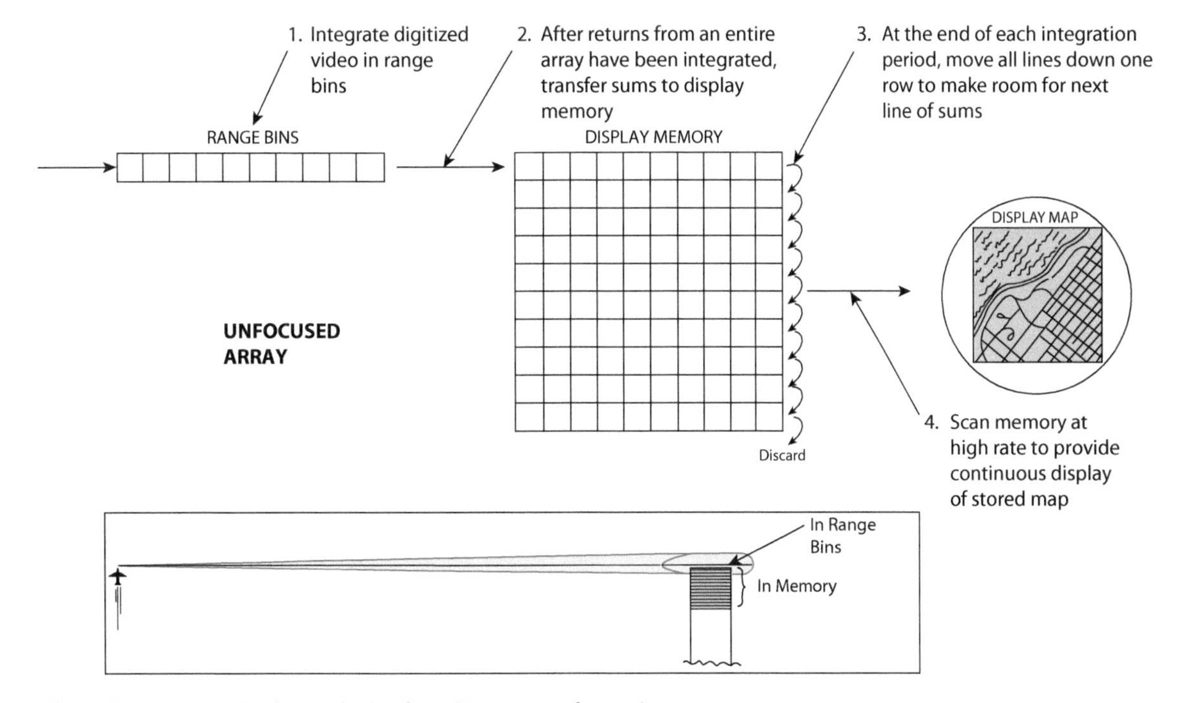

Figure 34-7. Shown here are steps in the synthesis of a rudimentary unfocused array.

the formation of a new aperture, the beam of which will still span the 2 km wide swath but is immediately ahead of the row of cells that have just been imaged. Since the image is formed a line at a time, this method of SAR signal processing is often referred to as *line-by-line processing*.

As the echoes from each new row of cells are received, the stored bins (pixels) are shifted down one row on the display to make room for the new image data. The operator is thus provided with a map that scrolls (usually) from top to bottom of the display screen in real time as the aircraft advances (rather like looking out the window of an aircraft and seeing the ground go past).

Signal Processing for Unfocused Array

THE SIGNAL PROCESSING REQUIRED TO SYNTHESIZE AN UNFOCUSED ARRAY antenna can be summarized mathematically as follows:

Inputs: For each resolvable range, R_r, N successive pairs of numbers are supplied.

$$x_n, y_n \; n = 1, 2, 3, \ldots\ldots N$$

Each pair represents the I and Q components of the return, received from range R_r by a single array element.

Integration: To form the beam (azimuth processing), the I and Q components are summed.

$$I = \sum_{n=1}^{N} x_n \qquad Q = \sum_{n=1}^{N} y_n$$

Magnitude detection: The magnitude or the vector sum of I and Q is computed and stored in memory prior to display.

$$S = (I^2 + Q^2)^{1/2}$$

S is the amplitude of the total return from a single-resolution cell on the boresight line at range R_r. The following figure shows how these steps are applied to process echoes from each range bin.

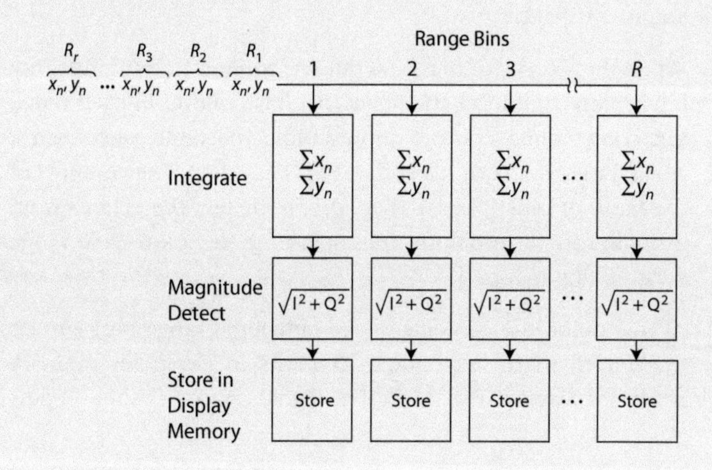

An intermediate step not shown here is the scaling of the detected magnitudes to the values of intensity (gray levels) that are to be displayed (sometimes in dBs).

Signal Processing Required. In the foregoing discussion, the inputs to the SAR processor were referred to only as the digitized radar returns. These inputs are the digitized I and Q components (x_n, y_n) of echoes after translation to baseband. The sums that build up in the range bins are the vector sums of the accumulated complex values of x_n and y_n (i.e., they include phase). The quantities are transferred from the range bins to the display memory as the magnitudes of the vector sums. The signal processing required to synthesize a simple array of this type is summarized in the blue panel above. Note: the equations are identical to those that must be solved to form a Doppler filter tuned to zero Doppler frequency.

Limitation of Unfocused Array. An unfocused array is short compared with the range to the swath being mapped. This

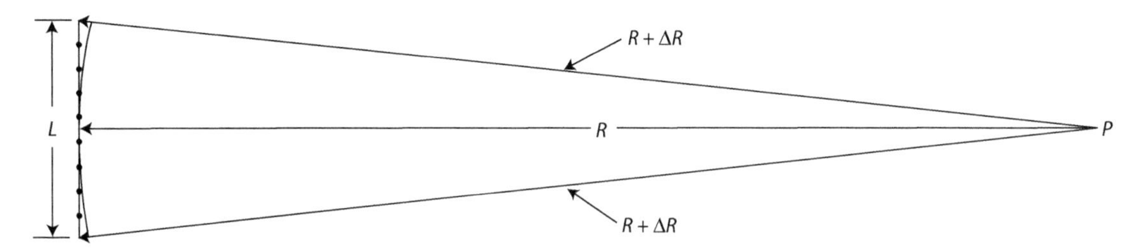

Figure 34-8. If the array length, L, is an appreciable fraction of the range, R, from a point, P, the distances to the end elements will be appreciably greater than the distances in the central element.

results in the lines of sight from any one point in the swath to the individual array elements being more or less parallel. However, the shorter the array, the less fine the along-track resolution.

Now, if the array length is an appreciable fraction of the swath's range, the lines of sight from a point at that range to the individual elements will diverge slightly because the wave front is curved. Then, even if the point is on the boresight line, the distances to the elements will not all be exactly the same (Fig. 34-8). If the wavelength is fairly short, very small differences in these distances can result in considerable differences in the phases of the echo returns that the individual array elements receive from a point, P.

Because these phase errors are not compensated in the unfocused array, image quality will be reduced because there will be a loss of contrast. While increasing the length of the array can initially improve the azimuth resolution (at a given range), a point is soon reached beyond which any additional increase in length only degrades performance due to the changes in phase becoming appreciable.

Degradation begins with a gradual increase in gain of the sidelobes relative to the mainlobe and a merging of the lower-order sidelobes with the mainlobe (Fig. 34-9).

This effect continues increasingly and is accompanied by a progressive falloff in the rate at which the mainlobe gain increases with array length.

The reason for the falloff in gain can be seen in examining Figure 34-8. It shows the distance from point, P, on the boresight line to each element of an array of length, L. For elements near the array center, there is very little difference in this distance. But for elements farther and farther removed from the center, the difference grows increasingly. As the array is lengthened, the phase of the returns received by the endmost elements falls increasingly behind the phase of the sum of the returns received by the other elements. As a consequence, the efficiency of the coherent summation is diminished.

The effect of this progressive phase rotation on the mainlobe gain of the unfocused synthetic array is illustrated in more

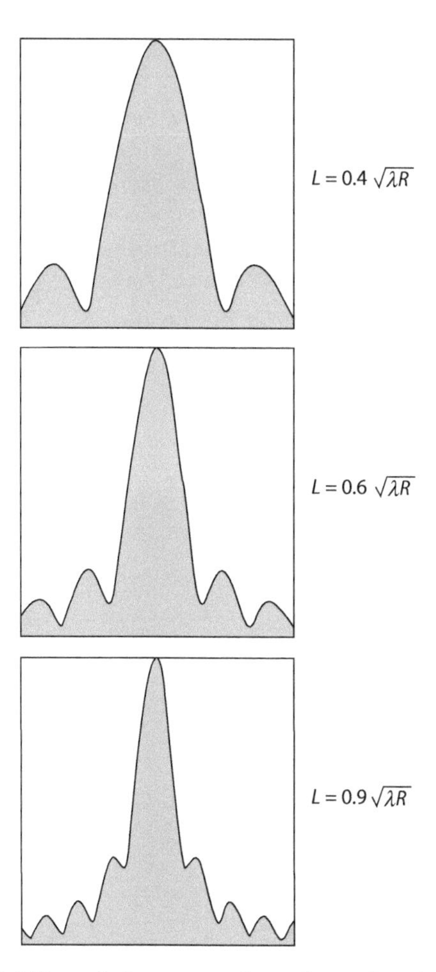

Figure 34-9. The radiation pattern of an unfocused synthetic array shows the increase in relative gain of sidelobes and the merging of sidelobes with the mainlobe as the array length, L, is increased.

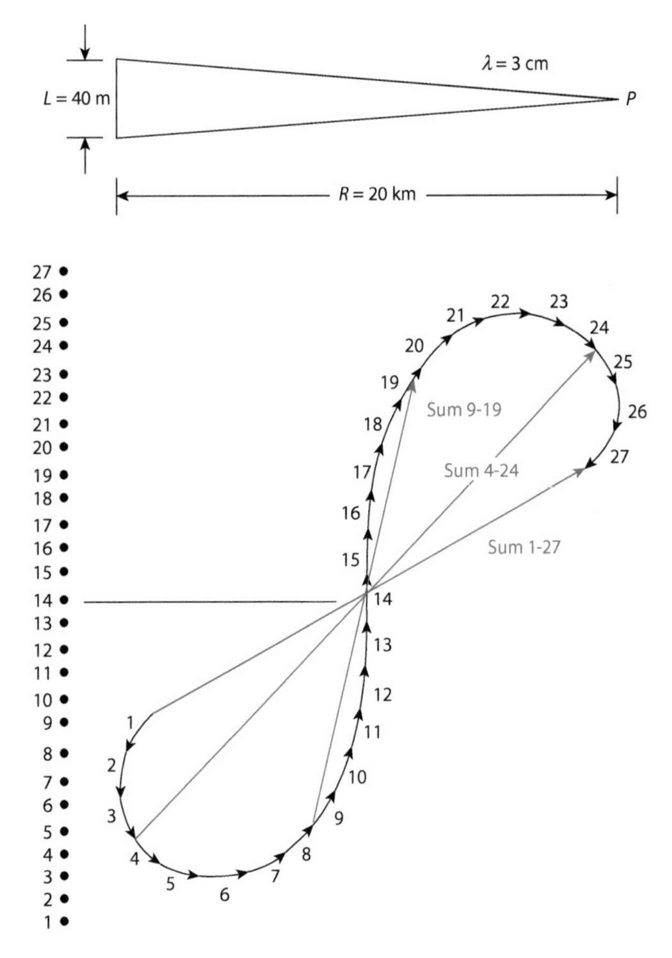

Figure 34-10. The degradation in the gain of an unfocused array is shown here. Phasors represent echo returns received from a distant point, *P*, by individual array elements. The gain is given by the sum of the phasors. In this case, decreasing the length of the array could increase the gain.

detail in Figure 34-10. The phasors show the returns received by the individual elements of a 27-element array from a distant point, *P*, on the boresight line. The phase of the return received by the middle element (14) is taken as the reference. The gain in the boresight direction corresponds to the sum of the phasors.

The returns received by the central elements (9 through 19) are so close to being in phase that their sum is virtually undegraded by the lack of focus, and coherent combining is almost 100% efficient. However, the phases of the returns received by elements farther and farther out are increasingly rotated. The returns received by elements 4 and 24 are nearly 90° out of phase with the sum of the returns received by the elements closer in and hence contribute negligibly to the coherent sum. The returns received by elements 1, 2, and 3 and 25, 26, and 27 are subtractive and will diminish the resulting sum. Under the conditions for which Figure 34-10 was drawn, the gain would have its maximum value if the array were only about 21 elements long (elements 4 through 24).

34.2 Focused SAR

The limitation on array length may be largely removed by focusing the array. This allows the length of the array to be increased in proportion to the range such that virtually the same resolution may be obtained at any desired range.

How Focusing Is Done. In principle, to focus an array all that needs to be done is to apply an appropriate phase correction (rotation) to the returns received by each array element. As illustrated in Figure 34-11, the phase error for any one element, and hence the phase rotation needed to cancel the error, is proportional to the square of the distance of the element from the center of the array, for example,

$$\text{Phase correction} = -\frac{2\pi}{\lambda R}\delta_n^2$$

where

δ_n = distance of element n from array center

λ = wavelength (same units as δ)

R = range to area being mapped (same units as λ)

In some cases, it may be possible to *pre-sum* the returns received by *blocks* of adjacent elements without impairing performance. By rotating only the phases of the sums, computing and storage requirements may be eased.

To simplify the description of focused SAR image formation, pre-summing is omitted here. It is also assumed that the combination of array length, PRF, and range is such that the along-track resolution distance, d_a, is roughly equal to the spacing between array elements.

To focus an array when no pre-summing is done, there must be as many rows of storage positions in the bank of range bins as there are array elements (Fig. 34-12). As the echoes from any

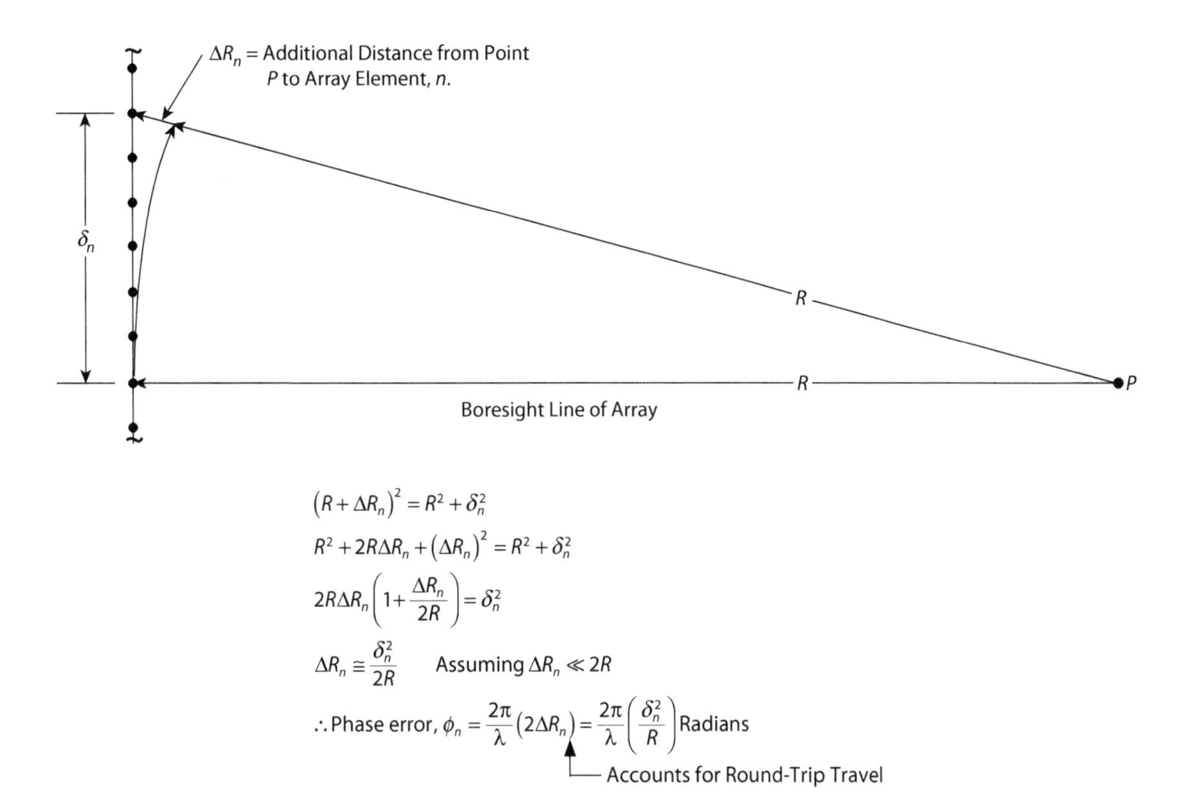

Figure 34-11. The phase error for a return received by any single array element, n, is proportional to the square of the distance, δ_n, from the element to the array center. The factor of two by which ΔR_n is multiplied accounts for the phase error being proportional to the difference in *round-trip* distance from the element to point *P*.

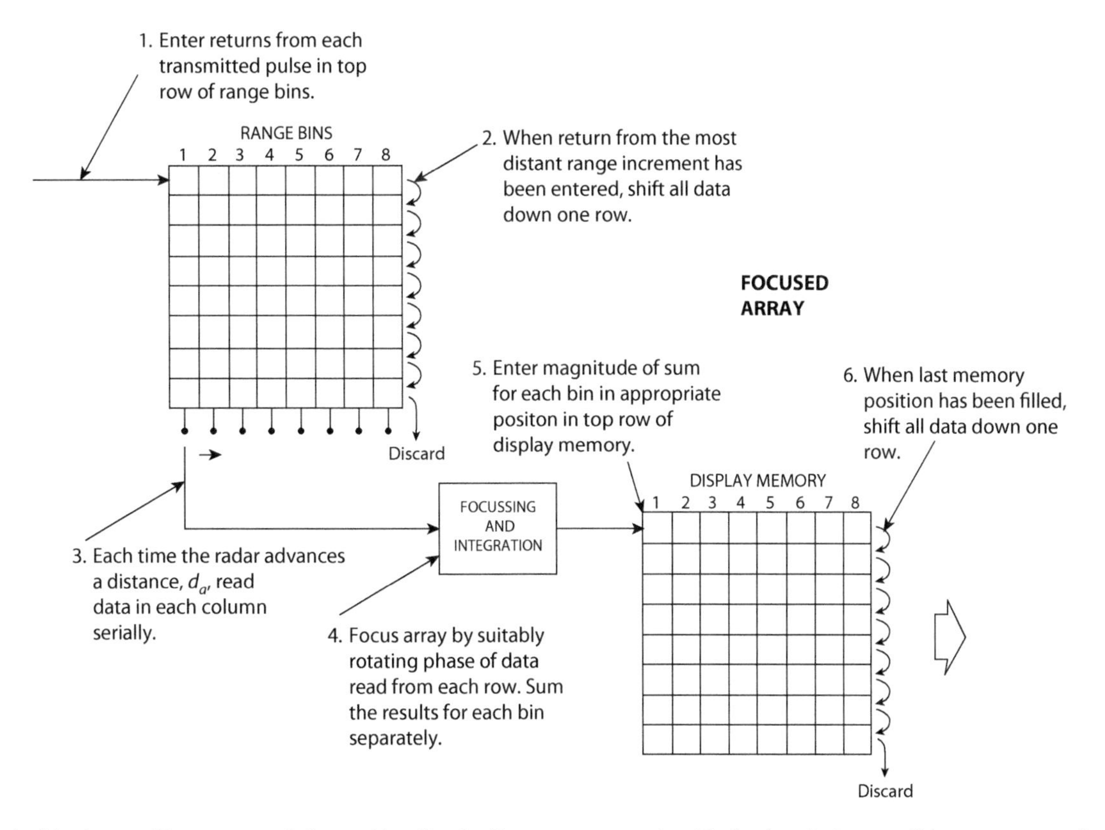

Figure 34-12. In this picture of how an array is focused in a line-by-line processor, to simplify the description, conditions are assumed to be such that the resolution distance, d_a, equals the spacing between array elements. Hence, every time a pulse is transmitted a new array must by synthesized.

one transmitted pulse are received (i.e., for one array element), they are stored in the top row. When the echo from the most distant range increment has been received, the contents of every row are shifted down to the row below it to make room for the incoming returns from the echo of the next transmitted pulse (i.e., the next synthetic aperture element). The contents of the bottom row are discarded.

The compressed returns from a single transmitted pulse form a vector of complex-valued data samples. Each sample of the vector is associated with a range resolution cell. The number of range bins (the length of this complex vector) is dependent on the swath width of the SAR image. The number of range bins is equal to the swath width divided by the range resolution, d_r.

Each successive transmitted pulse (array element) therefore results in a new range-bin vector of complex data samples. It is helpful to consider this graphically such that a two-dimensional array of data is formed (Fig. 34-12). In this 2D data array, the horizontal index indicates range bin, while the vertical index indicates synthetic array element (i.e., transmit pulse number).

This data array can also be viewed as a collection of column vectors, with one vector for each resolved range bin. The elements of these column vectors hold the measured synthetic array elements or data associated with that range bin. Properly focusing this synthetic aperture data leads to the desired azimuthal resolution, d_a, and to the scattered intensity from that resolution cell.

Specifically, a finite segment of each column vector is selected, corresponding to a specific synthetic array. The length of this synthetic array (the number of elements of the column vector that are selected) depends on the desired azimuthal resolution, d_a (a smaller d_a requires a longer synthetic aperture). For fine azimuth resolution, this synthetic aperture will be so long that the phase errors in Figure 34-11 must be compensated prior to summation. This process of phase correction, followed by summation of the corrected synthetic array data, is called *azimuth compression* and results in a focused array.

This process is repeated for the corresponding column data segment for each of the range bins, and the result is precisely one row of a SAR image. This row or image is in the across-track (i.e., range) dimension, from one side of the image swath to the other. However, this image is for just one along-track location (the location at the boresight of the synthetic aperture). In other words, it is for the along-track location directly opposite the center element of the synthetic array.

To determine the next row of the SAR image in the along-track dimension and to determine the image pixels for the next line of range resolution cells, the synthetic array is simply "shifted" a distance, d_a, in the along-track direction. Since, for this example, it is assumed that the synthetic array elements are spaced at a distance d_a, this shift in the synthetic array means adding one more element at one end of the array and discarding an element at the other.

For the 2D data array, this shift requires that the finite segment of column data is shifted downward one element. A column element is added at one end, and a column element is discarded at the other. This new synthetic array needs to be focused to provide azimuth resolution. This is repeated for each range bin, and a second row of the SAR image is created, again providing an image in the cross-track dimension. This time it is for scatterers at a location displaced a resolution distance, d_a, further away in the along-track dimension.

If this processing is progressively repeated (shifting the synthetic array by one element each time), the rows of the overall swath-mapping SAR image are progressively created, and the operator is presented with a moving or scrolling display of the SAR imagery as the aircraft flies along. The duration of the imaged swath is limited only by the length or duration of the radar flight track.

34.3 SAR Processing

To focus an array for every range bin, the signal processor must mathematically perform the equivalent of rotating the phasor, A, representing the echo received by each successive array element, n, through the required phase angle, ϕ_n (Fig. 34-13).

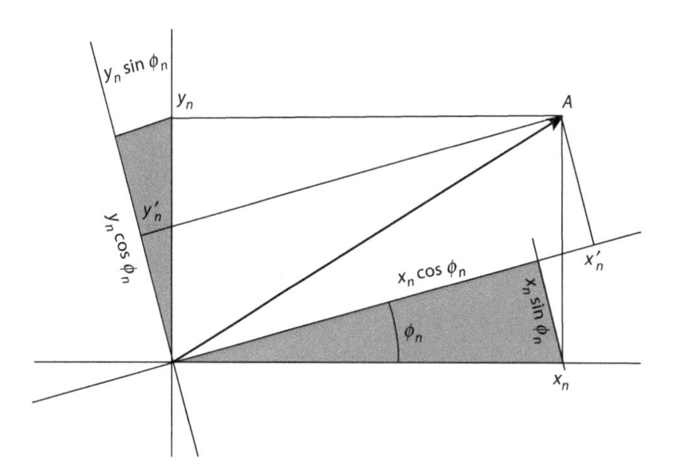

Figure 34-13. The I and Q components of the phasor before rotation are the inputs, x_n and y_n. After rotation, the components are represented by x'_n and y'_n.

To perform the rotation we use

$$x'_n = x_n \cos \phi_n + y_n \sin \phi_n$$
$$y'_n = y_n \cos \phi_n + x_n \sin \phi_n$$

The values of x'_n and y'_n for the total number of array elements, N, must then be summed separately

$$X = \sum_{n=1}^{N} x'_n \qquad Y = \sum_{n=1}^{N} y'_n$$

and the magnitude of the vector sum of X and Y must be calculated.

$$S = \sqrt{(X^2 + Y^2)}$$

Signal Processing Required. For very fine azimuth resolution, each individual element will be a part of many successive synthetic arrays. First, an element will appear at one end of the array. Sometime later it will become the center element of the synthetic array, and then finally it will be the last element. For each synthetic array, the *complex-valued measurement* associated with this single element remains the same. However, the phase *correction* applied to this element will be different for each focused array. For example, when the element is at the center of the synthetic aperture, no phase correction is applied, whereas, when at either end of the synthetic array, the phase correction is a maximum. Thus, a phase-rotated measurement value will be used only in SAR image synthesis.

Why the Element-to-Element Phase Shift is Double in a Synthetic Array

A LINEAR-ARRAY RADAR ANTENNA ACHIEVES ITS DIRECTIVITY BY VIRTUE OF THE progressive shift in the phases of the returns received by successive array elements from points off the antenna boresight line. For any one angle off boresight, this shift is twice as great for a synthetic array as for a real array having the same interelement spacing. The reason can be seen by considering the returns received from a distant point, *P*, displaced from the boresight line by a small angle, *θ*.

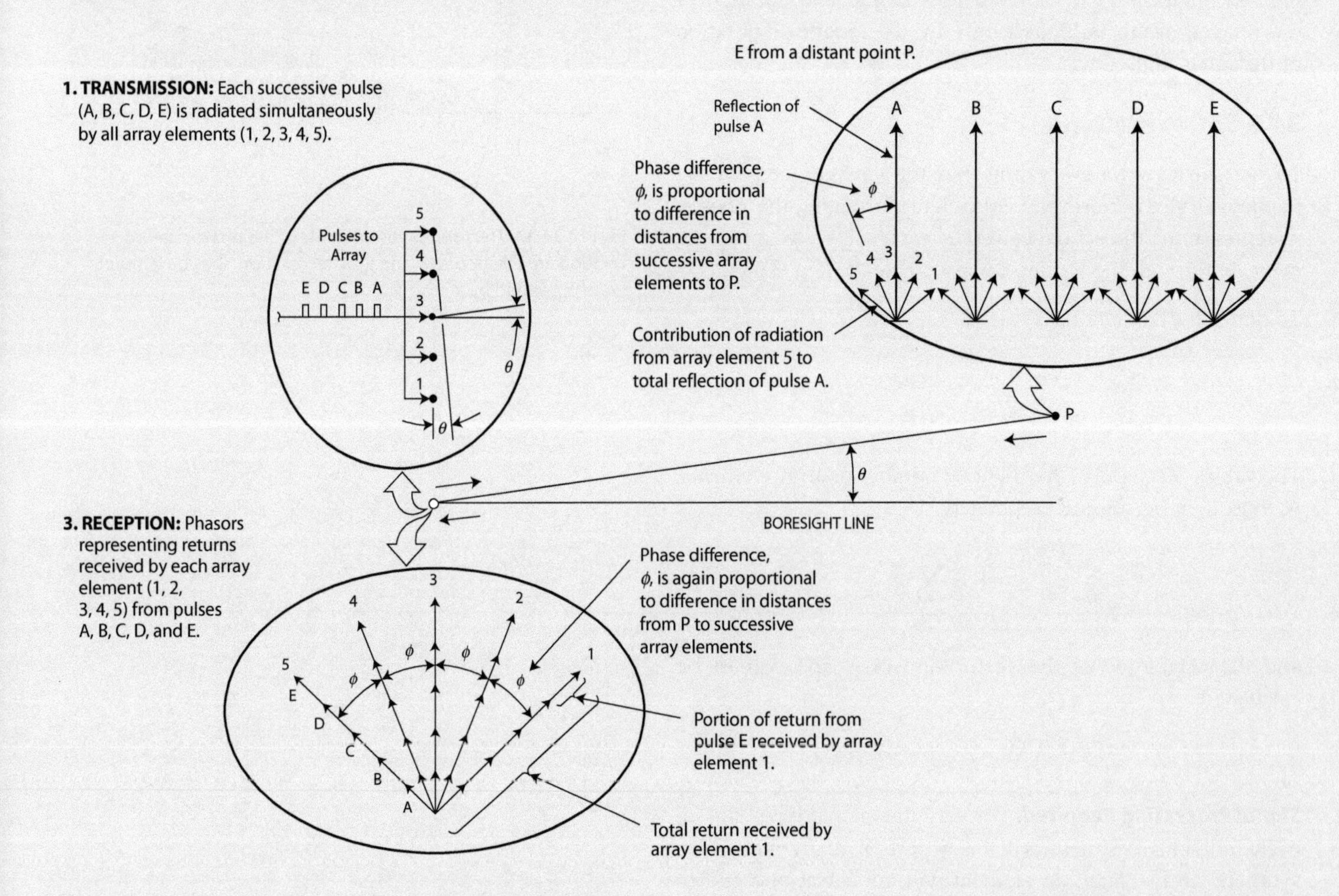

Real Array. In the case of a real array, transmission from all array elements is simultaneous. Every time a pulse is transmitted, the radiation from all elements arrives at *P* simultaneously (albeit staggered in phase as a result of the progressive differences in the distances from successive elements to *P*). The phase differences naturally reduce the amplitude of the sum of the radiation received at *P* from the individual elements. This reduction gives the one-way radiation pattern its sin *x*/*x* shape. But for each pulse the phase shift of the sum is fixed by the distance from the central element to a distant point, *P*. Therefore, if the position of the antenna is not changed, all of the pulses reflected by *P* have the same phase.

Why the Element-to-Element Phase Shift is Double in a Synthetic Array *continued*

The portions of each reflected pulse that are received by the individual array elements similarly differ in phase as a result of the progressive difference in the distances from *P* to the elements. As with transmission, the phase differences reduce the amplitude of the sum of the outputs the received pulse produces from the individual elements. This reduction, compounded with the reduction in the amplitude of the pulses reflected from *P*, gives the two-way radiation pattern its *sin x/x* shape. But the phase differences are again due only to the differences in the *one-way* distances from *P* to the elements.

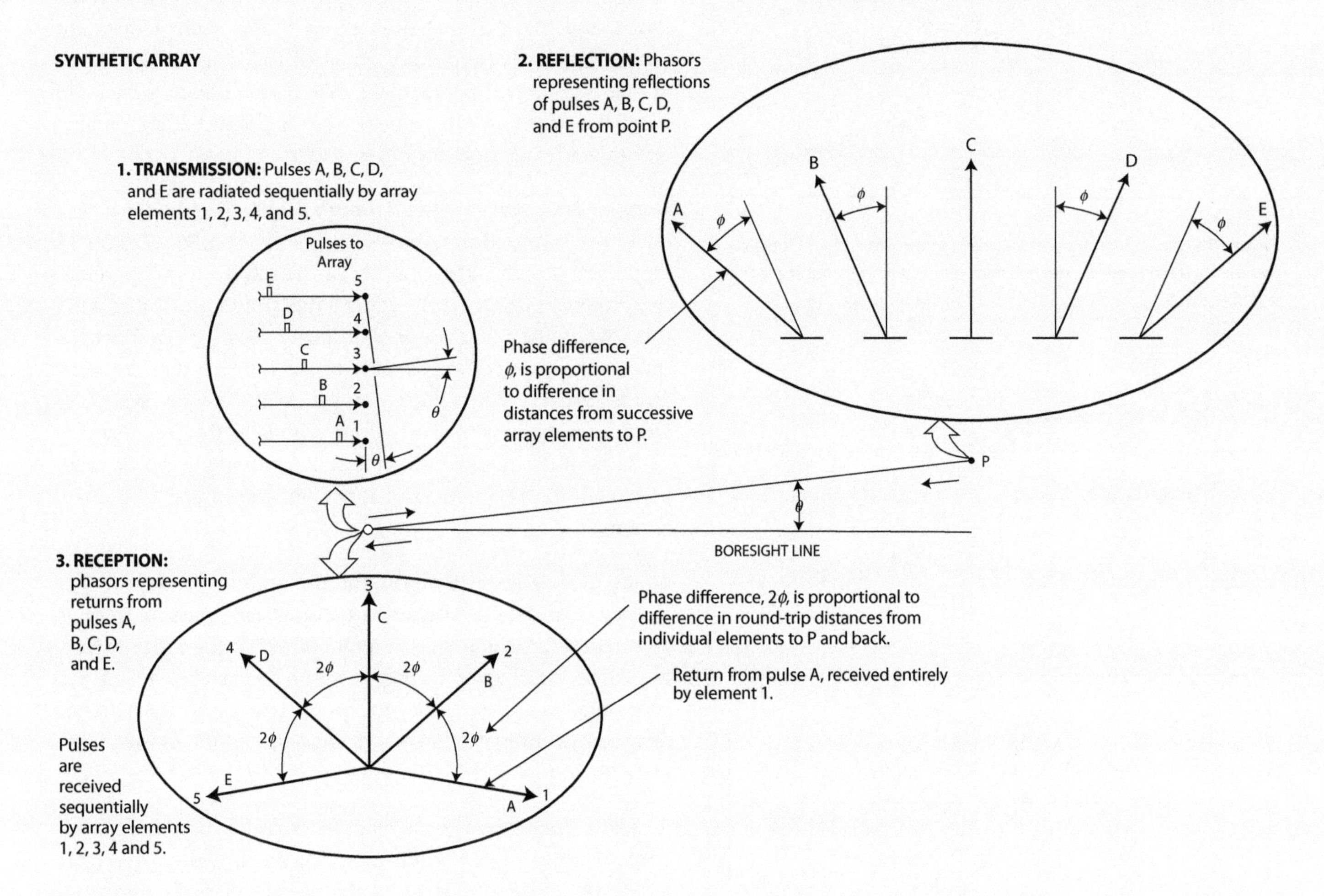

Synthetic Array. In the case of a synthetic array, transmission from the individual array elements is sequential. The first pulse is transmitted and received entirely by the first element, the second pulse entirely by the second element, and so on. Consequently, the returns received by successive elements differ in phase by amounts proportional to the differences in the *round-trip* distance from each element to point *P* and back to the element again.

Thus, for any one angle off the boresight line, the progressive shift in the phases of the returns received by successive array elements is twice as great for a synthetic array as for a real array. (The doubling of phase shift gives the beam of the synthetic array its sin $2x/2x$ shape.)

The doubling of the phase shifts must, of course, also be kept in mind when calculating such factors as the phase corrections necessary to focus an array and the angles at which grating lobes will occur.

Reducing the Computing Load: Doppler Processing. If the array is very long, a significant amount of computing is required for line-by-line focused processing. Every time the radar transmits a pulse, it must perform the phase correction all over again for every pulse that has been received over an entire array length

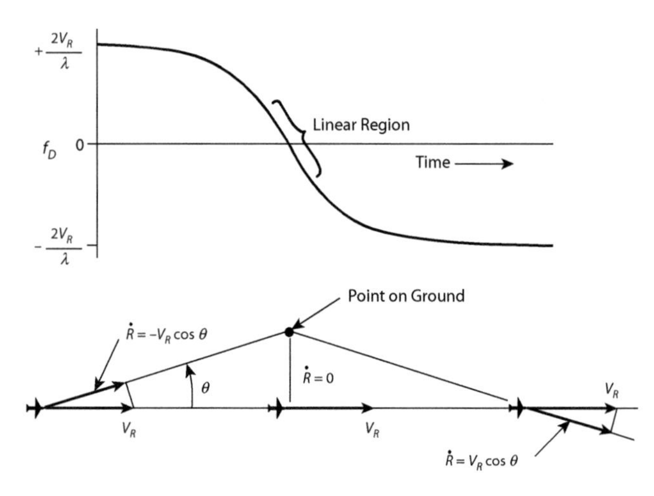

Figure 34-14. As a radar passes a point on the ground, its Doppler frequency decreases at a nearly linear rate, passing through zero when the point is at an angle of 90° to the radar's velocity.

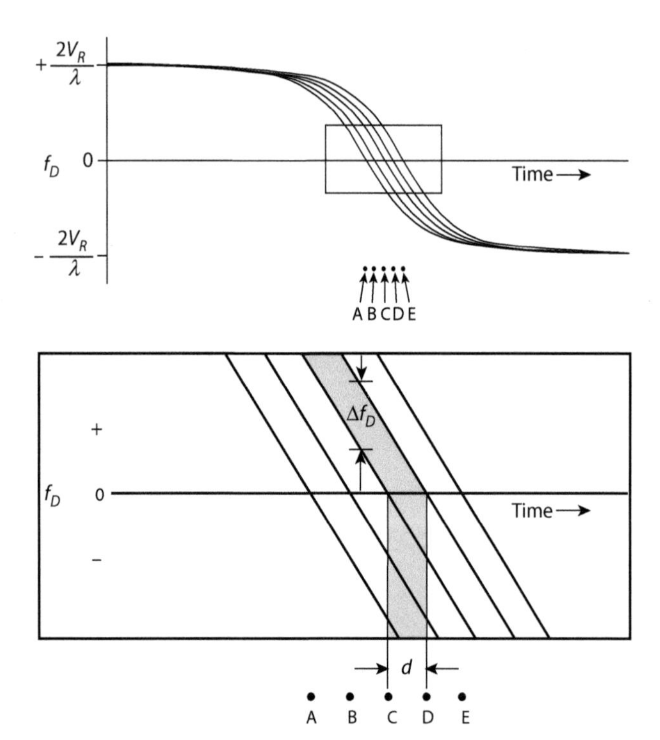

Figure 34-15. This depicts Doppler histories of evenly spaced points on the ground, the instantaneous frequency difference, Δf_D, is proportional to the azimuthal distance between points, *d*.

and then sum all of the results. Put another way, if there are *N* elements in a synthetic array, every time the radar advances a distance equal to the array length, it must correct and sum $(N \times N)$ phases for every range bin. This load may be reduced somewhat by pre-summing, but the processing challenge can still be formidable.

Processing the data in parallel can dramatically reduce the computing load. Here many lines of the map are produced at one time rather than serially, that is, one line at a time. For parallel processing, the returns from different azimuth angles are isolated with Doppler filters.

Doppler Frequency versus Azimuth Angle. We saw that SAR could be described in terms of Doppler in Chapters 32 and 33. Figure 34-14 shows the Doppler of the return from a point on the ground offset from the radar's flight path as a function of time. When the point is a great distance ahead, its Doppler frequency corresponds very nearly to the full speed of the radar and is positive. When the point is a great distance behind, its Doppler frequency similarly corresponds to the full speed of the radar but is negative. Note that the extreme Doppler values eventually become suppressed by the beamshape of the real antenna.

As the radar goes by the point on the ground, its Doppler frequency decreases at virtually a constant rate, passing through zero when the point is at an angle of 90° to the radar's velocity. If the radar antenna has a reasonably narrow beam and is looking out to the side, the point will be in the antenna beam only during this linearly decreasing portion of the point's Doppler history.

A plot of this portion of the Doppler histories of several evenly spaced points at the same offset range is shown in Figure 34-15. As can be seen, the histories are identical; the frequency decreases at the same constant rate except for being staggered slightly in time. Because of this stagger, at any one instant the echoes from different points have slightly different frequencies. The difference between the frequencies for adjacent points corresponds to the azimuth separation of the points. Thus, the return received from each point can be isolated by virtue of this difference in Doppler frequency.

Implementation. Although more advanced processing concepts have been developed for generating SAR imagery,[2] the method described here highlights the key steps. The block diagram of Figure 34-16 represents the general steps required for SAR processing.

At the outset a phase correction is made to the echo returns received from each pulse to remove the linear slope of the Doppler histories, converting the echo from each point on the

2. Indeed, many methods of efficient SAR processing have been reported, and perhaps a technique known as the *phase gradient algorithm* is the most popular. For a description of this and other approaches, see Further Reading.

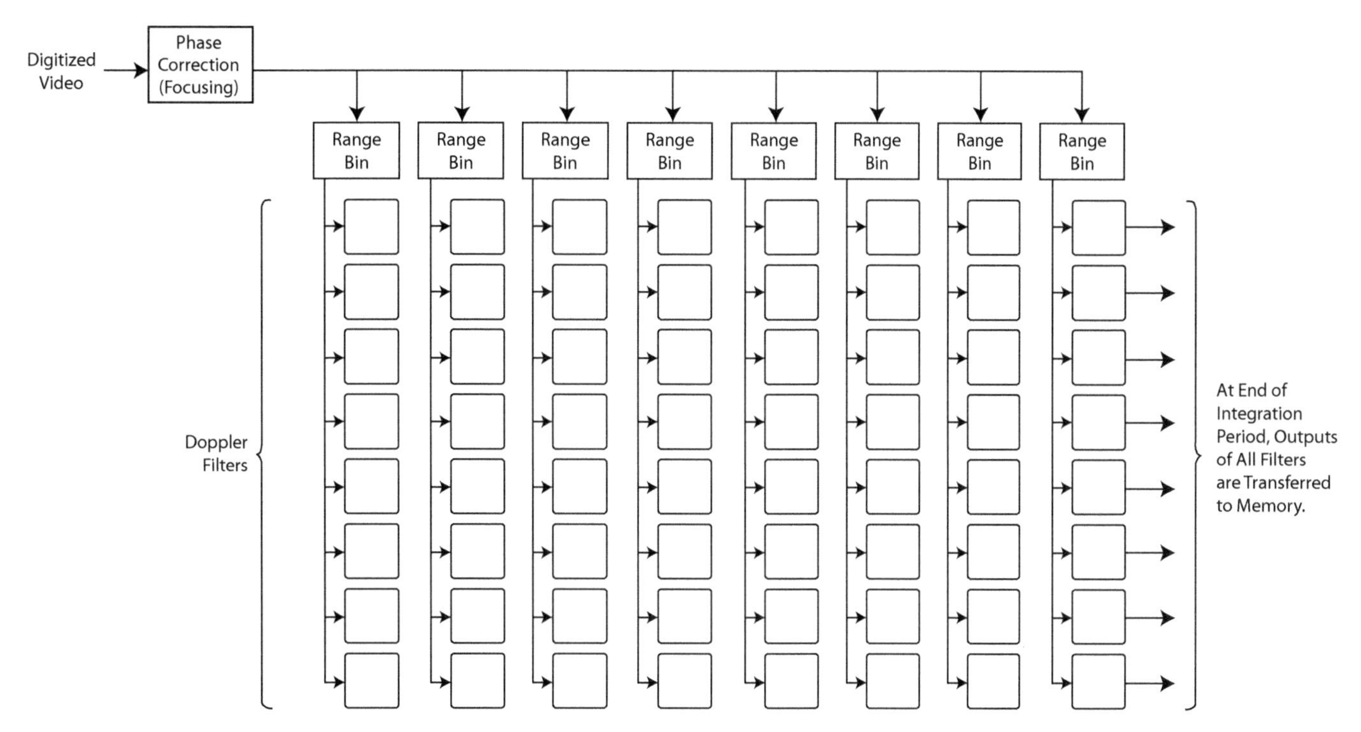

Figure 34-16. This figure shows how SAR processing is done. After focusing corrections have been made, the echo returns are sorted by range. When echoes from a complete array have been received, a separate bank of filters is formed for each range bin.

ground to a constant Doppler frequency (Fig. 34-17). This frequency corresponds to the azimuth angle of the point, as seen from the center of the segment of the flight path over which the return was received.

Every time the aircraft traverses a distance equal to the length of the array that is to be synthesized, the returns are phase corrected (for each range bin) and applied to a bank of Doppler filters. Thus, for every array length, as many banks of filters are formed as there are range bins. The integration time for the filters is the length of time the aircraft takes to fly the array length. The number of filters included in each bank correspondingly depends on the length of the array. The greater it is (the longer the filter integration time), the narrower the filter passbands (and the higher the azimuth resolution) and the greater the number of filters required to span a given band of Doppler frequencies.

Since the frequencies to be filtered are relatively constant over the integration time and (for uniformly spaced points on the ground) are evenly spaced, the fast Fourier transform (FFT) can be used to form the filters, greatly reducing the amount of computation. The filters are formed at the end of the integration period, that is, after the radar has traversed an entire array length. The outputs of each filter bank represent the echoes from a single column of resolution cells at the same range (the range of the range bin for which the

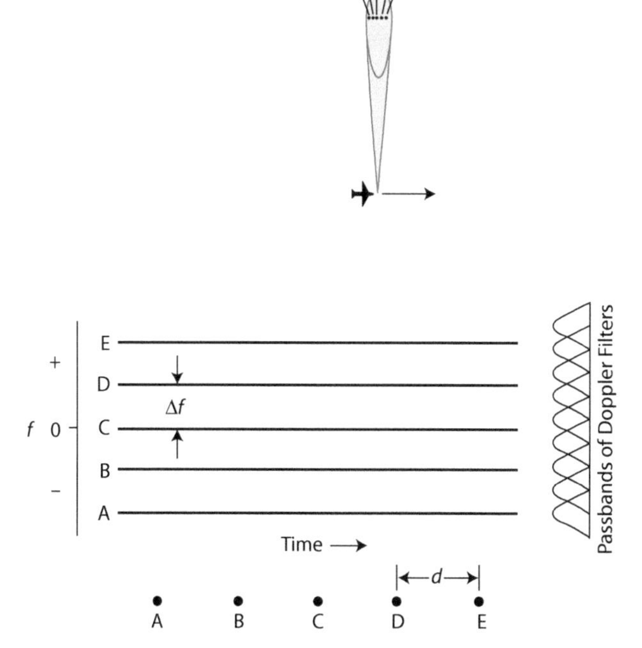

Figure 34-17. A phase correction converts the return from each point on the ground to a constant frequency, enabling the Doppler filters to be formed with the FFT.

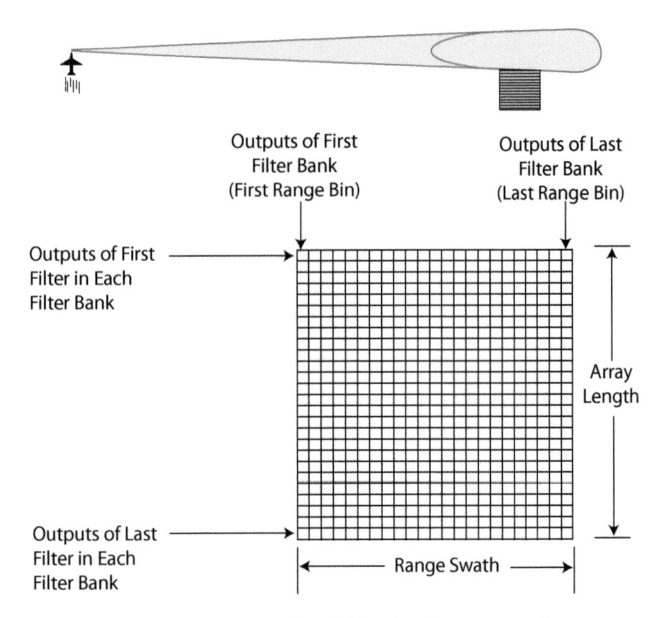

Figure 34-18. The outputs of each filter bank represent the return from a single column of range/azimuth resolution cells.

bank was formed; Fig. 34-18). The outputs of all of the filter banks therefore can be transferred as a block, directly to the appropriate positions in the display memory. The radar, meanwhile, has traversed another array length, thereby accumulating the data needed to form the next set of filter banks, and the process is repeated.

Reduction in Arithmetic Operations Achieved. Having gained a clear picture of the Doppler-filtering method of SAR azimuth compression, let us see what kind of saving in arithmetic operations it actually provides. To simplify the comparison, it is again assumed that no pre-summing is done.

In the Doppler processor, phase rotation takes place at two points: (1) when the return is focused; and (2) when the Doppler filtering is done. For focusing, only one phase rotation per pulse is required for each range bin. As explained in Chapter 21, in a large filter bank the number of phase rotations required to form a filter bank with the FFT is $0.5N \log_2 N$, where N is the number of pulses integrated. The total number of phase rotations per range bin is $N + 0.5N \log_2 N$. For line-by-line processing, the number of phase rotations per pulse per range gate is N^2.

Processing	Phase Rotations
Line by line	N^2
Parallel (Doppler)	$N(1 + 0.5 \log_2 N)$

To get a feel for the relative sizes of the numbers involved, let us take as an example a synthetic array having 1024 elements. With line-by-line processing a total of 1024 × 1024 = 1,048,576 phase rotations would be required. With parallel processing, only $1024 + 512 \log_2 1024 = 6{,}144$ would be required. The number of additions and subtractions would similarly be reduced, and the computing load would be reduced by a factor of roughly 170.

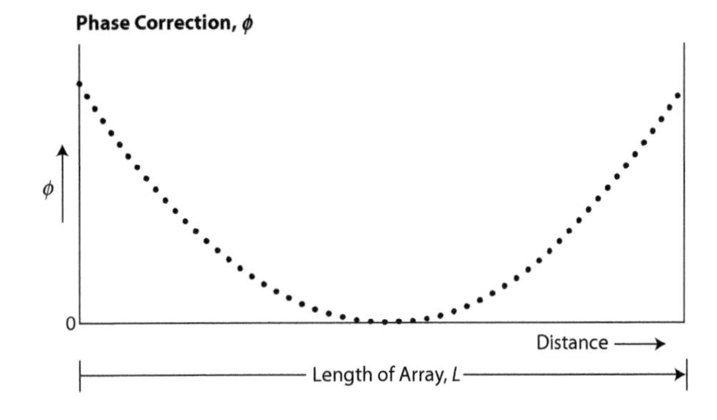

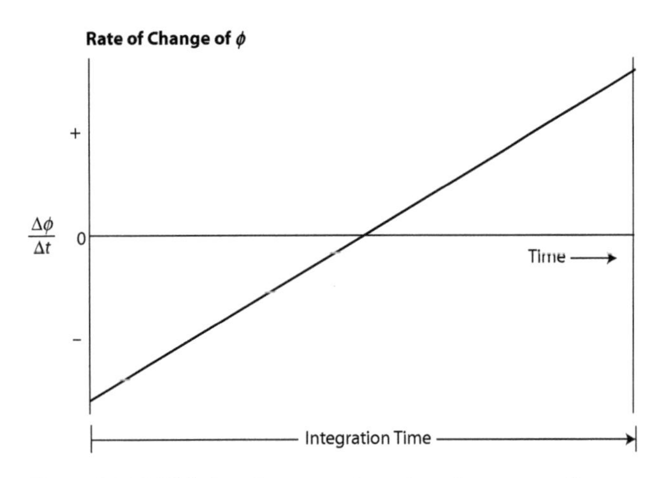

Figure 34-19. With focusing corrections, ϕ, made to returns from successive blocks of pulses, the rate of change of ϕ has the same slope as the Doppler history of a point on the ground, but it is rising rather than falling.

Correspondence to Conventional Array Concepts. Superficially, Doppler processing may seem like a fundamental departure from conventional array concepts, but it is not. A Doppler frequency is nothing more than a progressive phase shift. To say that a signal has a Doppler of 1 hertz is to say that its phase is changing at a rate of 360° per second. If the PRF is 1000 Hz, the pulse-to-pulse phase shift is 360°/1000 = 0.36°. Viewed in this light, the Doppler histories that have been considered are, more correctly, just phase histories.

The phase corrections used to remove the slope of the Doppler history curves are exactly the same as the corrections used to focus the array in the line-by-line processor. This is illustrated by the graphs in Figure 34-19. The U-shaped curve is a plot of the focusing corrections applied to the returns received by successive array elements in line-by-line processing. The straight diagonal line is a plot of the rate of change of these corrections. Its slope, you will notice, is identical to the slope of the Doppler history of a point on the ground, but it is rising rather than falling. The same focusing correction that is used by the line-by-line processor, therefore, converts the linearly decreasing frequency of the return from each point on the ground to a constant frequency.

The beams synthesized by the two processors are virtually the same; the only difference is in their points of origin (Fig. 34-20).

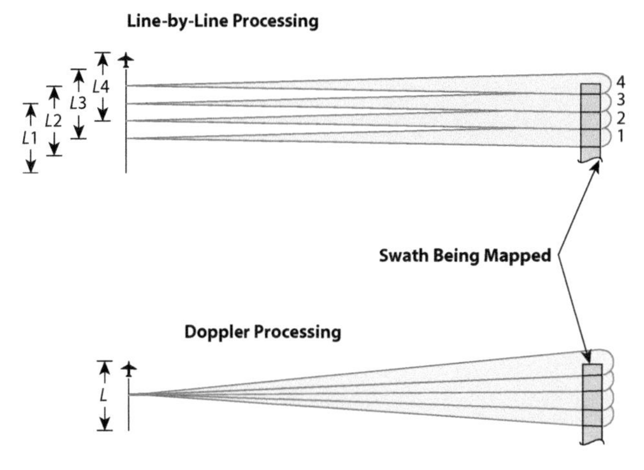

Figure 34-20. Synthetic array beams formed with line-by-line processing and Doppler processing differ only in their points of origin.

With the line-by-line processor, every time the radar advances one azimuth resolution cell, d_a, a new beam is synthesized. On the other hand, with the Doppler processor, every time the radar advances one array length, each Doppler filter bank synthesizes a new beam.

The beams formed by the line-by-line processor all have the same angle (90° in our example). However, because of the radar's advance, at the range being mapped they overlap only at their half-power points. The beams formed by the Doppler filters, on the other hand, all originate at the same point (center of the array) but fan out at azimuth angles such that they overlap at their half-power points.

And how do the along-track resolutions provided by the two processors compare? The 3 dB bandwidth of the Doppler filters is roughly equal to one divided by the integration time ($BW_{3\,dB} \cong 1/t_{int}$). As shown in Figure 34-21, the difference between the Doppler frequencies of two closely spaced points on the ground at azimuth angles near 90° is twice the radar velocity times the azimuth separation of the points, divided by the wavelength.

$$\Delta f_D = \frac{2V_R\Delta\theta}{\lambda}$$

Equating Δf_D to $BW_{3\,dB}$ and substituting $1/t_{int}$ for it, we obtain the following expression for the width of the beam synthesized by the Doppler processor.

$$\Delta\theta = \frac{\lambda}{2V_R t_{int}}$$

where

$\Delta\theta$ = beamwidth

V_R = radar velocity

t_{int} = integration time

The product of the radar's velocity and integration time, $V_R t_{int}$, is the distance flown during the integration time. As illustrated

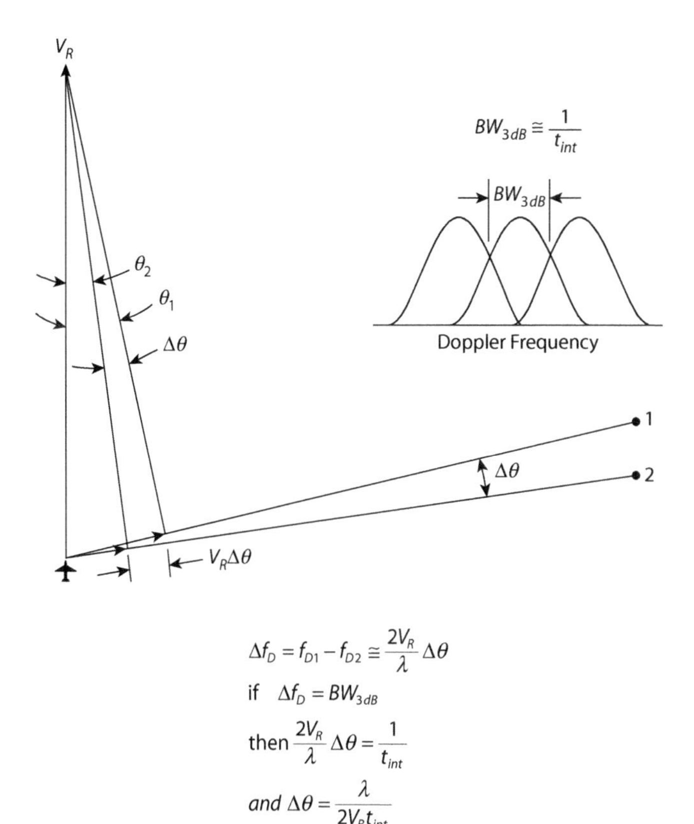

Figure 34-21. The 3dB bandwidth of a Doppler filter is one divided by the integration time. The difference in Doppler frequency, Δf_D, of the returns from two points on the ground is proportional to their angular separation, $\Delta\theta$. Equating BW_{3dB} to Δf_D yields an expression for the angular resolution.

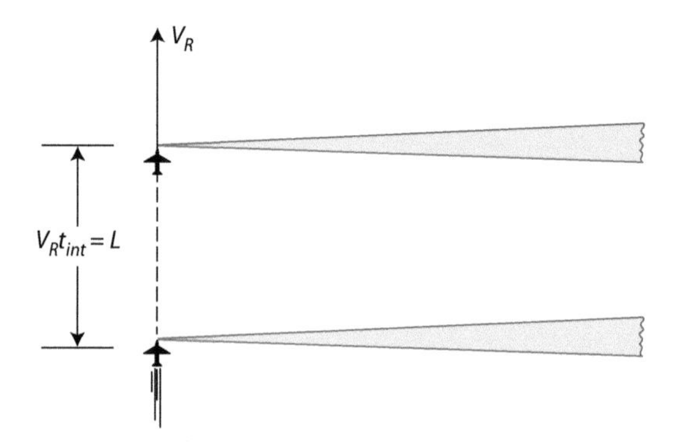

Figure 34-22. The length of the array synthesized with Doppler processing is the distance flown during the integration time of the Doppler filters.

in Figure 34-22, this is the length of the array, L. Substituting in the expression for $\Delta\theta$ and multiplying by the range, R, we find the azimuth resolution distance to be

$$d_a = \frac{\lambda}{2L} R$$

This is exactly the same as the azimuth resolution distance for the line-by-line processor. So both the Doppler processing and the conventional array concept are equivalent and merely represent two ways of achieving the same thing.

34.4 Motion Compensation and Autofocus

Throughout this chapter it has been assumed that the aircraft or spacecraft carrying the SAR instrument flies at a constant speed in a perfectly straight line. But this is virtually never the case. Aircraft are particularly susceptible to atmospheric turbulence, and spacecraft can experience drift and other orbital uncertainties. Since the whole SAR concept revolves around the effect of very slight differences in the phases of microwave signals received over a comparatively long period of time (e.g., 1 to 10 s), it is essential that any random perturbation or alteration of the flight track be measured and corrected for. These measurements may be made using an inertial navigation system or other navigational tracking systems, such as GPS.

On the basis of this measured flight track, accurate phase corrections are computed. These are applied to the received signals at virtually any point in the radar system, from local oscillator to final integration. Where pre-summing is employed, the corrections may, for example, be applied on a sample-by-sample basis after the pre-summing. Too many motion compensation and autofocus techniques have been developed to describe all of them here. Further Reading contains a number of excellent texts for delving more deeply into this topic. In this section the common sources of phase errors and their effects on image quality are listed.

Table 34-1. Phase Errors

Common Sources	Effects
Unmeasured velocity error	Increased sidelobe levels
Unmeasured acceleration along line of sight	Degraded resolution
Nonlinear motion of aircraft	Reduced antenna peak gain
Equipment imperfection	Beam wander (which can cause nonuniform gain across an image)
Processing approximations	
Atmospheric disturbances	

Limit on Uncompensated Phase Errors. The common sources of phase errors and their effects on the radar's performance are listed in Table 34-1. The significance of these errors cannot be overstressed. For example, it can be shown (mathematically) that an uncorrected frequency phase error of only 114° from the center to the ends of an array will result in a 10 percent spreading of the synthetic beam. At X-band, 114° corresponds to a distance just less than 1 cm. In other words, the synthetic array deviates from an ideal array across its entire length by just 1 cm. The predominant effect of uncompensated phase errors is increased sidelobe levels and decreased resolution. As a rule, the total random (high-frequency) phase error must be held to within 2° to 6° for the loss in resolution and growth in sidelobe levels to have minimal impact on image quality. However, at X-band, 6° of phase is equivalent to an antenna motion deviation of only 0.025 cm. The synthetic array might be 1 km or longer, and hence compensation to this extremely fine level is, to say the least, challenging. The phase-error sources listed here will always lead to imperfect compensation and consequently imperfect focusing of SAR data. However, the resulting errors in the SAR image can be somewhat mitigated by a process known as autofocus.

An improperly focused SAR image is very similar to an unfocused optical one. The image appears blurred, with a reduction in both sharpness and contrast. If the image were on the display of your camera, you would adjust the focal length of its lens until a sharp, focused image is revealed. Most digital cameras have an autofocus feature where a software algorithm instantaneously evaluates the display image. This provides a numeric indicator of image sharpness and contrast. The camera then automatically adjusts the focal length until this image quality measure is maximized.

For a SAR image, however, the "photograph" is already taken. There is no going back and adjusting the hardware to provide a better image. Instead, autofocus is accomplished by first viewing the imperfect SAR image as a perfect image that has been corrupted by *convolution* (Chapter 6) with some unknown blurring filter. If this distorting filter were known, its effects could theoretically be removed by a process known as *deconvolution*. Here, the deconvolution attempts to reverse the process that blurred the image. However, the form of the filter is not known since it represents the effects of unknown and uncompensated phase errors.

These phase errors are primarily due to motions such as along-track velocity and across-track acceleration perturbations. Nevertheless, it can be approximated by modeling as a relatively simple function with a limited set of unknown mathematical parameters. Its inverse (the deconvolution filter) similarly depends on these same unknown parameters. Thus, the general approach to autofocus invokes a procedure that applies an estimate of the correct deconvolution filter to the complex-valued SAR image. Subsequently, the deconvolution filter parameters are adjusted until one or more numeric indicators of image quality are maximized.

For example, contrast could be chosen as the image quality parameter. The correct deconvolution filter is the one that maximizes the image contrast. This may be selected on a trial-and-error basis. The resulting deconvolved SAR image is then accurately focused. Autofocusing algorithms cannot make SAR images perfect but have been shown to improve SAR focusing

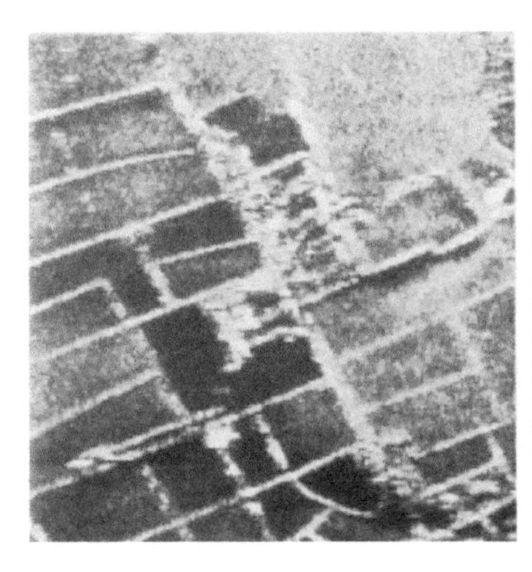

Figure 34-23. This is an example of SAR autofocusing. Contrast has been used as the image quality metric. The improvement in the sharpness of image features is clear, which makes it a much more suitable candidate for interpretation and exploitation. (Crown copyright DERA.)

and image quality substantially, as shown in Figure 34-23. Here the contrast is used as the image quality metric to determine the parameters for phase correction. The improvement in the sharpness of image features is clear, making it a much more suitable candidate for interpretation and exploitation.

34.5 SAR Image Interpretation

Figure 34-24. In this SAR image map, note the long radar shadows of a plant's stacks (center). With knowledge of the height of the radar, the height of the smoke stacks can be determined and the role of the industrial plant can begin to be inferred.

At first glance, the image in Figure 34-24 could easily be mistaken for a photograph of a section of terrain, taken from a high-altitude aircraft flying directly overhead. Moreover, it looks to be a cloudless day, and judging from the long shadows cast by the twin smokestacks it appears to have been taken at either sunset or sunrise, with the sun shining from the direction at the bottom of the image.

This interpretation would be quite plausible if the image was in fact an optical photograph, but a SAR created this image. It may have been formed from measurements made at night and through an overcast sky. Instead of passing directly overhead, the radar flew by this location at a significant offset distance, perhaps 100 km or more. More specifically, the flight track of the SAR evidently passed by the imaged area at an offset in the direction toward the bottom of the image. We can tell this to be the case from the shadows cast by the twin smokestacks. These shadows are not created by sunlight but instead by the electromagnetic waves propagating from the SAR transmitter. Just as with sunlight, the direction of these electromagnetic shadows points away from the source, in this case a source located at a long distance in the direction toward the bottom of the image.

Certainly, just like optical images, a fine-resolution SAR map allows for identification of structures such as roads, buildings, and vehicles as well as natural features such as lakes and mountains. However, SAR images are not affected by the time

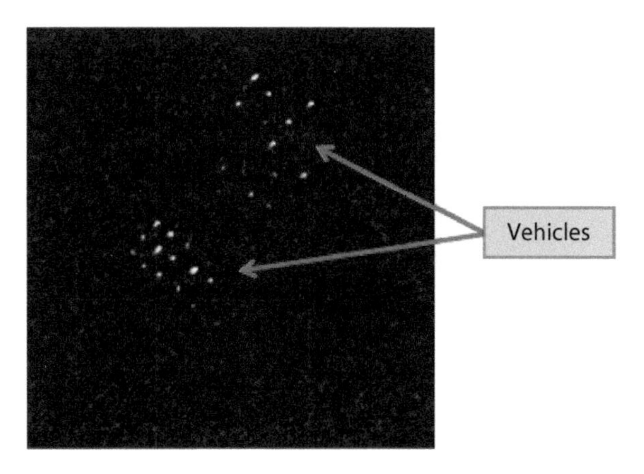

Figure 34-25. This VHF SAR image shows vehicles parked under a tree canopy that would not be detectable in the visible wavelengths.

of day and usually not by atmospheric conditions. Moreover, the long wavelengths of lower frequency SAR (below L-band) can often penetrate camouflage and vegetation, allowing for identification of objects that are otherwise visibly obscured. An example of a group of vehicles parked under a tree canopy, not detectable using visible wavelengths, is shown in Figure 34-25. The targets stand out very clearly when a SAR image is formed using very high frequencies (VHF) that can penetrate the tree canopy.

It is clear then that interpreting a SAR image is different from interpreting an optical image (photograph).

Since a swath-mapping SAR is side-looking, shadows will always be cast in a direction away from the radar and aircraft. This is often helpful in image interpretation; the length of the shadow cast can help infer the height of objects. For example, the height of the twin smokestacks in Figure 34-24 could easily be estimated from knowledge of the SAR's altitude and location.

SAR images formed at different times can contain valuable information as to changes that have occurred. The process of detecting such differences is known as *change detection*. This can be done using just the amplitude images (noncoherent change detection) or the amplitude and the phase (coherent change detection). Figure 34-26 shows an example of noncoherent change detection where targets detected in the second recorded image but not the first are shown in blue. Targets detected in the first recorded image but not the second are shown in red. This can be remembered with the phrase "red, fled, and blue, new."

Figure 34-26. Noncoherent change detection showing targets detected in the first image but not the second are shown in blue. Targets detected in the second image but not the first are shown in red. (Courtesy Air Force Research Laboratory, Sensors Directorate, Gotcha Radar Program, Public Release Number: AFRL/ WS- 88ABW-2013-5108.)

Chapter 33 revealed that polarimetric SAR improves the utility of SAR images for identifying and understanding illuminated objects. The polarization response of a scatterer is very dependent on its structure and orientation, so some physical characteristics can be inferred from polarimetric SAR images. For example, polarimetric SAR is often used to classify different

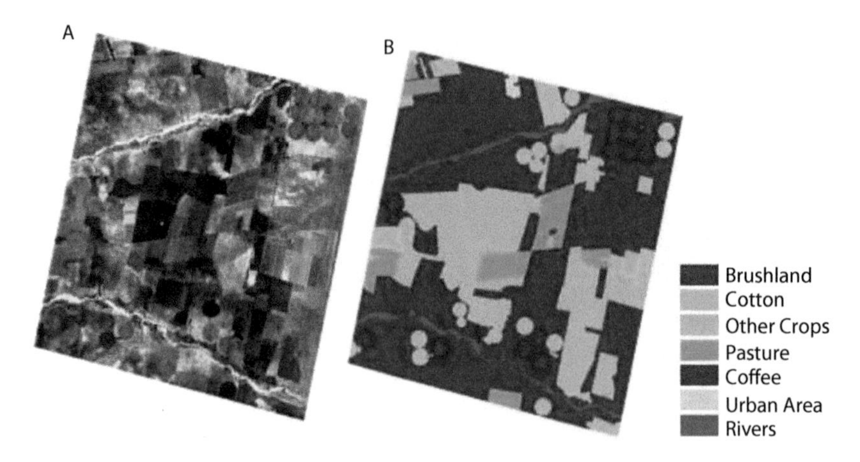

Figure 34-27. In polarimtric SAR image exploitation the different colors represent different polarimetric responses that enable crops and other image characteristics to be recognized. (From: http://www.scielo.br/scielo.php?pid = S0103-90162012000300005&script = sci_arttext.)

types of vegetation and ground cover, terrain that would otherwise all appear as green in an optical image. This is illustrated in Figure 34-27, where the different colors represent different polarimetric responses that enable crops and other image characteristics to be recognized. Man-made objects also often have characteristic polarimetric responses that can be used to differentiate them from natural features.

One difficulty associated with SAR image interpretation is the problem of *speckle*. Unlike visible light, which is incoherent electromagnetic energy spread across a wide bandwidth, a radar system is designed to be coherent so that phase plays a key role (indeed, the SAR process relies on the measurement of phases). As discussed in Chapter 33, the coherent scattering that results from different elements within a resolution cell can add either constructively or destructively, creating a large scattered intensity or no scattering at all (or any intensity in between).

The overall result is a SAR image with a grainy appearance that looks as if it has been corrupted by noise. Whereas speckle is a manifestation of this interference between scatterers, it cannot be diminished by increasing transmit power. Instead, speckle is reduced by incoherently averaging independent SAR images. Dividing a long synthetic aperture up into smaller segments and then forming a SAR image for each generally accomplishes this. The resulting images have coarser along-track resolution, but they can be averaged together to reduce the variance in intensity caused by fading. Similarly, the transmitter bandwidth can be partitioned into smaller blocks, resulting in multiple SAR images, albeit with degraded range resolution. Figure 34-28 compares an example of a single look SAR image and a thirty two look SAR image. To provide a more direct comparison between a single and multiple looks, the resolution in both cases is the same. Not only is the thirty two-look SAR image smoother, but also more structure of the underlying scattering can be discerned. This is because the averaging across the

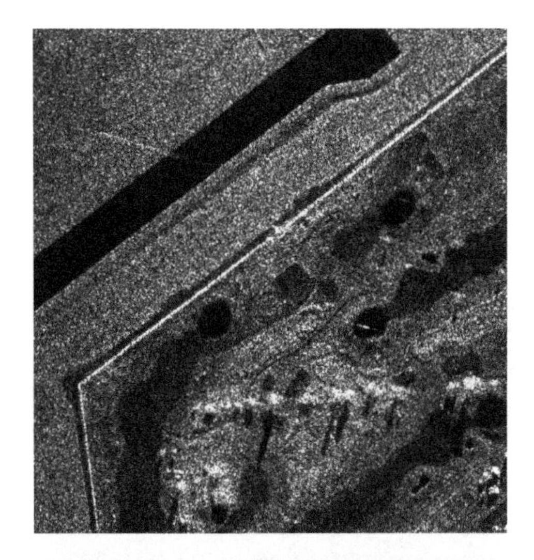

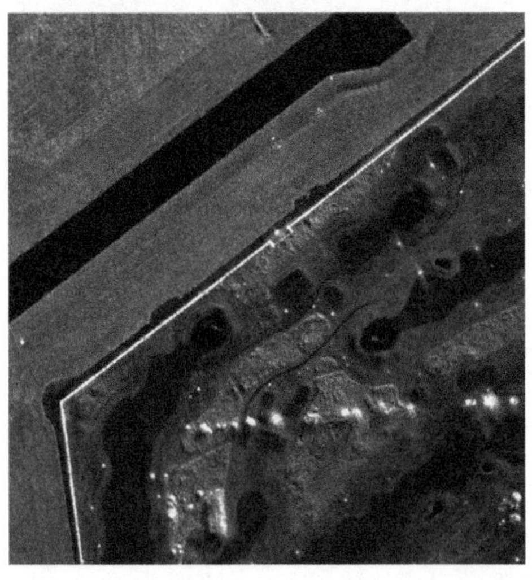

Figure 34-28. Pictured here are single and thirty two-look SAR imagery of the same scene. The thirty two-look image is much smoother (has less speckle), and detail can be more easily discerned. (Courtesy of MDA.)

eight images reduces the variance of the speckle noise, giving a better estimate of the mean value of the scattering and thereby revealing more of the physical features.

A SAR image made from incoherently averaging (summing) a set of images is known as a *multilook* SAR image. For multilook SAR, speckle reduction is usually achieved at the proportional cost of degraded resolution. Other processing approaches have attempted to reduce or even remove speckle. These are based on estimating a local average from a small patch of image amplitude values and assigning this average on a running basis.

34.6 Summary

Fine azimuth (or cross-range resolution) may be obtained by pointing a small radar antenna out to one side of an aircraft, storing the echo returns received over a period of time and integrating them to synthesize the equivalent of a long array antenna. This is SAR. The points at which successive pulses are transmitted can be thought of as the elements of this array.

Phase errors due to the greater range of a point on the ground from the ends of the array than from the center limit its useful length. The limitation may be removed through phase correction, a process called autofocusing.

With focusing, azimuth resolution can be made virtually independent of range by increasing the effective synthetic array length in proportion to the range of the region being mapped.

Computation may in some cases be reduced by pre-summing groups of echo returns and applying the phase corrections for focusing only to the resulting sums. In any event, computation may be substantially reduced by integrating the phase-corrected returns in a bank of Doppler filters, with the FFT.

The production of a well-focused SAR image requires extreme precision in the manufacture of the radar and the measurement of its flight track during the SAR data acquisition. Phase errors due to imprecise measurements can be partially mitigated with autofocus processing algorithms.

SAR image interpretation should proceed with care as radar imagery is different in a number of ways from optical imagery.

Further Reading

D. R. Wehner, *High-Resolution Radar*, Artech House, 1995.

C. V. Jakowatz, D. E. Wahl, P. H. Eichel, D. C. Ghiglia, and P. A. Thompson, *Spotlight-Mode Synthetic Aperture Radar: A Signal Processing Approach*, Springer, 1996.

M. Soumekh, *Synthetic Aperture Radar Signal Processing with MATLAB® Algorithms*, Wiley, 1999.

C. J. Oliver and S. Quegan, *Understanding Synthetic Aperture Images*, SciTech-IET, 2004.

Test your understanding

1. Explain the formation of a synthetic aperture for an unfocused array, describing the way phases add constructively to provided a response in a preferred direction.
2. What ultimately limits the maximum useful length of an unfocused synthetic array?
3. What adjustments to phase must be made to focus a synthetic array?
4. Outline the key steps in Doppler processing that result in a focused swath-mapping SAR image.
5. A stationary radar system trains its beam on a ship. Compare and contrast the form of the ISAR image resulting when (a) the ship is moving perpendicularly to the line of sight of the radar and is "rolling" about its short axis; and (2) the ship is moving toward the radar and "pitching" about its long axis.
6. How does interpretation of optical imagery differ from radar imagery?

Northrop Grumman RQ-4 Global Hawk (1998)

The RQ-4 Global Hawk is an unmanned aerial vehicle (UAV) used for high-altitude reconnaissance, surveillance, and security. It carries the Hughes Integrated Surveillance & Reconnaissance (HISAR) sensor system, which integrates a SAR-MTI system operating in X-band. In 2001, a Global Hawk flew non-stop from Edwards Air Force Base to RAAF Edinburgh in Australia and became the first unmanned aircraft to cross the Pacific Ocean.

35

SAR System Design

Thales I-Master radar, stripmap SAR image, 3 m resolution, showing a golf course (Courtesy of Thales UK.)

Designers of synthetic aperture radars (SAR) as with any radar system, must correctly select the best combination of the fundamental parameters to meet operational goals. The choice of these parameters cannot be made in isolation, and they are tightly linked to the characteristics of the aircraft (or spacecraft) on which the radar is carried (e.g., velocity and altitude). Particularly important are the choice of transmitter power, transmitter bandwidth, real aperture size, and pulse repetition frequency (PRF).

35.1 SAR Radar-Range Equation

In Chapter 11, the radar-range equation was developed. The radar-range equation relates the received signal energy to the transmitted signal power.

$$\text{Received signal energy} = \frac{P_{avg} G \sigma A_e t_{int}}{(4\pi)^2 R^4 L}$$

where

P_{avg} = average transmitted power

G = antenna gain

σ = radar cross section of target

A_e = effective area of antenna

t_{int} = integration time

L = losses

A similar expression can be derived for SAR and is usually referred to as the *SAR radar-range equation*. Its numerous differences from the usual radar case are now examined.

First, let us examine just the received signal energy from a single pulse that is, just one element of the synthetic array.

The transmitted energy, E_t^n, associated with this single pulse is given by the product of the average transmit power and the integration time (for a single pulse the integration time is the pulse duration):

$$E_t^n = P_{avg}\, t_{int}$$

The radar cross section (RCS) of a single resolution cell for the single pulse is expressed in terms of the backscattering coefficient, sigma-naught or σ^0 (see Chapter 22). The units of σ^0 are *meters squared per meter squared*. Therefore, the physical area illuminated must multiply this value to obtain the total amount of backscatter, σ. Thus, the total backscatter for a single pulse and a single resolution cell is given by

$$\sigma = \sigma^0 d_a d_r$$

Where d_a is the along-track resolution and dr is the across-track or range resolution. Substituting for $P_{avg}t_{int}$ and σ in the form of the radar-range equation leads to an expression for the received signal energy from a single pulse for a single resolution cell:

$$E_r^n = \frac{E_t^n G \sigma^0 (d_a d_r) A_e}{(4\pi)^2 R^4 L}$$

Note that the values of antenna gain, G, and effective aperture, A_e, are for the *real* antenna of the SAR.

If the received pulses of a synthetic array are fully focused to the location of the resolution cell, then the coherent summation of these phase corrected responses will have energy equal to the sum of the received energy of each individual pulse:

$$E_r = \sum_{n=1}^{N} E_r^n$$

If the antenna gain, the resolution cell backscatter coefficient, and range, R, are all assumed to be approximately identical for each pulse (a reasonable assumption), the total received energy from a SAR image resolution cell is increased by the number of pulses, N, used to synthesize the array, given by

$$E_r = \frac{E_t^n G \sigma^0 (d_a d_r) A_e N}{(4\pi)^2 R^4 L}$$

The SAR range equation has a number of different guises and can be expressed in terms of other SAR parameters. For example, Chapter 33 showed the azimumthal resolution to be

$$d_a = \frac{\lambda}{2L} R$$

where L is the length of the synthetic array. Inserting this into the SAR range equation, the received energy is found to be inversely proportional to the range raised to the third power:

$$E_r = \frac{\lambda}{2L} \frac{E_t^n G \sigma^0 d_r A_e N}{(4\pi)^2 R^3 L}$$

Likewise, the array length, L, is equal to the product of the radar velocity and the processing integration time, $L = V_R t_{int}$. The number of pulses, N, is related to the integration time by the pulse repetition frequency, $N = PRFt_{int}$.

Inserting these relationships reveals yet another form of the range equation. This one is explicitly dependent on PRF and radar velocity:

$$E_r = \frac{\lambda}{2V_r} \frac{(PRF)E_t^n G \sigma^0 d_r A_e}{(4\pi)^2 R^3 L}$$

Finally, inserting the relationship between antenna gain and effective aperture, $A_e = (\lambda^2 G/4\pi)$ also provides a commonly cited expression for the SAR radar-range equation:

$$E_r = \frac{\lambda^3}{2V_r} \frac{E_t^n G^2 \sigma^0 d_r}{(4\pi)^3 R^3 L}$$

From this result it might be concluded that the received signal energy is proportional to the wavelength raised to the third power, inversely proportional to the radar velocity, and independent of the azimuthal (along-track) resolution. Strictly speaking, this is true. However, it is important also to consider the balancing effect of the other parameters in any particular form of the SAR range equation, and it would be wrong to assume this is a fundamental property of a SAR system.

In addition, caution must be exercised because the parameters of this equation are often coupled in the way they relate to the particular design of a SAR. For example, the velocity of the aircraft carrying the real beam radar will affect the choice of PRF, wavelength will alter the antenna gain (for a fixed real antenna length), and azimuthal resolution will limit the real aperture size (in strip-mapping mode), which also reduces the gain, G, of the real beam.

Noise-Equivalent $\boldsymbol{\sigma^0}$**.** This measure of the sensitivity of a SAR system is often used instead of setting a required signal-to-noise ratio (SNR) as was the case for real beam radar in Chapter 13. Noise-equivalent σ^0 describes the strength of system noise with respect to the equivalent backscatter power in a pixel that would result from an idealized distributed scatterer of the same reflectivity value. In other words, it is the value of the backscatter coefficient at maximum imaging range that would be equal to system noise. Therefore, smaller noise-equivalent σ^0 values indicate greater SAR sensitivity.

The noise-equivalent σ^0 is normally set with respect to low backscatter features such as short grass, which ensures which ensures that they will be seen in the image above system noise. To accomplish this, values in the region of –20 dB are chosen as a design goal.

35.2 SAR Ambiguities

Ambiguous echoes in SAR imagery should be avoided because they result in false targets that can be confused with real ones, much as is the case for any radar system. Certain critical aspects of SAR design, if not properly attended to, may seriously degrade the quality of the images or perhaps even render them useless. These include selecting the optimum PRF, transmit bandwidth, and antenna beamwidth. Ideally, the PRF must be set low enough to avoid range ambiguities yet high enough to avoid Doppler ambiguities. In terms of antenna theory, the PRF must be high enough to have adequate spatial sampling of the synthetic array, thus avoiding problems with grating lobes. The transmit bandwidth must be large enough to allow for sufficient across-track (down-range) resolution, and the antenna beamwidth must be large enough to generate the desired along-track resolution.

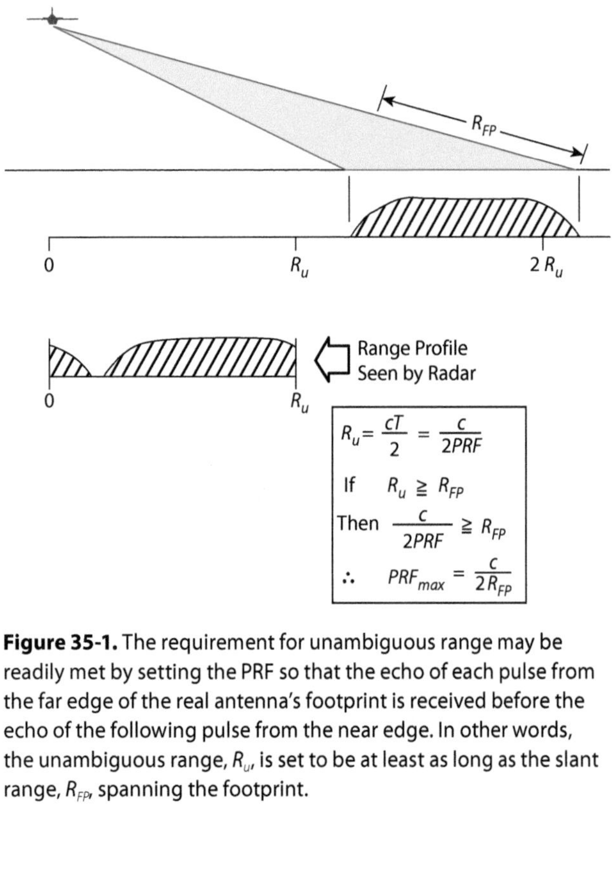

Avoiding Range Ambiguities. In general, to avoid range ambiguities the maximum value of the PRF for SAR imaging is limited by the maximum specified operating range, just as it is for standard pulsed radar (Chapter 27). More typically, this requirement is met by setting the PRF so that the echo of each pulse from the far edge of the real antenna's footprint is received before the echo of the following pulse from the near edge. In other words, the unambiguous range, R_u, is set to be at least as long as the slant range, R_{FP}, spanning the footprint (Fig. 35-1).

Range Profile Seen by Radar

$$R_u = \frac{cT}{2} = \frac{c}{2PRF}$$

If $R_u \geq R_{FP}$

Then $\frac{c}{2PRF} \geq R_{FP}$

$$\therefore \quad PRF_{max} = \frac{c}{2R_{FP}}$$

Figure 35-1. The requirement for unambiguous range may be readily met by setting the PRF so that the echo of each pulse from the far edge of the real antenna's footprint is received before the echo of the following pulse from the near edge. In other words, the unambiguous range, R_u, is set to be at least as long as the slant range, R_{FP}, spanning the footprint.

This criterion will usually be satisfied if the PRF is less than

$$PRF_{max} = \frac{c}{2R_{FP}}$$

where c is the speed of light (3×10^8 m/s), and R_{FP} is the range footprint of the real antenna.

This result can also be expressed in terms of the total illuminated swath width of a side-looking SAR. This width is simply the projection of R_{FP} onto the ground. So, from Figure 35-1,

$$\text{Swath width} = \frac{R_{FP}}{\sin\theta_i}$$

where angle θ_i is the incidence angle of the illuminating wave, describing its direction of propagation with respect to the ground. Therefore, the PRF must be less than

$$PRF_{max} = \frac{c \operatorname{cosec}(\theta_i)}{2(\text{swath width})}$$

If only a small segment of R_{FP} is being mapped, a higher PRF can be used. Within a narrow segment in the center of R_{FP}, ambiguities can be avoided, even with PRFs approaching twice the maximum given by the previous expression.

Sample Computation of PRF_{max}

Find the maximum PRF that a SAR system can have while still avoiding range ambiguities:

- The range segment being mapped may lie anywhere within footprint of antenna beam.
- Slant range, R_{FP}, spanning footprint = 50 km
- Speed of light = 3×10^8 m/s

The calculation is as follows:

$$PRF_{max} = \frac{c}{2R_{FP}} = \frac{3 \times 10^8}{2 \times 50000} = 3000 \text{ Hz}$$

Avoiding Doppler Ambiguities. In Chapter 34 we saw how the synthetic aperture is synthesized by processing a sequence of echo returns that have a changing Doppler frequency. In swath-mapping SAR, the Doppler bandwidth of the echo returns is determined by a combination of the angular size of the real beam and the along-track velocity of the aircraft for a given transmitter frequency. The PRF must sample this Doppler bandwidth at least at twice the maximum Doppler value to satisfy the Nyquist criterion. In other words, the PRF must exceed the maximum spread between the Doppler frequencies for points on the ground corresponding to the leading and trailing edges of the mainlobe of the real antenna.

The minimum PRF is given by

$$PRF_{min} = f_{D_L} - f_{D_T}$$

where f_{D_L} and f_{D_T} are the Doppler frequencies at the mainlobe's leading and trailing edges, respectively (Fig. 35-2).

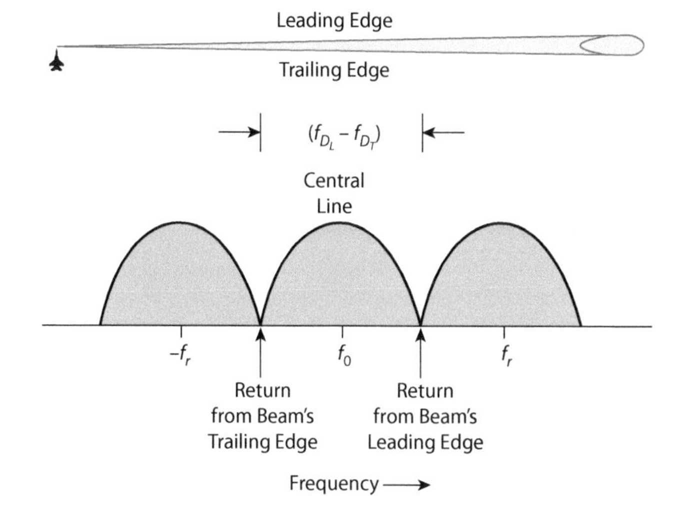

Figure 35-2. To avoid Doppler ambiguities, the PRF must exceed the difference between the Doppler shifts at the leading and trailing edges of the real antenna's mainlobe.

In the case of a narrow azimuth beamwidth, the Doppler spread is approximately equal to $2V_R\theta_{NN_a}/\lambda$ times the sine of the look angle, ε (Fig. 35-3). So the minimum PRF is

$$PRF_{min} \cong \frac{2V_R\theta_{NN_a}}{\lambda} \sin \varepsilon$$

where

V_R = velocity of the radar

θ_{NN_a} = null-to-null beamwidth of the real antenna in radians

ε = azimuth look angle

λ = wavelength, units consistent with V_R

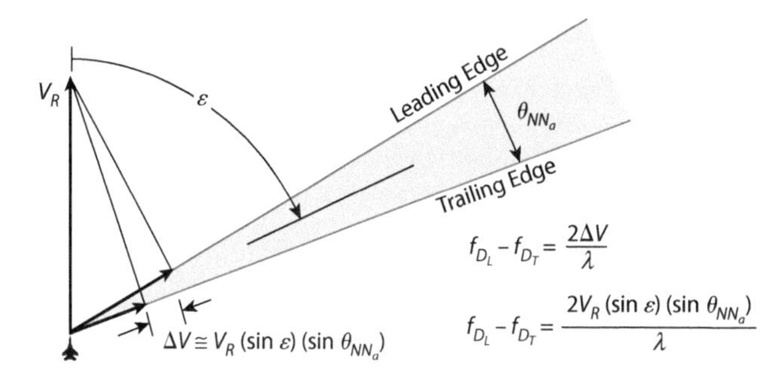

Figure 35-3. Geometric relationships determining the spread between the Doppler frequencies of echo returns from the leading and trailing edges of the real antenna's mainlobe.

1. If it is assumed that $\theta_{NN_a} = 2\lambda/l$ where l is the length of the real antenna, then at an azimuth look angle of 90°, $PRF_{min} = 4V_R/l$.

In typical airborne applications, V_R is between 250 and 500 m/s.[1]

$$\text{If} \quad PRF > f_{D_L} - f_{D_T}$$

$$\text{Then} \quad PRF > \frac{2V_R(\sin\varepsilon)(\sin\theta_{NN_a})}{\lambda}$$

$$\therefore \quad PRF_{min} \cong \frac{2V_R(\sin\varepsilon)(\sin\theta_{NN_a})}{\lambda}$$

$$\cong \frac{2V_R\theta_{NN_a}}{\lambda}\sin\varepsilon$$

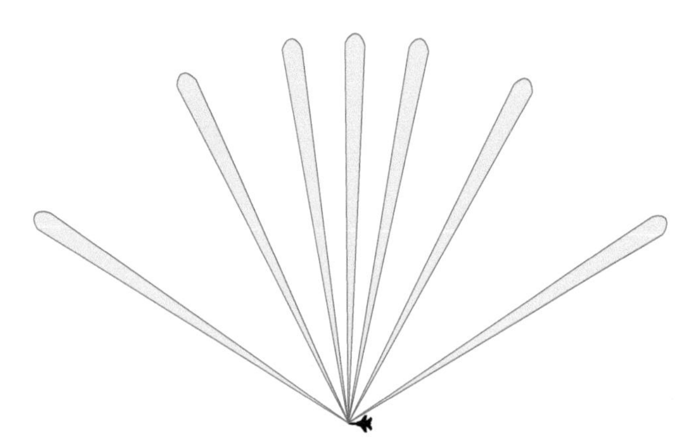

Figure 35-4. Grating lobes are replications of an array's mainlobe occurring at increasingly large intervals on either side of the mainlobe.

Grating Lobes. Another way of viewing the limitation on minimum PRF is in terms of antenna array theory (Chapter 9). From antenna array theory the determinant of the minimum PRF for unambiguous aperture synthesis is the distance, d_e, between successive array elements. Now, d_e equals the radar's speed times the interpulse period, $1/f_r$. If d_e is greater than half a wavelength, *grating lobes* will be produced, just as they would be for a real array antenna. These are replicas of the mainlobe, occurring at increasingly large intervals on either side of the mainlobe (Fig. 35-4).

Although grating lobes are not unique to synthetic arrays, they are more of a problem than in real arrays. There are two reasons for this. First, because of the restrictions on maximum PRF, the array elements generally cannot be placed as close together in a synthetic array as in a real array. Second, as we saw in the previous chapter, for a target at any one angle off boresight, the difference in the phase shift of the returns received by successive array elements is twice as great in a synthetic array as it is in a real array.

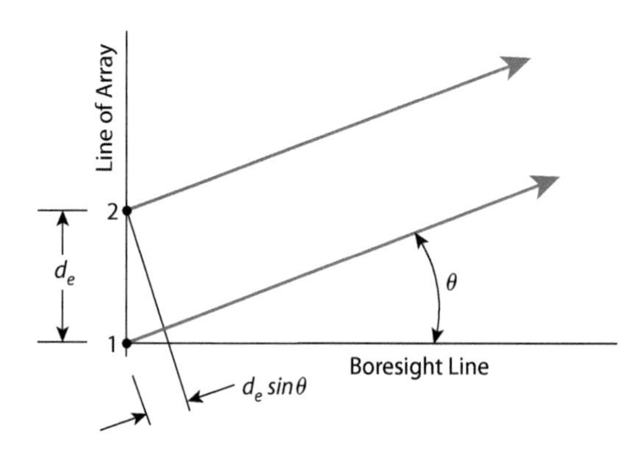

Figure 35-5. There are certain conditions under which grating lobes are produced. If spacing between array elements, d_e, times sine of angle, θ, off boresight is a multiple of half a wavelength, returns received from angle, θ, by successive elements will be in phase.

The lobes are produced in the following way. If a reflector is gradually moved away from the bore sight line of the array, a phase difference develops between the returns received by successive array elements (Fig. 35-5). This difference is proportional to twice the spacing of the elements times the sine of the azimuth angle, θ, of the reflector relative to the boresight line. The familiar pattern of nulls and sidelobes is thus observed.

$\Delta\phi$ = the difference between the phases of echoes received by array elements 1 and 2

$$\Delta\phi = \frac{2d_e\sin\theta}{\lambda} \times 360^\circ$$

$$\text{If } \Delta\phi = n \times 360^\circ$$

$$\text{Then} \quad \frac{2d_e\sin\theta}{\lambda} = n \quad (n = 1, 2, \ldots..)$$

$$\text{and} \quad d_e = n\frac{\lambda}{\sin\theta}$$

However, if the spacing of the array elements is much greater than half a wavelength, as the angle with respect to the reflector is increased, a point is soon reached where the phase shift is 180°. Beyond this point, the amplitudes of successive lobes start increasing ($\sin[180 + \theta] = -\sin\theta$). The increase continues until the element-to-element phase shift reaches 360°. At this point, the returns received by all of the elements add up, once again, exactly as they did when the reflector was in the center of the mainlobe. This "replica" of the mainlobe is the first grating lobe. If the azimuth of the reflector is increased further, the process repeats, and successive grating lobes appear.

In a real antenna, the gains of grating lobes fall off gradually from that of the mainlobe with increasing azimuth angle. But in a synthetic array, the fall off is much greater. The reason for this is that the synthetic array is formed from returns received through the real antenna. The strength of the returns received from any one direction is proportional to the real antenna's two-way gain in that direction. In general, the gain decreases rapidly as the azimuth angle increases. If the azimuth angle of the first grating lobe can be made sufficiently large, the amount of energy received through the grating lobes can be reduced to negligible proportions. SARs are generally designed to achieve this.

The restriction imposed on the PRF by the requirement that range be unambiguous is reasonably loose. This means that the PRF can usually be set high enough to place the first grating lobe well outside the mainlobe of the real antenna. If this isn't possible, it is placed in a null between adjacent sidelobes (Fig. 35-6). However, it must not be placed closer to the mainlobe than the second null.[2]

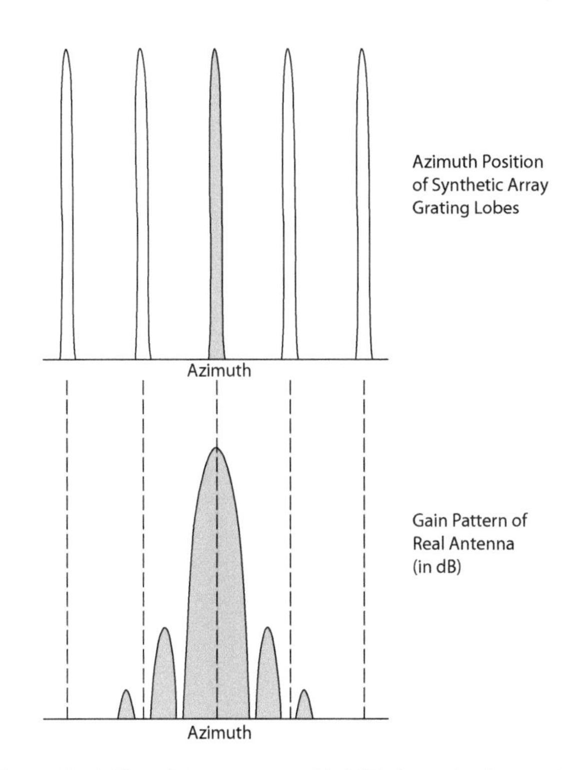

Figure 35-6. The minimum acceptable PRF places the first grating lobe between the first and second sidelobes of the real antenna.

2. Since the angle, θ_N, between this null and the boresight line equals θ_{NNa}, the PRF that places the first grating lobe here is the same as PRF_{min}.

To maintain a constant desired spacing between array elements and to keep the number of pulses that must be processed in a given array length constant, a common practice is to adjust the PRF to the effective ground speed of the aircraft while accounting for wind resistance.

35.3 Bandwidth and Cross-Track Resolution

The range resolution provided by radar is $\Delta r = c\tau/2$, where τ is the compressed pulse width, a value approximately equal to the inverse of the transmit signal bandwidth, B. Thus, alternatively, the range resolution is inversely proportional to this bandwidth:

$$\Delta r = \frac{c}{2B}$$

For SAR, this expression provides the resolution in the slant-range direction, as opposed to d_r, the resolution on the ground. The cross-track ground resolution, d_r, is simply the projection of the distance Δr onto the ground surface (Fig. 35-7):

$$d_r = \frac{\Delta r}{\sin\theta_i}$$

Therefore, like conventional real beam radar systems, the cross-track resolution of a SAR depends on the transmit bandwidth, and very fine resolution requires a wide bandwidth. In many practical airborne cases θ_i is near to 90° and Δr is approximately equal to d_r. This is not the case for space-based SAR where θ_i is of the order of 45° and significantly reduces (ground) range resolution.

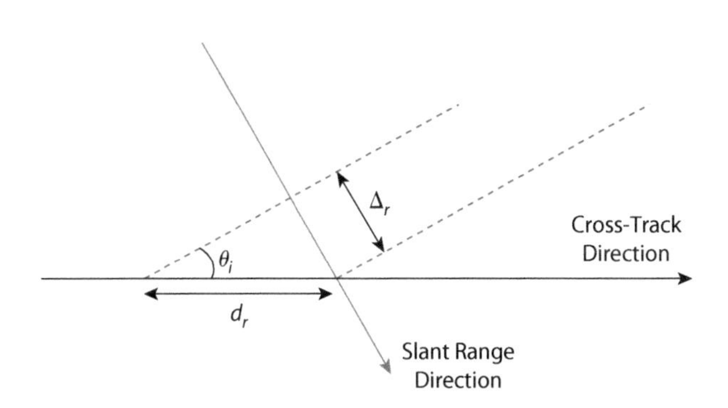

Figure 35-7. The cross-track resolution, d_r, is simply the projection of the distance, Δr onto the ground surface.

Thus, a fundamental parameter of a SAR sensor is the bandwidth, B. For a required cross-track resolution of d_r, the required transmit bandwidth must be in excess of

$$B_{min} = \frac{c}{2\Delta r} = \frac{c \text{ cosec } \theta_i}{2d_r}$$

35.4 Beamwidth and Along-Track Resolution

Chapter 34 showed that the along-track resolution of a strip mapping SAR sensor is given by $d_a = \lambda R/2\,L$, where L is the length of the synthetic aperture. The maximum value of L that the synthetic aperture can have is given by

$$L_{max} = \theta_a R$$

where θ_a is the azimuthal beamwidth of the real antenna in radians.

Thus, d_a depends on the azimuthal beamwidth, θ_a, of the real antenna. Hence, as we have seen in earlier chapters on SAR, finer azimuth resolution requires a wider real beam. In other words, to increase resolution in the along-track direction in a strip mapping SAR, a shorter real antenna length, l, is required. The shorter the real aperture, the larger the Doppler span of echo signals used to form the synthetic aperture (for a given aircraft velocity).

To achieve a required along-track resolution, d_a, the minimum real-aperture beamwidth, θ_a^{min}, is therefore

$$\theta_a^{min} = \frac{\lambda}{2d_a}$$

Thus, just as the across-track resolution is inversely proportional to transmit bandwidth, the along-track resolution is inversely proportional to the transmit (real antenna) beamwidth.

35.5 Minimizing Sidelobes

The performance of a synthetic array radar system may be degraded by both range sidelobes due to pulse compression and azimuth sidelobes in the formation of the synthetic aperture. Moving targets present in an image can also cause sidelobes in both range and azimuth. These sidelobes affect the radar imagery in two different ways. First, the peaks of the stronger sidelobes may cause a string of (progressively) weaker false targets to appear on either side of a strong target (Fig. 35-8). This is because the sideobes represent ambiguities, and a large enough target echo in a mainlobe can simultaneously appear in all the sideobes (which get progressively weaker the farther they are from the mainlobe).

Second, the combined power of all sidelobes (called the *integrated sidelobe return*) together with receiver noise tend to fog or wash out smaller details in SAR imagery.

The effect of the integrated sidelobe return can be visualized by imagining an area of ground the size of a resolution cell

Figure 35-8. In this high-resolution SAR image of the famous Ohio State University horseshoe stadium, note the effect of unsuppressed lower-order sidelobes when a strong point target is imaged. (Courtesy of Air Force Research Laboratory, Sensors Directorate, Gotcha Radar Program, Public Release Number: AFRL/WS-06-0592.)

that produces no return, for example, a smooth-surfaced pond, in the middle of a region of uniform backscattering, such as a grassy field. In an ideal image this would appear as a "black" area representing no echo, contrasted with a "white" area, representing echo from the surrounding grass. However, the signal output when the pond is being mapped is the sum of the simultaneously received powers of the range sidelobes and the azimuth sidelobes plus the receiver noise. If this power is comparable to that received from the surrounding terrain, the "hole" in the map corresponding to the pond will start to be filled in. If nothing is done to reduce the sidelobes, the integrated sidelobe return alone may contain up to 10 percent as much power as the mainlobe return has. This results in a loss of contrast and can be considerable. As an example, in Figure 35-9 the runway looks gray rather than black.

Figure 35-9. This SAR image demonstrates how integrated sidelobe return can wash out detail. The runway is smooth and reflects little echo signal. The gray appearance is due to a combination of system noise and integrated sidelobes. (Courtesy of DERA.)

The sidelobes of a synthetic array, like those of a real array, are produced as a result of the finite length of the synthetic array. Consequently, a synthetic antenna's sidelobes can be reduced via illumination tapering in the same way as those of a real antenna (Chapter 9). This is accomplished by weighting the returns received at the individual array elements (i.e., the returns from successive transmitted pulses). The returns received by the end elements are deemphasized relative to the returns received by the central elements. The cost of this reduction, of course, is a slight loss of along-track resolution. The weighting can be conveniently accomplished when the focusing corrections are applied to the stored returns.

35.6 SAR Design Examples

Since SAR relies on the movement of the radar to produce fine-resolution images, the radar itself must of course be deployed on some moving vehicle. In the early days of SAR, aircraft were exclusively used. High-altitude, high-speed aircraft, such as the U-2 and SR-71, allowed for the creation of SAR images with a wide swath-width, and over a large area. However, the limited altitude of aircraft means that ultimately the earth's curvature limits their range and coverage.

The next logical step was to place SAR systems on spacecraft, and in June 1978 NASA launched the Seasat satellite, which carried the first spaceborne SAR system. Since that time, a progressively increasing number of other SAR imaging radars have been deployed in space, and SAR has become an indispensible surveillance and remote sensing tool. For example, in 2006 the Canadian Space Agency launched the RADARSAT-2 satellite. Two European systems, TerraSAR-X and COSM-SkyMed, followed in 2007. RADARSAT-2 orbits the earth at an altitude of approximately 800 km and at a velocity of 7.5 km/s. It is a complex SAR system with a number of operating modes, including a swath width of 108 km (at an incidence angle of about 45°), a single-look along-track resolution of 7.9 m, and ground-range (across-track) resolution of 25 m. This system transmits at C-band (5.4 GHz) with a wavelength of 5.6 cm. At an incidence of 45°, the slant range to the center of the swath is roughly 1100 km. As of January 2013, RADARSAT-2 entered its fifth year of operational service (Fig. 35-10).

Figure 35-10. Pictured here is RADARSAT-2 orbiting Earth.

From these specifications of RADARSAT-2, a number of technical parameters can be inferred. For example, its transmit bandwidth can be calculated using the formula presented earlier in this chapter. This gives an approximate bandwidth of 8.5 MHz, for example,

$$B_{min} = \frac{c\,\text{cosec}\,q_i}{2d_r} = \frac{c\,\text{cosec}\,45°}{2(25)} \approx 8.5\,\text{MHz}$$

Likewise, the beamwidth of the real aperture can be computed to be approximately 0.2°:

$$\theta_a^{min} = \frac{\lambda}{2d_a} = \frac{0.056}{2(7.9)} \approx 0.2°$$

Consequently, the length of the real aperture must be no more than around 16 m (in fact it is 15 m). Note the change from degrees to radians to perform the calculation.

$$l_{max} = \frac{\lambda}{\theta_a^{min}} = \frac{0.056}{3.54 \times 10^{-3}} \approx 15.8\,\text{m}$$

Contrast this with the length of the synthetic array necessary to provide the stated along-track resolution of 7.9 m, of the order of 3.9 km! For example,

$$L = \theta_a^{min} R = 3.45 \times 10^{-3}\,(1100) \approx 3.9\,\text{km}$$

Moreover, at a velocity of 7.5 km/s, the satellite takes about half a second to form this synthetic array.

The PRF to avoid range ambiguities can also be computed and has to be less than 2 kHz, for example,

$$PRF_{max} = \frac{c\,\text{cosec}\,\theta_i}{2(\text{swath width})} = \frac{c\,\text{cosec}\,45°}{2(108 \times 10^3)} \cong 2.0\,\text{kHz}$$

However, assuming that the null-to-null beamwidth, θ_{NN_a}, is approximately twice that of the beamwidth, θ_a, the PRF must be in excess of 1.9 kHz to avoid Doppler ambiguities:

$$PRF_{min} = \frac{2V_R\theta_{NN\,a}}{\lambda}\sin\varepsilon = \frac{2(7.5\times10^3)2(3.45\times10^{-3})}{0.056}\sin 90^\circ$$
$$\approx 1.9\,\text{kHz}$$

This leaves only a small margin for error and illustrates the importance of careful design. These results show that for this operating mode RADARSAT-2 should have a PRF of around 1950 Hz. Since the synthetic aperture time was determined to be 0.5 s, the synthetic array will have slightly less than 1000 elements (PRF multiplied by the aperture synthesis time).

Although many more modern SAR systems are planned for launch, they are not exclusively spaceborne. A migration can be seen from piloted aircraft to unmanned air vehicles (UAVs). These systems are ideal for sensing and surveillance because they can sustain very long missions and the absence of a pilot allows for more flexible sensing payloads. An example of a UAV SAR system is the U.S. Air Force Global Hawk (Fig. 35-11), an unmanned aircraft that can fly at an altitude of nearly 20 km with a loiter velocity of approximately 175 m/s.

Figure 35-11. The USAF Global Hawk UAV, built by Northrop-Grumman, is an example of a UAV SAR system.

Unlike RADARSAT-2, which is used primarily as an Earth observation tool, the Global Hawk is a military surveillance system. As such, its SAR spatial resolution of 1 m in swath mapping mode and 30 cm in spotlight mode is much finer than RADARSAT-2. However, in swath mapping mode the swath width is much narrower (e.g., 9.25 km) reflecting its application requiring fine detail, achieved at the expense of a more restricted imaged ground area. Global Hawk uses a shorter wavelength (3 cm) than most satellite SARs (TerraSAR-X is an exception to this, also using a wavelength of 3 cm).

Using the same analysis as for RADARSAT-2, for an incidence angle of 45°:

$$B_{min} = \frac{c\,\text{cosec}\,q_i}{2d_r} = \frac{c\,\text{cosec}\,45^\circ}{2(1)} \approx 212\text{ MHz}$$

$$\theta_a^{min} = \frac{\lambda}{2d_a} = \frac{0.03}{2(1.0)} = 0.015\text{ rad} \approx 0.86^\circ$$

$$l_{max} = \frac{\lambda}{\theta_a^{min}} = \frac{0.03}{0.015} = 2.0\text{ m}$$

$$L = \theta_a^{min}R = 0.015(\sqrt{2})20000 \approx 4.2\text{ km}$$

$$PRF_{max} = \frac{c\,\text{cosec}\,\theta_i}{2(\text{swath width})} = \frac{c\,\text{cosec}\,45^\circ}{2(9.25\times10^3)} \approx 23\text{ kHz}$$

$$PRF_{min} = \frac{2V_R\theta_{NN_a}}{\lambda}\sin\varepsilon = \frac{2(175)2(0.015)}{0.03}\sin 90^\circ = 350\text{ Hz}$$

Compared with the velocity of RADARSAT-2, the Global Hawk is a much slower platform, which results in a lower minimum PRF requirement. Although the synthetic aperture length of Global Hawk is comparable to that of RADARSAT-2, the lower velocity also produces a far longer aperture time of more than 20 s. Even at the lowest PRF of 350 Hz, this long aperture time will result in a synthetic aperture of over 7000 elements. Compensating for unknown motions due to the air vehicle being buffeted by the atmosphere over such an extended period becomes increasingly important as well as difficult. However, the routine use of systems like Global Hawk show just how far SAR imaging has come. The more recent trend has been for SAR systems to come more and more miniaturized. This allows them to be carried on micro UAVs, much smaller than Predators or Global Hawks.

Test your understanding

1. Explain the importance of the SAR radar-range equation in designing SAR systems.
2. How does the avoidance of range ambiguities in SAR differ from the case of a real beam pulse-Doppler radar system?
3. What is the maximum spacing that synthetic aperture array elements can have if grating lobes are to be avoided?
4. In the RADARSAT example covered in this chapter, what would the effect on the real aperture length of setting the range and along-track resolution to 5 m?

35.7 Summary

The designers of synthetic aperture radar must correctly choose a combination of fundamental radar parameters to meet operational goals. These decisions cannot be made in isolation but are tightly linked to the characteristics of the aircraft on which the SAR resides.

Certain aspects of SAR design may seriously degrade the imaging and hence require careful consideration. Among the more important are choice of PRF, transmit bandwidth, antenna beamwidth, and sidelobe reduction.

Two primary factors influence the choice of PRF:

1. The *maximum* value is limited by the requirement that no range ambiguities occur within the span of ranges from which mainlobe return is received.
2. The *minimum* value is limited by the requirement that there be no Doppler ambiguities within the band of frequencies spanning the central spectral line of the mainlobe return.

In terms of antenna theory, this same minimum PRF places the first grating lobe between the first and second sidelobes of the real antenna.

A synthetic array's stronger sidelobes may cause weak, false targets to appear on either side of strong targets. The combined sidelobe return from all targets and the integrated sidelobe return may wash out detail. Such sidelobe corruption could be reduced through amplitude weighting.

Further Reading

J. C. Curlander and R. N. McDonough, *Synthetic Aperture Radar: Systems and Signal Processing*, John Wiley & Sons, Inc., 1991.

W. G. Carrara, R. S. Goodman, and R. M. Majewski, *Spotlight Synthetic Aperture Radar: Signal Processing Algorithms*, Artech House, 1995.

M. I. Skolnik (ed.), "Synthetic Aperture Radar," chapter 17 in *Radar Handbook*, 3rd ed., McGraw Hill, 2008.

C. V. Jakowatz, D. E. Wahl, P. H. Eichel, D. C. Ghiglia, and P. A. Thompson, *Spotlight-Mode Synthetic Aperture Radar: A Signal Processing Approach*, Springer, 2011.

PART VIII

Radar and Electronic Warfare

Boeing EA-18G Growler (2006)

The Growler is a variant of the two-seat Super Hornet, equipped with electronic warfare equipment developed by Northrop Grumman. It has an AN/APG-79 EASA radar and an ALQ-218 wideband receiver, and provides standoff and escort jamming protection to strike aircraft and ground forces with ALQ-99 jamming pods. When development of the next-generation jammer (NGJ) pod family is complete, it will replace the ALQ-99 pods on the Growler.

36

Electronic Warfare Terms and Concepts

AN/ALQ-131 Self-Protection Jammer Pod

This chapter provides an overview of electronic warfare, covering the basic terms, definitions, and concepts. More detailed information about each of the major subfields will be covered in the chapters that follow.

36.1 EW Definitions

Electronic warfare (EW) is the art and science of denying an enemy the benefits of the *electromagnetic* (EM) spectrum while preserving those benefits for friendly forces. This includes the whole EM spectrum, from just above dc to well above daylight. EW is practiced on land, on the sea, in the air and in space. Although all of these realms will be considered, the main focus of the material in the EW chapters will be on the impact of EW on airborne radars and the protection of aircraft against radars which threaten them.

The field of EW deals primarily with all types of hostile devices which transmit or receive electromagnetic signals. This includes radars, all types of communication, weapons which guide on optical, infrared or ultraviolet signatures, and a wide range of devices which locate potential targets using emissions in all of those frequency bands. It also deals with enemy attempts to interfere with the use of the electromagnetic spectrum by friendly forces to conduct military operations.

36.2 EW Subfields

There are three major electronic warfare subfields: *electronic warfare support* (ES), *electronic attack* (EA), and *electronic protection* (EP). These are shown in Figure 36-1. There were earlier subfield names that are still used occasionally. ES was called *electronic support measures* (ESM). This is still used in some applications, typically those in which high accuracy outputs are required. An important example is Naval EW systems in

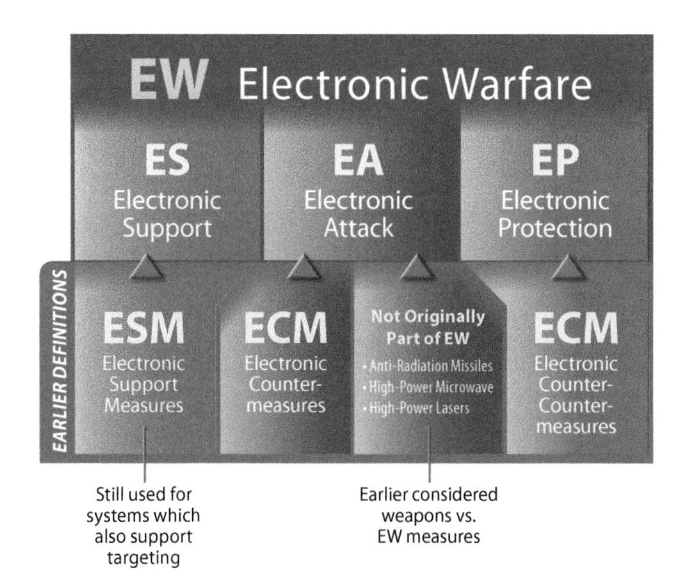

Figure 36-1. EW is divided into three subfields: ES, EA and EP. Earlier designations and some other types of activities are now collected into these subfields.

which the ES system is used to support weapon targeting. EA was previously called *electronic counter measures* (ECM). EW techniques involving lethal attacks were not considered part of EW but were designated as weapons. Now anti-radiation missiles, high power lasers, and high power microwave are included in EA. EP was previously called *electronic counter-countermeasures* (ECCM).

36.3 Electronic Warfare Support

Electronic Warfare Support (ES) involves receiving hostile signals anywhere in the EM spectrum. The purpose of ES is to acquire the information about hostile signals necessary to enable the other EW fields to be effective. Before countermeasures can be taken, it is usually important to know what hostile signals are present, where the associated emitters are located relative to friendly assets and the operating modes of those emitters. This field differs from *signal intelligence* (SIGINT) in several important ways. Both fields involve receiving hostile signals but do so for different reasons that dictate sharp differences in the type of equipment used and the way that the signals are processed. Figure 36-2 is a generalized block diagram of an ES or SIGINT system.

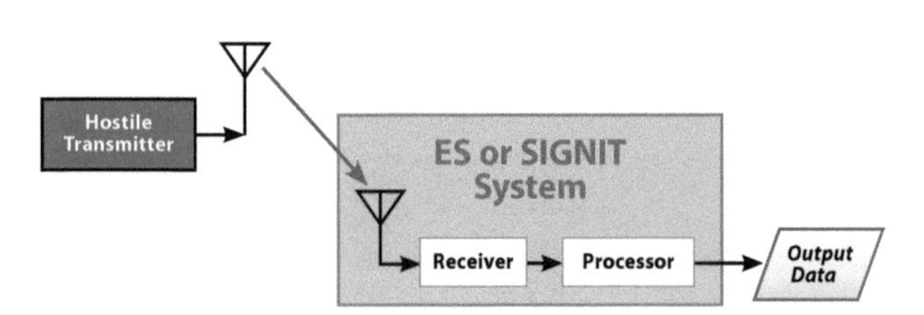

Figure 36-2. Both ES and SIGINT systems intercept and process signals from hostile transmitters, process the intercepted signals to extract required information, and output data to other assets. The differences are in the types of antennas, receivers, and processing.

In general, SIGINT systems collect and analyze hostile signals for strategic reasons. SIGINT is divided into *communications intelligence* (COMINT) and *electronic intelligence* (ELINT). COMINT is generally considered to involve listening to enemy communications to determine the enemy's situation and intentions. ELINT collects detailed information about enemy radars and other noncommunications signals to determine enemy capabilities and to support the development of countermeasures.

ELINT. The purpose of ELINT is to determine what the enemy's capabilities are. Thus, detailed parametric measurements are made and analyzed to determine whether collected signals come from previously known transmitter types or new types of transmitters. For a new type of transmitter, the operating specifications of the system must be determined. If it is a radar, what is its range and resolution? What is its angular tracking rate, and what are its vulnerabilities? This means that data are collected and stored to the maximum practical resolution. The analysis of the stored data for new threats can be conducted over an extended period of time, using multiple intercepts to resolve ambiguities. When confidence in the conclusions is high enough, the new threats are entered into data bases that are used in the design of EW equipment and tactics. Figure 36-3 shows the operational relationship between ELINT and radar ES systems.

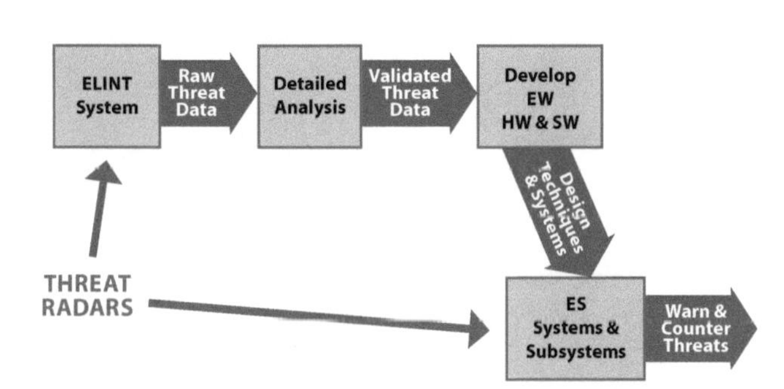

Figure 36-3. ELINT systems support detailed analysis of non-communications (typically radar) signals to support the development of EW hardware and software (including threat parameter tables), while radar ES systems perform short time frame (typically automatic) analysis to support tactical decisions.

ES versus ELINT System Capabilities. ES is different from ELINT in that it is used tactically and thus is time critical. Radar ES involves collecting and analyzing hostile radar signals to determine which known enemy threat system is a possible threat to a protected platform. A radar ES system also locates threat radars and determines their current operational modes. Analysis results are typically required within very few seconds of the time the data is collected. Data are collected and stored only to the resolution required to determine which of an enemy's known types of assets is being deployed in an immediate tactical situation. It is often said that ES has a proclivity toward action. Its mission is to counter the threat before an enemy can damage friendly assets or cause casualties among friendly forces. While some ES systems include the capability to record hostile signals, there are requirement differences that make it impractical to have combined ES and ELINT systems.

ELINT systems support the design of ES systems by collecting and storing signals which are analyzed to determine the parameters of each type of threat signal. These signals are also analyzed to determine the capabilities of the weapons that are controlled by the sensors and links that generate the intercepted signals.

As shown in Table 36-1, there are significant differences between ELINT and ES systems. In general, ELINT systems are designed to receive signals from distant transmitters that are not aimed at the friendly assets being protected by EW systems. This means that the antennas must be larger to allow additional gain and the receivers need to have greater sensitivity. Both of these increase the intercept range. ES systems, on the other hand, generally are designed to receive signals directly related to the tracking of protected platforms and guidance of weapons to those platforms. In general, this means that the emitters are close to the protected platform, and transmit toward the targeted platforms. In general, this means that the received signals are much stronger. This allows the use of smaller antennas with lower gain, and receivers which can have less sensitivity.

Table 36-1. ES versus SIGINT Data Collection

ES Data Collection	ELINT Data Collection
Collect only enough data to determine radar type and operating mode.	Collect enough data to support detailed analysis.
Data range of known threats.	Data range limited by practical range of any future threat.
Adequate parametric resolution to resolve identification ambiguities.	Adequate parametric resolution to determine capabilities of any future threat.

Another difference is the timing of intercepts and analysis. ELINT systems collect signals for non-real time analysis, so receivers can collect data over an extended time period. One real life example is an ELINT system in a satellite that searched for six months to collect a few seconds of data from a hostile radar that had never before been intercepted; it was considered very successful. This means that the antennas and receivers can search narrow angular and frequency segments to optimize sensitivity and precision of parametric measurements at the expense of delayed availability of output data.

An ES system, on the other hand, is trying to identify and/or locate an enemy weapon in time to prevent it from effectively engaging the protected platform. The engagement time can be minutes or even seconds from target identification to impact of a weapon. Thus, ES systems must be optimized for speed. This means that antennas or antenna arrays must rapidly cover larger angular areas and receivers must cover larger frequency ranges. This implies either larger instantaneous ranges or ability to move rapidly through those ranges.

The analysis of collected data is also different in ELINT and ES systems. ELINT systems must determine and store the parameters of intercepted signals to as much resolution as practical and over wide parametric ranges. In general, there is typically little or no available information about the parameters of new types of intercepted signals. We want to know *what* the enemy has: what capabilities the associated weapons have and the subtle ways that newly intercepted signals differ from known signals. This means that much data must be collected and thorough analysis must be performed—usually over extended time.

ES systems, on the other hand, just need to know *which* of the enemy's known weapons is a threat to the protected platform *right now*. Thus, only enough data needs to be collected and analyzed to differentiate between the parameters of the sensors associated with each of the enemy's weapons. This means that less data need to be collected but that the processing must be completed in a very short period of time (typically just a few seconds).

Situational Awareness. Situational awareness is a general term used in various contexts. In a mobile platform (e.g., aircraft, ship, tank) it provides the location and status of friendly and enemy assets relative to the platform. In modern systems, this usually includes information derived from optical and electronic sensors and other available information from off-board resources. The operator of the mobile platform uses this information to optimize maneuvers, weapon engagements, and defensive strategies against enemy weapons.

In a strategic or nonmobile tactical situation, situational awareness gives the responsible commander the necessary information, usually on a map display, of the identification, location and status of all enemy and friendly assets.

Radar Warning. Radar ES includes the radar warning mission, which involves the detection, identification, and location of hostile radars that acquire targets and guide weapons to those targets. Although the weapons are the actual threats, it is common to refer to the signals associated with such weapons as "threats": red threats are enemy signals associated with weapons; blue threats are friendly signals similar to enemy signals; and gray threats are signals from neutrals that look like enemy signals.

One of the ES tasks is to properly identify these types of signals and to remove the blue and grey threats from the data base of signals against which electronic attack can be made. Obviously, it is not desirable to attack friendly or neutral assets, but the presence of these signals complicates the identification of hostile signals both by increasing the number of signals that must be analyzed but also by causing increased levels of processing to distinguish the sometimes subtle differences between these three classes of "threat" signals.

Radar Warning Receivers. *Radar warning receivers* (RWRs) are receiving systems that intercept and analyze hostile radar emissions to determine the type and location of enemy radar controlled weapons that potentially threaten protected assets. The RWR compares the parameters of intercepted signals with a table of parameters of all (or many) types of enemy radar signals. It determines which type of radar is present, its operating mode, and its location relative to the protected platform. The RWR displays this information in real time to aircrew or ship crew personnel and typically also cues EA equipment to make appropriate responses. Figure 36-4 shows a top level block diagram for an airborne RWR.

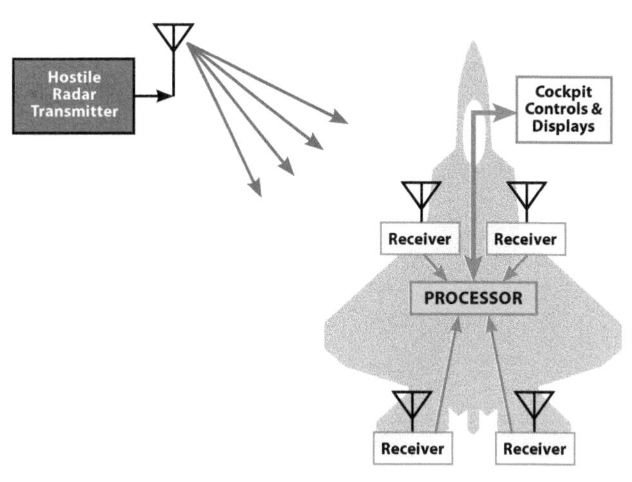

Figure 36-4. A radar warning receiver intercepts signals from hostile radars in multiple wide coverage antennas, and determines the types and locations of the radars from which the signals are received. It then displays thyreats to the aircrew.

Specific Emitter Identification. Specific Emitter Identification (SEI) systems determine not only the type of emitter from which signals are received, but also the specific enemy asset, for example the specific enemy ship on which a radar is mounted. As shown in Figure 36-5, the system makes very fine scale measurements of some of the radar signal's parameters. Then, it compares those parametric values to a table of the same parametric values on radars carried by a large number of enemy platforms. When the best fit is determined, the system outputs the identification of the platform carrying the identified radar.

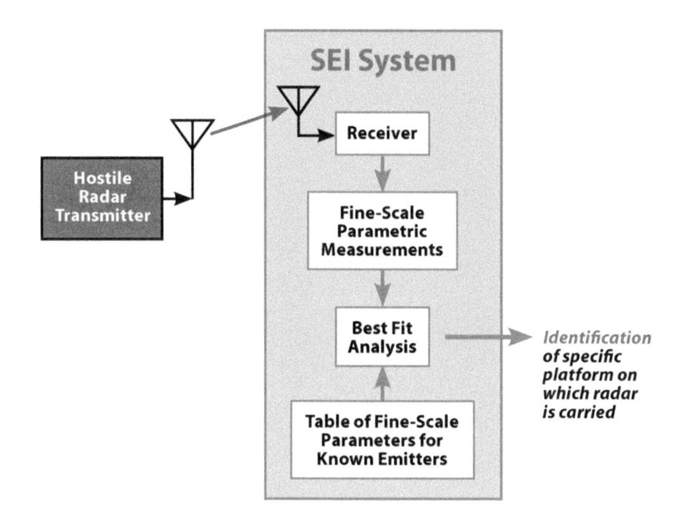

Figure 36-5. A specific emitter identification system analyzes fine scale parameters to allow determination of a specific radar transmitter.

Radar Detection Range versus Target Acquisition Range. ES systems detect hostile radars at a range dependent on the radar's parameters and the sensitivity of the ES system, including its antenna gain. Radars detect targets at a range defined by the radar-range equation. An important operational consideration is the ratio of these two ranges. *Low probability of intercept* (LPI) radars can detect a target before an ES receiver mounted on the target can detect the presence of the radar, as shown in Figure 36-6. An LPI radar that meets this criterion is also called a *quiet radar*, and the difference between the radar detection range and the ES receiver detection range is called the *quiet range*.

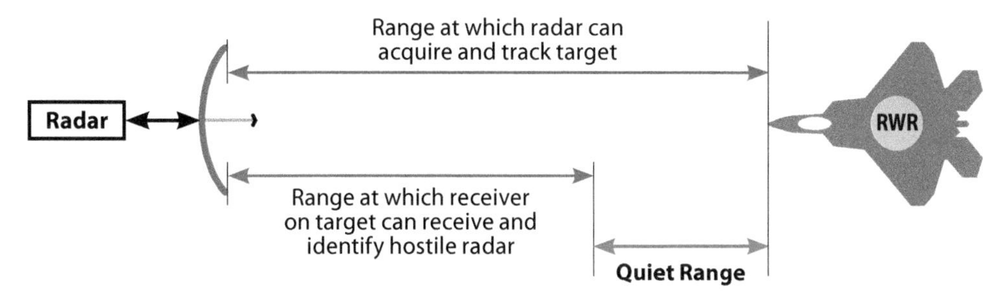

Figure 36-6. A radar that can acquire and track a target at a greater range than a receiver on that target can detect the presence of the radar is called a *quiet radar*, and the difference between the two ranges is called the *quiet range*.

Table 36-2. Ways That Jamming is Differentiated

Type of Jamming Target	RADAR	*Prevents RADAR from detecting or tracking targets*	
	Communications	*Prevents communication receiver from recovering information from desired signals*	
Location of Jammer	On RADAR Target	*Self Protection Jamming*	*Can be cover or deceptive*
	Remote from Target	*Stand Off Jamming*	*Jamming A/C beyond lethal range*
		Stand In Jamming	*Jamming closer to RADAR than Target*
		Escort Jamming	*Jamming aircraft; fly with strikers*
		Modified Escort Jamming	*Jamming aircraft; stay in hostile radar beam*
Jamming Approach	Cover Jamming	*Reduces RADAR's ability to detect target*	
	Deceptive Jamming	*RADAR thinks return is good, but calculated range or angle is incorrect*	

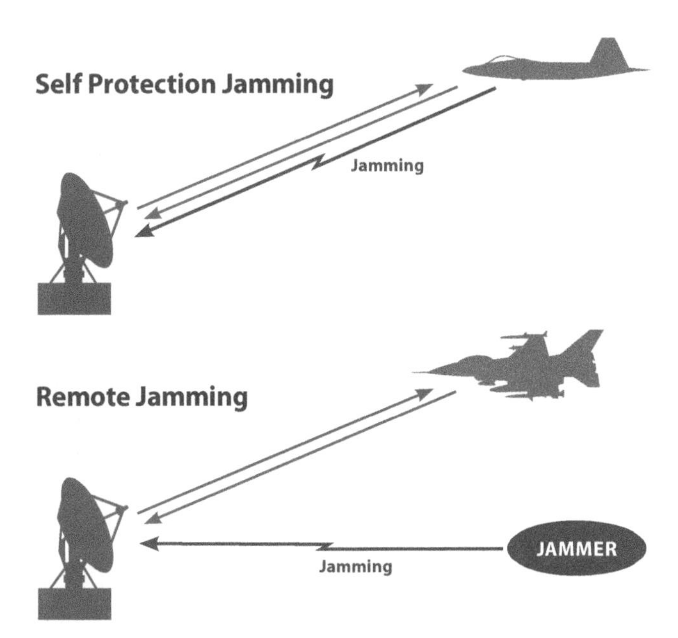

Figure 36-8. A self-protection jammer is mounted on the target, while a remote jammer transmits from some other location.

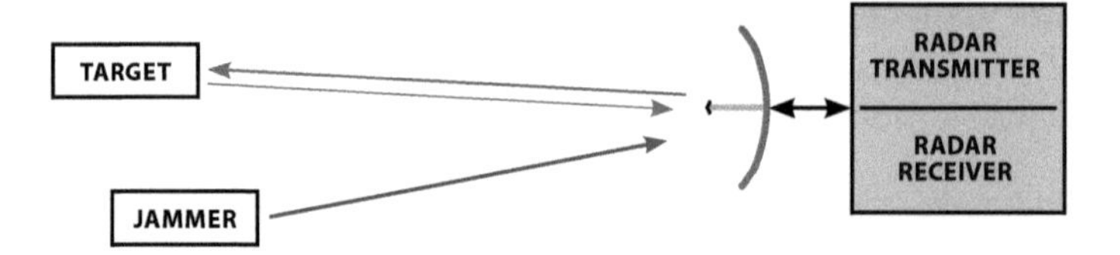

Figure 36-7. A radar jammer transmits to a radar's receiver to prevent it from properly receiving its own signal that is reflected from the target.

36.4 Electronic Attack

EA includes all of the ways that action is taken to reduce the effectiveness of hostile military assets by electronic means. This includes attacks on all types of enemy radars and communications. Methods include: jamming, chaff, flares, anti-radiation weapons, and high power radiation.

Jamming. Jamming involves the transmission of undesired signals into enemy receivers with power levels and modulations such that the ability of the receiver to receive and process its desired signals is degraded (Figure 36-7). Table 36-2 shows the way that jamming is described in terms of its target, geometry and approach to denying the enemy information.

The jamming geometry approaches are as follows.

- **Self-protection versus remote jamming** (Figure 36-8)
  - Self-protection involves transmission of jamming signals from a friendly asset to protect itself from detection, tracking or fusing by a hostile radar.
  - Remote jamming involves the transmission of jamming signals from some location other than that of the target of the hostile radar. This can be either stand-off or stand-in jamming.
- **Stand-off jammers** are farther way from the radar than the target. Typically, stand-off jammers are carried by special types of aircraft which are kept beyond the lethal range of weapons being controlled by the radars being jammed as shown in Figure 36-9.

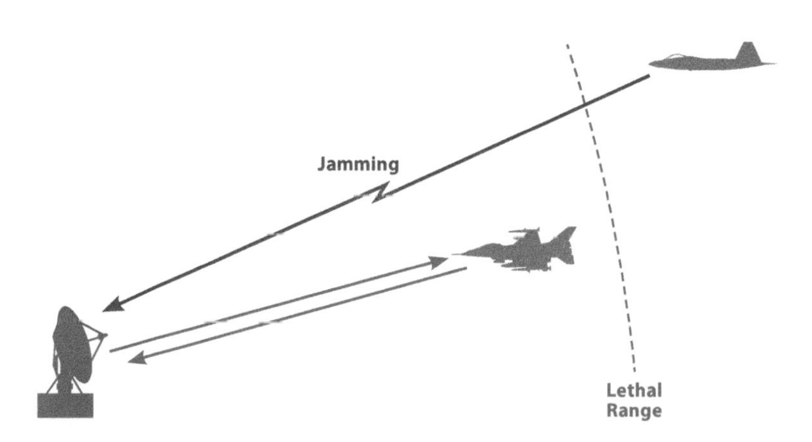

Figure 36-9. Stand-off jamming is transmitted from a dedicated jamming aircraft which is located beyond the lethal range of the aircraft weapon system controlled by the radar threatening the targeted aircraft.

- **Stand-in jammers** are closer to the radar than the target. These can be emplaced near the hostile radar or carried by unmanned vehicles which are remotely controlled to optimum jamming locations (Figure 36-10).
- **Escort jamming** is performed by special jamming aircraft that fly in with the strike forces they are protecting (Figure 36-11a).
- **Modified escort jamming** is performed by special jamming aircraft that fly-in behind the strike force aircraft they protect, but hold station so that they are within the main beam of specific, particularly dangerous radars to achieve more effective jamming. The jamming aircraft stay far enough from the threat radars that they remain beyond the lethal range of the associated weapons (Figure 36-11b).
- **Cover versus deceptive jamming**
  - Cover jamming involves the transmission of signals which prevent hostile receivers from receiving their desired signals as shown in Figure 36-12. This figure shows a plan position indicator (PPI) radar display with cover jamming. Cover jamming signals are typically modulated with noise, and are used in both radar and communication jamming applications.
  - Deceptive jamming involves the transmission of signals which are intended to be received as valid signals by a hostile receiver, but causing the associated radar processing circuitry to make incorrect conclusions (such as the range or angle to a target) as shown in Figure 36-13 or Doppler frequency.

Any of these techniques can use either noncoherent or coherent jamming. Coherent jamming is required for effectiveness against coherent radars.

Jamming Concepts. First, it is important to understand that jamming involves the transmission of unwanted signals to a target receiver to interfere with its ability to receive desired signals. Two important EA concepts are jamming-to-signal ratio and burn-through range. Both deal with the effectiveness of jamming.

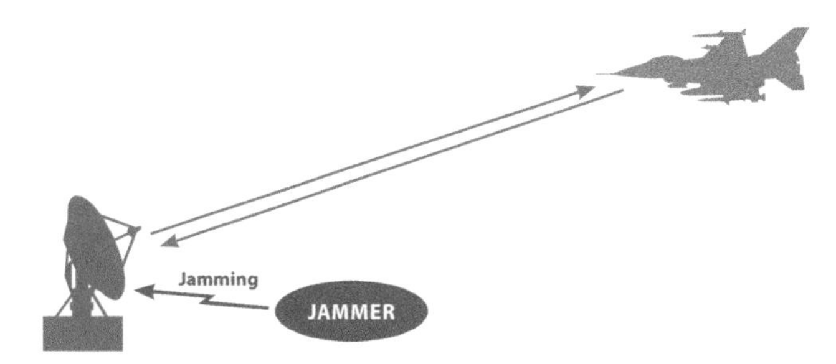

Figure 36-10. A stand-in jammer is emplaced or maneuvered closer to the radar than the target, to increase its jamming effectiveness.

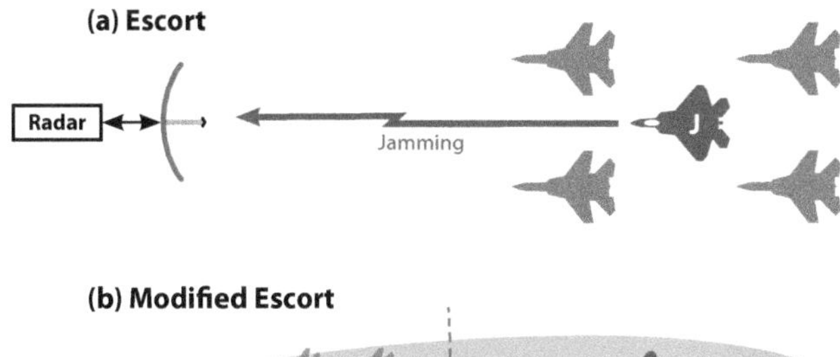

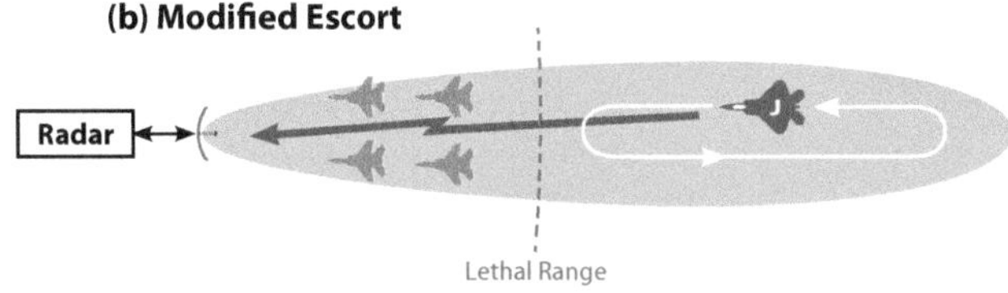

Figure 36-11. (a) In the escort role, a dedicated jammer aircraft flies in with a strike force, protecting the other aircraft. (b) In the modified escort role, the jamming aircraft keeps station within the threat antenna beam, but stays beyond the lethal range.

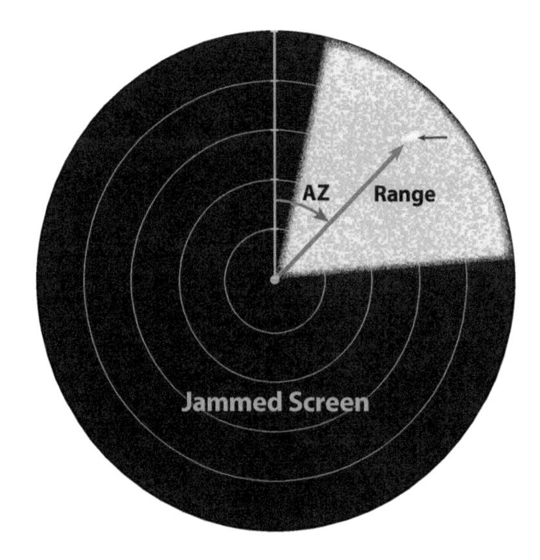

Figure 36-12. Cover jamming adds undesired signal energy, usually noise, to the radar receiver so that processing circuits will have difficulty acquiring or tracking the target. This is a PPI display of a target and jamming.

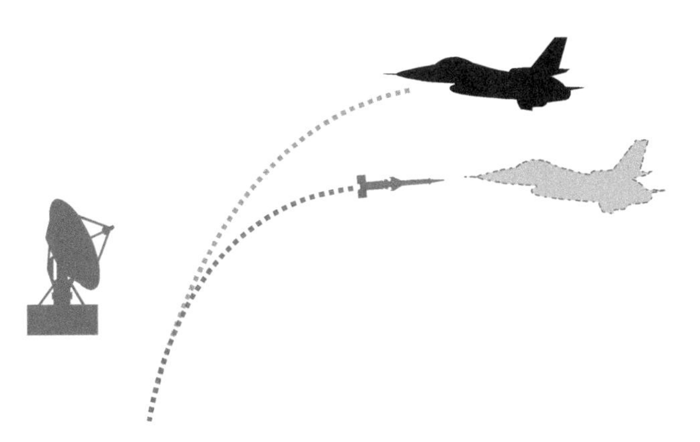

Figure 36-13. Deceptive jamming causes the radar processing circuitry to believe that it has a valid return signal, but the range or angle or frequency calculations are incorrect.

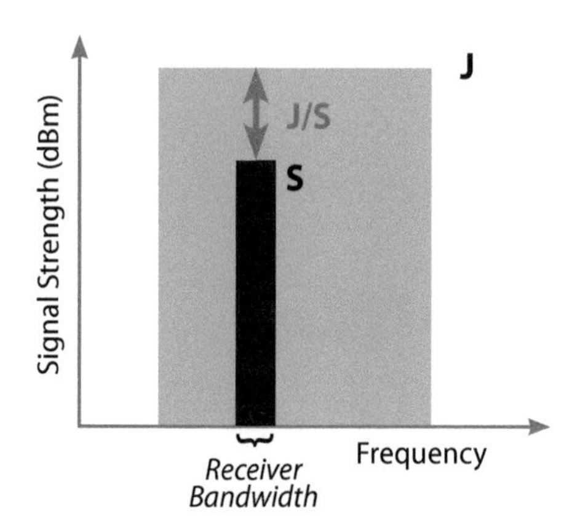

Figure 36-14. The jamming-to-signal ratio is the ratio between the jamming signal received by the target receiver and the desired signal received by the target receiver.

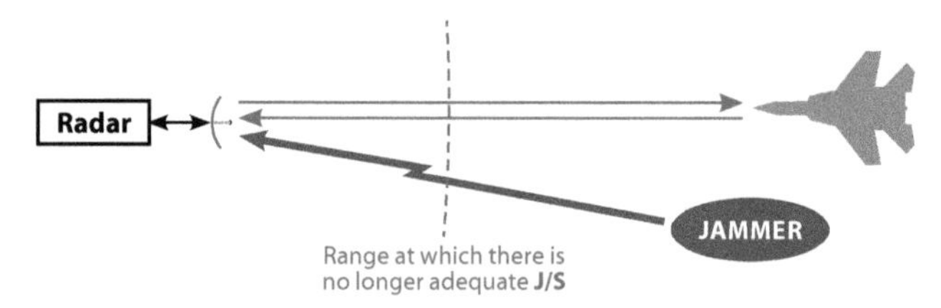

Figure 36-15. The J/S ratio reduces as the target approaches the radar. The range from the radar at which the J/S ratio is low enough that the radar can reacquire the target in the presence of jamming is the burn-through range.

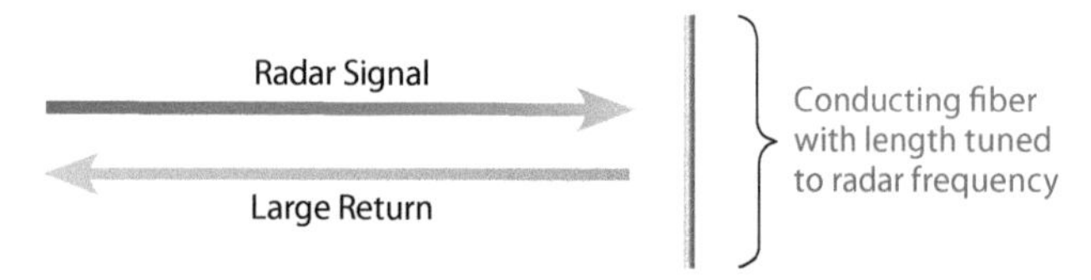

Figure 36-16. Chaff dipoles are cut to lengths that optimally re-radiate radar signals at a specific frequency when they are perpendicular to the incident signal direction and oriented with the polarization of the radar signal.

Jamming-to-Signal Ratio. The jamming-to-signal ratio (J/S) is the ratio between the receive power of the jammer and the desired signal in the target receiver as shown in Figure 36-14. Note that in this figure, the frequency range of the jamming (in blue) is wider than the bandwidth of the receiver (in black). It is only the jamming power that gets into the receiver that supports the J/S calculation.

Burn-Through Range. The burn-through range is the distance from a radar at which the radar can reacquire a target which has been protected by jamming as shown in Figure 36-15. At this range, the radar is said to "burn-through" the jamming.

Chaff. Chaff refers to small pieces of aluminum foil or metalized fiberglass that are cut to lengths which allow them to optimally rebroadcast radar signals as shown in Figure 36-16. Chaff was originally carried in packages that were opened and manually deployed from chutes in aircraft. Now it is fired from cartridges on aircraft and spread by turbulent air flow around the aircraft; or fired from ship mounted rockets and spread by explosive charges in the chaff rounds. It is also deployed by "string cutter" devices in aircraft or pods that cut the chaff material into proper lengths and blow it out into the air.

Chaff is used to create areas in which radars cannot find targets, or to cause radar returns that look like targets to capture the attention of radars and pull the focus of their tracking circuitry away from real targets.

Flares. Flares are expendable devices which protect aircraft from heat seeking missiles. They are fired from dispensers on Aircraft to create strong infrared false targets which move away from the protected target. To the infrared sensor in a heat seeking missile they look more like the aircraft than the aircraft does. The flare captures the attention of the missile tracker and leads it away from its intended target as shown in Figure 36-17.

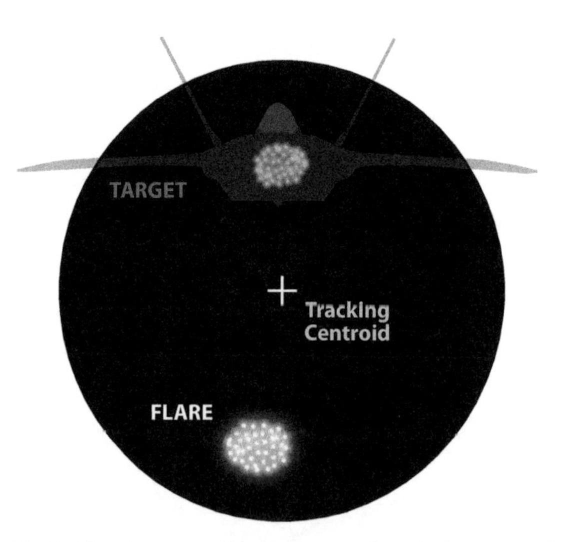

Figure 36-17. A flare has more thermal energy than the hot parts of the target against which a heat seeking missile is tracking. Therefore, the flare captures the missile's tracker, which then seduces the missile away from the target.

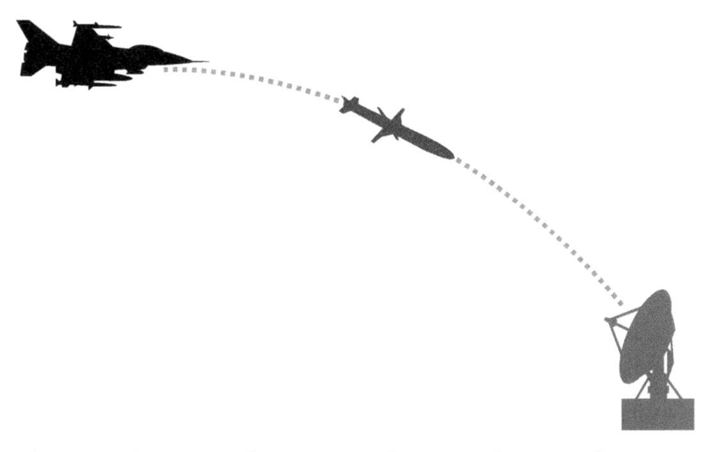

Figure 36-18. An anti-radiation weapon homes on the energy from an emitter. As long as the emitter is radiating, the missile guides itself to the radiating antenna with very great accuracy.

Anti-radiation Weapons. Anti-radiation weapons home on signals from hostile emitters. They have rather large war heads to destroy the hostile antenna and anything that is nearby. Although these weapons can be used against any type of transmitter, they are normally thought of as useful against hostile radars, either in the air or on the ground. Because they home on the hostile radar, they are capable of very high precision as long as the hostile emitter transmits. Figure 36-18 shows the action of an anti-radiation missile against a ground based anti-aircraft radar.

High Power Radiation. High power radiation can be either from lasers or radio frequency transmitters. This technique involves extremely high levels of transmitted power; three or more orders of magnitude greater than would be required to jam a receiver.

The purpose of high power radiation attack is either to destroy a hostile platform or, more commonly, to destroy an enemy sensor as shown in Figure 36-19. Jamming only temporarily defeats an enemy receiver; high power radiation permanently defeats an enemy sensor or directly damages an enemy platform. It is also significant that high power radiation weapons attack at the speed of light, while kinetic weapons attack at small multiples of the speed of sound.

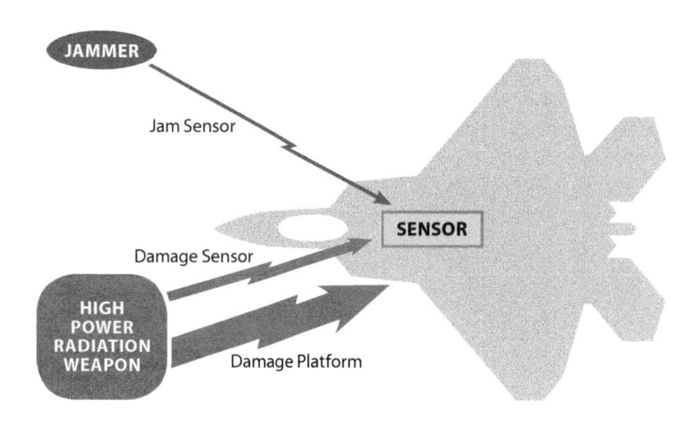

Figure 36-19. High power lasers can damage either enemy sensors or platforms. High power microwave can attack sensors. It takes about 3 orders of magnitude greater power to damage a sensor than to jam it, and it takes about 3 additional orders of magnitude of power to damage a platform.

36.5 Electronic Protection

Electronic protection (EP) does not protect platforms, but rather protects sensors against enemy electronic attack. Unlike ES and EA, EP seldom involves separate subsystems; it is a spectrum of design features of radars and communication systems that reduce the effectiveness of jamming. EP techniques involve special modulations, antenna design features, cancellation of jamming by addition of phase shifted replicas of jamming signals, and special operating modes. Table 36-3 summarizes the various general EP techniques.

Table 36-3. Electronic Protection Techniques

Radar EP Techniques	**Reduce Sidelobe Jamming**	*Auxiliary Antennas and Cancellation Circuits*
		Antenna Beam Shaping
	Reduce Deceptive Jamming	*Special Processing Circuits*
		Special Modulations
	Reduce All Jamming	*Special Operating Modes*
Communication EP Techniques	*Spread Spectrum Techniques*	

Table 36-4. Decoy Types, Missions, and Platforms Protected

Decoy Type	Mission	Platform Protected
Expendable	Seduction & Saturation	Aircraft & Ships
Towed	Seduction	Aircraft
Independent Maneuver	Detection, Seduction & Saturation	Aircraft & Ships

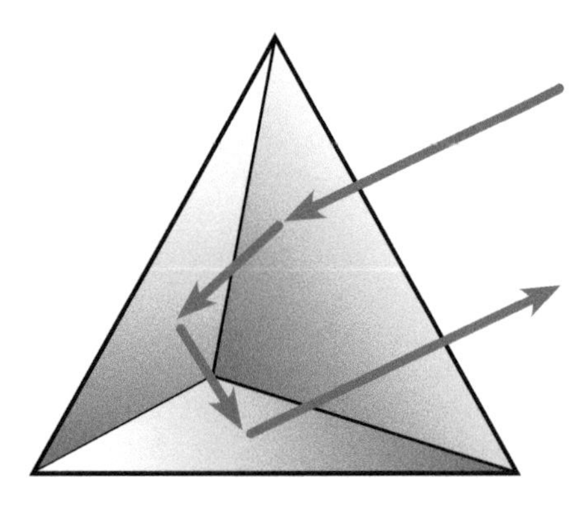

Figure 36-20. A corner reflector is a very efficient signal reflector and is therefore widely used in passive decoys to protect ships or other assets.

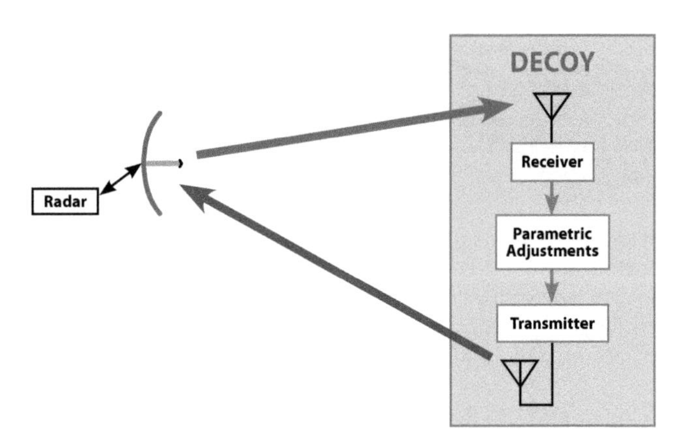

Figure 36-21. An active decoy creates the effect of a radar cross section by receiving a radar signal, amplifying it, and then rebroadcasting it to the radar.

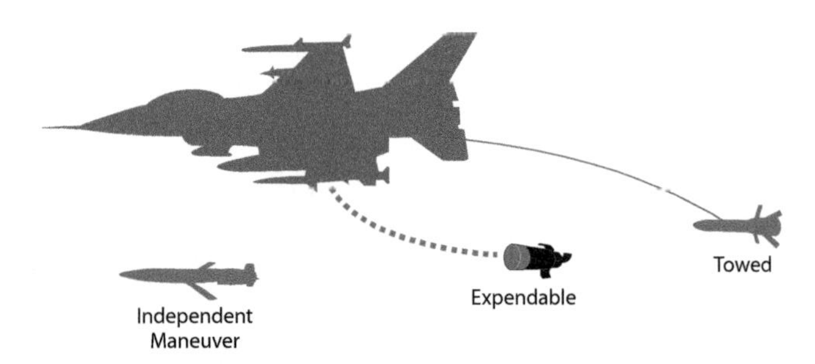

Figure 36-22. Decoys can be towed, expended from launchers, or mounted in independently maneuvering platforms.

36.6 Decoys

Decoys are devices designed to look like targets to enemy sensors. Flares, described above, are actually (thermal) decoys. Physical models (sometimes inflated) can be used to fool an enemy into believing that a particular piece of hardware is present in a location that will cause the enemy to draw false assumptions about planned military activities. However in EW, the term *decoy* is normally used to refer to devices which create radar returns that imitate the target returns in enemy tracking circuits. Decoys either give the enemy many false targets to complicate target acquisition or seduce the weapon's target tracker away from its intended tracker.

Table 36-4 shows the types of EW decoys, differentiating them by the way they are deployed, the way they protect a platform, and the type of platform protected. Seduction decoys capture an enemy tracker and lead weapons away from the target. Saturation decoys create false targets to overload the ability of a weapon system to acquire the target. A detection decoy causes an integrated defense system to believe that a dangerous weapon is arriving, thus causing the enemy system to turn on tracking radars which can then be attacked by anti-radiation weapons.

Decoys can be passive, relying on shapes like corner reflectors as shown in Figure 36-20 to create large radar returns; or they can be active, electronically creating large artificial radar returns as shown in Figure 36-21. Decoys can be launched from aircraft or ships. They can also be towed behind aircraft or can maneuver near ships or aircraft to steal the tracking focus to themselves rather than the intended targets. Figure 36-22 shows the various decoy deployment approaches.

36.7 Summary

EW is divided into three subfields: ES, which listens to enemy signals associated with threats; EA, which takes action to defeat or reduce the lethality of those threats; and EP, which protects friendly sensors against enemy EA. ELINT and radar ES are similar in that both collect hostile signals, however SIGINT collects as much detail of signals as possible to support detailed analysis to determine the capabilities and vulnerabilities of enemy threat systems and their associated signals. Radar ES, on the other hand, is more tactical, collecting only enough data to determine what threats are currently threatening a platform, supporting short term actions to protect the platform. Another role of ES is to support the development of an electronic order of battle, a description of enemy forces and capabilities derived from intercepted signal externals.

RWRs are special receiving systems which quickly intercept and analyze hostile radars and report the type of threats with which they are associated, their locations, and their current operating modes.

EA takes action against hostile signals. This includes jamming: the transmission of special signals to hostile receivers so that they cannot receive their desired signals or derive incorrect information from them. Self protection jamming is done from a target of an enemy radar to protect that asset. Stand-0ff jamming is done from a special aircraft which stays beyond the lethal range of weapons while protecting other aircraft which are within lethal range of those weapons. Cover jamming transmits noise to keep a hostile radar from seeing targets; deceptive jamming usually breaks the lock of an enemy radar on a target in range, angle or both. J/S is the ratio, in a hostile receiver, of the jamming signal to the desired signal (typically the skin return from a target). The J/S reduces as the square of the reducing range to an approaching target. The burn-through range is the range at which a hostile radar can reacquire its target in the presence of jamming. Other EA measures are chaff, flares, high power radiation, and anti-radiation weapons.

EP refers to special features of sensors to reduce the effectiveness of enemy EA; it protects sensors rather than platforms.

Decoys are designed to look more like a target than the target does, as perceived by hostile radars. Their mission is to saturate enemy weapons with false targets, lure enemy weapons away from their intended targets, or cause an enemy to turn on its tracking radars so they can be eliminated.

Test your understanding

1. Name and briefly describe the three major subfields of EW.
2. How does the role of ELINT differ from the role of ES?
3. Define the J/S ratio.
4. Define the burn-through range.
5. What is the specific role of an RWR?
6. Define Self-protection and stand-off jamming.
7. Does EP protect platforms?
8. How is a decoy different from a jammer that is remote from a target aircraft?

Further Reading

D. C. Schleher, *Electronic Warfare in the Information Age*, Artech House, 1999.

D. Adamy, *EW 101: A First Course in Electronic Warfare*, Artech House, 2001.

D. Adamy, *EW 102: A Second Course in Electronic Warfare*, Artech House, 2004.

D. Adamy, *Introduction to Electronic Warfare Modeling and Simulation*, SciTech-IET, 2006

F. Neri, *Introduction to Electronic Defense*, 2nd ed., SciTech-IET, 2007.

A. Graham, *Communications, Radar and Electronic Warfare*, John Wiley & Sons, Ltd., 2011.

A. De Martinio, *Introduction to Modern EW Systems*, Artech House, 2012.

DOD, Chairman of the Joint Chiefs of Staff, Joint Publication 3-13.1, *Electronic Warfare*, January 25, 2007, available at: https://www.fas.org/irp/doddir/dod/jp3-13-1.pdf

Boeing RC-135 RIVET JOINT (1961)

Rivet Joint provides intercept and accurate location of enemy radar and communication emitters. Onboard analysis allows near real-time reporting of reconnaissance information and ES support to theater commanders and combat forces. Key avionics include BAE Systems E&IS radar warning receiver (RWR) AN/ALR-94, AN/AAR-56 Infra-Red and Ultra-Violet MAWS (Missile Approach Warning System), and the Northrop Grumman AN/APG-77 Active electronically scanned array (AESA) radar. The AN/ALR-94 is a passive radar detector that is composed of more than 30 antennas blended into the wings and fuseIage and provides all-around coverage.

37

Electronic Warfare Support

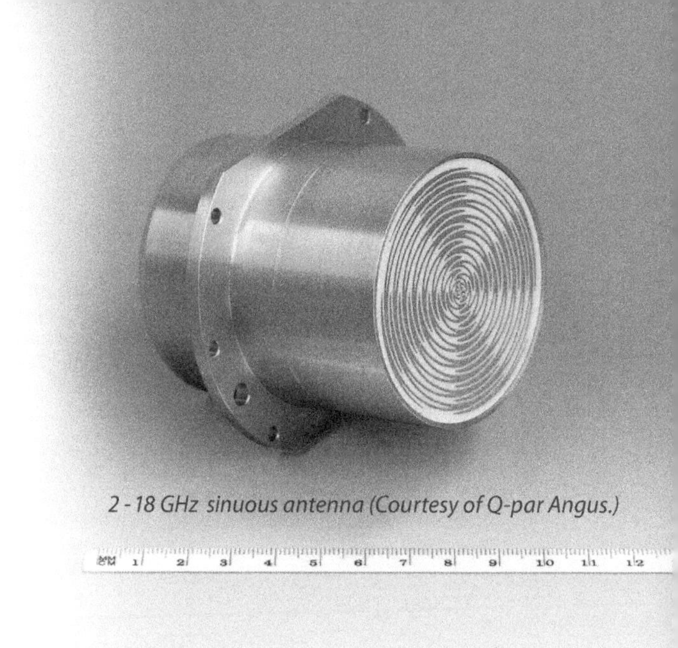

2 - 18 GHz sinuous antenna (Courtesy of Q-par Angus.)

Electronic warfare support (ES) is the listening part of electronic warfare. It is usually necessary to know something about the signals enemy radars are emitting before effective countermeasures can be taken against them. ES involves receiving hostile emissions (radars, communication and data links), analyzing their parameters, and displaying the appropriate information to the aircrew.

By receiving, identifying, and locating hostile emitters, ES supports the cueing of countermeasures and the development of electronic order of battle models of hostile forces. Although ES against hostile communications and data links is important, the primary emphasis in this chapter is on ES against radars.

Because ES is a receiving function, it is convenient to look at the building blocks of receiving systems at this point.

37.1 EW Antennas

Radar antennas are described in detail in Chapters 8, 9, and 10. This section discusses different types of antennas which are required for EW functions. These antennas either receive radar signals or transmit jamming signals. Their requirements can be significantly different from those of the earlier discussed radar antennas.

The *gain* of an antenna is the factor by which it increases the signal strength of a transmitted or received signal. It is caused by the angular concentration of energy.

An antenna that transmits or receives equally in all directions (i.e., over 4π steradians) is called an *isotropic* antenna. Its gain is defined as unity, which is 0 dB. Antenna gain is normally stated in dBi, which is dB above the gain of an isotropic antenna.

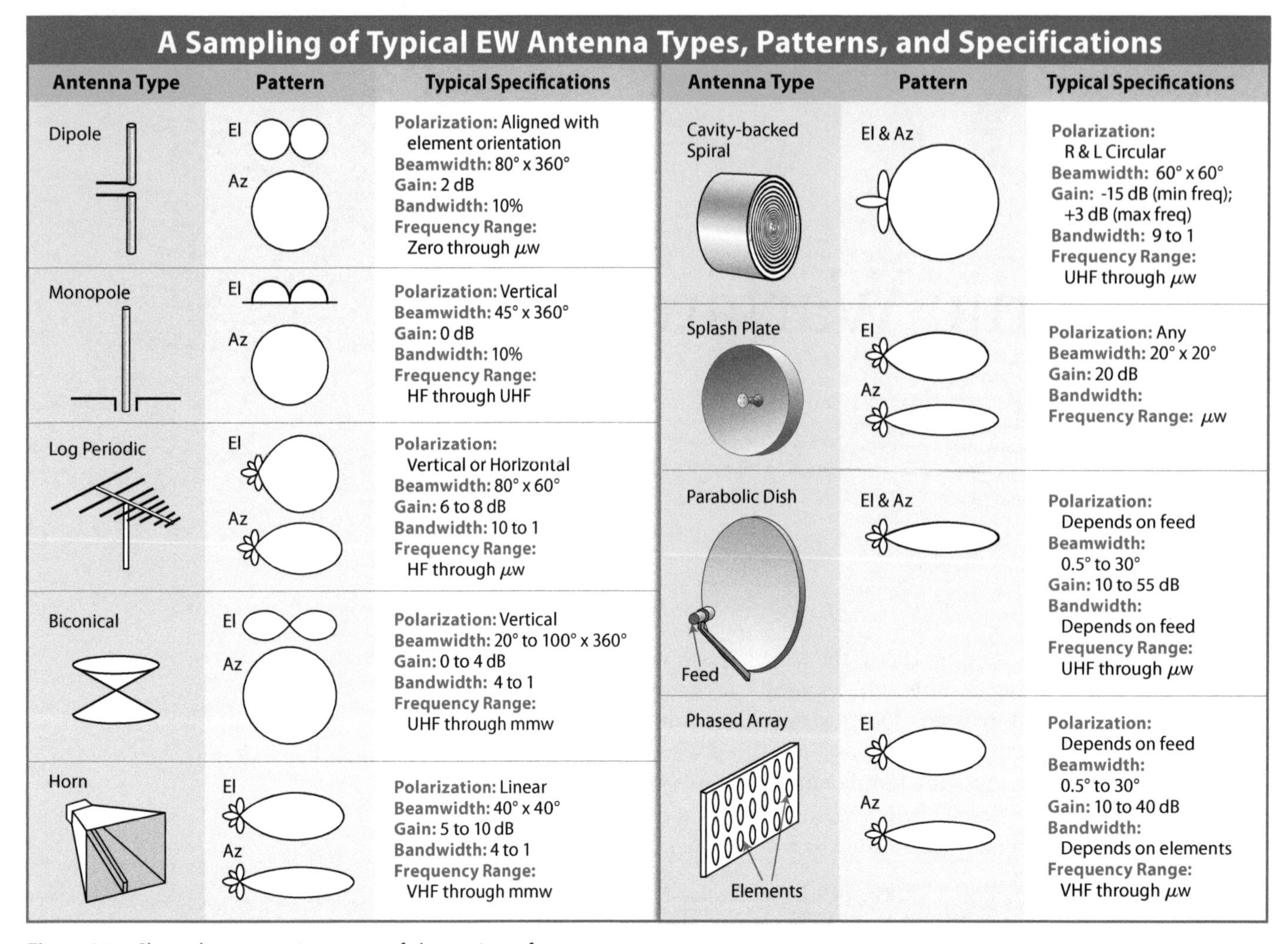

A Sampling of Typical EW Antenna Types, Patterns, and Specifications

Antenna Type	Pattern	Typical Specifications
Dipole	El, Az	**Polarization:** Aligned with element orientation **Beamwidth:** 80° x 360° **Gain:** 2 dB **Bandwidth:** 10% **Frequency Range:** Zero through μw
Monopole	El, Az	**Polarization:** Vertical **Beamwidth:** 45° x 360° **Gain:** 0 dB **Bandwidth:** 10% **Frequency Range:** HF through UHF
Log Periodic	El, Az	**Polarization:** Vertical or Horizontal **Beamwidth:** 80° x 60° **Gain:** 6 to 8 dB **Bandwidth:** 10 to 1 **Frequency Range:** HF through μw
Biconical	El, Az	**Polarization:** Vertical **Beamwidth:** 20° to 100° x 360° **Gain:** 0 to 4 dB **Bandwidth:** 4 to 1 **Frequency Range:** UHF through mmw
Horn	El, Az	**Polarization:** Linear **Beamwidth:** 40° x 40° **Gain:** 5 to 10 dB **Bandwidth:** 4 to 1 **Frequency Range:** VHF through mmw
Cavity-backed Spiral	El & Az	**Polarization:** R & L Circular **Beamwidth:** 60° x 60° **Gain:** -15 dB (min freq); +3 dB (max freq) **Bandwidth:** 9 to 1 **Frequency Range:** UHF through μw
Splash Plate	El, Az	**Polarization:** Any **Beamwidth:** 20° x 20° **Gain:** 20 dB **Bandwidth:** **Frequency Range:** μw
Parabolic Dish (Feed)	El & Az	**Polarization:** Depends on feed **Beamwidth:** 0.5° to 30° **Gain:** 10 to 55 dB **Bandwidth:** Depends on feed **Frequency Range:** UHF through μw
Phased Array (Elements)	El, Az	**Polarization:** Depends on feed **Beamwidth:** 0.5° to 30° **Gain:** 10 to 40 dB **Bandwidth:** Depends on elements **Frequency Range:** VHF through μw

Figure 37-1. Shown here are various types of electronic warfare antennas.

Figure 37-1 shows the important types of antennas used for EW functions. For each of these antenna types, the left column in the table shows a sketch of the antenna, the middle column has a sketch of its gain pattern, and the right column lists its typical specifications. It is important to note that these are *typical* specifications, because in EW systems it is often necessary to push their limits to provide extremely wide bandwidths to fulfill the EW mission.

Dipole Antenna. The dipole is used in phased arrays and is also widely used in *direction-finding* (DF) arrays. In DF applications the dipole is often operated over multiple octaves and must feed receivers through matching networks that greatly reduce its gain, particularly at the lower frequencies. The 360° pattern of the dipole, which is oriented perpendicular to the dipole elements, makes it ideal for ES receivers and for non-directional jammers. The nulls off the ends of the elements can be employed to significantly reduce interference between arrayed elements.

Monopole Antenna. Monopole antennas are used in receiving arrays and as jamming antennas. The monopole must operate over a ground plane and is thus ideal for mounting on the top or bottom of an aircraft fuselage for communication, signal intercept, or jamming applications. Like the dipole, it provides

360° azimuthal coverage. Although it has optimal gain when it is tuned in a narrow frequency band, it can provide wide frequency coverage with an appropriate matching network. When used over multiple octaves, its gain is down significantly at the lower frequencies.

Log Periodic Antenna. The log periodic (LP) antenna is used in low frequency phased arrays and can also serve as the feed antenna for a parabolic dish antenna. This antenna type is an array of dipoles of different lengths with different spacing to provide a directional gain pattern over an extended frequency range. The LP antenna is linearly polarized in the angle of its elements, so it can be oriented to provide any required linear polarization. Crossed dipole antennas can also be placed in a single array that can be fed through a phase shifting network to create a circular polarization.

Biconical Antenna. The biconical antenna is used as an airborne jamming antenna. Placed on a horizontal area of an aircraft or unmanned aerial vehicle (UAV) fuselage, it provides 360° azimuthal coverage with vertical polarization. Its gain pattern is similar to that of a vertical dipole, but the vertical angular dimension of the pattern is compressed, which provides additional gain. A polarizer can be used to change the polarization. This type of antenna can be ideal for transmitting jamming signals.

Cavity-Backed Spiral Antenna. Cavity-backed spiral (CBS) antennas are widely used in radar warning receivers (RWRs). This type of antenna has a gain pattern with a cosine^2 shape that looks like a big ball tangent to the surface of the antenna. This pattern has excellent front-to-back ratio. When a CBS antenna is used in an RWR for direction finding, the difference in output power from each antenna, expressed in dB, between neighboring antennas reduces linearly as a function of the true spherical angle from the antenna boresight.

By knowing the slope (dB/deg), which is stored as a part of the system data versus frequency, and the measured dB difference in receive signal strength from an off-board radar, an estimate of the radar angle relative to the aircraft may be obtained. Note that this linear gain variation can be significantly distorted by multipath from the platforms on which the antennas are mounted (Fig. 37-2). CBS antennas typically cover many octaves with significantly lower gain at the lower frequencies. However, at any covered frequency, the linear variation of gain applies. These antennas are circularly polarized, either right-hand or left-hand circular. Articulated antennas are capable of either sense circular polarization.

In aircraft RWRs, CBS antennas are typically mounted around the aircraft with 45° orientation relative to the aircraft's roll axis and approximately 15° depression below the aircraft's yaw plane (Fig. 37-3). The actual mounting locations are typically on the leading and trailing edges at the ends of the wings or in the fuselage at the wing roots and in the rear away from the jet engine outlet.

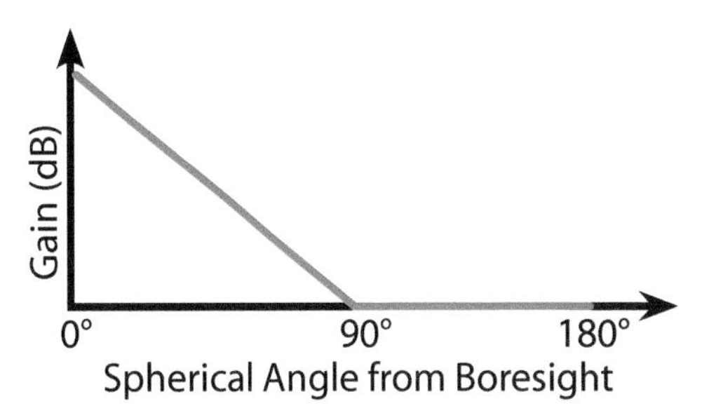

Figure 37-2. The cavity-backed spiral antenna has a gain pattern that varies approximately linearly (in dB) as a function of the spherical angle from the boresight.

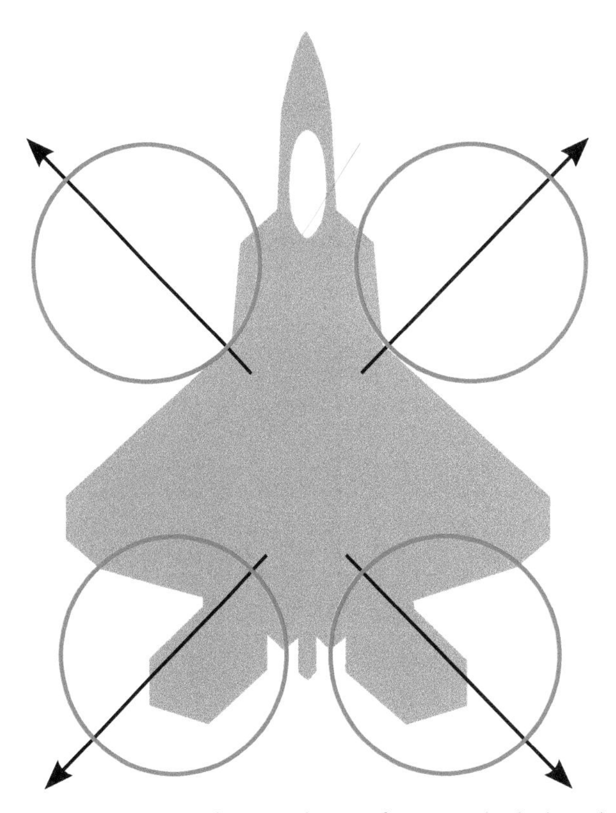

Figure 37-3. In a typical RWR application, four cavity-backed spiral antennas are mounted around the aircraft 45° from the roll axis and depressed 15° below the yaw plane.

Figure 37-4. These five cavity-backed spiral antennas provide vertical and horizontal baselines for an interferometer DF system.

Figure 37-5. Splash-plate antennas, like the one above, are placed at the forward and aft ends of jamming pods carried by standoff jammers.

Table 37-1. Important Types of EW Receivers

Receiver Type	Typical Sensitivity	Typical EW Application	System Performance Impact
Crystal Video	Low	RWR	*Wide frequency coverage; fast response time; one signal at a time*
Instantaneous Frequency Measurement	Low	RWR	*Up to octave frequency coverage, one signal at a time; measures only frequency*
Super-heterodyne	Med to High	RWR, ELINT, ES, and Targeting	*Selects one of multiple signals; recovers any modulation*
Channelized	Med to High	RWR, EW & Recon systems for complex signals	*Multiple simultaneous signals; can recover modulation*
Digital	Med to High	Every type of EW and Recon Application	*Wide frequency coverage; fast response time; one signal at a time*

CBS antennas can also be mounted in an array on the side of an aircraft for input to an interferometer DF system. The antenna placement for a two-dimensional DF system is shown in Figure 37-4.

Horn Antenna. Horn antennas are widely used in phased arrays and in jammers. An example application is the AN/SLQ-32 shipboard EW system, which uses sectoral horns in linear multibeam arrays for both ES receive and electronic array (EA) jamming purposes. Horn antennas provide gain and directivity with excellent front-to-back ratio. They can also operate over the wide frequency ranges required for EW applications. Although they can be used in ES applications, they are more commonly used in jamming systems. Horns can be rectangular or round and can be fed to provide any required polarization. Like many other antenna types, their gain will vary if used over wide frequency ranges with maximum gain at the upper frequencies and minimum gain at the lower frequencies.

***Splash-Plate* Antenna.** Splash-plate antennas have been long used in jamming pods carried by standoff jamming aircraft such as the EA-6B or the EF-18G. These antennas provide the gain required for long range jamming. The feed antennas can be horns or other antenna types that transmit from the edge of a tilted plate directing the power in the desired direction. Figure 37-5 shows a splash-plate antenna like that in a typical jamming pod.

Phased Array and Parabolic Dish Antennas. Phased array and parabolic dish antennas are covered extensively as radar antennas in other parts of this book. However, in EW applications, they are often required to cover much wider frequency ranges than those necessary for radars. Their achieved efficiency in wideband EW applications is significantly lower with the highest gain at the top frequency and the lowest gain at the lowest frequency.

The use of electronically steered arrays is an increasing trend in modern jammers. These antennas will provide multiple beams aimed at jammed radars to increase the achievable jamming-to-signal ratio, particularly in standoff jamming applications.

37.2 EW Receivers

Several types of receivers are used in ES system, each of which has specific advantages and disadvantages. Thus, most ES systems have multiple receiver types that are computer controlled in response to the encountered threat signals. Table 37-1 shows the most common receiver types used in electronic warfare along with typical sensitivity, applications, and impact on system performance.

Crystal Video Receiver. The crystal video receiver (CVR) is most commonly found in RWRs, where high probability of intercept is essential and low sensitivity can sometimes be tolerated. This receiver type is principally suited to the recovery of fairly strong pulse signals. As shown in Figure 37-6, the CVR has a diode detector that amplitude demodulates the incoming signals and passes them to a log video amplifier. The output of the amplifier is the video modulation of any signals received by the detector. Diode detectors operate in the square law region so that the amplitude of video output signals is proportional to the power of received radio frequency (RF) signals. The detectors instantaneously cover wide frequency ranges that are usually limited by input band-pass filters. The filters can be narrow but now usually cover large portions (typically 1/4) of the frequency range of interest (e.g., 2 to 18 GHz). All of the signals present at the detector are output together with no indication of the frequency of each signal.

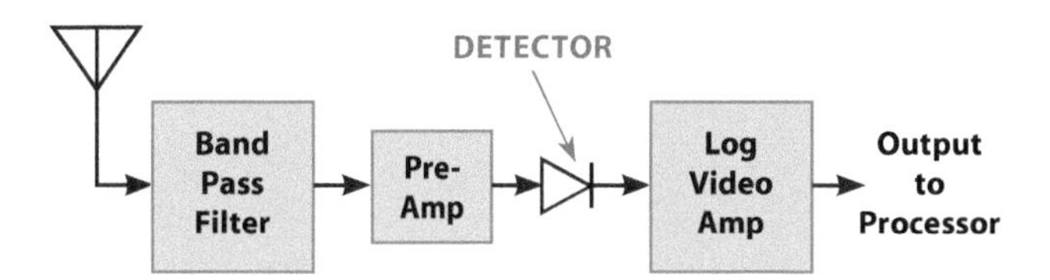

Figure 37-6. The crystal video receiver is widely used to receive pulsed signals in radar warning receivers.

These receivers are ideal for pulsed signals with relatively low duty factor because when two or more signals are present (e.g., when two pulses overlap) the output voltage is distorted.

CVRs have low sensitivity (approximately −40 dBm at the detector) but can be increased to approximately −65 dBm sensitivity with preamplification. They can receive signals with pulse widths as short as 50 to 100 nsec and as long as several milliseconds.

Instantaneous Frequency Measurement Receiver. An instantaneous frequency measurement (IFM) receiver measures only the RF frequency of received signals, producing a parallel digital output. The IFM has approximately the same sensitivity as the CVR and is often used in combination with multiple CVRs in RWRs. It can measure frequency in as little as 50 nsec, so it is very useful in isolating pulses from a single radar in a dense signal environment. It can also make multiple frequency measurements during a pulse, which is useful in detecting chirp or in specific emitter identification (SEI). The IFM receiver cannot handle more than one simultaneous signal. During the time that two or more signals overlap, the IFM outputs random bits.

An IFM can cover up to an octave of bandwidth and provides frequency resolution of about a tenth of a percent of bandwidth. Thus, an IFM covering a 4 GHz frequency range would provide frequency resolution to 4 MHz.

Superheterodyne Receiver. Superheterodyne receivers (SHRs) are used in EW systems to receive radar and communication signals in a dense signal environment. As shown in Figure 37-7, the SHR has a preselector that is tuned to select one signal or a narrow frequency range from all of the signals output by the antenna. The preselector output is passed to a mixer along with the output of a local oscillator (LO). The LO is tuned to a frequency a fixed amount higher or lower than the preselector frequency. The mixer output has both input signals and a number of other signals at the sums and differences of multiples of all input signals. An intermediate frequency (IF) amplifier filters the mixer output to select the difference frequency between the preselector and LO frequencies. It also provides a significant amount of gain. The output of the IF amplifier is passed to a detector that demodulates

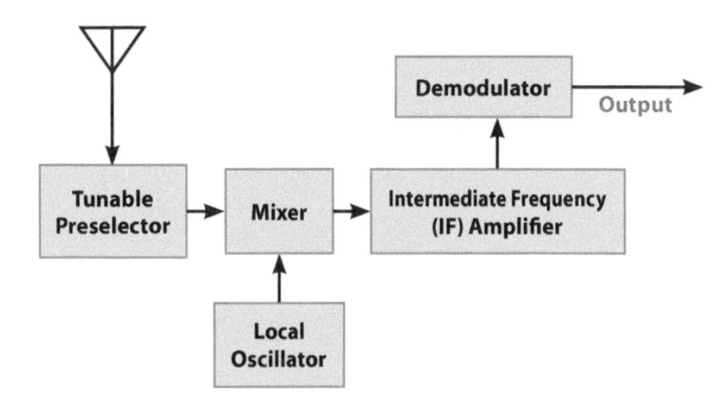

Figure 37-7. The superheterodyne receiver can isolate single signals or narrow frequency ranges in a high density environment.

the IF signal to recover the video modulation, which can be pulses or other waveforms such as continuous wave (CW). In other ES applications the instantaneous IF bandwidth may be broader, such as that as used for broad IF bandwidth digital channelized receivers.

The SHR can receive and demodulate any type of signal modulation and has sensitivity that is determined by the factors discussed in Section 37.3. In general, the SHR is significantly more sensitive than CVRs or IFMs.

Frequency converters that allow a receiver to selectively serve a wide frequency range operate on the heterodyne principle used in the SHR.

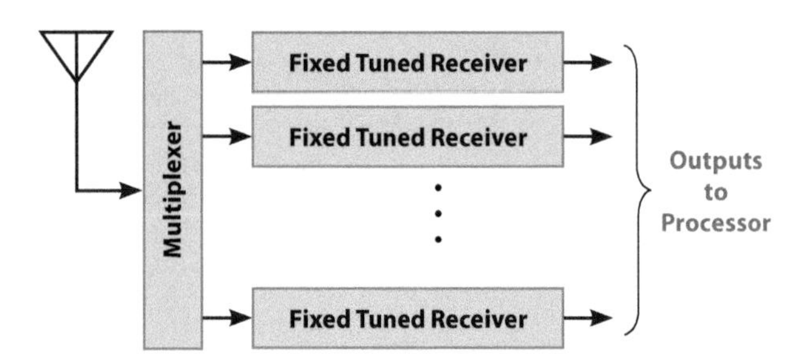

Figure 37-8. A channelized receiver can intercept multiple simultaneous signals.

Analog Channelized Receiver. The channelized receiver comprises a number of parallel fixed tuned receivers that cover different frequency ranges. These are typically designed with contiguous bands. As shown in Figure 37-8, an output from a single antenna is multiplexed into the receiver channels. Each channel can contain a complete receiver, producing a video output. Sometimes each channel has only the front end of a receiver, outputting a frequency range to a second set of channels, each of which has the output stages of receivers. In this case, when there is activity in one of the first stage channels, the second stage channels are assigned to handle the individual signals that are present.

The purpose of the channelized receiver is to recover simultaneous signals in a dense environment.

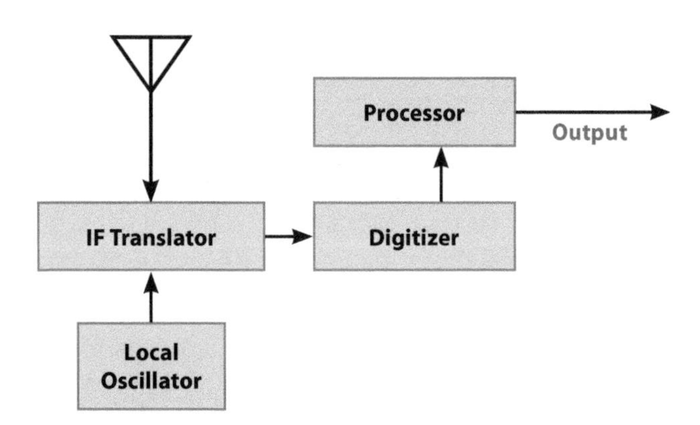

Figure 37-9. Digital receivers digitize the output of an IF amplifier and then perform the receiving and signal processing functions in a computer.

Digital Receiver. The digital receiver is growing in importance in EW systems. The typical block diagram of a digital receiver is shown in Figure 37-9. An analog front end usually includes a frequency converter to bring a portion of the input frequency range to a single wide IF band to be digitized. The digitizer or analog-to-digital (A/D) converter alters the entire IF band into a digital signal. Then focal plane arrays (FPAs), digital signal processors (DSPs), and a computer perform the receiver functions and various analysis functions that are difficult to perform on analog signals. In modern ES architectures, FPAs perform the channelization, detection, pulse, and CW parametric measurements, and DSPs perform such tasks as pulse sorting, DF estimates, and pulse train parametric calculations. A computer performs such functions as classification, identification, and emitter tracking.

The digitizer must sample its input bandwidth at the Nyquist rate. That is, it must sample the input band at a rate twice the sampled bandwidth. For example a 1 MHz bandwidth must be sampled at two million samples per second. The number of bits generated for each sample determines the sensitivity and dynamic range of the digital receiver. It is important that the sensitivity and dynamic range of the analog front end and the digital back end of the receiver be matched.

37.3 Receiver System Sensitivity and Dynamic Range

Sensitivity. Chapter 12 discusses the detection range for radars. Like the radar, an ES system must detect threat signals at adequate range. One of the contributing qualities to acquisition range is receiver system sensitivity, which is defined as the minimum signal a receiver system can take in and still perform its required tasks to specification. For an ES system, this usually means correctly identifying the type of threat signal received, determining the operating mode of associated weapons, and locating the threat emitter to the specified accuracy. The sensitivity is defined at the system input, that is, at the output of the antenna supplying signals to the receiving system.

Receiver system sensitivity is determined by the following formula:

$$S = kTB + NF + RFSNR_{RQD}$$

where S is the system sensitivity in dBm, kTB is the thermal noise in the system, NF is the system noise figure at the system input, and $RFSNR_{RQD}$ is the required predetection signal to noise ratio

Figure 37-10 shows the relationship of the three sensitivity contributors. The sensitivity is the signal level that allows full system performance. The minimum discernible signal (MDS) is the signal level at which the noise at the signal input is equal to the received signal from the antenna at that point.

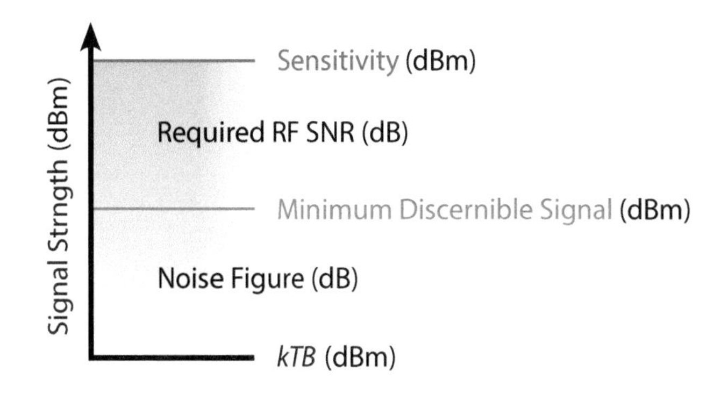

Figure 37-10. The receiver sensitivity is the sum of the thermal noise, the system noise figure, and the required prediction signal-to-noise ratio.

kTB. kTB is the internal thermal noise in the receiver system, caused by molecular motion and has three parts: k is Boltzmann's constant, which defines the thermal noise per Hertz at a given absolute temperature; T is the temperature in Kelvin; and B is the bandwidth in which the noise is measured.

There is a handy number for determining kTB within the Earth's atmosphere. This results from the definition of ambient temperature (for these purposes) as 290 K, and gives a value for kTB as

$$kTB = -114 \text{ dBm/MHz}$$

$$\text{or } -114 \text{ dBm} + 10 \log_{10}(B/1 \text{ MHz})$$

That means that the thermal noise in a system is −114 dBm if the effective system bandwidth is 1 MHz. This formula is also given as

$$kTB = -174 \text{ dBm/Hz}$$

which is the same value.

The temperature 290 K is 17 °C or 62.6 °F, but the −114 or −174 value for kTB is commonly used for applications within the atmosphere over a wide temperature range. Consider that the temperature would have to go up or down by 72.5 °C or 130.5 °F to change kTB by 1 dB.

For applications in outer space, sensitivity is not commonly calculated using *kTB* because of the extremes of temperature encountered. In that case, noise temperature is used.

***kTB* versus Sensitivity.** Since *kTB* is a part of the sensitivity calculation, the effective system bandwidth directly impacts the sensitivity. For example, reduction of the bandwidth by a factor of 10 will improve the system sensitivity by 10 dB.

Noise Figure. The system noise figure is the amount of noise that is added by the system above *kTB*. However, this noise is referenced to the system input. This noise level is often described as the amount of noise above *kTB* that would have to be input to the system to achieve the measured total output noise level if the system itself created no noise other than *kTB*.

Low-noise preamplifiers are added to systems to reduce the noise figure, thus improving the system's sensitivity. The amount by which the preamplifier improves the system noise figure depends on the specific system configuration, but it is common for the system noise figure to be reduced to only a dB or two above the preamplifier noise figure.

Table 37-2. Required Predetection Signal-to-Noise Ratio

Analysis Task	Required Predetection SNR
Analysis of signals by an expert operator	8 dB
Automatic analysis of signals by a computer	15 dB
Specific Emitter Identification	25 to 35 dB
Emitter Location	−15 to +15 dB

Predetection Signal-to-Noise Ratio. This is the way that the quality of received signals is quantified. The signal quality required depends on what information is being extracted from received signals. Some typical required prediction levels are shown in Table 37-2.

Analysis by a skilled operator may be required for *electronic signals intelligence* (ELINT) applications or cases in which signals must be extracted from a high noise environment.

The best example of automatic computer analysis is the RWR that very quickly identifies the threat radar type from its parameters and reports the type of associated weapon system and its location to a crew display.

Since SEI requires the detailed measurement of signal parameters, this process typically requires one or two orders of magnitude more signal-to-noise ratio to allow parametric measurements with adequate resolution.

Emitter location processing involves various approaches that may allow the effective bandwidth of the signal to be reduced through averaging and other techniques. This can create processing gains that improve the effective system throughput bandwidth and thuse the system sensitivity for this function.

Dynamic Range. The dynamic range of a receiving system is the difference (in dB) between the smallest signal the receiver must receive and the largest signal that can be present in the receiver bandwidth while the smallest signal is received. ES systems are required to have a relatively large instantaneous dynamic range. Unlike a radar, which usually must have *automatic gain control* (AGC), an ES system must be able to receive weak signals in the presence of much stronger signals. An AGC circuit reduces the throughput gain of the receiver so that reception of the strongest signal present is optimized. However, a lethal threat signal may be many dB weaker than the strongest signal present.

The required instantaneous dynamic range is typically 60 to 70 dB in an ES receiver system. For broadband receivers, the two-tone dynamic range defines the signal strength increase above the sensitivity level at which two signals will cause spurious responses at the sensitivity level.

The dynamic range of a digital receiver is determined by both the dynamic range of the analog front end and the digital circuitry. There is a trade-off between the sensitivity and dynamic range. In the digital part of the receiver, the dynamic range is determined by the number of bits with which each sample is digitized. The *effective number of bits* (ENOB) is determined from the actual dynamic range achieved and is typically a fractional number of bits less than the bits in the A/D converter output. The largest signal would be characterized by all bits being ones, whereas the smallest signal would be characterized by all zeros except a one in the least significant bit position. The formula for digital dynamic range is

$$DR = 20 \log_{10}(2^n)$$

where DR is the dynamic range in dB, and n is the number of digitizing bits.

37.4 One-Way Radio Propagation

One significant difference between radar and EW fields is that radar signals are typically received at the location of the transmitter after reflection from a target. In EW, transmissions are received by distant receivers—usually at locations hostile to the transmitter. In ES, systems are designed to intercept enemy radar or communication signals. Jamming systems generate signals designed for reception by enemy receivers and to interfere with their intended operation. Thus, the one-way link equation is central to the understanding of EW.

The one-way link equation gives the received signal strength in terms of everything else happening in the link. In its most basic form

$$P_R = P_T + G_T - L_P + G_R$$

where P_R is the received signal power in dBm, P_T is the transmitting antenna power in dBm, L_P is the propagation loss in dB, and G_R is the receiving antenna gain in dBi.

Another form of this equation combines the transmitter power and the transmitting antenna gain in the *effective radiated power* (ERP):

$$P_R = ERP - L_P + G_R$$

where ERP is the effective radiated signal power in dBm.

An important concept in EW is that the transmitting antenna gain and ERP are understood to be in the direction of the

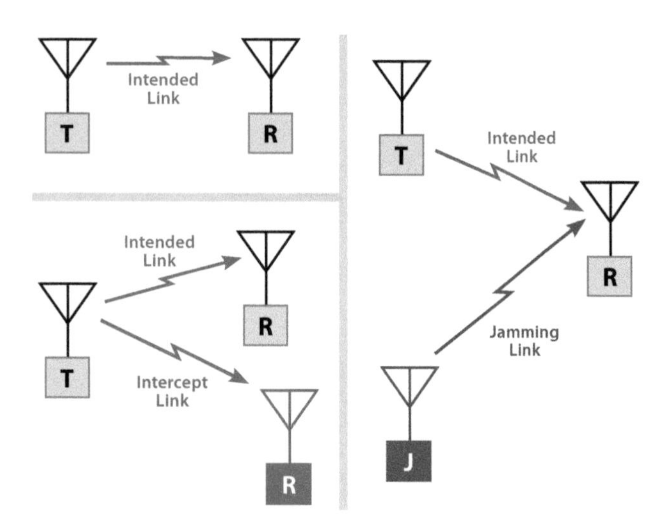

Figure 37-11. Communication links, intercept links, and jamming links all are described by the one-way link equation.

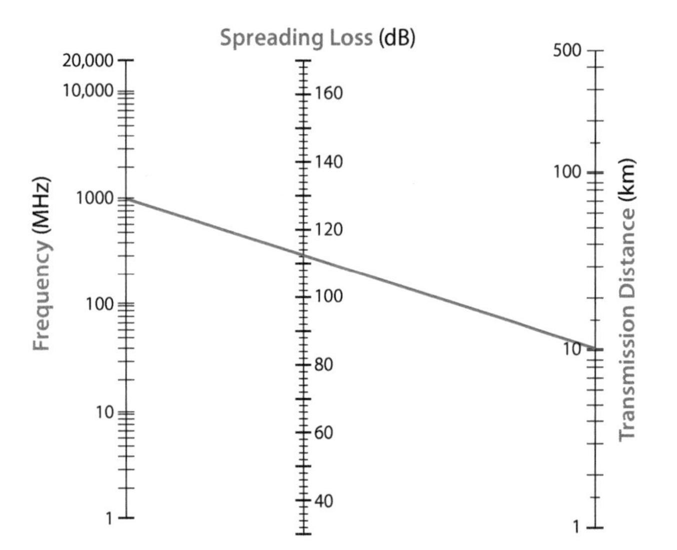

Figure 37-12. Free-space loss is determined from the link frequency and the link distance.

receiving antenna and the receiving antenna gain is understood to be in the direction of the transmitting antenna. This is important because the antennas are often not oriented toward each other. In ES, transmissions from hostile radars and communication transmitters are often intercepted from the sidelobes of transmitting antennas. Also, jammers often transmit into the sidelobes of enemy radar or communications receiving antennas.

Figure 37-11 shows several EW one-way communication links.

There are several models for radio propagation loss, but the model most appropriate to radars and radar-associated ES and EA is the *free-space propagation* model, which is also called *line of sight propagation* or *spreading loss*. The free space model is convenient although it neglects atmospheric attenuation, ducting, and other propagation factors that must be taken into account in detailed performance assessments. This is especially true at *millimeter wave* (MMW) frequencies. Note that this model is the basis of the radar-range equation presented in Chapter 13. The free-space propagation loss algebraic equation is

$$L_P = (4\pi)^2 d^2/\lambda^2$$

where L_P is the loss ratio, d is the distance between the two antennas in meters, and λ is the wavelength of transmission in meters.

There is also a convenient logarithmic form of this equation in which all of the values are in dB form:

$$L_P = 32.44 + 20\log_{10}(d) + 20\log_{10}(F)$$

where L_P is the loss in dB, d is the distance between the two antennas in kilometers, and F is the transmission frequency in megahertz

The constant, 32.44, allows the input of distance and frequency values in the most convenient units and avoids the requirement to determine the wavelength. This dB formula is also defined for distance in statute miles, but the constant must be changed to 36.52, or in nautical miles with the constant changed to 37.73.

This equation is often used to make calculations to 1 dB accuracy. In that case, the constants are normally written as 32, 37, and 38, respectively.

Figure 37-12 is a handy nomograph to determine the free-space loss between two isotropic antennas as a function of frequency and range. To use this nomograph, draw a line between the link frequency in MHz and the link distance in km. Read the link loss in dB where your line crosses the center scale.

37.5 Passive Emitter Location

One of the most important tasks of an ES system is to determine the location of noncooperative emitters. Ideally, the emitter location capability would cover all directions and provide high accuracy very quickly. The various approaches and techniques require trade-offs among these desired qualities. Unlike

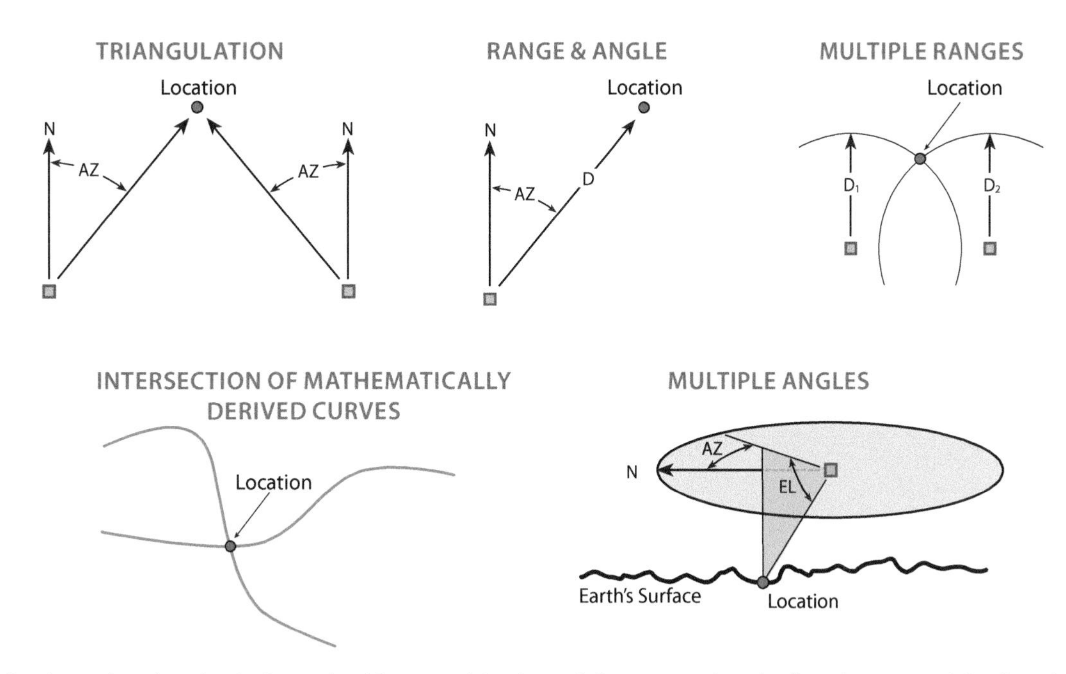

Figure 37-13. Passive emitter location is determined from combinations of distances and angles from known receiving locations or by the use of mathematically determined curves.

radars, ES systems do not deal with a known signal generated by a cooperating transmitter. They must determine the location of a wide variety of radar and communication transmitters. There are several basic approaches to emitter location as shown in Figure 37-13. This figure illustrates location in two dimensions, which is sometimes adequate. All techniques can, of course, be extended to emitter location in three dimensions. The approaches are as follows:

- Triangulation
- Range and angle from a known location.
- Distance from multiple locations
- Intersection of multiple mathematically determined loci
- Intersection of signal path with the earth

Triangulation requires that the direction of arrival (DOA) of a signal be determined from multiple known locations. Ideally, the locations will be 90° apart as seen from the target emitter. The equivalent operation can be accomplished with one moving platform, but the amount of time required to determine a location to required accuracy will depend on the range to the target and the speed of the platform.

Range and angle is the same technique used by a radar to locate its target. The difference in ES is that it cannot measure the time of transmission of the signal (at the speed of light) to determine the range because it does not know when the signal left the transmitter. Therefore, the range must be determined by measuring the strength of the arriving signal, estimating the ERP, and calculating the range from the equations in Section 37.4. This yields poor range accuracy because the ERP of a hostile transmitter cannot be accurately estimated. Radars often use emission control, which reduces their ERP to the minimum required optimum performance. Even without emission control, a radar's ERP varies with the location of the receiver relative to the bore sight of the radar's antenna.

Distance from multiple locations is used to calculate the target emitter location. However, it must still deal with the passive determination of range issue, so the location accuracy is limited.

Intersection of multiple mathematically determined loci is a technique used in the two precision emitter location techniques discussed shortly.

If an airborne ES system measures the azimuth and elevation to a hostile transmitter, it has defined a vector from the transmitter to the aircraft. If the transmitter is on the ground, the aircraft ES system can solve for the intersection of the signal path with the earth by use of a digital map and knowledge of the aircraft's elevation.

Statement of emitter location accuracy is the accuracy with which an ES system can determine the location of an emitter. It is commonly stated in terms of its circular error probable (CEP). The CEP is the radius of a circle around the measured location of the emitter that has a 50 percent chance of containing the emitter's true location. The smaller the CEP, the more accurate the location system is stated to be.

When the location is more accurately known in one dimension than another, the elliptical error probable (EEP) can be used. This is the ellipse with a 50 percent probability of containing the true location.

Direction of Arrival Techniques. In several of these approaches, it is necessary to determine the *direction of arrival* (DOA) of signals reaching an ES system. The accuracy of a DF system is stated as its root mean square (RMS) accuracy. This is determining by making DF measurements at many representative angles and frequencies. The DOA error is determined for each measurement, and those error values are squared. The squared errors are averaged, and the square root is taken. This is the RMS error. Common DOA techniques are (1) moving narrow-beam antenna, (2) amplitude comparison, (3) Watson-Watt direction finding, (4) Doppler direction finding, and (5) interferometer.

A *narrow-beam antenna* can be used to determine the DOA just as it docs in a radar. Thc antcnna is movcd through an angular search pattern, and the time at which signals are received determines the direction of arrival. This technique has the advantage of antenna gain and selectivity, which can be important in a dense signal environment. However, the time required to complete the search pattern delays the detection of lethal threats. This can be a significant problem to an ES system, which typically tries to detect and locate threat emitters within a very few seconds or less. A particular case of interest is to detect a threat missile having an active missile seeker and to apply EA against the missile to preclude missile lock-on to the EA protected platform. Breaklock is usually more difficult to achieve.

Using *amplitude comparison,* multiple antennas can be used to quickly determine the direction of arrival by amplitude comparison, as shown in Figure 37-14. The two pictured antennas receive the same signal with different amplitude depending on the antenna gain patterns and mounting orientation. The DOA is immediately calculated from the amplitude difference and can measure the DOA of every radar pulse received. Therefore, this approach is considered a monopulse direction finding technique.

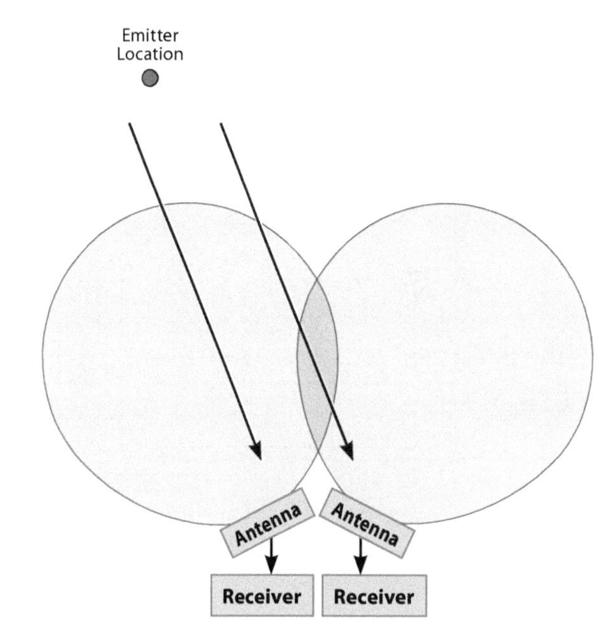

Figure 37-14. A signal's angle of arrival can be determined by comparing its amplitude as received by two antennas with overlapping coverage.

The accuracy achieved by this technique varies with the nature of the environment in which the antennas are mounted. In a very clean environment, it may achieve 3° to 5° RMS accuracy; in a highly complex environment, for example on a tactical aircraft, it may achieve accuracy only of the order of 10° RMS.

Figure 37-15 is the block diagram and amplitude versus angle plot of a *Watson-Watt direction-finding system.* This is a 360° coverage approach. It typically uses four dipole antennas in a circular array that is approximately one-quarter wavelength in diameter. It can also operate with a crossed pair of loaded coils. There are three receivers. Two are fed by opposite antennas in the circular array, and the third is fed by a reference antenna in the middle of the array. This receiver can also be fed by the summed outputs of all of the antennas in the circular array.

When the difference between the signal strengths from the array antennas is compared to the signal strength from the reference antenna, the result is a cardioid pattern as shown in the figure. At the instant when another opposite antenna pair is switched into the appropriate receivers, two cardioid patterns are defined. After rotating the array antenna selection a few times, the cardioid patterns are solved for the direction of arrival of the received signal.

The Watson-Watt approach can be used to receive all types of signal modulations. It is normally used in the very high frequency (VHF) and ultra high frequency (UHF) ranges and typically achieves about 2.5° RMS accuracy. Its sensitivity is such that there must be about +10 dB predetection SNR. It typically provides DOA in 1 to 2 s.

The *Doppler direction-finding approach* is based on the concept of one 360° coverage antenna rotating around another. As the moving antenna moves toward or away from a target emitter, there is a Doppler shift that increases or decreases the received frequency. As the moving antenna traces its circular path, the Doppler shift varies sinusoidally.

The direction of arrival of the signal is the DOA at which the negative going zero crossing of the Doppler sine wave occurs. In practice, the moving antenna is simulated by a circular array of antennas (three or more) around a fixed reference

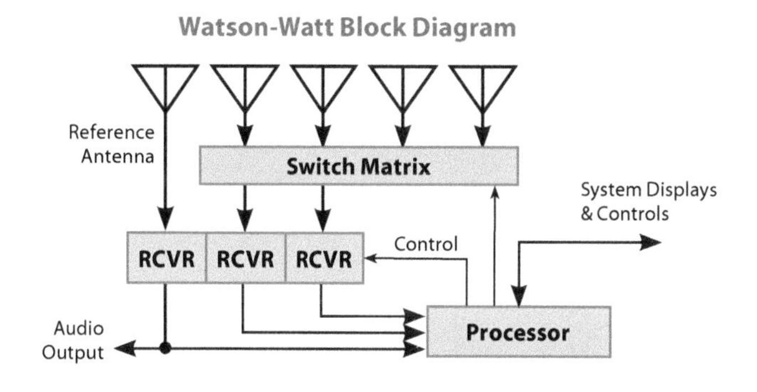

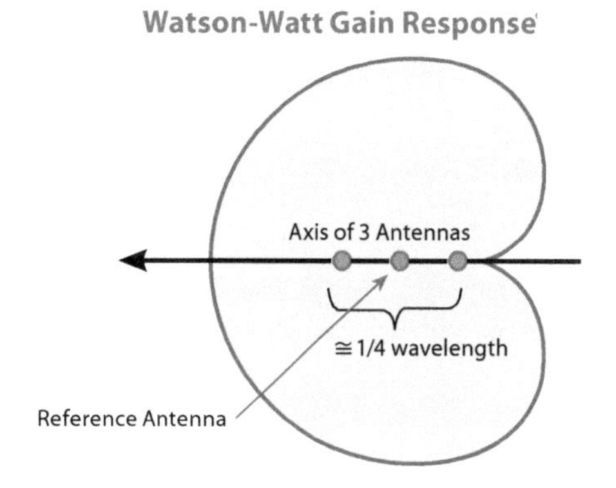

Figure 37-15. The Watson-Watt direction-finding approach measures direction of arrival over 360° of azimuth for all types of modulations.

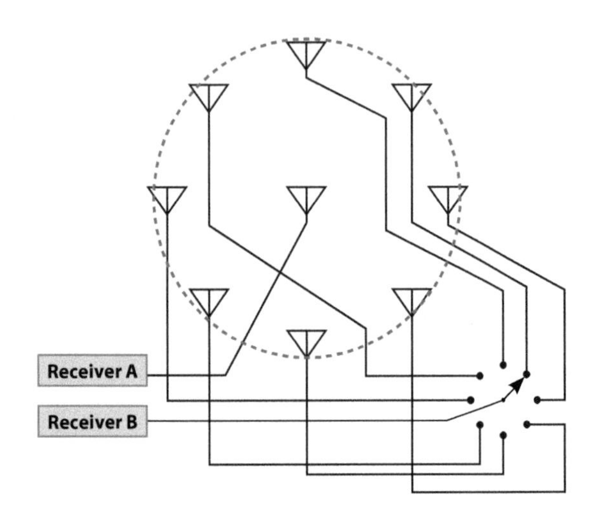

Figure 37-16. The Doppler direction-finding technique determines a signal's direction of arrival by commutating circularly arrayed antennas to one receiver and comparing signals at the "rotating" location to signals at a fixed reference location.

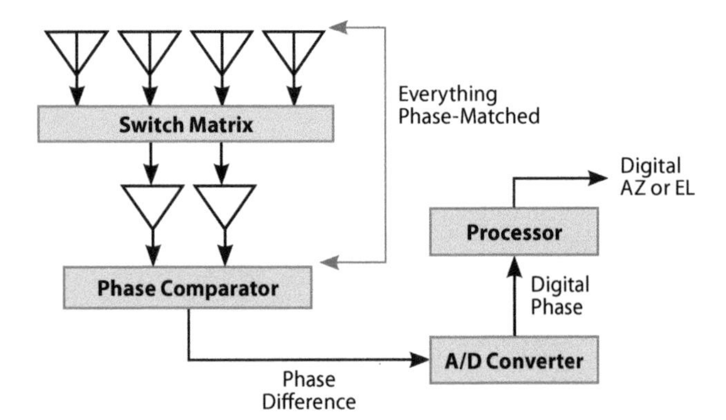

Figure 37-17. An interferometer calculates a signal's direction of arrival from the phase difference between the phase measured at two selected antennas.

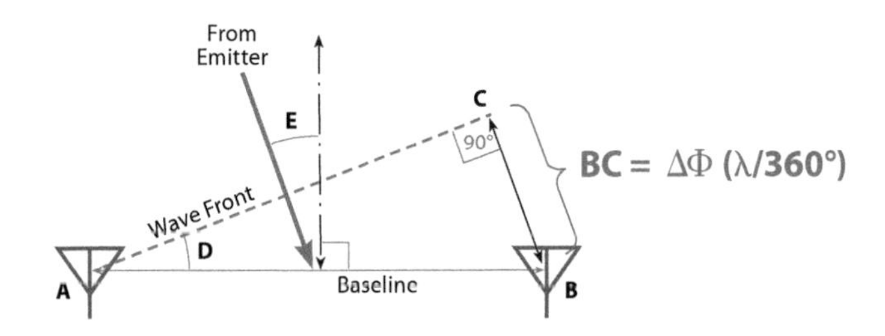

Figure 37-18. The interferometric triangle determines a signal's angle of arrival from its measured phase difference at two antennas forming a baseline.

antenna as shown in Figure 37-16. The antennas are sequentially switched into one receiver, and at the instant of switching a phase shift occurs. These phase shifts can be used to build the sine wave.

This technique has limited ability to handle frequency modulated signals since the Doppler frequency variation may be in the same parametric space with the signal modulation. It provides about 2.5° RMS accuracy over 360° of azimuth and produces DOA readings in 1 to 2 s.

When a high-accuracy DOA system is required, the *interferometric technique* is usually the best choice. It compares the phase of a signal at two antennas. Figure 37-17 shows the basic block diagram of an interferometer. Key to the use of this technique is matching the phase of the two signal paths from the antennas to the point where the phase is measured. Various approaches are taken to either measure the phase near the antennas or to calibrate the system in real time so that phase errors can be eliminated.

The two antennas form a baseline, and the DOA of the signal against this baseline is determined using the concept of the interferometric triangle as shown in Figure 37-18. The so-called wavefront of the signal is a line of constant phase of the arriving signal. Note that the phase at Point C is the same as the phase at point A. Therefore, the phase difference between the signal at the two antennas (A, B) is the same as the phase difference between points C and B. The wavelength of the signal is known from the formula:

$$\lambda = c/f$$

where λ is the wavelength, f is the frequency, and c is the speed of light.

There are 360° of phase in a wavelength, so the length of the line C–B is

$$BC = \Delta\Phi\ \lambda/360°$$

where $\Delta\Phi$ is the phase difference between points C and B (which equals the phase difference at the two baseline antennas), and λ is the wavelength of the received signal.

The angle between the wavefront and the baseline is

$$\text{Angle } D = \arcsin(BC/AB)$$

Note that the angle between the perpendicular bisector of the baseline and the DOA (angle E) is the same as angle D. It is angle E that is output from the process, because the perpendicular bisector of the baseline is the angle at which there are the most phase degrees per angle degree and hence the greatest DOA measurement accuracy.

A baseline is formed from each pair of antennas in an antenna array, and correlation of DOA answers from multiple baselines is used to resolve front–back ambiguities caused by the fact that the DOA and image signal arrival directions cause the same phase difference values.

To avoid further ambiguities, the baselines must be less than one-half wavelength long, and to have adequate accuracy it must be at least one-tenth wavelength. Therefore, a single array is typically useful over a one to five frequency range.

Aircraft-mounted interferometric arrays are often configured as shown earlier in Figure 37-3. In each axis, the longest baseline yields high accuracy with ambiguities, but the short baseline is not more than one half wavelength and can therefore resolve the ambiguities.

This description is called a *single-baseline interferometer* because it calculates from only one baseline at a time. It achieves 2.5° rms accuracy without calibration but is typically calibrated to achieve 1° RMS accuracy.

Other types of interferometers have baselines longer than one-half wavelength. One is the correlative interferometer, which uses many baselines longer than $\lambda/2$, each of which produces many ambiguities. These ambiguities are resolved statistically, because the correct DOA will have a higher correlation value. This approach achieves about the same accuracy as the single baseline interferometer.

Another type of interferometer is the multiple-baseline precision interferometer. It uses multiple baselines that are each several half wavelengths long. Note that at microwave and millimeter wave frequencies, the wavelengths may be shorter than the antenna diameters, so half wavelength spacing is not practical. The DOA is determined and ambiguities are resolved by solving the baselines simultaneously using modulo arithmetic.

This type of system can produce accuracies up to ten times higher than the single baseline interferometer. This technique is used against microwave signals because the short wavelengths allow antenna arrays of reasonable size.

Precision Location Techniques. Precision emitter location is used to determine the location of an emitter to adequate accuracy for weapon targeting. It also supports high accuracy passive ranging. There are two passive precision location techniques: *time difference of arrival* (TDOA) and *frequency difference of arrival* (FDOA).

TDOA involves measuring the time at which a signal from a single emitter reaches two widely spaced receivers. As shown in Figure 37-19, the difference in time of arrival determines the difference in the length of the two signal propagation paths. A specific difference in path length defines a hyperbolic surface in space.

Figure 37-20 shows the intersection of a family of these surfaces with a plane (or with the earth's surface). These hyperbolas are called *isochrones*. Since this time difference is measured to a few nanoseconds, the lines of the isochrones can be assumed

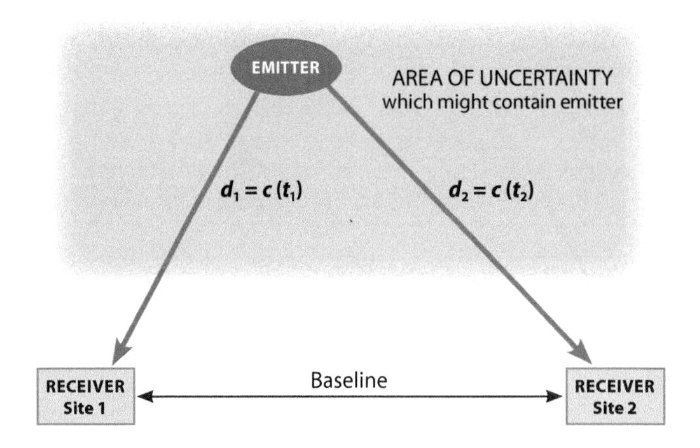

Figure 37-19. The difference in a signal's time of arrival at two receiving locations defines the difference in length of the two propagation paths.

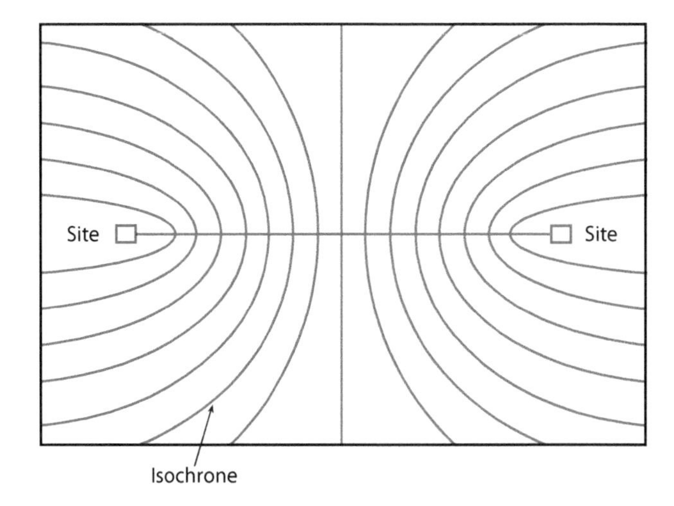

Figure 37-20. Isochrones fill all space. Each one is the locus of points that might be the location of an emitter received with a specific time difference of arrival.

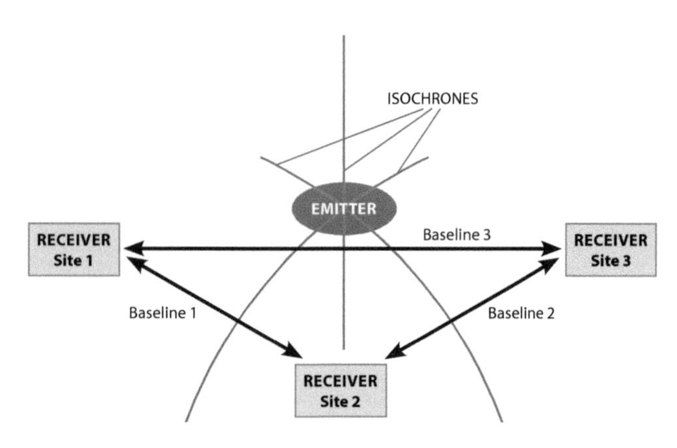

Figure 37-21. Three receivers form three baselines, each of which can determine an isochrones passing through the emitter location.

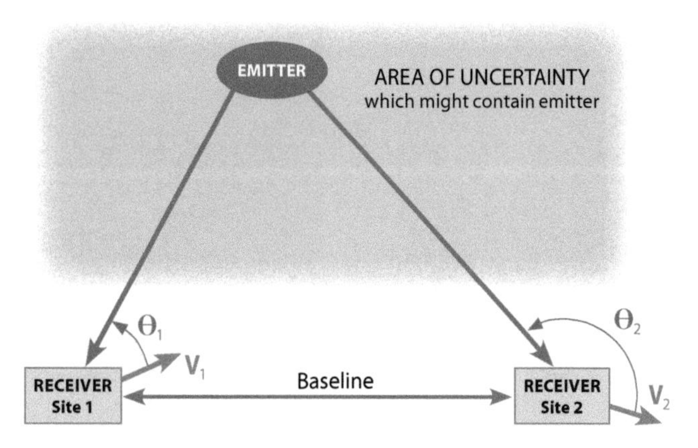

Figure 37-22. Two moving receivers will observe a single signal with different Doppler shifts so it will be received at different frequencies.

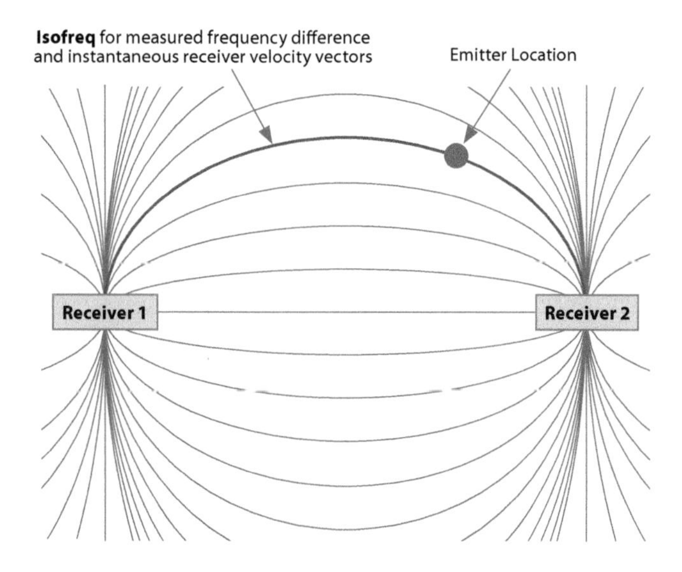

Figure 37-23. Each isodop describes the locus of points containing the emitter location if a specific frequency difference is observed.

to be very narrow. For example, if the time difference accuracy is 150 nsec, the line width will be 50 m. Thus, the emitter is known to be somewhere on a specific hyperbolic line.

If time of arrival is measured at three receivers as shown in Figure 37-21, three isochrones that pass through the emitter location are defined. Thus, the location is known to a few meters.

Precise time measurement requires that a highly accurate clock be available at each receiver location. This has become much more practical since GPS based time references have been available.

For pulsed signals, the leading edge of the pulse provides an excellent point of reference that can be timed. Since only a few pulses must be timed, the time information can be passed between locations over relatively narrow data links. However, for analog signals, it is necessary to digitize incoming signals at each receiver and pass that data to a single location for correlation. A series of delays are introduced at one or more receivers, and the correlation between the digitized signals is plotted. The differential delay that peaks the correlation is taken to be the time difference of arrival. Passing the digitized data between locations requires significant data link bandwidth.

FDOA is also called *differential Doppler* (DD). It involves measuring the received frequency of a single signal at two remote, moving receivers. As shown in Figure 37-22, a signal arriving at two moving receivers will have different offset angles between the signal path and the velocity vector of the platform. This will cause different Doppler shifts. The received frequency is measured at each receiver and passed to a single location for correlation. Only a narrowband data link is required.

Each frequency difference of arrival will describe a surface in space that contains the emitter location. The intersection of these surfaces with a plane is a family of curves called isofreqs or isodops. The curves shown in Figure 37-23 are for the special case in which the two platforms are traveling in the same direction at the same speed. Computer processing can handle data from any combination of velocity vectors. If the frequency is measured very accurately, the isofreq lines will be very narrow.

Like TDOA, FDOA requires three receivers to provide an emitter location. The three receivers form three baselines and hence three isofreqs that pass through the emitter location.

Fixed sites use only TDOA, but airborne platforms typically use both TDOA and FDOA. In this case, a single baseline provides an emitter location at the intersection of an isochron and an isofreq. Three receivers allow more operational accuracy by correlating multiple location calculations.

Multiple antennas on a single aircraft can form short baselines that can be used for both TDOA and FDOA. This will not provide the same accuracy as the longer baselines on multiple aircraft, but it will provide more accurate emitter location (including passive ranging) than the amplitude comparison or interferometer systems typical on RWRs.

37.6 Search

One of the significant challenges for ES systems is determination of the frequency of threat signals. This can be accomplished with a narrowband sweeping superheterodyne receiver, but the time it takes to complete a search will cause the probability of intercept (POI) to be relatively low. Another approach is to use a frequency measurement receiver. The IFM is very fast and covers an octave, but it cannot measure two simultaneous signals. The approach shown in Figure 37-24 is used to overcome this problem. When there is a CW or very high duty cycle signal (e.g., a pulse Doppler radar in high pulse repetition frequency [PRF] mode) present, the IFM cannot measure the frequency of pulsed signals in the same octave. The tunable band-stop filter removes the CW signal from the IFM input so that it can measure any low duty cycle signals present. A narrowband receiver (e.g., an SHR) sweeps the frequency range that is blocked by the band-stop filter (including its skirts).

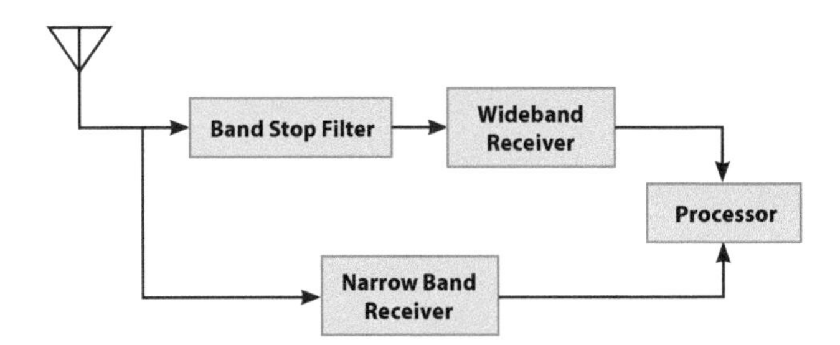

Figure 37-24. A tunable band-stop filter can remove CW or high duty factor pulsed signals from the input of a wideband IFM receiver to allow it to measure other signals. The SHR searches the frequency range covered by the band-stop filter.

Modern systems use digital channelized receivers to detect and process multiple time simultaneous signals and high PRF/CW signals, to measure internal modulations, and to perform precision DF and ranging. These receivers provide the response time and sensitivity necessary in a high pulse density and complex RF signal environments. As with other ES/EA systems, they must also support EA and operate in a less than ideal interference environment with other on platform signals and own platform EA.

37.7 Radar Warning Receivers

The mission of a RWR is to warn the crews of aircraft and other platforms of imminent attack by radar-controlled weapons. It must detect any threat radar signal quickly, in time for defensive measures to thwart the attack by the radar-controlled weapon. This means that the RWR must have antennas covering all directions from which threat radar signals may arrive and receivers covering the frequencies of all threat radars likely to be used against the protected platform. Ideally, the RWR would receive from all directions and at all frequencies all of the time. It should have enough sensitivity to receive all such signals to the range at which threat radars can detect or attack the platform. It should be able to correctly identify any threat type and determine its operating mode from its signal parameters.

As a practical matter, trade-offs must be made, but modern RWRs come reasonably close to the aforementioned set of requirements. Figure 37-25 is a block diagram for a typical RWR. It is deliberately a simple system to show the functionality. There are far more complex systems in use, and the growing complexity of threat radars is causing the complexity of RWR systems to increase.

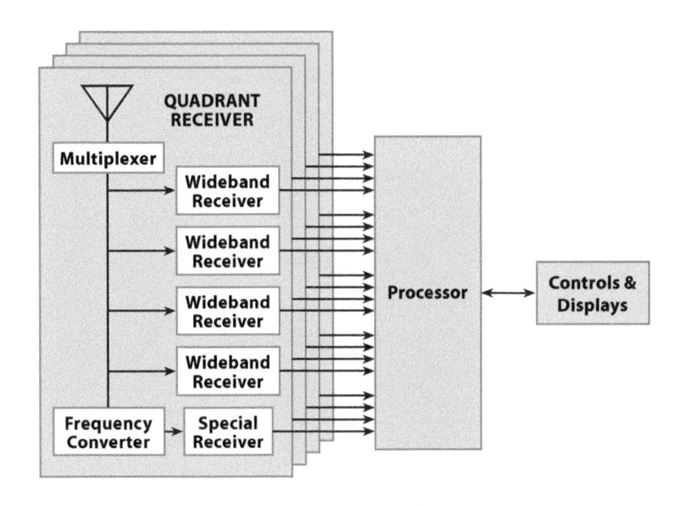

Figure 37-25. The radar warning receiver has four or more receiver channels, each covering an angular segment and each containing both wideband and special-purpose receivers.

The system shown has four antennas, each covering a quadrant plus half of the quadrants on either side. These are typically

CBS antennas covering a wide frequency range—often from 2 to 18 GHz. Sometimes a second part of the system covers MMW frequencies. Each antenna output is multiplexed into four channels: typically 2–6 GHZ; 6–10 GHz; 10–14 GHz; and 14–18 GHz. There is a receiver in each frequency channel; it may be a crystal video receiver or some other type, for example a tuned superheterodyne or digital receiver. There is normally a frequency converter that covers the whole 2–18 GHz range. The processor controls the frequency converter, which causes it to convert one of the four frequency ranges to a single range. The converter output is normally 6–10 GHz because that is less than an octave. The converter output feeds a special receiver. This receiver may be a superheterodyne receiver to receive CW or high duty factor pulse signals, a digital receiver, or an IFM receiver. Some configurations may locate the special receiver in an inboard location to save power and weight and to reduce the number of receivers and cost. When this is the case the signal may be downconverted to lower frequencies to reduce transmission line losses or RF over fiber may be used.

In the configuration shown earlier in Figure 37-25, the video outputs of all of these receivers go to the processor, which uses information from the wideband receivers for direction finding, pulse measurement, pulse train analysis, and selection and control of special receivers. The processor also performs signal analysis, accepts control inputs, and drives displays.

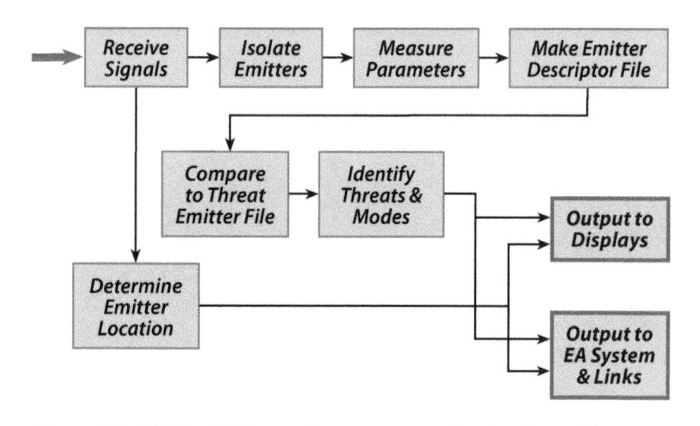

Figure 37-26. The RWR associates many received pulses with individual emitters, analyzes the signal parameters to form an emitter file, compares that file to a threat emitter database, and output the emitter type and location.

Figure 37-26 shows the RWR functional flow. There are two functional paths: signal analysis and emitter location. There are typically millions of pulses per second in the signal environment, and the processor must identify each pulse to one of the several signals that are received. The pulses are sorted by frequency (from an IFM receiver), direction of arrival (from the emitter location function), and timing (e.g., pulse interval). The signal parameters for each isolated signal are determined: antenna scan pattern; frequency; and several modulation elements. The parameters for each received signal are compared to a database of parameter values for each threat signal type called the threat identification (TID) file. A signal that matches the parameters of a threat radar is identified, and the weapon type associated with that radar is indicated on the RWR display.

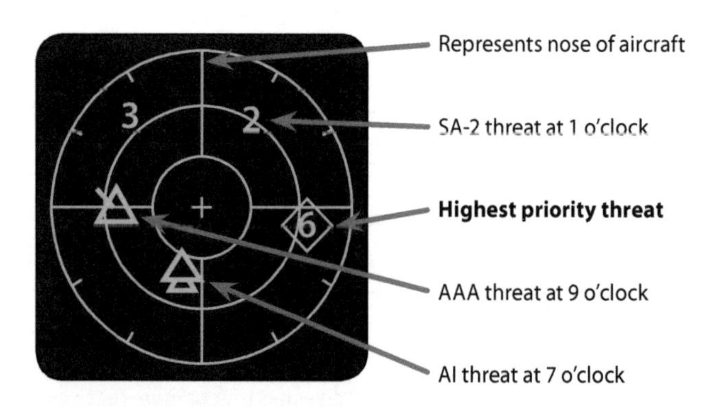

Figure 37-27. Radar warning receivers display threat types, operating modes, and emitter locations in real time.

The emitter location process involves comparing the amplitude of signals received from each of the RWR's four antennas. A signal is received in two antennas, and the direction of arrival of the signal is calculated from the amplitude ratio. The range to the emitter is calculated from the received signal strength using the received power equation described in Section 37.4.

Note that some RWRs use interferometric direction finding or one of the precision emitter location approaches rather than the amplitude comparison technique.

The location of the threat signal relative to the aircraft or other platform is displayed as shown in Figure 37-27. This is one of the types of displays used in RWRs, with a symbol for the kind of threat identified in a screen location indicating the threat location relative to the aircraft. These symbols are computer generated and can be anything the user desires. The symbols in Figure 37-27 are fairly common. The triangle with the diagonal line represents a radar-controlled gun known as an *automatic antiaircraft* (AAA), and the double triangle represents an enemy fighter plane. The associated radar is an *airborne intercept* (AI) radar. The numbers represent various types of antiaircraft missiles. The symbols can flash if a threat is in a launch mode, and various symbol enhancements can show such aspects as the highest (i.e., the most dangerous) threat.

37.8 Summary

ES is the listening part of EW, intercepting enemy radars and providing cues for effective EA.

EW antennas are different from radar antennas in that they usually cover much wider frequency ranges and have a large range of gain patterns. Important EW types include dipole, monopole, log periodic, biconical, cavity-backed spiral, horn, splash-plate, parabolic, and phased array antennas.

EW receivers are also different from radar receivers, in that they typically cover wider frequency ranges and have a great variety of functions. Some receive and demodulate individual signals. Some determine only the frequency of received signals. Some have low sensitivity and cover several octaves. Some accept multiple simultaneous signals. Some support exotic processing. Important types of EW receivers are crystal video, instantaneous frequency measurement, super-heterodyne, channelized, and digital.

The sensitivity of a receiver is the weakest signal it can receive and still perform its functions to full specification. Sensitivity is the sum (in dB) of three factors: thermal noise (kTB); noise figure (determined by the quality of components and system design); and the predetection signal-to-noise ratio. The former is −114 dBm per MHz of effective receiver bandwidth. Narrowing bandwidth increases receiver sensitivity proportionally. That is, narrowing bandwidth by a factor of 10 increases sensitivity by 10 dB.

The dynamic range of a receiver is the difference (in dB) between the weakest signal it can receive and the strongest signal that can be present. The dynamic range of a digital receiver is calculated from $DR = 20 \log_{10}$ (the number of bits of digitization)

One-way radio propagation produces a received signal defined by *Received Power* (in dBm) = *Effective Radiated Power* (in dBm) − *propagation loss* (in dB) + *receiving antenna gain* (in dB). The propagation loss in the frequency range of radars is typically line of sight loss. It is calculated by *Loss* (in dB) = 32.44 + 20 $\log_{10}$ (link distance in km) + 20 $\log_{10}$ (transmission frequency in MHz).

Passive emitter location is accomplished through triangulation, range and angle, multiple ranges, intersection of mathematically derived curves or multiple angles from an aircraft. Direction of arrival DF techniques are: moving narrow-beam antenna, amplitude comparison, Watson Watt, Doppler, and interferometer. DF accuracy is stated in RMS degrees, which is determined by squaring the errors at many angles and frequencies, averaging them and taking the square root. Precision location techniques include TDOA and FDOA; they can achieve location accuracy to a few meters.

Determination of the frequency of a threat emitter can be made by narrowband search or by use of a wideband frequency measuring receiver.

An RWR warns an aircrew to the presence of radar controlled weapons, identifies the associated weapons, indicates their operating mode, and shows their location relative to the aircraft. Threat identification is done by determining the parameters of each received signal and comparing them to a data base of the parameters of known threat emitters. Emitter location is typically by amplitude comparison or interferometer techniques.

Further Reading

J. B. Tsui, *Microwave Receivers with Electronic Warfare Applications*, SciTech-IET, 2005.

J. B. Tsui, *Digital Techniques for Wideband Receivers*, SciTech-IET, 2004.

R. G. Wiley, *ELINT: The Interception and Analysis of Radar Signals*, Artech House, 2006.

F. Neri, *Introduction to Electronic Defense*, 2nd ed., Chapter 4, SciTech-IET, 2007.

D. Adamy, *EW 103: Tactical Battlefield Communications Electronic Warfare*, Artech House, 2008.

J. B. Tsui, *Special Design Topics in Digital Wideband Receivers*, Artech House, 2009.

E. J. Holder, *Angle of Arrival Estimation Using Radar Interferometry*, SciTech-IET, 2014.

Test your understanding

1. How do EW antennas differ from normal radar antennas?
2. List three types of EW antennas not normally used by radar.
3. What kind of antenna is most commonly used in RWRs?

38

Electronic Attack

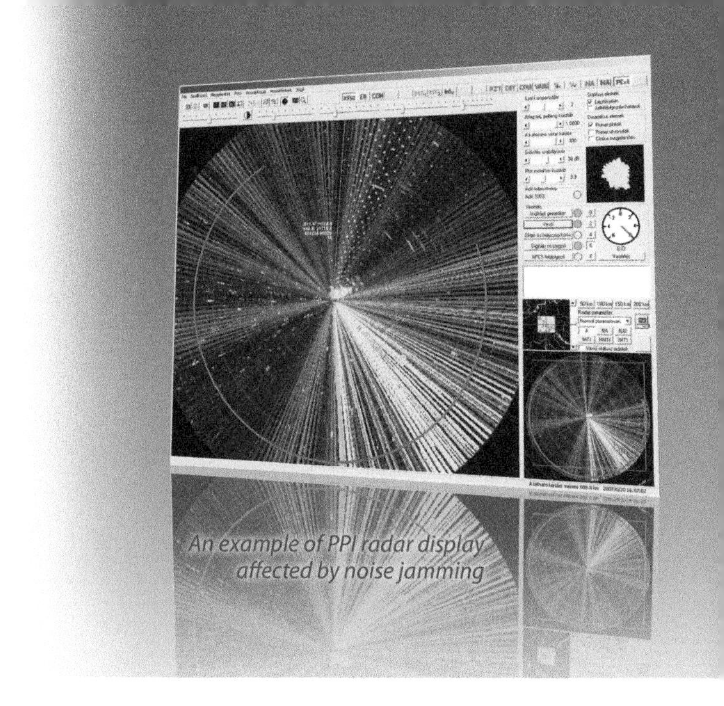

Electronic attack (EA) is the action part of Electronic Warfare. Actions are taken to reduce the effectiveness of enemy weapons. These actions include jamming deployment of chaff or flares, or transmission of energy to damage enemy sensors. This chapter will first cover jamming and then the other types of EA.

The effectiveness of jamming is impacted by the geometry, the types of signals being used by the enemy, and the strategy employed by the EA system. Although jamming is employed against enemy communications, the focus of this chapter is on EA against radars and the focus here will be on radar jamming.

Note that the electronic protection measures described in Chapter 39 are designed to reduce the effectiveness of all of these jamming techniques and require special tactics or EA system features to recover some or all of the jamming effectiveness.

38.1 Jamming Geometry

Jamming involves the transmission of undesired signals to disrupt an enemy receiver. The received signals must be received with sufficient signal strength to prevent the receiver from recovering the necessary information from the signals it is designed to receive. The jammer can be located either at the radar's target or at some other location. If the jamming signals are transmitted from the radar's intended target, this is self-protection or self-screening jamming. If the jamming signals are transmitted from any other location, this is either stand-off or stand-in jamming. Generally, it is called remote jamming. There is another geometrical approach – escort jamming. In this case, a dedicated jamming aircraft flies along with a group of other aircraft and protects them against enemy weapons by jamming.

Self-Protection Jamming. Figure 38-1 shows the propagation paths for *self-protection jamming* (SPJ). A jammer is located on the aircraft targeted by a radar. The radar transmits to its

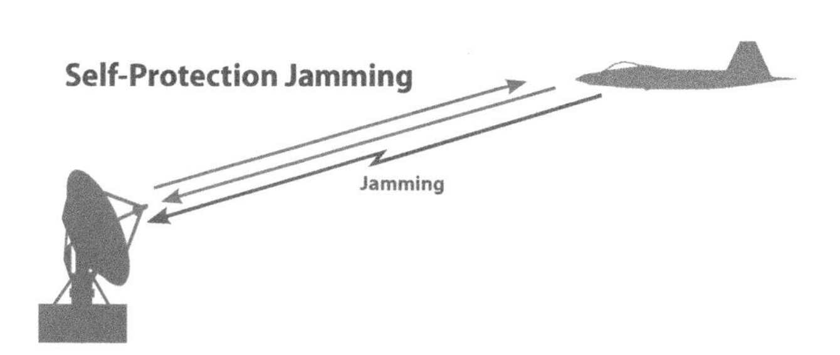

Figure 38-1. Self-protection jamming is performed by a jammer located on a target aircraft.

target, and the reflection from the target (called the *skin return*) is rebroadcast to the receiver in the radar. In a pulsed radar, the skin return is typically received in the same antenna from which the radar signal was transmitted. The self-protection jammer is located on the target, and the radar's antenna is pointed at the target at the moment the radar illuminates the target. Thus, the jamming signal is received by the radar receiver through the same antenna at its maximum gain (boresight) direction. In a CW radar, there must be separate transmit and receive antennas for isolation, but the radar receiver receives both the skin return and the jamming signal through the same receiving antenna.

The signal strength of the received jamming signal is increased by the high gain of the radar antenna, which increases the jamming effectiveness. Since a jammer transmits only one way (from the target to the radar), it normally experiences a propagation loss proportional to the square to the range from the target. The radar signal must propagate two ways (radar to target to radar) and thus has a propagation loss proportional to the fourth power of the range. This, along with the receiving gain of the radar antenna, allows a relatively high ratio of received jamming to received skin return signal strength (*jammer-to-signal ratio* or ***J/S***).

Because of the relatively high *J/S* achievable in SPJs, they can be used either to prevent an enemy radar from acquiring its target or can break the lock of a radar which is tracking the target. Section 38.2 will discuss cover and deceptive jamming techniques.

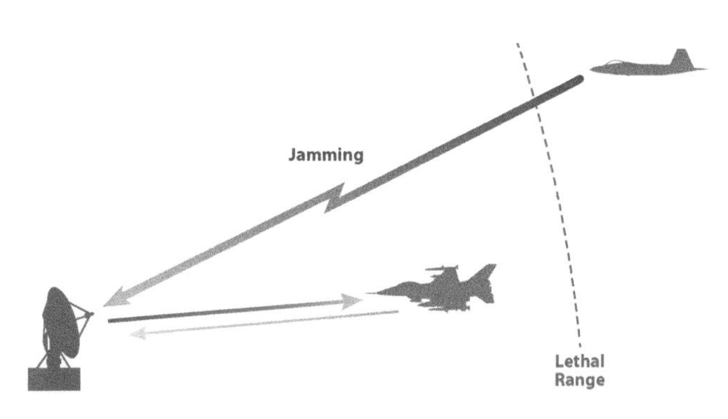

Figure 38-2. Stand-off jamming is transmitted from a dedicated jamming aircraft that is beyond the lethal range of weapons guided by the radar being jammed.

Stand-Off Jamming. Figure 38-2 shows the geometry of *stand-off jamming* (SOJ). The target aircraft is within the lethal range of a radar controlled weapon, and it is protected by a jammer in a second, specialized aircraft that flies beyond the lethal range of the weapon. The second aircraft carries a more powerful jammer than that used for self-protection jamming. The jamming aircraft is a high-value, low-inventory asset. Also, it typically has a very large radar cross section and is highly vulnerable to "home-on-jam" capabilities of enemy missile systems. Thus it is important that its exposure to enemy weapons be minimized.

The stand-off jammer is not on the target aircraft, so it is assumed that the radar antenna is pointed at the target and not at the jammer. This means that the stand-off jamming signal does not have the advantage of being received with the high gain of the radar antenna. Thus, in calculations of jamming to signal ratio in section 38.3 below, it is assumed that the jamming signal is received in a sidelobe of the radar antenna.

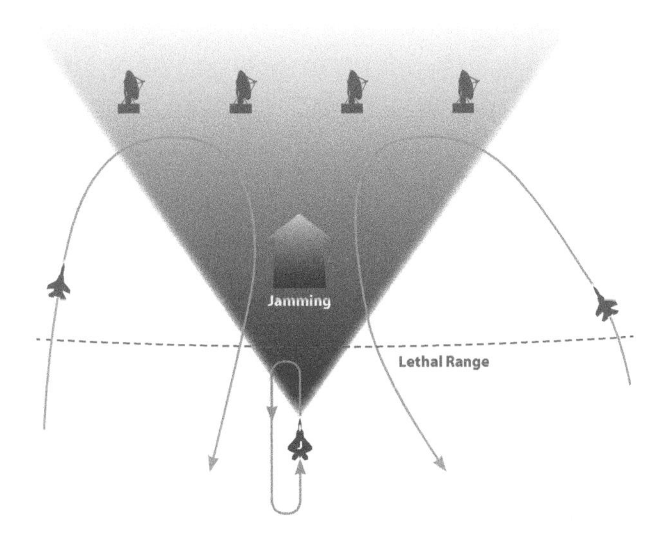

Figure 38-3. A stand-off jamming aircraft flies a pattern just beyond the lethal range of enemy weapons and jams multiple hostile radars to protect multiple strike aircraft assets.

As shown in Figure 38-3, a stand-off jamming aircraft typically flies a pattern as close as practical to the lethal range limit of the attacking weapon system, and holds this pattern from the time the protected aircraft enter the enemy's defended air space until they are beyond the range of enemy weapons. Note also in this figure, that the stand-off jammer protects against multiple enemy weapons. It can also protect multiple strike aircraft formations. Stand-off jammers typically have relatively wide antenna beams and can thus jam multiple radars over an angular range. This supports the assumption that the jammed enemy radars will receive the jamming signals in their sidelobes.

Because stand-off jammers are farther from the enemy radars than the aircraft they are protecting and because they transmit to the side lobes of those radars, it is difficult for the SOJ to achieve the high jamming-to-signal ratios (*J/S*) typically achieved by self-protection jammers. Since more *J/S* is required to break a lock than to prevent acquisition, SOJs are typically used to prevent the acquisition of targets by enemy radars. If it is necessary to break a lock, other EA techniques will be required.

In modern jammers which transmit from active electronically steered arrays (AESA), significantly more jamming power can be transmitted toward multiple threat radars than is available from broad coverage antennas. This allows a significant increase in the *J/S* achievable in stand-off jamming. This is shown in Figure 38-4.

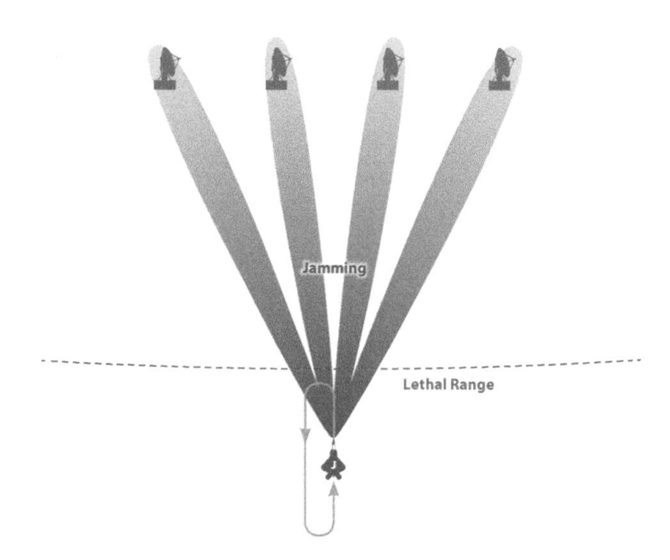

Figure 38-4. With an AESA, jamming signals can be directed at multiple threats in narrow beams that provide antenna gain to increase the *J/S* ratio.

Stand-In Jamming. A *stand-in jammer* (SIJ) is on an unmanned platform or is emplaced close to enemy radars by some other means. As shown in Figure 37-5, the stand-in jammer has a short transmission path. Because the propagation loss is a function of the square of the range to the radar, it is practical to generate a large *J/S* ratio in the jammed radar.

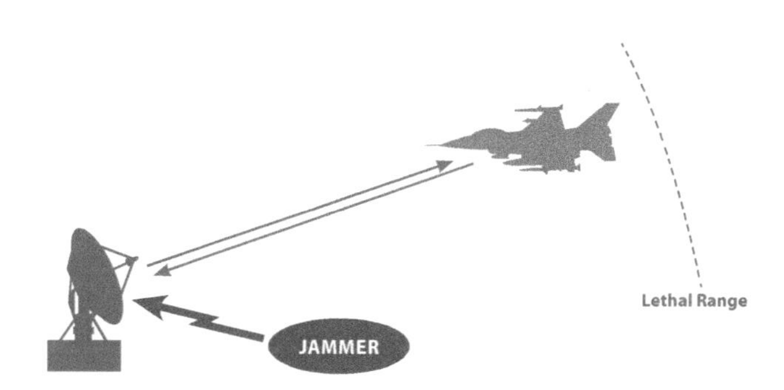

Figure 38-5. Stand-in jamming is transmitted from an unmanned device which is placed closer to the radar than the target threatened by the radar.

Since it is dangerous to be close to an enemy radar, stand-in jammers (SIJ) are typically unmanned. They can be flown in on unmanned aerial vehicles, fired in from artillery, or hand emplaced. An SIJ can be expected to transmit into the side lobes of the jammed radar as well as the main lobe depending upon the jamming platform. SIJs may radiate noncoherent or coherent jamming waveforms. Coherent techniques require a repeater or Digital RF Memory (DRFM) capability.

Escort jamming. Escort jamming is transmitted from a dedicated jamming aircraft that flies the whole mission as part of the formation of strike aircraft that it is protecting. This arrangement is shown in Figure 38-6. This gives the advantage of the higher jamming power typical of jamming aircraft. Since the jamming aircraft is collocated with the protected aircraft, it will be in the threatening radar's main beam, so it can achieve a very large *J/S* ratio. The disadvantage of escort jamming is that, like the stand-off jammer it is a high value asset and is put at risk. This is particularly true if modern enemy radars with home-on-jam capability are involved.

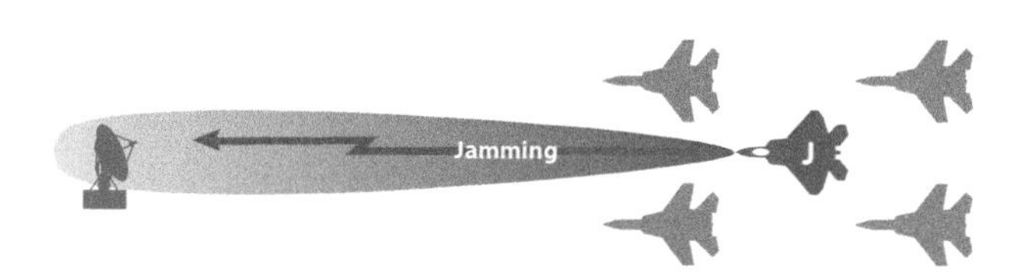

Figure 38-6. This picture illustrates escort jamming.

Thus use of air launched decoys with jamming capability is described in Chapter 40. These are also, in effect, escort jammers.

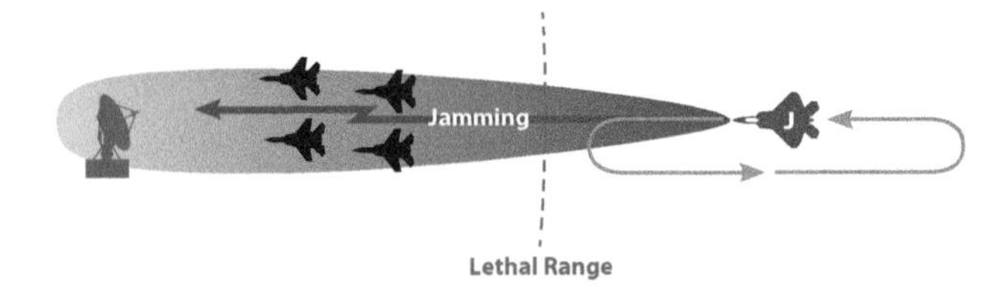

Figure 38-7. Modified escort jamming requires that a dedicated jamming aircraft files behind a strike force, staying beyond the lethal range, but holding station so that it can jam into the main beam of the radar which is threatening the strike force.

An alternate approach is called "Modified Escort Jamming." In this technique, the jamming aircraft is flown toward a particularly critical enemy radar to protect a critical aircraft or strike package. As shown in Figure 38-7, the jamming aircraft is farther from the jammed radar than the protected aircraft, but it flies so that it is in the main beam of a particularly dangerous radar that is illuminating the protected aircraft. The jamming aircraft flies in at the same speed, but turns around when it reaches the limit of the lethal range of the weapons guided by the jammed radar. With the advantage of jamming into the main beam of the jammed radar, the modified escort jammer can provide a significantly higher *J/S* ratio than a jammer which is jamming into the radar's sidelobes. Note that the modified escort jammer may also provide normal stand-off jamming protection for other strike packages by jamming into the sidelobes of other enemy radars.

Power of Noise Jamming on the Output of a Victim Radar's Receiver

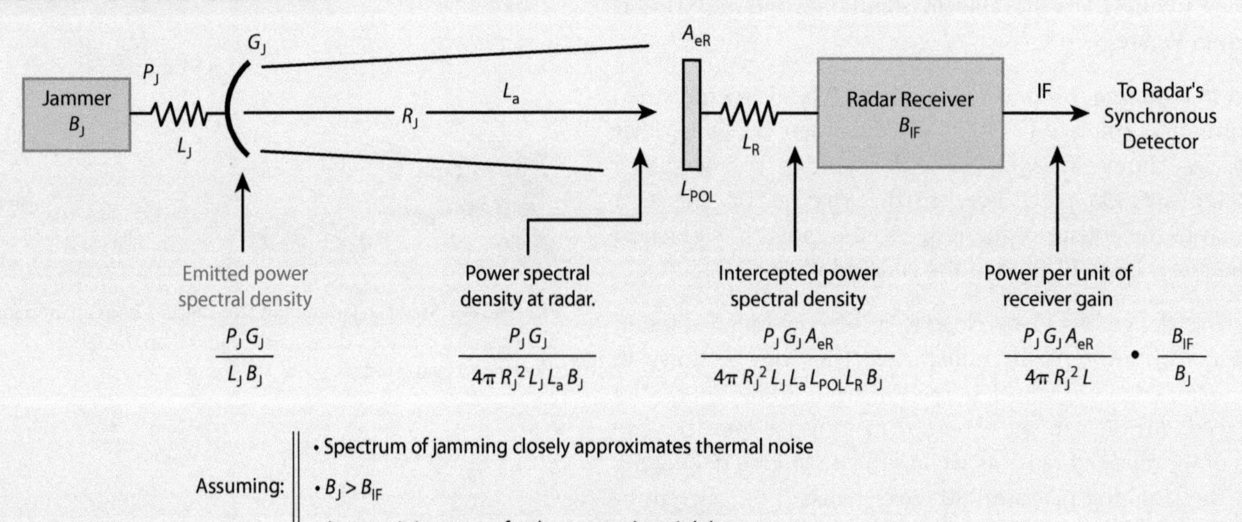

Mean power of the jamming in the receiver's output, per unit of receiver gain is

$$P_{JR} = \frac{P_J G_J A_{eR}}{4\pi R_J^2 L} \bullet \frac{B_{IF}}{B_J} \text{ Watts}$$

P_J = Power output of the jammer

G_J = Gain of jammer's antenna in radar's direction

A_{eR} = Equivalent area of radar antenna

B_{IF} = Bandwidth of receiver IF amplifier

R_J = Range from jammer to radar

L = Total losses: $L_J L_a L_{POL} L_R$

B_J = Bandwidth of Jammer's output

L_J = RF losses in jammer feed and antenna

L_a = Atmospheric loss (function of radar's operating frequency and R_J)

L_{POL} = Loss due to antenna polarization misalignment

L_R = RF losses in radar antenna & receiver front end

Note: If the radar antenna is not trained on the jammer, A_{eR} will be reduced by the ratio of the antenna gain in the jammer's direction to the gain at the center of the antenna's mainlobe.

38.2 Jamming Techniques

Cover Jamming. Cover jamming involves the transmission of signals which, when received by the radar, will degrade its ability to process the skin return signal. Typically, cover jamming uses noise modulation, although there are circumstances in which other modulations may be added to overcome specific features of a radar.

As shown in Figure 38-8, cover jamming fills the screen of a radar display with clutter so that radar returns from targets are obscured. This figure shows a plan position indicator (PPI) screen, but similar degradation of any display or other radar output is created by cover jamming.

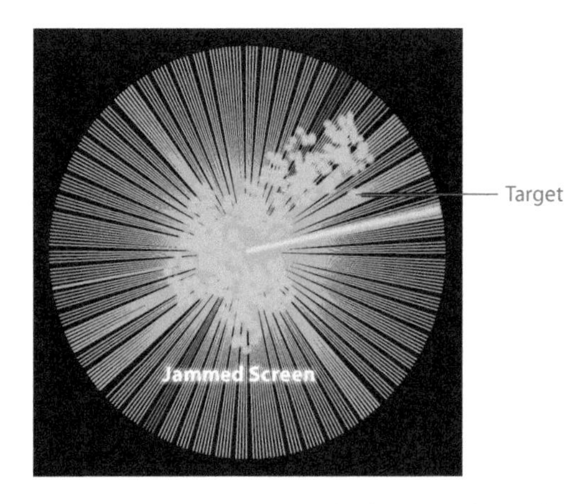

Figure 38-8. Cover jamming generates background clutter that makes it difficult or impossible for a radar to extract required information from received signals.

Barrage Jamming. The simplest form of cover jamming is "barrage" jamming. In this technique, noise is transmitted across a large frequency range which contains enemy radar operating frequencies. The advantage of barrage jamming is that it can be performed without any knowledge of the enemy radar parameters. The shortcoming is that jamming efficiency is low.

As shown in Figure 38-9, most of the jamming power is wasted because the jammed radar receives energy only within its bandwidth and ignores signals outside timing gates around its pulses. Jamming efficiency is defined as the percentage of transmitted jammer power that is actually received by a target radar.

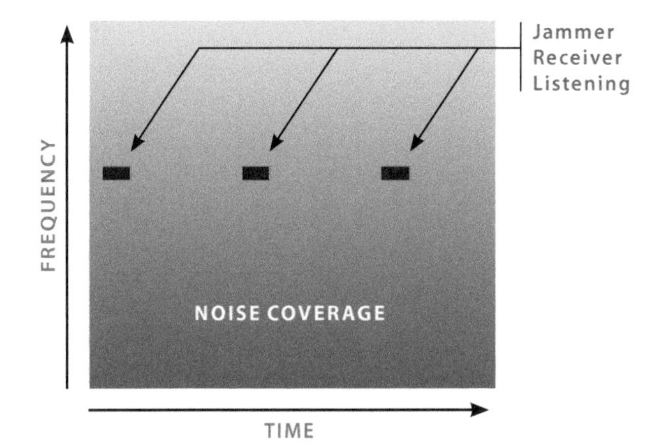

Figure 38-9. Barrage jamming spreads continuous signal power over a wide frequency range. It is inefficient because a jammed radar can see the jamming only within its bandwidth and accepts it only around the times that skin return signals are arriving.

Spot Jamming. If a noise jammer narrows its transmission to a small range around the target radar's operating frequency, it is a "spot" jammer as shown in Figure 38-10 and Figure 38-11. This technique yields better jamming efficiency, but it is necessary to "look through" the jamming (as described in Section 38-4) to be sure that the radar has not changed frequency.

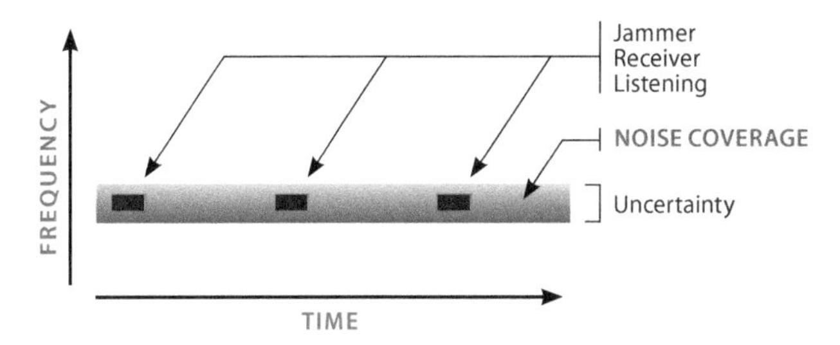

Figure 38-10. Spot jamming is broadcast in a narrow band containing the operating frequency of the radar being jammed.

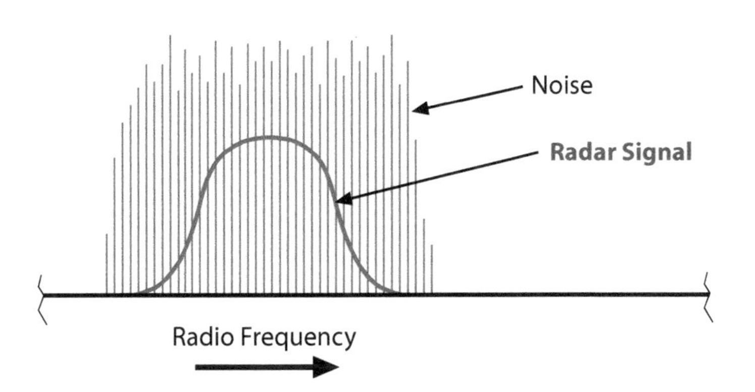

Figure 38-11. In spot noise jamming, maximum efficiency may be achieved by making the bandwidth of the jamming only slightly wider than the spectrum of the radar signal to be jammed. Because of mechanization limitations, however, the bandwidth is generally made much wider – between 3 and 20 MHz.

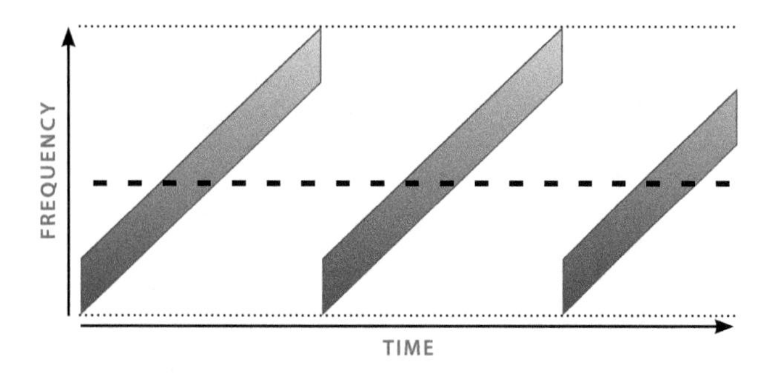

Figure 38-12. Swept spot jamming covers only a part of the frequency range of radars to be jammed but is swept over the complete range.

Swept Spot Jamming. "Swept spot" jamming involves tuning a narrow band of noise across the whole frequency range that might contain enemy radar signals, as shown in Figure 38-12. This jamming technique provides high jamming efficiency when it covers the operating bandwidth of a target radar, but has less than 100% jamming duty cycle. This means that some hostile pulses will not be jammed – or, for CW radars, some periods of reflected energy will return to the radar.

Figure 38-13 below compares several cover jamming techniques.

Deceptive Jamming. Deceptive jamming involves the transmission of signals which look like enemy radar returns, but cause the processor in the jammed radar to make false assumptions about the location or velocity of the target. Some jamming techniques associated with deceptive jamming can be used from off-board assets. These include the generation of false targets with threat signal parameters, as discussed in Chapter 40. However, these techniques are used to overcome the effects of various electronic protection (EP) techniques in threat radars rather than move the focus of the threat radar away from the true target in range, angle, or Doppler frequency. Therefore, it can be argued that deceptive jamming must be self-protection jamming because it requires precise (sub microsecond) information about radar signals at the target.

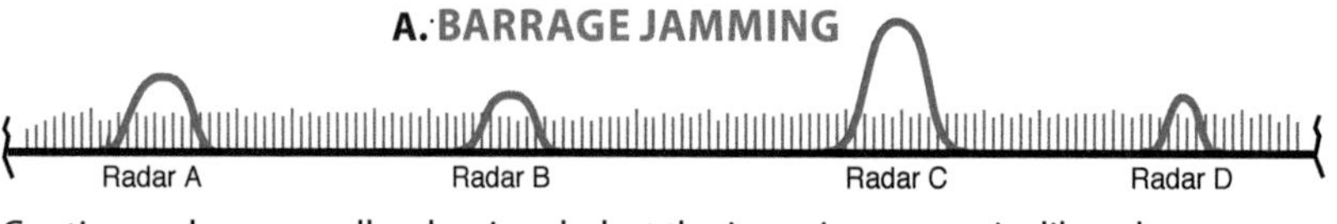

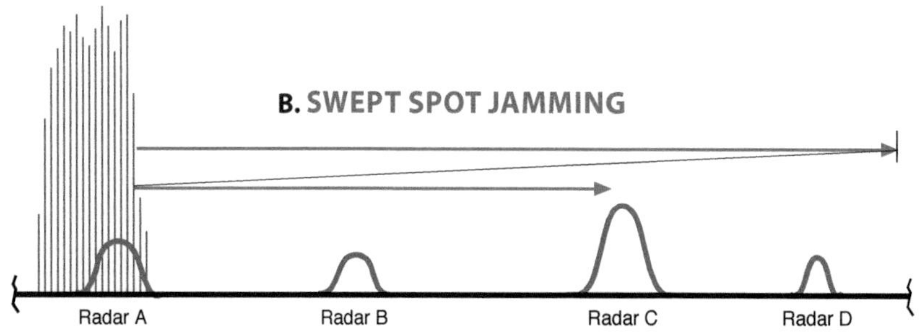

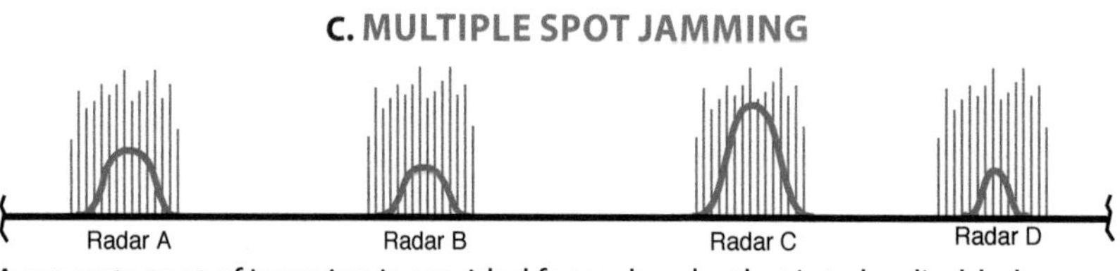

Figure 38-13. In these alternatives for jamming more than one radar operating on different frequencies, although requiring complex RF switching, multiple spot jamming is the most effective.

A related deceptive technique is the generation of false targets from remote jammers (stand-off or stand-in) is the generation of false targets. These false targets do not break the lock of tracking radars, but they can saturate a radar's processing and displays thus significantly reducing the capability of a hostile radar. False targets can have pulse compression modulations and Doppler signatures. They can also be synchronous with the radar or radar scan in some circumstances.

The first few techniques described below are not effective against monopulse radars. This is because monopulse radars get angular information from each pulse, and these techniques work against multiple pulses. Some will actually enhance the angular tracking performance in monopulse radars.

Range-Gate Pull-Off. Figure 38-14 shows the timing of radar pulses reflected from the target. *Range gate pull-off* (RGPO) involves rebroadcasting the enemy radar pulses with increased power and minimal delay. Then subsequent pulses are delayed by an increasing amount. The delay time is increased parabolically or exponentially. This delays the arrival of the return pulse at the enemy radar in a pattern that makes it appear that the target is turning away from the radar. As shown in Figure 38-15, the delayed pulses load up the radar's late gate, causing the radar-range tracking circuitry to conclude that the range to the target is greater than it actually is. The delay in the jamming pulses increases to a maximum and then snaps back to zero—repeatedly. This causes the radar to lose range track on the target. Another view of this process (called *range gate stealing*) is shown in Figure 38-16.

Note that if the radar changes to a mode in which it tracks on the leading edges of skin return pulses, the range tracker will ignore the delayed jamming pulses and continue to track the true skin return pulses, requiring that another jamming technique be used.

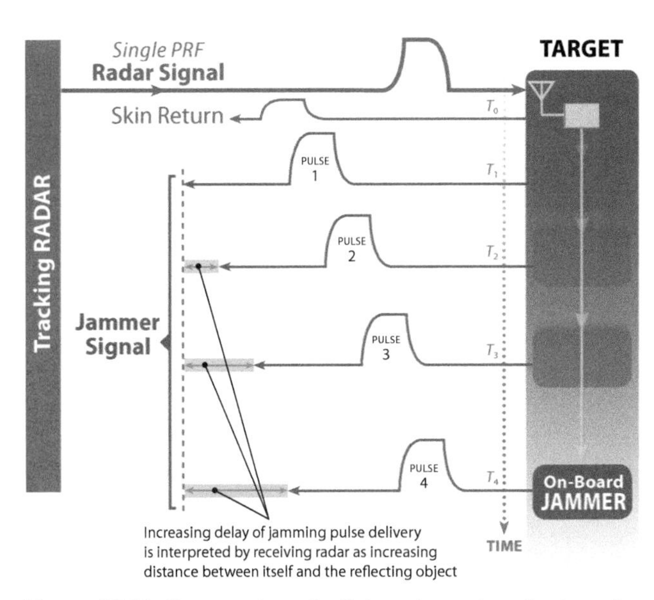

Figure 38-14. Range-gate pull-off jamming rebroadcasts radar pulses with increased power and delays subsequent pulses by an increasing amount to simulate movement of the target away from the radar.

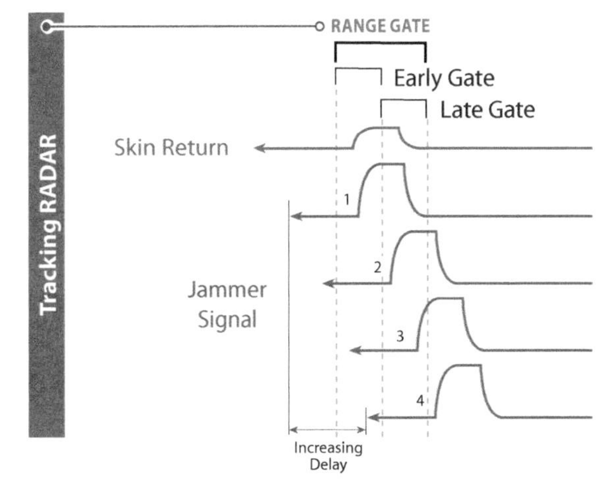

Figure 38-15. The delayed, amplified skin return pulses from range gate pull-off jamming increase the power in the radar's late gate, causing the radar to push out its range estimate, away from the actual target.

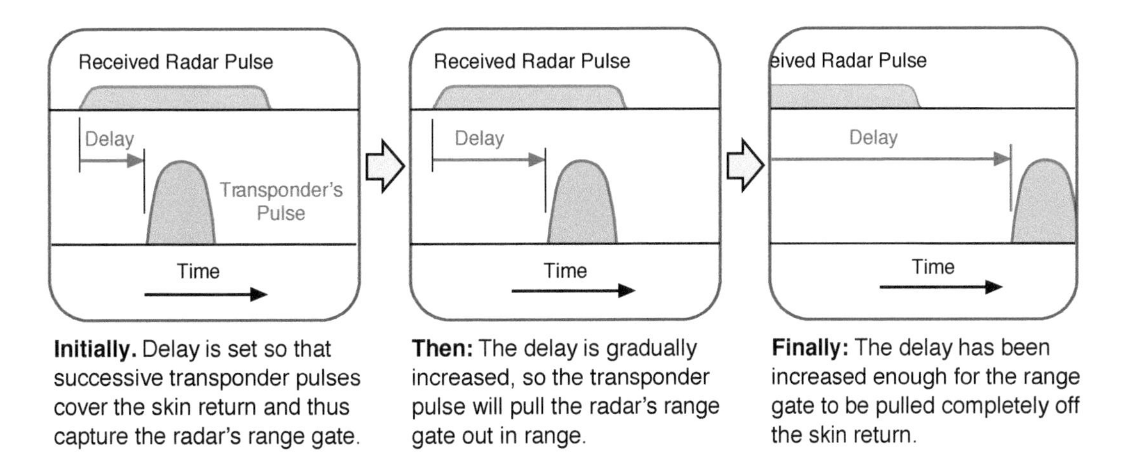

Figure 38-16. Against a noncoherent radar, the range gate stealer may be mechanized with a transponder. Upon receipt of each radar pulse, the transponder transmits a delayed RF pulse to the radar.

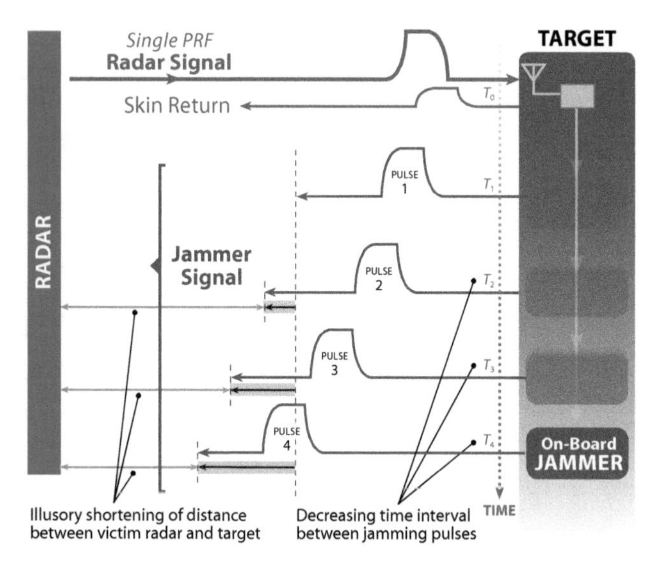

Figure 38-17. Range gate pull-in jamming generates high-power pulses that first correspond with the skin return pulse and then anticipate subsequent pulses by an increasing amount to simulate movement of the target toward the radar.

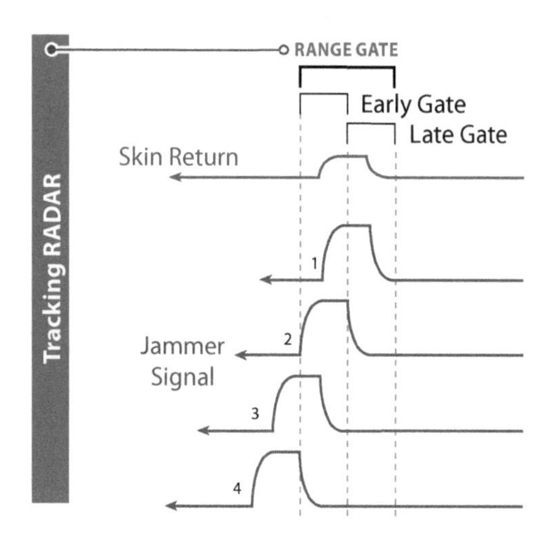

Figure 38-18. The strong jamming pulses from range gate pull-in jamming increase the power in the radar's early gate, causing the radar to pull in its range estimate away from the actual target.

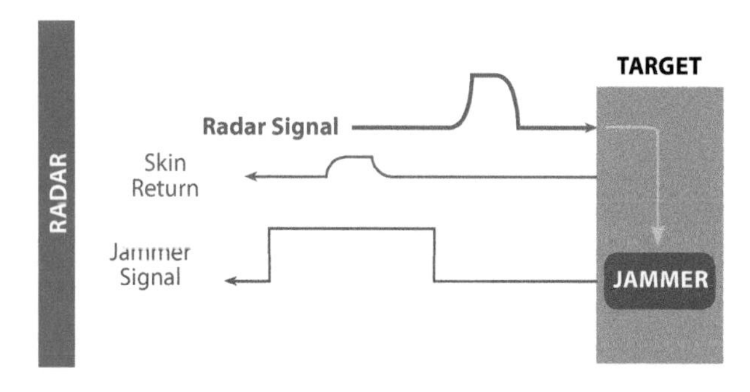

Figure 38-19. Cover pulses prevent the radar from determining the time of arrival of skin return pulses, thereby denying it range information.

Range Gate Pull-In. An alternate approach (which works against leading edge tracking) is to use a *pulse repetition rate* (PRF) tracker to anticipate the arrival time of each subsequent pulse and to transmit pulses with increased power before the skin return pulses leave the target as shown in Figure 38-17. This technique is called *range gate pull-in* (RGPI) or *inbound range gate pull-off.* The amount of lead time for the jamming pulses starts at zero and increases parabolically or exponentially to make it appear that the target is turning toward the radar. The radar's range tracking thus concludes that the range is shorter than it actually is. This loads up the early gate in the radar's tracking circuitry as shown in Figure 38-18. RGPI requires that the timing of future pulses be calculated. This can be accomplished for radars with fixed or staggered PRI, but not for radars which use randomly jittered PRI.

Cover Pulses. While not literally deceptive jamming, cover pulses are covered in this section because they require knowledge of the exact timing of radar pulses arriving at the target. As shown in Figure 38-19, cover pulses start before received radar pulses and end after those pulses. Figure 38-20 shows range bin masking which comprises multiple cover pulses and is useful against jittered pulses. This prevents the radar from determining the range to the target, while achieving increased jamming efficiency as compared to continuous jamming. This technique requires PRF tracking. For radars which use jittered PRI, the cover pulses must be expanded to cover the range of PRIs. This reduces the jamming efficiency.

Inverse Gain Jamming. Non-monopulse radars determine the azimuth and elevation to a target by observing the amplitude pattern (vs. time) of skin return pulses. For example,

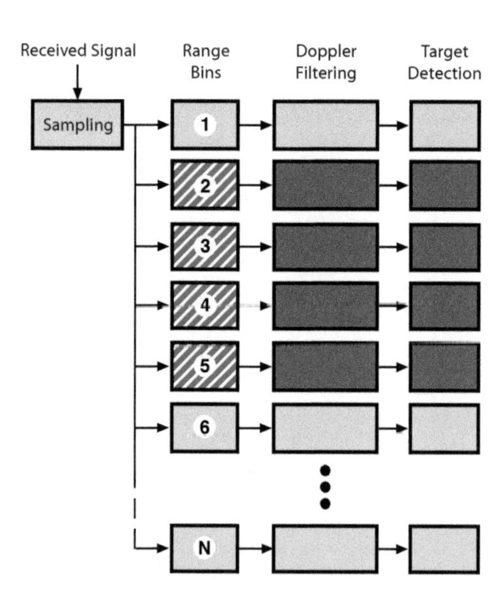

Figure 38-20. With range bin masking, the jamming is timed to fall within a block of range bins covering the range interval in which the aircraft to be screened may lie.

consider the time variation of skin return power from a conically scanned antenna as shown in Figure 38-21. The return power varies sinusoidally, with the maximum power occurring when the antenna beam is closest to the target and the minimum power when the beam is farthest from the target. The antenna is then steered to place the target in the center of the conical scan by steering in the direction of maximum pulse amplitude.

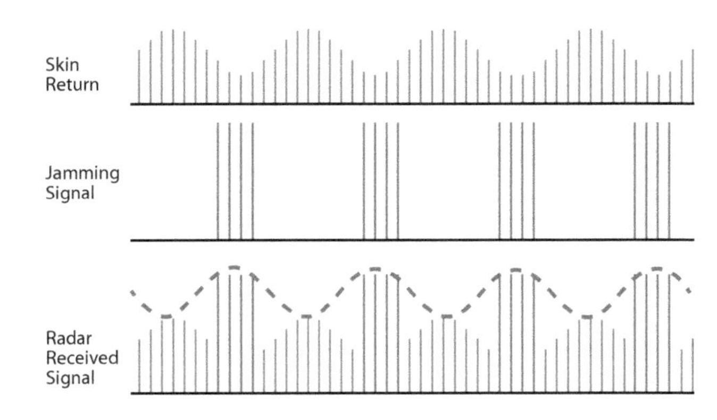

Figure 38-21. Inverse gain jamming broadcasts large pulses at time when the skin return pulses are at lower level because the radar antenna is not pointed directly at the target.

If a burst of synchronized pulses with increased power is transmitted at the low point of the sinusoid, the radar will receive a combined pulse amplitude pattern as shown in the bottom line of the figure. Since the radar must have a relatively narrow tracking filter to generate accurate guidance commands, the radar tracking circuit cannot see the sudden amplitude changes. Thus, it will react as though the phase of the sine wave is reversed. The blue dash line superimposed on the bottom line of the figure shows the received power pattern observed by the tracker. The result is that the angle tracking of the radar is broken by directing the centroid of the scan away from the target rather than toward it.

This technique can be used against any kind of radar antenna scan, but not against monopulse tracking radars.

AGC Jamming. Automatic gain control (AGC) jamming involves the transmission of strong, narrow, low duty cycle jamming pulses. A radar must have automatic gain control to handle its high required dynamic range. Further, the AGC must have the characteristic of fast-attack-slow decay. Therefore the jamming pulses will capture the radar's AGC, forcing the front end gain down to the point that the radar cannot detect the amplitude variation in the received skin return pulses caused by the radar antenna scan. In Figure 38-22, this is shown for a conically scanning radar. Note that the second line on this figure is an "under-exaggeration" in that the amount of reduction of the received signals would normally be sufficient to completely hide the sinusoidal pattern. It is presented this

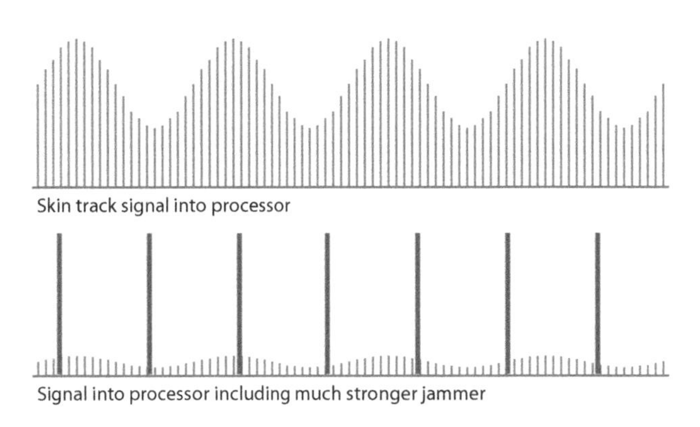

Figure 38-22. An AGC jammer broadcasts strong, narrow pulses which capture the AGC of the radar receiver, reducing front-end gain to compress the amplitude pattern of the skin return pulses from antenna scanning.

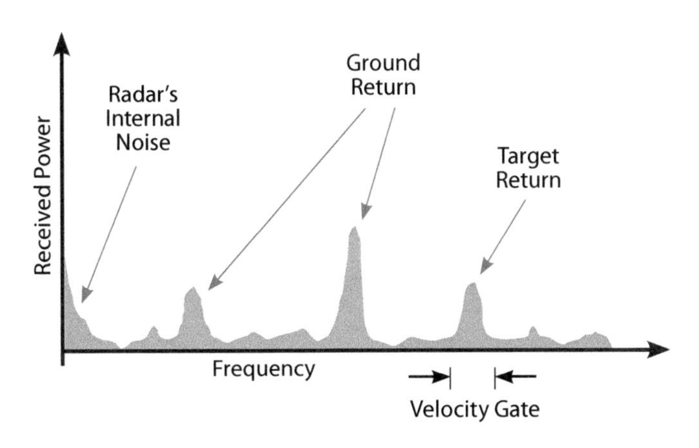

Figure 38-23. The response of a Doppler radar has Doppler frequency components corresponding to the relative radial velocity of terrain and of the target aircraft. A velocity gate is placed around the target being tracked.

Serrodyne Modulation

The time a signal takes to pass through a TWT depends to some extent on the velocity of the tube's electron beam, hence on the voltage applied to the anode of its electron gun. The phase, ϕ, of the tube's output, therefore, can be varied by modulating the anode voltage.

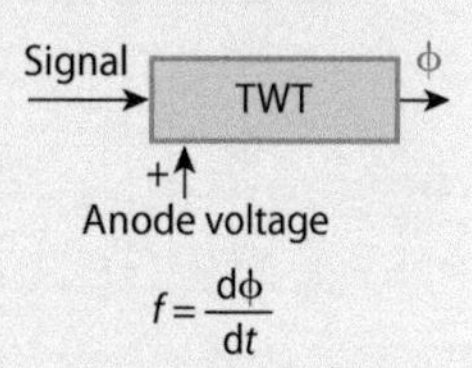

In essence, frequency, f, is a continuous phase shift, e.g., a phase shift of 360° per second is a frequency of 1 cycle per second.

By linearly advancing the phase of the TWT's output, therefore, the signal's frequency can be increased.

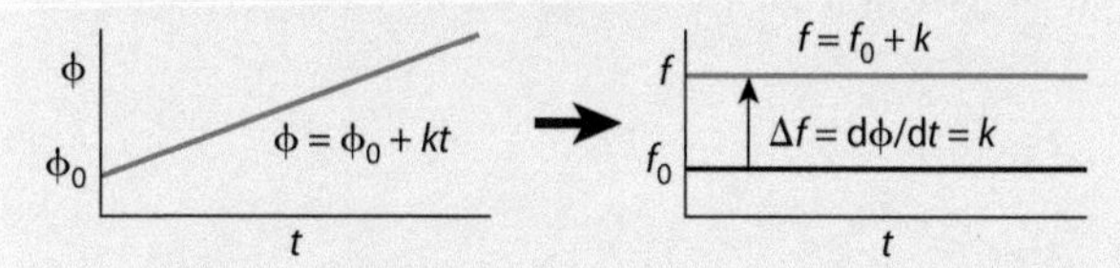

By advancing the phase at a geometrically increasing rate, the signal's frequency can be linearly swept through a band of frequencies.

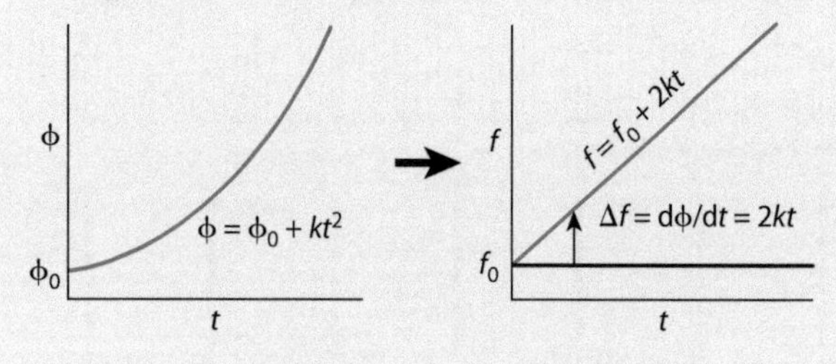

way in the figure to make clear that the scan amplitude variation is reduced.

Velocity Gate Pull-off. Figure 38-23 shows the power vs. frequency in a CW Doppler radar. Because of the relative velocity of terrain features, there are multiple frequency responses. A velocity gate is placed around the target return to allow tracking. If a strong signal is placed into the velocity gate it will capture the frequency tracking function and if it is moved away from the true target return frequency, the radar will conclude that the target velocity has changed from its true value, thus breaking the velocity track. This technique can also be used against a pulse Doppler radar.

38.3 Deceptive Jamming Techniques Effective against Monopulse Radars

Formation Jamming. If two aircraft fly formation within the resolution cell of a hostile radar as shown in Figure 38-24, the radar cannot determine that there are two targets in the cell. The radar will conclude that there is only one target, located between the two aircraft, with its location proportionally closer to the aircraft with the greater radar cross section.

At acquisition ranges, it is typical for the resolution cell to be larger in the cross-range dimension than it is in the down-range dimension. Therefore, it is easier to hold formation in angle (to the radar) than in range. If both aircraft jam the radar with approximately the same power, the radar will be denied range information, so the required station keeping is easier to maintain.

Blinking Jamming. Blinking jamming is shown in Figure 38-25. Two aircraft within the radar resolution cell with jammers can alternate transmissions at a rate that is too slow to average and too fast to track for a radar. In Figure 38-26, a missile is alternately aimed at each of the aircraft because the power of the jamming pulses dominates over the skin return pulse power. As the missile approaches the pair of aircraft, it must make increasingly high rate turns to move its aim point between the two aircraft. At some point, it will

Figure 38-24. If two aircraft hold formation within the resolution cell of a radar, the radar cannot distinguish two individual targets.

Figure 38-25. Cooperatively blinking the noise jamming from several closely grouped aircraft causes the centroid of the jamming as seen by the victim radar to oscillate erratically in angle.

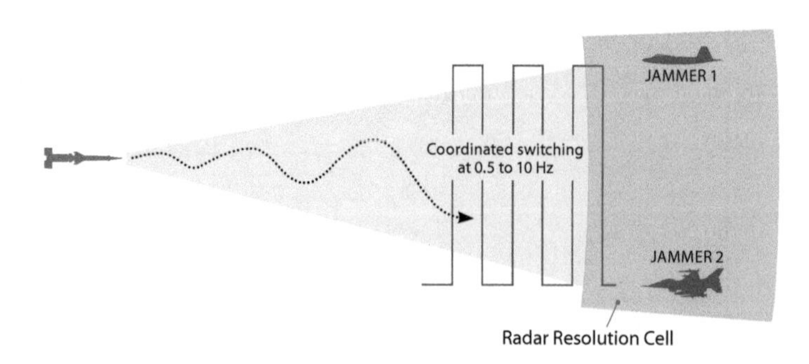

Figure 38-26. Blinking jamming requires that two aircraft within the resolution cell alternately jam to cause the jammed radar to direct a missile to one and then the other aircraft. As the range shortens, the missile will eventually fail to make one of the turns.

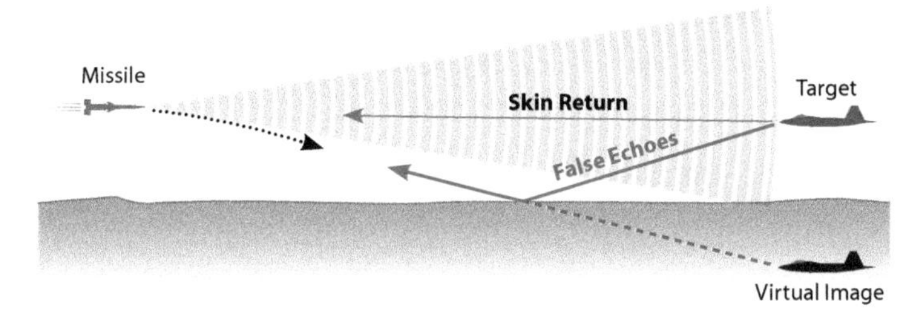

Figure 38-27. Terrain bounce: Downward deflected antenna in target bounces false echoes off terrain in front of missile, causing it to steer for a virtual image.

be unable to make one of the retargeting turns and will fly away from the targets.

Terrain Bounce Jamming. A terrain bounce jammer repeats pulses from a radar and retransmits them at very strong power down toward the ground or water as shown in Figure 38-27 abovr. This will cause the tracking radar to receive the vector sum of the skin return and ground bounce signals at an angle that is below the aircraft. Thus the missile or gun attacking the aircraft will be directed below its intended target.

Cross-Pol Jamming. *Cross polarization* (cross-pol) jamming takes advantage of edge effects in radar antenna beams. The forward geometry of a parabolic dish antenna, the curvature of a radome or reduced gain in edge modules of a phased array radar cause false lobes which are cross polarized to the main lobe of the radar' antenna. These are called "Condon lobes" and are small relative to the main lobe as shown in Figure 38-28. Figure 38-29 shows a three-dimensional view of the normal and cross polarized response in an antenna. However, if a jammer transmits a very large signal that is cross polarized to the skin return of the radar signal, these Condon lobes will dominate. A monopulse radar will direct one of its Condon lobes at the target causing the radar antenna (and hence the aiming point for weapons) to slew sharply away from the intended target. Figure 38-30 shows how a jammer can create

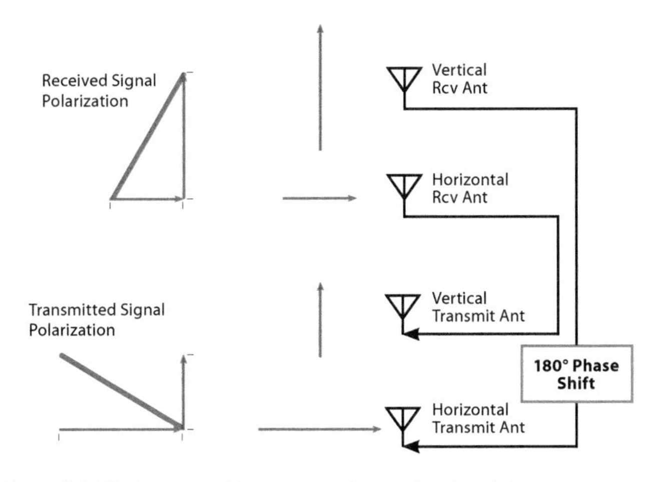

Figure 38-30. A cross-pol jammer receives radar signals in two, orthogonally polarized antennas and rebroadcasts both at the other polarization with a 180° phase shift in the horizontal transmission to create a cross-polarized jamming signal.

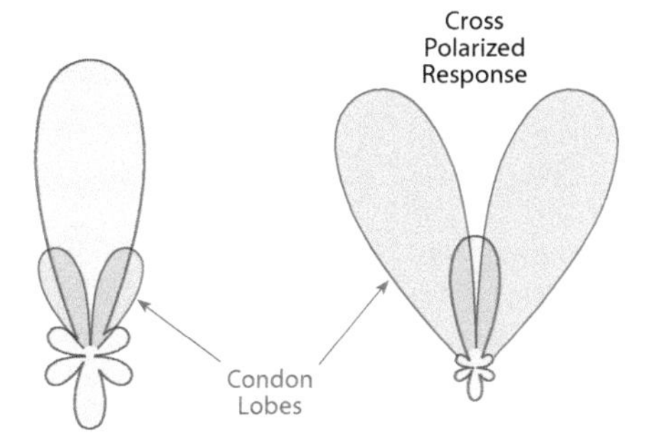

Figure 38-28. Radars with vulnerability to cross-pol jamming have cross-polarized lobes that become dominant to jamming signals that are cross polarized to the skin return signals from targets.

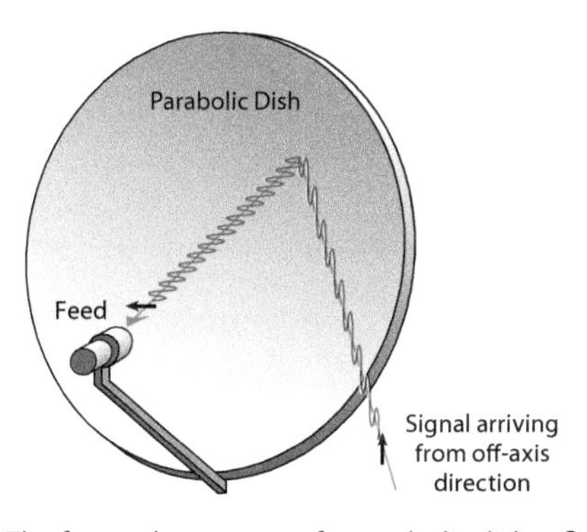

Figure 38-29. The forward geometry of a parabolic dish reflector causes off-axis signals to change polarization by 90° when reflected into the antenna feed.

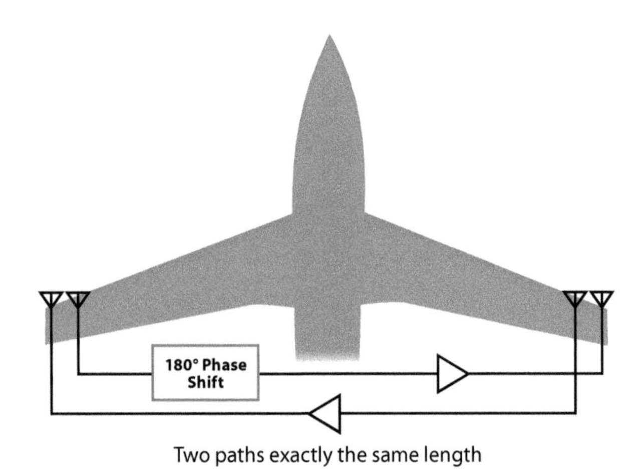

Figure 38-31. The cross-eye jammer receives signals in two widely-spaced antenna locations and rebroadcasts each (with significant gain) from the opposite location. One signal path has a 180° phase shift.

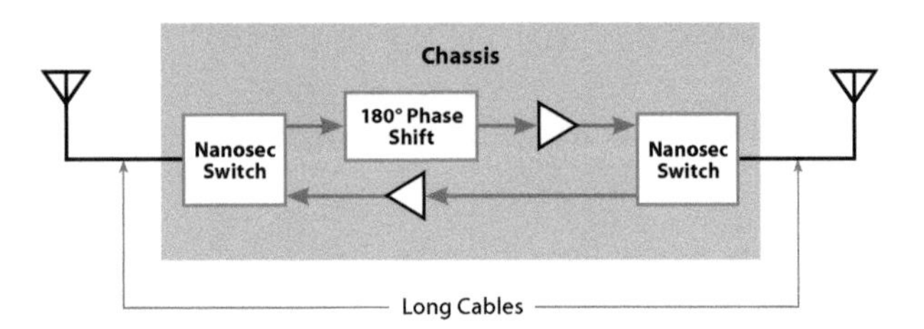

Figure 38-32. To avoid the significant challenge of phase matching long cables, the amplifiers can be placed in a chassis with nanosecond switches to reverse the signal flow through single cables to each of the widely spaced antennas.

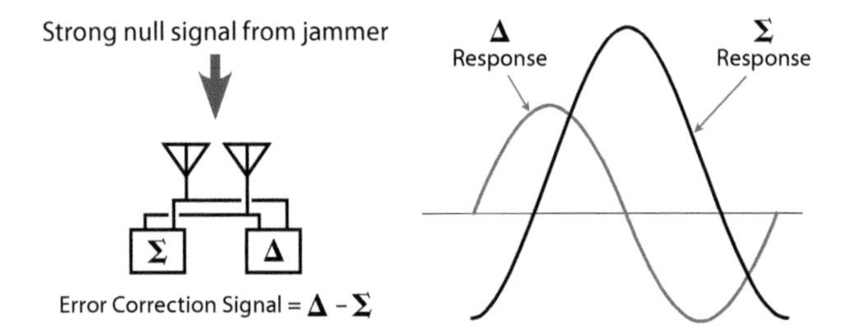

Figure 38-33. A monopulse tracker guides on the difference response normalized to the sum response of multiple antennas.

a cross polarized signal for any linearly polarized radar signal. A similar technique using circularly polarized antennas will produce a reverse sense cross polarized signal.

Cross-Eye Jamming. Cross-eye jamming involves two signal paths on the jamming aircraft as shown in Figure 38-31. Each receives a radar signal at one antenna and rebroadcasts it from a different antenna with significant gain. The receiving antenna for each path is collocated with the transmitting antenna for the other path. One of the paths has a 180° phase shift. The more widely spaced the two antenna locations, the greater weapon miss distance that can be created by the jammer. The two paths must be very closely phase matched; even tiny differences in the path lengths cause significant reduction in jamming effectiveness. The 180° phase difference between the two transmitted signals causes a null right at the sensors used by the monopulse radar to determine direction of arrival of the skin return signal. It is notoriously difficult to phase match long cables over temperature and dynamic range. One very elegant solution to this problem is to run one cable to one antenna at each transmit/receive location as shown in Figure 38-32. The amplifiers and the phase shifter are located in a chassis—where tight phase matching can be maintained. At each port of the chassis is an electronic switch that reverses the direction of transmission every few nanoseconds. Because the jammed radar cannot react to pulses orders of magnitude shorter than its pulse, it experiences two seemingly simultaneous signals, one from each antenna.

The effect of the cross-eye technique on the radar signals is described as causing a warp in the phase front of the radar signal, but it is more easily understood by considering the simplified two sensor monopulse receiving antenna in Figure 38-33. The actual antenna would have three or four sensors for two dimensional guidance. The monopulse radar determines the direction of arrival of the skin return by sensing the difference between the received power at its sensors (the difference response) divided by (i.e. normalized to) the sum response. The difference response should be linear across the 3 dB beamwidth of the sum response. By placing a null (from the phase cancellation of the two jamming signals) right on the receiving antenna, the sum response is made less than the difference response. Thus, the value of the difference response divided by the sum response reverses its sense and the antenna is forced sharply away from the target rather than toward the target.

Jamming Pulse Compression Radars. Radars with pulse compression have special modulations on them which reduce jamming effectiveness as discussed in Chapter 39. In order to effectively jam these radars, it is necessary to place matching modulations onto the jamming signals. This can be accomplished with various types of circuitry, but the most effective systems use direct digital synthesizers and *digital radio frequency memories* (DRFMs). The pulse compression waveforms

Digital RF Memories

THE *DIGITAL RADIO FREQUENCY MEMORY* (DRFM) IS AN IMPORTANT development supporting electronic countermeasures. It allows the rapid analysis of complex received waveforms and generation of countermeasure waveforms. It can increase the effectiveness of a jamming system against complex waveforms by many dB.

DRFM Block Diagram. As shown below, the DRFM down converts received signals to the appropriate *intermediate frequency* (IF) for digitization. Then it digitizes the bandwidth of the IF signal. The digitized signal is placed into a memory for transmission to a computer. The computer makes any necessary analysis of and modifications to the signal to support the jamming technique being employed. Then the modified digital signal is converted back to analog RF. This signal is frequency converted back to the received frequency using the same local oscillator used in the original frequency conversion. The use of a single oscillator maintains the phase coherence of the signal through the down conversion and up conversion process.

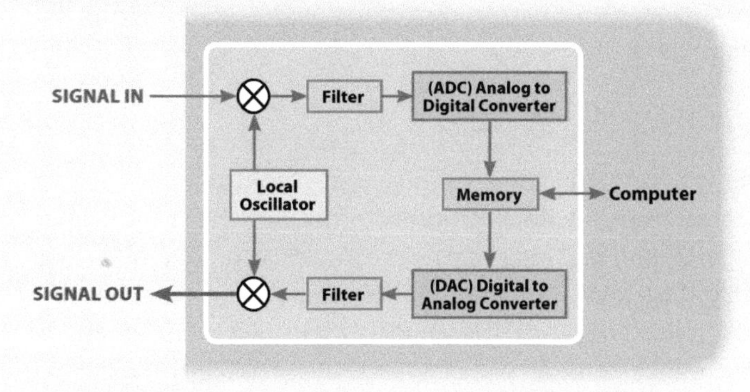

The DRFM digitizes a received signal, passes it to a computer for modification, and coherently regenerates the modified signal for rebroadcast.

The key element of the DRFM is the *analog-to-digital (A/D) converter* which is discussed in Chapter 14. It must support a digitization rate of about 2.5 samples per hertz of the frequency band it digitizes, and it must output an I & Q (in-phase and quadrature) digital signal. This captures the phase of the digitized signal. Note that the 2.5 samples per hertz is greater than the Nyquist rate of two samples per hertz that is required in a digital receiver because the DRFM reconstructs the signal. The digital signal must typically have several bits per sample, although there are cases in which one bit digitization or phase-only digitization is used.

The computer performs analysis of the captured signal, including determination of its modulation characteristics and parameters. The computer can typically analyze the first pulse received by the system and generate subsequent pulses with the same or systematically varied modulation parameters.

The *digital-to-analog (D/A) converter* that generates the RF output signal has more bits than the A/D converter to assure that the signal quality is not degraded in the reconstruction of the RF signal.

Wideband DRFM. A wideband DRFM digitizes a wide IF bandwidth that may include several signals. The jammer system tunes across the frequency range of threat signals it must jam and outputs an IF signal with the bandwidth the DRFM can handle. As shown in the figure below, the frequency conversion and the later reconversion to the received frequency are done using a single system local oscillator to preserve phase coherence. The DRFM bandwidth is limited by the sampling rate of its A/D converter. There can be multiple signals present in the bandwidth, which requires a significant spurious free dynamic range. Therefore, the A/D converter requires the maximum practical number of digitization bits. Wideband DRFMs are highly desirable, because they can handle signals with wide frequency modulations and frequency agile threats. The limitation is the state of the art in A/D converters.

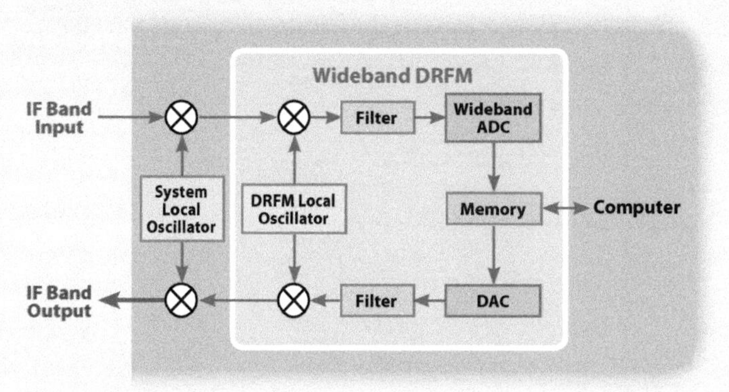

A wideband DRFM handles a frequency range containing multiple signals.

Narrowband DRFM. A narrowband DRFM need only be wide enough to capture the widest signal the jammer must handle. This means that a narrowband DRFM can operate with an A/D converter that is reasonably within the state of the art.

As shown below, the jammer system converts a frequency range of interest into the frequency range covered by multiple narrowband DRFMs. The DRFM input signal is power divided to the individual DRFMs. Each of the DRFMs is tuned to an individual signal and performs its function in support of jamming operation. Then, the analog RF outputs from the DRMFs are combined and converted (coherently) back to the original frequency range.

It should be noted that spurious responses are less of a problem in narrowband DRFMs because each contains only one signal.

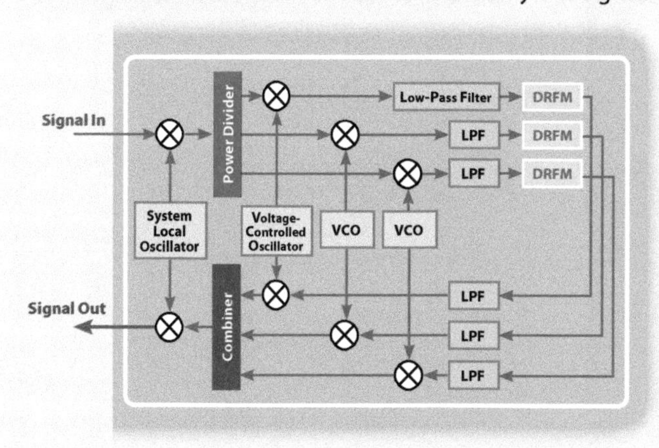

Digital RF Memories *continued*

A narrowband DRFM handles only one signal. Multiple narrowband DRFMs are required to handle a multiple signal environment.

DRFM Functions. DRFMs are particularly valuable in dealing with pulse-compressed radars. As described in chapter 16, radars which improve their range resolution through pulse compression do so by two techniques: "chirp" and "Barker code."

- *Chirp* frequency modulates each transmitted pulse. In the receiver, a compressive filter time compresses the signal into a small fraction of its transmitted pulse width before processing. If a jammer produces signals which do not have this frequency modulation, the effective jamming to signal ratio is reduced by the compression factor. By generating chirped jamming pulses, the DRFM maintains the full jamming to signal ratio.
- *Barker code* compression places a *binary phase shift keyed* (BPSK) modulation on each pulse in a coded pattern. A tapped delay line with phase shifters on the appropriate taps compresses the received pulse to the length of one bit of the code. Since a jamming pulse without the barker code is not compressed, the effective jamming to signal ratio is reduced by the number of bits in the code. The DRFM can create jamming pulses with the correct barker code, maintaining the full jamming to signal ratio.

DRFMs can also produce coherent jamming signals which occupy a single channel in a pulse-Doppler radar's processor. This denies the radar the ability to discriminate against jamming signals which spread across multiple Doppler channels.

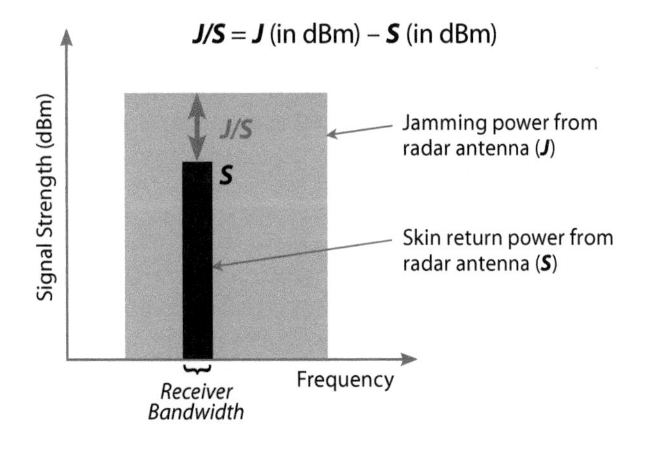

Figure 38-34. The jamming to signal ratio is the ratio of the received power from the jammer and the skin return within the bandwidth of the radar receiver.

are discussed in Chapter 16 and the jamming waveforms are discussed in Chapter 39.

38.4 Jamming Equations

The jamming equations in this section apply to any type of radar jamming. Following normal procedure they assume that the radar and jamming signals propagate according to the line of sight propagation model. At low frequencies and near terrain, different propagation models may apply. The equations do not include rain or atmospheric attenuation; these factors are usually added later if required in special circumstances.

Jamming to Signal Ratio. Figure 38-34 shows the received signal strength and frequency coverage of jamming and skin return signals in the jammed radar's receiver. Note that these are the signal levels from the radar antenna into the receiving system—considering the antenna orientation relative to the target and to the jammer.

Note also that the frequency coverage of the received jamming signal is shown as covering more bandwidth than the radar skin return. The jamming to signal ratio is the received power ratio within the radar receiver bandwidth. Ideally, the jammer would put all of its power into the radar receiver bandwidth, but as discussed earlier, this may not be the case. The ratio of in-band to total jamming energy defines the jamming efficiency. In the following equations, it is assumed that the jamming efficiency is 100%. The calculated jamming to signal ratio is then reduced by the efficiency factor.

Jamming is required only when the radar is illuminating the target, so that is the basis of the following jamming equations.

The jamming to signal ratio is the way that the effectiveness of the jammer is quantized. Each type of radar and each type of jamming requires the J/S ratio to be at or above some minimum level for the jammer to protect friendly aircraft.

Self-Protection Jamming. Figure 38-35 shows the geometry for self-protection (also called self screening) jamming. Since the jammer is located on the target, the radar antenna is pointed directly at the jammer. Thus, the jamming signal is received in the main lobe of the radar antenna. This allows simplification of the jamming formula. The jamming to signal ratio is given by the following equation:

$$J/S = \frac{ERP_J \ 4\pi R^2}{ERP_S \ \sigma}$$

where J/S is the jamming to skin return signal ratio, ERP_J is the effective radiated power of the jammer in watts, ERP_S is the radar effective radiated power in watts, R is the radar to target range in meters, and σ is the radar cross section of the target in square meters.

In dB form, this equation is

$$J/S = ERP_J - ERP_S + 71 + 20\log_{10} R - 10\log_{10} \sigma$$

where J/S is the jamming to skin return signal ratio in dB, ERP_J is the effective radiated power of the jammer in dBm, ERP_S is the radar effective radiated power in dBm, R is the radar to target range in km, and σ is the radar cross section of the target in square meters.

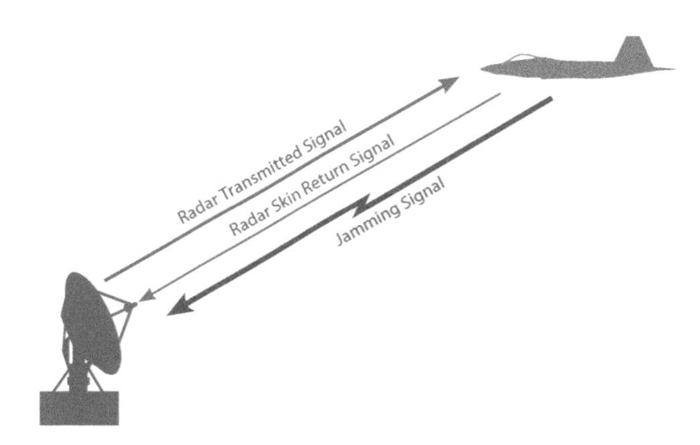

Figure 38-35. Self-protection jamming is transmitted from the target, so it is received in the radar main beam when the radar is illuminating the target.

Remote Jamming. Figure 38-36 shows the geometry for remote jamming. This can be stand-off or stand-in jamming. The only requirement is that jamming is conducted from some other location than that of the target. This means that the range from the radar to the jammer is different from the range to the target. It also means that the angle to the jammer from the radar is different from the angle to the target. This calculation assumes that the boresight of the radar antenna is pointed at the target and that the jammer is in an average gain sidelobe of the radar antenna. This means that the J/S formula has more terms than that for self protection jamming:

$$J/S = \frac{ERP_J \ 4\pi R_T^4 G_S}{ERP_S \ \sigma R_J^2 G_M}$$

where J/S is the jamming to skin return signal ratio in dB, ERP_J is the effective radiated power of the jammer in Watts, ERP_S is the radar effective radiated power in Watts, G_S is the average sidelobe gain of the radar antenna, G_M is the boresight gain of the radar antenna, R_J is the radar to jammer range in meters, R_T is the radar to target range in meters, and σ is the radar cross section of the target in square meters.

In dB form, this formula is

$$J/S = ERP_J - ERP_S + 71 - 20\log_{10} R_J + 40\log_{10} R_T - 10\log_{10} \sigma$$

where J/S is the jamming to skin return signal ratio in dB, ERP_J is the effective radiated power of the jammer in dBm, ERP_S is

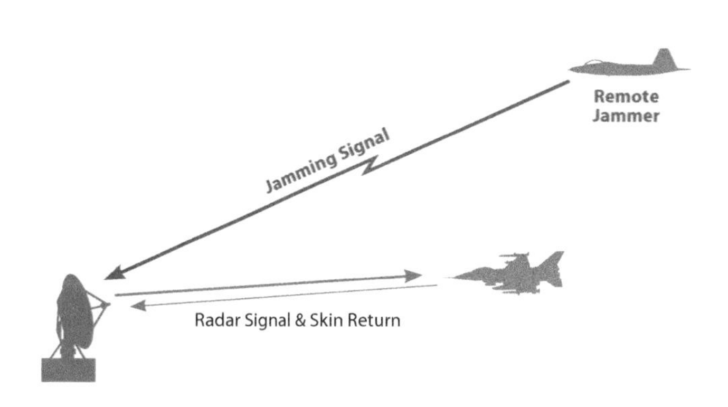

Figure 38-36. The remote jammer can be located anywhere but on the target. Thus, its range to the target will be different from that of the target and it is assumed to transmit into the radar antenna sidelobes.

the radar effective radiated power in dBm, G_S is the average sidelobe gain of the radar antenna in dB, G_M is the boresight gain of the radar antenna in dB, R_T is the radar to target range in km, and σ is the radar cross section of the target in square meters.

Burn-Through. Although the burn-through range is defined as the range at which the radar can reacquire the target in the presence of jamming, it is common practice in mission planning to set a *J/S* ratio at which the jammer could just barely protect the target. Thus, a minimum *J/S* ratio is used in burn-through calculations.

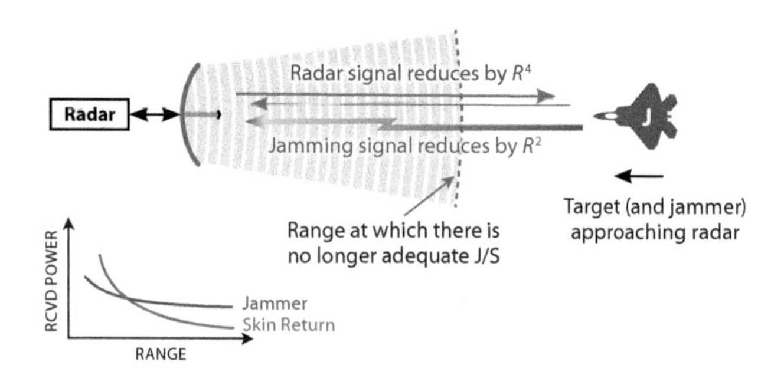

Figure 38-37. The *J/S* ratio decreases with reduced radar to target range. The range at which the radar can reacquire the target is the burn-through range.

Burn-Through of Self-Protection Jamming. As the jammer (mounted on the target aircraft) approaches the radar, the jamming signal in the radar receiver increases as the square of the reducing range. However, as the target approaches the radar, the skin return in the radar receiver increases as the fourth power of the reducing range (Fig. 38-37). Thus, the *J/S* ratio decreases as the square of the reducing target to radar-range. The following equation presents the range at which burn-through occurs:

$$R_{BT} = \sqrt{\frac{ERP_S \ \sigma \ J/S_{RQD}}{ERP_J \ (4\pi)}}$$

where R_{BT} is the radar to target range at burn-through in meters, ERP_S is the effective radiated power of the radar in Watts, ERP_J is the effective radiated power of the jammer in Watts, σ is the radar cross section of the target in square meters, and J/S_{RQD} is the jamming to skin return signal ratio, in dB, at which the jammer is no longer assumed to be able to prevent the radar from reacquiring the target.

In dB form, this calculation is often performed in two parts. The equation for the range term in the *J/S* formula is

$$20\log_{10} R_{BT} = ERP_S - ERP_J - 71 + 10\log_{10}\sigma + J/S_{RQD}$$

where R_{BT} is the radar to target range, in km, at burn-through, ERP_S is the radar effective radiated power of the radar in dBm, ERP_J is the effective radiated power of the jammer in dBm, σ is the radar cross section of the target in square meters, and J/S_{RQD} is the jamming to skin return signal ratio, in dB, at which the jammer is no longer assumed to be able to prevent the radar from reacquiring the target

The burn-through range is the range at which the required *J/S* value occurs. This range is determined from

$$R_{BT} = \text{antilog}\{[20\log_{10} R]/20\}$$

where R_{BT} is the radar to target range in km at burn-through.

Note that it is the numerical value of 20 $\log_{10} R$ that is entered into this equation. These two formulas can be combined, particularly if a calculation is being made in a computer, but are given separately in this discussion for clarity.

Burn-Through of Remote Jamming. In Figure 38-38, note that the jamming signal in the radar receiver increases as the fourth power of the reducing range *from the radar to the target.* The jamming signal in the radar receiver remains constant through the engagement because it is common practice to assume that the remote jammer does not move. Thus the J/S ratio decreases as the fourth power of the decreasing radar to target range. The following formula shows calculation of the burn-through range for remote jamming. This is the range from the radar to the target at burn-through.

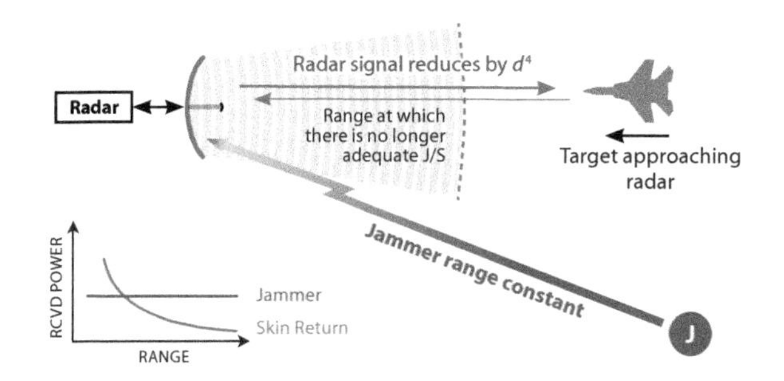

Figure 38-38. A remote jammer is assumed to stay in one location during the engagement, while the target approaches the radar. The J/S ratio decreases with the decreased radar to target range. The burn-through range is the range to the target at which the radar can reacquire the target.

$$R_{BT} = \sqrt[4]{\frac{ERP_S \ \sigma R_J G_M \ J/S_{RQD}}{ERP_J \ (4\pi) G_S}}$$

where R_T is the radar to target range in meters, ERP_S is the radar effective radiated power in watts, ERP_J is the effective radiated power of the jammer in watts, G_S is the average sidelobe gain of the radar antenna, G_M is the boresight gain of the radar antenna, R_J is the radar to jammer range in meters, and J/S_{RQD} is the radar cross section of the target in square meters to provide protection.

In dB form, this calculation can be performed in two parts. The equation for the range term in the J/S formula is

$$40\log_{10} R_T = ERP_S - ERP_J - 71 - G_S + G_M + 20\log_{10} R_J + \log_{10}\sigma + J/S_{RQD}$$

where R_T is the radar to target range in km, ERP_S is the radar effective radiated power in dBm, ERP_J is the effective radiated power of the jammer in dBm, G_S is the average sidelobe gain of the radar antenna. G_M is the boresight gain of the radar antenna, R_J is the radar to jammer range in km, and J/S_{RQD} is the radar cross section of the target in square meters to provide protection.

The actual radar to target range at which the minimum J/S for protection occurs is given by

$$R_{BT} = \text{antilog}\{[40\log_{10} R_T]/40\}$$

where R_{BT} is the radar to target range, in km, at burn-through

Note also that the numerical value of the $40\log_{10} R_T$ term is used in this equation.

38.5 Look-Through

Some jamming techniques require information about the radars being jammed. For example, spot jamming requires that any change in the radar's operating frequency be known so the jamming frequency can be changed. Jamming signals will also jam the friendly ES receiver, so jamming must be stopped

Figure 38-39. Chaff consists of thin metal-coated dielectric fibers, billions of which can be stored in a small space.

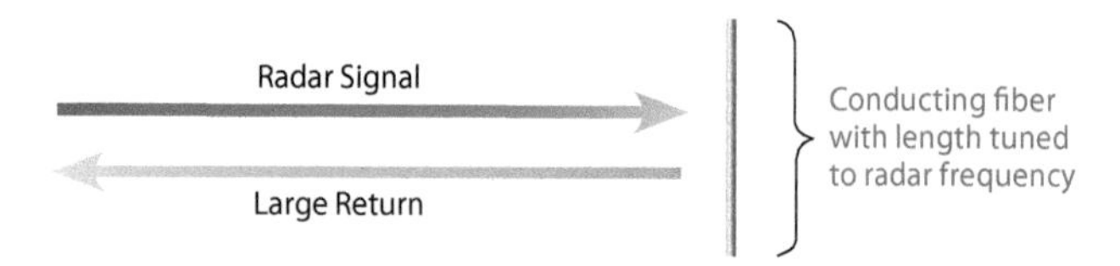

Figure 38-40. A chaff dipole is a stiff piece of metallized fiberglass or aluminum foil. It is a half-wavelength long at its resonant frequency to optimally re-radiate signals.

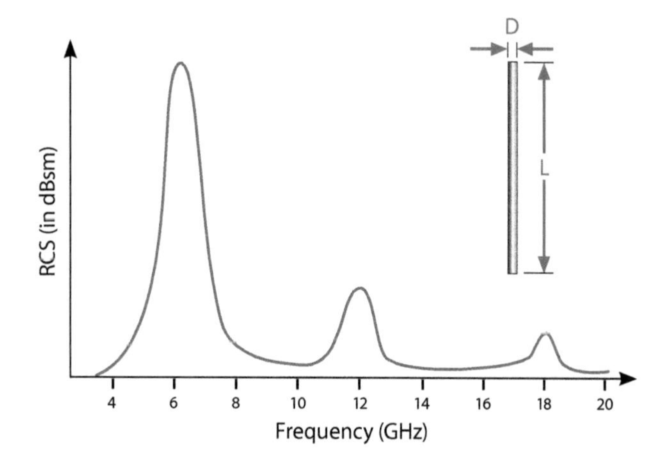

Figure 38-41. Bandwidth of chaff fibers one-half wavelength long at 6 GHz. The greater the fibers' ratio of length *L* to diameter *D*, the narrower the peaks will be. A wide band can be covered, however, by dispensing chaff of several different lengths.

or canceled in the ES system. This is can be done by adding a 180° phase shifted version of the jamming signal to the ES receiver input, but reflections from the aircraft and external stores makes this extremely challenging. Typically, look through involves stopping the jamming for short, nonperiodic "look-through" windows.

38.6 Chaff

Chaff is the name given for half wavelength dipoles which are launched into the air to interfere with hostile radars. Figure 38-39 shows an example of chaff. The chaff can either form a background of clutter to hide friendly aircraft or a false target to lure an enemy radar away from its intended target. As shown in Figure 38-40, each piece of chaff acts as a half wavelength shorted dipole. This means that if it is properly oriented (perpendicular to the angle of arrival of the radar signal and oriented at the correct polarization), it will optimally retransmit the radar signal.

Figure 38-41 shows the radar cross section caused by a single fiber of chaff.

Chaff is launched in various ways, typically in packets blown out of launchers or rockets. Each chaff cartridge will contain pieces of various length to cover specific threat frequencies or spaced to cover a frequency range as shown in Figure 38-42. As shown in this figure, each dipole length provides a radar cross section (RCS) peak and rolls off away from the resonant frequency.

The average RCS of a single dipole in a cloud of randomly oriented chaff pieces is given by the formula:

$$\text{RCS (in square meters)} = 0.15\ \lambda^2$$

where λ is the wavelength (in meters) associated with the resonant frequency.

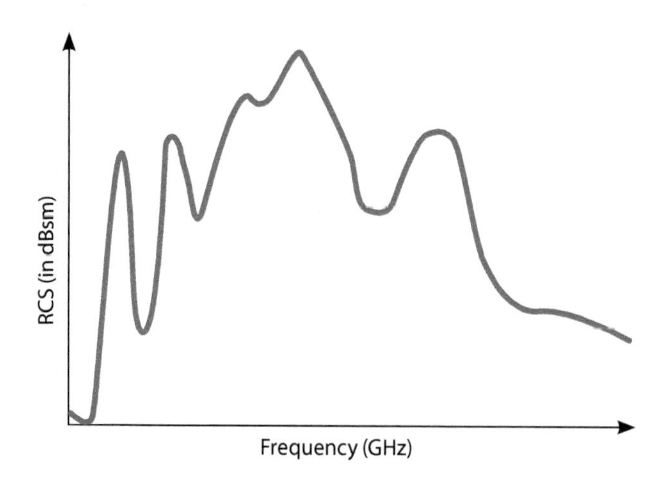

Figure 38-42. A packet of chaff will have several different dipole lengths, each causes a peak in generated RCS at its resonant frequency. The magnitude of each RCS peak depends on the number of dipoles at that wavelength.

Table 38-1. RCS of Chaff Cloud versus Dipole Spacing

Average Dipole Spacing	RCS of Cloud
One wavelength	0.925 $N \times$ RCS of one dipole
Two wavelengths	0.981 $N \times$ RCS of one dipole
Wide spacing	$N \times$ RCS of one dipole

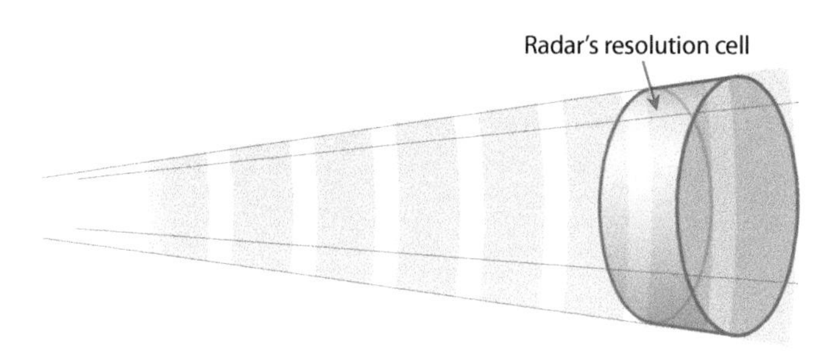

Figure 38-43. The RCS produced by chaff is the number of dipoles in the radar's resolution cell multiplied by the RCS per dipole with the average dipole separation.

Table 38-1 gives the formulas for the RCS of clouds of multiple dipoles with various average spacing of elements. Figure 38-43 is a sketch of a radar's resolution cell. If the number of dipoles within the volume of the cell is calculated, the RCS of the chaff can be calculated from the number of dipoles and their average spacing.

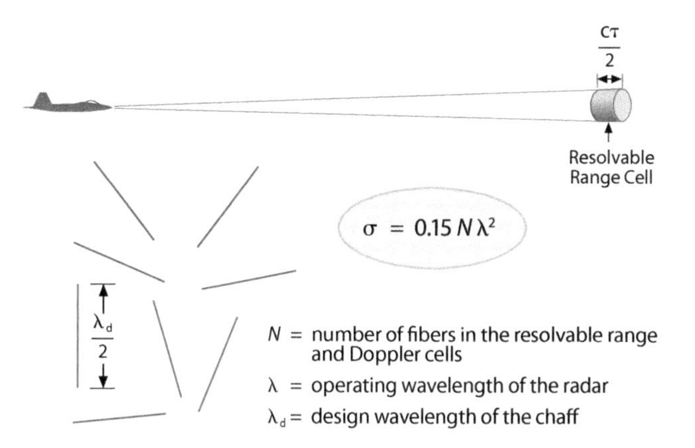

Figure 38-44. Radar cross section σ of randomly distributed and randomly oriented chaff fibers. Since the fibers are light and small, very large radar cross sections can readily be achieved.

Figure 38-44 summarizes the effect of chaff in the resolution cell. Movement of the chaff cloud will cause the radar illuminating the cloud to see a Doppler shift. When chaff is deployed from an aircraft, it will slow because of atmospheric drag, and thus will have different velocity from that of the aircraft – hence a different Doppler frequency. One solution to this is to illuminate the cloud with a jammer on the aircraft. Movement of the dipoles in the chaff cloud cause a spreading of the frequency response in an illuminating radar. Pulse Doppler radars can sense this spreading to allow discrimination against the chaff cloud.

38.7 Anti-Radiation Missiles

Figure 38-45 shows an anti-radiation missile attacking a ground-based radar. The missile is programmed to home on emissions from the target radar antenna. It can either be programmed to home on a specific radar signal at a specified general location, or can seek out a radar from a prioritized list. Then, it flies to that radar location by homing on sidelobe energy. As the missile approaches the target radar, its guidance accuracy increases until it can typically impact the actual antenna.

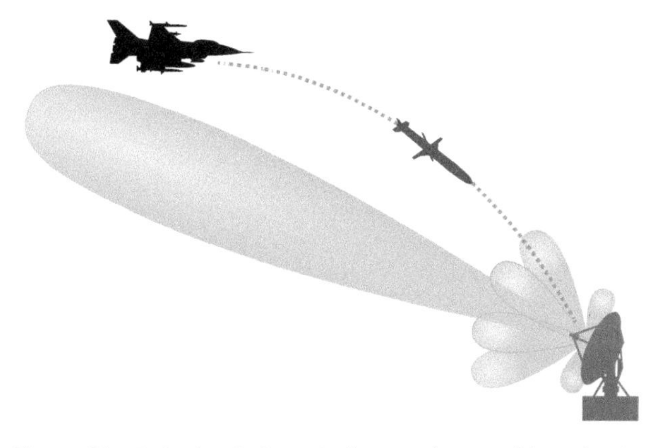

Figure 38-45. Anti-radiation missiles attack ground-based radars by homing on radiated energy from antenna sidelobes.

As shown in Figure 38-46, an anti-aircraft missile can be equipped with a radar homing guidance system that will home on an airborne radar. The missile is launched at a high angle so that it will fly beyond the range of its guiding radar. The missile attacks its radar equipped target from above.

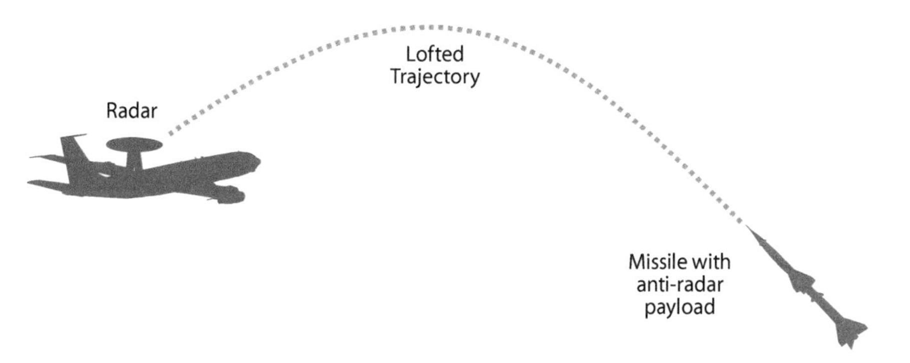

Figure 38-46. An airborne radar can be attacked by a missile with a payload that includes a radar homing capability.

38.8 High-Power Lasers

High-power lasers are electronic warfare weapons which can attack airborne radars from great distances. With adequate power, it is possible to destroy an airframe with a laser, but it requires far less power (of the order of three orders of magnitude) to damage sensors such as radars on the target aircraft. The range at which an attack can be prosecuted depends on the wavelength, the altitude and the weather.

38.9 High-Power Microwave

Strong microwave signals can be used to damage hostile sensors by burning out sensitive front end components. In general, it takes about three orders of magnitude greater power to temporarily disable components than the power required for jamming. It takes an additional three orders of magnitude more power to permanently damage components that required to temporarily disable them.

One complication to the application of high-power microwave tactics is that many radars have overload protection that will temporarily shut them down while very high levels of received signal strength are encountered. Therefore, attack with high-power microwave requires signal levels within the magnitude window between the maximum survivable received power level and the shut down level.

38.10 Summary

EA is the action part of EW, interfering with the effectiveness of hostile radars by jamming or the use of chaff, high-power lasers or high-power microwave.

Jamming involves the transmission of undesired signals into the receivers of enemy radars. This can be done as self-protection jamming from the protected platform or as remote jamming from some other location. Remote jamming can be standoff jamming from beyond the range of enemy weapons or stand-in jamming from locations closer to the radar than the target aircraft. Remote jamming transmits to the side lobes of its target radars. Jamming can also be performed by special aircraft that fly in with strike packages (escort jamming) or by special aircraft which align themselves with the strike package from outside the lethal zone and then turn back before entering the lethal zone.

Cover jamming transmits noise to blind enemy radars. It can be barrage jamming (over a large frequency range), spot jamming (about the enemy radar's operating frequency), or swept-spot jamming.

Deceptive jamming transmits signals that look like enemy radar signals, but cause the enemy radar to make false decisions about the location or velocity of protected aircraft. Although some deceptive techniques such as false targets are used off-board, deceptive jamming (to break the lock of a radar) must be self-protection jamming. However, some deceptive jamming techniques do not provide protection against mono-pulse radars. These include: range gate pull-off, range gate pull-in, cover pulses, inverse gain, AGC jamming and velocity gate pull-off. Techniques appropriate for use against monopulse radars include formation jamming, blinking, terrain bounce, cross-pol, and cross-eye.

Jamming to signal ratio is the ratio of received jamming power to skin return power in an enemy radar receiver. Burn-through range is the range at which an enemy radar can reacquire a target in the presence of jamming. In operation planning, this is calculated as the range at which a jammer can provide the minimum J/S to provide adequate protection against an enemy radar.

Chaff is a quantity of half wavelength dipoles which are launched into the air either to generate a large number of radar returns (to hide friendly aircraft) or to create a false target to lure an enemy radar away from its intended target. The RCS created by a chaff cloud depends on the number of dipoles (assumed to be randomly oriented) within the illuminating radar's resolution cell. A chaff cloud normally includes dipoles of various lengths to provide protection at important frequencies and across a frequency range.

Anti-radiation missiles attack ground based or airborne radars by homing on their sidelobe energy.

High power lasers and high power microwave can degrade the performance of radars, or with much more power can damage them.

Further Reading

R. N. Lothes, M. B. Zymanski, and R. G. Wiley, *Radar Vulnerability to Jamming*, Artech House, 1990.

F. Neri, *Introduction to Electronic Defense*, 2nd ed., SciTech-IET, 2007.

Test your understanding

1. List the different geometrical classifications for jamming – i.e. from where is the jamming performed.
2. Describe cover and deceptive jamming.
3. What J/S is provided by a self protection jammer if the jammer ERP is 100 Watts, the radar ERP is 10 Megawatts, the range from radar to target is 10 km and the target RCS is 10 m^2?
4. What J/S is provided by a stand-off jammer if the radar tube power is 10 kilowatts, the radar boresight gain is 30 dB, the jammer is located 30 km from the radar in a -20 dB side lobe, the jammer ERP is 200 kilowatts, the range from radar to target is 5 km and the target RCS is 10 m^2?
5. What is the burn-through range if a target is protected by a self-protection jammer with 100 watts ERP if the radar ERP is 10 Megawatts, the target RCS is 10 m^2, and the required minimum J/S is 2 dB?

General Dynamics F-16 (1978)

The Fighting Falcon is a small, lightweight aircraft designed for air superiority and has evolved into a successful all-weather multirole aircraft. Like all modern fighter aircraft, the APG-68 and APG-80 radar used in versions of the F-16 have significant electronic protection features to allow them to operate effectively in a dense electronic warfare environment.

39

Electronic Protection

144-element dual-polar UWB array of Vivaldi printed elements (Courtesy of UMass.)

39.1 Introduction

The third major subdivision of electronic warfare (EW) is *electronic protection* (EP). Unlike the first two parts of EW, it does not involve additional subsystems. Rather, it comprises technology and methods to make radar and communication systems less vulnerable to EW techniques, especially jamming. In this chapter we introduce and explore EP for radar systems. Table 39-1 shows the principal techniques and the counters provided to EW. Note that three of the "techniques" listed in this table are implemented in radars for some other purpose but that they also provide EP capability: monopulse, pulse compression, and pulse Doppler. Each of these radar characteristics is discussed in terms of its radar functionality in earlier chapters. In this chapter, it will be shown that each of these radar characteristics can have multiple EP advantages.

Table 39-1. Electronic Protection Features versus EW Techniques Countered

Electronic Protection Techniques	Electronic Warfare Techniques Countered
Ultra low sidelobes	All sidelobe jamming & threat signal detection
Sidelobe canceler	Narrow band sidelobe jamming
Sidelobe blanker	Wideband sidelobe jamming
Anti-cross pol	Cross polarization jamming
Monopulse	Range & angle gate pull-off, chaff, decoys
Pulse compression	Deceptive jamming without compression modulation
Pulse Doppler	Noncoherent jamming, chaff, separating targets, DRFM with spurs,
Leading edge tracking	Range gate pull-off
Dicke Fix	AGC jamming, wideband FM modulation
Burn-through modes	All jamming techniques
Frequency agility	Spot jamming
PRF jitter	Cover pulses
Home on jam	All jamming techniques
Adaptive arrays	Anti-radiation missiles
Adaptive waveforms	Threat signal identification
Processing	False targets, false tracks

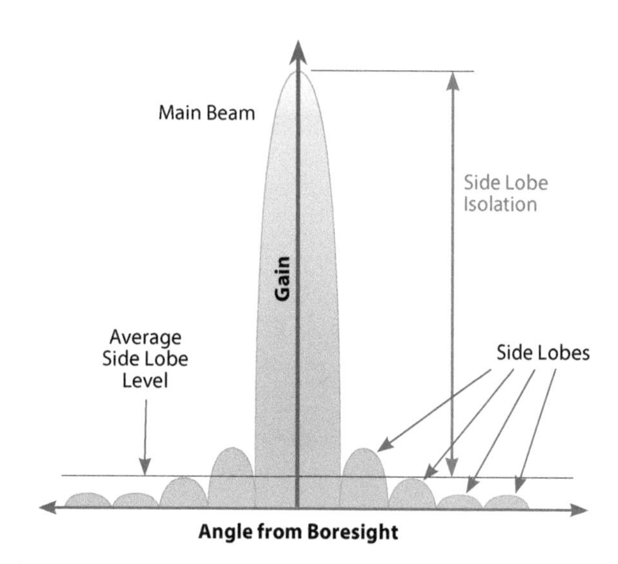

Figure 39-1. Sidelobe isolation is the difference in gain between the peak boresight gain and the average sidelobe level.

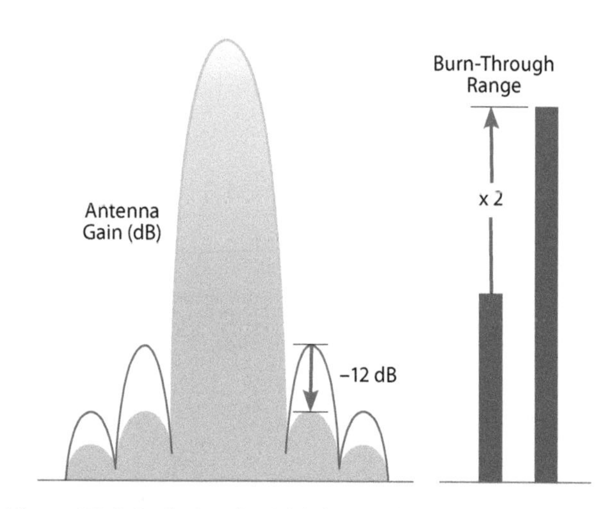

Figure 39-2. Reducing the sidelobe gain by 12 dB doubles target burn-through range.

In Chapter 8, the function of the radar's antenna beam in determining the azimuth and elevation of a target is covered. The beam is steered through angular segments and the amplitude of the return signal from a series of pulses increases sharply as the beam moves through the target this is the "skin return." The angular location of the target is determined by the time at which the skin return is detected. Another approach is to sense the angle of arrival of the skin return is by use of multiple, angularly offset antenna feeds (typically three or four feeds). This technique is called "monopulse" because the azimuth and elevation of the target is determined by comparing the outputs of the multiple feeds for each received pulse. As shown below, this angular measurement technique makes some types of jamming ineffective.

In Chapter 2, pulse Doppler (PD) radar is described. PD radars are coherent, allowing the determination of the frequency of each received pulse. As shown below, PD has significant impact on the effectiveness of some electronic attack (EA) techniques in addition to its enhancement of the basic radar performance.

Pulse compression is described in Chapter 16 in terms of the improvement of range resolution—that is the range increment within which the radar cannot detect multiple targets. Since some kinds of jamming depend upon a radar not being able to detect multiple targets, the reduction of the depth of the resolution cell (called pulse compression (PC) also reduces the effectiveness of decoys and some types of jamming.EP attempts to reduce the effectiveness of jammers in a variety of ways. Some are designed directly reduce the jamming to signal ratio (J/S) achieved by specific jamming techniques (for specific geometries). Other techniques are primarily designed for a different purpose, but also have the benefit of reducing jamming effectiveness. EP can even reduce the effectiveness of Electronic Support by making signal detection and identification more difficult and can counter anti-radiation missiles. Here, we examine these techniques.

39.2 Ultra-Low Sidelobes

As we saw in Chapter 38, jamming through the sidelobes of a radar system can render the radar useless, obscuring all possible targets. It follows that reduced sidelobe levels make it more difficult for ES and ELINT systems to detect, locate and jam a radar system. This has the beneficial effect of reducing the J/S achievable by stand-off or stand-in jammers and making the radar less vulnerable to electronic attack. Figure 39-1 shows the average sidelobe level for a radar antenna. Note how the nulls between the sidelobes are very narrow compared to the width of the peaks, effectively raising the average sidelobe level. Sidelobe isolation may be defined as the reduced average sidelobe gain relative to the main beam boresight gain.

Figure 39-2 shows the impact of reduced sidelobe levels on burn-through range.

Table 39-2. Reduced Sidelobe Levels

Level	Average Sidelobe Gain	Average Sidelobe Isolation
Ordinary Sidelobes	More than −3 dBi	More than 30 dB
Low Sidelobes	−3 to −10 dBi	35 to 45 dB
Very Low Sidelobes	−10 to −20 dBi	45 to 55 dB
Ultra-Low Sidelobes	Below −20 dBi	More than 55 dB

Table 39-2 lists some typical values for average sidelobe antenna gain for ordinary, low and ultra-low sidelobes. It also shows the amount of sidelobe isolation relative to the boresight gain of the radar antenna.

A further degree of protection can be achieved by lowering the sidelobes at just the discrete range of angles over which the jamming signal is being received. These techniques are known as *sidelobe cancellation* and *blanking*.

39.3 The Coherent Sidelobe Canceler

Coherent sidelobe cancelers can be incorporated into a radar system and are especially effective against narrowband sidelobe jamming signals. The most common narrowband sidelobe jamming modulation is CW noise.

For each sidelobe signal to be cancelled, an auxiliary antenna is required in addition to the main antenna. As shown in Figure 39-3, it must have more gain in the sidelobe direction than the gain of the sidelobes on the main radar antenna.

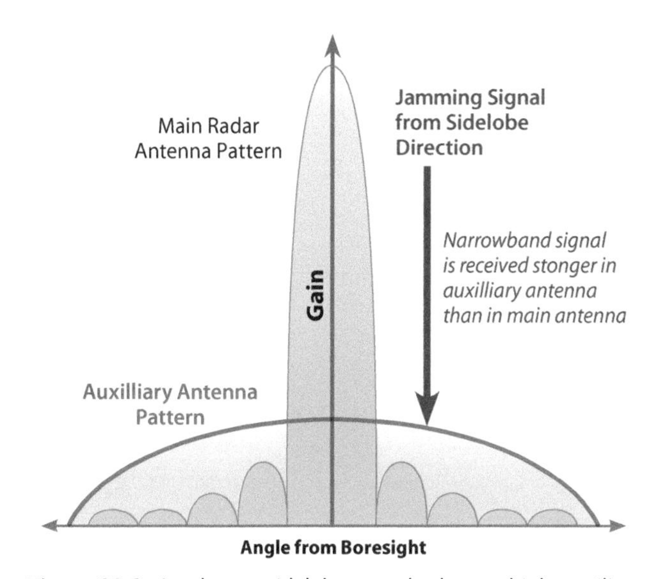

Figure 39-3. A coherent sidelobe canceler has multiple auxiliary antennas that have more gain than the main-lobe antenna in the sidelobes.

If a signal from an auxiliary antenna is stronger than that same signal from the main antenna, it is identified as a sidelobe jamming signal. This sidelobe jamming signal is used to generate a new version that is delayed by half a wavelength or a phase of 180°. When this phase delayed signal is added to the output of the main antenna (Fig. 39-4), it results in cancellation of the signal entering the sidelobes of the main antenna. In other words it, effectively, creates a null at that angle as far as the main antenna is concerned, thus making it much less sensitive to the incoming jamming signal.

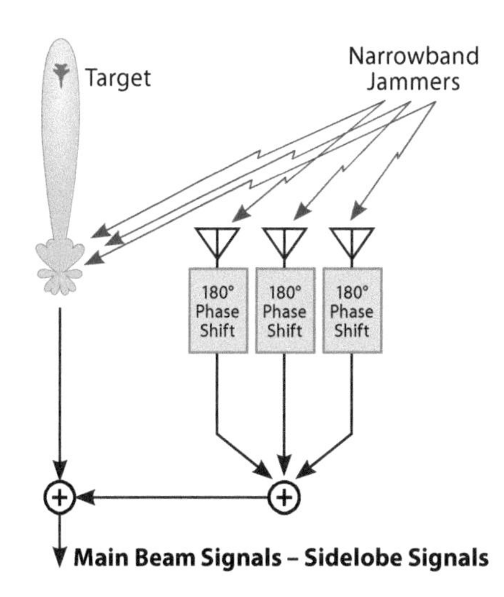

Figure 39-4. If a signal is stronger in the auxiliary antenna than the main antenna, a 180° phase delayed version of that signal is added to the output of the main antenna to cancel the sidelobe response.

Phase cancellation is inherently dependent on wavelength and therefore this process is only effective against narrow bandwidth signals (like noise modulated FM). Each jamming signal to be cancelled requires a separate phase canceler on the output of a separate auxiliary antenna.

The frequency spectrum of a pulsed signal has multiple spectral components that look like narrowband signals. Thus, a single pulsed signal can occupy several coherent cancelers (the addition of pulses to FM noise jamming signals is a way to counter coherent sidelobe cancelers) and in this way the job of the sidelobe canceler can become more and more challenging.

How Sidelobe Jamming Is Canceled

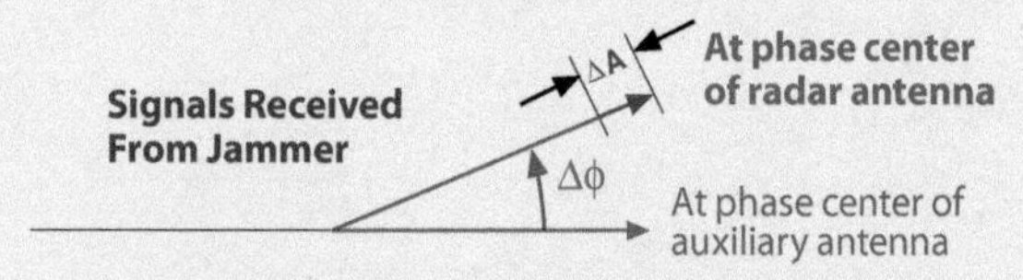

Amplitude difference, ΔA, is due to difference in gains of auxiliary antenna and main antenna in jammer's direction.

Phase difference, Δφ, is due to difference in distance from jammer to the two phase centers.

Amplitude Adjustment: By adjusting the gain of the auxiliary receiver, the amplitude difference is removed.

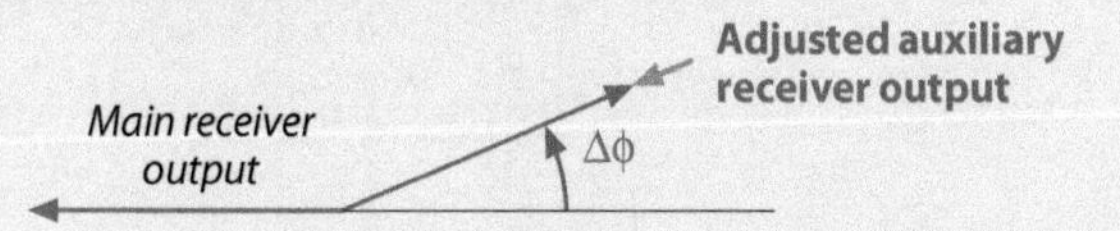

Note: In this example, the jammer is assumed to be in the radar antenna's first sidelobe. So the phase of jamming is reversed in the output of antenna.

Phase Adjustment: By adjusting the phase shift in the output of the auxiliary receiver, Δφ is removed.

Result: Because the jammer's signal in the output of the auxiliary receiver is now equal to and 180° out of phase with the jammer's signal in the output of the main receiver, they cancel when the outputs combine.

Another way of looking at this: a notch has been produced in the radar antenna's sidelobe pattern in the jammer's direction.

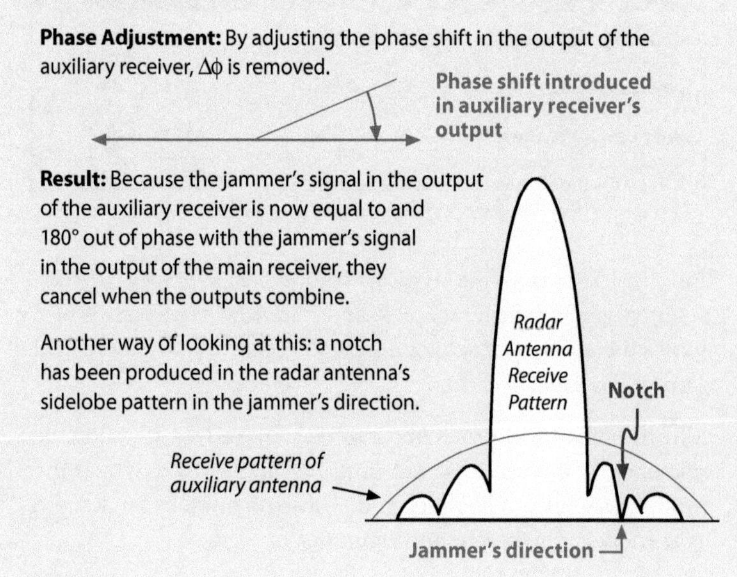

Figure 39-5. If a signal entering the radar system is stronger in the sidelobe blanking auxiliary antenna than in the main radar antenna, it is identified jamming signal and the main radar antenna is blanked during that pulse.

39.4 Sidelobe Blanker

The purpose of the blanker is to remove pulsed jamming signals from the radar receiver input. A sidelobe blanker requires the same type of auxiliary antenna shown in Figure 39-4 for the sidelobe canceler. The difference is that only one such antenna is required to handle multiple jammers.

Sidelobe blanking was covered in some detail in chapter 29 where we saw that when the signal strength in the auxiliary antenna is greater than that from the main antenna, the signal is again identified as a sidelobe signal. The radar receiver input may be blanked during a sidelobe pulse (Fig. 39-5). If there are multiple pulsed signals entering through the sidelobes from multiple directions, a single blanker is able to handle all of them.

If a jammer creates "cover" pulses over the radar's pulses, the side lobe blanker will disable the radar by having it blank its own pulses. This requires tracking the threat radar's PRF and careful timing of jamming pulses if the jammer is remote from the target. If the cover pulse is too wide, the coherent sidelobe canceler (if the radar has one) may be activated.

39.5 Anti-Cross Pol

Cross-polarization (cross-pol) jamming transmits cross-polarized signals towards a monopulse radar in a way designed to cause the radar to move its antenna away from the target it is tracking. This jamming technique depends on the presence of cross-polarized Condon lobes that occur in the antenna of a radar system (Chapter 38). Anti-cross pol electronic protection aims to eliminate or minimize these Condon lobes to make cross-pol jamming ineffective (or to significantly increase the level of jamming power required to make it work). Figure 39-6 shows the size of the Condon lobes with and without anti-cross pol EP. Note that an actual radar antenna will typically have four Condon lobes around its antenna while here only a two-dimensional representation is shown.

This form of this EP is implemented, typically, in two principal ways using either in planar phased arrays or by polarization filtering. Recall from Chapter 38 that Condon lobes are caused by differential gain in phased array elements located near the edge of the antenna. This means that flat, evenly illuminated phased arrays either have no or only very weak Condon lobes. A polarization filter matched to the polarization of the radar echo signal will therefore eliminate or minimize the cross-polarized jamming signals entering the radar. This can be difficult to achieve in practice depending on the polarimetric echo response of any given target.

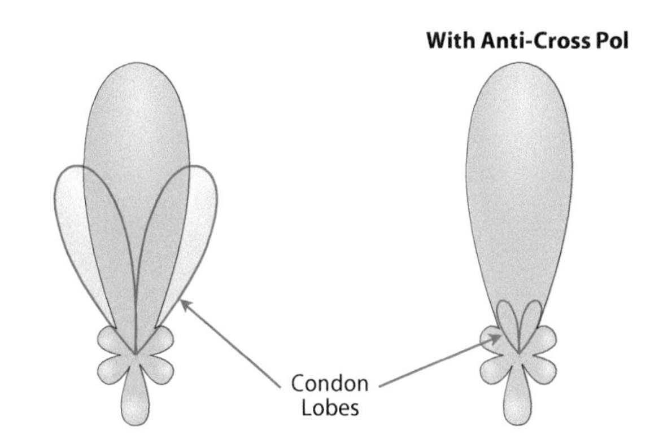

Figure 39-6. Reduced Condon lobes make cross-pol jamming less effective by reducing the cross-polarized response.

39.6 Monopulse Radars

Monopulse radars estimate range, range-rate and angle using each received echo return. This makes them less vulnerable to some jamming techniques and can provide them improved ability to track targets that are protected by certain types of jamming.

As an example, consider range gate pull off, range gate pull in, or inverse gain jamming as described in Chapter 38. These techniques create strong signals to be transmitted from the target to the radar system. These signals will, in fact, *improve* the angle tracking capability of monopulse radar systems, as these signals are usually a more reliable source of target location than the echo return. While monopulse radar systems are still vulnerable to range deception, their improved angle tracking will still allow guidance to the target.

39.7 Pulse Compression

Pulse compression is a technique often used to improve range resolution (Chapter 16). However, pulse compression has an impact on EP performance. Pulse compression is accomplished using techniques such as linear frequency modulation on pulse (LFMOP), commonly called "chirp," or binary phase modulation on pulse (BPMOP) using code sequences such as "Barker codes." Chirp can also employ non-linear or stepped frequency modulation and BPMOP can also use other codes.

In either case, a modulation is placed on the transmitted pulse and the receiver processing has a special circuit that compresses the received pulse power into a reduced duration.

For example, in the case of chirp, the receiver has a filter that delays each part of the received pulse by an amount proportional to its instantaneous frequency. In this way the slope of the delay versus frequency is matched to the time versus frequency of the transmitted pulse and compression results. The

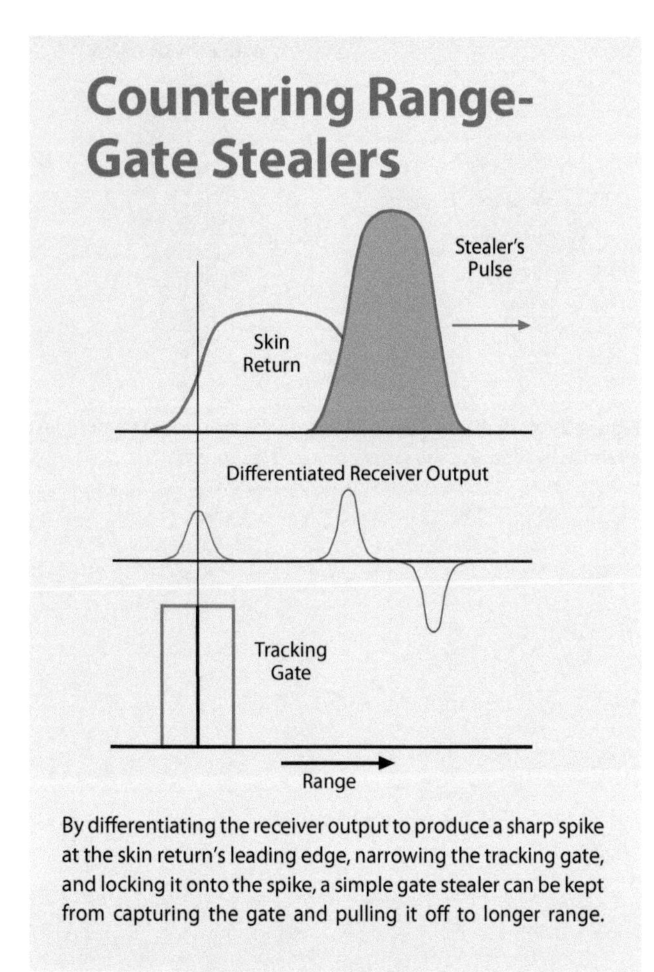

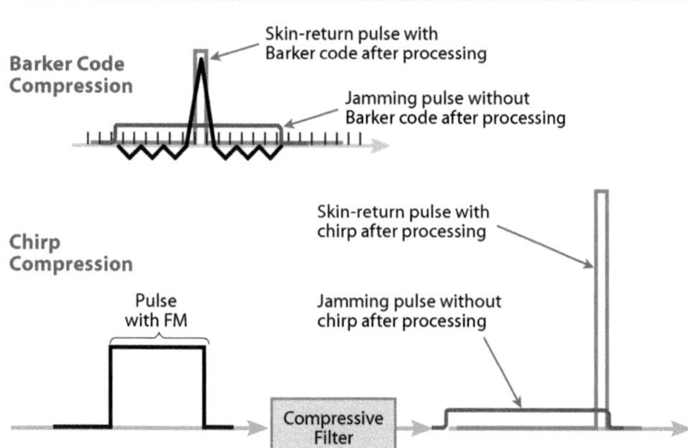

Figure 39-7. Pulse compression technique reduced the J/S if the jammer is not "matched" to the transmit pulse modulation used by the radar.

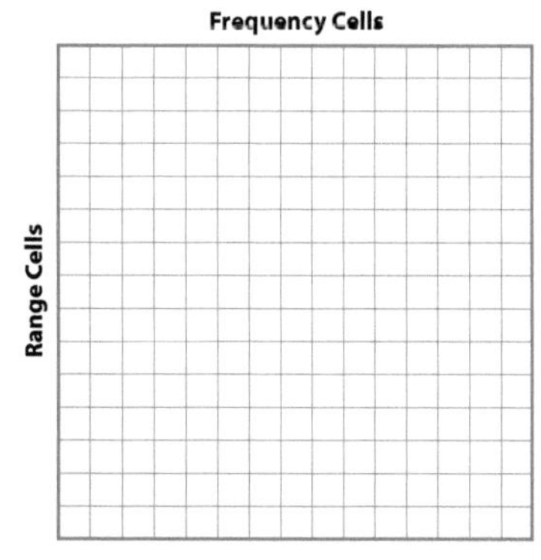

Figure 39-8. A pulse Doppler radar processor creates a "range-Doppler map."

frequency slope can be either positive or negative. The extent of the frequency excursion is the modulation bandwidth.

For a BPMOP Barker code, the transmitted pulse is a binary phase coded signal. Approximately half of the bits in the code are phase shifted 180° with respect to the other half. Recall that, in the pulse compression process, the receiver convolves (Chapter 6) the received echo with a replica of the transmitted pulse. When the convolution lag is zero, such that the two signals entirely overlap, the phases in the Barker code sequence all add constructively and the pulse is compressed. The properties of the Barker code are such that the output of the convolution is very low at all other lag values. The received pulse is compressed into the time occupied by just one single "bit" of the code (the inverse of the duration of "one bit" determines the modulation bandwidth and hence range resolution).

As shown in Figure 39-7, the radar receiver processes the signal only during the time period occupied by the compressed pulse. If a jamming signal is received without the proper compression modulation, the power it injects into the radar receiver is reduced by the amount of the compression. As we saw in Chapter 16, the time-bandwidth product determines the amount of compression. This is the product of the uncompressed duration of the transmitted pulse and the bandwidth of the modulation. Typical time-bandwidth product values are of the order of 100 to 1000 (20 to 30 dBs). The effective J/S ratio is reduced by the amount of compression. To overcome this loss of jamming efficiency, a jammer must either modulate its jamming pulses with the correct compression waveform or significantly increase its jamming power to overcome this compression loss.

Pulse compression radar waveforms lower the peak power transmitted by the radar thus also making the ES detection problem more difficult. Thus pulse compression is driving the need for increased ES sensitivity. It also requires either the use of coherent EA when employing support jamming (i.e., SOJ, ESCORT, or SIJ) or increased ERP for noncoherent jamming so that the greater losses can be overcome (e.g., by using high gain AESA beams). The use of EA high gain (barrow) beams is driving the need for improved direction finding accuracy in order to point the transmitter beam in the correct threat direction.

39.8 Pulse Doppler Radar

The output of a pulse-Doppler radar processor has the form of a "range-Doppler map" as shown in Figure 39-8. The frequency cells represent the output of a bank of narrow filters; usually implemented in software as a frequency channelization via a fast Fourier transform (FFT).

The range cells record the time after pulse transmission for each received pulse. This, of course, shows the range to each target from which an echo is received.

This range-Doppler map allows the pulse-Doppler radar to counter jamming in several ways. For example:

- Pulse Doppler radar can detect separating targets. This allows it to detect both valid target echoes as well as artificially delayed pulses from a range-gate pull-off jammer. The pulse Doppler radar computes the Doppler shift, and by examining this for each of the separating targets, the radar can determine which are valid targets and which can be rejected as false (i.e., jamming generated) targets.
- Since the transmitted signals of pulse Doppler radar are coherent, echoes from a target will generally fall into a single frequency channel. If a jamming signal consists of noncoherent noise, it will likely spread into multiple frequency channels and hence is diluted. In addition by detecting an increase in its noise floor, the radar system can know it is being jammed.
- The scintillation of chaff also causes radar echo returns to spread in frequency. A pulse Doppler radar can again determine this from the energy increase in multiple channels and thus detecting the presence of chaff. This enables the rejection of a chaff cloud as a false target.
- If a decoy target is separating from a true target aircraft, the pulse Doppler radar can determine the aerodynamic slowing of the decoy (unless it is frequency modulated to mimic the Doppler shift from the velocity of the target aircraft). This also applies to a chaff burst deployed from the aircraft.

39.9 Leading Edge Tracking

Range gate pull-off (RGPO) jamming involves sequentially delaying received pulses and re-broadcasting them with increased power. The jammer signal pulse in Figure 39-8 moves out to the right (i.e., increasing in delay) for each subsequent pulse, causing the radar to conclude that the range to the target is increasing. This causes the radar to lose range track. If the radar tracks on the leading edges of the target echo pulses rather on the energy of the full pulses (also in Fig. 39-9), the radar will not see the delayed jamming pulse because it is initially delayed by the latency in the jammer's delay circuitry. Thus, the radar continues to track on the real target echoes.

By changing to range gate pull-in (RGPI) jamming mode, the jammer creates false pulses that lead the true target echo return by an increasing amount, again causing the radar to lose range track.

Another attribute of leading edge tracking is that it can counter terrain bounce jamming as shown in Figure 39-10. The greater length of the terrain bounce signal path allows the radar to reject it if leading edge tracking is used.

39.10 Dicke Fix

In AGC jamming, a series of low duty cycle short pulses are transmitted. These "capture" the radar's automatic gain control (AGC) causing the signal sent to the radar's processor to reduce.

How Ground-Based Radars Counter Jamming

- **Increased ERP:** Use higher antenna gain and/or higher transmitted power.
- **Vertical Triangulation:** Angle track on jamming; compute range on basis of elevation angle, estimated target altitude, and earth curvature charts.
- **Multiple Radar Triangulation:** Simultaneously track jamming in angle with one or more widely separated radars; compute range on basis of measured angles and radars' known locations.
- **Second Radar Assist:** Track jamming in angle with main radar; briefly transmit on another frequency with a co-located second radar to determine range of target in noise strobe

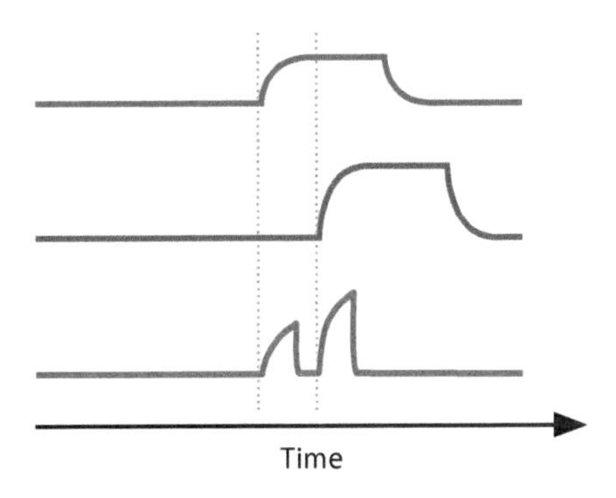

Figure 39-9. When a radar system uses leading edge tracking, it will ignore a delayed version of the true target echo.

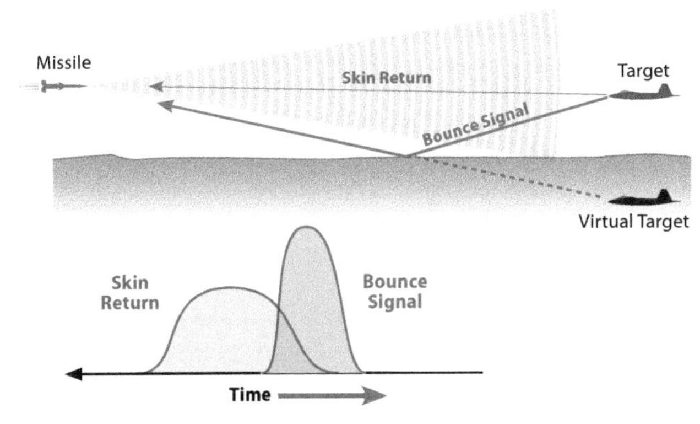

Figure 39-10. *Countering Terrain Bounce:* Because of the greater distance the bounce signal travels, it arrives at the radar a fraction of a pulse width behind the leading edge of the target echo return. As a result deception can be avoided by leading-edge tracking.

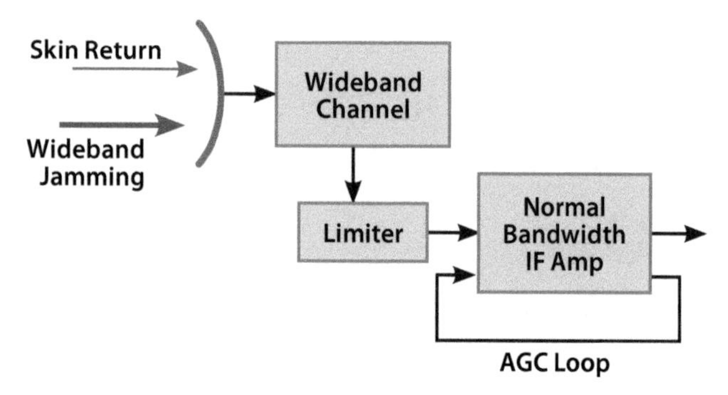

Figure 39-11. The "Dicke-fix" EP technique has a wide-band channel followed by a limiter that reduces the power of the wide-band signals. This allows the radar receiver's AGC to respond to normal bandwidth signals.

This continues until the target echoes cannot be observed. These short jamming pulses also occupy a wide bandwidth.

The "Dicke fix" is a receiver design in which received pulses are passed into a wideband amplifier as shown in Figure 39-11. The output of this amplifier is limited, clipping the strong jamming pulses. The signal stream is then passed to a decreased bandwidth channel matched to the radar signal. The AGC function is performed in this channel and this prevents the strong, narrow jamming pulses from capturing the AGC.

39.11 Burn-Through Modes

The J/S achieved by a jammer decreases as the range from a radar to its target decreases.

The burn-through range is the radar to target range at which the radar can re-acquire the target in the presence of jamming.

Radar burn-through modes are designed to increase the burn-through range. One approach is to increase the transmit power of the radar system. Typically this means eliminating the power reduction associated with normal emission control practice in which radiated power is reduced to the minimum level necessary for target echoes of sufficient strength for reliable detection.

Another approach is to increase the radar's duty cycle. Radar performance is a function of *energy* returned in the target echo pulses. Increasing the duty cycle increase target return energy and will thus increase the burn-through range against a jammer.

39.12 Frequency Agility

Frequency agility is a technique used in some radar systems to improve target detection whereby the radar changes the transmitted frequency on a pulse-by-pulse or burst-to-burst basis, as shown in Figure 39-12. Frequency agility also reduces the effectiveness of a jammer in multiple ways. If the frequency is changed in a pseudo-random pattern from pulse to pulse, the jammer never knows what the frequency of the next pulse will be (but the radar does). As a consequence, all potential frequencies must be jammed. This requires either that the jamming power be spread over the full frequency range or that each of the radar's frequencies be individually jammed. Since the radar only uses one frequency for each pulse, the effective J/S is reduced either by the number of frequencies or by the frequency range divided by the radar's bandwidth.

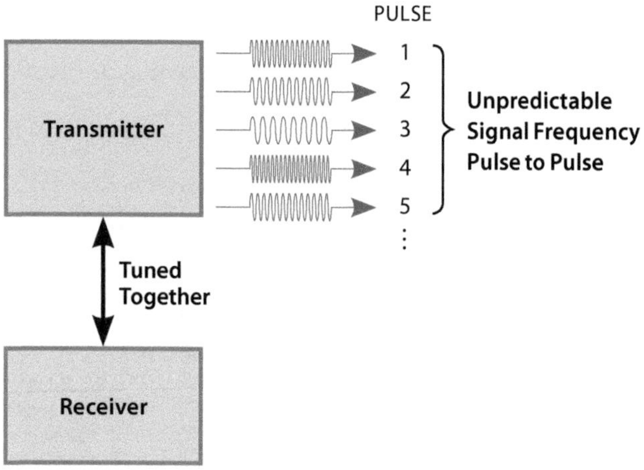

Figure 39-12. If a radar system employs frequency agility, the jamming power must be spread over a wide frequency range, reducing the effective J/S.

A second difficulty is caused for a jammer. This is because cover pulses or deceptive jamming waveforms that require anticipating future pulses will not know the frequency at which those pulses must be transmitted.

39.13 PRF Jitter

When a radar system has a pseudo-random pulse repetition interval, it is said to have a "jittered" PRF (Fig. 39-13). This means that a jammer cannot anticipate the time of arrival of the next pulse. This prevents the use of range gate pull in jamming and if cover pulses are generated, they must be extended to cover the full range over which the pulses are jittered.

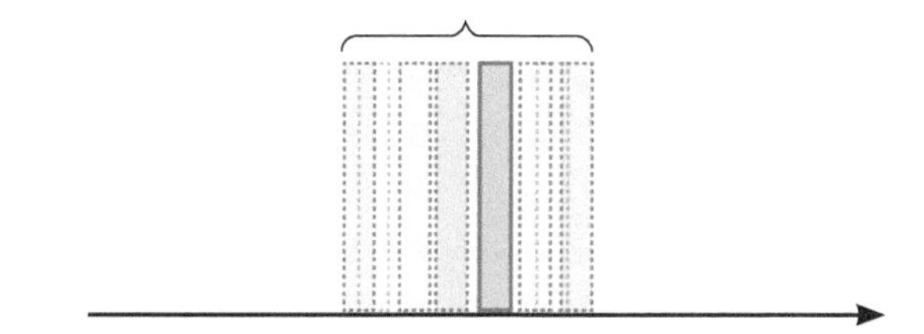

Figure 39-13. PRF jitter causes pseudo random shift in the pulse transmission time and prevents the prediction of the timing of subsequent pulses.

39.14 Home-on-Jam Modes

Many modern missiles have an operating mode that can be employed if a radar processor detects that it is being jammed. Pulse Doppler radars can process echoes to allow them to detect most kinds of jamming.

To enable a home-on-jam mode, the missile is equipped with homing guidance that can guide the missile to a jammer. As shown in Figure 39-14, the missile can be fired in an elevated trajectory that allows a significantly longer range than it would be the case if it were guided by a radar system.

Home-on-jam modes can make the use of self-protection jammers inappropriate, and they can also allow stand-off jamming aircraft to be attacked at great range.

Ground based radars can also have a "track-on-jam" capability in which the direction of a received jamming signal is determined, as shown in Figure 39-15. When this is done at multiple, cooperating radar sites, it can be used to triangulate the location of a jamming aircraft.

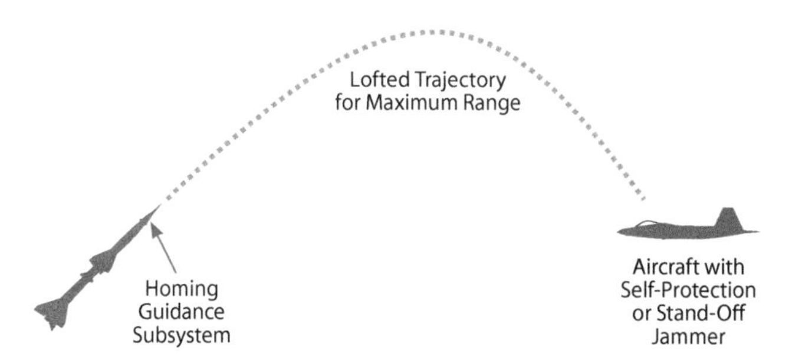

Figure 39-14. Missiles with home-on-jam capability can be commanded to guide themselves to an aircraft with an active jammer.

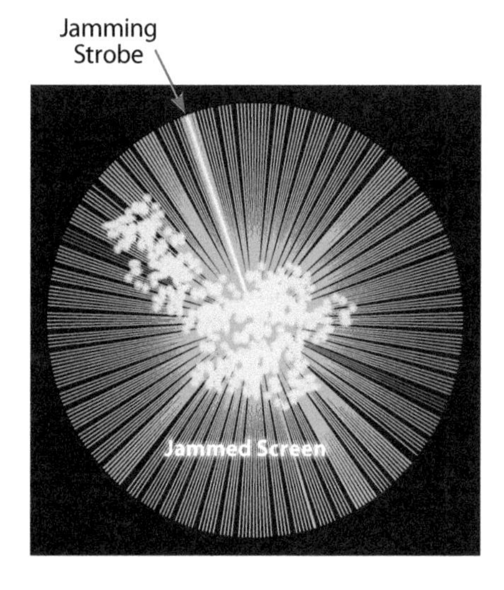

Figure 39-15. In angle-on-jamming, as the radar beam scans across a jamming aircraft in search, the jamming produces a bright line (strobe) on the radar display n the jammers direction. The radar can home-on-jam against this signal.

39.15 Adaptive Arrays

An adaptive phased array radar applies phase shifts to individual antennas in its array. These phase shifts can add for one angle to form a single, electronically steered beam in the direction of a target. They can alternately add in multiple directions to form beams toward multiple targets. In addition to forming beams toward targets, the phase shifts can be adjusted so that signals from some direction (or directions) will cause signals from that direction cancel each other to create a null in the direction of a jammer to reduce the J/S it can achieve.

If a jamming signal is received by multiple radars, their adaptive arrays can determine the direction of arrival of the jamming signal at the individual radars. This allows an integrated air defense system to determine the location of the jammer at great range by triangulation.

39.16 Adaptive Waveforms

Electronic support (ES) subsystems analyze received waveforms from radars in order to determine the type of radar illuminating the protected platform. This, in turn, allows the aircrew and automatic countermeasures subsystems to know the type of weapon with which it is under attack so that proper countermeasures can be employed. By altering the waveform of a radar signal, it is possible to cause an ES system to misidentify the illuminating radar. The altered waveform can imitate another type of radar (for example a "friendly" rather than hostile radar). will not be selected.

39.17 Processing EP

Scan-to-scan analysis can identify false tracks that are generated in response to jamming. These false tracks can then be eliminated from radar displays and other outputs. This allows more effective handling of valid target tracks.

39.18 Summary

Electronic protection techniques reduce the J/S achievable by a jammer. They also can increase the burn-through range of a radar system.

Ultra-low sidelobes make it more difficult to detect or jam a radar system. Sidelobe cancelers and blankers reduce the effectiveness of side lobe jamming with narrow bandwidth and pulsed modulations respectively. Anti-cross pol EP exploits antennas with weak or no cross-polarized "Condon" lobes.

Monopulse radar systems get angle estimate target parameters from every echo received. Therefore, inverse gain jamming does not reduce their angle tracking ability. In fact, any self-protection jammer that broadcasts strong signals from the target enhances their angle tracking.

Pulse compression reduces the J/S if the jamming signals do not have pulse compression modulation. The J/S is reduced by an amount determined by the time-bandwidth product.

Pulse Doppler radar processing enables detection of separating pulses, jamming that is wider than the radar receiver bandwidth, and chaff. This allows the radar to ignore most non-coherent jamming and chaff.

An anti-AGC jamming technique, called the "Dicke fix", prevents a jammer from capturing the radar's AGC by limiting wideband signals before the AGC is applied.

Increasing a radar system's ERP or duty cycle increases the burn-through range at which it can reacquire a target in the presence of jamming.

Frequency agility allows a radar system to tune to different frequencies in a pseudo-random pattern. This prevents a jammer from predicting the frequency of subsequent pulses, resulting in it having to cover a greater frequency range.

PRF jitter prevents a jammer from predicting the time of subsequent pulses. This prevents the use of range-gate pull-in and causes cover pulses to use an increased duty cycle.

A missile system with home-on-jam capability can command missiles to home on aircraft carrying active self-protection or stand-off jammers. In home on jam mode, a missile can be launched into an elevated trajectory that will carry in well beyond the range of missiles guided from the ground.

Further Reading

F. Neri, *Introduction to Electronic Defense*, 2nd ed., Chapter 6, SciTech-IET, 2007.

Test your understanding

1. Name two advantages of ultra-low sidelobes.
2. What kinds of modulations are vulnerable to side lobe cancelers and sidelobe blankers?
3. What is the impact on J/S of pulse compression in radars?
4. What two parameters do Pulse Doppler radars collect for every received pulse?
5. What is anti-AGC jamming EP commonly called?
6. Why can a missile's lethal range increase if it has home-on-jam capability?

Douglas EB-66 (1966)

The EB-66 was the electronic warfare version of the USAF EB-66, modified from the Navy A-3 Sky Warrior. Its modifications from the A-3 included removal of the folding wings, addition of ejection seats, and Allison J71 engines. With a crew of seven, it had four radar receivers and nine (later 20) jammers. It performed both ES and EA missions int he Vietnam war.

40

Decoys

40.1 Introduction

Decoys have three basic missions: to saturate a radar or integrated air defense system with targets; to seduce a weapon away from its intended target; and to cause an enemy to take some desired action.

When the mission is of a saturation type, as shown in Figure 40-1, the decoy needs to look, radiate, and reflect sufficiently like a real target. This will cause the enemy radar to spend significant resources and time to differentiate between real targets and decoys. Ideally, the enemy is unable to distinguish decoys and must deal with decoys as though they were real targets.

Figure 40-1. Decoys can saturate an air defense system because of the time and processing power required to differentiate true targets form the decoys.

When the mission is of a seduction type, the decoy must look as much like a true target as possible. One way of accomplishing this is for the decoy signal to be larger than the echo from a true target. In this case, the illuminating radar will track the decoy, leading the radar away from the true target as shown in Figure 40-2.

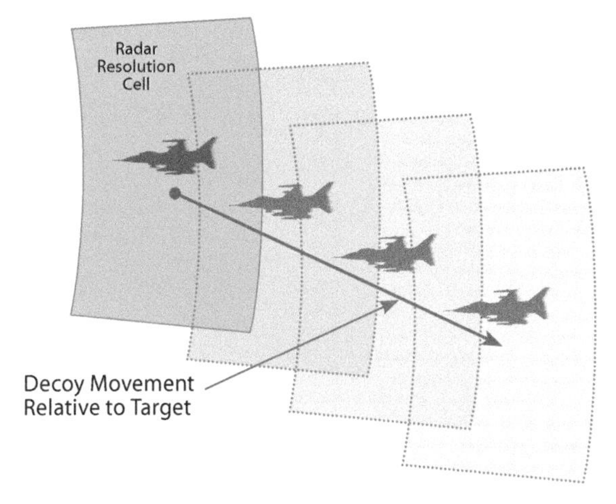

Figure 40-2. If a decoy has a larger RCS than a true target and both are within the resolution cell of the radar, the radar will track the decoy.

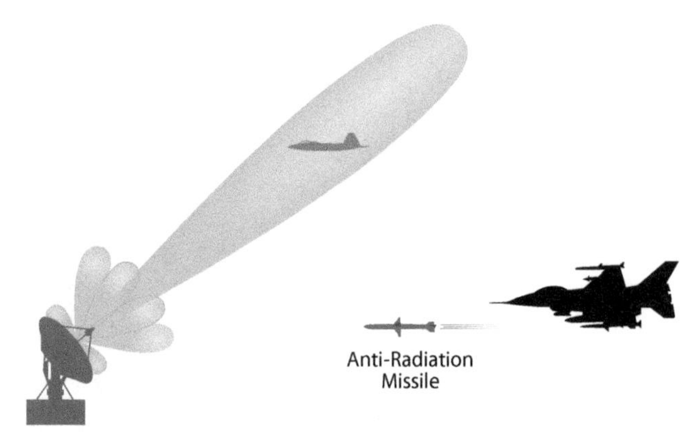

Figure 40-3. If a decoy has the RCS characteristics of an attack aircraft, it can cause an acquisition radar to identify it as a threat and activate a tracking radar. The tracking radar is then vulnerable to attack by antiradiation missiles.

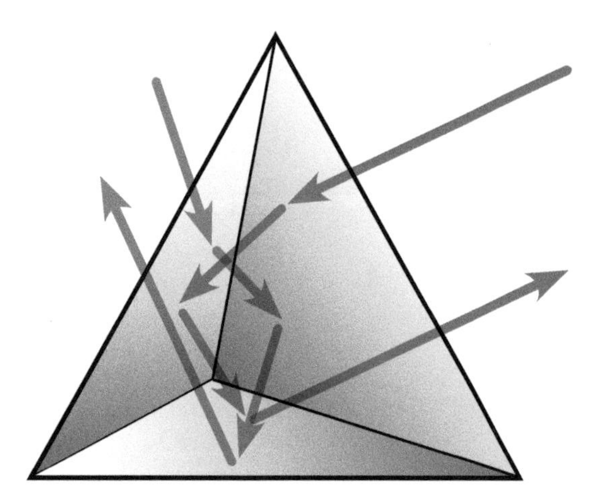

Figure 40-4. A passive decoy, such as a corner reflector, has a large RCS over a wide range of angles.

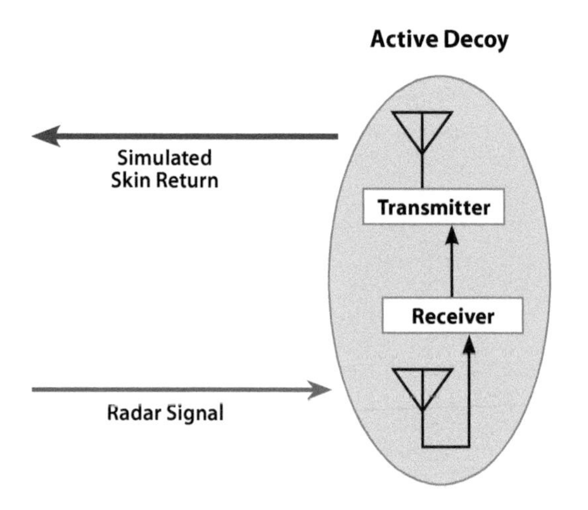

Figure 40-5. An active decoy develops a larger RCS than its physical size by rebroadcasting received signals with significant gain.

Unmanned aerial vehicles (UAVs) with decoy payloads can be used to fly into enemy airspace. The decoy payloads must be realistic enough such that when enemy acquisition radars see them, the UAVs will be processed as real aircraft targets and the enemy will employ tracking radars. When the tracking radars are active, they can be detected and located (and then destroyed). This is illustrated in Figure 40-3.

40.2 Active and Passive Decoys

Passive decoys have the materials, size and shape to create a large radar cross section (RCS). One favored decoy shape is the corner reflector, as shown in Figure 40-4. A corner reflector is a retro-reflector that sends signals back in the direction from which they came (i.e., toward the transmitter) and does this over a wide angular area with significant gain. Passive decoy design chooses the material, surface quality and shape to create the largest possible RCS. While passive decoys can have a large RCS, they may be discriminated in Doppler and possibly in range extent, hence negating their effectiveness.

An active decoy creates the effect of an RCS much larger than its physical size by use of an internal amplification circuit as shown in Figure 40-5. The receiver receives the signal from a radar system and rebroadcasts it as a jamming transmitter at a desired frequency, modulation and amplification to emulate the return of a large reflector. The transmitted signal frequency can be offset from that of the received signal to simulate Doppler shift, if desired.

If the active decoy is a (straight-through) repeater it will have a fixed throughput gain. If it is a primed oscillator decoy, it will rebroadcast the received signal at a fixed power. In either case, the throughput gain is the transmitted signal strength divided by the received signal strength (including antenna gains).

A straight-through repeater jammer can, if necessary, rebroadcast signals from multiple radars, while a primed oscillator can repeat only one signal.

The radar cross section simulated by the active decoy is determined from

$$\sigma \approx \frac{G\lambda^2}{4\pi}$$

where σ is the generated radar cross section in m^2, λ is the radar transmission wavelength in m, and G is the throughput gain of the receiving antenna, the transmitting antenna and any amplifiers.

This same formula in dB form is

$$\sigma = 38.55 + G - 20\log_{10}(f)$$

where σ is the generated radar cross section in dBsm; 38.55 is a constant to allow frequency (rather than wavelength) to be input (in MHz) and is often rounded to 39; G is the throughput gain in dB; and f is the radar operating frequency in MHz.

The gain of the decoy is the difference between the received signal and the transmitted signal. If the decoy has constant gain, the RCS generated is constant as the decoy to radar-range diminishes, i.e., until the amplifier in the decoy saturates. At this range, the generated RCS will roll off as a function of the square of the diminishing distance.

If the decoy is a primed oscillator, its transmit power is constant. Therefore, the gain diminishes with the square of the diminishing range from the radar to the decoy. These two scenarios are illustrated in Figure 40-6. The numerical values in this figure are for a specific application, but the nature of the variation with range to the target is general. Note that the horizontal scale in this figure has the range to the target diminishing to the right. This follows the nature of an engagement in which a weapon is approaching its target.

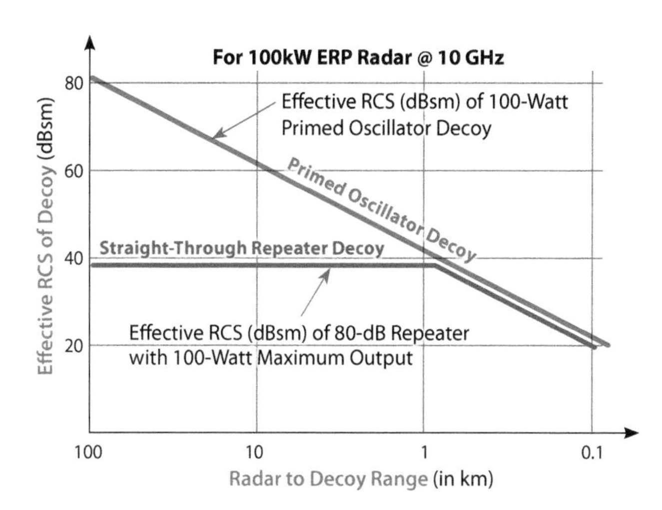

Figure 40-6. A "straight-through repeater" produces a constant RCS until its amplifier saturates. The RCS produced by a primed oscillator decreases with reduced range through the whole engagement.

40.3 Decoy Deployment

Decoys are deployed in many ways. The most common three when protecting aircraft are (1) expendable decoys, (2) towed decoys, and (3) independently maneuvering decoys.

In each case, it is desirable for the decoy to start in or enter the radar's resolution cell. As shown in Figure 40-7, the radar cannot determine that there are two targets in a single resolution cell. The radar sees only one target, which it locates as being between the aircraft and the decoy. This false target is perceived by the radar to be proportionally closer to the larger RCS of the two.

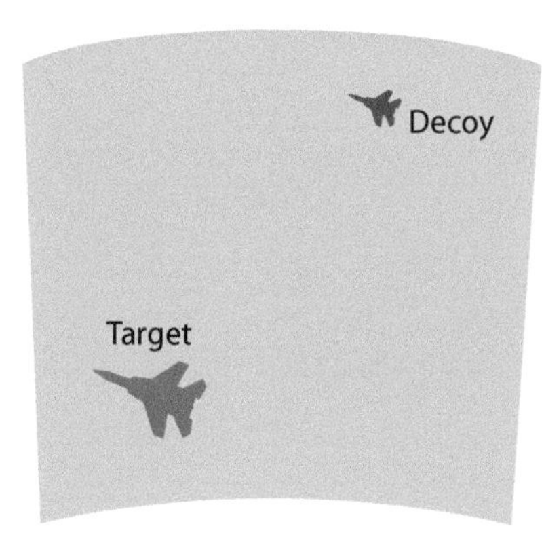

Figure 40-7. If a target and a decoy are both in the same range resolution cell, the radar will detect a single target located between the two and proportionally closer to the larger RCS.

When a decoy is within the resolution cell, it can capture a tracking radar's tracking gate. If it is outside the resolution cell, it can be acquired by a radar operating in search mode, but cannot break the lock of a tracking radar.

Expendable Decoys. The expendable decoy has the same basic block diagram as an active decoy (Fig. 40-5). However, it has a short operational lifetime so it can use a thermal battery to provide high power for its duration. It can be either a straight-through repeater or a primed oscillator decoy. Figure 40-8 shows the POET decoy, which is an example of a primed oscillator decoy. It is activated within the radar's resolution cell and captures the radar's tracking to carry the resolution cell away from the target as the decoy separates from the aircraft.

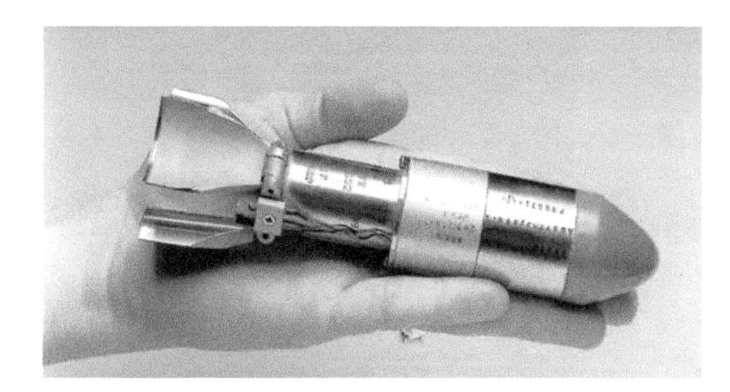

Figure 40-8. This active expendable decoy is used by the US Navy and the RAF.

The GEN-X decoy is a small, one-shot, expendable RF threat countermeasure that receives a signal from a recognized threat, such as airborne or land-based semi-active radar guided missile, and then transmits RF power to counter that threat. The GEN-X decoy can be launched from the AN/ALE-39 or AN/ALE-47 countermeasure dispensers using a CCU-63/B or CCU-136/A impulse cartridge. GEN-X has been designed and cleared

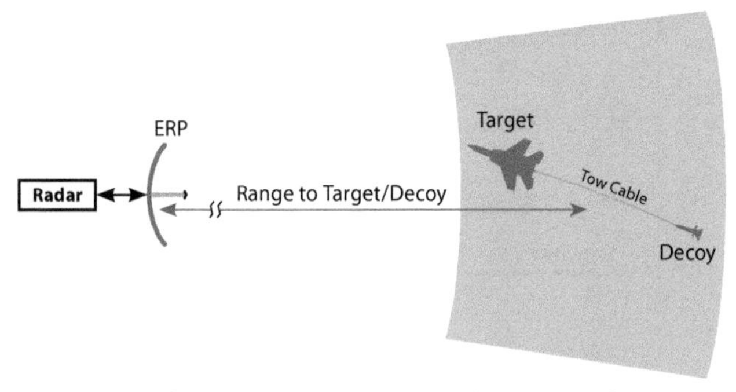

Figure 40-9. If a towed decoy is initially in the same resolution cell as the target aircraft, the radar will see only one target. As the range to the radar reduces, the decoy's greater RCS will capture the resolution cell away from the aircraft.

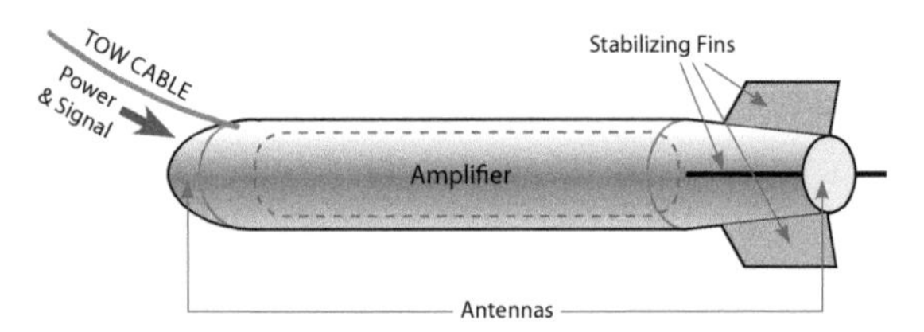

Figure 40-10. The towed decoy is composed of only an amplifier and antennas. Prime power and jamming signals are brought down the tow cable.

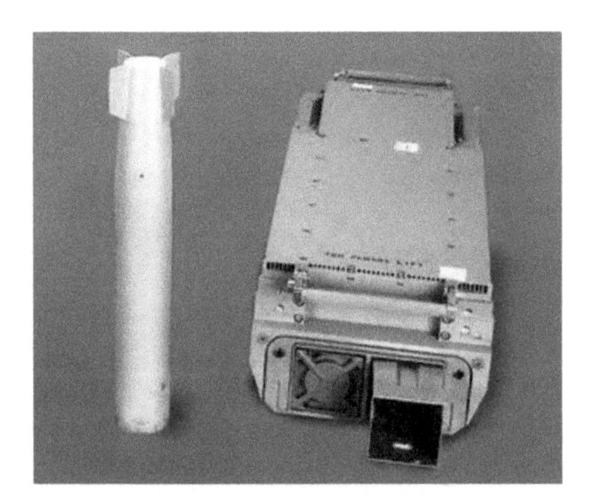

Figure 40-11. A towed decoy with its launcher as used in the F-16. The decoy is packaged in a sealed canister that also contains the payout reel.

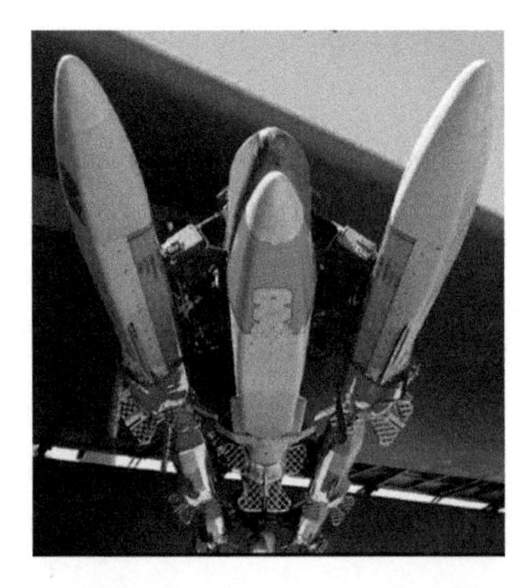

Figure 40-12. A miniature air launched decoy (MALD) is launched from an aircraft to seduce radar systems away from their intended target.

for flight on US Navy tactical aircraft. It is also used by the Royal Air Force. RT-1489 GEN-X measures 14.7 cm long with a 3.4 cm diameter and weighs 0.5 kg. It is activated within the radar's resolution cell and captures the radar's tracking to carry the resolution cell away from the target as the decoy separates from the aircraft.

Towed Decoys. As shown in Figure 40-9, a towed decoy is towed on a cable behind an aircraft. The cable is long enough that if a missile hits the decoy, the aircraft will be safe. If a towed decoy is initially in the same resolution cell as the target aircraft, the radar will see only one target. As the range to the radar reduces, the decoys greater RCS will capture the resolution cell away from the aircraft. Radar signals are received in the aircraft and jamming signals are passed over a fiber-optic path in the tow cable to the decoy for rebroadcast.

When a radar system uses pulse compression, the range dimension of the resolution cell will be reduced. Since the decoy needs to be in the resolution cell with the aircraft to capture the radar's tracking circuits, acquisition can be challenging.

As shown in Figure 40-10, the towed decoy has only an amplifier and antennas. Placement of the amplifier in the decoy (after the tow cable) optimizes the RCS of the decoy. Prime power for the amplifier and the modulated signals to be rebroadcast are sent down the tow cable from the aircraft. The signal is carried on the fiber optic path and the power is carried on a wire.

Towed decoys, like the one shown in Figure 40-11, are deployed from the aircraft and turn on when they reach the ends of their tow cables. Modulation, power and built-in-test signals are carried on the towline. To prevent self-oscillation between the transmitter and receiver, items such as gain control circuitry and special antenna technology are used. When they are no longer required in the mission, the cables are cut to discard the decoys. Later models have retractable cables so the decoys can be reeled up for redeployment. These models also allow selection of the length of tow cable for mission optimization. Expendable towed decoy have the same basic attributes as the standard expendable decoys described above.

Independently Maneuvering Decoy. Independently maneuvering decoys can be carried as UAV payloads or can be launched from aircraft. In either case, the decoy payload needs to simulate the radar cross-section, thermal and other relevant profiles of the aircraft it represents. Figure 40 12 shows the miniature air launched decoy (MALD), an example of an operational independently maneuvering decoy. If launched in a terminal engagement with an enemy missile, it looks more like the protected aircraft than the aircraft does, so that it will attract the missile to itself, protecting the aircraft.

Multiple MALDs can be launched at long range accurately duplicating the combat flight profiles and signatures of U.S.

and allied aircraft. MALDs appear to be aircraft as seen by multiple hostile radars in an Integrated Air Defense System. The MALD-J version has the ability to perform the decoy role and be used as a SIJ. MALD and MALD-J are long endurance vehicles that weigh less than 130 kg and have a range of approximately 900 km.

40.4 Chaff as a Decoy

Since chaff creates a false target with an RCS similar to true targets, it can be deployed from an aircraft. Chaff can be used as a decoy to draw hostile radar systems away from their intended targets (Chapter 36). Chaff can be deployed from multi-round dispensers on the aircraft or fired forward in rockets as shown in Figure 40-13. One characteristic of aircraft chaff is that it slows quickly when it is deployed and thus will present a Doppler shifted radar return quite different from that of the protected aircraft. A solution to this problem is to illuminate the chaff with a jammer. The jammer can provide a correcting Doppler shift in some engagement scenarios.

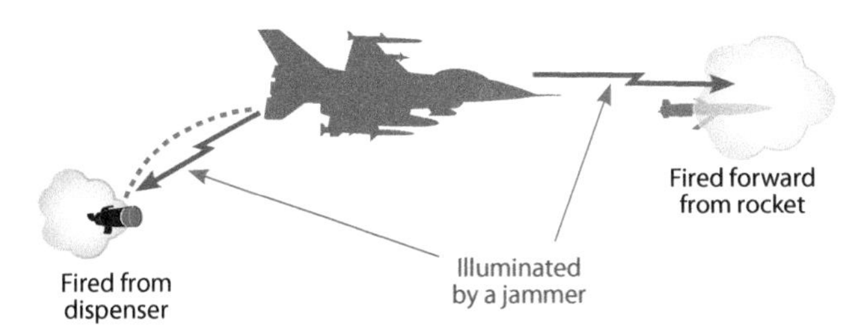

Figure 40-13. Chaff can act as a decoy to break the lock a radar system has on a target. It can be illuminated by a jammer to provide the correct Doppler frequency.

40.5 Summary

Decoys can fulfill three types of missions: They can saturate an enemy radar or air defense system; seduce an enemy radar away from its intended target; or cause the enemy to take some desired action (e.g., activating tracking radars so they can be located and attacked).

A decoy can be passive, developing an RCS from its size, shape, and surface characteristics. A decoy can be active; creating an RCS by amplifying and rebroadcasting received radar signals.

Active decoys can be either straight-through repeaters with fixed throughput gain or can be primed oscillators that broadcast an amplified copy of a received signal. In either case, the throughput gain (which determines the RCS) is the ratio of the power of the rebroadcast signal to the received signal.

A straight-through repeater produces a constant RCS until the radar is close enough to saturate its receiver; the RCS produced by a primed oscillator decreases through the whole engagement as the range to the radar reduces.

Aircraft protection decoys can be expendable, towed, or independently maneuverable.

Chaff can be deployed to move a radar away from tracking an aircraft, but its Doppler shift frequency will not match the aircraft. This can be avoided by illuminating the chaff with a jammer.

Further Reading

See the references provided in the previous EW chapters.

Test your understanding

1. How do active and passive decoys generate RCS?
2. What are the three decoy missions?
3. How does the RCS developed by a straight-through repeater vary with range to the radar?
4. How does the RCS developed by a primed oscillator decoy vary with range to the radar?
5. How many radar systems can a primed oscillator decoy operate against?
6. How can the Doppler shift frequency of a chaff cloud be made to match the Doppler frequency of the aircraft that deployed it?

Lockheed Martin F-35 Lightning II

The F-35 Lightning II is a family of single-seat, single-engine, fifth generation multi-role fighters under development to perform ground attack, reconnaissance, and air defense missions with stealth capability. The F-35 has three main models; the F-35A is a conventional take-off and landing variant, the F-35B is a short take-off and vertical-landing variant, and the F-35C is a carrier-based variant. Because stealth is a central part of its combat effectiveness, LPI radar is expected to be a requirement of all variants.

41

Low Probability of Intercept (LPI)

F-16 AN/APG-80 Radar

In this chapter, the term *low probability of intercept* (LPI) is used to indicate that there is a low probability that an intercept receiver will detect the emissions from a radar system. Other terms with essentially the same meaning are *low probability of detection* (LPD) and *low probability of exploitation* (LPE). The objective is to make a radar system whose transmitted signal is below the level of threshold of detection of an opposing electronic support (ES) system while, at the same time still being able to detect targets at useful ranges: "to see without being seen." In the case of LPE, the signal may be strong enough to be detectable but it is too weak to identify. Consider the particular case of a warning receiver carried by the radar's target where the range from the radar to the intercept receiver is the same as the range from the radar to the target. In addition, both the intercept receiver and the target are in the same part of the radar's antenna beam. The goal of the LPI radar is to detect the target at a range greater than the range at which the receiver can detect the radar's signal. Further discussion of the LPI situation may be found in Further Reading at the end of this chapter.

For the air battle of the future, LPI is essential. In conventional aircraft the most important need for LPI is to avoid electronic countermeasures. In low observable aircraft, LPI additionally enhances the element of surprise and denies the enemy use of radar intercept cueing of its fighters. In aircraft of both types, LPI may reduce the chance of a successful attack by an antiradiation missile.

In Chapter 37, various types of intercept receivers were described along with techniques for locating emitters, for example, angle of arrival (AOA) with triangulation, time difference of arrival (TDOA), and frequency difference of arrival (FDOA). In this chapter the various strategies used to defeat such receivers are considered. Next, specific design features that may be incorporated in a radar system to ensure a low probability of intercept will be presented. Finally, there is a brief assessment of LPI cost and a consideration of possible future trends in LPI design. Bear in mind that a discussion of LPI radar must include specifying the characteristics of the intercept receivers in use by the opposing force and also the scenario. A radar system may be LPI against one type of intercept receiver and *not* LPI against another type of intercept receiver.

41.1 Operational Strategies

The most effective LPI strategy is not to radiate at all. This strategy may be approached by limiting radar "on" time and operating with power no higher than absolutely necessary to achieve mission goals.

On stealthy interdiction missions, wherever possible the aircrew should use collateral intelligence and reconnaissance information. Through careful mission planning, they may be able to conduct an entire mission with only a few minutes, or even seconds, of radar operation.

In air-to-air combat situations, where continuous situation awareness is essential, the aircrew should use onboard passive sensors, such as an RWR or ESM system, an IR search-track set, or a forward-looking IR set. When a potentially hostile aircraft is detected, the radar may be used to measure range and possibly precise angle, which the passive sensors may not have provided. But it should be operated only in short bursts, and then only to search the narrow sector in which the passive sensors indicate the target to be.

41.2 Design Strategies

Since the range at which a radar system can detect a given target varies as the one-fourth power of the emitted signal power, whereas the range at which an intercept receiver can detect the radar varies only as the square root of the emitted power, the interceptor has a huge advantage over the radar. However, since signals from a multitude of other radars and electronic systems are inevitably present in a tactical environment, the radar designer has several opportunities to overcome this advantage.

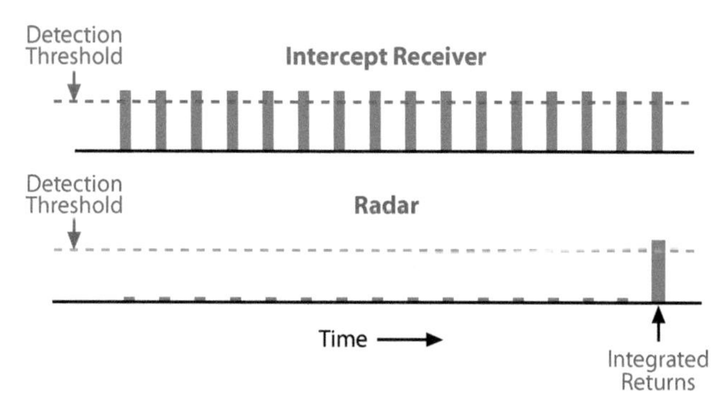

Figure 41-1. Because most of today's intercept receivers were designed without regard to an LPI radar threat, they generally detect individual pulses. Therefore its sensitivity is determined by the radar's peak power. Because radar receivers can coherently integrate the returns from the target over several pulses, radar performance depends on average power. Coherent integration time can be traded for reduced peak power.

Integration Considerations for Reduced Peak Power (Essentially Pulse Compression). For a signal to be usefully detected by an intercept receiver, its source must be identified on the basis of such parameters as angle of arrival, radio frequency, PRF (obtained from times of arrival), and pulse width. To satisfy this requirement, present intercept receivers generally detect individual pulses. Consequently, it can employ little or no signal integration; it is sensitive primarily to peak emitted power. The radar, on the other hand, is subject to no such requirement. By coherently integrating the echoes it receives over long periods, the peak power needed to detect a target can be greatly reduced, thereby reducing the detectability of the radar's signals (Fig. 41-1). Future intercept receivers designed with LPI radar in mind need to make use of integration to partially make up for this aspect of LPI radar design.

Bandwidth Considerations for Reduced Peak Power. This is the idea behind "spread spectrum" communications systems. In that case, a "spreading code" known at both the transmitter and the receiver is used to spread the power over an arbitrarily large bandwidth. An interceptor without knowledge of the spreading code may find that the spectral density of the communications signal is below that of thermal noise and not notice the presence of the signal at all. The same considerations apply to radar, but the dependence of range resolution on signal bandwidth must also be considered. The wider the coherently processed bandwidth, the finer the range resolution:

$$\Delta R = \frac{c}{2B}$$

where c is the speed of propagation, and B is the bandwidth of the signal during the coherent processing interval (also called its instantaneous bandwidth). This means that a radar system cannot use an arbitrarily wide bandwidth in the same way that communications systems do.

For example, to distinguish between two fighters in tight formation 30 m apart in range, B must be about 5 MHz. Figure 41-2 shows the relationship of range resolution to bandwidth.

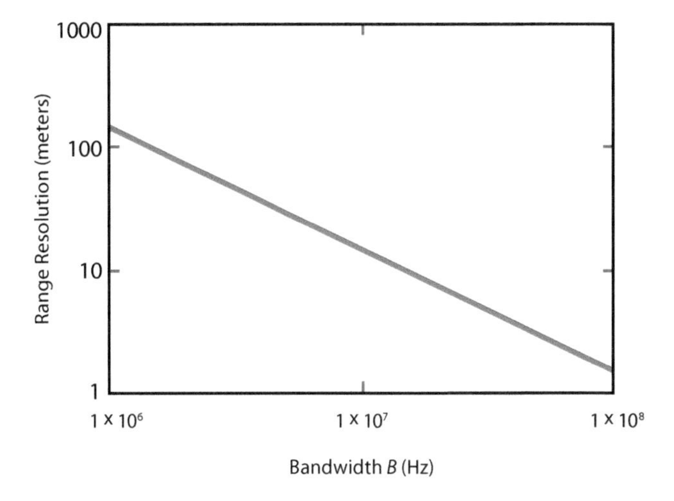

Figure 41-2. Range resolution related to radar coherent bandwidth.

Assuming a value of $B = 1$ GHz, the radar has a range resolution of 15 cm. This means that the target echoes are resolvable in 15 cm range increments called range cells. The echoes from a target extending 75 m in range would be spread across 500 range cells.

This spreading of the echoes across a multiplicity of range cells reduces the apparent radar cross-section (and thus reduces the SNR available) in a single range cell. For this and other reasons, radar designs use range resolution appropriate for their function. This often leads to choosing coherent bandwidths of 10 MHz or less (10 MHz corresponds to range resolution of 15 m). In this sense, today there is no such thing as a "spread spectrum" radar; what is transmitted is also received and the resulting range resolution is determined by the bandwidth. What this means for the LPI radar designers is that the coherent bandwidth of radar signals is likely to remain about the same as it is now, provided the radar performs the same task (Table 41-1).

Table 41-1. Instantaneous Bandwidth Required for Selected Applications

Range Resolution Required	Resolution (m)	Bandwidth (MHz)
1. Count aircraft in attack formation	30	5
	60	2.5
2. Detect missile separation at launch	15	10
3. Imaging of ships, vehicles & aircraft	0.5 - 1	150 - 300
4. High-resolution mapping	0.15	1000

Tactical radars (early warning, fire control, target acquisition, target track, airborne intercept) have bandwidth requirements typical of applications 1 and 2 in Table 41-1, that is, under 10 MHz. Some tactical radars and multifunction radars may have modes for imaging (application 3 in Table 41-1).

In modern-day intercept receivers, the instantaneous bandwidth of each of its channels is often chosen to pass the *shortest* pulses it can reasonably be expected to receive and measure their time and angle of arrival. On the other hand, radar receivers are designed with a bandwidth which "matches" the bandwidth of the transmitted signal. As intercept receivers adopt digital signal processing, it is likely that different channel bandwidths will be used in parallel to detect different types of threat signals. While the coherent bandwidth of radars is generally determined by its function, frequency agility (changing the carrier frequency from one coherent processing interval to the next) can be used to make the radar signal occupy a wider band and hence make the intercept receiver's task more difficult.

Antenna Gain Considerations for Reduced Peak Power. Against an RWR, the radar has the advantage of being able to employ a large directional antenna, which the RWR cannot. During transmission, of course, the high gain of this antenna benefits the RWR as much as it benefits the radar. But during reception, the radar antenna's large intercept area provides high gain, enabling the same detection sensitivity to be obtained with much lower peak power than if a small low gain antenna were used by the radar.

Against those intercept receivers which depend on sensing the radar antenna's sidelobe emissions. Besides having a high gain and large intercept area, the radar antenna has the advantage of a very large difference between mainlobe and sidelobe gains. This complicates detection for those intercepr receivers which depend on sensing the radar antenna's sidelobe emissions, i.e., EA systems receivers, ground-based DOA and ELINT systems and ARMs. All of these antenna gain characteristics can be traded for reduced peak power. While antenna size is generally limited by the dimensions of the aircraft, sidelobe reduction is less so. For LPI, the peak sidelobe gain should be down at least 55 dB, relative to the peak mainlobe gain (Fig. 41-3).

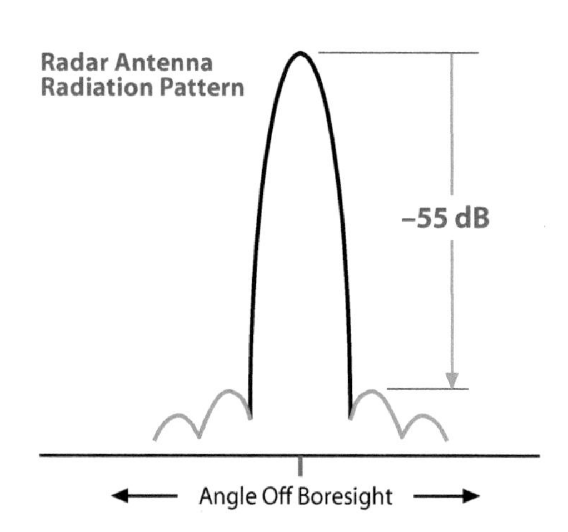

Figure 41-3. Since many intercept receivers must rely upon detecting the radar's side-lobe emissions, for LPI the peak sidelobe gain should be as low as possible, in the extremely low category and of the order of –55 dB.

Other Ways to Reduce Peak Power. Other features normally included in a radar to increase detection range that can correspondingly enable peak power to be reduced without reducing range include:

- High duty factor (but note that a CW radar system reaches the limit of 100%)
- Low receiver noise figure (but note that a similar noise figure may also be available for the intercept receiver).
- Low receive losses

Low transmit losses, it might be noted, are of no advantage for LPI. For unless the radar is operating at maximum range, the peak emitted power can be set to the desired level for LPI regardless of these losses.

41.3 Special LPI-Enhancing Design Features

Special features that may be used to further enhance LPI include power management, use of frequency agility (changing the carrier frequency from one coherent processing interval to the next), transmission via multiple antenna beams on different frequencies, randomizing waveform parameters, and mimicking the enemy's waveforms.

Power Management. The role of power management is to reduce the radar's peak radiated power to the absolute minimum needed to detect targets of interest at the *minimum* acceptable range, with *minimum* margin. As the radar's targets close to shorter range, the power management system must correspondingly reduce the emitted power

Superficially, it seems impossible for a radar system to avoid being detected by a target that the radar can detect. For the peak power that the radar must transmit to detect the target, P_{det}, is proportional to the fourth power of the target's range.

$$P_{\text{det}} = k_{\text{det}} R^4$$

where k_{det} is a constant of proportionality. Yet the peak power, P_{int}, which will enable an intercept receiver in the target to detect the radar is proportional only to the square of the target's range.

$$P_{\text{int}} = k_{\text{int}} R^2$$

where k_{int} is another constant of proportionality. The advantages of power management can best be appreciated by considering a simple example.

Avoiding Detection. By setting the radar's peak power just below a level corresponding to the range at which the two detection curves intersect (R_{dmax} in Figure 41-4) and progressively reducing it as the target closes to shorter ranges, the radar can avoid being detected by the intercept receiver. This is true when tracking a target but when searching out to a given range, the radar must radiate the power appropriate for that range

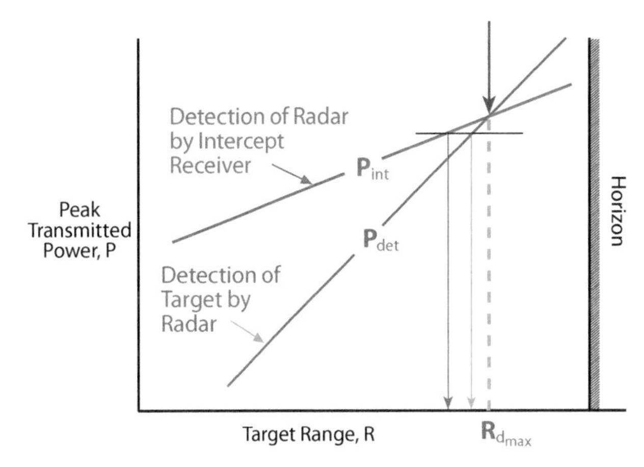

Figure 41-4. This graph depicts the radar transmit power versus target (and intercept receiver) range.

However, by increasing integration time, frequency agility, antenna gain, duty factor, and receiver sensitivity to reduce the peak-emitted power, the factor k_{det} can be made very much smaller than k_{int}. Figure 41-4 shows a plot of P versus R for detection of the target by the radar, together with a plot for detection of the radar by the intercept receiver, at a tactically useful range. The intercept receiver designer, when faced with the LPI radar threat, can try to improve the sensitivity of the intercept receiver in similar ways.

The range at which the two plots intersect is the *LPI design range.* Using the one-way range equation for signal interception and the two-way range equation for radar target detection, we can compute the *LPI design range.* For the RWR case (intercept receiver on the radar target), the power level available at each receiver is

$$\text{Power at the radar to detect target } P_{\text{det}} = \frac{P_T G_T G_R \lambda^2 \sigma}{(4\pi)^3 R^4}$$

Power Management

PROBLEM 1

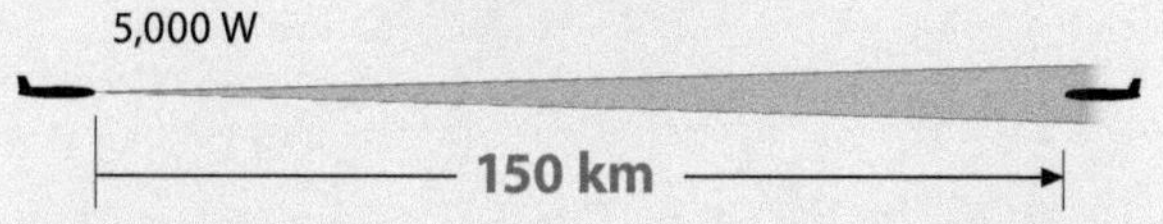

Conditions: A certain radar can detect a given target at range **R = 150 km** by emitting a peak power **P = 5,000 W.**

Question: How much power need the radar emit to detect the same target at **10 km?**

Solution: The required peak power varies as the fourth power of the desired detection range. Therefore, the required transmit power is reduced to 5000 (10/150) raised to the fourth power) or 0.099 W.

$$P_2 = P_1 \left(\frac{R_2}{R_1}\right)^4$$

$$P_2 = 5{,}000 \left(\frac{5}{150}\right)^4 = 0.099 \text{ W}$$

Power Management

PROBLEM 2

Conditions: When emitting a peak power of **5,000 W,** the radar of Problem 1 can be detected by a given intercept receiver at **550 km**

Question: At what range can the radar be detected by the same intercept receiver, when emitting a peak power of only **0.076 W**?

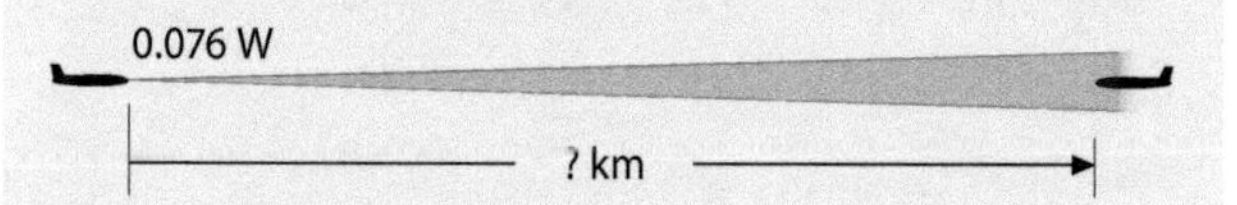

Solution: Since the signal travels only one way, the intercept range varies as the square root of the peak emitted power. Therefore,

$$R_2 = R_1 \left(\frac{P_2}{P_1}\right)^{0.5}$$

$$R_2 = 550 \left(\frac{0.076}{5{,}000}\right)^{0.5} = 2.1 \text{ km}$$

Power at the Intercept Receiver to detect the radar signal $P_{\text{int}} = \dfrac{P_T G_T G_{\text{int}} \lambda^2}{(4\pi)^2 R^2}$

If we solve each equation for the range and require that the intercept range be the same as the target detection range, the result for the LPI design range is

$$R = \left[\frac{P_{\text{int}}}{P_{\text{det}}} \frac{G_R \sigma}{4\pi G_{\text{int}}}\right]^{0.5}$$

If the power received is as small as possible for each receiver to perform its function, then the LPI design range then the sensitivity of each receiver can be inserted into the equation to estimate the LPI design range in free space. We can also find the ratio of the sensitivities needed to provide a given LPI range:

$$\frac{P_{\text{int}}}{P_{\text{det}}} = R^2 \frac{4\pi G_{\text{int}}}{G_R \sigma}$$

For example, suppose the LPI design range to detect a 10 square meter target is 10 km. Suppose the intercept receiver uses an omnidirectional antenna with unity gain (0 dBi) and the radar receive antenna has a gain of 1000 (30 dBi). Then, according to the equation above, the radar receiver must be more sensitive than the intercept receiver by a ratio of 125,636 or 51 dB.

Going back to Figure 41-4 the maximum LPI design range is the range at which the two curves cross. At that level of radar transmit power, the radar-range and EW interception range are equal. If the power is increased beyond that point, the EW detection range will exceed the range at which the radar detects its target, and the LPI property will be lost.

Consider the situations described in the blue boxes labeled problem 1 and problem 2. Suppose now that, when the radar was emitting full power, a given intercept system could detect it at 600 km. If the radar transmit power were reduced from 5000 Watts to 0.076 Watts, that same intercept system could detect the radar at only 2.4 km. This is because the ES detection range decreases in proportion to the square root of the ratio by which the power is reduced. The radar's target detection range is 10 km because the reduction in radar-range is reduced in proportion to the fourth root of the ratio by which the power is reduced. This illustrates a true LPI situation: the radar target detection range (10 km) is greater than the range at which the intercept receiver can detect the radar signal (2.4 km). To make up for this, the sensitivity of the intercept receiver needs to be improved by

$$10 \log_{10}\left(\frac{10}{2.4}\right) \text{ dB} = 13 \text{ dB}$$

In this equation the logarithm is multiplied by 20 because the required sensitivity improvement is proportional to the square of the ratio of the ranges.

Clearly, power management is essential for LPI in the case of a tracking radar. Note that until the radar-range is reduced to a relatively short range, the intercept receiver is able to receive the radar signal at range that is greater than the range at which the radar detects its target. LPI radar designs are usually short range systems. Also it is clear from the foregoing example that the power management system must be able to reduce the emitted power in small, precisely controlled steps over a very wide range—in this example, nearly 50 dB.

One point to bear in mind: the interception of the signal from a given radar depends upon its mode of operation and on the capabilities of the intercept receiving system. Both may vary within any one mission, as well as from mission to mission.

In searching a narrow sector for a designated target at a given range, for example, the peak power may be set so that the radar detects the target without being detected by the target's intercept receiver. Yet in conducting broad area surveillance with the same power setting, the radar may be detected by the intercept receiver of a target of the same type before it closes sufficiently to be detected by the radar. Detection range in this case is reduced because the radar's beam cannot dwell as long in the target's direction. Or a radar system might operate at a low enough peak power that its signals might be below the detection threshold of an RWR in a target aircraft. Yet, with that same power setting, the radar might be detected by a ground-based receiver with a large directional antenna and a more sensitive receiver.

Pulse Compression. A radar's power can be spread uniformly over a band of frequencies simply by transmitting short pulses. But, with the low peak power that is needed for LPI operation, this would result in such low average power that the radar could not detect targets at useful ranges.

A convenient solution to this dilemma is to transmit reasonably long duration pulses and to phase or frequency modulate the transmitter with pulse-compression coding.

The 3 dB bandwidth of the central spectral line of a pulsed signal is

$$BW_{3\,dB} = \frac{1}{\tau} \times \text{(pulse compression ratio)}$$

where τ is the uncompressed pulse width and the pulse compression ratio is the ratio of the duration of the transmitted pulse (that is, before compression) to the width of the pulse after compression. With 1 ms wide pulses and 2000-to-1 pulse compression coding, for example, a bandwidth of 2 MHz may be obtained. By selecting a pulse compression ratio of 2000, the emitted signal's duty factor can approach 100% with a PRF of 1000 Hz while maintaining range resolution of 75 m (which corresponds to the coherent bandwidth of 2 MHz using equation 41-1) needed to perform its functions. Suppose, instead, that a compression ratio of 125,000 is needed for LPI but the bandwidth of 2 MHz is appropriate for the radar's function. To completely

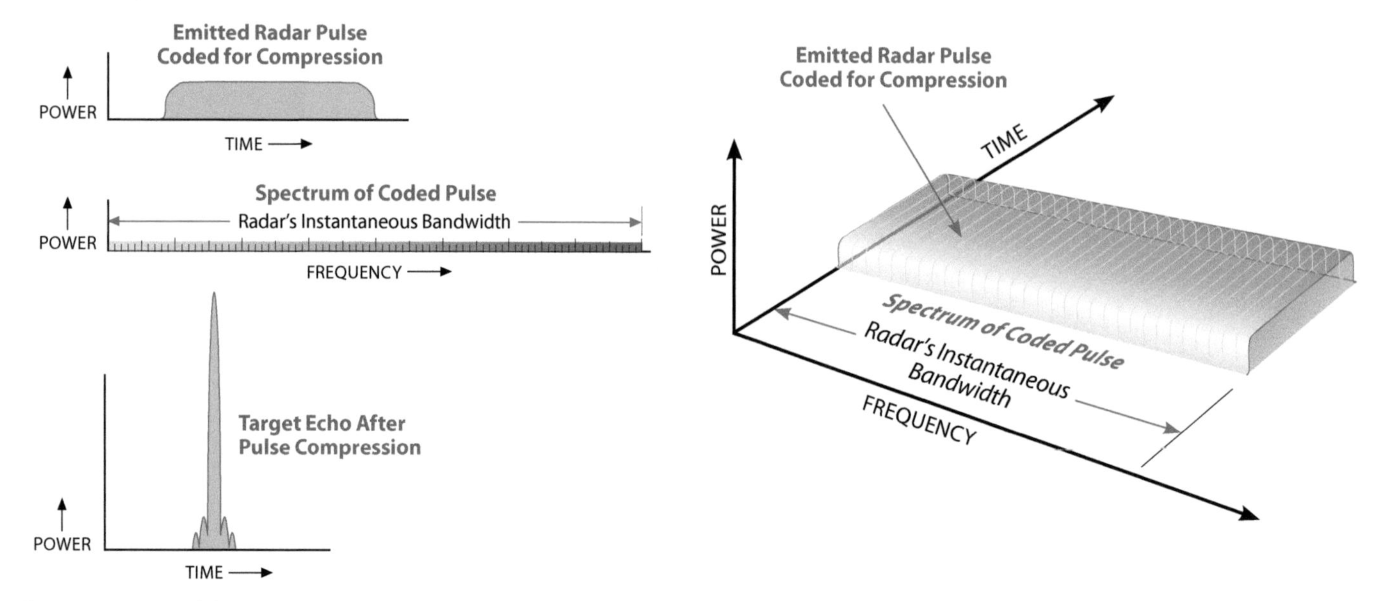

Figure 41-5. By modulating the radar's emitted pulses with pulse-compression coding, their power may be spread over the radar's entire instantaneous bandwidth and provide the range resolution suitable for the radar's function. When the radar echoes are decoded, they are compressed into narrow pulses containing virtually all of the received energy.

compress the pulse, the integration time must be equal to the uncompressed pulse duration. This means the integration time must be increased by a factor of (125,000/2000) = 62.5 to 62.5 ms. The range of a target closing at Mach 1 (about 350 m/s) would change by 21.7 m in that time. Such target motion during the pulse compression interval could drastically alter the echo's character and limit the ability of the radar to compress the signal. This illustrates that the amount of pulse compression is limited by the target's motion and size.

Upon being received by the radar and decoded, target echoes are compressed into short duration pulses providing fine range resolution, and containing virtually all of the received energy (Fig. 41-5). Yet not knowing the pulse compression code used, an interceptor cannot similarly compress the radar's emitted pulses. What the interceptor *can* do is use noncoherent integration to approximately the same extent that the radar uses coherent integration. Noncoherent integration has losses relative to the radar's coherent integration however the advantage of the radar can be approximately reduced from the pulse compression ratio to the square root of the pulse compression ratio. If the interceptor adopts this strategy, the fourth root of the compression ratio reduces the LPI design range. An explanation of the approximate effect of noncoherent integration on the intercept range is provided in the blue panel on integration.

Multiple Beams on Different Frequencies. For any mode of operation in which the radar must search a solid angle of space, the ability to reduce peak power by increasing the coherent integration time is limited by the acceptable scan frame time. Within this limit, however, dwell times may be substantially increased by transmitting multiple beams on different radio frequencies.

Suppose, for example, that a volume, V, expressed in multiples of an angle equal to the radar's 3 dB beamwidth, is to be searched

Noncoherent Integration

As shown in Chapters 12 and 16, using matched filter processing the radar can achieve integration gain nearly equal to the time bandwidth product. For ES systems, a matched filter is not practical because there are many signals and their characteristics are not known in advance. Also, while the radar is mainly interested in the time the echo returns and with what Doppler shift, the ES user is interested in the signal's characteristics so that it can be identified. The ES system could achieve some integration gain by using noncoherent processing. This is used in radar as well. If a number of pulses is transmitted with random phase and returned from the target, each single pulse can be processed by a matched filter and then those outputs are added. Similarly, an ES system could add multiple samples of the envelope of an LPI radar signal to improve the chances of detecting the signals.

When a large number of samples are added, the distribution approaches a normal distribution and the integration loss relative to coherent integration is approximately the square root of the number of samples integrated. For coherent integration, the improvement in the signal-to-noise ratio is proportional to the number of independent samples integrated, or, in dB, 10 $\log_{10}(N)$. For noncoherent integration the signal-to-noise ratio improves by about (5 $\log_{10}(N)$ + 5.5 dB) for values of probability of detection and false alarm used in practice. For example, for N = 100, the coherent or matched filter gives an output SNR which is 10 $\log_{10}(100)$ = 20 dB greater than the input SNR, whereas the noncoherent integrator gives an output SNR which is approximately 5 $\log_{10}(100)$ + 5.5 = 15.5 dB greater than the input SNR. In this example, the loss relative to a matched filter is approximately 4.5 dB. In summary, for noncoherent integration:

$$\text{Gain} \cong 3.55\sqrt{(B\tau)} \quad \text{or} \quad 5.5 + 5\log_{10}(B\tau)\,\text{db}$$

The ES system might choose to integrate noncoherently for as long as the radar integrates. Then the loss relative to coherent integration is

$$\text{Loss} \cong 5\log_{10}(B\tau) - 5.5\ \text{db}$$

By using noncoherent integration, the ES system can extend the range at which it may detect LPI radar signals. Because the signal strength decreases as the square of the range, the ES range increases as approximately the fourth root of the radar's time-bandwidth product.

in the time, T. If the search were done with a single beam, the maximum allowable dwell time would equal T/V (Fig. 41-6a).

On the other hand, if this same volume (V) were subdivided into N sectors and every sector were simultaneously searched by a different beam using a different radio frequency

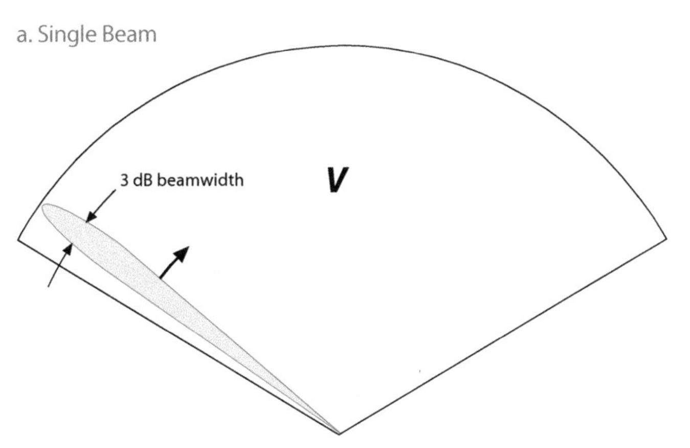

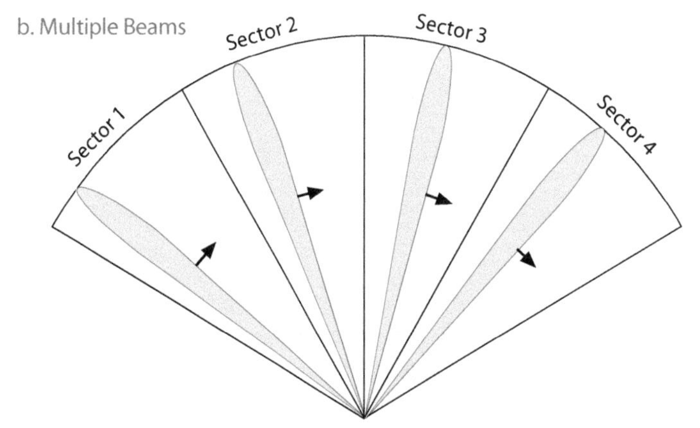

Figure 41-6. Increase in dwell time achievable by radiating multiple beams on different frequencies. For the same detection sensitivity, as the number, N, of beams is increased, peak power can be reduced by a factor of $1/N$. It is important to note that a moving target will move N times as far during the increased integration time.

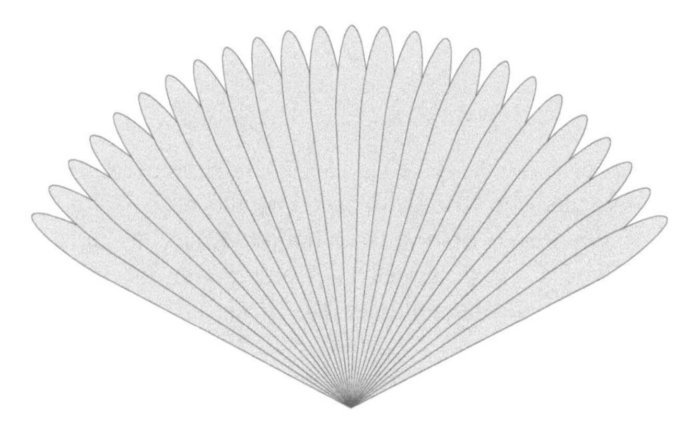

Figure 41-7. Enough beams might be provided to completely fill the scan volume. Then, no scanning would be needed, and the coherent integration time would equal the frame time.

(Fig. 41-6b), the dwell time in each beam direction could be increased by a factor of *N*. Then, if the coherent integration time were increased to match the dwell time, now equal to NT/V, the peak power emitted in any given beam direction could be reduced by the factor $1/N$.

In the extreme, provided adequate processor throughput is available, enough beams might be emitted to completely fill the scan volume (Fig. 41-7). No scanning would then be needed. Consequently, the coherent integration time could be made equal the total frame time, *T*. However note that an interceptor may also increase the amount of noncoherent integration it uses by a factor of *N*, reducing the radar's advantage from $\sim N$ to $\sim\sqrt{N}$.

Pseudo-Random Pulse Compression Codes

THESE ARE BINARY PHASE CODES THAT APPEAR TO BE ENTIRELY RANDOM IN virtually every respect, except for being repeatable. Their advantages:

- A great many different codes can be generated easily and conveniently
- Codes can be made almost any length and hence provide extremely large compression ratios.

The codes are commonly generated in a shift register having two or more feedback paths.

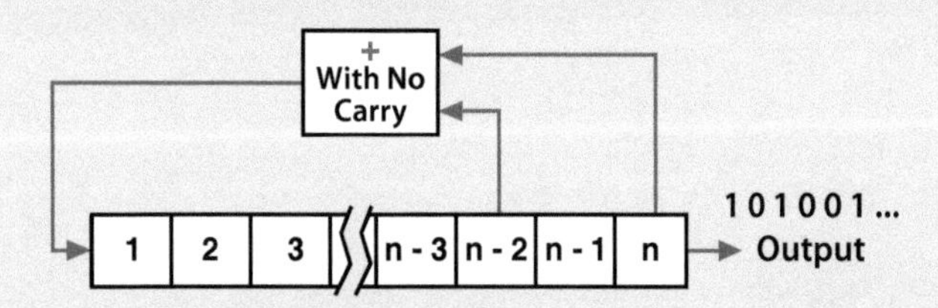

Filled initially with 1s or 1s and 0s, the register produces a code of 1s and 0s of length

$$N = 2^n - 1$$

where *N* is the number of bits output before the code repeats, and *n* is the number of bits the register holds. For an 11-bit register, $n = 11$, and therefore, $N = 2^{11} - 1 = 2047$

An 11-bit register with the 9th and 11th digits fed back to the input, for example, produces a code 2,047 digits long. By changing the feedback connections, 176 different codes of that length can be produced. These are called Maximal Length codes because they repeat only after *N* bits. Other feedback connections produce codes that repeat after a fewer number of bits. Changing the particular bit pattern may or may not affect the performance of an intercept receiver. Note also that pseudorandom codes are rather easy to predict. If one knows the length of the shift register and the feedback connections from collected intelligence, observing a sequence of ones and zeros equal to just twice the length of the shift register is sufficient to predict the entire sequence. More secure code generation methods are readily available.

The 0s and 1s in the code specify the relative phases – 0° and 180° – for successive segments of the radar's transmitted pulse.

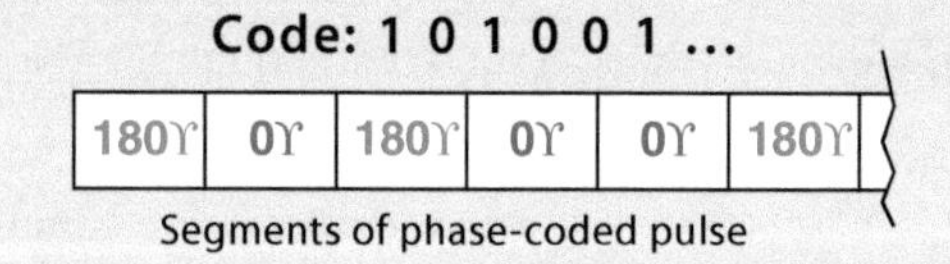

By shifting the register at intervals equal to the desired length of the segments, successive output bits can directly control the phase modulation of the radar signal.

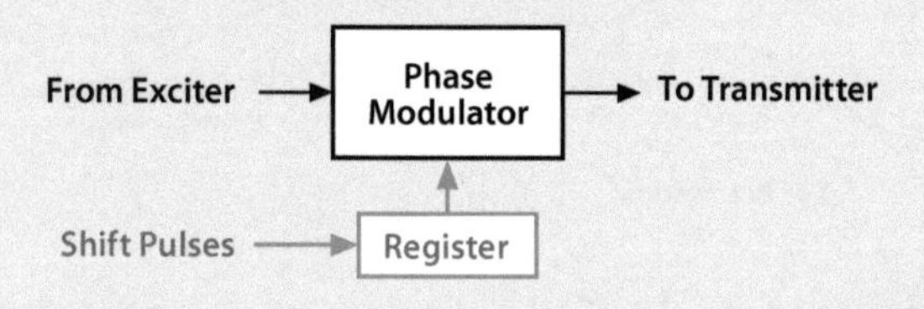

When the received pulse is decoded, the segments are superimposed, producing a pulse roughly *N* times the amplitude of the uncompressed pulse and only a little wider than the segments. The code generated by the 11-bit register of this example would thus yield a pulse compression ratio of roughly 2,000 to 1. Note that the time sidelobes of the compressed pulses vary somewhat randomly but are typically below the peak of the compressed pule by about a factor of the square root of the code length. Truly random codes may sometimes have undesirably high time sidelobes; however the Barker codes are too short for LPI designs. (Barker codes are discussed in Chapter 16.)

Multiple beams may also be employed to advantage in other ways. They may, for example, be used to selectively search different portions of the total scan volume. Or, each beam may be used to scan the entire volume on a different frequency, thereby increasing detection sensitivity through frequency diversity rather than through increased integration time.

Random Waveform Parameters. For all practical purposes, in a dense signal environment a signal has not been *usefully* intercepted unless it has been successfully de-interleaved (sorted) parameterized and classified or perhaps identified (Table 41-2). Consequently, besides reducing the probability that the radar's signals will be detected by an interceptor, the radar designer has opportunities for confounding the de-interleaving and identification processes, as well.

Table 41-2. Basic Intercept Receiver Functions

Detection	Detect single pulses (peak power), with little or no integration and detect CW Signals
De-interleaving	Separate pulses of individual emitters, in a dense signal environment
Parametric Analysis (Pulse and CW) Classification & Identification	Analyze pulse, pulse train and CW parameters for use in classification and identification Identify emitters by type; possibly even identify specific emitters to a specific platform

41.4 Further Processing of Intercepted Signals by the ES Receiver

Among the waveform parameters typically used for both de-interleaving and classification are as follows:

- Angle of arrival
- Radio frequency
- Pulse width
- PRF

Among those parameters typically used for classification alone are:

- Scan rate
- Intrapulse modulation
- Interpulse modulation
- Beamwidth
- Signal polarization

Except for angle of arrival, all of the above-listed parameters can be varied randomly from one coherent integration period to the next. If successive echoes from each pulse are not coherently integrated, pulse to pulse frequency agility can be used.

Variations can be achieved without reducing detection sensitivity by taking advantage of the waveform agility available in modern airborne radars. Moreover, with two or more aircraft operating cooperatively (i.e., alternately providing target illumination for each other; Fig. 41-8) even angle of arrival can be varied.

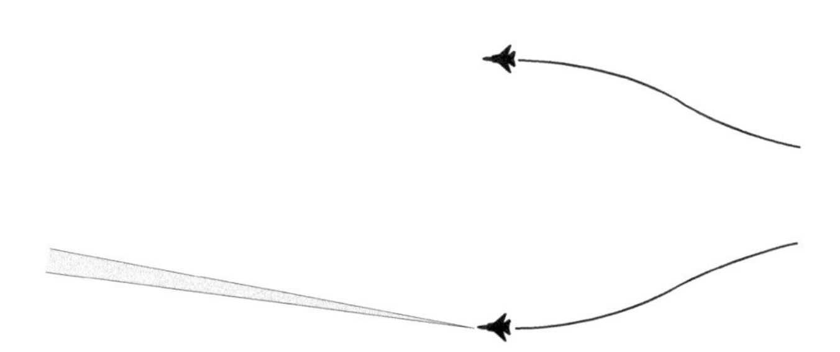

Figure 41-8. By cooperatively shifting radar transmission randomly between them, two or more aircraft can even vary the angle of arrival of their emissions.

Randomizing any of the parameters can confuse the classification process. That is particularly true for those intercept systems which classify signals by comparing their parameters with parameters stored in threat tables.

Mimicking Enemy Waveforms. Mimicking may also confuse signal classification. To be able to mimic an enemy's waveforms, though, the radar must not only have considerable waveform agility, but be able to operate over the full range of radio frequencies the enemy employs while still performing its required functions.

41.5 Cost of LPI

LPI techniques are not free; each of the LPI-enhancing features adds to the radar's cost. In this chapter, cost does not necessarily refer to monetary cost but also to increased signal processing and reduced time to perform it. Most increase the costs of both software and hardware. The main exception is that some savings are possible due to the lower peak power required by the transmitter.

But by far the greatest cost of LPI is in digital processing throughput. As instantaneous bandwidth is increased, for instance, the required throughput goes up proportionately because of the increased number of range bins whose contents must be processed.

For, to the extent that bandwidth is increased through pulse compression coding, the wider the bandwidth, the narrower the compressed pulses will be. Hence more range bins are required to cover the same range interval. Throughput similarly goes up with the number of simultaneous beams radiated.

To support a wide instantaneous bandwidth and a few simultaneous beams, the required throughput is large (until the 1990s these features were not deemed practical). With the dramatic advances being made in digital processor technology, however, the costs of these features are rapidly decreasing.

Moreover, in those situations where the advantages of maximum detection range and situation awareness outweigh the advantages of LPI, the operator always has the option of overriding power management, operating the radar continuously, and searching the antenna's entire field of regard.

41.6 Possible Future Trends in LPI Design

Looking to the long-term future, one thing is certain: competition between radar designer and intercept receiver designer will never be static. For every improvement in LPI, improvements in intercept receiver design can be expected. LPI designers will continue to exploit coherent processing, which the intercept receiver cannot duplicate. And designers of intercept receivers will continue to exploit the R^2 advantage of one-way versus two-way propagation as well as making more use of noncoherent integration.

Probably, the most spectacular gains in both LPI and intercept receiver design will occur in signal processing, which is the subject of the next chapter.

41.7 Summary

Operational strategies for LPI include limiting radar "on" time, using collateral intelligence and reconnaissance information wherever possible, relying heavily on onboard passive sensors, and searching only narrow sectors in which they indicate the target to be.

LPI design strategies capitalize primarily on the intercept receiver:

Having to detect individual pulses, so that it can de-interleave them and identify their sources

Consequently, trading long coherent integration time for reduced peak power and using frequency agility to further confuse the ES system can degrade the sensitivity of the ES system and enhance LPI capability.

High antenna gain, reduced sidelobe levels, high duty factor, and increased radar receiver sensitivity can likewise be traded for reduced peak power.

Several special features can further enhance LPI. First among these is power management—keeping the peak emitted power just below the level at which it can be usefully detected by an intercept receiver in an approaching aircraft, yet just above the level at which the radar can detect the aircraft.

Added to this feature are (a) using extremely large amounts of pulse compression; (b) simultaneously transmitting multiple beams on different frequencies to reduce the constraint imposed on integration time by limits on scan-frame-time; (c) randomly changing waveform characteristics to confound the intercept receiver's signal de-interleaving and identification process; and (d) mimicking enemy waveforms.

The principal cost of LPI is greatly increased signal processing throughput with cost reduction due to the lower peak power of the transmitter.

Test your understanding

1. How does the detection of signals by an intercept receiver differ from the detection of target echoes by a radar receiver?
2. If an LPI Radar were to be modified to double its detection range by increasing its average power, by how much would the new ES free space detection change? What does this tell you about designing a long range LPI radar?
3. Is it easier to create a long range LPI radar or a short range LPI radar?
4. List some ways the intercept receiver design might be changed to better detect LPI radars.
5. What roles does pulse compression play in the design of LPI radar?
6. What are some of the problems associated with CW radar?

Further Reading

P. E. Pace, *Detecting and Classifying Low Probability of Intercept Radar*, Artech House, 2004.

R. G. Wiley, ELINT: The Interception and Analysis of Radar Signals, Artech House, 2006.

Lockheed F-117 Nighthawk (1983)

The F-117A was developed as the result of a program in the late 1970s at the Lockheed Skunk Works under the codename HAVE BLUE. This program led to the development of a prototype that was both aerodynamically flyable and of low radar signature. These depended on substantial simulations and computing power, which were not feasible. The F-117 was a single-seat, twin-engined design for ground attack. It first flew in 1981, came into service in 1983, and was retired in 2008.

PART IX

Special Topics and Advanced Concepts

Joint Strike Fighter

The Joint Strike Fighter program is directly tied to the development of the F-35 Lightning II, a stealth multi-role fighter. The purpose of the program was to develop aircraft, weapons, and sensors that could allow services to use a single airframe to replace multiple families of aircraft. Multiple countries (United States, United Kingdom, Canada, Australia, and The Netherlands) are involved in the program.

42

Antenna Radar Cross Section Reduction

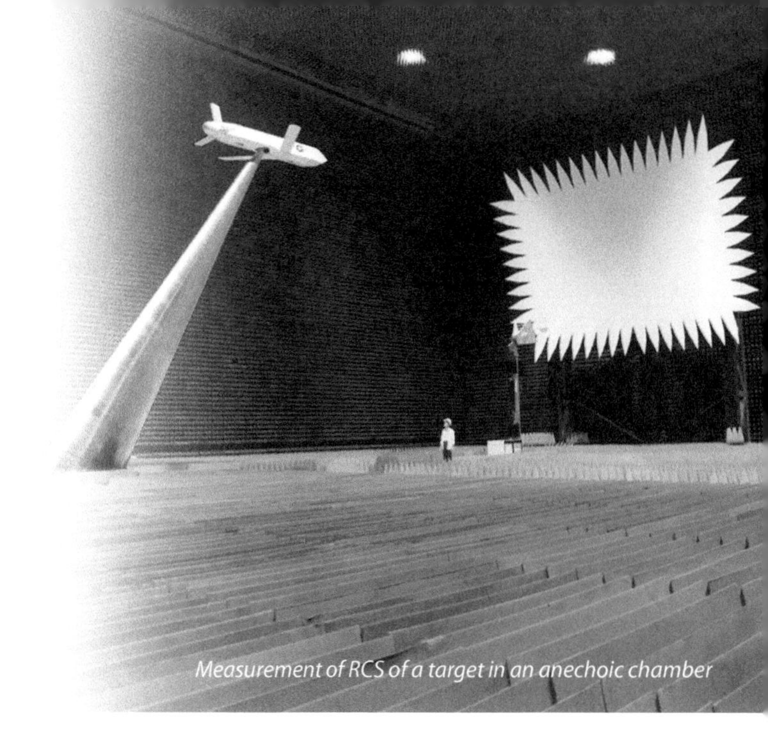
Measurement of RCS of a target in an anechoic chamber

Viewed nose-on, a typical fighter aircraft has a radar cross section (RCS) of the order of 1 m^2. A similarly viewed low observable aircraft may have an RCS of only 0.01 m^2. Unless special RCS reduction measures are employed, even a comparatively small planar array antenna can have an RCS of up to several thousand square meters when viewed from a broadside direction. Since an aircraft's radome is transparent to radio waves, if stealth is required, steps must be taken to reduce the RCS of the installed antenna.

In this chapter, we will introduce the sources of reflections from a planar array antenna, learn what can be done to reduce or render them harmless, and see why these steps are facilitated in an ESA. We will then take up the problem of avoiding so-called Bragg lobes, which are retrodirectively reflected at certain angles off broadside if the radiator spacing is too large compared with the radar's operating wavelength. Finally, we will very briefly consider the critically important validation of an antenna's predicted RCS.

42.1 Sources of Reflections from a Planar Array

For our purposes here, a planar array antenna, regardless of whether it is a mechanically steered array (MSA) or an electronically steered array (ESA), can conveniently be thought of as consisting of a flat plate—referred to as the *ground plane*—containing a lattice of radiating elements (Fig. 42-1).

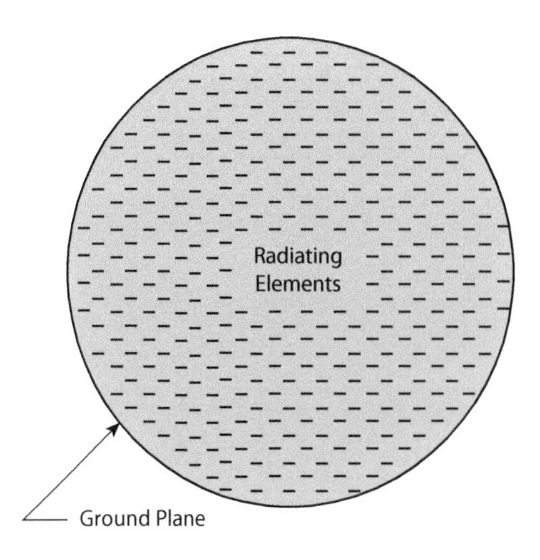

Figure 42-1. A planar array antenna, regardless of whether it is an MSA or an ESA, can conveniently be though of as a flat plate, termed the ground plane, containing a lattice of radiating elements.

The backscatter from the antenna when illuminated by a radar in another aircraft—*threat radar*, we'll call it—is commonly

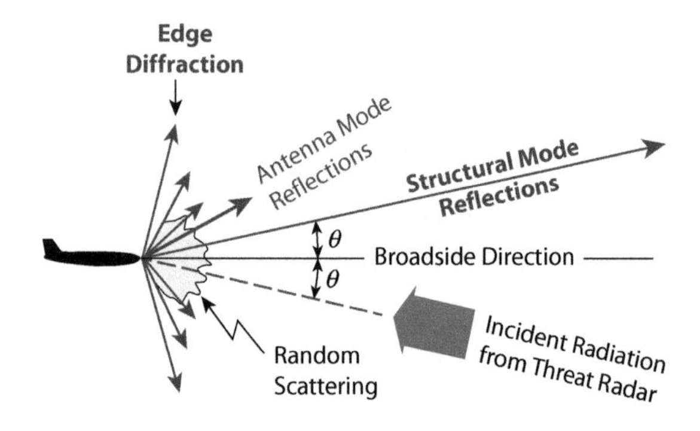

Figure 42-2. There are four basic components of backscatter from a planar array antenna. Random scattering is the sum of the random components of the structural mode and antenna mode reflections.

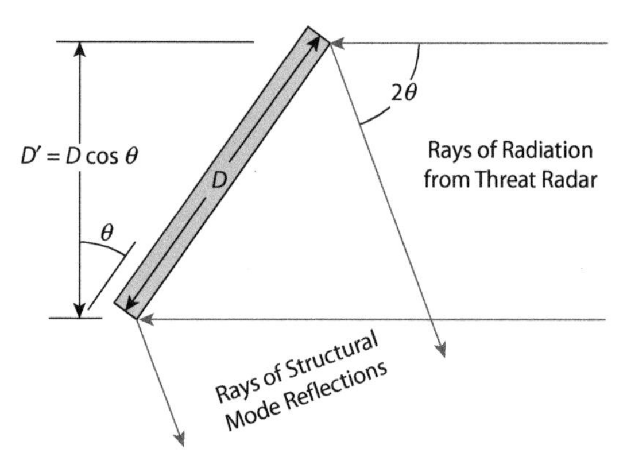

Figure 42-3. Structural mode reflections may be rendered harmless by tilting the array. The tilt reduces the effective aperture somewhat, but that is a small price to pay for the huge reduction in detectability achieved.

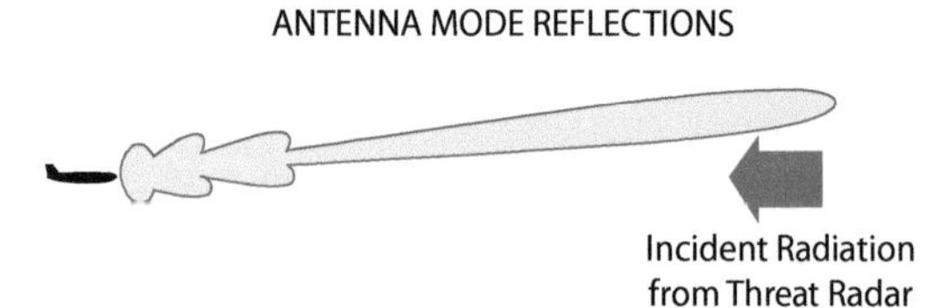

Figure 42-4. The radiation pattern of these reflections is similar to that of the transmitted signal. Since their direction is determined by internal phase shifts as well as by the angle of incidence of illuminating waves, they are not necessarily rendered harmless by antenna tilt.

categorized as being composed of four basic components (Fig. 42-2):

1. Specular (mirrorlike) reflections from the ground plane, called *structural mode* reflections
2. Reflections of some of the received power by mismatched impedances within the antenna, reradiated by the radiating elements and called *antenna mode* reflections
3. Reflections due to the mismatch of impedances at the edges of the array (i.e., between the ground plane and the surrounding aircraft structure), referred to as *edge diffraction*
4. Random components of the structural mode and antenna mode reflections, called *random scattering*

In case you're wondering, there are two reasons for separately breaking out random scattering. First, with the random scattering removed, the structural mode and antenna mode reflections can be characterized more simply. Second, there is then a one-to-one relationship between the individual categories of reflections and the techniques for reducing or controlling them.

42.2 Reducing and Controlling Antenna RCS

By carefully designing and fabricating an antenna, each of the four components of backscatter may be acceptably minimized or rendered harmless.

Rendering Structural Mode Reflections Harmless. As may be seen from Figure 42-3, these mirrorlike reflections may be controlled by physically tilting the antenna so that they are not directed back in the direction from which the illuminating radiation came. Although the tilt does not reduce the reflections, it prevents the threat radar from receiving them.

With an ESA, which is mounted in a fixed position in the aircraft, the antenna ground plane can be permanently tilted so that the incident radiation will be harmlessly reflected in the same direction as the irreducible "spike" in the pattern of reflections from the aircraft structure. The tilt reduces the antenna's effective aperture area somewhat, reducing the gain and broadening the beam about the axis of the tilt. But this is a small price to pay for the huge reduction in detectability that is achieved.

Minimizing Antenna Mode Reflections. At the radar's operating frequency, antenna mode reflections have a radiation pattern similar to that of the transmitted signal: a mainlobe, surrounded by sidelobes (Fig. 42-4). The direction of the mainlobe is determined by the angle of incidence of the illuminating waves and the element to element phase shift occurring within the array. As is clear from Figure 42-4, these reflections are not necessarily rendered harmless by the tilt of the antenna.

They can be acceptably minimized, however, by employing well-matched microwave circuitry in the antenna and by paying extremely close attention to design detail. In wideband MSAs and passive ESAs, even reflections from deep within the antenna

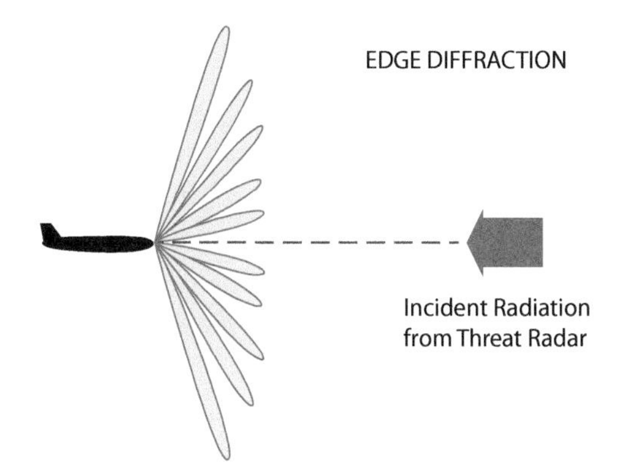

Figure 42-5. Backscatter due to edge diffraction is comparable to that from a loop the size and shape of the array's perimeter. Since its diameter is many times the operating wavelength, the backscatter fans out in many directions.

must be eliminated. This may be accomplished by inserting isolators, such as circulators, at appropriate points in the feed.

Minimizing Edge Diffraction. Edge diffraction produces backscatter comparable to what would be produced by a loop antenna having the same size and shape as the perimeter of the array. Since the dimensions of this loop are generally many times the operating wavelength of the radar, the radiation pattern of the loop typically consists of a great many lobes fanning out from the broadside direction (Fig. 42-5). Consequently, edge diffraction, too, is not rendered harmless by the antenna's tilt. Special measures must be taken to minimize it.

In some antenna installations, edge diffraction is rendered harmless by shaping the edge of the ground plane to disperse the diffracted energy so that it is beneath the threshold of detection of the threat radar.

In other installations, the diffraction is reduced by applying radar-absorbing material (RAM) around the edges of the ground plane so that its resistivity smoothly tapers to that of the surrounding structure. To be effective, the treatment must be at least four wavelengths wide at the lowest threat frequency (Fig. 42-6). Thus, it can seriously diminish the available aperture area and therefore reduce the radar's performance. Accordingly, careful trade-offs are necessary between radar performance and RCS performance.

In any event, the measures taken to reduce or render the diffraction harmless are greatly facilitated in an ESA, since it is permanently mounted in a fixed position on the aircraft structure.

Minimizing Random Scattering. The random components of structural mode and antenna mode reflections may be spread over a wide range of angles (Fig. 42-7). So they are not rendered harmless by the antenna's tilt. To reduce them to acceptable levels, the antenna's microwave characteristics must be highly uniform across the entire array. This requires exceptionally tight manufacturing tolerances.

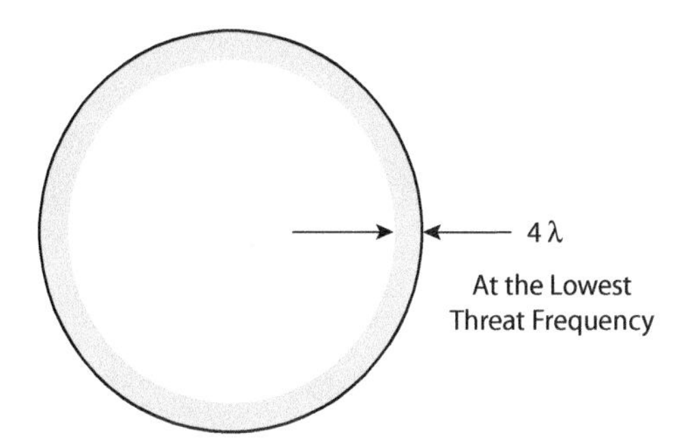

Figure 42-6. Edge treatment must be at least four wavelengths wide. Depending on the antenna's size, this can seriously diminish the effective aperture.

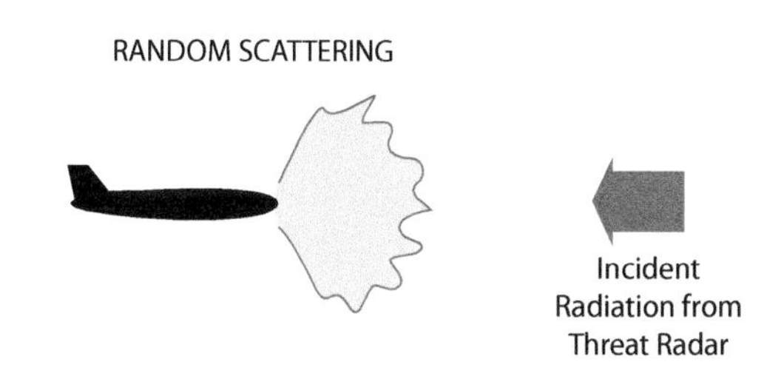

Figure 42-7. The random components of structural mode and antenna mode reflections are spread over a wide span of angles.

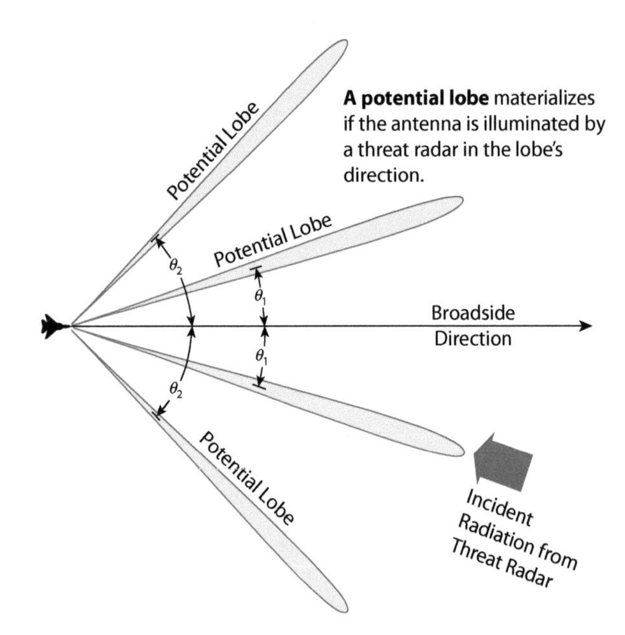

Figure 42-8. Bragg lobes are retrodirective reflections that may be received by an illuminating radar when it is a certain angle, θ_n, off broadside, if the spacing of the radiators is larger than half the wavelength of the illumination.

42.3 Avoiding Bragg Lobes

Bragg lobes are retrodirective reflections from the antenna's radiators, which may be received by an illuminating radar when it is in certain angular positions, θ_n, off broadside (Fig. 42-8). Depending on the antenna's design, besides direct reflections from the radiators, the lobes may also include energy reflected from within the antenna (i.e., antenna mode reflections).

The lobes are due to the periodicity of the radiator lattice. They occur at the angles for which the phases of the waves reflected in the illuminator's direction by successive radiators differ by 360° or multiples thereof and hence are all in phase and add up to a strong signal.

While for simplicity Figure 42-8 has been drawn for lobes in a single plane, bear in mind that for a two-dimensional array Bragg lobes occur about both lattice axes. As illustrated in the following panel, the directions of the lobes relative to the boresight direction are determined by the spacing of the radiators relative to the wavelength of the illumination. The greater the spacing or the shorter the wavelength, the closer the lobes will be to the broadside direction and the more lobes there will be.

Conditions Under Which a Bragg Lobe Will Be Produced

When adjacent radiators of an array antenna are illuminated by a threat radar, a Bragg lobe will be produced if the wave reflected in the radar's direction, θ_n, by the far radiator, *B*, is in phase with the wave reflected by the near radiator, *A*.

Assuming no regular radiator-to-radiator phase shift in reflections from within the antenna, that condition will occur if the additional round-trip distance, ΔR, traveled to and from radiator *B* is a whole multiple, *n*, of the incident radiation's wavelength, λ.

$$\Delta R = n\lambda$$

As is clear from the diagram,

$$\Delta R = 2\,d \sin \theta_n$$

where *d* is the spacing between radiators. Thus, the relationship between radiator spacing and Bragg lobe direction is

$$d = \frac{n\lambda}{2 \sin \theta_n}$$

To minimize the antenna's RCS, the first Bragg lobe, $n = 1$, must be placed 90° off broadside, $\sin \theta_1 = 1$. Substituting these values in the previous equation yields

$$d = \frac{\lambda}{2} \text{ for stealth}$$

Like grating lobes, Bragg lobes can be avoided by spacing the radiators close enough together to place the first lobe 90° off broadside. As the panel shows, if the illuminator's wavelength is the same as the radar's, this may be accomplished with a spacing of half the operating wavelength.

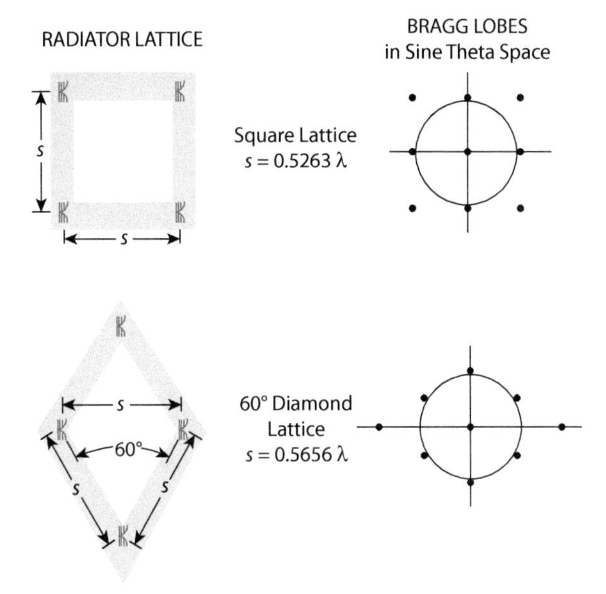

Figure 42-9. In the Bragg lobe patterns for square and 60° diamond radiator lattices shown here, despite the greater radiator spacing of the diamond lattice, all Bragg lobes except the central one are outside the boundary of visible (real) space. The central lobe is rendered harmless by the tilt of the antenna.

But if the illuminator's wavelength is shorter, the spacing must be proportionately reduced. Suppose, for instance, that the radar's wavelength is 3 cm and the illuminator is operating at 18 GHz, $\lambda = 1.67$ cm. To avoid Bragg lobes, the radiator spacing would have to be reduced to $1.67/2 = 0.84$ cm, which is little more than a quarter of the operating wavelength.

If such tight spacing is not economically feasible, the designer has three options. The first two are comparatively simple. One is to use a diamond lattice such as that illustrated in Figure 42-9. Despite the larger radiator spacing of this lattice, Bragg lobes may be rendered harmless.

The second option is simply to employ the tightest practical radiator spacing—at least along the axis of greatest concern.

The third and more costly option is to prevent any shorter wavelength radiation from reaching the array. One way of accomplishing this is to place a screen known as a *frequency selective surface* (FSS) in front of the array (Fig. 42-10). The screen is designed to pass all wavelengths in the radar's operating band with little attenuation yet reflect all out-of-band radiation. The screen may either be mounted externally or be built into the antenna face. As with structural mode reflections, because of the tilt of the antenna—hence also of the screen—radiation reflected by the screen will be directed in a direction in which it will not cause a problem.

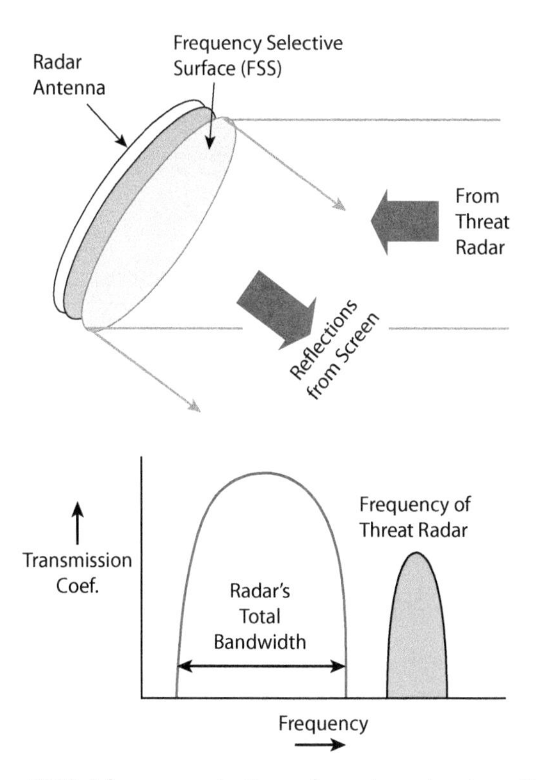

Figure 42-10. A frequency selective surface acts as a bandpass filter, rejecting radiation of high frequencies for which making the radiator lattice tight enough to avoid grating lobes would not be practical.

Frequency Selective Surfaces

FSSs ARE PRINTED PERIODIC SURFACES THAT CAN BE DESIGNED TO HAVE SPECIFIC transmission properties as a function of frequency, in the same way as filter circuits, and they can be designed in terms of equivalent circuits.

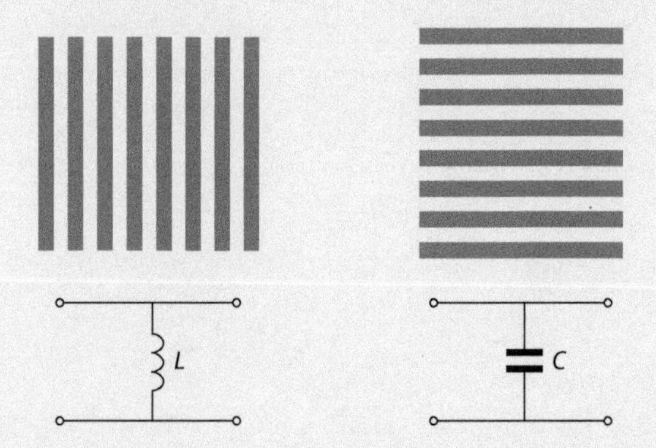

A grid of vertical metal strips behaves to a normally incident electromagnetic wave as an inductor. Likewise, a grid of horizontal strips behaves as a capacitor. The values in each case depend on the width and spacing of the strips compared with the wavelength.

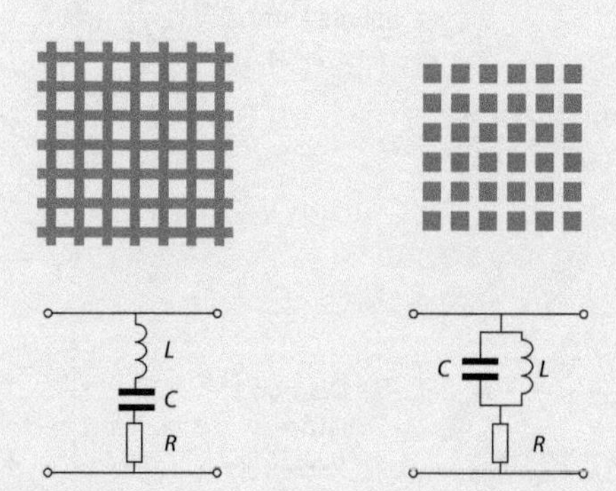

Combining these, the surfaces can be configured to have frequency selective properties corresponding to a series-tuned or parallel-tuned circuit.

Other types of FSS, with more elaborate frequency responses, may be built up from more complex shapes and with multiple layers.

In one possible implementation, the screen consists of a thin metal sheet containing a tight lattice of slots, mounted between two dielectric slabs. To be effective, the slots must be separated by no more than half the wavelength of the highest threat frequency.

Research and development are being undertaken into reconfigurable FSSs to provide tunable passband radomes. This would have the advantage of allowing the operation over tunable narrow bands without the need for a wideband FSS, which would make the platform vulnerable to enemy detection. Also being worked on are tunable FSS radomes that would act like an electromagnetic shutter; that is, they would allow the signal to either be passed (open state) or blocked (closed state). In the blocked state the radome becomes a mirror surface and the echo signal would be reflected away from the enemy direction, allowing the antenna to have a low cross section when the radar is not being used. This electronic shutter property has been demonstrated over a wide band (46 percent bandwidth centered at 650 MHz) and for over wide directions of the incoming signal (0° to 60° incidence angles).

42.4 Application of Radar Signature Reduction Techniques in Operational Active Electronically Scanned Arrays

The aforementioned techniques for the reduction of the radar cross section of active electronically scanned array (AESA) antennas have in the recent years been deployed or proposed for upgrades. As an early example, Figure 41-11 shows a low cross section approach for upgrading the F-111. It uses a tilted antenna to reduce backscatter toward hostile radars. It also shows the use of RAM and a shroud to reduce the scattering from the structures surrounding the AESA antenna. The first actual application for a deployed AESA with a tilted antenna is the Raytheon AESA APG-79 used on the F/A-18E/F Super Hornet. This radar allowed for the first time simultaneous air-to-air and air-to-ground modes. Figure 41-12 shows the antenna tilted up away from the threat direction. Aircraft from other countries that have used the same technique include the Russian MiG-35, PAK-PA and FLANKER fighter aircraft, and the Chinese J-10B.

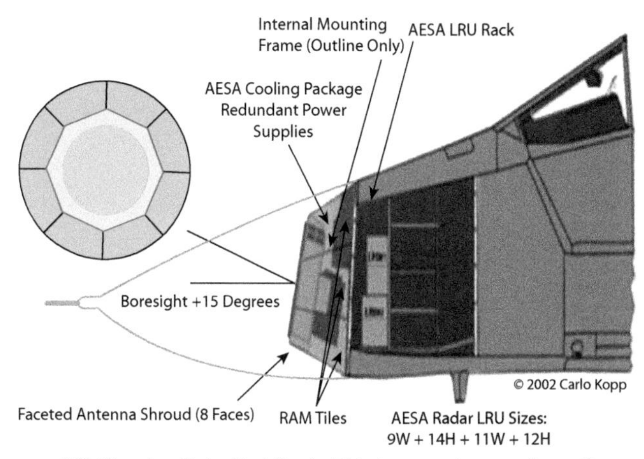

Figure 42-11. Illustrated here is the suggested AESA multimode radar upgrade of the F-111C Block C-4/C-5. (From Dr. Carlo Kopp, http://www.ausairpower.net/ ; click on F-111 tab.)

The B-2 stealth bomber (Fig. 42-13) has been upgraded with an AESA as part of a radar modernization program (RMP). Each antenna will require more than 2,000 of the two-channel modules. The AESA antenna design significantly improves the radar's performance and operational low radar signature. It boosts the radar's power, makes future upgrades easier, and was accomplished without making costly modifications to the 20-year-old B-2 platform. The AN/APQ-181 radar is a low probability of intercept, all-weather system that enables the B-2 Spirit stealth bomber to penetrate the most sophisticated air defenses.

Figure 42-12. This Raytheon APG-79 AESA is used for the F/A-18E/F Super Hornet. (Courtesy Raytheon.)

42.5 Cloaking and Stealthing Using Metamaterials

Research during the past decade has led to the development of *metamaterials*. These can be arranged to have the property of negative refractive index, which may allow electromagnetic waves to pass around an object rather than being scattered, thereby rendering the object invisible to radar. Physically they consist of periodic structures in a lattice

Figure 42-13. The B-2 Spirit stealth bomber can penetrate the most sophisticated air defenses. Its upgraded AN/APQ-181 AESA radar system enables the unique combination of stealth, range, payload, and precision weapons delivery capabilities.

configuration (Fig. 41-14), similar in some ways to FSSs. They may be incorporated in the designs of antennas to minimize their RCS. Current challenges are to develop designs that are physically thin, that can fit conformally over arbitrarily shaped objects, and that maintain their performance over wide bandwidths.

Figure 42-14. This negative index metamaterial array configuration is constructed of copper split-ring resonators and wires mounted on interlocking sheets of fiberglass circuit board. The total array consists of 3 by 20 × 20 unit cells with overall dimensions of 10 × 100 × 100 mm. (Courtesy of Jeffrey D. Wilson, NASA Glenn Research.)

Figure 42-15. The predicted RCS of the antenna of the radar of a fighter aircraft is verified in an anechoic chamber. The antenna in its radome is mounted on a low RCS test body.

42.6 Validating an Antenna's Predicted RCS

Because of the complexity of the factors contributing to an antenna's installed RCS, a key step in developing a low RCS antenna is validating the antenna's predicted RCS.

For this, one or more physical models of the radiating aperture are generally built. These are called *phenomenology models*, or *phenoms*. Typically, they include not only the radiators and any covering that goes over them but also the first few stages of internal circuitry. If the schedule allows, the phenoms may even be used to interactively refine the design.

Measurements made on the phenoms and on the complete antennas include the following:

- Closed-circuit measurements at the radiator level to isolate and quantify the complex reflections from each radiator and its internal circuitry—commonly referred to as *look-in measurements*.
- Angular *cuts* of the reflection pattern of the total array.
- Very high-resolution inverse synthetic aperture radar (ISAR) images of the antenna, made to isolate individual reflection "hot spots" and to determine the effectiveness of the edge treatment.

To realistically evaluate the installed antenna's RCS, a full-scale model of the nose section of the aircraft including the phenom is generally tested in a large anechoic chamber (Fig. 42-15).

42.7 Summary

Unless special measures are taken to reduce the reflections from a planar array, its RCS may be several thousand square meters. The reflections are of four basic types, which may be reduced or rendered harmless as follows:

- Mirrorlike reflections from the back plane (structural mode reflections)—may be rendered harmless by tilting the antenna
- Reflections due to mismatched impedances within the antenna (antenna mode reflections)—may be reduced by minimizing the mismatches
- Reflections due to mismatched impedances at the edges of the array (edge diffraction)—may be reduced by tapering the impedances with radar absorbing material or shaping the edges of the ground plane to disperse the diffracted energy
- Random components of structural and antenna mode reflections (random scattering)—may be reduced by holding to extremely tight manufacturing tolerances

To avoid retrodirective reflections from the radiator lattice—Bragg lobes—the radiator spacing must be less than half the wavelength of the illumination. If illuminators may be encountered whose wavelengths are shorter than the radar's, either the radiator spacing must be further reduced or a frequency-sensitive screen must be placed over the array to keep out the shorter wavelength radiation.

Because of the complexity of factors contributing to antenna RCS, RCS predictions are validated with physical models (phenoms) and a full-scale model of the nose section is usually tested in an anechoic chamber.

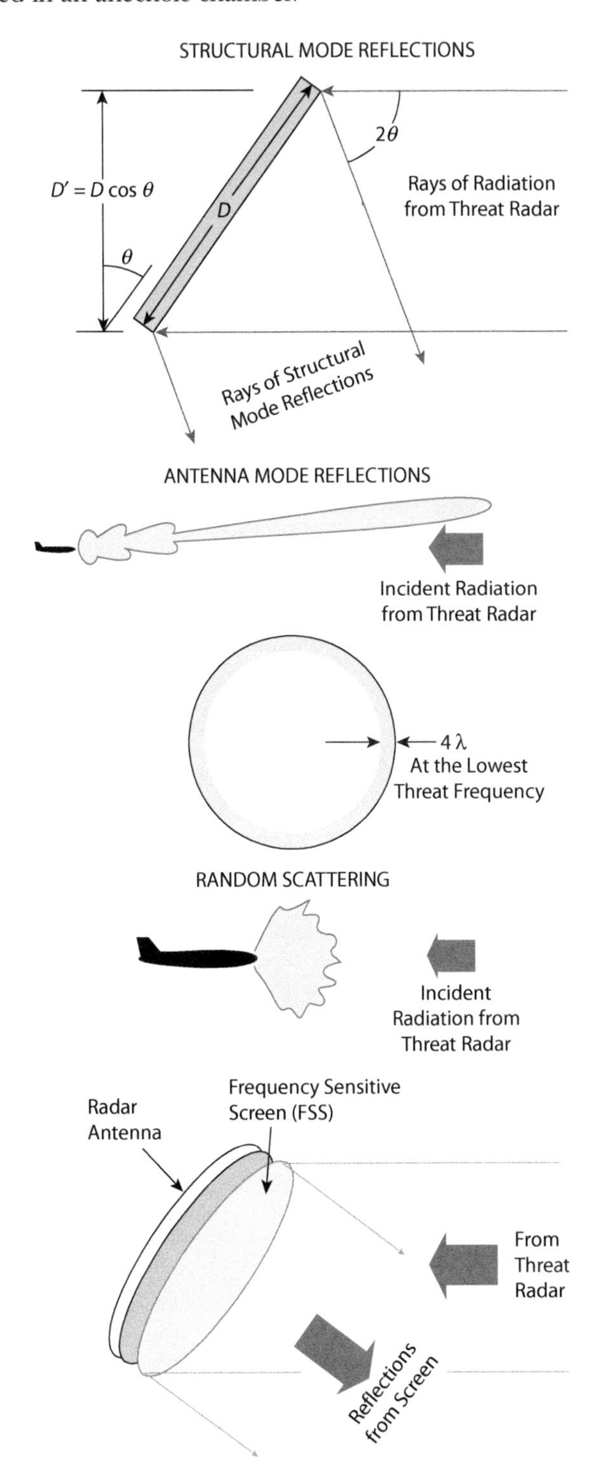

Further Reading

E. F. Knott, J. F. Shaeffer, and M. T. Tuley, *Radar Cross Section*, SciTech-IET, 2005.

D. C. Jenn, *Radar and Laser Cross Section Engineering*, AIAA Education Series, 1995.

B. A. Munk, *Frequency Selective Surfaces: Theory and Design*, Wiley, 2000.

B. A. Munk, *Finite Antenna Arrays and FSS*, Wiley, 2003.

D. Lynch Jr., *Introduction to RF Stealth*, SciTech-IET, 2004.

E. F. Knott, *Radar Cross Section Measurements*, SciTech-IET, 2006.

J. B. Pendry, D. Schurig, and D. R. Smith, "Controlling Electromagnetic Fields," *Science*, Vol. 312, pp. 1780–1782, 2006.

J. B. Pendry, "Time Reversal and Negative Refraction," *Science*, Vol. 322, pp. 71–73, 2008.

E. F. Knott, "Radar Cross Section," chapter 14 in M. Skolnik (ed.), *Radar Handbook*, 3rd ed., McGraw Hill, 3rd ed., 2008.

Test your understanding

1. List the four contributions to the RCS of an array antenna.
2. Why does ESA or AESA provide better stealth performance than a MSA?
3. What features can be incorporated into a radome to reduce the radar signature?

43

Advanced Processor Architectures

The 71620 module is a multichannel, high-speed data converter XMC that is designed for connection to HF or IF ports for communications, radar, and telemetry.

In light of the many existing advanced radar techniques that have been discussed to this point, it may seem like processor architectures are of little import. In fact, though, most of the advanced capabilities of airborne radars to date have been made practical only by substantial increases in digital processing throughput (Fig. 43-1).

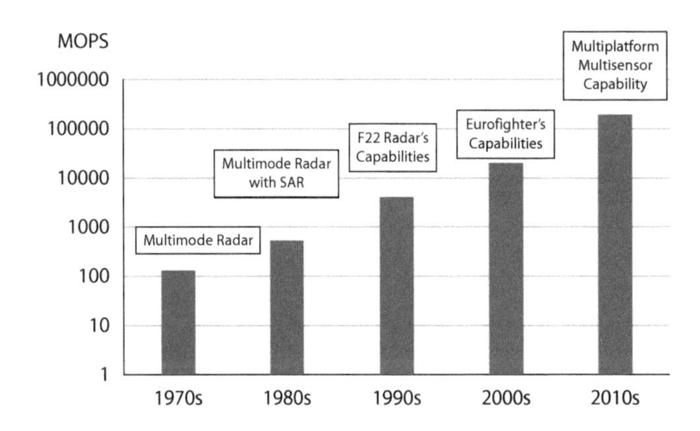

Figure 43-1. This graph shows 40 years of exponential growth in radar processing driven by Moore's law.

In the 1970s, multimode operation was made possible in fighters by replacing the hardwired fast Fourier transform (FFT) processor with a programmable signal processor (PSP) with 130 million operations per second (MOPS). Up to the mid-1990s these used custom designed digital signal processors based on discrete linear system-invariant (LSI) and later custom integrated circuit (IC) (application-specific integrated circuit [ASIC]) hardware. In the 1980s the development of high-speed ASICs increased throughput again by a factor of four, which enabled the addition of real-time synthetic aperture radar (SAR) processing. Over the course of the 1990s the processing power of microprocessors increased by a factor of at least 30. This increased the throughout enormously while extending the performance and functionality of modern fighter radars such as the F-22 and Eurofighter.

By the late 1990s, commercial microprocessors and digital signal processors for radar signal processing, which are based on commercial off-the-shelf (COTS) processing modules, were available. Scalable and modular radar processing solutions could then be implemented because of the increase in processing power, availability of standard processing modules, and acceptance of COTS for avionics use.

The availability of scalable, modular, high-power processing has made advanced radar capabilities practicable and enabled their combination with other radio frequency (RF) functions including electronic counter-countermeasures (ECCM) and communications electronic warfare (EW) functions within integrated RF systems. The demand for better radar and sensor system processing performance continues to increase at least to match the power available. Among the sources for this are escalations in resolution and area for mapping functions and in digitized bandwidths, new demands for multifunctionality and sensor fusion, and the use of sensors on unmanned platforms.

43.1 Basic Processing Building Blocks

We will start by considering the architectural features of processing. In architectural terms, the basic building blocks of a stored program computer are Control, Datapath, Memory, Input, and Output.

Control includes fetching from memory of instructions and data and storing data to memory. Control is also responsible for interpreting and decoding fetched instructions and controlling what the Datapath does with the instructions. Datapath is responsible for routing data to calculating units, performing calculations, and transferring data and to and from memory.

Most processors are synchronous and depend on a clock, which is normally a square wave. All operations and data transfers take place on a rising or falling edge of the clock and may take one or several cycles. The clock cycle rate is measured in units or multiples of hertz, with most modern processors having a cycle rate measured in gigahertz (10^9 Hz).

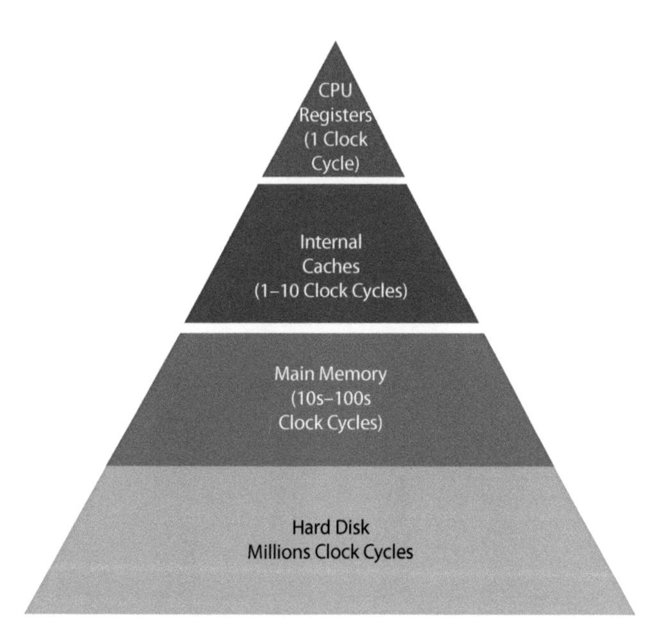

Figure 43-2. Memory hierarchy shows that the size and speed of memory can vary by a factor of a million or more from 100s of bytes and 100s of picoseconds at the top to terabytes and 100s of microseconds at the bottom.

Memory is used for storage of instructions and data. The memory system of a processor is hierarchical as shown in Figure 43-2, with small and fast memories at the top of the hierarchy and large and slow memories at the bottom. The fastest data storage is the processor register set that normally can be accessed in one clock cycle. Much of the complexity and circuitry used within modern processors is devoted to minimizing the effects of the low speed of large memories, mainly by using caches. Caches are small fast memories within the processor that mirror the contents of main memory. Caches depend on the principle of locality of reference, which states that the next data or instruction required has a high probability of being stored at a memory address near the last data or instruction accessed. Processor hardware and operating system software ensure that the internal caches contents reflect what is likely to be needed in the future and that main memory is updated to reflect any change to cache contents.

Input and output provide the familiar user input functions of keyboards, mice, and displays but also include high-speed communications channels that allow data and program code to be transferred to and from memory.

Processor Instruction Cycle. A simple sequence for a processor executing a program is as follows:

1. The processor fetches an instruction from an address in memory. The address is calculated by the control unit.
2. The instruction is decoded by the control unit, which determines what the processor does next.
3. The decoded instruction is executed. Some options for this instruction are:
   a. Address memory from a processor register OR
   b. Perform an arithmetic operation or logic operation using the contents of one or more of the processor registers OR
   c. If the instruction is a branch instruction then calculate a new value for the program counter if some condition is met.
4. The Write Phase. Some options for this phase are:
   a. Write the result of an arithmetic logic unit (ALU) calculation performed in the execute phase to a register OR
   b. Write from a register to main memory using an address set up in the execute phase OR
   c. Read from a main memory location addressed in the Execute phase and write to a processor register.

Architecture Classification. This simple processor instruction cycle is much more complicated in modern processors. It is for a single instruction performing a single operation at a time. In a real system the programmer works with a programming model of the processor, which usually involves parallelism. It is helpful to have basic terminology to describe architecture parallelism. One common classification of architectures is based on how the instructions and data are accessed and is known as Flynn's taxonomy. This provides four basic architecture classifications:

1. **Single Instruction, Single Data (SISD):** In any one clock cycle the processor executes one instruction on a stream of data; for example, it adds two values.
2. **Single Instruction, Multiple Data (SIMD):** In any one clock cycle the processor executes one instruction on a multiple streams of data; for example, it adds two vectors.
3. **Multiple Instruction, Multiple Data (MIMD):** In any one clock cycle the processor executes multiple instructions on a multiple streams of data; for example, it performs a scalar multiply on one vector and adds two other vectors together.
4. **Multiple Instruction, Single Data (MISD):** In any one clock cycle the processor executes multiple instructions but only on one data stream. For example, consider multiplication of two matrix elements: multiple instructions might be used to obtain the operands required by addressing multiple memories, while another instruction may direct the arithmetic unit to perform a multiplication operation.

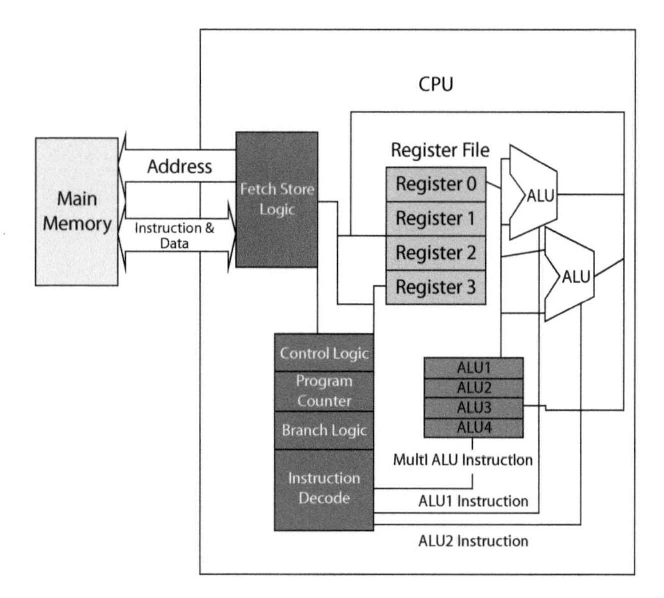

Figure 43-3. This illustration shows a single instruction, multiple issue superscalar architecture with a single instruction, multiple data coprocessor. The control logic is capable of determining if multiple instructions can be executed simultaneously and then decoding and issuing up to 3 instructions in a clock cycle.

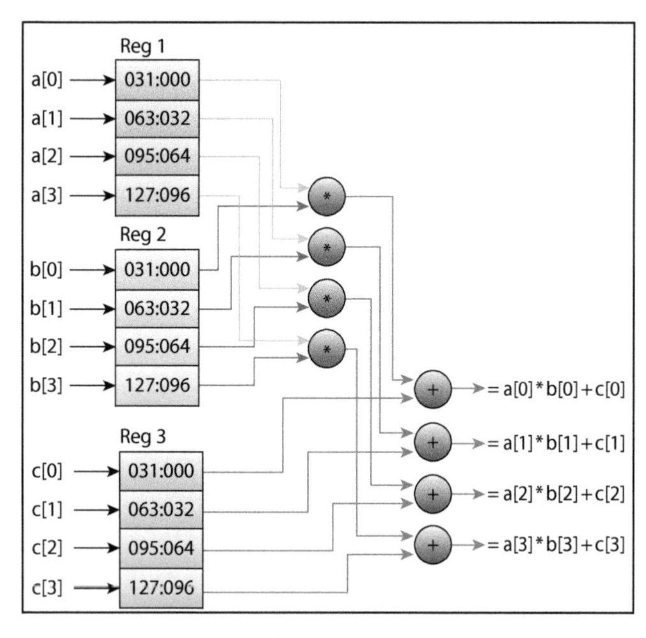

Figure 43-4. This example shows an SIMD co-processing unit performing a vector multiply and add being perfomed on 3 vectors of 4 floating point elements in a single clock cycle. Note the need for wide registers with multiple read access to support the simultaneous calculation.

There may be a difference between the architecture the high-level programmer sees and the actual hardware. The differences can be hidden by the high-level language compiler or hardware implemenation.

For example, the hardware may have an instruction pipeline that it checks to see if instructions can be executed or dispatched in parallel using parallel calculating units. This processor hardware technique is known as *superscalar dispatch*. If the processor is capable of launching N instructions at a time, then it is known as an N issue superscalar architecture (Fig. 43-3). However, from the programmer's point of view the complexity of this parallelism can be hidden and the architecture can appear SISD.

The Flynn taxonomy also applies to multiple processors. Most processor devices on the market today are multicore processors that operate on multiple instruction streams with multiple data streams and so may be considered MIMD. However, it is more usual for a programmer to write a single program with threads for each processor and synchronization between the processors when they need to exchange data. Therefore, it is more accurate to describe multiprocessors operating in this way as single program, multiple data (SPMD).

The addition of SIMD vector processing hardware to standard commercial microprocessor design has been especially important for radar signal processing. An additional bank of ALUs is controlled by a common instruction, each ALU getting its own stream of data from the register file (Fig. 43-4).

Number Representation. Computers store and manipulate their numerical data in binary format (strings of 1s and 0s). Numbers have various binary representations, the most common being unsigned integers, twos complement numbers, and floating points.

- **Unsigned integers:** These represent nonnegative integers. N bits can represent integers from 0 to 2^N.
- **Twos complement numbers:** These represent positive and negative integers. N bits can represent the integers from -2^{N-1} to $2^{N-1}-1$. A 32-bit twos complement number can represent integers between −4,294,967,296 and +4,294,967,295.
- **Floating points:** These are binary representations and approximations of real numbers. The may be considered to be the binary equivalent of scientific notation. A 32-bit floating point number (single precision) has a range from $\sim\pm10^{-38}$ to $\sim\pm10^{+38}$, and a 64-bit floating point number (double precision) has a range from $\sim\pm10^{-308}$ to $\sim\pm10^{+308}$.

Bit Growth, Rounding, and Dynamic Range. As arithmetic operations are performed, the number of bits needed to store results may grow. If the result cannot be expressed in the available number of bits, then the most significant bits of a result may be kept and the bottom bits discarded or rounded. This is called *rescaling* or *renormalization and rounding*. For integer operations, this must be managed manually. In field-programmable gate array (FPGA) implementations, rescaling and rounding hardware needs to be implemented and controlled explicitly. The methods used for rescaling and rounding may significantly impact results.

A floating point processing unit will handle renormalization automatically, thus greatly reducing the burden on the programmer. Most of us who have used scientific calculators or spreadsheets experience this when we are forced to select scientific notation format to display very large or very small results.

Today, much radar processing is now performed in floating point, preferably double precision. Double precision is useful where algorithm stability is important, for example, in constant false alarm rate (CFAR) calculations, Kalman filtering, and matrix inversion. It offers greater precision and dynamic range than a 32-bit integer does. However, the downside is that a double precision calculation is at least 1.6 times slower than a single precision calculation.

Moore's Law and Complementary Metal Oxide Semiconductors

Gordon Moore of Intel gave Moore's law in 1965. It predicted that the number of transistors on an integrated circuit would double every year. This was based on the observation that this had been happening every year since 1955, and Moore thought that the trend might continue until 1975. In 1975 Moore revised his law to a doubling every two years, and this trend has continued until the present day.

The main engine driving the exponential growth of processing power for the last 30 years has been the use of complementary metal oxide semiconductor and CMOS.

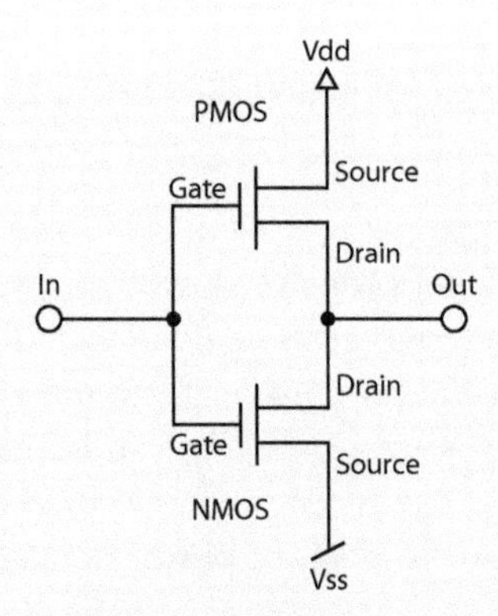

Basic CMOS inverter

Introducing a high-speed CMOS into the processor design increases the speed (clock rates) and the number of gates. From 1985 to 2005, processor clock rates rose by about 40 percent per year and processor performance by 54 percent per year. From 2005 onward, maximum clock rates have flattened out, with almost all being under 4 GHz. The reason for this limit is that increases in processor power dissipation could no longer be managed.

Scaling of CMOS Circuitry. CMOS provides an excellent technology for scaling the size of circuitry. It is far less complex to redesign transistors between generations of CMOS than for other technologies. The metric used to characterize a generation of CMOS is the channel length of the transistors used. For example, the current generation is known as 22 nm technology.

The linear scaling factor between generations is about ×0.7 to ×0.8. For example, from 130 nm to 90 nm is ×0.69 scaling, and 28 nm to 22 nm is ×0.78 scaling. These factors provide an approximate doubling of transistor density, such as $1/(0.7^2)$ ~ = ×2 scaling for transistor density.

Almost all of the basic structures needed to construct processors are made from the transistors, which are mainly CMOS. The normal CMOS design elements are supplemented by more compact transistors to create high-density random access memory, and specialized cells are used to create non-volatile memory cells. The Intel 4004 processor used about 2,500 transistors. The Intel 8086 used about 29,000 transistors.

A current multicore processor uses 3 to 4 billion transistors.

Integer processing is still employed in the early stages of processing, where FPGAs are frequently used to provide high-throughput, low-latency calculation such as in digital generation of in-phase quadrature (I/Q) signals, digital downconversion, filtering, and FFTs.

43.2 Low-Level Processing Architectures

A couple of methods can be used at the lower levels of implementation to speed up processing. They are implemented in FPGA design and in digital signal processors (DSPs) and microprocessors. These techniques often recur at a higher level where many processors may be used together to provide a radar or sensor processor. The methods we will look at are parallel calculation and pipelining. We will explore these methods using a single example, performing a complex multiply.

Pipelining and Parallelism. Pipelining has been used since the very earliest days of computing and radar processing to increase arithmetical throughput. The technique accepts a longer latency (how long it takes to get the first result) to achieve a higher calculation throughput. The pipeline may internally perform operations in parallel to gain speed, but this is not essential. For instance, an integer multiply may be split into eight stages of 5 ns for a 40 ns latency but may have a new pair of operands and new result every cycle of 5 ns.

There are three simple requirements for pipeline implementation:

1. The operation to be performed can be broken down into small pieces that can be performed faster than the overall operation.
2. The operations to be performed are expected to be performed multiple times consecutively.
3. The extra clock cycles for the latency do not cause any problems.

Today there are a number of options for implementing processors to create modules. Which one is selected will depend on the needs of the application and the costs and the benefits of the option chosen.

Full-Custom IC Design. This design offers the ultimate combination of highest performance, lowest power, and smallest size. However, the design costs of implementing large custom IC designs with highest available performance in the latest CMOS technologies is very high and can be justified only for volumes of millions of parts per year, or when no other digital logic provides adequate speed or power performance.

Uncommitted Logic Arrays, ASICs, and Gate Arrays. A method of creating semicustom ICs was developed in the late 1970s that allowed designers to implement their processing circuit in IC form without incurring the full nonrecurring costs of a custom integrated circuit. The initial versions of these semicustom circuits were based on *uncommitted logic arrays* (ULAs). In these devices

Measures of Processor Performance

THE MOST MEANINGFUL MEASURE OF PROCESSING PERFORMANCE IS HOW LONG it takes to run the desired application on the target hardware under real-world conditions. However, this measure is seldom of use before the application is developed. A number of metrics are commonly used, and all of these, the most common of which are now described, should be viewed with some degree of caution.

Clock Speed (MHz, GHz). Probably the least accurate of the common metrics. Performance can depend far more on architecture or the nature of the application. In one example, processor clock speed on a specific microprocessor was doubled from 500 MHz to 1,000 MHz; however, overall runtime for the application decreased by less than 15 percent (see Amdahl's law).

Millions of Instructions per Second. This measure of how many millions of instructions a second a processor is capable of performing is misleading in two ways. First, most processors take a different number of clock cycles per instruction for different instructions. Performance can vary widely depending on the instruction mix of the application. Second, many processors support instructions that execute many operations in a single instruction. For example, a single SIMD instruction may perform 16 arithmetic operations or more in a single cycle, but this would count as only 1, not 16, instructions.

Millions of Floating-Point Operations (MFLOPs). Usually, this is the number of millions of floating-point operations a processor can perform in a second. This is sometimes calculated by assuming the instruction with the maximum number of floating-point operations that can be executed on every cycle. For example, if a SIMD unit is capable of 16 floating-point multiply-accumulate operations in one cycle of a 3 GHz clock then the MFLOP figure would be $16 \times 3 \times 10^3$ MHz or 48,000 MFLOPs. As this assumes that no instruction fetches are taking place and all operands are in the appropriate registers, this figure can never be achieved in practice. In addition, some applications make very little use of floating point, and therefore a MFLOPs measure gives no guidance about performance.

Benchmarking Programs. Timing the execution of a variety of programs to characterize the performance of a computer is generally considered the best way of giving a general measure of a processor's performance. Sets of programs defined for this purpose are known as benchmark suites. Originally many of the programs were artificial examples, such as performing matrix multiplications or testing for prime numbers. Computer manufacturers have been known to optimize compilers or hardware to perform well on the synthetic benchmarks. To avoid this, today's benchmark suites are a combination of real-world and synthetic examples. The Standard Performance Evaluation Corporation (SPEC) provides a useful set of benchmarks. At the time of writing, the most commonly used general-purpose benchmark suite is SPEC CPU2006. Results from the individual benchmarks in the suite are available to allow assessment of performance, in particular application areas.

Amdahl's Law. This defines how much an application or hardware can be sped up depending on what proportion of the application or hardware is being improved. Software and hardware examples follow. In general terms, Amdahl's law can help guide where to speed up processing and supports the view that the most frequently performed operations should be made as fast as possible.

Example 1. Assume that a program spends 40 percent of execution time performing an FFT. It is proposed to make the FFT run 10 times faster. What program speed up can be expected? Assume initial execution time is 1 s.

$$\text{Execution time after improvement} = \frac{\text{Execution time affected by improvement}}{\text{Amount of improvement}} + \text{Execution time unaffected}$$

Execution time = 0.4/10 + 0.6 = 0.64 s ~ = 56 percent speed improvement. Note that even if the FFT were to take no time at all, the execution could not be less than 0.6 s.

Example 2. A processor takes 10 sec to run an application. To speed the application up, a proposal is made to increase the clock frequency for the processor from 500 MHz to 1,000 MHz. However, the processor spends 80 percent of its time accessing main memory, and the speed of these accesses will not change. What improvement in run time can be expected?

As before

$$\text{Execution time after improvement} = \frac{\text{Execution time affected by improvement}}{\text{Amount of improvement}} + \text{Execution time unaffected}$$

Execution time = (10 s × 20 percent)/2 + 8 s = 9 s, or a 10 percent improvement in performance.

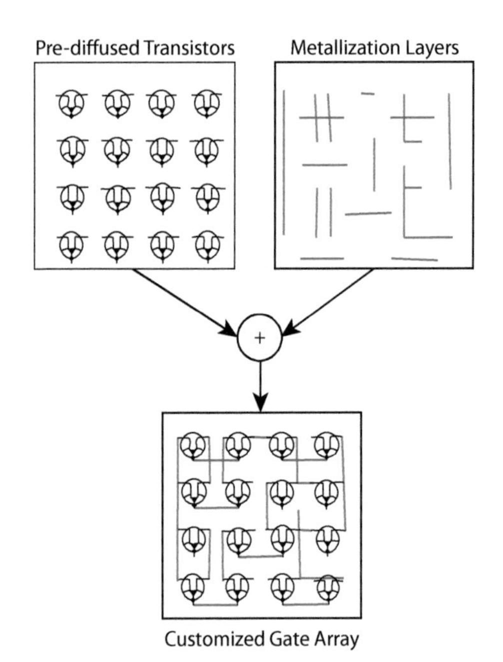

Figure 43-5. In this customization of a gate array by user defined metal track, the designer's circuit schematic defines the masks for the metallization layers. The metallization layers provide the connection to transistors that have been pre-diffused in a standardized pattern onto silicon wafers.

the transistors or gates are pre-diffused onto wafers and held in storage. Designers create their circuit design, which is then used to specify metal interconnections between the available gates. Metallization layers are fabricated to join the pre-diffused transistors or gates to create the desired circuit (Fig. 43-5).

Over time, the range of circuitry in the base array (pre-metallization) increased from simple gates to include more complicated devices such as arithmetic elements and small memories. Depending on the details, these are also referred to as ASICS or gate arrays.

The nonrecurring engineering (NRE) cost for these devices, although still high, may be less than a tenth that of a full-custom design.

FPGA. The next development in the evolution of customized processing circuitry was the electrically programmable and erasable gate array. Instead of having circuit interconnections that were determined by an expensive mask set, these devices have programmable interconnect with the programming information for the interconnect held in memory on the device. The inherent reprogrammability of these devices allowed reprogramming of circuitry in the field. Hence, the devices became known as FPGAs. As process geometries for CMOS have shrunk, the complexity of mask sets to manufacture custom or semicustom ICs has increased from perhaps $100,000 in the early 1990s to millions of dollars today. This has increased the competitive advantage of FPGAs for all but high-volume devices.

FPGAs offer many of the same advantages as custom IC design. By exploiting the hardware parallelism inherent in the FPGA structure (Fig. 43-6), designers can implement fast integer

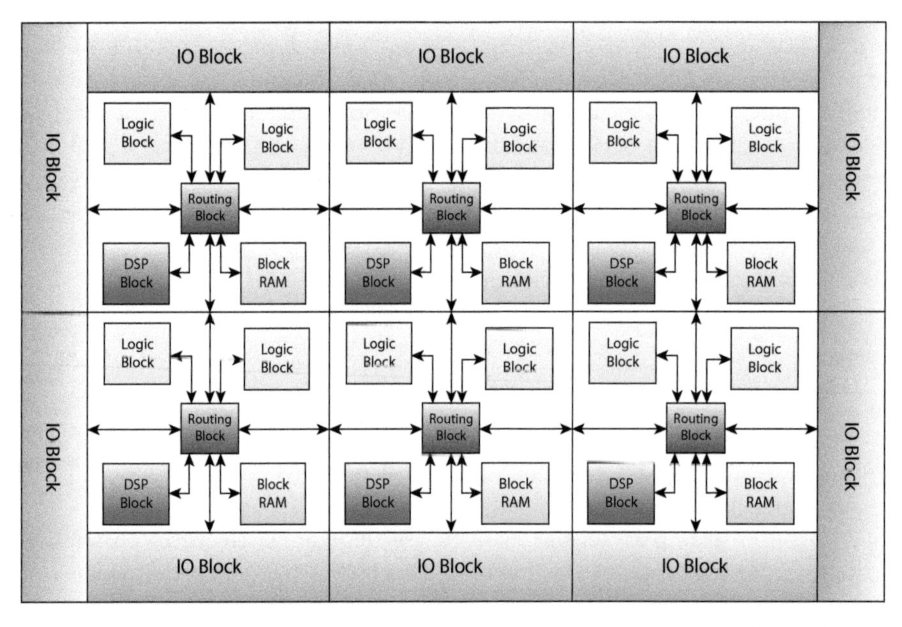

Figure 43-6. FPGA internal structure. In simplified form this shows many replications, in a regular pattern of pre-defined elements such as logic, memory and digital signal processing arithmetic blocks. Custom interconnections are created by programming electrical routing blocks. A real FPGA could have many thousands of all the basic elements.

processing, which can cope with hard real-time, high-speed streaming data. For example, if there is a requirement to process a digitized stream of radar data as it arrives, then FPGA can provide the only low-latency solution other than ASIC/ custom IC design. Other advantages of the FPGA solution are the low NRE, the inclusion of useful intellectual property (IP) or complex circuit blocks by the FPGA vendor, and the flexibility of being able to make modifications to the design cheaply. FPGAs also offer the capability to include embedded microprocessors within the design. The disadvantages of FPGA are that the clock speed at which arithmetic is performed is about five times slower for integer and about ten times slower for single-precision floating point than for a microprocessor or DSP at the same technology node (currently 28 nm). Because of the number of transistors devoted to programmable routing resource and the requirement to replicate circuit elements, whether or not they are required, an FPGA may be less efficient than an ASIC in terms of gates used by a factor of 10 or more, and unit cost and power dissipation can reflect this.

Digital Signal Processors. Offering a halfway house between FPGA and microprocessor implementation, DSPs provide multiple cores with integer and floating point with clock speeds comparable to microprocessors. DSPs usually provide multiple SIMD ALUs and address generators that support addressing multidimensional arrays and gather/scatter operations and separate memories for instructions and data. These facilities together with extensive use of DMA permit DSPs to achieve high operations per cycle count. To achieve the high performance, DSP devices require careful programming and optimization. Although specialized optimizing compilers are available, device simulators are still used to check and improve operation. The software effort to capitalize on the operation of the device is normally higher than for a microprocessor design, and generally the code is not as portable. An additional advantage of DSPs is that they can provide a lower power solution than a multicore microprocessor or a large, heavily utilized FPGA.

Embedded Processors. These microprocessors are embedded within other ICs to provide programmability. They are available in 32-bit and 64-bit versions and support multicore and SIMD acceleration. These devices are the most pervasive form of processing and provide the processing in mobile phones, tablets, and most home entertainment devices. More than 8 billion ARM-embedded processor cores were shipped in 2012 alone. Embedded processors are available in multicore format, eight being the largest number of cores at the time of writing. The performance of the devices can be less than a tenth of a high-end, multicore, general-purpose microprocessor, with the embedded core having only 2 percent of the power dissipation. One of their most promising applications currently is the use of embedded processors within FPGAs and DSPs allowing both these processing solutions to have a conventional processor interface programmable in a high-level language.

Microprocessors. The modern microprocessor device has multiple, identical, microprocessor cores with individual and shared caches. At the time of writing, devices with 12 cores are available, and this number will increase in the future. To support these cores a large percentage of the chip area is devoted to cache, and as the number of cores increases this percentage of chip devoted to cache must also increase.

Each core has multiple ALUs, including SIMD coprocessors that provide signal processing capability and a flexible way to perform radar signal processing. Optimizing and vectorizing compilers are available to extract good performance from these devices. The best signal processing performance is usually obtained using highly hand-optimized vector signal processing libraries. However, due to operating system cache management, and the multi-threaded nature of microprocessor operating sytems it can be difficult to ensure that hard real time requirements are met. In addition, to maintain the highest performance, the data has to be held in cache, which can make signal processing implementations sensitive to problem size. Another problem with recent and current generation devices is thermal power dissipation, which can limit the performance or require liquid cooling.

General-Purpose Computing on Graphic Processing Units. Graphic processing units (GPUs) are an example of massively parallel processing implemented in a single IC. They can have hundreds to thousands of ALUs operating in parallel and can be considered at the lower level of their hierarchy as SIMD devices and at a higher level as MIMD. In fact, the hardware and software support for different threads of calculation is what distinguishes programming a GPU from single or multicore processor core. At present, most multicores support two hardware supported threads (hyperthreads) per core, and at the time of writing a state-of-the-art GPU will support over 16,000 threads.

Figure 43-7 shows a block diagram of a GPU with eight building blocks, each with two streaming multiprocessors per building block and eight streaming processors per multiprocessor. GPUs currently need an X86 family CPU to act as a host.

If a GPU can sustain a high utilization on its streaming processors, then it is capable of exceeding the performance of a multicore microprocessor by a factor of between 10 and 100. To achieve this, the problem has to be highly parallel, the data accesses to the GPU's global memory have to permit large sequential transfers and transactions with the host processor involvement kept to a minimum. A drawback of GPUs is their high power consumption, which can be two to three times that of the largest microprocessors.

43.3 Meeting Real-Time Data Density Requirements

Most airborne radar systems need to complete data processing from a pulse or a burst of pulses for most modes in real time.

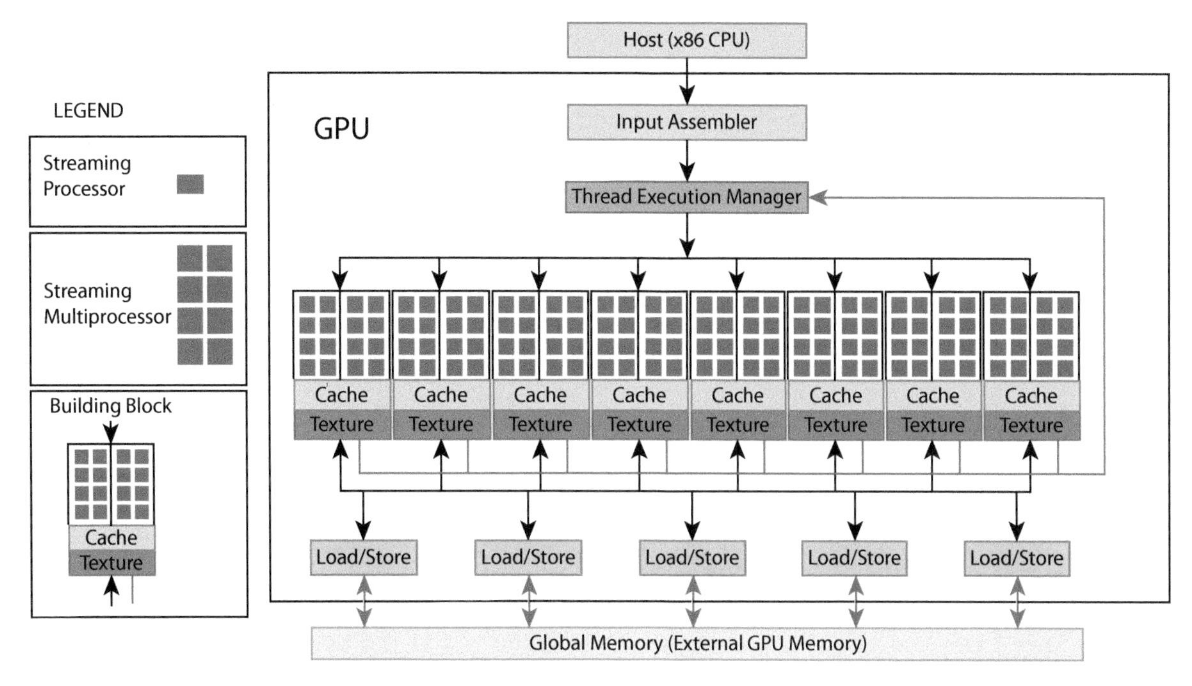

Figure 43-7. In this general-purpose GPU, the hierarchical parallel architecture is scaled down. Recent GPUs have hundreds of building blocks and thousands of streaming processors.

In other words, the time it takes to process the data should be less than the time it takes to collect it. That is not to say that the latency of the processing needs to meet the same requirement. It may take several pulse burst times before the first processed data appear at the output, but after that the output rate will match the input rate. There are exceptions of course. For example, for some high-resolution spotlight modes SAR it may take longer to process the data for an image than was required to collect the data, so that it is not possible to continuously image regions of interest.

The concepts of pipelining and parallelism that were discussed earlier in the context of individual calculations can now be applied at the level of entire radar processing chain. The underlying rules that apply at the individual calculation level apply equally here, and the overall aim is to keep data moving through the pipeline without creating bottlenecks. The first principle is to provide sufficient processing power where it is needed. However, this aim cannot be achieved perfectly. Differences in workload at the various stages of the processing pipeline mean the processor architecture must provide memory buffering to allow these throughput discontinuities to be smoothed out.

In addition, modern airborne radars require rapid interleaving of radar modes, which can have significant differences in workload, such as SAR and ground moving target indication (GMTI). This requires the processor to have the ability to rapidly reconfigure and to have sufficient memory buffering to smooth out the differences in data rate. Such mode interleaving can be made to appear seamless to the operator, but this

Table 43-1. Examples of Processing in a Generation 2 Radar Processor

Unit	Name	Functions	Processing Power
RGC	Receiver Gain Control	Channel and IQ formation, saturation detection, scaling, clutter frequency measurement	600 MoPs
TDP	Time-Domain Processor	Digital downconversion and decimation, clutter frequency offsetting and filtering, pulse compression, power, log power, or amplitude calculation	2,400 MoPs
FFP	Fast Fourier Processor	Doppler processing, filtering, power calculation	1,800 MoPs
CFP	CFAR Processor	Area averaging, thresholding, generation of detections	1,600 MoPs
COP	Correlation Processor	Space and time correlation of detections, track filtering	675 MoPs
DRM	Display Refresh Memory	Scan and range-to-plan conversion, display storage and generation	200 MoPs
QBM	Quad Bulk Memory	Interprocessor buffer memory	Not Applicable
DAP	Data Processor	Multitarget tracking, target recognition, control of radar, interfacing to external aircraft systems, distribution of navigation data, sensor integration	400 MoPs

requires careful design and attention to the timing details of the radar.

Table 43-1 shows stages of a radar pipeline for a radar processor from the mid-1990s, distribution of functions, and some indication of the processing power required at each stage. A block diagram and an example processing flow diagram is given in Figure 43-8.

Several generations of design of digital radar processor can be distinguished.

Generation 1 (Pre-1980s). Processors consisted of long-latency, pipelined processors with little flexibility in their programming. Control was hardwired, and radars generally supported a very restricted number of functions.

Generation 2 (1980s to Mid-1990s). Processor in this generation were able to take advantage of higher speed ASIC technology

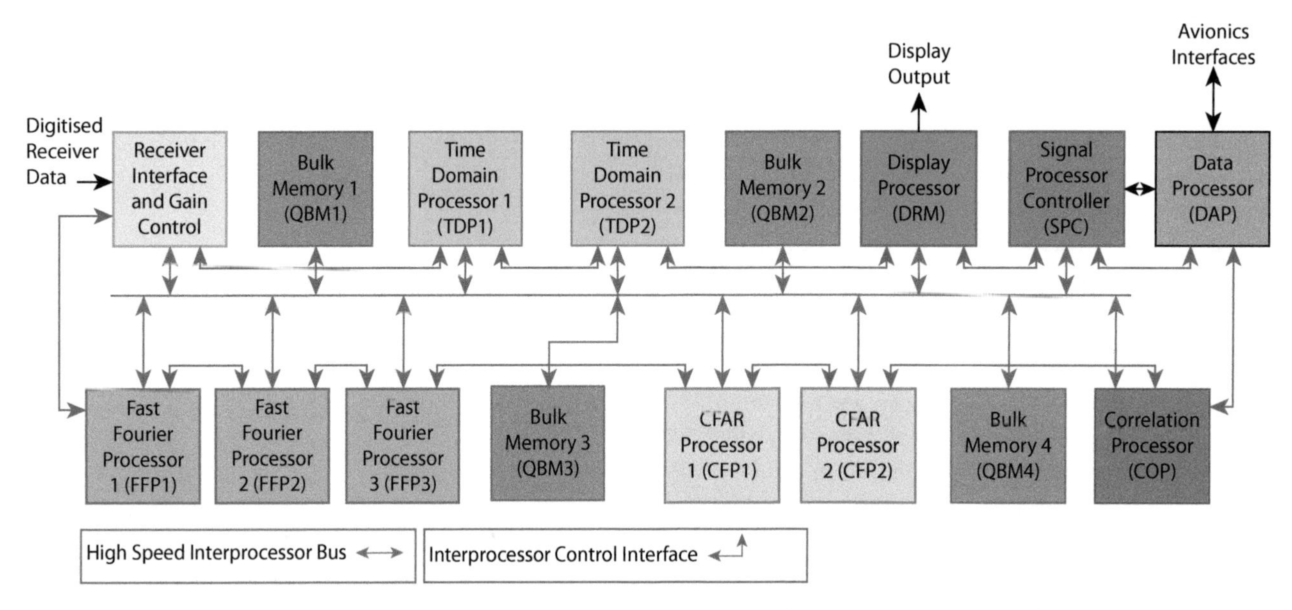

Figure 43-8. A modular second-generation pipelined radar processor with function specific processing modules conforming to a common module specification and wired to a common bus.

to shorten pipeline latency and offer more flexibility in programming by using microprocessors for control. For this generation most components were still custom designed, and to obtain high performance, custom backplanes and interconnect were required. This generation of processors enabled the first generation of true multimode pulsed-Doppler radars.

In this generation, designs could be highly modular, with the same processing module design being used many times. The operations of modules could be tailored to perform specific pipeline functions within the signal processing pipeline by using customized DSP ASICs.

Memory modules would typically be used to buffer data transfers between processors, and the address generators on these boards could support matrix reorganization (transposing, corner turning, and other functions) during data transfers. This removed the scatter/gather overhead from the individual processing modules.

The high-speed inter-processor bus was designed to support simultaneous multichannel data transfers with extremely low latency on a common wired bus. The most common COTS bus standard interfaces did not have the performance required for a radar processor.

Although this architecture is adequate for the applications for which it was designed, it has architectural shortcomings. The main shortcoming is that the functionally specialized nature of the ASICs restricts redeployment of resources to heavily loaded stages of the pipeline. For example, if the fast fourier processor (FFP) stage of the processing is overloaded, then unused resource elsewhere in the processor, such as the constant false alarm rate processor (CFP), cannot be used to boost performance. A second shortcoming arises from the use of a common bus, which limits simultaneous data transfers.

Generation 3: (Mid-1990s to Early 2000s). This generation of processor reduced the custom hardware requirement still further. Use of commercially available components, including general-purpose processors with SIMD signal processing units and FPGAs, replaced ASICs. Switching components developed for the communications market allowed for the implementation of fast and flexible data networks without using custom components and removed the need for a common bus. Larger memories, higher processing rates, and modules that were not function specific improved multifunction performance and lowered software development costs.

Figure 43-9 shows a third-generation architecture. Its processing module consists of three identical processing elements each with a data connection to a crossbar switch. The crossbar switch permits bidirectional data paths to be established between any two processing elements without blocking any other data transfers in the system.

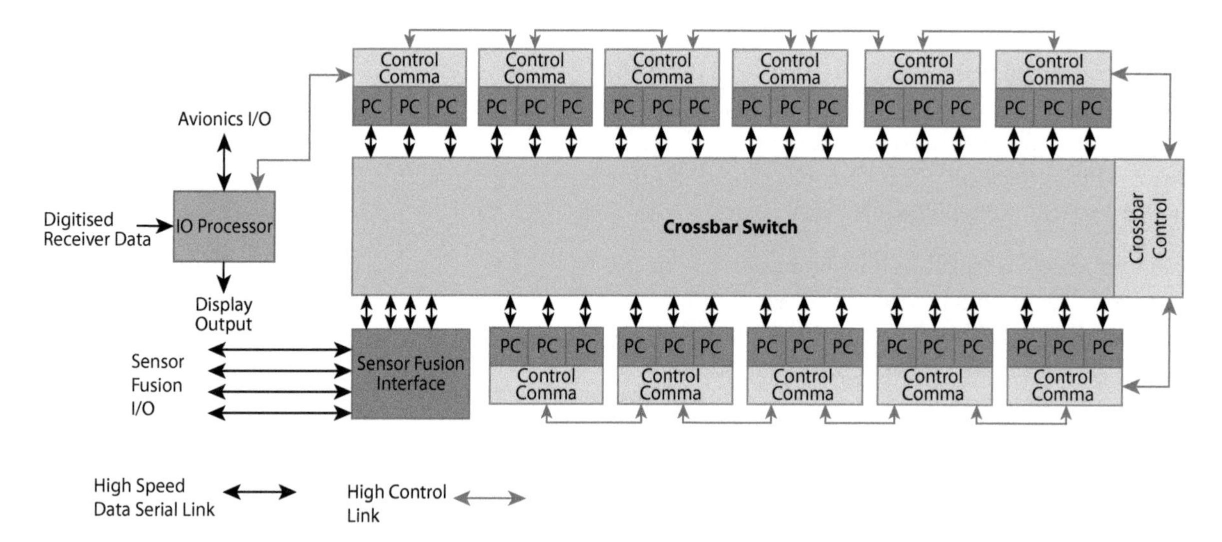

Figure 43-9. This third-generation radar processor architecture uses the same processing module for almost all of the processing stages. Higher speed communications are supported by replacing the common bus of previous generations with a crossbar switching structure.

This generation of design used ruggedized versions of commercial processor, memory, and switch components to create a custom architecture implementation. This architecture shows a much more regular structure with improved modularity overcoming the main defects of a Generation 2 design. Now the interconnections could be configured as required, and the processing was general purpose and not pipeline stage specific.

Generation 4: (Early 2000s Onward). The transition from third- to fourth-generation processing does not show as significant of a change as between previous generations. The performance advantages of this architecture come from rapid tracking of commercial processing developments and the use of commercial standards for high-speed communications.

The main challenges in using this architecture are in partitioning the software to make most efficient use of the multiple processor cores within the processing elements. This design activity can be performed either manually using conventional high-level language or with the aid of high-level and graphical design tools.

Apart from increases in speed and performance, a main difference between fourth-generation processors and previous ones is their use in military off-the-shelf (MOTS) modules that implement the sensor processor. The latest generation of processors also makes it possible to extend the processing between many line replaceable units (LRUs). Because the main benefits of the fourth-generation architecture derive from its modularity, it is discussed in greater detail in the next section.

43.4 Modular Design and Fault Tolerance

Modular design offers great operational, logistical, and cost benefits for an avionics system. Even within the scope of a single radar system, for a specific aircraft type it is possible to implement modular design and obtain many of the benefits.

For example, the second-generation architecture from the early 1990s displayed many of these benefits.

With the increase in processing power of very high-performance, general-purpose microprocessors, the need for function specific hardware has been reduced. Today, except for the most demanding applications, only the very front end of a radar processor needs to have custom-designed hardware. Commercial standards for processor modules have evolved to support high-speed serial interfaces, switching fabric standards, and multiple options for meeting military environmental specifications. The definitions of these modules are available as a set of American National Standards/VMEBus International Trade Association (ANSI/VITA) standards for VPX modules, which supplement the earlier Versa Module Eurocard (VME) standards for processing modules.

These new standards have been developed to support more mechanically rugged modules and to provide support for higher temperature modules and liquid cooling (VITA 46, 48, and 65). Basing processor designs on these modules allows rapid development, uniformity of design, multiple sources of modules, and reduced spares holdings for logistical support.

The disadvantages of these modules are that the underlying processor technology varies rapidly and that the specifications allow a great deal of flexibility in the choice of serial protocols. There is no guarantee that the MOTS or COTS module vendor will support the same interface configuration or processor configuration over two or more generations of module design.

Figure 43-10 shows an example of a fourth-generation radar processor designed with MOTS modules based on the VPX

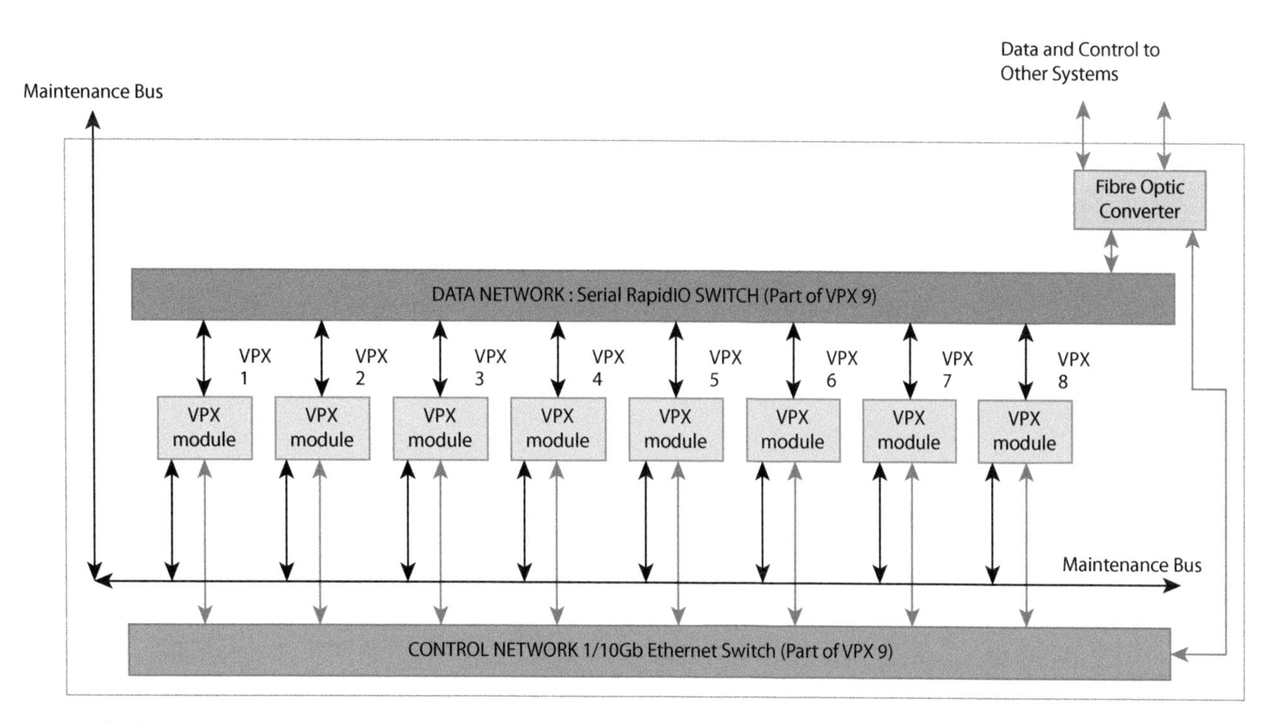

Figure 43-10. This fourth-generation radar processor uses VPX COTS standard modules for all processing elements and open standards compliant inter-module communication based on COTS switches.

standards. The system shown would occupy 10 slots in a VPX chassis: VP1 to VP8 for VPX processing modules; VPX9 for a data and control switching network; and VPX10 for a power supply, which is not shown.

The data network in this processor is using a standard called serial rapid IO, which allows flexible interconnect topologies together with strong support for redundancy in communications. The same VPX module also supports a multiport 10 GB Ethernet switch for a control network. With a sufficient number of spare ports these switches can support scalable processing outside the processor LRU chassis.

Figure 43-11 shows the block diagram of one of the eight VPX processing modules shown in Figure 43-10. The processing module consists of two quad core Intel microprocessors that are connected by PCIexpress and SRIO interfaces within the module and by Serial RapidIO and 10 GB Ethernet via the chassis backplane and the switches in VPx9.

Fault Tolerance. The mission-critical nature of airborne radar demands that the radar processor have high reliability and be capable of operating even in the presence of faults. If the processor is being used to support other sensors then this requirement becomes even more vital.

The main method used within a sensor processor to ensure correct operation is built-in test (BIT), whose purpose is twofold: (1) to detect errors when they occur; and (2) to locate the fault as accurately as possible. All parts of the processor will perform BIT on power-up (PBIT) and supplement this by a periodic or continuous, uninitiated, built-in test (CBIT). Additionally the operator can initiate a built-in test sequence (IBIT).

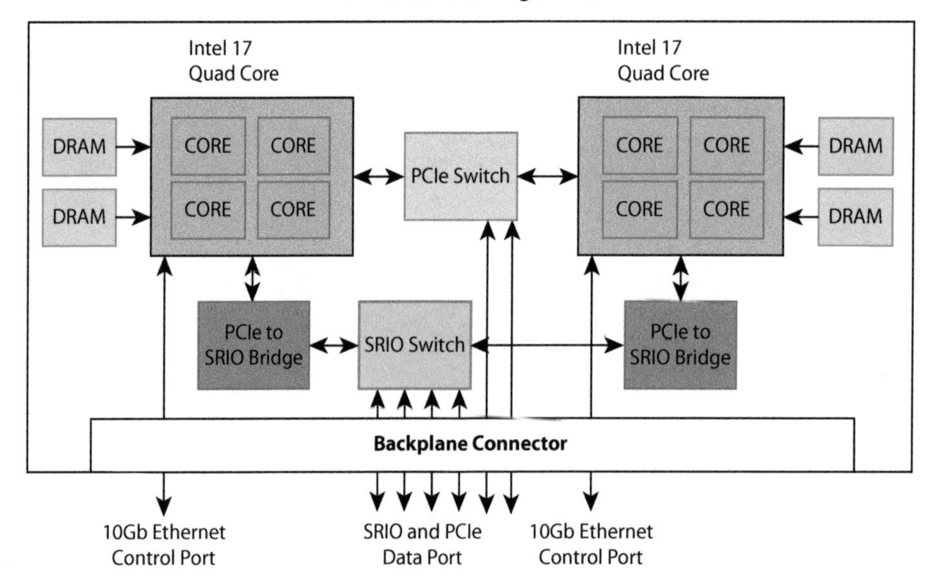

Figure 43-11. This VPX-based processor module consists of two quad core Intel microprocessors connected by PCIexpress and SRIO interfaces within the module and by Serial RapidIO and 10 GB Ethernet via the chassis backplane and the switches in VPx9.

In earlier generations of processor design, when custom hardware was used, the design would include hardware features specified to meet fault detection and location requirements. One weakness of COTS-based design is that hardware support for BIT is not normally available to the radar designer. At the software level the COTS module manufacture normally provides software BIT routines for PBIT and IBIT. Writing CBIT software that can reliably check out the operation of a billion gate microprocessor, while it continues to perform its radar processing, is no easy task. However, there is some microprocessor support for CBIT. For example, most systems support temperature monitoring, perform error detection and correction on memories and communications packets, and operate watchdog timers to detect programs or processors that are behaving abnormally.

Given that fault detection is challenging, modern modular processors do offer advantages for fault tolerance and isolation. Fault tolerant features of modular COTS designs include the following:

1. Identical hardware modules and flexible switching structures that allow faulty modules to be isolated and standby spares to be switched in.
2. Serial protocols, such as Serial RapidIO, that support fault detection and message rerouting. This can be supplemented by the inclusion of redundancy in the switching structures. For the VPX example, shown in Figure 43-10, this could involve including an additional switching module to provide redundancy in the routing resource.
3. Processing elements that have sufficient storage and processing power to store detailed information about their own structure down to the component level and to perform detailed diagnostics and to devise intelligent reconfiguration of the equipment.
4. Modern processors and their operating systems support operation of virtual machines. Virtual machines isolate executing programs from the real hardware and thus minimize the consequences of a program error or other malfunction. The failure can be limited to the virtual machine and not cause a more general systems failure. The supervising operating system can then perform appropriate recovery actions.

43.5 Future Challenges in Processing

What does the future hold for radar and sensor processing?

The next 10 years will be an interesting period in avionics computing and for computing in general. For airborne radar traditional tasks will continue to have increasing demands for processing power. These demands will arise from wide-area long endurance surveillance, wideband operation, sensor fusion, platform networking, cognitive sensing, autonomy, and artificial intelligence. All of these increases in scope are interrelated, and many of them arise from the requirements for operating sensors in unmanned air vehicles.

For computing in general, significant changes are needed to meet the challenge of making effective use of the ever-increasing number of parallel cores available. Technological limitations threaten to cause failure of Moore's law, and innovations will be required at all levels to find ways to extend the growth in processing capacity. The ever-present problem of removing waste heat from the ever increasing number and density of transistors will force new methods of actively cooling devices within packages.

Wideband multifunction RF operation arises from the requirement and opportunity to maximize the use of existing apertures and antennas for multiple-sensor applications. This creates a need for wideband frequency agility and an opportunity for much closer coupling between radar, ECM, ECCM, and signal intelligence. The implications for processing are for much higher bandwidth, low-latency processing near the front end of the radar, and this is likely to rely on much improved FPGA performance or the use of custom silicon.

Fusing the results from multiple sensors is already being performed to some degree on many platforms. The quantitative and qualitative aspects of sensor fusion are likely to change significantly. This will be due, in part, to increased use of data coming from networked platforms, cognitive aspects of the sensor suite, and requirements to provide data to autonomous or artificially intelligent systems.

Cognitive sensing will use what is known about the operating environment from the radar and other sensors and platforms to modify and optimize the operation of sensors including radar. There is a large overlap between cognitive sensing and sensor fusion, but one distinction from the processing perspective may be the need for rapid access to large nonvolatile memories to inform processing of previously known data. A simple example of this would be to perform real-time coherent change detection between previously collected data and a current SAR image, but many more opportunities and challenges will arise.

Unmanned air vehicles will operate in hostile environments with potentially limited control and data communications with their base. They will require the capability to achieve their mission objectives in the absence of continuous direct control. This implies that they must be able to fly their mission and optimally control their sensors autonomously. This capability has significant implications for all avionics processing systems.

Technological Changes. Moore's law continues to be valid for now, but at the time of writing it is predicted that this will no longer be the case by about 2022. There is agreement that fundamental quantum limits of CMOS gates will occur around 5 nm. At about this length, quantum effects cause currents to flow through gates regardless of the state of the gate voltage.

In practice, fabrication may become difficult at about the 8 nm technology point.

If the 8 nm node is reached, the number of transistors should be about 10 times the number available today (Table 43-2). Assuming a linear scaling from today's CPUs and GPUs, this would give 100–120 cores in a multicore CPU and 25,000 CUDA cores in a GPU. Conservative estimates based suggest a useable speed-up of only ×10 is likely.

Assuming that these devices can be programmed to achieve high utilization cores, they will require increased I/O bandwidth, probably by a factor of at least 10 times greater than today's rates. This increased bandwidth requirement will need advances in silicon photonics.

Reliability Concerns at Smaller Geometries. As a result of the reducing scale of CMOS geometries, there are concerns about what reliability issues may arise because of the use of these technologies:

- The intrinsic reliability of the very thin layers used
- The potentially extremely high thermal dissipation locally within a device, such as a 28 mm GPU with a transistor density of ~12 million transistors/mm^2
- The increased probability of malfunction due to single-event upsets (SEUs), which are caused by cosmic radiation passing through a chip and generating charged particles that trigger errors in behavior, of which smaller device geometries are more susceptible because many more transistors can be affected (lower switching energy and higher transistor density).

43.6 Advanced Developments

Three-Dimensional Device Fabrication. The process of extending device fabrication in the vertical axis has already begun with Intel's introduction of 3D transistors and multiple vendors mounting dies directly on top of each other. Apart from increasing density of components, a future main driver for this may be reducing the delay (and power) required to access off-processor (main) memory. Until Silicon photonic interfaces become commonplace on devices, there may be a period of time where memory interface performance will be improved by direct coupling of memory devices on processors to memories.

Three-dimensional assembly of chips also be used to mount active cooling layers (microfluidic, heat pump, or thermoelectric cooling) directly within the 3D device stack. This may be combined with direct 3D printing of cooling and packaging structures on silicon.

Silicon Photonics. Silicon photonics is a technology that enables fabrication of optical devices on silicon, potentially alongside standard CMOS gates. This technology may initially be used to transfer data on and off chip more quickly than is currently

Table 43-2. Extrapolation of Moore's Law

Year of Introduction	Feature Size (nm)	Number of Transistors
2012	28	7 billion
2014	22	11 billion
2016	16	20 billion
2018	11	45 billion
2020	8	80 billion

possibly. Data rates per optical link of 100 gigabits per second per link by 2020 are realistic possibilities. This would provide the ×10 speed-up required by increased clock rates and the ×10 increase in the number of cores previously predicted.

With sufficient miniaturization it may be possible to replace transistors with optical switches and electrical paths with optical waveguides. Another potential qualitative change would be a return to direct optical processing where waveform generation, pulse compression, filtering, up- and downconversion, convolution, and FFTs could all be performed as optical functions.

Optical systems are capable of very high storage densities and have the potential for high access speed and multiple simultaneous accesses. If this potential can be realized then the bottlenecks caused by the memory speed hierarchy may be eased. As the number of cores is increased the problem of simultaneous access to a large shared cache memory becomes more severe and optical systems may be able to provide a solution or easement of this problem.

A combination of these changes could permit an order of magnitude increase in performance due to improvement in switching speeds and propagation delay.

43.7 Summary

Processing power, driven by Moore's law, has increased enormously, and extremely powerful systems using only MOTS or COTS standard hardware modules are now feasible. This offers great benefits in modularity, flexibility, and scalability of processing.

Processing throughput can be increased at all levels by the use of parallelism and pipelining. At the calculation level care must be taken to ensure that data can be provided without gaps to the pipeline and that pipeline stages are regular. At the level of radar processor pipelining, flexibility must be available to redeploy processing power within the pipeline, and sufficient memory and bus bandwidth must be made available to compensate for internal throughput variations.

FPGA, DSP, and microprocessor solutions all have different benefits. Continuous high-speed data that needs integer processing in a pipelined fashion is usually suitable for FPGA implementation. Microprocessors can offer a uniform system that is programmable in a high-level language with single and double precision floating point, but operating system and cache issues can cause problems if very deterministic time performance is required. DSPs provide a good performance compromise between FPGA and microprocessor, but programming them can be more difficult than for a microprocessor.

The most reliable way of determining if a processing solution has sufficient processing throughput to work is to implement and test it on the target hardware conditions under real-world conditions. Failing this, predictions, based on benchmark program suites that are as close as possible to the final application, are the most reliable guide to processing performance. Moving data to and from processing units in the most efficient order can take longer than performing the calculations themselves. In particular, making effective use of high-speed memory such as processor data caches can have a critical impact on performance.

When calculating processor performance or processor speed-up, the time to taken to perform all the steps should be taken into account, and Amdahl's law should be used to target those areas where the greatest savings can be made.

The main challenges impeding future improvements processor performance are in managing the growth in power dissipation with increasing transistor density and scaling input/output and memory speeds to the number of cores within processors. The use of ever-shrinking technologies presents a risk to equipment reliability, whereas the use of commercially available microprocessors reduces the ability to detect and locate faults.

The eventual quantum limits to CMOS scaling will probably force a radical change in the transistor types used before 2022, and this will probably be accompanied by supporting developments in silicon photonics to overcome the challenges covered in this chapter.

Further Reading

J. Stokes, *Inside the Machine, an Illustrated Introduction to Microprocessors and Computer Architecture*, No Starch Press, 2007.

D. Liu, *Embedded DSP Processor Design*, Newnes-Elsevier, 2008.

J. L. Hennessy and D. A. Patterson, *Computer Architecture, a Quantitative Approach*, 5th ed., Morgan Kaufmann, 2011.

J. L. Hennessy and D. A. Patterson, *Computer Organisation and Design, The Hardware/Software Interface*, rev. 4th ed., Morgan Kaufmann, 2012.

Test your understanding

1. A processor design is shrunk from a 130 nm process to a 90 nm process. The original 130 nm process has an area of 1 cm^2. (a) Assuming all circuit elements scale linearly in both dimensions with the process shrink, what is the area of the device in the 130 nm process?
(b) By what factor would the number of transistors increase in a 1 cm^2 device in 90 nm?

2. A signal processing chain consists of a matrix transpose that takes 16 msec and an FFT that takes 32 msec for a total processing time of 48 msec. An FFT speed-up of ×10 is proposed. Use Amdahl's law to calculate by what factor the processing time has been reduced.

3. The front end of a radar processor is fed three channels of continuous data at 250 million 16-bit integers per second per channel. The data arrive continuously, and digital filtering must be performed on each channel. Which type of processing (GPU, microprocessor, DSP, or FPGA) would you use, and why?

Chengdu J-20 Black Eagle (2011)

The Black Eagle is reported as a fifth-generation, stealth fighter being developed in collaboration by the Chengdu Aircraft Corporation and Shenyang Aircraft Corporation. It is intended to enter service before 2020.

44

Bistatic Radar

PPI display of early UCL bistatic radar experiments (1982)

44.1 Basic Concepts

A conventional radar with a single antenna shared by transmitter and receiver is known as *monostatic*. In contrast, a radar in which the receiver is in a different location from the transmitter is *bistatic*. Although such an arrangement introduces a number of technical complications, particularly in synchronization between transmitter and receiver, and may be significantly costlier, there are several potential advantages.

Bistatic radar may improve the detection of stealthy targets that are shaped to scatter energy in directions away from the monostatic. A bistatic radar receiver is passive, which means that it is impossible to locate via electronic support ES methods.

It is difficult to deploy countermeasures against bistatic receivers because their location is not known. Therefore any jamming has to be spread over a range of angles, diluting its effectiveness. In the same way, a bistatic receiver will not be vulnerable to attack by anti-radiation missiles (ARMs). Bistatic operation is attractive especially in systems based on unmanned aerial vehicles (UAVs) because they can carry just the receiver, and the heavy, complex, and power-hungry transmitter can be located elsewhere.

Some of the earliest airborne radar experiments in the 1930s were bistatic, since at least initially it was not possible to generate high-power radar pulses in an aircraft-borne system.

An example of a bistatic radar system from the late 1970s to early 1980s is SANCTUARY, a U.S. bistatic air defense radar concept that used an airborne illuminating source at standoff range and passive, ground-based receivers (Fig. 44-1).

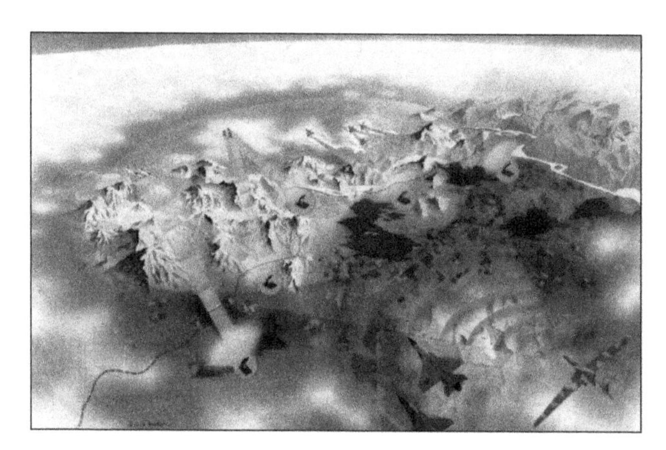

Figure 44-1. SANCTUARY was a U.S. bistatic air defense radar concept from the late 1970s to early 1980s. It used an airborne illuminating source at standoff range and passive, ground-based receivers.

This chapter will describe some of the properties of bistatic radar, showing how many of these are a consequence of the bistatic geometry. Examples are given of some practical bistatic radar systems and their results. A set of techniques known as *passive bistatic radar* are also given in which broadcast, communications, or radio navigation transmitters are used as the source of illumination instead of a dedicated radar transmitter.

Klein Heidelberg

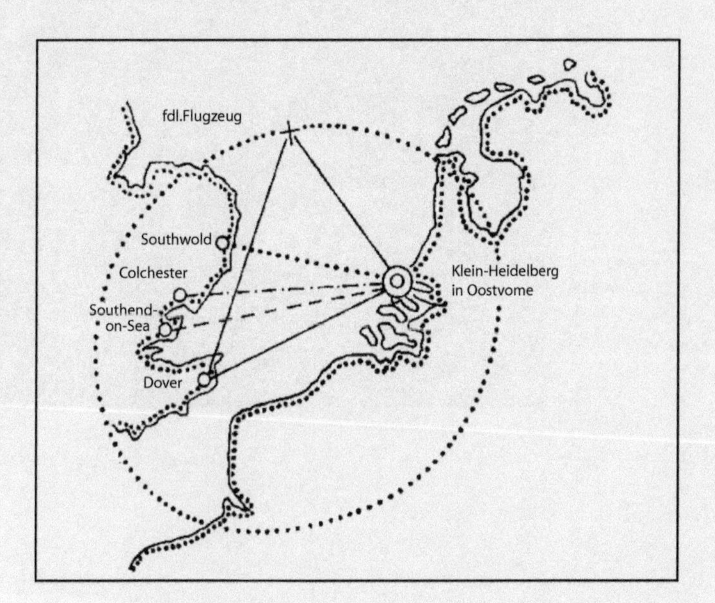

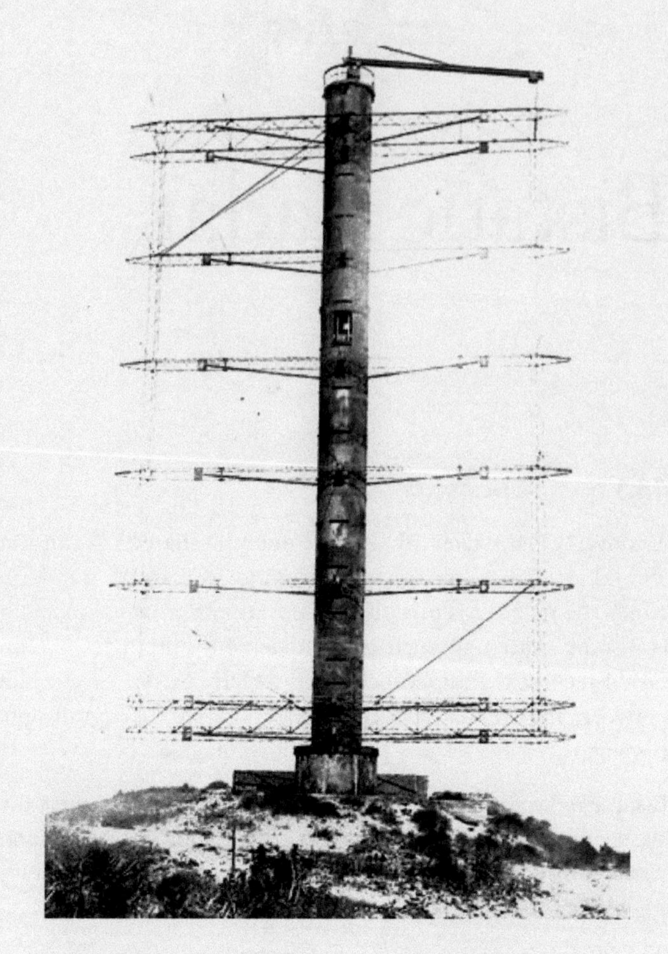

THE FIRST OPERATIONAL BISTATIC RADAR, IN THE MODERN SENSE, WAS THE German WW2 bistatic system *Klein Heidelberg*. This exploited transmissions from the British Chain Home radar stations on the east coast of England, which made it completely undetectable and therefore immune to jamming and countermeasures. In fact the Allies did not find out about it until November 1944. It achieved reported detection ranges of Allied aircraft of more than 400 km and was undoubtedly decades ahead of its time.

44.2 Properties of Bistatic Radar

The bistatic triangle formed by the transmitter, target, and receiver is shown in Figure 44-2. The distance of the receiver from the transmitter is known as the baseline, L. The angle at the target subtended by the transmitter and receiver is the bistatic angle, β. The transmitter-to-target range is R_T, and the target-to-receiver range is R_R. In most arrangements, the bistatic receiver measures the difference in delay between the direct pulse from the transmitter and the target echo, which, if L is known, gives the bistatic range sum $(R_T + R_R)$. Such a measurement defines an ellipse, with the transmitter and receiver as the two focal points. This is exactly the same as the trick you may have done when you were young: sticking two pins into a board with a loop of string and a pencil and then drawing an ellipse.

Target
β
R_T
R_R
Transmitter
Receiver
L
Bistatic Baseline

Figure 44-2. In bistatic radar geometry, a measurement of bistatic range, $R_T + R_R$, for the target defines an ellipse with the transmitter and receiver at the two focal points.

Usually, directional beams at the transmitter or receiver (or both) will be used, and these will allow the target's location on the ellipse to be determined. However, this introduces a complication if the transmitter signal is pulsed (which it usually will be), since the receive antenna beam must point instantaneously in the direction from which a target echo will come, following the propagation of the pulse through space. This process is known as *pulse chasing*. Because the change in the beam-pointing direction is both nonlinear as a function of time and is very rapid, the receive beam has to be steered electronically, which will certainly be more complex and costly than a mechanically steered approach.

Remember that the bistatic receiver has to be synchronized to the transmitter. More specifically, the processing at the bistatic receiver needs the following information: (1) the locations of the transmitter and receiver; (2) the instant of transmission of each pulse; (3) (if the transmitter has a directional beam) the pointing direction of the transmit beam; and (4) (if coherent processing is to be used) the phase of the transmitted signal.

If the transmit source is cooperative, then it may be possible to use a landline link. Otherwise, and particularly if one or both of the platforms is airborne, it will be necessary to obtain the information by reception of the direct signal. Synchronization and geolocation are made substantially easier using global positioning system (GPS).

The Bistatic Radar Equation. The radar equation for a bistatic radar is derived in exactly the same way as for a monostatic radar (Chapter 12). For comparison, the two equations are

$$\frac{S}{N}=\frac{P_{avg}G^2\lambda^2\sigma t_{ot}}{(4\pi)^3R^4\ kT_0BF} \quad \text{(monostatic)}$$

$$\frac{S}{N}=\frac{P_{avg}G_TG_R\lambda^2\sigma_Bt_{ot}}{(4\pi)^3R_T^2R_R^2\ kT_0BF} \quad \text{(bistatic)}$$

The key differences are (1) the antenna gain, G^2, is replaced by the separate transmit and receive antenna gains, G_T, and G_R; (2) the $1/R^4$ factor in the denominator is replaced by the factor $1/(R_T{}^2R_R{}^2)$, a consequence of which is that the signal-to-noise ratio for a given target is lowest when the target is equidistant between the transmitter and receiver and highest when the target is either close to the transmitter or close to the receiver; and (3) the target monostatic radar cross section (RSC), σ, is replaced by the bistatic RCS, σ_B.

Ovals of Cassini. From the bistatic radar equation it is evident that contours of constant signal-to-noise ratio are defined by R_TR_R = constant, which can be illustrated as *ovals of Cassini* (Figure 44-3). For small values of *c* these tend to circular regions centered on the transmitter and receiver; for large values they tend to ellipses, and for very large values of *c* to circles.

It is important to realize, though, that these figures are appropriate only for omnidirectional transmit and receive antenna patterns; if the patterns are directional, the contours are weighted by the radiation patterns and may be completely different in shape.

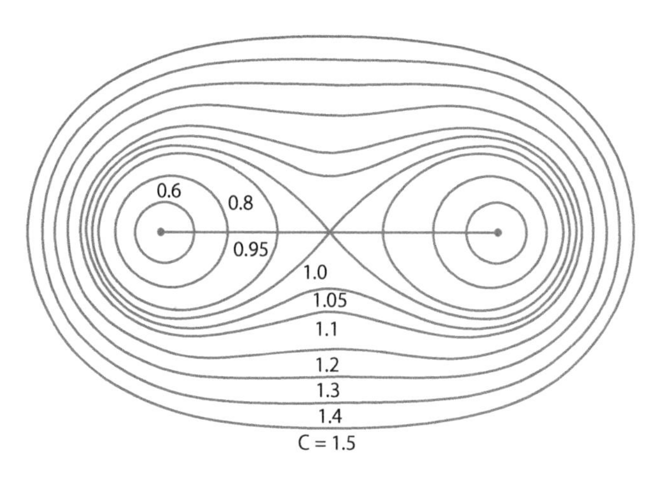

Figure 44-3. These ovals of Cassini are plotted for different values of the constant *c*, normalized to the baseline.

Forward Scatter. For a target on or close to the baseline, the echo will arrive at the receiver at the same time as the direct signal no matter where on the baseline the target lies. Furthermore, for a target crossing the baseline the Doppler shift will be zero, because at that point the transmitter-to-target range is changing in an equal and opposite way to the target-to-receiver range. These considerations show us that the range and Doppler resolution properties depend not only on the radar waveform but also on the location of the target with respect to the transmitter and receiver.

Target Bistatic RCS. As noted already, one of the advantages of bistatic radar is that it may offer a counter-stealth capability, since targets shaped or treated to minimize their monostatic RCS may nevertheless have higher bistatic RCS. This is not easy to verify, though, both because bistatic measurements of targets are difficult to make and because the values—certainly those of military targets—are likely to be classified.

For a nonstealthy target, the bistatic RCS will generally be comparable to its monostatic RCS. Early theoretical work on bistatic electromagnetic scattering from targets led to the *bistatic equivalence theorem*. This states that the bistatic RCS of a given target at a bistatic angle, β, will be the same as the monostatic RCS measured at the bisector of the bistatic angle, and at a frequency that is higher by a factor sec $\beta/2$ than that at which the bistatic RCS is desired. This depends on a number of assumptions: (1) the target is sufficiently smooth; (2) no part of the target shadows any other part; and (3) retroreflectors persist as a function of angle. In practice these conditions are not always met, so the theorem should be used with some care, particularly for complex targets and at large values of β.

Just as with monostatic signatures, bistatic target RCS will be enhanced at frequencies where the dimensions of target features are comparable to the radar wavelength (typically at VHF or HF frequencies for aircraft targets). This resonance effect occurs when contributions from different scatterers comprising the target (e.g., an aircraft's nose, cockpit, tailplane, engine intakes) add in phase at a particular radar frequency and geometry.

However, *forward scattering* can substantially enhance RCS, even for stealthy targets. This occurs when the target lies on or close to the baseline, and the effect can be understood using Babinet's principle from physical optics. Imagine that an infinite screen is placed between the transmitter and receiver so that the signal received is zero. Now suppose that a target-shaped hole is cut in the screen between the transmitter and receiver. Babinet's principle states that the signal that would be diffracted through the target-shaped hole must be equal and opposite to the signal diffracted around the target, since the two contributions must add to zero (Fig. 44-4).

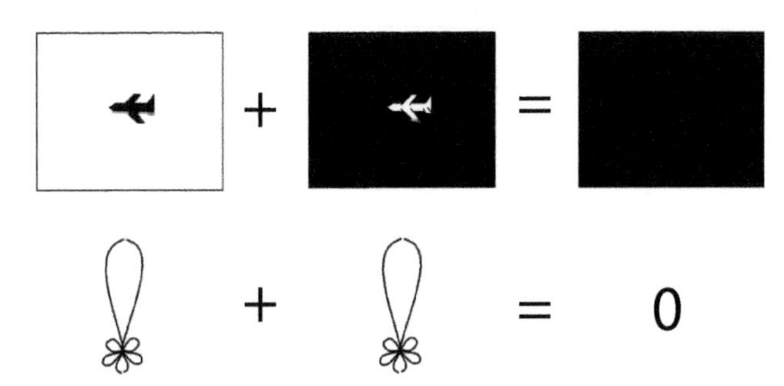

Figure 44-4. Babinet's principle is applied to determine the forward scatter radar signature of a target.

Determining the signal diffracted through an aperture of a given size and shape is a standard problem in electromagnetics, and the results for simple apertures are well-known. Babinet's principle means that these results at the same time give the forward scatter RCS of the target, and evaluation shows that this can be potentially several tens of dB greater than the monostatic RCS. While at first sight this appears attractive, the range and Doppler resolution for a target on or close to the baseline will both be poor. Therefore, forward scatter may be good for target detection, but using it for localization and tracking will be much more difficult.

Another issue with forward scatter is that the clutter cell area (because of poor range resolution) and clutter scattering coefficient will both be large, which means the clutter echo signal is likely also to be large. This means that target detection in forward scatter is likely to be clutter-limited rather than noise-limited.

Bistatic radar may also have the effect of reducing target *glint* and hence improving the performance of tracking radars. Glint is the angular displacement of the apparent phase center of a target return at a missile seeker. It is caused by phase interference between two or more dominant target scatterers lying within the seeker's resolution cell. Therefore, as the target aspect angle changes, the apparent phase center changes, increasing seeker angle tracking errors and thus miss distance. Conventional mitigation approaches include reducing the size of the seeker resolution cell so that individual scatterers are resolved, smoothing returns via noncoherent integration or operating bistatically.

For any equations or results derived for bistatic radar, it is sensible to exercise the sanity check $\beta \to 0$ or $L \to 0$, in which case they should reduce to those for monostatic radar. If not, an error has likely been made.

44.3 Examples of Systems and Results

Despite bistatic radar's advantages, there have been rather few examples of practically deployed airborne bistatic radar systems, though particularly in the past few years many experimental systems and trials have been conducted, some of which have been reported in the open literature. The following paragraphs and pictures give some examples.

Semiactive Missile Homing. A major development in bistatic radar during the 1950s and 1960s was the semiactive homing missile seeker, in which the large, heavy, and costly transmitter could be offloaded from the small, expendable missile onto the launch platform. While these seekers are clearly a bistatic radar configuration, missile engineers have developed a different lexicon to describe their technology and operation, for example, *semiactive* versus *bistatic*, *illuminator* versus *transmitter*, *rear reference signal* versus *direct-path signal*. The missile and radar communities continue to go their separate ways with only occasional technical interchanges. One such interchange can significantly reduce the missile's endgame miss distance by approaching the target at bistatic angles >20–30°, which in turn reduces the target's glint, a major contribution to miss distance.

Figure 44-5 illustrates this concept using the U.S. Navy's TALOS air defense system. In the picture, it is launched at elevation angles from 25 to 50°. After the boost phase, the rocket automatically separated from the missile, and the ramjet engine ignited. Midcourse beamriding then controlled the missile, which

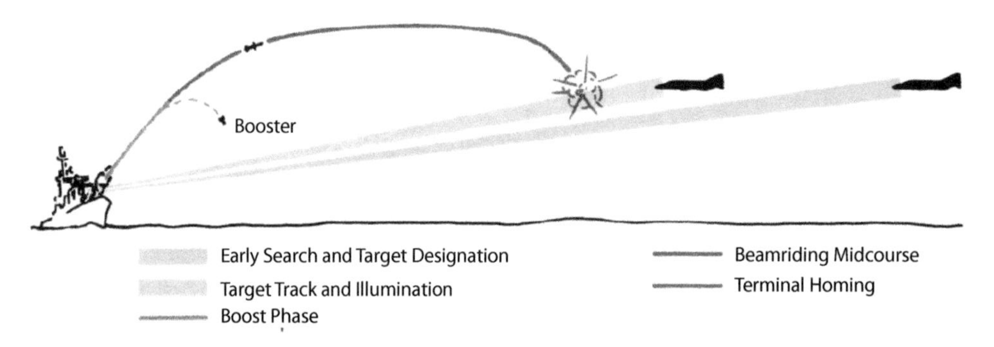

Figure 44-5. Pictured here is the operational concept of the U.S. Navy's TALOS air defense system. (Reprinted from *Johns Hopkins APL Technical Digest*, Vol. 3, No. 2, © The Johns Hopkins Applied Physics Laboratory.)

Figure 44-6. This photo shows TALOS missiles on the *USS Galveston* (CLG-3), which was the first TALOS cruiser in 1958. (Courtesy of APL/JHU.)

cruised at an altitude of 60–70 kft for long-range engagements. Toward the end of the missile's midcourse, it received a signal that activated its semiactive homing system and armed its warhead. Since aircraft flew at significantly lower altitudes, TALOS dived on its target, which generated a large bistatic angle, an exaggerated ~60° in the figure. This geometry in turn reduced target glint and therefore lowered the miss distance, often to the point of causing direct hits. Ironically, the concept of target glint reduction was unknown at the time, and engineers and operators attributed the improvement to good design, maintenance, and operation. Figure 44-6 shows the missile itself.

Bistatic Synthetic Aperture Radar. Almost all of the imaging radar techniques described in Chapters 32–35 can used bistatically via many different configurations: the transmitter and receiver can be fixed, airborne, or spaceborne, and the synthetic aperture can be formed by motion of the transmitter, the receiver, or both.

Figure 44-7 shows an example of an experimental airborne X-band bistatic synthetic aperture radar (SAR) system. Here the transmitter is carried by an aircraft and the receiver by a helicopter. The target scene is a small village in the southwest of England, and the bistatic angle is about 50°. Atomic clocks were used for synchronization between transmitter and receiver.

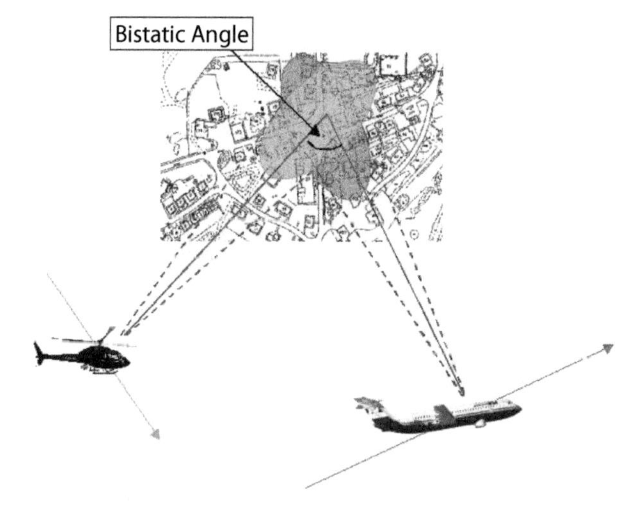

Figure 44-7. Pictured here is the geometry of an airborne bistatic SAR experiment with the transmitter carried by the aircraft and the receiver by the helicopter. (Courtesy of QinetiQ.)

Bistatic Synthetic Aperture Radar for Ground Attack

In this artist's rendition of the *Tactical Bistatic Radar Demonstration (TBIRD)* concept, a standoff SAR-equipped aircraft (F-4, upper right) detects and illuminates a target (lower right) while designating an attack aircraft (A-10, left) equipped with a bistatic receiver to the target area. The A-10 acquires and attacks the target directly on its velocity vector in radio frequency silence.

A 10 foot resolution image of Demonstration Hill within Fort Huachuca, Arizona, is pictured below:

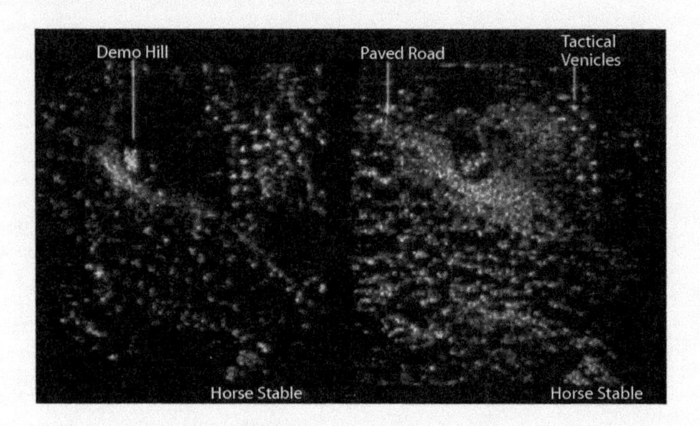

An example of a bistatic SAR image is depicted in Figure 44-8. The two targets at the bottom of the image have two shadows: one in the direction of look of the transmitter, and one in the direction of look of the receiver. Thus, if the shadow length and shape are used as part of the target classification process (as discussed in Chapter 47), the two shadows provide two separate pieces of information.

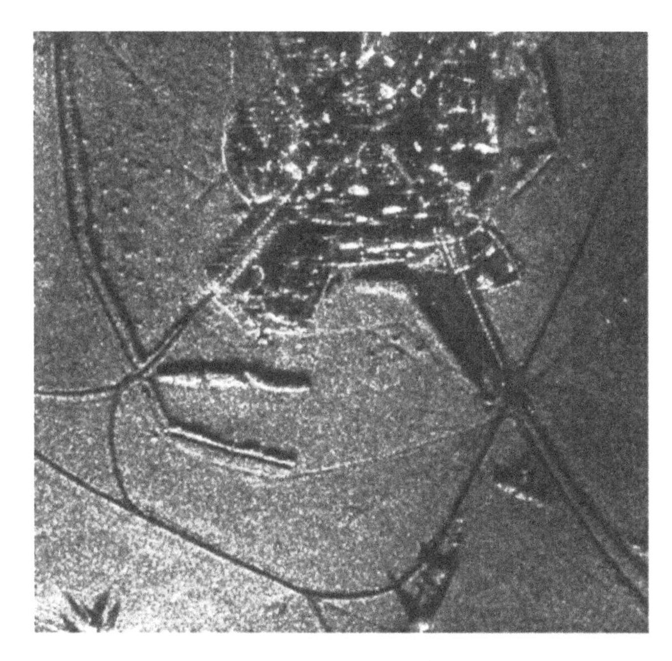

Figure 44-8. Pictured here is an example of a bistatic SAR image. The two targets at the bottom left and bottom center can both be seen to have two shadows (QinetiQ). (Courtesy of QinetiQ.)

44.4 Passive Bistatic Radar

So far we have discussed using a dedicated cooperative radar transmitter as the illuminating source. However, it is also possible to exploit other kinds of transmission, such as broadcast, communications, or radio navigation signals. These are known as *illuminators of opportunity*, and they have a number of potential advantages. For example, they tend to be high power and sited to give wide coverage. Most are terrestrial, but it is also possible to use various kinds of satellite-borne illuminators.

A key attraction of using these other signals is that the resulting radar is potentially completely covert. Passive bistatic radar may allow parts of the electromagnetic spectrum to be used, particularly at VHF and UHF, which are not normally available for radar, and where in defense applications, there may be an advantage against stealthy targets compared with conventional microwave radar frequencies. Since the transmitter already exists, cost and licensing are not an issue. Since passive bistatic radar does not cause any additional pollution of the electromagnetic spectrum, it has been described as *green radar.*

However, since the waveforms of broadcast, communications, and radionavigation sources are not explicitly designed for radar use, they may be far from optimum for radar purposes. It is therefore important to understand the effect of the waveform on the performance of the passive bistatic radar so that we can choose the most appropriate illuminating source, and to process the waveform in the optimal way.

Analog modulation formats, such as VHF frequency modulation (FM) and analog TV, are not so favorable because their performance as radar signals is time-varying and depends on the nature of the modulation (i.e., whether it is speech or music and whether the music has broad spectral content—thus a cacophony of rock music will be better than a solo instrument). Digital modulation formats such as digital audio broadcasting (DAB) and digital video broadcasting (DVB) are much better in this respect since the signals are more noise-like and do not depend on the instantaneous modulation. Of course, in many countries digital modulation formats are already replacing analog in broadcast and communications applications.

Another issue with passive bistatic radar is that because the signals are usually continuous in time (100 percent duty cycle) and because they are high-power, the environment of direct signals, multipath, and other co-channel interference will be severe, especially in urban locations, and quite sophisticated processing techniques are necessary to suppress these signals to allow targets to be detected.

Virtually all passive bistatic radar systems built and evaluated to date have used fixed, ground-based receivers, exploiting FM radio, television, or their more modern digital equivalents. Several of these have demonstrated detection and tracking of air targets at ranges of 100 km or more. It is interesting to consider how the same techniques might be used with an aircraft-borne receiver for applications such as airborne early warning (AEW) or air-to-ground surveillance.

Figure 44-9 shows one of the first results of this kind, from University College London. Here, a two-channel VHF receiver was mounted in a Piper PA 28-181 light aircraft with simple dipole antennas for the reference channel and signal channels taped to the inside of the window, flying from Shoreham airport on the south coast of England. The transmit sources are FM radio stations at Wrotham, Crystal Palace, Guildford, and Oxford, with transmit powers as high as 250 kW. An issue with such a system is that the receivers will certainly be line of sight to the transmitters, so the degree of suppression of the direct signal in the radar receiver may need to be even higher than for a terrestrial system.

The figure shows the results of detection of civil aircraft targets, giving in each case (from the differential time difference of arrival [TDOA]) an ellipse on which the target must lie. The target velocity vector is also computed from the measured Doppler shift and the known velocity of the aircraft carrying the receiver, and these are shown around the locus of the

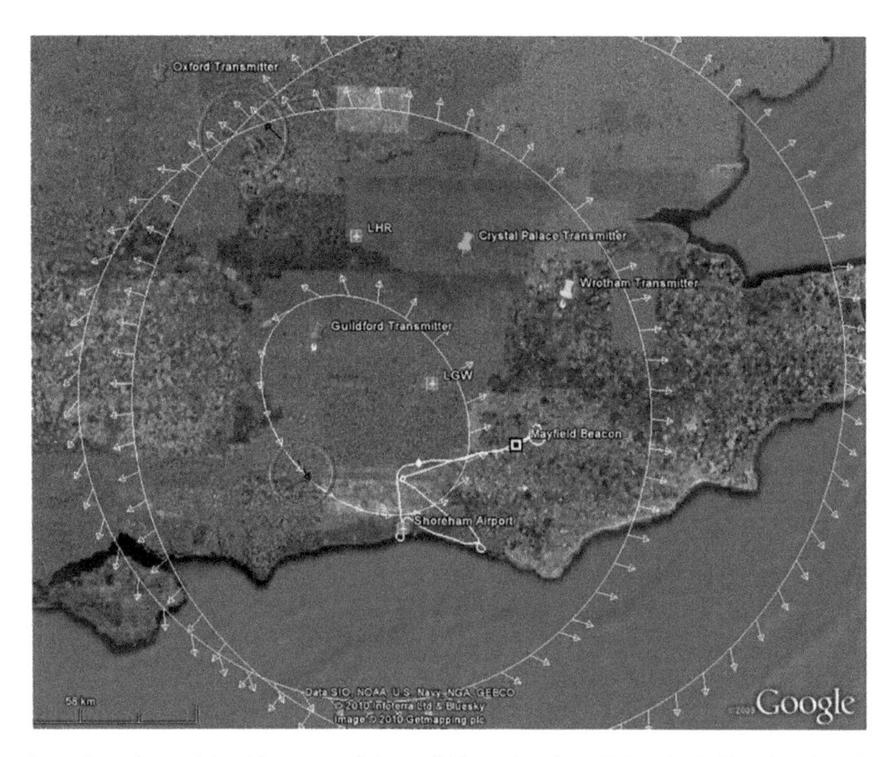

Figure 44-9. In the results shown here from trials with an aircraft-borne PBR receiver from University College London, the ambiguities associated with the intersection of the differential range ellipses may be resolved using the target velocity vectors.

ellipses. This information can be used to resolve the correct target location (i.e., to excise the ghosts) since the velocity vectors agree at one intersection point but not the other.

Although these results are very preliminary, they indicate the potential of the approach, and we can look with confidence toward its use in the development of real, practical systems in the future.

44.5 Summary

Although bistatic radar is more complex than monostatic, it has a number of potential advantages, the main one being that the receiver is covert. Although its principles have been known since the earliest days of radar and many experimental systems have been built and evaluated, there have been rather few operational systems.

Many of the properties stem from the geometry of the bistatic triangle formed by the transmitter, target, and receiver. The radar equation for a bistatic radar is derived in the same way as that for a monostatic radar. The key differences are (1) the antenna gain, G^2, is replaced by the separate transmit and receive antenna gains, G_T, and G_R; and (2) the $1/R^4$ factor in the denominator is replaced by the factor $1/(R_T^2 R_R^2)$. The target RCS in the bistatic configuration can be significantly increased over its monostatic RCS, particularly in forward scatter geometry.

Bistatic systems are starting to be introduced in real, fielded applications. Passive bistatic radar, using broadcast, communications, or radio navigation signals rather than dedicated radar transmitters, have some big attractions, and a recent defense business report has estimated that the market for passive bistatic radar over the next decade will be worth $10 billion.

Practical passive bistatic radar systems are often multistatic but can be treated as a set of bistatic systems.

Further Reading

M. C. Jackson, "The Geometry of Bistatic Radar Systems," *IEE Proceedings, Part F*, Vol. 133, No. 7, pp. 604–612, December 1986.

N. J. Willis, *Bistatic Radar*, 2nd ed., SciTech-IET, 2005.

N. J. Willis and H. D. Griffiths, *Advances in Bistatic Radar*, SciTech-IET, 2007.

H. D. Griffiths and N. J. Willis, "Klein Heidelberg—The First Modern Bistatic Radar System," *IEEE Transactions on Aerospace and Electronic Systems*, Vol. 46, No. 4, pp. 1571–1588, October 2010.

H. D. Griffiths and C. J. Baker, "Passive Bistatic Radar," chapter 11 in W. Melvin (ed.), *Principles of Modern Radar, Vol. 3, Applications*, SciTech-IET, 2013.

Test your understanding

1. List the potential advantages and disadvantages of bistatic radar compared with monostatic.
2. Why might a commercial ship target, which includes right-angled dihedral and trihedral features in its superstructure, be expected to exhibit a larger monostatic RCS than bistatic?
3. A target lies equidistant between transmitter and receiver such that $R_T + R_R = 10$ km. The baseline $L = 8$ km. Referring to the figure below, assuming that the transmit and receive antennas are omnidirectional and all other parameters remain unchanged, what is the echo signal power relative to that when the target is at (a), when it is instead at (b) or at (c), both points lying on the constant range sum ellipse?

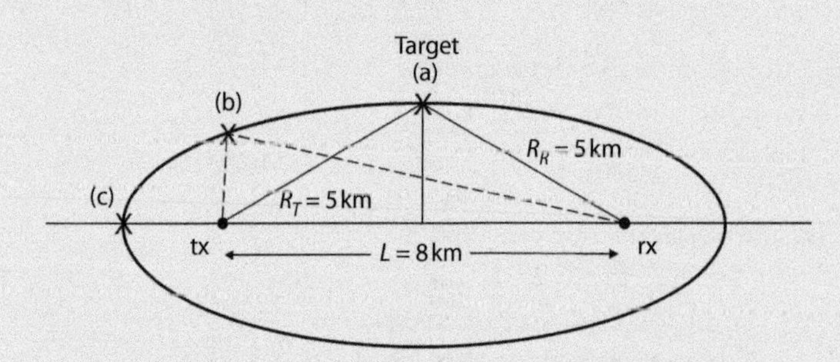

4. Why are digital modulation formats preferred over analog for passive bistatic radar sources?

45

Distributed Radar and MIMO Radar

In the previous chapter we saw that *bistatic* radar is a system where the receiver is in a different location from that of the transmitter. In a *distributed* radar system there will be more than one transmitter or more than one receiver positioned at locations that may be the same or different. The generic distributed radar network of Figure 45-1 is composed of one monostatic radar system (i.e., a transmitter collocated with a receiver) and a number of spatially distributed receivers. Together these make a single coherent network. Each transmitter or receiver is called a *node*. Pairs of nodes will form either bistatic or monostatic configurations.

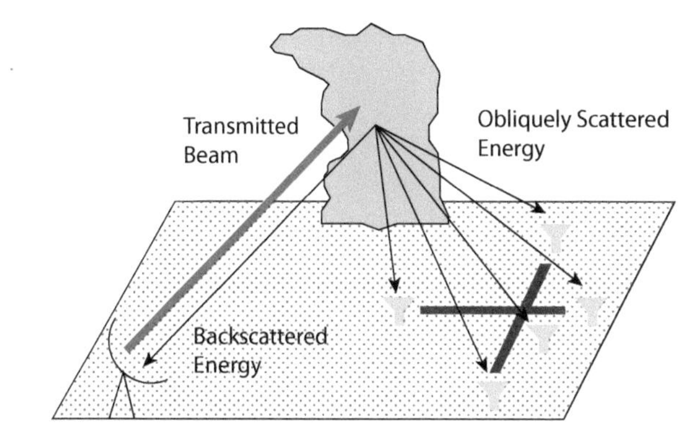

Figure 45-1. This generic distributed radar with one transmitter/receiver monostatic radar node and multiple distributed bistatic nodes operates as a single, synchronized, coherent whole.

45.1 Basic Concepts

Many of the concepts introduced in the previous chapters may be applied to distributed radar but in a combinatorial form that makes things a little different. Indeed, a network of geographically separated transmitters and receivers introduces a number of complications, most notably the synchronization between transmitters and receivers. However, distribution of the radar system over a geographical area also brings several potential advantages:

- It offers a performance improvement in system sensitivity enabling longer-range detection of targets, which is accomplished by collecting the scattered electromagnetic radiation more efficiently.
- It can also lead to a performance improvement in detection of stealthy targets by collecting energy otherwise deflected away from a monostatic radar receiver.
- If the receivers are not collocated with the transmitters, they are passive and thus electronic support methods cannot locate them.
- A distributed radar system exhibits *graceful degradation,* as physical or electronic loss of any one part of the system

does not mean a complete loss of all performance (as it would in monostatic radar).

- It can locate targets more accurately using *long baseline* techniques akin to large sparse arrays.
- It results in a target being illuminated from a number of different aspects that can be used to aid target classification performance.

In principle, distributed systems can be ground, air, space, or sea based (or any combination of these). They can operate over different frequency bands and widely varying distances. Here, we introduce a few fixed ground-based examples to illustrate the key concepts and modes of operation.

However, first, a word or two should be said about *multiple input, multiple output* (MIMO) radar systems. MIMO has become a topic of intense research. Its origins are in communications, where it was introduced as a technique to exploit severe multipath conditions. MIMO comes in two forms: *coherent MIMO* and *statistical MIMO.* In coherent MIMO, multiple input and output paths are formed via antenna array elements. Thus, they are essentially embedded into a single antenna manifold. This enables improvements in angular accuracy though aperture synthesis (Chapter 46).

Statistical MIMO also uses multiple input and output paths but within a much greater geographical distribution. In other words, statistical MIMO and distributed (or netted) radar are very close cousins, separated, somewhat arbitrarily, by the form of processing applied. In broad terms MIMO systems more typically use noncoherent processing, whereas distributed systems use coherent techniques. Distributed systems may also carry out part of the processing locally at receiver nodes and part may be carried out in a central processor. These distinctions are, to say the least, ill defined, reflecting the relative immaturity of the topic area.

45.2 Properties of Distributed Radar

The properties of distributed radar systems are a consequence of geometry (Fig. 45-2). In a distributed system, each transmitter is a node and each receiver is a node.

The distance from a receiver to a transmitter is called a *node baseline*, L_{mn}. The transmitter node and receiver node are denoted as m and n. Pairs of nodes can be collocated (i.e., a monostatic pair) or spatially separated (a bistatic pair). The angle at the target as subtended by a transmitter node and a receiver node is the *node angle.* The node angle is equivalent to the local bistatic angle, β_{mn} formed by the particular bistatic pair of transmitter and receiver nodes. If $\beta_{mn} = 0$, the geometry is that of a monostatic radar system. A transmitter-to-target range is R_{Tm}, where m denotes the transmitter node location, and a target-to-receiver range is R_{Rn}, where n denotes the

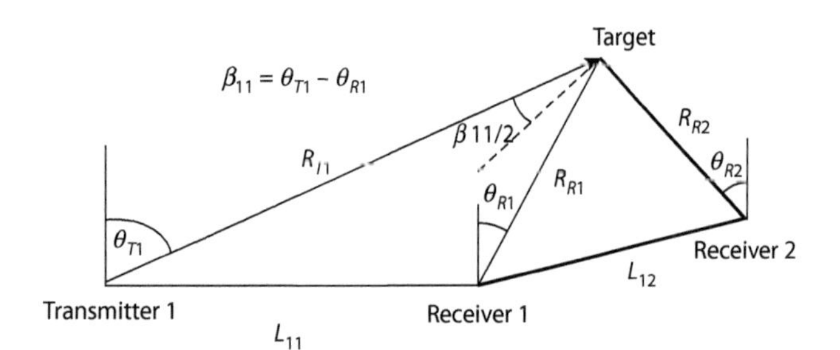

Figure 45-2. This simple distributed radar geometry is composed of a single transmitter and two receivers.

receiver location. A distributed network might be composed of a group of separated transmitters and receivers or a group of monostatic radars or some hybrid of both.

Targets scatter energy in all directions. By locating receivers at multiple sites, more of this scattered energy can be collected. It is this inherent characteristic that leads to improved sensitivity and detection performance. Every transmitter/receiver pair can be treated as a radar system. Where the transmitters and receivers are collocated they are monostatic radars, and where they are not collocated they are bistatic radar systems.

Figure 45-3 illustrates a coherent distributed radar system where three monostatic radars make up the network. By a coherent network, it is meant that a synchronization or clock signal is distributed across the network so that, to any given point in space, echoes are returned to each receiver in phase with one another. In other words, the nodes in the distributed network are synchronized, in time and space, with one another. This is the ideal case but can be difficult to implement in practice. Nevertheless, we will assume full coherency to introduce the basic principles of the distributed radar concept.

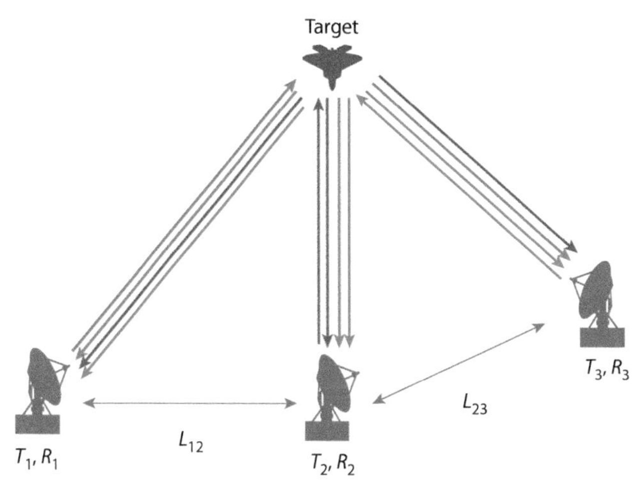

Figure 45.3. This simple distributed radar system is composed of three coherently synchronized monostatic radar systems. Each receiver can receive from all three transmitters as shown by the colored transmitter/receiver paths.

For the case shown in Figure 45-3 each transmitter results in two bistatic radar paths and one monostatic path by which echoes scattered by a target can be received. Thus, with three transmitters, there are a total of nine paths by which scattered energy can be collected. In other words, this distributed system is equivalent to three monostatic radars and six bistatic radars all viewing the same target at the same time. Note how the distributed radar system contrasts starkly to the case where there are three independent monostatic radars, each working in isolation (i.e., three paths only). In this way the distributed network of transmitters and receivers is able to collect scattered energy much more effectively: This translates into an overall improvement in system sensitivity pushing out the maximum detection range compared with monostatic radar or even a number of monostatic radars (say, having the same number of nodes).

To obtain these improvements the processing at each receiver needs the following information for each monostatic and bistatic pair:

- The locations of the transmitter and receiver pair
- The instant of transmission of each pulse
- If the transmitter has a directional beam, the pointing direction of the transmit beam
- If coherent processing is to be used, the phase of the transmitted signal

For collocated transmitters and receivers, synchronization is done locally as for any coherent monostatic radar. However, for the whole distributed radar to be fully coherent, a common local oscillator signal must be supplied to all nodes of the network. For fixed installations or small area networks, a landline link such as a fiber optic cable may supply such a synchronization signal. Alternatively, a wireless-distributed synchronization signal can be supplied via a global positioning system (GPS) or a local equivalent. A GPS, at its core, is an accurate timing signal and is available almost universally.

We saw in the previous chapter that passive bistatic radar is usually formed as a network. Coherency in a passive radar system is achieved using a technique known as *coherent-on-receive*. This results in each receiver node being locally coherent, but unless reception is synchronized across receivers the timing of phase measurements will vary from receiver to receiver and the network will not be fully coherent. This is an example of a partially coherent distributed radar system. The performance of a partially coherent distributed radar system is inferior to that of the coherent case but superior to that of the monostatic equivalent. However, the complexity and cost of the hardware are considerably eased.

Just as in the bistatic case, when a target crosses a baseline, the Doppler shift is zero because the transmitter-to-target range is changing in an equal and opposite way to the target-to-receiver range. However, in a network, there are multiple baselines, so this applies only to a local transmitter/receiver pair. Consequently, the impact on performance can be very different. Not only do range and Doppler resolution properties depend on the location of the target with respect to the position of the transmitters and receivers, but also radar performance will be a function of relative network geometry as well as target position and velocity. This adds additional complication but also presents new design freedoms that can be exploited.

Overall, distributing the transmitter and receiver nodes geographically brings about a range of parametric variables whose choice has a huge influence on the form of the network and ensuing performance. This provides an enormous increase in design freedoms and options. Here, we explore just a few of those design freedoms to illustrate the types of difference that can cause a wide range of resulting system behaviors. The range of possibilities is so vast that we first examine methods by which they can be categorized into broad types.

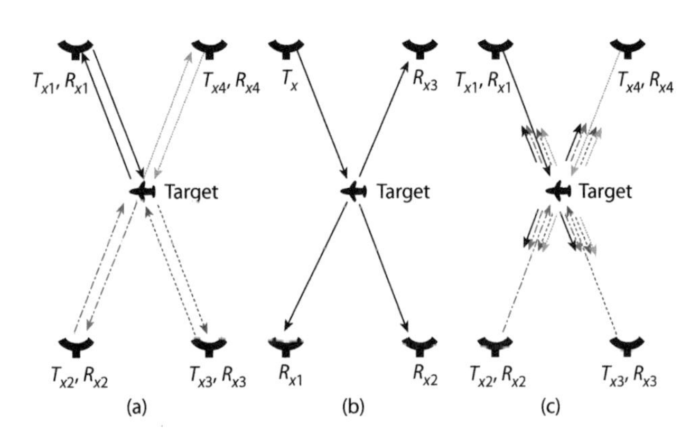

Figure 45-4. These diagrams give three types of distributed geometry: (a) multiple monostatic radar systems; (b) multiple bistatic radar systems, consisting of a single transmitter and three receivers; and (c) a fully coherent distributed system where all transmitters and receivers are interconnected and synchronized. T_{xm} is transmitter m, and R_{xn} is receiver n.

45.3 Categorization of Distributed Radar Systems

There are many ways distributed radar systems could be categorized. For example, they can be categorized by the composition of the system (Fig. 45-4). Figure 45-4a shows a system made up of four monostatic radar systems, each operating as an independent entity. The echo signals received by each of the monostatic radars can be combined in post processing to improve overall sensitivity.

In Figure 45-4b the distributed radar system consists of a single transmitter and three receivers, thus making three separate coherent bistatic radars. Their outputs can also be postprocessed to improve overall system sensitivity. Alternatively they could also be processed at the raw level (i.e., without preprocessing) if the system is coherently synchronized.

Figure 45-4c consists of four monostatic radar systems all operating coherently with one another. Each of the four transmissions (T_1 to T_4) is received monostatically as well as bistatically by each of the four receivers. Thus, there are 4 monostatic radar paths and 12 bistatic radar paths, all of which can act in phase to maximize system sensitivity. To achieve this, all the echoes arriving at each receiver must be combined coherently.

The degree of coherence within the distributed radar system is also a characteristic that can be used to categorize the type of distributed radar system. Two types of coherence can be considered: (1) the degree of coherence across the system; and (2) the level at which data are combined within the system.

Netted radar can be fully spatially and temporally coherent or locally coherent. In a fully spatially coherent distributed system, the radio frequency (RF) signal frequency and phase at each node of the system are known and stable over long time periods (i.e., greater than the duration of observation). This means that to any given point in space, the phase received at each receiver is the same. This allows the information contained in the scattered signals to be integrated with maximum efficiency thus, maximizing sensitivity.

The drawback is increased system complexity and cost. To provide the synchronization a distributed clock signal has to be available across the network. Additionally, but also the processing has to account for the location of each receiver so that each signal is phase corrected for the network geometry. This is because a fully coherent distributed system is responsive to the RF phase rather than that at IF as is normally the case for monostatic pulse Doppler radar systems. It is the exact location of the receivers that determines phase and an adjustment is required to ensure each node is in phase with the others in the network.

Consider the case where two echoes from two receivers are combined coherently together. If one of the receivers is located one quarter of a wavelength apart from the other, the two phases will add destructively and cancel. A phase shift must be added to account for the precise location of all receivers in the distributed network so that their received signals add up in phase.

An alternative to the fully coherent distributed radar system is one that is locally coherent. Locally coherent distributed radar systems exhibit coherence either monostatically or bistatically at each transmitter/receiver pair but are not coherent from receiver to receiver. Their signals can still be combined incoherently across transmit/receive pairs. The resulting system is less efficient, and sensitivity is reduced but remains above that of a monostatic equivalent. However, it reverts to being phase coherent at IF, and the implementation is much simpler.

A distributed radar can also be categorized according to the level at which data is combined within the system, partly determined by the form of coherence. Distributed systems may be categorized according to four levels of data combination that represent categories with lesser and lesser amounts of processing but also implying increasing levels of system complexity. The four levels in order of high to low are (1) tracks, (2) plots, (3) detections, and (4) raw data.

Tracks represent target information derived after processing raw (unprocessed) data into detections, associations of detection into plots and subsequently the formation and maintenance of tracks. As we saw in Chapter 31, tracks are derived from combining measurements and estimates of the changing position of a target in space and time. Tracks can be created at each of the transmit/receive pairs in the network. *Fusing* the tracks from each receiver site forms an overall integrated track. This seems straightforward, but in practice the different viewing geometries and radar parameters of the multiple nodes introduce errors that have to be understood and controlled, especially when there are many targets present in the observable space.

At the next level, plot data are formed from a series of detections such that a decision criterion is exceeded and a plot is declared. The decision criterion might be to detect a target in M of N processing intervals. Plot formation is a key step for forming tracks. Each node in the network sends plot information (e.g., position, velocity, time) to a central processor where they are combined with the plots from all the other nodes. From the combined plots, over time, tracks are formed. In this way, the distributed radar network fundamentally has more design freedoms in the plot association process, and thus overall tracking performance can be enhanced.

Detections are subject to even less preprocessing, thereby yielding further design freedoms and therefore further scope for enhancing performance. The presence or absence of a target is determined for each transmit/receive pair. These detections are subsequently sent to a central processor where they are combined to form plots and then tracks. In this case, the individual node probability of false alarm can be used as a variable such that, although the detection performance at each node may not be individually optimized, when the plot extraction and track formation stages are completed the overall tracking performance is superior.

At the lowest level, raw data are echoes received from each individual transmit/receive pair before being subject to any processing. All the raw data are sent to a central processor with a tag identifying its transmit/receive pair with other characteristics and subsequently combined. This processing stage might be optimized for detection or for tracking. At the raw level, data can be combined coherently and system sensitivity is maximized.

In general, the lower the level of the data to be combined, the more demanding the specification for communicating the data between nodes and the central processor. At the raw data level the system should be fully coherent necessitating a distributed and synchronization clock signal. The complexity and cost of the system increase rapidly increases further still.

Distributed radar can also be classified in other ways, such as *active* or *passive*. An active system is one where at least one transmitting station is included. A passive system has only receiving stations to detect target emissions. Naturally, it follows that systems cam be formed from combinations of both active and passive components. Passive radar systems that use multiple illuminators of opportunity are an area of distributed radar that has great future potential.

Transmitters and receivers can be stationary or moving, ground based, or airborne (or spaceborne). In general, the more parts of the distributed system that are in motion, the more complex it will be.

Table 45-1 is a guide to the various categorizations and complexity levels of radar networks and illustrate the distinctions between the different forms. It shows the increasing complexity of the different types of distributed radar, with the green boxes suggesting a simpler form, amber of intermediate complexity, and red requiring the most complex designs of all.

In *Case 1,* the distributed radar is composed of *N* fixed-position, monostatic radars. This simplest of all cases could be developed from combining existing individually coherent monostatic systems. Each of the radars operates independently, with its processed tracks sent to a central processing unit where they are combined. In this distributed system, both complexity and communication requirements are low, but so is performance improvement.

Table 45-1. Categories of Distributed Radar Systems[a]

	Case 1	Case 2	Case 3	Case 4	Case 5	Case 6
Location	Fixed	Fixed	Fixed	Fixed	Fixed and moving platforms	Nodes on moving platforms
Data Level	Tracks	Tracks	Detections	Detections	Raw	Raw
Coherency	Incoherent	Incoherent	Incoherent	Coherent	Coherent	Coherent
Operation Mode	N Tx, N Rx monostatic	1 Tx, N Rx multi-static	1 Tx, N Rx multi-static	M Tx, 1 Rx multi-static	M Tx, 1 Rx multi-static	M Tx, N Rx multi-static
Distribution	De-centralised	De-centralised	Semi De-centralised	Centralised	Centralised	Centralised
Assessment	Straight-forward	Multiple bistatics	Challenging	Complex	Very complex	Extremely complex

[a] Green = simplest complexity. Amber = intermediate complexity. Red = most complex.

In *Case 2,* there is one transmitter and N receivers. Each receiver is individually coherent with the transmitter. This is also a decentralized system that locally processes data to form tracks. These tracks are then communicated to a central processor where they are fused together to form the final track. Both complexity and communication requirements remain low, but separation of transmitters from receivers affords additional design freedoms.

In *Case 3,* the distributed radar has the same constituents as for Case 2. However, the balance between local processing and central processing differs. Here, detections are formed locally. Subsequently, these are sent to a central processor where they are used to form tracks. This affords additional flexibility so that, for example, dynamic adjustment of individual false alarm rates can be used to improve overall tracking performance.

In *Cases 4 and 5,* system topologies are fully distributed and fully coherent. Case 5 introduces the additional freedom of moving transmitters or receivers. As might be expected, this leads to superior overall performance that has to be balanced against significant additional system complexity.

In *Case 6,* the radar nodes are all mounted on mobile platforms. Data from each transmit/receive pair are coherently combined at the raw level followed by processing for detection and tracking. It provides the most performance flexibility and potentially best performance but creates the most severe technical challenges in maintaining system coherency. For example, raw (unprocessed) data must be transferred from the nodes to a central processing unit at high data rates.

The previous categorization should only be used as a guide and is in no way definitive. However, the categories do illustrate the vast range of options and new design freedoms that come about as a result of distributing multiple transmitters and receivers to form a radar system.

45.4 The Distributed Radar Equation

Both the monostatic and the bistatic forms of radar equation were introduced in Chapters 12 and 44, respectively, and are very similar in their construction. They both represent the power transmitted, the scattering from a target, and the power subsequently received. To enable a comparison with the distributed radar case, the monostatic and bistatic forms of the radar equation are restated here. The monostatic radar equation is given by

$$\frac{S}{N} = \frac{P_{avg} G^2 \lambda^2 \sigma t_{dt}}{(4\pi)^3 R^4 L k T_0 F_n}$$

where

S = received signal energy

N = noise energy

P_{avg} = average transmitted power (watts)

G = transmit and receive antenna gains

λ = wavelength (m)

σ = radar cross section of the target

t_{ot} = dwell time (s)

R = range to target (m)

L = losses

K = Boltzmann's constant

T_0 = ambient temperature (Kelvin)

F_n = noise figure of the receiver

In the bistatic case the different paths from the transmitter to the target and from the target to the receiver have to be identified separately, and the target radar cross section is a function of geometry. Thus, the bistatic form of the radar equation is

$$\frac{S}{N} = \frac{P_{avg} G^2 \lambda^2 \sigma_{Bi} t_{dt}}{(4\pi)^3 R_T^2 R_R^2 L k T_0 F_n}$$

where

R_T = range from the transmitter to the target (m)

R_R = range from the target to the receiver (m)

σ_{Bi} = bistatic radar cross section of the target

The distributed radar equation has to include all possible path pairs from each combination of transmitter to target to receiver paths. For example, there are nine of these path pairs shown in the three-monostatic radar example illustrated in Figure 45-3. The total signal-to-noise ratio is computed by summing across all possible path pairs (assuming a fully coherent system). Thus, the distributed radar equation for a geometry comprising M transmitters and N receivers is given by

$$\frac{S}{N} = \sum_{i=1}^{M} \sum_{j=1}^{N} \frac{P_{avgTi} G_{Ti} G_{Rj} \lambda^2 \sigma t_{dtij}}{(4\pi)^3 R_{Ti}^2 R_{Rj}^2 L_{ij} k T_0 F_{nj}}$$

where

i = the i-th transmitter

j = the j-th receiver

P_{avgTi} = average power transmitted by the i-th transmitter (Watts)

G_{Ti} = gain of the i-th antenna (associated with a transmitter)

G_{Rj} = gain of the j-th antenna (associated with a receiver)

λ_i = wavelength of signal emitted by the i-th transmitter

σ_{ij} = radar cross section of a target as seen by a transmitter/receiver pair

t_{dtij} = dwell time for at the j-th receiver for the i-th transmitter signal

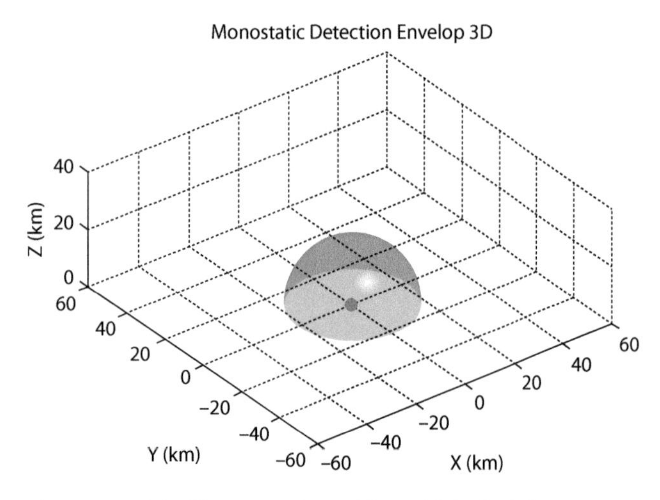

Figure 45-5. This illustration shows 3D sensitivity limit for a ground-based monostatic radar system as defined by a cutoff of 13 dB signal-to-noise ratio.

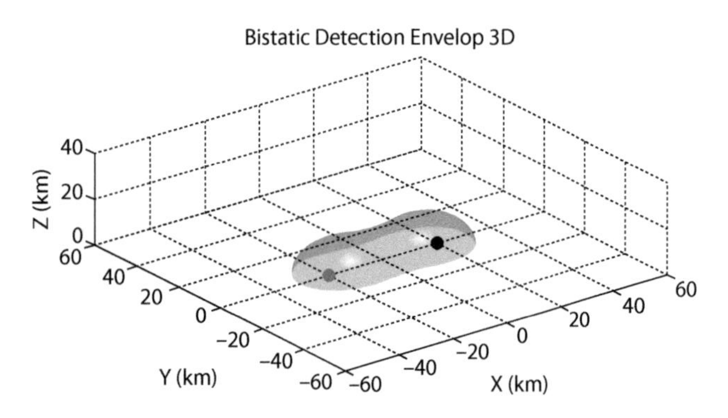

Figure 45-6. The 3D sensitivity limit for a bistatic radar system, as defined by a cut-off of 13 dB signal-to-noise ratio. This is equivalent to relocating the transmitter and receiver of the monostatic radar case shown in Figure 45-5, with all other parameters being the same.

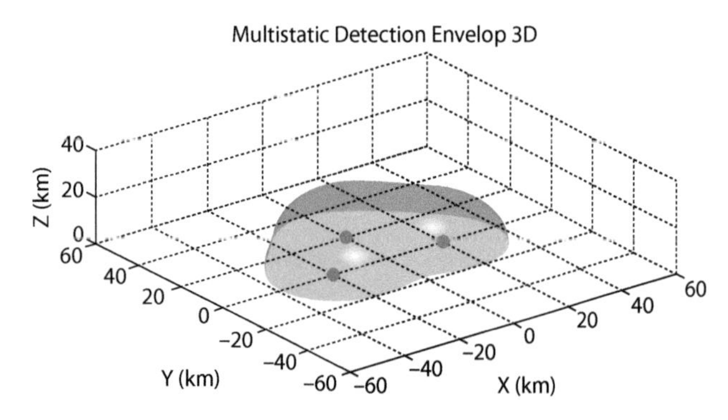

Figure 45-7. This 3D sensitivity limit for a distributed radar system is defined by a cut-off of 13 dB signal-to-noise ratio. The total emitted power is the same as for the monostatic and bistatic cases.

R_{Ti} = range from the i-th transmitter to the target (m)

R_{Rj} = range from the target to the j-th receiver (m)

L_{ij} = losses associated with a transmitter/receiver pair

F_{nj} = receiver noise figure at the j-th receiver

The distributed radar equation shows that the total power gain over a monostatic or bistatic system is equal to the number of transmitter/receiver path pairs in the distributed system (assuming the total transmit power to be the same and all other parameters to be equal). If the distributed system is made up of a number of monostatic radars, then the total power gain over a single monostatic radar is equal to the square of the number of individual monostatic systems. This is easily verified by referring to the example shown in Figure 45-3 with three monostatic systems leading to nine path pairs.

When computing detection performance for any radar, careful consideration has to be given to the 3D sensitivity of the system. This is straightforward in the monostatic case where the radar is at the center of a sphere. Figure 45-5 shows the hemispheric 3D plot of the signal-to-noise ratio, cut-off using a limit of 13 dB for a ground-based monostatic radar.

Now consider placing the transmitter and receiver of the monostatic system in separate locations, with all other parameters remaining the same (Fig. 45-6). In other words, the total emitted power (combination of transmitter power and antenna gain) is, again, kept constant, together with the receiver gain and noise figure.

Separating the transmitter and receiver considerably changes the shape of the sensitivity "bubble." When viewed from above, it takes the form of *ovals of Cassini* as described in the previous chapter. The interception of the bubble with the ground plane occurs over a larger area but at the expense of height coverage. The total volume represented by the bubble remains constant, as the total power emitted by the systems is constant. The positions of the transmitter and receiver determine the shape of the bubble. As they get closer and closer together, the height coverage extends more and more until it reaches a maximum as the monostatic is reached.

In a fully distributed example, the transmit and receive antennas of the monostatic case are divided into three separate monostatic radars located in differing positions. Again, the total emitted power is the same as in the monostatic and bistatic cases previously discussed. This system is a fully coherent network. (Fig. 45-7).

In Figure 45-7, the shape of the sensitivity bubble is defined by the geometry of the network. However, the network uses the nine transmit/receive paths to more efficiently collect scattered radiation from a target.

In our fourth and final example, the original monostatic radar is configured as three separate transmitters and three separate receivers. While they could be distributed anywhere, Figure 45-8 presents an example of a rectangular geometry with regular spacing between nodes.

Here, the limits of sensitivity are distributed over a larger surface area but at the expense of height coverage. Inspection of Figures 45-7 and 45-8 illustrates how the shape of the sensitivity volume is a function of the distributed geometry. When further combined with the freedom to choose the radar parameters at each transmitter and receiver, the possibilities are almost limitless.

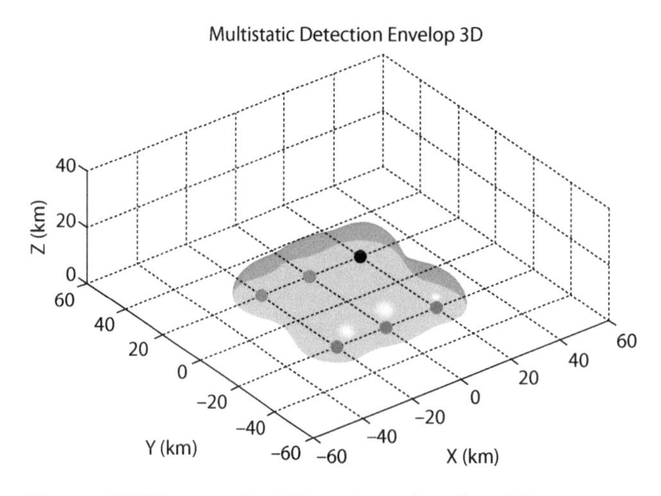

Figure 45.8 This example of 3D sensitivity limit for a fully distributed radar system is defined by a cut-off of 13 dB signal-to-noise ratio. The total emitted power is, again, the same as for the previous monostatic and bistatic cases.

45.5 Examples of Systems

Distributed systems go back to the earliest days of radar and have been reconsidered periodically from time to time. Both Russia and the United States made significant contributions to the early development of distributed radar systems. Other countries have also participated in advancing these types of concept.

In Russia, the earliest distributed radar, Vega, was composed of one transmitting and five receiving nodes. It was constructed in 1936 to detect aircraft. However, this system was not developed further and eventually fell out of use. In 1957, a distributed radar system with several spatially separated monostatic radars was used to track the first Sputnik satellite. Subsequently, there was a period of considerable development in both passive and active/passive, distributed radar systems made in Russia. The book by Victor Chernyak provides more details for the interested reader (see Further Reading).

In the United States, distributed radar systems played an important role in precision measurement of missile trajectories. These systems were composed of a ground-based transmitting station and several spatially separated, precisely located receiving nodes. One example is the continuous wave (CW) interferometric distributed radar, Azuza, which entered operation in the 1950s. Azuza was made up of one transmitter and nine receivers.

Navspasur (Navy Space Surveillance System) is another example of a CW system that began operation in 1960. It detected orbiting objects as they pass through an electronic fence positioned over continental United States. This system comprises three groups of stations with each group containing one transmitting node and two receiving nodes. In 1977, MIT Lincoln Laboratory began developing a netted radar program to improve battlefield surveillance, target acquisition, and battle management. The experimental system demonstrated the distributed concept to be highly effective. Lincoln Laboratory also developed a multistatic measurement system (MMS) in 1978–1980 at the Kwajalein Missile Test Range in Marshall Islands to collect bistatic signature data and to track reentry targets with extremely high accuracy.

Figure 45-9. Part of the JORN radar system located in Australia. (Supplied by DSTO (Stuart Anderson), Australia.)

Further development and implementation of distributed radar systems has been carried out in many countries. The Jindalee over-the-horizon Operational Radar Network (JORN) has been implemented in Australia (Fig. 45-9). This system provides long-range detection and tracking of aircraft and ships. It contains two cooperative but spatially incoherent high-frequency (HF) bistatic radars with a centralized control center, known as the JORN Coordination Center (JCC). An extensive network of beacons and sounders is also installed as part of the frequency management system. The network parts are located at widely separated sites around the northern Australian coastline, islands, and national offshore territories.

The Norwegian Defense Research Establishment developed a prototype distributed radar system, composed of several bistatic pairs. It has developed improved methods of detecting and extracting parameters for classifying targets. Testing measurements of a helicopter were made with experimental CW radar.

Distributed radar concepts have also been developed for civil applications. One example is the surface movement guidance and control system (SMGCS) utilized at airports. These are short-range radar networks, and depending on the local topology of the specific airport, multiple modules are employed for satisfactory surveillance, with each single module having at least three monostatic radar stations installed at separated sites. This is a good example of a distributed system exploiting the increased probability of obtaining line of sight to a target from a transmit/receive pair, while others are shadowed, thus improving overall detection and tracking performance of the system.

Distributed radar systems with similar functions have been proposed by the University of Rome. The subsystem architecture employed a network of short-range radars, known as *mini radars* (for their small dimensions and weight). The number of radar nodes depends on the actual airport topology, with typical numbers being between two and four. This system boasted high range resolution, elimination of shadowing, and the inclusion of image processing techniques in the track-while-scan function.

In another application, distributed radar with closely spaced sensors has been developed for adaptive automotive cruise control and collision avoidance. With the advent of low-cost vehicular radar technology, most cars will employ multiple radar sensors to operate as a distributed collective. This application may well become the largest and most significant of any of the distributed radar system concepts.

Longer-term initiatives to integrate multiple sensors, mainly developed by national military sectors, have been considered. One example is the Cooperative Engagement Capability (CEC). CEC is a United States Navy project for theater air defense. Functions include composite tracking, precision cueing, and coordinated cooperative engagements. Advanced features, such as cruise missile defense and tactical ballistic missile defense, are being added. Network-enabled capability (NEC) concepts have been developed typically aimed at the integration of sensors to achieve enhanced information sharing, improved situation awareness, collaborative decision making, and synchronization of actions.

A space-based example of a distributed concept is TechSat21 (Technology Satellite of the 21st Century). Figure 45-10 shows an artist's impression of a TechSat21 satellite cluster. This U.S. research program explored the technical challenges and benefits of replacing large single satellites with formations of microsatellites to perform the same mission.

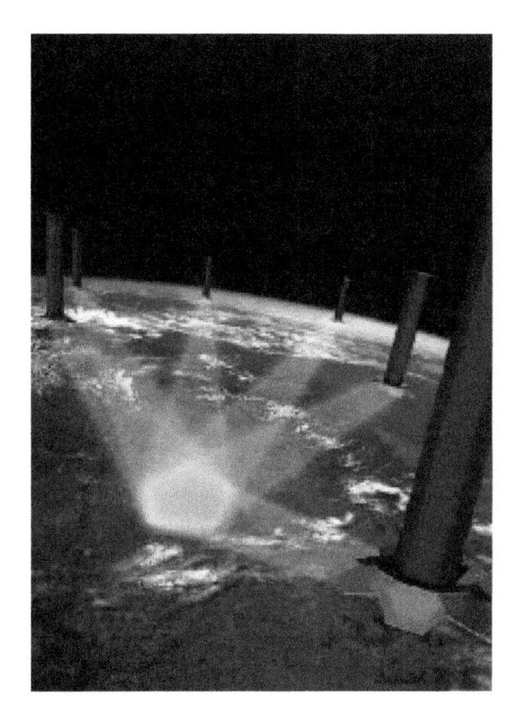

Figure 45-10. This artist's impression shows the TechSat 21 space-based distributed radar concept.

The TechSat21 concept was conceptualized as a cluster of free-floating satellites, each transmitting its own orthogonal signal and receiving all reflected signals. The satellites operate coherently at X-band. The cluster forms a multi-element interferometer with a large number of grating lobes and significant ground clutter. A novel technique for pattern synthesis in angle-frequency space with thinned arrays has been proposed, and a full evaluation of this technique and its effectiveness in clutter suppression has been completed using simulators. Although only a concept, it illustrates the potential and power of distributed radar sensing.

45.6 MIMO Radar

As discussed at the beginning of this chapter, the MIMO radar concept has two very different forms: coherent and statistical. Here we will describe statistical MIMO only. There are no strict definitions of what constitutes MIMO radar. Here, the main operating principles underlying the MIMO concept are introduced via a generic description. This will help show that MIMO is a part of the more general case of distributed radar and in some cases amounts to the same thing.

MIMO systems employ multiple nodes on both transmit and receive. They transmit independent data (e.g., x_1, x_2, $x_3, \ldots$) on differing transmit elements. At the receiver, a MIMO decoder operates on each of the nodes. Each radar receive node in the distributed system may receive signals coming from any of the transmit nodes, and these signals may or may not contain multipath. For a generic distributed MIMO system, we can write down a set of equations in matrix form:

$$r_1 = h_{11}x_1 + h_{12}x_2 + h_{13}x_3 \ldots\ldots\ldots\, h_{1N}x_N$$
$$r_2 = h_{21}x_1 + h_{22}x_2 + h_{23}x_3 \ldots\ldots\ldots\, h_{2N}x_N$$

$$r_N = h_{N1}x_1 + h_{N2}x_2 + h_{N3}x_3 \ldots\ldots\ldots\, h_{NN}x_N$$

where h_{ij} are the channel weights. These equations effectively treat the received data as a set of *channels* represented by the matrix equation. The challenge is to recover the individual data streams, x_i. To do this the channel matrix **H** must be estimated and inverted so that the individual data streams from the vector **r** can be recovered. This is equivalent to solving N simultaneous equations with N unknowns.

The MIMO concept can be applied to radar in a variety of ways with significant advantages. It can:

- Reduce multipath fading by recovering the original signal via matrix inversion
- Reduce fading effects caused by multibounce from a target that results in wildly fluctuating and difficult to detect signatures
- Separate out scatterers that cause glint signatures, well known for upsetting tracking systems
- Separate target signals from clutter signals in severely clutter-limited conditions

Note that both MIMO and fully distributed radar systems use all possible bistatic signal paths and can be thought of as equivalent. However, there can be some differences in the way signals are treated. Both require a set of transmitted signals or waveforms that can be individually identified and separately processed in the receiver to gain maximum benefit. This necessitates waveforms with *good orthogonality* properties. This means that the co-ambiguity function (one waveform correlated with itself as described in Chapter 16) shows a sharp peak at the center of the range-Doppler surface.

However, the cross-ambiguity function (i.e., one waveform correlated with another) should show no significant gain at any part of the range-Doppler surface. The easiest way to generate such waveforms is for them to be at differing operating frequencies with no bandwidth overlap. The search for suitable waveforms at the same operating frequency and bandwidth is ongoing. Overall, MIMO and distributed radar (and netted radar and multistatic and multisite radar) are variants around the same theme. Fundamentally, distributing radar in space brings new design freedoms and these are only just beginning to be explored.

45.7 Summary

Distributed radar systems offer the radar designer new and additional design freedoms by employing multiple transmitters and receivers positioned in different locations. A number of significant advantages can accrue: improved detection, tracking, and even classification performance. However, there is an attendant increase in system complexity, inevitably accompanied by an increase in cost. Some systems have been developed and deployed operationally, but this has only happened sporadically so far. The need for radar systems to maintain a competitive edge, the potential shown by networking and the rapid on-going developments in signal processing suggest that distributed radar will further mature until it is in routine use.

Further Reading

V. S. Chernyak, *Fundamentals of Multisite Radar Systems*, Gordon and Breach Scientific Publishers, 1998.

J. Li and P. Stoica, *MIMO Radar Signal Processing*, John Wiley and Sons, Inc., 2008.

W. Wang, *Multi-Antenna Synthetic Aperture Radar*, CRC Press, 2013.

Test your understanding

1. In a distributed radar system composed of four spatially separated monostatic radar nodes, how many transmit/receive paths can be integrated?
2. What is the difference between a *globally coherent* and a *locally coherent* distributed radar system?
3. Why is it easier to combine tracks rather than detections in a distributed radar system?
4. Describe the differences between a *statistical* and *coherent* MIMO radar system.

Lockheed Martin F-22 Raptor (2005)

The F-22 Raptor was designed as an air superiority fighter to replace the F-15 Eagle and F-16 Fighting Falcon. It is a stealth aircraft currently in service with the US Air Force with additional roles including ground attack, signal intelligence, and electronic warfare. The F-22 includes the AN/ALR-94 passive radar detector, composed of more than 30 antennas blended into the wings and fuselage, which provides all-around coverage.

46

Radar Waveforms: Advanced Concepts

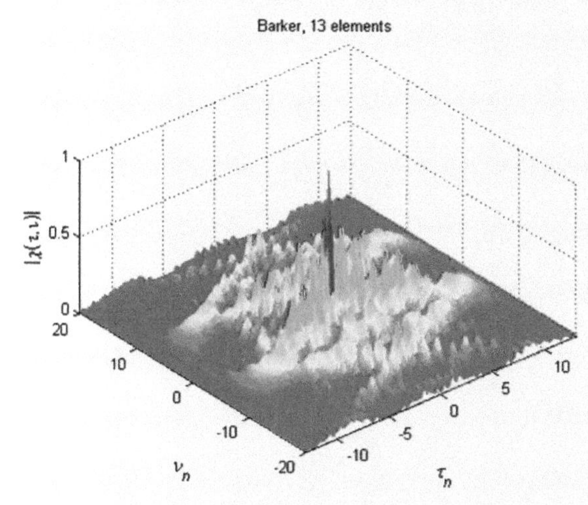

Ambiguity function of Barker code of length 13

The concept of pulse compression was introduced in Chapter 16 as a means of achieving both fine range resolution and high energy on target for long-range operation. It was observed that modulating the transmitted pulse and filtering the received echoes achieves this goal. Here, some practical aspects of pulse compression are considered, as well as new capabilities arising from advances in waveform generation technology and adaptive signal processing.

46.1 Practical Considerations

Pulse compression uses a long pulse, typically transmitted at peak power but modulated in frequency or phase. This modulated pulse, or *waveform*, is designed so that receive filtering enables the requisite sensitivity for detection and resolution along with robustness to factors such as target motion and external interference. The choice of waveform parameters such as pulse width, bandwidth, and modulation structure requires the consideration of additional factors dictated by hardware. Here these issues are addressed, including the impact of the transmitter, susceptibility to electromagnetic interference (EMI), and self-masking due to the finite pulse length (*pulse eclipsing*).

Transmitter Distortion. The radio frequency transmitter both amplifies and distorts the radar waveform. It is important to understand the nature of this distortion and correct for it in the radar receiver. There are many types of transmitters, and transmitter selection is dictated by the radar application, system architecture, and the specific components to be employed. The transmitter amplifies the generated waveform. The means of waveform generation and the design of the transmitter ultimately determine the specification of the emitted waveform (see Fig. 46-1).

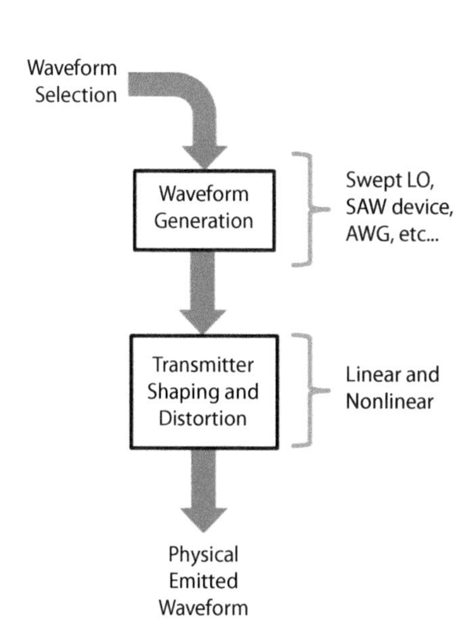

Figure 46-1. There are different schemes for generating a given waveform, which will subsequently experience spectral shaping and distortion in the radar transmitter.

The most common ways to generate a chosen waveform are

- a swept local oscillator (LO), which is often used to produce a linear frequency modulated (LFM) chirp;
- surface acoustic wave (SAW) devices, which generate both linear and nonlinear FM waveforms (see Section 46.3); and
- digital arbitrary waveform generators (AWGs), which are becoming increasingly popular because of their tremendous flexibility.

Transmitter power efficiency directly impacts the amount of energy the radar can emit and hence determines detection performance. However, maximizing power efficiency can cause transmitter nonlinearity. The combination of the waveform generation method and the transmitter can cause two forms of distortion effects on the transmitted waveform: linear distortion resulting in spectral shaping and nonlinear distortion.

Linear distortion is caused by the finite bandwidths of individual transmitter components. The passband of each of these components is not flat and exhibits amplitude ripple, causing amplitude distortion. In addition, dispersion (where different frequencies propagate through the system at different speeds) also introduces frequency (or phase) distortion. One approach that minimizes these effects is *predistortion*, which compensates for the subsequent linear distortion, leading to a transmitted waveform that has the desired specifications.

The finite transmitter bandwidth requires the waveform to also be band limited, at least to the degree possible for a finite pulse width. For this reason, phase-coded waveforms are generally implemented in such a way as to avoid abrupt transitions between code phase values (chips), as shown in Figure 46-2.

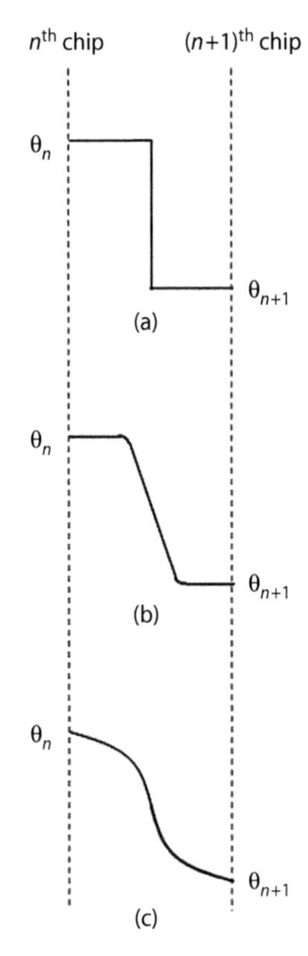

Figure 46-2. Phase transition between chips for coded waveforms where (a) is the idealistic implementation that will experience band-limiting distortion in the transmitter, (b) uses interpolated phase transitions, and (c) employs a code implementation scheme that avoids abrupt phase changes.

Nonlinear distortion is primarily caused by the radar power amplifier (the final component the waveform encounters prior to being launched from the antenna into free space). To maximize efficiency the power amplifier is generally operated at maximum gain (saturation), with the result that any waveform amplitude modulation is distorted nonlinearly.

A by-product of nonlinear distortion is the generation of *intermodulation products* resulting from the pairwise multiplication of the different frequency components in the waveform. These intermodulation products cause spreading of the spectral content of the waveform, resulting in *spectral leakage* that increases the likelihood of mutual interference between different spectral "users." Collectively, intermodulation products and transmitter-induced noise contribute to an increased spectral footprint of the radar, leading directly to interference in adjacent spectrum regions (see Fig. 46-3). This effect is also referred to as *spectral regrowth*. In an increasingly congested electromagnetic spectrum, such interference is not permissible. Therefore, the transmitted waveform should be designed such that spectral regrowth is avoided or at least minimized.

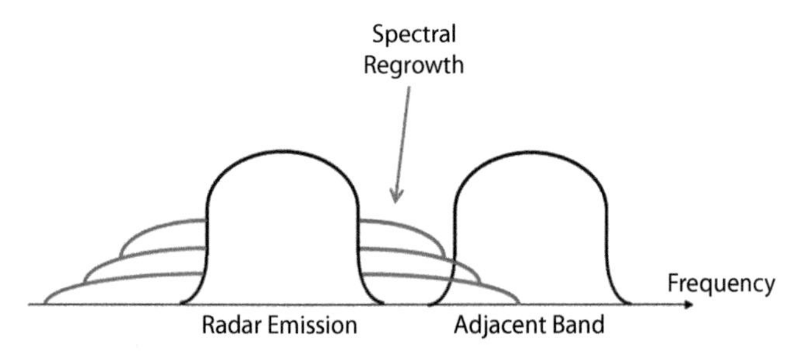

Figure 46-3. Radar spectral regrowth can induce interference in adjacent bands that may contain other spectral occupants.

Electromagnetic Interference. A radar system can be both the victim and the cause of EMI and both aspects must be considered in the design of waveforms. Indeed, EMI has always been an important consideration in both frequency selection and waveform design. However, the importance of waveform design for minimal EMI is growing as a result of the increasing usage of the electromagnetic spectrum.

A notable example of EMI causing problems for radar systems occurs at low frequencies, such as those used for foliage penetration (FOPEN). To achieve penetration through dense foliage, such as forests and jungles, frequencies between 20 MHz and 1200 MHz are used. FOPEN radars are imaging systems operating with high spatial resolutions. To obtain high range resolution, a bandwidth of several hundred MHz is necessary. However, these FOPEN bands are likewise used by many other systems, including broadcast radio, communications, and television. As a consequence, FOPEN systems employ strategies for avoiding those parts of the electromagnetic spectrum allocated to other users.

Avoidance of EMI is becoming ever more challenging. The electromagnetic spectrum is becoming increasingly congested due to the proliferation of commercial wireless communications, and particularly the demand for wireless streaming video. As a result, radar will have to contend with new sources of interference. Increasing spectral congestion will further complicate the design of waveforms and may lead to waveforms being modified on a pulse-by-pulse basis as the spectral environment changes.

Radars also have to comply with emission standards as determined by individual countries. These emission standards dictate the nature of the *in-band* and *out-of-band* spectral requirements. The out-of-band spectral requirements dictate the allowable powers that can be transmitted at frequencies within the allocated transmission band. Out-of-band spectral requirements are stated in terms of a spectral roll-off rate of decibels per decade, where a decade refers to a factor of 10 in frequency. Currently a 20 dB/decade roll-off is the common standard applied to radar systems. However, growing spectral congestion could result in changes that increase the roll-off rate to 30 or even 40 dB/decade. Thus it is very likely that future radar system designs and associated waveform designs will have to achieve a far more stringent degree of spectral containment.

Pulse Eclipsing. When a radar system transmits its waveform the receiver is switched off to avoid damage. Thus for a short period of time the radar is "blind", which can lead to the phenomena known as *pulse eclipsing*. Pulse eclipsing occurs at very short ranges and is repeated at range intervals determined by the pulse repetition frequency (PRF). This repetition makes pulse eclipsing particularly problematic for high PRF operation, as it creates many blind regions. It also has an effect on

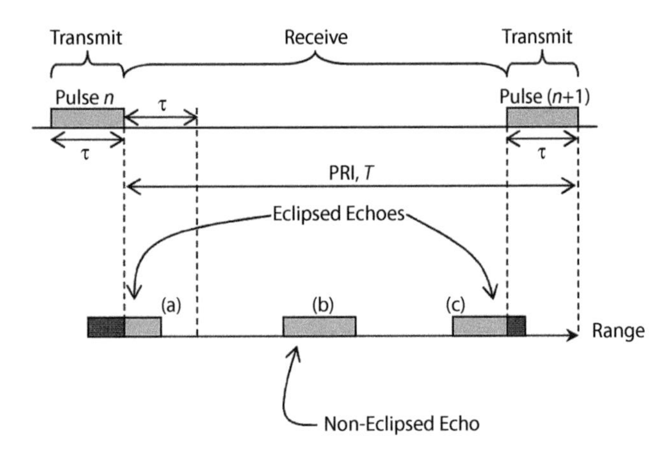

Figure 46-4. The echoes (a) and (c) are eclipsed because they arrive at the receiver during a time that overlaps with the transmission of a pulse (i.e., when the receiver is switched off). If the ratio of pulse width to PRI (the duty factor τ/T) is increased, more echoes will be eclipsed.

pulse compression and therefore on detection performance, as only a portion of the waveform is available for processing on receive.

Figure 46-4 illustrates how echoes from the n-th pulse are eclipsed within a single range interval. For the eclipsed echoes (a) and (c), a portion of each echo is missed as the receiver is switched off.

Figure 46-5 shows the impact of eclipsing on pulse compression when the waveform is an LFM chirp. Figure 46-5 compares the matched filter responses from (a) a complete LFM echo and (b) an LFM echo that has been eclipsed by 50%, where only half of the waveform has been received. Because only half the waveform is present, the amplitude gain from the pulse compression ratio (see Chapter 16) is halved.

The degradation of range resolution is also clearly evident from the increased mainlobe width shown in Figure 46-5. The 50% eclipsed LFM echo only possesses half the bandwidth of the original waveform, resulting in a range resolution degradation by a factor of 2.

46.2 Mismatch Filtering

In Chapter 16 the matched filter was introduced. The matched filter maximizes the receiver signal-to-noise ratio (SNR). In a matched filter, the magnitude of the range sidelobes is determined solely by the characteristics of the waveform. In this section, forms of receive filtering that are different from the matched filter are examined. These filters are termed *mismatched filters*. Their performance is a function of both the waveform and the filter thus providing more design freedom.

Frequency Weighting. In Chapter 6 we saw that the power spectral density of a signal is the Fourier transform of the

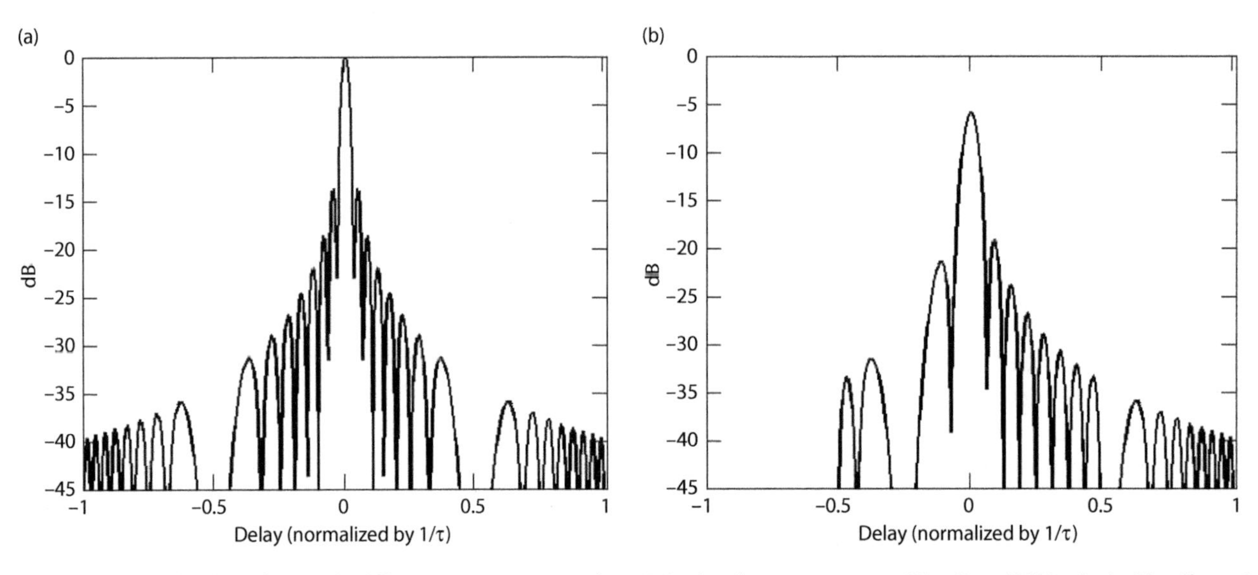

Figure 46-5. Compared with (a) the matched filter response to a complete LFM echo, the response to a 50% eclipsed LFM echo in (b) suffers a 6 dB loss at the matched point and a degradation factor of 2 in range resolution.

signal's autocorrelation function. Thus a waveform with low (autocorrelation) range sidelobes can be generated by weighting the spectral content of the waveform to reduce the power at the edges of the bandwidth. This effect is accomplished using a standard weighting function such as the Hanning, Hamming, or Blackman-Harris (and there are others).

For an LFM waveform, weighting simply deemphasizes the power at the beginning and end of the transmitted pulse, which corresponds to frequencies that are further from the center of the bandwidth. In this case, the receive filter is still a matched filter. While a significant reduction in range sidelobes can be achieved (see Fig. 46-6), there is an overall loss of sensitivity at the match point on reception, as energy has been removed at the band edges (relative to a constant envelope LFM). There is also some reduction in range resolution (generally measured as the mainlobe width 3 dB below the peak power). Both a reduction in detection range and in range resolution is tolerated, as there are benefits from having reduced sidelobes. For example, the reduced sidelobes help avoid ambiguities and also reduce clutter.

Optimized Mismatch Filtering. Another way to reduce pulse compression sidelobes is to design an optimized mismatched digital filter. The goals of this filter are to produce high range resolution, low sidelobes, and a low mismatch loss so that detection range is not reduced. Ideally, for a single delay shift the optimized mismatch filter should yield a perfect match, while for all other delays the filter output should be zero (no sidelobes). The filter design attempts to get as close to this ideal as possible. A digital filter (Fig. 46-7) can only approximate the impulse condition, thereby resulting in some degree of mismatch loss. Likewise, in practice, the sidelobes can be reduced but not eliminated.

One way to design such a filter is to use the frequency response, $S(F)$, of the time domain signal, $s(t)$ (i.e., the Fourier transform of the continuous waveform). The coefficients of the digital filter are obtained from the inverse of the frequency response, $1/S(F)$. This approach is referred to as an *inverse filter*. There are also other ways to solve for this optimized mismatch filter.

Pulse Compression with Adaptivity. As high-speed computing capabilities have advanced there has been an increasing trend towards performing pulse compression digitally (Fig. 46-7). However, digital implementation of pulse compression goes beyond simply replicating an analog filter with a digital counterpart. As processing speeds continue to increase, the application of adaptive signal processing techniques can be implemented that further improve sensitivity and robustness to interference.

As we have already seen, range sidelobes are a limiting factor in the use of pulse compression. If a large echo is received during an interval that overlaps with the reception of a smaller

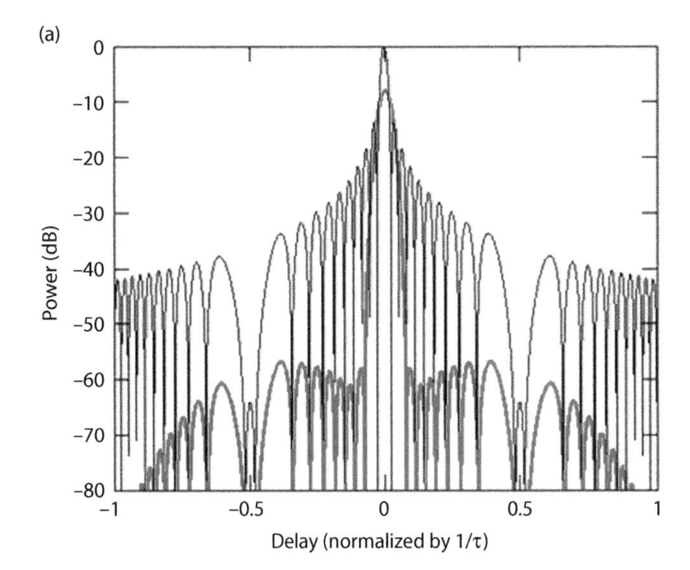

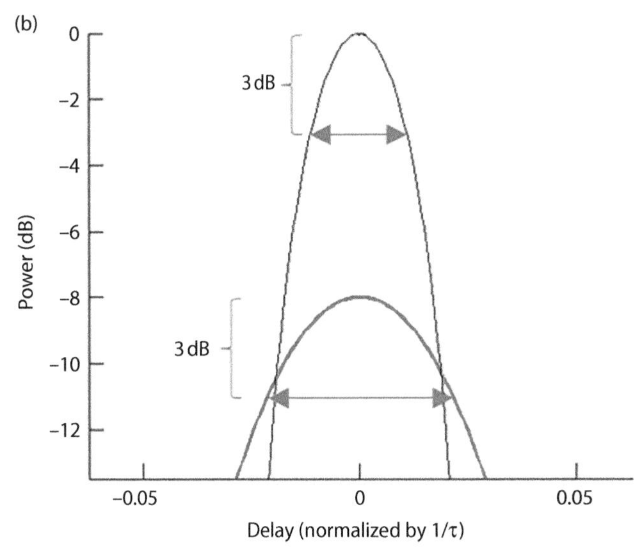

Figure 46-6. Compared with the standard LFM (in black), the matched filter response to a Hamming weighted waveform (in red) significantly reduces the range sidelobes (top panel) at the expense of range resolution (degraded by a factor of approximately 2) and roughly an 8 dB reduction in amplitude (lower panel). The overall detectability loss is significantly less than 8 dB because the weighting also reduces the competing noise signal.

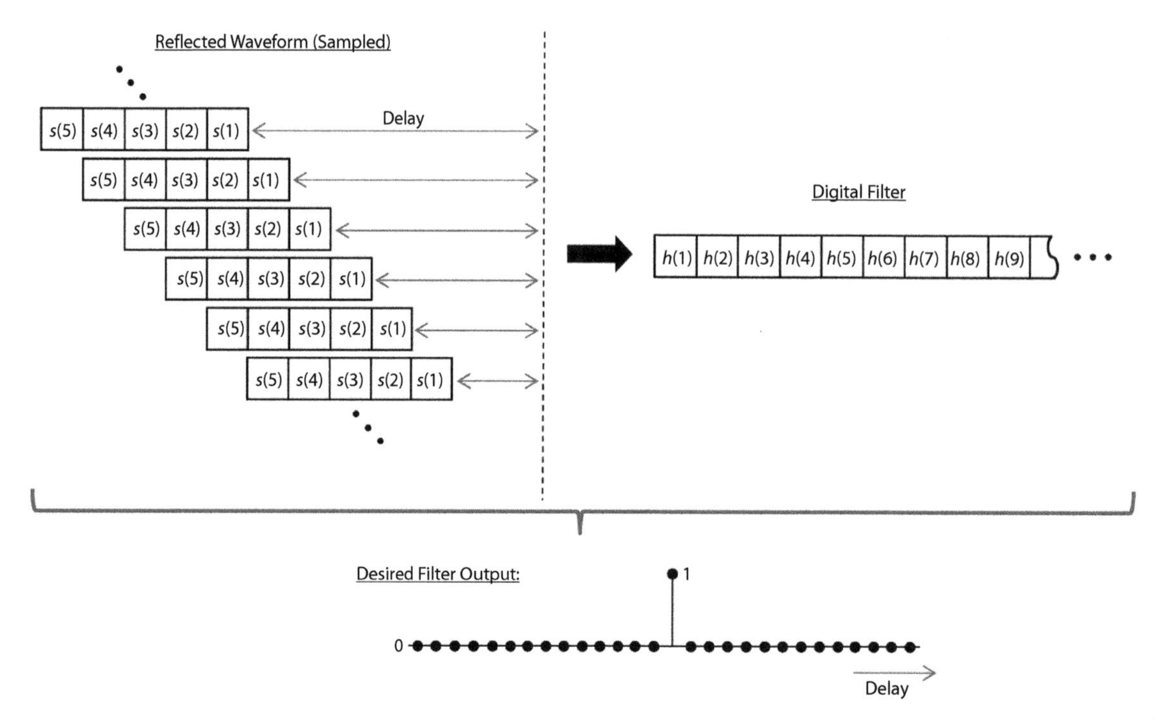

Figure 46-7. Based on a sampled representation of the reflected waveform for different delay shifts, the resulting digital filter response is designed to closely approximate a delayed impulse.

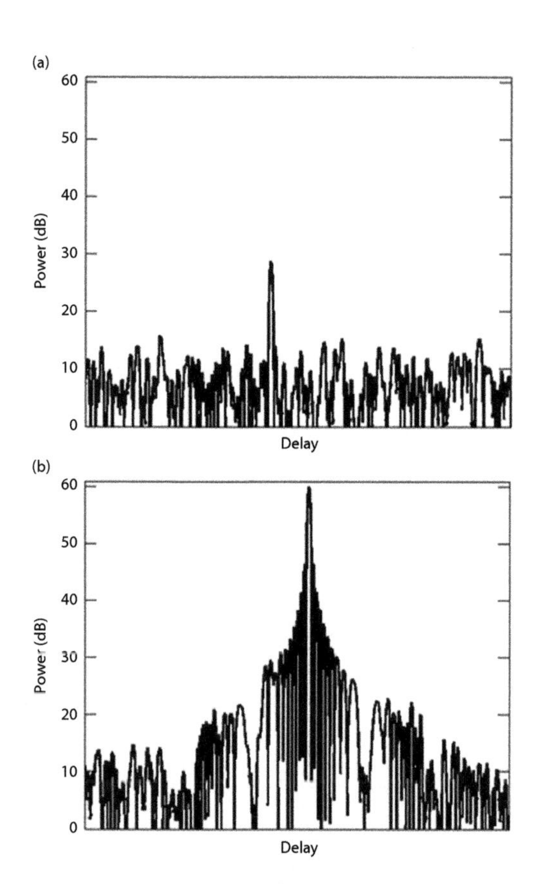

Figure 46-8. Pulse compression of large echoes can desensitize the radar to small echoes. (a) Depicts the matched filter response when an LFM chirp is used to illuminate a relatively small scatterer in the presence of noise. When an additional large scatterer is present, as shown in (b), the small echo is masked by the range sidelobes of the large echo.

echo, then the range sidelobes can mask the presence of the smaller echo. For example, Figure 46-8 depicts the pulse-compressed echo from a small scatterer alone (top panel) that is completely masked by range sidelobes if a large echo is also present nearby (bottom panel).

While careful waveform design and optimized mismatch filtering can be used to reduce these sidelobes, there is always a limit due to the selection of particular waveform design parameters and filter degrees of freedom. However, some of these limitations may be overcome through the incorporation of adaptivity into the receiver pulse compression stage.

Pulse compression adaptivity uses feedback (recursion). The output of the pulse compression filter is used to modify the filter, as shown in Figure 46-9. This approach is generally referred to as adaptive pulse compression and is analogous to adaptive beam forming used in electronically scanned antennas.

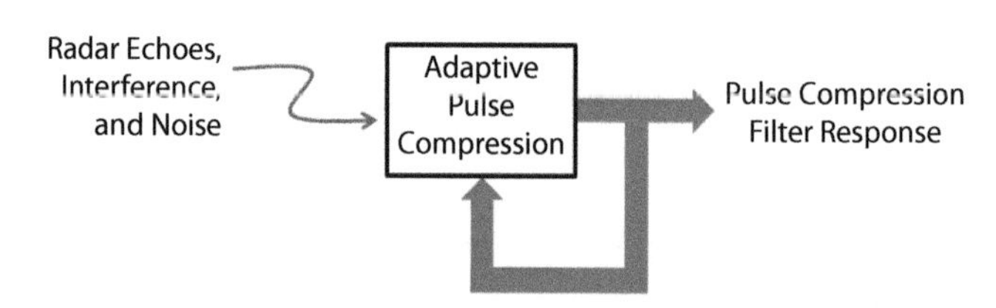

Figure 46-9. Incorporating adaptivity into pulse compression to cancel some forms of interference, including the self-interference of range sidelobes, involves feedback of the measured pulse compression output in a recursive manner.

46.3 Nonlinear FM Waveforms

Reducing range sidelobes can also be achieved with a frequency-swept waveform that does not spend equal time in each frequency, as is the case for LFM. In other words, the frequency is made to change nonlinearly with time. Here the structure of these nonlinear FM waveforms and their properties are considered.

A Different Approach to Frequency Weighting. To compare nonlinear FM (NLFM) with a standard LFM chirp, Figure 46-10 shows the instantaneous frequency of the two types of waveforms as a function of time. Both waveforms have the same pulse width τ and bandwidth ΔF, yet, where the LFM has a constant chirp rate, the NLFM has a time-varying chirp rate. Consequently the NLFM spends less time in the frequencies at the edges of the pulse, which results in deemphasizing the frequencies at either end of the spectrum.

As an example, given a 1 μs pulse and a time–bandwidth product of $\tau\Delta F = 64$, Figure 46-11 shows the power spectral density of an LFM waveform and a typical NLFM waveform. The NLFM waveform illustrates a rounding off of the spectral content relative to the much flatter spectral response of LFM.

When the NLFM waveforms are compressed in a matched filter the result is much lower range sidelobes, as shown in Figure 46-12. This result is a direct consequence of the NLFM energy being concentrated towards the center of the spectral band. There is some degradation of range resolution relative to LFM (i.e., broadening of the mainlobe). This degradation is also caused by the rounding off of the waveform spectral content. An additional advantage is that the NLFM matched filter output has no loss in sensitivity, as all the energy in the pulse is used and therefore there is no reduction in the detection range.

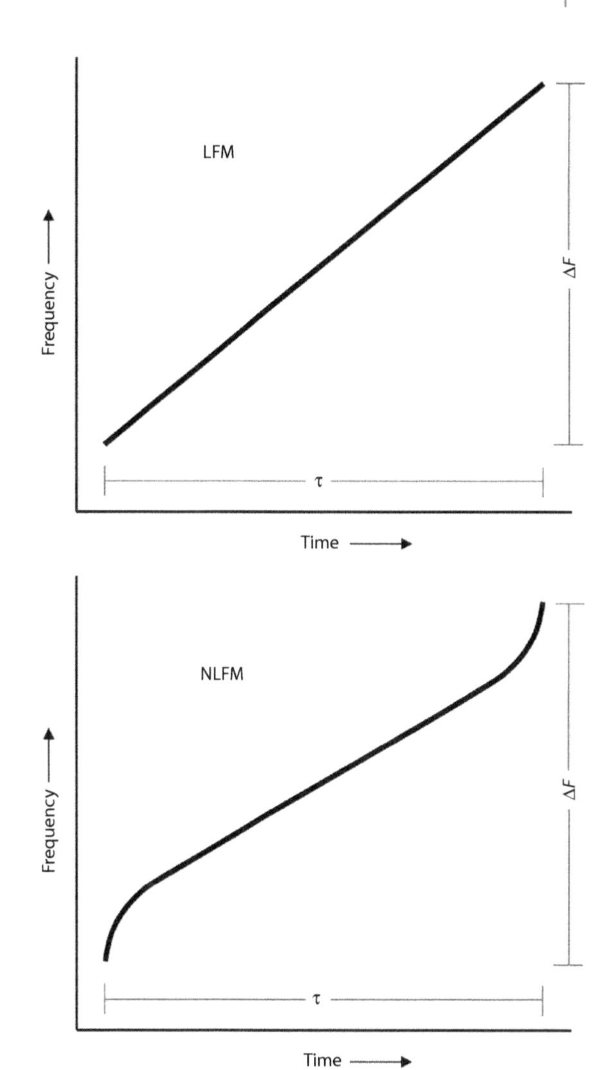

Figure 46-10. Time–frequency relationship for an LFM waveform and a generic NLFM waveform. While LFM has a constant chirp rate (a straight line), NLFM has a time-varying chirp rate so it spends less time in the frequencies at the edges of the pulse.

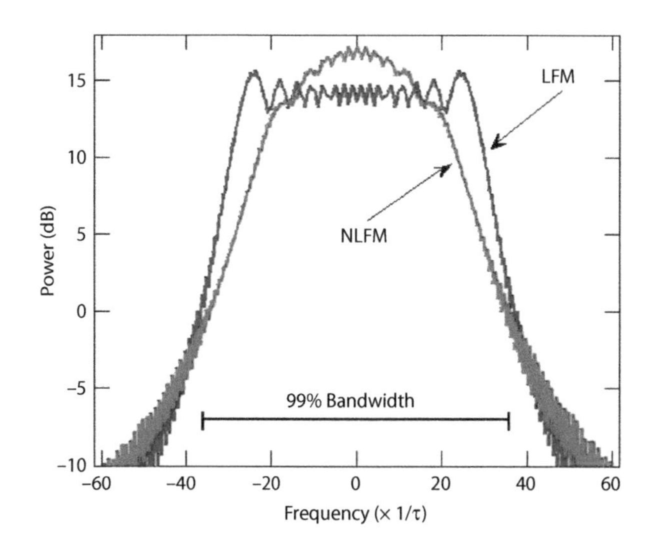

Figure 46-11. The power spectral density of LFM and NLFM waveforms having the same bandwidth and time-bandwidth product of 64. The NLFM spectrum energy is more concentrated in the middle and rounded at the edges of the band.

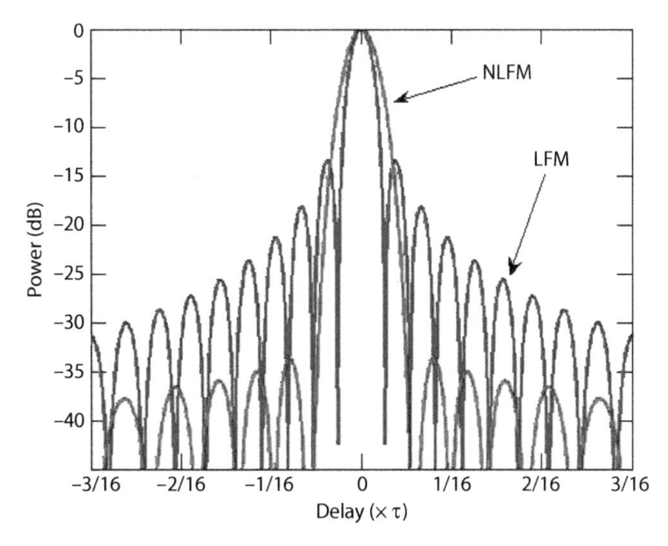

Figure 46-12. Matched filter responses for NLFM and LFM waveforms. NLFM provides lower range sidelobes , some degradation in range resolution, and no mismatch loss.

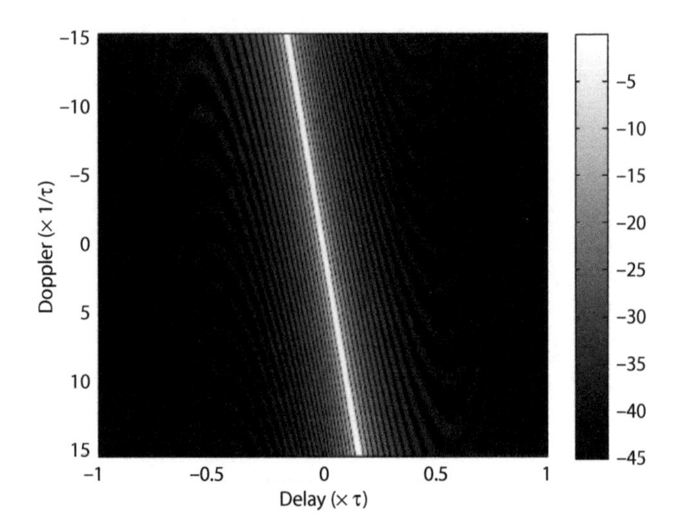

Figure 46-13. The ambiguity function for the LFM waveform with a time–bandwidth product of 64.

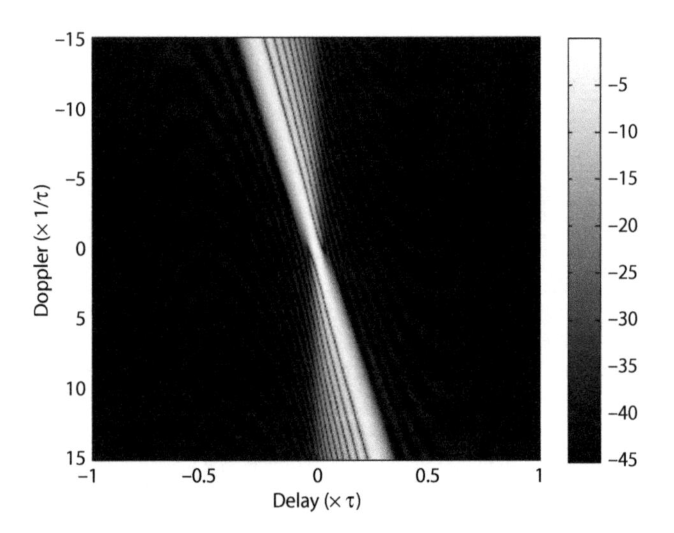

Figure 46-14. The ambiguity function for the NLFM waveform with a time–bandwidth product of 64. Fresnel lobes fan out from the main delay–Doppler ridge.

Doppler Effects. In Chapter 16 we saw that the ambiguity diagram of LFM chirp shows a linear relationship between range and Doppler (Fig. 46-13). This relationship means that the LFM waveform is tolerant of different Doppler frequencies as a result of the ambiguity between the delay shift and the Doppler frequency shift. For NLFM, however, the relationship between delay and Doppler depends on the form of the nonlinear frequency sweep.

Both LFM and NLFM waveforms exhibit *Fresnel lobes*. These are more pronounced in the case of NLFM and for nonzero Doppler frequencies; the ambiguity function fans out from the main delay–Doppler ridge (Fig. 46-14).

Waveform Design. The design of an NLFM waveform requires the determination of a nonlinear function that specifies how the frequency changes with time. The most common forms of nonlinear time–frequency relationships use polynomial phase functions or the concatenation of different piecewise LFMs to create the waveform. Good performance can also be achieved by combining NLFM with receiver mismatch filtering.

46.4 Waveform Diversity

Waveform diversity simply means using multiple different waveform designs to accomplish a radar function. This can be the selection of a single optimum waveform for a particular application or it might mean on-the-fly optimization with the specification of the waveform being continually altered. Here, some of the pertinent topics in waveform diversity are briefly summarized.

Enabling Technology. The most important technology development for waveform diversity is wideband digital waveform generation. Programmable digital waveform generators are commercially available with bandwidths exceeding 1 GHz. These allow almost any waveform to be emitted with design parameters that can be changed pulse by pulse. This technology is increasingly being incorporated into radar systems and provides the flexibility to tailor a waveform to any given application and operating environment. For example, it enables adaptive placement of nulls in range, thus helping to avoid unwanted EMI.

Combined with digital beam forming, waveform diversity provides additional and new design freedoms. Digital waveform generation, together with beam forming, is leading to improvements in the extraction of ever more useful information from radar echoes. For example, clutter, interference, and noise, which compete with target echoes, can be more effectively reduced. Waveform diversity is also enabling the operation of new processing concepts, such as multiple-input, multiple-output (MIMO) radar.

MIMO Radar. MIMO radar has two forms. In one form, spatially separated transmitter and receiver antenna elements form a distributed radar, as discussed in Chapter 45. This form

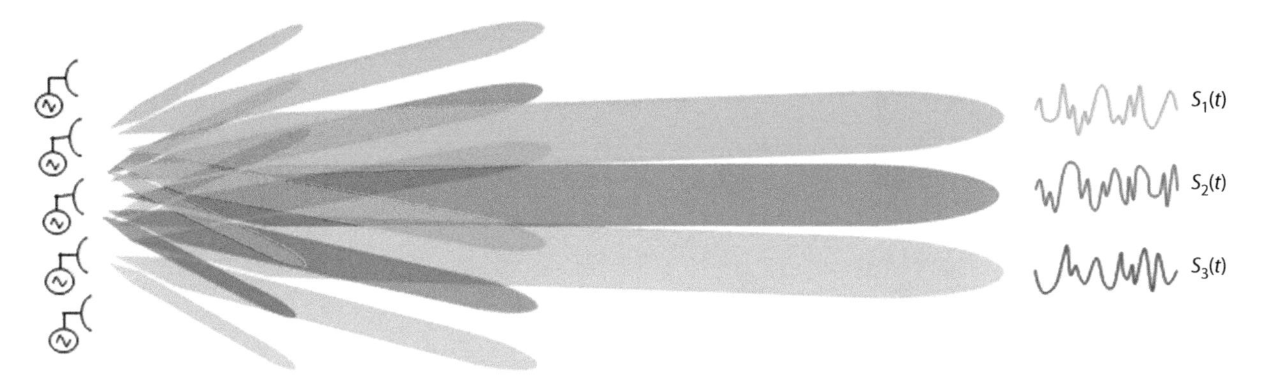

Figure 46-15. MIMO radar involves the emission of different waveforms in different spatial directions to provide additional separable channels of information in the radar receiver.

is sometimes referred to as *statistical MIMO*. Here, the other form of MIMO is considered, where the antenna elements are constrained within a single manifold (much as they are in an active electronically scanned array [AESA] radar). This form is sometimes called *coherent MIMO*. However, instead of forming a beam where each element transmits or receives the same waveform, coherent MIMO radar uses different waveforms, as illustrated in Figure 46-15.

For simultaneous emission of the different waveforms, there are essentially two ways they may be separated on receive. The waveforms can be separated into distinct frequency bands (i.e., frequency waveform diversity) or the waveforms can occupy the same frequency band but be coded so that they can be differentiated from one another (code waveform diversity). This is an area of ongoing research.

Let $s_j(t)$ be the jth waveform and $h_j(t)$ be the pulse compression filter function. As has already been seen, $s_j(t)$ must possess a matched filter response that provides the desired range resolution and has sufficiently low range sidelobes.

For MIMO radar, there is also the additional requirement that the response of the kth waveform to the jth matched filter be as small as possible for all delay shifts (and vice versa for the jth waveform and the kth filter). In other words, when the waveforms are cross-correlated they must have a consistently low output. Figure 46-16 illustrates this arrangement for a coded waveform diversity scheme where the coding is simply either an up-chirp or a down-chirp LFM. Consequently each of these waveforms is filtered by an up-chirp or down-chirp matched filter.

Pulse Agility. Pulse agility uses different waveforms at different times (pulses) within a coherent processing interval (CPI). This arrangement may be thought of as waveform time diversity. Pulse agility provides an alternative way to extend the unambiguous range of the radar.

As discussed in Chapter 15, range ambiguity results when echoes from an earlier pulse arrive during the receive interval

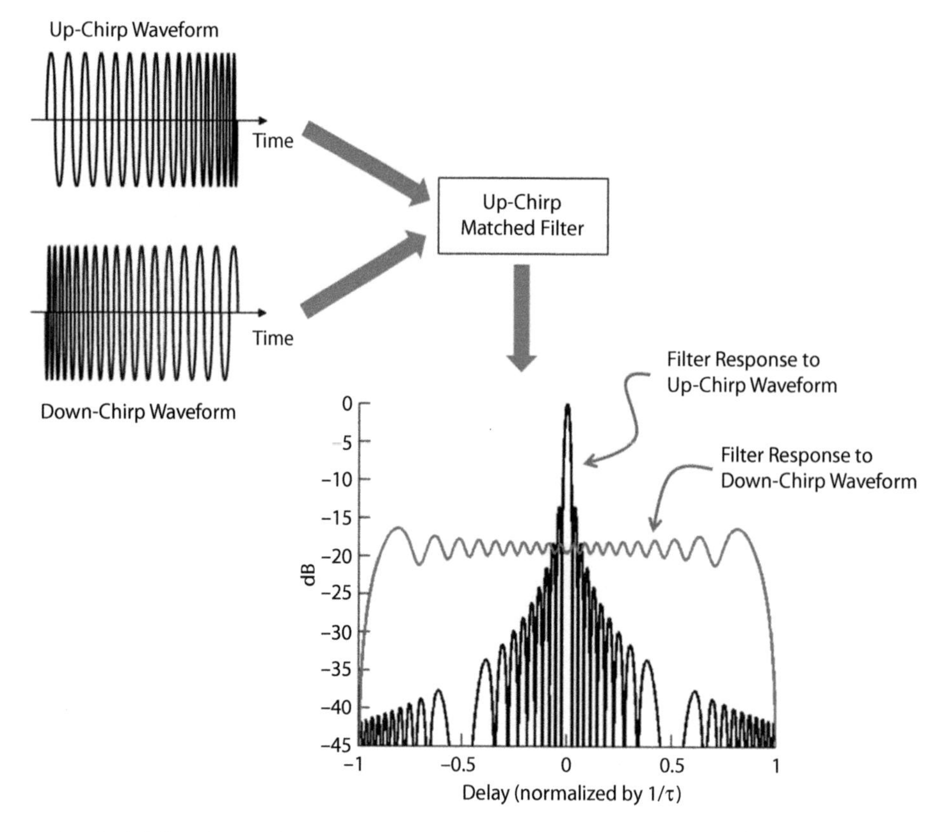

Figure 46-16. The separability of a MIMO radar waveforms depends on the waveforms used. This example depicts the up-chirp LFM matched filter responses for up-chirp and down-chirp waveforms.

for a later pulse (see Fig. 46-17). Because it is common for every pulse in a CPI to employ the same waveform, there is ambiguity as to which pulse the echo corresponds.

As illustrated in Figure 46-18, each agile pulse is modulated with a completely different waveform. This could be different frequencies or different modulation codes. The goal is that the waveforms be sufficiently separable after matched filtering. Hence, it is necessary to obtain a set of waveforms that, individually, possess the requisite range resolution and range sidelobes while also having a low cross-correlation.

Waveform Optimization. The two most prominent metrics used for waveform optimization are the *peak sidelobe level* (PSL) and the *integrated sidelobe level* (ISL) of the matched filter

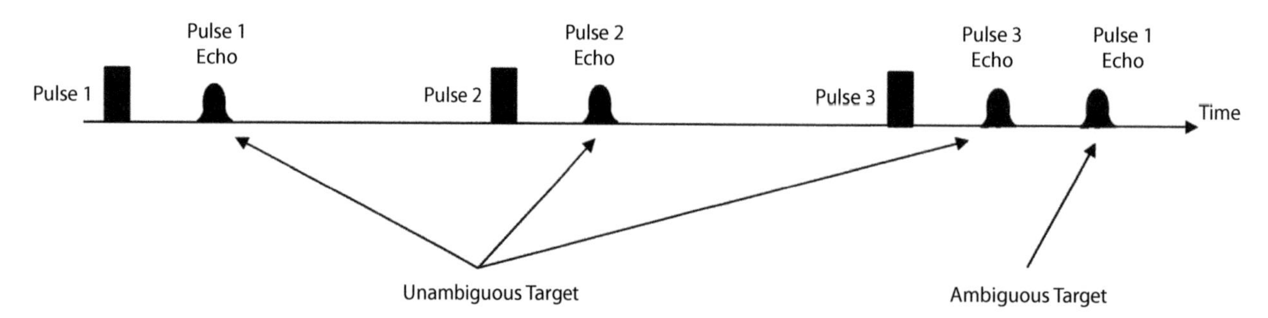

Figure 46-17. When the same waveform is modulated onto each pulse in a CPI, the maximum ambiguous range is determined by the interval between successive pulses. (Note: Ambiguous targets can be removed using PRF jittering.)

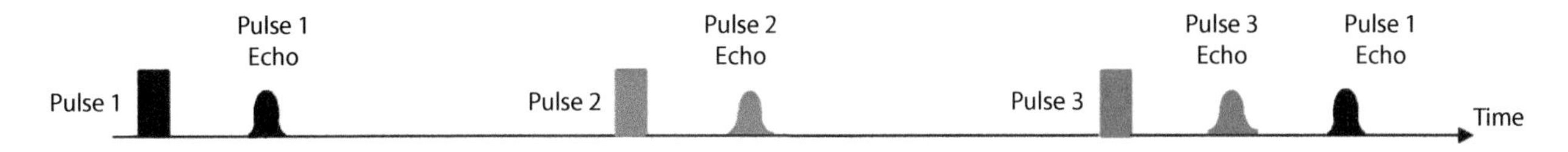

Figure 46-18. When a different waveform is used on each pulse in a CPI, the maximum ambiguous range depends on the number of pulses before the same waveform is used again. If the waveforms are sufficiently separable, the distant echo from pulse 1 can be identified accordingly.

output. Figure 46-19 shows that the PSL is the value of the largest sidelobe relative to the mainlobe, while the ISL is the area under all the sidelobes relative to the area under the mainlobe. Although both metrics measure aspects of the range sidelobes, the PSL can be viewed as a measure of the worst-case sidelobe response, while the ISL provides an aggregated measure. The ISL is particularly useful when ascertaining waveform suitability for diffused echoes such as those generated by distributed sources of clutter.

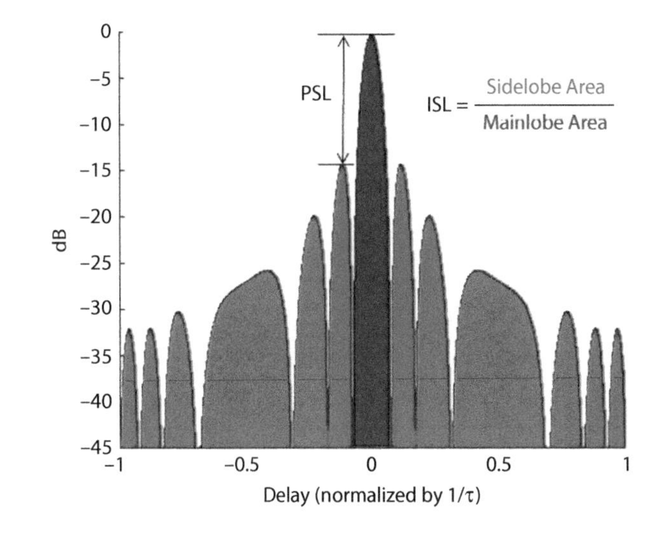

Figure 46-19. PSL and ISL metrics of the waveform matched filter response. These metrics are commonly used for waveform optimization.

When using these metrics to assess a waveform, it is important that the waveform structure closely resemble the actual physical waveform that the radar would transmit. In other words, transmitter distortion effects should be taken into account.

46.5 Summary

The evaluation of suitable waveforms and receive processing for pulse compression requires the method of waveform generation, the effects of the transmitter, and the nature of the spectral environment to be taken into account.

The transmitter induces both linear and nonlinear distortion that must be understood and corrected.

Radar waveform design has to be both tolerant to increasing EMI while also avoiding being a source of interference to other users of the electromagnetic spectrum.

Mismatched filters can reduce range sidelobes by reducing the influence of the frequencies at the ends of the pulse spectrum. However, this is at the expense of reducing maximum detection range. This limitation can be overcome by employing NLFM waveform designs that simultaneously result in maximal detection ranges and low range sidelobes.

Future radar systems will employ waveform diversity. This requires dynamic waveform design tailored to changing environmental and mission requirements. These systems may also utilize diverse operating modes, such as MIMO and pulse agility, that maximally exploit all available degrees of freedom.

Further Reading

N. Levanon and E. Mozeson, *Radar Signals*, John Wiley & Sons, Inc., 2009.

M. A. Richards, J. A. Scheer, and W. A. Holm (eds.), *Principles of Modern Radar: Basic Principles*, SciTech-IET, 2010.

M. Wicks, E. Mokole, S. Blunt, R. Schneible, and V. Amuso (eds.), *Principles of Waveform Diversity and Design*, SciTech-IET, 2011.

W. L. Melvin and J. A. Scheer (eds.), "Advanced Pulse Compression Waveform Modulations and Techniques," chapter 2 in *Principles of Modern Radar: Advanced Techniques*, SciTech-IET, 2012.

W. L. Melvin and J. A. Scheer (eds.), "Optimal and Adaptive MIMO Waveform Design," chapter 3 in *Principles of Modern Radar: Advanced Techniques*, SciTech-IET, 2012.

W. L. Melvin and J. A. Scheer (eds.), "MIMO Radar," chapter 4 in *Principles of Modern Radar: Advanced Techniques*, SciTech-IET, 2012.

F. Gini, A. De Maio, and L. K. Patton, *Waveform Design and Diversity for Advanced Radar Systems*, IET, 2013.

Test your understanding

1. Recall the tapped delay line filtering process described in Figure 16-18 of Chapter 16. Define the tapped delay line for application to a length 11 Barker code (from Figure 16-21 of Chapter 16), then apply the code segments below to the tapped delay line that would arise when pulse eclipsing occurs as depicted in Figure 46-4.
    a. [+ − − + −] (last five digits of the Barker 11 code)
    b. [+ + + − −] (first five digits of the Barker 11 code)
2. For each of the eclipsed tapped delay line responses from Problem 1, determine the resulting PSL.

47

Target Classification

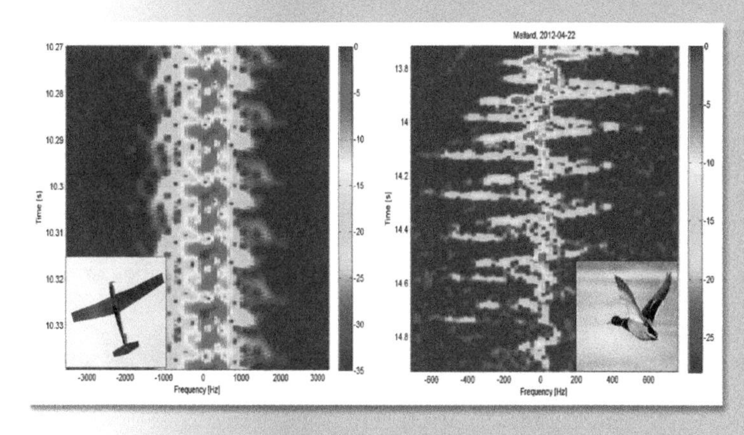

Micro-Doppler signatures at K-band of a small UAV (Left) and a mallard duck (right). (Boerge Torvik, FFI, Norway.)

The ability to detect and locate targets by day or night, over wide areas, regardless of weather conditions has long made radar a key sensor in military and civilian applications. However, the ability to reliably distinguish different targets such as an airliner from a fighter aircraft has proved much more elusive and has conventionally relied on skilled operators and on separate devices such as identification friend or foe (IFF) that require the cooperation of the target. To be able to do this automatically and without cooperation of the target represents a much more difficult challenge, although steady progress has been made over the past couple of decades.

In this chapter we discuss the principles of noncooperative target classification, using the example of the detection and identification of vehicles in high-resolution synthetic aperture radar (SAR) images. Note, though, that radar imagery is not the only means to achieve radar target classification. For example, one-dimensional high-range resolution profiles (HRRPs), especially for air targets, can provide operationally useful levels of performance. Another important technique for air target classification is jet engine modulation (JEM) of the radar signal, as illustrated in the blue panel on the following page. For maritime targets, a key consideration can be characterizing the maritime clutter environment against which targets are to be detected.

47.1 Introduction

Radar target classification is a huge topic, but many of the key ideas can be concisely highlighted using the example of SAR imagery, which will be the approach adopted here.

A typical SAR image is shown in Figure 47-1, including a highlighted helicopter. Remember that radar signals typically have wavelengths of the order of centimeters, so the scattering

Figure 47-1. This picture shows a typical high-resolution SAR image with a helicopter highlighted.

Jet Engine Modulation

A PARTICULARLY ELEGANT TECHNIQUE CAN BE USED IN THE RADAR IDENTIFICATION of air targets, through the modulation imposed on the radar echo due to the rotating compressor blades of jet engines. This is known as *jet engine modulation*.

A jet engine will have a number of compressor stages, each with a different number of blades.

The rotating compressor blades modulate the radar echo in a characteristic way. For a single rotor with N blades and a rotation period T, the spectrum of the echo has the form shown as follows.

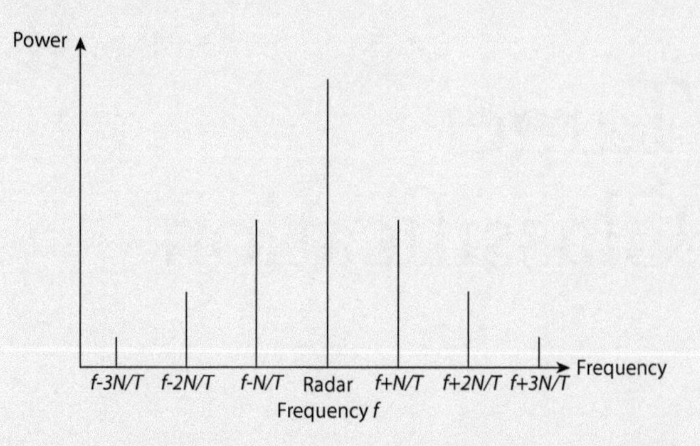

When the contributions from the other rotors are included, the spectrum provides a unique *fingerprint* that can allow the engine type, and hence the aircraft, to be recognized. Of course, the situation with multi-engine aircraft is slightly more complicated.

A similar process can be used to analyze and recognize the modulation of the radar signatures of helicopters, due to the rotation both of the main rotor and the tail rotor.

behavior is quite unlike that of visible light for which wavelengths are in the range of around 400 to 800 nm.

After a brief overview of classification terminology, this chapter will look at the impact of radar target phenomenology on classifier design, the detection and classification processing chain, the construction of databases to support classification, and performance assessment considerations for target classification systems. Finally, some of the key remaining challenges for radar target classification will be highlighted.

47.2 Classification Terminology

The terminology used for target classification can be very confusing. However, the North American Treaty Organization (NATO) AAP-6 Glossary of Terms and Definitions provides some precise definitions of concepts. The process is viewed as a hierarchy in which the targets are categorized into more and more precise subclasses. Six major steps are identified:

- Detection, separating targets from other objects in the scene
- Classification, giving the target a metaclass such as aircraft or wheeled vehicle

- Recognition, specifying the class of the targets such as fighter aircraft or truck
- Identification, giving the subclass of the target such as MIG29 fighter aircraft or T72 tank
- Characterization, taking into account the class variants such as MIG29 PL or T72 tank without fuel barrels
- Fingerprinting, leading to an even more precise technical analysis such as MIG29 PL with reconnaissance pod

Of course, the boundaries between these steps cannot be fixed for all problems and targets, so it is important to remember that wider usage is made of these terms. In particular, these definitions lead to the word *classification* being reserved solely to describe the process of metaclass separation, but it is more often used by engineers (and in this chapter) to describe the general process of assigning objects to categories.

It should also be noted that the classification of air targets is generally referred to as noncooperative target recognition (NCTR), whereas the classification of ground targets is generally referred to as automatic target recognition (ATR).

47.3 Target Phenomenology

Figure 47-2 shows a 10 cm resolution SAR image containing several vehicles. The individual targets have different characteristics, so in principle it should be possible to classify them. However, they are very unlike targets in optical images. For example, the highlighted vehicle displays some complex, periodic scattering structure. This vehicle is a type of pickup truck, and the multipath reflections from its open load-carrying section cause the observed structure.

Figure 47-3 provides another example of multipath. It shows the image of a tank with the gun barrel pointing to the side. The barrel appears three times in the image at three different ranges. The brightest image of the barrel is from the direct reflection and is at the closest range. The next barrel image results from the reflection off the ground onto the barrel and then back to the radar (or vice versa). The third barrel image results from the reflection off the ground onto the barrel and then back onto the ground before returning to the radar.

Another radar-specific phenomenon is the extreme sensitivity to small changes in imaging geometry. In Figure 47-4, the illumination angle (shown by the arrow) varies by only a small amount over the sequence, yet particular features of the target, such as the back edges of the aircraft wings, appear very differently throughout the sequence. This aspect angle dependency can be so strong that a target may look completely different when seen from directions separated by only few degrees.

Yet another consideration is that, unlike an optical image, radar explicitly measures range, which gives some particular effects (the blue panel on the following page). The lesson is

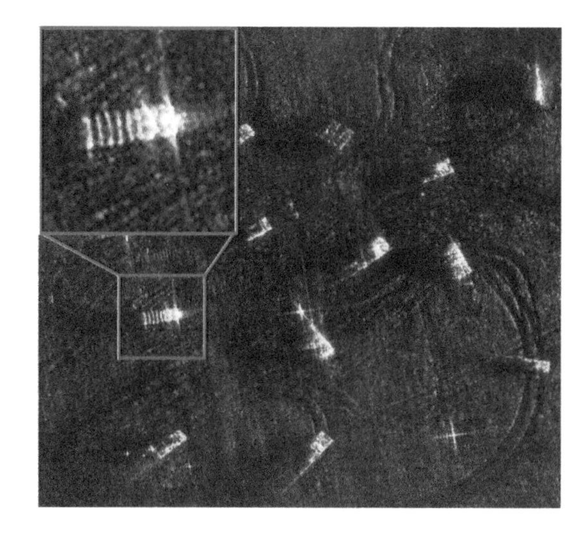

Figure 47-2. This example shows a 10 cm resolution spotlight SAR image and complex target phenomenology. The periodic structures are a result of multipath reflections.

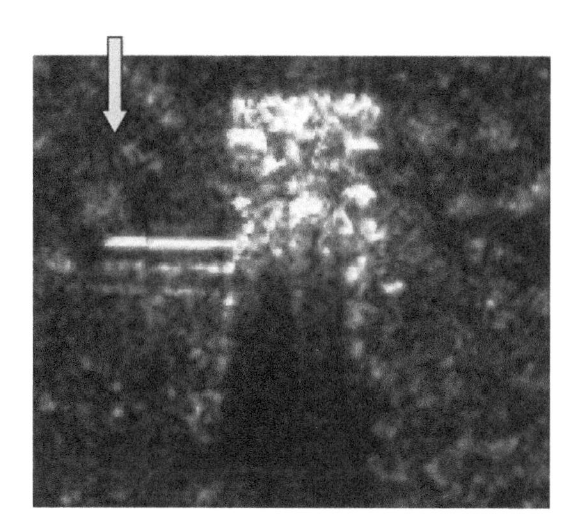

Figure 47-3. This illustration shows multipath reflections from the barrel of a tank, with direct-path (brightest), double-bounce, and triple-bounce barrel images.

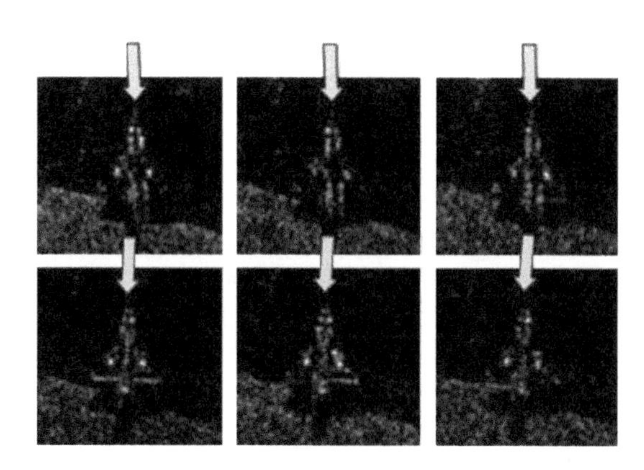

Figure 47-4. Sensitivity to imaging aspect angle causes the back edge of the aircraft wings to appear and disappear in the space of a few degrees of change.

not to expect radar images to look like optical images since they contain different information. SAR images are the result of complex phenomenology resulting from the imaging geometry, radar parameters, and radar scattering mechanisms. It is essential to account for this in classifier design to achieve a robust ATR system.

Layover and Shadowing

RADAR IS AN ACTIVE IMAGING SYSTEM THAT MEASURES RANGE. THIS HAS specific consequences for the appearance of 3D targets in 2D SAR images. It is possible to see through simple examples how the radar energy interacts with the target and ground, resulting in layover effects and self-shadowing.

The following figure illustrates that the radar image is formed by taking into account the range of the object from which the radar energy is scattered. In particular, point B on the target and point B′ on the ground are at the same range, so reflections from these points will be placed at the same position in the image.

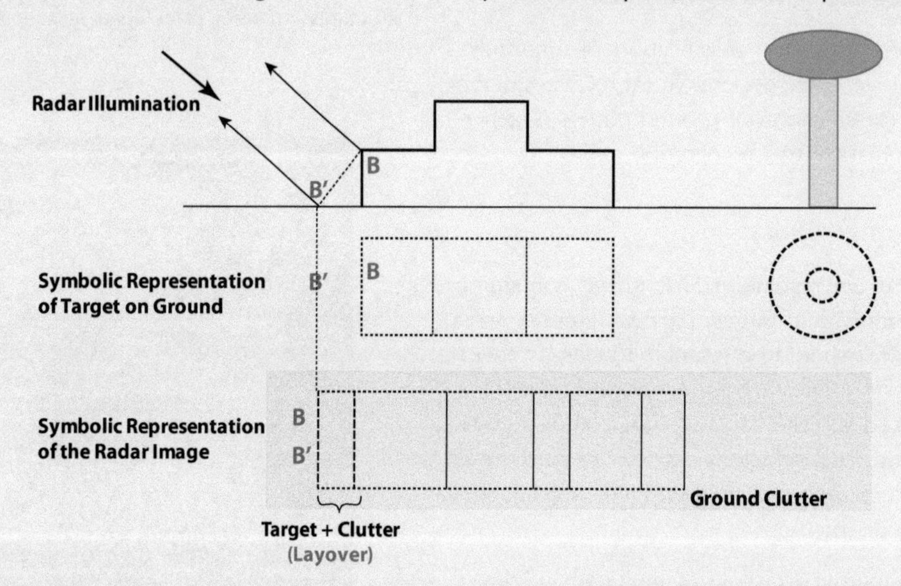

In the following image, all the reflections from the ground between B′ and A, and from the target between B and A, will occupy the same region. This is known as *layover*, and it results here in a superposition of target and clutter returns. Further layover occurs involving the target sections between B and C, between C and D, and between D and E. This produces a complex superposition of returns from various parts of the target in the final image.

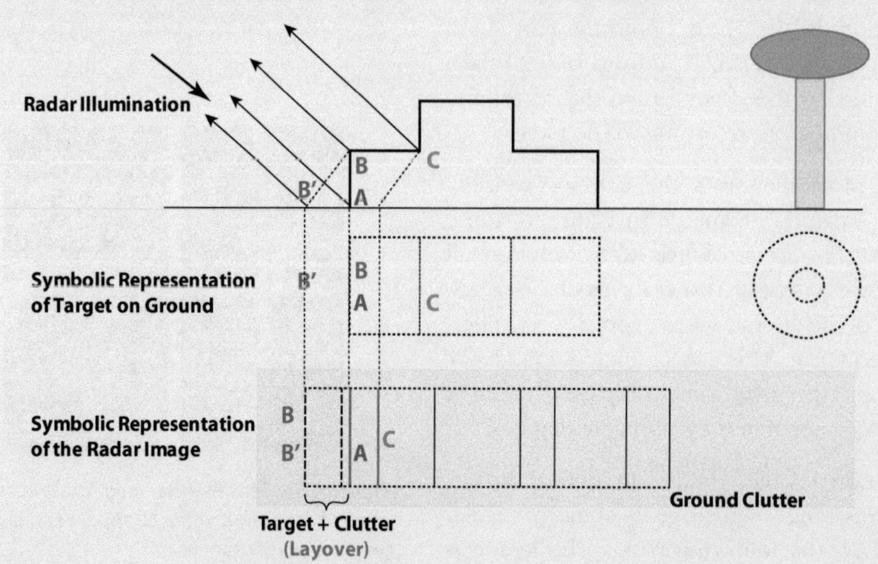

Layover and Shadowing *continued*

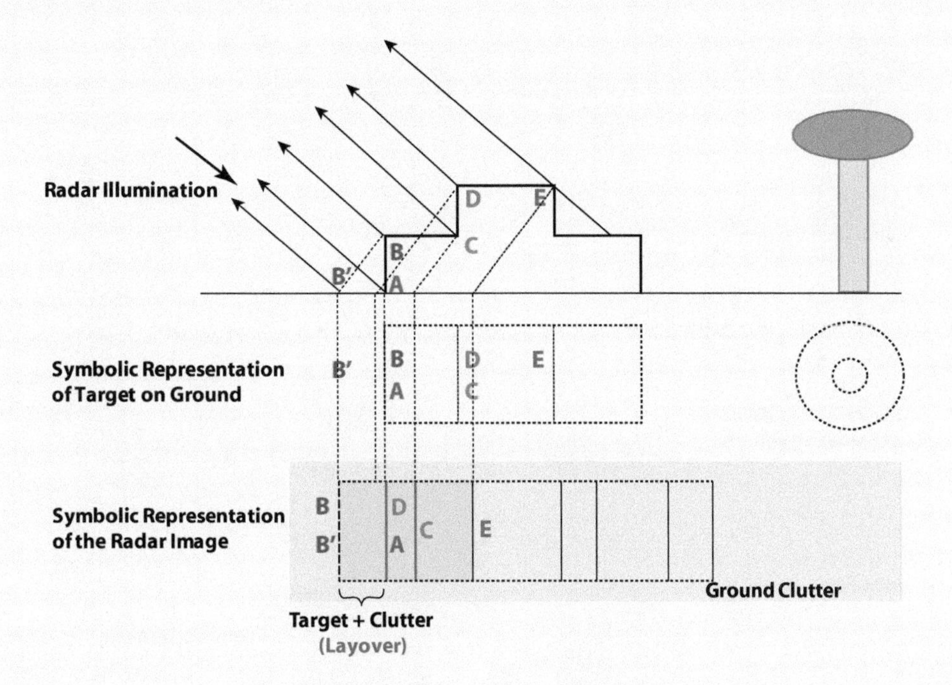

Another feature of radar imaging is *self-shadowing,* caused by an absence of line of sight from the radar to the target between points E and G. This inner shadow will obscure parts of the target itself, but the parts obscured will vary depending on the orientation of the target relative to the radar. There is also a shadow behind the target, again caused from regions that are not in the radar's line of sight. In particular, the target up to point H blocks the radar energy from reaching the ground.

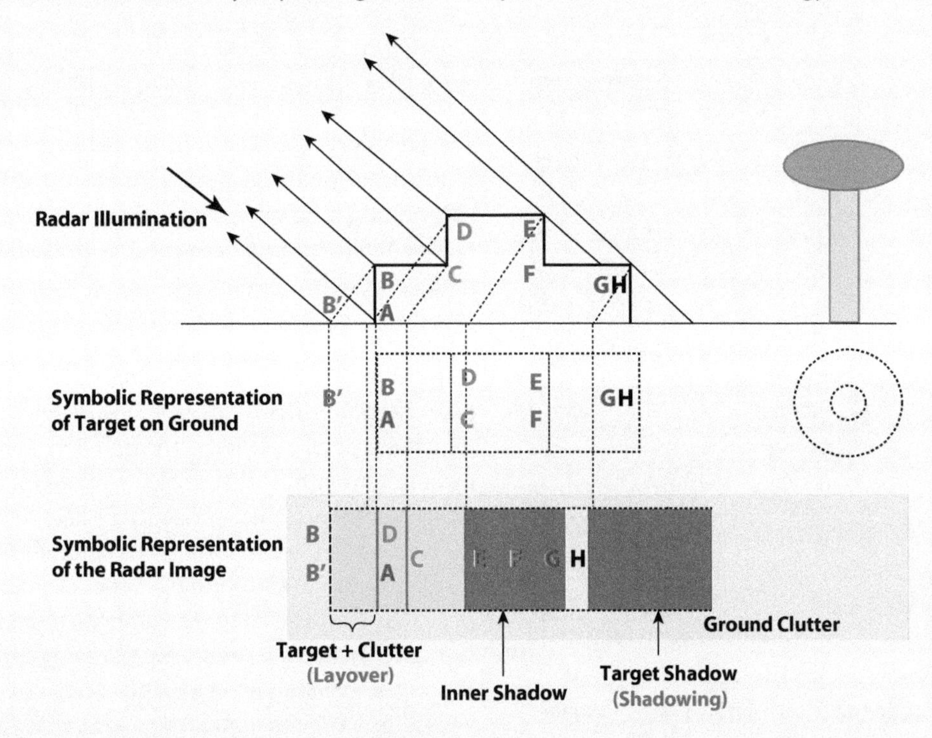

The shadow of the target can provide information about the shape of the target, so shadow properties can be used in the target recognition process. However, it should be noted that point H is subject to layover in the image; therefore, the extent of the shadow does not translate directly to the height of the object.

Another consideration is that objects separated from the target on the ground can interact with the target in the image. The following illustration shows that the reflections from the tree canopy will prevent the full target shadow being observed, which will corrupt any target shape information being measured from the shape of the shadow. In a more extreme case, the reflections from the tree canopy may actually lay over onto the target returns, so corrupting the target image and posing additional difficulty for target recognition.

Layover and Shadowing *continued*

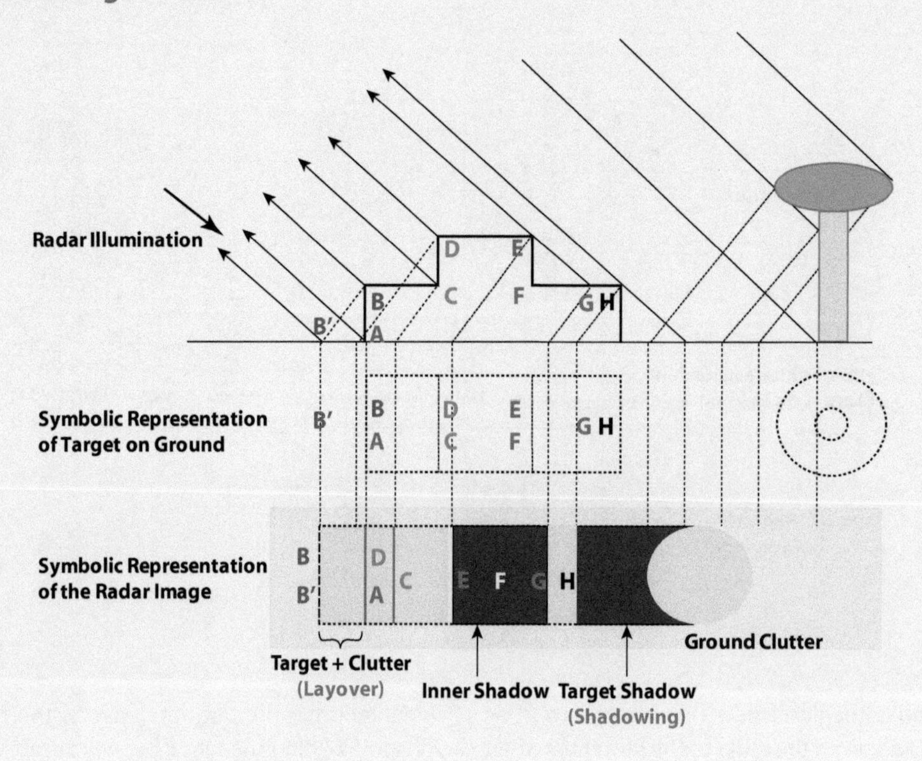

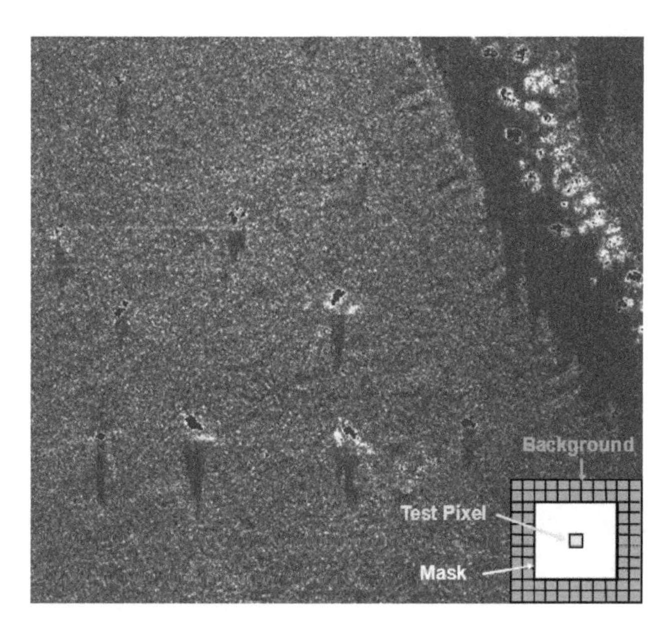

Figure 47-5a. A sliding window (inset) is passed over the image, and the central pixel is flagged up (blue) if it is significantly brighter than the surrounding background clutter.

47.4 The Target Classification Processing Chain

Prescreening. The first task in the target recognition process is to detect potential targets in the scene. This task in itself can contain a number of stages in which candidate detections are identified and filtered to reject those that do not meet the criteria for being a potential target. This process is often called *prescreening.*

Detection. The first prescreening stage is to perform a single pixel detection that flags pixels brighter than their neighboring background pixels. Fortunately in radar images the many metallic structures that make up targets give bright radar returns. Conventionally, a sliding window is used with a statistical test to flag up when the central pixel is significantly brighter than the surrounding clutter. A guard ring (or mask) of pixels is often excluded when characterizing the surrounding clutter to avoid contamination from target pixels. Figure 47-5a shows the detections (in blue) that result when this detection algorithm has been applied to a synthetic aperture radar (SAR) image of a number of vehicles in an open field. The inset illustrates the detection algorithm.

Clustering. A number of single pixel detections will be obtained from the detection stage. A clustering algorithm is then used to associate detections that belong to the same

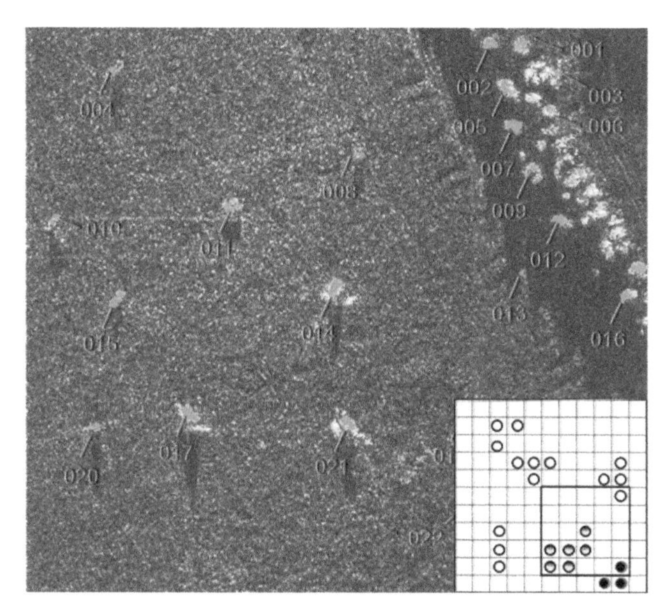

Figure 47-5b. The labeled clusters are shown that result from a clustering algorithm, based on simple cluster growing rules (inset).

potential target. One approach is to grow clusters using simple rules, as illustrated in the inset of Figure 47-5b. The detections shown in white have already been assigned to clusters such that four clusters have been identified. The shaded detection at the center of the bold square has initiated the next cluster. All the other shaded detections have been assigned to this cluster. The bold square indicates the predefined cluster size limit has been reached. The algorithm will now proceed to the next nonclustered (black) detection. Figure 47-5b shows the enumeration of distinct (green) clusters obtained by clustering.

Rejection of Clutter Discretes. The final prescreening stage is to examine each cluster on the basis of simple measures such as cluster size and average power. It is possible then to reject a number of candidate targets as being more likely to be discrete clutter objects such as trees rather than man-made objects. The remaining clusters are then passed forward to the classification stage. In Figure 47-5c, the potential targets have been marked with red crosses. It can be seen that clusters from vegetation and a decoy object (shown with a yellow arrow) have been rejected.

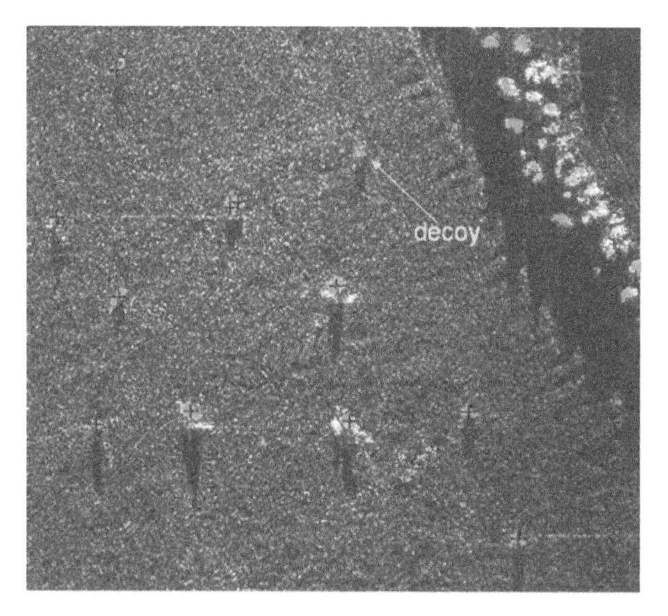

Figure 47-5c. Simple measures are used to reject clutter discretes. The red crosses flag up potential targets, and vegetation clusters and a decoy target (yellow arrow) are rejected.

Classification Techniques

ONCE FEATURES HAVE BEEN EXTRACTED, MANY CLASSIFICATION TECHNIQUES from the pattern recognition world can be used. While these are not radar specific, it is important to understand how they work.

Conceptually, the measured feature values can be plotted in a feature space with as many dimensions as there are features. For example, there are two features in the following illustration. The classification task is simply to draw the best decision

Classification Techniques *continued*

boundary to separate the classes, with *best* meaning maximizing correct and minimizing incorrect classifications.

If the density of the feature values can be modeled by a probability density function (e.g., using 2D Gaussian distributions as shown by the contours in the figure below), then an optimum statistical decision can be defined. This gives rise to the Bayesian classifier.

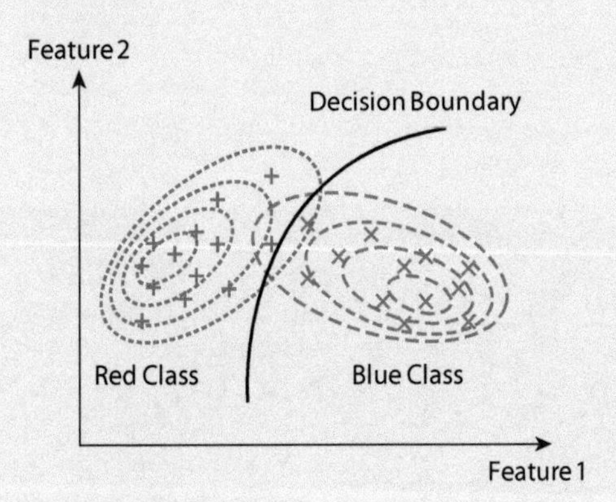

However, accurate statistical models can be difficult to formulate. An alternative is simply to declare the class of the target under test to be the same as the class of the closest training example in feature space, that is, the *nearest neighbor* (NN). In the following figure, the regions of feature space that will be declared red or blue, based on the training data (dots), are shown. Outliers can cause anomalous results (e.g., rightmost blue segment), although this can be mitigated by using the *k* nearest neighbors (kNN). In this approach, the class of the target under test is declared to be the dominant class among the *k* closest training examples, where *k* is a suitable chosen integer.

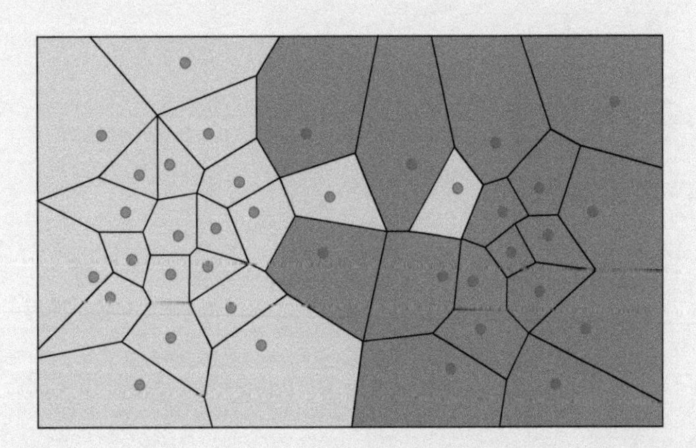

An alternative is to "learn" the decision boundary using a *multilayer perceptron* (MLP) neural network. An MLP consists of a number of layers of nodes or *neurons*, as illustrated in the following example. The first layer will have as many nodes as there are features, and the final layer will have as many nodes as there are classes. All nodes in one layer are connected to all nodes in the next layer and transmit their value via a weighting and a nonlinear function. Examples from the training set are presented to the first layer, and the MLP is "trained" (i.e., the weights are adaptively adjusted), such that only the output node corresponding to the correct class "fires" (i.e., has a value close to one).

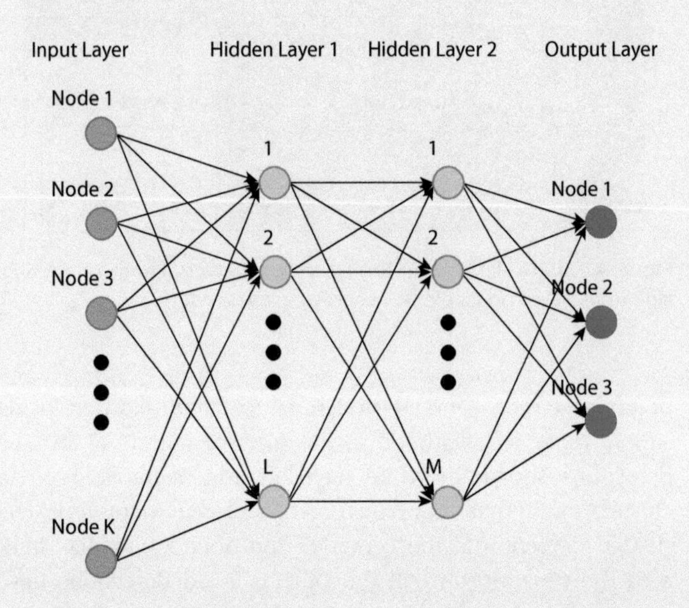

The basic operation of a *support vector machine* (SVM) is to maximize the margin between two classes by finding two linear boundaries with maximal separation as illustrated as follows. This assumes that the classes do not overlap, but there are generalizations to "soft" margin approaches to cope with overlapping classes.

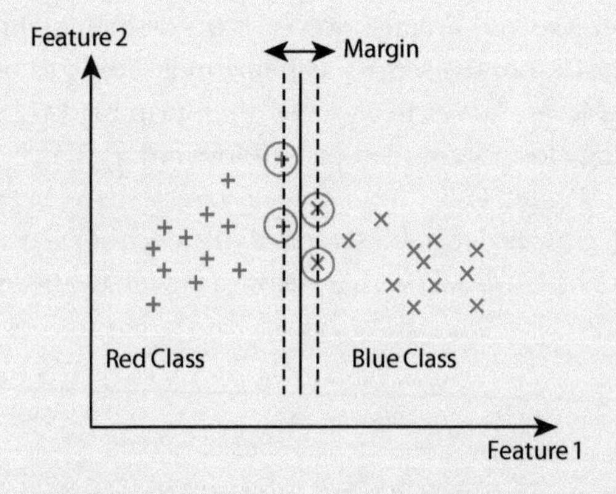

The choice of classifier will depend on the complexity of decision boundaries, the required flexibility to introduce new classes and the computational cost. However, the choice of features is the key to successful feature-based classification.

Template Matching. Once the candidate targets have been identified via prescreening, the classification process can begin. One approach is to attempt to match each candidate target with a database of images (i.e., templates) of the various targets known to the system. This approach is illustrated in Figure 47-6.

Since radar images are very variable as a function of imaging geometry, the database must contain example images of the targets at all possible geometries. The inset in Figure 47-6 shows a target database containing target images over 360° of aspect angle variation. In general, elevation angle and many other degrees of freedom would need to be included. The target under test has failed to match with the yellow database entries, but a match has been found with the red entry.

Although template matching is conceptually simple, the problem is that the required databases can be huge when there are many target classes and many potential degrees of freedom. Thus, template matching has an important role for classification problems that are relatively constrained, but an alternative approach is required for less constrained problems.

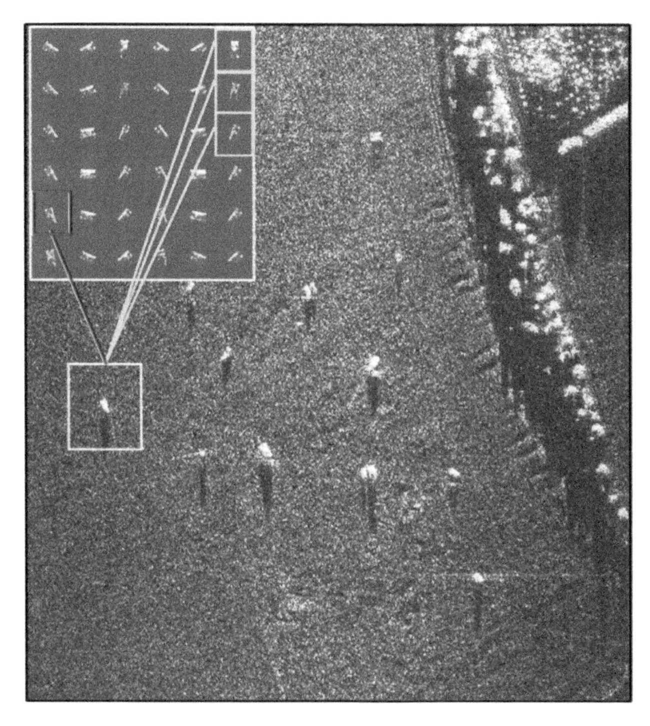

Figure 47-6. In template matching, the target is matched against previously gathered examples contained in a database. In this case a good match is found with the red database entry.

Feature-Based Classification. This alternative approach avoids the need for huge databases of imagery by characterizing the unique properties of the target class in terms of measured *features*. The choice of features is the key to classification performance. Standard features can be grouped into three main categories: geometrical features; texture-based features; and contrast-based features.

Geometrical features include the length, width, area, ratio of target length to width, and moments of inertia about the center of mass. Fourier coefficients may also be included in this category because they essentially characterize the edges of the target in their high-frequency components. Figure 47-7 illustrates how the use of length and width measurements can separate the SAR images of *main battle tanks* (MBT), *armored personnel carriers* (APCs), and support vehicles.

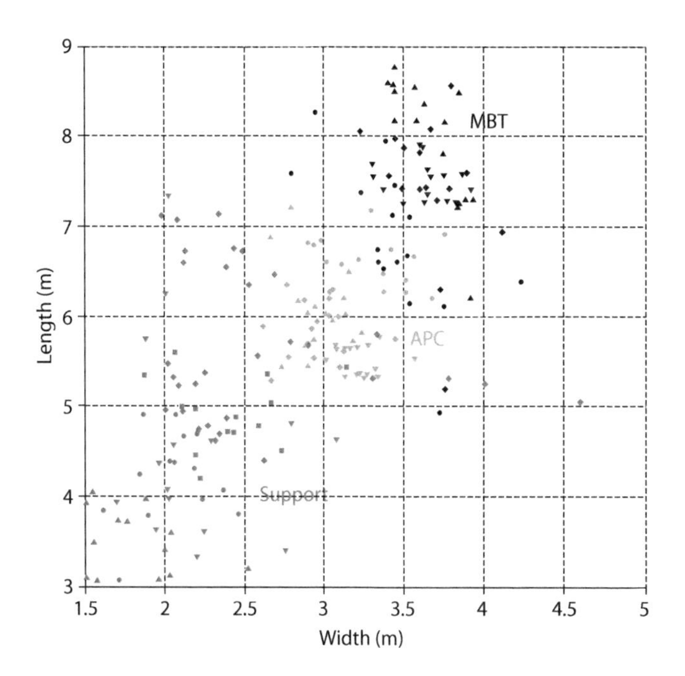

Figure 47-7. Features such as length and width can be used to characterize target classes. In this case three types of military vehicle can be separated reasonably well using these features.

Examples of texture-based features are the standard deviation and spatial correlation lengths of the pixel intensities and the entropy, which can characterize the randomness of the pixel intensities of the target. Contrast-based features include the *fractal dimension*, which measures the spatial distribution of the top scatterers within the target extent, and the *weighted rank fill ratio*, which measures the extent to which the power is concentrated within a few bright scatterers. Once a set of features has been defined, standard techniques from the pattern recognition area can be used to perform classification.

A reasonable level of classification can be obtained using the aforementioned standard features. However, in terms of finding features that provide the greatest robustness to target variability, the features that relate specifically to the underlying physical structure of the target, such as the positions of dominant scatterers, are likely to be the most robust.

47.5 Databases and Target Modeling

SAR Image Database Options. It is clear that a large database of example images of the targets of interest is needed for target classification. However, building such a database represents a great challenge due to the variability of radar imagery. For example, multiple expensive imaging flights will be needed to obtain enough data to characterize the variation of target signatures with imaging geometry.

The Defense Advanced Research Projects Agency (DARPA) undertook the first large data collection in the 1990s, which provided the moving and stationary target acquisition and recognition (MSTAR) data set. Figure 47-8 shows some examples of military vehicles and their corresponding SAR images from this data set. Images from all aspect angles and multiple elevation angles were obtained, and the target set included military variants and civilian vehicle to capture the huge variability of radar imagery.

Despite the extensive nature of the MSTAR database, given the totality of real-world variability it is clearly not feasible to collect real data to cover all the operating conditions that may be encountered. Thus, it is essential that databases are populated to a large extent by other means.

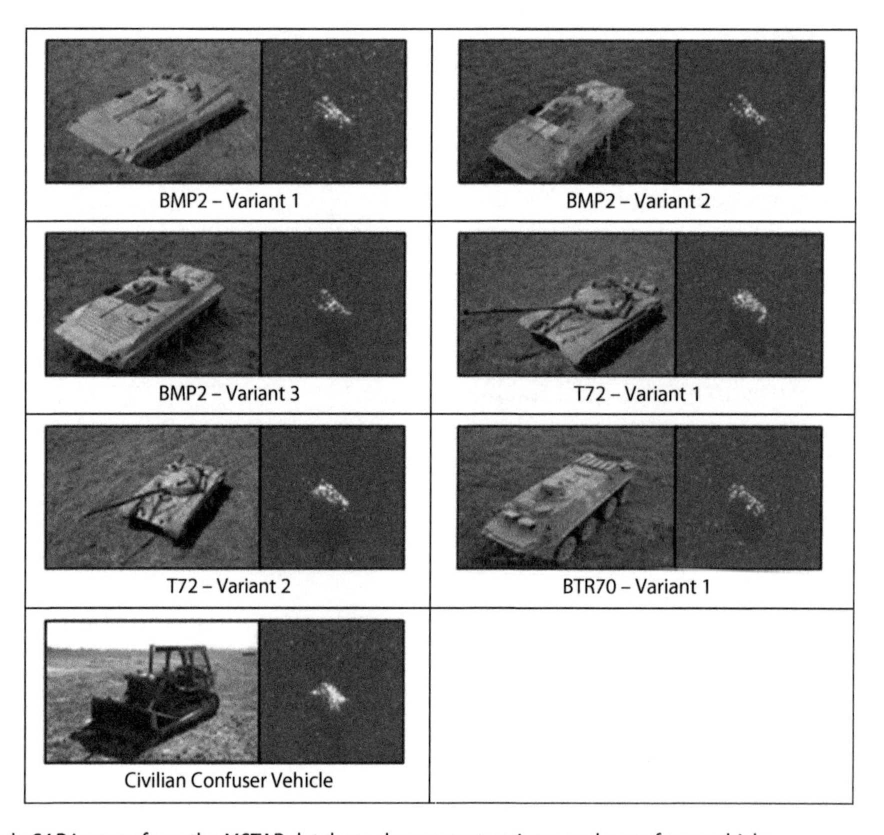

Figure 47-8. These example SAR images from the MSTAR database show target variants and a confuser vehicle.

One option is to place vehicles of interest on a turntable. This provides a very controlled situation for image formation in which the radar is stationary and relative motion is introduced by rotating the turntable. Image formation is then performed using inverse SAR techniques. It is important to be aware that such imagery may not be entirely representative of target imagery taken in the field. In particular, the background clutter against which the target appears, and hence ground–target interactions may not be representative.

A second option is to image scale models of the vehicles of interest using an appropriately scaled radar frequency to imitate the radar interactions that would be obtained at full scale. This is particularly relevant for air target recognition because the targets of interest tend to have a small number of scattering centers and few variants. Therefore, the modeling requirements are achievable.

However, it is less effective for ground targets, which will typically be much more complex structures from a radar imaging perspective and will also require target articulations (e.g., different turret positions) and target variants to be captured. These factors mean that the modeling requirements tend to become prohibitive.

SAR Image Simulation. An alternative to the collection of real data is to use simulation techniques that predict the radar signature from a computer-aided design (CAD) model of the target and thus allow images to be simulated. The propagation of radar signals and their interaction with a target in general require solution of Maxwell's equations, which can be challenging.

However, radar target classification will typically use radar systems that operate in a high-frequency regime; that is, the wavelength is small compared with the scattering structures. In this regime, approximate solutions to Maxwell's equations are valid based on geometric optics (GO) and physical optics (PO) giving rise to the ray-tracing approach.

Figure 47-9 illustrates the ray-tracing concept in which a bundle of rays propagates toward a target that is described by a CAD model consisting of a large number of flat facets. For each ray, the path of reflection based on geometric optics is calculated until the ray stops interacting with the scene or some maximum number of bounces is reached.

This is illustrated in Figure 47-10. For the final bounce, physical optics rather than geometric optics are used, so the scattering is diffuse and rays propagate in all directions. In particular, the ray that propagates back to the radar is taken into account. This is repeated for all rays, thus constructing the received signal.

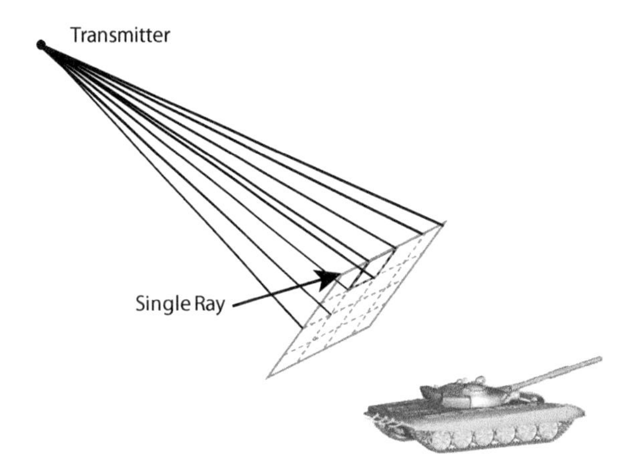

Figure 47-9. A bundle of rays is propagated toward the CAD model. The path of reflection based on geometric optics is calculated until the ray stops interacting with the target.

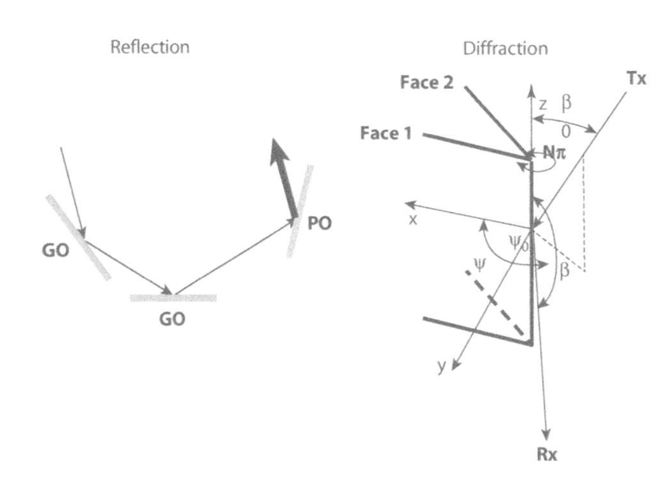

Figure 47-10. Interactions with the facets making up the CAD model are calculated based on multiple reflections and diffraction.

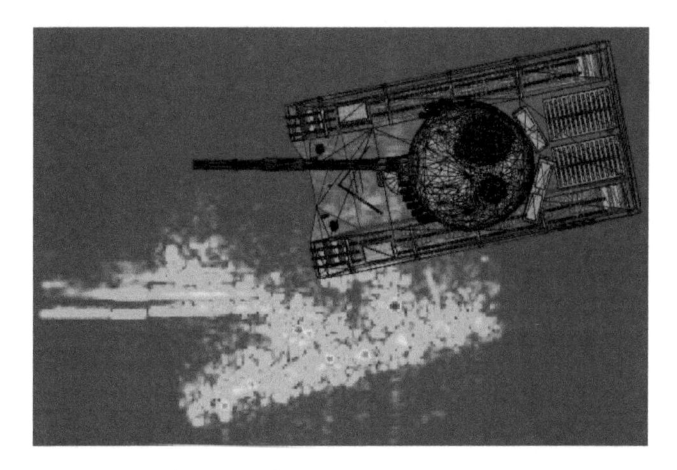

Figure 47-11. The returns for an entire synthetic aperture are calculated and passed through the SAR processor to obtain a simulated SAR image of the target.

More sophisticated schemes may also include edge diffraction, as illustrated in Figure 47-10. Figure 47-11 shows a simulated SAR image together with the CAD model used. The multipath effects that give rise to multiple images of the barrel have been incorporated in the figure.

A key consideration is how the accuracy of the CAD model affects the accuracy of the simulation and hence classification performance. If the performance is critically dependent on having a very accurate CAD model, then this is unlikely to be a robust solution since the actual targets in the field may easily vary from the CAD model representation.

Also, there is a question as to how the CAD models are obtained. If the vehicle is available, then laser-scanning techniques may be used to give an accurate CAD model. However, if the vehicle is not freely available, it may be necessary to produce a CAD model from a limited number of photographs. There is a need to understand the impact of CAD model fidelity on classification performance and how this relates to the accuracy of CAD models that can be obtained from, for example, photographs.

Performance Assessment: The Confusion Matrix

TARGET CLASSIFICATION PERFORMANCE IS GENERALLY ASSESSED USING A confusion matrix. The following illustration shows an example of a confusion matrix for a four-class problem in which there are two classes of friendly vehicles (blue) and two classes of enemy vehicles (red). It is initially assumed that there are 100 examples of targets from each class.

	ATR System Output				
	Declaration				
Truth	Blue1	Blue2	Red1	Red2	PCC
Blue1	0.78	0.03	0.06	0.13	0.78
Blue2	0.01	0.95	0.02	0.02	0.95
Red1	0.10	0.03	0.83	0.04	0.83
Red2	0.02	0.01	0.03	0.95	0.95
PCL	0.86	0.93	0.88	0.83	

The entries in the main body show the proportion of targets of a given true class that are declared as a given declaration class. For a perfect classifier, the entries down the main diagonal will all be unity (i.e., 100 percent classification), while all off-diagonal entries will be zero.

Let i denote the true target classes, and let j denote the declared target classes. Also let T_i be the number of targets with true class i, and let N_{ij} be the number of targets of true class i that are declared as class j. Then

$$P_{ij} = N_{ij}/T_i$$

is the proportion of targets of true class i that are declared as class j. This is the entry in the confusion matrix. The *probability of correct classification* (PCC) for true class i is then P_{ii}, which is shown on the right-hand side of Figure 47-12.

The number of targets declared as class j is given by

$$D_j = \sum_i P_{ij} T_i$$

so the proportion of targets declared as class j that have true class i is given by

$$Q_{ij} = N_{ij}/D_j$$

A more operationally relevant performance metric is the *probability of correct label* (PCL), that is, the probability that a target declared to be of a given class is truly a member of that class, which is given by Q_{jj}. The PCL is shown at the bottom of the confusion matrix and clearly tells a different story. For example, Red2 targets are declared as such 95 percent of the time. However, a target declared as Red2 is actually only a member of that class 83 percent of the time, that is, $100 \times 95/(13 + 2 + 4 + 95)$.

This distinction between PCC and PCL becomes more evident when the targets under test contain vehicles of two classes unknown to the classifier, that is, confusers. Typical results in this case are shown in the following example.

Performance Assessment: The Confusion Matrix *continued*

	ATR System Output				
	Declaration				
Truth	Blue1	Blue2	Red1	Red2	PCC
Blue1	0.78	0.03	0.06	0.13	0.78
Blue2	0.01	0.95	0.02	0.02	0.95
Red1	0.10	0.03	0.83	0.04	0.83
Red2	0.02	0.01	0.03	0.95	0.95
Conf1	0.24	0.28	0.22	0.27	N/A
Conf2	0.28	0.16	0.31	0.25	N/A
PCL	0.55	0.65	0.56	0.57	

It can be seen that the PCC is unaltered, but the PCL is dramatically affected. Now a target declared as Red2 will belong to the Red2 class only 57 percent of the time with a similar effect for all classes, that is, $100 \times 95/(13 + 2 + 4 + 95 + 27 + 25)$.

This highlights an important issue. The classifier is forced to classify every target as one of four classes even though it is exposed to targets outside its database. However, if an unknown class is used, the performance in the following figure results.

	ATR System Output					
	Declaration					
Truth	Blue1	Blue2	Red1	Red2	Unknown	PCC
Blue1	0.70	0.01	0.00	0.12	0.17	0.70
Blue2	0.00	0.95	0.01	0.00	0.04	0.95
Red1	0.09	0.01	0.79	0.02	0.09	0.79
Red2	0.01	0.01	0.02	0.95	0.02	0.95
Conf1	0.04	0.01	0.00	0.03	0.93	N/A
Conf2	0.04	0.00	0.07	0.04	0.85	N/A
PCL	0.80	0.96	0.89	0.82	N/A	

It can be seen that the PCC values are reduced as previous low-confidence correct declarations have now been declared as unknown. However, the PCL is significantly improved over the forced decision case. For example, the PCL for Red2 is now 82 percent, that is, $100 \times 95/(12 + 0 + 2 + 95 + 3 + 4)$.

Another salient factor is that some declaration errors can have more serious consequences than others. For example, if Red is classified as Blue, then an enemy attack may not be averted; however, if Blue is classified as Red, then fratricide may occur. Thus, it is important to include the probability of critical error (PCE) for a declaration of class j

$$PCE_j = \sum_{i \in CE_j} N_{ij} \Big/ D_j$$

where CE_j are the critical errors for a declaration of class j. The PCE is shown in the following illustration to assess the impact of a false declaration. For example, if Red2 is declared, then a critical error will result if the true class is one of the Blue classes or one of the confuser classes. The PCE is thus 16 percent, that is, $100 \times (12 + 0 + 3 + 4)/(12 + 0 + 2 + 95 + 3 + 4)$.

	ATR System Output					
	Declaration					
Truth	Blue1	Blue2	Red1	Red2	Unknown	PCC
Blue1	0.70	0.01	0.00	0.12	0.17	0.70
Blue2	0.00	0.95	0.01	0.00	0.04	0.95
Red1	0.09	0.01	0.79	0.02	0.09	0.79
Red2	0.01	0.01	0.02	0.95	0.02	0.95
Conf1	0.04	0.01	0.00	0.03	0.93	N/A
Conf2	0.04	0.00	0.07	0.04	0.85	N/A
PCL	0.80	0.96	0.89	0.82	N/A	
PCE	0.11	0.02	0.09	0.16	N/A	

Another consideration is the order of battle, that is, the probable number of units present. In this example, it is assumed that there are 10 times as many Red units as Blue units and confusers. The confusion matrix entries in the following illustration have been adjusted to give the performance figures if there were 1,000 of each Red unit and 100 of each other unit.

	ATR System Output					
	Declaration					
Truth	Blue1	Blue2	Red1	Red2	Unknown	PCC
Blue1	0.70	0.01	0.00	0.12	0.17	0.70
Blue2	0.00	0.95	0.01	0.00	0.04	0.95
Red1	0.09	0.01	0.79	0.02	0.09	0.79
Red2	0.01	0.01	0.02	0.95	0.02	0.95
Conf1	0.04	0.01	0.00	0.03	0.93	N/A
Conf2	0.04	0.00	0.07	0.04	0.85	N/A
PCL	0.40	0.81	0.97	0.96	N/A	
PCE	0.57	0.17	0.01	0.02	N/A	

This does not change the PCC but has a drastic impact on the PCL and PCE. For a Red2 declaration, the PCL is now 96 percent, that is, $100 \times 950/(12 + 0 + 20 + 950 + 3 + 4)$, and the PCE is now 2 percent, that is, $100 \times (12 + 0 + 3 + 4)/(12 + 0 + 20 + 950 + 3 + 4)$. In this situation, there would be a lot more confidence acting on a Red declaration, but there would be little confidence in a Blue declaration.

It has been seen that the confusion matrix provides a powerful means of representing classifier performance, but it is important to take all factors into account. A few key considerations have been introduced here, but this has by no means been exhaustive.

Despite these notes of caution, simulation from CAD models is an essential part of ground target database generation given the huge amount of variability that needs to be captured. However, real radar data should also be used when available with simulations filling the inevitable gaps in real data coverage.

47.6 Summary

Radar reflections from vehicles exhibit specific phenomenology, quite distinct from optical characteristics, that must be exploited. SAR image target classification proceeds through prescreening to identify potential man-made objects and then assignment to classes either through template matching or based on radar target features and using standard pattern recognition techniques. Target classification must use extensive databases that rely on signature predictions from CAD models to supplement necessarily limited real data. Performance can be assessed using confusion matrices, but there are many considerations for operational assessment.

Accommodating the huge variability of radar images dictated by imaging geometry and ground vehicle options, such as articulation of the turret or attachment of external fuel tanks, is an enduring challenge for radar target classification and links to the problem of robustly populating databases of example imagery.

Operational performance assessment is another key area for future development. It is very expensive to build and fly a radar classification system. Is it possible to assess potential performance based on a parametric description of the radar, the imaging geometry, the targets of interest, and the scene? Such a theory of target classification could have huge cost savings but is extremely challenging.

Radar target classification typically concentrates on the use of radar image chips. However, the scene context, such as the proximity of lines of communication, can provide additional information. Also, known military behavior as well as information from other sources, for example, other sensors or geospatial information systems, can contribute to the classification declaration. The exploitation of such contextual information is also an area where significant further development is required.

Finally, as target classification techniques mature, there will inevitably be a push to deceive such systems, so the whole area of denial and deception of target classification systems needs to be considered in the future.

Further Reading

P. M. Woodward, *Probability and Information Theory, with Applications to Radar*, Pergamon Press, 1953 (reprint Artech House, 1980).

R. O. Duda, P. E. Hart, and D. G. Stork, *Pattern Classification*, 2nd ed. Wiley-Blackwell, 2000.

V. N. Vapnik, *The Nature of Statistical Learning Theory*, 2nd ed., Springer, 2000.

C. J. Oliver and S. Quegan, *Understanding Synthetic Aperture Radar Images*, SciTech-IET, 2004.

P. Tait, *Introduction to Radar Target Recognition*, IET, 2005.

D. Blacknell and H. D. Griffiths (eds.), *Radar Automatic Target Recognition and Non-Cooperative Target Recognition*, IET, 2013.

Test your understanding

1. List the three steps in the target pre-screening process.
2. A target is known to belong to one of three classes. Each class contains targets that can be well approximated by cuboids. The length and width are the same for each class, but the heights are (a) 1 m, (b) 2 m, and (c) 3 m. A SAR image is taken at 45° incidence with the range direction being along the length of the target. A shadow of length very close to 2 m is observed. To which target class is the target likely to belong?
3. In the fourth confusion matrix in Panel 47-4 the PCE for a declaration of Red2 is 16%. If 100 new targets are introduced from a third confuser class and 10 of these are declared as Red2, what is the new PCE?

Thales WATCHKEEPER WK450 (2010)

The Thales Watchkeeper System is an unmanned aerial vehicle (UAV) for all-weather, intelligence, surveillance, target acquisition, and reconnaissance (ISTAR) delivered to the British Army.

48

Emerging Radar Trends

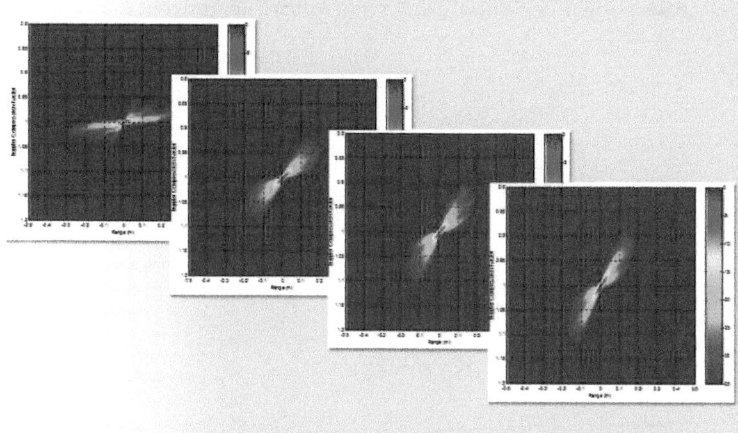

The bat species Pipistrellus pygmaeus *emits acoustic signals whose ambiguity functions are varied with time in a cognitive manner, according to its perception of the target scene. Future radar systems will mimic this behavior.*

48.1 Introduction

To date, much of the emphasis in improving radar performance has been on enhancing sensitivity and resolution. Sensitivity provides for better detection range, while resolution enables more detail to be observed, such as for target classification. These have been achieved, as we have seen throughout this book, with considerable success. Long-range detection and high-resolution imaging are now routine. However, as a consequence, discrimination between targets of interest and other objects is becoming more important than the problem of target detection. At the same time, advances in digital technology continue to be made such that almost every radar parameter, particularly the radar waveform, can be altered on a pulse-by-pulse basis. This is the basis of the subject of waveform diversity, and is creating many new and exciting possibilities not only for improving radar performance but also in opening up new applications.

Here we touch on just a few trends that are emerging from research to development. Specifically, we explore some of the technology advances that are already beginning to have an impact on radar design. We then go on to see how they facilitate what may well become a revolution in radar through the application of closed loop processing, much of which takes its cue from natural echolocation systems. Indeed, echolocation is a technique employed in the natural world to great effect by mammals such as the bat, whale, and dolphin. Although their sensing is rooted in acoustics, these mammals can effectively "see with sound" and exhibit many of the characteristics that would be highly desirable in radar systems.

48.2 Technology Trends

Perhaps the most significant of all technology trends that is impacting and will continue to impact the design of radar systems is that of digital technology. The trends in digital technology are as follows:

1. Increasing dynamic range and higher speed analogue to digital convertors
2. Increasing capability for processing at faster and faster rates
3. Increasing memory capacity and speed of accessibility
4. Reducing costs

The effects of improvements in digital technology are already evident in the design of both transmit and receive radar sub-systems.

For example, on transmit, waveforms can be programmed with almost limitless freedoms. The frequency, modulation, amplitude, bandwidth, and PRF can all be chosen with high accuracy and programmed into a digital waveform generator. Furthermore, these design parameters can be changed dynamically so that each pulse emitted can have a quite different specification.

On receive, the location of the analog-to-digital (A/D) converters is becoming progressively closer and closer to the antenna. It is now almost normal for digitization to take place at intermediate frequency (IF) at rather than at baseband, as was the case previously. Thus, the much greater flexibility that comes with digital processing is being used to overcome limitations of analog circuitry. This is already resulting in both greater versatility and overall improved system performance.

Research systems have begun to appear that are termed "software defined," By software defined it is meant that the radar operating modes can be programmed in software and, at least in principle, can be varied nearly instantaneously, such as on a pulse-by-pulse basis). This trend towards digital control and parameter programmability is likely to lead to the advent of all-digital radar. Lower operating frequency systems are again being reported in the literature, and higher frequency systems seem likely to follow.

However, A/D converter dynamic range is still often insufficient for many radar applications. At wide bandwidths high A/D converter speeds are required. It is more difficult to obtain high A/D converter dynamic range at fast A/D converter speeds. High range resolution demands very fast A/D converter speeds, and dynamic range requirements are less easy to meet. This necessity for both high dynamic range and high speed A/D converters is likely to slow the onset of all-digital, high frequency, wideband radar.

Another technology trend, yet one that is nearly as old as the invention of radar, is electronic scanning. Despite its longevity, electronic scanning is far from universal, most usually only appearing in complex and expensive military systems. Electronic scanning has been employed successfully to create beam patterns with very low sidelobes. Electronic scanning also enables adaptive beam forming that can simultaneously view targets with high gain while rejecting sources of interference. Indeed, multiple modes of operation such as search and track can be supported simultaneously. However, the adoption of electronically scanned systems has progressed relatively slowly. The main reasons for this are a combination of complexity, cost, and an inability to easily operate over wide instantaneous bandwidths. Some of today's operational systems use techniques such as sub-arraying and even mechanical scanning to help reduce complexity and cost. Nevertheless, such obstacles are steadily being eroded, and much of the technology to enable very wide band all-digital arrays with digitization at the element level already exists. Design drivers such as the need to use the electromagnetic spectrum resource more efficiently may eventually play a part in tipping the cost balance towards electronically scanned systems.

Electronic scanning coupled with advanced digital technology is moving radar into a new era. Electronic scanning means that the radar "beam" can be pointed anywhere at any time. Also almost all radar design variables can be changed as the beam moves from one position to another. It is the combination of both of these that allows electronic scanning to support multiple different tasks leading to true "multifunction radar systems." Figure 48-1 illustrates schematically the vast range of tasks that a multifunction radar might have to carry out.

However, this poses the question as to where and when to point the radar beam and how to optimize the radar design parameters to best carry out a given task. This simple question raises significant and fundamental issues that are influencing the direction of radar research that is likely to continue indefinitely.

Electronically scanned radar systems have led to the emerging topic of "resource management" that has the objective of

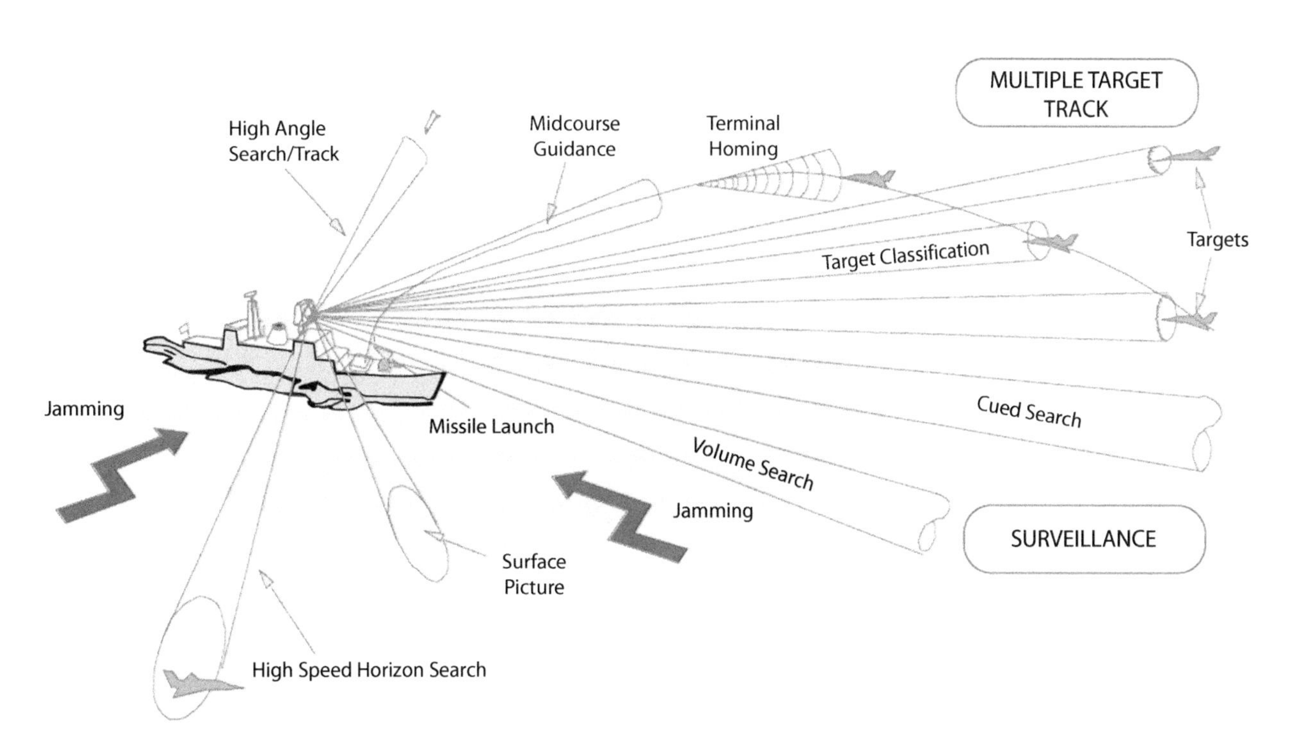

Figure 48-1. A multifunction phased array radar can perform a large number of tasks. The resource manager must prioritize the different tasks so that the most critical are performed first and the radar resource is used in the most efficient manner.

deploying radar resources to most effectively accomplish a task or group of tasks. This is largely through the design of waveforms and the direction of beam pointing. Much effort has been expended on how to schedule radar transmissions together with making decisions as to which radar resources to deploy. This notion of the radar making its own decisions is embedded in electronic scanning and digital processing. In the future adaptive feedback based on interrogation of received echoes will inform the design of future transmissions. We touch upon just a few aspects of electronically scanned digital radar systems in the next sections.

48.3 Radar Resource Management

Radar resource management is the efficient allocation of finite radar resources to accomplish a task or set of tasks in an optimal manner. In fact, electronically scanned radars will be able to fulfill all of their potential only if they are able to optimally deploy their resources within the finite amount of time that a radar system has available. In other words, if the radar has to send a series of pulses in multiple different directions at the same time it may not have enough time to send pulses for a tracking task. Thus a decision as to when to look and where has to be made, and task priorities have to be set. Most approaches to radar resource management have divided this topic into the following three separate categories:

1. Adaptive track updating,
2. Adaptive search scanning and
3. Scheduling.

More recently, attempts have been made to use measures of tracking and search performance as a basis for optimization and hence as a basis for determining and allocating radar resources.

Radar Resource Management Tasks. Radar resource management plays a central part in carrying out the mission to which the radar system contributes. This mission might be set at a high level by defining tasks such as the detection and tracking of all targets in a designated area or volume. These tasks may then be further refined as sub-tasks, such as the requirement that all tracked targets must fall within a set error bound. As shown in the box to the left, there are a large number of possible tasks that can be carried out by an electronically scanned radar system.

A mission can have inherent time dependence and may have to be redefined as a task unfolds. Targets may enter and leave the coverage area or volume and so the number of targets that must be detected and tracked will vary with time. Thus, there is a large number of radar variables, a large number of tasks to be carried out, and potentially a large number of targets. This means that the optimization problem faced by radar resource management is one that is extremely challenging.

Modes in a Multi-function Radar System

Surveillance

- Detection
- Volume search
- Verification
- Handover to tracker

Tracking

- Initiation, maintenance, termination
- Position estimation
- Track file updates

Missile guidance

- Mid-course correction

Signature interpretation

- Detection in clutter
- Classification

The previous blue panel illustrates the radar modes that have a requirement for a waveform design to be transmitted and a direction in which the waveform is emitted. Once the resource manager has made these decisions, it must allocate a time slot at which the waveform is to be transmitted and received. Each waveform requires different radar parameters such as modulation, power, frequency, bandwidth, and PRF to be selected. The selection and allocation of waveform parameters as a function of time can be accomplished in a wide variety of different ways. One approach is to trade off the radar design parameters as a function of the task to be carried out. For example, the following steps might be used:

1. Decide the functions to be carried out (e.g., detection, tracking or classification)
2. Ensure that targets can be detected at an adequate signal to noise ratio
3. Allocate radar resources to functions
4. Check the resource allocations against the available timeline
5. Trade off the allocations and
6. Make sure that nonviable modes are eradicated

However, this doesn't take into account the fact that different tasks may have different priorities and different coverage regions may be prioritized differently. In fact, these priorities will most likely dictate the detailed allocation and timing of radar resources such that the highest priority, time-critical tasks are carried out first. There can be occasions where so many tasks have to be carried out that they cannot all be accommodated. In this case the lowest priority, least time critical tasks would be set aside. Naturally, any tasks that cannot be accomplished must be flagged to the radar operator.

Radar Resource Management Categories. There are three broad categories by which radar resources are allocated:

1. *Rule based*, where a pre-determined behavior is programmed into the radar system as a function of a small range of scenario types and missions. This might pre-define separate waveforms for search and track but otherwise not change their specification. It may also predefine time spent searching versus time spent tracking.
2. *Self-organizing*, where the radar itself evaluates the scenario against requirements set by an operator. The waveform selection and beam pointing parameters, together with their allocation to the radar timeline, are determined on the fly by the radar system. Such an approach may continually review and respecify the radar waveform, the amount of time spent searching or tracking, etc.
3. *Hybrid*, where some system parameters are pre-determined while others are selected on the fly by the radar. Thus the hybrid approach is a combination of (1) and (2) that is part rule-based and part self-organizing.

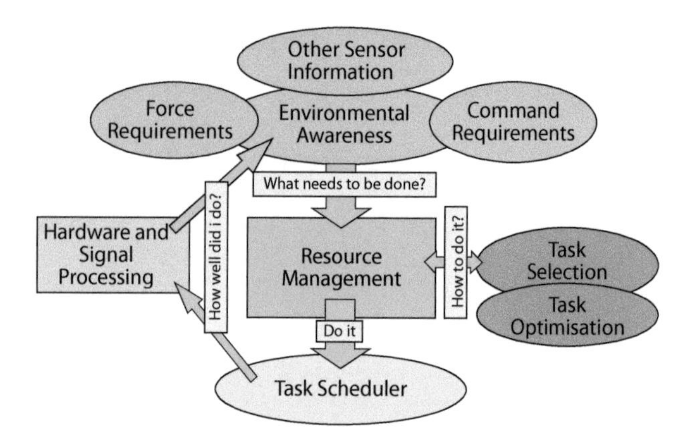

Figure 48-2. The scheduler allocates each radar task to be executed at a particular time, making sure that the highest priority time critical tasks are assigned first. The system then reassesses the scenario to examine how well the tasks have been, and are being, accomplished in preparation for the next resource management cycle.

Most current electronically scanned radar systems are very prescriptive in the way they allocate their time to different modes. Future systems will learn to adapt their performance as a function of their own understanding of their sensed environment. They will use this to modify waveform parameters and beam pointing to optimize performance in a given task. This is a challenging and vibrant area of electronically scanned radar research and one that is set to continue for some time into the future.

Radar Resource Management Activities. Figure 48-2 shows the flow of activities and information that are a part of radar resource management. There are a number of inputs that help set the tasks to be carried out. The job of the resource manager is to turn these into activities that will be carried out by the radar system. The figure shows that there are many and varied components that form part of resource management and consequently just how complex and unwieldy the whole process is. In addition to sensing the environment, the radar may receive information from a variety of other sources such as geographic information systems (GIS), databases, as well as other sources. This has to be assimilated into as accurate a picture of the scenario as possible so that designated tasks can be carried out. The resource manager then works out what tasks to do and how best to do them.

Once the radar tasks have been identified and specified, they have to be scheduled into a queue. This is the job of the task scheduler (Figure 48-2).

Role of Radar Resource Manager. Figure 48-3 shows a system architecture for an electronically scanned radar system highlighting the role of the resource manager. There is a tight

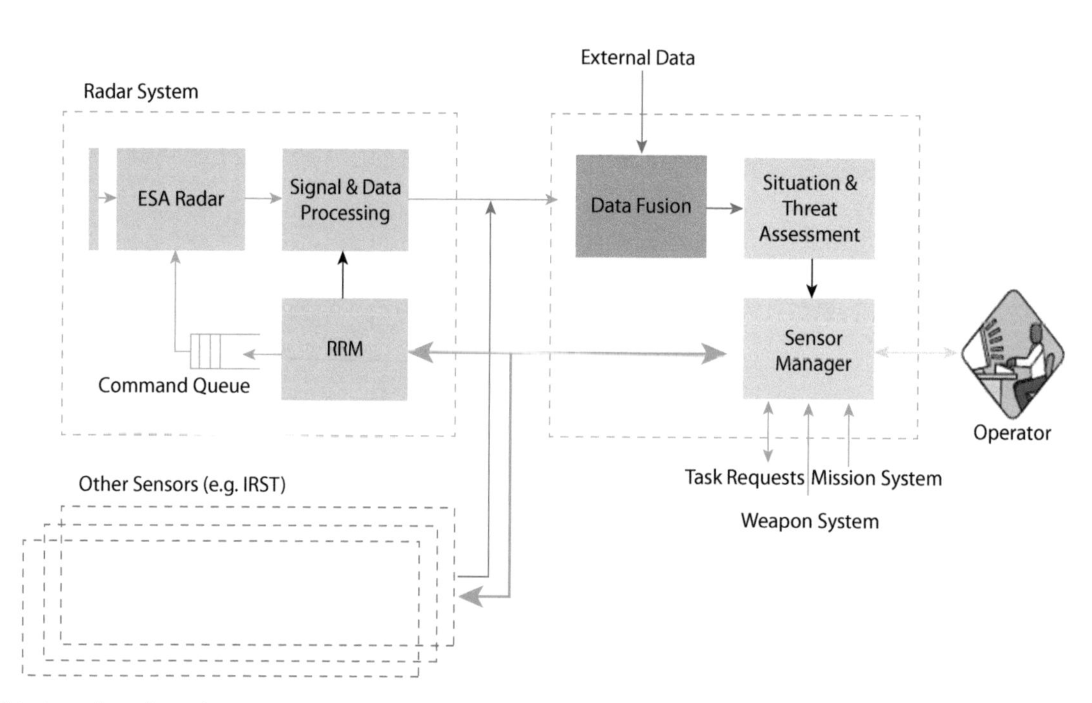

Figure 48-3. This shows how the radar resource manager (RRM) is set in the context of the overall electronically scanned radar system architecture.

coupling between the radar hardware, signal processing, and the resource manager that reflects the rate at which radar parameters can be re-assigned. There is a looser coupling with external data sources and the setting of requirements because they occur on a slower timescale.

Overall, resource management is an ongoing area of research, and there remains much to be done before optimal systems can be said to truly exist. Perhaps the most significant challenge is for the radar to correctly and accurately interpret echoes so that it has the best possible information for re-setting the parameters for future emissions. This can only be accomplished by the radar system itself as the radar parameters can be changed in times of a millisecond (i.e., well inside human decision-making and action). Accurate and full interpretation of radar echoes provides the motivation for the adaptive and cognitive sensing that we shall examine in the remainder of this chapter.

48.4 Echolocation in Nature

The bat, whale, and dolphin are well known for their ability to use echolocation as everyday activities key to their very survival as a species. Indeed, they have been successfully using echolocation for over fifty million years and have honed the technique to carry out a remarkable range of tasks. Perhaps less well known is the fact that humans can also echolocate, and there are some blind people who have become extremely expert. Bats are able to forage for food in dense clutter environments; some of their "targets" are stationary and some are moving. Their acrobatics are nothing short of extraordinary, often taking place in complete darkness. Dolphins are able to use echolocation underwater with such acuity that they have been "recruited" by the US Navy to search for mines. Although they are using acoustic rather than electromagnetic signals, their ability to achieve tasks beyond radar makes them a suitable subject for study with the aim of inspiring new and improved forms of concepts and processing. For these reasons "bio-inspired" radar research is a very active area and may well yield new insights that influence the design of future radar systems. Here we examine some of the characteristics of the echolocating bat to illustrate the techniques used.

Waveforms. Bats, dolphins, whales, and humans all use some form of *tongue-click* to generate their signals. The waveforms are generally pulsed with pulse lengths around 1 ms. The pulse repetition frequency is of the order of 100 Hz. The emitted waveforms are used as a tool for both navigation and target classification (i.e., selection and acquisition of prey).

One technique routinely employed that makes great sense from a radar perspective is to adjust the PRF such that range is always unambiguous. This is shown in Figure 48-4, in which the black dots are the position of a bat flying close to a hedge. The circles surrounding the black dots represent the extent of

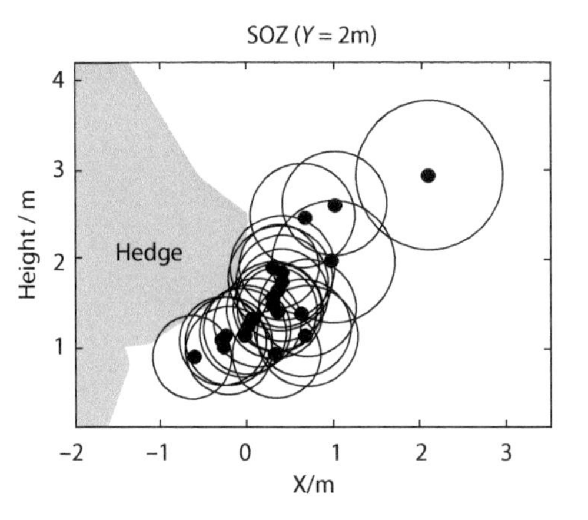

Figure 48-4. This shows how the bat dynamically adjusts the pulse repetition frequency of its calls so that the signal overlap zone (SOZ) is no more than the range to the hedge.

the first range zone that is adjusted to be no more than the range to an obstruction, in this case a hedge along which the bat is flying. The bat will also adjust its PRF to the wing beat of flying insects as a means of identification. This continuous adjustment of PRF as a function of the prevailing environment is not something routinely used in radar. Instead, either one of a small number of PRFs is selected or the unambiguous range is extended by using multiple PRFs.

Bats modulate their waveforms but with a structure not used in radar. Furthermore, the modulation is varied as a function of time throughout an activity, such as selection and acquisition of prey. The modulation itself varies considerably from one bat species to another, and within a species according to the particular task that the bat is undertaking. Bats will emit anything from a pure tone of a near constant frequency to a structure that has a time frequency profile close to a hyperbolic modulation. Even more elaborately, bats often emit, within a single pulse, a series of harmonically related modulations. These can be a group of tones or a group of hyperbolic modulations or a group combination of tones and hyperbolic modulations, as shown in Figure 48-5. Another feature of these waveforms is that the power emitted in the different components is not the same. The reasons for these waveform structures are far from clear and are the subject of current research. However, since they are the result of evolutionary optimization over millions of years, there must be some reason for them, and it therefore seems likely the waveforms have these structures in order to aid the bat in its navigation and recognition tasks.

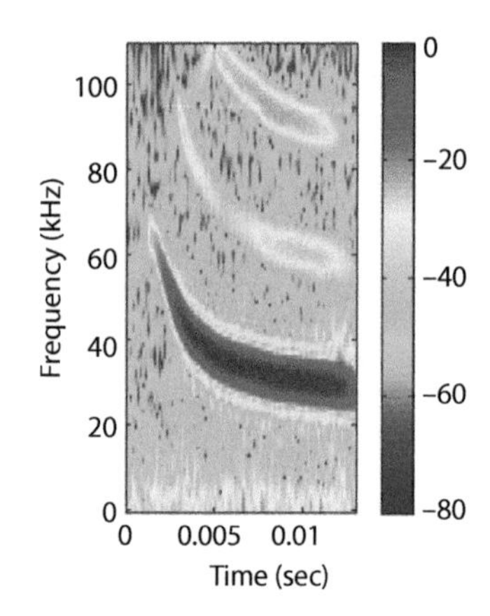

Figure 48-5. A spectrogram plot of the acoustic signal produced by a bat. The spectrogram shows three harmonically related components each exhibiting a hyperbolic component and a constant frequency component.

Another feature of these waveforms is that the different harmonic components are sufficiently separated in frequency that they will be emitted with different beam widths. The higher frequencies have narrower beam widths (shorter wavelength divided by a fixed "aperture" size). Perhaps this helps to distinguish between targets that have different reflective properties at the different frequencies? Alternatively, it might be a way of distinguishing between targets at different angles? We can only speculate, but nevertheless it shows how analysis of biological systems can help stimulate new thinking, thereby generating ideas for future radar research.

The waveform parameters used by bats are continually being changed as they fly their "mission." For example, the PRF for "general surveillance" is typically around three or four times lower than that used to intercept prey. Furthermore, the modulation that the bat applies to the waveform is continually being changed. In the final phase just prior to intercepting and capturing prey, the constant frequency component (Figure 48-5) is all but gone, leaving behind three much more steeply raked hyperbolic harmonics as shown in Figure 48-6.

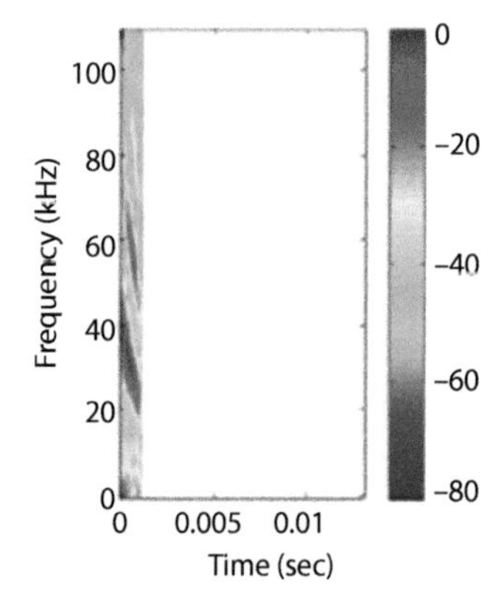

Figure 48-6. A spectrogram plot of the acoustic signal produced by a bat, showing three harmonically related components in the final phase of attacking prey. Note the absence of the constant frequency component that is evident in the latter part of the pulse shown in Figure 48-5.

Figure 48-6 also shows that the duration of the waveform has greatly reduced. One reason for this is that the bat expends energy when emitting a waveform. Thus, as it closes in on a target, the signal strength required for detection reduces, and consequently a shorter pulse can be used. A second reason is that the bat avoids “eclipsing” losses. As we have seen in Chapter 45, this is very desirable in airborne radar. Modulation of pulse width offers an additional variable that could be used to minimize the effects of eclipsing.

Thus the bat is adjusting its PRF and waveform modulation as it goes about the selection and acquisition of prey. In addition the bat is also continuously adjusting its trajectory. Rarely does a bat go straight towards its prey. More typically, it will circle the target, appearing to probe for information by using a number of pulses emitted at different orientations. Could it be gleaning additional information to confirm that this is truly a source of nourishment? Further research is needed to answer such questions, but it is clear that the bat is using echolocation to see the world and can do so with great effectiveness.

Overall, the bat is constantly adjusting many, if not all, of the waveform parameters under its control, including the orientation between it and the object being interrogated. It would be reasonable to assume it does this on the basis of information extracted from predecessor pulses. In other words, it is interrogating its environment and making changes to improve its ability to carry out its task within that environment. In this way the bat is an excellent example of an adaptive and cognitive echolocating system. Research is now beginning to examine cognition in echolocation mammals that may provide further important clues as to how radar systems can be improved and even extended into new applications.

48.5 Fully Adaptive Radar

The technology now exists for all radar parameters to be varied on a pulse-by-pulse basis. We have seen in the previous section that this is something done in natural echolocation systems to great effect. It seems intuitive that there are better and perhaps optimal parameters that should be collectively selected for a given radar function. Indeed, we have seen cases throughout this book where this is done for a particular application. One example where this occurs is in the setting of the radar PRF. This might be set to low, medium, or high depending on the application. However, this is fixing one set of parameters so that they are as close as optimal for one application. The likelihood of the selected parameters being optimal in all cases, let alone multiple applications, is very low. However, an adaptive approach, in which the parameters can be varied as a function of the information gleaned from predecessor pulses, provides a basis for continually iterating towards a more optimal parameter set.

Hence, the question immediately arises about the selection (and reselection) of the best parameters. This also has some quite fundamental implications. First, this has to be done by the radar itself, since the PRFs are too high for humans to intervene on a pulse-by-pulse basis. Secondly, it implies the radar system itself has to have knowledge of the scene it is illuminating as well as a clear understanding of the task or function to be carried. Only armed with this information can the radar choose the best parameters. Continuous optimization of the radar parameters is the objective for fully adaptive radar.

Consider an example based upon a single target tracking. The task of the radar system is to track an already detected target. It does this by continually pointing the peak gain of the illuminating radar beam onto the target in order to maintain the most accurate track. A maximum difference between the location as measured by the radar and the predicted location can be set. Radar measurement errors are largely determined by detection performance, which in turn is largely determined by the signal to noise ratio. Thus, adaptive radar might have a control loop (Figure 48-7) where the radar parameters are adjusted to maintain a desired signal-to-noise ratio (i.e., target detection performance). So if the target echo were to begin to fade, the radar system would interpret this as a reduction in signal to noise ratio. It can then react to automatically redress this, perhaps by increasing transmitter power (or pulse length or PRF or some combination of parameters). There may also be constraints that have to be observed. Constraints may be due to limits on transmitter duty cycle or the need for unambiguous ranging, and these also have to be built into the adaptive feedback concept.

Adaptive radar might then have a control loop (Figure 48-7) where the radar parameters are adjusted to maintain a desired signal-to-noise ratio (i.e., target detection performance).

A fully adaptive approach also allows parameters such as the PRF to be reduced when, for example, a target is flying a

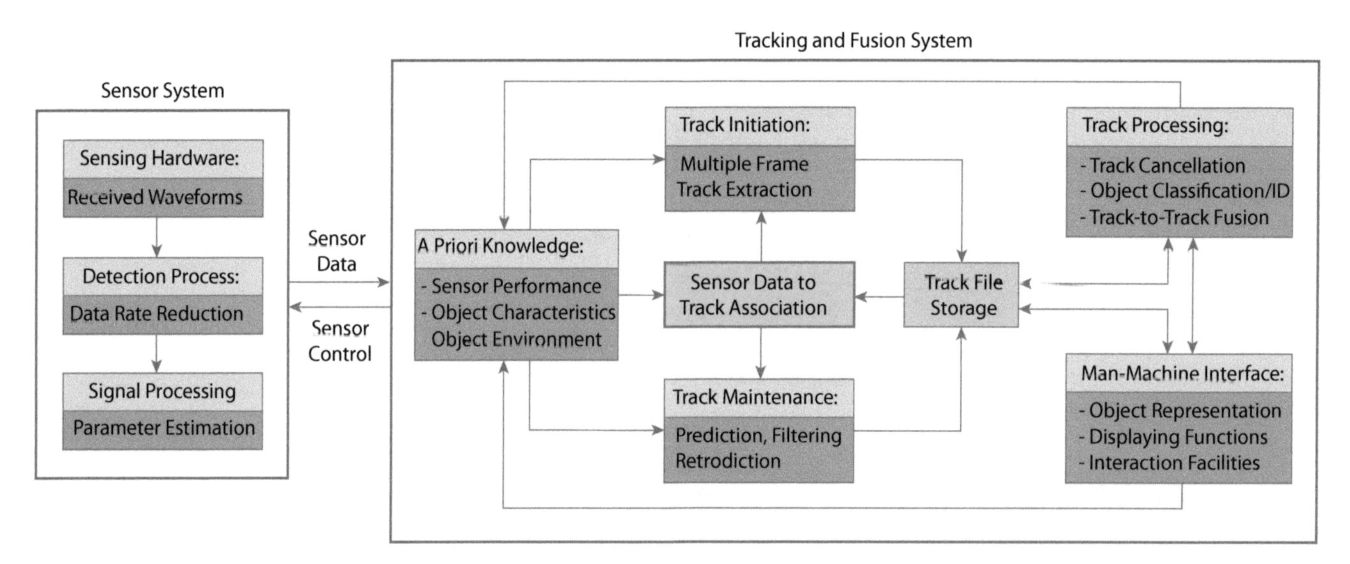

Figure 48-7. This shows a schematic example of the processing for an adaptive radar tracking system in which the radar parameters are dynamically and continually adjusted to optimize the radar performance, also making use of a priori knowledge of the sensor and the target scene.

straight and steady course and the track updates can be less frequent with little or no loss in performance. In this way radar resources can be deployed more flexibly. In other words, fully adaptive radar processing can also play an integral role within radar resource management. Indeed, there is an increasing interaction between radar performance and functions facilitated by software control of radar parameters.

Adaptive radar waveform design and optimization is also beginning to find its way into target classification. The principles are similar to those described in the tracking radar example except the criterion for success is now classification performance rather than detection. In detection a good waveform design is one that has an appropriate range and Doppler resolution with low sidelobes. The goal of the waveform design is to achieve a signal to noise ratio that ensures confident detection performance. This can be subject to feedback where the design is optimized until the desired detection performance is reached. In adaptive classification the feedback aims to select the design of waveform that optimizes classification performance, and this is not necessarily the one that results in best detection.

In some ways this might be thought of as counterintuitive since waveforms with high resolution and low sidelobes should provide an HRRP or imagery with best parameters. However, this assumes that a radar system operating as a coherent sensor using electro-magnetic radiation in the RF spectrum will see the world as we are used to seeing and interpreting it in the visual part of the spectrum using noncoherent imaging. Our current knowledge of these issues is poor, but conventional radar thinking is now being challenged, and simple demonstrations showing improved performance have already been reported. Certainly, the relationship between the way a radar illuminates a target and the ability to extract information that characterizes the target requires an improved understanding. Equally, the way in which received echoes can be processed to extract information characteristic to target type also has great scope for improvement. Again, it can be concluded that there are significant benefits to be gained from an adaptive approach.

Overall, fully adaptive radar is only just emerging as a research topic but is an exciting one that has great potential for improving radar performance in many ways and to improve the range of tasks that a single radar can carry out. Coupled with electronic scanning, the possibilities seem limitless for enhanced radar capability, but there remains much and challenging research before this potential can be realized.

48.6 Cognitive Radar Sensing

Cognitive sensing is an emerging strand of radar research that aims to tackle the challenges of interpreting and exploiting echoes. It builds on the ideas of fully adaptive radar but extends them by drawing on human and other animal forms of cognition. Indeed, cognitive computing and more general sensing and signal processing are making rapid advances in other related disciplines. Radar is but one example where a cognitive approach has the potential to both enhance performance and also may open up new applications area, especially those requiring autonomy.

Cognitive sensing has embedded within it the notion of the "perception-action" cycle that is at the heart of the cognitive process. By *perception* we mean the picture of the world viewed by the radar that is created internal to the radar processor. This is akin to our own artificial perception of the world that we create in our brain. Thus a radar perception might be the mapping out of aircraft in the sky.

Once a sufficiently accurate perception has been created, it can be used as a basis for decision-making. In air traffic control this might be relocating aircraft so that they remain at appropriately safe separations. The *action* part is provided through an instruction that repositions the aircraft to the desired locations. This perception-action activity is a task currently performed by air traffic controllers. It is they who supply the necessary cognition. However, it is a task that, in principle could be carried out via the radar processor issuing commands to pilots on an automatic basis. Whether or not a cognitive component of this type becomes part of air traffic management systems depends on many factors (not least that of public acceptability), but it does illustrate that radar systems are not so far away from being able to operate on a more autonomous basis.

Cognitive radars sensing goes hand in hand with adaptive feedback. It includes received echoes providing instructions for the design of future waveforms. It also includes the repositioning of the radar platform and direction of illumination. Furthermore, it implies slightly more subtle requirements, such as the explicit generation and exploitation of memories.

Memory is a fundamental component of cognition, but memory is not generally used in radar signal processing. The generation of memories can occur at many different levels. For example, predecessor pulses and subsequent detections can confirm a detection in a surveillance radar. Long-term memories based upon prior experience or missions can be used to create a database of target signatures to aid classification. Indeed, such a library could be subject to continuous updating. Perhaps future radar systems will be subject to *life-long learning*! Third-party databases, including Internet sources, can also provide a source of memory. For example, the maps in a geographical information system (GIS) can be used to plot an optimal trajectory for airborne radar that maximizes some aspect of performance.

It can also be seen that there is a need for the radar itself to be able to understand and respond to its perception of its environment. Potentially, it must do this on a timescale down to the milliseconds at which pulses are transmitted.

Cognitive radar sensing is also taking its cue from observations of cognition in natural systems. The cognitive processing architecture could look like that shown in Figure 48-8. This architecture has a number of features not seen in the examples discussed elsewhere in this book, such as multiple feedback loops, multiple parallel lines of processing, the dynamic creation and exploitation of memories, the link between sensing and action (the perception-action cycle), and the requirement for perpetual training and learning.

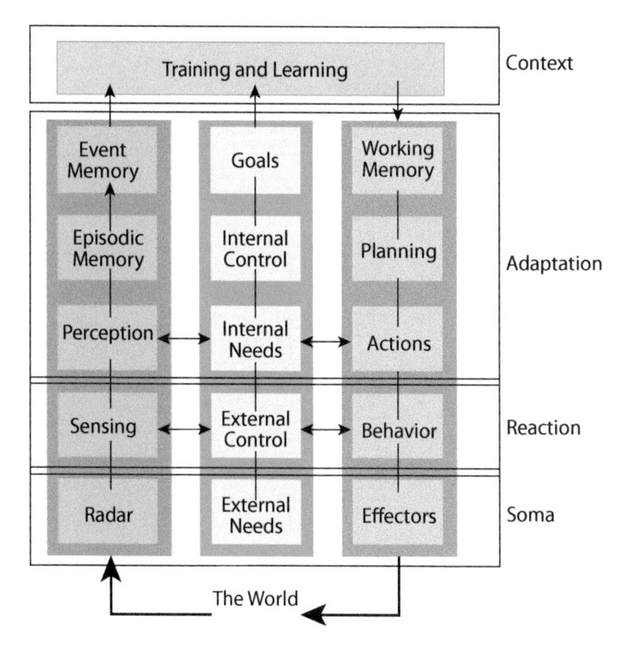

Figure 48-8. This shows a bio-inspired architecture for cognitive radar sensing, including the different stages of the processing and the multiple feedback loops within the overall process.

Cognitive sensing offers both enormous potential for improved performance but at the same time presents significant challenges. In the space available here we can only present just an outline description of cognitive sensing. This barely touches upon the myriad of options for processing echoes and turning them into a well understood picture for reliable decision-making. It remains to be seen how much progress will be made on this topic, but the potential makes for exciting possibilities. If a radar system has an accurate picture and understanding of its surroundings then, for example, it could lead to much greater autonomy. Perhaps a miniature aircraft, no larger than a small bird and guided by a combination of GPS and radar, can carry out a mission to survey the inside of a building too damaged to allow human investigation. Perhaps radar will play a vital role in the goal of collision-less automobiles. Are these and other examples beyond the realms of possibility? Perhaps, perhaps not.

48.7 Other Trends?

Taking these ideas still further, and combining them with the distributed radar concepts of Chapter 45, we can imagine a cognitive radar network using platforms operating multistatically, exploiting ground-based transmitters of opportunity as well, and configuring itself dynamically according to the target scene and its mission. There are numerous challenges, including geolocation and synchronization, communication between the nodes of the network, and the control and management of such a network.

48.8 Summary

Comparing a modern multifunction airborne radar with the very first examples, more than seventy-five years ago, we can see that the sophistication and performance are immeasurably greater, and we can confidently expect that trend to continue. Although we are always constrained by the laws of physics, the only other limit is our imagination.

Further Reading

S. Haykin, "Cognitive Radar: A Way of the Future," *IEEE Signal Processing Magazine*, Vol. 23, No. 1, January 2006.

J. Guerci, *Cognitive Radar: The Knowledge-Aided Fully Adaptive Approach*, Artech House, 2010.

G. Capraro, A. Farina, H. D. Griffiths and M. C. Wicks, "Knowledge-Based Radar Signal and Data Processing: A Tutorial Introduction," *IEEE Signal Processing Magazine*, Vol. 23, No. 1, January 2006.

PART

Representative Radar Systems

In this section we briefly describe the essential features and characteristics of a number of operational exemplar systems. In these increasingly security conscious and commercially competitive times information openly available about advanced radar systems is relatively sparse. Our descriptions are mainly compiled from knowledge distilled from the research literature and from open sources, especially, of course the Internet. Many of the concepts introduced throughout this book are embodied in these systems and we hope that you will find this useful context. If you come across more detailed open source of information, do please send them to us, and we'll attempt to incorporate updates into subsequent editions. The chapters comprising this section are:

Boeing E-3 SENTRY (1977)

With its distinctive rotating radome, the E-3 Sentry (more commonly known as AWACS) is based on the Boeing 707 airframe and is used to provide commanders with information to gain and maintain control of a battle. It uses the Westinghouse AN/APY-1 and AN/APY-2 passive electronically scanned array radar system which can provide surveillance over land and water. The information collected by the radar system can be sent to multiple command and control centers and includes positioning and tracking information of enemy and friendly aircraft and naval vessels.

49

Airborne Early Warning and Control

AEW Sea King

Airborne early warning and control (AEW&C) comprises two main components:

i. An airborne radar system: This is used to detect aircraft, ships and vehicles at long ranges

ii. A command and control system: This directs fighter and attack aircraft strikes.

The command and control function has a close similarity to the role performed by an Air Traffic Controller, managing the local air traffic to ensure mission success. AEW&C aircraft are used for both defensive and offensive air operations having the advantage of high mobility together with an advanced and extremely capable radar system. Together the air platform and radar system form a combination able to direct fighters to their target locations and direct counterattacks on enemy forces. The great advantage of operating at high altitudes is the very large ranges at which targets can be detected. For example, the United States Navy uses AEW&C aircraft from its large aircraft carriers to protect their on-board Command Information Centers (CICs).

AEW&C is also known by the terms "airborne early warning" (AEW) and "airborne warning and control system" (AWACS).

Modern AEW&C pulse Doppler radar systems have detection ranges of the order of 400 km. This means they can detect surface-to-air missiles sufficiently early to deploy countermeasures. A single AEW&C aircraft flying at an altitude of 9 km can cover an area of greater than 312,000 km^2. Indeed, three such aircraft in overlapping orbits can cover an area the size of central Europe. In air-to-air combat, AEW&C systems can cooperate with friendly aircraft via communication links. This

Figure 49-1. E-3 SENTRY system.

Figure 49-2. The Japanese E-767 AWACS flying over Mount Fuji.

Figure 49-3. The Swedish Air Force Saab 340 AEW&C system has a fixed side-looking antenna on top of the fuselage.

can effectively extend the sensor range of friendly aircraft and give them added stealth, since they no longer need to switch on their own radar systems.

Figures 49-1, 49-2 and 49-3 show some examples of AEW&C systems currently in operational use. The USAF E-3 Sentry and the Japanese Self Defense Force E-767 AWACS both exhibit a distinctive circular radome carried above the aircraft fuselage in which the antenna is housed. This placement ensures maximum all-round line of sight for the radar system giving the best possible coverage. The Swedish Air Force Saab 340 AEW&C system shows an elongated radome structure housing a side-looking antenna. Although the radar system has a more restricted range of azimuth viewing angles, the agility of the aircraft still enables a comprehensive picture of aircraft activity to be produced. Other types of AEW&C system include the UK Royal Navy's Sea King AEW series that dates back to an urgent operational requirement in 1982, and is still in operational service.

49.1 E-3 AWACS Radar

The APY-2 is the radar for the US Air Force E-3 Airborne Early Warning and Control System (AWACS). From an operational altitude of 9 km the radar is able to detect low altitude and sea-surface targets out to 400 km, co-altitude targets out to 800 km, and targets beyond the horizon at still greater ranges. Figure 49-4 indicates the air and surface coverage zones in which AWACs can detect targets.

Implementation. The radar frequency is in S-band (nominally 3 GHz). The radar employs an 8 m by 1.5 m planar-array antenna, steered electronically in elevation, and housed in a rotating radome (*rotodome*) that scans at a rate of 6 rpm (Fig. 49-5).

Besides phase shifters for elevation beam steering, phase shifters are also provided for offsetting the beam for reception during elevation scanning, to compensate for the time delay

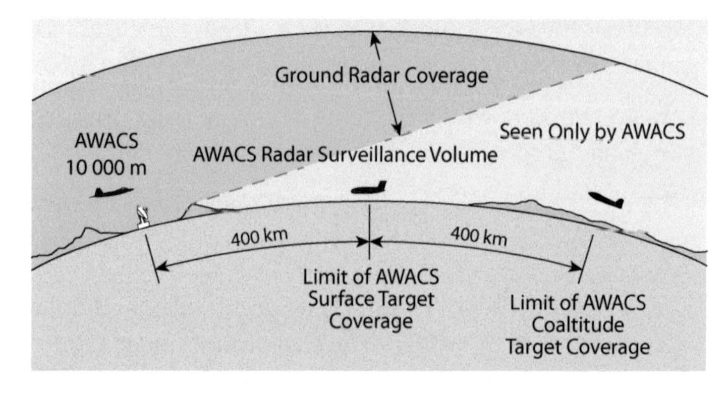

Figure 49-4. From an altitude of 9 km, AWACS can detect sea and low-altitude targets out to 400 km and co-altitude targets out to 800 km.

between transmission of a pulse and reception of returns from long-range targets. The antenna has an extremely narrow azimuth beamwidth and its excitation uses an amplitude taper to reduce the sidelobe level sidelobe.

The transmitter chain consists of a solid-state pre-driver (whose output power is increased as a function of antenna elevation angle) a traveling wave tube (TWT) intermediate power amplifier, and a high-power pulse-modulated dual-klystron amplifier. For reliability, dual redundancy is employed throughout.

Following an extremely low-noise (HEMT) receiver preamplifier, two separate receive channels are provided: one for range-gated pulse-Doppler operation and another for simple pulsed-radar operation.

Digital processing is performed by a signal processor, employing 534 pipeline gate arrays operating at 20 MHz and a data processor, employing four reduced instruction set computing (RISC) central processing units (CPUs).

Figure 49-5. The AWACS antenna consists of a stacked array of 28 slotted waveguides, plus 28 reciprocal ferrite elevation-beam-steering phase shifters and 28 low-power nonreciprocal beam offset phase shifters.

Modes of Operation. The radar has four primary modes of operation:

- *High-PRF pulse-Doppler range-while-search*, for detecting targets in ground clutter;
- *High-PRF pulse-Doppler range-while-search*, plus elevation scanning for additional elevation coverage and measurement of target elevation angles;
- *Low-PRF pulsed radar search* with pulse compression, for detecting targets at long ranges beyond-the-horizon, where clutter is not a problem;
- *Low-PRF pulsed radar search* for detecting surface ships, featuring extreme pulse compression and adaptive processing that adjusts for variations in sea clutter and blanks land returns on the basis of stored maps.

These modes can be interleaved to provide either all-altitude long-range aircraft detection or both aircraft and ship detection. A passive mode for detecting ECM sources is also provided.

Each 360° azimuth scan can be divided into up to 32 different sectors, in each of which a different operating mode and different conditions can be assigned or changed from scan to scan.

ASTOR Sentinel R-1 (2005)

SENTINEL (originally ASTOR – Airborne Stand-off Radar) is a British SAR/GMTI platform that utilises the Dual Mode Radar, a derivative of the Raytheon ASARS-2, on board a converted Bombardier Global Express aircraft. It flew its first operational sortie over Afghanistan in February 2009. With its crew of 5, including 2 trained Image Analysts, it can provide near-real-time, long-range, all weather, day-night battlefield intelligence to ground commanders for wide area surveillance, reconnaissance, target-imaging and tracking.

50

Reconnaissance & Surveillance

THALES I-Master radar (Courtesy of Thales UK.)

50.1 Manned Systems

Joint STARS is a long-range, long-endurance, air-to-ground surveillance and battle management system carried aboard the US Air Force E-8C aircraft. Operating at altitudes up to 12.8 km, the system's high-power pulse Doppler radar is capable of looking deep behind hostile borders from a stand-off position and monitoring fixed and moving targets with a combination of high-resolution SAR mapping and moving-target indication (MTI) vehicle detection and tracking.

The radar employs an 8 m long, roll-stabilized, slotted-waveguide, side-looking passive ESA, housed in a radome nearly 9 m long carried under the forward section of the fuselage (Figure 50-1). The antenna is steered electronically in azimuth and mechanically in elevation. In fact the antenna can be "rolled" all the way from one side to the other allowing illumination to the right or left of the aircraft's flight path.

Multiple signal processors perform the digital processing. The radar data and signal processors are controlled by an onboard distributed processing system that includes individual digital processors at each of 17 operator workstations and one navigator/operator workstation. All are readily accommodated in the E-8C's 43 m long cabin (Figures 50-2 and 50-3).

To separate targets having very low radial speeds from the accompanying mainlobe clutter, the displaced phase center technique described in Chap. 26 is used. To enable the targets' angular positions to be precisely determined, the antenna is subdivided lengthwise into three segments, also as described in Chap. 26.

The Joint STARS radar has three primary modes of operation:

- High-resolution SAR imaging mode, for detecting and identifying stationary targets

Figure 50-1. Joint STARS. The three-segment passive ESA, electronically steered in azimuth and manually steered in elevation, is housed in the 9 m long radome under the forward section of the fuselage. (Courtesy of US Air Force.)

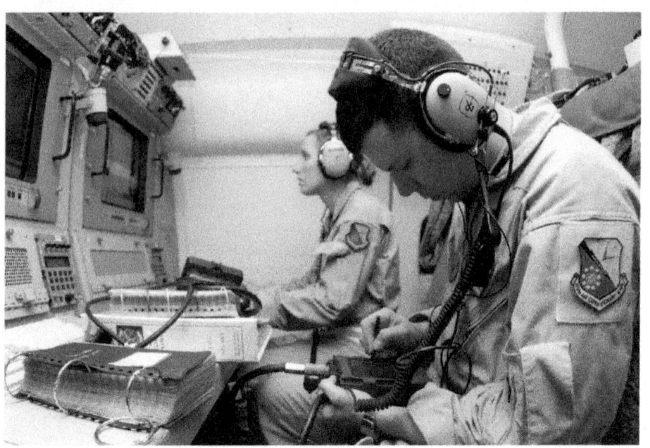

Figures 50-2, 50-3. The navigator's workstation (top) is one of 18 operator workstations. Each is equipped with a digital processor included in Joint STARS' distributed processing system.

- Wide-area MTI surveillance mode, for situational awareness
- Sector MTI search mode, for battlefield reconnaissance.

As the name implies, the MTI modes are used to locate, identify, and track moving targets. When a vehicle that is being tracked stops, the radar can almost instantly produce a high-resolution SAR image of the vehicle and its surroundings.

All three modes may be flexibly selected or interleaved.

Targets detected with MTI are displayed as moving images. These can be superimposed on digitally stored maps or on the radar's SAR images, and they can be stored and replayed at selectable speeds.

An operator can individually distinguish the vehicles in a convoy, and even determine which vehicles are wheeled and which are tracked.

The radar data is encrypted and may be relayed by a highly jam-resistant data link to an unlimited number of Army ground control stations.

SENTINEL R1 is a British SAR/GMTI system carried by a modified Bombardier Global Express aircraft, which is an ultra-long range, high-flying twin-engined business jet. It was formerly known as ASTOR (Airborne Stand-Off Radar). The crew of five consists of two pilots, a Mission Commander, and two image analysts. The radar is a derivative of the Raytheon ASARS-2 carried by the U-2,[1] with SAR and GMTI modes, and uses a 4.6 m long active scanned array mounted in a canoe-shaped radome under the front of the fuselage (Figures 50-4 and 50-5).

The SAR provides both swath and spotlight modes. The spotlight mode, which is used to identify and track specific targets, has a resolution of 0.3 m. The GMTI mode is able to track moving vehicle targets over a wide area.

By operating at high altitude, and at considerable long-range standoff distances, the radar platform is able to remain over safe territory while providing a favorable "look-down" angle of the target area. The SAR/GMTI radar identifies the location and number of hostile forces, giving information as to their speed and direction. Image data is transmitted in real time via a secure data link to a ground-based processing station. The radar signal processors transform the data into images that can be displayed and exploited, and subsequently transmitted to other areas. The system has directional and broadcast data links which are interoperable with existing U-2Rs, JSTARS and command and control networks.

1. HISAR is lightweight derivative of the same ASARS-2 radar, flown on the Global Hawk UAV.

Figure 50-4. SENTINEL: the radar system is housed in the long canoe-shaped radome beneath the fuselage of the aircraft. The radome just visible on the top of the aircraft houses the SATCOM antenna.

50.2 Unmanned Systems

TESAR (Tactical Endurance Synthetic Aperture Radar) is a strip mapping Ku-band SAR providing continuous 0.3 m imagery. The focused imagery is formed to a resolution of 30 cm, on-board the Predator and Army Gnat aircraft. It is compressed and sent to the Predator Ground Control Station over a Ku-band data link. The imagery is reformed and displayed in a scrolling manner on the SAR workstation displays. As the imagery is scrolling by, the operator has the ability to select square image patches (approximately 800 m × 800 m at an altitude of 5,000 m) for exploitation at a workstation. The imagery is also recorded continuously for further, off-line review.

There are 2 modes of operation. Mode 1 provides a non-centered strip map. That is the map center moves with respect to the aircraft motion. Mode 2 is the classic stripmap mode. Mapping occurs over a predetermined scene centerline, irrelevant to the aircraft motion. The radar is designed to map while squinting up to ±45 degrees off the velocity vector. At ground speeds from 25-35 m/s, the swath width is 800 meters. At speeds beyond 35 m/s, the swath width decreases proportionally with the increase in ground speed.

The TESAR subsystem functions autonomously by executing a series of preplanned mission commands loaded prior to operations, which are capable of being altered while inflight. Compressed and continuous SAR imagery is not available in the LOS/UHF modes of operation. It can only be transmitted over the 1.5 Mb/s Ku-band wideband data link via satellite relay to the ground control station (GCS) for decompression and display.

STARlite is a lightweight, compact SAR/GMTI radar system designed and built by Northrop Grumman, and compatible with a range of UAV systems. The radar has both stripmap and spotlight modes of operation, and is a successor to the TESAR system.

STARLite's radar offers four flexible modes for tactical reconnaissance:

- Synthetic Aperture Radar (SAR)
- Ground Moving Target Indicator (GMTI)
- Dismount Moving Target Indicator (DMTI)
- Maritime Moving Target Indicator (MMTI)

The radar provides two SAR modes: Strip and Spot. In Strip mode, the radar imagery is either parallel to the aircraft flight vector or along a specified ground path independent of the aircraft flight path. In Spot mode, the radar produces a high-resolution image at a specific geographic patch. In the GMTI mode, the radar provides moving target locations overlaid on a digital map. The MMTI mode performs a similar function for targets over water. The DMTI mode provides detection of personnel movement on the ground. STARLite is designed to be compatible with a standard ground control station. The ground station has the hardware and software tools necessary to control the radar and record and display the downloaded SAR imagery and GMTI/DMTI targets for increased situational Awareness and battlefield management.

Figure 50-5. The Sentinel aircraft provides a long-range, battlefield-intelligence, target-imaging and tracking radar for the Royal Air Force.

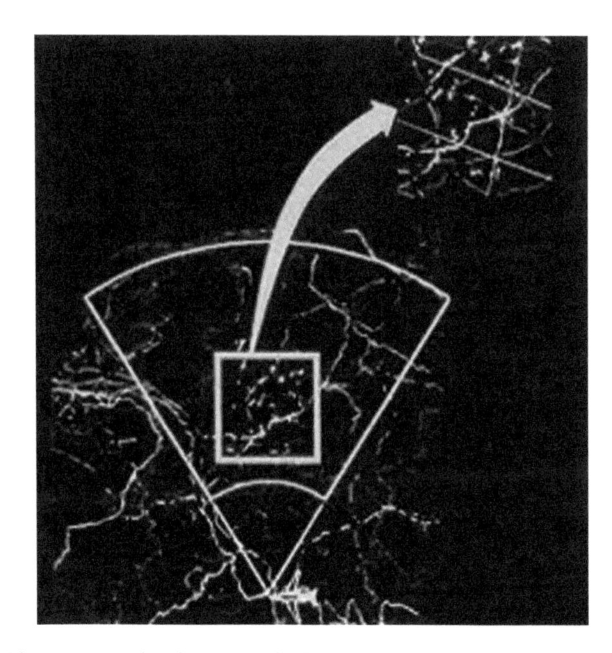

Figure 50-6. This display mode shows ground moving targets.

Figure 50-7. The TESAR radar encased in the Predator UAV. (Courtesy of US Air Force.)

Figure 50-8. STARlite undergoing flight test.

NASA UAV SAR is an L-band radar mounted in a pod that is slung under an aircraft such as a Gulfstream-III and can be carried by UAVs. The system is capable of producing polarimetric imagery. In addition, by overflying the same area twice it can produce three-dimensional imagery via an interferometric mode of operation.

The primary objective of the side-looking UAVSAR instrument is to accurately map crustal deformations associated with natural hazards, such as volcanoes and earthquakes. Topographic information is derived from phase measurements that, in turn, are obtained from two or more passes over a given target region. The frequency of operation, approximately 1.26 GHz, results in radar images that are highly correlated from pass to pass. Polarization agility facilitates terrain and land-use classification.

Figure 50-9. The NASA UAVSAR is mounted in a pod underneath a Gulfstream-III aircraft.

The UAVSAR radar is designed from the beginning as a miniaturized polarimetric L-band radar for repeat-pass and single-pass interferometry with options for along-track interferometry and additional frequencies of operation. By designing the radar to be housed in an external unpressurized pod, it has the potential to be readily ported to other platforms such as the Predator or Global Hawk UAVs. Initial testing has been carried out with the NASA Gulfstream III aircraft, which has been modified to accommodate the radar pod and has been equipped with precision autopilot capability developed by NASA Dryden Flight Research Center.

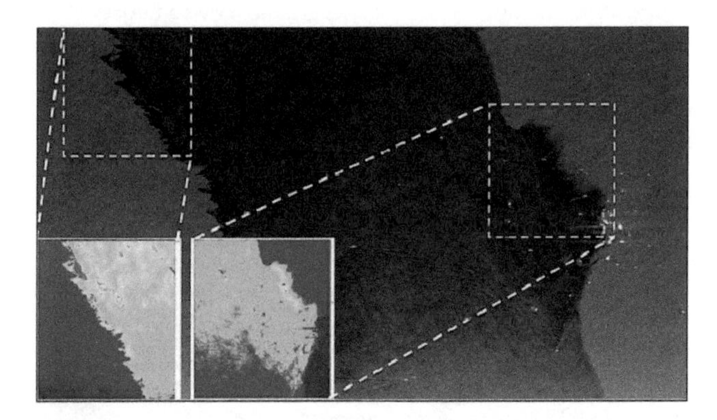

Figure 50-10. NASA UAVSAR image of the Deepwater Horizon oil spill, collected June 23, 2010.

UAVSAR characterizes an oil spill by detecting variations in the roughness of the sea surface and, for thick slicks, changes in the electrical conductivity of the surface layer (Figure 50-10). Just as an airport runway looks smooth compared to surrounding fields, a radar "sees" an oil spill at sea as a smoother (radar-dark) area against the rougher (radar-bright) ocean surface because most of the radar energy that hits the smoother surface is reflected in a specular manner away from the radar antenna. UAVSAR's high sensitivity and other capabilities allowed the difference between thick and thin oil to be distinguished by a radar system for the first time.

51

Space Based Radar Systems

ERS-1 SAR image of Straits of Gibraltar, showing internal waves (Courtesy of European Space Agency.)

Until recently Earth Observation (EO) from space was mainly for scientific and defense purposes. Today data and information products, specifically SAR imagery, are increasingly used for a variety of important scientific and commercial applications. Today, food and natural resource industries, insurance companies, government agencies and many more exploit SAR imagery from satellites. New systems have appeared that are capable of generating data of enhanced quality, fuelling both public and private sector exploitation of EO imagery. New SAR systems are producing imagery that is more detailed and is delivered faster and more reliably whilst, of course, being independent of weather conditions and cloud coverage.

51.1 RADARSAT-2

RADARSAT-2 (Figure 51-1) is a Canadian low Earth orbit (LEO) satellite-borne Earth observation SAR system developed as a collaboration between the MacDonald, Dettwiler and Associates (MDA) company and the Canadian Space Agency (CSA), and launched in December 2007. It follows on

Figure 51-1. The RADARSAT-2 satellite. Visible are the solar array, with the SAR antenna underneath. (from Canadian Space Agency website: http://www.asc-csa.gc.ca/eng/satellites/radarsat2/)

Figure 51-2. RADARSAT-2 image of Sendai, Japan, prior to the tsunami that devastated Fukoshima. (from Canadian Space. Agency website: http://www.asc-csa.gc.ca/eng/satellites/radarsat2/)

Table 51-1. Comparison of parameters of RADARSAT-1, RADARSAT-2 and RADARSAT missions (adapted from information from Canadian Space Agency http://www.asc-csa.gc.ca/eng/satellites/radarsat2/

	RADARSAT-1	RADARSAT-2	RADARSAT constellation
radar center frequency	5.3 GHz	5.405 GHz	5.405 GHz
radar bandwidth	30 MHz	100 MHz	100 MHz
SAR antenna dimensions	15 m × 1.5 m	15 m × 1.5 m	6.75 m × 1.38 m
polarization	HH	HH, VV, HV, VH	HH, VV, HV, VH, compact polarimetry
polarization isolation	>20 dB	>25 dB	>28 dB
mass	679 kg	750 kg	400 kg approx.
orbit altitude	793–821 km	798 km	592.7 km
orbit inclination	98.6°	98.6°	97.74°
orbit duration	100.7 min	100.7 min	96.4 min

from the highly successful RADARSAT-1 mission which was launched in November 1995 and which lasted till March 2013, well beyond its designed lifetime. Although the RADARSAT systems have global coverage, there is emphasis on maritime surveillance, ice monitoring, disaster management, environmental monitoring, resource management and mapping of polar regions. RADARSAT-2 typifies the development of Earth observation SAR systems, from NASA's SEASAT in 1978, though the European Space Agency's ERS-1, ERS-2 and ENVISAT, to the present. Many of these exploit the flexibility of electronically scanned arrays to give wide-swath and/or high-resolution modes, as well as polarimetric operation.

Like RADARSAT-1, RADARSAT-2 carries a C-band (5.3 GHz) SAR. Chapter 35 of this book has used the RADARSAT-2 SAR as a design example, and that section should be consulted to understand how the radar parameters are determined from the desired performance parameters.

The RADARSAT-2 SAR has several modes of operation, including standard (with a swath width of 100 km and a resolution of 25 m), fine (swath width 50 km, resolution 10 m) and ScanSAR (swath width 500 km, resolution 100 m). Figure 51-2 shows an example of a RADARSAT-2 image of Sendai, Japan, prior to the tsunami of March 2011 that devastated Fukoshima.

Its successor, the RADARSAT constellation, is planned for launch in 2018. Table 50-1 summarizes the parameters of RADARSAT-1, RADARSAT-2 and RADARSAT constellation, and shows the development of the polarimetric capability and multiple modes of operation.

51.2 TerraSAR-X

TerraSAR-X (Figure 51-3) is a German Earth-observation satellite. Its primary payload is an X-band (9.65 GHz) Synthetic Aperture Radar, with a range of different modes of operation, allowing it to record images with different swath widths,

Figure 51-3. Artist's impression of the two satellites forming the TanDEM-X mission, orbiting the Earth.

resolutions and polarizations. In this way TerraSAR-X provides space-based observation capabilities that were hitherto unavailable. The overall objective of the mission is to provide value-added SAR data at X-band, for research and development purposes as well as scientific and commercial applications. The system has image resolutions of 1, 3 and 18.5 m. The high-resolution modes allow specific areas to be targeted. The lower resolution modes are compatible with larger imaged areas. Image polarization can be switched rapidly.

The satellite is in a near-polar orbit around the Earth, at an altitude of 514 km. It employs a 5 m long active array antenna and the radar beam can be electronically tilted within a range of 20 to 60 degrees perpendicular to the flight direction, without having to move the satellite itself. This has the advantage of allowing the radar to zoom in on many more ground targets from the satellite's orbit than would be possible using a conventional non-steerable antenna beam. TerraSAR-X has been fully operational since 7 January 2008 and Fig. 51-4 shows an example image of Mount Bromo in Indonesia, produced by this system.

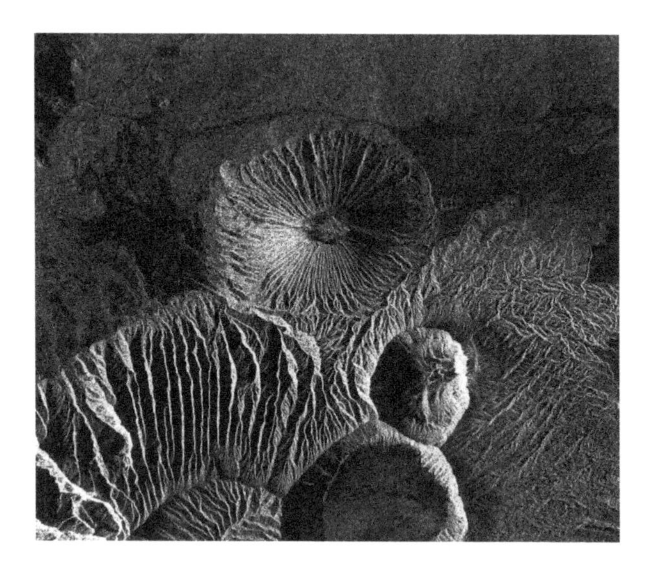

Figure 51-4. TerraSAR-X image of Mount Bromo, Indonesia.

The X-band SAR operates in three different modes:

- Spotlight mode: An area 10 km long and 10 km wide is imaged at a resolution of 1 to 2 m,
- Stripmap mode: A 30 km-wide strip is imaged at a resolution between 3 and 6 m,
- ScanSAR mode: A 100 km-wide strip is imaged at a resolution of 16 m.

Table 51-2 shows the main radar operating parameters.

Table 51-2 TerraSAR-X operating parameters.

Radar carrier frequency	9.65 GHz
Pulse Repetition Frequency (PRF)	2 kHz to 6.5 kHz
Signal bandwidth	150 MHz, 300 MHz (Advanced Mode)
Polarizations	HH, HV, VH, VV
Antenna height	0.7 m
Antenna length	4.8 m
Data access incidence angle range	15° to 60°
Incidence angle range for stripmap/ScanSAR modes	20° to 45°
Incidence angle range for spotlight mode	20° to 55°
Maximum achievable resolution in range	0.65 m at 300 MHz bandwidth (Advanced Mode)
Azimuth resolution	1 m to 16 m depending on imaging mode, incidence angle, and number of polarizations

If two antennas separated by a fixed distance are used, each antenna will be located at a slightly different distance from a given point on the ground. From the difference between these two distances, the elevation of the observed point above a reference plane can be calculated. Doing this for a large number of points can create a digital elevation model (DEM). The difference in distance is determined by measuring the time difference between the arrivals of the radar echoes. The accuracy of the elevation measurements can be improved significantly to sub-cm level by analyzing the phase differences between the backscattered signals.

This technique is known as radar interferometry (described in Chapter 33). Since TerraSAR-X only has one antenna, two passes in slightly different orbits are required in order to obtain a three dimensional view of the Earth. There is a drawback to using two images recorded at a fixed interval of time apart to determine elevation: any changes that take place at the Earth's surface during this interval (such as rainfall, plant growth, or even motion due to wind) can lead to decorrelation of the two images and hence inaccuracies. The TanDEM-X project however bypasses this limitation by flying TerraSAR-X in close formation with a practically identical second satellite. As the maximum distance between the satellites is 600 m, two images are recorded with a maximum time lag of just 0.08 seconds (i.e. almost simultaneously). This means that the accuracy of the digital elevation model no longer depends on any changes that could occur over time.

The main objective of the TanDEM-X mission is to generate an accurate topographic image of the Earth. At present, the elevation models available for large parts of Earth are of low resolution, inconsistent or incomplete. In addition, they are commonly based on different data sources and survey methods. TanDEM-X is designed to close these gaps and deliver an elevation model for many scientific and commercial applications. Orbiting Earth at an altitude of around 500 km, the two near-identical radar satellites have begun mapping its surface. Fig. 51-4 shows an example TerraSAR-X image of Mount Bromo in Indonesia.

51.3 COSMO-SkyMed

COSMO-SkyMed (*COnstellation of small Satellites for Mediterranean basin Observation*) is an Italian space system for Earth Observation, commissioned and funded by the Italian Space Agency (ASI) and the Italian Ministry of Defense (MoD), conceived as a dual-use (civilian and defense) end-to-end Earth Observation System supplying data, products and services relevant to a wide range of applications, such as risk management, scientific, commercial applications, as well as defense and intelligence use.

The system consists of a constellation of four Low Earth Orbit mid-sized satellites, each equipped with a multi-mode high-resolution Synthetic Aperture Radar (SAR) operating at X-band, and fitted with data acquisition and transmission

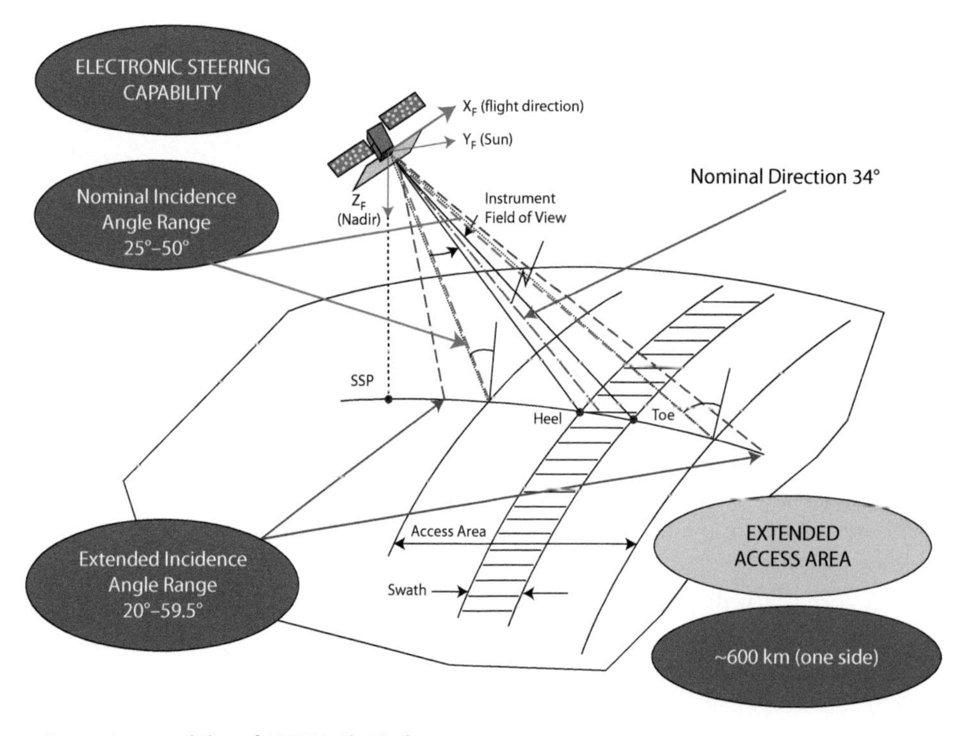

Figure 51-5. The electronic steering capability of COSMO-SkyMed.

equipment. Due to differing requirements for image size and spatial resolution the SAR sensor has multiple operating modes (Figure 51-5):

- A spotlight mode, for metric resolutions over small images
- Two stripmap modes with meter-level resolutions; one of these is polarimetric with images acquired in two polarizations
- Two ScanSAR modes for medium to coarse (100 m) resolution over a large swath

The four satellites allow for an interferometric orbiting configuration (Fig. 51-6). This is able to produce three-dimensional SAR images by combining two radar measurements of the same point on the ground interferometrically, in tandem configuration.

Figure 51-6. An artist's impression of one of the four COSMO-SkyMed satellites in orbit.

The sensor imaging modes (summarized in Table 51-3) are:

Spotlight Mode in which the antenna is steered (both in the azimuth and the elevation planes) during the data acquisition to illuminate the required scene for a time period longer than that of the standard strip side view. This increases the length of the synthetic antenna and therefore the azimuth resolution (at the expense of the azimuth coverage). In an "Enhanced Spotlight mode", illuminated spot extension is achieved by electronic steering so that the center of the beam is located beyond the center of the imaged spot. Figure 51-7 shows an example image of the Cape Town stadium, Green Point, South Africa.

Figure 51-7. An example of a COSMO-SkyMed enhanced spotlight mode SAR image of the Cape Town Stadium, Green Point, South Africa.

Stripmap Mode the antenna footprint covers a strip of the illuminated surface as the satellite platform moves. The acquisition is virtually unlimited in the azimuth direction, except for the limitations due to the SAR instrument duty cycle (about 600 s), allowing a strip length of over 4500 km. There are two different implementations of this mode, known as "Himage" and "PingPong".

In the Himage mode, the radar transmit/receive configurations are time invariant, allowing echoes to be received from each ground point over the full Doppler bandwidth (determined by the antenna azimuth beamwidth). The Himage is characterized by a swath width of approximately 40 km, corresponding to an acquisition time of about 6.5 seconds.

Table 51-3 COSMO-SkyMed modes of operation.

Mode	Image size (km)	PRF (kHz)	pulselength (μs)	pulse bandwidth (MHz) near range – far range
Spotlight	11 × 11	3.148 – 4.117	70 – 80	18.2 – 400
Stripmap Himage	40	2.906 – 3.874	36 – 40	65.64 – 138.6
PingPong	30	2.906 – 3.632	30	14.77 – 38.57
ScanSAR WideRegion (3 adjacent subswaths)	100 × 100	2.906 – 3.632	30 – 40	32.74 – 86.34
HugeRegion (6 adjacent subswaths)	200 × 200	2.906 – 3.632	30 – 40	8.86 – 23.74

The PingPong mode image acquisition is performed using strip mapping, alternating the polarization between two of possible types i.e. VV, HH, HV and VH. In this polarimetric burst mode only a part of the synthetic antenna length is available in azimuth, and consequently the azimuth resolution is reduced. This mode is characterized by a swath width and azimuth extent of approximately 30 km, corresponding to an acquisition time of about 5.0 sec.

ScanSAR Mode allows a larger swath in range but with lower spatial resolution. It is obtained by periodically stepping the antenna beam to neighboring sub-swaths. Since only a part of the synthetic antenna length is available in azimuth, the azimuth resolution is reduced.

In this configuration the acquisition is performed as a set of strip maps and hence is virtually unlimited in the azimuth direction (apart from the limitations of the SAR instrument duty cycle, about 600 s). There are two different implementations: "WideRegion" and "HugeRegion".

In the WideRegion mode, image acquisitions are grouped over three adjacent subswaths, allowing ground coverage of about 100 km in the range and azimuth directions, corresponding to an acquisition time of about 15.0 sec.

In the HugeRegion mode image acquisitions are grouped over six adjacent subswaths allowing ground coverage of about 200 km in the range and azimuth directions, corresponding to an acquisition time of about 30 seconds.

52

Fighter & Attack

52.1 AN/APG-76

The AN/APG-76 (Fig. 52-1) is a multimode Ku-band pulse-Doppler radar originally developed by Westinghouse Norden Systems for Israel's F-4 Phantom 2000 fighters for air-to-air and air-to-ground precision targeting and weapon delivery. Extended capability variants have been evaluated in simulated combat in wing tanks on the US Navy S-3 and US Air Force F-16.

Figure 52-1. The AN/APG-76 radar system in the nose of an F-16.

Capabilities. The radar is capable of simultaneous SAR mapping and ground moving target detection and tracking. Employing a three-segment mechanically steered planar array antenna and four low-noise receiver and signal processing channels, it features:

- Long-range multi-resolution SAR mapping
- All-speed ground moving target detection over the full width of the forward sector
- Automatic tracking of ground moving and "did-move" targets
- Automatic detection and location of rotating antennas

The antenna has seven receive ports: sum, azimuth difference, elevation difference, guard, and three interferometer ports. In air-to-air modes, the sum, azimuth difference, elevation difference, and guard outputs are processed in parallel through the four receive channels. In air-to-ground modes, the sum signal is processed through one channel, and the three interferometer signals through the remaining three channels.

GMTI and GMTT. The radar system employs interferometric notching and tracking techniques to detect and precisely track ground moving targets having radial velocities from 2 to 30 m/s anywhere within the radar's ±60° azimuth field of view.

Ground clutter is suppressed by subtracting the echo returns received by one interferometer antenna segment from the weighted returns received by another. This is done in the outputs of all of the Doppler filters passing frequencies within the mainlobe of the two-way antenna pattern.

Adaptive CFAR detection thresholds are independently determined for clutter and clutter-free regions. Those targets satisfying an M-out-of-N detection criterion are displayed as moving target symbols superimposed at the correct range and azimuth positions over the simultaneously produced SAR map.

As initially implemented, the radar employed five parallel operating vector pipeline processors and two scalar data processing elements. The radar also provides a wide selection of ground-map resolutions ranging from real beam, to Doppler beam sharpening and 3 m resolution SAR imaging. 1 m and 30 cm resolution SAR modes have also been developed and tested together with a wide-area surveillance mode that combines high-resolution SAR maps in a mosaic to facilitate continuous monitoring and tracking of moving targets.

52.2 AN/APG-77

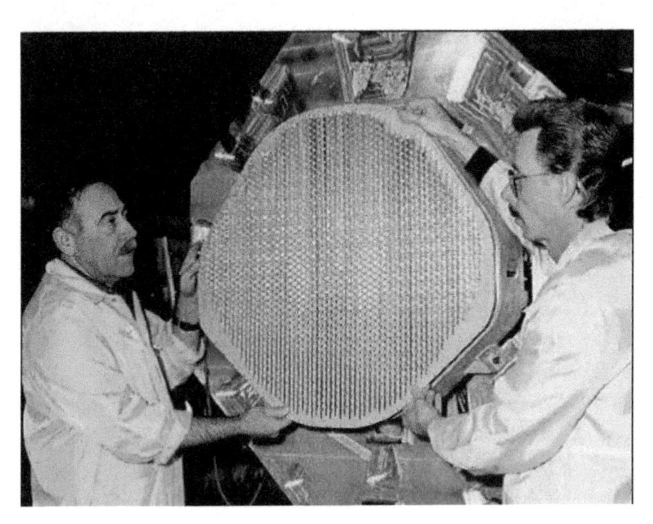

Figure 52-2. The active ESA employed by the APG-77 to meet low RCS requirements provides extreme beam agility and supports numerous growth features.

The AN/APG-77 (Fig. 52-2) is the F-22 fighter aircraft radar and employs an airborne electronically scanned phased array (AESA) antenna. The radiating modules are solid state enabling much greater flexibility as well as improving reliability. There are approximately 2300 modules and the system is said to be able to track a 1 m^2 target out to a range of 240 km.

The AN/APG-77 AESA radar is designed for air-superiority and strike operations. The AN/APG-77 changes frequencies more than 1,000 times per second to reduce the chance of being intercepted.

The AN/APG-77 is a multimode pulse-Doppler radar meeting the air dominance and precision ground attack requirements of the F-22 stealth dual-role fighter. The F-22 may be armed with six AMRAAM missiles or two AMRAAMs plus two 1,000-pound GBU-33 glide bombs, two sidewinder IR missiles, and one 20-mm multi-barrel cannon. All of these armaments are carried internally to maintain a low RCS. Four external stations are also available to carry additional weapons or fuel tanks.

The AN/APG-77 radar has a Non-Cooperative Target Recognition (NCTR) mode. Forming fine beams and generating a high-resolution image of a target using ISAR accomplish this. The pilot can compare the target with an actual picture radar image stored in a database. The radar also incorporates extensive LPI features. Signal and data processing requirements are met using a common integrated processor (CIP).

Two CIPs perform the signal and data processing for all of the F-22s sensors and mission avionics, with processor elements of just seven different types. One serves the radar, electro-optical, and electronic warfare subsystems; the other, the remaining avionics. Both have identical back planes and slots for 66 modules. Initially only 19 slots were filled in CIP 1 and 22 in CIP 2, leaving room for 200% growth in avionics capability.

52.3 CAPTOR-M

The Captor-M radar is the primary sensor for the multirole combat aircraft, the Eurofighter Typhoon. The radar detects, identifies, prioritizes and engages targets beyond the effective range of the enemy weapon systems, at the same time remaining resistant to severe electronic jamming.

The EuroRADAR consortium manufactures the Captor-M radar. SELEX Galileo is the lead contractor for this consortium, partnered with Cassidian (Germany) and Indra (Spain).

The Captor-M radar (Fig. 52-3) is an electronically scanned X-band radar system and is integrated with the Typhoon weapon system. It provides:

Figure 52-3. The CAPTOR-M radar mounted in the nose of the Eurofighter Typhoon aircraft.

- Long range detection and tracking
- RAID assessment and target identification
- Flexible, powerful and effective ECCM
- Designed for Beyond Visual Range (BVR) weapons
- Simultaneous multiple target engagement
- Decreased pilot workload through intelligent automation
- Close integration with other avionics sensors

The radar has multiple modes of operation:

- Simultaneous/interleaved A/A and A/G radar modes
- Air-to-Air search and track/search while track
- Air-to-Ground real beam ground map as well as high-resolution modes for surveillance and reconnaissance
- Ground moving target indication search and track
- Sea surface search
- Pilot workload reduction by efficient radar resource management

For long (beyond visual) range combat Captor automatically selects an appropriate mode depending on the current situation. Long-range look-up detection will typically find the system selecting a Low Pulse Repetition Frequency (LPRF). However, for lookdown situations a high pulse repetition (HPRF) will generally be used. For situations where both look-up and lookdown need to be covered simultaneously or where range and velocity data is required, a medium rate would be used. In addition Captor can automatically initiate Track While Scan (TWS) for a list of targets. The system employs Data Adaptive Scanning (DAS) to improve tracking of its selected targets while minimizing unnecessary movement of the antenna. For close-in combat situations Captor will automatically adjust its mode for a high precision single target track.

52.4 AN/APG-81

The AN/APG-81 (Fig. 52-4) is a highly advanced radar system developed for the F-35 aircraft, and is the successor radar to the F-22's AN/APG-77. The features of the AN/APG-81 include the same air-to-air modes of the AN/APG-77 plus advanced air-to-ground modes for high resolution mapping, multiple ground moving target detection and track, combat identification, electronic warfare and ultra high bandwidth communications. The antenna is an AESA operating at X-band and the array face is composed of 1200 solid-state modules allowing coverage of up to ±70° in both elevation and azimuth.

Figure 52-4. The AN APG-81 AESA radar in the nose of the F-35.

The use of solid-state technology and the elimination of mechanical moving are part of a strategy to improve reliability. The radar system features "replaceable assemblies" so that repairs or upgrades to hardware and software modules are easier. In this way the life-cycle costs of this AESA are expected to be significantly lower than those of its predecessors. The active arrays on the F-35 are expected to have almost twice the expected life of the airframe.

52.5 AH-64D Apache Helicopter (Longbow Radar)

Figure 52-5. Lurking behind cover with only the radome of its millimeter wave radar showing, Longbow can quickly detect, classify, and prioritize more than 100 moving or stationary targets.

Longbow is a fast-reaction, low-exposure, high-resolution, millimeter-wave fire-control radar designed for the AH-D Apache attack helicopter (Fig. 52-5). Mounted atop the main rotor mast to take advantage of terrain masking, the radar can pop up and, in seconds scan a 90° sector; then, drop down out of sight.

During that brief interval, it can detect, classify, and prioritize more than 100 moving and stationary ground targets, fixed wing aircraft, and both moving and hovering helicopters—discriminating between closely spaced targets of the same type with an extremely low false-alarm rate.

It then displays the 10 highest priority targets to the aircrew (Fig. 52-6) and will automatically cue either an RF or a semi-active laser guided fire-and-forget Hellfire missile to the first

Figure 52-6. Flat, fully interchangeable color displays are provided in both cockpit positions (courtesy of U.S. Army).

target (Fig. 52-7). Immediately after its launch, the system cues the next missile to the next priority target, and so on.

The radar also provides obstacle warning to alert the pilot to navigation hazards, including man-made structures, towers, etc.

Radar data is displayed on the pilot's night-vision helmet-mounted display and on two color-coded flat general-purpose displays in each cockpit.

A derivative of the Longbow radar will be forthcoming for the RAH-66 Comanche helicopter. Using the same millimeter-wave radar and the same Hellfire missiles as Apache, it will include a number of advanced features, such as a smaller antenna.

Figure 52-7. The AH-64D carries up to 16 RF or semi-active laser-guided Hellfire missiles and 76 70-mm folding fin aerial rockets or a combination of both, and up to 1,200 rounds of 30-mm ammunition.

Eurofighter TYPHOON (2003)

The Typhoon is a single-seat, twin-engined, canard-delta wing, multirole fighter. It is designed and manufactured by a consortium of BAE Systems, Airbus Group and Alenia Aermacchi. Its first flight was in 1994, and it is currently in service with the Royal Air Force, the Austrian Air Force, the Italian Air Force, the German Air Force, and the Royal Saudi Air Force.

Test Your Understanding: Numerical Answers

Ch. 1 Basic Concepts

3. 99.9 km

Ch. 4 Radio Waves and Alternating Current Signals

6. 10 cm
7. 0.33 ns

Ch. 5 A Nonmathematical Approach to Radar

4. 6.28×10^{10} rad/s
10. 20 dB

Ch. 6 Preparatory Math for Radar

1. 1.414, 45°

Ch. 7 Choice of Radio Frequency

2. 20%
3. 127%
4. 40 dB

Ch. 8 Directivity and the Antenna Beam

1. 63.5%
2. 9.5°, 2.9°, 0.95°
3. Approximately 11.5°, 3.5°, 1.15°

Ch. 9 Electronically Steered Antenna Arrays

4. 127°

Ch. 10 Electronically Steered Array Design

4. 0.54λ

Ch. 11 Pulsed Operation

1. (a) 300 km, (b) 500 Hz
2. 30 m
3. 6.67 ns
4. (a) 0.005 or 0.5%, (b) 50 W, (c) 0.1 J
5. 3 km

Ch. 12 Detection Range

1. 13 dB
3. 4×10^{-4} W/m^2
4. 4×10^{-7} J
5. 24 dB

Ch. 13 The Range Equation

1. 0.99
2. 19%, 41%, 19%
3. 250, 5 s
4. 81 km
5. 10^4 Wm^2

Ch. 14 Radar Receivers and Digitization

1. 2.3 dB
2. (a) 7.5 GHz (low side) or 12.5 GHz (high side); 5 GHz (low side) or 15 GHz (high side)
3. 100 MHz, because the I and Q A/D converters effectively provide a sampling rate of 200 MHz, so the Nyquist frequency is half of this.

Ch. 15 Measuring Range and Resolving in Range

2. 375 km or 200 nmi
3. $(75n + 17)$ km, $n = 0, 1, 2, \ldots$
4. (a) 10.2 km, 11.4 km; (b) 54 km

Ch. 16 Pulse Compression and High-Resolution Radar

1. 1.5×10^{13} Hz/s, 6000
2. (a) [−1, 0, −1, 0, −1, 0, −1, 0, −1, 0, +11, 0, −1, 0, −1, 0, −1, 0, −1, 0, −1]

   (b) [+1, +2, +1, −2, −3, −2, +1, −2, −3, +2, +11, +2, −3, −2, +1, −2, −3, −2, +1, +2, +1]
3. (a) mainlobe = $|+11| = 11$, largest sidelobe = $|-1| = 1$ so PSL = 1/11

(b) mainlobe = |+11| = 11, largest sidelobe = |−3| = 3 so PSL = 3/11

4. Four-phase Frank code: [0°, 0°, 0°, 0°, 0°, 270°, 180°, 90°, 0°, 180°, 0°, 180°, 0°, 90°, 180°, 270°]
   Tapped delay line output: [+j, −1 + j, −1, 0, −1, +1 − j, −j, 0, −j, −1 − j, +1, 0, +1, 1 + j, +j, +16, −j, 1 − j, +1, 0, +1, −1 + j, +j, 0, +j, 1 + j, −1, 0 −1, −1 − j, −j]

Ch. 17 Frequency-Modulated Continuous Wave Radar

1. 1.5 m
2. 4×10^{-7} s, 40 kHz
3. 75 m; 10 m/s
4. 26

Ch. 18 The Doppler Effect

4. 20 microns
5. 20 kHz
6. 20.2 kHz
7. 5 ms

Ch. 19 Spectrum of Pulsed Signal

2. 1 MHz
3. 2 Hz

Ch. 20 The Pulsed Spectrum Unveiled

3. 10 cm
6. 1 GHz

Ch. 21 Doppler Sensing and Digital Filtering

2. 1 Hz
3. ≥ 2 MHz
4. the null-to-null width = 2 × 3 dB width

Ch. 22 Measuring Range-Rate

1. 33.3 kHz
2. 67 kHz
3. 3 km

Ch. 23 Sources and Spectra of Ground Return

1. 1 m^2
3. (a) −20 dBm^2/m^2; (b) −7 dBm^2/m^2

Ch. 24 Effect of Range and Doppler Ambiguities on Ground Clutter

1. 150 km, 1 kHz
2. 200 km ± multiples of 150 km; i.e., 50 km, 200 km, 350 km, etc.
3. The sidelobe clutter returns will extend from ±2 $v/\lambda = \pm 2 \times 90/.03 = \pm 6000$ Hz, so the PRF must be at least 12 kHz; the air targets must have Doppler shifts greater than 6 kHz or less than −6 kHz.
4. The bandwidth required is 1 kHz because the frequencies of any signals with Doppler frequencies outside this bandwidth will be aliased into this bandwidth.

Ch. 26 Separating Ground Moving Targets from Clutter

1. Because even without any Doppler shifts induced by the motion of the radar platform, the Doppler shift of slow moving ground targets is often very similar to that of the surrounding clutter.
2. The antenna should move by 0.5 m between successive pulses; this will take 0.5/200 = 0.0025 s, so the PRF is 1/0.0025 = 400 Hz.
4. $\pm \frac{2v}{\lambda} \sin\left(\frac{\theta}{2}\right)$ Hz
5. STAP processing does not assume any prior knowledge of the relationship between the clutter Doppler spectrum and radar look direction, but is able to continuously adapt to changing clutter conditions and aircraft motion.

Ch. 27 PRF and Ambiguities

3. 490 kHz
4. 400 kHz

Ch. 28 Low PRF Operation

4. 5 kHz
5. $(2/30)^4$ times smaller

Ch. 29 Medium PRF Operation

1. 1.5 km and 15 m/s

Ch. 32 Radar and Resolution

3. 1500 m

Ch. 37 Electronic Warfare Support

2. The sensitivity (without a preamplifier) is normally about −40 dBm.
4. 112.4 dB
6. Three are required.

Ch. 43 Advanced Processor Architectures

1. The area would decrease to 1 $cm^2 \times (0.13/0.09)^2 =$ 0.48 cm^2. The number of transistors would be the inverse of this or ×2.08.

2. Execution time after improvement

$$= \frac{\text{Execution time affected by improvement}}{\text{Amount of improvement}}$$

$$+ \text{Execution time unaffected}$$

So Execution time after improvement = 32/10 +16 = 19.2 microseconds

48/19.2 = 2.5. So the processing time has been reduced by a factor of ~2.5

3. FPGA. The data is continuously streaming, involves integer calculations and the filtering operation is a strong candidate for a pure processing pipeline.

Ch. 44 Bistatic Radar

3. +4.6 dB, +8.9 dB

Ch. 46 Radar Waveforms: Advanced Concepts

1. (a) [−1, 0, −1, 0, −1, +1, +1, +3, −1, −3, +5, −2, −1, +2, −1]

   (b) [−1, −2, −1, +2, +5, +3, −1, −3, +1, −1, −1, 0, −1, 0, −1]

2. (a) mainlobe = |+5| = 5, largest sidelobe = |−3| = 3 so PSL = 3/5

   (b) mainlobe = |+5| = 5, largest sidelobe = |−3| = 3 so PSL = 3/5

Ch. 47 Target Classification

2. (i) 1 m

3. 23%

Index

Note: page numbers are followed by *f* or *b* or **t** (indicating figures or boxes or tables).

A

B

C

D

E

F

I

J

K

L

Q

R

S

T

U

V

W

Z

Radar Data and Relationships

Linear ⟵ ⟶ dB Scale

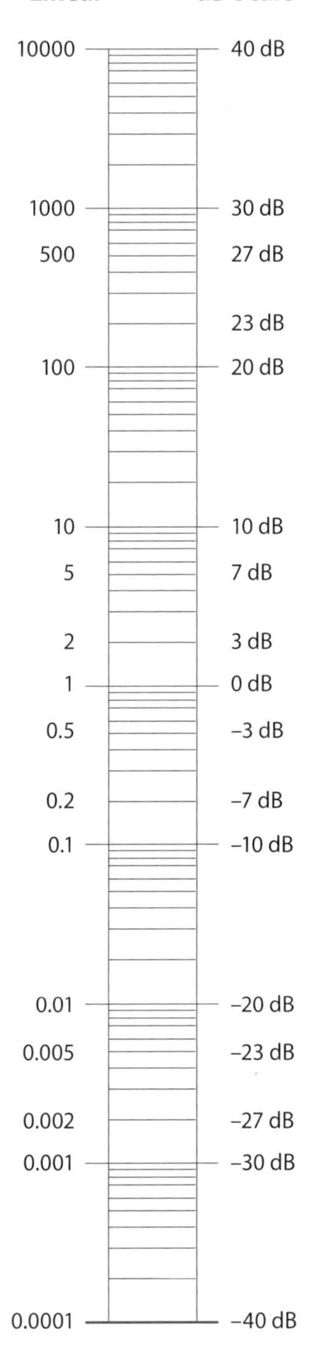

Radar Bands

Band	Frequency Range	Value
High Frequency (HF)	3–30 MHz	
Very High Frequency (VHF)	30–300 MHz	138–144 MHz 216–225 MHz
Ultra High Frequency (UHF)	300 MHz –1 GHz	420–450 MHz 890–942 MHz
L	1–2 GHz	1.215–1.400 GHz
S	2–4 GHz	2.3–2.5 GHz 2.7–3.7 GHz
C	4–8 GHz	5.250–5.925 GHz
X	8–12 GHz	8.500–10.680 GHz
Ku ("under" K-band)	12–18 GHz	13.4–14.0 GHz 15.7–17.7 GHz
K	18–27 GHz	24.05–24.25 GHz 24.65–24.75 GHz
Ka ("above" K-band)	27–40 GHz	33.4–36.0 GHz
V	40–75 GHz	59.0–64.0 GHz
W	75–110 GHz	76.0–81.0 GHz 92.0–100.0 GHz
mm	100–300 GHz	126.0–142.0 GHz 144.0–149.0 GHz 231.0–235.0 GHz 238.0–248.0 GHz

Atmospheric Attentuation (one-way)

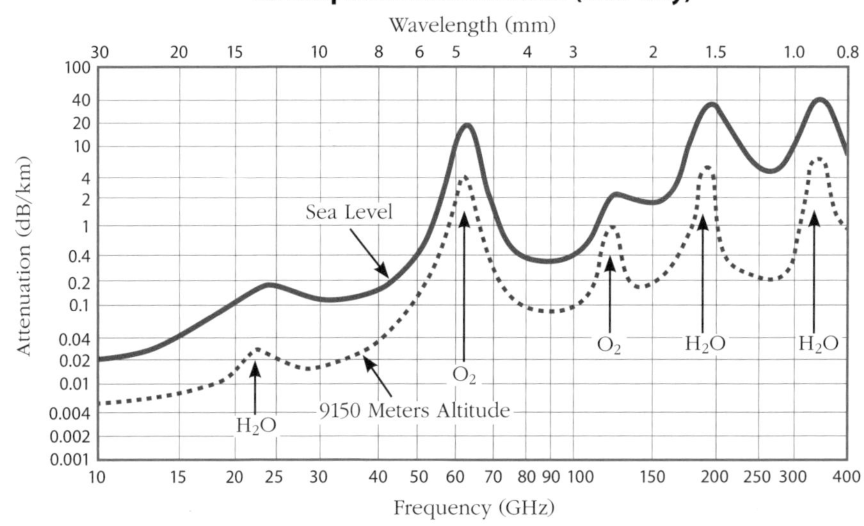

Table of Constants

Constant	Symbol	Value
Speed of light	c	2.99792458×10^8 m/s $\approx 3 \times 10^8$ m/s
Permittivity of free space	ε_0	8.85×10^{-12} F/m
Permeability of free space	μ_0	$4\pi \times 10^{-7}$ H/m
Impedance of free space	η	377 Ω
Bolztmann's constant	k	1.38×10^{-23} J/K

Time Delay

A time delay of...	...is approximately equivalent to a radar range of...
1 nanosecond (ns)	0.15 meters (m)
	15 centimeters (cm)
	0.5 feet (ft)
	6 inches (in)
1 microsecond (μs)	0.15 km
	150 meters (m)
	0.1 (0.093) miles
	500 (492) feet (ft)